History of Mechanism and Machine Science

Volume 42

This bookseries establishes a well-defined forum for Monographs and Proceedings on the History of Mechanism and Machine Science (MMS). The series publishes works that give an overview of the historical developments, from the earliest times up to and including the recent past, of MMS in all its technical aspects.

This technical approach is an essential characteristic of the series. By discussing technical details and formulations and even reformulating those in terms of modern formalisms the possibility is created not only to track the historical technical developments but also to use past experiences in technical teaching and research today. In order to do so, the emphasis must be on technical aspects rather than a purely historical focus, although the latter has its place too.

Furthermore, the series will consider the republication of out-of-print older works with English translation and comments.

The book series is intended to collect technical views on historical developments of the broad field of MMS in a unique frame that can be seen in its totality as an Encyclopaedia of the History of MMS but with the additional purpose of archiving and teaching the History of MMS. Therefore. the book series is intended not only for researchers of the History of Engineering but also for professionals and students who are interested in obtaining a clear perspective of the past for their future technical works. The books will be written in general by engineers but not only for engineers. The series is promoted under the auspices of International Federation for the Promotion of Mechanism and Machine Science (IFToMM).

Prospective authors and editors can contact Mr. Pierpaolo Riva (publishing editor, Springer) at: pierpaolo.riva@springer.com

Indexed by SCOPUS and Google Scholar.

Raffaele Pisano
Editor

A History of Physics: Phenomena, Ideas and Mechanisms

Essays in Honor of Salvo D'Agostino

Springer

Editor
Raffaele Pisano
Department of Physics
HOPAST Team at IEMN, Lille
University–CNRS
Villeneuve d'Ascq, France

ISSN 1875-3442 ISSN 1875-3426 (electronic)
History of Mechanism and Machine Science
ISBN 978-3-031-26173-2 ISBN 978-3-031-26174-9 (eBook)
https://doi.org/10.1007/978-3-031-26174-9

This Springer imprint is published by the registered company Springer Nature Switzerland AG
The registered company address is: Gewerbestrasse 11, 6330 Cham, Switzerland

Paper in this product is recyclable.

Acknowledgments

I want to express my gratitude to all Authors, invited (Parts 1 and 4) and contributors (Parts 2–3) for their efforts to produce papers of interest and of high quality worthy of this special volume. The result is excellent.

I take this opportunity to address my friendly warm appreciation to Prof. Stavros Katsanevas, Director of the *European Gravitational Observatory*—EGO (Italy) and Word Distinguished Scholar for his acceptance to write a *Foreword* for this Springer book, offering the readers his authoritative standpoint in the field and on the subject. Unfortunately, he passed away at the age of 69 on 27 November 2022 before delivering his manuscript. He was a known international interdisciplinary—intellectual scientist. Katsanevas was professor at the Université Paris Denis Diderot (*Paris Cité*). He led the prestigious *European Gravitational Observatory* (EGO), producing fundamental and extraordinary scientific developments in gravitational astronomy, including his personal interests in interdisciplinary topics like History of Physics.

Just as significantly, I address my acknowledgments to Dr. Caterina D'Agostino (Salvo's daughter) who was so generous to donate and remove all restrictions and authorizing for Salvo D'Agostino's images published in this volume.

I thank President Dr. Fausto Barbagli—and—Executive Secretariat Mrs Francesca Contini of the *Museo di Storia Naturale/Sezione di Zoologia La Specola* (University of Firenze and *Associazione Nazionale Musei Scientifici*—ANMS, Italy), Mrs Claudia Favaron, Responsible Librarian (*Fondazione Giorgio Cini ONLUS*, Venezia, Italy), *The National Library Service* (SBN, Polo in Venice, Italy) and Dr. Roberto Renzetti (*Fisicamente-Blog*) for their collaboration in bibliographic research of some Salvo D'agostino's less accessible papers, published in the past century.

Sincere gratitude also go to my friends and colleagues Prof. Enrico Giannetto (University of Bergamo, Italy), Prof. Arcangelo Rossi (University of Salento, Italy), Prof. Pasquale Tucci (University of Milano, Italy), Dr. Caterina D'Agostino (BnF-Gallica, France) and Dr. Roberto Mantovani (University of Urbino, Italy) for their friendly cooperation and enthusiasm that they have so kindly showed me during the preparation of this huge editorial celebrative project. Particularly, Giannetto, Rossi, Tucci and D'Agostino, on invitation, wrote four special papers *In Perpetuam Memoriam* of Salvo.

My thankfulness to Dr. Julie Robarts (University of Melbourne, Australia) for her observations on English language; eventual remaining English mistakes are mine. The English proofreading funding was kindly supported by IEMN, Lille University—UMR CNRS 8520 & Lille University, France.

In addition, my gratitude go to my friend and co-authors Paolo Bussotti (Udine University, Italy) for his friendly and professional help.

I regret, with much sadness, to communicate that during the final preparation of this volume my friends and colleagues, Prof. Joseph Agassi (Tel Aviv University, Israel and York University, Canada) and Dante Roberto Salatino (National University of Cuyo, Mendoza, Argentina), passed away. However, they completed their essays which were peer reviewed and revised in the usual manner. Michael Segre (University of Chieti Gabriele d'Annunzio, Italy) and Prof. Guillermo Alberto Cuadrado (University of Cuyo, Mendoza, Argentina) kindly accepted my invitation to write notes *In Memoriam* (Part IV) respectively of Joseph (1927–2023) and Dante (1949–2022). Thank you.

Finally, of great importance, I address my acknowledgments to Prof. Marco Ceccarelli (University of Roma Tor Vergata, Italy) Springer Editor-in-Chief book Series for his fine work and continuous assistance and collaboration during this extensive research–editorial project; also to the Springer reviewers for their excellent suggestions.

May 2024

Raffaele Pisano
The Editor

Remarks for the Reader

In order to celebrate Salvo D'Agostino (1921–2020), Professor in History of Physics and related Physics-topics at the University of Roma La Sapienza, Italy, the research for this major entirely original project aims to present *Phenomena, Ideas & Mechanisms within the History of Physics and Mathematics.*

This volume consists of *Preliminary Matters* and five main parts with respect to the above-cited aim.

The first part presents the *Introduction* to the volume and to the papers, and four invited essays on Salvo's *Life* and *Works.*

In the second part, the essays deal with Salvo's main research subjects/topics.

The third part comprises studies-papers in the History of Physics and Mathematics.

The fourth part is composed of two standalone essays *In Memoriam* of Prof. Joseph Agassi (1927–2023) and Prof. Salatino (1949–2022) who after completing their papers, suddenly passed away during the writing of this book.

The fifth part consists of a collected list of Salvo D'Agostino's publications since 1964.

References and Index end the book.

The essays appear in alphabetical order.

The Book attends to different individual ideas, and scientific, historical and epistemological accounts. Therefore, the editor is not responsible for the entire content. Each of the eminent authors is responsible for his/her/their own opinions, which should be regarded as reflecting their personal scientific background and expertise. They are also individually responsible for the images, reprints, quotations, acknowledgments, English proofreading and all related permissions/approvals displayed/not displayed in their papers.

All chapters as well as the whole volume have been prior peer-reviewed and revised, English proofread, then they were also independently blind peer-reviewed following *Springer*'s policy.

May 2024

Raffaele Pisano
The Editor

Contents

Editor and Contributors

Editor Project Leader

Raffaele Pisano Full Professor, Head History of Physics and Applied Science & Technology Team (HOPAST), IEMN, Department of Physics, Lille University, Lille, France

Contributors

Joseph Agassi (1927–2023), Emeritus Professor, Tel Aviv University, Tel Aviv, Israel;
York University, Ontairo, Canada

Javier Anta [see Carl Hoefer] PhD Candidate, University of Seville, Seville, Spain

Cécile Barbachoux Researcher, INSPE, University of Côte d'Azur, Nice, France

Franco Blezza Full Professor, Chieti & Pescara Gabriele d'Annunzio University, Chieti, Italy

Mariano Bonaccorsi [see Giuseppe Ruta] Research Associate to Department of Structural and Geotechnical Engineering, University of Roma La Sapienza, Roma, Italy

Paolo Bussotti Associate Professor, Udine University, Udine, Italy

Mauro L. Condé Professor, Federal University of Minas Gerais, Belo Horizonte, Brazil

Guillermo Alberto Cuadrado Professor, National University of Cuyo, Mendoza, Argentina

Thomas Cuff [see Jean-Marc Ginoux] Student, Frederick Community College, The United States of America, Frederick, USA

Giovanna D'Agostino Salvo D'Agostino's Family Private Archive, Italy, Italy

Caterina D'Agostino BnF-Gallica France & Salvo D'Agostino's Family Private Archive, Italy, Italy

Nicola D'Agostino Salvo D'Agostino's Family Private Archive, Italy, Italy

Antonino Drago Associate Professor, University of Napoli Federico II, Naples, Italy

Andrea Durlo [see Raffaele Pisano] PhD Candidate, IEMN, Lille University, Lille, France

Giovanni Ferraro Associate Professor, University of Molise, Campobasso, Italy

Enrico R. A. C. Giannetto Professor, University of Bergamo, Bergamo, Italy

Jean-Marc Ginoux Associate Professor, University of Toulon, Nancy, France; Associate to Archives Henri Poincaré, Lorraine University, France

Carl Hoefer Professor, University of Barcelona, Barcelona, Spain

Olga Kosheleva Associate Professor, University of Texas at El Paso, El Paso, TX, USA

Joseph Kouneiher Associate Professor, University of Côte d'Azur, Nice, France

Vladik Kreinovich Full Professor, University of Texas at El Paso, El Paso, Texas, USA

Jean-Marc Lévy-Leblond Emeritus Professor, University of Côte d'Azur, Nice, France

Montgomery Link Associate Professor, Suffolk University, Boston, USA

Quintino Lopes Researcher, NOVA University of Lisbon, Lisbon, Portugal

Donatella Marmottini [see Raffaele Pisano] Teacher, Campus dei Licei, Cisterna di Latina, Italy

Peeter Müürsepp Associate Professor, Tallinn University of Technology, Tallinn, Estonia

Nicla Palladino Associate Professor, Università del Molise, Italy

Emilio Marco Pellegrino Research Associate to ISEPP, Portland State University, Portland, USA

Raffaele Pisano Full Professor, HOPAST at IEMN, Lille University, Villeneuve d'Ascq, France;
Associate to Archive Henri Poincaré, Lorraine University, France

Jan Potters Heinrich Hertz Fellow, University of Bonn, Bonn, Germany; Postdoctoral Fellow Flanders, University of Antwerp, Antwerp, Belgium

Arcangelo Rossi Professor, University of Salento, Italy

Giuseppe Ruta Associate Professor, University of Roma La Sapienza, Italy

Dante Roberto Salatino (1949–2022) Professor and Physician, National University of Cuyo, Mendoza, Argentina

John A. Schuster Emeritus Professor, Campion College, Sydney, Australia

Michael Segre Professor, University of Chieti Gabriele d'Annunzio, Chieti, Italy

Galina I. Sinkevich Associate Professor, Saint Petersburg State University of Architecture and Civil Engineering, St. Petersburg, Russian Federation

Pasquale Tucci Professor, University of Milano, Milan, Italy

Part I
Introduction and Four Invited Essays in Honour of Salvo D'Agostino

Chapter 1
Homage to Salvo D'Agostino (1921–2020)

Raffaele Pisano

Abstract As this entire book stands in homage to Salvo D'Agostino, the following section is designed as an initial introduction of Salvo to the reader. I have only covered some recent events and images of Salvo, leaving, instead the next three devoted chapters to present him in more depth, both the essays—published in this volume in his honour–and to tell of, in as much detail as possible, and in chronological sequence, his extremely long life—and— career as a University professor in Physics and History of physics. I call to the reader's attention that at the end of this volume is also found the long list of publications produced by Salvo.

Keywords Salvo D'Agostino · Salvo · Physics · Mathematics · History of physics · History of instruments · Epistemology of science · Roma 2006 · Roma 2012 · Pergamenus

1 An Overview

Salvo D'Agostino passed away in the night of the 1st January 2020 at the age of 99 years.

Salvo D'Agostino (hereafter Salvo) was professor in History and Epistemology of Physics and related Physics-topics at the department of physics, University of Roma La Sapienza, Italy. From the end of the 1960s, he was one of the field's main scholars throughout the second half of the last century, he astonishingly had a specific research—&—teaching program in physics and the history of physics. Salvo was a notable scholar as well as an appreciated teacher in Italy and abroad (Fulbright Scholar in U.S.A., "Accademia dei Lincei, & Royal Society", at Cambridge, UK, etc.). He introduced to the reader both the criticalities of physics and the unknown aspects of historical scientific thought in order to present the history of science and its foundations as an academic discipline of research; including the different

R. Pisano (✉)
HOPAST at IEMN, Lille University, Villeneuve d'Ascq, Lille, France
e-mail: raffaele.pisano@univ-lille.fr

R. Pisano (ed.), *A History of Physics: Phenomena, Ideas and Mechanisms*,
History of Mechanism and Machine Science 42,
https://doi.org/10.1007/978-3-031-26174-9_1

epistemological interpretations and conflicts in context of science into the history. Research on the original documents, in as much detail as possible, was the priority of research for his method of inquiring: from mechanics to electromagnetism, relativity, quantum mechanics to the history of scientific instruments, comprising the teaching of physics, as well.

Salvo was one of the founders and active members of the *Gruppo Italiano degli Storici della Fisica* (*Italian Group of the Historians of Physics*) *at the CNR*, nowadays evolved into *Società Italiana degli Storici della Fisica e dell'Astronomia* (*Italian Society of the Historians of Physics and Astronomy*). He wrote 237 publications (or maybe more, this is the number I located), between articles and books, listed at the end of this volume. The first one (known) was an article published in 1964, entitled *Un insegnamento non dogmatico della Scienza* (*A non-dogmatic teaching of Science* (D'Agostino 1964)). I also note his early series of publications in notable journals such as *Giornale di Fisica, Scientia* and *Physis* et al. (D'Agostino 1964, 1965a, 1965b; D'Agostino 1966, 1967a, 1967b, 1968a, 1968b, 1968c, 1971, 1973; D'Agostino and Cametti 1970).

From the beginning, Salvo's interests both in physics and in history of physics were always intellectually complemented by deep epistemological debates, such as those, in 1992, on Koyré (Fig. 1) (Cfr. D'Agostino 1995).

Fig. 1 Salvo as speaker at the *La Concezione del Mondo in Europa. Religione, Scienza e Modernià attraverso L'opera di Alexandre Koyré 1892–1964* (Acquasparta, Terni, 1992, 30 September–3 October, Italy (Cfr. D'Agostino 1995)) *Source* Courtesy D'Agostino Family Private Archive

I recall his famous book published in 2000 (Fig. 2), *A history of the Ideas of Theoretical Physics: Essays on the Nineteenth and Twentieth Century Physics*

(D'Agostino 2000) for the prestigious *Boston Studies in History and Philosophy of Science* Kluwer-Springer Book series. Salvo's last publication is an article, *Schrödinger 1950: The Cat's Situation on the Background of the New Statistics and Second Quantization* (D'Agostino 2015). He submitted it in the end of 2019 and it has now been published posthumously (D'Agostino 2022).[1] It develops a prior article on the same topics (*A Good Chance for the Cat's Life: Erwin Schrödinger's New Statistics and a Second Quantization Theory of Quantum*) published in 2015 (D'Agostino 2015).

Fig. 2 Salvo D'Agostino in Amsterdam (journey with her daughters Caterina and Giovanna, as well, The Netherlands, October 1999) and his famous book *A History of the Ideas of Theoretical Physics* (D'Agostino 2000) *Sources* Courtesy D'Agostino Family Private Archive

I remark that, taking into account the historical period at that time, Salvo's early papers (1960s) established one of the first important Italian attempts to elevate and spread the *History of Physics* as a university discipline of research and within the intellectual and political academic environment.

I would also like to recall a charming small book of his, *Letture di Storia della Fisica* (*Lectures in History of Physics*, 2018 (D'Agostino 2018)). This book consists of seventeen academic lectures in the history of physics. They describe his historical research (from Galileo's *Two New Sciences* to Einstein's *Relativity*) later developed in teaching programs at the department of physics, University of Roma La Sapienza. They represent an outstanding collection of lectures from his long life's work, as well. In *Letture di Storia della Fisica* Salvo explained to the reader (Premise) the incredible genesis of his book. He wrote that he had encountered one of his former students from the 80 s in Frascati (near Roma, Italy) who had a detailed transcription of several D'Agostino's lectures from La Sapienza, Rome. The former student then sent these lectures (in hard copy only) to his old and dear professor. I had the privilege to be invited by *Metascience* (Springer) to review his book (Pisano 2021).

[1] At the request of the publisher and in collaboration with Arcangelo Rossi, we provided editing and proofreading of this publication.

2 Spotlighting Facts and Events

As announced above, I will be underlining some facts and events concerning Salvo's scientific activities.

Since the 2000s, and my entry into the *Italian Society of the Historians of Physics and Astronomy* (hereafter SISFA), I have had a very friendly and professional relationship with Salvo. We often spoke by telephone and he always sent me his works in the post.

I co-organized two SISFA Congresses both in Roma (2006 and 2012 and their related proceedings (Giannetto et al. 2009; Pisano et al. 2013b)). They were the first SISFA congresses to have an international environment, i.e., both for the organizing and scientific committees, financial support and for the publication of the proceedings. That to say, that scholars from all around the world appreciated – *in person* – Salvo's talks, both as *Keynote speaker* (2006 and 2012) and also as *Special speaker* (2012).

In 2006 he presented *Albert Einstein: science as self-criticism* (Giannetto et al. 2009) (Fig. 3):

Fig. 3 Salvo D'Agostino as Speaker at the XXVI SISFA Congress (University of Roma La Sapienza, 2006, 15–17 June, Italy (Giannetto et al. 2009)) *Source* Author (RP) Private Archive

The other friend and keynote speaker was Luigi Cerruti (1941–2021) who dealt with *The Count Amedeo Avogadro from Quaregna and Cerretto, Scientist and Citizen* (Fig. 5). I remember that both presentations greatly enriched the historical and epistemological debate during the SISFA event.

In 2007, due to their interest in Science and Religion, as well, I invited Salvo and Luigi to an international colloquium that I co-organized with Rafael Pascual (Fig. 4) in Latina (near Roma, Italy). At the event, several scholars from various scientific disciplines of research and traditions discussed the role and forms of language in relation to sciences–religions. The colloquium also became a proceedings book (Pascual and Pisano 2008) (Figs. 4 and 5) with significant contributors in the history of sciences, philosophy of science, science in society, all around the interplay between sciences–religions; obviously also science-faith and/or science-theology, evolutionism and other topics by means of historical, mathematical and logical, dimensions were addressed. Each contribution was freely expressed, according to the researchers' personal and professional convictions without interposing limits or obstacles to the dialogue and the general understanding.

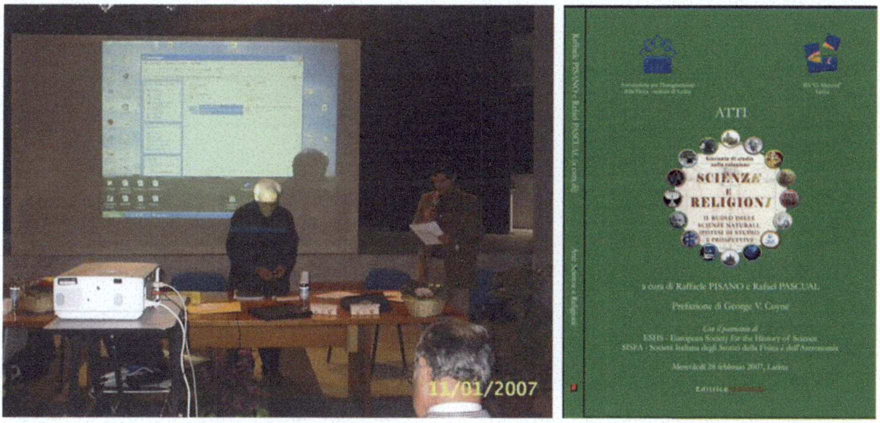

Fig. 4 Salvo D'Agostino as speaker at the *Sciences and Religions Colloquium* (IIS G. Marconi, 2007, 28 February, Latina, Italy) and the book of the Proceedings (Pascual and Pisano 2008) *Source* Author (RP) Private Archive [N.B.: the camera-date of photo was not updated]

INDICE

Fig. 5 The Content of the Proceedings of *Sciences and Religions Colloquium* (Pascual and Pisano 2008, preliminary matters) *Source* Author (RP) Private Archive

Salvo's talk was entitled *Religions and Sciences: Harmonies and Discordant Notes*, which became *Science and Religion: for a new alliance* (Fig. 5) in the proceedings.

In his abstract we read:

> By following the main theses of Juergen Habermass and Georg Henrik von Wright post-modern philosophies, I support the point of view of a collaborative approach between scientific and religious world views, because I believe that this approach and the ensuing dialogue between scientists and religious men and women is to-day necessary to meet the difficulties of the western world. I note that a collaborative approach is denied by religious fundamentalists on one side, and, on the other, by those scientists who believe in the absolute truth of science. After a short summary of their historical-philosophical critical analysis on the present development of science and technology, I conclude by quoting what I consider very important pronouncement of the founders of modern physics on the role which they assign to the future collaboration between physics and epistemology. I argue that the growing interests in epistemological studies in post-modern European cultures is to be considered as a welcome indication of a novel world-view. (Pascual and Pisano 2008, p. 107)

In 2009, as usual for SISFA members, we met again at the SISFA Congress held in Firenze. In the following, I would like to offer some happy and collegial images some from the event (Figs. 6, 7, 8 and 9).

Fig. 6 Salvo D'Agostino at the XXIX SISFA Congress (Galileo's Museum and Galileo's House (in Arcetri), Firenze, 2009, 14–18 September (Giannetto et al. 2014)) *Source* Author (RP) Private Archive

Fig. 7 Salvo D'Agostino who takes a photo in a photo at the XXIX SISFA Congress (Galileo's Museum and Galileo's House (in Arcetri), Firenze, 2009, 14–18 September (Giannetto et al. 2014)) *Source* Author (RP) Private Archive

Fig. 8 Salvo D'Agostino at the XXIX SISFA Congress (Galileo's Museum and Galileo's House (in Arcetri), Firenze, 2009, 14–18 September (Giannetto et al. 2014)) *Source* Author (RP) Private Archive

Fig. 9 Salvo D'Agostino at the XXIX SISFA Congress (Galileo's Museum and Galileo's House (in Arcetri), Firenze, 2009, 14–18 September, Italy (Giannetto et al. 2014)) with Stefano Bordoni (left) and Raffaele Pisano (right). The photo has been taken at Galileo's house *Source* Author (RP) Private Archive

In 2012, I co-organized another SISFA congress. The international environment were even more rich, and Salvo was once again invited to speak. His talk was entitled *Maxwell's dynamical approach to electromagnetism and kinetic theory*. Among the other invited scholars, I underline Walter Noll (USA), Frank A. J. L. James (UK), Jean Dhombres (France) and Marco Cecarelli (Italy). Some months before, on July, 4th, the ATLAS experiment presented a preview of its updated results on the search for the Higgs Boson (Higgs 1964, 1966). The results were shown at a seminar held jointly at CERN and via video link at the *International Conference for High Energy Physics* (ICHEP) in Melbourne, Australia (ATLAS Collaboration 2012; CMS Collaboration 2012). At CERN, preliminary results were presented to scientists on site and via webcast to colleagues located in hundred of institutions around the world. Aleandro Nisati (I.N.F.N.–CERN, Italy–Switzerland) was of the main founding physicists of one of the two main experiments at the *Large Hadron Collider* (LHC), *A Toroidal*

LHC ApparatuS (ATLAS) where he was *Physics Experimental Coordinator*: scientific program and the project on muon detection and spectrometer trigger system. Therefore, we took the opportunity to invite[2] Nisati to the SISFA 2012 congress in Roma (27–29 September) and Salvo, due to his expertise in modern physics, as well, was designated chairperson for his talk (Figs. 10 and 11).

Fig. 10 Salvo D'Agostino at the XXXII SISFA Congress (University of Roma La Sapienza, 2012, 27–29 September, Italy (Pisano et al. 2013b)) *Source* Author (RP) Private Archive

[2] I organized another meeting with Aleadro Nisati. It was an interview-workshop to (2012, December 12th) at the SEMM-Service Enseignement et Multimédia, UFR Physique, University of Lille 1, France. He and his colleague, Steven Goldfarb (CERN-University of Michigan, USA) kindly answered questions on the ATLAS detector, LHC machine and CERN-ATLAS laboratories proposed by the participants and scientific journalists (Pisano 2013a).

Fig. 11 Salvo D'Agostino designated Chairperson for Aleandro Nisati's (CERN) talk at the XXXII SISFA Congress (University of Roma La Sapienza, 2012, 27–29 September, Italy (Pisano 2013b)) *Source* Author (RP) Private Archive

As above cited, at the XXXII SISFA Congress in Roma, Salvo also was celebrated in *D'Agostino Tribute Session Special Event* with the following scheduled:

2012, 28 September

Tribute to: Salvo D'Agostino

Moderator–organising: Raffaele Pisano

Chairperson: Arcangelo Rossi

Communication: *Salvo D'Agostino Physics and History, Tradition and Innovation* by Stefano Bordoni

Special Speaker: Salvo D'Agostino

Free debate

The event ended with a gift presentation and a memorial plaque (Fig. 12):

In Honour of Your Achievements in Attendance, Deportment and the Impact of Your Distinguished History and Epistemology of Physics Studies in all the World

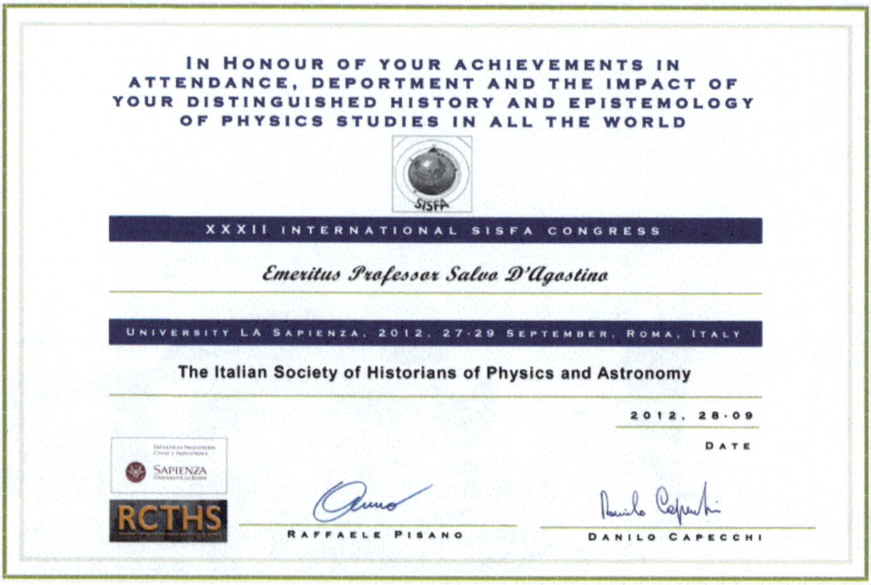

Fig. 12 Salvo D'Agostino *Pergamenus* at the XXXII SISFA Congress (University of Roma La Sapienza, 2012, 27–29 September, Italy (Pisano 2013b)) *Source* Author (RP) Private Archive

3 Concluding Remarks

Physics deals with observation & data, experimentation & measuring, modelling and theorizing. It is also a science understood through its history, such as the integrant part of its scientific foundations. *How does one speak about this fascinating integration*? Moreover, *how can one teach its results, historical foundations and epistemological interpretations?* Consequently, conflicts, anomalies and revolution arose building both crises and developments. These crises drove scientists to follow different research programs, producing intellectual, historical and epistemological reflections useful for university interdisciplinary lecturing.

This is what Salvo did for more than 50 years.

Salvo's publications and teachings aimed to arouse interest in students and scholars of humanistic culture, as well, in the scientific endeavours of the past. They represent a significant contribution to the history of physics literature.

Salvo's long life (Figs. 13, 14 and 15) and career—his last manuscript was completed at the age of 97 years—are also an expression of dignity, passion and commitment to the academic community, particularly to the Italian historians of physics. The *story* tells us that the international scientific community will always appreciate and benefit from the achievements of Salvo D'Agostino's research!

Fig. 13 Salvo D'Agostino in Bolsena (Viterbo, Latium, Italy), August 2012 *Source* Courtesy D'Agostino Family Private Archive

Fig. 14 Salvo D'Agostino in Bolsena (Viterbo, Latium, Italy), August 2014 *Source* Courtesy D'Agostino Family Private Archive

Fig. 15 Salvo D'Agostino in Piancastagnaio (Amiata Mountain Area, Siena, Tuscany, Italy), July 2012 *Source* Courtesy D'Agostino Family Private Archive

Acknowledgements I thank the anonymous reviewers for their careful reading of our manuscript and their many insightful comments and suggestions. I also express gratitude to both Dr. Julie Robarts (University of Melbourne, Australia) for her observations on English language (eventual remaining English mistakes are mine own) and referees.

References

ATLAS Collaboration (2012) Observation of a new particle in the search for the standard model Higgs Boson with ATLAS detector at the LHC. Physics Letter B 716/1:1–29. http://www.sciencedirect.com/science/article/pii/S037026931200857X

CMS Collaboration (2012) Observation of a new boson at a mass of 125 GeV with the CMS experiment at the LHC. Physics Letter B 716:30–61. http://www.sciencedirect.com/science/article/pii/S037026931200858

D'Agostino S (1964) Un insegnamento non dogmatico della Scienza. Scuola e Città:514–515.

D'Agostino S (1965a) Il contributo di Newton allo sviluppo dell'ottica. Giornale di Fisica VI/3:188–196.

D'Agostino S (1965b) Traduzione di Zadou-Naisky Georges. L'insegnamento delle scienze fisico-matematiche. La Nuova Italia, Firenze.

D'Agostino S (1966) Alcune considerazioni sul calcolo dei campi magnetici nei regimi quasi stazionari. Giornale di Fisica 7/3:179–184.

D'Agostino S (1967a) Maxwell e la scoperta della Teoria Elettromagnetica della luce. Archimede 4/5:276–282.

D'Agostino S (1967b) Alcune osservazioni su teorie elastiche e prime teorie del campo elettromagnetico nella prima metà dell'Ottocento. Scientia 102:166–171.
D'Agostino S (1968a) Il pensiero scientifico di Maxwell e lo sviluppo della teoria del campo elettromagnetico nella memoria "On Faraday's lines of force". Scientia 103:291–301.
D'Agostino S (1968b) I vortici dell'etere nella teoria del campo elettromagnetico di Maxwell: la funzione del modello nella costruzione della teoria. Physis X/3:188–202.
D'Agostino S (1968c) Per la preparazione all'insegnamento della fisica nella Scuola Media. Le Scienze 1:33–36.
D'Agostino S (1971) Hertz e Helmholtz sulle onde elettromagnetiche. Scientia 106:622–636.
D'Agostino S (1973) L'elettrodinamica di Lorentz sino alla soglia della teoria speciale della relatività di Einstein. Physis XV:260–279.
D'Agostino S (1995) Alcune considerazioni critiche sulla storiografia dell'universo della precisione di Koyré. In: Vinti C (ed). Alexandre Koyré. L'avventura intellettuale. Edizioni Scientifiche Italiane, Napoli, pp. 617–634.
D'Agostino S (2000) A history of the Ideas of Theoretical Physics: Essays on the Nineteenth and Twentieth Century Physics. Kluwer Academic Publishers, Dordrecht.
D'Agostino S (2015) A Good Chance for the Cat's Life: Erwin Schroedinger's New Statistics and a Second Quantization Theory of Quantum. Advances in Historical Studies 4/1:1–7.
D'Agostino S (2018) Letture di Storia della fisica. Da Galileo a Newton: Edizioni Accademiche Italiane, Beau Bassin–Berlin.
D'Agostino S ([submitted end 2019, published posthumous]) (2022) Schrodinger 1950: The Cat's Situation on the Background of the New Statistics and Second Quantization. Advances in Historical Studies 11/2:81–85.
D'Agostino S, Cametti C (1970) Esperienze didattiche di diffrazione con microonde. Giornale di Fisica 11:36–48.
Giannetto E, Giannini G, Capecchi D, Pisano R (2009) (eds) Da Archimede a Majorana: La fisica nel suo divenire. Proceedings of XXVI SISFA Congress (Roma, 15–17 June, 2006). Guaraldi Editore, Rimini.
Giannetto E, Ricciardo S, Antonello E, Mazzoni M (2014) (eds) Cielo e Terra: Fisica e Astronomia, un antico legame. Proceedings of the XXIX SISFA Congress, (Firenze, 14–18 September, 2009). Rimini, Guaraldi.
Higgs PW (1964) Broken symmetries and the masses of Gauge Bosons. Physics Review Letter 13:508–509.
Higgs PW (1966) Spontaneous symmetry breakdown without Mass less Bosons. Physics Review Letter 145:1156–1163.
Pascual R, Pisano R (2008) (eds) Scienze e Religioni. Il ruolo delle scienze naturali, ipotesi di studio e prospettive. Uniservice, Trento.
Pisano R (2013a) On the New Boson Higgs's Studies at the CERN-ATLAS Experiment. The Emergency of a Historical Discovery. AHS 2/1:3–5.
Pisano R (2013b) Introduction to the International 32nd Congress SISFA. In: Pisano R, Capecchi D, Lukešová A (eds). Physics A, Engineering. Critical Problems in the History of Science and Society. International 32nd Congress for The SISFA–Italian Society of Historians of Physics and Astronomy (Roma, 27–29 September). The Scientia Socialis UAB & Scientific Methodical Centre Scientia Educologica Press, Šiauliai University, Lithuania, pp xi–xxi.
Pisano R (2021) Book Review. Seventeen lectures in the history of physics. Metascience 30(2):207–210.

Chapter 2
Phenomena, Ideas & Mechanisms: Introducing the Essays in the History of Physics and Mathematics

Raffaele Pisano and Paolo Bussotti

Abstract Based on the previous celebratory chapter on Salvo D'Agostino (1921–2020), this section aims to introduce the research essays written in his honour. In this chapter the reader will find summaries of the papers in this book (accordingly with its five parts and in alphabetical order) which have been considered as an homage to Salvo D'Agostino. We note that in the end of the book the reader will also find the long list of publications by Salvo D'Agostino.

Keywords Epistemology of science · Essays · History of physics · History of instruments · Honour · Mathematics · Physics · Salvo D'Agostino · Salvo

1 Introducing Selected Papers

A History of Physics: Phenomena, Ideas & Mechanisms—Essays in Honour of Salvo D'Agostino is an editorial research project devoted to celebrating Salvo D'Agostino (1921–2020; hereafter Salvo), one of the main scholars in the field in the second half of the past century.

A phenomenon is an observable measurable fact, including data modelling, assumptions/laws. After a complex process of measuring and recording data, early results will be included in a modelling process in order to state assumptions and laws. It may be related to matter, motion, time, energy, etc. Generally speaking, a mechanical phenomenon is associated with equilibrium/motion, as well. Then, the conceptualisation of a mechanism is also devoted to understanding of causations of disequilibrium itself. But *what about in thermodynamics, electromagnetism and modern*

R. Pisano (✉)
HOPAST at IEMN, Lille University, Villeneuve d'Ascq, Lille, France
e-mail: raffaele.pisano@univ-lille.fr

P. Bussotti
DIUM, University of Udine, Udine, Italy
e-mail: paolo.bussotti@uniud.it

R. Pisano (ed.), *A History of Physics: Phenomena, Ideas and Mechanisms*,
History of Mechanism and Machine Science 42,
https://doi.org/10.1007/978-3-031-26174-9_2

physics with respect to the ideas within cultures and lives? Are all mechanisms really mechanisms of some phenomenon? For example, in the past, mechanics—and—mechanisms basically assumed planetary and terrestrial motions. Then, it aimed to understand causation of disequilibrium (shock, gravitational, attraction/repulsion, etc.) up to the apparent incoherence/problems of integrating of *Quantum Mechanics* and theories of *Relativity*.

This book analyses mechanisms, ideas, quantities and phenomena as a complex human product both of regular changes/interaction of a number of parts/components etc., and in the framework of irregular cases, e.g., when a classical mechanism cannot explain a physical phenomenon in modern physics. The implications and limits of this non-homogeneous framework for thinking about science and scientific practice are explored. Therefore, the essays in honour of Salvo present new insights in the *History of Physics*—and—*Mathematics*. They also deal with changes in human ideas over time within the history of human knowledge/thought and thinkers in certain cultures, lives and historical contexts. They have been written by leading scholars in the *arena*, ranging across the entire field and related in most instances to the works of Salvo D'Agostino.

The book is composed of the following:

- Part 1 consists in chapters devoted to celebrating and introducing Salvo's life and works to the reader.
- Part 2 presents historical research-essays on Salvo's research-topics, including history of ideas, historiography and epistemology of science.
- Part 3 is composed of papers on important subjects pertaining to physics, mathematics and geometry applied to different aspects of science, accurately presented.
- Part 4 is composed of two standalone chapters *In Memoriam* of Prof. Joseph Agassi (1927–2023) and Prof. Salatino (1949–2022), dear—colleagues, authors of two chapters in this book, who passed away few months ago. They finished their papers.
- Part 5 presents a collected list of Salvo D'Agostino's publications since 1964.
- The references section (concerning the essays) and an Index conclude the book.

2 Part 1: Salvo D'Agostino and Invited Essays in His Honour

Raffaele Pisano (France) in *Homage to Salvo D'Agostino (1921–2020)* offers a touching reminder of Salvo D'Agostino. Several episodes of Salvo's personal and scientific biography are reported. A panorama of his long and profound activity as a historian of science is referred to. Pisano enriches his homage with his own memories concerning his discussions with D'Agostino, the co-organization of scientific conferences, the cooperation in scientific research. The text also presents some photos as a tribute to Salvo.

Caterina D'Agostino (France) Giovanna D'Agostino (Italy) and Nicola D'Agostino (Italy) in *Three Sicilian Tales by Salvo D'Agostino* offer a tribute to their father. Salvo was both a great historian of science and man with an artistic temperament. Among his unpublished manuscript–sheets preserved among the family papers, there are also some short stories. Three of them are presented here in English translation and in the original Italian version. The first one is a story entitled "The Christmastime Fig Tree". The second one is a family-story concerning Salvo's youth. It is dedicated to the *Saie*, which are canals used to water the lemon trees. They characterize the Sicilian countryside. Finally, the last story concerns mount Etna.

Enrico R. A. C. Giannetto (Italy) in *Salvo D'Agostino and the History of Physics: Relativity and Quantum Physics* the author describes a broad panorama of the historiography of physics in Italy. Afterwards he highlights the role of D'Agostino within such a historiography. Even more in detail, Giannetto focuses on Salvo's contributions to the history and epistemology of relativity and quantum mechanics. He uses published works and personal dialogues with D'Agostino, which offer an original picture of the role of geometry within relativity and of Einstein's conception of geometry. With regard to quantum mechanics, the author explains some aspects of the relations Bohr-Einstein and expounds the difference conception of indeterminacy developed by Bohr and Heisenberg as conceived by D'Agostino.

Arcangelo Rossi (Italy) in *Salvo D'Agostino: Life and Work, from Physics to History of Physics* offers a precious picture of Salvo's life and work. In the first section, the scientific biography of D'Agostino is proposed: his early studies, his scientific interest, his career as a teacher and as a professor. The milieu of the faculty of Mathematical, Physical and Natural Sciences of Roma *La Sapienza* between the 1950s and 1970s, where D'Agostino worked, is presented. In the second section the very broad production of Salvo as an historian of physics is explained. No words can express D'Agostino's scientific attitude better than Rossi's: "His historiographical approach was based on a direct and in-depth comparison with the classics of physics. In fact, while not ignoring and indeed paying close attention to the great debates of the 60's and 70's on the epistemology and historiography of science, D'Agostino preferred an independent comparison with the classics of the history of physics".

Pasquale Tucci (Italy) in *History of Scientific Instrumentation and History of Science* analyses the contributions made by D'Agostino to the history of physics in a particular perspective: Salvo was convinced that the role of instruments within the development of theoretical physics is usually underestimated. The author refers to historians and philosophers of science such as Hanson and Amsterdamski, who criticized as too strong the distinction between the history of theoretical science and of scientific instruments. After that, Tucci analyses Salvo's conception: he explains the original elements in D'Agostino's train of thought. In particular, his analysis of Hertz's approach to physics is examined focusing on the central role of the instruments for Hertz's theoretical conceptions.

3 Part 2: Essays on Salvo D'Agostino's Research Topics

Joseph Agassi (Israel and Canada) in *Versions of Determinism* expounds Popper's view of determinism and indeterminism. He begins analysing Laplace's conception and the early criticism it attracted, such as those of Poincaré. After that, he explains Popper's variant of Laplace's thesis showing that determinism cannot be scientific. Finally, the relations between science and metaphysics within Popper's theory are analysed. In the Conclusion the author clarifies, summarizing his preceding arguments, that Popper's fault was his conviction to have proved that every kind of determinism is refuted as unscientific.

Mauro L. Condé (Brazil) in *Ideas, Phenomena and Mechanisms: The Role of Modern Physics in the Historical Epistemology of Fleck and Canguilhem* distinguishes between traditional epistemology, which is basically grounded on physics, and historical epistemology which also includes biology, life sciences and medicine. Fleck and Canguilhem are two scholars who have made fundamental contributions to historical epistemology. The step from traditional epistemology to the historical one was favoured by changes that occurred in physics, in particular by the foundation of relativity theory and quantum mechanics. In this paper Condé analyses the concept of phenomenon and mechanism in physics and in the other sciences up to and including biology. He then focuses on the epistemological contributions of Fleck and Canguilhem. Later, the author identifies the tradition of thought to which Fleck and Canguilhem tried to affiliate. Finally, an important panoramic section on historical epistemology is proposed.

Antonino Drago (Italy) in *When was Electromagnetic Theory Completed? An Alternative Formulation to that of Maxwell* deals with a discussion on the traditional idea concerning the completion of electromagnetic theory by Maxwell which—according to the author—can only be considered complete when the symmetric explanation of electromagnetic induction between moving bodies is obtained. Drago inserts the particular case of electromagnetism into his more general view: there are two dichotomies, one concerning mathematics and one logic. Newtonian physics is characterized by one kind of mathematics and logic. Formulations of physics alternative to Newton's are based on the opposite kind of mathematics and logic. Maxwell's electromagnetism enters Newton's tradition. Whereas a theory alternative to Maxwell's, described by Drago, is based upon the opposite couple mathematics/logic. The author describes this net of relations and dichotomies between physics and mathematics/logic.

Olga Kosheleva (USA) and Vladik Kreinovich (USA) in *Can Physics Attain Its Goals: Extending D'Agostino's Analysis to the 21st Century and Beyond* starts from the analysis of D'Agostino's book *The History of Ideas of Theoretical Physics: Essays on the Nineteenth and Twentieth Century Physics* (2000), where he offers a profound analysis of the physics developed in those two centuries. The authors wonder whether Salvo's analysis can be extended to the physics of the twenty-firstcentury. First of all, they examine what kinds of prediction could enter science and how difficult it was to recognize them. This process lasted from the beginning of the seventeenth century to the Laplace epoch. The consequences for the structure of physics due to Laplace's determinism are analysed. After that a weaker kind of determinism, the statistical one, is analysed as an important element of twentieth century physics. Finally, some considerations of the future of physics are proposed.

Jean-Marc Lévy-Leblond (France) in *Einsteinless Pathways into Spacetime. A Gedankenexperiment in Science History* imagines that Einstein never existed. What would have been our conception of inertia and, in particular, of the relation between mass and energy? What about principles of conservation? What about the conceptions of space and time, and their terminology? By analysing the ideas developed between the end of the nineteenth century and the first decade of the 20th, the author constructs a possible scenario of physics without Einstein, an alternative scenario to the effective development of special relativity. He suggests that the construction of a counterfactual history may have: (1) Educational aims because alternative approaches are formative; (2) Epistemological goals: how would science and its language have been? (3) Cultural purposes: what about the history of culture without the myth of Einstein?

Montgomery Link (USA) in *Rotating Universes* deals with many fundamental concepts of modern physics, particularly of general relativity. Link highlights that our universe is not rotating, but that an inquiry on the features of a rotating universe can be useful to understand some characteristics of our universe. In particular, in a rotating universe closed time-like curves exist, which are coherent with general relativity, but which destroy our usual causal structure. Thus, the author speaks of the Gödel's solution of Einstein's equation which gives rise to a stationary rotating universe. After the analysis of Gödel's universe, Link examines the relation between time and matter if closed like-time curves exist or if they do not. This leads to a discussion on the notion of physical time.

Raffaele Pisano (France), Donatella Marmottini (Italy) and Andrea Durlo (France) in *Lagrange's Method and Lagrangian's Mechanism in Maxwell's A Treatise on Electricity and Magnetism (1873)* clarify how/when Maxwells' electromagnetism came to be based on the concepts of electric charge and energy rather than the Newtonian mass and force. Thus, the magnetic field appeared as a kinetic energy system whose features might be explained without resorting directly to forces. The Lagrangian formalism was a fundamental step which allowed Maxwell to develop his program alternatively to the Newtonian one. This implies a new approach to physics which the authors name Physics-Mathematics; no similar historical analysis exists so far. After that, they examine the physical meaning of Faraday's work and Maxwell's interpretation, through which they arrive at discussing the most important conceptual, historical and epistemological aspects of Maxwell's *Treatise*. The next step is the presentation of Lagrange's variational principles (the concept of Lagrangian). Finally, Maxwell's applications of Lagrange's formalism are fully explained. A comparison between Faraday's and Maxwell's conceptions closes this chapter essay.

Raffaele Pisano (France) and Emilio Marco Pellegrino (USA) in *Clausius's Laws, Boltzmann's H-Theorem & Entropy: Mathematical Frameworks into a Kinetic Mechanism (1850–1901)* analyse the relation between Boltzmann's H-theorem, the concept of entropy and Clausius' second law. The paper begins with an outline concerning the concept of entropy and the laws of thermodynamics. After that, a picture of the authors who founded statistical mechanics is presented. Later on, the kinetic model of gas as well as the H-theorem are introduced. Pisano and Pellegrino focus on the discoveries of Clausius and Boltzmann. A profound analysis of the Clausius' laws is then proposed. The results obtained by Boltzmann as well as his methods are afterwards explained in comparison to Clausius'. Finally, a discussion of the different aspects regarding the concept of entropy is proposed, correlating this notion with the H-theorem.

Jan Potters (Germany and Belgium) in *Going Beyond the World of Atoms: Heinrich Hertz and the Electromagnetic Worldview* analyses the influence exerted by Hertz's *Principles of Mechanics* on some aspects of the following fundamental physics. The author claims that this influence has been underestimated. After some introductory considerations Potters expounds the basic concepts of Hertz's mechanics. In what follows, the author analyses Lorentz's conception of electromagnetism and shows the way in which Hertz's conception influenced Wilhelm Wien's and Max Abraham's formulation of the electromagnetic worldview. All these conceptions were strictly tied to the notion of ether and to the different features the physicists attributed to this mysterious intermediary of electromagnetics actions. Within this context Poincaré, Abraham and Lorentz played a remarkable role, up to arriving at a reformulation of the notion of inertia. Hertz's *Principles* inspired Abraham's electromagnetic worldview in several ways explained by Potters. Einstein's and Planck's contributions to this subject are analysed. The discussion of the different viewpoints is finally proposed.

Dante Roberto Salatino (Argentina) in *Maxwell's Electromagnetism: A Matter of Symmetry* aims at showing that the analogies used by Maxwell reveal the symmetries behind electromagnetism. The author resorts to *transcursive logic*. This means looking at the events from the observer's point of view, which allows valuing analogies as a source of knowledge. This approach is consistent with Maxwell's who pointed out that a single phenomenon can be subject to various explanations. To expound his own epistemology, Salatino clarifies that the different analogies used to explain a phenomenon create a whole explanatory pattern with the phenomenon itself. The essential features of a phenomenon can be regarded as a group of transformations. Having clarified the epistemological and mathematical bases of his interpretation based on *transcursive logic*, the author applies this apparatus to Maxwell's electromagnetism.

4 Part 3: Essays in History of Physics—and—Mathematics

Javier Anta (Spain) and Carl Hoefer (Spain) in *Rethinking the Intellectual Genesis of Landauer's Principle* analyse the relation of the concept of irreversibility in thermodynamics and in the theory of computing machines. In this context they examine Landauer's principle formulated in 1961: the physical implementation of a logically irreversible operation implies an entropy increase in the system which computes the operation. The authors trace a story of the Principle and its reception. The role of John Swanson is explained as well as of Leon Brillouin, the advisor of Landauer in Harvard, who made fundamental contributions to the physical meaning of information. The concept of *negentropy* is explained. The thermal effect of information is then expounded. The role of Von Neumann in the story of Landauer's paradox is also clarified. Finally, the Principle is explained in all it nuances.

Cécile Barbachoux (France) and Joseph Kouneiher (France) in *Foundations of the Operators Approach in Mathematics and Physics* analyse the concept of operator, which has been fundamental in mathematics and physics. In this paper the authors deal with the origin of this notion. Descartes' algebrization of geometry is the first important example concerning the extensive use of operators within a classical branch of mathematics. After that, the authors focus on infinitesimal calculus, the *locus classicus* of operators. They begin by analysing Leibniz's basic concepts which represent the first stage of the history of operators in mathematical analysis. The second stage is their use in the theory of differential equations. Boole was important. With the analytical theory of heat, operators entered physics. The next step concerns the history of matrices up until arriving at spectral theory. The role of operators in integral equations is then examined. With the fifth section, the second part of this paper begins with an explanation of the use of operators in physics, focusing in particular on quantum mechanics.

Franco Blezza (Italy) in *Physics as a Fundamental Medical Humanity* stresses the importance of a training in physics for future surgeons: this is useful not only because several parts of physics are connected to physiology, but also because of the rigour imposed by a training in physics. Therefore, this discipline should belong to the human culture of a surgeon. After some considerations on the philosophy of science, the author argues that thermodynamics is a branch of physics which is particularly useful for surgeons. After that, the connection between quantum mechanics and virology is shown. The distinction between the systemic approach, which is a synthetic one, in contraposition to the analytical is highlighted. The author explains the advantages of a synthetic approach to medicine and, generally, to science.

Giovanni Ferraro (Italy) in *Between Natural Philosophy and Mathematics: Euler's Investigation of Planetary Perturbations in the 1740s* explains that Euler's inquiries of planetary perturbations are preceded by three hypotheses on their probable origin, one of which rests on the possibility that the inverse square law holds only approximately. Euler's doubts are derived in the argument on action at a distance and some empirical evidence. The author explains Euler's ideas on these subjects. After that, he broadens the picture to include D'Alembert's, Clairaut's and Buffon's opinions. Later, Ferraro deals with Euler's view of the second law of dynamics.

Mariano Bonaccorsi (Italy) and Giuseppe Ruta (Italy) in *The Creator only may destroy vis viva: Natural Philosophy in Early William Thomson* connect Lord Kelvin's spiritualistic *Weltanschauung* with his physical conviction on the conservation of *vis viva.* The authors narrate that Kelvin worked in Paris and encountered Carnot's and Clapeyron's research on thermal phenomena. They introduced the caloric fluid, which, falling through a difference of temperature, produces mechanical work. This means that caloric contains *vis viva*, which must be conserved, otherwise a *perpetuum mobile* would be constructible. But this was absurd because, according to Kelvin, only God can create *vis viva*, not natural phenomena. The authors trace a picture of English science in the Victorian Age, the scientific biography of Kelvin, the way he reached his philosophical convictions, his attempt to prove God's existence as well as his interest towards Carnot's-Clapeyron's theory. The core of the article concerns Kelvin's account of such a theory and his new heat theory.

Paolo Bussotti (Italy) and Raffaele Pisano (France) in *Details in Galileo's Mechanism of Inertia as New Intellectual Historical Insights* the authors offer new insights on the complex theme concerning Galileo's concept of inertia. In the first part of the paper, they present a panorama of the physical concepts having a connection with inertia from ancient times to the nineteenth century. After that, they focus on Galileo. Bussotti and Pisano analyse the nuances behind Galileo's views of uniform circular motion up to clarifying in which circumstances he regarded such a motion to be inertial or not. The same analysis is developed for uniform rectilinear motion. The authors clarify that Galileo had a clear view of uniform rectilinear motion, whereas his conception of circular motion was anything but clear. Finally, a broad discussion with the secondary literature is developed where the authors show that no previous interpretation of Galileo's concept of inertia is completely satisfactory.

Thomas Cuff (USA) and Jean-Marc Ginoux (France) in *Deciphering Enigma* describe the mechanism and the functioning of the Enigma machine. The authors present several clear photos of the machine and of its different parts. Relying upon such photos they describe the different parts of the machine. In this way, the reader can have a clear and visual picture of Enigma and its functioning. The Encryption/Decryption mechanism is well explained. The alphabet used by the machine is expounded. After the mechanical part, the electric one and its functioning is explained. The global aim of this paper can be summarized through the words written by the authors in the Conclusion: "With this document, we hope to provide the interested layperson or professional with enough background, including citations from the original source material, so that they can discover for themselves both the anatomy (structure) and physiology (function) of this man made mechanism".

Quintino Lopes (Portugal) in *Physics between the Wars: Mobility and Exchange, Lisbon Physics Laboratories and Institutions beyond the Iberian Peninsula* deals with physics in Portugal in the period between the two World Wars. This research is developed through the analysis of the relations between the laboratories of the faculty of Science of Lisbon University, the High Technical Institute of Lisbon on the Portuguese side, and the *Laboratoire Curie*, the *Collège de France* and other prestigious scientific European institutions on the other side. This contribution also offers a picture of Portugal's scientific policy in the period analysed, focusing particularly on the public financing of research in physics. It is, therefore, a contribution to the history of scientific policy.

Peeter Müürsepp (Estonia) in *Physics as Science Proper: A New Historical Justification* starts from the observation that physics represents a model for science. The other sciences can, therefore, be divided as physics-like and not physics-like. In this respect, chemistry is the most interesting topic because it has many aspects which are physics-like and others which are not physics-like. Thus, the author develops a comparison between physics and chemistry and their respective histories. The argumentation begins with the distinction between hypothetico-deductive and constructive-hypothetico-deductive disciplines as well as with an inquiry concerning their respective philosophies. In the third section a comparison between chemistry and physics and their histories is developed, focusing, particularly, on the notion of "arrow of time". The Conclusion follows.

Nicla Palladino (Italy) in *Giambattista Suardi's Machine for Drawing Plane Curves and other Instruments for Cycloids* describes the instrument invented by Gianbattista Suardi to draw many curves obtained through the combination of different rotary motions. In his treatise Suardi considered four topics: (1) description of curves by points; (2) construction of a mechanism able to draw such curves; (3) use of this mechanism; (4) use of similar instruments made up of gears. The concepts and the names of the parts which compose Suardi's instruments are drawn from Ptolemaic theory. For, the main parts of the mechanism are named *Primo mobile* (First Moved) and *epiciclo* (epicycle). Palladino enriches his paper with beautiful images of the different parts of Suardi's instrument. She also explains how it is used to draw curves such as the trochoids, Archimedean and elliptic spirals. Finally other similar instruments are described.

John A. Schuster (Australia) in *The Laws of Collision at the Royal Society, 1668–1669: A Case Study in the Organization of Inquiry in Early Classical Physics* takes a new approach to an event that has been well studied in internalist history of physics. He brilliantly examines the Royal Society's inquiry into the laws of collision at the intersection of two interpretative axes: (1) the Royal Society's organizational patterns of management of investigations, (2) the emergence of classical mechanics, between Galileo and Newton, out of the larger field of natural philosophizing to become the hegemonic discipline within the new ecology of the sciences. In particular he incisively shows how, in cognitive and institutional terms, consensus was actively managed by leaders within the Society and how Fellows with strong natural philosophical proclivities were diplomatically marginalized. As a result of this approach he is able to offer novel historiographical criticisms of well known previous attempts to deal with the collision inquiry—or with experimental research in general—at the Royal Society, offered by Steven Shapin, Dana Jalobeanu and Peter Anstey.

Galina I. Sinkevich (Russian Federation) in *Historia Matheseos: from the History of Phenomena to the History of Ideas. Early Formation of a History of Mathematics* focuses on the historiography of mathematics. For, the author analyses a series of scholars who, starting from the fourth century B.C., can be considered historians of mathematics *tout court* or who, nonetheless, made important contributions to the history of mathematics. For each scholar, Sinkevich presents a picture of his cultural milieu, of his education and of his contributions to mathematics and to its history as well his explicit or implicit philosophy of mathematics. The following authors are analysed: Eudemus of Rhodes, Pappus, Thabit ibn Qurra, Johannes Buteo, Pierre de la Ramée, Joseph Justus Scaliger, Dionisius Petavius, Bernardino Baldi, Joseph Blancanus, Gerardus Johann Voss, ClaudeFrancois de Chales, Jean Mabillon, John Wallis, Edward Bernard, Joseph Raphson, Christian von Wolff and Johann Heilbronner.

5 Part 4: *In Memoriam*

Michael Segre (Italy) in *In Memoriam: Prof. Joseph Agassi (1927–2023)* offers a brief scientific biography of Joseph Agassi. Particularly, the author refers to Agassi's science & social science, metaphysics, logics, analysis of scientific method and politics.

Guillermo Alberto Cuadrado (Argentina) in *In Memoriam: Prof. Dante Roberto Salatino (1949–2022)* offers a brief scientific biography of Dante Roberto Salatino. Particularly, the author refers to Salatino's contribution to *transcursive logic*.

6 Part 5: Appendix. Salvo D'Agostino's Publications

Raffaele Pisano (France)—in collaboration with Pasquale Tucci (Italy), Arcangelo Rossi (Italy), Caterina D'Agostino (France), Roberto Mantovani (Italy) and Roberto Renzetti (Italy)—provided a collected list of Salvo D'Agostino's publications from 1964 to 2019–[posthumous 2022]: 237 Works.

Avvertimento

The Book's authors attend to different individual ideas, scientific, historical and epistemological accounts. Therefore, the Editor is not responsible for the entire contents. Each of the eminent Invited Contributors and Authors are responsible for his/her/their own opinions, which should be regarded as reflecting their personal scientific background and expertise. They are also individually responsible for the images, reprints, quotations, acknowledgments, and all related permissions/approvals displayed/not displayed in their papers.

Acknowledgements We thank the anonymous reviewers for their careful reading of our manuscript and their many insightful comments and suggestions. We also express gratitude to Dr. Julie Robarts (University of Melbourne, Australia) for her observations on English language; eventual remaining English mistakes are our own.

Chapter 3
Three Sicilian Tales by Salvo D'Agostino

Caterina D'Agostino, Giovanna D'Agostino, and Nicola D'Agostino

Abstract *Three Sicilian stories* by Salvo D'Agostino, published here for the first time—except for the simple and beautiful handmade limited edition by *Biblioteca Plautilla*—belong to those familiar and informal, non-academic manuscripts that our father irregularly produced in the last years of his long life (Cfr. Figs. 1, 2, 3, 4, 5, 6, 7 and 8). Starting with the birth of the grandchildren, over time they became a sort of periodic but irregular correspondence: *Letter to my children and grandchildren, Strange stories, Tolomeo the cat and a frog named Filadelfa* [...] inspired by his moralizing eighteenth-century spirit and his passion for teaching lessons, they were also animated, in their various forms, by the desire of communicating his indisputable knowledge and, *last but not least*, by the fatherly desire to amend our poor intellectual training. But these stories can also be seen as faded photographs of the home land he left behind in his youth and which reveal, with evident nostalgia, the yearning for a religious communion with Nature (*The Christmastime fig three*), the passion for its extreme manifestations (*Mount Etna*) and also his lifelong interest in the history of technology (the *Saie*). In the end, the reader will find the three tales in original Italian versions.

Keywords D'Agostino's family · Literature · Salvo D'Agostino · Tales

---

C. D'Agostino (✉) · G. D'Agostino · N. D'Agostino
Salvo D'Agostino's Family Private Archive, Paris, France and Rome, Italy
e-mail: caterina.dagostino@wanadoo.fr

R. Pisano (ed.), *A History of Physics: Phenomena, Ideas and Mechanisms*,
History of Mechanism and Machine Science 42,
https://doi.org/10.1007/978-3-031-26174-9_3

1 The Christmastime Fig Tree

When the fig season starts, I am always reminded of the story of the Christmastime fig tree and now I want to tell it to you. As I remember it, the Christmastime fig tree was a tree hidden somewhere, way back in the lemon tree garden, in Sicily, at Acireale, hidden among other trees in a depression which was very shaded because little light filtered through the brown leaves of the lemon trees. Trees of various species that I do not remember now hid it. One had to look for it, then, scrambling down the low, mossy walls, one-step after another, on slippery steps carved out in the black lava rock so many years ago by industrious hands. That is why we children were the only ones to go down those narrow stairs. The light was scarce, filtered through the brown leaves of the lemon trees, and the hollow was filled with a shadowy silence that contrasted with the dazzling light of a Sicilian summer. There was still, stagnant water all around it, and the grass and moss surrounded the Christmastime fig tree in dark green. I remember eating its plump fruit with its brownish-green skin and pinkish-white juicy pulp. It must have been close to Christmas, because the fig season was over and there were only dried figs, hanging in clusters along the beams in the cellar.

Sometimes it seems that I dreamt it, since so many years have gone by and no one has ever been able to tell me whether fresh figs can be found at Christmas time. I am sure there is a Christmastime fig tree behind that screen of greenery. I remember, too, that sometimes the farmer brought us a gift basket for Christmas, and everyone greeted it with wonder, because of the lovely display of walnuts, hazelnuts, and dried figs on the table, and no one expected that plump fruit with its brownish-green skin and pinkish-white juicy pulp. Everyone exclaimed: how can there be such ripe figs at this cold Christmas time? But we children knew there was the Christmastime fig tree that retained its leaves and fruit until Christmas.

The years went by, and the memory of the fig tree was lost. One day, wandering through the streets of a Tuscan town near the sea, Castiglione della Pescaia, I perceived a special smell and saw a basket of plump figs, with a brownish-green skin and pinkish-white pulp. It was August, and Christmas was still a long way off, but hope guided my hand, and I bought a kilo of them. I did not find the same taste in them, and even the name they were given by the fruit vendor disappointed me.

But I have always looked for the Christmastime fig tree. You know I have always had a special love for fig trees, and I can recognize the smell of its leaves even in the midst of a country road. When I travel through the countryside, in the South or the North, I always ask peasants whether they have eaten fresh figs at Christmas. It was a fable, they told me, a childhood dream. But I still search for it. I hope you too will help me find it among the fig trees, wherever you live.

Fig. 1 Salvo (standing, on the right) with his brother Nemi (sitting, on the right) and friends. Acireale (Sicilia, Italy) 1930–1935 *Source* Courtesy D'Agostino Family Private Archive

Fig. 2 Hare hunting with dog Giulio near S. Cosmo (Acireale, Sicily, Italy), 1936 *Source* Courtesy D'Agostino Family Private Archive

2 The *Saie*

The canals that are used to water the lemon trees in Sicily are called *Saie*, or *Sahie*. The word is derived from the Arab, and the canals are made of terra cotta, with brick borders, like tiles. They say that in Sicily, the Arabs built a system of canals to irrigate the arid land where no river or springs were to be found. In the part of Sicily where I was born, water was even scarcer, since the fields are what remains of the lava descended from the Etna long ago, then dried up. My grandparents said that no fields are more fertile than those composed of ancient lava. But large trees like lemon and orange trees need large amounts of water. The water came from melted snow from the Etna, but it ran deep under the layers of lava, and had to be pumped up. The pumps were near the sea, and belonged to the rich masters who charged large sums for the water to be taken to my lemon tree garden. Therefore, you understand why the water arrived during the night, when the electricity cost less.

The water spurted out unexpectedly from a big black hole, which was an artesian canal. It was greeted with some apprehension by us children, because it came from underground and surfaced like an unexpected miracle. We were allowed to go out at night, in the dark, with a peasant carrying an oil lamp, to watch the water flow down the *Sahie*. When there was no moon you could not see much around the trees in the hollows, and we groped our way among the unexpected rustling of sparrows disturbed in their sleep. Sometimes a rabbit or a hare ran out, or one of those little animals who lived in the fields. For us, it was an excursion into the mysterious nighttime world, when all the grown-ups were sleeping in their beds and only revolutionaries and bandits roamed the countryside. But for the farmer, this was the beginning of the hard work of using his hoe to divert the turbulent current so that it would enter the ditches that circled the lemon tree trunks. The soil was parched and dry, and the water contrasted with that aridity, and we dipped our hands and feet in it. The farmer worked his way around, among the trunks and the branches, in the feeble light of the lantern which barely pierced the darkness, his skillful hoe opening and closing the gullies and canals. We were almost ashamed of our playfulness, in contrast to that man, gaunt and burnt by the sun, who would continue to work the following morning after the night's labor. But what surprised us was that during his nocturnal fatigue, he never raised his eyes to the sky where the stars could be seen shining more or less brightly in the clear, dark country sky, and was unable to locate the motionless Pole Star. We too were surprised by that intimidating display of stars.

Sometime later we returned to the *Sahie* in daytime. The lemon trees were all in bloom covered with those pungent flowers the Arabs call "zahara".

Fig. 3 Soldier (right side) in the 3rd Artillery Regiment, Milan, March 1941 *Source* Courtesy D'Agostino Family Private Archive

3 Mount Etna

When I was twelve I called it The Mountain, not knowing it was a volcano. From the patio of my house in Acireale you could see its three high peaks reaching out to the west, the highest, middle one which I was told was the main crater, often smoking, the other two smaller ones on the right and left like the outreaching arms of a quietly resting giant. Lower down were many small hills and cone-shaped forms which had been the mouths of small volcanoes many centuries ago, the children or grandchildren of the great mountain. I gazed at them from the patio in full daylight or in the evening shadows, illuminated by the lights of Zafferana.

You cannot imagine how much I wanted to see that mountain from close up, so one day I took off from home on my bicycle and after much hard pedaling I arrived at Zafferana. A path led up from the square, climbing along a rocky trail and after the vineyards I entered an orchard densely populated with large brown-leaved chestnut trees. Then the land grew steeper, and the chestnut trees smaller and fewer. Unexpectedly, a large valley known as the *valle del Bove*, opened up, deep with cone-shaped black rocks, craters formed by the mouths of ancient volcanoes, now extinct. An icy wind was blowing, chasing the clouds, and the boundless horizon took your breath away. It was a landscape that, to my imagination, must have looked like the moon. At the other end of the valley a steep wall of rock was the flank of the highest volcano, which I was told culminated in a large crater. This was the Etna described in Goethe's poem: "Do you know the mountain and its clouded paths?".

But I saw it again one night, in the midst of an eruption. After walking in the dark among many curious people, we arrived in a clearing facing a black stone wall, with glittering flames, which revealed its true nature: it was the fire coming out of the belly of the great volcano. Suddenly a hellish gale of death and destruction was unleashed, all around us, trees were downed, burning like lit matches.

We all raced away, dismal spectators defenseless in the face of the tremendous might of the Mountain.

Fig. 4 Soldier (left side) in Sestriere Vecchio (Piemonte, Italy). Mounted Artillery Regiment, September 11, 1942 *Source* Courtesy D'Agostino Family Private Archive

4 Concluding Remarks

With the publication of these three tales, we wished to emphasize another aspect of Salvo D'Agostino intellectual activity. Poetic, literary, musical and artistic passions were increasingly dominant during the last period of his almost century-long life. Over all, his extremely close ties to Dante's Divine Comedy, entire passages of which he was able to quote by heart. Anyone who spent time with him in those last years will remember the very frequent quotation of Paradise, *Canto XXXIII*:

> Even thus the snow is in the sun unsealed,
> Even thus upon the wind in the light leaves
> Were the soothsayings of the Sibyl lost.[1]

5 Italian Original Versions

I Tre racconti siciliani di Salvo D'Agostino, pubblicati qui per la prima volta, eccezion fatta per la semplice e bellissima edizione fatta a mano con tiratura confidenziale della *Biblioteca Plautilla*, fanno parte di quegli scritti meno accademici e più familiari che nostro padre è andato alimentando irregolarmente negli ultimi anni della sua lunga vita. Cominciati con la nascita dei nipoti e diventati con il tempo una sorta di periodico ma irregolare carteggio: *Lettera ai figli e ai nipoti, Storie strane, Il gatto Tolomeo e la rana Filadelfa* [...] prendevano spunto dal suo moraleggiante spirito settecentesco e dalla passione per la lezione didattica, ed erano animati, nelle loro varie forme, dal desiderio di comunicare le sue indiscutibili conoscenze, ma anche, *last but not least*, dall'aspirazione paterna di correggere la nostra precaria formazione intellettuale. Ma questi racconti possono anche essere letti come sbiadite fotografie di una terra abbandonata in gioventù dove trapelano, con evidente nostalgia, l'anelito per una comunione religiosa con la natura (*Il fico natalino*), la passione per le sue manifestazioni anche estreme (*L'Etna*) e non ultimo l'interesse di tutta una vita per la storia della tecnica (*Le Saie*).

[1] Translated by Henry Wadsworth Longfellow (1867) [Wikisource]; see below for Italian version.

Fig. 5 Original Handmade Cover of the *Three Sicilian Tales Source* Courtesy Monteverdelegge & D'Agostino Family Private Archive

Fig. 6 Pen Drawing, Assisi, June 20, 1947 *Source* Courtesy D'Agostino Family Private Archive

5.1 *Il Fico Natalino*

Ora che arriva la stagione dei fichi mi viene sempre in mente questa storia del fico natalino e ve la voglio raccontare. Nei miei ricordi il fico natalino era un albero nascosto in fondo al giardino dei limoni, ad Acireale in Sicilia, nascosto tra gli alberi in una conca molto scura per la poca luce che passava attraverso le foglie brune dei limoni. Alberi di varie specie che ora non ricordo lo nascondevano. Bisognava allora cercarlo scendendo giù dai muretti di pietra coperti di muschio, un passo dietro l'altro, su gradini scivolosi che mani laboriose avevano scolpito tanti anni addietro sulle rocce nere di lava. È per questo che su quelle strette scale ci andavamo solo noi bambini. La poca luce filtrava attraverso le foglie brune dei limoni e nella conca c'era un silenzio ombroso che contrastava con la luce accecante dell'estate siciliana. L'acqua ristagnava tutt'intorno ed erbe e muschio circondavano di un verde scuro l'albero del fico natalino. Ricordo di averne mangiato i frutti, grossi, di un verde marroncino nella buccia e bianco rossiccio nella polpa succosa. Doveva essere il tempo vicino a Natale, perché era finita la stagione dei fichi e solo i fichi secchi pendevano a grappoli dalle travi della cantina.

A volte mi sembra di averlo sognato perché tanti anni sono passati e nessuno mi ha mai saputo dire se ci sono in giro fichi freschi intorno a Natale. Ma sono sicuro che dietro la verde parete c'era l'albero del fico natalino. Ricordo anche che qualche volta il contadino ci portava un cesto per Natale e tutti l'accoglievano con meraviglia, perché sulla tavola facevano bella mostra le noci, le nocciole e i fichi secchi e nessuno si aspettava i frutti grossi di un verde marroncino nella buccia e bianco rossiccio nella polpa succosa. Ma com'è possibile avere fichi maturi in questo freddo Natale, dicevano tutti? Ma noi piccoli sapevamo che c'era l'albero del fico natalino che teneva foglie e frutti sino a Natale.

Passarono gli anni e si perse la memoria del fico. Un giorno, girando per le vie di un paese della Toscana vicino al mare, a Castiglione della Pescaia, sento un certo odore e vedo un cesto di fichi grossi, la buccia di un verde marroncino e l'interno di bianco rossiccio. Era il mese di agosto e Natale era ancora lontano, ma la speranza mi prese la mano e ne comprai un chilo. Non mi sembrava che avessero lo stesso sapore e anche il nome che gli diede il fruttarolo mi aveva deluso.

Ma io l'ho sempre cercato il fico natalino. Lo sapete che ho sempre avuto un amore speciale per gli alberi di fico e sento l'odore delle sue foglie anche in mezzo a un viottolo di campagna. In giro per le campagne del sud e del nord ho sempre chiesto ai contadini se avevano mangiato fichi freschi intorno a Natale. Era una favola, mi dicevano, un sogno infantile. Ma io continuo a cercarlo. Spero che anche voi mi aiuterete a trovarlo fra gli alberi di fico delle vostre parti.

Fig. 7 A mountain-climbing excursion with his Colleague and Friend Emma Castelnuovo, (both at University La Sapienza, Roma) early 1960s *Source* Courtesy D'Agostino Family Private Archive

5.2 *La Saie*

I canali che servono per innaffiare le piante di limoni in Sicilia si chiamano *Saie* o *Sahie*. È una parola che deriva dall'arabo, e sono di terracotta con i bordi di mattone come le tegole. Si dice che gli Arabi in Sicilia avessero costruito un sistema di canali per irrigare le terre aride dove non ci sono fiumi e sorgenti d'acqua. In quella parte della Sicilia dove io sono nato, l'acqua era ancora più scarsa perché i terreni sono i resti di antiche lave discese dall'Etna che poi si sono disseccate. I miei nonni dicevano che non c'era terreno più fertile di quello delle antiche lave. Ma grandi piante come i limoni e gli aranci abbisognavano di molta acqua. L'acqua veniva dallo scioglimento delle nevi dell'Etna, ma era profonda sotto gli strati di lava e bisognava pomparla. Le pompe stavano vicino al mare ed erano proprietà di ricchi signori che si facevano pagare cara l'acqua per portarla nel mio giardino di limoni. Così avete capito perché l'acqua arrivava di notte, quando la corrente elettrica costava di meno.

L'acqua spuntava all'improvviso da un grosso buco nero che era un canale artesiano, accolta con trepidazione da noi bambini perché veniva da sotto terra e affiorava di colpo come un miracolo improvviso. Avevamo il permesso di uscire di notte al buio, assieme a un contadino che portava una lanterna a petrolio, per vedere scorrere l'acqua nelle *Sahie*. Nelle notti senza luna si vedeva poco sotto il giro delle conche degli alberi e si camminava a tentoni fra l'improvviso fruscio dei passeri disturbati nel sonno notturno. Qualche volta scappava un coniglio o una lepre o uno dei piccoli animali abitanti dei campi. Per noi era come entrare nel mondo misterioso della notte, quando tutti i grandi dormivano nei loro letti e nella campagna camminavano soltanto i rivoluzionari e i briganti. Ma per il massaro cominciava la dura fatica di far deviare a colpi di zappa il fiotto irruente per farlo entrare nei cerchi di terra, le conche che circondavano i tronchi dei limoni. Il terreno era arido e secco e l'acqua contrastava con tutta quella arsura e noi vi tuffavamo le mani e i piedi. Il massaro si aggirava fra tronchi e rami, alla fioca luce della lanterna che appena interrompeva quel buio e con la zappa sapiente apriva e chiudeva conche e canali. Noi, quasi vergognosi del nostro giocare, davanti a quell'uomo scarno e bruciato dal sole, che domani avrebbe ancora lavorato dopo aver vegliato la notte. Ma eravamo sorpresi che nella sua notturna fatica non avesse mai rivolto gli occhi al cielo, dove le stelle si mostravano a strati degradanti di splendore nel limpido e buio cielo della campagna e non sapesse dove collocare l'immobile stella polare. Sorpresi anche noi da quella incombente stellata.

Dopo qualche tempo ritornavamo di giorno sulle Saie. Gli alberi di limone erano tutti fioriti di quel fiore odoroso che gli arabi chiamano "zahara".

Fig. 8 With his three children in a park, Rome, 1965 *Source* Courtesy D'Agostino Family Private Archive

5.3 L'Etna

Quando avevo dodici anni la chiamavo la Montagna, senza pensare che fosse un vulcano. Dal terrazzo della mia casa di Acireale si stendeva a ponente con tre grandi cime, una centrale più alta che spesso fumava e mi dicevano era il cratere centrale, e le due altre più piccole a destra e a sinistra come le braccia distese di un gigante in tranquillo riposo. Più in basso tanti piccoli monti a forma di cono che tanti secoli prima, erano state bocche di piccoli vulcani, figli o nipoti della grande montagna. Li guardavo dal terrazzo nella luce piena del giorno o con le ombre della sera, quando apparivano scintillanti di luci del paese di Zafferana.

Non vi dico che voglia di vederla da vicino quella montagna, così un giorno scappai da casa in bicicletta e dopo una gran pedalata arrivai a Zafferana. Dalla piazza saliva un sentiero che si inerpicava lungo un sentiero sassoso e dopo le viti s'incontrava il castagneto, fitto di grandi alberi dalle foglie brune. Poi il terreno saliva e i castagni diventavano sempre più piccoli e radi. All'improvviso si apriva una grande vallata, detta la valle del Bove, uno sprofondo di rocce nere a forma di cono, crateri di antiche bocche del vulcano ormai spente. Tirava un vento gelido che trasportava le nuvole, il panorama vastissimo lasciava senza fiato. Un paesaggio che nella mia fantasia doveva rassomigliare alla luna. In fondo alla vallata una ripida parete di roccia era il fianco più alto del vulcano, che mi dicevano terminava con un grande cratere. Questa era l'Etna dei versi di Goethe: "... conosci la montagna e il suo cammino alle nuvole vicino?".

Ma la rividi una notte in piena eruzione. Dopo una camminata al buio in mezzo a tanta gente curiosa si arrivava a uno spiazzo davanti a una parete nera di pietre, con qualche barbaglio di fiamma, che rivelava la sua vera natura: era il fuoco uscito dalla pancia del grande vulcano. All'improvviso si scatenò un inferno di morte e distruzione, all'intorno crollavano gli alberi bruciando come fiammiferi.

Siamo tutti scappati, miserabili spettatori inermi davanti a questa immensa forza della Montagna.

6 Note Conclusive

Pubblicando qui questi tre racconti, abbiamo voluto mostrare un altro aspetto della vita intellettuale di Salvo D'Agostino. Le passioni poetiche, letterarie, musicali e artistiche in senso lato hanno preso man mano, negli ultimi tempi della sua vita quasi centenaria, sempre più spazio. Non ultimo, lo strettissimo legame con la Commedia dantesca di cui citava a memoria canti interi. Chiunque l'abbia frequentato in questi ultimi anni ricorderà l'ostinata e ripetuta citazione del *Canto XXXIII del Paradiso*:

> Così la neve al sol si disigilla,
> così al vento ne le foglie levi
> si perdea la sentenza di Sibilla.

Chapter 4
Salvo D'Agostino and the History of Physics: Relativity and Quantum Physics

Enrico R. A. C. Giannetto

Abstract Here I would like to discuss Salvo D'Agostino's perspective on the History of Physics, which he pursued in his writings and which he left us as a cultural heritage. Through the memory of various personal conversations and the analysis of some of his writings, I will try to bring out the fundamental cultural value that he attributed to the History of Physics. I will then analyze some of his works on the theory of relativity and quantum physics, aimed at bringing out the philosophical, gnoseological and ontological aspects of physics.

Keywords Epistemology · History of physics · Physics ontology · Quantum physics · Relativity · Salvo D'Agostino

1 Introduction: The History of Physics in Italy

It would be impossible to adequately remember and discuss all the contributions of Salvo D'Agostino to the History of Physics (Rossi 2019). Salvo studied among others the works of Newton, Ampère, Boltzmann, Weber, Maxwell, von Helmholtz, Hertz, Planck, Einstein, Bohr, Schrödinger, Dirac, Bartoli, Fermi, Giorgi, Enriques (D'Agostino 2000a). The history of electromagnetism (D'Agostino 1975), the history of physics from Galileo to Einstein (D'Agostino 2018a), the history of theoretical physics, the history of the theory of relativity (D'Agostino 2018b, c) and the history of quantum physics are the areas in which he has worked the most.

Here, first of all, I would like to discuss the perspective of the History of Physics that he elaborated, which he pursued in his writings and which he left us as a cultural heritage. Through a quick analysis of some of his writings, I will try to bring out the fundamental cultural value that he attributed to the History of Physics.

E. R. A. C. Giannetto (✉)
Department of Letters, Philosophy, Communication, Bergamo University, Bergamo, Italy
e-mail: enrico.giannetto@unibg.it

R. Pisano (ed.), *A History of Physics: Phenomena, Ideas and Mechanisms*,
History of Mechanism and Machine Science 42,
https://doi.org/10.1007/978-3-031-26174-9_4

Salvo had trained abroad, through scholarships, in the United States, and had also followed the lessons of Mary Hesse: this American training always characterized the breath of his approach to the history of science, strongly linked to the philosophy of science and epistemological reflection (D'Agostino 1991a). When he returned to Italy, in the 1970s, the school of Ludovico Geymonat was developing in Milan, which from a more orthodox Marxist point of view valued well science, and, on the other hand, a critical current of science was developing especially in Rome, related to Western Marxism's criticism of science as a non-neutral and bourgeois activity, aimed at the technical domination of Nature as an integral part of Western capitalism.

It was a great season of commitment and cultural ferment for the whole of Italian society,—"formidable those years" wrote Mario Capanna (Capanna 1988)—a period in which student and worker movements were born in 1977 which brought about winds of revolution. In this context, a historical–critical study of science and physics was born in Italy, and which saw many historians of physics united in a common research that found space, an Italian phenomenon, in an institutional practice within the Departments of Physics and not in the humanities, and which, unfortunately, is now over, at least from the point of view of University structures.

The originally Marxist perspective of Lello Rossi and others (Baracca and Rossi 1974, 1976, 1977, 1978; Baracca et al. 1979; Ciccotti et al. 2011), opened up a broader and more relevant horizon of reflection: for me, the lesson "of Angel (Baracca) and of Archangel (Rossi)", ironically cited by Paolo Rossi in a polemical analysis against the Marxist critique of science (Rossi 1986), was valid and to follow, beyond the sharing of some questionable results, of the specific form it had assumed for them in individual cases. Recently a re-evaluation of this perspective was proposed by the editors of a new translation of a 1930 text by Hessen (Hessen [1930] 2017).

This is the root of the history of Italian physics, this is our history. It was no longer a question of hagiography and historical anecdotes, of ornamental illustrations in physics school books, of a self-delight or in any case of an exaltation of the scientific community in contrast to the historicistic humanistic idealism that had dominated the Italian tradition, but rather a historical critique of physics. This materialistic historical critique of physics, however, had evident defects: the critique was at times general and therefore vague, it was not anchored to the study of experiments and mathematical formalism, but moved on an a priori defined ideological level.

2 Salvo D'Agostino and the Historiography of Physics

Already Edwin Arthur Burtt since 1924, in showing the metaphysical foundations of modern physical science, had studied the cases of Copernicus, Kepler, Galileo, Descartes, More, Newton and many others (Burtt [1924, 1932] 2003).

In 1938 Tarner Lectures, Arthur Stanley Eddington presented the ideas then published in 1939 in his book, *The Philosophy of Physical Science* (Eddington 1939). Here he stated that also modern physics was part of philosophy, philosophy of Nature, and ontology and epistemology (gnoseology) were parts of physics. After the so-called scientific revolution, physics conserved its philosophical nature, but without great new inquiries about it. Ontology and epistemology were re-elaborated with the scientific revolutions of relativity and quantum physics and we have to understand their ontology and epistemology to deeply understand the new physics. This was the work also for history of physics, to which Salvo D'Agostino mostly dedicated his research.

What was history of physics for Salvo? Of course, the fundamental aspect that could be shared with Italian trends was that (1) one could *not limit at an internal history of the discipline*, but having to look at the internal–external relationships of physics with the rest of knowledge and with human life in its concreteness, but (2) *avoiding a purely external story*. A history of physics which therefore does not neglect the dimension of historiography, and which is confronted with all types of documents and all types of objects in physics. In addition, in this context, we must understand Salvo's work and, for me, our work still to do.

Salvo understood that not only could one not limit at an internal history, but also that he could *not limit at a history of ideas*, privileged by the humanistic tradition.

It was need for a history of physics that does not neglect (3) *the history of experiments*, or, as we say now, of experimental practices, and (4) *the history of instruments*. A history of physics that is therefore intertwined with the (5) *history of technology*. A history of physics that is *not a mere history of ideas*: *no longer a mere history of scientific theories but also of experimental practices*.

Salvo has devoted various reflections and writings to the problem of instruments: *The instruments of electromagnetism* (D'Agostino 1989), *The scientific instruments and science* (D'Agostino 2005), which also deals with the problems of quantum measurement. Salvo not only considers them philosophically as "conditions of possibility of the experiment", but also, from the theoretical point of view, as physical constraints.

In his 1986 essay on the uncertainty principle he poses the problem of the ontology of classical physics and its difference from the ontology of quantum physics, also in relation to the different approaches of Heisenberg and Bohr. That is, the philosophical problem par excellence of physics is posed, and the problem of the relationship between language and reality is also posed. Salvo thus implicitly delineates (6) *a history of physics which is therefore intertwined with the history of philosophy* to illuminate the gnoseological and ontological dimension of physics in its development. In other essays, such as those on Helmholtz and Einstein, the concrete problem arises of the comparison with some philosophers, in particular with Kant and his gnoseology. *A history of physics that is intertwined with the epistemology of physics*. It is on this front that, according to Salvo, the eventual recognition of the ideological role of science is played: whether the ontology of a physicist is idealistic or ideologically generated.

However, fundamental, in this analysis, is the differential knowledge of the mathematical languages in which the various authors base the theoretical reconstruction of the experiments: in the 1986 essay, he highlights the difference between Heisenberg's mathematical language, based on the calculation of statistical errors starting from the probabilistic theory of the Schrödinger wave function, and that of Bohr, based on the algebra of the Einstein-De Broglie relations, and again the mathematical language of classical physics, and finally the ordinary language. Here, Salvo also discusses the correspondence principle and finds a broader understanding of a continuity in changing theories, which he will explore more in other essays, notably on Einstein and Bohr. The same importance is given to mathematical languages by Salvo in his studies on Weber, Ampère, Maxwell. Salvo thus indicates (7) *a history of physics that is intertwined with the history of mathematics*. A history of physics which, however, is not limited to a merely descriptive technical and formal analysis of mathematics, but also pays attention to the ontological implications of mathematics. And it indicates, also through the study of Bohr's problem of ordinary language, (8) a history of physics that is not a mere disciplinary history, *a history of physics that is inserted in the broader history of science, in the broader history of culture, in the larger history*.

Salvo then focused his attention, for many German-language authors, on their perspective of physical theory as *Bild*, or as representation or image (see the essays contained in the 2000 volume, published in Boston Studies). It was not just an investigation into the epistemology of physicists like Hertz, Boltzmann, Schrödinger, Dirac and others. It was (9) *a history of physics as a culture* that allows us to understand the image of the world implicit in physics and is itself aimed at understanding the world and the Nature in which we live. I believe in the value of these characterizations of the history of physics given by Salvo's work, not in an abstract historiographical perspective, but in the concreteness of his historical investigations.

3 The History and Philosophy of Relativity: Albert Einstein

The first concrete occasion to meet Salvo and to discuss with him was that of the *Representation and object workshop. From physics to other sciences*, held at the Physics Department of Rome in October 1986, where Salvo spoke on *The uncertainty principle and the transition from the ontology of classical physics to that of quantum mechanics* (D'Agostino 1987). In the fall of 1987, there was another workshop in Rome, *Images Languages Concepts*, where Salvo spoke about *Einstein and Kant: theoretical physics against the* "logical but empty scheme of concepts" (D'Agostino 1991b). These two conferences held by Salvo are related to his wider researches on the history of quantum mechanics and special and general relativity theories.

First of all, Salvo tried to understand the historical process of theory change. Beyond recognizing the conceptual incommensurability among theories, related to the so-called meaning variance, Salvo focused his attention on the particular factors of continuity and discontinuity in theory change. In his book *A History of the Ideas of Theoretical Physics,* Salvo studied the role of the "correspondence principle" as an epistemic tool used by Albert Einstein to formulate general relativity (since 1907–1949) (D'Agostino 2000b) and by Niels Bohr to formulate quantum mechanics (D'Agostino 2000c). Salvo discussed the various formulation of the correspondence principle explaining how it worked. Constructing a new theory implies to contain, in a specific mathematical limit, the mathematical relations that one can consider valid in a certain physical approximation or in a more limited physical context. Even if there is meaning variance of mathematical symbols, mathematical relations as approximations must be preserved.

This mathematical limit allows for the preservation also of the experimental content of the previous theory. In this way, Einstein could have constructed general relativity establishing the mathematical limits that give back special relativity and Newtonian gravitation theory with a pluri-correspondence principle, even if general relativity as field theory is conceptually completely different from Newtonian action at a distance. In the same way, Bohr could have constructed quantum theory that at a mathematical limit in relation to macroscopic world reduced to Newtonian classical mechanics.

Even if mathematical symbolism radically changes in relation to a new ontological commitment—as in quantum mechanics or in quantum-relativistic field theory within which the mathematical symbols of wave functions or of fields no more denote properties of physical objects but physical entities themselves (for example, electrons)—epistemic meta-theoretical operations of mathematical limits can ensure an element of continuity in the history of physics (D'Agostino 2002).

Thus, Salvo pointed out how even the most revolutionary theory, like general relativity or quantum mechanics, presents some factors of continuity with the previous theory in the mathematical formulas and experimental contents. Salvo was able to show how an epistemic principle, correspondence principle, is a constitutive principle of a physical theory: this epistemic principle is very important not only within the so-called context of discovery, but it is constitutive within the context of legitimation, determining the mathematical structure and the experimental content of any physical theory. Field equations, space-time intervals are structurally defined by specific applications of correspondence principle. Epistemology is a deep part of physics.

Outlining Einstein's revolution of general relativity, Salvo studied historically the great epistemological problem of the relationship between geometry and physics (D'Agostino 1999). He started his analysis from Descartes and considered the development of this problem in Kant, Helmholtz, Poincaré, Einstein till quantum physics, comparing Einstein's perspective with Cassirer's and Reichenbach's one. Salvo pointed out the originality of Einstein's solution. Poincaré and Reichenbach considered the possibilities to write physical laws by different geometrical languages. Poincaré allows for the use of different conventions (conventionalism): conventions can be also mathematical abstractions with no physical content, like Euclidean geometry as an abstract mathematical language. Reichenbach allows for what he called different coordinative definitions and concluded, on one hand, the relativity of geometry in respect to different definitions, and, on the other hand, the experimental, anti-Kantian, foundation of general relativity. Einstein rejected the possibility of an abstract mathematical definition of geometry: geometry must be physically defined, geometry is a physical science. Thus, there is no relativity of geometry, no possibility of different conventional geometrical languages. Geometrical structure of the world is not a Kantian or neo-Kantian (Helmholtz, Cassirer) knowledge as an a priori synthesis nor a logical neo-empiricism's result (Reichenbach): space is a physical concept and physical geometry is constituted as a posteriori synthesis. Space is not a subjective form of knowledge: space–time is always determined as curvature-variable by matter or field, and for this reason general relativity presents a four-dimensional, pseudo-Riemannian chrono-geometry. Absolute space and absolute time were eliminated by an epistemic principle of rejection of non-observable quantities. The same general covariance principle, related by the correspondence principle to the special relativity covariance principle, is an epistemic principle about the fact that physical laws must be invariant in form for all the reference frame systems.

However, physical laws have always an epistemic nature. There is always a subjective, Kantian element in physical knowledge, because theoretical principles cannot be experimentally deduced: theory presents a free creation of mind element of imagination which is not reducible to pure experiments, whereas experiments only verify or falsify a theory.

Thus, Einstein's conception of physics gives to epistemology a very fundamental, structural role: physical laws or principles are intrinsically epistemic, that is "transcendental laws" and one can give them an objective reality (they are "objective transcendentals", a posteriori synthesis) as long as they are partially experimentally verified.

Salvo noted also that in the last period, Einstein was involved in a field unification theory and spoke about the incompleteness of general relativity, where the material instruments constituting the "physical subject" and matter particles were not explained in terms of the field.

According to Salvo, this epistemic nature of physics can explain Einstein's rejection of quantum mechanics in respect to the epistemic requirement of completeness of a theory. If there is some kind of experiment that can give an exact measurement of the position of the electron, if the indeterminacy of position has to be related to the disturbance of measurement as presented by Heisenberg, then exact position must be considered an element of reality which must have a correspondent term in the theory: that is, the ignorance of the exact position of the electron must be considered epistemic and not due to an ontological indeterminacy. Thus, according to Einstein, quantum mechanics is epistemically incomplete. Certainly, Einstein did not consider the possibility that indeterminacy could not depend from measurement disturbance, but from a real intrinsic change. Einstein did not consider the fact that exact measurability of position is possible only in particular cases and not in other ones. Anyway, independently from the fact that Einstein's rejection of quantum mechanics is right or wrong, Einstein's criterium is an epistemic conception of physical theories, in which completeness is a constitutive principle. Einstein looked for a Kantian *Anschaulichkeit* (theoretical, space-time visualisability) of physical reality: theory, following Hertz, must be a *Bild* (image) of physical reality. Bohr had a different perspective: quantum theory must have only a symbolic character with an abstract mathematical diagrammatic visualisation.

4 The History and Philosophy of Quantum Physics: Niels Bohr

Salvo continued to analyse Bohr's work and to show the emergence of epistemological aspects of physics (D'Agostino 2000c). Bohr too, since 1913, used correspondence arguments to develop quantum theory: in the first papers correspondence is considered as a form of analogy between classical physics and the new quantum theory of atom. In 1920, Bohr used the term correspondence and after 1925 Heisenberg paper on quantum mechanics, his correspondence principle started to be more formally specified in relation to frequency and intensity of radiation formulas and the observables of the new mechanics. Salvo showed that Bohr's complementarity principle is not a mere philosophical interpretation of Heisenberg's 1927 indeterminacy principle, without any true physical relevance as stated by Karl Popper. Indeed, complementarity is a sophisticated form of the correspondence principle.

In 1927, Bohr presented the complementarity principle before Heisenberg's indeterminacy relations. Heisenberg's indeterminacy principle will be only a case of illustration of the complementarity principle. Bohr deduced equations like but different from indeterminacy relations from the so-called Einstein-De Broglie formulas $E = h\nu$ and $p = h/\lambda$. Bohr's equations are not in terms of statistical errors but in terms of effective change in variables related to wave spectrum $\Delta\nu = 1/(\Delta t)$ and $\Delta\lambda$.

Bohr's indeterminacy relations are different from Heisenberg's indeterminacy relations. These new Bohr's equations (inequalities) are not concerning epistemic errors due to measurement interactions, but ontological changes related to the motion as wave propagation and to the operative definition of physical reality. Instruments and experimental procedures of measurements are involved in the definition itself of physical reality: ontology is re-defined epistemically in relation to the operations of physical subjects. There is an inseparability of the observed system and of the observing/defining instrument (observing/defining physical subject). There is a pragmatic turn of Kantian transcendentalism. Discontinuity of physical variables and the finitude of the Planck constant as quantum of action or of angular momentum were not due to measurement interaction changes, but to the intrinsic structure of a general wave packet which is a propagation of a finitely space-time extended form. The particle velocity is the group velocity of a spreading wave packet of a limited extension and a limited time rhythm of propagation. These relations connect wave (frequency and wave length) and corpuscular aspects (energy and momentum), which are correlated variables at quantum level. Classically these variables are not correlated, therefore the correspondent models have only a symbolic character of mathematical abstraction. The non-exhaustiveness of any model, according to Bohr, made it symbolic, metaphoric. Formal metaphors (particle energy is wave frequency modulo the Planck constant, particle momentum is the inverse of wavelength modulo the Planck constant) constitute epistemically quantum physics.

After 1927, complementarity allows for the correspondence with various classical models: superposition principle and conservation laws, corpuscular model (interaction phenomena of matter and light) and wave model (interference phenomena of propagation of matter and light), causality (energy, momentum) explanation and space–time description, and also operative definition of a physical variable and observation of its variability. In this way, Bohr rejected mechanical pictures of quantum reality, but allowed for a limited space–time visualisability and Kantian *Anschaulichkeit.*

Correspondence has a complementary nature: Bohr's *Bild* of Nature was influenced by Kierkegaard's philosophy and criticism of Hegel's synthesis (Giannetto 2019). Nature can no more be understood in any unitary, synthetical frame. Following Kierkegaard's philosophy of paradoxes/contradictions (Petruccioli 1988), which, for example, did not reduce the dual (human and divine) nature of Christ to unity, quantum physics must accept the dual nature of matter and light, of electrons and photons as corpuscle and wave. There is a double irreducible analogy of quantum physics with classical physics, a double complementary correspondence. There is also complementarity of physical laws: physical laws are no more universal but complementary-contradictory and in some way are never exact (indeterminacy of any variable is never exactly zero). Therefore, complementarity is more than a mere extension of correspondence principle: it does not imply a simple inclusion of classical physics into quantum physics: the double contradictory correspondence to different contradictory classical models means an irreducibility of quantum physics to classical physics.

5 Conclusion

Salvo studied also the late theoretical contributions by Erwin Schrödinger in fifties to the development of quantum physics (D'Agostino 2000d). He noted that second quantisation, that is quantum-relativistic field theory, explained indistinguishability (non-individuality) of quantum particles leaving the corpuscular model, and overcame the many-particle multi-dimensionality of wave function, the wave packet spreading problem and Born's subjectivist probability interpretation of wave function, making it possible to restore in a new form his first interpretation of quantum reality in terms of physical wave fields. The analysis of Schrödinger's work given by Salvo showed the complexity of contemporary physics in its relation with epistemology and philosophy.

Acknowledgements I would like to thank Pasquale Tucci and Arcangelo (Lello) Rossi for the possibility given to me to read their papers on Salvo. I have learned many things from Salvo, who is continuing to be source of inspiration and guide for my work in history and philosophy of physics. I would like to continue to thank Salvo forever.

References

Baracca A, Rossi A (1974) 1789: prassi e organizzazione della scienza. Sapere 775:46–52.
Baracca A, Rossi A (1976) Marxismo e scienze naturali. De Donato, Bari.
Baracca A, Rossi A (1977) Scienza e produzione nel '700. Guaraldi, Firenze.
Baracca A, Rossi A (1978) Materia e energia: antologia di testi. Feltrinelli, Milano.
Baracca A, Ruffo S, Russo A (1979) Scienza e Industria, 1848–1915. Laterza, Bari.
Burtt EA ([1924, 1932] 2003) The metaphysical foundations of modern science. Doubleday, Dover, New York.
Capanna M (1988) Formidabili quegli anni. Rizzoli, Milano.
Ciccotti G, Cini M, De Maria M, Jona-Lasinio G, Donini E ([1976] 2011) L'Ape e l'Architetto. Paradigmi Scientifici e Materialismo Storico. Feltrinelli, Milano.
D'Agostino S (1975) L'elettromagnetismo classico (with an anthology of texts, translated by Maria Grazia Ianniello). Sansoni, Firenze.
D'Agostino S (1987) Il principio d'indeterminazione e la transizione dall'ontologia della fisica classica a quella della meccanica quantistica. In: La Forgia M, Petruccioli S (eds) Rappresentazione e oggetto. Dalla fisica alle altre scienze. Theoria, Roma, pp 23–34.
D'Agostino S (1989) (ed) Gli strumenti dell'elettromagnetismo. Università La Sapienza di Roma Press, Roma.
D'Agostino S (1991a) Immagini della scienza. Per una cultura della riforma. Quaderni del Centro di Metodologia delle Scienze Sociali, Serie I – Saggi. L.U.I.S.S., Roma.
D'Agostino S (1991b) Einstein e Kant: la fisica teorica contro il "logico ma vuoto schema dei concetti. In: Petruccioli S (ed). Immagini Linguaggi Concetti. Theoria, Roma, pp 195–217.
D'Agostino S (1999) Alcune considerazioni sulle teorie dello spazio e sulle geometrie da Descartes ad Einstein. In: Conti L, Mamone Capria M (eds). La scienza e i vortici del dubbio. Edizioni scientifiche Italiane, Napoli, pp 193–211.
D'Agostino S (2000a) A history of the ideas of theoretical physics. Essays on the nineteenth and twentieth century physics, Boston studies in the philosophy of science. Kluwer, Dordrecht.
D'Agostino S (2000b) Ch. 10. Einstein's correspondence criterium and the construction of general relativity. Ch. 11. Einstein's life-long doubts on the physical foundations of the general relativity and unified field theories. A history of the ideas of theoretical physics. Essays on the nineteenth and twentieth century physics, Boston studies in the philosophy of science. Kluwer, Dordrecht, pp 219–252.
D'Agostino S (2000c) Correspondence and complementarity in Niels Bohr's papers 1925–1927. A history of the ideas of theoretical physics. Essays on the nineteenth and twentieth century physics, Boston Studies in the philosophy of science. Kluwer, Dordrecht, pp 253–271.
D'Agostino S (2000d) From the 1926 wave mechanics to a second-quantisation theory: Schrödinger's new interpretation of wave mechanics and microphysics in the 1950's. A history of the ideas of theoretical physics. Essays on the nineteenth and twentieth century physics, Boston studies in the philosophy of science. Kluwer, Dordrecht, pp 273–302.
D'Agostino S (2002) From rational numbers to Dirac's bra and Ket: Symbolic representation of physical laws. Phys Perspect 4:216–229.
D'Agostino S (2005) Gli strumenti scientifici e la scienza. Barbieri, Manduria.
D'Agostino S (2018a) Letture di Storia della Fisica. Da Galileo ad Einstein. Edizioni Accademiche Italiane, Beau Bassin.
D'Agostino S (2018b) An asymmetric evaluation of Einstein's epistemological contribution. Lambert, Beau Bassin.
D'Agostino S (2018c) The centennial recurrence of Einstein's general relativity. Lambert, Beau Bassin.
Eddington AS (1939) The philosophy of physical science. Cambridge University Press, Cambridge.
Giannetto E (2019) Something is rotten in the state of Denmarke: Søren Kierkegaard, Niels Bohr e la nascita della fisica quantistica. Altre Modernità 22:179–197. https://riviste.unimi.it/index.php/AMonline/article/view/12152

Hessen B ([1930] 2017) Le radici sociali ed economiche della meccanica di Newton. Saggio introduttivo di Gerardo Ienna e Giulia Rispoli. Castelvecchi, Roma.
Petruccioli S (1988) Atomi Metafore Paradossi - Niels Bohr e la costruzione di una nuova fisica. Theoria, Roma.
Rossi P (1986) Italo-marxismo, scienza società. I ragni e le formiche. Un'apologia della storia della scienza. Il Mulino, Bologna, pp 229–247, p 236.
Rossi A (2019) Ricordo di Salvo D'Agostino. Physis LIV (1/2):491–495.

Chapter 5
Salvo D'Agostino: Life and Work, from Physics to History of Physics

Arcangelo Rossi

Abstract This text reconstructs the figure of Salvo D'Agostino (1921–2020), eminent and passionate scholar of the historical development and epistemological roots of physics. Moreover, D'Agostino was a fervent and convinced supporter of the importance of a historical perspective for the purpose of a more fruitful training of physicists, both for those actively working in the field of scientific research and for those engaged in teaching activities and therefore in the formation of younger generations.

Keywords History of physics · History and epistemology of science · Salvo D'Agostino · Scientific instruments · Scientific education and popularization

1 Introduction: Salvo D'Agostino's Life

In the lively, stimulating and in some ways even difficult path of the History of Physics in Italy, the figure of Salvo D'Agostino (1921–2020) stands out. His undoubted qualities of vivacity, intellectual passion and vastness of cultural interests as well as his very strong critical sense led him to leave, still as a young man, the research in the field of physics, from which he had also started, to devote himself to the study of the historical development of physics and of its epistemological roots. Actually, he was convinced supporter of the fertility and, even more, of the need for a historical and philosophical perspective that would lead to an adequate training of physicists, both those actively working in the field of scientific research and those engaged in teaching and dissemination activities.

Salvatore D'Agostino, known as Salvo D'Agostino (S. D.) was born in Acireale, province of Catania, on February 3, 1921 (Antiseri and Tagliagambe 2008). In Catania he carried out his first studies. Like many young men of that age, he was initially

A. Rossi (✉)
Dipartimento di Matematica e Fisica "Ennio De Giorgi", Università del Salento, Lecce, Italy
e-mail: arcangelo.rossi@unisalento.it

R. Pisano (ed.), *A History of Physics: Phenomena, Ideas and Mechanisms*,
History of Mechanism and Machine Science 42,
https://doi.org/10.1007/978-3-031-26174-9_5

fascinated by the nationalist and warlike myth expressed by the fascism. Young avant-garde, he left as a volunteer at the outbreak of the war:

> [...] I will not tell you about my pain when I left my land and my trips under Etna, but I was young, I was a little fascist and Italy was after all my homeland. (D'Agostino 2019a, p. 5, Author's translation).[1]

He then went to Milan, in the horse artillery regiment, becoming a second lieutenant in the army of the Kingdom of Italy. After the armistice of September 8, 1943, he returned to Sicily, with an adventurous journey that lasted about a month, crossing the North and much of the Central Italy on freight trains, disguised as a railway worker (D'Agostino 2019a). In a nation half destroyed by bombing, the major railway lines were gone and the remaining railways followed secondary and winding lines, transporting soldiers and civilians displaced from large cities, seeking refuge in small towns. At the station of Cittaducale (RI), the train stopped and was surrounded by German soldiers hunting for soldiers of the Italian army:

> [...] I felt lost, that terrible fear was followed by a strange calm and I looked at my frightened comrades and told them: let's get off the train. We got off and sat at the side of the train waiting for them to pick us up. But after a while the train left with all the Germans and the passengers who had not gotten off. We were safe! (D'Agostino 2019a, p. 6, Author's translation)[2]

When they arrived in the village they met a marshal of the *Carabinieri* (one of the Italian army corps) originating from Sicily as well as the small group of soldiers and displaced people who had gathered around D'Agostino. The marshal advised them to avoid vehicle accessible roads and to return home by foot, following the ancient ways of transhumance:

> [...] These routes were not indicated on the maps [...] but there was no need for geography because the marshal's advice had been followed by hundreds of disbanded southern soldiers who all wanted to return home [...]. Days and days of walking with my companions of misfortune, I no longer remember how many [...]. Many memories now come to mind, but what marks a strong change in our journey was the passage of the front, that is of that no man's land that separated the land occupied by the Germans and the fascists from the one that was slowly conquered by the victorious American, British, Canadian, Australian armies, almost half the world that wanted to defeat the Germans and Mussolini's Italians [...]. I would like to tell you my ideas of how ugly and stupid wars and armies are that pass through the countryside, the mountains and the hills which belong to all men of good will [...]. We were in liberated Italy, there were no Germans but the landscape was the same: ruins and ruins and we risked starving. But there were also the new Italian soldiers of the king who had fled to Bari, we found a military truck [...]. (D'Agostino 2019a, pp. 7-8, Author's translation)[3]

---

[1] "[...] Non vi dico con che dolore ho lasciato la mia terra e le mie gite sotto l'Etna, ma ero giovane, ero un poco fascista e l'Italia era dopo tutto la mia patria." (D'Agostino 2019a, p. 5).

[2] "[...] Mi sono sentito perduto, alla terribile paura succedette una strana calma e guardando i miei compagni spaventati dissi loro: scendiamo dal treno. Scendemmo e ci sedemmo a lato del treno aspettando che ci prendessero. Ma dopo un poco il treno ripartì con tutti i tedeschi e i passeggeri che non erano scesi. Eravamo salvi!" (D'Agostino 2019a, p. 6).

[3] "[...] Queste vie non erano indicate nelle carte geografiche [...] ma non c'era bisogno di geografia perché il consiglio del maresciallo era stato seguito da centinaia di soldati sbandati meridionali

Thanks to some rides on military trucks, long stretches on foot and an adventurous crossing of the strait of sea, S. D. finally arrived in Messina:

> [...] I held back the emotion also because anguished by sad omens. I had a relative in Messina and I went to his house but instead of the door there was an upside-down piano hanging from the strings outside the window. Later I thought it was a symbol of the unhappy time that was awaiting me in Sicily after my return. (D'Agostino 2019a, p. 9, Author's translation)[4]

At the end of the war, D'Agostino resumed his studies at the University of Catania, graduating in Physics. Shortly afterwards he moved to Rome, together with his brother and part of the family. In Rome, driven by his multiple interests and his attraction to the figurative arts, he enrolled in the Faculty of Architecture of the University of Rome *La Sapienza* with the intention of acquiring a second degree. However, his "imprinting" as a physicist made him dissatisfied with some approaches that he considered too qualitative, so he decided to interrupt his studies in architecture and to resume the path of physics.[5]

Thus, in the early 1950s, he began his activity as a teacher of mathematics and physics, first at the classical high school of Orvieto and then in that of Viterbo. These experiences gave rise to his first reflections on the methodology, teaching and history of physics. Reflections that he had the opportunity to develop in his activity at the Institute of Physics of the University *La Sapienza*, initially as a volunteer assistant and later as an ordinary assistant, delivering courses of Experiments of Physics and Complements of General Physics. Those years he met Germana Moech, an English teacher and translator of several important texts, who later became his wife.

Both winners of Fulbright fellowships, in 1959 they went to the USA and they got married there. S. D. attended the course of History of Science at Harvard University held by James Bryant Conant (Antiseri and Tagliagambe 2008; D'Agostino 2009a). Moreover, he signed a contract as a professor of Experimental Physics at the Union College (Schenectady) and he delivered a course in Theoretical Physics at Brandais University. In this manner he had the opportunity to deepen his interests in teaching and his knowledge of theoretical physics (Antiseri and Tagliagambe 2008; D'Agostino 2009a). This stay in the USA was decisive for the subsequent evolution

che tutti volevano tornare a casa [...]. Giorni e giorni di cammino assieme ai miei compagni di sventura, non ricordo più quanti [...]. Tanti ricordi mi tornano ora in mente, ma quello che segna un forte cambiamento nel nostro cammino è stato il passaggio del fronte, cioè di quella terra di nessuno che separava la terra occupata dai tedeschi e dai fascisti da quella che lentamente veniva conquistata dagli eserciti vittoriosi americani, inglesi, canadesi, australiani, quasi mezzo mondo che voleva sconfiggere i tedeschi e gli italiani mussoliniani [...]. Mi verrebbe la voglia di raccontarvi le mie idee di come sono brutte e stupide le guerre e gli eserciti che passano distruggendo e odiando su campagne, monti e colline, che appartengono a tutti gli uomini di buona volontà [...]. Eravamo nell'Italia liberata, non c'erano tedeschi ma il paesaggio era lo stesso: rovine e rovine e rischiavamo di morire di fame. Ma c'erano anche i nuovi soldati italiani del re che se ne era fuggito a Bari, trovammo un camion militare [...]." (D'Agostino 2019a, pp. 7–8).

[4] "[...] Trattenevo la commozione anche perché angosciato da tristi presagi. Avevo un parente a Messina e mi avviai alla sua casa ma invece del portone c'era un pianoforte rovesciato appeso alle corde fuori dalla finestra. Più tardi pensai che fosse un simbolo di quel tempo non felice che mi avrebbe aspettato in Sicilia dopo il mio ritorno." (D'Agostino 2019a, p. 9).

[5] Private communication.

of his interests. In particular, he greatly appreciated some aspects of the American universities that were linked to the variety of curricula and interdisciplinary openness, something that was poorly existent in the rather rigid Italian training course of scientific degrees.

Back to Rome, still at the University *La Sapienza*, he was entrusted with courses of Theoretical Physics and Experimental Physics as part of the didactic address of the degree course in Physics:

> [...] I was happy to be back with Italian physicists in the Institute of Physics that was still under the direction of illustrious physicists of Fermi's time, among whom the names of Edoardo Amaldi and Enrico Persico stood out. I was grateful to Amaldi for his generous hospitality and for his letters of presentation [...]. (D'Agostino 2009a, p. 5, Author's translation)[6]

On the other hand, once back to Rome, another change awaited him: the birth of his first daughter, Caterina. Later on, the family expanded with the birth of two more children, Giovanna and Nicola.

In those years of teaching in Rome, S. D. developed the need for a substantial renewal of teaching both in the overall organization of the courses and in the individual teaching, and this certainly met some resistance among his colleagues of the Institute of Physics, who were particularly engaged in an intense scientific research activity and therefore for the most part reluctant to spend time discussing didactic-methodological aspects (D'Agostino 2009a). However, with the progress of the 60's the social-political and cultural climate was changing and the universities, not only the Italian ones, entered a period of intense and swirling movement of ideas, even radical ideas, not without contrasts with the institutions and not only with them, a movement that, as it is known, had its climax in 1968–69, continuing for most of the 70's. This climate obviously reverberated within the Institute of Physics too, involving young students and many physicists of the new generation in particular:

> [...] I am left with a rewarding memory of those years which were very alive for the ferment of ideas. (D'Agostino 2009a, p. 5, Author's translation)[7]

Nevertheless, while appreciating and sharing the cultural effervescence of those years, S. D. remained somewhat wary and skeptical about some more radically ideologized positions (D'Agostino 2009a). Maybe the hard-experienced disillusionment of his youthful ideological enthusiasm led him to more cautious positions.

Topics such as the relationship between the two cultures, the neutrality/non-neutrality of science, the moral responsibility of scientists and other themes linked to the cultural and social role of science, were the subjects of numerous reflections and lively debates, both in Italy and abroad, as widely reported in a vast literature (Snow 1962; Capocaccia et al. 1965; Cerroni 1970; Cini and Maccacaro 1975;

---

[6] "[...] Ero felice di essere ritornato fra i fisici italiani, nell'Istituto di Fisica ancora sotto la direzione di illustri fisici del tempo di Fermi, fra i quali spiccavano i nomi di Edoardo Amaldi ed Enrico Persico, grato ad Amaldi per la sua generosa ospitalità e per le sue lettere di presentazione [...]." (D'Agostino 2009a, p. 5).

[7] "[...] Di quegli anni molto vivi per il fermento delle idee, mi rimane una gratificante memoria." (D'Agostino 2009a, p. 5).

Baracca et al. 1979; Guerraggio 2010). In particular, with regard to the Roman environment in the Faculty of Mathematical, Physical and Natural Sciences, we should remember the contribution given by various physicists and mathematicians such as Marcello Cini, Giovanni Ciccotti, Michelangelo De Maria, Giovanni Jona-Lasinio, Lucio Lombardo Radice, Giorgio Israel, Carlo Bernardini and many others (Ciccotti et al. 1976; Lombardo Radice 1962, 1981; Battimelli et al. 1984). On the other hand, we must not forget the role and attention to the history of physics of Edoardo Amaldi and Enrico Persico, getting a prestigious tradition and a consolidated memory at the university of Rome *La Sapienza* (it would be enough to mention the role of Federigo Enriques). S. D. himself wrote (D'Agostino 2009a):

> [...] In those years my interest matured towards training physicists in a way that would give prominence to historical and philosophical knowledge, as a premise to a more appropriate function for the modern society and to a greater attention to the problem of the diffusion of scientific culture in our country. The Institute gave me great freedom in teaching programs, a merit that must be recognized to the intellectual freedom of physicists, today not frequent in other sectors of school and university. The CNR (National Research Council of Italy, author's note) financed my research with small funds and I was very happy to be able to teach students who dedicated themselves to teaching, because the problems of the school in general had always interested me. Many of my ideas about the relationship between theory and experiment were formed in those years in a modest microwave laboratory. (D'Agostino 2009a, p. 6, Author's translation)[8]

However, 1970 was the year that marked a turning point in S. D.'s life, definitively pushing him towards the history of physics. In that year he won a fellowship for the history of physics, funded by the *Accademia dei Lincei* and the Royal Society. This fellowship gave him the opportunity of a long stage at the University of Cambridge (UK), having as research advisor Mary Hesse, the eminent scholar of the Maxwellian science.

Back in Italy in 1971, S. D. decided to leave the teaching positions of physics and pure physics researches, to devote himself to teaching the course of History of Physics that had just been established in the degree course in Physics of the University *La Sapienza*. In fact, the reform of 1961, strongly supported by Giovanni Polvani, president of the Italian Physical Society (SIF) until 1961 and president of the CNR since 1960, introduced the teaching of History of Physics in the study plans of the Italian degree courses in Physics (AAVV 2021). Actually, D'Agostino would keep this teaching job until his retirement.

---

[8] "[...] In quegli anni era infatti maturato il mio interesse ad una formazione del fisico che desse rilievo a conoscenze storiche e filosofiche, come premessa ad una sua funzione più adeguata alla moderna società e a una maggiore attenzione al problema di una diffusione nel nostro paese della cultura scientifica. L'Istituto mi dava ampia libertà nei programmi di insegnamento, un merito che va riconosciuto alla libertà intellettuale dei fisici, oggi non frequente in altri settori della scuola e dell'università. Il CNR finanziava con piccoli fondi le mie ricerche ed io ero ben felice di poter insegnare agli studenti che si dedicavano all'indirizzo didattico, perché i problemi della scuola mi avevano sempre interessato. Molte delle mie idee sul rapporto fra teoria ed esperimento si sono formate in quegli anni in un modesto laboratorio di microonde." (D'Agostino 2009a, p. 6).

Consequently with this choice, he began to establish various contacts with various Italian and foreign researchers of history of physics, history of science and epistemology. Some significant moments in the development of these contacts were represented by the participation of S. D. in some conferences and in some important international courses on those topics that took place in those years. In particular, the course in Varenna, in 1972, organized by the SIF, namely the *International School of Physics,* on the subject of *History of 20th Century Physics,* which had among its speakers eminent historians as Martin Klein and John Heilbron as well as illustrious physicists such as P. A. M. Dirac, H. Casimir, E. Amaldi, V. F. Weisskopf and others (Weiner 1977). Other important courses took place at the Center for Scientific Culture *Ettore Majorana* in Erice, for example the course of History of Physics directed by Giorgio Tabarroni and the course of History of Science directed by Vincenzo Cappelletti.

Always in that years S. D. organized a series of history of physics courses at the *Deutsches Museum* in Munich, together with Jürgen Teichmann and Maria Grazia Ianniello, at that time young researcher of the Institute of Physics of *La Sapienza.* Meanwhile, at the end of the 70's, S. D. separated from his wife, with whom in any case he maintained good relations until her death in 2010.

On the other hand, in 1980 two meetings were organized in Pavia and Milan, dedicated to the role of history of physics in teaching, to which S. D. participated with enthusiasm. The following year, the need to develop a coordination on a national basis was felt by various Italian researchers including D'Agostino. This led to the organization of two conferences in Pavia (AAVV 2021) which resulted in the establishment of the National Coordination Group for the History of Physics and, following contacts with the CNR, the establishment of the GNSF, National Group of History of Physics of the CNR in 1982 (AAVV 2021). This group consisted of various units based in Italian universities: initially Bologna, Genoa, Lecce, Milan, Naples, Palermo, Parma, Pavia and Rome, which were soon joined by Catania, Florence, Padua, Turin and Urbino. Guido Tagliaferri was elected as the first president of the group in 1982 (AAVV 2021) while S. D. was a member of the Scientific Council of the group.

The collaboration that D'Agostino had with the National Academy of Sciences called "Dei Quaranta" was also intense and fruitful, thanks also to the work of its presidents, Giovanni Battista Marini Bettolo Marconi and Gian Tommaso Scarascia Mugnozza (Antiseri and Tagliagambe 2008).

Other stimuli to his research activity came from his assiduous participation in the periodic meetings of the Roman Epistemological Circle directed by Dario Antiseri, where lively debates took place with the participation, among others, of Vittorio Somenzi, Massimo Carrea and Ernst Hutten, an illustrious German physicist who was also Einstein and Schrödinger's student (Antiseri and Tagliagambe 2008).

In the same years, D'Agostino began collaborating with the editorial staff of the *Dizionario Biografico degli Italiani*, by the *Istituto della Enciclopedia Italiana Giovanni Treccani.* Our personal relationships were consolidated there, since for almost twenty years we edited the entries dedicated to Italian scientists in the physical and mathematical fields (Rossi 2019). In fact, this activity lasted until the end of 2005, when health problems forced its end.

Among the many initiatives and assignments that saw him involved in those years, we must certainly point out his role of head of the Museum of Scientific Instruments of the Institute (later Department) of Physics of Rome *La Sapienza* and of scientific consultant, held in 1988 during the making of the film "I ragazzi di via Panisperna" under the direction of Gianni Amelio.

In the second half of the 80's the Faculty of Mathematical, Physical and Natural Sciences of *La Sapienza* decided to announce a call for a full professor position in the field of Complements of Physics, a discipline which at that time also included History and Didactics of Physics. S. D. was at that time an associate professor and he believed he could deservedly aspire to the role of full professor. However, the final result of the selection was for him a great disappointment. In that occasion, he was probably penalized by his own age since he was not far from retirement, which actually took place in 1992. This episode was a burning wound for S. D., who froze his relationships with the Department of Physics (as the old Institutes were by now called according to a law of 1980), yet this disillusion did not stop his passion for teaching and research activities.

Actually, from this point of view, D'Agostino spent almost thirty years after his retirement in a very vital and productive way, and the unofficially given motivation for the outcome of the selection, namely his "old age", resounded more and more like a bitter joke to him.

As a demonstration of his strong commitment to research even after retirement, in the second half of the 90's S. D. undertook an intense collaboration with the electrical engineer Adriano Paolo Morando of the Polytechnic of Milan, who was a cultured and passionate scholar of the history of his discipline. This collaboration, in which I too participated, led him to deal with the developments of electrotechnical science in Italy, an activity that resulted among other things in the organization of two international congresses dedicated to Galileo Ferraris (Rome, 1997) (D'Agostino and Rossi 1998; Morando 1998) and Giovanni Giorgi (Rome, 2001) (D'Agostino and Rossi 2004; Morando 2004) with the support of the *Istituto della Enciclopedia Italiana.* Moreover, always with the support of the same institute, we edited the reprinting of the entries written by Enrico Fermi for the *Enciclopedia Italiana* (D'Agostino and Rossi 2001).

In that period, it was published one of the most important D'agostino's works for deepness and vastness of the overall historical vision, namely the book *A History of the Ideas of Theoretical Physics* within the Boston Studies series, which analyzes the advent of theoretical physics as a manifestation of a new way of doing physics (D'Agostino 2000a).

The fruitfulness of D'Agostino's research activity is documented by about two hundred publications on history and teaching of physics and more generally on history of science, in national and international journals, books by several authors, contributions to conferences and seminars as well as a dozen monographs (among these last D'Agostino 2009a, 2000a, 1975, 2005a, 2013, 2017, 2018a, 2018b).

In recognition of his activity and his scientific value he was invited to give lectures in various prestigious international institutions such as the *Deutsches Museum* in

Munich (invited by Jürgen Teichmann), the *Max Planck Institut für Wissenschaftsgeschichte* in Berlin (invited by Jürgen Renn), the *London School of Economics* (invited by T. Heinze), *La Sorbonne* (invited by C. Chevalley), the *University of South Carolina* (international congress on Heinrich Hertz), as well as in various Italian institutes and universities.

He was a member of the scientific council of the National Coordination Group for the History of Physics and then a member of the board of the Italian Society of Historians of Physics and Astronomy, SISFA. As already mentioned, he was part of the editorial staff of *Dizionario Biografico degli Italiani* by the *Istituto della Enciclopedia Italiana Treccani* as well as part of the editorial boards of various journals, including *Physis, Rivista Internazionale di Storia della Scienza* and *Nuova Civiltà delle Macchine.*

From the end of 2005, some health problems progressively reduced his movements but did not weaken his interests, his *verve,* his curiosity and his passion for research. Only in recent years his uninterrupted study left increasing spaces for drawing and painting (his old vocations also linked to his early studies in architecture) and listening to music, a passion that accompanied him throughout his life. In the first hours of 2020 he passed away serenely.

2 Salvo D'Agostino's Work

Scrolling through the long list of works published by Salvo D'Agostino (reported here only partially) immediately we realize how vast and deep his research interests were: from the study of the great physicists, such as Newton (D'Agostino 1965, 1976, 1983a), Ampère (D'Agostino 2000a, 1975, 1983b, 1990, 1991a), Helmholtz (D'Agostino 2000a, 1986a, 2004a, 2005b, 2011, 2014), Hertz (D'Agostino 2000a, 1975, 1971, 2000b), Maxwell (D'Agostino 2000a, 1975, 2000b, 1968, 1968a, 1978, 1981, 1986b, 1996), Boltzmann (D'Agostino 2000a, 1990a, 1990b), Einstein (D'Agostino 2000a, 2017, 2018b, 1973, 1995, 1998b), Bohr (D'Agostino 2000a, 1985a, 1987b), Schrödinger (D'Agostino 2000a, 1991b, 1991c, 1992, 1992a, 2009b; D'Agostino et al. 1992), Dirac (D'Agostino 2000a, 1985b, 2002, 2003b), Fermi (D'Agostino and Rossi 2001, 1995, 1996) and many others (Antiseri and Tagliagambe 2008; D'Agostino 2009a, 2000a, 1975, 2005a, 1998a, 2013, 1998a; D'Agostino and Rossi 1998, 2004), to the reflections on the different epistemological, methodological and didactic aspects involved in the study of the history of physics and the history of science and, more generally, in the development of a scientific culture as widely spread as possible (Antiseri and Tagliagambe 2008; D'Agostino 2009a, 2018a, 1964, 1968c, 1979, 1980a, 1983c, 1983d, 1984, 1986c, 1987a, 2003a, 2004b; D'Agostino and Cametti 1970; D'Agostino and Ianniello 1977; D'Agostino et al. 1979, 1980). His historiographical approach was based on a direct and in-depth comparison with the classics of physics. In fact, while not ignoring and indeed paying close attention to the great debates of the 60's and 70's on the epistemology and historiography of science (Holton 1973; Kuhn 1962; Cassirer 1970;

Popper 1970, 1972), S. D. preferred an independent comparison with the classics of the history of physics. His scientific skills were a useful starting point that allowed an interpenetration with the research and historiographical analysis. When presenting his research activity, he himself described it as:

> [...] His historiography was inspired by the reading of the historical text and by a sometimes difficult immersion in it, and when the thought went back to the knowledge of the physicist of the twentieth and twenty-first centuries, for a comparison on an equal footing with the "scientists" of the past, a comparison that did not presuppose a progress of today's physics compared to past science. The immersion reserved some difficulties in understanding the ideas underlying the argumentative procedures of the "scientists" (their "logic of research") and sometimes caused striking surprises in finding procedures and methods of research that are unusual today, to say the least, but which led to the great theoretical and experimental discoveries that preceded today's physics. (Antiseri and Tagliagambe 2008, p. 141, Author's translation)[9]

Another subject to which S. D. paid a lot of attention concerns the modalities of change and of historical mutation of the ideas of science. His reflections led him to elaborate a thesis that:

> [...] it did not imply a Kuhn-like "incommensurability" among the theories, because mutations were not totalizing, as components such as mathematics for theories (or instrumentation for experiments) caused the new confronted with the old by accepting it partially or totally and therefore justifying it. These components constituted stratifications of continuity among different physical theories. (Antiseri and Tagliagambe 2008, p. 141, Author's translation)[10]

As already mentioned, one of D'Agostino's most important historiographical contributions is the book, published in year 2000 in the collection of Boston Studies (D'Agostino 2000a). According to the same author, in this book:

> [...] a research is presented to identify elements characterizing the change, both on the level of ideas and on that of experiments. It is argued that only through a characterization of the change of ideas in the new theoretical physics, and even through an observation of their partial failure in the so-called post-relativistic theories of "unified fields", was it possible to understand the "conceptual turn" advanced with quantum theories. (Antiseri and Tagliagambe 2008, p. 142, Author's translation)[11]

---

[9] "[...] La sua storiografia prendeva lo spunto da una lettura del testo storico e da un'immersione non sempre facile nello stesso, e soltanto dopo il pensiero risaliva alle conoscenze del fisico del ventesimo e ventunesimo secolo, per un confronto alla pari con gli "scienziati" del passato, un confronto cioè che non presupponesse un progresso della fisica odierna rispetto alla scienza trascorsa. L'immersione riservava difficoltà non lievi nel comprendere le idee sottostanti ai procedimenti argomentativi degli "scienziati" (la loro "logica della ricerca") e procurava sorprese a volte eclatanti nel ritrovare procedimenti e metodi di ricerca oggi inusuali, a dir poco, ma che pure sboccavano nelle grandi scoperte teoriche e sperimentali che hanno preceduto la fisica di oggi." (Antiseri and Tagliagambe 2008, p. 141).

[10] "[...] non comportava una "incommensurabilità" alla Kuhn fra le teorie, perché le mutazioni non erano totalizzanti, in quanto componenti come la matematica per le teorie (o la strumentazione per gli esperimenti) facevano sì che il nuovo si confrontasse con il vecchio accettandolo parzialmente o totalmente e quindi giustificandolo. Queste componenti costituivano stratificazioni di continuità fra fisiche diverse." (Antiseri and Tagliagambe 2008, p. 141).

[11] "[...] si presenta una ricerca per individuare elementi caratterizzanti il cambiamento, sia sul piano delle idee che su quello degli esperimenti. Vi si sostiene che soltanto attraverso una caratterizzazione

Another relevant subject in D'Agostino's research concerns the history of scientific instrumentation and its interconnection with the development of science (D'Agostino 2005a). A subject closely linked to the mutual interaction between theories and experiments, a problem that intrinsically crossed all of S. D.'s research (D'Agostino 1980, 1982, 1983, 1985c, 1986, 1989a, 1990, 1991d, 2004c; D'Agostino and Ianniello 1979, 1980).

It is still necessary to underline another aspect of S. D.'s thought that supported and motivated a large part of his work, his life and his choices, namely the conviction of the need to link scientific and historiographical research with the didactic training of the new recruits of researchers and teachers in order to develop a critical and dynamical vision of the scientific knowledge and to contribute to a wider dissemination of the scientific culture.

To the development of this thought certainly contributed the interaction, in the first years at the Institute of Physics of Rome, with some personalities like Amaldi and Persico, whom he considered among his dearest teachers (D'Agostino 2018a), together with Russel Mc Cormach, Mary Hesse, Don Howard and Giuliano Toraldo di Francia. In particular, he wholeheartedly agreed with Persico's definition and interpretation of scientific culture:

> [...] Enrico Persico said that the scientific culture concerns the relationships among the scientific knowledge, their contextualization in unitary views, the evolution over time of such knowledge and their active and passive relationships through technology with the various aspects of civilization [...]. According to Persico, scientific culture is therefore not identified with either scientific erudition or specialized research. It also concerns the epistemological roots of science and its active and passive relationships with philosophy: we must recognize that in the scientific field, between the superficial knowledge of the results of scientific research and the specialized technique of such a research, there is a vast field of knowledge and mental dispositions that precisely constitute the scientific culture. So he concluded with an accusation against Italian scientists and science scholars: it is a serious fault not to have set up adequate institutional and cultural structures for the development of a scientific culture in universities and academies in Italy. (D'Agostino 2019b, p. 4, Author's translation)[12]

On the other hand, after year 2000, S. D. developed a more optimistic view of the situation:

---

del mutamento delle idee nella nuova fisica teorica, e persino attraverso una constatazione del loro parziale fallimento nelle cosiddette teorie post-relativistiche dei "campi unificati", si riusciva a comprendere la "virata concettuale" che è stata avanzata con le teorie quantistiche." (Antiseri and Tagliagambe 2008, p. 142).

[12] "[...] Enrico Persico sosteneva che la cultura scientifica riguarda i rapporti fra le varie conoscenze scientifiche, il loro inquadramento in vedute unitarie, l'evolversi nel tempo di tali conoscenze e i loro rapporti attivi e passivi attraverso la tecnica con i vari aspetti della civiltà [...]. Secondo Persico la cultura scientifica non si identifica quindi né con l'erudizione scientifica né con la ricerca specializzata. Riguarda anche le radici epistemologiche della scienza e i suoi rapporti attivi e passivi con la filosofia: si deve riconoscere che nel campo scientifico, tra la conoscenza superficiale dei risultati della ricerca scientifica e la tecnica specializzata di tale ricerca, è situato un campo vastissimo di conoscenze e disposizioni mentali che costituisce appunto la cultura scientifica. E concludeva con una accusa ai cultori della scienza italiani: è una grave colpa non aver predisposto presso le università e le accademie, strutture istituzionali e culturali adeguate allo sviluppo di una cultura scientifica in Italia." (D'Agostino 2019b, p. 4).

> [...] Fortunately, a different, more optimistic picture is presented to me today regarding the ideas of scientific researchers and, in particular, of secondary school teachers. Nowadays, the idea of an essential need to rethink the physics and the science of the past as essential for an effective interpenetration of today's science in its problems and hopes for a better future, just this idea is the subject of a debate at the level of physicists and scientific researchers themselves, and the lessons in the history of science are part of several training courses for secondary school teachers. There is also a conviction of the sterility of the contrast between science and philosophy compared to the wealth of interests, including scientific ones, deriving from their harmonization [...]. Unfortunately, the conviction that a modern training for researchers and teachers cannot do without some courses of history of physics and science included as fundamental courses in the first years of university preparatory courses, has not completely penetrated into the organizers of the *curricula* at ministerial level and, in general, into many Italian scientific specialists. For what concerns the last point, the situation has even worsened, a condition more and more anomalous if we consider that these needs had a past and a tradition in the Italian culture, and Federigo Enriques as well as Ludovico Geymonat's activities and ideas remain an emblematic example. (D'Agostino 2009a, p. 8, Author's translation)[13]

Still with regard to the important social role of the scientific culture, intended as aforementioned, S. D. fully shared many of sociologist Luciano Gallino's theses (D'Agostino 2009d; Gallino 1992). In particular, he complained about:

> [...] fragmentation of the cognitive domain that is the result of today training of specialists, inadequate models of the world are nowadays in the repertoire of the majority of specialists or experts in the various fields of science. Since specialists are framed as experts in advisory commissions in which political and economic decision-making bodies participate, the world's inadequate models are at the root of erroneous decisions. (D'Agostino 2009a, p. 15, Author's translation)[14]

D'Agostino, as well as Gallino, considered the need to correct this cognitive fragmentation as a priority, obviously they did not imagine an impossible abolition

---

[13] "[...] Un quadro diverso più ottimistico, per fortuna, mi si presenta oggi riguardo alle idee dei ricercatori scientifici e, in particolare, degli insegnanti di scuola secondaria. L'idea di un essenziale bisogno di ripensare la fisica e la scienza del passato come imprescindibile per una effettiva compenetrazione della scienza di oggi, nei suoi problemi e nelle speranze di un migliore futuro, è oggi argomento di dibattito a livello degli stessi fisici e dei ricercatori scientifici e lezioni di storia della scienza fanno parte di parecchi corsi di formazione dei docenti di scuola secondaria. Si fa anche strada la convinzione della sterilità di una contrapposizione fra scienza e filosofia di contro alla ricchezza di interessi anche scientifici derivanti da una loro armonizzazione [...]. Purtroppo non ha fatto del tutto breccia fra gli organizzatori dei *curricula* a livello ministeriale e, in generale, fra molti specialisti scientifici italiani, la convinzione che una formazione moderna del ricercatore e dell'insegnante non può fare a meno di corsi di storia della fisica e della scienza inseriti come corsi fondamentali nei primi anni dei corsi propedeutici universitari. Per quanto riguarda questo ultimo punto, la situazione è peggiorata, una condizione tanto più anomala se si considera che esigenze del tipo di quelle qui espresse hanno avuto un passato e una tradizione nella cultura italiana, nella quale restano come esempio emblematico l'attività e le idee di Federigo Enriques e di Ludovico Geymonat." (D'Agostino 2009a, p. 8).

[14] "[...] frammentazione del dominio cognitivo che è l'esito appunto della formazione odierna degli specialisti, modelli inadeguati del mondo sono oggi nel repertorio della maggioranza degli specialisti o esperti dei vari campi delle scienze. Poiché gli specialisti vengono inquadrati come esperti nelle commissioni consultive a cui partecipano gli organi decisionali politici ed economici, i modelli inadeguati del mondo sono all'origine delle decisioni errate." (D'Agostino 2009a, p. 15).

of specialisms, actually they wanted to act on university *curricula* in such a way as to at least allow a comparison and a constructive dialogue between them (D'Agostino 2009e).

S. D. identified the aspect linked to the substantial marginalization of scientists by the ruling class as a critical element in the relationship between scientific culture and society. In this regard, he came to the following conclusions:

> [...] my theses can be summarized as follows: if it is true that the persistence of paleo-humanistic cultures that oppose scientific culture constitutes a phenomenon of cultural backwardness of which our country is certainly a victim, it is also true that the positions of some scientists with their ostracism at every introduction of history and philosophy in scientific curricula contribute to weakening any innovation attempts (Israel 2008). Aversion to philosophical problems in science only demonstrates a lack of appreciation of the great dignity of humanistic-philosophical problems (ultimately this is a lack of scientific culture) (Lèvy-Leblond 2007). In the absence of access to modern scientific culture, scientists contribute to their marginalization from important decision-making processes. What is meant here is that the just need for the inclusion of scientific and technical skills in political decisions must be accompanied - for its very effectiveness - by a reform aimed at modifying the formation of the scientific community, so that it can be considered an indispensable counterpart to the positions of reformation of the current recruiting processes of political power. (D'Agostino 2009a, p. 25, Author's translation)[15]

According to S. D., another crucial and very current topic in the relationship between scientific culture and society consisted in the image of the science in the contemporary culture. In that regard, he stated:

> [...] I think the positions that discredit science have a good grip if all we do to oppose them is to preserve outdated images that, arisen in different eras and for historically limited purposes, have unnaturally lived until today outside their historical scope [...]. The risks associated with these images derive from the underestimation of needs that are significant for today's people and that are variously expressed in the idea of freedom, spontaneity, moral law ("the human world"). These demands, not satisfied in the outdated images of science and scientific thought, are then transferred to the realm of the irrational, thus delegating to it a formidable power of image, or, at best, they are considered as the prerogative of the other "culture", the so-called humanistic one, which thus presents itself as another "world" without connections or in antagonism with the world of science. (D'Agostino 2009a, p. 51, Author's translation)[16]

---

[15] "[...] le mie tesi si possono così riassumere: se è vero che il perdurare di culture paleo-umanistiche che si oppongono alla cultura scientifica costituisce un fenomeno di arretratezza culturale di cui il nostro paese è certamente vittima, è anche vero che le posizioni di alcuni scienziati con il loro ostracismo ad ogni immissione di storia e filosofia nei curricula scientifici contribuiscono ad indebolire ogni tentativo di innovazione (Israel 2008). L'avversione alla problematica filosofica nella scienza dimostra solo mancanza di apprezzamento della grande dignità delle problematiche umanistico-filosofiche (in definitiva questa è mancanza di cultura scientifica) (Lèvy-Leblond 2007). In mancanza di un accesso alla cultura scientifica moderna, gli scienziati contribuiscono alla loro emarginazione da importanti processi decisionali. Ciò che si vuol dire è che la giusta esigenza di un inserimento di competenze scientifiche e tecniche nelle decisioni della politica, deve essere accompagnata per la sua stessa efficacia da una riforma tendente a modificare la formazione della comunità scientifica, in modo che essa possa essere considerata una controparte indispensabile delle posizioni di riforma degli attuali processi reclutativi del potere politico." (D'Agostino 2009a, p. 25).

[16] "[...] Io penso che le posizioni screditanti la scienza abbiamo buona presa se per opporsi ad esse si conservano immagini sorpassate che, sorte in epoche diverse e per scopi storicamente limitati,

3 Conclusion

Therefore, if the various issues related to the meaning and role of the scientific culture in the society were subject of so many of D'Agostino's in-depth reflections, it is easy to understand how relevant and permanent his attention and commitment were on all the issues of didactic training both at university and school level. In fact, these themes were the subject of numerous studies and publications throughout his life (D'Agostino 2009a, 2018a, 1964, 1968c, 1979, 1980a, 1983c, 1983d, 1984, 1986c, 1987a, 2003a; D'Agostino and Ianniello 1977; D'Agostino et al. 1979, 1980).

Acknowledgements The author warmly thanks Caterina, Giovanna and Nicola D'Agostino for their great availability and kind collaboration in finding various kinds of information concerning the figure of S. D.

References

AAVV (2021) SISFA: 40 Years of History of Physics in Italy. Percorsi 37(1–2):37–48.
Antiseri D, Tagliagambe S (2008) Storia della Filosofia. Filosofi Italiani contemporanei, vol. 14. RCS Libri, Bompiani, Milano, pp. 138–151.
Baracca A, Ruffo S, Russo A (1979) Scienza e industria, 1848–1915. Laterza, Bari.
Battimelli G, De Maria M, Rossi A (1984) (eds) La Ristrutturazione delle scienze fra le due guerre mondiali, Atti del Congresso, Firenze e Roma, 1980. Editrice Universitaria La Goliardica, Roma, vol. 1 e 2.
Capocaccia A, Eco U, Geymonat L (1965) Verso una nuova cultura. Esigenze, aspetti e problemi di una cultura unitaria. Civiltà delle Macchine XIII(1).
Cassirer E (1970) Determinismo e indeterminismo nella fisica moderna. La Nuova Italia, Firenze.
Cerroni U (1970) Tecnica e libertà. De Donato, Bari.
Ciccotti G, Cini M, De Maria M, Jona Lasinio G (1976) L'ape e l'architetto. Paradigmi scientifici e materialismo storico. Feltrinelli, Milano.
Cini M, Maccacaro GA (1975) (eds) Scienza e Politica. Scienza e potere, AA. VV. Feltrinelli, Milano.
D'Agostino S (1964) Un insegnamento non dogmatico della Scienza. Scuola e Città, pp. 514–515.
D'Agostino S (1965) Il contributo di Newton allo sviluppo dell'ottica. Giornale di Fisica VI (3):188–196.
D'Agostino S (1968a) I vortici dell'etere nella teoria del campo elettromagnetico di Maxwell: la funzione del modello nella costruzione della teoria. Physis 10:188–202.
D'Agostino S (1968b) Il pensiero scientifico di Maxwell e lo sviluppo della teoria del campo elettromagnetico nella memoria "On Faraday's lines of force." Scientia 103:291–301.

si sono innaturalmente prolungate sino ad oggi al di fuori delle loro portate storiche […]. I rischi connessi con queste immagini derivano dalla sottovalutazione di esigenze che sono significative per l'uomo di oggi e che vengono variamente espresse nell'idea di libertà, spontaneità, legge morale ("il mondo umano"). Queste esigenze, non soddisfatte nelle immagini sorpassate della scienza e del pensiero scientifico, vengono allora trasferite nel regno dell'irrazionale, delegando così ad esso un formidabile potere di immagine, o, nel caso migliore, vengono considerati come appannaggio dell'altra "cultura", quella cosiddetta umanistica, che si presenta così come un altro "mondo" senza nessi o in antagonismo con il mondo della scienza." (D'Agostino 2009a, p. 51).

D'Agostino S (1968c) Per la preparazione all'insegnamento della fisica nella Scuola Media. Le Scienze 1:33–36.
D'Agostino S (1971) Hertz e Helmholtz sulle onde elettromagnetiche. Scientia. Rivista di Scienza 106:622–636.
D'Agostino S (1973) L'elettrodinamica di Lorentz sino alla soglia della teoria speciale della relatività di Einstein. Physis 15:260–279.
D'Agostino S (1975a) Hertz's researches on electromagnetic waves. Historical studies in the physical sciences 6:261–323.
D'Agostino S (1975b) L'elettromagnetismo classico. Sansoni, Firenze.
D'Agostino S (1976) Il problema del moto assoluto e della distinzione tra forze impresse e forze insite nel "Philosophiae naturalis principia mathematica" di Isaac Newton. Giornale di Astronomia, Rivista di Informazione Cultura e Didattica della Società Astronomica Italiana 4:275–290.
D'Agostino S (1978) Esperimento e teoria nell'opera di Maxwell: Le misure per le unità assolute elettromagnetiche e la velocità della luce. Scientia 113(5–8):453–468.
D'Agostino S (1979) Il rinnovamento dei musei scientifici. A.N.M.S., Notiziario dell'Associazione Nazionale Musei Scientifici.
D'Agostino S (1980a) Attività scientifica e didattica del gruppo di storia della fisica dell'Istituto di fisica, 1974–1979. Bollettino della Facoltà di Scienze matematiche, fisiche e naturali 1979–1980 4:5–25.
D'Agostino S (1980b) Quale museo per gli strumenti scientifici? risultati e proposte di una ricerca sulla storia degli strumenti scientifici. In: D'Agostino S, Ianniello MG (eds) Storia della scienza e della tecnica: problemi di ricerca storica e didattica nella scuola e nei musei scientifici. Atti del Seminario di studio "Storia della scienza e della tecnica e insegnamento scientifico" Università di Roma, Centro Europeo dell'Educazione, Villa Falconieri, Frascati, 3–7 novembre 1980, Centro Europeo dell'Educazione, Frascati. pp. 183–191.
D'Agostino S (1981) Weber and Maxwell on the discovery of the velocity of light in 19th century electrodynamics. In: Grmek MD, Cohen RS, Cimino G (eds) On Scientific Discovery, The Erice Lectures 1977. Reidel, Dordrecht, pp. 281–293.
D'Agostino S (1982) Empirismo e strumenti scientifici. Physis XXIV(2):281–283.
D'Agostino S (1983a) On the problem of the presumed redundancy of absolute motion in Newton's principia. Physis XXV(1):167–169.
D'Agostino S (1983b) Symmetry in Ampère electrodynamics. In: Doncel MG, Hermann A, Michel L, Pais A (eds) Symmetries in physics (1600–1980). Proceedings of the 1st International Meeting on the History of Scientific Ideas, Sant Feliu de Guixols, Catalonia, Spain, September 20–26. World Scientific Publishing, Singapore, pp. 93–103.
D'Agostino S (1983c) L'importanza della storia delle scienze nel rinnovamento dell'insegnamento scientifico, in AA. VV. L'insegnamento della matematica e delle scienze sperimentali nella scuola secondaria superiore. Parte seconda, Atti del Convegno Internazionale presso la Fondazione Cini, Fondazione Cini, Venezia, p 263.
D'Agostino S (1983d) Conceptions and concepts in physics teaching. In: Kennedy PL (ed). Using history of physics in innovatory physics education. Atti del Seminario Internazionale Centro Studi per la Didattica, Pavia, pp 173–182.
D'Agostino S (1983e) Strumenti scientifici e teorie fisiche: considerazioni storico-critiche sulla storia della strumentazione scientifica. In: Tarozzi G (ed) Gli strumenti nella storia e nella filosofia della scienza. Istituto Beni Artistici Culturali Ambientali Regione Emilia Romagna, Bologna, pp. 173–182.
D'Agostino S (1984) Alcune annotazioni sulla posizione del pensiero scientifico nel pensiero contemporaneo. Cultura e Scuola 90:239–243.
D'Agostino S (1985a) The problem of the link between correspondence and complementarity in Niels Bohr's papers: 1925–1927. Proceedings of the International symposium on Niels Bohr. Università "La Sapienza", Rome. Rivista di Storia della Scienza 2(3):369–390.

D'Agostino S (1985b) La matematica e la fisica nelle idee di P. A. M. Dirac. Memorie di scienze fisiche e naturali. Rendiconti della Accademia nazionale delle scienze detta dei XL, serie V, IX(II):287–290.
D'Agostino S (1985c) Strumenti elettromeccanici e concezioni meccaniciste nella storia dell'elettrodinamica. Epistemologia 8:119–138.
D'Agostino S (1986a) Scienza e cultura nella Germania dell'800. Le basi filosofiche della fisica-matematica nell'opera di Helmholtz. Cultura e Scuola 99:265–282.
D'Agostino S (1986b) Maxwell's dimensional approach to the velocity of light. Centaurus. International Journal of the History of Science and Its Cultural aspects 29:178–204.
D'Agostino S (1986c) La formazione dello storico della scienza: problemi e prospettive. Il Nuovo Saggiatore 3:8–10.
D'Agostino S (1986d) Strumenti scientifici e beni culturali: alcune considerazioni. Insegnare II:10–12.
D'Agostino S (1987a) Federigo Enriques (1871–1946) and History of Science in Italy in the Pre-War Years. Organon 22–23:17–24.
D'Agostino S (1987b) Le relazioni di indeterminazione e la fondazione della meccanica quantistica nell'opera di Bohr ed Heisenberg, 1925–1927. Atti del VII Congresso nazionale di Storia della fisica - Padova, 2–7 ottobre 1986. Gruppo nazionale di coordinamento per la storia della fisica del C.N.R, Milano, pp. 63–85.
D'Agostino S (1989a) La classificazione degli strumenti scientifici. Nuncius 4(1):87–92.
D'Agostino S (1989b) Pourquoi Hertz et non pas Maxwell a-t-il découvert les ondes électriques? Centaurus. International Journal of the History of Science and Its Cultural Aspects 32:66–76.
D'Agostino S (1990a) Boltzmann and Hertz on the bild-conception of physical theory. History of Science 28:380–398.
D'Agostino S (1990b) Da Maxwell a Boltzmann: la costituzione della fisica teorica. Memorie di Scienze Fisiche e Naturali, Rendiconti della Accademia Nazionale delle Scienze detta dei XL, Serie V, XVI(2):343–357.
D'Agostino S (1990c) Matematica ed elettrodinamica nell'opera di Ampère: lo sviluppo delle teorie dell'elettromagnetismo. Cultura e Scuola 116:261–271.
D'Agostino S (1990d) Teorie sistematiche e strumenti precisi: per una storia degli strumenti dell'elettromagnetismo. In: Dragoni G (ed) Instrumenta, il patrimonio storico-scientifico italiano: una realtà straordinaria. Soprintendenza per i Beni Culturali della regione Emilia Romagna, Bologna, pp. 247–254.
D'Agostino S (1991a) Ampère e l'elettrodinamica: leggi matematiche e ipotesi fisiche. Giornale di Fisica 32 (2-3):127-135.
D'Agostino S (1991b) Erwin Schrödinger: la teoria fisica come "bild" e l'integrazione di scienza e cultura. Nuova Civiltà delle Macchine 9(34):151–157.
D'Agostino S (1991c) Il secondo Schrödinger. La riformulazione del concetto di teoria fisica. In: Cattaneo G, Rossi A (eds) I Fondamenti della Meccanica Quantistica. Analisi storica e problemi aperti, EditEI, Commenda di Rende, pp. 91–103.
D'Agostino S (1991d) Gli strumenti dell'elettromagnetismo. Collezioni del Museo del Dipartimento di Fisica, "La Sapienza". Istituto Tecnico Nautico "Colonna", Roma. Università degli studi di Roma "La Sapienza", Roma.
D'Agostino S (1992a) Schrödinger on continuity and completeness in physical theory. Physis. Nuova Serie 29(2):539–561.
D'Agostino S (1992b) Continuity and completeness in physical theory: Schrödinger's return to the wave interpretation of quantum mechanics in the 1950's. In: Bitbol M, Darrigol O (eds) Erwin Schrödinger, Philosophy and the Birth of quantum mechanics. Editions Frontières, Gif-sur-Yvette, pp. 339–362.
D'Agostino S (1993) Hertz's researches and their Place in nineteenth Century theoretical Physics. Centaurus. International Journal of the History of Science and Its Cultural aspects 36:46–82.

D'Agostino S (1995) Einstein's life-long doubts on the physical foundations of the general relativity and unified field theories. In: Garola C, Rossi A (eds) The Foundations of Quantum Mechanics - Historical Analysis and Open Questions. Kluwer Academic Publishers, Dordrecht, pp. 167–178.
D'Agostino S (1996) Absolute systems of units and dimensions of physical quantities: a link between Weber's electrodynamics and Maxwell's electromagnetic theory of light. Physis 33(1–3):5–51.
D'Agostino S (1998a) J. J. Thomson al tempo della scoperta dell'elettrone. Il metodo di Thomson spiegato in una lettera inedita di Thomson a Righi. Memorie di scienze fisiche e naturali, Rendiconti della Accademia nazionale delle scienze detta dei XL, serie V, XXII(2):317–336.
D'Agostino S (1998b) La relatività generale nel dibattito degli anni venti fra neo-kantiani ed empiristi logici: annotazioni su recenti studi einsteiniani. Physis. Nuova Serie 34(3):643–658.
D'Agostino S (2000a) A history of the ideas of theoretical physics. Essays on the nineteenth and twentieth century physics, Boston studies in the philosophy of science n. 213. Kluwer Academic Publishers, Dordrecht.
D'Agostino S (2000b) On the difficulties of the transition from Maxwell's and Hertz's pure-field theories to Lorentz's electron. Physics in Perspective 2:398–410.
D'Agostino S (2002) From rational numbers to Dirac's bra and ket: symbolic representation of physical laws. Physics in Perspective 4:216–229.
D'Agostino S (2003a) La riflessione critica in Europa e la fisica teorica. Arché, Rivista di Filosofia 5:217–230.
D'Agostino S (2003b) Dirac: la ragionevole potenza della matematica. Nuova Civiltà delle Macchine XXI(3):92–105.
D'Agostino S (2004a) The bild conception of physical theory: Helmholtz, Hertz and Schrödinger. Physics in perspective 6:372-389.
D'Agostino S (2004b) La formazione umanistica è indispensabile per un efficace insegnamento della scienza? Nuova Civiltà delle Macchine XXI I(1):144–146.
D'Agostino S (2004c) Un problema delle teorie classiche e quantistiche: esiste una logica della strumentazione? In: Tassani I (ed). Quanti Copenaghen? Bohr, Heisenberg e le interpretazioni della meccanica quantistica: Atti del Convegno, Cesena, 17–19 ottobre 2002. Il ponte vecchio, Cesena, pp. 135–148.
D'Agostino S (2005a) Gli strumenti scientifici e la scienza. Barbieri Selvaggi editore, Manduria.
D'Agostino S (2005b) Il difficile recupero dell'Anschaulichkeit di Goethe nell'opera di Helmholtz. Nuncius 20(2):401–414.
D'Agostino S (2009a) Fra storia e memoria. Ricerca e insegnamento nella fisica italiana. Barbieri Selvaggi Editori, Manduria.
D'Agostino S (2009b) Il "gatto" di Schroedinger in una lettura del testo originale del celebre saggio. Physis. Nuova Serie 46(1–2):235–252.
D'Agostino S (2011) Hermann von Helmholtz e il problema delle due culture. Nuova Civiltà delle Macchine 27(2):207–218.
D'Agostino S (2013) La scienza dell'elettricità da Volta ad Einstein. Una prospettiva storico-didattica. Booksprint Editore, Romagnano al Monte (SA).
D'Agostino S (2014) Fenomenologia e matematica nell'opera di Helmholtz e di Kirchhoff. Physis. Nuova Serie 49(1/2):243–258.
D'Agostino S (2017) The centennial recurrence of Einstein's general relativity. Lambert Academic Publishing, Saarbrücken.
D'Agostino S (2018a) Letture di storia della fisica. Da Galileo a Newton. Edizioni Accademiche Italiane, Beau Bassin (Mauritius).
D'Agostino S (2018b) An asymmetric evaluation of Einstein's epistemologic contributions. Lambert Academic Publishing, Saarbrücken.
D'Agostino S (2019a) A Piedi dalle Alpi verso il mare. In: Complementi ai quaderni, Unpublished.
D'Agostino S (2019b) La cultura scientifica. In: Quaderni più o meno scientifici per amiche, amici e conoscenti. Unpublished.
D'Agostino S, Cametti C (1970) Esperienze didattiche di diffrazione con microonde. Giornale di Fisica 11:36–48.

D'Agostino S, Fabbri B, Ianniello MG, La Forgia M, Tarsitani C, Veronese L, Violini G (1979) La termodinamica nella secondaria superiore. Riforma della Scuola 5:51–56.

D'Agostino S, Ianniello MG (1977) Il Deutsches Museum tra storia e scienza. La Fisica Nella Scuola 4:179–181.

D'Agostino S, Ianniello MG (1979) Catalogo della collezione di strumenti elettrici e magnetici del museo dell'Istituto Fisico dell'Università di Roma, Roma.

D'Agostino S, Ianniello MG (1980) Elettrometri e galvanometri dell'Ottocento. Evoluzione strumentale e contesti teorici. Annali dell'Istituto di Storia della Scienza di Firenze 5(2):69–81.

D'Agostino S, Ianniello MG, Violini G (1980) L'esperimento storico nella didattica della fisica, iniziative e proposte. Fisica nella Scuola XII I(1):19–21.

D'Agostino S, Rossi A (1995) Enrico Fermi e l'origine della fisica teorica in Italia. In: Rossi, A (ed) Atti del XIV e del XV Congresso nazionale di storia della fisica, Udine, 27 settembre - 2 ottobre 1993–Lecce, 26 settembre - 1 ottobre 1994. Conte, Lecce, pp. 467–474.

D'Agostino S, Rossi A (1996) Fermi, Enrico. Dizionario Biografico degli Italiani, Istituto della Enciclopedia Italiana, Roma. 46:329–343.

D'Agostino S, Rossi A (1998) (eds) Galileo Ferraris e il suo tempo. Physis XXXV(2).

D'Agostino S, Rossi A (2001) (eds) Enrico Fermi e l'Enciclopedia Italiana. Istituto dell'Enciclopedia Italiana, Roma.

D'Agostino S, Rossi A (2004) (eds) Giovanni Giorgi nella realtà del suo tempo. Physis XLI(2).

Gallino L (1992) L'incerta alleanza. Modelli di relazioni tra scienze umane e della natura. Einaudi paper backs, Torino.

Guerraggio A (2010) (ed) Il '68 e la scienza in Italia. Pristem–Storia, Note di Matematica, Storia, Cultura 27–28:1–167.

Holton G (1973) Thematic origins of scientific thought, Kepler to Einstein. Harvard University Press, Harvard.

Israel G (2008) Chi sono i nemici della scienza? Lindau s.r.l, Torino.

Kuhn T ([1962]1969) La struttura delle rivoluzioni scientifiche. Einaudi, Torino.

Lèvy-Leblond JM (2007) La Velocità dell'ombra. Ai limiti della scienza. Codice Edizioni, Torino.

Lombardo Radice L (1962) L'educazione della mente. Editori Riuniti, Roma.

Lombardo Radice L (1981) L'Infinito. Numeri, mondi, Dio, spazio, tempo. Da Pitagora a Cantor itinerari filosofici e matematici di un concetto di base. Editori Riuniti, Roma.

Morando AP (1998) Galileo Ferraris e la nascita dell'ingegneria elettrica moderna. Physis XXXV(2):291–339.

Morando AP (2004) Giovanni Giorgi, fisico matematico ed ingegnere elettrico. Physis XLI(2):391–447.

Popper KR (1970) Logica della scoperta scientifica. Einaudi, Torino.

Popper KR (1972) Conoscenza oggettiva, un punto di vista evoluzionistico. Rossi A (ed), Armando Editore, Roma.

Rossi A (2019) Ricordo di Salvo D'Agostino. Physis. Nuova Serie LIV(1–2):491–493.

Snow CP (1962) The two cultures and the scientific evolution. Feltrinelli, Milano.

Weiner C (1977) (ed) History of 20th century physics, Atti della LVII Scuola Internazionale di Fisica Enrico Fermi, Varenna, 1972. Academic Press, New York.

Chapter 6
History of Scientific Instrumentation and History of Science

Pasquale Tucci

Abstract The central point in D'Agostino's reflection about the history of instrumentation was to leave the area of antiques and collecting to become an integral part of research in the History of Science, guided by specific historiographical thesis. For a long time, 'theoreticism' had characterized historian's interest. In the second half of the twentieth century alongside a naïve empiricism widespread especially in scientific circles, the theory-ladenness of observables had driven research in the history of science. D'Agostino's analysis of Hertz's experiment on electromagnetic waves was paradigmatic of a new historiography. From one hand Maxwell had considered the numerical equality between the speed of light and the ratio of absolute units of electrical quantities, obtained with electromechanical instrumentation, to be fundamental for his theory. Hertz, on the other hand, had succeeded in his purpose of detecting electromagnetic waves only by abandoning electromechanical instrumentation, and giving centrality to the resonator. In recent years Hanson and Amsterdamski provided a historiographical background for D'Agostino's analysis about the relationships between History of science and History of the instrumentation.

Keywords Amsterdamski · D'Agostino · History of Instruments · Hanson

1 Introduction

Since his first papers Salvo D'Agostino (hereafter Salvo; D'Agostino 1976, 1978a, b, 1979a, b, 1980, 1981, 1982, 1983, 1985, 1989a; D'Agostino and Ianniello 1979, 1980) scattered his considerations of the history of the scientific instruments among his papers. He was convinced that the criteria of the historical analysis should be founded on the examination of structural features which belong to the instrument *per se* and the modes in which it has been used in meaningful experiments. In this way he addressed from a methodological point of view a specific theme that had to

P. Tucci (✉)
University of Milano, Milan, Italy
e-mail: pasquale.tucci@unimi.it; pasquale1943@gmail.com

R. Pisano (ed.), *A History of Physics: Phenomena, Ideas and Mechanisms*,
History of Mechanism and Machine Science 42,
https://doi.org/10.1007/978-3-031-26174-9_6

do with the relationships between theory and experiment in the empirical sciences. The subject was addressed in the paper on Hertz's electromagnetic wave researches (D'Agostino 1975) and in other papers related to electromagnetism[1] (D'Agostino 1981, 1968, 1986, 1989b, c, 1990, 1996, 2000). At the end it was systematically addressed in a dense and handy-format book published in the 2005 (D'Agostino 2005).

In the book Salvo highlighted in detail how scientific instruments are not merely "data providers" unrelated to theoretical processes but they influenced the formulations of the theory. The comparison then took place between a theoretical theory and a theory of instrumentation in its historical evolution, thus defining a historiography of scientific instruments that was not limited to describing their constructive aspects.

The theoretical context dated back to the interpretation of the papers of Helmholtz, Hertz, Boltzmann that Salvo had dealt with in several publications, while the historiographical context was that of Hanson and Amsterdamski.

In the following paragraphs, I will examine the historiographical context and, subsequently, the original contributions of Salvo.

2 Norwood Russell Hanson

In the 1958 Hanson published a book which emphasized, among many other interesting topics, the theory-ladenness of observations carried out with experimental apparatuses (Hanson 1958). According to the author, the scientific instruments, and the experiments and observations that were carried out with them, were imbued with theory. Secondo Salvo

> [...] the "theoreticism" of the present historiography has been even more accentuated with the researches that, starting from Duhem to the most recent Hanson, highlight the theory-ladenness of the observables.[2] (D'Agostino 1989a, p. 88)

Hanson questioned the neo-positivistic view that theory and observations were clearly separated as if they were part of different vocabularies: the theoretical one, to which theories belong and the observational one, to which the observational reports belong. While the observational vocabulary is stable, the theoretical one is subject to change. If the predictions of a theory (which belonged to the observational vocabulary) proved to be true (i.e. they coincided with the observational report of the experiment written by a scientist) then the theory was accepted. It was known to neo-positivists that the acceptance of a theory often passed through other criteria, but the analysis of such events, according to them, did not belong to the logic but to psychology; acceptance of a theory had nothing to do with the testing of the theory, which was essentially a problem of logic. If this was true, such a point of view

---

[1] I will limit myself to list some of those that I consider most interesting.

[2] "[...] il "teoreticismo" della presente storiografia si è poi ancora più accentuato con le ricerche che, a partire da Duhem sino al più recente Hanson, pongono in risalto la *theory-ladenness* degli osservabili." (D'Agostino 1989a, p. 88).

was unable, according to Hanson, to give an account of the ways of discovering the theories developed by great scientists: Kepler, Newton, Maxwell, Einstein, etc.

H. Putnam reviewed the book and claimed that "[...] this is the most exciting book in the philosophy of science to appear in the last 10 years" (Putnam 1959, p. 1666). Two reasons justified such a so enthusiastic judgment: Hanson was endowed with both the technical equipment of a first rank philosopher and with a deep knowledge of the physics. So Hanson could speak about science and not about "[...] the papier-mâché constructions that replaced science in the writings of philosophers and logicians of science" (Putnam 1959, p. 1666). In addition to the theory-ladenness Hanson introduced also the theory-loaded words discussing the concept of causal explanation.

3 Stefan Amsterdamski

The polish philosopher of science Amsterdamski published his book in the 1975 (Amsterdamski 1975) when the neo-positivistic point of view was still diffused, above all among the scientists, in the form of naïve empiricism. The author devoted the majority of his book to the critics of neo-positivism epistemology and to the alternatives to it (Popper, Kuhn, Feyerabend, Polanyi).

Amsterdamski tried to replace both the idea of one-sided impact of facts on theoretical thinking (as the neo-positivistic inductivism suggested) and the idea that facts are shaped by theoretical thinking, as Hanson and many other historians and philosophers of science after him had done (Amsterdamski 1975, p. XVII). His primary objective was to question the radical empiricist conception of science. He claimed:

> [...] In any case, neither statements of scientists nor reasonings of philosophers provide sufficient confirmation of the thesis, that in posing experimental questions about nature, and reading the answers by means of measuring instruments, scientists were free from hypotheses and presuppositions and that in declaring the glory of radical empiricism they were not subject to any philosophy or metaphysics. (Amsterdamski 1975, p. 21)

Amsterdamski dedicated a section of the book (Amsterdamski 1975, pp. 69–86) to the 'measuring instrument'. Actually, he was referring to the 'man' who, according to the empiriocriticists, was a passive measuring instrument, able to react to the states of the outside world as well as to its own states through his senses. Some kind of coordination between these reactions and the states that the instrument reacted to existed. According to the polish philosopher of science, the epistemology underlying this approach was not unambiguous. Depending on the answer to the following questions: to what signals was the instrument sensitive?; what was its resolving power? Did it transform the signals it received? Different epistemologies could be arranged (Amsterdamski 1975, p. 70). These questions dealt with the measuring instruments. But even for man, considered a measuring instrument, it would have been possible to ask the question: In what way do the internal states of the man, such as memory, as well as the interaction with other similar man affect its functioning?

(D'Agostino and Ianniello 1980, p. 70); and even in this case the answers could have given rise to multiple epistemologies.

Amsterdamski questioned the concept of an 'ideal observer', capable of acquiring information about an object by means of a measuring instrument without changing the state of the object. The measuring instrument was not simply a passive extension of our senses, but it mediated some energetic and informational relationships between the knowing subject and the object.

> [...] Any act of experience, even that which previously was called a passive observation, disturbs to some degree the state of the object and the information we receive about this state is at the same time information about the disturbance which the act of observation has caused, regardless of whether it was a direct or indirect observation. (Amsterdamski 1975, p. 71)

Reports of experimental data were established by means of measuring instruments, constructed always on the basis of some theory, which therefore determined the meaning of observational statements which had their impact upon our accepting or refuting observational statements, so the experimental reports could not be considered as purely empirical (Amsterdamski 1975, p. 79).

As we will see in the next paragraph these considerations were particularly important for Salvo who placed them at the origin of his analysis.

4 Salvo D'Agostino

Both Hanson and Amsterdamski agreed that the rules of the scientific method were not permanent through the centuries and that therefore one could not speak of a historically stable phenomenon. In particular, Hanson had argued that the twentieth-century sciences—Theory of Relativity, Quantum Mechanics etc.—could not be regarded as paradigms of physical inquiry and that some features that are identified in them cannot be transferred to sciences such as planetary mechanics, optics, electromagnetism, classical thermodynamics that are no longer, according to Hanson, research disciplines (Hanson 1958).

On the other hand Amsterdamski wrote:

> [...] By what right do we use the name *science* for such dissimilar systems of belief as, for example, Greek cosmology and the theory of relativity, the theory of preformation and contemporary molecular biology, the atomistic theory of Democritus and the contemporary theory of the structure of matter, in other words, for all that different period was considered as science? (Amsterdamski 1975, p. 39)

Kuhn had attempted to solve the problem by stating that

> [...] Rather than seeking the permanent contributions of an older science to our present vantage, they [the historians of science] attempt to display the historical integrity of that science in its own time. (Kuhn 1962, p. 3)

Geymonat (Amsterdamski 1975, p. 38) argued a similar view in the book *Filosofia e Filosofia della scienza* (Geymonat 1960). The scholar who is dedicated to the historical approach will state, that rather than search for permanent, normative definitions of science, we should instead search for

> [...] what in different periods men had in mind when they called a specific field of human activity as science, on what basis they judged activity as scientific, what changes did the notion of science undergo from antiquity to modern times, and what changes did the moving of the scientific method undergo. (Geymonat quoted by Amsterdamski; Amsterdamski 1975, pp. 26-27)

Salvo elaborated an original theory of instrumentation borrowing the theory-ladnness from Hanson, at the time of Salvo endorsed by many other historians and philosophers of science, and by Amsterdamski the thesis that not only the instrument is imbued with theory but that the instrument perturbed the state of the examined object.

Salvo emphasized the importance of Amsterdamski's point of view about the revolution that was at the origin of physics as an empirical science in the 17th and 18th Centuries. According to Salvo this revolution was supported by a philosophy whose fundamental tenet was

> [...] the conception of an *ideal observer*, i.e., an observer who reacts by his feelers (the senses) to the states of the outside world without interfering in any way with the observed object and/or the measuring instrument. (D'Agostino 2004b, p. 130)

It's worth to stress that these considerations were included in a contribution that Salvo had written for a conference on the Foundation of Quantum Mechanics, where the concept of a clear separation between the measuring instrument and the investigated object had been abandoned. However they are found, more or less explicitly, in all his papers that had to do with the history of the instrumentation.

Amsterdamski's point of view reversed, according to Salvo,

> [...] the traditional understanding of the process of the origin and the growth of physics as an empirical science, because, in his view, these processes were not originated by the mere transition from speculative to empirical thinking, as usually intended, but this transition was conditioned by the assumptions of an *ideal observer* and of a *theory neutral instrument*. (D'Agostino 2004b, pp. 130–131)

Even when we do not use any measuring instrument in the course of an experiment, the result is interpreted, and the observational statement is accepted or refuted, on the grounds of different assumptions and theories concerning the structure of the world and the mechanism of our cognitive functions.

The thesis that the thinking man is conceived as a passive receiver of data through the senses was, according to Salvo,

> [...] a very fruitful idea because it legitimated [...] the use of the tools as passive data receptors (secondary receptors). (D'Agostino 2005, p. 18)[3]

---

[3] "[...] un'idea fecondissima perché legalizzava [...] l'impiego degli strumenti come ricettori passivi di dati (ricettori secondari)." (D'Agostino 2005, p. 18).

The use of the instruments as an expansion of our senses, together with the mathematical analysis of data, characterized the scientific revolution of the eighteenth and nine-tenth centuries. According to Salvo, therefore, it was possible to develop

> [...] Amsterdamski's thesis states that the *empirical hypothesis* was fundamentally associated with a conception of the physical object, as shown among other things by the Galilean postulation of an objective nature of geometric shapes. (D'Agostino 2005, p. 18; Author's translation)[4]

Later on the statistical theory of errors was formulated by Laplace (1749–1827) and Gauss (1775–1855) in order to obtain the "true" measurement from *perturbations* that the instruments introduced into the actual measurements. This was the same as to say that the "true" measurement was the result of a theory. We are however within the framework of a classical conception of the science—what Salvo called "instrumental empiricism"—according to which scientific theories had to deal with entities that had an existence instrument-independent.

Not only the instrument was imbued with theory, but it introduced a new metric, as the Galilean experiment of the inclined plane already showed. Actually it provided for a new correlation between space and time, quantities that were not homogeneous, and therefore not correlated, according to ancient science. The Theory of Relativity and of Quantum Mechanics made even more explicit that every measurement involved a new metric.

In the second half of the nineteenth century, thanks also to the elaboration of Kantian philosophy by great scientists, instrumental empiricism began to be questioned in favour of a conception according to which the instrument as well as the thinking subject was a mediator between the subject and the object.

The subject/object dualism was first questioned by the success of the Thermodynamics. According to this theory a Maxwellian demon, able to select the molecules of a gas according to their temperature, would have been impossible, as the entropy would decrease, so violating the first principle. In addition, definitively, the dualism was destroyed by Heisenberg's uncertainty principle.

Salvo addressed these issues in the analysis of difference between Helmholtz and Hertz regarding the "parallelism of laws".[5] For Helmholtz the measurements of physical quantities that provided the properties of the objects examined were not objective, but they were only the result of an appropriate instrumental choice that introduced its own metric into the fields of investigation. For him there was a close correlation between external physical agents and their effects on our sensations or perceptions, observable or directly or through some physical instrument. A simple representation of physical facts and laws by a system of differential equations was the right way to approach the natural phenomena and their relationships with our theories (D'Agostino 2004a, p. 377). Hertz, one of Helmholtz's favourite pupils, criticized the idea of a uniqueness of parallelism. For Hertz, different ways of measuring the

[4] "[...] la tesi di Amsterdamski affermando che l'*ipotesi empirica* si associò fondamentalmente a una concezione dell'oggetto fisico, come viene rivelato fra l'altro dalla postulazione galileiana di una natura oggettiva di forme geometriche." (D'Agostino 2005, p. 18).

[5] In Italian "parallelismo delle legalità" (D'Agostino 2005, p. 25).

same phenomenon led to different metrics; so it did not determine unambiguously a physical theory.

For Hertz a physical theory was a system of equations that did not change over time, it is the "undying part" of it. But modes of representation of a theory could change over time: the same system of equations could correspond to various ways of representations of them. Salvo, in the analysis of Hertz's Preface to his *Electrical waves*, noted that there was a difference between Helmholtz and Hertz regarding the relationships between the mathematical structure of a theory and its modes of representation.:

> [...] by separating the mathematical structure of a theory from its modes of representation he [Hertz] has profoundly challenged the conception of a physical theory as an indivisible unity of the two [...]. (D'Agostino 2004a, p. 379)

Hertz's position thus was in conflict with Helmholtz's parallelism of laws between concepts and perceptions. The propagation of electric force could have been explained by the contiguous action of Hertz but also by Helmholtz's action-at a-distance theory; experiments alone were unable to decide between the two (D'Agostino 2004a, p. 379). More generally, the thesis of an absolute differentiation between instrumental theories and theories concerning an autonomously definable system regardless of observations and measurements was not sustainable (D'Agostino 2005, p. 26).

But an instrument not only "observes" through a theory but is built according to a theory; indeed, it often stratifies various theories that together constitute its history. The Coulomb balance, for example, is built according to the mechanical theory of torsion. But when it is used by swinging the suspended arm, the theory of oscillations comes into play. Moreover, under certain conditions, it is necessary to take into account air resistance. And so on. But the balance cannot control the mechanical laws of torsion, otherwise we would fall into a vicious circle.

On the other hand, an instrument built according to some theoretical rules cannot control a theory that has nothing to do with the theory according to which it was built. It would be impossible to control the law of falling of a body by means of a spectrometer, whose constructive theory is basically based on laws that do not belong to Mechanics. It is necessary, therefore, that between the theory according to which an instrument is built and the theory that one wants to control there is some form of connection. Salvo called this connection "correspondence". In addition, there is no doubt that the linguistic allusion to the principle of correspondence in quantum mechanics is not at all accidental. In Quantum Mechanics, in fact,

> [...] the measuring instrument (and more generally every device) behaves in a classic way in that process of interaction between the observation system and the measuring apparatus on which the measurement is based. (D'Agostino 1983, p. 182; Author's translation)[6]

[6] "[...] lo strumento di misura (e in generale ogni apparecchio) si comporta in modo classico in quel processo di interazione fra sistema di osservazione e apparato di misura in cui consiste la misurazione." (D'Agostino 1983, p. 182).

Coming back to the Coulomb balance we can give to the masses an electrical charge in order to check if the law of the inverse of the square between electrical charges applies. There is therefore an area of correspondence between the mechanical and electrostatic theory that needs to be checked. Indeed, on closer analysis, the torsion theory represents only a first stratification in the area of correspondence of the electrostatic theory.

The same can be said of Hertz's experiment on the propagation of electromagnetic waves. It was carried out using generator instruments (Ruhmkorff's coil) and electromagnetic radiation detector instruments (electromagnetic oscillators and resonators). A first stratification can be placed in the circuit electrodynamics, while Maxwellian electromagnetism gives the second stratification.

The history of instrumentation, made in Salvo's way, is therefore an integral part of the history of science and any hierarchy that privileges the theoretical or instrumental aspect ought to be banned from the historical reconstruction.

5 Conclusion

If you scroll through the list of almost 200 publications of Salvo we realize that the topics covered range from the history of ancient science to that of quantum mechanics. The core of Salvo's research activity concerns the history of science in the second half of the nineteenth century and, in particular, the history of electromagnetism from Maxwell to Hertz. In the papers dedicated to this research Salvo shows all his ability to highlight the interplay between the physics, the history of physics, the history of instrumentation, epistemology, philosophy with a surprising continuity of argumentation. Salvo was able to balance the detailed and meticulous analysis of the memoirs with an analysis of the epistemological and philosophical context in which the authors he treated acted.

From the contributions that Giannetto and Rossi have also given in this volume it is clear that for Salvo the history of science is a discipline in which some research topics cannot be isolated. This was the undying contribution that Salvo made to the discipline he loved and to which he dedicated over half a century of his life.

References

Amsterdamski S (1975) Between experience and metaphysics/Philosophical problems of the evolution of science. Reidel, Dordrecht–Boston.

D'Agostino S (1968) Il pensiero scientifico di Maxwell e lo sviluppo della teoria del campo elettromagnetico nella memoria "On Faraday's lines of force". Scientia 103:291–301.

D'Agostino S (1975) Hertz's Researches on Electromagnetic Waves. Historical Studies in the Physical Sciences 6:261–323.

D'Agostino S (1976) La scoperta di una velocità quasi uguale alla velocità della luce nell'elettrodinamica di Wilhelm Weber (1804-1891). Physis XVIII (3-4):297–318.

D'Agostino S (1978a) Esperimento e teoria nell'opera di Maxwell. Scientia 113:453–468.

D'Agostino S (1978b) Esperimento e teoria nell'opera di Maxwell: Le misure per le unità assolute elettromagnetiche e la velocità della luce. Scientia 113(5-8):453–468.

D'Agostino S (1979a) Modelli ed oggetti della teoria fisica. La nuova critica. Rivista di scienze dell'uomo e di filosofia delle scienze XIII(49):5–19.

D'Agostino S (1979b) Storia della strumentazione scientifica e sviluppo della scienza: i problemi posti dall'uso conoscitivo della techne. Medicina nei secoli 1:13–29.

D'Agostino S (1980) Quale museo per gli strumenti scientifici?: risultati e proposte di una ricerca sulla storia degli strumenti scientifici. In: D'Agostino S, Ianniello MG (eds). Storia della scienza e della tecnica: problemi di ricerca storica e didattica nella scuola e nei musei scientifici. Atti del Seminario di studio "Storia della scienza e della tecnica e insegnamento scientifico" Università di Roma, Centro Europeo dell'Educazione, Villa Falconieri, Frascati, 3-7 novembre 1980. Centro Europeo dell'Educazione, Frascati, pp. 183–191.

D'Agostino S (1981) Weber and Maxwell on the Discovery of the Velocity of light in 19th Century Electrodynamics. In: Grmek MD, Cohen RS, Cimino G (eds). On Scientific Discovery: The Erice Lectures 1977. Reidel, Dordrecht, pp. 281–293.

D'Agostino S (1982) Empirismo e strumenti scientifici. Physis XXIV(2):281–283.

D'Agostino S (1983) Strumenti scientifici e teorie fisiche: considerazioni storico-critiche sulla storia della strumentazione scientifica. In: Tarozzi G (ed). Gli strumenti nella storia e nella filosofia della scienza. Istituto per i Beni Artistici Culturali Ambientali della Regione Emilia Romagna, Bologna, pp. 173–182.

D'Agostino S (1985) Strumenti elettromeccanici e concezioni meccaniciste nella storia dell'elettrodinamica. Epistemologia 8:119–138.

D'Agostino S (1986) Maxwell's Dimensional Approach to the Velocity of Light Centaurus. International Journal of the History of Science and its Cultural Aspects 29:178–204.

D'Agostino S (1989a) La Classificazione degli strumenti scientifici. Nuncius 4(1):87–92.

D'Agostino S (1989b) Pourquoi Hertz et non pas Maxwell a-t-il découvert les ondes électriques? Centaurus. International Journal of the History of Science and its Cultural Aspects 32:66–76.

D'Agostino S (1989c) The bild-conception of physical theory and the foundation of Boltzmann's mechanics. X SISFA Proceedings, Cagliari, 28 settembre - 2 ottobre 1989. Gruppo Nazionale di Coordinamento per la Storia della Fisica del CNR, Milano, pp. 95–110.

D'Agostino S (1990) Boltzmann and Hertz on the Bild-Conception of Physical Theory. History of Science 28:380–398.

D'Agostino S (1996) Absolute systems of units and dimensions of physical quantities: a link between Weber's electrodynamics and Maxwell's electromagnetic theory of light. Physis 33(1-3):5–51.

D'Agostino S (2000) On the Difficulties of the Transition from Maxwell's and Hertz's Pure-Field Theories to Lorentz's Electron. Physics in Perspective 2:398–410.

D'Agostino S (2004a) The Bild Conception of Physical Theory: Helmholtz, Hertz and Schrödinger. Physics in Perspective 6:372–389.

D'Agostino S (2004b) Theory-Laden instruments and quantum mechanics. In Garola C, Rossi A, Sozzo S (eds). The Foundations of Quantum Mechanics/Historical Analysis and Open Questions-Cesena 2004. World Scientific Press, Singapore, pp. 130–141.

D'Agostino S (2005) Gli strumenti scientifici e la scienza. Barbieri, Manduria.

D'Agostino S, Ianniello MG (1979) Catalogo strumenti. Catalogo della collezione di strumenti elettrici e magnetici. Museo dell'Istituto di fisica dell'Università di Roma, Roma.

D'Agostino S, Ianniello MG (1980) Elettrometri e galvanometri dell'Ottocento. Evoluzione strumentale e contesti teorici. Annali dell'Istituto di Storia della Scienza di Firenze 5(2):69–81.

Geymonat L (1960) Filosofia e Filosofia della scienza. Feltrinelli, Milano.
Hanson NR (1958) Pattern of Discovery/An inquiry into the conceptual foundations of science. Cambridge University Press, Cambridge.
Kuhn T (1962) The structure of scientific revolutions. University of Chicago Press, Chicago.
Putnam H (1959) Review of Pattern of Discovery/An inquiry into the conceptual foundations of science. Science 129(3364):1666–1667.

Part II
Essays on Salvo D'agostino's Research Topics

Chapter 7
Versions of Determinism

Joseph Agassi

Abstract Karl Popper's "Indeterminism in Quantum Physics and in Classical Physics" suffers unjust neglect. He judged determinism false: the future is open. In principle, replacing in Laplace's variant of predictable predetermination renders "scientific" determinism scientific and so refutable. Popper claimed that he had refuted it. Now a metaphysical system may have an extension—in the mathematical sense—that may render it explanatory and testable. If it exists, then it is not unique but may have many alternative extensions. Popper's proof is then inconclusive.

Keywords Determinisms · Extension · Fault · Indeterminism · Laplace's metaphysical determinism · Popper · Quantum physics · Variant

1 An Outline

The attraction of determinism is in its avoidance of judgment. Heinrich Heine notes that Friedrich the Great had expelled Voltaire from his court for his view that soldiers do not deserve flogging, as they are mere automata. Popper responded to that story, saying, the monarch expelled Voltaire because he had no response to him, but there is a simple one: if they are automata, then I may flog them to my heart's content! The last word against determinism is that of Alfred Landé: it renders miraculous the possibility of any meaningful activity (like writing) sine the laws of physics fully determines it and yet it also follows its rules (namely, grammar). This is over-determination. The simplest example is from elementary algebra: the values of n variables is fully determined by n independent equations; adding another independent equation makes it insoluble and the likelihood that the additional one will depend on the other is very slim.

J. Agassi (✉)
Tel Aviv University, Tel Aviv, Israel
e-mail: agass@tauex.tau.ac.il

York University, Ontairo, Canada

R. Pisano (ed.), *A History of Physics: Phenomena, Ideas and Mechanisms*,
History of Mechanism and Machine Science 42,
https://doi.org/10.1007/978-3-031-26174-9_7

2 Popper: *Indeterminism* and *Determinism*

Karl Popper's *Indeterminism in Quantum Physics and in Classical Physics* (Popper 1950, 1957, 1982; Agassi 1975) suffers unjust neglect. It discusses a variant of Laplace's determinism. Laplace offered a solution to a metaphysical problem: he wrote about probability that seems to clash with the determinism that he took for granted; is the clash real? The answer of Laplace is the subjective interpretation of the axioms of probability. All his life Popper argued against this interpretation and for a realist one.[1]

The preface to Laplace's *Philosophical Essay on Probability* of 1814 introduces an intellect—the literature refers to it as to "Laplace's demon"—armed with Newtonian mechanics, an image of the universe in any single moment, and an unlimited ability to compute. That intellect, Laplace declared, can know every past and every future event in the whole universe. This claim is Laplace's determinism; it is a version of "scientific" determinism. The discussion of determinism that followed the publication of his book centered on his version of it: obviously received opinion deemed it the best version. First, the versions of "scientific" determinism are all superior to religious fatalism—to the mere declaration that the future is predetermined—in that they appeal to science. Second, Laplace's determinism is a more detailed version "scientific" determinism than that of Spinoza, as it appeals to a specific, highly corroborated scientific theory.

The first attack on the version of determinism of Laplace was mathematical; Poincaré (Poincaré [1902] 1905) and Jacques Hadamard refuted some assumptions that Laplace had made about the power of mathematics. They refuted his assumption that in principle mathematics offers solutions to all the problems that it raises. Their just and historically very significant mathematical arguments are not relevant to this essay, that concedes the supposition of Laplace that in principle mathematics is complete—in order to examine the rest of his assumptions.

Popper judged determinism false: the future is open, he declared. His interest in determinism was to discredit it as much as he could, and mainly but not only for moral and political reasons: he viewed the thesis of historical inevitability a corollary to "scientific" determinism and he found it responsible for much political iniquity and moral irresponsibility.[2] He conceded that both the thesis of historical inevitability

---

[1] Popper ((1935), end of §27) declared that his methodology is open to both the subjectivist and the objectivist interpretation, and that his preference for the objectivist one is personal. In his preface to its 1959 English translation, he declared a change of mind: he viewed his position as objectivist and anti-subjectivist. The most challenging part of this reading was the chapter on probability. He made two great contributions here that he worked on for the rest of his life: he offered the first autonomous axiom system for probability, namely, a system that is open to all interpretations. Moreover, he developed the objectivist one, namely, his propensity interpretation of the axioms of probability.

[2] Popper ((1945), Ch. 13) rightly emphasized that it is possible to hold a metaphysical version of determinism and consider one's behavior as if determinism is false—as Spinoza did quite successfully and as Einstein did. Nevertheless, Popper also observed in that chapter, determinism did influence conduct, at least that of Karl Marx.

and “scientific” determinism irrefutable and hence[3] possibly true. Replacing mere predetermination with a stronger claim, namely predictability in principle, and so of testability in principle, renders “scientific” determinism scientific and so refutable and then, Popper claimed, he had refuted it by the observation that already Henri Poincaré has made: it is impossible to know our future predictions while keeping them in the future. Clearly, then, this observation of Poincaré (Poincaré [1902] 1905) proves inconsistent any Laplacian system that contains a predictor. It describes a universe containing a predictor able to predict every future event yet unable to predict at least one future event, namely, at least one item out of its own future behavior, namely, any future prediction of its own.

One might object to this line of thought: although you cannot predict your own future prediction, I can. This objection is easy to refute: a pair of predictors comprise a predictor, as is a community of predictors—since no limitation on the space that the predictor occupies are irrelevant here. True, when you and I try to predict the outcome of each other’s prediction of each other prediction, the outcome is the same as any Popper-type short-circuit. Watkins (1970) has adduced non-trivial and informative instances of Popper-type short-circuits, both in history and in game theory.[4] This sound too sophisticated; it seems we may exclude ad hoc self-prediction at least in the early stages of the discussion, as hardly possible anyway and so as scarcely relevant to the discussion of the impact of science on philosophy, and then ignore Popper’s discussion. Not so: in his discussion of our inability to predict our own predictions Poincaré referred to our inability to predict the course of science; since science has tremendous and unforeseeable influence on our lives, no significant prediction of human future is possible.[5] Yet, Popper admitted, determinism irrefutable. It is thus possibly true. This situation he found disagreeable. There is one technique for handling such situations, already repeatedly illustrated in Plato’s early dialogues: the disagreeable idea is too thin to be applicable; for this it needs strengthening, and its

[3] Modal logic validates the inference from “x is irrefutable” to “x is possibly true” (Dummet 2011, 19). The verification principle denies meaning to the irrefutable and the possibly true. Its popularity at the time explains the neglect of Popper’s discussion of determinism.

[4] Cfr. Watkins. Predictions that players make about each other’s prediction about each other’s conduct, refute the idea that full knowledge precludes risk (Watkins (1970), pp. 197–198). Extending this to the prisoners’ dilemma refutes a basic assumption of some versions of traditional game theory (Watkins (1970), p. 206): it creates the situation—known from tense international borders—of an undesired Nash equilibrium of mutual distrust where the desired one is of mutual trust.

[5] Two examples. First, Marx used the fact that the efficiency of a steam engine increases with its size to predict the concentration of wealth due to competition, consequently the inability of the middle classes to compete with big capital, and thus their disappearance. Soon after he died, the new applicability of the electric dynamo and motor enabled the rise of the Edison Electric Company and its likes, and these enabled the creation of a new middle class of small entrepreneurs whose production depended on the available electric current. Second, Keynesian economic proposals prevented the allegedly ever increasing misery due to economic crises that Marx deemed unstoppable, thus leading inexorably to the socialist revolution. Instead, this led to the rise of the welfare state that Marx had wrongly deemed impossible. It may be appropriate to mention science fiction here: Isaac Asimov’s *Foundation* (Asimov 1951) trilogy has its hero, Hari Seldon, made people in power forget his own prediction—to avoid defeating it.

strengthened version is open to criticism. Popper undertook the task of enriching[6] Laplacian determinism so as to render it open to criticism and then to try to criticize it. The assumption that it is possible to perform this task raises the problem, then, as to whether other adequate extensions of the system are available. Still, let us first go over Popper's presentation of the situation. Popper's extension of the Laplacian system adds to it a Laplace-predictor, namely, the intellect of Laplace's initial description. Is this system, Popper asks, allowing the assumption of Laplace that our universe is determinist, is it still so after assuming that it contains predictors like you and me? (This addition of the Laplace-predictor to Laplace's system is the view that the intellect in his initial description of his system is an idealized version of Laplace himself.) The Laplace-predictor cannot perform every prediction, not the prediction that one Laplace-predictor should make about what another Laplace-predictor will predict. A Laplace-predictor thus cannot predict its own future prediction. This is a short circuit. To be able to prove his thesis, Popper replaced predetermination with predictability (within agreed limits of accuracy), which may be testable.

3 Popper's *Variant*

Popper claimed that his *variant* of the Laplace thesis is scientific, since it is inconsistent. This is an error: by his demarcation of science, a scientific system of statements must be:

(a) Consistent
and
(b) inconsistent with at least one conceivable[7] observation report.

Hence, by Popper's own criterion of demarcation of science, his variant of the system of Laplace is unscientific. Hence, at most Popper has shown that "scientific" determinism is impossible; he has not refuted the metaphysical thesis extended to be

[6] Diverse synonyms name enriching or increasing the content or the force of a system, or extending it in the mathematical sense of the word. Popper notices in his *The Logic of Scientific Discovery* (Popper 1959, §15, Note *1), that the enrichment of a theory is self-understood. Incidentally, this he had to state since the reading of his book as a theory of scientific language—his protestation notwithstanding—renders highly problematic this rather intuitively admissible process. The extensions under discussion here are rather organic: otherwise, the mere conjunction of any two theories extends them. The reverse is also significant: of given variants of a given theory, Popper's methodology recommends the preference of the less informative but equally testable one (e.g., Mach's version of Newton's mechanics minus its assumption of the existence of absolute space).

[7] The metaphysics of Parmenides is refutable, as already Antisthenes has claimed (by moving back and forth). Parmenides would not admit the refuting observation, and Zeno tried to prove him right on this. To meet this, Popper replaced the requirement for tests with the requirement of admissible ones: he allowed for the conventionalist's refusal to admit the possible refutation of a significant theory, and offered in opposition to it the convention to avoid the apologetic rescue of a theory. This convention will nullify Parmenides theory as scientific but not as a metaphysics. This is important since that theory led to the atomism whose importance is beyond contest (Popper 1952, 1963).

scientific *à la* Popper, as he had not constructed one. (His extended version adds too much.)

A metaphysical system may have an extension—in the mathematical sense—that may render it explanatory and empirically testable. The paradigm case for this is the case of atomism, ancient and modern, as these are metaphysical and scientific respectively.[8] All sorts of writers[9] have taken for granted the ability to render a metaphysics scientific, including those most hostile to metaphysics. Thus, when members of the Vienna Circle dismissed theology as meaningless on the ground that the Holy Name does not designate clearly, they explained that had it clearly and unequivocally designated, say, the column of fire that supposedly went before the Children of Israel in the desert (*Exodus*, 13:21–22), then they would consider assertions about the Divine false, not meaningless. The method of extension, however, is in great neglect. The characterization of metaphysics remained for long unstudied, although repeatedly some commentators took it for granted that a metaphysics is a system of general suppositions for science, and thus a general outlook or a general framework or a worldview of sorts.[10] Ludwig Wittgenstein declared high-handedly that all efforts to state a metaphysics in sufficiently clear statements is impossible. He thereby blocked all effort to characterize metaphysics.[11] Popper's early publications characterized scientific systems (not sentences); (ignoring here logic and mathematics, as the status of these is uncontested) he advised to disregard all non-science, including systems of superstition, of religion, of metaphysics, and of pseudo-science, calling them all "metaphysical" even though they clearly differ from each other in many respects. He noted that metaphysical systems such as ancient atomism have heuristic value (Popper (1935), §4, §78, §85 and Appendices) but while insisting there on leaving aside all heuristic. After his famous clash with Wittgenstein, he discussed the problems that philosophy (metaphysic) comes to answer as rooted in scientific theories (Popper 1952). These problems appeared there as cosmological, namely, in search of a worldview, and in this study of his ancient atomism played a major role.

---

[8] Admittedly, a theory can have metaphysical and scientific readings. Thales theory does, since water is decomposable and since Einstein has demolished the substance theory.

[9] The literature on this item is immense. It is all elaborations on Wittgenstein "[...] whenever someone else wanted to say something metaphysical, to demonstrate to him that he had failed to give a meaning to certain signs in his propositions" (Wittgenstein ([1921] 1922), §6.53). For, they all took for granted his "what can be said at all can be said clearly, and what we cannot talk about we must pass over in silence" (Preface). This is a false assertion and an objectionable demand.

[10] A metaphysics is of science in general and, by courtesy, of any specific science (Agassi (1977), Ch. 1, note 21).

[11] Nevertheless, Carnap characterized metaphysics as the confusion between language and metalanguage. Atomism refutes this proposal.

4 Popper's *Extension*

Some metaphysical systems may be not open to "natural" (see below) extension, much less to scientific ones. What is necessary for it? Popper has constructed one trivial extension of an obviously untestable system to a testable one: a system containing one purely existential statement—there exists a mermaid—may become testable by the addition to it of space–time specification—a mermaid is now present in the neighborhood aquarium. Rudolf Carnap has suggested that this cannot hold for statements with universal and existential quantifiers (as if this makes any difference for this example to be a refutation of his system). Watkins called these "all and some statements" and refuted Carnap's suggestion by an example. His example is, "everyone has a soul-mate": it is untestable and a specification of it, "everyone is married to one's soul-mate", that is a scientific extension of it, and that is refutable (and refuted). Now "every event has a cause" is metaphysical determinism; Popper's *extension* of it to "every event is predictable from its cause" is refutable. It is, but, by Popper's own magnificent theory of scientific character, to be scientific a theory must be both consistent and empirically refutable; yet, as he has claimed, his extension of determinism is inconsistent. Popper's *extension* of determinism by specifying decidability to predictability achieves its aim of making it unattractive. His view of it as scientific was a technical error. Still, this leaves open the question, is there a possible "natural" extension of determinism (not by rendering it predictable but) by adding to it another qualification that would enable it to depict some event as (erroneously) predictable? If there were such an extension, then there would be no reason to assume that it is unique. Moreover, the determinism could be subject to many alternative extensions. Then there would be no reason to assume that this holds only for determinism and not for any other metaphysics. Indeed, ancient atomism underwent extensions into diverse scientific theories–all of them refuted by now.[12]

Traditionally, metaphysics was not just the presuppositions for (some) science but also what claims some comprehensiveness in some intuitive sense of the word. Take the metaphysical system of Newton's *Principia*, namely, Euclid's space populated with point-masses interacting with central forces (namely, with forces that obey his famous three laws). It is metaphysical and it is untestable. Adding to it his law of gravity (or Coulomb's law of electricity) will yield a testable system *par excellence,* and this is an extension in the obvious sense of specification (of the laws). It is not necessarily an extension in the strict ("natural") sense of being comprehensive. The view, once quite popular, of it as possessing one and only one force (the idea that all forces are reducible to one) is comprehensive and so it is more "natural" an extension of the theory that renders it a metaphysics proper. Another possible extension of it is would be a list the forces governing the system of the world, plus the claim that the list is complete, namely that all known phenomena are in principle

[12] To be precise, the initial version of atomism is not as thin as the enriched one. This, however, is understandable. We do not have the full ancient story; quite possibly Democritus distinguished between the thinnest version of his doctrine and the thickest version that he could (and intended to) generate.

thus explicable. (An alternative is to list all the possible forms of energy; deciding that the list is complete makes the law of conservation of energy comprehensive and thus refutable, as Poincaré noted ([1902] 1905, Ch. 5) he advised against this—to escape testability.) Any scientific theory, such as Newtonian theory of gravitation that conforms to Newton's metaphysics, is a possible part of the "natural" extension of it. This proves that if Newton's metaphysics can undergo extension its extension is not unique, even if its "natural" extension may be. The claim that an extension need not be unique is common sense. Hence, the onus of the proof is on anyone who claim uniqueness. Hence, in principle, what extension is "natural" or comprehensive is unclear.

5 Conclusions

The fault with Popper extension of Laplace's metaphysical determinism into scientific version is the tacit but very clear claim for uniqueness: Popper took it for granted that he had refuted all possible scientific versions of determinism. Now, supposing Newtonian metaphysics were extendable to a scientific system, and suppose, with Laplace, that all problems within it have unique solutions, then, clearly, each such extension is already both deterministic in Laplace's original sense, predictable in principles and scientific as refutable in principles.

To put it generally, when a metaphysics has a set of scientific theories that conform to it, and we add to that set the claim of completeness, then the metaphysics evolves into a scientific system (Agassi 1964). Here scientific character is considered a refutable explanation; if one wants refutability alone, then it is much easier to extend a metaphysics that has some scientific theory conforming to it: one can simply claim completeness anyway and have the completeness claim trivially refuted. Also, metaphysics is here taken in the traditional sense, not in the (much broader) Wittgenstein-style sense or Vienna Circle-style sense, in which "non-science" and "metaphysics" are synonyms. In this sense, determinism is not a metaphysics proper. It is nonetheless metaphysical, as it is a character of some metaphysical systems.

To put all this most generally, there is still no consensus about the traditional dispute within philosophy between intellectualism and empiricism, although most philosophers of science are empiricists. Historically, this was the verdict of Laplace: the scientific choice between the Cartesian and the Newtonian systems of the world was also the methodological choice between the intellectualists and the empiricists; and when Newton won, he won on both fronts. Nevertheless, the metaphysics of Descartes fascinated Laplace sufficiently to try a system of the world (with *fluide gravidique*) that should comply with both Descartes' and Newton's systems. He never tried his hand with the parallel system that should comply with both Descartes' and Newton's methodologies. The idea, that a theory has both an empirical base and an a priori justification looks either encumbered with redundancy or a matter of course. For, logically, the idea of redundancy is that the redundant item follows from the rest. It is easy to see that neither intellectualism nor empiricism entails each other.

At most both justify the same system, yet not very likely. This would be a serious impediment for classical methodology that requires proof; not for Popper's system, as it requires openness to criticism. In this sense, Popper's extension of Laplace's "scientific" determinism is in line with his methodology, but this does not mean that his extension is the only one possible; hence, we have so far no refutation of all possible, reasonable versions of "scientific" determinism.

References

Agassi J (1964) The Nature of Scientific Problems and Their Roots in Metaphysics. In: Bunge M (1964) (ed). The Critical Approach Essays in Honor of Karl Popper. Free Press of Glencoe, NY., pp. 189–211.

Agassi J (1975) Determinism: Metaphysical versus Scientific. Memoirs, 5th International Congress of Logic, Methodology and Philosophy of Science. Kluwer, Dordrecht, pp. V 1–4.

Agassi J (1977) Towards a Rational Philosophical Anthropology. Nijhoff. The Hague.

Asimov I (1951) Foundation Series. Gnome Press, NY.

Dummet M (2011) Metaphysical Necessity. In: DeVidi D, Hallett M, Clarke P (2011) (eds) Logic, Mathematics, Philosophy, Vintage Enthusiasms: Essays in Honour of John L. Bell, Springer, Dordrecht, pp. 19–34.

Popper K (1935) Logik de Forschung. Springer, Berlin.

Popper K (1945) The Open Society and Its Enemies. Routledge, London.

Popper K (1950) Indeterminism in Quantum Physics and in Classical Physics. British Journal for Philosophy of Science 1:117–195.

Popper K (1952) The Nature of Philosophical Problems and their Roots in Science. [Republished in Popper 1963].

Popper K (1957) The Poverty of Historicism. Routledge, London.

Popper K (1959) The Logic of Scientific Discovery. Routledge, London.

Popper K (1963) Conjectures and Refutations. Routledge, London.

Popper K (1982) The Open Universe, An Argument for Indeterminism: From the Postscript to The Logic of Scientific Discovery. Routledge, London.

Poincaré H ([1902] 1905) Science and Hypothesis. Walter Scott Publishing Co., NY.

Watkins J (1970) Imperfect Rationality. In: Borger R, Cioffi F (1970) (eds) Explanation in the Behavioral Sciences. Cambridge University Press, Cambridge, pp. 167–218.

Wittgenstein L ([1921] 1922) Tractatus Logico-Philosophicus. Routledge, London.

Chapter 8
Ideas, Phenomena and Mechanisms: The Role of Modern Physics in the Historical Epistemology of Fleck and Canguilhem

Mauro L. Condé

Abstract Although historical epistemology is frequently understood in opposition to the traditional (ahistorical) epistemology grounded mainly in classical physics (from the epistemology of modern science to logical positivism), physics also developed an essential role in the rise of historical epistemology. To a certain extent, the inclusion of biology, life sciences, and medicine—until then an "art of cure"—as one of the crucial parameters of historical epistemology (Fleck, Canguilhem) was due to the changes in the "new science" of physics (Relativity and Quantum Mechanics) occurred from the beginning of the twentieth century. This chapter aims to demonstrate that, in the 1930s and 1940s, ideas and theories presented by modern physics boosted an epistemological development in the sciences of life. In turn, biology and life sciences were vital in consolidating historical epistemology and understanding the history of science. This analysis is established from the works of two seminal authors in the historiography of science, Ludwik Fleck and Georges Canguilhem. When reconsidering the phenomenon of life inspired by ideas from modern physics, these two physicians allowed us to understand it as a new type of mechanism much more dynamic, including biological and social aspects.

Keywords Canguilhem · Fleck · Historical epistemology · Ideas · Life sciences · Mechanism · Modern physics · Phenomena

1 Introduction: Ideas, Phenomena and Mechanisms, from Physics to Biology

At the beginning of the twentieth century, physics developed different ideas and theories, analysing new phenomena or re-evaluating old ones. In this context, the very idea of the mechanisation of the world (Dijksterhuis 1961)—essential in the

M. L. Condé (✉)
Federal University of Minas Gerais, Belo Horizonte, Brazil
e-mail: mauroconde@ufmg.br

R. Pisano (ed.), *A History of Physics: Phenomena, Ideas and Mechanisms*,
History of Mechanism and Machine Science 42,
https://doi.org/10.1007/978-3-031-26174-9_8

rise of classical mechanics—is resized, and the understanding of what a mechanism is will change throughout the twentieth century (Craver and Tabery 2015). The epistemological deployments arising from these modern physics ideas performed an essential role in expanding contemporary scientific rationality. In other words, new ideas, conceptions and theories in the different fields of physics—thermodynamics, relativity and quantum mechanics—considerably extended the physical knowledge consolidated by classical mechanics. Furthermore, it drove science to rethink its epistemology, markedly positivist or idealistic. Consequently, the historiography of science started to be written considering the new epistemological perspective.

In *Le Nouvel Ésprit Scientifique*, Gaston Bachelard was one of the first authors to comprehend that the theories of modern physics presented new epistemological possibilities that would drive us to have other eyes for history of science. From what quantum mechanics, relativity and non-Euclidean geometries demonstrated, according to Bachelard, it was necessary to formulate "a new scientific spirit" to constitute a new non-Archimedean, non-Cartesian and non-Newtonian epistemology (Bachelard [1934] 2013). With his approach, Bachelard outlined the foundations of historical epistemology[1] to understand the history of science.

However, Bachelard's innovative proposal was not immediately considered. In the twentieth century, classical physics epistemology remained for an extended period as a model for the historiography of the sciences. This traditional epistemology still underlined the history of science conception of key authors from Alexandre Koyré to Thomas Kuhn. Koyré's influential "epistemology of precision" (Koyré [1948] 1971) helped to understand the classical idea of mechanisation of the world, but it was not adequate to think of modern physics or biology. And no matter how many innovations were brought by Kuhn's *The Structure of Scientific Revolutions* (Kuhn [1962] 1970), this seminal work in a certain sense still reflects the epistemological model of classical physics as presented in Koyré's history of science. And Kuhn recognised Koyré's influence on his book (Kuhn 2000, p. 285). Paradoxically, Kuhn's decisive influence in the history of science still reinforced aspects of the epistemology of precision or the epistemology of classical physics. With effect, partially, Kuhn encouraged historical epistemology adopting the social elements of science. Still, he somewhat also restrained it, affirming his history of science close to the epistemology of classical physics. The difficulties of Kuhnian concepts like paradigm and incommensurability arose from the affirmation of their "precision".[2]

---

[1] In his genealogy of the expression "historical epistemology", Yves Gingras informs us that this expression was first used by Dominique Lecourt, in 1969, in his book *L'epistemologie Historique de Gaston Bachelard* (Gingras 2010, p. 442). Although Bachelard did not use the term, he laid the foundations for what would be consolidated as a historical epistemology throughout the twentieth century by authors such as Fleck and Canguilhem.

[2] It is worth noting that Kuhn, in his last writings, turns to biology as an epistemological model and no longer to classical physics or even modern physics. Unfortunately, the later Kuhnian ideas, already closer to biology, did not significantly impact the historiography of science. Kuhn's promised book – in which he would develop an evolutionary theory of the history of science – was ever published. In *The Road since Structure* (Kuhn 2000, pp. 90–104), he outlines his new epistemological proposal in which the idea of evolution (speciation, niche, evolutionary epistemology, lexical taxonomy,

Meanwhile, since the 1930s, the new ideas in modern physics have influenced other fields of knowledge, especially biology. At that time, biology sought its epistemological foundations. Thus, the new physics also became a parameter for thinking about epistemological issues in explaining the phenomenon of life. This epistemological approximation between these fields of knowledge occurred in different manners, as a direct influence or a simple analogy between the two areas of knowledge. However, establishing an analogy between these fields does not mean reducing one to the other.

On the contrary, analogy seeks to bring out meaning (paradoxically the "logical") not by affirming identity but by noting differences and similarities. Thus, analogy aims to demonstrate that something "is" through what "is not", that is, by differentiation in similarity. When one proposes an influence or inspiration from the epistemology of physics to think about biology, one does not intend the relationship between the two domains to be one of identity but similarity or analogy. In other terms, modern physics has been appropriated as epistemological inspiration by biology—sometimes as an analogy or metaphor—where comparisons are made between two domains: the target (epistemology of the life sciences) and the vehicle (ideas from modern physics). The known vehicle is used to reach the unknown target to explain it. Thus, from this analogy, we have a rich heuristic instrument that makes it possible to broaden the understanding of the life sciences. But, of course, without establishing identities between these two domains. Thus, for example, Fleck made an interesting analogy between physics and bacteriology, characterising the "revolution" represented by the theory of relativity as a "mutation" in the scientific thought style. "The transformation in physics and in its thought style brought about by relativity theory represents such a mutation, as does the adjustment in bacteriology resulting from the theories of variability and cyclogeny" (Fleck [1935] 1979, p. 26).

One can see that using "analogies and metaphors" to understand different sciences can be a helpful tool. Of course, although methodologically useful and didactic, analogies and metaphors have their limits. For example, we can approximate the dynamics of biology to the dynamics of historical and cultural events. Still, we cannot understand this approximation as reductive because one phenomenon is not entirely reduced to the other even though one may be analogous to the other. Indeed, although the dynamics of biology—especially when compared to classical mechanics and its epistemology of precision—may be a closer parameter to the dynamics of social life, even so, one model is not reduced to the other. Although they may be analogous, biological, physical, and social phenomena are not identical. They are different mechanisms with "family resemblances".

According to Fleck, ideas (or pre-ideas coming from the past, like the atom, syphilis, etc.) are guidelines for developing any scientific discovery (Fleck [1935]

etc.) gains a more significant meaning. According to Graver and Tabery (Craver and Tabery 2015), Kuhn and other historians of science in 1960/1970 had a relevant influence shaping the new idea of mechanism as something much more dynamical plus social aspects in the production of scientific knowledge. Indeed, one can attribute to Kuhn these historical and social conceptions in understanding the history of science as an influence in the new idea of mechanism, but not to Kuhn's later biological approaches.

1979, p. 23). These ideas circulate among different thought styles or different sciences. There is communication among various sciences or, in Fleck's terms, communication of thoughts both within a "thought collective" and among "thought collectives" (*Der intra- und der interkollektive Denkverkehr*) (Fleck [1935] 1979, p. 98, pp. 108–109). Modern physics' ideas circulated in other collectives of sciences like biology and medical sciences. Of course, in their reception, these ideas are transformed. The new fields create other possibilities different from the initial purposes of these ideas.

Therefore, drawing inspiration from physics, biology sought to broaden its epistemological repertoire, which enabled a better understanding of the history of the life sciences. More than that, this epistemological history of the life sciences developed by authors such as Fleck and Canguilhem allowed a broader understanding of the history of the sciences. However, the main aim here is not directly to compare Fleck and Canguilhem but to demonstrate as they are a product of the same "thought style" that rose influenced by the new transformation in physics and biology.[3] Moreover, based on these epistemological family resemblances between physics and biology, Fleck and Canguilhem directly or indirectly drew inspiration to develop their conceptions of the history of biology and biomedical sciences. Thus, the tradition coming from biology—especially when opposing the epistemology of classical physics to the modern physics ideas—seems to have allowed Fleck and Canguilhem to understand in greater detail that natural sciences are also historical products.[4] Consequently, epistemology is also a "historical epistemology" like all human knowledge is social and historical. In a certain sense, modern physics, or the new scientific spirit stated by Bachelard, pointed out results very close to those provided by an epistemology of biology. In other words, modern physics, biology and life sciences have developed a robust epistemology applicable to various areas of knowledge.

From a particular viewpoint, we can understand the change of the traditional epistemology to historical epistemology as a product of the new concepts of phenomena and mechanisms presented by modern physics and biology. When modern physics started to consider the idea of mechanism as a much more dynamical system in understanding the physical phenomena, biology found room to use this dynamical conception of mechanism to analyse life phenomena. These interactions between physics and biology will stem to a more complex and valuable notion of a mechanism. In other words, according to the idea of mechanism and mechanisation of the

[3] I first attempted to understand these authors as a turning point in the historical epistemology (Condé 2016), and now I am looking for the role of physics in this process. Before, Braunstein (Braunstein 2008) and (Braunstein 2009) compared the philosophy of Fleck and Canguilhem. Then, Sinding (Sinding 2009) approximated Fleck and Canguilhem to point out that medicine is based on an epistemology of uncertainty.

[4] From an epistemological perspective, the idea that knowledge has a solid social and historical character had already been presented, with different nuances, by authors such as Mannheim, Durkheim, and Jerusalem. However, these authors failed to extend the sociological approach to the natural sciences. Certainly, Reichenbach's distinction (Reichenbach 1938) between the "context of discovery" and the "context of justification" restrained these historical and sociological approaches. However, in seeking to understand the history of science, Fleck and later Kuhn understand that science is a genuinely historical and social product.

world present in classical physics, biology could not be adequately understood as a science. However, as modern physics reconfigure the notion of mechanism as a complex and dynamical "system of reference" to understand the behaviour of nature, biology comes to be understood as a particular type of mechanism in understanding the phenomenon of life.

Consequently, biology and life sciences allowed a new horizon for boosting the idea of epistemology. Moreover, allied to this conception, another essential tenet to construct the historical epistemology, as mentioned, was provided by a new role for social and historical aspects in producing scientific knowledge. Fleck and Canguilhem were the first to comprehend these new possibilities from biology and social elements to think of the phenomenon of life as an innovative mechanism.

For Graver and Tabery (Craver and Tabery 2015), the contemporary conception of a mechanism as a complex and dynamical system has different features depending on the authors. However, one can see "family resemblances" among them. Thus, according to them, mechanisms must have four primary characteristics: they are (1) "phenomena", they have (2) "parts", they present (3) "causing", and they are (4) "organisation". Furthermore, it is worth noting for our specific topic that "not all mechanisms are machines", and "a mechanism is not just fiction or metaphor" (Canguilhem 1977). Still, one can use analogies and metaphors comparing mechanisms. As we will see, it looks like Fleck and Canguilhem followed these basic features by analysing the phenomena of life. Thus, they can be seen as forerunners of the contemporary notion of the mechanism because they consider the phenomenon of life as a peculiar type of mechanism. Moreover, they include the social and historical aspects in this process. A historical epistemology is a product of this complex.

Thus, starting from new ideas of modern physics where the concept of a mechanism is reconsidered, Fleck and Canguilhem reflected on the epistemology of the life sciences. In following this direction, they deepened the path towards a historical epistemology[5] when thinking about the status of the life sciences. To accomplish their epistemological projects, both authors emphasised the social and historical aspects of the sciences. So, they expanded the idea of historical epistemology. The following aim is to demonstrate how they established their approaches of scientific knowledge as historical epistemology.

---

[5] It is important to say that, although Fleck and Canguilhem were pioneers in featuring the historicity of scientific knowledge, they did not use the term "historical epistemology". Additionally, historical epistemology's idea is multifaceted, with different nuances and terminology, for example, "social epistemology" or "historical ontology". For the problem of the constitution of "historical epistemology", see Rheinberger (Rheinberger 2010) and Carrier (Carrier 2012); for related approaches to "historical ontology", see Hacking (Hacking 2002) and for "social epistemology", see Fuller ([1988] 2002) and Kusch (2011). However, there seems to be a convergence point among these conceptions stating that scientific knowledge is more than nature's answer to our questions. It is also the product of a given historical context. Science has a history, and its history is a constitutive element of the very results it achieves. An assumption of what is here called historical epistemology.

2 Fleck, Canguilhem and Historical Epistemology

In approaching Fleck's and Canguilhem's ideas, the aim is not to address the peculiarity of their works or even compare one with the other, but, as aforementioned, to try to understand them in the context or the "thought style" in which they appear. Both thinkers are fruits of a particular historic moment in sciences, with the transformations in physics and biology creating a new thought style that will enable historical epistemology. Therefore, my focus is not on the extension of the thought of these authors. I do not want to analyse their intellectual trajectories but understand the significance of the rise of these works in the context of the 1930s/1940s, in a specific niche that made possible the birth of historical epistemology.

It was not a coincidence that the rise of an epistemology of life sciences guided the history of life sciences right after the epistemological changes in classical physics. Until then, classical physics was the primary reference, and biology had no room in this epistemological project. Nor is it a coincidence that Fleck and Canguilhem started their careers in medicine. As aforementioned, a few years before Canguilhem epistemologically inquired about the relationship between the "normal" and the "pathological", Bachelard had already outlined the path of the future historical epistemology (Bachelard [1934] 2013). Even though Canguilhem was dedicated to the philosophical and historical understanding of the life sciences and not physics like Bachelard, this physician and philosopher of science will clarify his social and historical inspiration from Bachelard. "It is hardly necessary to say that by linking so closely the development of epistemology to the elaboration of historical studies, we are inspired by the teaching of Gaston Bachelard" (Canguilhem 1977, pp. 19–20; my translation). In 1955, Canguilhem succeeded Bachelard at the Sorbonne and took over the Institute for the History of Science and Technique direction.

Fleck's interest in the epistemology presented by the new physics was no different. His reaction to quantum mechanics is even more direct. He inserts himself into the epistemological controversy brought about by the "new science". In one of his first articles, "On the Crisis of 'Reality'" (Fleck [1929] 1986), he addresses the question of the observation of a natural phenomenon, starting from Niels Bohr's quantum postulate (Bohr 1928) that states that, in quantum phenomena, the measuring instrument interferes with the measurement. Fleck endorses Bohr's position by saying that "to observe, to know (*erkennen*) is always to test and thus literally to change the object of investigation" (Fleck [1929] 1986, p. 53). Faced with this new framework brought about by physical science, Fleck and Canguilhem found a more promising ground to develop the epistemological consequences of the tradition they are affiliated with, that is, the life sciences tradition. This tradition goes back a long time, but only in this new context of the classical physics epistemological crisis does it seem to find room to flourish its epistemology. In other words, this tradition finds an epistemological foundation to transform medicine from an "art of cure" into a science.

Belonging to different cultures and without reciprocal or unilateral influences, in the 1930s/1940s, Fleck and Canguilhem reacted against the old conception of the history of science legitimated by a positivist or idealistic epistemology based on the model of classical physics. In 1935, the physician Ludwik Fleck published his book *Genesis and Development of a Scientific Fact*. Only eight years later, in 1943, appeared *The Normal and the Pathological*,[6] by Georges Canguilhem, also a physician. In these books, the authors established their respective innovative positions constituting a new "thought style"—to use a concept precious to Fleck—to understand science and its history. These two books start from a similar perspective: the reaction against positivism / logical positivism, then hegemonic in epistemology. They sought to demonstrate the need for a new epistemological model for understanding the history of the life sciences in this rich context of new ideas in physics.

Considering the medical and biological problems, Fleck and Canguilhem question the rigid limits of the idea of "precision" of rationalist epistemology and the conception of fact as something given and fixed as postulated by positivists. They lead us to think of new parameters for knowledge in a scenario in which facts are also social constructions, and the precision of knowledge is established not from the formalisation of a pre-existing essence but considering the adequacy among the empirical and social changes as elements of a "system of reference". Our authors set up a new epistemology that operates in a universe of uncertainties and fluctuations by analysing biology and understanding historical aspects as an element of scientific knowledge. In effect, they abandon the "universe of precision", as formulated by Koyré's history of science, to embrace a type of "epistemology of uncertainty". Therefore, something much closer to Bohr's principle of complementarity or Heisenberg's uncertainty.

Both Fleck and Canguilhem perceived that biology would operate similarly to what the new epistemology of modern physics offered. Biology and the life sciences seemed to need an interpretative system without ultimate and absolute references. What is *normal* and what is *pathological* should be established by a "system of reference" defining these values. Even if we knew what is alive and what is dead, the limit between these two parameters would not have a "precision" along the lines of classical physics. Nevertheless, this limit could be established similarly to the location of an electron, for example. More than that, these malleable criteria seemed to show the very historicity in the interpretation of a natural phenomenon. The subject producer of knowledge makes hermeneutic choices, and these choices are an essential part of the production of knowledge in the interaction between humankind and nature. This approach helps to affirm the historical nature of knowledge, while it is a criticism of positivist epistemology.

Establishing a criticism of Comte's positivism, Canguilhem's conception of the history of science tells us that history is a fundamental element in understanding science because sciences integrate history. History, culture and sciences are integrated into a whole. According to Canguilhem,

[6] This work appeared in 1943 with the title, *Essay on Some Problems Concerning the Normal and the Pathological*. In 1966, extending its reflections, it was reissued under the title, *The Normal and the Pathological*. Approaching the rise of the historical epistemology context in the 1930s/1940s, my analysis concentrates on the first part written in 1943.

> [...] as scientists lead their lives as men in an environment and social setting that is not exclusively scientific, the history of science cannot neglect the history of ideas. In following a thesis to its logical conclusion, it could be said that the modifications it undergoes in its cultural milieu can reveal its essential meaning. (Canguilhem [1943] 1991, p. 46)

For the French thinker, the very historical condition of human knowledge resides in this social set. The scientist, for example, when producing science, is an individual located in space and time acting in a peculiar way to obey specific circumstances. Consequently, we could not understand science from an abstract and ahistorical epistemology. On the contrary, it would be necessary to assume this historical condition of science. Finally, formulated from this perspective, the knowledge problem effectively takes the character of a "historical epistemology".

In opposition to the Vienna Circle ideals, for Fleck, epistemological understanding of science is indelibly marked by historical and social aspects. Asserts the Polish thinker, "epistemology without historical and comparative investigations is no more than an empty play on words or an epistemology of the imagination (*epistemologia imaginabilis*)" (Fleck [1935] 1979, p. 21). If knowledge production takes place essentially in a historical process, it is not enough to understand the production mechanisms of that knowledge by themselves. It becomes necessary also to know this historical process in which such knowledge develops. In fact, for Canguilhem and Fleck, the history of science could only be made considering the marks of the time and space in which it is produced. Science has a history, which is a fundamental part of understanding science.

Aiming to understand the importance of Fleck and Canguilhem for the rise of historical epistemology, we need to guide by these questions: (1) Why, in such close periods and independently, do these authors affirm the importance of the historical and biological perspectives to understand scientific knowledge? (2) In what sense do their historical and biological approaches constitute the basis of a historical epistemology? Initially, I aim to demonstrate that, starting from the history of medical knowledge, our authors insert themselves in a long tradition of knowledge far from the epistemological conceptions present until the beginning of the 1930s, mainly based on the physical sciences. In this way, inspired by the new epistemological perspectives brought by the transformations in the new physical sciences, our authors will conceive a new epistemological proposal to reflect on the life sciences. Then, after approaching aspects of the history of medical knowledge in which our authors are inserted, I aim to demonstrate the guidelines of this new epistemological perspective presented by them. These guidelines contributed to the rise of historical epistemology.

With these guidelines, Fleck and Canguilhem established a criticism of the traditional conception of mechanism to understand the phenomenon of life from an expanded notion of mechanism, that is, mechanism as a dynamic "system of reference" (*Bezugssystem*) (Fleck [1935] 1979, p. 50) or "system of regulations" (*Système du regulations*) (Canguilhem [1943] 1991, p. 260). For both physicians, the disease is something more complex than a mere physiological mechanism, for it includes social, historical and geographical aspects. Thus, one can read their works as a new type of mechanism.

3 Clues of a Thought Style

To understand this tradition to which Fleck and Canguilhem sought to affiliate, I base myself on the classic text by the eminent Italian historian Carlo Ginzburg, "Clues: Roots of an Evidential Paradigm (*Paradigma Indiziario*)"[7] (Ginzburg [1978] 1983). In Ginzburg, one can find an extensive mapping of this tradition in which, among other knowledge marginalised by classical epistemology, both medical and historical knowledge is inserted. Such types of knowledge were discarded by the classical epistemology constructed from classical physics. For example, the epistemology of the philosopher-scientists (primarily physicists) of the Vienna Circle (Fleck [1935] 1979, p. 50) does not include history or medicine. These traditions of medicine and history were neglected when Koyré aligned "Plato and Galileo" (Koyré [1943] 1973) to produce his epistemology of precision. The works of Fleck and Canguilhem give voice to this tradition of marginalised knowledge of history and medicine that remained in the background over the centuries.[8] In the Italian thinker's words, this tradition was "(…) implicit (crushed by the prestigious—and socially higher—model of knowledge elaborated by Plato)" (Ginzburg [1978] 1983, p. 70; my translation).

Because they are constituted from medical and historical knowledge, Fleck and Canguilhem's books can be included in what Ginzburg calls the "evidential paradigm" (*paradigma indiziario*). However, even before addressing what an evidential paradigm is, it is essential to note that in the context of the 1970s, with the strong influence of *The Structure of Scientific Revolutions*, Ginzburg uses the Kuhnian concept of "paradigm" (Ginzburg [1978] 1983, p. 55). Perhaps the expression paradigm, at that time, loaded with a solid argumentative appeal and a great methodological potential, was even inevitable. However, time has shown us that evidential knowledge, as explained by Ginzburg, as something flexible, adaptive, and without rigidly defined limits, seems to be much more a "thought style", in Fleck's claim, than a rigid paradigm circumscribed to its incommensurability, as defined by Kuhn. Of course, this in no way diminishes the value of Ginzburg's text. However, if the great Italian historian would allow us, I would argue that it is much more an "evidential thought style" than an "evidential paradigm".

What would be this evidential thought style (evidential paradigm)? According to the Italian historian, it would be the search for the details, clues, and elements that do not occupy the foreground but can lead us to a more detailed understanding

[7] It is hard to translate the Italian term *indiziario* into English. Perhaps, the word "evidential" is not the best translation because, according to Ginzburg, the aim is not to demonstrate "evidence" but find "clues" or "indications" that reveal a hidden process or system. Therefore, due to the difficulty in finding a better translation, I am following the English edition (Ginzburg [1978] 1986) for this term. But all citations of Ginzburg's text are my translations from the Spanish edition (Ginzburg [1978] 1983).

[8] Of course, it is not the case to affirm that this tradition did not come from a long line in the time since Aristotle, but to say it was neglected as a central point in the rise of modern science. Thus, this lapse delayed the advancement of biology during the scientific revolution of the seventeenth century. Biology and medicine are not sciences inserted into a "precision" thought style (Koyré 1971).

of the whole. As an exemplary model of this methodology, Ginzburg points to the painting treatise by an Italian doctor named Giovanni Morelli. According to this method, when analysing an original work of art, it is necessary to observe the details, the minutiae and subtleties marked out by the painter. From these aspects, we can establish the identification of fake works; that is, by focusing on the details, we perceive "characteristics present in the originals but not in the copies" (Ginzburg [1978] 1983, p. 57; Author's translation).

The search for detail, for clues, according to Ginzburg, is also what characterises the method of Sherlock Holmes, a character created by Conan Doyle, also a physician. More than the mere similarity between the methodology of clues and details practised by Sherlock Holmes, Ginzburg believes that Conan Doyle had, through his uncle, knowledge of Morelli's method (Ginzburg [1978] 1983, p. 59). As if these two doctors' closeness was not enough, even if they had professions far from medicine, Morelli, according to Ginzburg, was also very close to another famous doctor, Sigmund Freud. The technique of psychoanalysis would be directly linked to this tradition of clues and details. And, according to Ginzburg, Freud himself, with his work, indicates this perspective when he conceives "the proposal of an interpretative method centred on the leftovers, the marginal data, considered as revealing. In this way, details usually considered unimportant, or directly trivial, 'vulgar', provided the key to access the highest products of the human spirit" (Ginzburg [1978] 1983, p. 63; Author's translation).

In *The Moses of Michelangelo* (Freud [1914] 1946), Freud, who had in his library a copy of Morelli's book *Della Pintura Italiana: Studii Storico Critici*, gives us an accurate synthesis of the Italian art critic's method in his relationship with anatomy to distinguish a masterpiece from a copy. According to Freud, details of the masterpiece are not present in the copy.

> The characteristic importance of secondary details, of insignificant minutiae such as the conformation of the fingernails, the earlobes, the areola, and other elements generally go unnoticed, and the copyist fails to imitate. While instead, every artist follows these details precisely to imprint his uniqueness. (p. 185, quoted in Freud [1914] 1946, p. 60; Author's translation)

After explaining the method of clues, mapping its tradition and showing its close relationship with medicine through the analogy and approximation between Morelli, Holmes and Freud, Ginzburg asks:

> How does one explain this threefold analogy? The answer, at first sight, is straightforward. Freud was a doctor; Morelli graduated in medicine; Conan Doyle had been a doctor before turning to literature. In all three cases, we see the model of medical symptomatology: the discipline that allows diagnosing illnesses inaccessible to direct observation based on superficial symptoms. (Ginzburg [1978] 1983, p. 65; Author's translation)

However, perhaps more importantly, this similar methodological approach of these three authors would not be just a fortuitous resemblance but a type of response to a spirit of the times. According to Ginzburg, "by the end of the nineteenth century—more precisely in the decade 1870/1880—an evidential paradigm based precisely on symptomatology began to assert itself in the human sciences, but its roots were much older" (Ginzburg [1978] 1983, p. 65; Author's translation). It comes from other times; this was the knowledge of "doctors, historians", in short, a "conjectural knowledge" (Ginzburg [1978] 1983, p. 70). This knowledge "would be characterised for its capacity to go back to a complex reality not directly experienced, starting from apparently negligible data" (Ginzburg [1978] 1983, p. 66; Author's translation).

In his conclusion, although not premeditatedly emphasising this point, Ginzburg unites the knowledge of the physician and the historian precisely as present in Fleck and Canguilhem. Ginzburg states that the knowledge of the "historian is comparable to that of the physician who uses nosographic charts to analyse the specific illness of the singular patient. And like that of the physician, historical knowledge is indirect, indicative, conjectural" (Ginzburg [1978] 1983, p. 72; Author's translation). Of course, I am not assuming for historical epistemology the uses of the method of the evidential paradigm but only characterizing a similar type of reasoning between these two traditions based on the common origins, that is, medicine and its epistemology of uncertainty (Sinding 2009). In this sense, the works of Fleck and Canguilhem have similar origins. These works are thus located in a well-defined historical context. They emerge from the tradition of medicine and after the epistemological crisis of classical physics. This new context finds a good ground for flourishing a unique biological thought style with a new epistemology. As the Italian historian points out, "the group of disciplines that we call evidential [*indiziario*] (including medicine) does not fit at all into the scientific criteria deducible from the Galilean paradigm" (Freud [1914] 1946, p. 71; Author's translation).[9]

More than outlining the adequate ground to affirm the new legitimacy of this old knowledge by clues, Ginzburg concludes his text by demonstrating the importance that this event will have for epistemology. According to him, "in the discussions on the 'uncertainty' of medicine, the future epistemological knots of the human sciences were formulated" (Ginzburg [1978] 1983, p. 84; Author's translation). Of course, history is at the epicentre of this issue.

It is not difficult to find in Fleck's or Canguilhem's words a kind of affiliation to this tradition of the clues and details, although they never used this terminology. For Canguilhem, "medicine seemed to us and still seems to us like a technique or art at the crossroads of several sciences, rather than, strictly speaking, like one science" (Canguilhem [1943] 1991, p. 34). According to the author of *The Normal and the Pathological*, medicine has never conformed to the prevailing idea of science: "clinical practice is not and will never be a science even when it uses means whose effectiveness is increasingly guaranteed scientifically" (Canguilhem [1943] 1991, p. 226).

In Canguilhem's conception, certainly, scientific knowledge was necessary for the development of medicine. However, for the French philosopher, "despite so many laudable efforts to introduce methods of scientific rationalisation, the essential lay

[9] Of course, the Galilean paradigm had a critical contribution to the rise of modern science. However, as recalled by Husserl, in his famous sentence, Galilei "is at once a discovering and a concealing genius" (Husserl [1936] 1970, p. 52) of modern science. He "discovers" science affirming its "primary properties", but, at the same time, he "conceals" the knowledge of the "secondary qualities" essential for biology and human sciences.

in the clinic and therapeutics, that is, in a technique of establishing or restoring the normal which cannot be reduced entirely and simply to a single form of knowledge" (Canguilhem [1943] 1991, p. 34). And it would be exactly life, in all its complexity, that never allowed medicine to be reduced to the scientific rationality of "exactness" or even positivity of facts. Thus, unlike physics, Canguilhem concludes: "we can say that in biology it is the pathos which conditions the logos because it gives it its name. It is the abnormal which arouses theoretical interest in the normal" (Canguilhem [1943] 1991, pp. 208–209).

It is no different for the author of *Genesis and Development of a Scientific Fact*. Fleck is aware that even if medicine could be considered from a scientific point of view, it is linked to the many cultural, social and historical aspects present in evidential [*indiziario*] knowledge. It would be precisely from this confluence that a new epistemology would emerge. It is in this sense that, for Fleck, "a scientific fact taken from the annals of medicine is particularly suitable for our observations, because it is very rich in history and content but has not yet been thoroughly examined epistemologically" (Fleck [1935] 1979, p. xxiii). By excellence, medicine is a knowledge constituted in its historicity. Naturally, at the same time, this medical knowledge seeks to refine its scientific criteria. With this understanding, Fleck establishes medical fact as a privileged fact to have a broader notion of science. According to him, "a medical fact, the importance and applicability of which cannot be denied, is particularly suitable because it also appears to be very rewarding historically and phenomenologically" (Fleck [1935] 1979, p. xxviii).

4 Guidelines of the Historical Epistemology

Beyond the historical and social presence in the production of knowledge—which already carries the criticism of positivism / logical positivism—our authors developed other fundamental guidelines in formulating historical epistemology. These characteristics appear in their works partially in a more developed form and others more embryonic. From these guidelines, Fleck and Canguilhem see the phenomenon of life as a particular mechanism that enables us to understand knowledge as a whole. I want to discuss at least three of these guidelines: (1) "Biological matrix" or the idea that the matrix for understanding knowledge is biology. The development of knowledge is something that takes place in evolutionary terms; (2) "Social and historical perspective" or the understanding of knowledge as the result of a collective and its social interactions situated in time; (3) "self-referential system" or the idea of the of rationality as a product of a dynamic and open "system of reference" or "system of regulations". This system has autonomy and mobility to constitute, suppress or exchange its postulates. This system conception demonstrates that Fleck and Canguilhem were forerunners of the contemporary idea of mechanism. As a whole, all these guidelines are close to the features of the new concept of mechanism, and, at the same time, they are essential elements of historical epistemology.

4.1 The Biological Matrix

As I pointed out since the beginning of this chapter, thinking about science from the medical and biological perspectives is a characteristic that brings our two authors together. I name the importance of biology to epistemology as the biological matrix. Although the differences between biology and medicine may bring different implications in the construction of historical epistemology, they are both understood in the same life sciences group. Before thinking about the history of science from the standpoint of historical epistemology, Fleck and Canguilhem were physicians. Of course, this was not a mere coincidence if we think from the idea of evidential thought style. Not only do they have a practice and a discourse based on the biological matrix, but they understand the need to incorporate this matrix as an epistemological referential. In a certain sense, a historical epistemology was designed from a biological or evolutionary epistemology.

By contrasting the biological matrix with the old matrix of classical physics, Fleck points out that what often seems fixed, objective, and absolute to us, in reality, goes through a long process of natural, social, cognitive, and linguistic evolution. Strongly present in Fleck's thought, the idea of evolution shows his affiliation to a Darwinian perspective. He transposes the conception of evolution to epistemology; that is, scientific development is seen in a way analogous to Darwinian evolution. Fleck tells us: "biology taught me that a field undergoing development should always be investigated from the viewpoint of its historical development (*entwicklungsgeschichtlich*)" (Fleck [1935] 1979, p. 20). Science has a history and an evolution, and in this process, its transformations are much more "mutations" than revolutions. Knowledge evolves from one thought style to another. Thus, there are "mutations in the thought style" (Fleck [1935] 1979, p. 26) and not drastic ruptures, as postulated by the Kuhnian idea of paradigm (Kuhn [1962] 1970).

Similarly, for Canguilhem, the life sciences compose a special kind of knowledge, unlike the physical sciences, because they start from a more primordial need than knowing the physical world, that is, that of knowing the very life of those who know. The life of those who live and suffer. For the author of *The Normal and the Pathological*, "it is first and foremost because men feel sick that a medicine exists" (Canguilhem [1943] 1991, p. 299) and not because the doctors tell men about their illnesses (Canguilhem [1943] 1991, p. 93). "Biological pathology exists, but there is no physical or chemical or mechanical pathology" (Canguilhem [1943] 1991, p. 127).[10]

[10] As we can see, Canguilhem understands pathology beyond a traditional mechanism idea. There is no mechanical pathology, for pathology goes beyond the conventional conception of mechanism. However, he did not comprehend his pathology approach as a new type of mechanism. Nevertheless, this possibility emerged years later after contemporary philosophy developed the new idea of a mechanism. In a sense, even if they did not claim that label, Canguilhem and Fleck were pioneers of the new mechanism idea.

As the pathos of human suffering, the pathological problem becomes a medical referential. However, for Fleck and Canguilhem, it is also an epistemological referential to think about the limits of science itself. This pathological problem is the central question of Canguilhem's doctoral dissertation, becoming his book *The Normal and the Pathological*. For Fleck, it is also a significant issue that appears as the theme of his first article, published in 1927, entitled "Some specific characteristics of the medical way of thinking". In this article, the Polish thinker develops his first epistemological considerations around the peculiarity brought by the normal and the pathological question. Fleck established the system of reference from which he will extend his ideas, contrasting the medical tradition's knowledge with the physical sciences' tradition. The physical science tradition can look for the "normal" in structures that repeat themselves and, thus, identify and label natural phenomena. In medical knowledge, this would not be possible. As Fleck states, "[…] even the very subject of medical knowledge differs in principle from that of scientific knowledge. A scientist looks for the typical and normal phenomenon, while a medical man studies precisely the atypical, abnormal, morbid phenomena" (Fleck [1927] 1986, p. 39). This specificity of medical knowledge inserts it in Ginzburg's evidential tradition and opposes it to the exactitude of the facts desired by the epistemology of classical physics. For Fleck, medicine faced a critical challenge when faced with the question of the normal and the pathological.

> There exists no strict boundary between what is healthy and what is diseased, and one never finds exactly the same clinical picture again. But this extremely rich wealth of forever different variants is to be surmounted mentally, for such is the cognitive task of medicine. How does one find a law for irregular phenomena?—This is the fundamental problem of medical thinking. (Fleck [1927] 1986, p. 39)

Thus, Fleck concludes that, although the specificity of medical knowledge also seeks regularities—"laws, relationships, some types of higher-order" (Fleck [1927] 1986, p. 40)—and thus constitutes itself as scientific knowledge, it situates in another tradition. In this tradition, "medicine has its own motivations which, however, do not lie along the line of classical theory, but require a change in mental attitude" (Fleck [1927] 1986, p. 43). In his *magnum opus*, it will be no different. Fleck opens his book, *Genesis and Development of a Scientific Fact*, precisely by drawing attention to this question of the peculiarity of medical knowledge (Fleck [1935] 1979, p. xxiii).

For Canguilhem, who seeks to understand the scientific aspects of medicine by analysing the normal and the pathological question, the difference between these states is not given by quantity differentiation. According to the French thinker, "the pathological state is not a simple, quantitatively varied extension of the physiological state, but something else entirely" (Canguilhem [1943] 1991, p. 89). Each of these conditions has its own rules. "There are healthy biological norms and there are pathological norms, and the second are not the same as the first" (Canguilhem [1943] 1991, p. 129). It is up to medicine to understand these proper rules that are much more dynamic than a mere positivist description. "The anomalous is not the pathological. […] But the pathological is indeed abnormal" (Canguilhem 1991, p. 137). We need to understand all these rules peculiar to medical knowledge when considering medicine as a science.

As in Fleck, for Canguilhem, the medical understanding of the transition between normal and pathological is a process conditioned by cultural and historical issues, besides biological ones. We cannot understand this relationship through the eyes of the epistemological model of classical physics but through the specificity that the phenomenon of life holds in each individual in particular. Something much more akin to the epistemology that emerged with modern physics.

> If the normal does not have the rigidity of a fact of collective constraint but rather the flexibility of a norm which is transformed in its relation to individual conditions, it is clear that the boundary between the normal and the pathological becomes imprecise. But this in no way leads us to continuity between a normal and a pathological identical in essence save for quantitative variations, nor to a relativity of health and disease so confusing that one does not know where health ends and disease begins. The borderline between the normal and the pathological is imprecise for several individuals considered simultaneously but it is perfectly precise for one and the same individual considered successively. (Canguilhem [1943] 1991, p. 182)

4.2 *The Social and Historical Perspective*

The social and historical aspects have a role in physiology. What is normal and abnormal in Canguilhem's physiology is conditioned by biological, geographical, historical, and cultural factors. This presupposition is because medical interpretations are conditioned by history, where medical conceptions are conditioned by the ideas of the time (Canguilhem [1943] 1991, p. 103) and because men and women in their geographical distributions are embedded in history and culture. Both biological and cultural environments influence human beings. Consequently, they react differently from context to context. Therefore, the pure observation of a medical phenomenon does not exist. Instead, these different factors always affect it.

> […] biological functions are unintelligible as observation reveals them to us, if they express only states of a material which is passive before changes in the environment. In fact, the environment of the living being is also the work of the living being who chooses to shield himself from or submit himself to certain influences. We can say of the universe of every living thing what Reininger says of the universe of man: "*Unser Weltbild ist immer zugleich ein Wertbild*" our image of the world is always a display of values as well. (Canguilhem [1943] 1991, pp. 178–179)

We see the world from different values. But, as the environment in which we live, the world is dynamic and diverse.

> Nothing happens by chance, everything happens in the form of events. Here is how the environment is inconstant. Its inconstancy is simply its becoming, its history. For the living being life is not a monotonous deduction, a rectilinear movement, it ignores geometrical rigidity, it is discussion or explanation (what Goldstein calls *Auseinandersetzung*) with an environment where there are leaks, holes, escapes and unexpected resistances. (Canguilhem [1943] 1991, p. 198)

In effect, for Canguilhem, medicine cannot conform to a vision of static science because the phenomenon of life is not a stable event identical to itself. On the contrary, life is evolution, variation of forms and invention of behaviour. In short, the structure of life is historical. "Physiology would then tend toward history, which is not, no matter what you do, the science of nature" (Canguilhem [1943] 1991, pp. 203–204). Thus, as physiology is historical, the traditional mechanism of physiology must be seen through a new enlarged conception of mechanism.

As pointed out, Fleck does not use the expression historical epistemology. Instead, to characterise science as a historical phenomenon, he uses many other expressions linked to the idea of producing scientific knowledge as a historical phenomenon. Thus, throughout his book, *Genesis and Development of a Scientific Fact*, he makes references to scientific knowledge using expressions such as: "history", "historical", "cultural and historical conditioning", "history of thought", "history of knowledge", "historical development", "historical links", "history of the concept", "historical relations", "context of historical ideas", etc. In short, a set of expressions to characterise that "scientific experience in particular derives from special conditions established by the history of ideas and by society" (Fleck 1979, p. 48).

Therefore, for the Polish thinker, a scientific fact can only have a "genesis" (and a development) from a defined historical and cultural context, or a thought style, as he calls it. A fact will never be something purely raw.

> This is how *a fact* arises. *At first, there is a signal of resistance in the chaotic initial thinking, then a definite thought constraint, and finally a form to be directly perceived.* A fact always occurs in the context of the history of thought and is always the result of a definite thought style. (Fleck [1935] 1979, p. 95)

Emphasising the social and the historical in the construction of knowledge, Fleck defends a position opposite to logical positivism. In fact, for the Polish thinker, more than being a mere description or observation,

> Every knowledge is a social act, not only it actually requires cooperation, because it is always based on knowledge and skill handed down from many others. [...] knowledge must be considered as a function of three components: it is a relation between the individual subject, the certain object and the given community of thought (*Denkkollektiv*) within which the subject acts; it works only when a certain thought style (*Denkstil*), originating in the given community is used. (Fleck [1960] 1986, p. 154)

Therefore, knowledge is a process of interaction between men (as a collective) and nature. Although this interaction process takes place in time, it is also a historical and social process. According to Fleck, "at least three-quarters if not the entire content of science is conditioned by the history of ideas, psychology, and the sociology of ideas and is thus explicable in these terms." (Fleck [1935] 1979, p. 21). Or again: "knowledge is therefore not an individual process of any theoretical 'particular consciousness'. Rather it is the result of social activity since the existing stock of knowledge exceeds the range available to any one individual" (Fleck [1935] 1979, p. 38).

Thus, for Fleck, in the history of syphilis, it is "not so-called empirical observations that led to the construction and fixation of the idea. Instead, special factors of deep psychological and traditional significance greatly contributed to it" (Fleck [1935] 1979, p. 3). Syphilis, which as a whole is much more than the etiological agent, exists only as a social product that also brings together scientific conceptions and practices (also the conflict between them), experiments, public health policies, popular beliefs, etc. In short, the "thought collective" constructs its "thought style".

4.3 Self-Referential System

Canguilhem presents a systemic perspective in his view of health and disease. For example, in explaining what diabetes is, he clarifies that this is not a "disease because of glycosuria, nor a pancreatic disease because of hypoinsulinemia, nor a disease of the pituitary; it is the disease of an organism all of whose functions are changed […] moreover, it can strike man or woman" (Canguilhem [1943] 1991, p. 88). Similarly, every symptom has to be evaluated from a background; the "part" has to be considered with the "whole". For the French thinker,

> When an isolated symptom or a functional mechanism is termed pathological,[11] one forgets that what makes them so is their inner relation in the indivisible totality of individual behaviour. The situation is such that if the physiological analysis of separated functions is known in the presence of pathological facts, this is due to previous clinical information, for clinical practice puts the physician in contact with complete and concrete individuals and not with organs and their functions. (Canguilhem [1943] 1991, p. 88)

We should not punctually see a medical problem because we arrive at the "cell by moving backwards, starting from the total organism. […] The solution to a problem posed by the entire organism, first to the sick man, later to the clinician, has been sought in the tissue or cell […] To look for disease at the level of cells is to confuse the plane of concrete life" (Canguilhem [1943] 1991, p. 223). Canguilhem conceives life, health and disease (the normal and the pathological) from multiple systemic relations of the individual and his environment. For him,

> The species is the grouping of individuals, all of whom are different to some degree, whose unity expresses the momentary normalisation of their relations with the environment, including other species, as Darwin had clearly seen. Taken separately, the living being and his environment are not normal: It is their relationship that makes them such. For any given form of life, the environment is normal to the extent that it allows it fertility and a corresponding variety of forms such that, should changes in the environment occur, life will be able to find the solution to the problem of adaptation […] in one of these forms. (Canguilhem [1943] 1991, pp. 143–144)

Thus, we perceive in Canguilhem a systemic perspective of life that can be seen as a new conception of mechanism. The phenomenon of life presents interactions between the "parts" and "whole", integrating the being and its environment. Therefore, the

[11] We can see that, as for Fleck [1935] 1979, p. 62), for Canguilhem, physiology is understood as a mechanism. However, these authors try to show physiology as a much broader mechanism. They include social, psychological and even geographic aspects into the composition of physiology. In other words, it is a much more comprehensive and more dynamic mechanism or a new type of mechanism.

knowledge of life is not a mere description, but more than that, it is the understanding of its evolution and its systemic interactions in its "causing" and "organisation". In other words, the phenomenon of life presents the characteristics of the new conception of mechanism.

Fleck carries this evolutionary and systemic perspective into his theory of knowledge. If Canguilhem finds physiology as a way to think about science, in his turn, Fleck makes an explicit comparison between physiology and epistemology, which he calls "physiology of knowledge". For the author of *Genesis and Development of a Scientific Fact*,

> The physiology of knowledge *[Erkenntnisphysiologie]* is analogous to the physiology of movement. To move a limb, an entire so-called myostatic system must be immobilised to provide a basis of fixation. Every movement consists of two active processes; namely, motion and inhibition. The corresponding features in the operation of knowledge are purposive, directed determination and cooperative abstraction, which complements one another. (Fleck [1935] 1979, p. 30)

Fleck conceives the scientific knowledge of an area and the process of knowledge production as a social whole, as a system or an articulated unit of practices, actions, and languages (system of reference).

> The tenacity of systems of opinion shows us that, to some extent, they must be regarded as units, as independent, style permeated structures. They are not mere aggregates of partial propositions, but as harmonious holistic units exhibit those particular stylistic properties which determine and condition every single function of knowledge. The self-contained nature of the system as well as the interaction between what is already known, what remains to be learned, and those who are to apprehend it, go to ensure harmony within the system. (Fleck [1935] 1979, p. 38)

Although it is an open system for the polish thinker, the thought style constitutes, as a self-organised structure, a "system of reference" from which we understand what a fact is. Fleck clarifies: "natural scientists in their philosophising commit the opposite and also very typical error. They are aware that there are no 'solely objective features and conditions' but only relations governed by a more or less arbitrary system of reference (*Bezugssystem*)" (Fleck [1935] 1979, p. 50).

In criticising the positivist notion of fact, Fleck thus reaffirms the idea of fact as the product of a social system of objects, actions, and linguistic denominations interacting in its "parts" and "causing" reciprocally to create an "organisation". Thus, one can also see in Fleck a new type of mechanism. The scientific fact is seen as the fruit of a system, an interwoven network that forms multiple and varied connections. For him, science is an extensive system that is established in the interwoven of theories and experiments, objects and actions that are processed interacting in a natural, social, and historical network in which "there is probably no such thing as complete error or complete truth" (Fleck [1935] 1979, p. 20). Therefore, this notion of system is not restricted to an abstract methodological system but extends to a natural and social system that brings, necessarily in the historical weave, a fundamental factor influencing the final result of science.

We thus realise that Fleck's systemic conception of science considering natural, social, and historical dimensions is not only concerned with the influences that science receives from society as a cohesive block. It is not a mere correlation between science and society but a complex connection between social and science. Even before any social factor influences science externally, it has already been interpenetrated by social attitudes and rules that have shaped cognitive practices. This social dimension is shown, for example, in Fleckian conception of "active connection", which is human's ability to produce, as a social being, the orderings of objects, called by Fleck "passive connections" (Fleck [1935] 1979, p. 10). In other words, for Fleck, more than describing nature, human, as a social being, interferes in this process. As already pointed out, "to observe, to know (*erkennen*) is always to test and thus literally to change the object of investigation" (Fleck [1929] 1986, p. 53).

From a systematic viewpoint, an excellent exercise in understanding a particular knowledge is to use a comparison device. Both Fleck and Canguilhem make use of this resource. Thus, Canguilhem speaks of a "comparative physiology".

> As far as man and his permanent physiological characteristics are concerned, only a comparative human physiology and pathology—in the sense that there exists a comparative literature—of the various ethnic, ethical or religious, and technical groups and subgroups, which would take into account life's intricacy and its kinds and social levels, could furnish a precise answer to our hypotheses. (Canguilhem [1943] 1991, p. 163)

As noted, Fleck, in his turn, extends this perspective of comparison to epistemology. Hence, for him, "it is one of the most important tasks in comparative epistemology to find out how conceptions and hazy ideas pass from one thought style to another, how they emerge as spontaneously generated pre-ideas" (Fleck [1935] 1979, p. 28). We can use this idea of comparative epistemology to compare Fleck's and Canguilhem's works to the new concept of mechanism.

At its origin, the phenomenon of life in its complexity was understood as something very distant from the traditional idea of mechanism. However, the new notion of mechanism allows us to realise that Fleck and Canguilhem's interpretation of the normal and the pathological establishes a peculiar type of mechanism. For Fleck, it is not a matter of understanding "the mechanism of the disease" (Fleck [1935] 1979, p. 62) in an isolated viewpoint, but the whole context in which it presents itself. In turn, for Canguilhem, in the same way, one should not consider the mechanism of the disease itself but also its expression, including all its biological and social aspects (Canguilhem [1943] 1991, p. 79). In other words, the phenomenon of life should not be understood only by physiological mechanisms—physiology of the liver, for example—but we must "regard it as an event involving the living organism taken as a whole" (Canguilhem [1943] 1991, p. 80). This is the systemic character of the phenomenon of life.

Indeed, in this sense, we can conceive Fleck and Canguilhem expanded the traditional idea of a mechanism to understand it in more complex interactions of a system of reference. In this system, to understand the disease or the relationship between normal and pathological, different phenomena intersect, that is, the physiological, psychological and sociological phenomena. The phenomenon of illness can only be understood at this crossroads of factors. Only along the lines of this system of reference do we know the phenomenon of life as a peculiar mechanism. Since, in addition to biological phenomena, historical and social phenomena enter into this process, any epistemology to understand it will be a historical epistemology.

Fleck and Canguilhem brought their contributions from the life sciences. Still, as pointed out, they were also influenced—directly in Fleck's case and indirectly for Canguilhem—by the new epistemological implications of modern physics. Therefore, even for them, it is not a matter of reducing epistemology to biology, but of thinking it from biology, without detriment to the new conceptions of the physical sciences, especially quantum mechanics. Considering various biological, physical and social models of knowledge expands our analogies, helping us better understand the relationship between humankind and nature. All this process enables us to conclude that epistemology necessarily is historical.

5 Conclusion

In this chapter, although I have made several approximations between Fleck and Canguilhem, the central point was not to perceive similarities between these two authors but to demonstrate how they independently established their conceptions of historical epistemology. In the 1930s/1940s, the two authors set themselves in the context of the transformations of sciences, primarily modern physics with quantum mechanics and relativity, where they thought about the epistemological status of the life sciences. In other words, by observing modern physics ideas and opposing them to classical epistemology, Fleck and Canguilhem affirmed an epistemology of biology and life science. We also saw that the historical perspective adopted by them—which states that history interferes in the final product of science—and the biological matrix, or life sciences, were essential references for developing the historical epistemology. Besides this strong emphasis on history's role in understanding science and using biology as an epistemological reference, we saw that these authors established a systemic perspective to understand the production of sciences and its history. Consequently, one can see in Fleck and Canguilhem a new mechanism to understand the "phenomena of life" considering the integration of its "parts", "causing", and "organisation". Moreover, one can conclude that historical epistemology has its "genesis" from this new mechanism idea.

References

Bachelard G ([1934] 2013) Le Nouvel Ésprit Scientifique. PUF, Paris.

Bohr N (1928) Das Quantenpostulat und die neuere Entwicklung der Atomistik. Die Naturwissenschaften. No. 15.

Braunstein J (2008) Deux Philosophies de la Medecine: Canguilhem et Fleck. In: Fagot-Largeault A, Debru C, Morange M (eds). Philosophie et médecine. En hommage à Georges Canguilhem. Vrin, Paris.

Braunstein J (2009) Fleck, Canguilhem, Foucault et le "Style Français" en Philosophie des Sciences. In: Fehr J, Jas N, Löwy I. (eds). Penser avec Fleck—Investigating a Life Studying Life Science. Collegium Helveticum, Heft 7, Zurich.

Canguilhem G ([1943] 1991) The Normal and the Pathological. Zone Books, New York.

Canguilhem G (1977) Idéologie et Rationalité dans l'Histoire des Sciences de la Vie. Vrin, Paris.

Carrier M (2012) Historical epistemology: On the diversity and change of epistemic values in science. Ber Wiss 35(3):239–251.

Condé M (2016) Entre o normal e o patológico: Ludwik Fleck, Georges Canguilhem e a gênese da epistemologia histórica. Intelligere, Revista de História Intelectual 2(1):51–67.

Craver C, Tabery J (2015) Mechanisms in Science. *The Stanford Encyclopedia of Philosophy* (Summer 2019 Edition), Zalta EN (ed). Via: https://plato.stanford.edu/archives/sum2019/entries/science-mechanisms. last accessed 2021/11/21.

Dijksterhuis EJ (1961) The mechanisation of the world picture. Clarendon Press, Oxford.

Fleck L ([1927] 1986) Some specific features of the medical way of thinking. In: Cohen R, Schnelle T (eds). Cognition and fact: Materials on Ludwik Fleck. Reidel, Boston.

Fleck L ([1929] 1986) On the crisis of 'Reality'. In: Cohen R, Schnelle T (eds). Cognition and fact: Materials on Ludwik Fleck. Reidel, Boston.

Fleck L ([1935] 1979) Genesis and development of a scientific fact. The University of Chicago Press, Chicago.

Fleck L ([1960] 1986) Crisis in sciences. In: Cohen R, Schnelle T (eds). Cognition and fact: Materials on Ludwik Fleck. Reidel, Boston.

Freud S ([1914] 1946) Der Moses des Michelangelo. Gesammelte Werke. Band X. Imago, London.

Fuller S ([1988] 2002) Social epistemology. Indiana University Press, Bloomington.

Gingras Y (2010) Naming without necessity: On the genealogy and uses of the label historical epistemology. Revue de Synthèse 131(3):439–454.

Ginzburg C ([1978] 1983) Señales: Raíces de un paradigma indiciario. In: Gargani A (ed). Crisis de la Razón: Nuevos modelos en la relación entre saber y actividad humana. Siglo Veintiuno, Mexico.

Ginzburg C ([1978] 1986) Clues: Roots of an evidential paradigm In: Clues, Myths and the historical method. (trans) Tedeschi J, Tedeschi A. The Johns Hopkins University Press, Baltimore.

Hacking I (2002) Historical ontology. Harvard University Press, Cambridge–MA.

Husserl E ([1936] 1970) The Crisis of the European Sciences and Transcendental Phenomenology. An Introduction to Phenomenological Philosophy. Northwestern University Press, Evanston.

Koyré A ([1943] 1973) Galilée et Platon. In: Koyré A. Étude d'histoire de la pensée scientifique. Gallimard, Paris.

Koyré A ([1948] 1971) Du Monde de l'"à-peu-près" à l'univers de la précision. In: Koyré A (ed). Étude d'histoire de la pensée philosophique. Gallimard, Paris.

Kuhn T ([1962] 1970) The structure of scientific revolutions. The University of Chicago Press, Chicago.

Kuhn T (2000) The road since structure. The University of Chicago Press, Chicago.

Kusch M (2011) Social epistemology. In: Bernecker S, Pritchard D (eds). The routledge companion to epistemology. Routledge, London.

Reichenbach H (1938) Experience and prediction: An analysis of the foundations and the structure of knowledge. The University of Chicago Press, Chicago.

Rheinberger H (2010) On historicizing epistemology: An essay. Stanford University Press, Stanford.

Sinding C (2009) De Fleck a Canguilhem: La medicine comme epistémologie de l'incertain. In: Fehr J, Jas N, Löwy I (eds). Penser avec Fleck—Investigating a life studying life science. Collegium Helveticum, Heft 7, Zurich.

Chapter 9
When Was Electromagnetic Theory Completed? An Alternative Formulation to That of Maxwell

Antonino Drago

Abstract Here the usual historical account on the completion of electromagnetic theory is contested; electromagnetic theory has to be considered a completed theory not before a symmetric explanation of electromagnetic induction between moving bodies was fulfilled. The notion of incommensurability between different approaches explains why this case-study constitutes a hard subject for the historians of physics and hence why in the past they contented themselves to consider as the decisive event of this history a partial achievement, i.e. the birth of Maxwell's equations. After the introduction of Lorentz's group into theoretical physics, some scholars assumed on the foundations of electromagnetism a retrospective theoretical view, which suggested new formulations of this theory. Among such new formulations. I recognize a recent formulation, which is alternative to Maxwell's in the following sense. I interpreted the foundations of theoretical physics as constituted by two basic dichotomies: one about two kinds of mathematics, and another about two kinds of logic. Every theory relies on a couple of choices on the two dichotomies. I characterize the couple of choices of Maxwell-Hertz-Lorentz' formulation as that of Newton's mechanics; the alternative electromagnetic theory is based upon the couple of opposite choices. By improving this theory in a formal way according to the couple of alternative choices I obtained the extreme polarities of all possible formulations of electromagnetism. An overview of the history of EM according to the new foundational viewpoint constitutes the conclusion.

Keywords Alternative formulations of electromagnetism · Couple of the alternative choices · Dominant couple of choices · Lorentz' group · Maxwell equations · Two dichotomies

---

A. Drago (✉)
University of Napoli "Federico II", Naples, Italy
e-mail: drago@unina.it

R. Pisano (ed.), *A History of Physics: Phenomena, Ideas and Mechanisms*,
History of Mechanism and Machine Science 42,
https://doi.org/10.1007/978-3-031-26174-9_9

1 Introduction

The end point of the history of Electromagnetic Theory (EM) is not clear in both textbooks (Jenkin 1873) and the historical accounts of this theory; e.g. the excellent Olivier Darrigol's book, terminating his account exactly at the date of 1905, characterizes it by means of the following words:

> In the history of electrodynamics, 1905 was a year of abundance, not a year of conclusion. (Darrigol 2000, p. 394)

The question of the date of completion of this theory will be settled on the base of a previous recognition of the foundations of theoretical physics (Drago 2012, 2013, 2017). This base will be proved as adequate to EM by suggesting a formulation of EM which is formally alternative to James Maxwell's one; this new formulation was anticipated by a recent paper which independently suggested a new formulation of EM by assuming Hendrik Lorentz' group of transformation (Diener 2013).

In Sect. 2 an interpretation of the foundations of theoretical physics as constituted by two dichotomies is suggested and then in Sect. 3 they are applied to the history of EM. Section 4 introduces Lorentz' group (LG) as an essential component of the foundations of EM. Section 5 offers a first list of different formulations of EM. In Sect. 6 an alternative formulation, based on the alternative choices to the dominant ones, is looked for among the already presented new formulations derived from the viewpoint of the transformations groups. It is recognized as substantially anticipated by a recent one, which then is re-formulated in a detailed way according to the two alternative choices to Maxwell's ones. An overview of the history of EM (e.g., Maxwell 1873) according to the new foundational viewpoint constitutes the conclusion.

2 Definition of the Foundations of Theoretical Physics

The above question of the completeness of Maxwell's theory brings us to ask what the foundations of EM are. No previous historian of EM claimed to be informed by a definition of the foundations of theoretical physics. The present paper represents a new investigation on the subject at the extent that it applies a general conception of these foundations of physics; through their definition, the above question will be settled.

As a matter of fact, several scientific theories have been presented by the respective authors in a non-axiomatic way and by making use of elementary mathematics. Previous studies mutually comparing their structures have suggested that the foundations of theoretical physics are constituted by two basic dichotomies (Drago 2012, 2013). One dichotomy concerns the kind of *infinity*, either actual infinity (AI) or potential infinity (PI). In the last century these two notions have been formalized as respectively:

- Classical mathematics (including notions—e.g. infinitesimals, or principles e.g. Zermelo's principle—relying on the notion of actual infinity. This formulation of mathematics is commonly misinterpreted as indispensable, although it applies idealistic techniques to physical reality);
- Constructive mathematics (making use of only mathematical objects which have been constructed by algorithms (Bishop 1967, pp. 1–10).

One more dichotomy concerns the kind of *organization of a theory*, either an entirely deductive organization from some axiom-principles (AO) (this organization is commonly misinterpreted as the unique rigorous organization) or a problem-based organization (PO) (i.e., a theory aimed at discovering a new scientific method for solving a basic problem. These two organizations of a theory make use of two different ways of reasoning, respectively, the deductive one or the inductive one; or, in formal terms, classical logic (which is commonly misinterpreted as the unique valid one for the applications to the reality) and non-classical logic, first of all the intuitionist one, where the law of double negation fails. The ideal model of the latter organization presents six steps:

(i) The declaration of a problem, which the theory is aimed at solving;
(ii) The use of doubly negated propositions whose corresponding affirmative ones lack of evidence, hence the latter ones are not equivalent to the former ones; since here the double negated law fails, the propositions belong to intuitionist logic (DNPs);
(iii) A methodological principle leading the search for discovering a new method for solving the given problem;
(iv) An argumentation composed by *ad absurdum* proofs;
(v) The conclusion of the entire argumentation is a doubly negated predicate;
(vi) By applying the principle of sufficient reason to it an affirmative proposition is obtained; it represents a new hypothesis, from which one derives all possible conclusions; they are then tested with reality in order to at last establish whether the new hypothesis solves the given problem or not. Instances of theories exhibiting such organization are Sadi Carnot thermodynamics (Drago and Pisano 2000), Lobačevskij's non-Euclidean geometry (Drago 2011) and Einstein's 1905 paper on quanta (Drago 2013).

In sum, the foundations of theoretical physics are constituted by two dichotomies which are philosophical in nature because they are represented by philosophical notions (infinity, organization), and at the same time are scientific in nature, because they are formalized through scientific theories. They are different from previous suggestions of foundations because they are more general than a single physical notion (e.g. force, field of forces); or group of notions (e.g. space and time), of a system of equations (e.g., Maxwell's), and at the same time the dichotomies are more adequate to physical theories than the philosophical ideas of causality, determinism, holism, or even a system of ideas (e.g. Kant's). Moreover, being the two choices antagonistic in nature, also historical conflicts can be represented by them.

Each couple of choices characterizes one out four Models of Scientific Theory (MSTs). Notice that the choice of PI cannot be combined with the choice of AI since the latter one ignores any limitation (e.g. undecidabilities) generated by PI. Moreover, the choice PO is incompatible with the choice AO because so are their respective kinds of logic, which are essentially different owing to the validity or not of the law of the double negation. Given the antagonism of the two choices on each dichotomy, an accurate notion of incommensurability (Drago 1987) between two theories is derived: Two physical theories are said mutually incommensurable in the case they rely on different couples of choices on the two dichotomies. This phenomenon is manifested by radical variations in meanings of the basic notions shared by such two theories; a variation is caused by the different choices on which the two theories rely.

3 A New History of EM

The double nature of the two dichotomies, when considered as interpretative categories of the history of physics gives to a historian a great interpretative power; actually, they cover all formal aspects of the history of physics as well as the philosophical aspects, conflicts of ideas too.

Under the light of the two dichotomies, we easily recognize that the basic choices of Ampère's program of research were AI (owing to his use of the most sophisticated mathematics, i.e. infinitesimals) and AO (owing to his search of general laws to put as principles of an axiomatic-deductive theory). The choices of Faraday's program of research were instead PI (owing to his minimum recourse to mathematics, however an elementary mathematics) and PO (owing to his theorizing was aimed at solving a problem, i.e. whether electricity is equivalent to magnetism). This characterization of the two programs of research gives full meaning to the conflict between them: being the two couples of choices one the opposite of the other no greater foundational distance between these programs was possible. Moreover, the basic choices of Maxwell's theory are recognized as AI (owing to the mathematics of his four local laws expressed by differential equations), and AO (because his equations were considered as axioms from which one deduces the resolutions of all EM laws).

It is well-known that in order to include Faraday's results into his mathematical framework Maxwell's hard worked; actually, in order to achieve his theory he had to overcome both the radical variations in meanings of his notions with respect to those pertaining to the incommensurable Faraday's theoretical approach, and the radical difference between his mathematics and that of the latter one. I can hint that owing to this incommensurability phenomenon and the consequent radical variations in meanings of many notions most historians of EM avoided to deepening their historical appraisals and have contented themselves of little more than chronicles of a historical course that they saw as inexorably leading to a final result, i.e. Maxwell's equations. Notice that Maxwell's choices were the same of Newton's (i.e. AI&AO) (Drago 1988). The foundational convergence of Maxwell's theory, concerning an at all new field of phenomena, to the old paradigm confirmed and renewed it, or better its basic choices, within theoretical physics. This fact gives reason of the inertia of historians to change the common view on the discovery of Maxwell's equations as the conclusive step of the history of EM.

4 The Revolution Caused by the Introduction of Lorentz' Group into Theoretical Physics

In the history, EM emerged through a physical research performed within exceptional circumstances: at all new phenomena, surprising notions and laws which often have been discovered not before the inventions of new specific instruments, surprising mathematical novelties, pervasive influence of the Newtonian paradigm. Surprisingly, after the short time of only two generations Maxwell discovered his equations concerning a field of phenomena, which no long ago were almost completely unknown. His indubitable great success made euphoric the physicists so much that the underlying problems of the foundations of the new theory have been almost ignored. EM physicists mistakenly assumed as foundations of their theory a single notion, the Newtonian space or its mere specification, field of forces (Hesse 1965; Landau and Lifschitz 1975). Owing to this naïve view, the hazardous hypothesis of the ether (in the time chemistry emerged as a science exactly because it was capable to foresee new elements of matter) appeared as unavoidable. Yet, the hypothesis resulted to be a "fiasco".

At present time we think that the failure of ether hypothesis and Einstein's introduction of Lorentz' group (LG) constituted a revolution within theoretical physics. Yet, Einstein's paper gives any relevance to the notion of group of transformations (only one time he made use of the word "group" in the sense of a grouping). Later, he prolonged his work on special relativity theory by founding the theory of general relativity again on the notion of space studied by means of analytical tools, not groups. Therefore, the revolution elicited by his 1905 paper (Einstein 1905a, see also 1905b) was not completed by a change in the basic mathematical tools, i.e. groups instead of differential equations.

Yet, in the year 1960 symmetries emerged as constituting an essential component of theoretical physics. A re-visitation of *past history* of physics was necessary. In particular, a question has arisen from this re-visitation of the 1905 revolution, which constituted the first celebrated introduction of symmetries into theoretical physics, i.e. the application of the group of transformation of EM, LG, to the theory of mechanics: Which is the theoretical role of LG within the foundations of EM? A subsidiary role or a constitutive role?

The former answer agrees with the appraisals given by past EM theorists. Indeed, we see that past textbooks of EM present Maxwell theory as it was autonomous from LG, i.e. as if this theory could be considered independently from its group of symmetry; but, if for the theory of mechanics the approximation $v << c$ gives a valid theory, in EM no such approximation makes sense (Jammer and Stachel 1980). Therefore I think that it is time to abandon any reference to Newton's space and hence to the Galilei's group and so to present the birth of LG as a decisive improvement of the foundations of EM.

The latter answer (the proposed constitutive role of LG within the foundations of EM) has to be proved by means of a definition of the foundations of theoretical physics. Unfortunately, in past time there was no common agreement on what they are.

A scholar contested the usual historical account on the end of the development of electromagnetic theory on the same base of the problem presented by the introduction of Einstein's 1905 paper. In his opinion Maxwell's theory can be considered a complete theory not before a symmetric explanation of the electromagnetic induction of moving bodies was given (Drago 2003) and hence the introduction of LG into EM. Einstein's 1905 paper may be severed into two parts (presented by Einstein in the reverse order): his proof of LG invariance of EM and his astonishing new formulation of mechanics obtained by applying LG to Newton's mechanics. It was the former part that put an end to the construction of EM (whereas the latter part constituted a dethroning of Newton's mechanics from its bi–secular paradigmatic role and therefore received much more attention). The relevance of this point is clear from the above illustrated viewpoint on the foundations of a physical theory. LG is not derived as a specific deduction from the axioms of a theory—as it occurs in an AO theory—and it cannot play the role of an a priori axiom either. It is rather included within a search for solving (as all PO theories do) a problem, the problem of what are the "invariants", as Lazare Carnot stated in the first introduction of a transformation group into a physical theory which is a formulation of mechanics, based on the impact of bodies. Indeed, the main idea of L. Carnot's mechanics was the following one:

> Therefore there exists [...] in each percussion or communication of movement [...] a quantity which is not altered by the impact. (Carnot [1783] 1786, p. 44; Author's translation)

Then he defined a group of transformations (called by him "geometrical motions"). In fact, this formulation of mechanics followed Leibniz' program: "Our minds look for in-variants" (Leibniz 1698, p. 215). Being the result of the application of a transformation group expressed by a DNP (in-variant), this mathematical technique is essentially linked to a PO.[1] In the same way Einstein looked for the invariant laws under changes of the reference systems.

In sum, the introduction of LG within EM changed the kind of its theoretical organization from AO to PO. This fact gives reason of the cultural inertia of the physicists (and in previous times also the mathematicians) in receiving this innovation; it was difficult to realize the theoretical change because the cultural framework of a transformation group, i.e. to look for solving the problem of the invariants, at last a PO, was ignored owing to the prejudice that only AO is a correct organization of a theory. Yet, no revolution in theoretical physics would have been possible if LG was a mere addition to an already accomplished theory, Maxwell's. Therefore, Lorentz' group (LG) has to be considered an essential component of the foundations of EM.

[1] Lazare Carnot wanted to found his physical theory on an "empirical" mathematics, i.e. whose nature is operative, constructive; hence, his choice was PI.

5 Alternative Approaches to the Electromagnetic Theory

The history of EM includes some radical, theoretical conflicts: between Volta's laws (Drago 2003) and Newton's mechanics, between Ampère's program and Faraday's program, between Faraday's mathematics and Maxwell's. Under this light, one can ask: Do Maxwell's equations constitute the one possible formulation of EM? Moreover, by viewing EM as a theory, which was completed by the introduction of LG, one can ask the following question converging with the previous one: in a retrospective view, at present time can the EM theory be founded on different grounds?

Just few years after Einstein's paper Leigh Page suggested a new way to found part of EM theory through LG (Page 1912): by starting from Coulomb law of electrostatic, he derived Lorentz force and magnetostatic from LG. (This approach is presented by many textbooks: Purcell and Morin 2013; Landau and Lifshits 1975 and also Feynman 1967). Much more was suggested by some Feynman's notes, discovered half a century ago, but interpreted and commented few years ago (De Luca 2019). He derived:

> Lorentz force and the transformation laws of electric and magnetic fields only by requiring charge invariance (coming from experiments), relativity and linearity, but without stating Coulomb's law. Relativity played a pivotal role also in later courses, in particular in the Hughes Lectures (Feynman 1967-1968), where he followed another path, based on a relativistically invariant generalization of the least action principle, where a 4-vector potential is added to the free particle relativistic action. A comparison of the resulting equations of motion with the expression of Lorentz force gives the link between the electric and magnetic fields and the components of the 4-potential. Finally, by means of well-known vector analysis identities, homogeneous Maxwell equations are recovered. (De Luca 2019, p. 238)

A further paper (Diener 2013) completed the application of Feynman's viewpoint by deriving the inhomogeneous Maxwell's equations. The same paper listed some similar approaches:

> Starting from the pioneering Page's paper in 1912 (Page 1912), it became clear that magnetic fields could be understood as relativistic effects due to relative motion with respect to electric charge. In particular magnetic field of a uniformly moving point charge (and in fact, the whole of magnetostatics if the superposition principle is assumed) can be obtained by boosting the Coulomb field of a static charge (Purcell et al. 2013; French 1968; Rosser 1991). Also, the fields of accelerating charges, including radiation, can be derived from the knowledge of the Coulomb field and relativity, with a few more assumptions, i.e. the action and reaction principle, charge conservation and the request that the electromagnetic interaction propagates with the speed of light (Tessman 1966). Once the general fields produced by a moving charge are known, the whole set of Maxwell's equations can be derived using the superposition principle (Page et al. 1940; Rosser 1997). An alternative approach starts from the fundamental equation of electrostatics, namely Gauss' theorem (i.e. the first Maxwell equation), which in turn can be derived from the Coulomb force plus the superposition principle, and to boost it (Lorrain et al. 1988), or to *covariantize* it (Schwartz 1972). More sophisticated and abstract approaches exist as well (Landau el at. 1975; Susskind et at. 2017), in which the superposition principle is assumed while the Coulomb law is not. Rather, an action for the electromagnetic field is constructed by assuming the latter principle, plus relativistic and gauge invariance, whose variation gives the non-homogeneous Maxwell equations. In such an approach, the Coulomb law is derived as the solution of the equations for a point static source. (Diener 2013, p. 2)

In conclusion, we see that by starting from the viewpoint of LG one can recover the entire Maxwell's theory within a consistent and well-founded framework of theoretical physics (e.g. in Feynman's approach that of the least action). This fact gives additional evidence for considering Maxwell's theory as no more than a specific theoretical system of EM theory.

6 Two Carnotian Formulations of EM

In order to recover the deeper aspects of the history of EM, let us investigate whether there exist a complete, alternative formulation to Maxwell's theoretical system whose choices are the same of Newtonian paradigm (Kuhn 1969), i.e., AI and AO; the wanted formulation has to be therefore based on the alternative choices PI and PO, i.e. the choices of the Carnotian model of a scientific theory.

First, let us remark that *EM may be represented in the same way of phenomenological thermodynamics, i.e. through its phenomenological laws without translating them, as Maxwell did, into local equations.*[2] This translation is necessary in order to obtain the wave equation, which however does not pertain to the foundations of EM. This phenomenological formulation, represented by extended laws, e.g., is presented by the book (Rossi 1957, p. 365).

Let us characterize the specific choice of this formulation on the kind of mathematics. We have to test through constructive mathematics the formulas representing its global laws. They all are space integrations plus some first derivatives. In constructive mathematics, the first derivatives on essentially continuous functions are constructive. Moreover, the result of an integration operation is assured in the case the integrand function is uniformly continuous (Bishop 1967); this condition is surely satisfied by electromagnetic laws. In sum, the choice of this formulation on the kind of mathematics is PI. In addition, we have to test the kind of organization of this formulation. To my knowledge, no previous scholar looked for a PO of these laws; we have to test whether the development of the theory presents the six steps listed in Sect. 2: (*i*) The declaration of a problem, which the theory is aimed at solving; the basic problem of EM is easily recognized as expressed by the following doubly negated proposition (DNP): "Is electricity non-different (equivalent) to magnetism, and vice versa?" (*ii*) The previous proposition—which is a DNP—introduces the use of DNPs. (*iii*) the same proposition, without the interrogative mark is the methodological principle: "It is not true that electricity is not magnetism; and vice versa. The next three steps—(*iv*) *Ad absurdum* proofs, (*v*) to obtain a final DNP and (*vi*) to apply the principle of sufficient reason—are usually left as understood; to explicit them

[2] Let us recall that the phenomenological formulations of the physical theories are usually depreciated. "Like every other physical theory, there is an inductive [phenomenological] approach to electrodynamics, and there is also a deductive one. Sommerfeld [1952] prefers deductive approach, and following a great paper of Hertz [1890; 1893] takes the Maxwell's equations themselves as the postulate of the theory calling the traditional inductive approach "awkward, incoherent, and by no means self-contained." (Barut 1974, p. 478).

in our case is an open problem. When these *ad absurdum* proofs will be suggested, a complete EM formulation which is alternative in the basic choices to Maxwell's formulation, will be obtained.

In addition to this possible formulation, let us consider the several new approaches of EM obtained by applying LG or more in general groups of transformations to a minimum of EM laws. A recent approach (Di Mauro 2021) is the most attractive, because, the same title of this paper—like the title of Einstein's 1905 paper on quanta, whose organization is clearly PO (Einstein 1905b)—, claims have obtained "heuristically" its result. Even more interesting is its presentation, which includes *ad absurdum* arguments, in agreement with the model of a PO theory. The authors summarize their paper as follows:

> Starting from the experimental fact that a moving charge experiences the Lorentz-force and applying the fundamental principles of simplicity (first order derivatives only) and linearity (superposition principle), we show that the structure of the microscopic Maxwell equations for the electromagnetic fields can be deduced heuristically by using the transformation properties of the fields under space inversion and time reversal. Using the experimental facts of charge conservation and that electromagnetic waves propagate with the speed of light together with Galilean invariance of the Lorentz force, allows us to finalize Maxwell's equations and to introduce arbitrary electrodynamics units naturally. (Di Mauro 2021, p. 120)

Or, more shortly,

> [...] Maxwell equations do indeed equate quantities with the same transformation behavior under time reversal and space inversion. The idea presented here is that the transformation behaviors, together with the Lorentz force and the heuristic demand for simplicity and linearity, allows us to deduce the structure of the Maxwell equations. (Di Mauro 2021, p. 121)

Let us examine whether the entire organization of this formulation is a PO. Its problem is the same previous problem, i.e. whether electricity is equivalent to magnetism, and vice versa. Notice that the superposition principle is a DNP; it means neither that two states are equal, nor that they are distinct, because each state is not severed from the other state. Being a DNP, this "principle" makes impossible an AO of its theory. Like all DNPs beginning a PO theory, it represents a methodological principle rather than an axiom; i.e. it plays the role of orienting the search for a new method capable of solving the problem stated at the beginning of the theory. This principle plays a similar role to that played within past theoretical developments of mechanics and thermodynamics by the impossibility of a perpetual (= without an end) motion.

Next authors' illustration of their theory is composed by five steps; each of the first three steps makes an essential use of the word "only" (= nothing else), hence a DNP. Each proposition, including in an essential way the word "only", is proved by an *ad absurdum* proof. For instance, I quote the first one:

> [...] there is no experimental evidence for magnetic charges so there cannot exist any field-generating_pseudoscalars. The only pseudoscalar [...] that can be generated and obey linearity is to take the first-order spatial derivative of an axial field vector. Thus, we quickly conclude that [the scalar product of nabla with the magnetic vector is null]. (Diener 2013, p. 121)[3]

[3] One more "only" is used in an additional argument to prove this result.

The entire argumentation consists of the sequence of the three *ad absurdum* proofs; it obtains a theoretical framework including five undetermined constants; therefore, the final (implicit) proposition of this theoretical framework cannot be different from a DNP, of the following kind: "It is not true that [...] not [...]". Then, by imposing the fulfillment of three constraints (i.e. charge conservation, invariance of Lorentz' force and wave equation) the indetermination is reduced to two parameters. Owing to this residual indetermination, the (implicit) final proposition is again a DNP.

The last two undetermined parameters are decided through the choice of the system of units, as the authors state:

> For the five unknown constants we therefore have three independent equations, meaning there is a degree of ambiguity [caused by the two remaining undetermined parameters] in this system. This ambiguity will be resolved with the discussion of three frequently used systems of units, the metric [...]. (Di Mauro 2021, p. 122)

This decision cannot be a clear-cut one, because each system of units is a heuristic result of a theoretical tradition; hence the conclusion is again represented by a DNP. Only an application of PSR to this final DNP can give the wanted results that is the corresponding affirmative system proposition stating that this theoretical system obtains exactly the theory of EM. It is a common practice of theorists to leave understood this application, as a trivial step. In conclusion, this paper almost completely represents the model of a PO theory, notwithstanding their authors ignored the model of this kind of organization.

The kind of mathematics of the presentation of this formulation does not present constructive problems, because the operations are the usual ones of the first-order derivatives of the vector differential calculus, except for the wave equation; which however here only works as a test for proving the correct mathematical form of the theory, not as producing a set of solutions, which may be considered as non-constructive (Pour-El and Richards 2017). *Eventually we have obtained a new formulation of EM relying in a detailed way on the two choices PI and PO*; it is alternative to Maxwell's formulation, which is based on the opposite choices. In sum, we have represented through a suitable formulation of EM the two extreme polarities of all possible EM formulations.

7 Conclusion

The theorists of EM were unaware of the great foundational problem left by previous theoretical mechanics: the incommensurability between different formulations of this theory. They thus ignored that their new field of research allowed the birth of at all divergent theoretical approaches which could be mutually incommensurable and moreover incommensurable with those of previous mechanics theory. As a fact, they met extraordinary problems:

(i) How to decide Alessandro Volta's suggestion of the possibility of a perpetual motion in either electrical or magnetic phenomena;

(ii) How to resolve a conflict between different meanings of a basic notion: e.g. the interaction as an action at distance or a step-by-step action along a line of force;

(iii) How to resolve the conflict between: a priori notions (e.g. absolute space, universal force, etc.) and merely experimental notions.

(iv) The same for the different roles played by mathematics in theoretical physics: to assume Euclidean geometry, as the unavoidably basic mathematics, or every mathematical container of physical phenomena had to be rejected? How to solve the conflict between discrete mathematics and infinitesimal analysis?

(v) How to start? From a priori principles, or no more than methodological principles for developing a heuristic research?

(vi) Does a field of research allow only one formulation of a physical theory or alternative formulations are possible?

(vii) In the latter case which are their mutual relationships: Complementarity or divergence?

In particular, the last foundational problem played a crucial role since the birth of the first theoretical framework for electromagnetic phenomena. The divergence between the two programs of research of Ampère and Faraday constituted a very deep problem. Maxwell, who improved the set of discovered laws until to a consistent and almost complete mathematical system, solved it. It is not surprising that after Ampère's and Faraday's great theoretical efforts—elapsed in the first half of nineteenth century—and after the great Maxwell's work for obtaining his equations EM theorists wanted to mainly apply Maxwell's mathematical framework. He aimed to conquer a greatest number of new results without reflecting on both the performed historical path and the kind of obtained results in a similar way to what Mach did for the history of mechanics. This lack of reflections has made the theorists of that time blind to theory's incompleteness of Maxwell's theory; which eventually have been astonishingly revealed by Einstein's 1905 paper. It revealed that EM, although already highly perfectioned along the second half of the nineteenth century, yet was not completed owing to the asymmetry in the description of the magnetic induction in moving bodies, and hence had to introduce LG. Moreover, it revealed that EM equipped with LG surprisingly was a more general theory than Newton's mechanics, which had played the paradigmatic role for the entire theoretical physics. Hence, this lack of reflections obscured theorists' conscience of the foundations of EM and the entire theoretical physics, too. It was again Einstein's paper, which eventually gave an answer to an old foundational problem. It discussed whether the two theories, mechanics and EM, were "irreconcilable" (Einstein 1905a, p. 891) or not. This paper also gave birth to a new kind of study, i.e. the comparison of two different physical theories. For the first time Einstein showed that two "apparently irreconcilable" theories can be conciliated. He was successful because he introduced the group technique into theoretical physics. This innovation forced theorists to acquire a new level of knowledge of the foundations of physical theories. Unfortunately, in that time and also in the next times this new level of knowledge remained confined to on one hand improve the mathematics of the theoretical physics, and on the other hand take the distance from past philosophies of knowledge as essentially inadequate to the new scientific achievements, without seizing the opportunity of dissecting the problem of what are the foundations of physics and trying to solve it.

Acknowledgements I offer as a homage to Salvo D'Agostino this paper dealing with a subject to which he devoted great part of his studies, i.e. the intriguing history of electromagnetic theory.

References

Barut AO (1974) On the formulations of electrodynamics from a single principle. Foundations of Physics 24(4):477–485.
Bishop E (1967) Foundations of Constructive Mathematics. Mc Graw-Hill, New York.
Carnot L ([1783] 1786) Essai sur les machines en général. Defay, Dijon.
Darrigol O (2000) Electrodynamics from Ampère to Einstein. Oxford University Press, Oxford.
De Luca (2019) An unpublished approach to electrodynamics by Richard Feynman. European Journal of Physics 40(6):1–15.
Diener G (2013) Obtaining Maxwell equations heuristically. American Journal of Physics 81(2):120–123.
Di Mauro M (2021) Some insights into Feynman approach to electromagnetism. European Journal of Physics 42(2) 025206.
Drago A (1987) An effective definition of incommensurability. VIII LMPS, Moscow, 4, pt.1, 159–162 [see also: Cellucci C (1988) (ed) Temi e prospettive della logica e della filosofia della scienza contemporanea. II, CLUEB, Bologna, pp. 117–120].
Drago A (1988) A Characterization of Newtonian Paradigm. In: Scheurer PB, Debrock G (eds). Newton's Scientific and Philosophical Legacy. Kluwer Academy Press, Dordrecht, pp. 239–252.
Drago A (2003) Volta and the strange history of electromagnetism. In: Giannetto EA (ed) Volta and the history of Electricity. Hoepli, Milano, pp. 97–111.
Drago A (2011) The birth of the non-Euclidean geometries As the more significant crisis in the foundations of modern Mathematics, Logic and Philosophy of Science 9(1):103–110.
Drago A (2012) Pluralism in Logic. The Square of opposition, Leibniz's principle and Markov's principle. In: Béziau J-Y, Jacquette D (eds) Around and Beyond the Square of Opposition. Birkhaueser, Basel, pp. 175–189.
Drago A (2013) The emergence of two options from Einstein's first paper on Quanta. In: Pisano R, Capecchi D, Lukesova A (eds). Physics, Astronomy and Engineering. Critical Problems in the History of Science and Society. Scientia Socialis, Siauliai, pp. 227–234.
Drago A (2017) Dalla storia della Fisica alla scoperta dei fondamenti della Scienza. Aracne, Roma.
Drago A, Pisano R (2000) Interpretazione e ricostruzione delle *Réflexions* di Sadi Carnot mediante la logica non classica. Giornale di Fisica 41:195–215 [English translation: *Id.*, (2004) Atti Fondazione Ronchi 59:615–644].
Einstein A (1905a) Zur Elektrodynamik bewegter Koerper. Annalen der Physik 17:891–921.
Einstein A (1905b) Ueber einen die Erzeugung der Verwandlung des Lichtes betreffenden heuristisch Gesichtpunktî, Annalen der Physik 17:132–148 [reprinted: Stachel J (1989) (ed) Collected Papers of Albert Einstein. Princeton University Press, Princeton, II, pp. 149–16].
Hertz H (1890) Über die Grundgleichungen der Elektrodynamik für ruhende Korper. Göttinger Nachrichten, March, and Annaler der Physik 40 [Über die Grundgleichungen der Elektrodynamik für bewegte Korper. Annaler der Physik 41].
Hertz H (1893) Electric Waves. MacMillan, London.
Hesse M (1965) Forces and Fields. Littlefield & Adams, London.
Jammer D, Stachel J (1980) If Maxwell had worked between Ampere and Faraday. An historical fable with a pedagogical moral. American Journal of Physics 48:5–7.
Jenkin F (1873) Electricity and Magnetism. Appleton, New York.
Kuhn TS (1969) The structure of the scientific revolutions. Chicago University Press, Chicago.
Landau LD, Lifschitz EM (1975) The Classical Theory of Fields, III. Pergamon, London.
Leibniz GW (1698) Essai de dynamique sur les lois du mouvement. MS, VI, pp. 215–221.

Masterman M (1970) The Nature of a Paradigm. In: Lakatos I, Musgrave A (eds). Criticism and the Growth of Knowledge. Cambridge University Press, London, pp. 59–89.
Maxwell JC (1873) A Treatise on Electricity and Magnetism. Oxford University Press, Oxford.
Page N (1912) Derivation of the fundamental relations of electrodynamics from those of electrostatics. American Journal of Science 34:57–68.
Pour-El MB, Richards JI (2017) Computability in Analysis and Physics. Cambridge University Press, Cambridge.
Rossi B (1957) Optics. Addison Wesley, New York [Italian Translation: *Id.* (1979) Ottica. Masson Italia, Milano].
Sommerfeld A (1952) Electrodynamics, Lectures on Theoretical Physics. Vol. 111, Academic Press, New York.

Chapter 10
Can Physics Attain Its Goals: Extending D'Agostino's Analysis to the 21st Century and Beyond

Olga Kosheleva and Vladik Kreinovich

Abstract In his 2000 seminal book, Salvo D'Agostino provided the detailed overview of the history of ideas underlying nineteenth and twentieth century physics. Now that we are two decades into the twenty-first century, a natural question is: How can we extend his analysis to the twenty-first century physics—and, if possible, beyond, to try to predict how physics will change? To perform this analysis, we go beyond an analysis of what happened and focus more on why paradigm changes happened in the history of physics. To better understand these paradigm changes, we analyze not only what were the main ideas and results of physics, but also on how (and why) the objectives of physics changed with time.

Keywords Full cognition · History of physics · Modern approach · Laplace determinism · Paradigm · Statistical determinism · Unity of science

1 Introduction

In his 2000 seminal book (D'Agostino 2000), Salvo D'Agostino provided the detailed overview of the history of ideas underlying nineteenth and twentieth century physics; see, e.g., (Feynman et al. 2005; Thorne and Blandford 2017). Now that we are two decades into the twenty-first century, a natural question is: how can we extend his analysis to the twenty-first century physics—and, if possible—beyond, to try to predict how physics will change?

Of course, the progress of physics never stops: new theories appear all the time, theories that lead to a better and more accurate description of physical phenomena. However, from the foundational, philosophical viewpoint, what is important is

O. Kosheleva (✉) · V. Kreinovich
University of Texas at El Paso, El Paso, TX 79968, USA
e-mail: olgak@utep.edu

V. Kreinovich
e-mail: vladik@utep.edu

R. Pisano (ed.), *A History of Physics: Phenomena, Ideas and Mechanisms*, History of Mechanism and Machine Science 42,
https://doi.org/10.1007/978-3-031-26174-9_10

that not only theories change: sometimes, physics undergoes more fundamental changes—when what changes are the very objectives of physics, i.e., the very paradigm of what it means to do physics. To try to predict how physics paradigm will change, we need to go beyond an analysis of what happened and focus more on why paradigm changes happened in the history of physics. In other words, to better understand future changes, we need to analyze not only what were the main ideas and results of physics, but also how (and why) the objectives of physics changed with time.

In the beginning—way before the nineteenth century—physicists believed that the objective of science was full cognition: e.g., Kepler not only described how planets move, he also tried to explain why they are located at given distances from the Sun—and even what will be the fate of a person. This paradigm was replaced by Laplace Determinism, when initial conditions are left to Hokusai and other artists, but, based on these conditions, we can predict the world's future state. Starting with statistical physics—and emphasized by quantum theory—the new paradigm of Statistical Determinism appeared, when we can only predict probabilities of future events. In many areas of physics, this paradigm is being replaced with what we call Modern Approach, when often we can only predict partial information about the probabilities.

In this essay, (following ideas outlined in Finkelstein and Kreinovich 1989; Fox et al. 1998; Kreinovich et al. 2011; Kreinovich and Kosheleva 2021), we explain that this progression of paradigms—and even specific ideas within these paradigms—was natural: In each case, the main objective of the previous paradigm turned out to be not feasible and thus, a modification was needed. For example, the universal bound on interaction speed—introduced by Special Relativity—naturally comes from the need to reliably predict future events: Without such a bound, unknown faraway events in the potentially infinite Universe could affect the future events. This, in turn, explains the ubiquity of differential equations.

Our analysis shows that modern approach also has limitations, and there are already seeds of the new idea—which we call Unity of Science—when to predict a phenomenon, we need to consider other phenomena as well.

An interesting—and largely unexplored—twist is that while historical physics paradigm shifts decrease our ability to predict future events, the corresponding physical phenomena can be used to enhance computations: everyone heard about spectacular promises of quantum computing; we will also overview potential computational use of relativistic effects and of possible future physical paradigms.

2 Full Cognition: The Original Paradigm

2.1 *What is Full Cognition Paradigm*

One of the main objectives of science is to predict the future—and one of the main objectives of science and (generally understood) engineering is to find the best way to make this future as good for us as possible.

Often, people object to this description, saying that probably a more important objective of science is to provide understanding of the world phenomena. This is true, but what does it mean "to understand"? How do you check that a person understands calculus? By asking this person to solve problems—especially problems that go beyond routine exercises. How do you check that a person understands local weather patterns? If this person can, in many cases, successfully predict tomorrow's weather based on today's observations.

In the beginning, people did not know what can be predicted and what cannot be predicted, they did not know what information is needed for such predictions. So the most optimistic viewpoint was that everything can be predicted—so that, in principle, we can reach full cognition.

This was not just the viewpoint of a few optimists, it was a prevailing paradigm. People went to oracles, people asked prophets, people made future predictions based on the flights of birds and on tea leaves—and everything was the subject of such predictions: Results of future battles, individual histories, etc. People who consulted with the Delphi oracle did not perform any measurements or observations.

This is what kings and princes hired Wise Men for: To give them advice about the future. The famous astronomer Kepler, the first astronomer to discover that the planets follow elliptical orbits, was also routinely asked to predict future fate of different people. This is why people went to their spiritual leaders, to their priests and rabbis—to seek advice based on the leaders' ability as smart folks to know the future.

While some people continue to seek advice of spiritual seers, in general, this is now treated as a joke. In Russia—where we come from—in 1990s, right after the fall of communist dictatorship, there was a popular joke about an airplane designer who consults with a wise rabbi because at high speed, wings keep falling off the plane that he is trying to design. The rabbi without hesitation suggests where exactly to make holes in these wings and—magic!—the plane flies perfectly. Why? asks an engineer, expecting to hear that the rabbi consulted with the angels or even with God himself, but the rabbi's explanation is more mundane: During the communist times—the rabbi explained—there was Soviet-produced toilet paper where holes were places at exactly these locations, and it would never tear off at these holes, no matter how much you tried.

This is how we view this approach now.

2.2 Limitations of this Paradigm

History shows that many of the oracle's predictions turned out to be wrong—and even predictions that turned out to be correct were often so vague that they could easily be interpreted either way. There is a famous—probably apocryphal—example when a mighty king asked an oracle about the future outcome of a war with his equally mighty neighboring king, and the answer was: A great empire will be destroyed. And it was—except the oracle did not indicate which of the two fighting empires.

2.3 But There Were Successes

On the other hand, in many cases, when predictions were made not based on supposed divine inspiration but based on observations, many of these predictions worked well. In many cases, medical doctors were able to correctly predict what will happen to a sick patient, astronomers could predict eclipses, and there were many correct and useful weather predictions as well.

Eventually, people realized that while some predictions are useless, often, predictions based on measurements and observations work well. In other words, they realize that to make a good prediction, they need to first make some observations and/or measurements. This led to a new paradigm.

3 Laplace Determinism: The Second Paradigm

3.1 What is Laplace Determinism

As we have mentioned, people realized that to make good predictions, you need to first perform as many observations and measurements as possible. So, the optimistic view is that if you perform a sufficient number of observations and measurements, then you will be able to predict all future events.

In other words, if you know the state of the Universe at a given moment of time, then, based on this information, you can predict its state at any future moment of time. This idea—while prevailing for some time—was first explicitly formulated only reasonably recently, by a 18th/19th century French mathematician and physicist Pierre-Simon Laplace. This paradigm is therefore known as Laplace Determinism.

This paradigm is exactly the idea behind Newton's mechanics that we study at school: once we know the initial positions and velocities of all the bodies, we can predict their future positions and velocities. A classical example is celestial mechanics, where we can successfully predict the positions of all the planets centuries ahead, with spectacular accuracy. One of the main discrepancies between the actual planet positions and predictions of Netwon's theory—resolved later by General

Relativity theory—constitutes 0.43 angular seconds per year, something that only sophisticated telescopes can capture.

This is the main idea behind electrodynamics: Once we know the positions, velocities, and charges or all the particles, we can predict what will happen later. This is how most nineteenth century physical theories work: They were usually formulated in terms of differential equations, so that once we know the initial conditions, we can uniquely predict all future states.

This paradigm is not only about the future—we can similarly uniquely determine the past, as long as have a sufficient number of measurements and observations. This is the main idea behind the legend of Sherlock Holmes—that, if a person can observe as many features as possible, all past events can be uniquely reconstructed—and thus, the Great Detective can solve every crime. And one of the main reasons why Dr. Watson could not do it was that—as Sherlock Holmes mentions in almost every story—unlike the Great Detective, Dr. Watson did not observe seemingly minor things which turned out to be sometimes important, e.g., how many steps one needs to climb to get to Holmes's study.

3.2 Laplace Determinism Implies a Universal Limit on Speeds—i.e., in Effect, Implies Special Relativity

What we plan to show now is that this paradigm naturally leads to many ideas that, at first glance, seem to be completely independent of this paradigm—and even revolutionary in comparison with this paradigm. The first such idea is the universal limit on all the speeds—the idea that was firmly established in physics only after Einstein's special relativity.

Indeed, if there is no limit on speeds, then, to predict what will happen in one second at a current location, we need to know the current state not only at this location, but anywhere at the Universe. Indeed, maybe a very fast body—which is currently located in a different galaxy—can come here in one second and change the state at our location? The Universe is large—according to Newton's understanding, infinitely large—and we cannot know the initial state at all locations.

So, if we believe—as people did who believed in this paradigm—that we can predict the future state of the Universe, this means that there must be a universal limit on all the speeds. In Relativity Theory, this limit is usually described by saying that nothing can be faster than the speed of light, but, fundamentally, it does not have to be the speed of light. For example, it may turn out that photons—quanta of light—have rest mass (after all, this may be true for neutrinos, about which people also originally believed that they travel with the speed of light). This would imply their speed is slightly smaller than the upper bound c used in all the formulas of Special Relativity Theory—but these formulas will not change, the only thing that will change is our interpretation of c as the speed of light.

Please note that here—and in the following sections—we do not claim that Special Relativity could be derived simply from the idea of Laplace Determinism: all we claim is that the idea of a universal limit of speed follows from Laplace Determinism. There is a big distance between this simple idea and a sophisticated theory—just like there is a big distance between the Biblical idea that our world had a beginning (and that in the beginning it was light) and complex formulas of modern cosmology.

3.3 This Explains Why Physics Uses Differential Equations

The universal limit c on all the speeds means that if we want to predict the state at a given location x t seconds into the future, we only need to consider the current state in locations whose distance from x does not exceed c^*t—all other locations simply do not have enough time to affect what we will observe. For example, if we want to predict what will happen on Earth in 1 s, then knowing what astronauts are doing on the Moon will not help—the distance to the Moon is larger than 300,000 km (the distance that light travels in 1 s), so whatever they do will take longer than 1 s to affect us.

In other words, whatever will happen at location x t seconds from now depends only on the state of the small neighborhood of x—neighborhood of radius c^*t. The smaller t, the smaller the neighborhood. So, to find the changes corresponding to infinitesimally small t, we only need to know the values in the infinitesimal vicinity of x. This is exactly what we mean by differential equations: When a change in small vicinity of moment of time is affected only by values of the corresponding quantities in an infinitesimal vicinity of this location.

3.4 Laplace Determinism Naturally Leads to the Notion of Fields

How can objects interact? If I place a magnet close to my scissors, they start moving towards it—why? There is no action-at-a-distance. As we have argued, the motion of the scissors is determined only by the objects in the close vicinity of their location—and the magnet is not that close. So there must be something that is located close to the scissors and that affect their motion—and this something is what physicists call a field.

This is how the Sun's gravity affects the Earth—via a field. So, this notion of a field—a notion that did not exist in Newton's time—actually naturally follows from the main idea of Laplace Determinism.

3.5 But Why Not Full Cognition?

The absence of full cognition explains why many processes are irreversible while equations of physics are reversible; Laplace Determinism is not just saying that we can predict the future if we know the initial state of the word, it also claims that we cannot have full cognition—and thus, that we cannot describe this initial state of the world by any law.

What does that mean? For example, we can have regular sequences of 0s and 1s, e.g., a sequence 010101... We can also have sequences which are kind of lawless, they cannot be described by a regular law. For example, if we flip a coin and record when we get heads (0) and when we get tails (1), then we will get such a lawless sequence—and if we get a sequence like 010101... in this flipping experiment, then any observer would say that we were cheating. Such lawless sequences are called random.

In these terms, Laplace Determinism not only claims that we can predict the future based on the initial conditions, it also claims that the initial conditions are, in some sense, random. (Of course, randomness is a vague term: To describe it precisely, we need to specify a probability distribution on the set of all possible initial conditions.)

This sounds trivial, but it has important consequences. Namely, it is a known fact that equations of Newton's mechanics are reversible: If at any given moment of time, we change the direction of each velocity to the opposite one, the system will start moving kind of backwards in time, eventually returning to its initial state. This can be experimentally shown on the example of interacting balls. So, in principle, whatever processes we start with, if we reserve the velocities of all the atoms, we will eventually get back to the original state. In other words, in principle, all physical processes are reversible. (And, by the way, relativity and quantum physics do not change this reversibility of basic equations.)

However, we all know that some physical processes are irreversible. If a cup falls down and breaks, there is no way to make it whole. How can we explain this? This is the "arrow of time" problem that bothered (and continues to bother) philosophically inclined physicists (and physically inclined philosophers) for centuries. What we claim is that the solution—on the qualitative level—is simple. Physics is not just about describing the equations, it is also about claiming that initial conditions are random. If we reverse time and start with the future, equations remain the same, but the new "initial conditions"—which, in the original description, formed the future state—are no longer random. For example, if we consider the effect of gravity, and we start with the random distribution of matter, then locations with higher density (and thus, higher gravitational field) will attract matter from other locations, and we will see what we see now—clumps of matter separated by empty space. In other words, what we will see is definitely not a random state anymore.

3.6 Physical Ideas that Follow from the Possibility of Feasible Predictions: Atomism, Finite Universe, Quantization, Symmetries, Conservation Laws, Inequalities (Like Second Law of Thermodynamics)

Up to now, we considered ideas that directly follow from Laplace Determinism, ideas without which Laplace Determinism is simply not possible, i.e., without which it is not possible to have algorithms that predict future states.

But, as everyone now knows, there are algorithms and algorithms. Some algorithms are practically feasible, while others are not realistic: e.g., their running time is much larger than the lifetime of the Universe (10–20 billion years). If we have too many inputs—or if we need to compute too many outputs—the problem becomes algorithmically intractable. So, the only way to be able to have realistic predictions is to have a realistic number of inputs and a realistic number of outputs in the prediction problem.

In the Laplace Determinism paradigm, to predict the future state, we need to know the values of all the quantities at all the locations at the current moment of time. The more locations there are, the more possible values of each quantity, the more combinations of such values—the more inputs we have. So, if we believe—as we do in this paradigm—that predictions are possible, we conclude that the number of such locations is realistic, that the number of possible values of each quantity is realistic, etc.

In general, realistic means finite. So, we must have at most finitely many locations—i.e., we cannot have a continuous medium, we must have a finite number of point-wise particles. This conclusion can be split into two parts.

The first part is that in each spatial region, there should be only finitely many particles. This idea—called atomism—thus follows from the natural assumption that we can have feasible Laplace-Determinism-related prediction algorithms.

The second part is that, since we only have finitely many particles overall, this means that the part of the space where particles are—and everything else does not make any physical sense, since we cannot observe it—must be finite. In other words, the Universe must be finite in size. This idea—seriously explored only with the advent of General Relativity—thus also naturally follows from the requirement that feasible prediction algorithms are possible.

Similarly, for each quantity, we can only have finitely many different values. This is indeed observed in physics: Electric charge cannot be any number, it must be proportional to electron's charge (to be more precise, if we take quarks into account, proportional to 1/3 of electron's charge). For each atom, its energy level can only take discrete set of values, etc. This is what is called quantization and what is now viewed as one of the main features of quantum physics—but it was actually conjectured and then observed way before quantum physics; quantum physics just provided a quantitative explanation for the observed discrete values. We see that on the qualitative level, the idea of quantization naturally follows from the assumption that there exist feasible prediction algorithms.

Do we need to describe all the values of all the quantities for all the particles? There may still be too many of them, so a natural conclusion is that there is usually some a priori relation between some of these values. A simple example is, e.g., when the initial state is symmetric—e.g., spherically symmetric. Then, we do not need to describe the values at all possible locations—for each distance from the center, it is sufficient to describe the values at one single location at this distance—for all other locations at the same distance from the center, the values are the same. Because of such geometric example, physicists call all such relations symmetries. In this sense, Laplace Determinism naturally leads to the existence of (and need for) symmetries—which are indeed an important part of modern physics.

Finally, do we need to run complex algorithms—and prediction algorithms in physics are usually complex and time-consuming—to predict all the future values of all the particles? There may be too many of these future values. It is therefore reasonable to conclude that some of these values—or at least some combinations of these values—do not need to be complicatedly computed: For example, these combinations may have the same value as in the previous moment of time. Such combinations are known as conservation laws—energy conservation law, momentum conservation law, etc. —and, on the qualitative level, their existence indeed follows from the requirement that feasible prediction algorithms are possible.

For some combinations of values, the previous value does not uniquely determine the future value—as in conservation laws—but it should at least restrict possible future values by some appropriate inequality: e.g., the future value should be greater than or equal to the current value. An example of such an inequality is the Second Law of Thermodynamics, according to which the entropy can only increase.

3.7 Limitations of Laplace Determinism

Interestingly, discreteness—one of the natural conclusions of Laplace Determinism paradigm—leads to the need to go beyond this paradigm. Let us illustrate this need on the example of a radioactive decay. A radioactive atom, at some moment of time, emits some particle (a photon, an electron, or an alpha-particle: the Helium ion).

According to the Laplace Determinism paradigm, at each moment of time, the current state of the system must be uniquely determined by its current state. But in this case, such uniqueness is not possible. If we start with the state 2 s before the actual decay, then in the following second, the state remains the same, but if we start with the state 1 s before, in 1 s there will be a decay. This means that we sometime cannot predict the future state of the system, even if we know everything about the current state. This impossibility led to a different paradigm.

4 Statistical Determinism: The Third Paradigm

4.1 What is Statistical Determinism

As we have just mentioned, it turned out that it is not always possible to predict the exact future state. In this sense, in such cases, the future state is unpredictable—i.e., random. For random predictions, at best, we can predict the probabilities of different future events.

An optimistic viewpoint is—naturally—that, if we have a sufficient amount of information about the current state, then we can uniquely determine the probabilities of all possible future states. This viewpoint constitutes what is usually called Statistical Determinism.

This paradigm became prevailing with the advent of quantum physics, where, once we know the current quantum state, Schroedinger's equations indeed enable us to uniquely determine the future quantum state—and thus, for all future measurements, uniquely determine the probabilities of all possible measurement results.

In this paradigm, the main objective is thus to find the exact form of the equations that describe all the world's physical processes—what physicists call Theory of Everything. Once such a theory will be found, fundamental physics will, in effect, come to an end—a glorious end but still end. But will it?

Let us explain what is the problem with this paradigm.

4.2 Fundamental Limitations of Statistical Determinism

We started this essay by saying that while the main objective of physics is to predict future events, the main objective of engineering and decision making is to ensure that this future is as beneficial for us as possible. And herein lies a problem.

According to the Statistical Determinism, the probabilities of all future events are pre-determined by the initial state of the Universe: Nothing we do will change it. From this viewpoint, engineering makes no sense—although in reality, its successes are undeniable. Clearly, something is not right with this paradigm.

On a more philosophical level, this can be described as a freedom of will problem: We all know that we can make decisions that make changes in our lives and in lives of others—and thus, that the future is not uniquely pre-determined, and probabilities of different future states are also not uniquely pre-determined. Of course, we can claim that Statistical Determinism is correct, and freedom of will is an illusion—but this looks like too unorthodox an approach. We can as well claim that—like in the movie Matrix—all our life is an illusion, and in reality, we are harvested by robots. So if we dismiss this weird possibility, then we arrive at a different paradigm, which is reasonable to be called Modern Approach.

5 Modern Approach—the Fourth Paradigm

5.1 What is Modern Approach

The main idea of Statistical Determinism was that, once we know the current state of the world, we can uniquely determine the probabilities of all future states. Since this idea contradicts common sense, a natural new idea is that, given the current state of the world, we can determine some—but not all—information about the probabilities of the future states.

This means, in particular, that we cannot have an absolutely correct Theory of Everything: no matter how good is a currently proposed theory, eventually, new observations will surface that are not explained by the current theory, and a new modification will be needed. In this sense, physics is eternal, it will never come to an end. This seems to be, at present, the prevailing view among physicists.

5.2 But Aren't We Unnecessarily Pessimistic?

From one paradigm to another, we are becoming less and less optimistic about the possibility of predictions. At first, in the Full Cognition paradigm, we believed that we can predict everything without even bothering to determine the current state of the world. In the Laplace Determinism paradigm, we still believed that we can predict everything—although this time, we can only make these predictions after a hard work of measuring the current state in detail. In the Statistical Determinism paradigm, even after all this work, we can only predict probabilities of future events. And in the Modern Approach, we cannot even predict all the probabilities. This looks somewhat pessimistic. What can we do to make the situation better?

To answer this question, let us recall that physics is usually divided into sub-disciplines: Cosmology studies Universe as a whole, astrophysics studies stars and galaxies, biophysics studies living creatures, there are solid body physics, atmospheric physics, etc. In each of these sub-disciplines taken on its own, as we go from one paradigm to the next one, we are less and less optimistic about the possibility to predict the future. But good news is that this division into sub-disciplines is somewhat artificial. All of these sub-disciplines describe the same world, the world in which all the processes are interconnected. So while, if we only use facts from each sub-discipline alone, we cannot make good predictions about its object of study, maybe facts and laws from other sub-disciplines will help?

This is exactly what is happening, to the extent that we can consider the resulting emerging state of physics as the new paradigm—a paradigm that is natural to be called Unity of Science.

6 Unity of Science—An Emerging Paradigm

6.1 What is Unity of Science

On the methodological level, most physicists believe in unity of physics: No matter what are the applications, be it cosmology, particle physics, geophysics, biophysics—these are important parts of physics. However, from the practical viewpoint, physics has been mostly compartmentalized.

For example, to understand the cosmological structure and evolution of the Universe, physicists observe distance galaxies, 3K radiation, and other astronomical phenomena. It would have been weird to suggest that the observation of living creatures on Earth can somehow help understand cosmological forces.

To understand the interaction between elementary particles, physicists perform experiments in particle accelerators and measure cosmic rays. It would have been weird to propose that to better understand some parameters of the particles interaction, we need to observe animals on Earth or the distribution of galaxies.

This compartmentalization extended beyond physics. For example, to understand how people make decisions, a natural idea is to observe their decision making, to ask them to make decisions in some artificially designed situations, to measure their brain signals during their decision making. It would have been weird to suggest that to better understand how people make decisions, we need to observe how elementary particles interact—this seems as weird as astrology where somehow the observation of planets helps predict people's fate.

Interestingly, lately, these seemingly weird things are actually happening: It turned out that observing animals on Earth can help explain fundamental physical constants, that observing distance galaxies can help better estimate masses of elementary particles, and that observing interaction between elementary particles can help better predict human behavior.

We will provide more details in the next section, but now we want to emphasize that this emerging phenomenon means a new change in the objectives of physics, i.e., in effect, the emergence of the new paradigm. In the past, physicists (and other scientists) believed that to predict phenomena in each sub-discipline, the most important data is the data from this sub-discipline. For example, to predict human behavior, it is important to study human behavior and not where the planets are or how elementary particles interact. To estimate the mass of an elementary particle, it is important to observe these and other particles—and not distant galaxies or animals. In contrast to this belief, there are several case when it turns out that such seemingly unrelated observations can help make predictions—predictions that observing the phenomena within the same sub-discipline did not allow. In other words, the essence of this paragraph is that to predict future events (or their probabilities), it is often important to use facts and laws from all parts of physics—and even from outside physics—and not just from a related sub-discipline. So, the Unity of Science is not just a methodological claim, it is a very practical and helpful idea.

6.2 Examples

One of the first examples of using this paradigm was the Anthropic Principle. This example is related to the fact that equations of physics have constants that need to be experimentally determined. Newton's theory of gravitation had the gravitation constant—an empirical constant that describes how strong is the gravitational force between two bodies of unit mass separated by a unit distance. Maxwell's equations—that describe electromagnetic field—contain speed of light as an empirical constant, a constant that describes the speed with which electromagnetic waves propagate. Quantum physics contains Planck's constant—that described how strong are quantum effects. Many other equations contain other constants.

If we change the numerical values of these constants, the world will change. It turns out that even if we change these constants a little bit—e.g., by 10%—the world will change so drastically that it will no longer be possible to have life as we know it: Either the stars and planets will not be formed at all, or water will become impossible, etc. Thus, the very fact that we have life in the Universe severely restricts the values of the physical constants. One possible explanations—which is in line with multi-world interpretation of quantum physics—is that in reality, there are many universes, and we—the living creatures – Can only exist in those that allow life. Irrespective of the explanation, this is a clear example of how, contrary to the usual beliefs, facts and laws from one area of science helps make predictions in another area—namely, facts from biology and chemistry help to explain current values of fundamental physical constants.

Another example is how cosmology helps particle physics. A good example is neutrino physics. Neutrinos are notoriously difficult to study, since they practically do not interact with any other particles. Interestingly, we can find upper bounds on their rest mass by the fact that if they had a larger rest mass, the Universe would evolve differently from what we see now. In this case, to get a better understanding of elementary particle, we observe distant galaxies—something that until recently would be considered weird.

A different example is that quantum models—when only slightly modified—explain how people actually make economic decisions (Baaquie 2004; Busenmeyer and Bruza 2012; Haven and Khrennikov 2013; Kreinovich and Kosheleva 2023). In other words, observing how elementary particles interact can help better understand how people make decisions. This fits well with the famous general observation by a Nobelist Eugene Wigner that many mathematical theories—initially developed for a completely different purpose—are often surprisingly helpful in describing physical phenomena. This is also related to the computer science discovery that all complex computational problems (known as NP-complete) can be reduced to each other, so that any algorithm which helps solve some instances of one of the problems can be—almost automatically—used to solve several instances of all other complex problems; see, e.g., (Kreinovich et al. 1998; Papadimitriou 1994).

This last remark leads us to the final topic of our essay: How all these paradigms affect computations.

7 How Different Physical Paradigms Affect Computations

At first glance, the effect should be negative. For example, we want computations to always produce the same (correct) results, but if we take into account that in quantum physics, we can only make probabilistic predictions, this makes such deterministic computations difficult—if not impossible.

Good news is that researchers' ingenuity makes a tasty lemonade out of this sour-lemon situation. Many readers are familiar with successes of quantum computing (Cfr. Nielsen and Chuang 2010), where, e.g., Grover's algorithm allows us to find the desired element in an unsorted n-element array in time proportional to square root of n—while any non-quantum computer would require, in the worst case, at least n steps. For a database of 1 million elements, this means a thousand times speed-up.

An even more spectacular speed-up is achiever by Shore's algorithm. Once a universal quantum computer is designed, this algorithm will allow us to break all existing codes—something that now requires astronomically unrealistic computation time—and thus, will enable us to read all encoded messages in the world.

Similarly, the use of curved space–time can speed up computations (Morgenstein and Kreinovich 1995), effects of special relativity can lead to a square root speed-up (Kosheleva and Kreinovich 2020), and the Modern Approach, according to which no physical theory is final, can help solve almost all instances of NP-complete problems in reasonable time (Kosheleva et al. 2014).

8 Conclusion

Successes of the nineteenth century physics were so great and overwhelming that Max Planck, one of the founding fathers of quantum physics, was strongly advised not to go into physics—since physics at that time was perceived to be almost finished. In spite of these gloomy predictions, twentieth century became even more spectacular—with quantum physics and relativity. So spectacular it was, that some people—even some physicists—believed that now finally physics is coming to its end, and that anyone interested in great discoveries should move to, e.g., biology.

We hope that this essay—continuing the analysis (D'Agostino 2000) of the nineteenth and twentieth century physics performed by Professor D'Agostino—will convince you that 21 century physics will be no less spectacular than the two previous ones!

Acknowledgements This work was supported in part by the National Science Foundation grants 1623190 (A Model of Change for Preparing a New Generation for Professional Practice in Computer Science), and HRD-1834620 and HRD-2034030 (CAHSI Includes), and by the AT&T Fellowship in Information Technology. It was also supported by the program of the development of the Scientific-Educational Mathematical Center of Volga Federal District No. 075-02-2020-1478, and by a grant from the Hungarian National Research, Development and Innovation Office (NRDI).

The authors are greatly thankful to Prof. Dr. Raffaele Pisano for his encouragement and support, and to the anonymous referees for valuable suggestions.

References

Baaquie BE (2004) Quantum Finance: Path Integrals and Hamiltonians for Options and Interest Rates. Cambridge University Press, New York.

Busenmeyer JR, Bruza OD (2012) Quantum Models of Cognition and Decision. Cambridge University Press, New York.

D'Agostino S (2000) A History of the Ideas of Theoretical Physics: Essays on the Nineteenth and Twentieth Century Physics, Kluwer Academic Publishers, Dordrecht.

Feynman R, Leighton R, Sands M (2005) The Feynman Lectures on Physics. Addison Wesley, Boston, Massachusetts.

Finkelstein AM, Kreinovich V (1989) Can Science Attains its Goals? Informatika Technical Reports, St. Petersburg, Russia [in Russian].

Fox D, Schmidt M, Koshelev M, Kreinovich V, Longpre L, Kuhn J (1998) We must choose the simplest physical theory: Levin-Li-Vitanyi theorem and its potential physical applications. In: Erickson GJ, Rychert JT, Smith CR (eds). Maximum Entropy and Bayesian Methods, Kluwer, Dordrecht, pp. 238–251.

Haven E, Khrennikov A (2013) Quantum Social Science, Cambridge University Press, Cambridge, UK.

Kosheleva O, Kreinovich V (2020) Relativistic effects can be used to achieve a universal square-root (or even faster) computation speedup. In: Blass A, Cegielsky P, Dershowitz N, Droste M, Finkbeiner B (eds). Fields of Logic and Computation III, Springer, Cham, Switzerland, pp. 179–189.

Kosheleva O, Zakharevich M, Kreinovich V (2014) If many physicists are right and no physical theory is perfect, then by using physical observations, we can feasibly solve almost all instances of each NP-complete problem. Mathematical Structures and Modeling 31:4–17.

Kreinovich V, Kosheleva O (2021) Joule's 19th Century Energy Conservation Meta-Law and the 20th Century Physics (Quantum Mechanics and Relativity): 21st Century Analysis. Foundations of Science 26:703–725.

Kreinovich V, Kosheleva O (2023) Why quantum techniques are a good first approximation to economic phenomena, and what next. In: Nguyen Hoang Phuong, Kreinovich V (eds) Biomedical and Other Applications of Soft Computing. Springer, Cham, pp. 57–66.

Kreinovich V, Lakeyev A, Rohn J, Kahl P (1998) Computational Complexity and Feasibility of Data Processing and Interval Computations. Kluwer, Dordrecht.

Kreinovich V, Araiza R, Ferret J (2011) Analysis of Information and Computation in Physics Explains Cognitive Paradigms: from Full Cognition to Laplace Determinism to Statistical Determinism to Modern Approach. In: Dodig-Crnkovic G, Burgin M (eds). Information and Computation, World Scientific, Singapore, pp. 203–223.

Morgenstein D, Kreinovich V (1995) Which algorithms are feasible and which are not depends on the geometry of space-time. Geombinatorics 4(3):80–97.

Nielsen M, Chuang I (2010) Quantum Computation and Quantum Information. Cambridge University Press, New York.

Papadimitriou CH (1994) Computational complexity. Pearson, Upper Saddle City, New Jersey.

Thorne KS, Blandford RD (2017) Modern Classical Physics: Optics, Fluids, Plasmas, Elasticity, Relativity, and Statistical Physics. Princeton University Press, Princeton, New Jersey.

Chapter 11
Einsteinless Pathways into Spacetime. A Gedankenexperiment in Science History

Jean-Marc Lévy-Leblond

Abstract Suppose Einstein had not existed. How, in the early twentieth century, might our understanding of space–time physics have developed?

This paper proposes a reconstruction of history as it *could* have evolved, drawing on attested pre-Einsteinian works, and some little known post-Einsteinian ones, as well as introducing a few imaginary characters. The virtual rise of Minkowskian chronogeometry will be reviewed, focusing on: the discovery of the inertia of energy and the subsequent construction of velocity-dependent formulas for energy and momentum by Von Dida (1909) and Hunepierre (1909), and the application of the Erlangen program to the construction of possible space–time structures and classification of chronogeometrical groups, from von Ignatowsky (1910, 1911), Franck and Rothe (1911, 1912) to Blondeville and Prostov. Some secondary sources will also be discussed Zweistein (1905). Finally, the relevance of such an approach, on educational, epistemological and cultural grounds, will be highlighted and put in perspective within the ongoing debate about the contingency vs inevitability of science history.

Keywords Chronogeometry · Contingency · Einstein · History of relativity

J.-M. Lévy-Leblond (✉)
University of Côte d'Azur, Nice, France
e-mail: jmll@unice.fr

R. Pisano (ed.), *A History of Physics: Phenomena, Ideas and Mechanisms*,
History of Mechanism and Machine Science 42,
https://doi.org/10.1007/978-3-031-26174-9_11

1 Introduction. In Praise of Counterfactual History

Gedankenexperiments consist in asking counterfactual questions—"What if…?".[1] It all started when Galileo wondered "What if air resistance did not exist?" and discovered the law of free fall (Lévy-Leblond 2001). This strategy was later fruitfully extended and popularised in physics, in particular by Einstein. I propose to use it here, not *in* physics, but *on* physics. Understanding the unfolding of any specific stretch of history greatly benefits from imagining how it could have developed otherwise, if such and such crucial circumstance had been different. Counterfactual narrative attempts do play a significant role in modern historiography (Rodwell 2013; Singaravélou and Deluermoz 2016). Fictional literature has produced a host of uchronic novels, some of which are considered as classic works of great import (Dick 1962; Spinrad 1972; Silverberg 2003; Roth 2004). Such an endeavour has already been used in science history (Jammer and Stachel 1980; Davies 2002; Peter Bowler 2013). I will advance what I believe to be a plausible scenario of the history of space–time physics at the beginning of the twentieth century had Einstein not been there to make his justly celebrated 1905 breakthrough. I will rely on the thorough studies by Arthur Miller and Olivier Darrigol of pre-Einsteinian developments (Miller 1981; Darrigol 2022), while inventing a few characters and contributions (the latter usually corresponding to real later work).

2 The Inertia of Energy

By the end of the XIXth century, the development of electromagnetic theory was leading physicists to a fully electromagnetic worldview, according to which all phenomena could ultimately be reduced to the interactions of electrical charges through the ether. All properties of matter would be explainable in electromagnetic terms. Mass itself should then have its origin in the electromagnetic field, as first suggested by Thomson (1881). Thomson demonstrated that by electrically charging a conducting sphere, one would change its inertia. The physical reason was that the motion of body through the ether generated an electrical current that gave rise to a magnetic field acting back upon the charge; this self-induction effect would modify the inertia of the moving body (a primitive example of renormalization). Thomson found that the additional inertia was proportional to the energy of the electrostatic field, with a coefficient of proportionality $8/15\,c^2$, c denoting as usual the propagation velocity of the electromagnetic field. Heaviside (1889), using a somewhat different version of electromagnetic theory, found a coefficient $4/3\,c^2$.

At all events, whatever the value of the coefficient, due to the velocity dependence of the field and its energy, the total kinetic energy of the body could no longer show the

---

[1] The present work is a revised, corrected and extended version of a talk given at the 24th International Colloquium on Group Theoretical Methods in Physics (Paris, July 2002) (Lévy-Leblond 2003).

usual quadratic dependence on velocity. However, obtaining the exact formula proved rather difficult, as the calculations depended upon ad hoc assumptions concerning the cohesive forces holding the body against the internal repulsive Coulomb forces, as well as on its form and its possible deformations when set into motion. Thus ensued a decade of confusion, with various proposals as to the dependence of mass on the velocity, by Searle (1897), Wien (1900), Abraham (1903), Lorentz (1904), Langevin and Bucherer (1905); the trouble was compounded by the distinction between a "longitudinal mass" m_{L} and a "transversal mass" m_{T}.

Here is a small sample of the variety of the results obtained for a spherical "electron" with mass m_{e}, radius R and velocity v with $\beta = v/c$

$$\text{Abraham (1903):} \quad \begin{cases} m_{\mathrm{L}} = m_{\mathrm{e}}\beta^{-3}\left[\beta\left(1-\beta^2\right)^{-1} - \tanh^{-1}\beta\right] \\ m_{\mathrm{T}} = m_{\mathrm{e}}\beta^{-3}\left[\tfrac{1}{2}\beta\left(1+\beta^2\right)\tanh^{-1}\beta - \tfrac{1}{2}\beta\right] \end{cases}$$

$$\text{Lorentz (1903):} \quad \begin{cases} m_{\mathrm{L}} = m_{\mathrm{e}}\tfrac{2}{3}\left(1-\beta^2\right)^{-3/2} \\ m_{\mathrm{T}} = m_{\mathrm{e}}\tfrac{2}{3}\left(1-\beta^2\right)^{-1/2} \end{cases}$$

$$\text{Langevin-Bucherer (1904):} \quad \begin{cases} m_{\mathrm{L}} = m_{\mathrm{e}}\tfrac{2}{3}\left(1-\tfrac{1}{3}\beta^2\right)\left(1-\beta^2\right)^{-4/3} \\ m_{\mathrm{T}} = m_{\mathrm{e}}\tfrac{2}{3}\left(1-\beta^2\right)^{-1/3} \end{cases} \tag{1}$$

Meanwhile, Kaufmann's experiments (1902–1903) confirmed the reality of the variation of mass with velocity. However the confusion was exacerbated by the question of the agreement between the theoretical and the experimental results, the precision of which was clearly overrated at the time (Kaufmann claimed a 1% precision for $0.6 < \beta < 0.9$).

Many discussions on the nature and significance of this velocity dependence took place between Abraham, Lorentz, Kaufmann, Planck and Poincaré, who had written as early as 1900 that "Electromagnetic energy can be viewed as a fluid with inertia". Finally, Planck, at a famous Köln conference (1908) concluded to a general "law of inertia of energy" by showing that the flow of any sort of energy, whether thermal, chemical, elastic, gravitational, etc., could be associated with a momentum density. At the same period, in one of the first important American contributions, Lewis and Tolman (1909) stated that the variation of mass with velocity transcended electromagnetic theory, and must be a universal feature, independent of the electric charge of the body.[2]

Then, by the end of 1909, came the breakthrough. Von Dida, a student of Planck, asked himself what could be the simplest and most natural form for the inertia of a body as a function of its velocity.[3] Here is how he proceeded, starting from first

[2] All of the preceding is historically correct, as can be checked from the book by Miller (1981). In particular, for the discussions of the mass variation with velocity within electromagnetic theory, see Sects. 1.8 to 1.14. For Planck's law of inertia of energy, see Sect. 12.5.7. For Lewis and Tolman contribution, see Sect. 12.2, note 4.

[3] Here we depart from history as it unfolded. The following derivation in fact is due to Davidon (1975), who also gave an account of the historical developments sketched above.

principles. Von Dida first *defined* the (possibly variable) inertia I of a body as the coefficient of the velocity in the expression of the momentum:

$$p = Iv. \tag{2}$$

He then *assumed* the simplest expression for the variability of inertia, asking that any variation in the energy E of the body would entail a proportional variation of its inertia I:

$$dI = \chi dE \tag{3}$$

where χ should be a universal constant. Finally, by appealing to Hamilton's first equation

$$v = \frac{dE}{dp}, \tag{4}$$

he had three relationships between the four magnitudes v, p, E, I, enabling him to express for example the last three of them in terms of the first one. One now computes quite simply $dI = \chi v dp = \chi v(v dI + I dv)$, hence

$$\frac{dI}{I} = \frac{\chi v dp}{1 - \chi v^2} \tag{5}$$

so that, by a simple integration, one gets

$$I = \frac{I_0}{\sqrt{1 - \chi v^2}}. \tag{6}$$

Of course, considering the low-velocity limit leads one to identify I_0 with the mass m of the body. Since the constant χ has the dimensions of the inverse-square of a velocity, it is natural to set $\chi = c^{-2}$, defining a *universal* constant c with the dimension of a velocity, acting as a limit velocity for any massive body. Note that c at this stage is not necessarily the velocity of any specific physical agent, although of course it was ultimately to be identified with the velocity of electromagnetic waves. Von Dida finally obtained the following expressions for the dynamical quantities of a body:

$$I = \frac{m}{\sqrt{1 - v^2/c^2}}, \quad p = \frac{mv}{\sqrt{1 - v^2/c^2}}, \quad E_{\text{kin}} = \frac{mc^2}{\sqrt{1 - v^2/c^2}} - mc^2. \tag{7}$$

The last expression results from the fact that, because of (3), $E = \chi^{-1} I + \text{Cst}$, and the requirement that for the purely kinetic energy, one should have $E_{\text{kin}}(v)|_{v=0} = 0$.

A puzzling consequence of these formulas was that they put into question the very notion of velocity. Indeed, since the momentum is no longer proportional to the velocity, the conservation of the total momentum no longer entails the usual law of addition of velocities. But the trouble with velocity was still to deepen, as we are going to see.

3 The New Conservation Laws

At the very same time, a young Swiss physicist, Albert Hunepierre, hit on a different but related approach. Attending the 1908 Köln conference already alluded to, he was stimulated by conversations with Paul Langevin on the nature of conservation laws, and by a comment of Minkowski at the end of Planck' lecture: "In my view, the law of momentum is obtained from the energy law; namely, in Lorentz's theory, the energy law depends on the reference system. We write the energy law for every possible reference system, so that we have many equations and in those are contained the law of momentum".[4]

Hunepierre's idea was to exploit this clue in order to characterize the functions $E(u)$ and $p(u)$ giving the energy and momentum of a body in terms of its velocity u (the change with respect to the preceding section of the notation concerning velocity is purposeful). In fact, he simply revived, in very general terms, considerations going back to Huygens. For the sake of simplicity, let us confine ourselves to the one-dimensional case, so that E and p are an even and an odd function respectively.[5] Consider a system of interacting particles, with velocities $u_k (k = 1, 2, \ldots)$; it is ruled by the conservation laws for the total energy E_{tot} and total momentum p_{tot}:

$$\begin{cases} E_{\text{tot}} = \sum_k E(u_k) = \text{Cst} \\ p_{\text{tot}} = \sum_k p(u_k) = \text{Cst} \end{cases} \tag{8}$$

expressed in a particular inertial reference frame. Let us now describe the system in another inertial reference frame, moving with respect to the first one with velocity U. The crux of the argument is to *assume the additive law of combination for velocities*, so that the particles then have velocities

$$u'_k = u_k + U \quad (k = 1, 2, \ldots) \tag{9}$$

The conservation laws in this new reference frame then read

[4] While Hunepierre's character is a fiction, the remark by Minkowski is true to the facts; see Miller (1981, p. 367) and Darrigol (2022, p. 51). In fact, Minkowski referred to a rather precocious memoir by Schütz establishing the mechanical conservation laws by applying the (Galilean) relativity principle to collisions (Schütz 1897).

[5] The following derivation has been proposed by Lévy-Leblond (1976a). See also Comte (1986).

$$E\begin{cases} E'_{\text{tot}} &= \sum\limits_k E(u_k+U) = \text{Cst} \\ p'_{\text{tot}} &= \sum\limits_k p(u_k+U) = \text{Cst} \end{cases} \tag{10}$$

If we now require the two conservation laws to hold in *all* equivalent inertial frames, we see, by developing the expressions (10) in power series of U, that not only are the total energy E and momentum p conserved, but also their derivatives with respect to U, that is, all the quantities formed by adding up the successive derivatives of the individual energies and momenta,

$$\sum_k \frac{dE}{du}(u_k), \sum_k \frac{d^2E}{du^2}(u_k), \ldots, \sum_k \frac{dp}{du}(u_k), \sum_k \frac{d^2p}{du^2}(u_k), \ldots.$$

But we cannot have more than two independent conservation laws; otherwise the collision process would be overdetermined (as the case of two particle makes clear). The 'new' conservation laws must therefore be offshoots of the 'old' ones. Since the individual velocities are arbitrary, the derivatives of the energy and momentum functions must depend on the energy and momentum for each particle separately. Moreover, the additivity of the conservation laws requires this functional dependence to be linear. Finally, the even and odd nature respectively of E and p severely restricts the possibilities. For the first derivatives, indeed, the most general expressions are:

$$\begin{cases} \frac{dE}{du} = p \\ \frac{dp}{du} = m + \chi E \end{cases} \tag{11}$$

where we have chosen a coefficient unity in the first equation according to the usual dimensional convention, and where the constant term in the right-hand-side of the second equation is identified with the mass of the particle in order to recover the standard low-velocity Newtonian expressions. By so doing, we automatically take E to be the purely kinetic energy of the particle, since, in the case $\chi = 0$, one recovers from (11) the expressions $p = mu$ and $E = mu^2/2$. Of course, we know, with hindsight, that the coefficient χ is the same as in the previous Section, and we put $\chi = c^{-2}$. The solution of the system (11) of differential equations, given the even and odd nature of E and p respectively, is unique:

$$\begin{cases} E_{\text{kin}} = mc^2[\cosh(u/c) - 1] \\ p = mc\sinh(u/c) \end{cases}. \tag{12}$$

Of course, the comparison of the expressions (12) with those given by (7) compounded the problem concerning the very notion of velocity, to which we will return anon.

But Hunepierre's major accomplishment, in a second paper of the same year 1909, was to realize that the new form (12) of the conservation laws was inconsistent with the hitherto unquestioned additivity of mass. It suffices to consider two particles, with the same mass m. In the reference frame of their center of inertia, they have respective velocities $(u, -u)$; the internal energy of the system, that is, the kinetic energy of the particles in this very reference frame thus is

$$E_0 = 2mc^2[\cosh(u/c) - 1] \tag{13}$$

In another reference frame, moving with respect to the first one with velocity U, the particles have velocities $(u + U, -u + U)$, so that the total momentum is

$$p = mc\sinh(u/c + U/c) + mc\sinh(-u/c + U/c) = 2mc\cosh(u/c)\sinh(U/c). \tag{14}$$

However, on the other hand, the momentum of the system, moving with global velocity U, should read, according to the second of Eq. (12), $p = Mc\sinh(U/c)$, where M is its (total) mass. We are then led to recognize that the total mass is given by

$$M = 2m\cosh(u/c) = 2m + E_0/c^2. \tag{15}$$

Mass is no longer additive, and any change in the internal energy of the system entails a proportional change in its mass, $\Delta M = \Delta E_0/c^2$. Hunepierre then proposed to normalize the (hitherto arbitrary) zero of energies so as to include, for any M, its 'mass energy' Mc^2 in its internal energy, which enabled to write in full generality the now famous Hunepierre's equation

$$M = E_0/c^2 \tag{16}$$

which he jokingly transcribed as

$$\text{Mass} \propto \text{Innergy}\ .$$

4 Chronogeometry

By that time, that is, around 1910, it was becoming obvious that the changes in dynamics (i.e. in the expressions for the energy and momentum) required a parallel modification in kinematics (i.e. in the structure of space–time), as illustrated by the debate around the notion of velocity.

It had been known for several years that Maxwell's equations were invariant under a particular set of transformations, as noted by Lorentz (1904)—who left his name to them—, and by Poincaré (1905), who emphasized the group structure of this set. Minkowski had given in 1908 a neat mathematical description of this pseudo-Euclidean group. Still, for many, this invariance property was thought to be specific of electromagnetic theory, and its interpretation was unclear. Poincaré, for instance, up to the end of his life (1912), clung to an epistemology which required Lorentz's ether (Darrigol 2022, Chap. 6). However, Sommerfeld, who had been a student of Felix Klein, was familiar with the Erlangen program and the geometrical interpretation of transformation groups (Miller 1981, p. 181, note 41). He thus suggested to investigate the possible transformation groups in space–time, depending only on general and abstract requirements, irrespective of the specific physical phenomena (electromagnetic, gravitational or others) taking place on a supposedly universal spatio-temporal stage.

In the early 1910s, several independent works appeared resolving this question—relying, it must be said, on rather awkward approaches and complicated calculations, due to a lack of familiarity with group theory (and with infinitesimal Lie algebraic methods in particular). Various contributions by von Ignatowsky (1910–1911), Franck and Rothe (1911), van Rijn van Alkemade (1912), Hahn (1913) etc., finally converged towards the following conclusion.[6] Under the very general assumptions of:

- homogeneity of space–time
- isotropy of space
- existence of causal relationships,

the Lorentz-type transformations (written here in the one-dimensional case)

$$\begin{cases} x' = \frac{x-vt}{\sqrt{1-\chi v^2}} \\ t' = \frac{t-\chi vx}{\sqrt{1-\chi v^2}} \end{cases} \quad \text{with a constant } \chi \geq 0 \tag{17}$$

are the *only* possible ones. Of course, if these transformations are to be universal, and if Maxwell equations are indeed correct, one has again to identify $\chi = c^{-2}$.

Under the influence of the Kleinian point of view, the theory of space–time soon became known under the very apt name of "chronogeometry", or, more precisely, "Minkowskian chronogeometry", when necessary to distinguish it from the classical conception of space–time or "Galilean chronogeometry" (which, it must be stressed indeed is a space-with-time theory, as it already mixes space and time—or, rather, time with space, if not space with time as in the Minkowskian case). Once this generalized geometrical perspective was adopted, it was a simple matter to understand the puzzling phenomena so-called "length contraction" and "time dilatation"

[6] None of these names or contributions are invented. See Ignatowsky (1910–1911), Frank and Rothe (1911), van Rijn van Alkemade (1912), Hahn (1913). These papers were immediately neglected and forgotten, and their conclusions somewhat later rediscovered by Pars (1921), Esclangon (1936), Le Roy (1936), Lalan (1491), which were no more successful in propagating the idea. Later independent proofs, using simpler and more modern arguments, were proposed independently by Berzi and Gorini (1968) Lee and Kalotas (1975) and Lévy-Leblond (1976b). A general review of this long but occult line of argument has been made by Lecardonnel (1979). For recent rediscoveries or comments on this approach, see e.g. Mermin (1984), Pal (2003), Gannett (2007).

as *parallax* effects in space–time, quite analogous to the customary parallax effects in space. Langevin, in particular, was especially keen on this interpretation.[7]

5 Speeds

The confusion around the notion of speed was due to the existence of two equally natural derivations for the energy and momentum, yielding the two expressions stemming from (7) and (12) respectively, which we rewrite in the natural system of units already advocated by Planck, where $c = 1$:

$$\begin{cases} E = \frac{m}{\sqrt{1-v^2}} = m \cosh u \\ p = \frac{mv}{\sqrt{1-v^2}} = m \sinh u \end{cases} . \tag{18}$$

It was immediately clear that the two competing notions were related through

$$v = \tanh u \tag{19}$$

but it was generally held that the 'true' velocity was v, since, after all, it entered Hamilton's equation and was consistent with the old Galilean definition of velocity as the ratio of the span of space traveled by a mobile object to the travel time, i.e. $v = \Delta x/\Delta t$.

While most physicists maintained that u was merely a formal quantity, that is, a purely mathematical device, others insisted that the limitation $v \leq c$ and the non-additivity of the velocity parameter v required u to be given a primary physical meaning—as well as a name of its own, for which the term 'rapidity' was chosen.[8] Agreement was finally reached that a single notion in Galilean chronogeometry split into two different albeit related ones in Minkowskian chronogeometry. The situation, it was understood, is quite similar to that in ordinary geometry where slope, when small, is characterized by a single parameter, while one has to distinguish angle and tangent for larger values.

Finally, rapidity was given a direct physical meaning by Blondeville who discussed the operational definitions of speed; he stressed that an observer, if isolated within his own reference frame without any external clue, cannot measure the distance he covers (so that the Galilean definition is useless to him). However, he can measure his (proper) acceleration γ_0 as a function of his (proper) time t_0, and integrate it, which results, as a simple calculation shows, in the increment of his rapidity:

---

[7] Although Langevin did not explicitly use the idea of parallax, he nevertheless put forward a very modern view of space–time (Langevin 1911).

[8] This additive magnitude was introduced as early as 1910 by Varicak (1910), and named "rapidity" by Robb ([1911] 2015). Its common use however had to wait for several decades, and only gained acceptance first in high energy particle physics.

$$\int_A^B dt_0 \gamma_0(t_0) = u_B - u_A \tag{20}$$

The most natural *intrinsic* measurement of speed changes thus yields directly the variation of rapidity.[9]

Once the status of rapidity as a *bona fide* physical magnitude was established, it was used by Blondeville and Prostov for a much simpler derivation of the Lorentz group as the only possible chronogeometrical group (along with its singular Galilean limit) by relying on the elementary but deep lemma that any one-parameter continuous group can be additively parametrized.[10]

6 Another Route to Spacetime?

No account of the birth of modern chronogeometry would be complete without mentioning an unexpected historical discovery, which shows that another path could have been taken. Recently, in the archives of the famous journal *Annalen der Physik*, the manuscript was uncovered of a paper submitted in 1905 by a young physicist named Albrecht Zweistein. In this paper, which was rejected for reasons we will shortly discuss, Zweistein studied the electrodynamics of moving bodies. While sticking to the customary (at the time) electromagnetic worldview of Lorentz and Poincaré, he consistently pursued an earlier argument by Cohn (1901, 1902), who, following a Machian train of thought stressing 'scientific economy', had shown the superfluousness of the ether.[11] Zweistein then completely did away with the ether and its privileged frame. Instead, he proposed an interpretation of the Lorentz invariance of Maxwell equations based on two principles:

(1) the classical equivalence of reference frames in uniform motion, which he called "Principle of relativity", borrowing the terminology from Poincaré,
(2) the apparently weird idea that the velocity of light was the same in all these reference frame—his 'second postulate'.

Zweistein then proceeded to analyse time and space measurements, through detailed gedankenexperiments using the exchange of light signals to synchronize

[9] While this simple fact may have been known to quite a number of people, I do not know of any published mention before 1980 (Lévy-Leblond 1980).

[10] It remains surprising that such a simple mathematical result was recognized so lately as a solid founding stone for chronogeometry (Lévy-Leblond and Provost 1979). It may also be mentioned here that a still more general derivation of the (very few) possible chronogeometries, relaxing the linearity of the action of the group on space–time, has been given by Bacry and Lévy-Leblond (1968).

[11] The important contributions of Emil Cohn and his epistemological stand are barely known today. They probably exerted a definite influence on Einstein who thought highly of Cohn (as did the usually overcritical Pauli); see Miller (1981, pp. 181–182, note 42) and Darrigol (2022, pp. 195–198).

distant clocks. He was able to show that the time and space coordinates defined by such operational procedures, when compared in two equivalent reference frames, were linked by the standard Lorentz transformations.

Although this was quite a clever paper, it raised strong objections and was refused for publication. A first referee sternly condemned the dismissal of the ether as "unphysical", in line with Poincaré's position. A second one pointed to the ad hoc and fragile nature of the operational point of view taken by Zweistein: what, he asked, if non-electromagnetic signals, of a hitherto unknown nature, were discovered and used? Would we not have to face the possibility of a different description of space–time for these new phenomena, negating the idea of a universal spatio-temporal arena? And what if Maxwell's equations were only approximate?[12] The paper was resubmitted by the author a few years later, claiming priority after the development of chronogeometry as sketched above. It was then rebutted by Sommerfeld on more epistemological grounds. Here is what he wrote: "[The theory of space–time] is an *Invariantentheorie* of the Lorentz group. The name 'relativity theory' is an unfortunate choice: the relativity of space and time is not the essential thing, which is the independence of the laws of nature from the point of view of the observer".[13]

One may only wonder what the dominant interpretation and terminology of our theory of space–time would have been if Zweistein's paper had been published as early as 1905.

7 Conclusions

I will now abandon fiction and propose some conclusions.

But first, in order to ward off any accusation of blasphemy or lese-majesty, let me stress that Einstein himself was quite aware that his path towards what we now know as 'special relativity theory' was not the only possible one, and that others could have fathered the new space–time physics. Here is, for instance, what he wrote in his obituary to Langevin (1947): "It appears to me as a foregone conclusion that he would have developed the special relativity theory, had not that been done elsewhere" (Einstein 1947).

I leave it to specialists in science history to assess the degree of plausibility of the alternate account narrated here. Of course, as for any counterfactual historical attempt, a larger reworking of the context, epistemological, intellectual and even social, would be necessary in order to guarantee the credibility of the anachronisms here presented. The present story thus has to be taken with the proviso "all things being equal otherwise", which is obviously impossible; indeed, if Einstein had not

[12] In modern terms, think of the situation if the photon finally had a nonzero mass, however small; light would not travel with the invariant velocity, and any derivation of chronogeometry based on the exchange of light signals would become invalid.

[13] This statement (and other similar ones) by Sommerfeld is well documented—although it came historically much later, that is, in a 1948 review paper on philosophy and physics after 1900; see Miller (1981, p. 181, note 41).

been there, what can make us believe that Lorentz, Poincaré, Minkowski, etc., would have been there nevertheless?

What then is the purpose of this fictitious reconstruction? In fact, I see three areas of relevance for such considerations:

(a) *Educational.* The presentation of scientific notions as they unfolded historically is not the only possible one, nor even the best one. Alternative arguments and novel derivations should be pursued and developed, not necessarily to replace, but at least to supplement the standard ones. As a matter of fact, several recent textbooks on relativity offer, in place of the traditional, Einsteinian-like, derivation of the Lorentz transformations, the more modern approach to chronogeometry sketched in Sect. 4 above.
(b) *Epistemological.* As I have tried to show, if Einstein had not existed, we still would have 'his' theory—or would it really be the same? Apart from a few details perhaps, the formalism, that is, the notations and equations, would be very similar to ours today. But the language and, underneath, the words and the ideas we would use could be rather different. Many familiar terms, beginning with 'relativity' itself, shunned by Sommerfeld, might be absent from our vocabulary. My point here is to stress the polysemy of science, even for such a highly formalized one as physics. We need to actualise the variety of potential meanings behind our symbols—would it be only to keep open impredictable future paths of development.
(c) *Cultural.* Science is too often perceived by lay people as a mechanistic and inhuman endeavour. We should try to offer a less absolute and less deterministic view of its development, so that science would clearly appear for what it is, that is, a human venture, the solidity of which is no doubt based on its collective structure, but in which the role of the individuals cannot be neglected. Think for instance of the whole mythology of the twentieth century without the figure of Einstein (it is on purpose that in the above fiction the emergence of chronogeometry has been attributed to a largely collective effort).

In more general terms, my goal here was to advocate a view of the history of science which gives due credit to the notion of *contingency*, so forcefully stressed by Stephen J. Gould. In his work, and especially in his book *Full House*, he stressed the importance of the notion of contingency for the history of life (Gould 1996). If you 'rewind the film', as he was fond of saying, and let it unfold again starting a billion years ago, there is no reason the story would follow the same course. Chance events, it is now accepted, have played a crucial role in the history of life, events such as the crashing of a meteorite in Yucatan 65 million years ago, which tolled the knell of dinosaurs and paved the way for mammals. Why should the history of ideas keep blind to such a crucial element in the history of life itself?

Keeping to the history of physics, it would not be difficult to develop other examples of such gedanken experiments. For instance, and pursuing the fiction of an Einsteinless world, the construction of a theory of gravitation consistent with the new chronogeometry could have followed a way completely different from the geometrical picture set up by Einstein in what he called general relativity. Indeed, it has been shown, but is too little known, that one may work within the conventional classical theory of fields, and start writing a suitable Lagrangian with mass-energy of matter as the source of the field. Now the crux of the situation is that the field itself is endowed with energy, yielding a new term to be included in the source term, which in turn will modify the field, etc. In other words, the gravitational field is self-coupled, contrarily to the case of the electromagnetic field, which does not carry an electric charge. The resulting recurrence may be solved, resulting in nonlinear field equations which turn out to be identical with the standard Einstein equations. Nothing then is changed as concerns the predictions of the theory, except that the physical picture is completely different—and closer to the case of the other interactions (electromagnetic, strong, etc.). Indeed, instead of picturing gravitation as the effect of a curved spacetime, one may stick to the idea of a field living in the underlying but undetectable flat Minkowskian spacetime. This is not the place to dwell longer on this example, the interest of which is left to the reader to judge.[14] But consider the following statement by J. S. Schwinger: "It was Einstein's great ambition in his later years to construct a unified theory of gravitation and electromagnetism. In this he failed. The reason for that is quite clear once the problem is considered to be a physical one and not an exercise in pure mathematics. There is no known physical phenomenon in which the gravitational action of the electromagnetic field differs in any significant way from that of any other energy momentum bearing substance" (Schwinger 1976).

The present essay should be considered in the context of a wider attempt to put into question the too common view of the development of science as an inescapable one, with its successive theories and terminologies as inescapable signposts along a predetermined and unique road. Several recent works by historians of science have put forward the idea of a potentially pluralistic progress—and even possible regress (Ziman and Midgley 2001; Ziman 2003; Soler et al. 2015). Of course, as Gould repeatedly emphasised, this view is not akin to an irrational one: the general features of history, would it be natural, social or scientific, must and can be explained. But historical phenomena are so complex as to be extremely sensitive to a host of apparently secondary conditions. This should not come as a surprise to us physicists who today are familiar with deterministic chaos, sensitivity to initial conditions, etc.

Might it not be claimed, then, that Einstein was a Loren(t)z butterfly?

Acknowledgements It is a pleasure to thank Françoise Balibar and Bruno Latour for their comments on the first version of this paper (Lévy-Leblond 2003) and George Morgan for his most helpful linguistic advice. As concerns the present version, I am greatly indebted to Olivier Darrigol for his advice and critical remarks. I also benefitted from helpful indications by Thibaut Damour, Yves Gingras, Marc Lachièze-Rey.

References

Avishai Y, Ekstein H (1974) Einstein's Equivalence Principle and its Derivation from Special Relativity Presymmetrey. Communications in Mathematical Physics 37:193.

Bacry H, Lévy-Leblond JM (1968) Possible Kinematics. Journal of Mathematical Physics 9:1605.

[14] There have been quite a number of publications on the subject by distinguished authors since more than half a century (Birkhoff 1944; Gupta 1954; Thirring 1961; Sexl 1967; Deser 1970; Misner et al. 1973; Avishai and Ekstein 1974; Schwinger 1976; Di Mauro et al. 2021), culminating with Feynman's presentation (Feynman et al. 1995 in Di Mauro et al. 2021).

Berzi V, Gorini V (1968) Reciprocity Principle and the Lorentz Transformation. Journal of Mathematical Physics 10:1518 [These authors offer a very large list of the previous works, beyond the ones quoted here].

Birkhoff GD (1944) Flat Space-time and Gravitation. Proceedings of the National Academy of Sciences of the United States of America 30:324.

Comte C (1986) Leibniz aurait-il pu découvrir la relativité? European Journal of Physics 7:225.

Darrigol O (2022) Relativity Principles and Theories from Galileo to Einstein. Oxford University Press, Oxford.

Davidon WC (1975) Consequences of the Inertial Equivalence of Energy. Foundations of Physics 5:525.

Davies EB (2002) The Role of Astronomy in the History of Science. arXiv:physics/0207043

Deser S (1970) Self-Interaction and Gauge Invariance. General Relativity and Gravitation 1:9.

Di Mauro M, Naddeo A (2021) A Road Map for Feynman Adventures in the Land of Gravitation. arXiv:2102.11220v2

Dick PhK (1962) The Man in the High Castle. Putnam's Sons, New York [numerous editions following; in this remarkable example of counterfactual history, the plot takes place in America, twenty years after it has lost Second world war and is occupied by Nazi Germany and imperial Japan].

Einstein A (1947) Paul Langevin, La Pensée 12:13 [see also: Einstein A (1993) Œuvres choisies. Le Seuil-CNRS, Paris, Vol. 4, p. 183].

Esclangon E (1936) Sur les formules de Lorentz et la théorie de la relativité. Compte Rendu de Académie des sciences 202:1492.

Feynman RP, Morinigo FB, Wagner WG (1995) Feynman lectures on gravitation. Addison-Wesley, Massachusetts [For an introduction and assessment, see Di Mauro and Naddeo 2021].

Frank PH, Rothe H (1911) Uber die Transformation der Raumzeitkoordinaten von ruhenden auf bewegte Systeme. Annalen der Physik 34:825 [*Id.*, (1912) Zur Herleitung der Lorentz transformation, Physik Z 13:750].

Gannett JW (2007) Nothing but relativity redux. European Journal of Physics 28:1145.

Gould SJ (1996) Full House. Harmony Books, New York.

Gupta S N (1954) Gravitation and electromagnetism Physical Review 96:1683 [*Id.*, (1957) Einstein and Other Theories of Gravitation. Reviews of Modern Physics 29:334].

Hahn E (1913) Grundlagen zu einer Theorie der Lorentztransformationen. Archiv der Mathematik und Physik III:21/1.

Ignatowsky WV (1910–1911) Das Relativitätsprinzip. Archiv der Mathematik und Physik III:17:1–18; Physik Z 11:972 & 12:776–779.

Jammer M, Stachel J (1980) If Maxwell had worked between Ampère and Faraday: "An historical fable with a pedagogical moral". American Journal of Physics 48:5–7.

Lalan V (1936–1937) Compte Rendu de Académie des sciences 203:1491 [see also: Bulletin des Sciences Mathématiques 65:83].

Langevin P (1911) L'évolution de l'espace et du temps. Scientia 10:34.

Lecardonnel JP (1979) Un autre point de vue pour présenter les transformations de Lorentz. Bulletin de l'union des physiciens 615:1171.

Le Roy E (1936) Sur les postulats qui sont à la base des cinématiques. Compte Rendu de Académie des sciences 202:794.

Lee AR, Kalotas TM (1975) Lorentz Transformations from the First Postulate. American Journal of Physics 43:434.

Lévy-Leblond JM (1976a) What is So Special about Relativity? Group-Theoretical Methods in Physics. Janner A & al (eds). Lecture Notes in Physics n. 50. Springer-Verlag, New York.

Lévy-Leblond JM (1976b) One More Derivation of the Lorentz Transformations. American Journal of Physics 44:271.

Lévy-Leblond JM (1980) Speed(s). American Journal of Physics 48:345.

Lévy-Leblond JM (2001) Science's fiction. Nature 413:573.

Lévy-Leblond JM (2003) What if Einstein had not been there? In: Gazeau J-P & al (eds). Proceedings of the 24th International Colloquium on Group Theoretical Methods in Physics, Institute of Physics, London.
Lévy-Leblond JM, Provost JP (1979) Additivity, Rapidity, Relativity. American Journal of Physics 47:1045.
Mermin ND (1984) Relativity without light. American Journal of Physics 52:120.
Miller AI (1981) Albert Einstein's Special Theory of Relativity, Emergence (1905) and Early Interpretation (1905–1911). Addison-Wesley, Massachusetts.
Misner CW, Thorne KS, Wheeler JA (1973) Gravitation, Freeman & Company, NY, p. 424.
Pal PB (2003) Nothing but relativity. European Journal of Physics 24:315.
Pars LA (1921), The Lorentz transformation, Philosophical Magazine 42:249.
Peter Bowler PJ (2013) Darwin Deleted. Imagining a World without Darwin. Chicago Press, Chicago.
Robb A ([1911] 2015) Optical Geometry of Motion, a New View of the Theory of Relativity. Heffner & sons. Cambridge 9. Andesite Press, Warsaw.
Rodwell G (2013) Whose History? Engaging history students through historical fiction University of Adelaide Press, Adelaide. Chapter 7, Counterfactual Histories and the Nature of History, pp. 81–98.
Roth Ph (2004) The Plot Against America. Houghton Mifflin, Boston.
Schütz I (1897) Prinzip der absoluten Erhaltung der Energie. Königlich Preussische Akademie der Wissenschaften, Nachrichten, pp. 110–123.
Schwinger J S (1976) Gravitons and photons: The methodological unification of source theory General Relativity and Gravitation 7:251–256.
Sexl RU (1967) Theories of Gravitation. Fortschritte der Physik 15:269.
Silverberg R (2003) Roma Æterna. Eos Books, New York.
Singaravélou P & Deluermoz Q (2016) Pour une histoire des possibles. Analyses contrafactuelles et futurs non advenus, Seuil, Paris.
Soler L, Trizio E, Pickering A (2015) (eds) Science as it Could Have Been: Discussing the Contingency/Inevitability Problem. University of Pittsburgh Press, Pittsburgh.
Spinrad N (1972) The Iron Dream. Avon Books, New York.
Thirring W (1961) An Alternative Approach to the Theory of Gravitation. Annales of Physics 16:96.
van Rijn van Alkemade AC (1912) Über die Transformationsgleichungen der Relativitätstheorie. Annalen der Physik 38:1033.
Varicak Vladimir (1910) Anwendung der Lobatschefskijschen Geometrie in der Relativtheorie. Physikalische Zeitschrift 11:93–6.
Ziman J (2003) Emerging out of Nature into History: the Plurality of the Sciences. Philosophical Transactions of the Royal Society of London A 361:1617–1633.
Ziman J, Midgley M (2001) Pluralism in Science: a Statement. Interdisciplinary Science Reviews 26:153.

Chapter 12
Rotating Universes

Montgomery Link

Abstract This paper concerns the dispute about space, time, and motion, about whether space, time, and motion are absolute or relative. This paper focuses on one aspect of the dispute by investigating a kind of motion, the rotation of galaxies with respect to the inertial frame. In rotating universes, distant galaxies rotate with respect to the inertial frame. Our universe does not rotate. Nonetheless, rotating universes throw light on our own, or so this paper argues. First, the paper presents historical evidence of the fundamental theoretical use of the rotating solution in physics. Second, the paper identifies three consequences for time and the causal order.

Keywords Closed time-like curves · Geodesic completeness · Gödel · Kurt · Rotation · Rotating universe · Time

1 Introduction

One of the overarching themes of the volume dedicated to Salvo D'Agostino is the irregular case in the history of physics. Herein the present paper looks at an irregular case that arose in the history of general relativity, a case which sprang from a remarkable insight. The insight was to put in motion not some individual object or another but instead the galaxies themselves. The particular mechanism of motion is rotation, like the spinning of a top. Rather than a rotating top, however, this paper discusses the rotation of universes. It focuses on the issue that the laws of general relativity hold in a rotating universe, and yet, for fundamental reasons the rotating structure does not quite fit with the structure of our actual universe. The rotation disrupts the order of time in such a way that causation breaks down. This paper argues for an unexpected conclusion, that the objectivity of time depends on the motion of matter.

M. Link (✉)
Suffolk University, Boston, USA
e-mail: mlink@suffolk.edu

R. Pisano (ed.), *A History of Physics: Phenomena, Ideas and Mechanisms*,
History of Mechanism and Machine Science 42,
https://doi.org/10.1007/978-3-031-26174-9_12

The argument is as follows. Section 2 introduces the rotating universe, with a brief description of it. With the ground laid, Sect. 3 looks at some of the theoretical consequences and conceptual impact of rotating universes, including historical evidence that corroborates the fundamental use of rotating solutions in theoretical physics. Section 4 then discusses the implications for the independence, objectivity, and beginning of time. The final assessment is that there remains a standoff on time, because the existence of closed time-like curves is compatible with the theory of general relativity. Section 5 finishes quickly with a reminder that current theory determines our real commitments and sets the conceptual parameters.

Before turning to the rotation of universes, the paper introduces several thought experiments. The first is the water pail experiment.

If one is not familiar with the water pail experiment and the objects involved or is under the impression that it defies imagination, one might actually conduct it. It might be best to do outdoors, especially if one enlists the help of children. Get a pail with a handle, a length of cord, and some water. Tie one end of the cord to the handle of the pail. Use the other end of the cord to suspend the pail in air, from the limb of a tree, for example. Having secured the cord, spin the pail counterclockwise until the cord winds up tightly. The cord represents the axis of rotation. Holding the pail fast, fill it with water, but not too full. Then release it.

Now, observe the water in the bucket. The water surface, which had been flat, starts to become concave. One can observe the water climbing the walls of the bucket. One gauges the rotation by measuring how far up the pail the water rises, marking the measurement on its side, for example. This is a force to be reckoned with. The rotation of the water starts slowly but eventually begins to catch up to the clockwise rotation of the bucket. Once that happens, and the water comes to be at rest with respect to the pail, the measurable change in the water surface is not due to the relative motions of the water and the pail. The clockwise rotation of the pail continues to slow, then ceases. The water follows in its turn. The pail seems to pause before it begins to rotate the other way in a counterclockwise motion.

One who does not have a length of rope or a bucket on hand might try the following procedure. Go to a bar and order a cognac in a snifter, the kind without a stem or base. Place the glass on the bar and spin it. Such experiments resonate with personal experience. One does not seem to need to carry them out. Why waste good brandy? Whether it is a bucket at the beach filled with seawater, or the cognac in a snifter, there is no question about how things will turn out. One might even imagine that the walls of the bucket were a mile thick.

Two other experiments come to mind. The first begins with two spheres connected together by a cord. If spun about an axis at the center of their mass, the cord tightens between the two balls. The second uses a rotating disc, which will tend to bulge at its equator. Saturn is a good example, much larger than Earth and yet with a faster rotation. The surface of Saturn is more distant from its center at its equator than at its poles.

Let us now restate the idea of rotation. In three-dimensional space, imagine two cords both perpendicular to and laid across the horizontal axis x. We replace the cords with cylinders. Imagine that one of the cylinders C is at rest and the other D is in motion rotating through the y z plane. These might be two towers on the horizon, one at rest and the other rotating. Then, if the motion is relative, one can fix either as the frame of reference for the rotation. Yet, the experiences of two observers, one standing on *C* and the other on *D*, would differ.

For another example, hang a siren from the rotating tower *D*. For an observer standing in the plane of the axis of rotation, there is no Doppler effect. The Doppler effect is relative to the frame of reference. It is apparent in the plane of rotation as a change in frequency.

For one last example, growing up, there was a roundabout out back. I would run, jump on board, and off I went, spinning. On clear nights, I watched the stars, the Milky Way, often the moon, all the lights in the sky, spinning as one. By looking up at the stars, one can gauge the spin.

2 Galaxies Rotating with Respect to the Inertial Frame

The mechanical phenomenon of the rotating pail takes on a life of its own. The vortex of water that results evokes the idea that the change in the place of the water that occurred has an absolute frame of reference permanently fixed, that this one frame of reference is the Frame against which to measure motion.

Absolute motion is a direct consequence of absolute space and time on one view.[1] Those who disagree about absolute space and time agree that the bucket represents a true motion. They argue instead, for example, that a completely accurate account does not require absolute space to explain the motion. A leading argumentative strategy has been to use modus tollens reasoning to take down absolute space and time by arguing that true motion is not absolute. Perhaps, for example, the mass of distant galaxies is the origin of local inertia.

Ernst Mach made such a supposition:

> The possible rotation of the universe has a special significance in general relativity because one of the influences that led Einstein to the theory in 1915 was Mach's principle. The exact formulation of the principle is rather obscure, but it is generally interpreted as denying the

---

[1] Cfr. Rynasiewicz (2011, 6.2; 2018). The water pail experiment did not arise out of nothing. Isaac Newton introduced it to counter a competing theory in the Scholium to the *Principia* (1687) (Newton 1687; see also Hoefer et al. 2021, 3–4). The competing theory of motion, due to Descartes ([1644]1983, II.13, II.24–II.25), is inadequate to explain the measurable result of the rotation of the bucket. To add this historical context is not to diminish Newton's complete commitment to the reality of space as distinct from the bodies in it and to the uniformity of the passage of time independent of events of the world. For the debate with Leibniz, see the correspondence between Leibniz and Clarke (in Leibniz 1969, pp. 675–721). Was Newton refuted? Recent historical accounts do not see a single sustained historical debate over space and time as much as separate debates (Huggett et al. 2021, 2.4), but other accounts see more continuity (Earman 1995).

By engaging with this mechanical phenomenon, we have entered into the great "European tradition in which science and philosophy are often linked by the same aims" (D'Agostino 2000, p. 303). This linkage serves as a background hypothesis for our investigation of rotating universes. We argue neither for it nor against it but assume it as a coordinating principle.

> existence of absolute space. In other words, matter has inertia only relative to other matter in the universe. The principle is generally taken to imply that the local inertial frame defined by gyroscopes should be non-rotating with respect to the frame defined by distant galaxies. (Gödel 1990, p. 189)

Mach suggested that the Earth's rotation sets the time scale.

> [An] [...] investigator must feel the need of [...] knowledge of the immediate connections, say, of the masses of the universe. They will hover before him as an ideal [...] insight into the principles of the whole matter, from which accelerated and inertial motions will result in the same way".[2] (Ciufolini and Wheeler 1995, p. 269)

Einstein wrote to Mach that it "turns out that inertia originates in a kind of interaction between bodies, quite in the sense of your considerations on Newton's pail".[3] There is on Einstein's view no natural boundary to space. He rejected a space-infinite universe. In discussing his field equations, he stated:

> In my opinion the general theory of relativity can only solve this problem [of inertia] satisfactorily if it regards the world as spatially self-enclosed. (Ciufolini and Wheeler 1995, p. 271)

Both Mach and Einstein denied absolute space, time, and motion. Einstein endorsed Mach's thought. Yet, it is plausible that they would not have agreed on a single principle or on what they were rejecting.

I have not let go of the roundabout. I returned recently to the roundabout with two children. They climbed on the carousel. I said, "Penelope, Noah, hold on tight", and off we went. Let us develop some parameters for an accurate description of the motion. In one frame of reference, the carousel was spinning. In another, it was still and the world was turning. Relative to one frame of reference, I pushed my two companions on the merry-go-round. They felt free, but I plodded, turning them slowly. Relative to another frame of reference, the merry-go-round was not in motion, and the world beneath my feet was turning. Was I turning the world in that reference frame? No, certainly not. The rotation of the stars is due to gravity and a rotational field. I merely prevented the roundabout from taking part in this rotation, from the perspective of general relativity (Reichenbach 1958, p. 239).

"Then what is that rotation relative to?" The answer is that the rotation is relative to the inertial frame, a reference frame that includes a time scale. Begin by fixing a collection of objects, like points of mass or the stars in the Milky Way (Disalle 2020). This becomes a frame, against which one describes the geometry of motion. A dynamical account of the motion, which includes the relevant masses or forces, requires the addition of a time scale to the frame of reference. Such is an inertial frame. Inertial frames have a constant rectilinear motion relative to each other. A collection of inertial frames describes every region of space–time on the theory of

---

[2] Most "physicists nowadays would not accept Mach's principle, because they feel it makes an untenable distinction between the geometry of space–time, which represents the gravitational and inertial field, and other forms of fields and matter" (Gödel 1990, p. 189).

[3] Einstein, letter to Mach from Zurich dated 25 June 1913 (Ciufolini and Wheeler 1995, pp. 269–270). Circumstance and history impinge on scientific theory (Ryckman 2011).

general relativity. The space–time geometry and ultimately the flux and density of momentum and energy across the universe determine the local inertial situation.

General relativity at first seemed essentially bound up with the denial of absolute space. Who needs it for motion? On one interpretation of general relativity, mass-energy out there determines space–time geometry locally. The local geometry of space–time geometry guides matter and fields. To expand quickly upon that quotation, the Einstein field equations connect matter to the curvature of space–time (Einstein 1916). Space–time is a four-dimensional space $<M, g>$ specified with a manifold and a metric. The metric tensor g underlies the complex functional relation determining the curvature of space–time in the Einstein field equations as a function of mass-energy (Hawking and Ellis 1973, 3.4; see also Malament 1977). In sum, the mass of distant galaxies determines local inertia.

What does this mean for the water pail experiment? The measurable change in the surface of the water need not depend on an absolute frame. Motion is geodesic in general relativity. It presupposes not absolute space and time but the mass of the universe, which governs space–time locally and determines inertia. Does this refute absolute time and space?

In "An Example of a New Type of Cosmological Solutions of Einstein's Field Equations of Gravitation", Kurt Gödel, best known for his proofs of the completeness of first-order logic, the incompleteness of any formal system, and the relative consistency of the continuum hypothesis, introduced the possibility of absolute frames into general relativity theory[4] (Gödel 1930, 1931; see also Svozil and Baaz 2011, Gödel 1938, 1949a, b, 1952).

> [The] most conspicuous physical property distinguishing these solutions from those known at present is that the compass of inertia in them everywhere rotates [in the same direction] relative to matter, which in our world would mean that it rotates relative to the totality of galactic systems. These worlds, therefore, can fittingly be called "rotating universes". (Gödel 1990, p. 204; Author's original quotation marks)

The lack of rotation gives meaning to the denial of absolute space.

He explained these universes in a lecture:

> If we now consider the problem from the standpoint of relativity theory, the first question which arises is in what sense we can speak of a rotation of the whole universe in relativity theory, where we have no absolute space to which we could refer it. The answer, of course, is that in relativity theory, as a substitute for absolute space, we have a certain inertial field which determines the motion of bodies upon which no forces act. In particular, this inertial field determines the behaviour of the axis of a completely free gyroscope or of the plane of a pendulum, and it is with respect to the spatial directions defined in this way (by the gyroscope or the plane of a pendulum) that matter will have to rotate. In the usual terminology, this means that it will rotate relative to the compass of inertia. You see that this kind of rotation does not involve the idea of an axis around which the whole world rotates. The world may

[4] The Gödel metric is a two-term solution. For a full stability analysis of the Gödel universe (Barrow and Tsagas 2004), and for some of its particular properties (Rindler 2009). For the Gödel solutions and Newtonian cosmology (Heckmann and Schücking 1955, 1956; Raychaudhuri 1957). Mach had a profound influence on the Vienna school. Reichenbach was a prominent member, and Gödel was associated with the Vienna school for some time.

> be perfectly homogeneous and still rotate locally in every place, as is actually the case in the example I am going to give. Nevertheless, the world may be said to rotate as a whole (like a rigid body) because the mutual distances of any two material particles (measured by the orthogonal distance of their world lines) remain constant during all times. Of course, it is also possible and even more suggestive to think of this world as a rigid body at rest and of the compass of inertia as rotating everywhere relative to this body. (Gödel 1949c, pp. 270–271)

Earlier we discussed an issue about change in place. In this universe, every place locally is rotating and yet the orthogonal distance between the geodesics of any two particles is constant, so the whole place is a rigid body that is homogeneous and isotropic. Any rotation at the periphery would have to be at a low rate, to the degree that the background microwave radiation is isotropic, to conform to Hubble's observations of galaxies at a distance of half the radius from the center (Gödel 1949a, p. 450 and n. 16; see also Hubble 1934, p. 74).

Gödel provided in general relativity an alternative static universe that has rotational symmetry around any point. It is suggestive to think of a dust cloud of spinning particles. The particles are mass points.

> Evidently this state of affairs shows that the inertial field is to a large extent independent of the state of motion of matter. This contradicts Mach's principle but it does not contradict relativity theory. (Gödel 1949c, p. 271)

According to the Einstein field equations, the local inertial field depends on the state of motion of matter far away. Has Gödel slipped up? Probably not. When Gödel says the inertial field is largely independent, his point is that the connection is purely contingent, something about our actual world, and in no way a necessary connection. The state of motion of matter fixes the local inertial field. The inertial field, on the other hand, is independent of the state of motion of matter in the theoretical sense that if we interpret particles of Gödel dust as galaxies, we have a model of general relativity theory that rotates, while other models do not rotate.

The earliest cosmological models of the Einstein field equations were the Friedmann-Robertson-Walker solutions developed during the 1920s. These solutions present universes for which there is no rotation, in conformity with Mach's principle. Gödel solves the Einstein field equations in such a way that the local inertial frame rotates with respect to distant galaxies. The result is that Mach's principle and general relativity are independent issues. The cosmological equations of general relativity theory have solutions in spatially homogeneous, isotropic, and expanding models that do not rotate, as well as in homogeneous, isotropic, and expanding models that do rotate. Gödel

> [...] showed that it was possible to have solutions of the Einstein field equations in which the galaxies were rotating with respect to the local inertial frame. He therefore demonstrated that general relativity does not incorporate Mach's principle. (Gödel 1990, p. 189)

that rotation is an independent issue.

Even without an absolute frame of reference, a close relative is the cosmological expansion itself. The laws of physics remain the same in the cosmic microwave background frame even if this radiation is the closest approximation to an absolute frame of reference. The cosmic microwave background radiation does not appear to be isotropic. Instead, the dipole anisotropy is observable. Therefore, we are not at rest with respect to the cosmic microwave background or Hubble flow.

One talks of space–time without presupposing that space–time is a substance. Space–time is not a pre-theoretical container. "According to general relativity", "because of the dynamical nature of all space–time structures", "except for continuity, differentiability and local topology of the differentiable manifold" (Stachel 2005, p. 57):

> The points of space-time lack inherent haecceity; thus they cannot be used for individuation of other physical events in a general-relativistic theory of matter and/or non-gravitational fields. This is the purport of the "hole argument" [...]. The points of space-time have quiddity as such, but only gain haecceity (to the extent that they do) from the properties they inherit from the metrical or other physical relations imposed on them. In particular, the points can obtain haecceity from the inertia-gravitational field associated with the metric tensor.
>
> [...]
>
> This is better said in the language of fibred manifolds, in which particular dynamical physical fields are represented by cross-sections of the manifold: one can now define the base space as the quotient of the total space by the fibration. Thus, even the points of the base space (let alone its metric) are not defined a priori, and their individuation depends on the choice of a cross-section of the fibred manifold, which will include specification of a particular inertio-gravitational field. (Stachel 2005, p. 57; Author's original quotation marks)

If mappings from field to field need not preserve the same "don't cares", then the outcome of the hole argument is that space–time is not a manifold containing events (see also Stachel 2014; Norton 2019; Janssen 2004). Indeed, points in space–time are not identifiable individuals separate from one another. Their identity depends on their specific spatio-temporal properties.

There is no reason to think that the theory of general relativity preserves pre-theoretical entities. The energy–momentum tensor determines the metric field. Space–time on more than one view is rather a coordinate system than a substance in its own right, but the debate continues (Cfr. Hoefer 2000).

3 Closed Time-Like Curves

Before we go further, let us return to our mode of experimentation. Lay a cord from an earlier experiment horizontally and extend it to an infinite length in both directions. Imagine it is a cylinder (Fig. 1).

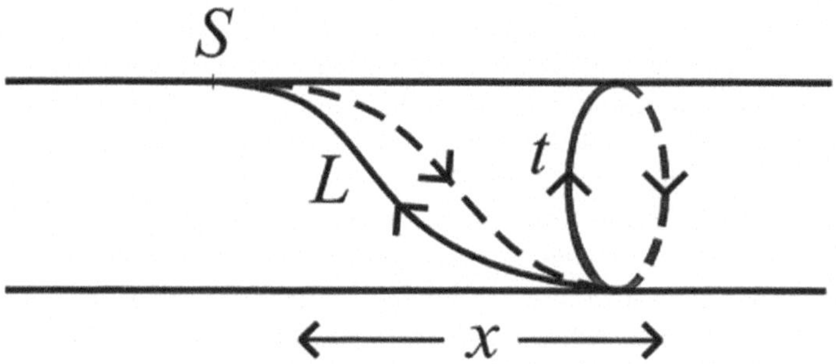

Fig. 1 A two-dimensional map of space (x) and time (t) onto the surface of a cylinder. The light signal L spirals (Reichenbach 1958, p. 272) *Source* Author, ML

Our subject is the surface of that cylinder in two dimensions, one spatial and one temporal. The spatial dimension x runs the length of the cylinder parallel to its axis. For the temporal dimension slice the cylinder and draw t on the edge of the surface of that slice so that it loops. The manifold has no curvature.

> Only as a whole it has the peculiarity that time lines may be closed. A light signal L will travel like a spiral to infinity, yet it may be reflected (at [some distant point on the surface of the cylinder] S) so as to return to its starting point.
>
> This simple model demonstrates strictly that closed timelike world-lines can exist in a world in which the special theory of relativity holds without singularities. (Reichenbach 1958, p. 273; see also Ryckman 2021, 2.4).

The reflection is like a return back in the sense of circling.

Gödel proposed closed time-like curves, world lines along which a material particle in space–time returns to its starting point. Again, travelling along these curves, a material particle ends at the same space point as it started. There is a "breakdown of causality" here because a line can be drawn that is everywhere time-like and yet returns back on itself (Raychaudhuri 1979, 5.4, in particular pp. 92–94). The picture of returning along this line is like going around in a circle, as in Fig. 2.

Fig. 2 Picture of the return. Closed time-like curve (Gödel 1949c, p. 286) *Source* Author, ML

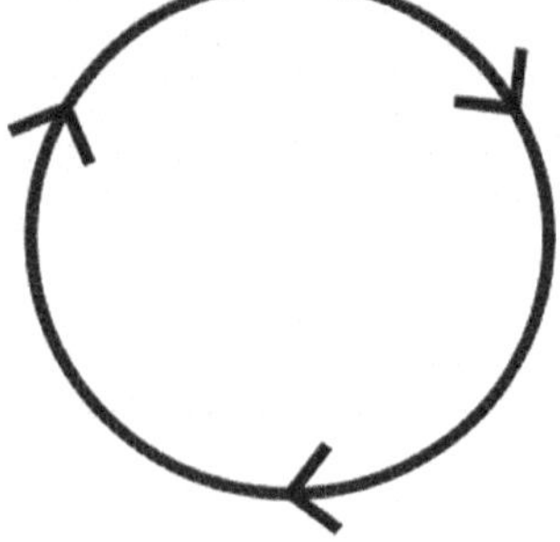

It does not "mean that on returning you would find your surroundings look like a film running in the wrong direction", as in Fig. 3.

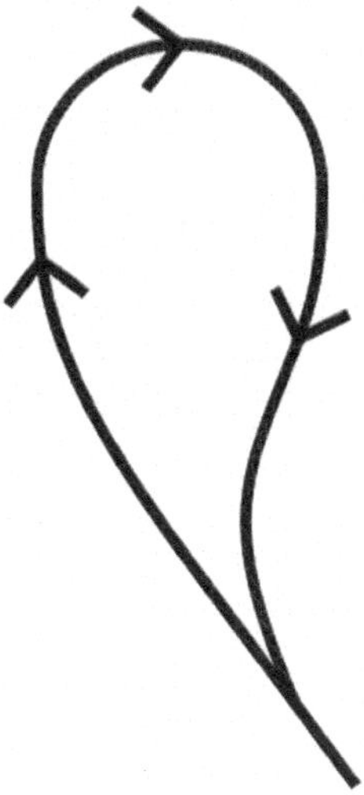

Fig. 3 Mistaken picture of the return. Upon returning, the surroundings do not look like a video rewinding (Gödel 1949c, p. 286)
Source Author, ML

Gödel continues:

> But in spite of the existence for the whole world of a uniform positive direction of time, there exists, on the other hand, no uniform time for the whole world. This assertion can be given a quite precise meaning. Namely, one may say that a coordinate x_0 is a time coordinate if x_0 always increases if one moves along any possible world line of matter in its positive direction. (Gödel 1949c, p. 286)

One does not return to see the film rewinding, but rather to see it replayed. This eternal return has an unforeseen consequence:

> The existence of closed time-like lines implies that such a time coordinate for the whole four-dimensional space cannot exist. For if you move along a closed time-like line in its positive direction, the time coordinate x_0 for the initial point must be the same as for the end point (because initial and end point coincide), but then x_0 cannot have been increasing along the whole line. (Gödel 1949c, p. 286)

So that the solution may have a physical meaning, "the existence of a time coordinate in this sense is not absolutely necessary" (Gödel 1949c p. 286). Still, a well-defined positive direction of time remains, since

> The direction of time determines, for example, what it means to send a light signal, since the light is to be found only in the positive half of the double light cone issuing from the space-time point from which the signal is sent. The positive direction of time also allows you to formulate the second law of thermodynamics, at least for closed systems; whether it could be applied to the whole universe I do not know.[5] (Gödel 1949c, pp. 286–287)

---

[5] Note that, without an admissible time function, there is no definable simultaneity relation in terms of a rotating space–time structure (Stein et al. 2009, pp. 436–437).

Einstein described the importance of the rotating solution "to the general theory of relativity, especially to the analysis of the concept of time" (Schilpp 1949, p. 687). If *P* is a world point, it has a light cone. On a time-like world line through *P* observe two points, *A* and *B*, separated by *P*. "Does it make any sense to provide the world-line with an arrow, and to assert that *B* is *before P*, *A after P*?" From the physical point of view, Einstein asked, could we point the arrow in the opposite direction (Fig. 4)?

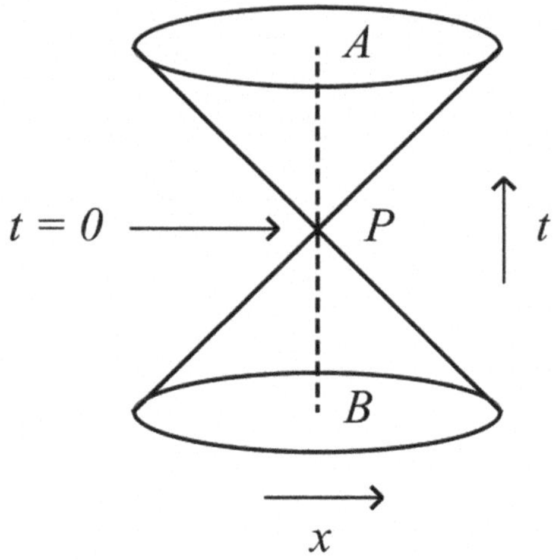

Fig. 4 The light cone for world point *P*. The dashed line is a time-like world line. *B* and *A* are world points (Cfr. Schilpp 1949, p. 687) *Source* Author, ML

Not in the sense of sending a signal, in which case time would be asymmetrical.

> What is essential in this is the fact that the sending of a signal is, in the sense of thermodynamics, an irreversible process, a process which is connected with the growth of entropy (whereas, *according to our present knowledge,* all elementary processes are reversible). (Schilpp 1949, pp. 687–688; Author's italic)

Where there is an irreversible process in the sending of the signal, "*B* is before *A*" makes sense. What if there is a closed series such that each world point precedes the preceding one?

> [In] that case the distinction 'earlier-later' is abandoned for world-points which lie far apart in a cosmological sense, and these paradoxes, regarding the *direction* of the causal connection, arise, of which Mr. Gödel has spoken.[6] (Schilpp 1949, pp. 687–688; Author's italic)

Scholarly opinion on Einstein's response is not unanimous. Einstein and Gödel met regularly in those days, so there would have been no need to record the details, and there was no interest to generate a public response. This proposal was not out of the blue for Einstein. Einstein had considered the introduction of closed time-like

[6] In Gödel (1949c), Gödel "makes several remarks that help one visualize the configuration of light-cones in Gödel space–time and visualize its time-like and null geodesics. [...] Free particles and light rays exhibit a kind of boomerang effect." (Gödel 1995, p. 269; Chandrasekhar and Wright 1961).

world lines in print already thirty-five years earlier (Einstein 1914, p. 1079). Einstein also said that the paths run "counter to my physical imagination most vividly". Much later, in 1954, Einstein will completely relent: "As a matter of fact, one should no longer speak of Mach's principle at all".[7] Indeed. Einstein had abandoned the principle long before, perhaps as early as June 1918.[8]

The closed time-like curves occur if space has donut-like holes, exhibiting a toroidal geometry. It is not clear at all that the actual universe looks like that, but it is currently unknown if general relativity forbids the curves for more realistic conditions. Gödel establishes a framework in which the expanding bubble, present in order to mimic cosmological expansion, may have holes that locally resemble a torus-like structure.[9] In the context of much shorter time scales, Enrico Fermi had rejected the possibility of time travel in lectures on the quantum theory of radiation delivered to the Symposium for Theoretical Physics at the University of Michigan during the summer of 1930, at which time his reasoning was as follows:

> Two atoms, A and B, are separated by a distance R. At time $t=0$, A is assumed to be in [an] excited state and B in its ground state, with no photons present. Atom A will decay into its ground state under the emission of a photon. This photon can then, with a small probability, be absorbed by atom B. Fermi asked the question at what earliest time atom B will "notice" the decay of atom A with its accompanying photon. He expected that B moves out of its ground state only after a time $t=R/c$, in accordance with the speed of light. And indeed, this is what he found by his calculations. (Hegerfeldt 1997, p. 2; see also Fermi 1932; Hegerfeldt 1994, p. 596; Maddox 1994, p. 509)

Fermi's calculations were based on second-order perturbation theory—a technique still common today—and on the approximation of an integral over positive frequencies by an integral over positive and negative frequencies ranging from $-\infty$ to ∞ instead of 0 to ∞. The mistaken approximation translates into the impossibility of time travel. Fermi's calculations were incorrect in the context of standard quantum mechanics.

Time travel appears to have a connection to teleportation. Since both require the transportation of information at least at the speed of light, the actual mechanisms behind them may be similar. Investigations into experimental quantum teleportation using correlated pairs of particles began in the early 1990s[10] (Bennett 1993; Einstein et al. 1935; see also Bell 1964, pp. 195–200). The correlation is entanglement.

---

[7] Letter to Felix Pirani of 2 February 1954 (The Albert Einstein Archives of Jerusalem 17; via: http://www.albert-einstein.org/.index3.html). This letter is quoted by, among others, Abraham Pais in *Subtle is the Lord...: The Science and the Life of Albert Einstein* (Clarendon, Oxford, 1982, p. 288); and Gerald Holton (1968, Mach, *Einstein, and the search for reality*, Daedalus 97, pp. 636–673, on p. 667).

[8] John Stachel, personal communication. For Einstein's interaction with Willem de Sitter, which began in the fall of 1916 (Barbour and Pfister 1995). The de Sitter world provides a counterexample to a version of Mach's principle.

[9] Raul de la Fuente Marcos helped me with this paragraph and the next two. Two time travel mechanisms proposed to exploit that spatial configuration are wormholes and cosmic strings (Morris et al. 1988; Gott 1991).

[10] Some experimental results for photons and atoms are: Barrett et al. (2004), Riebe et al. (2004), Ursin et al. (2004). For teleportation and quantum computation (Brassard 1996).

The immediate response of the rest of the scientific community to rotating universes in its day was almost negligible. This is not surprising since interest in general relativity among physicists was very low at that time.[11] Yet, beginning in 1955 the rotating solution appears in fundamental papers on cosmology, some of which we discuss briefly below over the next few paragraphs.[12] We identify in particular two influences on general relativity theory.

The first influence is through the causality violations Amal Raychaudhuri identified, which sharpen the geodesic properties of space–time.[13] A space–time geodesic is a line on the surface of space–time. If a universe is geodesically complete, one can give a rule mapping the tangent vector space at p to the manifold, which for each vector $\boldsymbol{X}$ in the space selects the point in the manifold on the geodesic λ_X that is a unit parameter distant from p. Once the map is defined for all the vectors in the space, then the geodesic λ_X (v) is defined for all v, in which case the geodesic λ_X (v) is a *complete geodesic.* If all the geodesics on the manifold are complete, then the manifold is *geodesically complete.* In our universe, the connection does not converge at every point.

Our universe does not in general satisfy geodesic completeness. A geodesic that is not complete ends in an undefined point: a singularity.[14] $<M, g>$ is *b-complete* if there is an endpoint for every curve of a special sort that includes both geodesic curves and curves that are not geodesics. B-completeness implies geodesic completeness, or g-completeness. Geodesic completeness distinguishes a rotating universe from our actual universe, the feature mentioned above in Sect. 2. A space–time is singularity free if it is b-complete. Our universe violates geodesic completeness. Causality breakdowns in general do not prevent the occurrence of singularities, which leads us to the second influence.

The second influence is through the Raychaudhuri fundamental equation of gravitational attraction, which Roger Penrose and Stephen Hawking used in their essential work on singularities. Raychaudhuri cited and discussed Gödel's work. In a model of the universe that does not rotate, singularities are ineliminable even under the absence of homogeneity and isotropy. Rotation is the specific property that led to Raychaudhuri's fundamental equation. When Hawking (1967) formalized cosmic time, he showed it to be exactly that which rotating universes exclude.

The big bang, for instance, is a singularity. Cosmologists have information about the universe just fractions of a second after it began, yet the beginning itself is not testable or observable in the same way. Observation of the background microwave radiation confirms big bang theory and the expansion of the universe, but the big bang itself did not occur at a specifiable space–time coordinate. As such, scientific

---

[11] Daniel Kennefick, personal communication.

[12] For a comprehensive and insightful account (Ellis 2001).

[13] Raychaudhuri (1955, II; see also Hawking and Penrose 1970, p. 540; Hawking and Ellis 1973, pp. 170, 30–33, 259–263). This influence is obvious in the work of Chandrasekhar and Wright (1961).

[14] Penrose and Hawking developed these ideas in a series of papers in the middle to late 1960s, among which (Hawking and Penrose 1970; Penrose 1965).

investigation of the big bang, as of any singularity, is somewhat asymptotic. There also exists a singularity at the end of time, the final state.

How then are we to understand time, given galactic rotation with respect to the inertial frame? Time is a venerable topic in philosophy. Is there time without change? Is time just the events that occur within time? What about branches of time? We will not even broach many of the basic questions about this vast topic (Arntzenius 2012; Emery et al. 2020). By an absolute stretch of time in what follows, I mean one that is independent of the events that occur in that stretch. Does an absolute stretch of time consist of points in time that themselves have no temporal duration? If we are to answer the first question of this paragraph, all other such questions must await another time and place.

A well-known and often discussed consequence of the rotation of universes is time travel. Time travel into the future does not seem to pose any problems; however, there has been much consideration of the philosophical implications of the closed time-like curves in connection to time travel into the past.[15] Might one travel back into the past and prevent one's own parents from having any children? This provocative but unlikely scenario is distinctly separate from more penetrating questions about the physical operation of time machines and time travel. The paradoxes of time travel into the past point to the issue that we cannot define one absolute world time in terms of the local times of observers in rotating space–time structures[16] (Gödel 1949b, pp. 203–207). Anyone of them can step into a region of the universe that is past.

In rotating universes, change is not objective. The "decisive point is this: that for *every* possible definition of a world time one could travel into regions of the universe which are past according to that definition. This again shows that to assume an objective lapse of time would lose every justification in these worlds". His concluding paragraph begins by raising the question of whether an appeal to possible worlds can be illuminating:

> It might, however, be asked: Of what use is it if such conditions prevail in certain *possible* worlds? Does that mean anything for the question interesting us whether in *our* world there exists an objective lapse of time? I think it does. (Gödel 1949b, p. 207)

There are two reasons to consider prevailing conditions in certain possible worlds. The first reason is that the redshift observed in light emitted from distant galaxies is not due to the Doppler shift but to the expansion of space itself. Gödel had at first produced solutions for universes that are not expanding. While photons travel from the other side of the universe to Earth, the actual distance from the source to Earth

[15] For Gödel and time travel (Earman 1995, ch. 6; see also Thorne 1994, p. 499, n. 8; Stein 1991) discusses the openness of the future (Malament 1985; Stein 1970) provide two early and cogent keys to the debate. For the physical operation of time machines and time travel (Earman et al. 2009; Smeenk and Wüthrich 2011).

[16] In arguing against the reality of time, Gödel stands in the tradition of Parmenides (McTaggart 1908; see also Dorato 2002; Earman 1995, pp. 194–200). More generally, many might agree that time in general relativity theory becomes simply space or space–time, for which (Yourgrau 1999, pp. 51–52; see also Savitt 1994; Stachel 2007). That has not finished the debate about the reality of time. For questions raised by eternalism and presentism (Putnam 1967; Savitt 2000) and on becoming (Stein 1968). On the question of a preferred notion of time (Belot 2005).

changes over time, increasing (Eddington 1933). Light experiences a redshift as it travels through expanding space.

As for the possibility of rotating solutions that are also expanding: it "is not impossible that our world is a universe of this kind" (Gödel 1949b, p. 206). In fact, in his lecture at the Institute for Advanced Study on 7 May 1949 that we examined earlier in this section (Gödel 1949c), he gave a new solution for the cosmological equations that is rotating and expanding, the result published in 1952 (Gödel 1952). This universe fulfills the cosmological principle, expands, and rotates. It is like the mixmaster universe. It is unstable early in its past, with chaotic oscillations that "rattle back and forth"[17] (Misner 1969, p. 1074).

Let us turn to the second reason:

> [The] mere compatibility with the laws of nature of worlds in which there is no distinguished absolute time, and [in which], therefore, no objective lapse of time can exist, throws some light on the meaning of time also in those worlds in which an absolute time *can* be defined. For, if someone asserts that this absolute time is lapsing, he accepts as a consequence that whether or not an objective lapse of time exists (i.e., whether or not a time in the ordinary sense of the word exists) depends on the particular way in which matter and its motion are arranged in the world. This is not a straightforward contradiction; nevertheless, a philosophical view leading to such consequences can hardly be considered as satisfactory. (Gödel 1949b, pp. 206–207)

If the laws of nature are compatible with a world with no objective lapse of time, then an absolute lapse of time depends on mass out there. Yet, absolute time was to be independent of material conditions and particular events.

4 Antinomy of Time

The purpose of this section is to assess the unsatisfactory situation that the objective lapse of absolute time depends on the arrangement of matter and its motion in the world. The method is to respond to Gödel. The response is to accept the basic philosophical premises that Gödel saw as involved in a satisfactory philosophy, including actualism and realism, and then, by lowering his standards, to maintain the reality of time. If an objective stretch of absolute time independent of the motion of matter is a clear concept that is not the reality, one might call it an ideal that is not mental. The reality is that the order of time is lost with rotation. There are consequences for the independence, objectivity, and beginning of time and of the universe.

Actualism is the "one-world assumption" that "all that can properly be said to exist must exist somewhere in the actual world and that therefore all that we can meaningfully speak about must in the analysis be denizens of the actual world".[18] It is the position "that all our meaningful concepts can in the last analysis be analyzed

[17] This was one in a series of articles by Misner in 1969 opening up the field of quantum cosmology (Coleman et al. 1991).

[18] Hintikka argued that Gödel is an actualist (Hintikka 1998, pp. 13, p. 20; Hintikka 2000, pp. 46–53; see also Kreisel 1980; Feferman 1984). If we distinguish two varieties of actualism (following

so as to [pertain] to our one and only actual world". "What else could there be that we would be interested in talking and thinking about?".

In contrast, "other thinkers believe that even when our aim is to deal with the actual world, we can—and sometimes must—use for this purpose concepts whose meaning involves alternatives to the actual state of affairs or courses of events". Possibilism involves "many-worlds thinking", that thinking in terms of worlds apart from our own. One approach to possible scenarios, for we do not have to imagine an entire world, is to treat them as instruments of thought, while another approach is to treat them as conceptually real.

From almost any perspective, the model that expands marks an advance over the earlier model that did not expand. Unlike the first solution, this one works for our actual universe, if the rate of rotation of our universe is quite slow. Freeman Dyson reported that Gödel continued to check observational data about our universe for a rotational tilt to distant galaxies, which would imply that his model was actually realizable.

Gödel comes to call himself a "mathematical and conceptual realist"[19] (Gödel 1975). Conceptual realism features not only numbers. "Classes and concepts" may "be conceived as real objects ... existing independently of our definitions and constructions".

> It seems to me that the assumption of such objects is quite as legitimate as the assumption of physical bodies and there is quite as much reason to believe in their existence. They are in the same sense necessary to obtain a satisfactory system of mathematics as physical bodies are necessary for a satisfactory theory of our sense perceptions.[20] (Gödel 1951, p. 137)

The reality of objects, classes, and of an independent conceptual order, and the rejection of the objectivity of time, taken together, elicits the obvious question about our access to this reality. A crux for Gödel's philosophy is the dependence on intuition and perception to access mathematical reality (Atten and Kennedy 2003). With actualism, the commitment to large-scale intuition becomes more pressing, otherwise, "how are we supposed to come to know the abstract Platonic region of our world?" (Hintikka 2000, p. 50).

The position of Kant's First Critique is that space and time are pure forms of intuition, transcendentally ideal and empirically real. That is the starting point, a basis from which the organization of the First Critique proceeds with two constitutive conditions, first that space is a form of outer intuition and second that time is a form of inner intuition. These are to explain synthetic a priori knowledge of space and time

---

Contessa 2010), his actualism is "hardcore". We mentioned that a rotating structure for space–time does not have an objective temporal order globally (Barrett et al. 2004; Gödel 1952). For the conditions for the theoretical existence of the Gödel universe (Clifton and Barrow 2006, I–III; see also Hintikka 1998, p. 14; Hintikka 2000, p. 47).

[19] The Grandjean material (Gödel 2003, pp. 441–450). For conceptual realism, see the Gibbs lecture (1951). He approved of Russell's realism (Gödel 1944; see also Martin 2005, pp. 207–213; Parsons 1995; Belot 2005). In particular (Gödel 1947, p. 520; 1951, pp. 128–130; 1964, p. 264; 1972; compare also Frege 1891, p. 13; 1892, p. 34).

[20] For apt statements of his philosophy, Gödel will recommend his own (Gödel 1964) and Wang's (1974; see also Feferman 1984, 1986).

and of objects in space and time.[21] One wants to ask how to reach the goal like that, how to get to the infinite from the finite. For a physical analogue, a rotating universe is akin to the idea of reducing infinite space to finite space. The possibility of the endless division of space *in indefinitum,* of continuing indefinitely, is not enough to secure a priori knowledge of the concept of infinity *in infinitum.* That gap in knowing jeopardizes the synthetic method of the First Critique; of course, in many cases, a rule will allow one to proceed constructively; however, this is not adequate for every case.

"Gödel cannot adhere to Kant's view" (Gödel 1949b, p. 200):

> [...] so far as Kantian philosophy is concerned, the principle analogy that Gödel has in mind between it and relativity theory concerns the strong sense in which temporal properties become (in general) well defined only relative to certain structures within the world: the world-lines of bodies.[22] (Gödel 1990, pp. 200–201)

What is our interface with the rotation of universes? Do we not just need a bigger, better telescope?[23] Some have used the strong telescope metaphor to argue that possible worlds are not real but instrumental, but it also poses a challenge for the actualist. "Yes, we need a much bigger telescope" is not a convincing answer in this scenario. A better response is to point out that we do not need a bigger telescope to settle the question of universal rotation. "It can in fact be settled by present means of observation" [...] "of mean radial velocities of galaxies located in different regions of the sky" (Gamow 1946, p. 549).

Appeals to possibility and necessity are popular, but unnecessary. Even "if our Universe were the interior of a black hole existing in a bigger universe" (Popławski 2010, p. 113), that still does not require a possible state of affairs. It requires experiment by an observer entering or residing in the interior of a black hole to determine the radial geodesic motion of a massive particle as it falls in. Due to the laws of general relativity, an observer under these circumstances can never send word out. Even so, the motion of a massive particle as it falls into a black hole is determined at a distance due to the Raychaudhuri fundamental equation.

If one does not appeal to a possible state of affairs in order to explain the rotation of universes, then one seems to make a deeper claim about the objectivity of time. At least one philosopher has been.

---

[21] Kant ([1781–1787] 1956, A38–39/B55–56; A508–515/B536–543; A523–527/B551–555), Ryckman (2011) see also Parsons (1964, p. 109).

[22] The two versions of "Some Observations about the Relationship between Theory of Relativity and Kantian Philosophy" (Gödel 1995, pp. 202–260), which Gödel wrote in preparation for his (Gödel 1949b), throw added light on the issues of temporality and causation. For the causal relation and that which it relates (Schaffer 2016). We are not discussing causation or its metaphysics, but are instead repeating the well-known and often discussed connection between the concept of causality and general relativity without attempting to solve any of the well-known problems.

[23] The speed of light restricts our observations in the actual universe. Parallel universes and counterfactual situations are not entirely like possible worlds, although there are working assimilations. Lewis reduces possibilia to counterfactuals (Lewis 1968, 1986, ch. 1), while Kripke reduces counterfactuals to possible worlds (Kripke [1972]1980, p. 6, pp. 76–78). Possibilities are instruments of thought for Kripke but are real for Lewis.

> [...] moved to laughter at the thought of how *presumptuous* it would be to reject mathematics for philosophical reasons. [...] Can you tell them, with a straight face, to follow philosophical argument wherever it may lead? If they challenge your credentials, will you boast of philosophy's other great discoveries: that motion is impossible [...] that time is unreal. (Lewis 1991, p. 59)

An implicit suggestion in this passage is that it would be presumptuous to reject the objectivity of time. Physics itself does not bar other ways of looking at motion and time but to the contrary takes these into account (Wang 1995). Above, we have seen a series of fundamental cosmological discoveries predicated upon the studied rejection of rotating solutions. As with chiaroscuro in painting, heightened contrast clarifies differing aspects. One understands the contrast with an alternative state of affairs of geodesic completeness via *negativa*, as one understands the hill and the dale. Proving theorems gives rise to cosmological knowledge of geodesic completeness and violation of causality conditions. That constitutes a strict minimum case. Knowledge of singularities includes at the minimum the singularity theorems themselves, what they state, what they prove. The proofs of these theorems, while ideal, are empirical objects, although obviously not in Mill's sense. Raychaudhuri, once again, studied the rotating solution in order to understand gravitation.

There are no breakdowns in causality in our world, and, in that sense, any stretch of time is objective. There is no travel to regions that are past for us, a claim which depends on the fact that our universe does not rotate. If the motion of matter is a characteristic of our universe, then a change in the motion of matter changes the universe. That means that time depends on a particular material distribution and its motion, on whether or not the universe rotates. If this summarizes our best current understanding, we conclude that time depends on the motion of matter. A stretch of time is, to put it crudely, material.

Put more carefully, there is an unresolved standoff in the debate over the reality of an objective stretch of absolute time. A rotating universe is g-complete. It entails a breakdown in the causal order, time-like lines that return back again to a previous space–time point. We are unable to take such a time sequence of events and uniquely assign an order to it, unlike in the actual universe, which is g-incomplete. If one assumes there is no causality other than that which the laws of nature provide, then a precondition for every occurrence in the world is a prior situation from which the occurrence follows by a law of nature. This prior situation had no earlier permanence. If it had, then the consequent state would not have just occurred. A precondition for the prior situation, then, is a situation prior to it. According to the laws of nature, there is no first beginning; yet, the universal expansion is a visible sign of the big bang. Here we have a cause that is itself not caused. It cannot be, because there is no point in the temporal order that precedes it.

On the other hand, assume a cause that has no cause, which itself functions as the kind of causality from which effects in the world would indeed follow, that is, the capacity to initiate a worldly state and a causal series of effects that arise solely through this beginning. The series of causes begins absolutely on this assumption. Such a beginning of a series implies that a situation arises that is in no way caused by the previous one. Nature provides the example of the big bang, and yet such "freedom" as this in turn depends on the laws of nature insofar as big bang explains universal expansion. The big bang provides evidence that, while nature everywhere shows us causes that, in turn, are themselves caused, further investigation reveals that the causal network has a peculiar feature. There are singularities in space–time, but these do not lead to breaks in the sequence of events from our position in the middle of things, because the backwards chain of causation is closed by the big bang but unbounded as it approaches that limit. Scientists now have information about very early times in universal history. The big bang is evidence of an earlier uncaused state that leads to our current universal state, and yet this state is not reachable by reversing the causal process (Kant [1781–1787] 1956, A444/B472–A449/B477). Singularities, which are not themselves somehow linked in the causal chain, are characteristic of our universe. Our universe avoids breakdowns in causality, but there are particles in freefall for infinite lengths of time. In a g-complete universe, there are no particles in infinite freefall. We conclude that causation is independent of the theory of general relativity.

What about the beginning of the universe? Concerning it, we have a mathematical standoff. There are homogeneous, isotropic, expanding models of the universe that satisfy general relativity yet disagree on rotation. Some of them avoid it while others do it. The ones that do it have no singularities, so no big bang, no final state, no beginning or end to these universes. Universes that are not b-complete do not rotate. The question about the universe having a beginning or not having a beginning depends on whether distant galaxies are rotating with respect to the inertial frame, a question that observation will answer.

In other words, the sudden discovery that distant galaxies are rotating with respect to the inertial frame, after all, would change our view of time, at least in the narrow sense of time that we have explored. We have looked into the motion of matter and rotation. If it is the actual motion of matter that determines whether the universe had a beginning, then that leaves much to wonder about time.

5 Conclusion

Earman argued the dispute between two accounts of space, time, and motion remains undecided. I have not repeated that argument, but wanted to add something to it. The theory of general relativity does not rule out rotating structures in which there are causal breakdowns. Instead, it is compatible with them, as it is with the structure of our actual universe, which does not rotate. The result is that the existence of an absolute stretch of objective time is independent of the theory of general relativity. This is so whether the rotating structures are real or instrumental. Since some of these rotating structures satisfy the cosmological principle and expand, the motion of matter determines time's objectivity.

Clearly, theoretical solutions to physical problems do not reveal what was essential about an original purported entity. Instead, a theoretical formulation reveals the state of our knowledge. The theory tells us what we know. It guides and informs us about the world. Even so, if the theoretical entities did not exist, then it would be hard to make sense of any claim one might make to have a theory that provides knowledge of them.

Acknowledgements For an earlier related project, thanks to Raul de la Fuente Marcos, John Stachel, Kip S. Thorne, Walter Johnson, Daniel Kennefick, Akihiro Kanamori, Charles Parsons, John D. Barrow, Wolfgang Rindler, Gregory H. Moore, Michael Stoeltzner, Karl Svozil, and Gerhard Heinzmann. Thanks also to Richard H. Jandovitz, Alex Taylor, Azul Pinochet-Barros, and Nicole T. Russell. For this paper, thanks to Raffaele Pisano, Trevor Link, Brian Kiniry, and Joseph X. Donovan. Thanks to Aaron Brode for the figures. Errors are mine.

References

Arntzenius F (2012) Space, Time, and Stuff. Oxford University Press, New York.

Atten van M, Kennedy J (2003) On the philosophical development of Kurt Gödel. The Bulletin of Symbolic Logic 9:425–476.

Barbour J, Pfister H (eds) (1995) Mach's Principle: From Newton's Bucket to Quantum Gravity. Birkhauser, Boston.

Barrett MD, Chiaverini J, Schaetz T, Britton J, Itano WM, Jost JD, Knill E, Langer C, Leibfried D, Ozeri R, Wineland DJ (2004) Deterministic quantum teleportation of atomic qubits. Nature 429:737–739.

Barrow JD, Tsagas CG (2004) Dynamics and stability of the Gödel universe. Retrieved 26 January 2022 from Cornell University Library Web site: http://arxiv.org/abs/gr-qc/0308067.

Bell JS (1964) On Einstein Podolsky Rosen paradox. Physics 1:195–200.

Belot G (2005) Dust, time, and symmetry. British Journal for the Philosophy of Science 56:255–291.

Bennett CH (1993) Teleporting an unknown quantum state via dual classical and Einstein-Podolsky-Rosen channels. Physical Review Letters 70:1895–1899.

Brassard G (1996) Teleportation as a quantum computation. Retrieved 26 January 2022 from Cornell University Library Web site: https://arxiv.org/abs/quant-ph/9605035.

Chandrasekhar S, Wright JP (1961) The geodesics in Gödel's universe. Proceedings of the National Academy of the Sciences of the USA 47:341–347.

Ciufolini I, Wheeler JA (1995) Gravitation and Inertia. Princeton University Press, Princeton.

Clifton T, Barrow JD (2006) The existence of Gödel, Einstein, and de Sitter universes. Retrieved 27 January 2022 from Cornell University Library Web site: https://arxiv.org/abs/gr-qc/0511076.

Coleman S, Hartle JB, Piran T, Weinberg, S (1991) (eds) Quantum Cosmology and Baby Universes. World Scientific. River Edge, NJ.

Contessa G (2010) Modal truthmakers and two varieties of actualism. Synthese 174:34–353.

D'Agostino S (2000) A History of the Ideas of Theoretical Physics. Essays on the Nineteenth and Twentieth Century Physics. Kluwer, Boston.

Descartes R ([1644] 1983) Principles of Philosophy. Translation by Miller VR and Miller RP. Reidel, Dordrecht.

Disalle R (2020) Space and time: Inertial Frames. Stanford Encyclopedia of Philosophy, Zalta EN (ed). Retrieved 23 February 2022 from https://plato.stanford.edu/entries/spacetime-iframes/.

Dorato M (2002) On becoming, cosmic time and rotating universes. Royal Institute of Philosophy Supplement 50:253–276.

Earman J (1995) Bangs, Crunches, Whimpers, and Shrieks. Singularities and Acausalities in Relativistic Spacetimes. Oxford University Press, New York.

Earman J, Smeenk C, Wüthrich C (2009) Do the laws of physics forbid the operation of time machines? Synthese 169:91–124.

Eddington A (1933) The Expanding Universe. Macmillan, New York.
Einstein A (1914) Die formale Grundlagen der allgemeinen Relativitätstheorie. Königlich Preussische Akademie der Wissenschaften Berlin. Sitzungsberichte:1030–1085. Reprinted as Document 9 [Einstein 1996, pp. 72–130].
Einstein A (1916) Die grundlage der allgemeinen relativitätstheorie. Annalen der Physik 49:769–822. Document 30 [Einstein 1996, pp. 283–339].
Einstein A (1996) The Collected Papers of Albert Einstein. Volume 6: The Berlin Years: Writings, 1914–1917. In: Klein MJ, Kox AJ, Schulman R (eds). Princeton University Press, Princeton.
Einstein A, Podolsky B, Rosen N (1935) Can quantum-mechanical description of physical reality be considered complete? Physical Review 47:777–780.
Ellis GFR (2001) Contributions of K. Gödel to relativity and cosmology. In: Hájek (ed). Gödel '96. Logical Foundations of Mathematics, Computer Science, and Physics—Kurt Gödel's Legacy. AK Peters, Natick, pp. 34–49.
Emery N, Markosian N, Sullivan M (2020) Time. Stanford Encyclopedia of Philosophy. Zalta EN (ed). Retrieved 8 February 2022 from https://plato.stanford.edu/entries/time/.
Feferman S (1984) Kurt Gödel: Conviction and caution. Philosophia Naturalis 21:546–562.
Feferman S (1986) Gödel's life and work [Gödel 1986, pp. 1–36].
Fermi E (1932) Quantum theory of radiation. Reviews of Modern Physics 4:87–132.
Frege G (1891) Funktion und begriff. [Reprinted Frege 1967, pp. 125–142].
Frege G (1892) Über sinn und bedeutung. Zeitschrift für Philosophie und philosophische Kritik (n.s.) 100:25–50. [Reprinted Frege 1967, pp. 143–162].
Frege G (1967) Kleine Schriften. Angellini (ed.). Georg Olms, Hildesheim.
Gamow G (1946) Rotating Universe? Nature 158:549.
Gödel K (1930) The completeness of the axioms of the functional calculus of logic. [Translation Gödel 1986, pp. 100–123].
Gödel K (1931) On formally undecidable propositions of the Principia Mathematica and related systems I. [Translation Gödel 1986, pp. 144–195].
Gödel K (1938) The consistency of the axiom of choice and of the generalized continuum hypothesis. [Reprinted Gödel 1990, pp. 26–27].
Gödel K (1944) Russell's mathematical logic. 3rd ed. [Reprinted, Gödel 1990, pp. 102–141].
Gödel K (1947) What is Cantor's continuum problem? American Mathematical Monthly 54:515–525; errata 55:151. [Reprinted, Gödel 1990, pp. 176–187].
Gödel K (1949a) An example of a new type of cosmological solutions of Einstein's field equations of gravitation. Reviews of Modern Physics 21:497–450. [Reprinted, Gödel 1990, pp. 189–198].
Gödel K (1949b) A remark about the relationship between relativity theory and idealistic philosophy. [Reprinted, Gödel 1990, pp. 199–207].
Gödel K (1949c) Lecture on Rotating Universes. [Reprinted, Gödel 1995, pp. 261–289].
Gödel K (1951) Some basic theorems on the foundations of mathematics and their implications. The Gibbs lecture [Gödel 1995, pp. 290–323].
Gödel K (1952) Rotating universes in general relativity theory. Proceedings of the International Congress of Mathematicians: Cambridge, Massachusetts, U.S.A. August 30–September 6, 1950, I. American Mathematical Society, Providence, pp. 175–181. [Reprinted, Gödel 1990, pp. 208–216].
Gödel K (1964) Revised and expanded version of Gödel 1947. [Gödel 1990, pp. 254–270].
Gödel K (1972) On an extension of finitary mathematics which has not yet been used. [Gödel 1990, pp. 271–280].
Gödel K (1975) Letter to Grandjean of 19 August 1975 and Grandjean's questionnaire. [Gödel 2003, pp. 443–449].
Gödel K (1986) Collected Works, Volume I: Publications 1929–1936. Feferman S, et al. (eds). Oxford University Press, New York.
Gödel K (1990) Collected Works, Volume II: Publications 1938–1974. Feferman S, et al. (eds). Oxford University Press, New York.

Gödel K (1995) Collected Works, Volume III: Unpublished Essays and Lectures. Feferman S, et al. (eds). Oxford University Press, New York.
Gödel K (2003) Collected Works, Volume IV: Correspondence A-G. Feferman S, et al. (eds). Oxford University Press, New York.
Gott III JR (1991) Closed timelike curves produced by pairs of moving cosmic strings: exact solutions. Physical Review Letters 66:1126–1129.
Hawking SV (1967) The occurrence of singularities in cosmology. III. Causality and singularities. Proceedings of the Royal Society of London. Series A 300:187–201.
Hawking SV, Ellis GFR (1973) The Large-Scale Structure of Space-Time. Cambridge University Press, Cambridge.
Hawking SV, Penrose R (1970) The singularities of gravitational collapse and cosmology. Proceedings of the Royal Society of London. Series A 314:529–548.
Heckmann O, Schücking E (1955) Bemerkungen zur Newtonschen Kosmologie. I. Zeitschrift für Astrophysik 38:95–109.
Heckmann O, Schücking E (1956) Bemerkungen zur Newtonschen Kosmologie. II. Zeitschrift für Astrophysik 40:81–92.
Hegerfeldt G (1994) Causality problems for Fermi's two atom system. Physical Review Letters 72:596–599.
Hegerfeldt G (1997) Problems about causality in Fermi's two-atom model and possible resolution. Retrieved 27 January 2022, from Cornell University Library Web site: https://arxiv.org/abs/quant-ph/9707016.
Hintikka J (1998) On Gödel's philosophical assumptions. Synthese 114:13–23.
Hintikka J (2000) On Gödel. Wadsworth, Belmont.
Hoefer C (2000) Kant's hands and Earman's pions: Chirality arguments for substantival space. International Studies in the Philosophy of Science 14:237–256.
Hoefer C, Huggett N, Read J (2021) Absolute and relational space and motion: Classical theories. Stanford Encyclopedia of Philosophy. Zalta EN (ed). Retrieved 26 January 2022 from https://plato.stanford.edu/entries/spacetime-theories-classical/.
Hubble E (1934) The distribution of extra-galactic nebulae. Astrophysical Journal 79:8–76.
Huggett N, Hoefer C, Read J (2021) Absolute and relational space and motion: Post-Newtonian theories. Stanford Encyclopedia of Philosophy. Zalta EN (ed). Retrieved 26 January 2022 from https://plato.stanford.edu/entries/spacetime-theories/
Janssen M (2004) Relativity. New Dictionary of the History of Ideas. Horowitz MC et al. (eds). Charles Scribner's Sons, New York.
Kant I ([1781–1787] 1956) Kritik der reinen Vernunft. Felix Meiner, Hamburg.
Kreisel G (1980) Kurt Gödel, 28 April 1906–14 January 1978. Biographical Memoirs of Fellows of the Royal Society 26:148–224. Corrections 27:697; 28:718.
Kripke S ([1972]1980) Naming and Necessity. Harvard University Press, Cambridge.
Leibniz GW (1969) Philosophical Papers and Letters, Loemker LE (ed). 2nd ed. Reidel, Boston.
Lewis D (1968) Counterpart theory and quantified modal logic. The Journal of Philosophy 65:113–126.
Lewis D (1986) On the Plurality of Worlds. Blackwell, New York.
Lewis D (1991) Parts of Classes. With an appendix by Burgess J, Hazen A, Lewis D. Basil Blackwell, Oxford.
Maddox J (1994) Time machines still over horizon. Nature 367:509.
Malament D (1977) The class of continuous timelike curves determines the topology of spacetime. Journal of Mathematical Physics 18:1399–1404.
Malament D (1985) Time travel in the Gödel universe. In Asquith PD, Kitcher P (eds). PSA: Proceedings of the 1984 Biennial Meeting of the Philosophy of Science Association. Philosophy of Science Association, East Lansing, pp. 91–100.
Martin DA (2005) Gödel's conceptual realism. The Bulletin of Symbolic Logic 11:207–213.
McTaggart JME (1908) The unreality of time. Mind (n.s.) 17:457–474.
Misner C (1969) Mixmaster universe. Physical Review Letters 22:1071–1074.

Morris M, Thorne K, Yurtsever U (1988) Wormholes, time machines, and the weak energy condition. Physical Review Letters 61:1446–1449.
Newton I (1687) Philosophiae Naturalis Principia Mathematica. Autore IS. Newton, Trin. Coll. Cantab. Soc. Matheseos Professore Lucasiano, & Societatis Regalis Sodali. Imprimatur. S. Pepys, Reg. Soc. Præses. Julii 5. 1686. Londini, Jussu Societatis Regiæ ac Typis Josephi Streater. Prostat apud plures Bibliopolas. Anno MDCLXXXVII.
Norton JD (2019) The hole argument. Stanford Encyclopedia of Philosophy, Zalta EN (ed). Retrieved 26 January 2022 from https://plato.stanford.edu/entries/spacetime-holearg/.
Parsons C (1964) Infinity and Kant's conception of the "possibility of experience". [Reprinted] Mathematics in Philosophy: Selected Essays. Cornell University Press, Ithaca, pp. 95–109.
Parsons C (1995) Platonism and mathematical intuition in Kurt Gödel's thought. The Bulletin of Symbolic Logic 1:44–74.
Penrose R (1965) Gravitational collapse and space-time singularities. Physical Review Letters 14:57–59.
Popławski NJ (2010) Radial motion into an Einstein-Rosen bridge. Physics Letters B 687:110–113.
Putnam H (1967) Time and physical geometry. The Journal of Philosophy 64:240–247.
Raychaudhuri A (1955) Relativistic cosmology. I. Physical Review 98:1123–1126.
Raychaudhuri A (1957) Relativistic and Newtonian cosmology. Zeitschrift für Astrophysik 43:161–164.
Raychaudhuri A (1979) Theoretical Cosmology. Clarendon, Cambridge.
Reichenbach H ([1928] 1958) The Philosophy of Space and Time. Translation by Reichenbach M and Freund J. Dover, New York.
Riebe A, Häffner H, Roos CR, Hänsel W, Benhelm J, Lancaster GP, Körber TW, Becher C, Schmidt-Kaler F, James DF, Blatt F (2004) Deterministic quantum teleportation with atoms. Nature 429:734–747.
Rindler W (2009) Gödel, Einstein, Mach, Gamow, and Lanczos: Gödel's remarkable excursion into cosmology. American Journal of Physics 77(6):498–510.
Ryckman TA (2011) What does history matter to philosophy of physics? Journal of the Philosophy of History 5:496–512.
Ryckman TA (2021) Early philosophical interpretations of general relativity. Stanford Encyclopedia of Philosophy. Zalta EN (ed). Retrieved 29 January 2022 from https://plato.stanford.edu/archives/fall2021/entries/genrel-early/.
Rynasiewicz R (2011) Newton's views on space, time, and motion. Stanford Encyclopedia of Philosophy. Zalta EN (ed). Retrieved 26 January 2022 from https://plato.stanford.edu/entries/newton-stm/.
Rynasiewicz R (2018) Newton's Scholium on time, space, place and motion. [Preprint]. Retrieved 11 March 2022 from http://philsci-archive.pitt.edu/id/eprint/15525.
Savitt S (1994) The replacement of time. Australasian Journal of Philosophy 72:463–474.
Savitt S (2000) There's no time like the present (in Minkowski spacetime). Philosophy of Science 67–Proceedings:S563–S574.
Schaffer J (2016) The metaphysics of causation. Stanford Encyclopedia of Philosophy. Zalta EN (ed). Retrieved 3 February 2022 from https://plato.stanford.edu/entries/causation-metaphysics/.
Schilpp PA (ed.) (1949) Albert Einstein, Philosopher-Scientist. Library of Living Philosophers, Evanston.
Smeenk C, Wüthrich C (2011) Time travel and time machines. In Callender (ed). Oxford Handbook of Philosophy of Time. Oxford University Press, New York, pp. 577–630.
Stachel J (2005) Structure, individuality, and quantum gravity. Retrieved 27 January 2022 from Cornell University Library Web site: https://arxiv.org/abs/gr-qc/0507078.
Stachel J (2007) Review of Yourgrau, A World Without Time: The Forgotten Legacy of Gödel and Einstein. Notices of the American Mathematical Society 54:861–868.
Stachel J (2014) The hole argument and some physical and philosophical implications. Living Reviews in Relativity 17(1). Retrieved 26 January 2022 from https://doi.org/10.12942/lrr-2014-1.

Stein H (1968) On Einstein-Minkowski space-time. The Journal of Philosophy 65:5–23.
Stein H (1970) On the paradoxical time-structures of Gödel. Philosophy of Science 37:589–601.
Stein H (1991) On relativity theory and the openness of the future. Philosophy of Science 58:147–167.
Stein H (2009) "Definability," "conventionality," and simultaneity in Einstein-Minkowski Space-Time. In Myrvold WC, Christian J (eds). Quantum Reality, Relativistic Causality, and Closing the Epistemic Circle (Essays in Honor of Abner Shimony). Springer, Dordrecht, pp. 403–442.
Svozil K (2011) Physical unknowables. In Baaz M, et al (eds). Kurt Gödel and the Foundations of Mathematics. Cambridge University Press, Cambridge, pp. 213–251.
Thorne KS (1994) Black Holes and Time Warps: Einstein's Outrageous Legacy. Norton, New York.
Ursin R, Jennewein T, Aspelmeyer M, Kaltenbaek R, Lindenthal M, Walther P, Zeilinger A (2004) Quantum teleportation across the Danube. Nature 430:849.
Wang H (1974) From Mathematics to Philosophy. Humanities, New York.
Wang H (1995) Time in philosophy and physics: From Kant and Einstein to Gödel. Synthese 102:215–234.
Yourgrau P (1999) Gödel Meets Einstein. Time Travel in the Gödel Universe. Expanded ed. Open Court, Chicago.

Chapter 13
Lagrange's Method and *Lagrangian*'s Mechanism in Maxwell's *A Treatise on Electricity and Magnetism* (1873)

Raffaele Pisano, Donatella Marmottini, and Andrea Durlo

Abstract James Clerk Maxwell advanced Electromagnetic theory, while also mathematically enlarging Michael Faraday's works. He declined Newtonian mechanics for the formulation of new conceptual frameworks because more adequate for describing a field of new phenomena such as electromagnetism; so he also went beyond the Newtonian paradigm introducing a novelty in the relationship physics–mathematics based on new magnitudes, e.g., electric charge and energy instead of mass and force. However, through exactly which terms were these advances made? In particular, in *A Treatise on Electricity and Magnetism, II, Part IV, Chapter IV–VII* (1873) Maxwell aimed to formulate a dynamical justification for field equations; he focused on the fact that magnetic field appeared a complete kinetic energy system. In this paper, we analyse and describe both: (a) what he called, Lagrange's *method of reducing the ordinary dynamical equations* in *Mécanique analytique* (1788)–-including—three methods used by Maxwell for *expressing the kinetic energy—detecting the existence of the terms of the form* $T_{me,}$ and (b) the application of the Lagrangian formulating through the *equations of motion of a connected system.*

Keywords Faraday · Lagrange · Lagrange's Method · Lagrangian · Maxwell · Maxwell's Equations · *Mécanique analytique* · Relationship Physics–Mathematics

---

R. Pisano (✉) · A. Durlo
HOPAST at IEMN, Lille University, Villeneuve d'Ascq, France
e-mail: raffaele.pisano@univ-lille.fr

D. Marmottini
Campus dei Licei, Cisterna di Latina, Italy

R. Pisano (ed.), *A History of Physics: Phenomena, Ideas and Mechanisms*,
History of Mechanism and Machine Science 42,
https://doi.org/10.1007/978-3-031-26174-9_13

1 An Outline

1.1 Looking for a Suitable Mathematics in Electromagnetism

Both in *A Dynamical Theory of the Electromagnetic Field* (read 1864 and published 1865; Maxwell 1865; Maxwell [[1865] 1996][1] and in *A Treatise on Electricity and Magnetism* (1873) (Maxwell 1873, II, Part IV, Chaps. IV–V) James Clerk Maxwell (1831–1879) advanced in his Electromagnetic theory also enlarging—from a mathematical standpoint—Michael Faraday's (1791–1867) experimental works (without the use of mathematics; D'Agostino 2000; Faraday 1839–1855; Pisano 2014; Pisano et al. 2021a; Fig. 1). This new scientific approach gives rise to reasonable questions: *what was the original concept of energy that Maxwell had to deal with? What kind of mathematics did he use for inscribing his physics into electromagnetic theory?*

Despite holding numerous scientific interests evidenced by several publications (Maxwell 1860, 1871, 1874, 1879, 1881, [1876–1877] 1920; D'Agostino 2013), *A Treatise on Electricity and Magnetism* is most probably Maxwell's masterpiece. Maxwell formulated an electromagnetic theory that included new mathematical models beyond a Newtonian approach thereby achieving an innovation in the relationship physics–mathematics based on new quantities, such as e.g., electric charge and energy instead of mass and force. In fact, he avoided any fundamental argument on the physical–mathematical nature of force, focusing on the concept of energy (which is a result of an integral). He proposed theoretical arguments to describe the basic phenomena of electricity and magnetism. He directed attention to the formulation of the potential and mutual energy of two interacting currents. His theory in electricity and magnetism has a double formulation: a mechanical and an electromagnetic one. Within the "physics mathematics" domain (see below) Maxwell formulated Lagrangian mathematics in electromagnetic theory (Maxwell 1873, II) in order to clarify the complex structure of a natural phenomenon, e.g., the pattern of a field. In particular, he also used analogies in the mechanics of connected systems described by Joseph-Louis Lagrange (1736–1813) in his equations (Lagrange 1788, 1892; Lagrange 1870–1873). However, through what exact scientific—and—methodological relations were these changes described and effected? In the following, we present an intellectual path taken by Maxwell from Faraday's studies to a Lagrangian formulation in his electromagnetic theory.

[1] The paper was received on October 27, 1864, and read on December 8, 1864. In 1865, the abstract was published in The *Proceedings of the Royal Society of London* (Maxwell 1864) and the paper in the *Philosophical Transactions of The Royal Society of London* (Maxwell 1865). On the subject and in the field see Pisano's works (Pisano and Marmottini 2017; Pisano 2013; Pisano, *Faraday and Maxwell, Two Lives, One Theory, pre-print*).

1.2 Looking for Working Hypotheses—and—Objectives

Both Faraday (1831) and Joseph Henry (1797–1878) (1832) studied electromagnetic induction (Faraday 1839–1855, I, Series IX, XI). In *On the Induction of a Current on Itself* (Maxwell 1873, II, Part IV, Chap. IV, pp. 180–183) Maxwell reminded the reader of electromagnetic induction and self–induction as previously described by Faraday (Maxwell 1873, II, Part IV, Chap. III, pp. 172–173). He stated his dissatisfied evaluation of, and reasonings on the matter.

> 540.] The conception of such a quantity, on the changes of which, and not on its absolute magnitude, the induction current depends, occurred to Faraday at an early stage of his researches*[*Exp. Res.*, series i. 60.]. He observed that the secondary circuit, when at rest in an electromagnetic field which remains of constant intensity, does no shew any electrical effect, whereas, if the same state of the field had been suddenly produced, there would have been a current. (Maxwell 1873, II, Part IV, Chap. IV, p. 173; Author's "*" reference and italics)

Then he suggested a correlated analogy on the subject between a *current in a wire* and a *fluid flowing in a pipe* through the idea of inertia[2] and momentum. Analysing in detail the two cited phenomena in his authoritative work *A Treatise on Electricity and Magnetism* (hereafter *Treatise*; Fig. 2) Maxwell emphasized that this analogy–parallelism cannot survive when electromagnetic induction effects are considered (Maxwell 1873, II, Part IV). In fact, while in the motion of the fluid in a pipe only internal characteristics of the tube are involved, in the case of wires crossed by the same current, behaviour is changed both by changes in shape and by the presence of bodies outside the wire.

In the whole of Chap. IV (Maxwell 1873, II) Maxwell dealt with energy and made yet another interesting comparison between mechanical and electrical phenomena. As a conventional force increases the energy of the body to which it is applied, so a current leads to effects that produce a certain type of energy:

> 548.] These effects of the inertia of the fluid in the tube depend solely on the quantity of fluid running through the tube, on its length, and on its section in different parts of its length. They do not depend on anything outside the tube, nor on the form into which the tube may be bent, provided its length remains the same.
>
> [...]
>
> 549.] These results shew clearly that, if the phenomena are due to momentum, the momentum is certainly not that of the electricity in the wire, because the same wire, conveying the same current, exhibits effects which differ according to its form; and even when its form remains the same, the presence of other bodies, such as a piece of iron or a closed metallic circuit, affects the result. (Maxwell 1873, II, Part IV, Chap. IV, p. 181)

Hence, Energy & Momentum (mechanical phenomena) are inserted into the new Electricity (electrical phenomena). Maxwell, in support of his statement, described effects related to currents that lead to energy production in different forms—such as

---

[2] Cfr. On the subject, see the Chapter Bussotti and Pisano's *Details in Galileo's Mechanism of Inertia as New Intellectual Historical Insights* (Bussotti and Pisano 2024); below in this book, Part 3.

the Joule effect/mechanical work[3] (Gillispie and Pisano 2014; Joule 1845, [1849] 1850]; Pisano et al. 2021a, b)—resulting in the movement of magnets (Joule 1838, 1839, 1841a, b, 1842, 1843, 1884). At the end of Chap. IV, the working hypothesis—and—objectives were declared as established in agreement with Langrage's method in *Mécanique analitique* (1788; Fig. 3) (Lagrange 1788; Pisano and Marmottini 2017; Francaviglia 1991) and Faraday's studies in *Experimental Researches in Electricity* (1839–1855) (Figs. 1 and 2):

> 552.] What I propose now to do is to examine the consequences of the assumption that the phenomena of the electric current are those of a moving system, the motion being communicated from one part of the system to another by forces, the nature and laws of which we do not yet even attempt to define, because we can eliminate these forces from the equations of motion by the method given by Lagrange for any connected system. (Maxwell 1873, II, Part IV, Chap. IV, p. 183)

Therefore, he aimed to:

- Analyse the consequences of the hypothesis that the phenomena tied to the electric current are those of a system in motion in which the motion is communicated from one side to the other of the system through forces;
- Apply Lagrange's method and its formalisms for connected systems, making it effectively unnecessary to explain both the nature and the laws governing these forces, because the resulting contributions are eliminated in the formulation of the equations of motion deduced by Lagrange.

[3] James Prescott Joule's (1818–1889) bicentenary took place in 2018. The *Inter-Divisional Teaching Commission* (IDTC of DHST/DLMPST) commemorated (RP was the elected President of the IDTC) and organized *Symposium—James Joule's Bicentenary: Scientific and Pedagogical Issues Concerning Energy Conservation* at the European Society for the History of Science Conference (ESHS & BSHS), 14th–17th September, 2018 in London. Consequently, a Special Issue on *Joule's Bicentenary* was guest edited by one of us (RP) and published by Springer (Pisano et al. 2020).

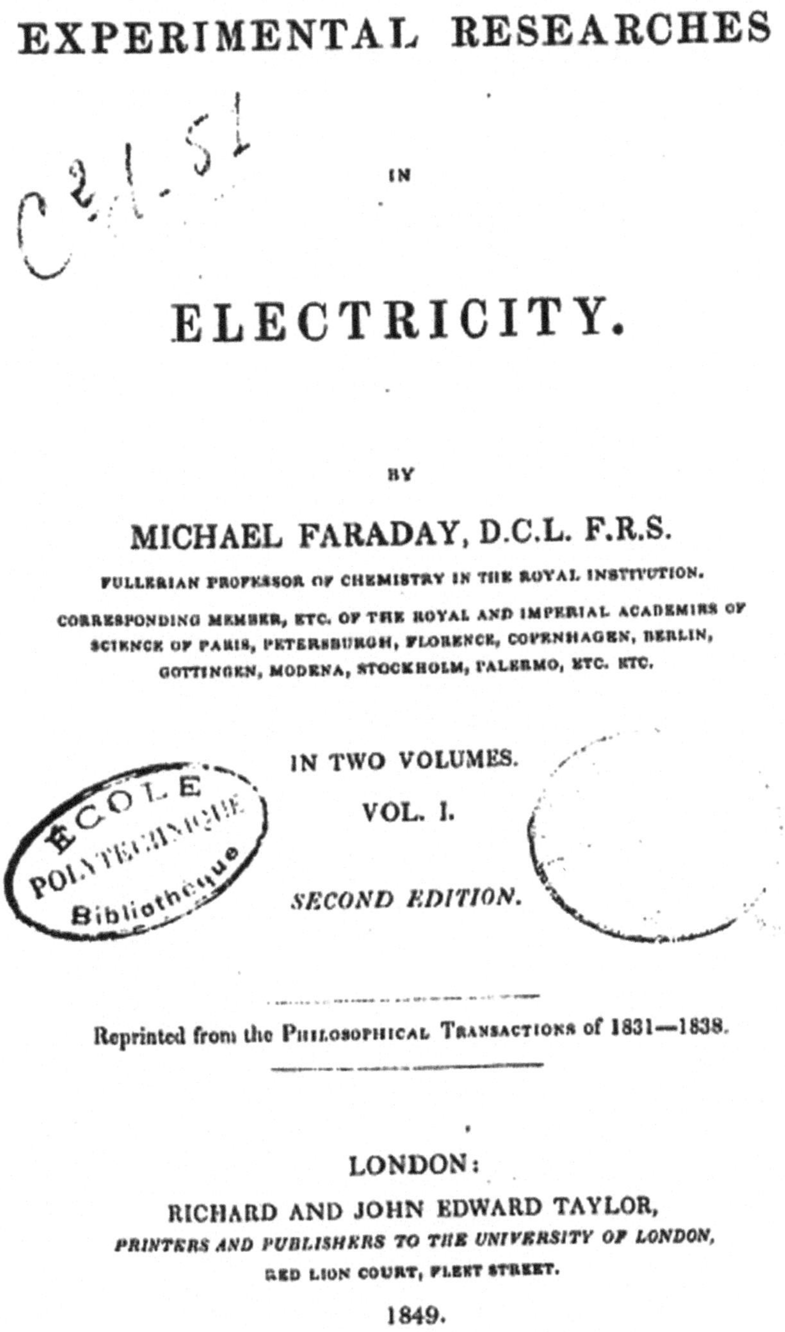
EXPERIMENTAL RESEARCHES

IN

ELECTRICITY.

BY

MICHAEL FARADAY, D.C.L. F.R.S.

FULLERIAN PROFESSOR OF CHEMISTRY IN THE ROYAL INSTITUTION.

CORRESPONDING MEMBER, ETC. OF THE ROYAL AND IMPERIAL ACADEMIES OF SCIENCE OF PARIS, PETERSBURGH, FLORENCE, COPENHAGEN, BERLIN, GOTTINGEN, MODENA, STOCKHOLM, PALERMO, ETC. ETC.

IN TWO VOLUMES.

VOL. I.

SECOND EDITION.

Reprinted from the PHILOSOPHICAL TRANSACTIONS of 1831—1838.

LONDON:

RICHARD AND JOHN EDWARD TAYLOR,

PRINTERS AND PUBLISHERS TO THE UNIVERSITY OF LONDON,

RED LION COURT, FLEET STREET.

1849.

Fig. 1 Frontispiece of Faraday's *Experimental Researches in Electricity*, I (3 Vols.) (Faraday 1839–1855) *Source* French National Library–Gallica. Public Domain

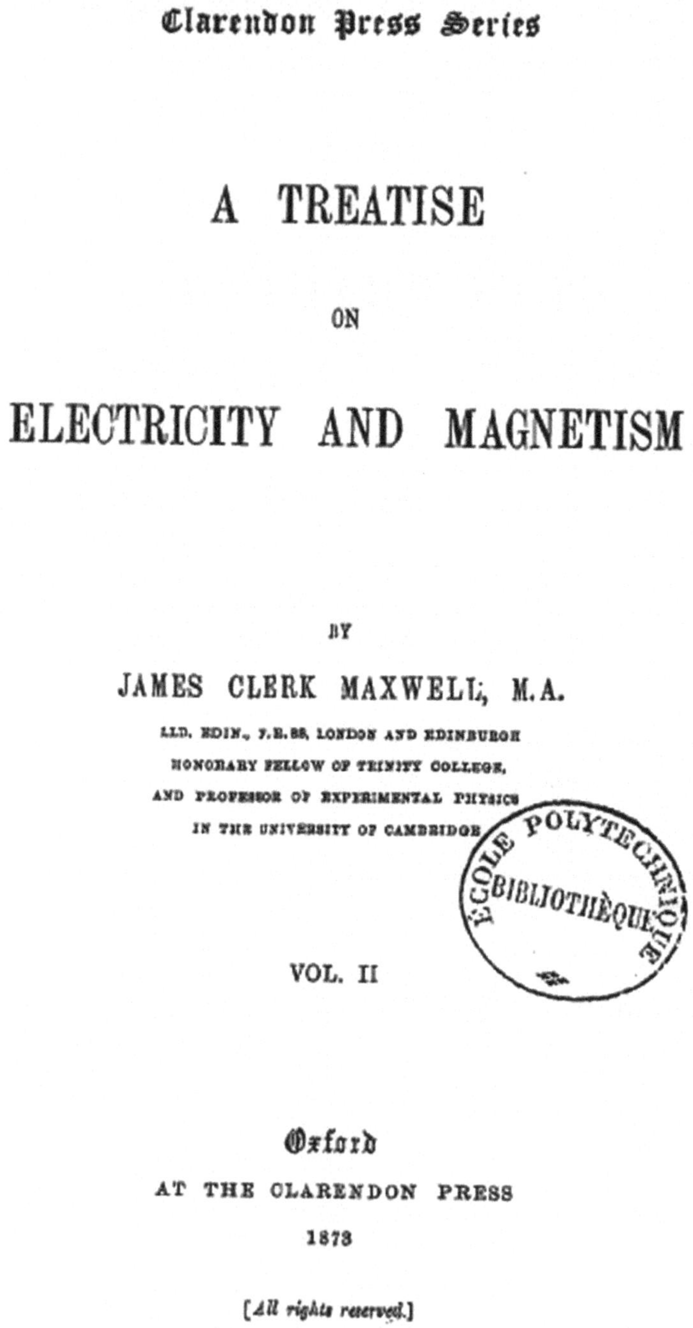

Clarendon Press Series

A TREATISE

ON

ELECTRICITY AND MAGNETISM

BY

JAMES CLERK MAXWELL, M.A.

LL.D. EDIN., F.R.SS. LONDON AND EDINBURGH
HONORARY FELLOW OF TRINITY COLLEGE,
AND PROFESSOR OF EXPERIMENTAL PHYSICS
IN THE UNIVERSITY OF CAMBRIDGE

VOL. II

Oxford
AT THE CLARENDON PRESS
1873

Fig. 2 Frontispiece of Maxwell's *A Treatise on Electricity and Magnetism* (Maxwell 1873, II) *Source* French National Library–Gallica. Public Domain

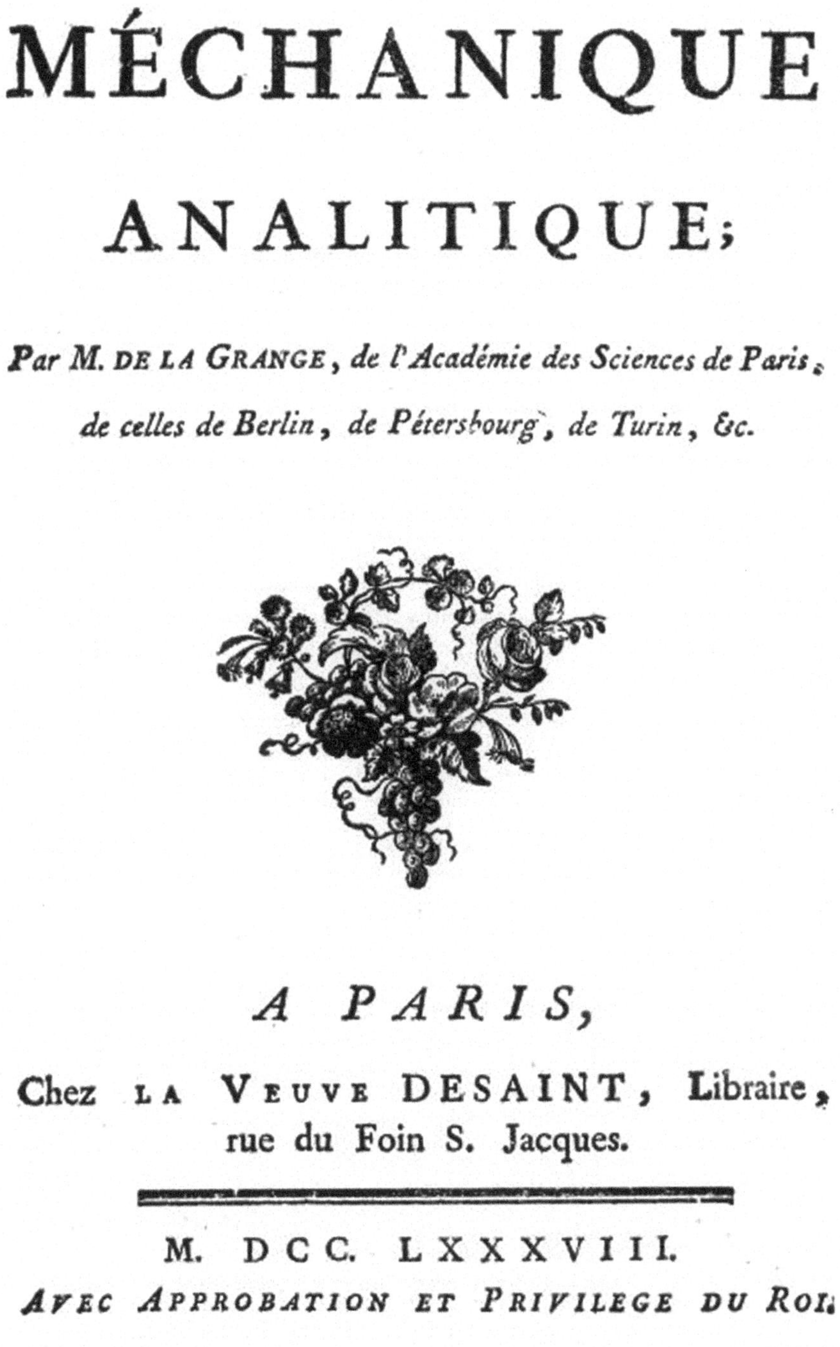

MÉCHANIQUE

ANALITIQUE;

Par M. DE LA GRANGE, de l'Académie des Sciences de Paris, de celles de Berlin, de Pétersbourg, de Turin, &c.

A PARIS,

Chez LA VEUVE DESAINT, Libraire, rue du Foin S. Jacques.

M. DCC. LXXXVIII.

AVEC APPROBATION ET PRIVILEGE DU ROI.

Fig. 3 Frontispiece of Lagrange's *Mécanique analitique* (Lagrange 1788). *Source* French National Library–Gallica. Public Domain

1.3 Introducing Langrage's Method

In *On the Equations of Motion of a Connected System* (Maxwell 1873, II, Part IV, Chap. V, pp. 184–194) Maxwell presented the relationships between motion and energy within a connected system, rather than in terms of laws of motion governing the actions of forces (Pisano 2024).
Maxwell wrote:

> 553.] In the fourth section of the second part of his *Mécanique analytique*, Lagrange has given a method of reducing the ordinary dynamical equations of the motion of the parts of a connected system to a number equal to that of the degrees of freedom of the system.
>
> [...]
>
> 554.] The aim of Lagrange was to bring dynamics under the power of the calculus. He began by expressing the elementary dynamical relations in terms of the corresponding relations of pure algebraic quantities, and from the equations thus obtained he deduced his final equations by a purely algebraical process. Certain quantities (expressing the reactions between the parts of the system called into play by its physical connexions) appear in the equations of motion of the component parts of the system, and Lagrange's investigation, as seen from a mathematical point of view, is a method of eliminating these quantities from the final equations. (Maxwell 1873, II, Part IV, Chap. V, p. 184)

In order to arrive at a suitable formulation of a dynamical theory of electromagnetism, Maxwell emphasized the possibility of writing kinetic energy in three ways. The first formulation presented in the *Treatise* is the one introduced by Hamilton in which kinetic energy is written in terms of variables and momenta.

> 560.] The instantaneous state of the system is completely defined if the variables and the momenta are given. Hence the kinetic energy, which depends on the instantaneous state of the system, can be expressed in terms of the variables (q), and the momenta (p). This is the mode of expressing T introduced by Hamilton. (Maxwell 1873, II, Part IV, Chap. V, p. 188)

Consequently, based on the stated method of Lagrange (see also below) Maxwell proposed a way to formulate a kinetic energy[4] in terms of velocities $\dot{q}$ & coordinates q ($T_{\dot{q}}$):

[4] The conceptualisation of field in relation to Maxwell's *processes of reasoning*, and also *analogies* has been addressed by one of us (RP) (Pisano 2013; Joule ([1849] 1850). Maxwell used figures and analogies in physics, e.g., in *On Faraday's Line of Force* he claimed: "[...] In order to obtain physical ideas without adopting a physical theory we must make ourselves familiar with the existence of physical analogies. By a physical analogy, I mean that partial similarity between the laws of one science and those of another which makes each of them illustrate the other. Thus all mathematical sciences are founded on relations between physical laws and laws of numbers, so that the aim of exact science is to reduce the problems of nature to the determination of quantities by operations with numbers." (Maxwell [read 1855 and 1856] 1855, p. 156). "[...] It is by the use of analogies [...] that I have attempted to bring before the mind, in a convenient and manageable form, those mathematical ideas which are necessary to the study of the phenomena of electricity." (Maxwell [read 1855 and 1856] 1855, p. 157). The relationship physics mathematics in term of dialectic would be expressed by means of his figures (Cfr. Maxwell 1873, I, Plate I), as well as geometry (one of Maxwell's great interest) playing an important role in this change of dialectic. The electric field would be *static* in the diagram described by means of an elastic medium under strain (in mechanics). The passage to a kinetic field (Maxwell 1873, II, Part IV) is an evident expression of the development of mathematical inquiry in physics (see also Part III on the magnetic shell as

- *A system consists in several coordinates* q_i;
- dT/dq_i *describes the dependence of the kinetic energy* T *on the variables* q_i.

Maxwell used Lagrange's mechanics as a new approach to physical theory on broader terms about the connected mechanical system. Since Lagrange's equation of motion is expressed in terms of rates of change, one arrives at the form of a differential equation. Particularly, he referred to it as a "third method":

> 563.] There is a third method of expressing the kinetic energy, which is generally, indeed, regarded as the fundamental one. By solving the Eqs. (3) [$\dot{q} = \frac{dT_p}{dp}$] we may express the momenta in terms of the velocities, and then, introducing these values in (13) [$T_{p\dot{q}} = \frac{1}{2}(\dot{q}_1 p_1 + \dot{q}_2 p_2 + \ldots)$] we shall have an expression for T involving only the velocities and the variables. When T is expressed in this form we shall indicate it by the symbol $T_{\dot{q}}$. This is the form in which the kinetic energy is expressed in the equations of Lagrange. (Maxwell 1873, II, Part IV, Chap. V, p. 191)
>
> [...]

In this way, Maxwell, first presented Lagrange's method and then used his Langangian formulation to describe his studies on electromagnetic phenomena (Maxwell 1873, II, Part IV, Chap. V). We anticipate (see more below) that because in Langrage's formulation the field is a continuum—in order to formulate a dynamical justification of field equations (Cfr. D'Agostino 2014)—Maxwell worked to obtain a magnetic field with an energy that was totally kinetic (Maxwell 1873, II, Part IV, Chap. V, pp. 189–192). In fact, at the end of the chapter, he wrote:

> 567.] In this outline of the fundamental principles of the dynamics of a connected system, we have kept out of view the mechanism by which the parts of the system are connected. We have not even written down a set of equations to indicate how the motion of any part of the system depends on the variation of the variables. Our only assumptions are, that the connexions of the system are such that the time is not explicitly contained in the equations of condition, and that the principle of the conservation of energy is applicable to the system. (Maxwell 1873, II, Part IV, Chap. V, pp. 193–194)

Hence Maxwell, introducing Lagrange's method, fixed the following two main subjects in his electromagnetic theory:

- System connections are such that time does not appear explicitly in equations;
- A principle of energy conservation is applicable to the electromagnetic system.

Below more details are presented.

1.4 Work and Energy

Maxwell used a mathematical formalism (from Lagrangian mechanics) to conceptualize dynamic frameworks in his theory of electromagnetism (Maxwell 1873, II,

---

a mathematical physical figure). In some figures, mathematical construction and physical image coincided in the aims (Maxwell 1873, I, Plate V, pp. 143–145; p. 403), which Maxwell also used to call *eye–knowledge* (Pisano: *Faraday and Maxwell, Two lives, One Theory*, pre-print). See also below.

part IV, Chap. VI, pp. 195–205). However since Lagrange's method came from mechanics, a question arises: what about the relationship between work and energy in electromagnetic theory?

A current on a wire allows the possibility of mechanical work. The ability to do work means energy. This is a very important epistemological point, because energy is still defined as the possibility for a body to perform work, and it is very interesting to find in Maxwell the same definition. In the case of an electric current, *what kind of energy arises? Is it kinetic or potential?* [Cfr. Gillispie and Pisano 2014; Pisano 2011, 2013; Joule 1839; Pisano and Bussotti 2015]. According to Maxwell, an electric current is a kinetic phenomenon. A current is due to movement whose cause is called electromotive force, and this force is not to be confused, Maxwell pointed out (Maxwell 1873, II, Part IV, Chap. VI, p. 196)—with an ordinary force—that acting on a body can cause it to change its state of motion. Following the analogy of the flow of water in a pipe, the work produced by electricity, that is, by a current, is given by the product between the electromotive force and the amount of electricity flowing through the section of a conductor under the action of the electromotive force. Finally, the energy of the considered system then has to be totally kinetic:

> 568.] We have shewn, in Art. 552, that, when an electric current exists in a conducting circuit, it has a capacity for doing a certain amount of mechanical work, and this independently of any external electromotive force maintaining the current. Now capacity for performing work is nothing else than energy, in whatever way it arises, and all energy is the same in kind, however it may differ in form.
>
> […]
>
> 569.] The electric current cannot be conceived except as a kinetic phenomenon. Even Faraday, who constantly endeavoured to emancipate his mind from the influence of those suggestions which the words 'electric current' and 'electric fluid' are too apt to carry with them, speaks of the electric current as 'something progressive, and not a mere arrangement'*[*Exp. Res.*, 283.]. (Maxwell 1873, II, part IV, Chap. VI, p. 195; Author's "*" reference and italics)
>
> […]
>
> 570.] In the same way the work done by an electromotive force is measured by the product of the electromotive force into the quantity of electricity which crosses a section of the conductor under the action of the electromotive force. The work done by an electromotive force is of exactly the same kind as the work done by an ordinary force, and both are measured by the same standards or units. (Maxwell 1873, II, part IV, Chap. VI, p. 196)

Taking into account Joule's prior results (Joule 1838, 1839, 1841a, 1842, 1843; Joule [1849] 1850; Pisano 2014; Pisano et al. 2020) the work done by an electromagnetic force consists in three different parts (Maxwell 1873, II, Part IV, Chap. VI, pp. 195–205):

A part is spent to overcome the *resistance* of the circuit and it is converted to *heat*;	*Joule Effect* (Pisano et al. 2020; Joule 1841b, 1842, 1843)
A part is spent to produce electromagnetic phenomena in which conductors are placed in motion as a magnetic *force between two parallel wires carrying current*;	Cfr. Ampère's works (particularly Ampère 1822a, b, 1826; D'Agostino 2000)
A part is spent to increase the kinetic *energy of the current*.	Cfr. Faraday's *induced and self–induced currents* (Faraday 1839–1855, I, Series IX, Sect. 15, pp. 1048–118; Series XI, Sect. 18, pp. 1161–1306; see also Series I, Sect. 1, pp. 6–26)

1.5 The Structure of this Paper

In this paper we present how:

- The approach of Maxwell literally changed the current scientific dialectic present in the field over the course of the second half of nineteenth century: finding a way to mathematically describe Faraday's works in electricity. This also meant a change of perspective for the physical experimental sciences.
- Maxwell proposed a new field of applications by means of an already existing mathematical apparatus, that is, Lagrange's equations in mechanics; resulting in an astonishing *Treatise* on the electrical and magnetic phenomena to rigorously describe the different situations Faraday has previously observed, but in a completely different light.
- In detail, Lagrange's mechanisms are used in Maxwell's *Treatise*, and the role played by the *nature* of the forces in play disappeared in Lagrange's treatment.
- A new Physics Mathematics was approached in the nineteenth century and how the conceptualisation of *field* moved into electromagnetism.
- Conclusion

2 Introduction

2.1 Physics Mathematics *Conceptualisation*

Generally speaking, *Mathematical Physics* considers a set of mathematical tools, like formulas, methods or strategies, to be used to solve problems inspired by physics. It is known how Newton used his geometrical–mathematical calculus to describe motion, overcoming the problem of non–homogeneous quantities.

Physics Mathematics, instead, was a different and brand-new concept born in the nineteenth century (see sections below and related Pisano's works). Maxwell studied Faraday's experimental works (and brought together Lagrange's mechanics and equations). It is Maxwell himself that frames this new approach.

> When I had translated what I consider to be Faraday's ideas into a mathematical form, I found that in general the results of the two methods coincided, so that the same phenomena were accounted for, and the same laws of action deduced by both methods, but that Faraday's methods resembled those in which we begin with the whole and arrive at the parts by analysis, while the ordinary mathematical methods were founded on the principle of beginning with the parts and building up the whole by synthesis. I also found that several of the most fertile methods of research discovered by the mathematicians could be expressed much better in terms of ideas derived from Faraday than in their original form (Maxwell 1873, I, *Preface*, p. ix).

The fusion of these two aspects constitutes the deepest core of the physics mathematics system at that time; especially because electromagnetism was (like chemistry first, followed by thermodynamics) outside the Newtonian mechanical paradigm during the nineteenth century (Pisano et al. 2014). In Maxwell's mind, Lagrange's mechanics would be the perfect key to understanding, or at least get as close as possible to understanding, Faraday's physical problems expressed in a non–mathematical way. Mathematics is no longer a mere instrument to *solve* physics problems, but the way to *describe* the situations in order to *understand* them.

Newton had his system to solve problems through calculus, connecting quantities such as space and time, as two variables in the same function, so describing the behaviour of a system under the action of forces. Maxwell broke this paradigm founding his system on energy. As he explained in his *Treatise*, he dealt with kinetic energy in detail and the description of electromagnetic issues through energy is possible if the mathematical approach of Lagrange is applied. He found a way to express energy that he himself described as:

> [...] a third method of expressing the kinetic energy, which is generally, indeed, regarded as a fundamental one. (Maxwell 1873, II, Part IV, Chap. V, p. 191)

This was made through Lagrange's equations. In this way Maxwell, as we do nowadays, could *read* the physical situation by means of quantities that, at the same time, are both physical and mathematical. There is no break between physics and mathematics here, but rather a merging of the two intellectual plans. Moreover, taking measurements, a typical physical operation, is no longer important since it is not a value of a quantity that we are interested in, but the description and the comprehension of a physical fact itself.

Maxwell's genius and ability to synthesise, combined with his skill in mathematical calculation, led him to organize a new discipline characterised by a profound sense of cohesion between two subjects that had hitherto been conceived as separate, i.e. functional to the solution of each other's problems. With this insight from Maxwell, the approach to the physical problem shifts from laboratory observation and mathematical modelling to the highest level of understanding, achieved with objects whose nature becomes dual, through the fusion of the disciplines of physics and mathematics.

2.2 *Faraday and Maxwell: A* Physics Mathematics *Context*

The historical development of the theory of electromagnetism has made it possible to frame its complex structure through a series of fundamental works, prepared by those scientists who are credited among the brightest minds in this area of physics. As we have already pointed out, Maxwell's brilliant intuition, stated in his *Treatise*, was to overcome Newton's descriptive paradigm of physics, in favour of a new system based on the concept of energy rather than that of force. However, it was necessary to define the role played by the concept of force and how it involved the other concept, energy. The equations of motion are the tools that allow us to describe and predict the behaviour of a system as a function of space and time, as happens when we analyse the motion of a particle on which a force acts. The differential equations that govern the system, i.e., Newton's second law, Euler–Lagrange equations (Lagrange 1870–1873), the solutions of these equations are sometimes referred to as the equations of motion.

Maxwell recognised that Lagrange's mathematical formalism could prove an extremely effective tool for the new path of physics mathematics in electromagnetic theory. He established the foundations and relationship between electricity and magnetism, that is, a new scientific theory within the *physics mathematics* domain. The latter was a new discipline, physics mathematics and not mathematical physics. It was a new methodological approach where variables are physical (i.e., space and time) and mathematical (i.e., for calculating speed) at the same time; and the measurements—in this approach—are not a priority (Pisano et al. 2013, 2019, 2017; Pisano 2015, 2010, 2017; Pisano and Bussotti 2015, 2022, 2023; Pisano and Pellegrino 2023). For example, at beginning of the nineteenth century, Jean Baptiste Joseph Fourier (1768–1830) in his analytic heat rational theory wrote:

> It is recognised that the same principles regulate all the movements of the stars, their form, the inequalities of their courses, the equilibrium and the oscillations of the seas, the harmonic vibrations of air and sonorous bodies, the transmission of light, capillary actions, the undulations of fluids, in fine the most complex effects of all the natural forces, and thus has the thought of Newton been confirmed: *Quod tam paucis tam multa proestet geometria gloriatur.* But whatever may be the range of mechanical theories, they do not apply to the effects of heat. These make up a special order of phenomena, which cannot be explained by the principles of motion and equilibrium. [...] whereby every physical problem of this kind is brought back to an investigation of mathematical analysis. From these general facts I have concluded that to determine numerically the most varied movements of heat, it is sufficient to submit each substance to three fundamental observations. (Fourier 1822, pp. ij–iij; Author's Italic; Translation from Fourier 1878, p. 2)

> [...] The principles of the theory are derived, as are those of rational mechanics, from a very small number of primary facts, the causes of which are not considered by geometers, but which they admit as the results of common observations confirmed by all experiment. The differential equations of the propagation of heat express the most general conditions, and reduce the physical questions to problems of pure analysis, and this is the proper object of theory. They are not less rigorously established than the general equations of equilibrium and motion. (Fourier 1822, p. xj; Translation from Fourier 1878, p. 6)

Thus, the second half of the nineteenth century was characterised by the predominance of the mathematical approach, when several theories were included in mechanics, giving shape to new rational–analytical theories, highlighting the close relationship between physics and mathematics (Pisano 2013; Pisano and Capecchi 2016), where principles "[…] ne présuppose aucune loi physique" ("It does not presuppose any physical law") (Lamé 1852, pp. 21–22; Pisano and Capecchi 2009) and there were no experimental observations. Gabriel Lamé (1795–1870) proposed, in the wake of Fourier's work (Pisano and Capecchi 2009; Fourier 1822), concepts such as heat propagation, applied velocity and so on. He confirmed that the new trend was clearly called "physics mathematics":

> Physics Mathematics properly so–called is a completely modern creation, which belongs exclusively to the Geometers of our century. Today, this science really comprises only three chapters, variously extended, which are treated rationally; that is to say, which are based only on indisputable principles or laws. These chapters are: the theory of static electricity on the surface of conducting bodies; the analytical theory of heat; and the mathematical theory of the elasticity of solid bodies. The latter is the most difficult, the least complete; it is also the most useful, at a time period where we appreciated the importance of a mathematical theory by the results that it can immediately provide for industrial practice. […] really Physics Mathematics is a science as rigorous, as exact as rational mechanics.[5]
>
> The principle cited [above] at Sect. II [heat propagation] is therefore justified. It does not presuppose any physical law. The only fact that it derives from experience, or, if you will, the only hypothesis that one can admit, can be stated as follows: the quantity of heat exchanged [on heat and caloric]. This principle does not claim the nature [measurability/calculability in physics] of heat.[…]
>
> § LVI. Differential part and integral part
>
> Any branch of physics mathematics is subdivided into two distinct parts: the differential part and the integral part. The first includes the partial difference equation, and all the consequences that are deduced from the order and the various transformations of these equations. The second deals with the integration of the same equations in general or particular cases, and finally leads to the numerical discussion [also justification] of the theoretical results. The extent or relative importance of these two parts changes with the progress of science.[6]

[5] "La Physique mathématique proprement dite est une création toute moderne, qui appartient exclusivement aux Géomètres de notre siècle. Aujourd'hui, cette science ne comprend en réalité que trois chapitres, diversement étendus, qui soient traités rationnellement; c'est-à-dire qui ne s'appuient que sur des principes ou sur des lois incontestables. Ces chapitres sont: la théorie de l'électricité statique à la surface des corps conducteurs; la théorie analytique de la chaleur; et la théorie mathématique de l'élasticité des corps solides. Le dernier est le plus difficile, le moins complet; il est aussi le plus utile, à une époque où l'on veut apprécier l'importance d'une théorie mathématique par les résultats qu'elle peut fournir immédiatement à la pratique industrielle. […] la véritable Physique mathématiques est une science aussi rigoureuse, aussi exacte que la Mécanique rationnelle." (Lamé 1852, pp. V–VI; Author's translation).

[6] "Le principe posé au § II [heat propagation] est donc justifie. Il ne présuppose aucune loi physique. Le seul fait qu'il emprunte à l'expérience, ou, si l'on veut, la seule hypothèse qu'il admette, peut s'énoncer ainsi: la quantité de chaleur échangée [on heat and caloric]. Ce principe ne se prononce pas sur la nature [measurability/calculability in physics] de la chaleur. (Lamé 1861, pp. 21–22). […] § LVI. Parte différentielle et parte intégrale Toute branche de la physique mathématique se subdivise en deux parties distinctes: la partie différentielle et la partie intégrale. La première comprendre la mise en équation aux différences partielles, et toutes les conséquences qui se déduisent de l'ordre

Thus, it was necessary to establish a new theory, with new magnitudes to justify both the *new nature of ideas* and to avoid hiding physical reasoning with a "[...] terrific array of symbols [...]" (Maxwell [1892] 1954, p. 176).[7] The novelty consists in a new conception of physical quantities which were not solely physical, nor solely mathematical, but physics mathematics quantities since they, altogether, represent *relationships of thought* among *mathematical quantities* and *physical structures* in order to foster "[...] reducing these [experimental electric and magnetic] phenomena into scientific form [...]" (Maxwell 1865, p. 459; see also Maxwell 1855–1856; 1861–1862). Mathematical techniques of derivatives and integration are the instruments to link new mathematical results to the new physical *process of reasoning*. This new approach necessarily required Maxwell to go beyond Newton's mathematics, precisely through the new formalism—as he mentioned this in his *Treatise* when he devoted his work to mathematical mechanisms: *On the equations of motion of a connected system* (Maxwell 1873, II, Part IV, Chap. V, pp. 183–194). For his part, Maxwell claimed:

> The first process therefore is the effectual of study of the science, must be one of simplification and reduction of the results [...]. The results of this simplification may take the form of a purely mathematical formula or of a physical hypothesis [...]. In order to obtain physical ideas without adopting a physical theory we must make ourselves familiar with the existence of physical analogies. By a physical analogy I mean that partial similarity between the laws of one science and those of another which makes each of them illustrate the other. Thus all mathematical sciences are founded on relations between physical laws and laws of numbers, so that the aim of exact science is to reduce the problems of nature to the determination of quantities by operations with numbers. (Maxwell [read 1855 and 1856] 1855, pp. 156–157)

Maxwell's *Treatise* was preceded, historically, by the complex flourishing of a series of theoretical models that saw the light of day in the 18th Century (Pisano 2011; Pisano 2013; Joule 1839) in order to solve a variety of problems. We owe to Pierre–Simon de Laplace (1749–1827) the attempt to mathematically solve any kind of motion, through central forces and differential equations, presented in his *Traité de Mécanique céleste* (Fox 1974; Laplace 1805; Euler 1736). Again, differential equations were the tool used by Jean Baptiste Joseph Fourier (1768–1830), in a sophisticated way, to address problems related to heat, a quantity whose nature he did not consider (Pisano 2010). André–Marie Ampère (1775–1836) also approached the problems concerning the new fields of research such as electricity and the study of magnetic phenomena through a mathematical description (Ampère 1822a, b, 1826), the basis of which can be traced back to the experiences of Hans Christian

et des diverses transformations de ces équations. La seconde s'occupe de l'intégration des mêmes équations dans des cas généraux ou particuliers, et conduit finalement à l'évaluation numérique des résultats théoriques. L'étendue ou l'importance relative de ces deux parties change avec le progrès de la science." (Lamé 1861a, p. 95; Author's translation; see also Lamé 1861b; Lamé 1836).

[7] We refer to the famous debate (1886) between Tait and Ludwig Boltzmann (1844–1906) in *XX. On some Questions in the Kinetic Theory of Gases. Reply to Prof. Boltzmann* (Maxwell [1892] 1954, pp. 172–179); see more below. On Boltzmann (including Clausius, Gibbs and related references) we note Pisano and Pellegrino's *Details in Clausius' Laws, Boltzmann's H-Theorem & Entropy: Mathematical Frameworks into a Kinetic Mechanism (1850–1901)* (Pisano and Pellegrino 2023, see below Part 3 in this book).

Ørsted (1777–1851). Looking at the works of Sadi Carnot (1796–1832) (Gillispie and Pisano 2014), Ørsted (1822, 1823) and Michael Faraday, *Experimental Researches in Electricity* (Faraday 1839–1855), we note that there is no recourse to infinitesimal analysis. Faraday, in particular, was able to introduce the basis for the concept of contiguous actions in electromagnetic induction without resorting to particularly sophisticated mathematics (Pisano 2013). In spite of Maxwell's own opinion on his own *Treatise*, interpreted by the author as a mathematical description of Faraday's work, the approach could be said to be completely different, because new abstract concepts and new problems were involved in the development of the work (Maxwell 1873); see also (Maxwell [1876–1877] 1920). Maxwell always took Faraday's works into serious account:

> After taking his degree in 1854, Maxwell read through Faraday's Experimental researches, a course which he always recommended his students to follow. In Faraday he found a mind essentially of his own type. Thoroughly conversant himself with the Theory of Attractions as developed in Mathematical Treatises, and with the laws of electrical action as illustrated by Sir William Thomson in his paper "on the Uniform motion of heat in homogeneous solid bodies, and its connection with the Mathematical Theory of Electricity," a paper published in the Cambridge Mathematical Journal, February 1842, and "on a Mechanical representation of Electric, Magnetic, and Galvanic Forces," published in the Cambridge and Dublin Mathematical Journal, January 1847, Maxwell saw the connection between Faraday's point of view and the method of research adopted by the Mathematicians. He used to say that he had not a good nose to smell heresy, but whatever was good and true Maxwell would detect beneath the mass of misconception, or even falsehood, which had gathered round it, and which caused its rejection by nearly every one else without inquiry. Faraday's conception of a medium he adopted as a guide throughout his electrical researches. (Campbell and Garnett 1882, pp. 513–514; Author's italics)

In early writings from the Edinburgh period—before he moved to *Trinity College Cambridge* for his degree—Maxwell showed uncommon skills in handling calculus tools, in problem setting, and in focusing his attention on the continuous media that he would study through the partial differential equations, thus building the basis that would allow him to successfully address the theory of electromagnetism. Therefore, a breaking of the consolidated Newtonian paradigm based on the concept of force to endorse a new one based on the concept of energy allows him to deal deeply with new phenomena into another paradigm, the Lagrangian one.

Maxwell's idea of calling into question Lagrange's equations can be traced back historically to William John Macquorn Rankine's (1820–1872) definition of potential energy. Rankine defined potential energy as *the product of a change of state into the resistance against which that change is made* (Rankine 1888, pp. 203–208). This definition was related to Rankine's idea that the actually observable changes in complex systems should be traced back to changes in energy and that the set of internal mechanisms that acted as a junction could be ignored, in other words the forces within the system could be neglected (Cfr. Simpson 1970).

In a later paper, *Outlines of the Science of Energetics* (Rankine 1885; see also Maxwell 1860) published in 1855, Rankine refined these ideas and made explicit his concept of the structure of physical theory. According to Rankine, science has an objective, which is achieved by the following two paths:

1. The laws underlying physical phenomena are established through observation and experiment, in accordance, if possible, with the Galilean scientific method;
2. The laws underlying a class of phenomena are condensed into a physical theory.

In Rankine's mind, the structure of a physical theory:

> [...] consists of definitions and axioms as first principles, and propositions, their consequences [...] a definition states properties common to a class of existing objects, or real phenomena, and a physical axiom states a general law as to the relations of phenomena. (Rankine 1888, note 8, p. 209)

Again, regarding definitions and axioms their status is:

> [...] to a certain extent, *tentative* – that is to say involving the trial of conjectural principles, and their acceptance or rejection, according as their consequences are found to agree or disagree with the formal laws deduced immediately from observation and experiment. (Rankine 1888, p. 209)

Rankine was well aware that the traditional scientific process, as conducted by British physicists of his time, aimed to bring observable phenomena under a mechanical point of view, since mechanics was considered the only formally complete physical theory. Thus he reformulated this paradigm on the basis of the laws of energy, rather than those of mechanics.

> Energy, or the capacity to effect change, is the common characteristic of the various states of matter to which the several branches of physics relate; if then, there be general laws respecting energy, such laws must be applicable *mutatis mutandis* to every branch of physics, and must express a body of principles as to physical phenomena in general. (Rankine 1888, p. 213)

In the context of this scientific debate, two more authors should be taken into consideration: William Thomson—Lord Kelvin (1824–1907) and Peter Guthrie Tait (1831–1901). We refer to the *Treatise on Natural Philosophy* (1867) essentially published with the aim of unifying the different branches of physics under the common principle of energy. It was, to all intents and purposes, the first text which gathered together the laws of energy. Even the authoritative journal *Nature*, which first appeared in 1869, reviewed the volume in 1872 and cited it as "new physics" (Moyer 1977, p. 255). Thomson and Tait's work became affectionately known to the public as *T and T*^{1} (MacFarlane 1917; Fig. 4).

PETER GUTHRIE TAIT 43

that Tait I am sure could not conceive of any higher compliment. Maxwell had facetiously referred to Thomson as T and to Tait as T^1. Hence the *Treatise on Natural Philosophy* came to be commonly referred to as *T and* T^1 in the conversation of mathematicians.

It appears that the introduction of the quaternion method was a serious point of difference between the joint authors. Prof. Thomson, as you know, subsequently became Lord Kelvin and recently he wrote to Prof. Chrystal as follows with respect to the joint authorship of the Treatise. "I first became personally acquainted with Tait a short time before he was elected professor in Edinburgh; but, I believe, not before he became a candidate for the chair. It must have been either before his election or very soon after it that we entered on the project of a joint treatise of natural philosophy. He was then strongly impressed with the fundamental importance of Joule's work, and was full of vivid interest in all that he had learned from and worked at, with Andrews. We incessantly talked over the mode of dealing with energy which we adopted in the book, and we went most cordially together in the whole affair. He gave me a free hand in respect to names, and warmly welcomed nearly all of them. We have had a thirty-eight years' war over quaternions. He had been captivated by the originality and extraordinary beauty of Hamilton's genius in this respect, and had accepted, I believe, definitely, from Hamilton to take charge of quaternions after his death, which he has most loyally executed. Times without number I offered to let quaternions into Thomson and Tait, if he could only show that in any case our work would be helped by their use. You will see that from beginning to end they were never introduced."

In 1864 Tait published in the *North British Review* articles on "The dynamical theory of heat" and "Energy" which were afterwards made the basis of his *Sketch of Thermodynamics* published in 1868. The articles, mainly historical, are written from the British point of view, so much so, that he was accused of Chauvinism. To this charge he replied, "I cannot pretend

Fig. 4 "T" and "T^1" from Kelvin and Tait (MacFarlan 1917, p. 43) *Source* Archive.org. Public Domain

After Newton's *Principia*, mechanics advanced to include new approaches and results. Thomson and Tait set aside the structural principles of mechanics to look at Newton's more fertile ideas from their own perspective (Truesdell 1968, 1970). They focused on the distinctions, both for the science of motion and the science of machines, with the aim of making dynamics the basis of all physical science. Their basic idea was based on the conservation of certain macroscopic quantities: once this hypothesis was guaranteed, the conservation laws would allow the description of the changes in any complex system, regardless of its hidden microscopic structure. Thomson and Tait developed the generalised equations of motion, those of Lagrange and Hamilton, from conservation laws. The principle of conservation of energy became their cornerstone to ensure the absolute generality of dynamics because, according to their view, the conditions of equilibrium or motion of a system could immediately be derived from it (Moyer 1977, p. 257). A milestone in the development of Thomson's and Tait's ideas was a minimum theorem formulated by Thomson himself in 1849, which in principle referred to the motion of liquids: the variation of a system caused by impulsive forces is such that of the kinetic energy of the system is minimised. An impulsive force acts in an infinitesimal time interval, such that there can be no appreciable change in the potential energy of the system, so that the only appreciable energy changes are in the kinetic contribution, and—since the state of the system does not depend on how it was achieved—the use of impulsive forces is of a completely general nature. But, no specific studies were given to the hidden structure of a system (Moyer 1977, pp. 258–259).

In 1876–1877, Maxwell succeeded in improving, in terms of clarity of exposition, *T and T'*, through his work *Matter and Motion* (Maxwell [1876–1877] 1920). In order to understand how much the work of his predecessors was of fundamental importance to Maxwell, one can state that he gave the same definitions of physical quantities that we can find first in Rankine and then in Thomson and Tait, attributing the same fundamental importance to the principles of conservation. The contribution of these last two authors in particular is clear in the words of Maxwell's *Thomson and Tait's Natural Philosophy*:

> It is to Lagrange, in the first place, that we owe the method which enables us to answer this question without asserting either more or less than all that can be legitimately deduced from the observed facts. But though this method has been in mathematicians hands since 1788, when *Mécanique Analytique* was published, and though a few great mathematicians, such as Sir W. R. Hamilton, Jacobi etc., have made important contributions to the general theory of dynamics it is remarkable how slow natural philosophers at large have been to make use of these methods. Now, however, we have only to open any memoir on a physical subject in order to see that these dynamical theorems have been dragged out of the sanctuary of profound mathematics in which they lay so long enshrined, and have been set to do all kinds of work, easy as well as difficult, throughout the whole range of physical science. The credit of breaking up the monopoly of the great masters of the spell, and making all their charms familiar in our ears as household words, belongs in great measure to Thomson and Tait. The two northern wizards were the first who, without compunction or dread, uttered in their mother tongue the true and proper names of those dynamical concepts which the magicians of old were wont to invoke only by the aid of muttered symbols and inarticulate equations. And now the feeblest among us can repeat the words of power and take part in dynamical

discussions which but a few years ago we should have left for our betters. (Maxwell [1890] 2003, II, p. 782)

Dynamics became for Maxwell an extremely powerful tool to investigate the changes in complex systems without having to refer to some internal and hidden mechanics. The most significant fruit of the path traced by Rankine, Thomson and Tait and taken up by Maxwell was the dynamic theory of electromagnetism collected in the two volumes of *A Treatise on Electricity and Magnetism*. Starting from Faraday's idea of a state in electricity for a description of electromagnetic induction, a dynamic theory of electricity and magnetism was developed.

> 547.] He observes, however, that 'the first thought that arises in the mind is that electricity circulates with something like momentum or inertia in the wire'. (Maxwell 1873, II, Part IV, Chap. IV, p. 180)

Maxwell set out to investigate the motion of the system through dynamics.

> 550.] [...] The fundamental dynamical idea of matter, as capable by its motion of becoming the recipient of momentum and of energy, is so interwoven with our forms of thought that, whenever we catch a glimpse of it in any part of nature, we feel that a path is before us leading, sooner or later, to the complete understanding of the subject.
>
> 551.] Now an ordinary force, acting on a body in the direction of its motion, increases its momentum, and communicates to it kinetic energy, or the power of doing work on account of its motion. [...] But a conducting circuit in which a current has been set up has the power of doing work in virtue of this current, and this power cannot be said to be something very like energy, for it is really and truly energy. [...]
>
> 552.] It appears therefore that a system containing electric current is a seat of energy of some kind; and since we can form no conception of an electric current except as a kinetic phenomenon, its energy must be kinetic energy. [...] We are therefore led to inquire whether there may not be some motion going on in space outside the wire, which is not occupied by the electric current, but in which the electromagnetic effects of the current are manifested. [...] What I propose now to do is to examine the consequences of the assumption that the phenomena of the electric current are those of a moving system, the motion being communicated from one part of the system to another by forces, the nature and laws of which we do not yet even attempt to define, because we can eliminate these forces from the equations of motion by the method given by Lagrange for any connected system. In the next five chapters of this treatise I propose to deduce the main structure of the theory of electricity from a dynamical hypothesis of this kind [...]. (Maxwell 1873, II, Part IV, Chap. IV, pp. 181–183)

He then developed the generalised equations of motion, starting with the theory of impulsive forces. According to Maxwell, when applying dynamics to electromagnetism, no assumptions about the internal structure of the system would be necessary. As above cited:

> 567.] In this outline of the fundamental principles of the dynamics of a connected system, we have kept out of view the mechanism by which the parts of the system are connected. (Maxwell 1873, II, Part IV, Chap. V, p. 193)

Furthermore, dynamics is a consequence of the universal law of energy.

Table 1 A Partial Timeline on Electricity and Magnetism Laws (Ampère 1822a, b, 1826; Coulomb 1785; Ørsted 1822, 1823; Joule 1884)

Year	Electricity	Magnetism
1785	Coulomb law $F = K\frac{q_1 q_2}{r^2}$	
1800	Volta's battery	
1813	Poisson's equation electrostatic potential $\nabla^2 V = -4\pi\rho$	
1820	Ørsted's experiments	
1820	Biot–Savart's law: $dB = id\vec{l} \times \vec{r}/r^3$	
1821		Faraday's lines of forces
1825	Ampère's mathematization for the interaction between magnetism and electricity (motion of particles)	
1825–27	Ohm's law: $V = RI$	
1831	Faraday's law: $emf = -\frac{d\Phi(\vec{B})}{dt}$	
1832	Faraday's law for electrolytic conduction: $m = \frac{Mq}{Fz}$	
1834		Lenz's rule

> 567.] Our only assumptions are, that the connections of the system are such that the time is not explicitly contained in the equations of condition, and that the principle of the conservation of energy is applicable to the system. (Maxwell 1873, II, Par IV, Chap. V, p. 194)

We will look at the details of Maxwell's work later in this paper. It has often been assumed that the Newtonian mechanical paradigm[8] (Bussotti and Pisano 2013; Pisano and Bussotti 2014; Newton 1687) was capable of encompassing—without any consequences—every theoretical and deductive mechanics formulation of the eighteenth and nineteenth centuries (Fox 1974; Laplace 1805, 1822; Joule 1845). This kind of assumption led to the conclusion that the various formulations of mechanics are equivalent; and *new ideas* based on *actual facts* and alternatives to exclusive Newtonian mechanics were widely obscured (Mach 1883). In the following, a concise historical view of the scientific studies around Faraday–Maxwell is presented (Mottelay 1922) (Table 1).

Prior to Charles Augustine Coulomb's (1736–1806) discovery that the forces between electrically charged bodies follow a law quite similar to Newton's law of universal gravitation (Oliveira and Pisano 2022) the physics of electricity was conducted by empirical means. After Coulomb, other researchers including Pierre-Simon de Laplace (1749–1827) and Siméon-Denis Poisson (1781–1840) applied Newtonian-type laws to physical phenomena. However, three other illustrious names were responsible for a quantum leap in the field of physical sciences oriented towards the study of electricity.

[8] Different approaches to Newtonian physical theory mainly came from d'Alembert's dynamics (d'Alembert 1743), Lazare Carnot's mechanics (Gillispie and Pisano 2014; Carnot [1783] 1786, 1803; Pisano 2021b) and Lagrange's mechanics (Fraser 1983, 1985; Lagrange 1788, 1870–1873, 1892; Pisano et al. 2013; Joule 1839; Pisano and Bussotti 2016; Pisano and Capecchi 2009, 2013).

1. Alessandro Volta (1745–1827) used the battery, and was able to produce with greater intensity the effects historically observed by rubbing materials together.
2. Hans Christian Oersted (1777–1851) discovered the magnetic action of current in a conductor.
3. André–Marie Ampère (1775–1836) discovered that these actions are due to the sum or difference of current elements subject to the inverse–square law.

Ampère was referred to by Maxwell as the "Newton of electricity". However, at the time, a unifying framework for the phenomenology of electrical charges, magnets and currents had not yet been found. The likes of Georg Simon Ohm (1787–1854), Gustav Robert Kirchhoff (1824–1887), Franz Neumann (1798–1895), Hermann von Helmholtz (1821–1894) and Wilhelm Weber (1804–1891) tried to formalise the laws of electromagnetic induction, a phenomenon which had in turn been observed in depth by Michael Faraday (1791–1867). The mathematical theories successfully elaborated by these scientists were, however, still within a Newtonian and Laplacian framework of forces at a distance. This was so even if the introduction of Weber's view of forces dependent on the velocity and acceleration of the parts of the system involved had proposed an approach outside the conceptual framework established by Newton and Leibnitz. First, Faraday gave more importance to electric and magnetic forces considered in themselves, detached from their sources, which could be charges or currents, breaking the dichotomy between force and mass and the Newtonian paradigm from which this dichotomy arose. The impenetrability of matter, for example, can be interpreted as resistance to the approach of other matter, hence as a force that increases rapidly as the distance between interacting objects decreases. Inertia, on the other hand, can be interpreted as a force that opposes the driving force. Faraday focused on an interpretation that was complementary to the prevailing mechanism, emphasising the dynamic aspect of change in physical reality. Maxwell, in turn, turned to the contributions of English scientists and in particular, as noted in the previous paragraph, to the work of Lord Kelvin and Tait. Thomson, in turn, had rigorously developed the analogy between heat propagation and the theory of potential, including in his paper praise of Faraday's work.

> The general conclusions established in it show that the laws of distribution of electric or magnetic force in any case whatever must be identical to the laws of distribution of the lines of motion of heat in certain perfectly defined circumstances. [...] they constitute a full theory of the characteristic of lines of force, which have been so admirably investigated experimentally by Faraday [...]. (Thomson [1842] 1854, p. 1)

Still on the subject of Thomson's work taken into account by Maxwell for the development of his electromagnetic theory, one must bear in mind the equations that George Gabriel Stokes (1819–1903) had obtained to describe the equilibrium of a deformed elastic body. Thomson had observed that these laws are in fact analogous to those expressing the distribution of the electric field near a point charge and of the magnetic field near a magnet. Maxwell considered these analogies confirmed by Thomson as fundamental to the development of his work.

In the first memoir that Maxwell wrote for the *Cambridge Philosophical Society*, *On Faraday's Lines of Force*, one of the founding theories of electrodynamics was that proposed by Weber, based on a law of force at a distance between moving electric charges, as a function of their velocity and acceleration. Weber's theory was effective, as it was able to explain both electrostatic phenomena and actions between currents.

However, Maxwell set out to demonstrate how these phenomena could be explained in the light of a completely different approach. He took up Faraday's intuition that a conductor or insulator, in the presence of a current or a magnetic field, would change its condition and give rise to induced currents. Maxwell gave a mechanical interpretation of Faraday's state, as we can see from the original passage in his paper (Fig. 5).

Maxwell hypothesised that this "electro-tonic" (Maxwell 1855–1856; 1861–1862, p. 205) state introduced by Faraday could be compared to a mechanical state characterised point by point by certain quantities, such as displacements, velocities or forces. A little earlier in the same paper it is also possible to read Maxwell's definition of the system's self–potential and its mathematical description, obtained by applying the principle of conservation of energy: today this quantity is known as vector potential.

A few pages before, Maxwell wrote (Maxwell [read 1855 and 1856] 1855, p. 203) the integral (Fig. 6):

$$Q = \iiint \rho_1 p_1 \, dxdydz \tag{1}$$

where Q is the total potential of the system on itself, p_1 is the magnetic tension and ρ_1 represents a real magnetic density.

> The increase or decrease of Q will measure the work lost or gained by any displacement of any part of the system, and will therefore enable us to determine the forces acting on that part of the system. (Maxwell [read 1855 and 1856] 1855, p. 203).

Maxwell, in this way, opened a new path towards a field theory comprehensive of all the measures of electricity and magnetism.

In the years 1861–1862 Maxwell published in the Philosophical Magazine a work titled *On Physical Lines of Forces*, the importance of which lies in the fact that it contains the first statement of an electromagnetic field theory together with the electromagnetic theory of light. In this work (Fig. 7) conducting and insulating bodies are no longer described in the light of the classical conduction current, but rather from the new point of view of the new concept of displacement current, a current that can also propagate through the insulating space between conductors, which Maxwell assumes is filled by a substance called ether, behaving like a fluid.

ON FARADAY'S LINES OF FORCE. 205

It appears from experiment that the expression $\frac{da_0}{dt}$ refers to the change of electro-tonic state of a *given particle of the conductor*, whether due to change in the electro-tonic functions themselves or to the motion of the particle.

If α_0 be expressed as a function of x, y, z and t, and if x, y, z be the co-ordinates of a moving particle, then the electro-motive force measured in the direction of x is

$$\alpha_2 = -\frac{1}{4\pi}\left(\frac{da_0}{dx}\frac{dx}{dt} + \frac{da_0}{dy}\frac{dy}{dt} + \frac{da_0}{dz}\frac{dz}{dt} + \frac{da_0}{dt}\right).$$

The expressions for the electro-motive forces in y and z are similar. The distribution of currents due to these forces depends on the form and arrangement of the conducting media and on the resultant electric tension at any point.

The discussion of these functions would involve us in mathematical formulæ, of which this paper is already too full. It is only on account of their physical importance as the mathematical expression of one of Faraday's conjectures that I have been induced to exhibit them at all in their present form. By a more patient consideration of their relations, and with the help of those who are engaged in physical inquiries both in this subject and in others not obviously connected with it, I hope to exhibit the theory of the electro-tonic state in a form in which all its relations may be distinctly conceived without reference to analytical calculations.

Summary of the Theory of the Electro-tonic State.

We may conceive of the electro-tonic state at any point of space as a quantity determinate in magnitude and direction, and we may represent the electro-tonic condition of a portion of space by any mechanical system which has at every point some quantity, which may be a velocity, a displacement, or a force, whose direction and magnitude correspond to those of the supposed electro-tonic state. This representation involves no physical theory, it is only a kind of artificial notation. In analytical investigations we make use of the three components of the electro-tonic state, and call them electro-tonic functions. We take the resolved part of the electro-tonic intensity at every point of a

Fig. 5 In the last paragraph of this page the author compares Faraday's "electro-tonic" state to a mechanical state (Maxwell 1855–1856; 1861–1862, p. 205) *Source* https://en.wikisource.org. Public Domain

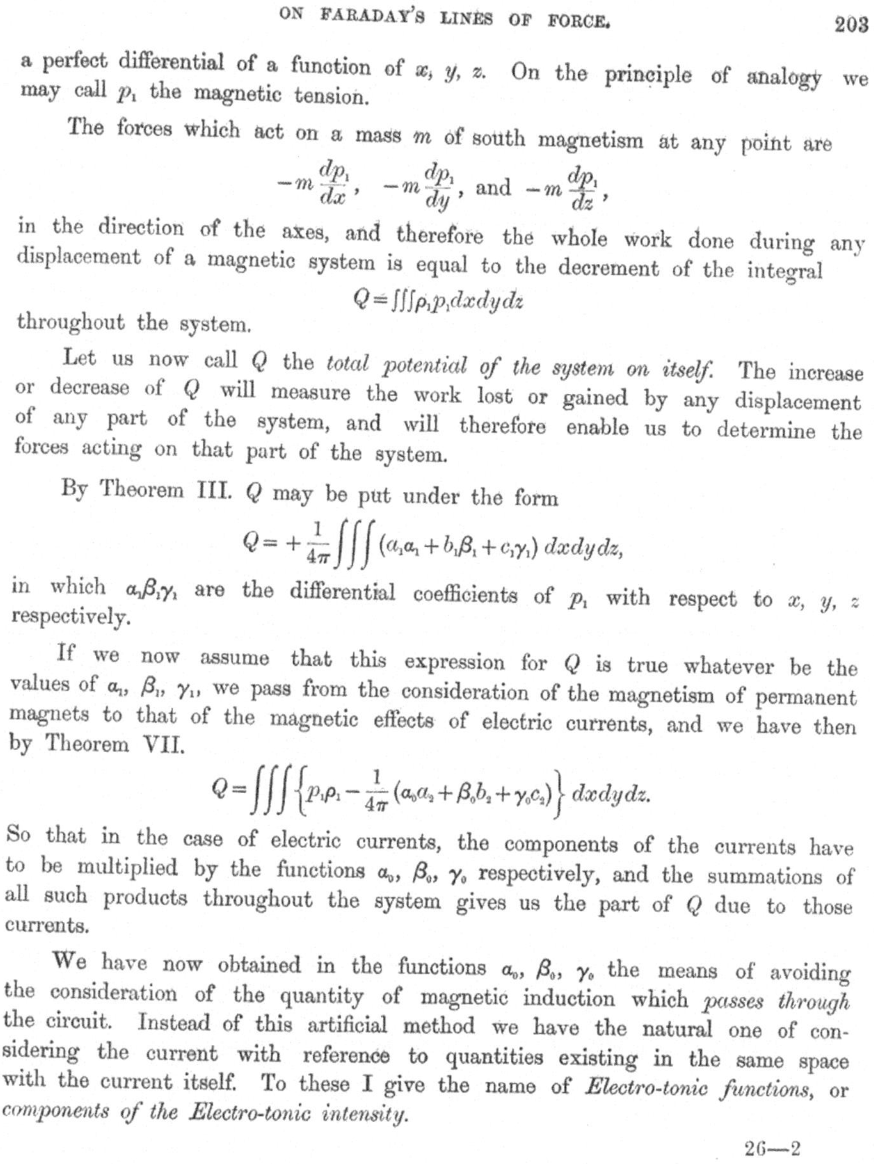

ON FARADAY'S LINES OF FORCE. 203

a perfect differential of a function of x, y, z. On the principle of analogy we may call p_1 the magnetic tension.

The forces which act on a mass m of south magnetism at any point are

$$-m\frac{dp_1}{dx},\quad -m\frac{dp_1}{dy},\ \text{and}\ -m\frac{dp_1}{dz},$$

in the direction of the axes, and therefore the whole work done during any displacement of a magnetic system is equal to the decrement of the integral

$$Q=\iiint \rho_1 p_1\,dx\,dy\,dz$$

throughout the system.

Let us now call Q the *total potential of the system on itself*. The increase or decrease of Q will measure the work lost or gained by any displacement of any part of the system, and will therefore enable us to determine the forces acting on that part of the system.

By Theorem III. Q may be put under the form

$$Q=+\frac{1}{4\pi}\iiint(a_1\alpha_1+b_1\beta_1+c_1\gamma_1)\,dx\,dy\,dz,$$

in which $\alpha_1\beta_1\gamma_1$ are the differential coefficients of p_1 with respect to x, y, z respectively.

If we now assume that this expression for Q is true whatever be the values of a_1, β_1, γ_1, we pass from the consideration of the magnetism of permanent magnets to that of the magnetic effects of electric currents, and we have then by Theorem VII.

$$Q=\iiint\left\{p_1\rho_1-\frac{1}{4\pi}(\alpha_0 a_2+\beta_0 b_2+\gamma_0 c_2)\right\}dx\,dy\,dz.$$

So that in the case of electric currents, the components of the currents have to be multiplied by the functions α_0, β_0, γ_0 respectively, and the summations of all such products throughout the system gives us the part of Q due to those currents.

We have now obtained in the functions α_0, β_0, γ_0 the means of avoiding the consideration of the quantity of magnetic induction which *passes through* the circuit. Instead of this artificial method we have the natural one of considering the current with reference to quantities existing in the same space with the current itself. To these I give the name of *Electro-tonic functions*, or *components of the Electro-tonic intensity*.

26—2

Fig. 6 Maxwell's integral describing the vector potential (Maxwell [read 1855 and 1856] 1855, p. 203) *Source* SICD Strasbourg University. Public Domain

while a dielectric is like an elastic membrane which may be impervious to the fluid, but transmits the pressure of the fluid on one side to that on the other.

As long as electromotive force acts on a conductor, it produces a current which, as it meets with resistance, occasions a continual transformation of electrical energy into heat, which is incapable of being restored again as electrical energy by any reversion of the process.

Electromotive force acting on a dielectric produces a state of polarization of its parts similar in distribution to the polarity of the particles of iron under the influence of a magnet*, and, like the magnetic polarization, capable of being described as a state in which every particle has its poles in opposite conditions.

In a dielectric under induction, we may conceive that the electricity in each molecule is so displaced that one side is rendered positively, and the other negatively electrical, but that the electricity remains entirely connected with the molecule, and does not pass from one molecule to another.

The effect of this action on the whole dielectric mass is to produce a general displacement of the electricity in a certain direction. This displacement does not amount to a current, because when it has attained a certain value it remains constant, but it is the commencement of a current, and its variations constitute currents in the positive or negative direction, according as the displacement is increasing or diminishing. The amount of the displacement depends on the nature of the body, and on the electromotive force; so that if h is the displacement, R the electromotive force, and E a coefficient depending on the nature of the dielectric,

$$R = -4\pi E^2 h;$$

and if r is the value of the electric current due to displacement,

$$r = \frac{dh}{dt}.$$

These relations are independent of any theory about the internal mechanism of dielectrics; but when we find electromotive force producing electric displacement in a dielectric, and when we find the dielectric recovering from its state of electric displacement with an equal electromotive force, we cannot help

* See Prof. Mossotti, "Discussione Analitica," *Memorie della Soc. Italiana* (Modena), Vol. xxiv. Part 2, p. 49.

62—2

Fig. 7 The page where Maxwell explains how to calculate the displacement current (Maxwell [1890] 2003, p. 491) *Source* French National Library–Gallica. Public Domain

In the ether, according to Maxwell's idea, there are *vortical circulatory* motions (Fig. 8) whose kinetic energy can be assimilated to the energy of the magnetic field; similarly, Faraday had proved that rotation phenomena could be associated with magnetism.

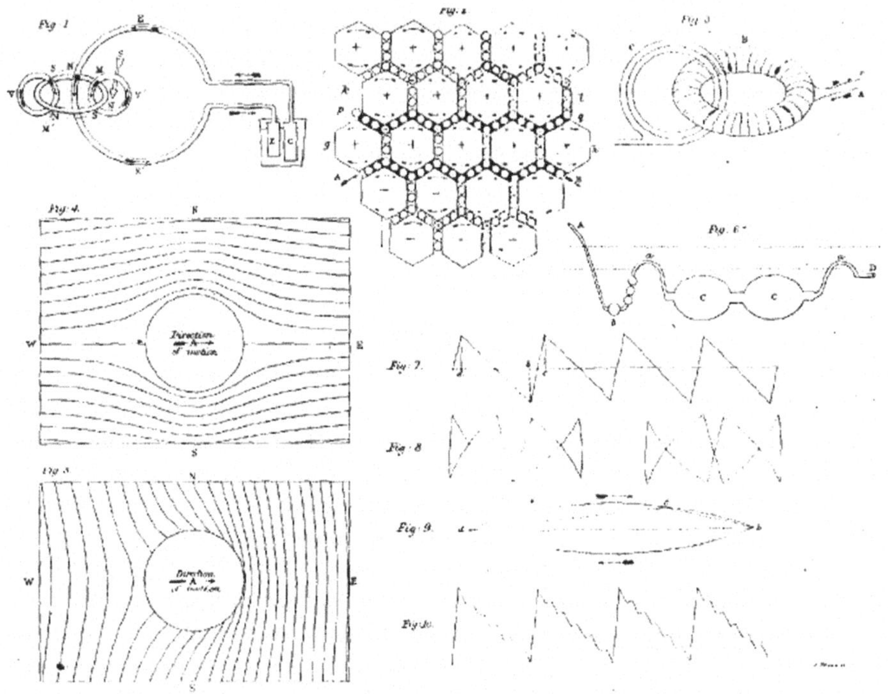

Fig. 8 Here, top centre, we can see the famous image where Maxwell depicts the vortexes of current (Maxwell 1855–1856; 1861–1862, III, Plate) *Source* French National Library–Gallica. Public Domain

The electrostatic energy corresponds, for Maxwell, to the elastic potential energy of the ether fluid and the conduction and displacement currents are represented by particles in tangential motion with respect to the vortices of the magnetic field. The model therefore takes on, as Faraday predicted, a physical meaning that goes beyond simple kinematic representation. Although the model was complicated, even by Maxwell's own admission, it was exploited by the author to obtain significant results such as, for example, the relationship between the induction of the current and the variation of the speed of rotation of that fluid called *magnetic field*.

Maxwell's third major work, preceding the extensive *Treatise* on electromagnetism, was *A Dynamic Theory of the Magnetic Field* read in 1864 and published in 1865 (Maxwell 1865). In this work the mathematical approach becomes predominant with respect to the preceding works, and in it we find Maxwell's intuition to consider electromagnetic energy as an intrinsic property of the field. This intuition was made possible by the analogies established by Maxwell between electromagnetic phenomena, such as the interactions between currents and magnets, and the behaviour of machines, described by mechanical devices such as gears, transmission belts and so on. These phenomena could be framed in the mechanical theory of *Reduced Momentum* (Maxwell 1864), which was particularly well suited to being addressed by the equations of analytical mechanics developed by Lagrange. This led

Maxwell to treat the phenomena of electromagnetism by means of this mathematical tool, as we shall see in detail in the *Treatise*. Actually, the mathematical instruments developed during the 18th Century would play a fundamental role in the development of completely new approaches to physical sciences (Fraser 1985, 1989, 1992, 1994; Fufay [1733] 1735; Pisano and Capecchi 2013; Joule 1843).

3 On Lagrange's Mechanics

In the field of classical Physics (e.g. in the study of Mechanics), one can physically describe a dynamic motion and thus obtain the mathematical equations of motion of free particles through a careful application (e.g. calculation of derivatives) of Newton's laws. If it is required to model some physical phenomena through mathematical physics, then an integral in time of a function called Lagrangian should be considered. What are the main features of this procedure? In general, a dynamical system is a mathematical concept in which a fixed implicit relation (i.e. a differential equation, a difference equation or another time scale) describes the time dependence of a point in a geometric space. It also concerns the state of a system only for a short period in the future. For example, to determine the state for all future times, an iterative process (such as solving/integrating the system) is required. By means of a deterministic process, given an initial point it is possible to determine all its future positions. That is, the set of known points represents the trajectory of the laws of motion used. However, some difficulties may arise, e.g. with respect to the ability to solve any kind of trajectory in any kind of system. Considering only deterministic time evolution, an ordinary differential equation can be used to represent the evolution of the system. For example:

$$\frac{dx(t)}{dt} = f(x(t), t) \tag{2}$$

where f is a mathematical function that defines the *dynamical system* with *n–degrees of freedom* (i.e., a n–dimensional vector field). Hence, a *dynamical system* whose equations of motion are established by means of the *Principle of least action*[9] is called a *dynamic Lagrangian system*. The *Lagrangian L* of *a dynamic Lagrangian* system is a mathematical function that describes the dynamics of the system. In

[9] Or, *Principle of stationary action*. At first glance, Pierre Louis Maupertuis (1698–1759) is usually considered prior for the principle due to his lectures at the *Académie des sciences* (1744, *Loi du repos des corps*; 1746, Accord de plusieurs lois naturelles qui avaient paru jusqu'ici incompatibles), and his paper *Les loix du mouvement et du repos déduites d'un principe métaphysique* (Maupertuis [1746] 1748). Euler also proposed his own formulation in the *Additamentum* 2 to his Methodus Inveniendi (Euler [1744] 1952). Analytical approach to motus is in (Euler 1736). The full account of the principle was stated by Lagrange in the transaction of the Accademia delle Scienze di Torino known as *Miscellanea philosophicomathematica Societatis privatae Taurinensis* (Lagrange [1766–1769] 1773, II). The work contains notations of the calculus of variations, as well as, presenting its application by deducing the principle of least action.

classical Mechanics, it is defined as the difference between the kinetic energy T and the potential energy V of the system:

$$L = T - V \tag{3}$$

If the Lagrangian, as expressed in (3), of a certain system is known, then the equations of motion of a *Langragian dynamical system* may be calculated by a direct substitution of the expression for the *Lagrangian* into what is nowadays called *Euler–Lagrange equation*:

$$\frac{d}{dt}\left(\frac{\partial L}{\partial \dot{x}_i}\right) - \frac{\partial L}{\partial x_i} = 0 \tag{4}$$

This is a differential equation whose solutions are the functions for which a given functional is stationary. Leonhard Euler (1707–1783) (Euler 1911) and Lagrange, in connection with their separate[10] studies (1750s) on the *tautochrone*,[11] analytically proposed a solution essentially related to (4). In 1755, Lagrange solved it and sent[12] the solution to Euler (Lagrange 1892, IV, 1755, Torino, 20 November, pp. 146–151) just after reception of Euler's letter on the Calculus of Variation[13] (Lagrange 1892, IV; 1755, Berlin, 6 September, pp. 144–146). Finally, both scholars developed Lagrange's method and applied it to their analytical mechanical problems (Fraser 1983). Today, the term Lagrangian[14] is generally introduced taking into account the reformulation of classical mechanics proposed by Lagrange in his *Mécanique analitique* (1788).

---

[10] *Methodus inveniendi* (1744) is the most significance work, which inspired Langrage. See also Euler's letter to Lagrange (Lagrange 1892, IV 1766, Berlin, 3, May, p. 209).

[11] A tautochrone or isochrone is a curve for which the time taken by an object sliding without friction in uniform gravity to its lowest point is independent of its starting point. The problem was anticipated by Johann Bernoulli (1667–1748) within his studies on the brachistochrone curve (*Problema novum*, 1696).

[12] In this letter, Lagrange signed himself as Ludovico de La Grange Tournier (Lagrange 1892, IV, p. 151). Other minor variations of his signature can also be read.

[13] The Lagrange–Euler correspondence (37 letters from 1754, 28th June to 1775, 23th March) (Lagrange 1892) are amazing documents in French and Latin. They produced interesting foundational reflexions and consecrated assumptions evidently concerning the novelty of the calculus of variation (Fraser 1985, 1992). On latter, early correspondences, essentially, begun on 1755, from Lagrange to Euler (Lagrange 1892, IV; 1755, 12th August, pp. 138–144): Lagrange revealed his new method on δ–symbolism for Euler's equation. Later in 1756, Euler himself coined the term *calculi variationum* for Lagrange's variation method when he entitled *Elementa calculi variationum* as his lecture for a meeting at the Berlin Academy (1756, 16th September). The work was published later (Euler [1764] 1766). If it is true that Euler established the main foundation and structure of the calculus of variations, Lagrange definitely performed its formalism and, above all, worked on the generalizations of the new calculus, no less important, for accrediting a new scientific theory. See also (Lagrange 1892, IV; Cfr. Euler [1750] 1752, 1775; Fraser 1989, 1994).

[14] We specify that the results within Langragian mechanics prescinds from difference T–V, and they are valid for the generalized form of the Lagrangian, as well.

Taking into account expression (3), an electric particle[15] of mass m, electrical charge q (coulombs) and a speed v (ms^{-1}) free to move in an electromagnetic field of scalar potential ϕ (volts) and vector potential A (Vsm^{-1}). The kinetic and potential energies of the particle are respectively:

$$T = \frac{1}{2} m \vec{v} \cdot \vec{v} \quad \text{and} \quad V = q\phi - q\vec{v} \cdot \vec{A} \tag{5}$$

It is not hard to show that from (5), the electrodynamics Lagrangian formulation[16] defined by the following equation can be obtained:

$$L = \frac{1}{2} m \vec{v} \cdot \vec{v} - q\phi + q\vec{v} \cdot \vec{A} \tag{6}$$

After the description given above of the main features of the Lagrangian conceptualization and its mathematical formulation, in the following, the methods and Lagrangian formulation of Maxwell's electrodynamics field are briefly presented. In "*On the induction of a current on itself*" (Maxwell 1873, II, Part IV, Chap. IV, pp. 180–183):

- Maxwell reminded the reader of some phenomena of electromagnetic induction and self–induction described by Faraday;
- and only then he created an analogy between a current in a wire and a fluid flowing in a pipe through the idea of inertia and momentum.

Analysing in detail the two phenomena, Maxwell—as above cited—remarked that this parallelism cannot persist during electromagnetic induction effects (Maxwell 1873, II, Part IV). While in the motion of the fluid in a pipe only internal characteristics of the tube are involved, so—in the case of wires crossed by the same current—the behaviour changes both for changes in shape and for the presence of bodies outside the wire. In Chap. IV (Maxwell 1873) dealt with energy and still made an interesting comparison between mechanical and electrical phenomena. As a conventional force increases the energy of the body to which it is applied so a current leads to effects that produce a certain type of energy. Maxwell, in support of his statement, described effects related to currents that lead to energy production in different forms such as the Joule effect or mechanical work resulting in movement of magnets. At the end of Chap. IV (Maxwell 1873, p. 183; Pisano et al. 2019, 2020) the working hypothesis and objectives are declared as established by Langrage's methods:

> 552.] What I propose now to do is to examine the consequences of the assumption that the phenomena of the electric current are those of a moving system, the motion being

[15] A modern description of a Lagrangian formulation concerns the interaction of an electromagnetic field with a charged particle.

[16] There is insufficient space here to present the covariant formulation of Lagrangian in Maxwell's electromagnetic theory, as well. It refers to ways of writing forms of (modern 4–expressions) Maxwell's equations and Lorentz's force as invariants under Lorentz's transformations (i.e., in special relativity using rectilinear inertial coordinate systems). One of the aims is to show that laws of classical electromagnetism assume *idem* form in any inertial coordinate system.

> communicated from one part of the system to another by forces, the nature and laws of which we do not yet even attempt to define, because we can eliminate these forces from the equations of motion by the method given by Lagrange for any connected system. (Maxwell 1873, II, Part IV, Chap. IV, p. 183)

In other words, to analyse:

(a) The consequences of the hypothesis that the phenomena tied to the electric current are those of a system in motion in which the motion is communicated from one side to the other of the system through forces;
(b) The nature and laws that govern these forces are not explained because they are eliminated in a formulation of the equations of motion deduced by the applying Lagrange method for connected systems.

Maxwell's theory became a field theory making use of Lagrange equations (Pisano 2014) of motion and without a priority of physical measurements to remain within the physics mathematics domain (Joule 1845). Maxwell formulated his momentum concept (Maxwell 1873, II, Part IV, V, pp. 190–191) as one of the basic concepts to build his *Dynamical theory of electromagnetism* (Maxwell 1873, VI, pp. 195–205). In order to proceed toward his dynamical project he needed to establish the mathematical conditions to write a form of the equations capable of describing a particular[17] physical system within electromagnetic theory. In his words:

> 567.] Such a description of the methods of pure dynamics is not unnecessary, because [of] Lagrange and most of his followers, to whom we are in debt for these methods [...]. (Maxwell 1873, II, Part IV, Chap. V, p. 194)
>
> As the development of the ideas and methods of pure mathematics has rendered it possible, by forming a mathematical theory of dynamics, to bring to light many truths which could not have been discovered without mathematical training, so, if we are to form dynamical theories of other sciences, we must have our minds imbued with these dynamical truths as well as with mathematical methods. (Maxwell 1873), II, Part IV, Chap. V, p. 194)

Maxwell looked for (Maxwell 1873, II, Part IV, Chap. VI) adequate coefficients to convert a previous general equation (Maxwell 1873, II, Part IV, Chap. VI, pp. 198–199). From a strictly mathematical point of view, he pointed out three categories among coordinates q_i (Maxwell 1873, pp. 197–198) obtaining the total energy T divided in three parts generating terms of three forms: $x_i x_j$, $y_i y_j$, and $x_i y_j$. Non-homogeneous coordinates are combined in terms of the form $x_i y_j$. From a physical point of view they are mechanical (x) and electrical (y) variables. These products give rise to three corresponding sources of kinetic energy when they appear in the equation for T.

> 571.] [...]. We may therefore divide T into three portions, in the first of which T_m, the velocities of the coordinates x only occur, while in the second, T_e, the velocities of the coordinates y only occur, and in the third, T_{me}, each term contains the product of the velocities of two coordinates of which one is x and the other y. Where:

$$T_m = \frac{1}{2}(x_1 x_1)\dot{x}_1^2 + \&c. + (x_1 x_2)\dot{x}_1 \dot{x}_2 + \&c.$$

[17] In the previous chapter V Maxwell had pointed out (from Lagrange) a general form of the equations of motion which was not able to describe a particular physical system.

$$T_e = \frac{1}{2}(y_1 y_1)\dot{x}_1^2 + \&c. + (y_1 y_2)\dot{y}_1 \dot{y}_2 + \&c.$$

$$T_{me} = \frac{1}{2}(x_1 y_1)\dot{x}_1 \dot{y}_1 + \&c.$$

These physics mathematics arguments dealt with three methods:

> […] of detecting the existence of the terms of the form T_m, none of which have hitherto led to any positive results. (Maxwell 1873, II, Part IV, Chap. VI, pp. 198–199).

Therefore, the equations from Chap. VI were carried to bear on a simple electromagnetic system consisting of only two circuits in *Theory of electric circuits* (Maxwell 1873, II, Part IV, VII).

In 1788, Joseph–Louis Lagrange (1763–1813) published his *Méchanique analitique.* Other studies mainly were by Alexis Clairaut (1713–1765), Jean–Baptiste le Rond d'Alembert (1717–1783), Pierre–Simon Laplace (1749–1827), Leonhard Euler (1707–1783), Johann Bernoulli (1667–1748) and Jacob Bernoulli (1655–1705). This is clear from his introduction: Lagrange shifted the geometrical Newtonian paradigm of the *Philosophiae Naturalis Principia Mathematica* to a new one founded on mathematical analysis (Fig. 9).

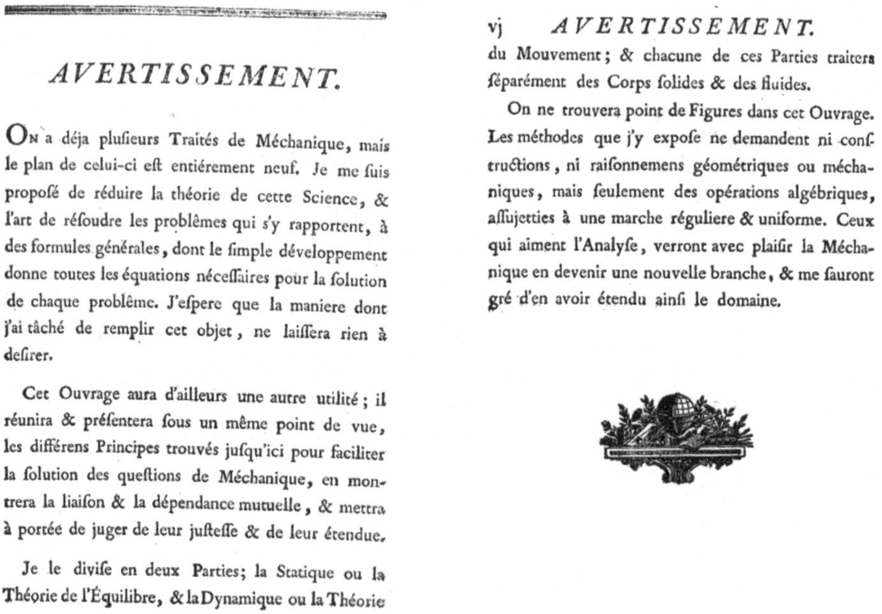

AVERTISSEMENT.

On a déja pluſieurs Traités de Méchanique, mais le plan de celui-ci eſt entiérement neuf. Je me ſuis propoſé de réduire la théorie de cette Science, & l'art de réſoudre les problêmes qui s'y rapportent, à des formules générales, dont le ſimple développement donne toutes les équations néceſſaires pour la ſolution de chaque problême. J'eſpere que la maniere dont j'ai tâché de remplir cet objet, ne laiſſera rien à deſirer.

Cet Ouvrage aura d'ailleurs une autre utilité; il réunira & préſentera ſous un même point de vue, les différens Principes trouvés juſqu'ici pour faciliter la ſolution des queſtions de Méchanique, en montrera la liaiſon & la dépendance mutuelle, & mettra à portée de juger de leur juſteſſe & de leur étendue.

Je le diviſe en deux Parties; la Statique ou la Théorie de l'Équilibre, & la Dynamique ou la Théorie

vj *AVERTISSEMENT.*

du Mouvement; & chacune de ces Parties traitera ſéparément des Corps ſolides & des fluides.

On ne trouvera point de Figures dans cet Ouvrage. Les méthodes que j'y expoſe ne demandent ni conſtructions, ni raiſonnemens géométriques ou méchaniques, mais ſeulement des opérations algébriques, aſſujetties à une marche réguliere & uniforme. Ceux qui aiment l'Analyſe, verront avec plaiſir la Méchanique en devenir une nouvelle branche, & me ſauront gré d'en avoir étendu ainſi le domaine.

Fig. 9 Left side: Lagrange explains in this introduction (right side) that his *Treatise* will bring together the different principles issued so far. Right side: Lagrange declares his new approach, different from the Newtonian one is going to be applied within the text so the reader will find a Mathematical Analysis–based on development of the different subjects (Lagrange 1788, I, pp. v–vj) *Source* French National Library–Gallica. Public Domain

In the following we analyse and explain Lagrange's dynamics formulation that describes his formulation (Lagrange 1788, II) then used by Maxwell (1873, II, Parts IV–V).

The Second part of *Méchanique analitique* (Lagrange 1788) deals with considerations on dynamics, and Lagrange offers the reader the following definition: dynamics is the science of accelerating and retarding forces and of the varied motions they can produce. We remark that Lagrange dealt with accelerating and retarding forces, instead of the single term acceleration, and in accordance with the common sense that sees objects gaining speed, and thus accelerating, or losing it, and thus slowing down, when they are under the effect of one or more forces.

In *General Formula for the Motion of a System of Bodies Animated by any Forces* (Lagrange 1788, Part II, Sect. II, § 7) Lagrange synthesizes all the results he obtained in a single formula that we quote:

$$S\left(\frac{d^2x}{dt^2}\delta x + \frac{d^2y}{dt^2}\delta y + \frac{d^2z}{dt^2}\delta z + P\delta p + Q\delta q + R\delta r + etc.\right)m = 0 \tag{7}$$

The first three terms in brackets refer, in particular, to the forces acting in preferential directions such as those of the three co–ordinates, while the other terms (Lagrange indicates etc. in order to maintain a condition of generality and not to mislead the reader by reporting only three terms) are the so–called accelerating forces each with its own direction of action.

In the Fourth Section of Part Two (Lagrange 1788, Part II, Sect. IV) Lagrange takes up this equation, which we observe contains no reference to the forces acting between the individual parts of the system. Considering each time the mechanical problem to be solved and therefore the conditions it dictates, reducing the different variables by replacing them with functions of different variables, it is possible to arrive at the solution of any mechanical problem. To show how to work with (2), the author proposed the following.

He broke the equation into two terms which he calls respectively Γ and Δ on which he worked on separately, in detail:

$$\Gamma = S\left(\frac{d^2x}{dt^2}\delta x + \frac{d^2y}{dt^2}\delta y + \frac{d^2z}{dt^2}\delta z\right)m \tag{8}$$

$$\Delta = S(P\delta p + Q\delta q + R\delta r + etc.)m \tag{9}$$

Consequently, this can be even more simply constructed as $\Gamma + \Delta = 0$. Then Lagrange considered the following expression:

$$d^2x\delta x + d^2y\delta y + d^2z\delta z \tag{10}$$

which consists, in fact, of the numerators of (8) and aims to find a way of expressing it through different variables. In this respect, he proceeds as in the following.

He completes (10) with three terms in order to obtain an *integrable* form such as:

$$d^2x\delta x + d^2y\delta y + d^2z\delta z + dxd\delta x + dyd\delta y + dzd\delta z \tag{11}$$

which is rewritten as follows:

$$d^2x\delta x + dxd\delta x + d^2y\delta y + dyd\delta y + d^2z\delta z + dzd\delta z \tag{12}$$

appears to be integrated through the first rule of differentiation to give this writing:

$$dx\delta x + dy\delta y + dz\delta z \tag{13}$$

Therefore, Eq. (10) becomes:

$$d^2x\delta x + d^2y\delta y + d^2z\delta z = d(dx\delta x + dy\delta y + dz\delta z) - (dxd\delta x + dyd\delta y + dzd\delta z) \tag{14}$$

Lagrange resorts to a result that he proved in Section Three (Lagrange 1788, Part II, Sect. III, § 12, pp. 209–211) namely that the writing dδ is equivalent to writing δd and therefore the second parenthesis of the right-hand member of (14) is rewritten as:

$$dx\delta dx + dy\delta dy + dz\delta dz \tag{15}$$

which integrates into the:

$$\frac{1}{2}\delta\left(dx^2 + dy^2 + dz^2\right) \tag{16}$$

Therefore:

$$\begin{aligned} d^2x\delta x + d^2y\delta y + d^2z\delta z \\ &= d(dx\delta x + dy\delta y + dz\delta z) \\ &- \frac{1}{2}\delta\left(dx^2 + dy^2 + dz^2\right) \end{aligned} \tag{17}$$

Therefore, the question posed by Lagrange is resolved by evaluating the two parentheses of the right–hand member of (17) that contain only first–order differentials (squared in the case of the second parenthesis). In order to replace the variables x, y and z with functions of variables ξ, ψ and φ expressions of the following form are constructed:

$$dx = Ad\xi + Bd\psi + Cd\varphi + etc. \tag{18}$$

$$dy = A^{'}d\xi + B^{'}d\psi + C^{'}d\varphi + etc. \tag{19}$$

$$dz = A^{''}d\xi + B^{''}d\psi + C^{''}d\varphi + etc. \tag{20}$$

$$\delta x = A_{*}d\xi + B_{*}d\psi + C_{*}d\varphi + etc. \tag{21}$$

$$\delta y = A^{'}{}_{*}d\xi + B^{'}{}_{*}d\psi + C^{'}{}_{*}d\varphi + etc. \tag{22}$$

$$\delta z = A^{''}{}_{*}d\xi + B^{''}{}_{*}d\psi + C^{''}{}_{*}d\varphi + etc. \tag{23}$$

where we stress that Eqs. (21), (22) and (23) are not given in the original article but are functional for the calculations that follow. The terms denoted by A, A', A'', B, B', B''... are functions of ξ, ψ and φ. If the latter expressions are substituted in (13), after elementary algebra calculations we obtain:

$$Fd\xi\delta\xi + G(d\xi\delta\psi + d\psi\delta\xi) + Hd\psi\delta\psi + I(d\xi\delta\varphi + d\varphi\delta\xi) + etc. \tag{24}$$

where F, G, H, I,... are functions of ξ, ψ and φ. The expression in brackets (16) becomes, by substitution of δ in d, the following one:

$$Fd\xi^2 + 2Gd\xi d\psi + Hd\psi^2 + 2Id\xi d\varphi + etc. \tag{25}$$

At this point, differentiating Eqs. (24) for d and (25) for δ, one obtains two expressions that Lagrange gave as follows:

$$\begin{aligned} d\cdot(Fd\xi)\times\delta\xi + Fd\xi d\delta\xi + d\cdot(Gd\xi)\times\delta\psi + d\cdot(Gd\psi)\times\delta\xi \\ +Gd\xi d\delta\psi + Gd\psi d\delta\xi + d\cdot(Hd\psi)\times\delta\psi \\ +Hd\psi d\delta\psi + etc. \end{aligned} \tag{26}$$

and:

$$\begin{aligned} \delta Fd\xi^2 + 2Fd\xi\delta d\xi + 2\delta Gd\xi d\psi + 2Gd\psi\delta d\xi + 2Gd\xi\delta d\psi \\ +\delta Hd\psi^2 + 2Hd\psi\delta d\psi + etc. \end{aligned} \tag{27}$$

Lagrange divided the (27) by 2 and subtracts it from (26) to finally construct the right–hand member of (17) in its final form with the substitution of the variables:

$$\begin{aligned} d\cdot(Fd\xi)\times\delta\xi - \frac{1}{2}\delta Fd\xi^2 + d\cdot(Gd\xi)\times\delta\psi + d\cdot(Gd\psi)\times\delta\xi \\ - \frac{1}{2}\delta Gd\xi d\psi + d\cdot(Hd\psi)\times\delta\psi - \frac{1}{2}\delta Hd\psi^2 + etc. \end{aligned} \tag{28}$$

and this is, for Lagrange, the value that assumes:

$$d^2x\delta x + d^2y\delta y + d^2z\delta z \tag{29}$$

As we can see, and as the author himself pointed out several times (Lagrange 1788, Part II, Sect. IV), reaching this result is a matter of pure calculation, a process that allows us to appreciate Lagrange's working methodology, which is made up of rigorous logical–mathematical speculation, without recourse to schemes that would not add substantial supports for understanding.

Following Lagrange (Lagrange 1788, Part II, Sect. IV, § 4, pp. 218–220), if one algebraically elaborates expression (27), i.e., by dividing all the terms by 2 and changing the signs of the terms that do not contain δd, then one obtains:

$$2Fd\xi\delta d\xi \rightarrow d\cdot(Fd\xi)\times\delta\xi \tag{30}$$

$$2Gd\psi\delta d\xi \rightarrow d\cdot(Gd\psi)\times\delta\xi \tag{31}$$

$$2Hd\psi\delta d\psi \rightarrow d\cdot(Hd\psi)\times\delta\psi \tag{32}$$

Idem if the characteristic δ operates on the last variable and characteristic d operates on the first difference. If we identify with the letter the function of the variables ξ, ψ, φ and dξ, dψ, dφ the function into which the writing is transformed:

$$\frac{1}{2}(dx^2 + dy^2 + dz^2) \tag{33}$$

Lagrange finally obtained the final form sought:

$$\begin{aligned} d^2x\delta x + d^2y\delta y + d^2z\delta z \\ &= \left(-\frac{\delta\alpha}{\delta\xi} + d\cdot\frac{\delta\alpha}{\delta d\xi}\right)\delta\xi + \left(-\frac{\delta\alpha}{\delta\psi} + d\cdot\frac{\delta\alpha}{\delta d\psi}\right)\delta\psi \\ &+ \left(-\frac{\delta\alpha}{\delta\varphi} + d\cdot\frac{\delta\alpha}{\delta d\varphi}\right)\delta\varphi + etc. \end{aligned} \tag{34}$$

The next paragraph (Lagrange 1788, Part II, Sect. IV, § 6, pp. 220–223) the sixth, is devoted to obtaining the same result by applying the principles of the method of variations, the one we know today as the Euler–Lagrange method.

The conclusion reached by Lagrange, therefore, is that to obtain the value of the first quantity Γ as a function of the variables ξ, ψ, φ it will be sufficient to determine the value of the quantity:

$$S\left(\frac{dx^2 + dy^2 + dz^2}{2dt^2}\right)m \tag{35}$$

which we easily recognise as having the structure of the quantity defined as kinetic energy, as a function of ξ, ψ, φ and the respective differentials, to obtain:

$$\Gamma = \left(d \cdot \frac{\delta T}{\delta d\xi} - \frac{\delta T}{\delta \xi}\right)\delta\xi + \left(d \cdot \frac{\delta T}{\delta d\psi} - \frac{\delta T}{\delta \psi}\right)\delta\psi + \left(d \cdot \frac{\delta T}{\delta d\varphi} - \frac{\delta T}{\delta \varphi}\right)\delta\varphi + etc. \tag{36}$$

The calculation of the second part of the $\Gamma + \Delta = 0$ is much less laborious because it is a matter of rewriting separately the different perpendicular distances of the forces from their pivotal points, what Lagrange simply calls "distances". Lagrange observes that the sum of the moments is indeed *intégrable* and therefore writes, in the tenth paragraph of the Third Section:

$$d\Pi = Pdp + Qdq + Rdr + etc. \tag{37}$$

Therefore, it will also be:

$$\delta\Pi = P\delta p + Q\delta q + R\delta r + etc. \tag{38}$$

and consequently we have:

$$\Delta = S\delta\Pi m = \delta \cdot S\Pi m \tag{39}$$

because the sign of sum does not depend on δ. The value of the expression $S\Pi m$, called V, allows the solution to be obtained in these terms:

$$\Delta = \frac{\delta V}{\delta \xi}\delta\xi + \frac{\delta V}{\delta \psi}\delta\psi + \frac{\delta V}{\delta \varphi}\delta\varphi + etc. \tag{40}$$

Finally, one obtains the general formula of motion according to Lagrange, a formula that gives a writing of the type $\Gamma + \Delta = 0$ becomes:

$$\Xi\delta\xi + \Psi\delta\psi + \Phi\delta\varphi + etc. = 0 \tag{41}$$

where:

$$\Xi = d \cdot \frac{\delta T}{\delta d\xi} - \frac{\delta T}{\delta \xi} + \frac{\delta V}{\delta \xi} \tag{42}$$

$$\Psi = d \cdot \frac{\delta T}{\delta d\psi} - \frac{\delta T}{\delta \psi} + \frac{\delta V}{\delta \psi} \tag{43}$$

$$\Phi = d \cdot \frac{\delta T}{\delta d\varphi} - \frac{\delta T}{\delta \varphi} + \frac{\delta V}{\delta \varphi} etc. \tag{44}$$

Moreover, if we suppose that:

$$T = S\left(\frac{dx^2 + dy^2 + dz^2}{2dt^2}\right)m \tag{45}$$

$$V = S\Pi m \tag{46}$$

$$\delta\Pi = P\delta p + Q\delta q + R\delta r + etc. \tag{47}$$

In this way Lagrange reduced the solution of the problem to the solution of as many equations as there are degrees of freedom in the system, reducing the variables to as few as possible.

4 Lagrangian Formulation in Maxwell's *Treatise* (1873)

4.1 Lagrange's Methods and Lagrangian in Maxwell's Treatise

As we pointed out earlier, Lagrange's work (1788) was based on a solid logical–deductive mathematical structure, leaving no room for geometrical digressions involving the use of graphs or figures. By developing step–by–step relationships between the algebraic quantities in the equations functional to the discussion, Lagrange demonstrated that a mechanical problem can be solved by means of a minimum number of equations equal to the number of degrees of freedom of the system. In his well–known work *A Treatise on Electricity and Magnetism*, Maxwell adopted Lagrange's approach and set out to bring to light a methodology that would establish as clear as possible a relationship between the mathematical approach to electromagnetism and dynamics, a science he himself described as fundamental (Maxwell 1873, p. vi; see also Maxwell 1875; Fig. 10).

vi PREFACE.

most important of these phenomena, to shew how they may be subjected to measurement, and to trace the mathematical connexions of the quantities measured. Having thus obtained the data for a mathematical theory of electromagnetism, and having shewn how this theory may be applied to the calculation of phenomena, I shall endeavour to place in as clear a light as I can the relations between the mathematical form of this theory and that of the fundamental science of Dynamics, in order that we may be in some degree prepared to determine the kind of dynamical phenomena among which we are to look for illustrations or explanations of the electromagnetic phenomena.

In describing the phenomena, I shall select those which most clearly illustrate the fundamental ideas of the theory, omitting others, or reserving them till the reader is more advanced.

The most important aspect of any phenomenon from a mathematical point of view is that of a measurable quantity. I shall therefore consider electrical phenomena chiefly with a view to their measurement, describing the methods of measurement, and defining the standards on which they depend.

In the application of mathematics to the calculation of electrical quantities, I shall endeavour in the first place to deduce the most general conclusions from the data at our disposal, and in the next place to apply the results to the simplest cases that can be chosen. I shall avoid, as much as I can, those questions which, though they have elicited the skill of mathematicians, have not enlarged our knowledge of science.

Fig. 10 Dynamics is "fundamental" science (Maxwell 1873, I, p. vi) *Source* French National Library–Gallica. Public Domain

In particular, in *On the Induction of a Current on Itself* (Maxwell 1873, II, Part IV, IV–VI) Maxwell quoted specific parts of Michael Faraday's work *Experimental Researches in Electricity* (Faraday 1839–1855) about the phenomena shown by the current in a solenoidal coil of an electromagnet. According to Maxwell, Faraday observes that current possesses properties comparable to momentum or inertia, i.e. mechanical characteristics. For this reason, Maxwell made an intellectual leap and uses, through analogy, the mechanical paradigm, and the fluid–dynamic one in particular, to give a description of the phenomena at the basis of his research, providing the reader with experiences of the behaviour of water inside a conduit described on the basis of the fluid's properties of inertia. However, Maxwell observed, water pipes and electrical wires are not exactly the same thing, electrical wires behave very differently and yet, again according to Maxwell, the idea of analogy is what will lead, sooner or later as he says, to a complete understanding of the subject. Thinking about the motion of a fluid in a conduit means focusing on the characteristics of the fluid and on the internal structure of the conduit, length and cross–section; if a current flows in a conduit it will be influenced by the shape of the conduit and by the presence of possible bodies outside, such as a piece of metal or a closed circuit close to the current conducting wire. From this point on, Maxwell made the fundamental paradigm shift, moving away from the concept of force as expressed by Newton, i.e. agents responsible for motion, embracing the concept of energy, kinetic energy in particular, and invoking the definition of energy as the power to do work.

> 551.] In the case of the electric current, we find that, when the electromotive force begins to act, it does not at once produce the full current, but that the current rises gradually. What is the electromotive force doing during the time that the opposing resistance is not able to balance it? It is increasing the electric current. Now, an ordinary force, acting on a body in the direction of its motion, increases its momentum and communicates to it kinetic energy, or the power of doing work on account of its motion. [...]
>
> 552.] It appears, therefore, that a system containing an electric current is a seat of energy of some kind; and since we can form no conception of an electric current except a kinetic phenomenon, its energy must be kinetic energy, that is to say, the energy which a moving body has in virtue of its motion. (Maxwell 1873, II, Part IV, Chap. IV, p. 182)

Maxwell had no necessity to distinguish among the different definitions of energy, kinetic and potential, because his idea was initially very clear in this respect. For Maxwell, a system in which electric current circulates is therefore a reservoir of energy through which work of some kind can be done: this type of energy must necessarily be kinetic, the system is in some kind of motion. It is now time for a paradigm shift: the phenomena of electromagnetism are those of a connected mechanical system.

> 552.] [...] What I propose now to do is to examine the consequences of the assumption that the phenomena of the electric current are those of a moving system, the motion being communicated from one part of the system to another by forces, the nature and laws of which we do not yet even attempt to define, because we can eliminate these forces from the equations of motion by the method given by Lagrange for any connected system. In the next five chapters of this treatise I propose to deduce the main structure of the theory of electricity from a dynamical hypothesis of this kind. [...] (Maxwell 1873, II, Part IV, Chap. IV, p. 183)

At the beginning of chapter five (Maxwell 1873, II, Part IV, V) the style of the text changes radically, first because Maxwell immediately refers to Lagrange and his *Méchanique Analytique*, meticulously describing the theory he is going to apply, then because in the following chapters he will engage in a real systematisation of the possible terms appearing in Lagrange's equations of motion for a connected system, placing them in electromagnetism where congruent. The experimental information with which Maxwell provides the reader is sufficient to support the statements made, and even sufficient to provide for further fundamental experiments to prove them. When Maxwell's equations are finally written down, they follow an orderly logical thread that makes it possible to realise what dynamic meaning they embody (Simpson 2005, p. 250).

The *Treatise* can be considered the work that concludes a path in Maxwell's scientific literature: *On Faraday's Lines of Force* (1855–6) (Maxwell [read 1855 and 1856] 1855), *On Physical Lines of Force* (Maxwell 1855–1856; 1861–1862) and *A Dynamical Theory of the Electromagnetic Field* (1864–1865) (Maxwell 1864), and each of these treatises takes a different approach to understanding electromagnetic phenomena. *On Physical Lines of Force* could be considered the paper that follows the conventional route, supported by the right experiment at the right time to test the hypotheses made. As above discussed, Maxwell dealt with matter and motion without including mass under a Lagrangian applied to an electromagnetic theory formulation (Cfr. Simpson 2005, p. 252).

In his *Treatise*, Maxwell presents and applies Lagrange's methods and equations, in particular, in Part IV (Maxwell 1873, II, Part. IV, Chap. V, pp. 189–192). His aim, in the field of mathematical physics, is to clarify the basic and complex physical structure of a natural phenomenon (e.g. the model of a field). To successfully carry out such a high–level scientific speculation, Maxwell needed a more dynamic and adequate mathematics, certainly far from the one employed by Newton (Newton [1687] 1803). Lagrange's equations of motion are expressed in terms of rates of change, so a form of a differential equation was adopted. Maxwell pointed out a physics mathematics conceptualization of the concept of field and tried to retrieve a geometrical and mathematical solution in the paragraph *On induction of electric current* (Maxwell 1873, II, Part IV, Chap. III, pp. 172–173). Thus, Maxwell.

> *firstly accepted Lagrange's method and, in a second moment only, he accepted his equations to describe electromagnetic problems* (Maxwell 1873, II, Part IV, V, pp. 184–186; see also pp. 191–194).

In the fifth chapter of his *Treatise* (Maxwell 1873, II, Part IV, Chap. V, pp. 184–194)—and contrary to the Newtonian paradigm – Maxwell expressed motion and energy relationships within the system as a whole, rather than in terms of laws of motion governing the actions of forces. Thus, after several methods proposed (Maxwell 1873, II, Part IV, Chap. V, pp. 184–186) he announced a *third method*[18] related to Lagrange (Maxwell 1873, II, Part IV, Chap. V, p. 191; Fig. 11):

[18] See also Chapter VI where he—only after presentation of the Lagrangian frameworks –passed to describing his physical system within electromagnetic theory.

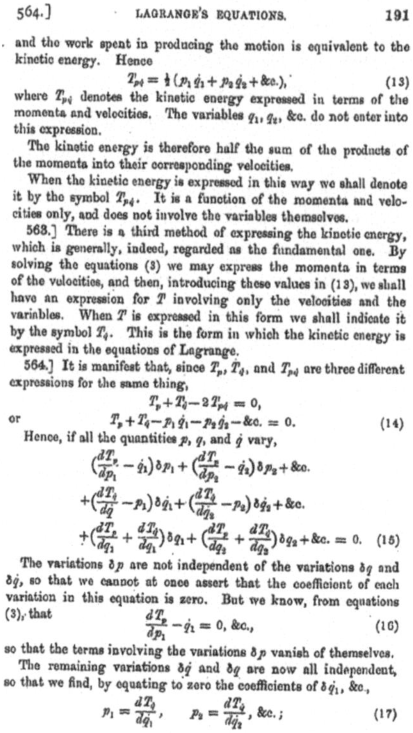

564.] LAGRANGE'S EQUATIONS. 191

and the work spent in producing the motion is equivalent to the kinetic energy. Hence

$$T_{p\dot{q}} = \tfrac{1}{2}(p_1 \dot{q}_1 + p_2 \dot{q}_2 + \&c.), \quad (13)$$

where $T_{p\dot{q}}$ denotes the kinetic energy expressed in terms of the momenta and velocities. The variables q_1, q_2, &c. do not enter into this expression.

The kinetic energy is therefore half the sum of the products of the momenta into their corresponding velocities.

When the kinetic energy is expressed in this way we shall denote it by the symbol $T_{p\dot{q}}$. It is a function of the momenta and velocities only, and does not involve the variables themselves.

563.] There is a third method of expressing the kinetic energy, which is generally, indeed, regarded as the fundamental one. By solving the equations (3) we may express the momenta in terms of the velocities, and then, introducing these values in (13), we shall have an expression for T involving only the velocities and the varinbles. When T is expressed in this form we shall indicate it by the symbol $T_{\dot{q}}$. This is the form in which the kinetic energy is expressed in the equations of Lagrange.

564.] It is manifest that, since T_p, $T_{\dot{q}}$, and $T_{p\dot{q}}$ are three different expressions for the same thing,

$$T_p + T_{\dot{q}} - 2T_{p\dot{q}} = 0,$$

or

$$T_p + T_{\dot{q}} - p_1\dot{q}_1 - p_2\dot{q}_2 - \&c. = 0. \quad (14)$$

Hence, if all the quantities p, q, and $\dot{q}$ vary,

$$\left(\frac{dT_p}{dp_1} - \dot{q}_1\right)\delta p_1 + \left(\frac{dT_p}{dp_2} - \dot{q}_2\right)\delta p_2 + \&c.$$
$$+\left(\frac{dT_{\dot{q}}}{d\dot{q}} - p_1\right)\delta\dot{q}_1 + \left(\frac{dT_{\dot{q}}}{d\dot{q}_2} - p_2\right)\delta\dot{q}_2 + \&c.$$
$$+\left(\frac{dT_p}{dq_1} + \frac{dT_{\dot{q}}}{dq_1}\right)\delta q_1 + \left(\frac{dT_p}{dq_2} + \frac{dT_{\dot{q}}}{dq_2}\right)\delta q_2 + \&c. = 0. \quad (15)$$

The variations δp are not independent of the variations δq and $\delta\dot{q}$, so that we cannot at once assert that the coefficient of each variation in this equation is zero. But we know, from equations (3), that

$$\frac{dT_p}{dp_1} - \dot{q}_1 = 0, \&c., \quad (16)$$

so that the terms involving the variations δp vanish of themselves.

The remaining variations $\delta\dot{q}$ and δq are now all independent, so that we find, by equating to zero the coefficients of $\delta\dot{q}_1$, &c.,

$$p_1 = \frac{dT_{\dot{q}}}{d\dot{q}_1}, \quad p_2 = \frac{dT_{\dot{q}}}{d\dot{q}_2}, \&c.; \quad (17)$$

Fig. 11 The Lagrangian, the third method on kinetic energy (Maxwell 1873, II, Part IV, Chap. V, p. 191. Please see also p. 192) *Source* French National Library–Gallica. Public Domain

With Lagrangian, the field is a continuum and it remains so in the whole and in substance undivided.
Since Maxwell aimed to formulate a dynamical justification to field equations, in this part, he focused on the fact that magnetic field appeared as a completely kinetic system.
Therefore, the energy of the system should be kinetic [Maxwell 1873, II, Part IV, Chap. V, pp. 189–1920].
Of course, a strict relationship between Lagrange's equations and the second law becomes evident for a single moving body. Particularly, dT/dq_i (in a system of multi–variables) that is the dependence of T on the variable qi, is not only a mathematical derivative operation.
It represents a physical variation of qi. In other words, the state of equilibrium of the system is unchanged while qi is varied (Simpson 2005, p. 65). This evidently represents, in effect, a sort of thought –experiment.

The derivative considered in this mathematical conceptualisation is today known to be the partial derivative of T with respect to q_i (Maxwell 1873, II, Part IV, Chap. V, p. 191). Subsequently, the theory essentially became a field theory making use of Lagrange's equations of motion. This proposal would be followed without a priority of physical measurements, that is, the idea was to remain within this new physics mathematics domain created (Pisano 2013). In order to proceed toward his dynamical project, Maxwell needed to establish the mathematical conditions to write the equations under a form suitable to describe a particular physical system within electromagnetic theory. By following differentiation of the energy function and by using Lagrange's equations of motion, Maxwell was able to formulate his concept of momentum (Maxwell 1873, II, Part IV, Chap. V, pp. 190–193), i.e., one of the basic concepts necessary to build his dynamical theory, which he also presented in Chap. VI (Maxwell 1873, II, Part IV, Chap. VI, pp. 195–205). Therefore, Maxwell arrived to formulate his mathematical field (and without discussing the concept of force or mentioning mass) concept through three phases (Fig. 12):

1. A geometrical study of Faraday's hydrodynamic analogical model, based on lines systems of forces—imagined by Faraday—and "[…] the collection of imaginary properties […]" (Maxwell 1855–1856; 1861–1862, p. 160) of the theory of motion of an incompressible fluid (Maxwell 1855);
2. A concrete mechanical model (Maxwell 1855–1856; 1861–1862, Part I), based on the production of magnetic forces from electric current, called molecular vortexes;
3. A dynamical justification to field equations (Maxwell 1865, 1873).

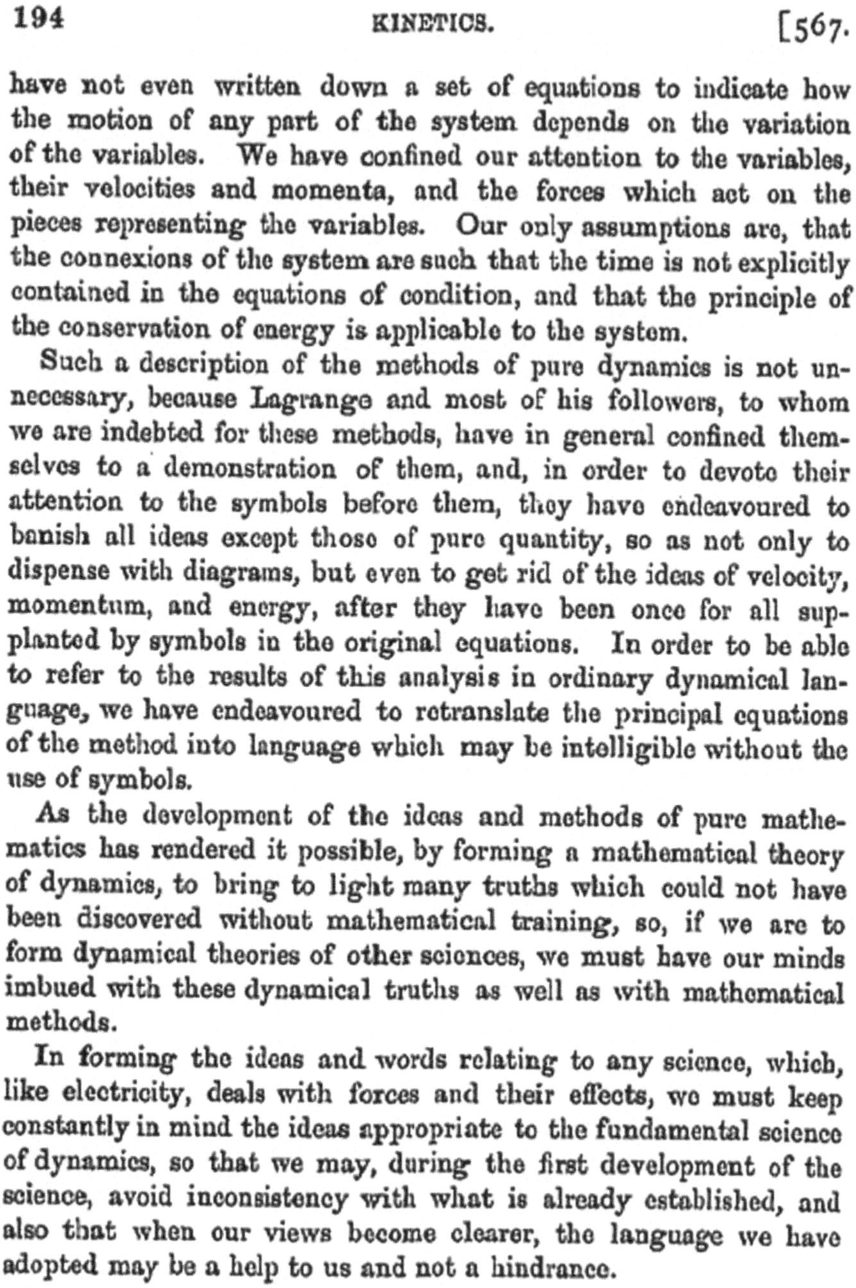

194 KINETICS. [567.

have not even written down a set of equations to indicate how the motion of any part of the system depends on the variation of the variables. We have confined our attention to the variables, their velocities and momenta, and the forces which act on the pieces representing the variables. Our only assumptions are, that the connexions of the system are such that the time is not explicitly contained in the equations of condition, and that the principle of the conservation of energy is applicable to the system.

Such a description of the methods of pure dynamics is not unnecessary, because Lagrange and most of his followers, to whom we are indebted for these methods, have in general confined themselves to a demonstration of them, and, in order to devote their attention to the symbols before them, they have endeavoured to banish all ideas except those of pure quantity, so as not only to dispense with diagrams, but even to get rid of the ideas of velocity, momentum, and energy, after they have been once for all supplanted by symbols in the original equations. In order to be able to refer to the results of this analysis in ordinary dynamical language, we have endeavoured to retranslate the principal equations of the method into language which may be intelligible without the use of symbols.

As the development of the ideas and methods of pure mathematics has rendered it possible, by forming a mathematical theory of dynamics, to bring to light many truths which could not have been discovered without mathematical training, so, if we are to form dynamical theories of other sciences, we must have our minds imbued with these dynamical truths as well as with mathematical methods.

In forming the ideas and words relating to any science, which, like electricity, deals with forces and their effects, we must keep constantly in mind the ideas appropriate to the fundamental science of dynamics, so that we may, during the first development of the science, avoid inconsistency with what is already established, and also that when our views become clearer, the language we have adopted may be a help to us and not a hindrance.

Fig. 12 Lagrangian and the 3–steps used by Maxwell to formulate his field (Maxwell 1873, II, Part IV, Chap. VI, p. 194) *Source* French National Library–Gallica. Public Domain

Maxwell looked then for adequate coefficients to convert the general equation (Maxwell 1873, II, Part IV, Chap. VI, pp. 198–199). From a strictly mathematical point of view, he pointed out three categories among coordinates q_i (Maxwell 1873, II, Part IV, Chap. VI, pp. 197–198) obtaining the total energy T divided in three parts generating terms of three forms: $x_i x_j$, $y_i y_j$, and $x_i y_j$ (Fig. 13).

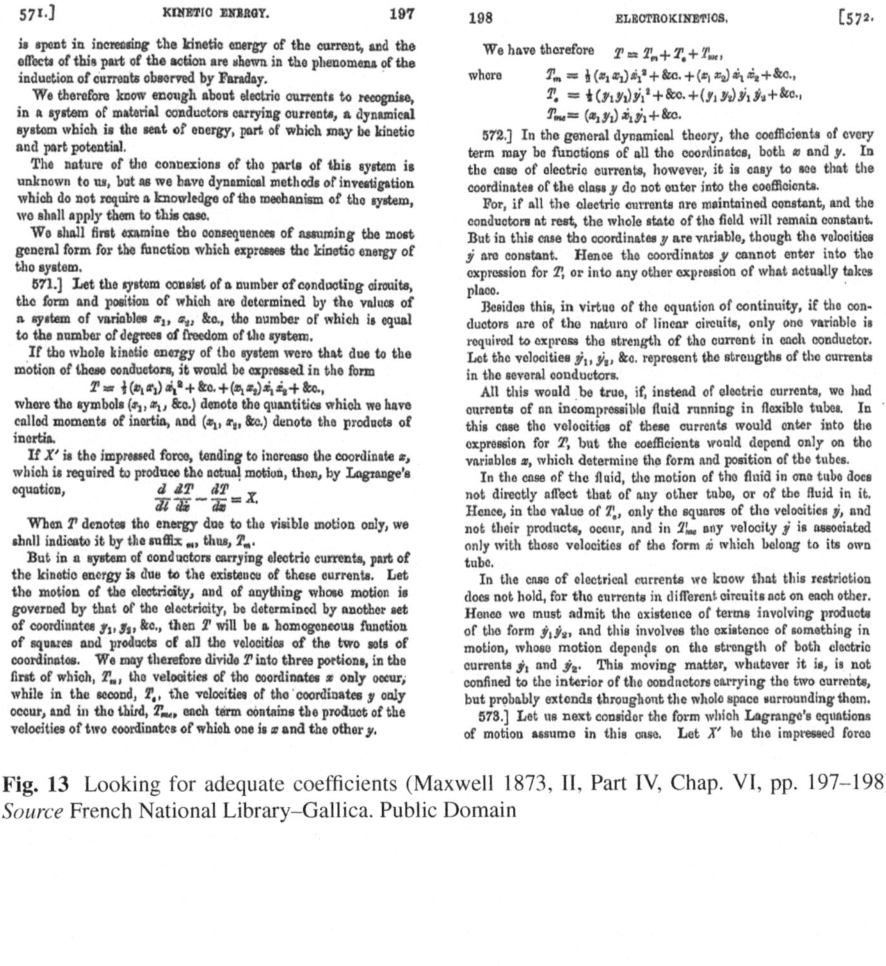

571.] KINETIC ENERGY. 197

is spent in increasing the kinetic energy of the current, and the effects of this part of the action are shewn in the phenomena of the induction of currents observed by Faraday.

We therefore know enough about electric currents to recognise, in a system of material conductors carrying currents, a dynamical system which is the seat of energy, part of which may be kinetic and part potential.

The nature of the connexions of the parts of this system is unknown to us, but as we have dynamical methods of investigation which do not require a knowledge of the mechanism of the system, we shall apply them to this case.

We shall first examine the consequences of assuming the most general form for the function which expresses the kinetic energy of the system.

571.] Let the system consist of a number of conducting circuits, the form and position of which are determined by the values of a system of variables x_1, x_2, &c., the number of which is equal to the number of degrees of freedom of the system.

If the whole kinetic energy of the system were that due to the motion of these conductors, it would be expressed in the form

$$T = \tfrac{1}{2}(x_1 x_1)\dot{x}_1^2 + \&c. + (x_1 x_2)\dot{x}_1 \dot{x}_2 + \&c.,$$

where the symbols $(x_1, x_1, \&c.)$ denote the quantities which we have called moments of inertia, and $(x_1, x_2, \&c.)$ denote the products of inertia.

If X' is the impressed force, tending to increase the coordinate x, which is required to produce the actual motion, then, by Lagrange's equation,

$$\frac{d}{dt}\frac{dT}{d\dot{x}} - \frac{dT}{dx} = X.$$

When T denotes the energy due to the visible motion only, we shall indicate it by the suffix $_m$, thus, T_m.

But in a system of conductors carrying electric currents, part of the kinetic energy is due to the existence of these currents. Let the motion of the electricity, and of anything whose motion is governed by that of the electricity, be determined by another set of coordinates y_1, y_2, &c., then T will be a homogeneous function of squares and products of all the velocities of the two sets of coordinates. We may therefore divide T into three portions, in the first of which, T_m, the velocities of the coordinates x only occur, while in the second, T_e, the velocities of the coordinates y only occur, and in the third, T_{me}, each term contains the product of the velocities of two coordinates of which one is x and the other y.

198 ELECTROKINETICS. [572.

We have therefore $T = T_m + T_e + T_{me}$,

where

$$T_m = \tfrac{1}{2}(x_1 x_1)\dot{x}_1^2 + \&c. + (x_1 x_2)\dot{x}_1 \dot{x}_2 + \&c.,$$
$$T_e = \tfrac{1}{2}(y_1 y_1)\dot{y}_1^2 + \&c. + (y_1 y_2)\dot{y}_1 \dot{y}_2 + \&c.,$$
$$T_{me} = (x_1 y_1)\dot{x}_1 \dot{y}_1 + \&c.$$

572.] In the general dynamical theory, the coefficients of every term may be functions of all the coordinates, both x and y. In the case of electric currents, however, it is easy to see that the coordinates of the class y do not enter into the coefficients.

For, if all the electric currents are maintained constant, and the conductors at rest, the whole state of the field will remain constant. But in this case the coordinates y are variable, though the velocities $\dot{y}$ are constant. Hence the coordinates y cannot enter into the expression for T, or into any other expression of what actually takes place.

Besides this, in virtue of the equation of continuity, if the conductors are of the nature of linear circuits, only one variable is required to express the strength of the current in each conductor. Let the velocities $\dot{y}_1$, $\dot{y}_2$, &c. represent the strengths of the currents in the several conductors.

All this would be true, if, instead of electric currents, we had currents of an incompressible fluid running in flexible tubes. In this case the velocities of these currents would enter into the expression for T, but the coefficients would depend only on the variables x, which determine the form and position of the tubes.

In the case of the fluid, the motion of the fluid in one tube does not directly affect that of any other tube, or of the fluid in it. Hence, in the value of T_e, only the squares of the velocities $\dot{y}$, and not their products, occur, and in T_{me} any velocity $\dot{y}$ is associated only with those velocities of the form $\dot{x}$ which belong to its own tube.

In the case of electrical currents we know that this restriction does not hold, for the currents in different circuits act on each other. Hence we must admit the existence of terms involving products of the form $\dot{y}_1 \dot{y}_2$, and this involves the existence of something in motion, whose motion depends on the strength of both electric currents $\dot{y}_1$ and $\dot{y}_2$. This moving matter, whatever it is, is not confined to the interior of the conductors carrying the two currents, but probably extends throughout the whole space surrounding them.

573.] Let us next consider the form which Lagrange's equations of motion assume in this case. Let X' be the impressed force

Fig. 13 Looking for adequate coefficients (Maxwell 1873, II, Part IV, Chap. VI, pp. 197–198) *Source* French National Library–Gallica. Public Domain

Non–homogeneous coordinates are combined in terms of the form$x_i x_j$. From a physical point of view, they are mechanical (x) and electrical (y) variables, and at the same time mathematical. These products give rise to three corresponding sources of kinetic energy when they appear in the equation for$T$. When establishing the electronic state as a state of momentum, Maxwell was driven by Faraday's studies. Particularly in Faraday's scientific panorama, Lagrange's method is not needed for electromagnetic theory. Therefore, in exploring of the field by means of the secondary circuit (Maxwell 1873, II, Part IV, Chap. VIII, pp. 211–226) Maxwell wrote:

> We have thus three methods of detecting the existence of the terms of the form *Tme*, none of which have hitherto led to any positive results (Maxwell 1873, II, Part IV, Chap. VI, p. 205. Author's italic).

Thus, from Chap. VI, the equations were proposed to bear on a simple electromagnetic system consisting of only two circuits in Maxwell (1873, II, Part IV, Chap. VII), *Theory of electric circuits*. Later, his reasoning would be extended to more complex systems which involve various conductors and a variety of mechanical motions and corresponding equations are interpreted in electrical terms following Faraday's experimental discoveries in electromagnetic induction (Pisano 2013). In Chap. VIII (Maxwell 1873, II, Part IV, Chap. VIII) Maxwell argued on the very mathematical structure of the electromagnetic field making extensive use of Faraday's studies and experimental data to accomplish this goal. In order to succeed in his endeavour, Maxwell needed:

> 594.] [...] to deduce from dynamical principles the expressions for the electromagnetic force acting on the conductor carrying an electric current through the magnetic field, and for the electromotive force acting on the electricity within a body moving in the magnetic field. (Maxwell 1873, II, Part IV, VIII, p. 217)

The tipping point lies in the fact that the Lagrangian formulation of mechanics is focused on the energy of the systems; no forces are doing work in them. The system of the differential equations of motion provided by a Lagrangian formulation are alternatives to Newton's laws of motion. Here a Lagrangian function, L, is involved. Also, we have to take into account the difference between the kinetic and potential energies of a system. L is known as a function of a set of independent coordinates of a system sufficient to fix the state of that system (generalized coordinates, and their time–derivatives, and generalized velocities). The forces that constrain the system without doing work do not need to be considered, and any coordinates on which the energy of the system does not depend on are ignored. As above cited, since his *Preface* (Maxwell 1873, I), Maxwell based his theoretical studies first on Faraday's experimental genius about how science works and its scientific methods, and then on Lagrange's mechanics (Lagrange 1788). Thanks to his formidable physical and mathematical integrated sense he built a new theory within the *physics mathematics* trend at that time as above discussed. The combination constituted the main physics mathematics apparatus that is Lagrangian in its entirety. The electromagnetic phenomena are very well connected with equations of motion and their forms generated the foundation for Maxwell's theory of the electromagnetic field. His reasoning is full of importance for physics mathematics, which we are going to describe in the next chapter, and for his related translation of Faraday's thought.

In order to proceed toward his dynamical project, Maxwell needed to establish the mathematical conditions to write the equations under a form suitable to describe a particular physical system within electromagnetic theory before formulating his physics mathematics field concept. The Maxwellian motion, in fact, can be imparted to the various parts of a machinery by means of ropes, and the number of degrees of freedom of the system equals the number of ropes. By following his analogies (Simpson 2005, pp. 23–26) Maxwell aimed to develop a Lagrangian formulation of electromagnetism in which the ether mechanism would be the analogue to the mechanism in a belfry, a bell tower; the same consideration he made for positions and velocities of the ropes, calculable charge and current distributions serving to determine the electromagnetic energy (Maxwell 1873; Simpson 2005, II, Part IV, Chap. IX, p. 231). Finally, Lagrangian mathematics methods themselves required that Maxwell returned to his main part (Maxwell 1873, I) of his *Treatise* and Faraday's science. All these considerations allow us to say that Maxwell's theory has a double formulation: the first is a mechanical one, and the other an electromagnetic one.

The primary sources of Maxwell's inspiration in his time were based on a dynamic conception of Nature, one side inspired by Faraday's ideas, the other side from the contributions of the physical mathematical debate in context at that time: dynamic aspect of phenomena, continuous medium, transmitting forces through space, etc. Maxwell's idea (1864–1865), was also to link up with Lagrange's *Méchanique Analitique* to research the equations of the electromagnetic field, effectively establishing a conceptual link between ideas relating to the phenomena of electromagnetism and the formalism of mechanics in order to describe, for example, to the reader in a simplified way the concept of *Reduced Momentum* (Maxwell 1864).

Maxwell examined the phenomenon of electromagnetic induction, proposing to study the Electromagnetic Momentum of a current and considering the state of the field that is generated in the vicinity of the electric current. Returning to the movement of matter referred to above, Maxwell poses the problem of considering what force must be applied to the medium in which the field is established, if its magnetic state depends on the motion of the medium, to increase or decrease the movement itself. It will then be possible to measure a momentum of some kind for the current when the motion of the medium is driven by this force: thanks to this momentum it is possible, for Maxwell, to give an interpretation to the interaction between field and current. To achieve this result (Maxwell 1865, p. 467) Maxwell models a body in motion with velocity w, which he denotes by C, connected to two other so–called driving–bodies, A and B, in motion with velocity u and v respectively. The relationship between the velocities of the objects is expressed by Maxwell with the formula:

$$\frac{dw}{dt} = p\frac{du}{dt} + q\frac{dv}{dt} \tag{48}$$

with p and q multiplying factors. Furthermore, in terms of the forces X and Y acting on A and B and the simultaneous displacements of the three objects, it is possible to write the following law of dynamics for the system (Fig. 14):

$$C\frac{dw}{dt}\delta z = X\delta x + Y\delta y \tag{49}$$

and:

$$\delta z = p\delta x + q\delta y \tag{50}$$

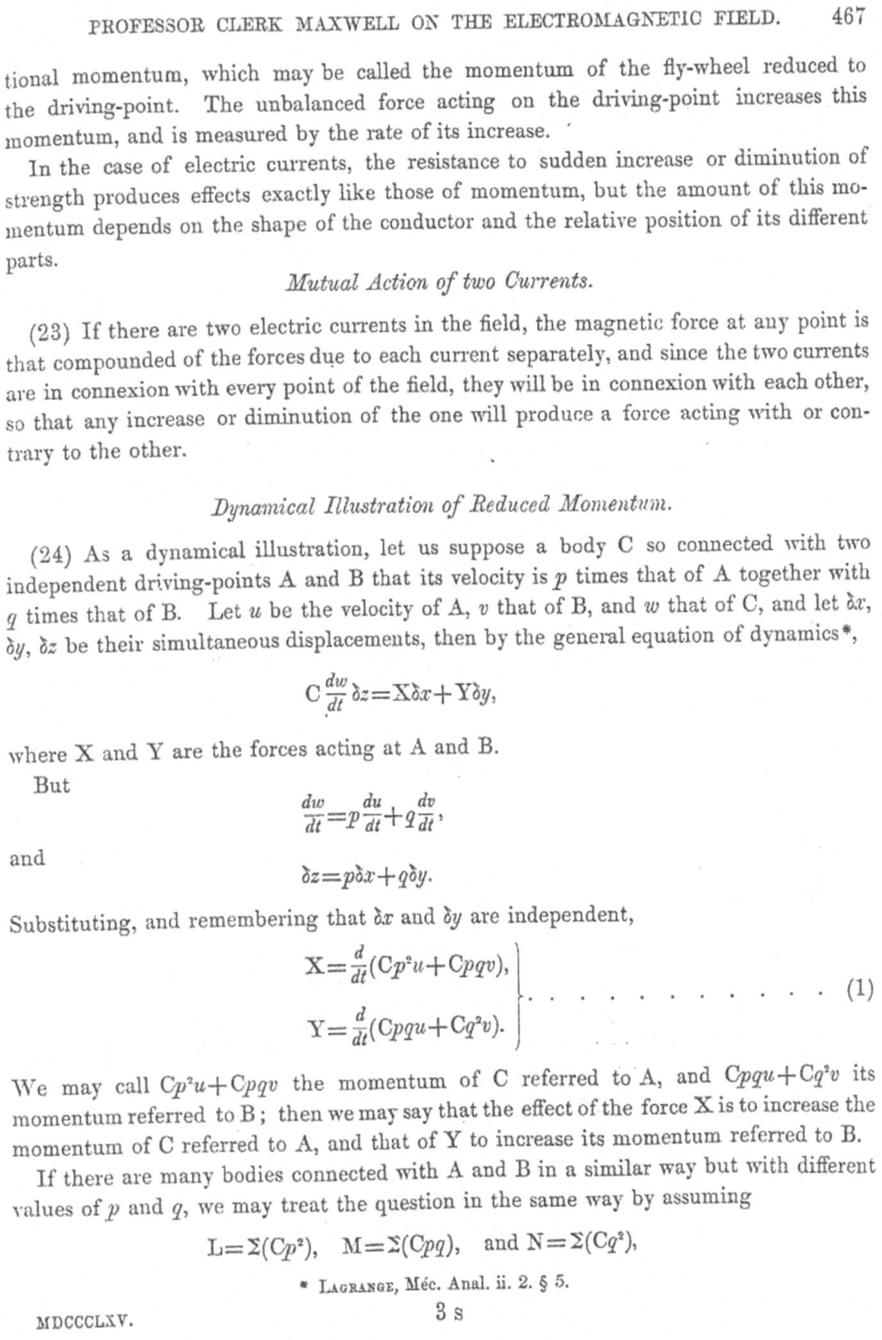

PROFESSOR CLERK MAXWELL ON THE ELECTROMAGNETIC FIELD. 467

tional momentum, which may be called the momentum of the fly-wheel reduced to the driving-point. The unbalanced force acting on the driving-point increases this momentum, and is measured by the rate of its increase.

In the case of electric currents, the resistance to sudden increase or diminution of strength produces effects exactly like those of momentum, but the amount of this momentum depends on the shape of the conductor and the relative position of its different parts.

Mutual Action of two Currents.

(23) If there are two electric currents in the field, the magnetic force at any point is that compounded of the forces due to each current separately, and since the two currents are in connexion with every point of the field, they will be in connexion with each other, so that any increase or diminution of the one will produce a force acting with or contrary to the other.

Dynamical Illustration of Reduced Momentum.

(24) As a dynamical illustration, let us suppose a body C so connected with two independent driving-points A and B that its velocity is p times that of A together with q times that of B. Let u be the velocity of A, v that of B, and w that of C, and let δx, δy, δz be their simultaneous displacements, then by the general equation of dynamics*,

$$C\frac{dw}{dt}\delta z = X\delta x + Y\delta y,$$

where X and Y are the forces acting at A and B.

But

$$\frac{dw}{dt} = p\frac{du}{dt} + q\frac{dv}{dt},$$

and

$$\delta z = p\delta x + q\delta y.$$

Substituting, and remembering that δx and δy are independent,

$$\left.\begin{aligned} X &= \frac{d}{dt}(Cp^2u + Cpqv), \\ Y &= \frac{d}{dt}(Cpqu + Cq^2v). \end{aligned}\right\} \quad . \quad . \quad . \quad . \quad . \quad (1)$$

We may call $Cp^2u+Cpqv$ the momentum of C referred to A, and $Cpqu+Cq^2v$ its momentum referred to B; then we may say that the effect of the force X is to increase the momentum of C referred to A, and that of Y to increase its momentum referred to B.

If there are many bodies connected with A and B in a similar way but with different values of p and q, we may treat the question in the same way by assuming

$$L = \Sigma(Cp^2), \quad M = \Sigma(Cpq), \quad \text{and } N = \Sigma(Cq^2),$$

* Lagrange, Méc. Anal. ii. 2. § 5.

MDCCCLXV. 3 S

Fig. 14 The Eq. (49) in Maxwell's work (Maxwell 1865, p. 467) prior to *Treatise* (1873), including quotation of Lagrange's *Mechanics* (II, 2, § 5) at the bottom of the page *Source* French National Library–Gallica. Public Domain

Here we can see the interplay between Lagrange's mechanics and Maxwell's electrodynamics. The latter dealt with electromagnetic interactions through mechanical consideration.

Back to the analytical discussion, considering that δx and δy are independent, substituting and decoupling the Eq. (49) we get the expressions for the forces:

$$X = \frac{d}{dt}\left(Cp^2u + Cpqv\right) \tag{51}$$

$$Y = \frac{d}{dt}\left(Cpqu + Cq^2v\right) \tag{52}$$

The expressions in brackets in Eqs. (51) and (52) are the momenta of C referred to A and B respectively. So X increases the momentum of C referred to A and Y increases the momentum of C referred to B. This is in the case of one body joint to two driving–ones. If there are multiple bodies connected to A and B with different values of p and q, Maxwell pictures three variables:

$$L = \sum\left(Cp^2\right) M = \sum(Cpq)\, N = \sum\left(Cq^2\right) \tag{53}$$

corresponding to coefficients of induction. Maxwell proves that in this case of multiple bodies the forces X and Y acting on A and B are:

$$X = \frac{d}{dt}(Lu + Mv) \tag{54}$$

$$Y = \frac{d}{dt}(Mu + Nv) \tag{55}$$

and where the terms in brackets have the same meaning as before. Now, this hypothesis is to be set: the motion of A is resisted by a force proportional to its velocity, say Ru, and the motion of B is resisted by a similar force, say Sv, where R and S are coefficients Maxwell calls "of resistance". The forces acting on the driving–bodies are then:

$$\xi = X + Ru = \frac{d}{dt}(Lu + Mv) + Ru \tag{56}$$

$$\eta = Y + Sv = \frac{d}{dt}(Mu + Nv) + Sv \tag{57}$$

If we suppose that A is moving and its velocity changes according to the ratio du/dt, if we want B to be standing and not–moving, a force:

$$\eta = \frac{d}{dt}(Mu) \tag{58}$$

must be exerted. In Maxwell's words:

> This effect on *B*, due to an increase of the velocity of *A*, corresponds to the electromotive force on one circuit arising from an increase in the strength of a neighbouring circuit. This dynamical illustration is to be considered merely as assisting the reader to understand what is meant in mechanics by Reduced Momentum. The facts of the induction of currents as depending on the variations of the quantity called Electromagnetic Momentum, or Electrotonic State, rest on the experiments of FARADAY*, FELICI†, &c. (Maxwell 1864, p. 468)

In this substantial text, too, he intended to apply to his field theory a dynamic philosophy and a dynamic approach, in order to follow the track that, as we have already had occasion to see, Lagrange had outlined in his *Méchanique Analitique*. Of course, following Maxwell, a theory can be defined as dynamic once it provides expressions for the kinetic energy and the potential energy of the system considered, without reporting in the equations the coordinates of the supposed particles that compose the system itself.

Maxwell wanted to develop his electromagnetic theory applying, ad hoc, Lagrange's method and fundamental equations of dynamics. Nowadays we can derive Maxwell's equations from variational calculus, as well. It was also observed (D'Agostino 2013) that, since the nature of the moving particles does not constitute a fundamental factor with respect to the formal aspects of the motion, the exposition of electromagnetism according to Maxwell and the mathematical theory formulated by Lagrange are in fact compatible because both focus on the motion itself (D'Agostino 2014, p. 124). This starting point also allowed us to state that, through his approach to Lagrange's equations, Maxwell succeeded in establishing a quasi–mechanical field theory, neglecting the internal mechanism of the motions responsible for the genesis of the electromagnetic field. We have seen in connection with the *Reduced Momentum* (Maxwell 1864) how the quotation of Lagrange makes its full appearance, so that in his dynamic approach to the theory of the electromagnetic field Maxwell draws from the *Méchanique Analitique* as a source of inspiration for the reasoning that will lead him to the equations he sought. Thus, briefly Maxwell:

- Neglected the mechanical concept of mass;
- Neglected characteristic quantities of motion such as the positions and velocities of all the objects and the binding forces acting on them;
- Considered the variables and the coordinates the motion of the system as dependent variables;
- Selected appropriate expressions to represent the energy of the magnetic field as kinetic energy and the energy of the electric field as potential energy.

In *On the Dynamical Evidence of the Molecular Constitution of Bodies* (1875) Maxwell made clear what he meant by the dynamic explanation of some phenomenon.

> When any phenomenon can be described as an example of some general principle which is applicable to other phenomena, that phenomenon is said to be explained. Explanations, however, are of very various orders, according the degree of generality of the principle which is made use of. [...] When a physical phenomenon can be completely described as a change in the configuration and motion of a material system, the dynamical explanation of

that phenomenon is said to be complete. We cannot conceive any further explanation to be either necessary, desirable, or possible, for as soon as we know what is meant by the words configuration, motion, mass, and force, we see that the ideas which they represent are so elementary that they cannot be explained by means of anything else. (Maxwell 1875, p. 357)

In 1929 the expression *Lagrangian function* first appeared in literature to define a function, for a holonomic dynamical system with n degrees of freedom, expressed through kinetic energy, written considering the independent variables determining the position of the system, and potential energy. The system being holonomic, both the physical constraints and the potential energy, say V, can be expressed by equations depending on generalized coordinates and also on time. Luther Pfahler Eisenhart (1876–1965) referred to the Lagrangian function giving it that formalism found in classical Mechanics books we know today (Eisenhart 1929, p. 591; Cfr. Simpson 1970, 2005; Thomson [1842] 1854):

$$L = \frac{1}{2} g_{\alpha\beta} \dot{q}^{\alpha} \dot{q}^{\beta} - V \tag{59}$$

where α and β are to be considered from 1 to $n+1$ and the dot means that the variable is differentiated with respect to the time. In the development of his calculus, Lagrange reduced the number of variables necessary to describe a system to what he calls generalised variables, whose number is equal to that of the actual degrees of freedom of a system and which correspond to the quantities for which we have information. These generalised co–ordinates are quite general, they need not be linear displacements, but they represent quantities sufficient to describe the state of the system. Maxwell precisely started from this point, considering a system made up of several moving parts interconnected by appropriate mechanisms without friction, inertia and non–deformable due to the actions acting on the system, and constructs the mathematical formulation with which to explain his approach.

4.1.1 Looking for a System Configuration in Motion

According to Maxwell (Maxwell 1873, II, Part IV, p. 185) q is the position of one of the moving objects, when all the positions of the single components of the system are known, the configuration of the latter at rest is determined. In order to know the configuration of the system in motion, however, it is necessary to enlarge the useful information, so that it is no longer sufficient to know all the initial positions of the constituents of the system, but also their respective velocities. The following quantities must be known:

$$\frac{dq}{dt} = \dot{q} \tag{60}$$

which Maxwell also reported for the reader in brackets (Maxwell 1873, II, Part IV, Chap. V, p. 186) with classical mechanical notation. The way in which the momentum increases gives, for Maxwell (1873, II, Part IV, Chap. V, p. 186) the motive force in the case in which the configuration of the body is always the same. Following Maxwell (1873, II, Part IV, Chap. V, p. 186) if F is the force and p the momentum, these two quantities are linked by the following relationships:

$$F = \frac{dp}{dt} \rightarrow p = \int F dt \tag{61}$$

Maxwell then defines (Maxwell 1873, II, Part IV, Chap. V, p. 187) the work function of a force as F_1 in a fully analytical manner:

$$W = \int F_1 dq_1 = \int F_1 \dot{q}_1 dt \tag{62}$$

We can note that—with respect to modern terms—it recalls the concept of a limit in a non–canonical manner establishing the maximum and minimum value of the work as a function of the maximum and minimum values of the system's velocities.

$$W_{max} = \dot{q}_1' \int F dt = \dot{q}_1'(p_1' - p_1) \tag{63}$$

$$W_{min} = \dot{q}_1'' \int F dt = \dot{q}_1''(p_1' - p_1) \tag{64}$$

and thus reducing the impulse, i.e. the difference between the maximum speed $\dot{q}_1'$ and the minimum speed q_1'', which will tend towards the common value $\dot{q}_1$. It is therefore possible to write the value of the infinitesimal work as (Maxwell 1873, II, Part IV, Chap. V, p. 188):

$$\delta W_1 = \dot{q}_1 \delta p_1 \tag{65}$$

where $\delta p_1 = p_1' - p_1$. The infinitesimal work of a force is obtained by multiplying the infinitesimal impulse by the velocity. Maxwell's later considerations on kinetic energy (Maxwell 1873, II, Part IV, Chap. V, p. 188) are inspired by Hamilton's theory. In 1874, he also applied it in his *On Hamilton's characteristic function for a narrow beam of light* (Maxwell 1874): it is a superstructure of Lagrange's theory, according to which the instantaneous state of a system can be described by its positions q and the momenta p.

We remark that—as above cited—Maxwell did not carry out all mathematical passages/calculations. In addition, in the *Preface* to the *Treatise* Maxwell declared that one of his aims was "[...] to trace the mathematical connexions of the quantities measured" (Maxwell 1873, I, Preface, p. vi). Therefore, we aim to stress the mathematical passages (including those lacking) and their correlated physical conceptualisations in order to determinate the force via momentum and kinetic energy (Maxwell 1873, II Part IV, Chap. V, p. 190). Thus, remaining calculations carried out by Maxwell to arrive at the result (75) and which are not reported in the *Treatise*, appear—in modern terms—as in the following.

The definition of Lagrangian as a function of generalised coordinates, generalised velocities and time, is:

$$L = L(q_j, \dot{q}_j, t)$$

The subscripts indicate the directions. Now let us consider an element of the system whose spatial coordinates are x_j on which acts a force derivable from a potential function $V(q_j)$. We can write:

$$L = T - V = \frac{1}{2} m \dot{q}_j \dot{q}_j - V(q_j) \tag{66}$$

where:

$$\frac{\partial L}{\partial \dot{q}_j} = \frac{\partial}{\partial \dot{q}_j}(T - V) = m\dot{q}_j = p_j \tag{67}$$

Thus we have:

$$\frac{\partial L}{\partial \dot{q}_j} = p_j \tag{68}$$

This result is obtained because the acting forces derive from a potential that is a function solely of position. Now, since the potential has this characteristic, we can write the relation:

$$\frac{d}{dt}\frac{\partial L}{\partial \dot{q}_j} - \frac{\partial L}{\partial q_j} = 0 \tag{69}$$

and thus we can also write:

$$\frac{d}{dt} p_j - \frac{\partial L}{\partial q_j} = 0 \rightarrow \dot{p}_j = \frac{\partial L}{\partial q_j} \tag{70}$$

If the symbol δ indicates an infinitesimal increment for any function of the variables q_j, p_j, t, or q_j, $\dot{q}_j$, t, they increase due to an infinitesimal variation of these variables:

$$\delta L = \frac{\partial L}{\partial q_j}\delta q_j + \frac{\partial L}{\partial \dot{q}_j}\delta \dot{q}_j + \frac{\partial L}{\partial t}\delta t \tag{71}$$

As δt is made to tend to zero, the last term of the right–hand member disappears.

$$\delta L = \frac{\partial L}{\partial q_j}\delta q_j + \frac{\partial (T - V)}{\partial \dot{q}_j}\delta \dot{q}_j \tag{72}$$

Multiplying and dividing the second term of the right–hand member by m we move from velocities to momentum:

$$\delta L = \frac{\partial (T - V)}{\partial q_j}\delta q_j + \frac{\partial (T - V)}{\partial p_j}\delta p_j \tag{73}$$

By considering that the partial derivatives are carried out on the respective variables of all the components of the system, we can move on to the sums:

$$\delta L = \sum \frac{dL}{dq}\delta q + \sum \frac{dL}{dp}\delta p \tag{74}$$

At this stage, Maxwell only discusses the following results (Maxwell 1873, II, Part IV, Chap. V, p. 188) that we make explicit in the following. By applying this result (74) to the kinetic energy—where Maxwell explicitly indicated with a subscript p the expression of the kinetic energy according to Hamilton (Maxwell 1873, II, Part IV, p. 188)—the total variation of the kinetic energy expressed according to the positions (q) and momenta (p) is:

$$\delta T_P = \sum \frac{dT_P}{dq}\delta q + \sum \frac{dT_P}{dp}\delta p \tag{75}$$

A result that immediately follows is that:

$$\dot{q} = \frac{dT_P}{dp} \tag{76}$$

If the system is in arbitrary motion and constrained by the conditions imposed by the connections between the parts, then it is possible to write the changes in p and q as:

$$\delta p = \frac{dp}{dt}\delta t \delta q = \dot{q}\delta t \tag{77}$$

consequently:

$$\frac{dT_P}{dp}\delta p = \frac{dT_P}{dp}\frac{dp}{dt}\delta t = \frac{dp}{dt}\dot{q}\delta t = \frac{dp}{dt}\delta q \tag{78}$$

Recalling the kinetic energy theorem, it follows that the work done by the forces exerted on the system results in an increase in the kinetic energy itself (Maxwell 1873, II, Part IV, Chap. V, p. 190) i.e.:

$$\delta T_P = \sum F\delta q \tag{79}$$

Since the variations δq are all independent of each other, we can write the following equality:

$$\sum\left[\left(\frac{dT_P}{dq} + \frac{dp}{dt}\right)\delta q\right] = \sum F\delta q \tag{80}$$

from which, by eliminating the signs of summation and equalising the coefficients, we obtain:

$$\left(\frac{dT_P}{dq} + \frac{dp}{dt}\right) = F \tag{81}$$

At this stage, we make explicit a calculus (Maxwell 1873, II, Part IV, p. 190) which is lacking in Maxwell's *Treatise* as in the following.

From (81) writing p_r for the momentum and F_r for the force associated with the coordinate q_r, we obtain the following equality:

$$F_r = \frac{dp_r}{dt} + \frac{dT_P}{dq_r} \tag{82}$$

We note that in the third edition (1892) of Maxwell's *Treatise*—edited and commented by Joseph John Thomson (1856–1940)—a footnote (added by Thomson, Fig. 15) claims that this result (82) is obtainable under the hypothesis that it is $\delta q = \dot{q}\delta t$.

In these two expressions the variations δq are all independent of each other, so that we are entitled to equate the coefficients of each of them in the two expressions (7) and (8). We thus obtain

$$F_r = \frac{dp_r}{dt} + \frac{dT_p}{dq_r}, \tag{9}$$

where the momentum p_r and the force F_r belong to the variable q_r*.

There are as many equations of this form as there are variables. These equations were given by Hamilton. They shew that the force corresponding to any variable is the sum of two parts. The first part is the rate of increase of the momentum of that variable with respect to the time. The second part is the rate of increase of the kinetic energy per unit of increment of the variable, the other variables and all the momenta being constant.

* {**This proof does not seem conclusive as δq is assumed to be equal to $\dot{q}\delta t$, that is to $\frac{dT_p}{dp}\delta t$, so that all we can legitimately deduce from (7) and (8) is**

$$\Sigma\left\{\left(\frac{dp_r}{dt} + \frac{dT_p}{dq_r} - P_r\right)\frac{dT_p}{dp_r}\right\} = 0.$$

Fig. 15 Thomson's specification about Hamilton's equation in Maxwell (Maxwell [1892] 1954, II, p. 205)

In addition, beginning in the *Preface*, Thomson also added his "considerable difficulties" to complete some Maxwell's calculus:

> I have endeavoured to add something in explanation of the argument in those passages in which I have found from my experience as a teacher that nearly all students find considerable difficulties; to have added an explanation of all passages in which I have known students find difficulties would have required more volumes than were at my disposal. I have attempted to verify the results which Maxwell gives without proof; I have not in all instances succeeded. In arriving at the result given by him, and in such cases I have indicated the difference in a foot-note. (Maxwell [1892] 1954, II, *Preface*, *Dec.* 5, 1891, p. XVI).

Thus, since (76) and (77) are known, by means of the following calculations (lacking in the *Treatise*) and considering that:

$$\dot{q} = \frac{dT_p}{dp} \rightarrow \dot{q}\delta p = \frac{dT_p}{dp}\delta p \rightarrow \dot{q}\frac{dp}{dt}\delta t = \frac{dT_p}{dp}\delta p \rightarrow \frac{dp}{dt}\delta q = \frac{dT_p}{dp}\delta p \tag{83}$$

we can obtain the previous (82) via the following passages:

$$\delta T_P = \sum\left(\frac{dT_p}{dp}\delta p\right) + \sum\left(\frac{dT_p}{dq}\delta q\right)$$

$$= \sum\left(\frac{dT_p}{dp}\delta p + \frac{dT_p}{dq}\delta q\right)$$

$$= \sum\left(\frac{dp}{dt}\delta q + \frac{dT_p}{dq}\delta q\right)$$

$$= \sum\left(\left(\frac{dp}{dt} + \frac{dT_p}{dq}\right)\delta q\right)$$

We remark that we are keeping the same notation Maxwell adopted (Maxwell 1873, II, Part IV, Chap. V, p. 190) by using two nested round brackets. Then, recalling the (79) and the (80) the result immediately follows. The (83) takes into account the hypothesis mentioned above. On the subject, some pages before, Maxwell wrote:

> 561.] […] *the velocity corresponding to the variable q is the differential coefficient of T_p with respect to the corresponding momentum p*. We have arrived at this result by the consideration of impulsive forces. By this method we have avoided the consideration of the change of configuration during the action of the forces. [...]. In other words, the variables, and the corresponding velocities and momenta depend on the actual state of motion of the system at the given instant, and not on its previous history. (Maxwell 1873, II, Part IV, Chap. V, p. 189; Author's italic)

Thus, the only writing possible with the calculations carried out leads to:

$$\sum\left[\left(\frac{dp_r}{dt} + \frac{dT_P}{dq_r} - F_r\right)\frac{dT_P}{dp_r}\delta t\right] = 0 \tag{84}$$

Maxwell then wrote (Maxwell 1873, II, Part IV, Chap. V, p. 191) kinetic energy as a function of momentum and velocity as follows:

$$T_{p\dot{q}} = \frac{1}{2}(\dot{q}_1 p_1 + \dot{q}_2 p_2 + \ldots) \tag{85}$$

indicating that kinetic energy can be expressed as the semi–sum of the products of momentum and velocity, a notation that is incorporated in the subscript of kinetic energy. Maxwell arrived (Maxwell 1873, II, Part IV, Chap. V, p. 191) at this result (85) by reasoning on two systems of momentum and velocity linked together by the following bonding conditions:

$$\mathrm{p}_i = np_i \, \dot{\mathrm{q}}_i = n\dot{q}_i \tag{86}$$

where n is a dimensionless term, the objects in italics refer to the first system and those in round font to the second system. Maxwell calculated (Maxwell 1873, II, Part IV, Chap. V, p. 190) the work of a force on the second system by varying n of an amount δn between 0 and 1 so as to eventually equalise the two systems, by writing:

$$F_1\delta\mathrm{q}_1 = \dot{\mathrm{q}}_1\delta\mathrm{p}_1 = n\dot{q}_1 p_1 \delta n \tag{87}$$

remembering Eq. (83) and applying it for an infinitesimal variation. Integrating:

$$W = (\dot{q}_1 p_1 + \dot{q}_2 p_2 + \ldots) \int_0^1 n dn = \frac{1}{2}(\dot{q}_1 p_1 + \dot{q}_2 p_2 + \ldots) \tag{88}$$

This writing, together with the Eq. (76), allows us to immediately write the:

$$T_p = \frac{1}{2}\left(p_1 \frac{dT_1}{dp_1} + p_2 \frac{dT_1}{dp_2} + \ldots \right) \tag{89}$$

This form in which Lagrange wrote kinetic energy in his equations, i.e. in terms of velocities and coordinates (Lagrange 1788, p. 237), is not derived by Maxwell; he only described–slightly explicated it (Maxwell 1873, II, Part IV, Chap. V, pp. 190–192). On our side we make it explicit as in the following.

$$p = m\dot{q} = \frac{2}{2} m\dot{q} = \frac{1}{2} md(\dot{q})^2 = \frac{dT_{\dot{q}}}{d\dot{q}} \tag{90}$$

and we can write:

$$T_{\dot{q}} = \frac{1}{2}\left(\dot{q}_1 \frac{dT_{\dot{q}}}{d\dot{q}_1} + \dot{q}_2 \frac{dT_{\dot{q}}}{d\dot{q}_2} + \ldots \right) \tag{91}$$

Consequently the components of the momentum are the differential coefficients, in Maxwell's words, of $T_{\dot{q}}$ compared to the corresponding speeds. At this point, Maxwell arrived (Maxwell 1873, II, Part IV, Chap. V, p. 191) at writing his kinetic energy, the same quantity, with three different expressions, which is why it is possible to construct the equation:

$$T_p + T_{\dot{q}} - 2T_{p\dot{q}} = T_p + T_{\dot{q}} - \dot{q}_1 p_1 - \dot{q}_2 p_2 - \ldots \tag{92}$$

from which (92) one obtains, by varying p, q and $\dot{q}$ and equating the coefficients of δq_i:

$$\frac{dT_p}{dq_i} + \frac{dT_{\dot{q}}}{dq_i} = 0 \tag{93}$$

The (93) means that the differential coefficient of kinetic energy with respect to q_i assumes the same numerical value when kinetic energy is expressed as a function of coordinates or velocities, but is of opposite sign in one case to the other. The equation of motion (82) then becomes:

$$F_i = \frac{d}{dt}\frac{dT_{\dot{q}}}{d\dot{q}_i} - \frac{dT_{\dot{q}}}{dq_i} \tag{94}$$

which is precisely the expression of the equations of motion according to Lagrange (1788, p. 237).

4.1.2 A Formal Derivation of Lagrange's Equations of Motion

For the sake of clarity, completeness and the fact that Maxwell will often refer to the concept of Lagrange equations, let us recall how they can be formally derived (Cfr. Maxwell 1873, II; Corben and Stehle [1950] 1994).

Let us suppose that a system of N parts described by $3N$ Cartesian coordinates that can be written as $x_{\rho,i} = x_{\rho,i}(q_1, q_2, \ldots, q_{3N}, t)$. By assuming that the system is also subject to a set of k constraints, however strictly less than $3N$. Cartesian coordinates can be expressed as a function of the so–called generalised coordinates q_m, for which the subscript m varies from 1 to $3N - k$. It is known that the Cartesian components of speed have the following expression:

$$\dot{x}_{\rho,i} = \frac{\partial x_{\rho,i}}{\partial q_m}\dot{q}_m + \frac{\partial x_{\rho,i}}{\partial t} \tag{95}$$

an equation obtained by obvious differentiation of the coordinates. Let us now consider that the variables q_m and $\dot{q}_m$ are independent so that the following relationship can be written:

$$\frac{\partial \dot{x}_{\rho,i}}{\partial \dot{q}_m} = \frac{\partial x_{\rho,i}}{\partial q_m} \tag{96}$$

In Cartesian coordinates and mechanical notation the equations of motion are written as:

$$F_{\rho,i} = m_\rho \ddot{x}_{\rho,i} \tag{97}$$

where F are the forces applied to the system. If we multiply both members of (97) by an arbitrary displacement that nevertheless respects the constraints and sum over all the indices, then we can calculate the work done by the forces as a result of the displacement:

$$W = \sum_{\rho,i} F_{\rho,i} dx_{\rho,i} = \sum_{\rho,i} m_\rho \ddot{x}_{\rho,i} dx_{\rho,i} \tag{98}$$

We can now process this expression algebraically, taking into account (95). Then we obtain:

$$W = \sum\nolimits_{\rho} m_\rho \ddot{x}_{\rho,i} dx_{\rho,i} \tag{99}$$

$$= \sum\nolimits_{\rho} m_\rho \ddot{x}_{\rho,i}\left(\frac{\partial x_{\rho,i}}{\partial q_m}dq_m + \frac{\partial x_{\rho,i}}{\partial t}dt\right) \tag{100}$$

Now, we add and subtract this quantity:

$$\dot{x}_{\rho,i}\frac{d}{dt}\frac{\partial x_{\rho,i}}{\partial q_m}dq_m \tag{101}$$

Inside the round bracket and break up the summation by grouping the terms that contain dq_m and isolating the term containing dt. We get:

$$\begin{aligned} W = &\sum\nolimits_{\rho} m_\rho\left[\frac{d}{dt}\left(\dot{x}_{\rho,i}\frac{\partial x_{\rho,i}}{\partial q_m}\right) - \dot{x}_{\rho,i}\frac{d}{dt}\frac{\partial x_{\rho,i}}{\partial q_m}\right]dq_m \\ &+ \sum\nolimits_{\rho} m_\rho \ddot{x}_{\rho,i}\frac{\partial x_{\rho,i}}{\partial t}dt \end{aligned} \tag{102}$$

$$\begin{aligned} = &\sum\nolimits_{\rho}\left[\frac{d}{dt}\frac{\partial}{\partial \dot{q}_m}\left(\frac{1}{2}m_\rho \dot{x}_{\rho,i}\dot{x}_{\rho,i}\right) - \frac{\partial}{\partial q_m}\left(\frac{1}{2}m_\rho \dot{x}_{\rho,i}\dot{x}_{\rho,i}\right)\right]dq_m \\ &+ \sum\nolimits_{\rho} m_\rho \ddot{x}_{\rho,i}\frac{\partial x_{\rho,i}}{\partial t}dt \end{aligned} \tag{103}$$

Equation (97) provides us with a means of calculating the work done for an arbitrarily small displacement in a coordinate–independent manner through the derivatives of a function calculated with respect to dq_m, $d\dot{q}_m$, and t. This function is the kinetic energy, which will become fundamental for Maxwell. In other words, we can write work as:

$$W = \left(\frac{d}{dt}\frac{\partial T}{\partial \dot{q}_m} - \frac{\partial T}{\partial q_m}\right)dq_m + \sum_{\rho,i} m_\rho \ddot{x}_{\rho,i}\frac{\partial x_{\rho,i}}{\partial t}dt \tag{104}$$

We can also rewrite (98) as:

$$W = Q_m dq_m + Q_t dt \tag{105}$$

where we have to define what are called generalised components of the force:

$$Q_m = \sum\nolimits_{\rho,i} F_{\rho,i}\frac{\partial x_{\rho,i}}{\partial q_m}\quad Q_t = \sum\nolimits_{\rho,i} F_{\rho,i}\frac{\partial x_{\rho,i}}{\partial t} \tag{106}$$

Equalizing (104) and (105) with the substitutions dictated by (59) we have:

$$\frac{d}{dt}\frac{\partial T}{\partial \dot{q}_m} - \frac{\partial T}{\partial q_m} = Q_m \tag{107}$$

and this writing expresses the equations of motion in the Lagrange form. The forces $F_{\rho,i}$, which are called applied forces are the only ones that contribute to the generalised forces. All forces due to constraints do not contribute to the generalised forces as they do not do any work, this is because the displacements are chosen so as not to violate the constraints. Under these assumptions, kinetic energy becomes a homogeneous quadratic function of the generalised components of velocity. The velocity will be written as:

$$\dot{x}_{\rho,i} = \frac{\partial x_{\rho,i}}{\partial q_m}\dot{q}_m \tag{108}$$

and therefore the kinetic energy becomes in turn:

$$T = \frac{1}{2}\sum\nolimits_{\rho,i} m_\rho \left(\dot{x}_{\rho,i}\right)^2 = \frac{1}{2} g_{mn}\dot{q}_m\dot{q}_n \tag{109}$$

where:

$$\frac{1}{2} g_{mn}\dot{q}_m\dot{q}_n = \sum\nolimits_{\rho,i} m_\rho \frac{\partial x_{\rho,i}}{\partial q_m}\frac{\partial x_{\rho,i}}{\partial q_m} \tag{110}$$

In order to complete the construction of the equations, we can consider the case where the forces acting on the system are derived from a potential, thus:

$$F_{\rho,j} = -\frac{\partial V}{\partial x_{\rho,j}} \tag{111}$$

Thus:

$$Q_m = \sum\nolimits_{\rho,i} F_{\rho,i}\frac{\partial x_{\rho,i}}{\partial q_m} = -\sum\nolimits_{\rho,i} \frac{\partial V}{\partial x_{\rho,j}}\frac{\partial x_{\rho,i}}{\partial q_m} = -\frac{\partial V}{\partial q_m} \tag{112}$$

Therefore, if the system is encompassed in a force field expressible as a gradient of a scalar function (111) we will obtain the final expression we are looking for:

$$\frac{d}{dt}\frac{\partial T}{\partial \dot{q}_m} - \frac{\partial T}{\partial q_m} + \frac{\partial V}{\partial q_m} = \frac{d}{dt}\frac{\partial T}{\partial \dot{q}_m} - \frac{\partial}{\partial q_m}(T - V) = 0 \tag{113}$$

4.2 *Maxwell's Application of Lagrange–Equations to Electromagnetism*

After having constructed the mathematical tool necessary for a dynamic approach to the development of his ideas, Maxwell proceeds (Maxwell 1873, II, Part IV, Chap. VI, pp. 195–205) to apply the concepts of dynamics to the theory of electromagnetism.

Chapter VI. In the beginning of *On the Dynamic Theory of Electromagnetism* (Maxwell 1873, II, Part IV, Chap. VI, p. 195) Maxwell returns to the capacity of the electric current in motion in a circuit to perform work independently from the presence or absence of an external electromotive force:

- One part of this energy is kinetic because it is linked to the actual motion of the charge carriers
- The other one is potentially kinetic energy because it consists in the possibility of the carriers being set in motion.

The carrier has kind of accounted energy, which, if released, becomes kinetic energy, a kind of energy–in–being. Maxwell called the second type of energy *potential* (Maxwell 1873, II, Part IV, Chap. VI, p. 195); it is due to the tendency of the carriers to oppose the variation of their relative position.

The electric current, according to Maxwell, can be nothing more than a kinetic phenomenon and this is why he recalls Michael Faraday's observation that the electric current is "something that progresses and not a mere arrangement" (Faraday 1839–1855). The velocity of the current in the conductor is not yet known, in fact according to Maxwell:

> 569.] [...] As to the velocity of the current we have shewn that we know nothing about it, it may be the tenth of an inch in an hour, or a hundred thousand miles in a second. So far are we from knowing its absolute value in any case, that we do not even know whether what we call the positive direction is the actual direction of the motion or the reverse. (Maxwell 1873, II, Part IV, Chap. VI, p. 196)

Maxwell applied Lagrange's theory to the dynamic description of electromagnetism in the following steps.

He considered a series of conducting circuits that compose a system, the form and the position of which are determined by a series of variables that he indicates with x_i whose number is equal to the number of the degrees of freedom of the system. The kinetic energy of the system will take the following form, taking into account the movement of the conductors:

$$T = \frac{1}{2}(x_1 x_1)\dot{x}_1^2 + \cdots + (x_1 x_2)\dot{x}_1\dot{x}_2 + \ldots \tag{114}$$

where the expressions $(x_1 x_1)$ indicate the momentum of inertia as the expressions $(x_1 x_2)$ are the so–called products of inertia. A force $X\prime$ necessary to modify the coordinates x and give rise to true motion is obtained through the solution of the following Lagrange equation:

$$\frac{d}{dt}\frac{dT}{d\dot{x}} - \frac{dT}{dx} = X' \tag{115}$$

By writing T_m the kinetic energy is calculable because of the actually visible movement of the system. However, according to Maxwell (Maxwell 1873, II, Part IV, Chap. VI, p. 198) in a system of conductors such as the one designed for these considerations, part of the kinetic energy is due to the presence of the currents themselves. Therefore, another set of coordinates should be defined in order to describe the movement of the electricity itself; this will be indicated by y_i. The overall kinetic energy will therefore be the resultant of the composition of three contributions, the one due solely to the coordinates x_i, that due solely to the coordinates y_i and a mixed contribution consisting of the products of these coordinates. So forth:

$$T = T_m + T_e + T_{me} \tag{116}$$

where:

$$T_m = \frac{1}{2}(x_1 x_1)\dot{x}_1^2 + \cdots + (x_1 x_2)\dot{x}_1\dot{x}_2 + \ldots \tag{117}$$

$$T_e = \frac{1}{2}(y_1 y_1)\dot{y}_1^2 + \cdots + (y_1 y_2)\dot{y}_1\dot{y}_2 + \ldots \tag{118}$$

$$T_{me} = (x_1 y_1)\dot{x}_1\dot{y}_1 + \ldots \tag{119}$$

According to Maxwell (1873, II, Part IV, Chap. VI, p. 198) the existence of mixed terms such as $\dot{y}_1\dot{y}_2$ is linked to the presence of something, which he does not define at this point in the discussion, and which depends on the strength of the electric currents $\dot{y}_1$ and $\dot{y}_2$. Whatever this moving matter is, it is not bound within the conductors but probably extends to the space around them. Maxwell derived another Lagrange equation of motion in this circumstance (Maxwell 1873, II, Part IV, Chap. VI, p. 199). He assumed (Maxwell 1873, II, Part IV, Chap. VI, p. 199) that the force imparted at the coordinate x, one of the co–ordinates identifying the shape and position of the conductor circuit, is itself composed of three contributions which it indicates as:

$$X\prime = X\prime_m + X\prime_e + X\prime_{me} \tag{120}$$

The contribution of $X\prime_m$ is neglected by Maxwell because it is related to considerations of pure dynamics. If we write Lagrange's equation for $X\prime_e$ we get:

$$\frac{d}{dt}\frac{dT}{d\dot{x}} - \frac{dT}{dx} = X'_e \rightarrow X'_e = -\frac{dT_e}{dx} \tag{121}$$

because T_e does not contain terms differentiable with respect to $\dot{x}$. This relationship indicates that the mechanical force that must be applied to a conductor to counteract the electromagnetic force is measured by the rate of decrease in kinetic energy due to the change in coordinate x only. The electromagnetic force X_e which calls into play this external mechanical force is equal in magnitude and opposite to $X\prime_e$.

The third Lagrange's method–equation is:

$$\frac{d}{dt}\frac{dT_{me}}{d\dot{x}} - \frac{dT_{me}}{dx} = X^{'}_{me} \tag{122}$$

The first term of the left–hand member depends on (as indicated by the derivative with respect to time) the rate of change of the force of the currents and indicates a mechanical force on the conductor that is zero if the currents are constant and takes on a positive or negative value depending on whether the currents increase or decrease in intensity. The second term depends on the actual intensity of the currents and changes sign when the currents themselves change sign.

A further application of Lagrange's equations can be found in Maxwell's calculation of electromotive forces (Maxwell 1873, II, Part IV, Chap. VI, p. 204): if Y is the electromotive force due to induction phenomena, the force that must act on the circuit from outside to counterbalance it is $Y^{'} = -Y$ and this immediately leads to the following equation of Lagrange:

$$Y = -Y^{'} = -\frac{d}{dt}\frac{dT}{d\dot{y}} + \frac{dT}{dy} \tag{123}$$

Since there are no terms in the expression of T which contain the coordinate y the second term of the right-hand member is zero and, as a fundamental consequence, there can be no electromotive forces in a system that is at rest, traversed by currents whose variation cannot be measured: constant, in other words. If it is:

$$Y = Y_m + Y_e + Y_{me} \tag{124}$$

as T_m does not contain $\dot{y}$ it is also $Y_m = 0$ and Lagrange's equation becomes what is known in the textbooks as Faraday's *Law of Induction* (Maxwell 1873, II, Part IV, Chap. VI, p. 204):

$$Y_e = -\frac{d}{dt}\frac{dT_e}{d\dot{y}} \tag{125}$$

Finally, the part of the kinetic energy that depends on the product of the velocities for the currents is derived as follows:

$$Y_{me} = -\frac{d}{dt}\frac{dT_{me}}{d\dot{y}} \tag{126}$$

Chapter VII. In the seventh chapter (Maxwell 1873, II, Part IV, Chap. VII, pp. 206–210) Maxwell devotes himself to the study of the *electrokinetic* energy involved in electrical circuits.

Before applying Lagrange's equations, Maxwell considers a set of conductive circuits A_i whose shapes and positions are expressed by variables x_i. Through y_i he then indicates the quantities of current that have passed through a section of the conductors A_i at the instant $t = 0$; the current intensity will be $\dot{y}_i$. At this point, it is interesting to observe how Maxwell uses the fundamental term Newtonian "fluxion" (Maxwell 1873, II, Part IV, Chap. VII, p. 206; see also Pisano and Bussotti 2023, 2022) to indicate the derivative of the quantities of current y_i. Maxwell called y_i the integral current and $\dot{y}_i$ the effective current (Maxwell 1873, II, Part IV, Chap. VII, p. 206). The *electrokinetic* energy that Maxwell refers to has the following analytical form (Maxwell 1873, II, Part IV, Chap. VII, p. 206):

$$T = \frac{1}{2}L_1\dot{y}_1^2 + \frac{1}{2}L_2\dot{y}_2^2 + \cdots + M_{12}\dot{y}_1\dot{y}_2 + \ldots \tag{127}$$

where the coefficients $L, M, \ldots$ are functions ofx_i. We note that the y_i variables that Maxwell defined as *electrical* (Maxwell 1873, II, Part IV, Chap. VII, p. 207) do not appear in the equation. The terms L_i are the electrical moments of inertia of the circuits, or self–induction coefficients in the language of electromagnetism, while the terms M_{ij} are the electrical inertia products of the circuit i and of the circuit$j$, or, in other words, the mutual induction coefficients. By differentiating (127) with respect to the$\dot{y}_i$, it is possible to get the so–called e*lectrokinetic* momentum:

$$p = L_1\dot{y}_1 + M_{12}\dot{y}_2 + \ldots \tag{128}$$

Now, let E be the electromotive force impressed on the circuit A, produced by a suitable source. If R is the resistance of the circuit, then, to overcome the internal resistance, an electromotive force must be applied whose value is $R\dot{y}$, so the momentum of the circuit can only be modified by the quantity $E - R\dot{y}$, so:

$$E - R\dot{y} = Y' = \frac{dp}{dt} - \frac{dT}{dy} = \frac{dp}{dy} \tag{129}$$

or:

$$E = R\dot{y} + \frac{dp}{dy} \tag{130}$$

Thus, the electromotive force is composed of two terms. The first, $R\dot{y}$, gives quantitatively the value of the force that must be applied to maintain the current against the resistance; the second term identifies the change in electromagnetic momentum

p. Maxwell, therefore, wrote Lagrange's equation for the mechanical force $X\prime$, i.e. the external cause that increases x (Maxwell 1873, II, Part IV, Chap. VII, p. 208):

$$X^{'} = \frac{d}{dt}\frac{dT}{d\dot{x}} - \frac{dT}{dx} \tag{131}$$

The expression of electro kinetic energy contains no $\dot{x}$ so the Eq. (125) becomes:

$$X^{'} = -\frac{dT}{dx} \tag{132}$$

but $X\prime$ is the external balancing force that depends on electrical causes and is equal and opposite to the electromagnetic force, which is indicated by X:

$$X = \frac{dT}{dx} \tag{133}$$

Thus, in Maxwell's words:

> [...] *the electromagnetic force tending to increase any variable is equal to the rate of increase of the electrokinetic energy per unit increase of that variable, the currents being maintained constant*. (Maxwell 1873, II, Part IV, Chap. V, p. 208; Author's italic)

Finally, the electromagnetic force that tends to increase any variable is equal to the rate of increase of *electrokinetic* energy (T) per unit of the variable considered (x), under the assumption that currents are constant.

In conclusion, Maxwell developed his theory by applying Lagrange's equations to the various and increasingly complex applications of electromagnetism. From this point of view, Maxwell established the concluding chapters of his work and came up with the twenty equations that were originally recognised as *Maxwell's Equations.*

5 Maxwell's Equations (1865)

Maxwell described the phenomena relating to electricity and magnetism through theoretical arguments, avoiding dwelling on the nature of the forces involved, but revolutionising the founding paradigm, and considering the concept of energy. He took care to detail the difference between kinetic energy and potential energy, the latter being used, for example, to describe two interacting currents. He pointed to this concept, fundamental to Faraday's physics, and sought a geometrical and mathematical solution in the paragraph *On induction of electric current* (Maxwell 1873, II, Part IV, pp. 172–173). The application, in Maxwell's *Treatise*, of the Lagrangian mathematical formalism would have led the author to develop a complex but fundamentally important theory. It is worth dwelling on the paragraphs that constitute the final part of the second volume, because from them it is possible to extrapolate those relations that today go under the name of "Maxwell's Equations". This expression actually creates some confusion, especially in the light of what we can read in part III of *A Dynamical Theory of the Electromagnetic Field*[19] entitled *General Equations of the Electromagnetic Field* (Fig. 16) (Joule 1841b, III, pp. 480–488). See also his abstract (Joule 1841a, pp. 531–536):

1864.] *Dynamical Theory of the Electromagnetic Field.* 531

ducts of the action of heat on the ethylphosphates, may aid in determining this question. Meanwhile the observation, already recorded, as to the stability of ordinary ethylphosphoric acid and its salts in the presence of permanganate of potassium requires a word or two of further comment. When argentic diethylphosphate is acted upon with iodine, the silver and one atom of ethyl may be removed, and after treatment with finely divided silver and a little oxide of silver to remove any iodine and hydriodic acid, and then with excess of carbonate of barium, an ethylphosphate of barium is obtained, which, unlike the ordinary salt, immediately reduces permanganate of potassium; perhaps the ethyl in this salt exists in a different and less intimate form of combination. I am inclined to think that the permanganates will afford, in some cases, criteria for the detection of slight differences in isomeric compounds, although it would be premature at present to hazard an exact interpretation of the phenomena to which they give rise. I may add that treatment of an ethylphosphate with strong nitric acid fails to decompose the ethylphosphoric acid; so that phosphoric acid cannot thus be separated from this remarkably stable body.

II. "A Dynamical Theory of the Electromagnetic Field." By Professor J. Clerk Maxwell, F.R.S. Received October 27, 1864.

(Abstract.)

The proposed Theory seeks for the origin of electromagnetic effects in the medium surrounding the electric or magnetic bodies, and assumes that they act on each other not immediately at a distance, but through the intervention of this medium.

The existence of the medium is assumed as probable, since the investigations of Optics have led philosophers to believe that in such a medium the propagation of light takes place.

The properties attributed to the medium in order to explain the propagation of light are—

1st. That the motion of one part communicates motion to the parts in its neighbourhood.

2nd. That this communication is not instantaneous but progressive, and depends on the elasticity of the medium as compared with its density.

The kind of motion attributed to the medium when transmitting light is that called transverse vibration.

An elastic medium capable of such motions must be also capable of a vast variety of other motions, and its elasticity may be called into play in other ways, some of which may be discoverable by their effects.

One phenomenon which seems to indicate the existence of other motions than those of light in the medium, is that discovered by Faraday, in which the plane of polarization of a ray of light is caused to rotate by the action

480 PROFESSOR CLERK MAXWELL ON THE ELECTROMAGNETIC FIELD.

will be limited sheets, terminating in the electric circuit as their common edge or boundary. The number of these will be equal to the amount of work done on a unit pole in going round the current, and this by the ordinary measurement $=4\pi\gamma$, where γ is the value of the current.

These surfaces, therefore, are connected with the electric current as soap-bubbles are connected with a ring in M. Plateau's experiments. Every current γ has $4\pi\gamma$ surfaces attached to it. These surfaces have the current for their common edge, and meet it at equal angles. The form of the surfaces in other parts depends on the presence of other currents and magnets, as well as on the shape of the circuit to which they belong.

PART III.—GENERAL EQUATIONS OF THE ELECTROMAGNETIC FIELD.

(53.) Let us assume three rectangular directions in space as the axes of x, y, and z, and let all quantities having direction be expressed by their components in these three directions.

Electrical Currents (p, q, r).

(54) An electrical current consists in the transmission of electricity from one part of a body to another. Let the quantity of electricity transmitted in unit of time across unit of area perpendicular to the axis of x be called p, then p is the component of the current at that place in the direction of x.

We shall use the letters p, q, r to denote the components of the current per unit of area in the directions of x, y, z.

Electrical Displacements (f, g, h).

(55) Electrical displacement consists in the opposite electrification of the sides of a molecule or particle of a body which may or may not be accompanied with transmission through the body. Let the quantity of electricity which would appear on the faces $dy.dz$ of an element dx, dy, dz cut from the body be $f.dy.dz$, then f is the component of electric displacement parallel to x. We shall use f, g, h to denote the electric displacements parallel to x, y, z respectively.

The variations of the electrical displacement must be added to the currents p, q, r to get the total motion of electricity, which we may call p', q', r', so that

$$\left.\begin{aligned} p' &= p + \frac{df}{dt}, \\ q' &= q + \frac{dg}{dt}, \\ r' &= r + \frac{dh}{dt}, \end{aligned}\right\} \quad \ldots \ldots \ldots \quad (A)$$

Electromotive Force (P, Q, R).

(56) Let P, Q, R represent the components of the electromotive force at any point. Then P represents the difference of potential per unit of length in a conductor

Fig. 16 On the left, the first page of Maxwell's "abstract" at the *Royal Society of London* (Maxwell 1864, p. 531). On the right, First page of *General Equations of the Electromagnetic Field* (Maxwell 1865, p. 480) *Source* French National Library–Gallica. Public Domain

Here eight (six plus two) equations appear, equations which, in turn, if projected along a system of Cartesian axes, expand the collection of relations to no less than twenty equations in as many unknowns (Pisano 2013).

For all intents and purposes, the term "Maxwell's Equations", which originally applied to a set of equations published in *A Dynamical Theory of the Electromagnetic Field* (Maxwell 1865), now refers to four equations that were brought together in 1884 by Oliver Heaviside in conjunction with similar work by Willard Gibbs (1839–1903) (Pisano et al. 2019; Pisano and Pellegrino 2023) and Heinrich Hertz. In this regard, it is significant from an epistemological point of view to observe that the publication of the twenty equations in 1865 and their subsequent condensation into the four equations we know today, was anticipated the previous year by a six–page abstract that Maxwell sent to the Royal Society in London. In the

[19] *A Dynamical Theory of the Electromagnetic Field* was received in October 27, 1864, and read by Maxwell in December 8, 1864 at *the Royal Society of London.*

document (Maxwell 1864, p. 534) Maxwell mentions the result of the twenty field equations he was going to write, but there is no reference to the use of partial derivative mathematics or, remarkably, to their being written down in any analytical form. The equations, which cover a wide range of physical situations, are described by a name: *Equations of Total Currents, Magnetic Force, Electric Currents, Electromotive Force, Electric Elasticity and Electric Resistance*. These, in particular, give rise to six sets of three equations, together with an equation of Free Electricity and one relating to Electrical Continuity (finally, eight main sets), bringing the total number of relations to twenty.[20] Their analytical formalisation will take place the following year only, with the publication of the full article. Only there the reader can find a conventional mathematical description of the equations in the way we would expect them to appear in a scientific report.

6 Faraday and Maxwell: Electromagnetic Field

The field concept—already assumed by Faraday (1839–1855, I, III; Joule 1845)—became a part of Maxwell's mathematical physical arguments and constituted a mere defining moment in *A Treatise on Electricity and Magnetism* (Maxwell 1873; Pisano 2014; Simpson 2005). In order to include Faraday's insight, Maxwell projected and realized one of the basic structures of his new physics mathematics: the field concept (Maxwell 1865, 1873) earlier and physically described by Faraday.

> Faraday and Thomson invented field theory: they introduced theoretical entities in space between electric and magnetic sources, and they elaborated powerful techniques for investigating the properties of these entities (Darrigol 2000, p. 134).
>
> Faraday had no mathematical or mechanical preconceptions and his theory mostly reflected patient experimental explorations (Darrigol 2000, p. 136).
>
> Proceeding from Faraday's and Thomson's writings, Maxwell reached the essentials of his electromagnetic field theory stepwise, in three great memoires. In *On Faraday's line of force* his aim was to obtain a mathematical expression of Faraday's field conception (Darrigol 2000, p. 172. Author's quotations marks).

In Maxwell's opinion (Table 2) the field was part of a new theory, a physics mathematics within the domain[21] of a physical theory. In the following, we report an excursus on the field concept from Faraday to Maxwell:

[20] (Cfr. Maxwell 1864, p. 534). In this abstract (Fig. 16, left side) no mathematical equations are reported. They officially appeared in the paper published in 1865 (Maxwell 1865, pp. 459–512). For more info on this abstract and the lack of Maxwell's equations in it, see (Pisano 2014). In addition, the abstract (Fig. 16, left side) was printed in the *Proceedings of the Royal Society of London* (Joule 1841a, pp. 531–536). The full paper appeared in 1865 (Fig. 16, right side) in the *Philosophical Transactions of The Royal Society of London* (Joule 1841b, pp. 459–512; Cfr. Simpson 1997, 2005, 2010).

[21] These were stressed within previous historical epistemology studies from Pisano (Gillispie and Pisano 2014; Joule 1845).

Table 2 A Timeline on Field conceptualization from Faraday to Maxwell

Main	
1831 *On field concept*	Faraday "114. By magnetic curves, I mean the lines of magnetic force, however modified by the juxtaposition of poles, which would be depicted by iron filings; or those to which a very small magnetic needle would form a tangent." (Faraday 1839–1855, I, §114, footnote *, p. 32)
1845 *On field concept*	"2149. But before I proceed to them, I will define the meaning I connect with certain terms which I shall have occasion to use: thus, by *line of magnetic force*, or *magnetic line of force* or *magnetic curve*, I mean that exercise of magnetic force which is exerted in the lines usually called magnetic curves, and which equally exist as passing from or to magnetic poles, or forming concentric circles round an electric current. By *line of electric force*, I mean the force exerted in the lines joining two bodies, acting on each other according to the principles of static electric induction (1161, &c.), which may also be either in curved or straight lines." (Faraday 1839–1855, III, § 2149, p. 2)
[1845] 1846 *On magnetic field concept*	"2247. Another magnet which I have had made has the horseshoe form. [...] the poles are, of course, 6 inches apart, the ends are planed true, and against these move two short bars of soft iron [...]. The ends of these bars form the opposite poles of contrary name; the magnetic field between them can be made of greater or smaller extent and the intensity of the lines of magnetic force be proportionately varied." (Faraday 1839–1855, III, § 2247, p. 29; see also: § 2252, p. 30 and related figure)
1848–1851 *On magnetic field concept*	See also (Faraday 1839–1855, III, §§ 2463–2475, pp. 85–89; §§ 2806–2810, pp. 203–205; § 2831, pp. 211–212; § 3171, pp. 367–368)
[1850] 1851 *On magnetic field concept*	"2806. I will now endeavour to consider what the influence is which paramagnetic and diamagnetic bodies, viewed as conductors (2797.), exert upon the lines of force in a magnetic field. Any portion of space traversed by lines of magnetic power, may be taken as subject to a field, and there is probably no space without them. The condition of the field may vary in intensity of power from place to place, either along the lines or across them; but it will be better to assume for the present consideration a field of equal force throughout, and I have formerly described how this may, for a certain limited space, be produced (2465.). In such a field the power does not vary either along or across the lines but the distinction of direction is as great and important as ever, and has been already marked and expressed by the term axial and equatorial, according as it is either parallel or transverse to the tile magnetic axis." (Faraday 1839–1855, III, §2806, p. 203)
[1851] 1852 *On field concept*	"3071. A line of magnetic force may be defined as that line which is described by a very small magnetic needle, when it is so moved in either direction correspondent to its length, that the needle is constantly a tangent to the line of motion; or it is that line along which, if a transverse wire be moved in either direction, there is no tendency to the formation of any current in the wire, whilst if moved in any other direction there is such a tendency; or it is that line which coincides with the direction of the *magnecrystallic* axis of a crystal of bismuth, which is carried in either direction along it. The direction of these lines about and amongst magnets and electric currents, is easily represented and understood, in a general manner, by the ordinary use of iron filings" (Faraday 1839–1855, III, §3071, p. 328)

(continued)

Table 2 (continued)

Main	
	Maxwell
[1864] 1865 *On electromagnetic field concept*	"(3) The theory I propose may therefore be called a theory of the *Electromagnetic Field,* because it has to do with the space in the neighbourhood of the electric and magnetic bodies, and it may be called a *Dynamical Theory,* because it assumes that in that space there is matter in motion, by which the observed electromagnetic phenomena are produced." (Maxwell 1865, p. 460) "(4) The electromagnetic field is that part of space which contains and surrounds bodies in electric or magnetic conditions" (Maxwell 1865, p. 460)
1873 *On electro-magnetic field concept*	"The electric field is the portion of space in the neighbourhood of electrified bodies, considered with reference to electric phenomena. It may be occupied by air or other bodies, or it may be so–called vacuum [...]." (Maxwell 1873, I, §44, p. 44) "476. It appears therefore that in the space surrounding a wire transmitting an electric current a magnet is acted on by forces dependent on the position of the wire and on the strength of the current, The space in which these forces act may therefore be considered as a magnetic field, and we may study it in the same way as we have already studied the field in the neighbourhood of ordinary magnets, by tracing the course of the lines of magnetic force, and measuring the intensity of the force at every point." (Maxwell 1873, II, §476, pp. 128–129)

Faraday's *Experimental Researches in Electricity* (1839–1855) were emblematical expectations; without formulas he introduced the basis for the concepts of field and vectors in electromagnetic induction theory. Otherwise, in late (electro–thermal and) electromagnetic theory an advanced use of mathematics was presented (1864–1873) by Maxwell and mechanically (*vortex*) explained Faraday's phenomena.

> I have confined my self almost entirely to the mathematical treatment of the subject, but I would recommend the student, after he has learned, experimentally if possible, what are the phenomena to be observed, to read carefully Faraday's Experimental Research in Electricity. (Maxwell 1873, I, *Preface*, p. xiii)

Finally, electromagnetic phenomena are very well connected with equations of motion, and their forms generated the foundation for Maxwell's physical mathematical (Pisano 2011) theory of the electromagnetic field. This is a discipline structured on *relationships of thought* among *mathematical quantities* and *physical structures* (including logic and language) thereby "[...] reducing [experimental electric and magnetic] phenomena into scientific form [...]" (Maxwell 1865), p. 459). It has its own hypotheses, methods of proofs, internal coherent logic, where a change of mathematics (i.e., from Newtonian to Langrangian) produces a change in both significant physical processes and interpretations of *physical quantities* (Joule 1845; Pisano 2014).

7 Concluding Remarks

This article analysed the original approach proposed by James Clerk Maxwell to deal with his theory of electromagnetism in context (Cfr. Pisano's Faraday and Maxwell, *pre-print*; Pisano 2015). The result was the elaboration of a coherent and complete self-evident system, which gained enormous success and was able to lay the foundations for a theoretical and practical development of knowledge, the consequences of which we still benefit from today. After our exposition, two main, closely connected aspects emerged which follow three strands of thought and can be described:

1. The first aspect concerns the historical context that allowed a new idea such as the one advocated by Maxwell to flourish. It was a context in which profound changes were taking place both from a theoretical point of view and from the point of view of the experimental approach in the physical sciences. The rise of physics mathematics was an evident cause—and—example.
2. The influence on Maxwell that was decisive, greatly inspiring his thought which was then translated so effectively into his *Treatise*, was that of Michael Faraday, through his collection of results on physics entitled *Experimental Researches in Electricity*. These results form the basis of Maxwell's work and he, as we have shown in our paper, often referred to a precise scientific relationship between Faraday's writings and his ideas. However, understandably, Faraday was not the only one Maxwell drew on for his work.
3. Maxwell's century was characterised by a paradigm shift in the physical sciences. This was a shift that led from a purely mechanical view of phenomena (mechanics being the only complete and accepted physical theory) consolidated over time since the distant foundations of Greek thought and definitively made to blossom by Newton's genius in the concepts of mass, inertia and force (Pisano and Bussotti 2014, 2016, 2022, 2023; Pisano 2015; Bussotti and Pisano 2023) to a view that accept an approach based on the concept of energy. An approach shared by several brilliant minds of the time.

In this context, we also analysed the role played by fundamental studies which aimed to go beyond the mechanistic concept then accepted, realising the importance of the conceptualisation of energy (Pisano and Pellegrino 2023) as a quantity able to produce both effective changes in a system, and also realising the possibility of organising an organic body of laws based on energy itself that could be applied to any branch of physics. We also précised William Thomson–Lord Kelvin and Peter Guthrie Tait's activities with respect to Maxwell's electromagnetic theory. These scientists tackled the question of unifying the different branches of physics under the common denominator of energy (Pisano et al. 2019). They sought to go beyond mechanics in favour of the energy approach, which for them was much more effective in making dynamics the basis of all science, through the conceptual cornerstone known as the principle of conservation of energy. Their ability to translate physical problems mathematically through Lagrange's and Hamilton's equations allowed them to identify physical quantities through the language of energy, developing dynamics in terms of conservation principles. In fact, this constituted the first step in Maxwell's theoretical construction.

The mathematical tool so fruitfully applied by Maxwell is due to the ingenuity of Joseph–Louis Lagrange, and is what is known as the Lagrangian formulation or method. It consists in a much more adequate and flexible mathematics than the mathematics used up to that time, based on Isaac Newton's approach. Maxwell found a way to rely on differential mathematics, using Lagrange's equations that describe motion in terms of ratios of variations. Until then, the dynamic conditions of a system could be described, effectively, according to Newton's idea based on an accurate application of the laws of dynamics, in fact on the forces acting on the system, be it particle or complex. Maxwell, in a very effective way:

- Succeeded in translating Faraday's thought and his experimental results into a mathematical language, through Lagrange's theory for connected systems.
- Gave a description of the founding hypotheses of his theory of electromagnetism, identifying how to write the expression of kinetic energy, which he guessed to be necessarily the only possible form of energy, in terms of coordinates and velocity, after having expressed the relations between the motion of a system as a whole and the energy that this system competes with.
- Reinforced the concept that the energy in a mechanical system (first) and in an electric circuit (secondly) is the capacity for doing some amount of work.

At the end of its extended formulation, Maxwell's theory had become a de facto new field theory (D'Agostino 2000; Maxwell and ([1865] 1996), described through the Lagrangian equations of motion. In this way, the whole formulation of the problems focuses on the energy of the system, going beyond the forces acting on it, by applying a system of differential equations of motion completely alternative to those proposed by the mechanical paradigm at that time. In addition, also the binding reactions that are present and act on the system can be neglected in the moment in which they do not perform work; in the same way it is possible to neglect any co-ordinate upon which the energy of the system does not depend.

Maxwell's works were based on the brilliant intuitions of Faraday and Lagrange, and supported by the extraordinary ability to integrate with great effectiveness the mathematical point of view with the physical one, opening the way to a completely new discipline, the mathematical physics, decisive for science and for the history of science. The equations of motion, for example, were connected with electromagnetic phenomena. The result was the foundation of Maxwell's modern electromagnetic field theory, a theory through which it is possible to return to Faraday's work with a renewed capacity for interpretation. Maxwell himself recommended students to go back to Faraday's method and experimental results, whose work was to be read with maximum attention after having learnt what is in the *Treatise* (Campbell and Garnett 1882, pp. 513–514).

Maxwell's definitive leap from the laws of mechanics based on masses and forces to dynamics based on the concept of energy thus constitutes a fundamental historical–epistemological turning point towards a new description of reality from a different point of view, through a completely new use of already known mathematics, towards a new discipline that we could call *physics mathematics* (Pisano 2015, 2010, 2017; Pisano et al. 2017a, b, 2019; Pisano and Bussotti 2015, 2023, 2022; Pisano and Pellegrino 2023). This is a completely new methodological approach to solve physical problems (origin), where the quantities can be physical and mathematical at the same time (first novelty), and measurements are not a priority or a prerogative (second novelty) that makes, however, a coherent and valid physical science. According to this thought, the emergence of physics mathematics (discipline) belonged to physics, not to an advanced use of mathematics to solve physical problems, since physics has changed its *face* because, within this new physical mathematical discipline, it changed its foundations. With regards to Maxwell:

> Maxwell's challenge was to expound a new doctrine and at the same time to establish a new standard in the treatment of current problems. In order to meet these two conflicting requirements, he carefully separated the basic mathematical and empirical foundations of the subject from more speculative theory. [...] Maxwell defined the basic physical quantities in a neutral manner that could be accepted both by fluid and field theorists. For example, he introduced the quantity of electric charge of a body by means of Faraday's hollow conductors [...] with these neutral definitions, Maxwell could conduct much of mathematical analysis without deciding the nature of electricity and magnetism. (Darrigol 2000, p. 167)

Finally, we are confronting a change in significant physical processes and interpretations of *physical quantities*. One of the possible scientific approaches to understanding the history of the foundations of science may be to combine historical and epistemological aspects (primary sources, historical hypotheses, shared knowledge and epistemological interpretations) by means of physical foundations and mathematical inquiry within a combined intellectual history of science—and—*historical epistemology of science* (Cfr. Pisano and Bussotti 2016; Pisano et al. 2017; Pisano 2014, p. 52; 104; et al. Pisano's works).

Acknowledgements We thank the anonymous reviewers for their careful reading of our manuscript and their many insightful comments and suggestions. We also express gratitude to Dr. Julie Robarts (University of Melbourne, Australia) for her observations on English language; eventual remaining English mistakes are our own.

References

Ampère AM (1822a) Recueil d'observation électrodynamiques (Collections electrodynamics of observations). Crochard, Paris.

Ampère AM (1822b) Extrait d'une lettre d'Ampère au Professeur de La Rive sur des expériences électro–magnétiques et sur la formule qui représente l'action mutuelle de deux portions infiniment petites de courants électriques –12 juin 1822. (Extract of Ampère's letter to Professor de la Rive on electromagnetic experiments and on formula representing the mutual action of two infinitely small portions of electric currents). Bibliothèque universelle des sciences, belles–lettres, et arts XX:185–192.

Ampère AM (1826) Théorie des phénomènes électro–dynamiques uniquement déduite de l'expérience (Theory electro–dynamic phenomena only inferred from experience). Méquignon–Marvis. Paris.

Bussotti P, Pisano R (2013) On the Conceptual Frames in René Descartes' Physical Works. Advances in Historical Studies 2(3):106–125.

Bussotti P, Pisano R (2024) Details in Galileo's Mechanism of Inertia as New Intellectual Historical Insights. In Pisano R (ed). A History of Physics: Phenomena, Ideas & Mechanisms. Essays in Honour of Salvo D'Agostino. Dordrecht, Springer, pp. 457–514.

Campbell L, Garnett W (1882) The life of James Clerk Maxwell with a selection from his correspondence and occasional writings and a sketch of his contributions to science by Lewis Campbell, M.A., LL.D. Professor of Greek in the university of St. Andrews and William Garnett, M.A. Macmillan and Co., London.

Carnot L ([1783] 1786) Essai sur les Machines en Général. Defay, Dijon.

Carnot L (1803) Principes fondamentaux de l'équilibre et du mouvement. Deterville, Paris.

Corben HC, Stehle P ([1950] 1994) Classical Mechanics. Dover, Mineola.

Coulomb CA (1785) Théorie des machines simples en ayant égard au frottement de leurs parties et à la roideur du cordage (Theory of simple machines with respect to the friction of parts and the stiffness of the rope), Mémoire de mathématique et de physique présentes à l'Académie Royal des Sciences par divers savants. Bachelier, Paris.

D'Agostino S (2000) On the Difficulties of the Transition from Maxwell's and Hertz's Pure–Field Theories to Lorentz Electron. Physics in Perspective 2(4):398–410.

D'Agostino S (2013) La Scienza dell'Elettricità da Volta ad Einstein. BookSprint Edizioni, Romagnano al Monte.

D'Agostino S (2014) Maxwell's Dynamical Philosophy: An Early Solution to the Problem of a Discrepancy between Particles and Waves. Advances in Historical Studies 3:123–129.

d'Alembert JLR (1743) Traité de Dynamique. David, Paris.

Darrigol O (2000) Electrodynamics from Ampère to Einstein. Oxford University Press, Oxford.

Eisenhart LP (1929) Dynamical Trajectories and Geodesics. Annals of Mathematics. Princeton University & Institute for Advanced Studies. NJ, pp. 591–606.

Euler L (1736) Mechanica sive motus scientia analytice exposita. Auctore Leonhardo Eulero academiae imper. scientiarum membro et matheseos sublimioris professore. Tomus I. Instar supplementi ad commentar. acad. scient. imper. Ex typographia academiae scientarum, Petropoli.

Euler L ([1744] 1952) Methodus inveniendi lineas curvas maximi minimive proprietate gaudens, sive solutio problematis isoperimetrici latissimo sensu accepti, Bousquet, Losanna e Ginevra In: Opera Omnia mathematica, s. I, Vol. XXIV.

Euler L ([1750] 1752) Découverte d'un nouveau principe de Mécanique. Mémoires de l'académie des Sciences et des Belles Lettres de Berlin, VI, Paris, pp. 185–217.

Euler L ([1764] 1766) Elementa calculi variationum. Novi Commentarii academiae scientiarum Petropolitanae 10:51–93 [See also: Opera Omnia mathematica, s.1, Vol. XXV, pp. 141–176].

Euler L (1775) Nova methodus motum corporum rigidorum determinandi. Novi Commentarii academiae scientiarum Petropolitanae, XX, pp. 208–238.

Euler L (1911–) Leonhardi Euleri Opera Omnia. Serie I: Opera Mathematica, 29 Vols. Serie II: Opera Mechanica et Astronomica, 29 Vols. Series III: Opera Physica Miscellanea 12 Vols. Series IVA: Commercium Epistolicum 9 Vols. Series IVB: Manuscripts. Euler Committee of the Swiss Academy of Sciences (eds). Petersburg Academy of Sciences in Basel. Springer Birkhäuser Basel.

Faraday M (1839–1855) Experimental Researches in Electricity. 3 Vols. Taylor, London.

Fourier J (1822) Théorie analytique de la chaleur. Firmin Didot, Paris.

Fourier J (1878) The Analytical Theory of Heat. Cambridge University Press, London–Cambridge.

Fox R (1974) The rise and fall of Laplacian physics, Historical Studies in the Physical Sciences 4:89–136.

Francaviglia M (1991) (ed) Mechanics, Analysis and Geometry: 200 years after Lagrange. Amsterdam, North–Holland.

Fraser CG (1983) J. L. Lagrange's early contributions to the principles and methods of mechanics. Archive for History of Exact Sciences 28:197–241.

Fraser CG (1985) Lagrange, J. L. changing approach to the foundations of the calculus of variations. Archive for History of Exact Sciences 32/2:151–191.

Fraser CG (1989) The calculus as algebraic analysis: Some observations on mathematical analysis in the 18th century. Archive for History of Exact Sciences 390:317–335.

Fraser CG (1992) Ipsometric Problems in the Variational Calculus. Historia Mathematica 19:4–23.

Fraser CG (1994) The origins of Euler's variational calculus. Archive for History of Exact Sciences 470:103–141.

Fufay CF ([1733] 1735) Quatrième mémoire sur l'électricité. De l'attraction et répulsion des corps électriques (Fourth memory on electricity. The attraction and repulsion of electric corps). Histoire et mémoires de l'Académie [royale] des sciences avec les mémoires de mathématiques et de physique, partie «Mémoires». L'Imprimerie Royale, Paris, pp. 457–476.

Gillispie CC, Pisano R (2014) Lazare and Sadi Carnot. A Scientific and Filial Relationship. 2nd ed. Springer, Dordrecht.

Joule JP (1838) Description of an electro-magnetic engine, by Mr. J. P. Joule. The Annals of Electricity, Magnetism, and Chemistry; and Guardian of Experimental Science. Conducted by William Sturgeon, Lecturer on Experimental Philosophy at the Honourable East India Company's Military Academy, Addiscombe, &c.&c.; Assisted by Gentlemen Eminent in these Departments of Philosophy, Vol. II (pp. 122–123). Published by Sherwood, Gilbert, and Piper, Paternoster Row, London.

Joule JP (1839) Investigations in Magnetism and Electro-magnetism. By James P. Joule, Esq. In two letters to the Editor. The Annals of Electricity, Magnetism, and Chemistry; and Guardian of Experimental Science. Conducted by William Sturgeon, Lecturer on Experimental Philosophy at the Honourable East India Company's Military Academy, Addiscombe, &c.&c.; Assisted by Gentlemen Eminent in these Departments of Philosophy. Published by Sherwood, Gilbert, and Piper, Paternoster Row, London, Vol. IV, pp. 131–135.

Joule JP (1841a) On a New Class of Magnetic Forces. Conversazione, February 16, 1841. Lecture at the Royal Victoria Gallery. In: The Annals of Electricity, Magnetism, and Chemistry; and Guardian of Experimental Science. Conducted by William Sturgeon, Superintendent of the Royal Victoria Gallery of Practical Science, Manchester; Formerly Lecturer on Experimental Philosophy at the Honourable East India Company's Military Academy, Addiscombe, etc. etc.; Assisted by Gentlemen Eminent in these Departments of Philosophy, Published by Sherwood, Gilbert, and Piper, Paternoster Row, London, Vol. VIII, pp. 219–224.

Joule JP (1841b) On the Heat evolved by Metallic Conductors of Electricity, and in the Cells of a Battery during Electrolysis. The London, Edinburgh, and Dublin Philosophical Magazine and Journal of Science XIX/124:260[–277].

Joule JP (1842) On the Effects of Magnetism upon the Dimensions of Iron and Steel Bars. The London, Edinburgh and Dublin Philosophical Magazine and Journal of Science, XXX, 76–87. [New and United Series of the Philosophical Magazine, Annals of Philosophy, and Journal of Science, January-June, 1847, Richard and John E. Taylor, Red Lion Court, Fleet Street, Printers and Publishers to the University of London].

Joule JP (1843) On the Calorific Effects of Magneto–Electricity, and on the Mechanical Value of Heat. The London, Edinburgh, and Dublin Philosophical Magazine and Journal of Science. Conducted by Sir David Brewster, K.H. LL.D. F.R.S.L.&E. &c. Richard Taylor, F.L.S. G.S. Astr.S. Nat.H.Mosc. &c. Richard Phillips, F.R.S.L.&E. F.G.S. &c. Robert Kane, M.D. M.R.I.A., Richard and John E. Taylor, Red Lion Court, Fleet Street, Printers and Publishers to the University of London, London, New and United Series of the Philosophical Magazine, Annals of Philosophy, and Journal of Science XXIII:263–276, 347–355, 435–443.

Joule JP (1845) On the Changes of Temperature produced by the Rarefaction and Condensation of Air. By J. P. JOULE, Esq. The London, Edinburgh, and Dublin Philosophical Magazine and Journal of Science [ser. 3, May 1845] 26/174:369–383. [*Id.*, (1884) The Scientific Papers of James Prescott Joule. Cambridge University Press New York, pp. 172–189].

Joule JP ([1849] 1850) On the Mechanical Equivalent of Heat. By James Prescott Joule, F.C.S., Sec. Lit. and Phil. Society, Manchester, Cor. Mem. R.A., Turin, &c. Communicated by Michael Faraday, D.C.L., F.R.S., Foreign Associate of the Academy of Sciences, Paris, &c. &c. &c. Received June 6, Read June 21, 1849. Philosophical Transactions of the Royal Society of London, 140, pp. 61–82.

Joule JP (1884) The Scientific Papers of James Prescott Joule. The Physical Society of London. Taylor and Francis, London.

Lagrange JL ([1766–1769] 1773) Sur la méthode des variations. Miscellanea philosophico-mathematica Societatis privatae Taurinensis IV:163–187. [See also: Œuvres de Lagrange, II, pp. 37–63.

Lagrange JL (1788) Méchanique analitique. Desaint, Paris [see also: Lagrange JL (1973) Œuvres de Lagrange. Seconde Edition. Courcier (ed.), I–XIV vols. (in X), Gauthier–Villars, Paris].

Lagrange JL (1870–1873) Œuvres de Lagrange (Lagrange works) 2 edn. Courcier Paris. XIV vols. (in X) Gauthier–Villars, Paris.

Lagrange JL (1892) Correspondance de Lagrange avec Condorcet, Laplace, Euler et divers Savants (Correspondence) In: Lagrange 1870–1873, Vol. XIV.

Lamé G (1836) Cours de physique de l'école polytechnique, I. Bachelier, Paris [see also: *Id.*, Cours de physique de l'école polytechnique, 2nd ed. Bruxelles].

Lamé G (1852) Leçons sur la théorie mathématique de l'élasticité des corps solides. Bachelier, Paris

Lamé G (1861a) Leçons sur la théorie analytique de la chaleur. Mallet–Bachelier, Paris.

Lamé G (1861b) Discours préliminaire. In: Lamé 1861a.

Laplace PS (1805) Traité de mécanique céleste. Courcier, Paris.

Laplace PS (1822) Notes sur la vitesse du son. Annales de chimie et de physique XX:266–268.

MacFarlane A (1917) Lectures on Ten British Physicists of the Nineteenth Century. John Wiley, Chapman and Hall.

Mach EWJW (1883) The Science of Mechanics. 4th ed. The Open Court Publishing Co, Chicago.

Maupertuis PLM ([1746] 1748) Les Loix du mouvement et du repos déduites d'un principe métaphysique. Académie Royale des Sciences et des Belles Lettres de Berlin. Histoire de l'Académie Royale des Sciences et des Belles Lettres. Haude. Libraire de la Cour & de l'Académie Royale, Berlin, pp. 267–294.

Maxwell JC ([read 1855 and 1856] 1855) On Faraday's lines of force. In Maxwell [see also: *Id.*, [1890] 2003], I, pp. 155–229.

Maxwell JC (1855–1856; 1861–1862) On physical lines of forces. Philosophical Magazine XXI, pp. 161–175, 281–291, 338–348; Philosophical Magazine XXII, pp. 12–24, 85–95.

Maxwell JC (1860) Illustrations of the dynamical theory of gases. Philosophical Magazine XIX:19–32; XX:21–37 [See also *Id.*, [1890] 2003, I, pp. 377–409].

Maxwell JC (1864) [abstract] A dynamical theory of the Electromagnetic field [The paper was received in October 27, 1864, and read in December 8, 1864] The Proceedings of the Royal Society of London. From November 19, 1863, to December, 1861, inclusive. XIII:531–536.

Maxwell JC (1865) A Dynamical theory of the Electromagnetic field. Philosophical Transactions of The Royal Society of London CLV:459–512.

Maxwell JC ([1865] 1996) Una Teoria Dinamica del Campo Elettromagnetico. Translated into Italian by Salvo D'Agostino. Edizioni Teknos, Roma.

Maxwell JC (1871) Theory of heat. Longmann Green, Roberts & Green, London.

Maxwell JC (1873) A Treatise on Electricity and Magnetism. 2 Vols. The Clarendon Press, Oxford.

Maxwell JC (1874) On Hamilton's characteristic function for a narrow beam of light. Proceedings of the London Mathematical Society s1–6 /1:182–190.

Maxwell JC (1875) On the Dynamical Evidence of the Molecular Constitution of Bodies. Journal of the Chemical Society London 28:493–508.

Maxwell JC ([1876–1877] 1920) Matter and motion. Larmor J (ed). Macmillan, London.

Maxwell JC (1879) On stresses in reified gases arising from inequalities of temperature. The Philosophical Transactions of the Royal Society of London 170:231–256.

Maxwell JC (1881) An elementary treatise on electricity. Garnett W (ed). The Clarendon Press, Oxford.

Maxwell JC ([1890] 2003) The scientific papers. Niven WD (ed). 2 Vols. Dover, New York.

Maxwell JC ([1892] 1954) A Treatise on Electricity and Magnetism. 2 Vols. Unabridged third edition. Dover Publications, Inc. New York.

Mottelay PF (1922) Bibliographical History of Electricity and Magnetism. Charles Griffin and Company Limited Publication, London.

Moyer DF (1977) Energy, Dynamics, Hidden Machinery: Rankine, Thomson and Tait, Maxwell. Studies in History and Philosophy of Science 8/3:251–268.

Newton I (1687) Philosophiae Naturalis Principia Mathematica. Imprimatur S. Pepys, Reg. Soc. Preses. Julii 5. 1686. Londini, Jussi Societatus Regiae ac Typis Josephi Streater. Prostat apud plures Bibliopolas. Anno MDCLXXXVII.

Newton I ([1687] 1803) The Mathematical Principles of Natural Philosophy, by Sir Isaac Newton. Translated into English by Motte A., Symonds, London, Vol. I.

Oliveira ARE, Pisano R (2022) Historical Details on Coulomb's Torsional Forces by Metal Wires (with Oliveira Agamenon, Polytechnic School of UFRJ, Brazil) In: Ceccarelli M, López-García R (eds). Explorations in the History and Heritage of Machines and Mechanisms. Vol. 40. Springer, Dordrecht, pp. 241–254.

Ørsted HC (1822) Instruments pour mesurer la compression de l'eau (Instruments to measure the compression of the water). Annales de chimie et de physique XXI:99–100.

Ørsted HC (1823) Sur la compressibilité de l'eau (On the compressibility of water). Annales de chimie et de physique XXII:192–19.

Pisano R (2010) On Principles in Sadi Carnot's Thermodynamics (1824). Epistemological Reflections. Almagest 2/1:128–179.

Pisano R (2011) Physics–Mathematics Relationship. Historical and Epistemological notes. In Barbin E, Kronfellner M, and Tzanakis C (eds). ESU–European Summer University History and Epistemology. Verlag Holzhausen GmbH–Holzhausen Publishing Ltd., Vienna, 457–472.

Pisano R (2013) History Reflections on Physics Mathematics Relationship in Electromagnetic Theory. In: Barbin E, Pisano R (eds). The Dialectic Relation between Physics and Mathematics in the 19th Century. Springer, Dordrecht, pp. 31–58.

Pisano R (2014) On Lagrangian in Maxwell's electromagnetic theory. Scientiarum VI. The Federate University of Rio de Janeiro Press, Rio de Janeiro, pp. 44–59.

Pisano R (2015) (ed) A Bridge between Conceptual Frameworks, Science, Society and Technology Studies. Springer, Dordrecht.

Pisano R (2017) A Development of the Principle of Virtual Laws and its Framework in Lazare Carnot's Mechanics as Manifest Relationship between Physics and Mathematics. International Journal for Historiography of science 2:166–203.

Pisano R (forthcoming) Faraday and Maxwell, two life one theory, pre-print.

Pisano R, Bussotti P (2014) On the Jesuit Edition of Newton's Principia. Science and Advanced Researches in the Western Civilization. Newton Sp. Issue: History and Historical Epistemology of Science. Advances in Historical Studies 3/1:33–55.

Pisano R, Bussotti P (2015) The Emergencies of Mechanics and Thermodynamics in the Western Society during 18th–19th Century. In Pisano 2015 (ed) pp. 399–436.

Pisano R, Bussotti P (2016) A Newtonian Tale Details on Notes and Proofs in Geneva Edition of Newton's Principia. Bulletin BJHM–Journal of the British Society for the History of Mathematics 31/3:160–178.

Pisano R, Bussotti P (2022) Conceptual Frameworks on the Relationship between Physics–Mathematics in the Newton Principia Geneva Edition (1822). Foundations of Sciences (Springer) 27/3:1127–182.

Pisano R, Bussotti P (2023) The Physics–Mathematics Interplay in Newton Principia Geneva Edition (1822): A New Case Study of the Three-body Problem, Proposition LXVI–Theorem XXVI. Submitted to *Philosophical Transactions A*.

Pisano R, Capecchi D (2009) La Théorie Analytique de La Chaleur. Notes on Fourier and Lamé. Bulletin de la Sabix 44:83–90.

Pisano R, Capecchi D (2013) Conceptual and Mathematical Structures of Mechanical Science in the Western Civilization around 18th Century. Almagest 4/2:86–121.

Pisano R, Frankowiak R (2014) Lavoisier and Sadi Carnot. Chemical–and–Physical Sciences as Dating Back to Two Survived Scientific Revolutions: 1789 And 1824. Bilek M (ed). Proceedings of the 9th International Organizing Science Technology Education – IOSTE symposium for Central and Eastern Europe. Hradec University, IX, Gaudeamus Králové, pp. 42–54.

Pisano R, Marmottini D (2017) Nature–of–Science Teaching: Notes on the Lagrangian Methods in Maxwell's electromagnetic Theory. XXXVI SISFA Proceedings. Pavia University, Pavia, pp. 263–268.

Pisano R, Pellegrino EM (2024) Clausius's Laws, Boltzmann's H-Theorem & Entropy: Mathematical Frameworks into a Kinetic Mechanism (1850–1901). In Pisano R (ed). A History of Physics: Phenomena, Ideas & Mechanisms. Essays in Honour of Salvo D'Agostino. Springer, Dordrecht, pp. 263–328.

Pisano R, Capecchi D (2016) Tartaglia's Science of Weights and Mechanics in the Sixteenth Century. Selections from Quesiti et inventioni diverse: Books VII–VIII. Springer, Dordrecht.

Pisano R, Anakkar A, Pellegrino EM, Nagels M (2019) Thermodynamic Foundations of Physical Chemistry. Reversible Processes and Thermal Equilibrium into History. Foundations of Chemistry 21:297–323.

Pisano R, Agassi J, Drozdova D (eds) (2017a) Hypotheses and Perspectives in the History and Philosophy of Science. Homage to Alexandre Koyré 1892–1964. Springer, Dordrecht.

Pisano R, Fichant M, Bussotti P, Oliveira ARE (2017b) (eds) The Dialogue between Sciences, Philosophy and Engineering. New Historical and Epistemological Insights. Homage to Gottfried W. Leibniz 1646–1716. The College Publications, London.

Pisano R, Cooppersmith J, Peake M (2021) Essay on Machines in General (1786). Text, Translations and Commentaries. Lazare Carnot's Mechanics–Vol. 1. Springer, Dordrecht.

Pisano R, Vincent P, Dolenc K, Ploj-Virtič M (2021) Historical Foundations of Physics & Thermodynamic foundations of physical chemp; Applied Technology as Dynamic Frameworks in Pre-Service STEM. Foundations of Science 26:727–756.

Pisano R, Vincent P, Mauricio P (2020) Introduction. Joule's Bicentenary: History of Science, Foundations and Nature of Science. Joule's Bicentenary Special Issue. Foundations of Science 26:531–551.

Rankine WJM (1885) Outlines of the Science of Energetics. Proceedings of the Royal Philosophical Society of Glasgow, Volume 3. Royal Society of Glasgow. Richard Griffin & Company, Glasgow and London, pp. 381–399.

Rankine WJM (1888) On the General Law of the Transformation of Energy. Miscellaneous Scientific Papers. Charles Strahan and Company, London.

Simpson TK (1970) Some Observations on Maxwell's Treatise on Electricity and Magnetism. Studies in History and Philosophy of Science 1/3:188–276.

Simpson TK (1997) Maxwell on the Electromagnetic Field: A Guided Study. Rutgers University Press, New Jersey.

Simpson TK (2005) Figures of Thought. A Literary Appreciation of Maxwell's Treatise on Electricity and Magnetism. Green Lio Press, Santa Fe–New Mexico.

Simpson TK (2010) Maxwell's mathematical rhetoric. Rethinking the Treatise on electricity and magnetism. Green Lio Press, Santa Fe, New Mexico.

Thomson W ([1842] 1854) On the Uniform Motion of Heat in Homogeneous Solid Bodies, and its Connection with the Mathematical Theory of Elasticity. The London Edinburgh and Dublin Philosophical Magazine and Journal of Science 7/48:502–515.

Truesdell CA (1968) Essays in the history of mechanics. Springer, Berlin.

Truesdell CA (1970) Reactions of late baroque mechanics to success, conjecture, error, and failure in Newton's Principia. In: Palter R (ed). The annus mirabilis of sir Isaac Newton 1666–1966. MIT Press, Cambridge–MA, pp. 192–232.

Chapter 14
Clausius' Laws, Boltzmann's H-Theorem & Entropy: Mathematical Frameworks into a Kinetic Mechanism (1850–1901)

Raffaele Pisano and Emilio Marco Pellegrino

Abstract Boltzmann's H-theorem (1872) was explicitly conceived in order to proof Clausius' Second Law (1850s) and for cyclic processes only (1854). Secondary literature used to join the *function E*, in Boltzmann's 1872–paper, to the development of *Entropy*. However, Clausius' Second Law for cyclical processes did not deal with Entropy; moreover, in a cyclic process, entropy variation, being a state function, is necessarily equal to zero. Later, Clausius introduced the Entropy quantity in order to settle the framework of non-cyclic processes mathematically (1865; 1872). In this paper we will shed light on this issue through the textual analysis of Clausius and Boltzmann's seminal works; a critical discussion of the main historiography of the H-theorem and its interplay with Entropy will be presented as well.

Keywords Boltzmann's H-theorem · Clausius · Cyclic processes · Entropy · Kinetic model · Mechanisms · Non-cyclic processes · Relationship physics–mathematics · Second law of thermodynamics · Statistical mechanics

1 An Outline

Still to this day Entropy, generally indicated as *S*, is ambiguously introduced, both historically and in the Nature of Science (Pisano et al. 2019, 2021b, c, 2020). Taking into account the evident intellectual complexity of its historical and scientific roots, we remark that, on the one hand it confusingly deals with an heuristic character of a *physical measure of the disorder*—spontaneous, increasing–decreasing, reversible–irreversible transformations, etc., on the other hand it is finally presented as a rigorous *mathematical–physical state function.* A competition between physical processes

R. Pisano (✉)
HOPAST at IEMN, Lille University–CNRS, Villeneuve d'Ascq, France
e-mail: raffaele.pisano@univ-lille.fr

E. M. Pellegrino
ISEPP, Portland State University, Portland, USA
e-mail: emiliomarco.pellegrino@chimicifisici.it

R. Pisano (ed.), *A History of Physics: Phenomena, Ideas and Mechanisms*,
History of Mechanism and Machine Science 42,
https://doi.org/10.1007/978-3-031-26174-9_14

and mathematical definitions is set up which causes confusion within the crucial relationship physics–mathematics in the theory (Pisano and Bussotti 2022; Pisano 2017; Pisano and Capecchi 2013); in addition, the textbooks usually deals with the definition of Second Law of thermodynamics defining the total entropy of a system and its surroundings increases for a spontaneous process (Gillispie and Pisano 2014).

1.1 Ideas and Mechanisms

Below (Pisano 2021), for our aims, we briefly outline the intellectual mechanisms of the physical and mathematical role played by the laws of thermodynamics and of entropy:

(1) The *First Law of Thermodynamics* states the definition of a new ad hoc quantity, internal energy U, for a thermodynamic system. This law, since it takes into account (friction and therefore) heat and temperature as new quantities, deals with the extension of the conservation of mechanical total energy law. The energy is conserved (spontaneous processes).

(2) The *Second Law of Thermodynamics* (in all its ideas–&–formulations) is concerned with the direction of a natural (irreversible/spontaneous) process; the Entropy of the universe increases (i.e., arrow of time). Thus, natural process runs only and always spontaneously in one sense. The other *versus* is considered not natural and also called reversible.

(3) The *Third Law of Thermodynamics* states that Entropy (ideal crystal substances) approaches zero as the temperature approaches absolute zero–K (Thomson 1848–1849a, 1851a, b, 1882–1911; see also Thompson 1798). One can calculate Entropy changes (increase) for a reaction using a summation law; but following the 3rd Law a question arises: what is the relationship between Enthalpy and Entropy?

In the appendix to his *ninth Memoir*, Clausius presented the mathematical expressions of the first and Second Laws as follows:

> We have therefore:
>
> $$dQ = dU + dw \qquad \text{(A)}$$
>
> $$dQ = Tds \qquad \text{(B)}$$
>
> The first of these two equations applies to reversible, as well as to non-reversible changes; the second, on the contrary, holds good for reversible changes solely. But in order to be able to bring the two equations into conjunction, we will suppose that they relate to one and the same reversible change of the body. In this case the thermal element dQ is the same in both equations, hence we can eliminate it from the equations, whereby we obtain
>
> $$Tds = dU + dw \qquad \text{(a) [Clausius 1867, p. 366]}$$

It is to be noted that Clausius does not use the symbol δ for partial differential, even if he considers heat and work as *path dependent* functions. Moreover, dw represents, in Clausius' writings, the infinitesimal work element expressed in the heat unit of measure, that is to say $dw = AdW$. In current thermodynamics, Clausius' laws are exploited by writing them in the form of the following fundamental equations:

- $\Delta U = Q \pm W$ (Planck's vs Clausius' conventions) Cfr. Clausius, 1850 (Clausius 1850)
- $dU = \delta Q - \delta W$ (closed, in modern terms infinitesimal quasi-static/reversible or irreversible) Cfr. Clausius (1850)
- $dU = TdS - PdV$ (closed, general, in modern terms infinitesimal quasi-static processes/reversible) Cfr. Clausius, 1866 (Clausius 1867, p. 366)

In the special case of ideal gases, the solutions of the foregoing fundamental equations of thermodynamicsin the case of adiabatic, isothermal, isobaric and isochoric transformationsare recalled in the following Table 1.

Table 1 Thermodynamic properties for Ideal Gases[1]

	Adiabatic $Q = 0$	Isothermal T constant	Isobaric P constant	Isochoric V constant
State equation	$PV^{\gamma} = K$ where $\gamma = C_P/C_V$	PV = constant	V/T = constant	P/T = constant
Mechanical work (W)	$-[1/(\nu - 1)] \cdot [p_2 V_2 - P_1 V_1]$	$nRT \ln(V_2/V_1)$	$P\Delta V$	$P\Delta V = 0$
Internal energy (ΔU)	$C_V \Delta T$	0	$C_V \Delta T$	$C_V \Delta T$

In this framework, the relationships among heat, entropy and temperature are so–usually defined, in general, by the following equation:

$$\delta Q \leq TdS$$

where the equality sign holds true in the case of reversibility. Then, for reversible processes, the entropy is given by:

$$dS = \delta Q/T$$

whilst mechanical work reduces to:

$$\delta W = PdV$$

[1] In the case of ideal gases, in general, we have: $PV = nRT$; $U = nc_v T$; $c_p - c_v = R$; $\gamma = c_p/c_v$. In particular, for a monatomic ideal gas: $c_v = (3/2)R$, and $\gamma = 5/3$.

Hence:

- For isolated system, it never decreases, at equilibrium, $S \rightarrow S_{max}$
- Following 2ndlaw, the Entropy (closed/constant volume) of the systems increases as
- $S_{max} = \log N$ (equilibrium states)
- Irreversible cases: free adiabatic expansion ($\delta Q \leq TdS$)
- Motion and friction and consequent heat production ($\delta W \leq PdV$)

In other words, we can write:

$$dS = dQ_r/T = dU/T + P\,dV/T = C_V(T)dT/T + P\,dV/T$$

where for an ideal gas the internal energy and entropy are so usually defined:

- $dU = C_v dT | PV = nRT$
- $\Delta S = C_V \int dT/T + nR \int dV/V = C_V \ln(T_2/T_l) + nR \ln(V_2/V_i)$
- It is a State function so it is independent of path (bounded integral-calculus)
- Way of the process for isolated system: versus state of maximum value of the entropy.
- Entropy may be negative for a portion of a composite system.

We remark that:

- An increase of entropy does not mean an increase of temperature (denominator of the fraction).
- An increase of temperature does not mean an increase of entropy (e.g., adiabatic compression).

On so–called *Effective Entropy* and 2nd Law of Thermodynamics, Arthur Stanley Eddington (1882–1944) wrote:

> The law that entropy increases—the Second Law of Thermodynamics—holds, I think, the supreme position among the laws of Nature. (Eddington [1928] 1948, p. 68)

In agreement with the above definitions and considerations, we can summarize that:

- Entropy S increases as disorder increases and energy decreases;
- Gases result as high disorder and low energy;
- Solids result as low disorder and high energy;
- Entropy S is studied as a state property and its variation calculated as $\Delta S = S_f - S_i$
- Entropy variation $\Delta S > 0 \rightarrow$ increase in entropy $\rightarrow$ disorder increased;
- Entropy variation $\Delta S < 0 \rightarrow$ decrease in entropy $\rightarrow$ disorder decreased

Finally, the following reflections arise:

- In order to define Entropy S, one starts from the 2ndlaw of thermodynamics. But which statement? Carnot's one? Clausius' one? Kelvin's one? Clausius-Kelvin's one? Planck's one? Caratheodory's one?
- What is disorder? Is one able to measure this quantity?
- Did S really measure something? If yes, what is the measuring tool? Or, did we calculate something?
- Entropy S is also a state function: can a mathematical entity (function) measure a physical (measurable) quantity? For example: is length (physical quantity) the measure of the meter (tool of measurement)?
- In the case of Entropy S, what is the physics-mathematics relationship between a measurable quantity (disorder?) and its mathematical description (Integral of state function)? And in turn what is the physics-mathematics relationship between a measurable quantity (randomness?) and its measurement (yet unknown) tool? (Pisano 2013)
- Finally, can a mathematical-physical term (e.g., state function S) measure a physical quantity (e.g., randomness? Disorder?) that is in turn measured by something else?

2 Introduction

The foundational roots of Statistical Mechanics (Cfr. Pisano and Capecchi 2013) are retraced within the development of the former kinetic models of gases. They are mainly authored by August Karl Krönig (1822–1879) (Krönig 1856), Rudolf Julius Emanuel Clausius (1822–1888) (Clausius 1867, 1850, 1854, 1857a, b, 1865, 1871, 1872; Pisano et al. 2019, 2021b), Ludwig Edward Boltzmann (1844–1904) (Boltzmann ([1865–1874] 1909), 1866, 1871,1872, 1877, 2003; Pisano et al. 2019, 2021b) and James Clerk Maxwell[2] (1831–1879). Some of these works can be framed within the search for a proof of the Second Law of Clausius' Mechanical theory of heat (Clausius 1872, pp. 132–146; Cfr. Pellegrino and Cerruti (2015, 2016), Pellegrino (2014)), as is made explicit in the introduction of Boltzmann's 1866 paper (Boltzmann 1866):

> The coincidence of the First fundamental Law of the Mechanical Theory of Heat with the mechanical principle of kinetic energy has long been widely acknowledged. Conversely, the Second Law represents an exceptional peculiarity: its proof is sometimes uncertain and definitely unclear. This paper's aim is to provide a truly analytic and general proof of the Second Law of the Theory of Heat, through a mechanical approach.[3] (Boltzmann 1866, p. 9; Authors' translation)

[2] On the subject and its interplay with new coming disciplines in the nineteenth century (D'Agostino 2014; Faraday 1839–1855; Lagrange 1788; Maxwell 1864, 1865, 1871) we direct the reader to Pisano's works (Pisano et al. 2013, 2023; Pisano 2014; Pisano and Marmottini 2017) and the recent one on Maxwell's and Lagrangian (Pisano et al. 2024) in this book above, Part 2.

[3] "Bereits längst ist die Identität des ersten Hauptsatzes der mechanischen Wärmetheorie mit dem Prinzip der lebendigen Kräfte bekannt; dagegen nimmt der zweite Hauptsatz eine eigentümliche exzeptionelle Stellung ein und wird sein Beweis auf hier und da nicht einmal sichern, keinesfalls aber klar vor Augen liegenden Umwegen geführt. Es soll nun der Zweck dieser Abhandlung sein, einen rein analytischen, vollkommen allgemeinen Beweis des zweiten Hauptsatzes der Wärmetheorie zu liefern, sowie den ihm entsprechenden Satz der Mechanik aufzufinden."

From the former kinetic models on, mechanical approaches—conceived to represent thermodynamic systemsevolved towards a plurality of theories and concepts that converged into a corpus of models that is currently known under the common name of Statistical Mechanics.

Frigg (2008) delivered an excellent and wide scientific production on Statistical Mechanics wherein he brilliantly outlines its current status as follows:

> Unlike quantum mechanics and relativity theory, Statistical Mechanics has not yet found a generally accepted theoretical framework, let alone a canonical formulation. What we find in Statistical Mechanics is a plethora of different approaches and schools, each with its own program and mathematical apparatus, none of which has a legitimate claim to be more fundamental than its competitors. (Frigg 2008, p. 6)

To handle this conceptual polymorphism, Frigg identifies two theoretical frameworks: one referring to Boltzmann's approach and one referring to Gibbs' approach. In Frigg's own words:

> In what follows I take my lead from the fact that all these different schools (or at any rate those that I discuss) use (slight variants) of either of two theoretical frameworks, one of which can be associated with Boltzmann (1877) and the other with Gibbs (1902), and can thereby classify different approaches as either 'Boltzmannian' or 'Gibbsian'. (Frigg 2008, p. 7)

In two recent works of ours in the field (Pisano et al. 2019, 2021b)—supported by the analysis of historical documents—we have disclosed some aspects of the statistical expressions variously related with the thermodynamic function that Clausius seminally defined as entropy—and that can be found in both Boltzmann's statistical approach [1865–1872; Boltzmann ([1865–1874] 1909, 1866) and Gibbs' geometrical approach (1873–1906 [posthumous]). Within this physical–chemical–mathematical framework, in particular, we have retraced previously unknown aspects of the intellectual process that led Boltzmann to formulate his H-theorem (Pisano et al. 2021b; Pellegrino 2020).

The H-theorem is presented as the climax of Boltzmann's kinetic model; starting in 1866 by means of a kinetic physical–mathematical approach, that allowed the expression of Clausius' Second Law (1850–1854) for cyclic processes (Clausius 1854, 1872) that is to say

$$\int \frac{dQ}{T} \geq 0 \tag{1}$$

as in the following further form (Fig. 1):

$$2\Sigma \log \int_{t_1}^{t_2} \frac{mc^2}{2} dt \geq 0$$

145

Genau dieselbe Formel entwickelt Tait in seinem *Sketch of Thermodynamics* S. 100, aber anstatt sie als meine Formel anzuführen, sagt er: *„This is Thomson's expression for the amount of heat dissipated during the cycle."* Dabei citirt er als den Ort, wo Thomson diesen Ausdruck gegeben haben soll, den schon oben erwähnten Artikel *„On a Universal Tendency etc."*[1]). In diesem Artikel befindet sich aber weder der oben angeführte, noch irgend ein ihm gleichbedeutender Ausdruck. Es kommen darin überhaupt nur vier Formeln vor, welche von der oben angeführten gänzlich verschieden sind. Ich mufs sagen, dafs mir die Behauptung des Hrn. Tait und das hinzugefügte Citat vollkommen unerklärlich sind.

In einer späteren Abhandlung[2]) habe ich die Betrachtungen noch weiter vervollständigt. Der obige Satz über uncompensirte Verwandlungen, welcher unter Fortlassung des Zeichens N auch so geschrieben werden kann:

$$\int \frac{dQ}{T} \geqq 0,$$

bezog sich nur auf Kreisprocesse. Es kam mir nun darauf an, eine Gröfse zu gewinnen, welche für jede beliebige Veränderung eines Körpers gilt, und ihren Werth immer nur in Einem Sinne ändern kann. Zu dem Zwecke fügte ich zu den bisher betrachteten beiden Verwandlungsarten (nämlich der Verwandlung von Arbeit in Wärme und umgekehrt, und dem Wärmeübergange aus einem wärmeren in einen kälteren Körper und umgekehrt), noch eine dritte Verwandlungsart hinzu, welche sich auf die Zustandsänderung eines Körpers bezieht, und durch eine Gröfse Z, welche ich die ***Disgregation*** des Körpers nannte, dargestellt wird. Mit Hülfe dieser Gröfse und der durch H bezeichneten, im Körper wirklich vorhandenen, Wärme

1) *Phil. Mag. Ser. IV, Vol. IV, p. 304* und *Proc. of the Edinb. R. Soc. 1852.*

2) Diese Ann. Bd. CXVI (1862) S. 73 und Abhandlungensammlung Th. I, S. 242.

Poggendorff's Annal. Bd. CXLV. 10

Fig. 1 Clausius' Second Law and entropy (Clausius 1872, p. 145; see also pp. 132–146) *Source* French National Library–Gallica. Public Domain

Thus, by expressing[4] (Eq. 1) in terms of the kinetic energy (i.e. *vis viva;* (Pisano and Capecchi 2013; Pisano and Bussotti 2017a) of the particles that constitute the microscopic structure of gases. We have made explicit that this achievement represented the necessary step towards attempting a demonstration of the Second Law, which would finally be accomplished in 1872 after the H-theorem. To be comprehensive, in the following we present a selection of the historical chronology in the field and on the subject from 1840s to 1906 (Table 2):

[4] In (1) the integral is extended over a closed curve. Clausius and Boltzmann did not use the symbolism instead currently used for cyclical integral (Cfr. Pisano et al. 2021b).

Table 2 From 2nd law, Heat–Work equivalence (H–W) to entropy arguments

Time period	Author, arguments and references
1786–1827	Lazare Nicolas Marguérite Carnot (1753–1823)'s *mechanical machines/mechanics* 1786. *Essai sur les machines en général* (Pisano et al. 2021a) Nicolas Léonard Sadi Carnot's (1796–1832)'s *heat machines/thermodynamics* 1824. *Réflexions sur la puissance motrice du feu* (Carnot (1824, 1878a, b) A new physical theory in thermodynamics by means of a book without mathematics (Gillispie and Pisano 2014; Pisano 2010) based on *impossibility of perpetual motion* and the second principle only 1786–1824. A filial (1786) and scientific project (1824) on *machines en general*—mechanical and heat—(Gillispie and Pisano 2014; Carnot 1824) 1819–1827. Unpublished manuscripts (Gillispie and Pisano 2014; Carnot 1986): *Recherche d'une formule propre à représenter la puissance motrice de la Vapeur d'Eau* (Carnot [1824] 1978, pp. 223–234; Gabbey and Herivel 1966; Carnot 1878a); *Notes sur les mathématiques, la physique et autres sujets* (Carnot (s.d.), 1878b)
1840s–1842	Julius Robert von Mayer (1814–1878) *Principle of equivalence* (Mayer 1842a, b) Sadi Carnot obtained the value of the heat *equivalent* 18–20 years before Mayer. He discarded the theory of conservation of *calorique* and introduced the *conservation* of ("puissance mécanique") energy (Pisano et al. 2021, p. 248; Gillispie and Pisano 2014, pp. 142–148)
1843–1850	James Prescott Joule (1818–1889) He experimentally confirmed *H–W Equivalence* (Joule 1843, 1845, 1884, 1887; [1849] 1850) 1847; 1849–1850. On the *mechanical equivalent* of heat, as determined by the heat evolved by the friction of fluids (Joule 1845)
1848–1911	William Thomson Lord Kelvin (1824–1907) *1*st *and 2*nd *Law of Thermodynamics* (historical inversion with respect to Carnot (Pisano et al. 2021a; Pisano 2010) 1848–1849. *It is impossible, by means of inanimate material agency, to derive mechanical effect from any portion of matter by cooling it below the temperature of the coldest of the surrounding objects* [Thomson (1848–1849a, 1848–1849b), see also Thomson (1851a, b)] 1882–1911. On *Carnot's theory* (Thomson 1882–1911)

(continued)

Table 2 (continued)

Time period	Author, arguments and references
1850–1872	Rudolf Julius Emanuel Clausius (1822–1888) 1850. $dQ = dU + AdW$. By means of new quantity U, Clausius defined the new kind of conservation law. The 1st Law was born (Clausius 1867) 1850. Equivalence H–W and Interior—&—Exterior Works (1867) (Clausius 1850) 1850–1867. The ancient and erroneous notions-ambiguities *heat-caloric* is surpassed by Clausius' new *Mechanical Theory of Heat* (Clausius 1867) "Energy of system" (Clausius 1867) 1854. First fundamental Theorem and Equivalence H-W (Clausius 1854) 1854–1867. He advanced cyclic (1854) and not-cyclic (1865) processes differently (Clausius 1854, 1865) Kinetic model of Gases (Clausius 1857a, b, 1865, 1871, 1872) 1867. Advanced form of the First fundamental Theorem and Equivalence H-W (Clausius 1867) 1865. Second Law for adiabatically isolated systems (Clausius 1865) "(1) *The energy of the universe is constant*. (2) *The entropy of the universe tends to maximum.*" (Clausius 1865, p. 400) 1865–1872. Second Law of Thermodynamic, non-cyclical processes towards *a Mathematical formulation of the Entropy* (Clausius 1865, pp. 390–391; Clausius 1872, pp. 145–146) See also 1872. "Entropy of a system" (Clausius 1865, p. 355)
1856	August Karl Krönig (1822–1879) The roots of Statistical Mechanics (Krönig 1856) He developed one of the former mathematical models sustained by the idea that the thermal properties of a gas were connected with to the motion of the particles, constituting the microscopic structure of the gaseous state (Krönig 1856)
1865–1877	Ludwig Eduard Boltzmann (1844–1906) *Mechanical model* (Boltzmann 1866) He dealt (for gases) with mathematical expressions of Heat (δQ) and Temperature (T), in terms of the *kinetic energy* of the constituent atoms (or molecules) (Boltzmann [1865–1874], 1909, 1866, 1871, 1872, 1877, 2003) Physical meaning of Boltzmann's *H-theorem* supported by on the mechanical expression of *Clausius' Second Law* for cyclic processes (Boltzmann 1866, 1872) Within his kinetic model (1866, 1872) it results a mathematical-statistical connection of the entropy with the kinetic energy of the constituent particles (Boltzmann 1872, 1877, 2003). The expression of this relation will be made explicit by Max Planck in 1900 (Planck 1901, p. 556)

(continued)

Table 2 (continued)

Time period	Author, arguments and references
1901–1917	Max Planck (1858–1947) *Planck's-Boltzmann Entropy equation* within a probability mathematical framework $S = k\log W + const.$ (Planck 1901, p. 556; Cfr. Planck 1914) currently known as Boltzmann's equation 1917. *Treatise on Thermodynamics* (Planck 1917; see also Planck 1914)
1873–1906	Josiah Willard Gibbs (1839–1903) 1873. *Équilibre des Système Chimiques* (Gibbs 1899) 1875. *Extension of Clausius'* 2nd *Law* Equilibrium of heterogeneous systems (Gibbs 1906) 1902. *Statistical mechanics* with a particular focus on the statistical expression of the entropy in the case of a canonical system (Gibbs 1902) 1906 (posthumous). Thermodynamics within a *geometrical formulation* of Clausus' Second Law (Gibbs 1906)
1909	Constantin Carathéodory (1873–1950) An Axiomatic approach to Thermodynamics by relying on a method that allows building up thermodynamics as a kind of extension of mathematics, hence delivering one of the first axiomatic approaches to a physical theory Carathéodory obtained a purely abstract formal system without any reference to phenomenological principles in contrast with the approach of Clausius

Taking into account the above Table 2, particularly Clausius' and Boltzmann's works, we can highlight three main final different methodological conceptualizations of entropy:

1. 1875

   In Gibbs' works we see a process of geometrization of entropy such that entropy becomes a steering variable, whose geometrical behaviour serves to define the spontaneous direction of any thermodynamic transformation (Gibbs 1892, 1899, 1902, 1906).
2. 1894

   In Wien's model, entropy mainly corresponds to a macroscopic quantity characterizing the radiation, this latter being modelled as a thermodynamic system made of a set of wavelengths (Wien 1894).
3. 1901–1902.

   In Planck's theory entropy corresponds to a statistical function, which plays a pivotal role in the microscopic modelling of the radiation itself (Planck 1901).

These three steps belong to almost 175+ years of research within an Intellectual History of Physics–Mathematics from Thermodynamics to Statistical Mechanics (Pisano 2021):

- 1780-1813. Lazare Nicolas Marguerite Carnot (1753–1823)
- 1824. *Nicolas Léonard Sadi Carnot (1796–1832)*
- 1842. Julius Robert von Mayer (1814–1878)
- 1843-1847. James Prescott Joule (1818–1889)
- 1848-1852. William Thomson Lord Kelvin (1824–1907)
- 1850-1872. *Rudolf Julius Emanuel Clausius (1822–1888)*
- 1856. *August Karl Krönig (1822–1879)*
- 1865-1870. Paolo Ballada Comte de St Robert (1815–1888)
- 1865-1872. *Ludwig Edward Boltzmann (1844–1904)*
- 1873-1906. Josiah Willard Gibbs (1839–1903)
- 1894. Wilhelm Wien (1864–1928)
- 1901-1917. Max Planck (1858–1947)
- 1909. Constantin Carathéodory (1873–1950)
- 1931. Lars Onsager (1903–1976)
- 1948. Claude Elwood Shannon (1916–2001) | Gibbs-Shannon
- 1955. Ilya Prigogine (1917–2003)
- ….. works in progress

For sake of brevity and acknowledging Sadi Carnot's heritage, for our aims, in this paper we will focus from 1850 to 1872, and on Clausius, Krönig and Boltzmann only (above in italics).

3 The Structure of the Paper

In this paper, we would like to further reveal the theoretical substrate whereon the foundational principles of the current Statistical Mechanics lay:

(1) Our first focus concerns an *ante litteram* on the Kinetic model and H-theorem as expounded by Boltzmann (Pisano et al. 2019, 2021b). Then we will analyse the *Mechanical Theory of Heat*, which Clausius delivered in several papers, from 1850 and 1865, and which were published, in 1867, in a unique volume (Clausius 1867)—entitled *The mechanical theory of heat with its applications to the steam-engine and the physical properties of bodies*—in the form of a collection of nine memoirs, translated into English. In particular, we will retrace some aspects of the intellectual path that lead Clausius to formulate his First Law, which intertwines heat and work, for cyclical processes. We will make explicit how the origin of Clausius' theory also relies on a model of matter implying the motion of the constituent particles. Then we will briefly present the Second Law for cyclical processes. We have mainly based this research on the First and Fourth memoirs (Clausius 1867); nevertheless, we have analyzed some parts excerpted from the 1850 paper (Clausius 1850*), Über die bewegende Kraft der Wärme und die Gesetze, welche sich daraus für die Wärmelehre selbst ableiten lassen,* of which we report our translation instead of the translation given in the First Memoir.

(2) Second, we will make explicit the construction of Clausius' kinetic model that allowed him to obtain the Second Law for reversible-cyclical processes in terms of the kinetic energy of the constituent particles. We will focus on two of Clausius' works which are not part of 1867–Memoirs: (a) the 1857 paper (Clausius 1857a)[5] entitled *Über die Art der Bewegung welche wir Wärme nennen*[6] and (b) the 1871–work (Clausius 1871a), *Über die Zurückführung des zweiten Hauptsatzes der mechanischen Wärmetheorie auf allgemeine mechanische Prinzipien.*

(3) Then, we will analyse Boltzmann's note (Boltzmann 1871), *Zur Priorität der Auffindung der Beziehung zwischen dem zweiten Hauptsatze der mechanischen Wärmetheorie und dem Prinzip der kleinsten Wirkung*, published in 1871

---

[5] We will rely on the English translation published by Brush in 1965 (Clausius 1857b).

[6] "On the kind of movement which we call heat". (Authors' translation).

to claim the priority of his kinetic model—delivered in 1866 in Boltzmann (1866)—with respect to the results of Clausius' research (Clausius 1871), published in 1871. Boltzmann's arguments claiming the priority depend on a comparative analysis between both models. To write this part of our contribution, besides the aforementioned paper (Boltzmann 1871), we worked on two other articles of Boltzmann: (a) 1866 paper (Boltzmann 1866), Über die mechanische Bedeutung des zweiten Hauptsatzes der Wärmetheorie and (b) 1872–work (Boltzmann 1872), entitled Weitere Studien über Wärmegleichgewicht unter Gasmolekülen. We would like to add a final clarification about our Boltzmann primary sources. Boltzmann's opera omnia is collected in three volumes, in German language, published by Barth (Leipzig) in 1909 under the title Wissenschaftlichen Abhandlungen von Ludwig Boltzmann (Boltzmann [1865–1874] 1909). We worked on the papers taken from this collection instead of the original papers published on Annalen der Physik und Chemie and Wiener Berichte.

(4) Finally, we will make explicit the origin of two conceptual aspects of entropy. First, we will focus on the former definition of entropy delivered by Clausius between 1862 and 1865 by analyzing, in particular, the Sixth and Ninth Memoirs (Clausius 1867). Second, we will retrace the scientific roots of the statistical expression of entropy as delivered by Planck in his 1901 work. To this end, we will consider: (a) Boltzmann's 1866 paper (Boltzmann 1866), *Über die mechanische Bedeutung des zweiten Hauptsatzes der Wärmetheorie*; (b) 1901 Planck's work (Planck 1901), *Über das Gesetz der Energieverteilung im Normalspectrum*; (c) 1903 edition of Planck (1901) Planck's *Vorlesungen über Thermodynamik* (Edited in English translation as *Treatise on Thermodynamics*).

4 *Ante Litteram*: Kinetic Model (1866) and H-theorem (1872)

In his 1872 paper (Boltzmann 1872), entitled *Weitere Studien über das Wärmegleichgewicht unter Gasmolekülen* (*Further Studies on the Thermal Equilibrium of Gas Molecules*) (Fig. 2) Boltzmann rebuilds his 1866–work (Fig. 3) kinetic model through a statistical perspective. He frames his reflections by considering that the macroscopic thermodynamic state of a system does not change even if the dynamical states of its constituent molecules vary continuously per effect of the continuous motions affecting those molecules. He goes further by observing that despite this variability at the microscopic level, the laws of behaviour of warm bodies remain completely definite. The explanation, which he outlines, recalls two intertwined occurrences: (a) the fact that the most random events, when they occur in the same proportions, give the same average value and that (b) the molecules of the body are indeed so numerous, and their motion is so rapid, that we can perceive nothing more than average values. It is from these premises that he suggests the use of probability theory to determine those average values.

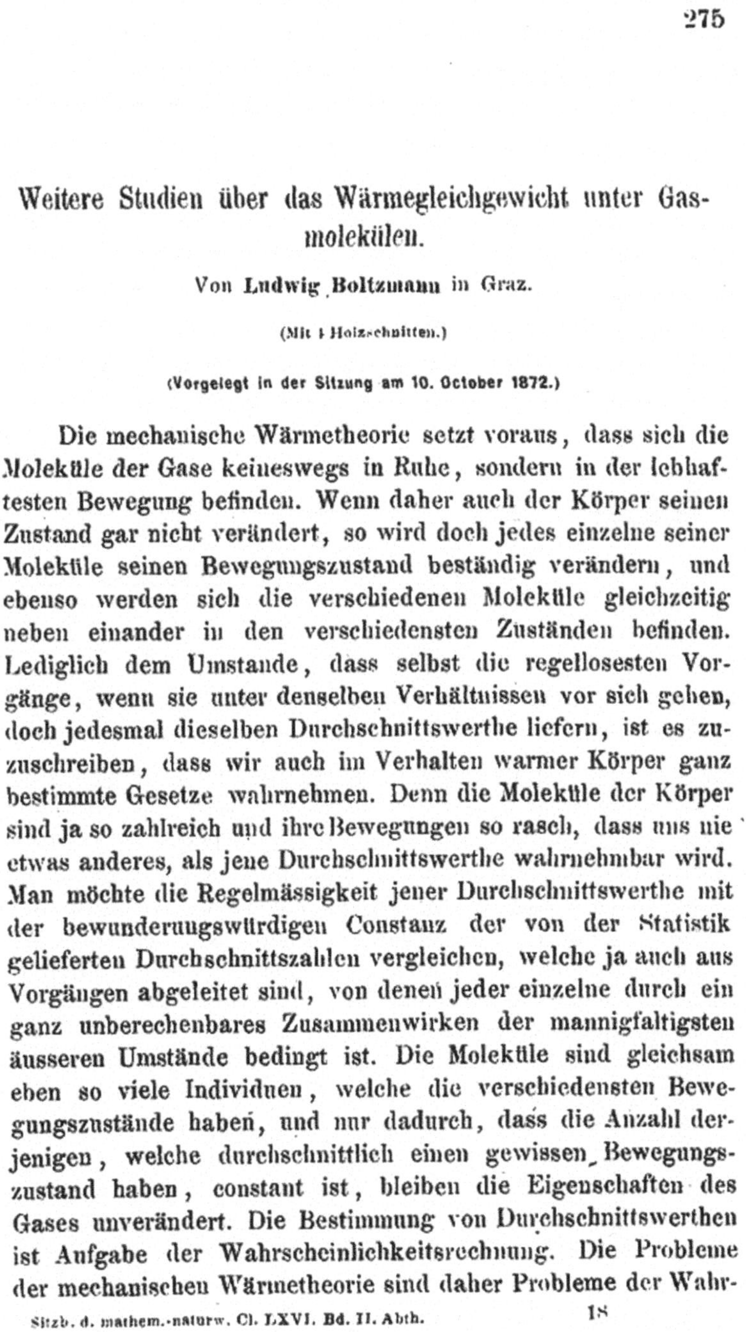

275

Weitere Studien über das Wärmegleichgewicht unter Gasmolekülen.

Von Ludwig Boltzmann in Graz.

(Mit 1 Holzschnitten.)

(Vorgelegt in der Sitzung am 10. October 1872.)

Die mechanische Wärmetheorie setzt voraus, dass sich die Moleküle der Gase keineswegs in Ruhe, sondern in der lebhaftesten Bewegung befinden. Wenn daher auch der Körper seinen Zustand gar nicht verändert, so wird doch jedes einzelne seiner Moleküle seinen Bewegungszustand beständig verändern, und ebenso werden sich die verschiedenen Moleküle gleichzeitig neben einander in den verschiedensten Zuständen befinden. Lediglich dem Umstande, dass selbst die regellosesten Vorgänge, wenn sie unter denselben Verhältnissen vor sich gehen, doch jedesmal dieselben Durchschnittswerthe liefern, ist es zuzuschreiben, dass wir auch im Verhalten warmer Körper ganz bestimmte Gesetze wahrnehmen. Denn die Moleküle der Körper sind ja so zahlreich und ihre Bewegungen so rasch, dass uns nie etwas anderes, als jene Durchschnittswerthe wahrnehmbar wird. Man möchte die Regelmässigkeit jener Durchschnittswerthe mit der bewunderungswürdigen Constanz der von der Statistik gelieferten Durchschnittszahlen vergleichen, welche ja auch aus Vorgängen abgeleitet sind, von denen jeder einzelne durch ein ganz unberechenbares Zusammenwirken der mannigfaltigsten äusseren Umstände bedingt ist. Die Moleküle sind gleichsam eben so viele Individuen, welche die verschiedensten Bewegungszustände haben, und nur dadurch, dass die Anzahl derjenigen, welche durchschnittlich einen gewissen Bewegungszustand haben, constant ist, bleiben die Eigenschaften des Gases unverändert. Die Bestimmung von Durchschnittswerthen ist Aufgabe der Wahrscheinlichkeitsrechnung. Die Probleme der mechanischen Wärmetheorie sind daher Probleme der Wahr-

Sitzb. d. mathem.-naturw. Cl. LXVI. Bd. II. Abth. 18

Fig. 2 Boltzmann *Weitere Studien über das Wärmegleichgewicht unter Gasmolekülen*–1872 (Boltzmann 1872, p. 275; see also Bolzmann [1865–1874], 1909, p. 316) *Source* Google Books, Public Domain

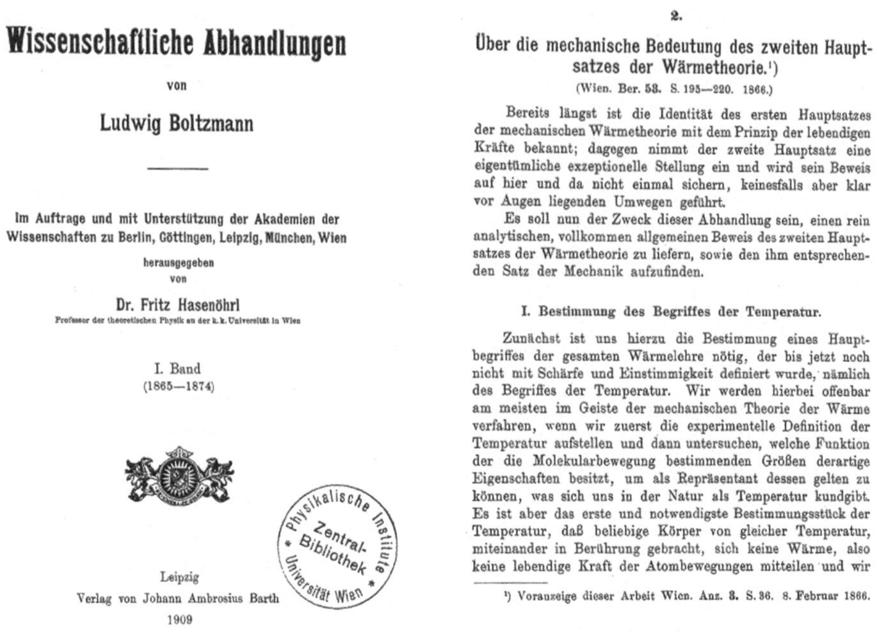

Wissenschaftliche Abhandlungen

von

Ludwig Boltzmann

Im Auftrage und mit Unterstützung der Akademien der Wissenschaften zu Berlin, Göttingen, Leipzig, München, Wien

herausgegeben von

Dr. Fritz Hasenöhrl

Professor der theoretischen Physik an der k. k. Universität in Wien

I. Band

(1865—1874)

Leipzig

Verlag von Johann Ambrosius Barth

1909

Physikalische Institute Zentral-Bibliothek Universität Wien

2.

Über die mechanische Bedeutung des zweiten Hauptsatzes der Wärmetheorie.[1])

(Wien. Ber. 53. S. 195—220. 1866.)

Bereits längst ist die Identität des ersten Hauptsatzes der mechanischen Wärmetheorie mit dem Prinzip der lebendigen Kräfte bekannt; dagegen nimmt der zweite Hauptsatz eine eigentümliche exzeptionelle Stellung ein und wird sein Beweis auf hier und da nicht einmal sichern, keinesfalls aber klar vor Augen liegenden Umwegen geführt.

Es soll nun der Zweck dieser Abhandlung sein, einen rein analytischen, vollkommen allgemeinen Beweis des zweiten Hauptsatzes der Wärmetheorie zu liefern, sowie den ihm entsprechenden Satz der Mechanik aufzufinden.

I. Bestimmung des Begriffes der Temperatur.

Zunächst ist uns hierzu die Bestimmung eines Hauptbegriffes der gesamten Wärmelehre nötig, der bis jetzt noch nicht mit Schärfe und Einstimmigkeit definiert wurde, nämlich des Begriffes der Temperatur. Wir werden hierbei offenbar am meisten im Geiste der mechanischen Theorie der Wärme verfahren, wenn wir zuerst die experimentelle Definition der Temperatur aufstellen und dann untersuchen, welche Funktion der die Molekularbewegung bestimmenden Größen derartige Eigenschaften besitzt, um als Repräsentant dessen gelten zu können, was sich uns in der Natur als Temperatur kundgibt. Es ist aber das erste und notwendigste Bestimmungsstück der Temperatur, daß beliebige Körper von gleicher Temperatur, miteinander in Berührung gebracht, sich keine Wärme, also keine lebendige Kraft der Atombewegungen mitteilen und wir

[1]) Vorauzeige dieser Arbeit Wien. Anz. 3. S. 36. 8. Februar 1866.

Fig. 3 Boltzmann's 1866–paper developing his former kinetic model (Boltzmann [1865–1874] 1909) *Source* Google Books, Public Domain

The conceptual core of *Weitere Studien über das Wärmegleichgewicht unter Gasmolekülen*[7]–1872 (Boltzmann 1872, [1865–1874] 1909), pp. 316–402; Fig. 4] concerns a function E that Boltzmann defines Boltzmann [1865–1874] 1909, p. 335) with reference to a gaseous system modelled as made of many material points (representing the constituent molecules) moving into a container. For the most of time, these points move following a linear uniform motion, with velocity x. When the molecules interact, they do it by couple as elastic spheres. The interactions between the molecules and the walls of the container are elastic too. Following Boltzmann's arguments, the function E has the following form:

$$E = \int_{0}^{\infty} f(x,t)\left\{\log\left[\frac{f(x,t)}{\sqrt{x}}\right] - 1\right\}dx \tag{2}$$

where the variable x is the velocity. Boltzmann demonstrates ([1865–1874] 1909, pp. 335–344) that the first derivative of E can never be positive and that zeroes when:

$$f(x,t) = C\sqrt{x}e^{-hx} \tag{3}$$

[7] The same 1872 paper can be found in the *Wissenschaftliche Abhandlungen von Ludwig Boltzmann* ([1865–1874] 1909), published by Barth in 1909. All the quotations and references of Boltzmann in our present paper, are excerpted from Boltzmann ([1865–1874] 1909).

where we recognize in the foregoing function—Eq. (3)—Maxwell's distribution of velocity (Cfr. Pisano et al. 2023).

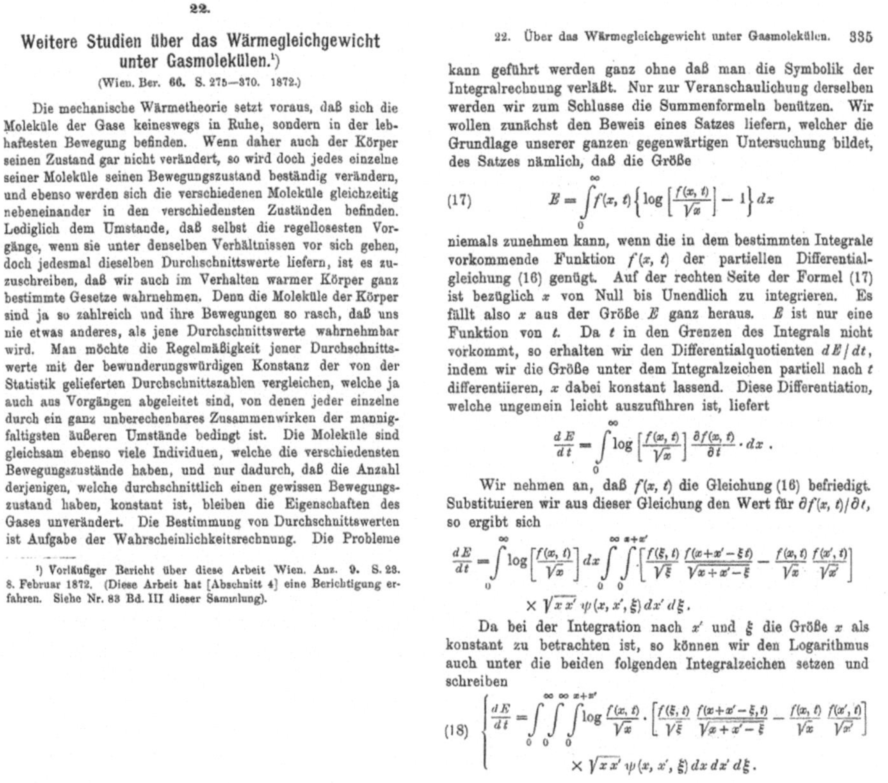

22.

Weitere Studien über das Wärmegleichgewicht unter Gasmolekülen.[1])

(Wien. Ber. 66. S. 275—370. 1872.)

Die mechanische Wärmetheorie setzt voraus, daß sich die Moleküle der Gase keineswegs in Ruhe, sondern in der lebhaftesten Bewegung befinden. Wenn daher auch der Körper seinen Zustand gar nicht verändert, so wird doch jedes einzelne seiner Moleküle seinen Bewegungszustand beständig verändern, und ebenso werden sich die verschiedenen Moleküle gleichzeitig nebeneinander in den verschiedensten Zuständen befinden. Lediglich dem Umstande, daß selbst die regellosesten Vorgänge, wenn sie unter denselben Verhältnissen vor sich gehen, doch jedesmal dieselben Durchschnittswerte liefern, ist es zuzuschreiben, daß wir auch im Verhalten warmer Körper ganz bestimmte Gesetze wahrnehmen. Denn die Moleküle der Körper sind ja so zahlreich und ihre Bewegungen so rasch, daß uns nie etwas anderes, als jene Durchschnittswerte wahrnehmbar wird. Man möchte die Regelmäßigkeit jener Durchschnittswerte mit der bewunderungswürdigen Konstanz der von der Statistik gelieferten Durchschnittszahlen vergleichen, welche ja auch aus Vorgängen abgeleitet sind, von denen jeder einzelne durch ein ganz unberechenbares Zusammenwirken der mannigfaltigsten äußeren Umstände bedingt ist. Die Moleküle sind gleichsam ebenso viele Individuen, welche die verschiedensten Bewegungszustände haben, und nur dadurch, daß die Anzahl derjenigen, welche durchschnittlich einen gewissen Bewegungszustand haben, konstant ist, bleiben die Eigenschaften des Gases unverändert. Die Bestimmung von Durchschnittswerten ist Aufgabe der Wahrscheinlichkeitsrechnung. Die Probleme

[1]) Vorläufiger Bericht über diese Arbeit Wien. Anz. 9. S. 23. 8. Februar 1872. (Diese Arbeit hat [Abschnitt 4] eine Berichtigung erfahren. Siehe Nr. 83 Bd. III dieser Sammlung).

22. Über das Wärmegleichgewicht unter Gasmolekülen. 335

kann geführt werden ganz ohne daß man die Symbolik der Integralrechnung verläßt. Nur zur Veranschaulichung derselben werden wir zum Schlusse die Summenformeln benützen. Wir wollen zunächst den Beweis eines Satzes liefern, welcher die Grundlage unserer ganzen gegenwärtigen Untersuchung bildet, des Satzes nämlich, daß die Größe

$$E = \int_0^\infty f(x,t)\left\{\log\left[\frac{f(x,t)}{\sqrt{x}}\right] - 1\right\}dx \tag{17}$$

niemals zunehmen kann, wenn die in dem bestimmten Integrale vorkommende Funktion $f(x, t)$ der partiellen Differentialgleichung (16) genügt. Auf der rechten Seite der Formel (17) ist bezüglich x von Null bis Unendlich zu integrieren. Es fällt also x aus der Größe E ganz heraus. E ist nur eine Funktion von t. Da t in den Grenzen des Integrals nicht vorkommt, so erhalten wir den Differentialquotienten dE/dt, indem wir die Größe unter dem Integralzeichen partiell nach t differentiieren, x dabei konstant lassend. Diese Differentiation, welche ungemein leicht auszuführen ist, liefert

$$\frac{dE}{dt} = \int_0^\infty \log\left[\frac{f(x,t)}{\sqrt{x}}\right]\frac{\partial f(x,t)}{\partial t}\cdot dx\,.$$

Wir nehmen an, daß $f(x, t)$ die Gleichung (16) befriedigt. Substituieren wir aus dieser Gleichung den Wert für $\partial f(x,t)/\partial t$, so ergibt sich

$$\frac{dE}{dt} = \int_0^\infty \log\left[\frac{f(x,t)}{\sqrt{x}}\right]dx \int_0^\infty\int_0^{x+x'}\left[\frac{f(\xi,t)}{\sqrt{\xi}}\frac{f(x+x'-\xi t)}{\sqrt{x+x'-\xi}} - \frac{f(x,t)}{\sqrt{x}}\frac{f(x',t)}{\sqrt{x'}}\right] \times \sqrt{xx'}\,\psi(x,x',\xi)\,dx'\,d\xi\,.$$

Da bei der Integration nach x' und ξ die Größe x als konstant zu betrachten ist, so können wir den Logarithmus auch unter die beiden folgenden Integralzeichen setzen und schreiben

$$\left\{\begin{aligned}\frac{dE}{dt} = \int_0^\infty\int_0^\infty\int_0^{x+x'} \log\frac{f(x,t)}{\sqrt{x}}\cdot\left[\frac{f(\xi,t)}{\sqrt{\xi}}\frac{f(x+x'-\xi,t)}{\sqrt{x+x'-\xi}} - \frac{f(x,t)}{\sqrt{x}}\frac{f(x',t)}{\sqrt{x'}}\right]\\ \times\sqrt{xx'}\,\psi(x,x',\xi)\,dx\,dx'\,d\xi\,.\end{aligned}\right. \tag{18}$$

Fig. 4 Boltzmann's 1872–paper where he conceptualised his H-theorem (Boltzmann [1865–1874] 1909, p. 316, p. 335) *Source* Google Books, Public Domain

The subtext of Boltzmann's theorem is explicit in the following excerpt:

> It has thus been rigorously proved that, whatever may be the initial distribution of kinetic energy, in the course of a very long time it must always necessarily approach the one found by Maxwell. The procedure used so far is of course nothing more than a mathematical artifice employed in order to give a rigorous proof of a theorem whose exact proof has not previously been found. It gains meaning by its applicability to the theory of polyatomic gas molecules. There one can again prove that a certain quantity E can only decrease as a consequence of molecular motion, or in a limiting case can remain constant. (Boltzmann 2003, p. 291)

That is to say that, whatever the initial distribution is of the velocity of a large number of moving particles, this distribution evolves with time and after a sufficiently long time it converges with Maxwell's distribution of velocity of the particles constituting a gaseous system in equilibrium (Pisano 2013). This result allowed Boltzmann to attain the following conclusions:

> One can also prove that for the atomic motion of a system of arbitrarily many material points there always exists a certain quantity which, in consequence of any atomic motion, cannot increase, and this quantity agrees up to a constant factor with the value found for the well-known integral $\int \frac{dQ}{T}$ in my paper[8] on the "Analytical proof of the 2nd law, etc." (Boltzmann 2003, p. 291; Author's quotations marks)

In other words, Boltzmann states that, up to a constant factor, the function E, that he has just proved cannot increase, agrees with the value of Clausius' integral that he obtained in terms of kinetic energy in his 1866–paper (Boltzmann [1865–1874] 1909, p. 28), that is to say:

$$2\Sigma \log \int_{t_1}^{t_2} \frac{mc^2}{2} dt + C \tag{4}$$

In our previous work (Pisano et al. 2021b), we have made this argument of Boltzmann explicit; we have shown the aforementioned agreement between the function E and the expression given by Eq. (4). This result allowed Boltzmann to argue that he had found a proof of Clausius' Second Law for cyclical processes as we read in the conclusion of his 1872 paper.

> We have therefore prepared the way for an analytical proof of the second law in a completely different way from those previously investigated. Up to now the object has been to show that $\int \frac{dQ}{T} = 0$ for reversible cyclic processes, but it has not been proved analytically that this quantity is always negative for irreversible processes, which are the only ones that occur in nature. The reversible cyclic process is only an ideal, which one can more or less closely approach but never completely attain. Here, however, we have succeeded in showing that $\int \frac{dQ}{T}$ is in general negative, and is equal to zero only for the limiting case, which is of course the reversible cyclic process (since if one can go through the process in either direction, $\int \frac{dQ}{T}$ cannot be negative). (Boltzmann 2003, p. 291)

In effect, Boltzmann's kinetic model—as developed in the 1866 paper (Boltzmann [1865–1874] 1909; Fig. 5)—serves to attain a mathematical expression for the temperature and the heat exchanged by a gaseous system in terms of the kinetic energy of the particles (atoms or molecules) constituting the system. This allowed the Second Law of Clausius to be written—in the case of cyclic processes—in terms of the kinetic energy of the constituent particles.

[8] Cfr. Boltzmann's *Über die mechanische Bedeutung des zweiten Hauptsatzes der Wärmetheorie*—1866 (Boltzmann 1866).

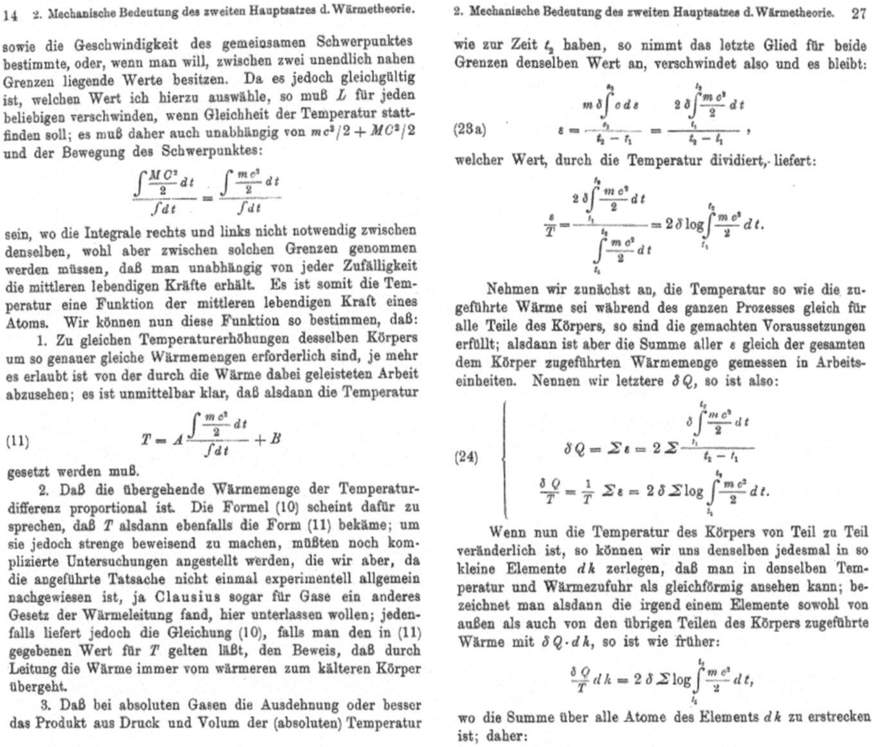

14 2. Mechanische Bedeutung des zweiten Hauptsatzes d. Wärmetheorie.

sowie die Geschwindigkeit des gemeinsamen Schwerpunktes bestimmte, oder, wenn man will, zwischen zwei unendlich nahen Grenzen liegende Werte besitzen. Da es jedoch gleichgültig ist, welchen Wert ich hierzu auswähle, so muß L für jeden beliebigen verschwinden, wenn Gleichheit der Temperatur stattfinden soll; es muß daher auch unabhängig von $mc^2/2 + MC^2/2$ und der Bewegung des Schwerpunktes:

$$\frac{\int \frac{MC^2}{2} dt}{\int dt} = \frac{\int \frac{mc^2}{2} dt}{\int dt}$$

sein, wo die Integrale rechts und links nicht notwendig zwischen denselben, wohl aber zwischen solchen Grenzen genommen werden müssen, daß man unabhängig von jeder Zufälligkeit die mittleren lebendigen Kräfte erhält. Es ist somit die Temperatur eine Funktion der mittleren lebendigen Kraft eines Atoms. Wir können nun diese Funktion so bestimmen, daß:

1. Zu gleichen Temperaturerhöhungen desselben Körpers um so genauer gleiche Wärmemengen erforderlich sind, je mehr es erlaubt ist von der durch die Wärme dabei geleisteten Arbeit abzusehen; es ist unmittelbar klar, daß alsdann die Temperatur

$$T = A \frac{\int \frac{mc^2}{2} dt}{\int dt} + B \tag{11}$$

gesetzt werden muß.

2. Daß die übergehende Wärmemenge der Temperaturdifferenz proportional ist. Die Formel (10) scheint dafür zu sprechen, daß T alsdann ebenfalls die Form (11) bekäme; um sie jedoch strenge beweisend zu machen, müßten noch komplizierte Untersuchungen angestellt werden, die wir aber, da die angeführte Tatsache nicht einmal experimentell allgemein nachgewiesen ist, ja Clausius sogar für Gase ein anderes Gesetz der Wärmeleitung fand, hier unterlassen wollen; jedenfalls liefert jedoch die Gleichung (10), falls man den in (11) gegebenen Wert für T gelten läßt, den Beweis, daß durch Leitung die Wärme immer vom wärmeren zum kälteren Körper übergeht.

3. Daß bei absoluten Gasen die Ausdehnung oder besser das Produkt aus Druck und Volum der (absoluten) Temperatur

2. Mechanische Bedeutung des zweiten Hauptsatzes d. Wärmetheorie. 27

wie zur Zeit t_2 haben, so nimmt das letzte Glied für beide Grenzen denselben Wert an, verschwindet also und es bleibt:

$$\varepsilon = \frac{m\delta \int_{s_1}^{s_2} c\, ds}{t_2 - t_1} = \frac{2\delta \int_{t_1}^{t_2} \frac{mc^2}{2} dt}{t_2 - t_1}, \tag{23a}$$

welcher Wert, durch die Temperatur dividiert, liefert:

$$\frac{\varepsilon}{T} = \frac{2\delta \int_{t_1}^{t_2} \frac{mc^2}{2} dt}{\int_{t_1}^{t_2} \frac{mc^2}{2} dt} = 2\delta \log \int_{t_1}^{t_2} \frac{mc^2}{2} dt.$$

Nehmen wir zunächst an, die Temperatur so wie die zugeführte Wärme sei während des ganzen Prozesses gleich für alle Teile des Körpers, so sind die gemachten Voraussetzungen erfüllt; alsdann ist aber die Summe aller ε gleich der gesamten dem Körper zugeführten Wärmemenge gemessen in Arbeitseinheiten. Nennen wir letztere δQ, so ist also:

$$\left\{ \begin{aligned} \delta Q &= \Sigma \varepsilon = 2\Sigma \frac{\delta \int_{t_1}^{t_2} \frac{mc^2}{2} dt}{t_2 - t_1} \\ \frac{\delta Q}{T} &= \frac{1}{T} \Sigma \varepsilon = 2\delta \Sigma \log \int_{t_1}^{t_2} \frac{mc^2}{2} dt. \end{aligned} \right. \tag{24}$$

Wenn nun die Temperatur des Körpers von Teil zu Teil veränderlich ist, so können wir uns denselben jedesmal in so kleine Elemente dk zerlegen, daß man in denselben Temperatur und Wärmezufuhr als gleichförmig ansehen kann; bezeichnet man alsdann die irgend einem Elemente sowohl von außen als auch von den übrigen Teilen des Körpers zugeführte Wärme mit $\delta Q \cdot dk$, so ist wie früher:

$$\frac{\delta Q}{T} dk = 2\delta \Sigma \log \int_{t_1}^{t_2} \frac{mc^2}{2} dt,$$

wo die Summe über alle Atome des Elements dk zu erstrecken ist; daher:

Fig. 5 Boltzmann's conceptualization of temperature and heat in 1866–paper (Boltzmann [1865–1874] 1909, p. 14, p. 27) *Source* Google Books, Public Domain

For the sake of clarity, we recall that, in the case of cyclical processes, Clausius' Second law corresponds to the following inequality (Clausius 1867, p. 219):

$$\int \frac{dQ}{T} \geq 0 \tag{5}$$

and that, in 1866, Boltzmann wrote (Boltzmann [1865–1874] 1909), p. 28) the integral in the left side of Eq. (5) as:

$$2\Sigma \log \int_{t_1}^{t_2} \frac{mc^2}{2} dt + C \tag{6}$$

where m is the mass of one particle and c its velocity. The sum is extended over all the particles.

In 1872, thanks to the so-called H-theorem, Boltzmann showed ([1865–1874] 1909), p. 344] that the quantity given by Eq. (6) can always decrease, except in the case of reversibility where it zeroes; that is to say that it can never be negative. From this result, he concluded to have found a proof of Clausius' Second Law for cyclical processes.

Let us consider more closely Boltzmann's intellectual process leading to these conclusions, by itemizing the main historical steps of his reasoning:

(a) In 1866 (Boltzmann 1866, 1871, 1872), Boltzmann obtained a mathematical expression of the following integral:

$$\int \frac{dQ}{T} \tag{7}$$

In terms of the kinetic energy of the constituent particles of a gaseous system. More precisely, he implemented a physical model wherein cyclical process were represented by considering that each particle returned to its initial position after a certain time, hence describing a closed trajectory.

(b) In 1872, with reference to a set of material points in motion, Boltzmann showed that a certain function *E*, defined as Eq. (2), can always only decrease, except when the distribution of velocity of the material points is equal to Maxwell's distribution. In that case, the first derivative of *E* zeroes.

(c) In the same 1872 paper, Boltzmann affirmed that Eq. (6) is mathematically equivalent to the function *E* (Eq. 2). The logical-mathematical steps leading to affirming this correspondence are not explicit in Boltzmann's works. We have made them explicit in one of our previous papers (Pisano et al 2021; see also Frigg 2008).

(d) The mathematical correspondence between the function *E* and the expression given by Eq. (6) allowed Boltzmann to state that this latter kinetic expression can always only decrease except at equilibrium, when it zeroes. This can be mathematically written through the following inequality:

$$2\Sigma \log \int_{t_1}^{t_2} \frac{mc^2}{2} dt \geq 0 \tag{8}$$

(e) As already recalled in the foregoing point (d), the left side of Eq. (8) represents the *kinetic expression* of Clausius' integral (see Eq. 7) written within the formalism that Boltzmann implemented in 1866. Hence, in a certain measure, the foregoing inequality (Eq. 8) can be seen as a proof of Clausius' Second Law for cyclical processes, as Boltzmann concluded in 1872:

> We have therefore prepared the way for an analytical proof of the second law in a completely different way from those previously investigated. Up to now the object has been to show that $\int \frac{dQ}{T} = 0$ for reversible cyclic processes, but it has not been proved analytically that this quantity is always negative for irreversible processes, which are the only ones that occur in nature. The reversible cyclic process is only an ideal, which one can more or less closely approach but never completely attain. Here, however, we have succeeded in showing that $\int \frac{dQ}{T}$ is in general negative, and is equal to zero only for the limiting case, which is of course the reversible cyclic process (since if one can

go through the process in either direction, $\int \frac{dQ}{T}$ cannot be negative). (Boltzmann 2003, p. 291)

In 1871, Clausius developed (Clausius 1871) a similar kinetic model of gases and he also found a mathematical expression for the cyclic integral:

$$\int \frac{dQ}{T} \tag{9}$$

in terms of the kinetic energy of the constituent particles. More specifically, he obtained the following expression:

$$\int \frac{dQ}{T} = A \sum 2mc \log(Ti) \tag{10}$$

where A is the mechanical equivalent of thermal units, m the mass of one particle, c its velocity, T the temperature and i the time that an atom takes to cover its closed trajectory. The sum is extended to all the constituent particles. This result appears to be equivalent to Boltzmann's kinetic equation (Eq. 9), except for the mathematical notation. Nevertheless, Clausius' approach leading him to the foregoing kinetic expression follows a different logical path. This is mainly for the syllogistic strategy he chose to establish the semantic connection between the kinetic expression (Eq. 10) and the Second Law for cyclical processes that he derived from Carnot's equation (Eq. 5). We will develop these epistemological junctures in the next sections of the present work.

4.1 Boltzmann (1866–1872) and Clausius (1857–1872) Methods to Model the Thermodynamic–Properties of the Gaseous Systems

Whilst secondary literature had widely commented on Boltzmann's kinetic model, at present, as far as we know, there are no detailed analyses and commentaries of Clausius' aforementioned methodological approach to model the thermodynamical properties of gaseous systems, in terms of the dynamical properties of their microscopic constituents. Based on the study of Clausius' *Über die Zurückführung des zweiten Hauptsatzes der mechanischen Wärmetheorie auf allgemeine mechanische Prinzipien* (1871) (Clausius 1871), an intellectual historical analysis is presented.

Just after the publication (1871) of Clausius' paper *Über die Zurückführung des zweiten Hauptsatzes der mechanischen Wärmetheorie auf allgemeine mechanische Prinzipien* (Clausius 1871), Boltzmann delivered (1871) his *Zur Priorität der Auffindung der Beziehung zwischen dem zweiten Hauptsatze der mechanischen Wärmetheorie und dem Prinzip der kleinsten Wirkung* (Boltzmann 1871, [1865–1874] 1909, pp. 228–236). In this work he claimed (Boltzmann 1871, [1865–1874]

1909, p. 228). The priority of his kinetic approach (delivered in 1866) with respect to Clausius' one (published in 1871), by developing a comparative analysis of both approaches. Below we itemize the main steps of this latter *intellectual exercise* of Boltzmann.

On Boltzmann's kinetic model, Cerciniani (Cercignani 1998) commented:

> Boltzmann not only showed that the equation bearing his name admits Maxwell's distribution as an equilibrium solution, but he also gave a heuristic proof that it is the only possible one. To this end he introduced a quantity, which he denoted by E and later (as here) denoted by H, defined in terms of the molecular velocity distribution. He then demonstrated that as a consequence of this equation, this function must always decrease in an isolated system or, at most, remain constant, the latter case occurring only if a state of statistical equilibrium prevails. His result is usually quoted as the "H-theorem" and indicates that H must be proportional to minus the entropy. (Cercignani 1998, p. 96)

It is worth focusing on the last sentence. Cerciniani, according to a general *idea* shared by several other scholars, associates the function E (that will later be denoted as H) to the entropy through a relation of proportionality. More specifically, E would be proportional to minus the entropy. We point out that—according to our historical-epistemological analysis of the original documents (see above and references listed below)—the function E that Boltzmann defined in 1872 (Boltzmann 1872) and that concerns the so-called H-theorem is not directly intertwined with entropy; and it cannot be. In fact, the H-theorem (1866) is developed with reference to a cyclical process, as can be seen in the following passage:

> We want now to assume that every arbitrary atom returns to the same place at the same velocity and direction of motion at every state of the body after a certain (albeit arbitrarily large) time, the beginning and end of which we want to name t_1 and t_2, thus describing a closed curve and repeating its movement after this time, albeit not in the same way, but in such a likely way that the average living force during time can be regarded as the average living force of the atom for any length of time [...]. (Boltzmann 1866, p. 24; Authors' translation)

In the case of a cyclical process, Clausius' Second Law (Clausius 1865) does not include any entropy. The entropy will be defined only to settle the more complicated case of the second law for non-cyclical processes. To this end, Clausius was searching for an integral path independent function. In 1862 (Clausius 1862), he introduced a new function he named *Disgregation (Z),* integral path independent, that he intertwined with another integral path independent function (H) to write the former expression of the Second Law. Later, in 1865 (Clausius 1865), he combined the foregoing functions by defining the entropy. The brand-new quantity was a mathematical entity, integral path independent: its variation for any cyclic processes is necessarily zero. Clausius' Second Law for cyclical processes does not include and cannot include entropy nor any other integral path independent functions.

5 Clausius' Laws (1850–1865)

Clausius' Mechanical theory of heat is the result of a long cognitive process that allowed for settling the complex issue of mathematically expressing the principles of Nicolas Léonard Sadi Carnot (1796–1832) (Carnot 1824)—by using calculus. Historically, Clausius developed his theory through a preliminary hermeneutical reading of Carnot's theory that allowed him to identify as the key conceptual element the principle stating that when work is obtained as an effect of a steam engine then a certain quantity of heat passes from *a warm body to a cold one*. More precisely he wrote:

> Carnot proves that whenever work is produced by heat and a permanent alteration of the body in action does not at the same time take place, a certain quantity of heat passes from a warm body to a cold one; for example, the vapour which is generated in the boiler of a steam-engine, and passes thence to the condenser where it is precipitated, carries heat from the fireplace to the condenser. (Clausius 1867, p. 15)

Nevertheless, an inciting incident that marked Clausius' research was the foundational question on the nature of heat (Fig. 6). We recall the following passage excerpted from Carnot's *Réflexions sur la puissance motrice du feu et sur les machines propres à développer cette puissance* (Carnot 1824; Gillispie and Pisano 2014; Pisano 2010):

> So the production of motive power, in steam engines, is not due to an actual consummation [that is a formal manner of its conservation] of *caloric but to its passage from a hot body to a cold one*. It is due, in other words, to a restoration of the equilibrium of caloric [...].[9] (Carnot 1824, pp. 10–11; Carnot's italic. Authors' translation)

Carnot assumed the re-establishment of equilibrium of caloric, that is that the *calorique* was conserved. We focus on Clausius' reading of Carnot's statement in the following:

> He says expressly, that *no heat is lost* in the process, that the quantity remains unchanged; and he adds, "This is a fact which has never been disputed; it is first assumed without investigation, and then confirmed by various calorimetric experiments. To deny it, would be to reject the entire theory of heat, of which it forms the principal foundation." (Clausius 1867, p. 15; see also Fig. 6 below)

---

[9] "La production de la puissance motrice est donc due, dans les machines à vapeur non à une consommation réelle du calorique, *mais à son transport d'un corps chaud à un corps froid*, c'est-à-dire à son rétablissement d'équilibre [...]." (Carnot 1824, pp. 10–11). The re-establishment of *equilibrium of caloric* with respect to thermodynamics transformations of Carnot's cycle has been object of wide and deep investigation by one of us. Thus for sake of brevity, we only remark that thermodynamic work of a heat machine was produced thanks to (nowadays we say infinitesimal) a difference of temperature of the two isotherms only, because the two adiabatic works are equals and opposites. Thus, at that time, one used to say that this difference of temperature was a *re-establishment of equilibrium of caloric* in turn (Pisano 2010). Furthermore, Lazare Carnot (Pisano et al. 2021a) already studied the matter—in the case of mechanical machines—and transferred ideas and methods to his son for heat machines (Cfr. Gillispie and Pisano 2014).

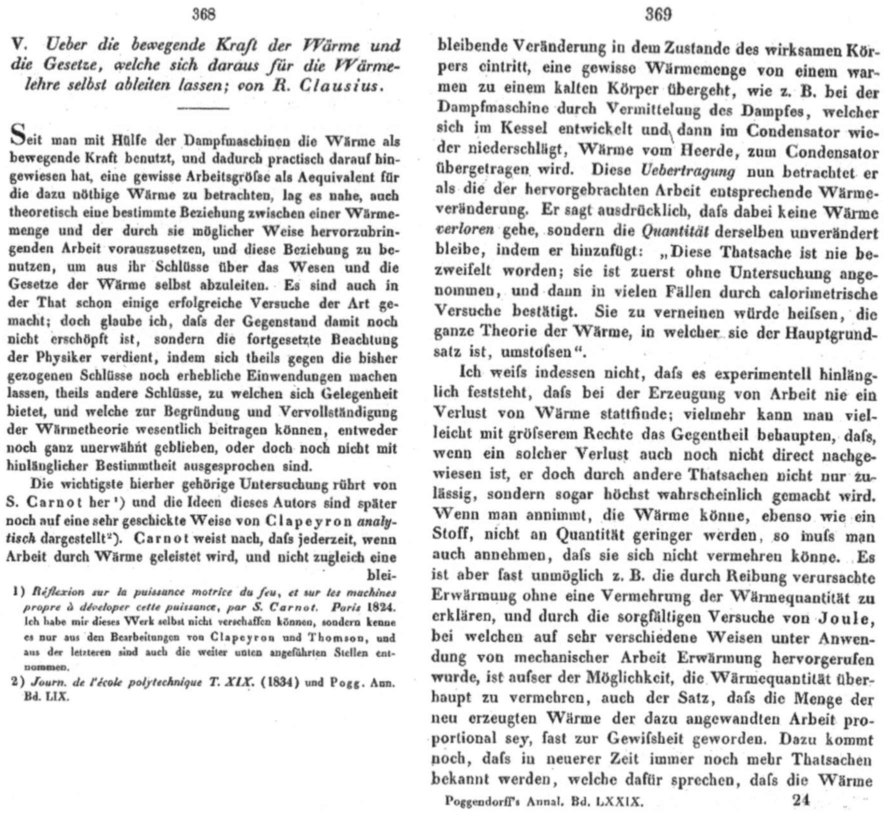

368

V. *Ueber die bewegende Kraft der Wärme und die Gesetze, welche sich daraus für die Wärmelehre selbst ableiten lassen; von R. Clausius.*

Seit man mit Hülfe der Dampfmaschinen die Wärme als bewegende Kraft benutzt, und dadurch practisch darauf hingewiesen hat, eine gewisse Arbeitsgröfse als Aequivalent für die dazu nöthige Wärme zu betrachten, lag es nahe, auch theoretisch eine bestimmte Beziehung zwischen einer Wärmemenge und der durch sie möglicher Weise hervorzubringenden Arbeit vorauszusetzen, und diese Beziehung zu benutzen, um aus ihr Schlüsse über das Wesen und die Gesetze der Wärme selbst abzuleiten. Es sind auch in der That schon einige erfolgreiche Versuche der Art gemacht; doch glaube ich, dafs der Gegenstand damit noch nicht erschöpft ist, sondern die fortgesetzte Beachtung der Physiker verdient, indem sich theils gegen die bisher gezogenen Schlüsse noch erhebliche Einwendungen machen lassen, theils andere Schlüsse, zu welchen sich Gelegenheit bietet, und welche zur Begründung und Vervollständigung der Wärmetheorie wesentlich beitragen können, entweder noch ganz unerwähnt geblieben, oder doch noch nicht mit hinlänglicher Bestimmtheit ausgesprochen sind.

Die wichtigste hierher gehörige Untersuchung rührt von S. Carnot her[1]) und die Ideen dieses Autors sind später noch auf eine sehr geschickte Weise von Clapeyron *analytisch* dargestellt[2]). Carnot weist nach, dafs jederzeit, wenn Arbeit durch Wärme geleistet wird, und nicht zugleich eine blei-

1) *Réflexion sur la puissance motrice du feu, et sur les machines propre à déveloper cette puissance, par S. Carnot. Paris* 1824. Ich habe mir dieses Werk selbst nicht verschaffen können, sondern kenne es nur aus den Bearbeitungen von Clapeyron und Thomson, und aus der letzteren sind auch die weiter unten angeführten Stellen entnommen.

2) *Journ. de l'école polytechnique T. XIX.* (1834) und Pogg. Ann. Bd. LIX.

369

bleibende Veränderung in dem Zustande des wirksamen Körpers eintritt, eine gewisse Wärmemenge von einem warmen zu einem kalten Körper übergeht, wie z. B. bei der Dampfmaschine durch Vermittelung des Dampfes, welcher sich im Kessel entwickelt und dann im Condensator wieder niederschlägt, Wärme vom Heerde, zum Condensator übergetragen wird. Diese *Uebertragung* nun betrachtet er als die der hervorgebrachten Arbeit entsprechende Wärmeveränderung. Er sagt ausdrücklich, dafs dabei keine Wärme *verloren* gehe, sondern die *Quantität* derselben unverändert bleibe, indem er hinzufügt: „Diese Thatsache ist nie bezweifelt worden; sie ist zuerst ohne Untersuchung angenommen, und dann in vielen Fällen durch calorimetrische Versuche bestätigt. Sie zu verneinen würde heifsen, die ganze Theorie der Wärme, in welcher sie der Hauptgrundsatz ist, umstofsen".

Ich weifs indessen nicht, dafs es experimentell hinlänglich feststeht, dafs bei der Erzeugung von Arbeit nie ein Verlust von Wärme stattfinde; vielmehr kann man vielleicht mit gröfserem Rechte das Gegentheil behaupten, dafs, wenn ein solcher Verlust auch noch nicht direct nachgewiesen ist, er doch durch andere Thatsachen nicht nur zulässig, sondern sogar höchst wahrscheinlich gemacht wird. Wenn man annimmt, die Wärme könne, ebenso wie ein Stoff, nicht an Quantität geringer werden, so mufs man auch annehmen, dafs sie sich nicht vermehren könne. Es ist aber fast unmöglich z. B. die durch Reibung verursachte Erwärmung ohne eine Vermehrung der Wärmequantität zu erklären, und durch die sorgfältigen Versuche von Joule, bei welchen auf sehr verschiedene Weisen unter Anwendung von mechanischer Arbeit Erwärmung hervorgerufen wurde, ist aufser der Möglichkeit, die Wärmequantität überhaupt zu vermehren, auch der Satz, dafs die Menge der neu erzeugten Wärme der dazu angewandten Arbeit proportional sey, fast zur Gewifsheit geworden. Dazu kommt noch, dafs in neuerer Zeit immer noch mehr Thatsachen bekannt werden, welche dafür sprechen, dafs die Wärme

Poggendorff's Annal. Bd. LXXIX. 24

Fig. 6 Incipit of Clausius' 1850–paper on new theory of heat (Clausius 1850, pp. 368–369) *Source* French National Library–Gallica. Public Domain

We remark that in *Notes sur les mathématiques, la physique et autre sujets* (Carnot s.d., 1878a, b, 1986), an unpublished manuscript (fl. between 1819 and March 1827; Gillispie and Pisano 2014), Sadi Carnot already—and—clearly introduced his *general thesis* ("thèse générale") on an embryonic conceptualisation of energy:

> Hence we may state, as a general assumption, that the quantity of motive power in nature is unchangeable and that, strictly speaking, motive power is neither produced nor destroyed.[10]

However, certainly, after the experiments (Joule 1843, 1845, [1849] 1850, 1884, 1887; Cfr. Pisano 2020, 2021c) of James Prescott Joule (1818–1889), the idea (caloric) that *no heat is lost* was no longer acceptable. In 1850 Clausius made it explicit (Fig. 6) in the following passage:

> If we assume that heat, like matter, cannot be lessened in quantity, we must also assume that it cannot be increased; but it is almost impossible to explain the ascension of temperature

[10] "On peut donc poser en thèse générale que la puissance motrice est en quantité invariable dans la nature, qu'elle n'est jamais à proprement parler ni produite, ni détruite." (Carnot s.d., 1986).

> brought about by friction otherwise than assuming an actual increase of heat. The careful experiments of Joule, who developed heat in various ways by the application of mechanical force, establish almost to a certainty, not only the possibility of increasing the quantity of heat, but also the fact that the newly-produced heat is proportional to the work expended in its production. (Clausius 1867, p. 15)

At the beginning of his *First Memoire* (Clausius 1867, pp. 14–69) Clausius wrote the following lines of reasoning which opened the way to a new conception of heat: it is conceived as being related to the motion of what Clausius called "ultimate particles". (Clausius 1867, p. 15):

> It may be remarked further, that many facts have lately transpired which tend to overthrow the hypothesis that heat is itself a body, and to prove that it consists in a motion of the ultimate particles of bodies. If this be so, the general principles of mechanics may be applied to heat; this motion may be converted into work, the loss of *vis viva* in each particular case being proportional to the quantity of work produced. (Clausius 1867, p. 15)

In other words, heat in its caloric terms, is no more a self-repellent fluid (also considered having a kind of form) that flows from hotter bodies to colder ones. These few lines above have a very high epistemic value to the intellectual history of science. Hence, we can identify three historical inquiring–keys, which mark completely the scientific production of Clausius (also see next section):

(1) A microscopic model of matter, that is considered to be made of minuscule particles;
(2) The idea of using the "general principle of mechanics" to build a theory of heat, being this latter conceived as a motion of particles;
(3) The idea that work can be obtained from heat, by converting the motion of the constituent particle into work: the work obtained in such a way is proportional to the loss of kinetic energy (*vis viva*).

Nevertheless, still the issue related to Carnot's assumption *no heat (caloric in Carnot's term) is lost* remains. Therefore, following Clausius as "To deny it [caloric idea], would be to reject the [current] entire theory of heat [...] [and so] should Carnot's theory be rejected?" (Clausius 1867, p. 15). Of course, the caloric was an erroneous assumption, a child of its time. The question was controversial, as witnessed by the position of the other great father of thermodynamics, William Thomson (1824–1907). Friedrich Wilhelm Ostwald (1835–1932) was among those historians sympathetic to Carnot. He was one of the first to observe (Ostwald 1892) that Carnot used two different words to indicate heat: "chaleur" and "calorique". Particularly, Ostwald affirmed that Sadi Carnot's, *calorique* indicated *heat changed without variation in temperature*. Thus, Carnot's *calorique* should be equivalent (due to the constant value of temperature) to entropy. Of course this was not the case: no energy and entropy conceptualisations are in Carnot's 1824–book. Rather, Carnot's simple language had left the appearance of an unsound historical hypothesis. Ostwald proposed replacing *calorique* with entropy, thereby giving Carnot's theory validity within modern thermodynamics. He translated Carnot's book into German, aiding its dissemination (Gillispie and Pisano 2014). Antoine Laurent Lavoisier (1743–1794), in his famous *Traité élémentaire de chimie* (Gillispie and Pisano 2014; Lavoisier [1789] 1937), had already defined *calorique* and *chaleur* according to their different scientific contexts (Lavoisier [1789] 1937, pp. 12–17, pp. 21–22). Lavoisier also discussed *chaleur* and *calorique*, attempting an early distinction between heat and temperature. He also dealt with the conceptualisation of *lumière* (Lavoisier [1789] 1937, pp 12–17). Thus Sadi Carnot, who knew and cited Lavoisier (Gillispie and Pisano (2014), pp. 394–398) considered the two words–quantities as already assumed by Lavosier (Cfr. Pisano's works; Gillispie and Pisano (2014).[11]

In the following, we read how Clausius summarized the debate with Thomson on the nature of heat.

> The difference between the two ways of regarding the subject has been seized with much greater clearness by W. Thomson, who has applied the recent investigations of Regnault, on the tension and latent heat of steam, to the completing of the memoir of Carnot. Thomson mentions distinctly the obstacles which lie in the way of an unconditional acceptance of Carnot's theory, referring particularly to the investigation of Joule, and dwelling on one principal objection to which the theory is liable. If it be even granted that the production of work, where the body in action remains in the same state after the production as before, is in all cases accompanied by a transmission of heat from a warm body to a cold one, it does not follow that by every such transmission work is produced, for the heat may be carried over by simple conduction; and in all such cases, if the transmission alone were the true equivalent of the work performed, an absolute loss of mechanical force must take place in nature, which is hardly conceivable. (Clausius 1867, p. 16)

Even if Thomson was aware of the obstacles which lay in the way of unconditional acceptance of Carnot's theory, at the same time he considered "insurmountable" the difficulties that could stem from Carnot's theory being rejected. His conviction was that in that case a completely new theory of heat would have to be formulated. On this matter, Clausius wrote (Fig. 7):

[11] On the subject, one can also see *Mémoires sur la chaleur* (Lavoisier and Laplace 1784) carried out by Lavoisier and Pierre Simon Laplace (1749–1827). Nevertheless, any deep discussion about the notion of *calorique* in general and in Carnot is out of the scope of this paper. On the ambiguity caused by the two words, Gillispie and Pisano (Gillispie and Pisano 2014) definitively detailed examples in which Sadi Carnot uses the term "chaleur" but (from a physical point of view) meant "calorique". The matter has already been masterfully analysed by one of us (Gillispie and Pisano 2014) and in recent outstanding works (Pisano et al. 2019; Pellegrino 2020; Pisano and Capecchi 2013; Pisano 2010), as well.

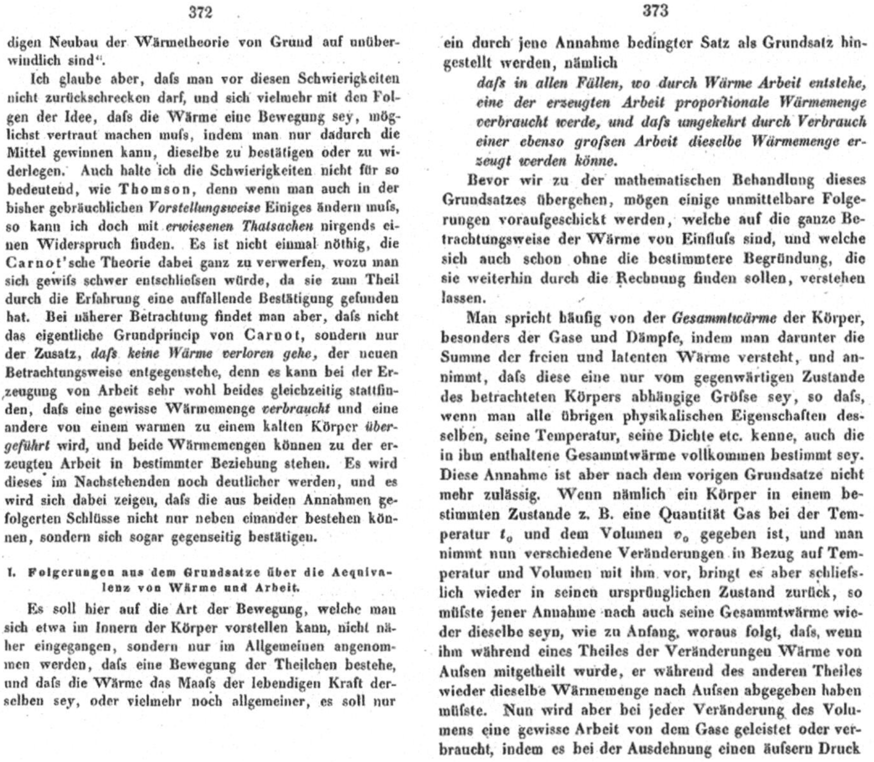

372

digen Neubau der Wärmetheorie von Grund auf unüberwindlich sind".

Ich glaube aber, dafs man vor diesen Schwierigkeiten nicht zurückschrecken darf, und sich vielmehr mit den Folgen der Idee, dafs die Wärme eine Bewegung sey, möglichst vertraut machen mufs, indem man nur dadurch die Mittel gewinnen kann, dieselbe zu bestätigen oder zu widerlegen. Auch halte ich die Schwierigkeiten nicht für so bedeutend, wie Thomson, denn wenn man auch in der bisher gebräuchlichen *Vorstellungsweise* Einiges ändern mufs, so kann ich doch mit *erwiesenen Thatsachen* nirgends einen Widerspruch finden. Es ist nicht einmal nöthig, die Carnot'sche Theorie dabei ganz zu verwerfen, wozu man sich gewifs schwer entschliefsen würde, da sie zum Theil durch die Erfahrung eine auffallende Bestätigung gefunden hat. Bei näherer Betrachtung findet man aber, dafs nicht das eigentliche Grundprincip von Carnot, sondern nur der Zusatz, *dafs keine Wärme verloren gehe*, der neuen Betrachtungsweise entgegenstehe, denn es kann bei der Erzeugung von Arbeit sehr wohl beides gleichzeitig stattfinden, dafs eine gewisse Wärmemenge *verbraucht* und eine andere von einem warmen zu einem kalten Körper *übergeführt* wird, und beide Wärmemengen können zu der erzeugten Arbeit in bestimmter Beziehung stehen. Es wird dieses im Nachstehenden noch deutlicher werden, und es wird sich dabei zeigen, dafs die aus beiden Annahmen gefolgerten Schlüsse nicht nur neben einander bestehen können, sondern sich sogar gegenseitig bestätigen.

I. Folgerungen aus dem Grundsatze über die Aequivalenz von Wärme und Arbeit.

Es soll hier auf die Art der Bewegung, welche man sich etwa im Innern der Körper vorstellen kann, nicht näher eingegangen, sondern nur im Allgemeinen angenommen werden, dafs eine Bewegung der Theilchen bestehe, und dafs die Wärme das Maafs der lebendigen Kraft derselben sey, oder vielmehr noch allgemeiner, es soll nur

373

ein durch jene Annahme bedingter Satz als Grundsatz hingestellt werden, nämlich

dafs in allen Fällen, wo durch Wärme Arbeit entstehe, eine der erzeugten Arbeit proportionale Wärmemenge verbraucht werde, und dafs umgekehrt durch Verbrauch einer ebenso grofsen Arbeit dieselbe Wärmemenge erzeugt werden könne.

Bevor wir zu der mathematischen Behandlung dieses Grundsatzes übergehen, mögen einige unmittelbare Folgerungen voraufgeschickt werden, welche auf die ganze Betrachtungsweise der Wärme von Einflufs sind, und welche sich auch schon ohne die bestimmtere Begründung, die sie weiterhin durch die Rechnung finden sollen, verstehen lassen.

Man spricht häufig von der *Gesammtwärme* der Körper, besonders der Gase und Dämpfe, indem man darunter die Summe der freien und latenten Wärme versteht, und annimmt, dafs diese eine nur vom gegenwärtigen Zustande des betrachteten Körpers abhängige Gröfse sey, so dafs, wenn man alle übrigen physikalischen Eigenschaften desselben, seine Temperatur, seine Dichte etc. kenne, auch die in ihm enthaltene Gesammtwärme vollkommen bestimmt sey. Diese Annahme ist aber nach dem vorigen Grundsatze nicht mehr zulässig. Wenn nämlich ein Körper in einem bestimmten Zustande z. B. eine Quantität Gas bei der Temperatur t_0 und dem Volumen v_0 gegeben ist, und man nimmt nun verschiedene Veränderungen in Bezug auf Temperatur und Volumen mit ihm vor, bringt es aber schliefslich wieder in seinen ursprünglichen Zustand zurück, so müfste jener Annahme nach auch seine Gesammtwärme wieder dieselbe seyn, wie zu Anfang, woraus folgt, dafs, wenn ihm während eines Theiles der Veränderungen Wärme von Aufsen mitgetheilt wurde, er während des anderen Theiles wieder dieselbe Wärmemenge nach Aufsen abgegeben haben müfste. Nun wird aber bei jeder Veränderung des Volumens eine gewisse Arbeit von dem Gase geleistet oder verbraucht, indem es bei der Ausdehnung einen äufsern Druck

Fig. 7 Incipit of Clausius' 1850–paper where he frames his new theory of heat (Clausius 1850, p. 372, p. 373) *Source* French National Library–Gallica. Public Domain

> He [Thomson] says, "If we forsake this principle, we stumble immediately on innumerable other difficulties, which, without further experimental investigations, and an entirely new erection of the theory of heat, are altogether insurmountable." (Clausius 1867, p. 17: Author's quotation marks)

Nevertheless, despite all these difficulties Clausius undertook the journey for a new theory of heat without completely rejecting Carnot's theory. On the contrary, the theory of heat on which he spent more than twenty years of his life was erected starting from the principle *heat cannot flow by itself from a colder to a warmer body* (Clausius 1872, p. 136). In 1872, he claimed to have derived his assumptions from Carnot's works, after having purified the caloric hypothesis, (*no heat is lost*) as we can read in his *Zur Geschichte der Mechanischen Wärmetheorie* (Clausius 1872):

> However, I have further shown that it is possible to solve the problem of consistency between Carnot's theorem and the First Law by modifying the theorem itself, and I have succeeded in formulating such modified theorem by adopting a new principle. This principle in its shortest form is: *heat cannot flow by itself from a colder to a warmer body.*[12] (Clausius 1867, p. 136; Authors' translation)

[12] "Dagegen habe ich weiter gezeigt, dass man durch seine gewisse Modifikation des Carnot'shen Satzes seine Übereinstimmung mit dem ersten Hauptsatze herstellen kann, und den so modifiert

Supported by these premises, Clausius provided a conceptual distinction of heat into two categories:

> On a nearer view of the case, we find that the new theory is opposed not to the real fundamental principle of Carnot, but to the addition "no heat is lost"; for it is quite possible that in the production of work both may take place at the same time; a certain portion of heat may be consumed, and a further portion transmitted from warm body to a cold one; and both portions may stand in a certain definite relation to the quantity of work produced. (Clausius 1867, p. 17; Author's quotation marks)

This dual conceptual nature of heat is at the base of Clausius' mathematical model developed in 1850 (Clausius 1850) where he distinguished heat as *freie Wärme* and *latente Wärme*—in English *free heat* and *hidden heat*, respectively. The discriminant between these two strands is the fact that *freie Wärme* (symbolically represented with the letter Q) corresponds to the quantity of heat that cannot be transformed into work and is *transmitted from a warm body* (higher temperature) *to a cold one* (lower temperature) hence being dispersed. The *latente Wärme* (designated by the letter H) corresponds to the quantity of heat that can actually be transformed into work.

A dual conceptual distinction also concerned work. In Clausius we find two entities:

(1) The *Innere Arbeit* (or interior work)—cannot be measured but only calculated inside a mathematical model.

(2) The *Äußere Arbeit* (or exterior work)—corresponds to the current notion of mechanical work: it is a measurable quantity and it can be calculated.

More specifically, Clausius made this distinction in the following way:

> The work produced is of a twofold nature. First, a certain quantity of work is necessary to overcome the mutual attraction of the particles, and to separate them to the distance that they occupy in a state of vapour. We name it interior work [*innere Arbeit*] [...]. (Clausius 1850, p. 375; Authors' translation)

The *innere Arbeit* is defined as the physical–mathematical quantity of work necessary to overcome the mutual attraction of the particles and to separate them to the distance they occupy in a state of vapour. Clausius represented it by script I and it can be considered an embryonic pre-modern thermodynamic quantity that—as well as the *latente Wärme* H—is no longer in use in current thermodynamics. In addition, both dI and dH are exact differentials, that is *latente Wärme* and *innere Arbeit* are integral path-independent.

Clausius defines *äußere Arbeit* as follows:

> Secondly, the vapour during its development must, in order to procure room for itself, force back an outer pressure. We name this latter exterior work [äußere Arbeit]. (Clausius 1850, p. 375; Authors' translation)

The physical–mathematical meaning of the *äußere Arbeit* is very close to the notion of macroscopic mechanical work (pdv) as used in current thermodynamics.

Satz habe ich durch Annahme eines neuen Grundsatzes bewiesen. Dieser Grundsatz lautet in seiner kürzesten Form: Die Wärme kann nicht von selbst aus einem kälteren in einen wärmeren Körper.".

Clausius represented it by W, and—as well as the *freie Wärme* (Q)—it is not an exact differential. Thus dW and dQ are integral path-dependent. We remark that Clausius did not use the script δ to distinguish non-exact from exact differentials. Hence in modern terms, we should read δW and δQ instead of dW and dQ.

This is the theoretical substrate whereon Clausius delivered the *First* and the *Second Fundamental Theorems* of his new *Mechanical Theory of Heat* (Clausius 1867). More specifically, the *First Fundamental Theorem* is deduced from the principle of the equivalence of heat and work and is formerly presented in 1850–paper (Clausius 1850). The 1850–paper is also included in the volume *The Mechanical Theory of Heat* (Clausius 1867), published in 1867, where Clausius' papers on the theory of heat, from 1850 till 1866, are translated in English and collected in the form of nine Memoirs. Clausius himself wrote the Preface of this collection. Many quotations of our present work are excerpted from this source, that is the reference (Clausius 1867). In particular, the 1850–paper constitutes a part[13] of Clausius' *First Memoir* (Clausius 1867, pp. 14–69).

In Clausius' *First Memoir* (1867), the *First Fundamental Theorem* is presented as follows:

> *Deductions from the principle of the equivalence of heat and work*
>
> We shall forbear entering at present on the nature of the motion which may be supposed to exist within a body, and shall assume generally that a motion of the particles does exist, and that heat is the measure of their *vis viva*. Or yet more generally, we shall merely lay down one maxim which is founded on the above assumption:
>
> *In all cases where work is produced by heat, a quantity of heat proportional to the work done is consumed; and inversely, by the expenditure of a like quantity of work, the same amount of heat may be produced.* (Clausius 1867, p. 18; Author's italic)

In the same 1850 work (Clausius 1850, 1867, pp. 14–69) Clausius aimed to obtain a mathematical formulation of this assumption; he attained a general relation intertwining the heat and the work under the form of:

> [...] the following complete differential equation of the first order [is] obtained
>
> $$d\,Q = d\,U + A\,p\,d\,v$$
>
> The function U, here introduced, is of great importance in the theory of heat; it will frequently come under discussion in the following memoirs. (Clausius 1867, p. 18)

In other words, this mathematical expression of the *First Fundamental Theorem* (Clausius 1867, p. 78) corresponds to the current first principle of thermodynamics which, including friction, temperature and heat enlarges the phenomenological field of the principle of conservation in mechanics. Let us consider more closely the three terms of Clausius' equation $dQ = dU + A\,pdv$.

(1) The term dU, as nowadays, corresponds to the differential of internal energy U, the *brand new* physical–mathematical quantity that Clausius introduced in 1850 (Clausius 1867, pp. 14–69; Clausius 1850). It has a mathematical character only: that is to say that the function U cannot be measured but one can only calculate it.

[13] We specify that each *Memoir* normally has some Appendixes.

(2) The term dQ, the *freie Wärme* or *free heat* in Clausius' writings, plays the conceptual role played by the heat, as used nowadays in thermodynamics. It is an integral path-dependent function or process function.

(3) To conceptualize the term $A\,pdv$, let us consider, as the first term pdv. In the case of a steam engine, it is $pdv = dW$, that is the *äußere Arbeit* (exterior work) expressed in the Joule unit. This term plays the same conceptual role played nowadays by mechanical work in thermodynamics. It is to be noted that W, as well as Q, is an integral path-dependent function or process function.

The letter A "denotes a constant which expresses the equivalent of heat for the unit of work [Joule]." (Clausius 1867, p. 27). Hence, the term $A\,pdv$ corresponds to the exterior work expressed in thermal unit. Further Clausius clarified the physical meaning of the term "$A\,pdv$" in the Appendix, of his *Sixth Memoir*[14] (Clausius 1867, pp. 250–266) entitled *On Terminology* where we read:

> Let heat and work continue to be measured each according to its most convenient unit, that is to say, heat according to the thermal unit and work according to the mechanical one. But besides the work measured according to the mechanical unit, let another magnitude be introduced denoting *the work measured according to the thermal unit*, that is to say, *the numerical value of the work when the unit of work is that which is equivalent to the thermal unit*. For the work thus expressed a particular name is requisite. I propose to adopt for it the Greek word *ergon*. (Clausius 1867, p. 253; Author's italic)

Hence the term "$A\,pdv$"—according to Clausius' terminology—represents the *äußere Ergon*, that is the exterior work expressed in thermal units. This terminology is explicit in the *Ninth Memoir* (Clausius 1867, pp. 327–365; Author's italic) where we can read:

> If for the sake of brevity, we denote the exterior ergon by a simple letter,
>
> $$w = AW,$$
>
> we can write the foregoing equation[15] as follows,
>
> $$dQ = dU + dw$$
>
> (Clausius 1867, p. 328)

This former general result (Clausius 1867, p. 78) allowed Clausius to obtain in thermodynamics, as the particular solution for cyclical processes, the mathematical expression of the extension of the principle of conservation in Mechanics, that he expressed as $Q = AW$.

In 1854 (Clausius 1854), Clausius attained this same result in a more advanced form. Based on his *Fourth Memoir* (Clausius 1867, pp. 111–135) we will pinpoint the path of reasoning that he followed. As first, he modelled the exchange of heat between a system and the outside by stating the following convention of the sign of heat:

---

[14] The *Sixth Memoir* (Clausius 1867, pp. 215–250) contains Clausius' 1862–paper (Clausius 1862).

[15] The aforementioned equation is $dQ = dU + AdW$, reported at page 328 of the *Ninth Memoir* (Clausius 1867, pp. 327–365).

> Let Q, therefore, be the quantity of heat which must be imparted to a body during its passage, in a given manner, from one condition to another, any heat withdrawn from the body being counted as an imparted negative quantity of heat. (Clausius 1867, p. 114)

In other words, if heat flows from the body towards the outside, then its sign will be negative. Then, he mathematically expressed Clausius' *First Law* in an alternative form, by using the interior quantities H and AI, that is to say *innere Wärme* (*interior heat*) and *innere Ergon* (interior ergon), respectively, this latter being the interior work in thermal units. More precisely he wrote the following mathematical formulation of the first law for a general process (i.e.: not necessarily a cyclical process):

$$Q = H + AI + AW$$

that he commented by writing:

> Then Q may be divided into three parts, of which the first is employed in increasing the heat actually existing in the body, the second in producing the interior work and the third in producing the exterior work. (Clausius 1867, p. 114)

Let us consider more closely Clausius reading of the above *First Law*. This mathematical form, in a certain measure, lead to Clausius to *mathematically resolving* the heat, exchanged by the system, into three components.

- The first term H, is the *innere Wärme*. It corresponds to the heat actually existing in the body and, according to Clausius theory of heat, is the part of heat that can be transformed into mechanical work. It is a mathematical entity and is an integral path-independent function.
- The second term, AI, is the *interior ergon*, that is to say the *innere Arbeit* expressed in heat units or, as Clausius specifies, the heat "producing the interior work". This idea, for which the interior work expressed in thermal units can be seen as the quantity of heat necessary to produce a corresponding portion of interior heat, is an expression of the principle of equivalence of heat and work whereon Clausius implemented his *Theory of Heat*. Also, this term is a mere mathematical expression and it is an integral path independent function.
- The third term, AW, is the exterior ergon, that is to say the exterior work expressed in thermal units. Following the same way of reasoning deployed for the *interior ergon*, AW can be seen as the heat "producing the exterior work", W. This term corresponds to the mathematical expression of a physical quantity, that is integral path-dependent.

Clausius exploited the fact that both the interior ergon and interior heat are integral path-independent functions to introduce a new integral path-dependent function, that he named U as the sum of the aforementioned two quantities. More precisely, he wrote:

> What was before stated of the second part also applies to the first—it is independent of the path pursued in the passage of the body from one state to another: hence both parts together may be represented by one function U, which we know to be completely determined by the initial and final states of the body. (Clausius 1867, p. 114)

In other words, the fact that both H and AI are integral path-independent allows gathering them in one function U, that is given by the following sum:

$$U = H + AI$$

We note the introduction of the function U is an alternative with respect to the former definition of U accomplished in 1850 (Clausius 1867, pp. 14–89; Clausius 1850). In particular, this current definition of the function U—that corresponds to the internal energy in today's thermodynamics—reveals the microscopic physical meaning that U holds within Clausius' earliest thermodynamic model. In fact, its physical meaning can be assigned only by considering the atomic or molecular structure of the matter. The *internal heat H*, that is the portion of heat that can be transformed into work, would be in some way related to the *kinetic energy* of the constituent molecules. The *internal ergon AI* served to overcome the intermolecular forces and to separate the constituent particles to the distance they occupy in a state of vapour. Finally, for the *third component of the heat* we read:

> The third part, however, the equivalent of exterior work, can, like this work itself, only be determined when the precise manner in which the changes of condition took place is known. If W be the quantity of exterior work, and A the equivalent of heat for the unit of work, the value of the third part will be AW, and the first fundamental theorem will be expressed by the equation Q = U + AW. (Clausius 1867, p. 113)

In other words, after the introduction of U, the *First Law* reduces to $Q = U + AW$ where AW is the exterior ergon or the exterior work expressed in "heat units". Starting from this general result (Clausius 1867, p. 113) Clausius considered the particular case of cyclical processes, hence obtaining these conclusions:

> When the several changes are of such nature that through them the body returns to its original condition, or when, as we shall in future express it, these changes form a cyclical process, we have
>
> $$U = 0,$$
>
> and the foregoing equation becomes
>
> $$Q = AW$$
>
> (Clausius 1867, p. 114)

This above latter equation, stemming from Clausius' new theory of heat, expresses the fact that in all cases "where work is produced by heat, a quantity of heat proportional to the work done is consumed; and inversely, by the expenditure of a like quantity of work, the same amount of heat may be produced." (Clausius 1867, p. 181). In other words, $Q = AW$ mathematically describes the *First Fundamental Theorem*. It is based on on this pivotal result that Clausius framed the *Second Fundamental Theorem* also called the *Theorem of the Equivalence of Transformation*. In his words:

> *Theorem of the equivalence of transformations.*
>
> Carnot's theorem, when brought into agreement with the first fundamental theorem, expresses a relation between two kinds of transformations, the transformations of heat into work, and the passage of heat from a warmer to a colder body, which may be regarded as the transformation of heat at a higher, into heat at a lower temperature. The theorem, as hitherto used, may be enunciated in some such manner as the following: - *In all cases where a quantity of heat is converted into work, and where the body effecting this transformation ultimately returns to its original condition, another quantity of heat must necessarily be*

> *transferred from a warmer to a colder body; and the magnitude of the last quantity of heat, in relation to the first, depends only upon the temperatures of the bodies between which heat passes, and not upon the nature of the body effecting the transformation.* (Clausius 1867, p. 116; Author's italic)

The intellectual process that leads Clausius to mathematically express this *Second Fundamental Theorem* took him almost fifteen years. In part, we have detailed this subject in our previous papers (Pisano et al. 2021b) and the matter is outside the scope of the present paper. Hence, we limit ourselves to itemize some aspects of the long story of Clausius' Second Law.

Historically we can distinguish two conceptual steps:

(1) 1854. Clausius delivered a mathematical expressions of the Second Law for cyclic thermodynamic processes (Clausius 1854).

(2) 1862–1865. Clausius solved the more complex issue of not-cyclic processes (Clausius 1865). In this context Clausius introduced the *brand-new* thermodynamic function that he named *entropy*.

The kinetic models that we will discuss in the next sections refer to cyclical processes. It is useful to briefly recall the mathematical expression of the *Second Law* only for these latter cases and to specify what Clausius' idea of a cyclical process was. To this end, we present below both his cycle (1850) to represent thermodynamics transformation in the wake of Sadi Carnot (Clausius 1850; Gillispie and Pisano 2014, pp. 199–203) and the model of a cyclical-working steam engine that he sketched out in 1856 and reported at page 157 of the *Fifth Memoir* (Clausius 1867, pp. 136–207) (Fig. 8):

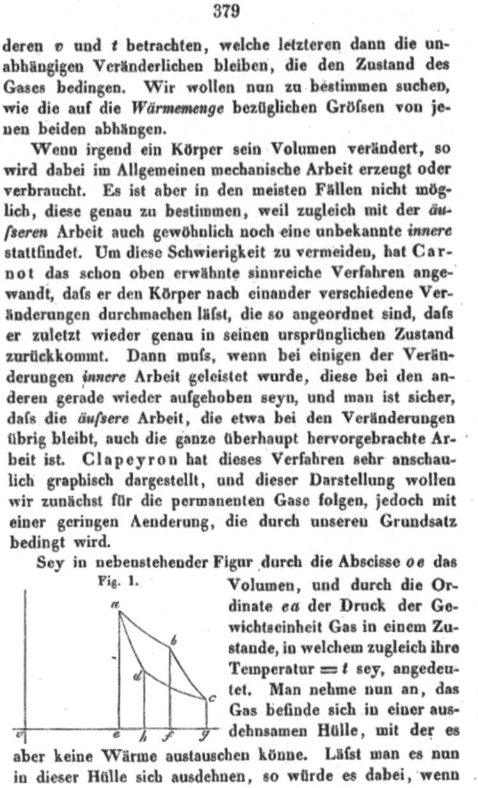
379

deren *v* und *t* betrachten, welche letzteren dann die unabhängigen Veränderlichen bleiben, die den Zustand des Gases bedingen. Wir wollen nun zu bestimmen suchen, wie die auf die *Wärmemenge* bezüglichen Gröfsen von jenen beiden abhängen.

Wenn irgend ein Körper sein Volumen verändert, so wird dabei im Allgemeinen mechanische Arbeit erzeugt oder verbraucht. Es ist aber in den meisten Fällen nicht möglich, diese genau zu bestimmen, weil zugleich mit der *äufseren* Arbeit auch gewöhnlich noch eine unbekannte *innere* stattfindet. Um diese Schwierigkeit zu vermeiden, hat Carnot das schon oben erwähnte sinnreiche Verfahren angewandt, dafs er den Körper nach einander verschiedene Veränderungen durchmachen läfst, die so angeordnet sind, dafs er zuletzt wieder genau in seinen ursprünglichen Zustand zurückkommt. Dann mufs, wenn bei einigen der Veränderungen *innere* Arbeit geleistet wurde, diese bei den anderen gerade wieder aufgehoben seyn, und man ist sicher, dafs die *äufsere* Arbeit, die etwa bei den Veränderungen übrig bleibt, auch die ganze überhaupt hervorgebrachte Arbeit ist. Clapeyron hat dieses Verfahren sehr anschaulich graphisch dargestellt, und dieser Darstellung wollen wir zunächst für die permanenten Gase folgen, jedoch mit einer geringen Aenderung, die durch unseren Grundsatz bedingt wird.

Sey in nebenstehender Figur durch die Abscisse *oe* das Volumen, und durch die Ordinate *ea* der Druck der Gewichtseinheit Gas in einem Zustande, in welchem zugleich ihre Temperatur = *t* sey, angedeutet. Man nehme nun an, das Gas befinde sich in einer ausdehnsamen Hülle, mit der es aber keine Wärme austauschen könne. Läfst man es nun in dieser Hülle sich ausdehnen, so würde es dabei, wenn

Fig. 1.

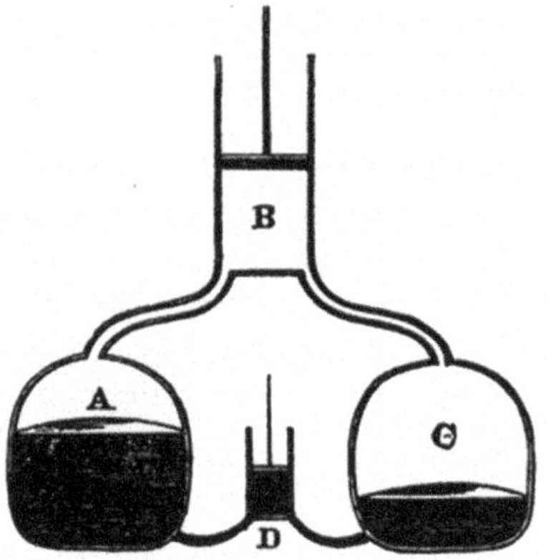

Fig. 8 Clausius' cycle (1850, left) (Clausius 1850, p. 379) and his model (1856 right) (Clausius 1867, p. 157) of a cyclically working steam-engine *Source* French National Library–Gallica. Public Domain

Clausius analytically described (Clausius 1850) Carnot's geometrical cycle (Carnot 1824, p. 118). We specify that previously, in 1834, Benoît Paul Émile Clapeyron (1799–1864) had already mathematically reworked Carnot's theory by means his own cycle known as the *PV-diagram* (Clapeyron 1834, Fig. 1, p. 190; see also Gillispie and Pisano 2014, pp. 199–203).

The container *A* (Fig. 8, right side) represents the boiler that is maintained at a constant temperature T_1. A part of the steam passes from the boiler to the cylinder *B*. The cylinder and boiler are disconnected; the vapour raises the piston by its own expansion. After this the cylinder is put in communication with the space C, which represents the condenser. It is supposed that the condenser is kept at lower temperature by external cooling, and not by additional water injected at a different temperature. The constant temperature of the condenser is T_0. During the connection of the cylinder with the condenser the piston tracks back its former path, and thus the vapour which did not immediately pass by itself into the condenser is driven into it, and there becomes condensed. Taking into account above both the Fig. 8, below we read how Clausius thought to close his cycle in order to produce—for cyclical processes—thermodynamic work:

> In order to complete the cycle of operations, it is now necessary to convey the liquid produced by condensation back again into the boiler. This is done by means of the small pump D, whose action is so regulated, that at every ascent of the piston just as much liquid is withdrawn from the condenser as entered it by the above condensation; and during the decent of the piston this same quantity of liquid is forced back into the boiler. As soon as this liquid is again raised in the boiler to the temperature T_1, everything is once more in its initial condition, and the same series of operation can commence again. Here, therefore, we have a complete cyclical process. (Clausius 1867, p. 157)

In Clausius' memoirs, we can read several mathematical formulations of the Second Law for cyclical processes.

In the following excerpt we report the mathematical description as published in *Über die Anwendung des Satzes von der Äquivalenz der Verwandlungen auf die innere Arbeit* (1862) (Clausius 1862) (Fig. 9) and that can be also found in the Mechanical Theory of Heat (Clausius 1867) as the *Sixth Memoir* (Clausius 1867, pp. 215–250).

> Let dQ be an element of the heat given up by the body to any reservoir of heat during its own changes (heat which it may absorb from a reservoir being here reckoned as negative), and T the absolute temperature of the body at the moment of giving up this heat, then the equation
>
> $$\int \frac{dQ}{T} = 0$$
>
> must be true for every reversible cyclical process, and the relation
>
> $$\int \frac{dQ}{T} \geq 0$$
>
> must hold good for every cyclical process which is in any way possible. (Clausius 1867, pp. 218-219)

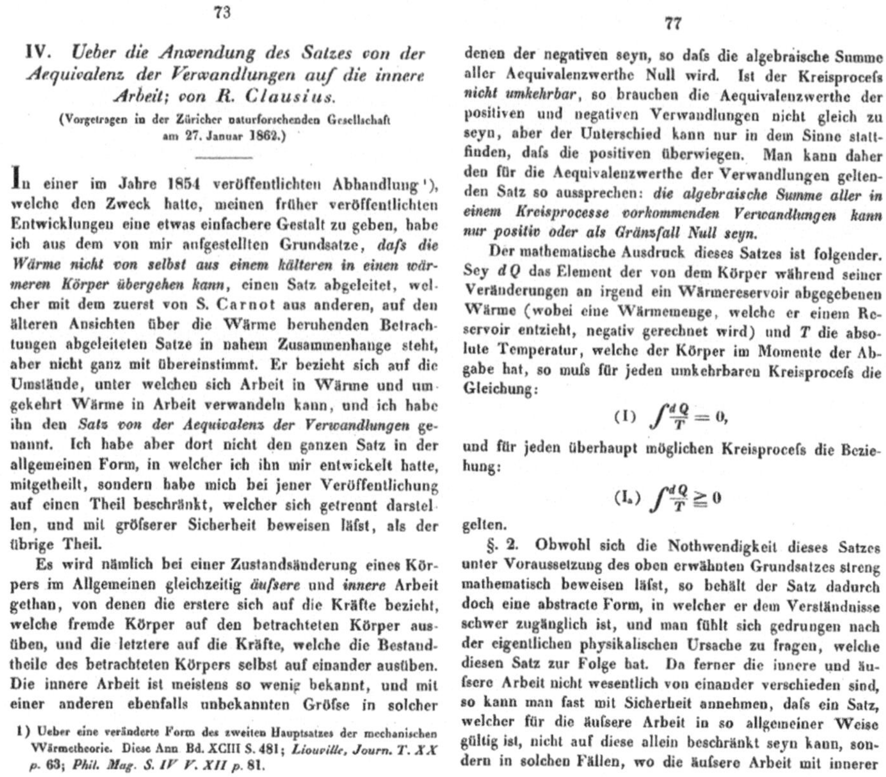

73

IV. *Ueber die Anwendung des Satzes von der Aequivalenz der Verwandlungen auf die innere Arbeit; von R. Clausius.*

(Vorgetragen in der Züricher naturforschenden Gesellschaft am 27. Januar 1862.)

In einer im Jahre 1854 veröffentlichten Abhandlung [1]), welche den Zweck hatte, meinen früher veröffentlichten Entwicklungen eine etwas einfachere Gestalt zu geben, habe ich aus dem von mir aufgestellten Grundsatze, *dafs die Wärme nicht von selbst aus einem kälteren in einen wärmeren Körper übergehen kann*, einen Satz abgeleitet, welcher mit dem zuerst von S. Carnot aus anderen, auf den älteren Ansichten über die Wärme beruhenden Betrachtungen abgeleiteten Satze in nahem Zusammenhange steht, aber nicht ganz mit übereinstimmt. Er bezieht sich auf die Umstände, unter welchen sich Arbeit in Wärme und umgekehrt Wärme in Arbeit verwandeln kann, und ich habe ihn den *Satz von der Aequivalenz der Verwandlungen* genannt. Ich habe aber dort nicht den ganzen Satz in der allgemeinen Form, in welcher ich ihn mir entwickelt hatte, mitgetheilt, sondern habe mich bei jener Veröffentlichung auf einen Theil beschränkt, welcher sich getrennt darstellen, und mit gröfserer Sicherheit beweisen läfst, als der übrige Theil.

Es wird nämlich bei einer Zustandsänderung eines Körpers im Allgemeinen gleichzeitig *äufsere* und *innere* Arbeit gethan, von denen die erstere sich auf die Kräfte bezieht, welche fremde Körper auf den betrachteten Körper ausüben, und die letztere auf die Kräfte, welche die Bestandtheile des betrachteten Körpers selbst auf einander ausüben. Die innere Arbeit ist meistens so wenig bekannt, und mit einer anderen ebenfalls unbekannten Gröfse in solcher

1) Ueber eine veränderte Form des zweiten Hauptsatzes der mechanischen Wärmetheorie. Diese Ann Bd. XCIII S. 481; *Liouville, Journ. T. XX p.* 63; *Phil. Mag. S. IV V. XII p.* 81.

77

denen der negativen seyn, so dafs die algebraische Summe aller Aequivalenzwerthe Null wird. Ist der Kreisprocefs *nicht umkehrbar*, so brauchen die Aequivalenzwerthe der positiven und negativen Verwandlungen nicht gleich zu seyn, aber der Unterschied kann nur in dem Sinne stattfinden, dafs die positiven überwiegen. Man kann daher den für die Aequivalenzwerthe der Verwandlungen geltenden Satz so aussprechen: *die algebraische Summe aller in einem Kreisprocesse vorkommenden Verwandlungen kann nur positiv oder als Gränzfall Null seyn.*

Der mathematische Ausdruck dieses Satzes ist folgender. Sey dQ das Element der von dem Körper während seiner Veränderungen an irgend ein Wärmereservoir abgegebenen Wärme (wobei eine Wärmemenge, welche er einem Reservoir entzieht, negativ gerechnet wird) und T die absolute Temperatur, welche der Körper im Momente der Abgabe hat, so mufs für jeden umkehrbaren Kreisprocefs die Gleichung:

$$\text{(I)} \quad \int \frac{dQ}{T} = 0,$$

und für jeden überhaupt möglichen Kreisprocefs die Beziehung:

$$(\text{I}_a) \quad \int \frac{dQ}{T} \geqq 0$$

gelten.

§. 2. Obwohl sich die Nothwendigkeit dieses Satzes unter Voraussetzung des oben erwähnten Grundsatzes streng mathematisch beweisen läfst, so behält der Satz dadurch doch eine abstracte Form, in welcher er dem Verständnisse schwer zugänglich ist, und man fühlt sich gedrungen nach der eigentlichen physikalischen Ursache zu fragen, welche diesen Satz zur Folge hat. Da ferner die innere und äufsere Arbeit nicht wesentlich von einander verschieden sind, so kann man fast mit Sicherheit annehmen, dafs ein Satz, welcher für die äufsere Arbeit in so allgemeiner Weise gültig ist, nicht auf diese allein beschränkt seyn kann, sondern in solchen Fällen, wo die äufsere Arbeit mit innerer

Fig. 9 Clausius' formulation of the Second Law for cyclical processes published in 1862 (Clausius 1862, p. 73, p. 77) *Source* French National Library–Gallica. Public Domain

Actually, Clausius recalled and summarized (Fig. 9) the conceptual frameworks and the main findings published in 1854 (Clausius 1854), which constituted the necessary premise of this 1862 work (Clausius 1862): to deliver the former mathematization of the Second Law for non-cyclical process.

6 Krönig (1856) and Clausius (1871) Kinetic Models

This section deals with different approaches to the *Theory of Heat*, which depend on the microscopic structure of gaseous systems, modelled as made of a great number of moving particles.

In 1856 Krönig firstly developed a mechanical hypothesis regarding the microscopic nature of heat, based on the idea that, with reference to a gaseous system, "[...] the heat of a system consists in nothing but a movement of its smallest parts." Krönig (1856) (Fig. 10).

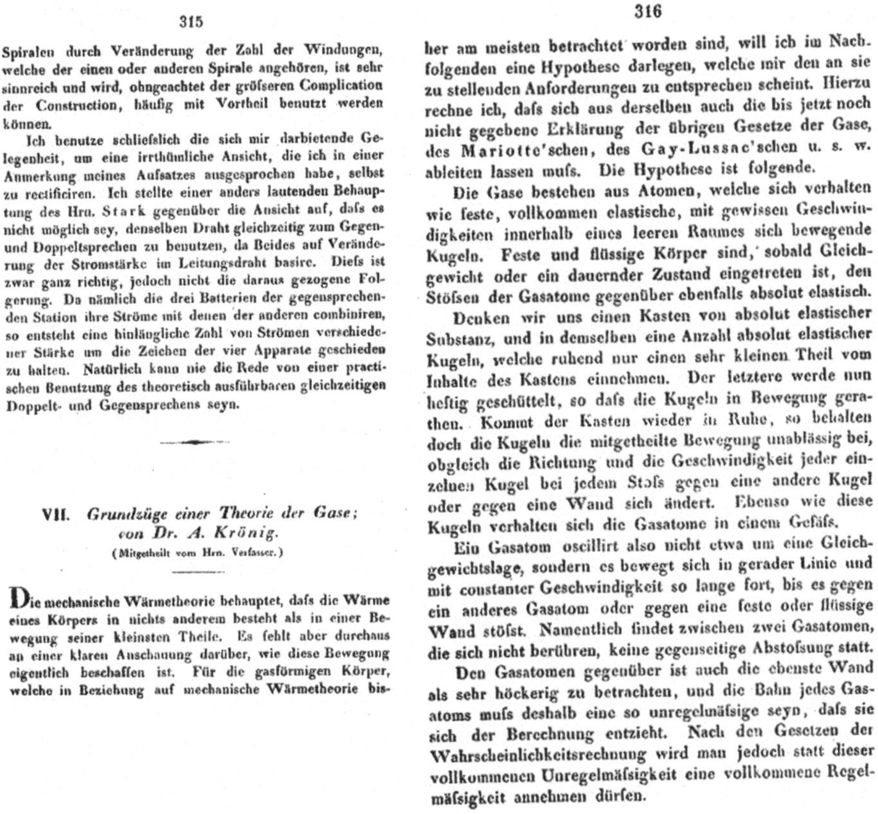

315

Spiralen durch Veränderung der Zahl der Windungen, welche der einen oder anderen Spirale angehören, ist sehr sinnreich und wird, ohngeachtet der gröfseren Complication der Construction, häufig mit Vortheil benutzt werden können.

Ich benutze schliefslich die sich mir darbietende Gelegenheit, um eine irrthümliche Ansicht, die ich in einer Anmerkung meines Aufsatzes ausgesprochen habe, selbst zu rectificiren. Ich stellte einer anders lautenden Behauptung des Hrn. Stark gegenüber die Ansicht auf, dafs es nicht möglich sey, denselben Draht gleichzeitig zum Gegen- und Doppeltsprechen zu benutzen, da Beides auf Veränderung der Stromstärke im Leitungsdraht basire. Diefs ist zwar ganz richtig, jedoch nicht die daraus gezogene Folgerung. Da nämlich die drei Batterien der gegensprechenden Station ihre Ströme mit denen der anderen combiniren, so entsteht eine hinlängliche Zahl von Strömen verschiedener Stärke um die Zeichen der vier Apparate geschieden zu halten. Natürlich kann nie die Rede von einer practischen Benutzung des theoretisch ausführbaren gleichzeitigen Doppelt- und Gegensprechens seyn.

VII. *Grundzüge einer Theorie der Gase; von Dr. A. Krönig.*

(Mitgetheilt vom Hrn. Verfasser.)

Die mechanische Wärmetheorie behauptet, dafs die Wärme eines Körpers in nichts anderem besteht als in einer Bewegung seiner kleinsten Theile. Es fehlt aber durchaus an einer klaren Anschauung darüber, wie diese Bewegung eigentlich beschaffen ist. Für die gasförmigen Körper, welche in Beziehung auf mechanische Wärmetheorie bis-

316

her am meisten betrachtet worden sind, will ich im Nachfolgenden eine Hypothese darlegen, welche mir den an sie zu stellenden Anforderungen zu entsprechen scheint. Hierzu rechne ich, dafs sich aus derselben auch die bis jetzt noch nicht gegebene Erklärung der übrigen Gesetze der Gase, des Mariotte'schen, des Gay-Lussac'schen u. s. w. ableiten lassen mufs. Die Hypothese ist folgende.

Die Gase bestehen aus Atomen, welche sich verhalten wie feste, vollkommen elastische, mit gewissen Geschwindigkeiten innerhalb eines leeren Raumes sich bewegende Kugeln. Feste und flüssige Körper sind, sobald Gleichgewicht oder ein dauernder Zustand eingetreten ist, den Stöfsen der Gasatome gegenüber ebenfalls absolut elastisch.

Denken wir uns einen Kasten von absolut elastischer Substanz, und in demselben eine Anzahl absolut elastischer Kugeln, welche ruhend nur einen sehr kleinen Theil vom Inhalte des Kastens einnehmen. Der letztere werde nun heftig geschüttelt, so dafs die Kugeln in Bewegung gerathen. Kommt der Kasten wieder in Ruhe, so behalten doch die Kugeln die mitgetheilte Bewegung unablässig bei, obgleich die Richtung und die Geschwindigkeit jeder einzelnen Kugel bei jedem Stofs gegen eine andere Kugel oder gegen eine Wand sich ändert. Ebenso wie diese Kugeln verhalten sich die Gasatome in einem Gefäfs.

Ein Gasatom oscillirt also nicht etwa um eine Gleichgewichtslage, sondern es bewegt sich in gerader Linie und mit constanter Geschwindigkeit so lange fort, bis es gegen ein anderes Gasatom oder gegen eine feste oder flüssige Wand stöfst. Namentlich findet zwischen zwei Gasatomen, die sich nicht berühren, keine gegenseitige Abstofsung statt.

Den Gasatomen gegenüber ist auch die ebenste Wand als sehr höckerig zu betrachten, und die Bahn jedes Gasatoms mufs deshalb eine so unregelmäfsige seyn, dafs sie sich der Berechnung entzieht. Nach den Gesetzen der Wahrscheinlichkeitsrechnung wird man jedoch statt dieser vollkommenen Unregelmäfsigkeit eine vollkommene Regelmäfsigkeit annehmen dürfen.

Fig. 10 Krönig's model of gases (Krönig 1856, pp. 315–316) *Source* French National Library–Gallica. Public Domain

More precisely, we can read:

> The mechanical theory of heat states that the heat of a body consists in nothing but a movement of its smallest parts. However, there is absolutely no clear view of how this movement actually is. For the gaseous systems, which so far have mostly been considered in relation to the mechanical theory of heat, in the following I will present a hypothesis, which seems to me to correspond to their characteristics.[16] (Krönig 1856, pp. 315–316; Author's translation)

One year later, in 1857, in *Poggendorff Annalen*, Clausius published *Über die Art der Bewegung, welche wir Warme nennen* (Clausius 1857a) where he developed his own kinetic model resting on almost the same hypotheses as Krönig (Krönig 1856) (Fig. 11).

[16] "Die mechanische Wärmetheorie behauptet, dass die Warme eines Körpers in nichts anderem besteht als in einer Bewegung seiner kleinsten Theile. Es fehlt aber durchaus en einer klaren Anschauung darüber, wie diese Bewegung eigentlich beschaffen ist. Für die gasförmigen Körper, welche in Beziehung auf mechanische Wärmetheorie bisher am meisten betrachtet worden sind, will ich im Nachfolgenden eine Hypothese darlegen, welche mir den an sie zu stellenden Anforderungen zu entsprechen scheint".

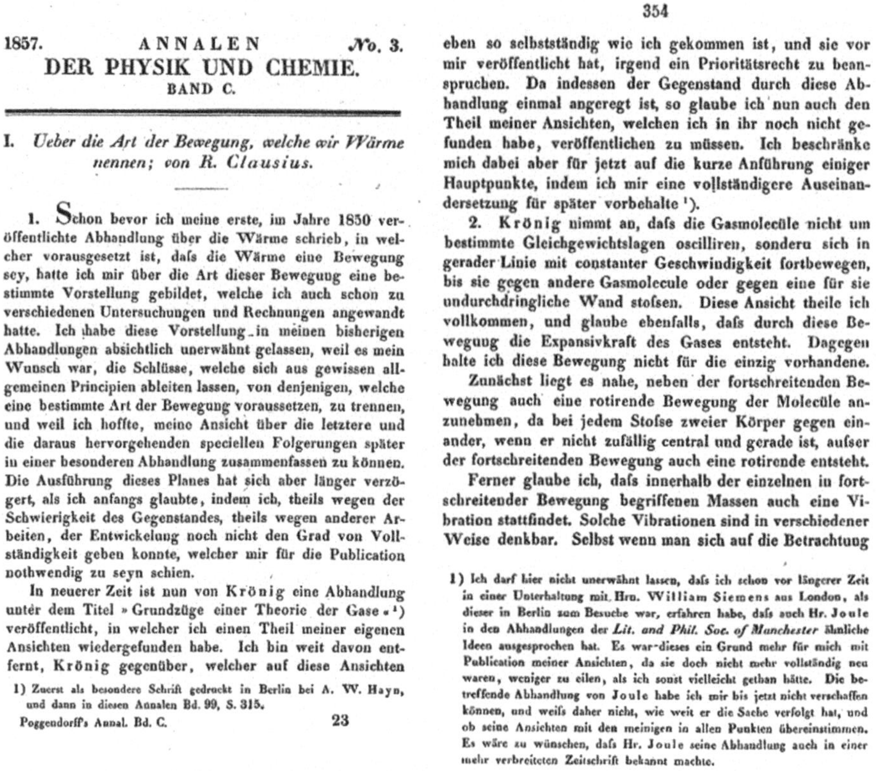

1857. ANNALEN *No.* 3.
DER PHYSIK UND CHEMIE.
BAND C.

I. *Ueber die Art der Bewegung, welche wir Wärme nennen; von R. Clausius.*

1. Schon bevor ich meine erste, im Jahre 1850 veröffentlichte Abhandlung über die Wärme schrieb, in welcher vorausgesetzt ist, dafs die Wärme eine Bewegung sey, hatte ich mir über die Art dieser Bewegung eine bestimmte Vorstellung gebildet, welche ich auch schon zu verschiedenen Untersuchungen und Rechnungen angewandt hatte. Ich habe diese Vorstellung in meinen bisherigen Abhandlungen absichtlich unerwähnt gelassen, weil es mein Wunsch war, die Schlüsse, welche sich aus gewissen allgemeinen Principien ableiten lassen, von denjenigen, welche eine bestimmte Art der Bewegung voraussetzen, zu trennen, und weil ich hoffte, meine Ansicht über die letztere und die daraus hervorgehenden speciellen Folgerungen später in einer besonderen Abhandlung zusammenfassen zu können. Die Ausführung dieses Planes hat sich aber länger verzögert, als ich anfangs glaubte, indem ich, theils wegen der Schwierigkeit des Gegenstandes, theils wegen anderer Arbeiten, der Entwickelung noch nicht den Grad von Vollständigkeit geben konnte, welcher mir für die Publication nothwendig zu seyn schien.

In neuerer Zeit ist nun von Krönig eine Abhandlung unter dem Titel »Grundzüge einer Theorie der Gase«[1]) veröffentlicht, in welcher ich einen Theil meiner eigenen Ansichten wiedergefunden habe. Ich bin weit davon entfernt, Krönig gegenüber, welcher auf diese Ansichten

1) Zuerst als besondere Schrift gedruckt in Berlin bei A. W. Hayn, und dann in diesen Annalen Bd. 99, S. 315.

Poggendorff's Annal. Bd. C. 23

354

eben so selbstständig wie ich gekommen ist, und sie vor mir veröffentlicht hat, irgend ein Prioritätsrecht zu beanspruchen. Da indessen der Gegenstand durch diese Abhandlung einmal angeregt ist, so glaube ich nun auch den Theil meiner Ansichten, welchen ich in ihr noch nicht gefunden habe, veröffentlichen zu müssen. Ich beschränke mich dabei aber für jetzt auf die kurze Anführung einiger Hauptpunkte, indem ich mir eine vollständigere Auseinandersetzung für später vorbehalte[1]).

2. Krönig nimmt an, dafs die Gasmolecüle nicht um bestimmte Gleichgewichtslagen oscilliren, sondern sich in gerader Linie mit constanter Geschwindigkeit fortbewegen, bis sie gegen andere Gasmolecule oder gegen eine für sie undurchdringliche Wand stofsen. Diese Ansicht theile ich vollkommen, und glaube ebenfalls, dafs durch diese Bewegung die Expansivkraft des Gases entsteht. Dagegen halte ich diese Bewegung nicht für die einzig vorhandene.

Zunächst liegt es nahe, neben der fortschreitenden Bewegung auch eine rotirende Bewegung der Molecüle anzunehmen, da bei jedem Stofse zweier Körper gegen einander, wenn er nicht zufällig central und gerade ist, aufser der fortschreitenden Bewegung auch eine rotirende entsteht.

Ferner glaube ich, dafs innerhalb der einzelnen in fortschreitender Bewegung begriffenen Massen auch eine Vibration stattfindet. Solche Vibrationen sind in verschiedener Weise denkbar. Selbst wenn man sich auf die Betrachtung

1) Ich darf hier nicht unerwähnt lassen, dafs ich schon vor längerer Zeit in einer Unterhaltung mit Hrn. William Siemens aus London, als dieser in Berlin zum Besuche war, erfahren habe, dafs auch Hr. Joule in den Abhandlungen der *Lit. and Phil. Soc. of Manchester* ähnliche Ideen ausgesprochen hat. Es war dieses ein Grund mehr für mich mit Publication meiner Ansichten, da sie doch nicht mehr vollständig neu waren, weniger zu eilen, als ich sonst vielleicht gethan hätte. Die betreffende Abhandlung von Joule habe ich mir bis jetzt nicht verschaffen können, und weifs daher nicht, wie weit er die Sache verfolgt hat, und ob seine Ansichten mit den meinigen in allen Punkten übereinstimmen. Es wäre zu wünschen, dafs Hr. Joule seine Abhandlung auch in einer mehr verbreiteten Zeitschrift bekannt machte.

Fig. 11 Clausius' theory of gases (Clausius 1857a, pp. 353–354) *Source* French National Library–Gallica. Public Domain

In the incipit (Clausius 1857a) Clausius notes his perspective with respect to Krönig's contribution by saying:

> A memoir has lately been published by Krönig, under the title *Grundzüge einer Theorie der Gase*, in which I have recognized some of my own views. Seeing that Krönig has arrived at these views just as independently as I have, and has published them before me, all claim on my part is of course out of the question; nevertheless, the subject having once been mooted in this memoir, I feel myself induced to publish those parts of my own views which I have not yet found in it. For the present, I shall confine myself to a brief indication of a few principal points, and reserve a more complete analysis for another time. (Clausius 1857b, p. 112)

We remark that the idea of modelling heat as a motion of the microscopic constituent of matter was already discussed by Joule. He worked on the notion developed by Benjamin Thompson, also known as count de Rumford (1753–1814) that heat resulted from the movement of particles and investigated how the changes in the pressure of a gas led to changes in temperature. He also developed the idea that heat is related to the *motion of particles* assuming *the velocity of rotation of atomic elements* and *electricity* (Joule 1884; Cfr. Pisano et al. 2021c). *On the Changes of*

Temperature produced by the Rarefaction and Condensation of Air (1845) (Joule 1845, 1884, 1887) and *On the Mechanical Equivalent of Heat* (1849) (Joule 1845, 1884, 1887) are perhaps two of the best examples on the subject (Pisano et al. 2021c) where Joule made several interesting remarks:

> These results [heat-work/mechanical equivalent of heat] are inexplicable if heat be a substance. [...] The results are, however, such as might have been deduced à priori from any theory in which heat is regarded as a state of motion among the constituent particles of bodies. It is easy to understand how the mechanical force expended in the condensation of air may be communicated to these particles so as to increase the rapidity of their motion and thus may produce the phenomenon of increase of temperature. (Joule 1884, pp. 185–186)
>
> [...]
>
> The beautiful idea of Davy, that the heat of elastic fluids depends partly upon a motion of particles round their axes, has not I think hitherto received the attention it deserves. I believe that most phenomena may be explained by adapting it to the great electro-chemical discovery of Faraday, by which we know that each atomic element is associated with the same absolute quantity of electricity. Let us suppose that these atmospheres of electricity, endowed to a certain extent with the ordinary properties of matter, revolve with great velocity round their respective atoms, and that the velocity of rotation determines what we call temperature [...]. (Joule 1884, pp. 187–188; see also Joule 1887)

In the end, through his experiments Joule showed (Pisano et al. 2021c) that heat was not a *substance* but probably a kind of motion of particles as it could be *destroyed*.[17]

> Believing that the power to destroy belongs to the Creator alone, I entirely coincide with Roget and Faraday in the opinion that any theory which, when carried out, demands the annihilation of force, is necessarily erroneous. (Joule 1845, pp. 382–383; Joule 1884, pp. 188–1890)

For his part, Clausius himself kept the reader informed about the Joule's affair:

> I must not omit to mention here, that some time ago Mr. William Siemens of London, when on a visit to Berlin, informed me that Joule had also expressed similar ideas in the *Memoirs of the Literary and Philosophical Society of Manchester*. My views being consequently no longer completely new, this was an additional reason why I should hasten their publication less than I otherwise probably should have done. (Clausius 1857b, Footnote, p. 112)

In effect, Clausius' kinetic model rests on the idea that a gaseous system is made of a number of moving molecules and that heat is associated to the *vis viva* of these motions. We recall that *vis viva* corresponds to the current notion of kinetic energy, as Clausius clarified in the following excerpt:

> In accordance with a practice lately become general, and with what I have myself done in former memoirs, I call the semi-product of the mass into the square of the velocity *vis viva*, because it is only with this definition of the notion that we can, without the addition of a coefficient, equate the expressions representing a quantity of work and the increase or decrease of *vis viva* which corresponds to the same. (Clausius 1857b, p. 129)

[17] Joule also discussed the interplay between chemical energy, electricity and mechanical motion thus paving the way to modern understanding of the transformations of energy, the principle of conservation of energy and the first law of thermodynamics.

In Clausius' model, constituent molecules can be monoatomic or polyatomic and, in general, can undergo three kinds of motions (Clausius 1857b):

(1) *Translatory motion*	Molecules moving at constant velocity in straight lines until they strike against other molecules or against some surface (Clausius 1857b, p. 113)
(2) *Rotatory motion*	It can concern only polyatomic molecules and it consist in the motion of the entire molecule around a rotational axis (Clausius 1857b, p. 113)
(3) *Vibrations motion*	It can concern only polyatomic molecules and it consists in the oscillation of the atoms with respect to each other (Clausius 1857b, p. 113)

The *vis viva* of a *translatory motion* is not enough to entirely model the heat properties of gases; the other two freedom degrees of the motion are necessary. In his words:

> By means of a mathematical investigation given at the end of the present memoir, it may be proved that the *vis viva* of the translatory motion alone is too small to represent the whole heat present in the gas; so that without entering into the probability of the same, we are thus compelled to assume one or more motions of another kind. (Clausius 1857b, p. 113)

We underline that the assumption of both *rotational* and *vibrational* degrees of freedom is one of the main novelties of Clausius' kinetic model with respect to Krönig's one. Clausius pinpointed it as in following:

> Krönig assumes that the molecules of gas do not oscillate about definite positions of equilibrium, but they move with constant velocity in straight lines until they strike against other molecules, or against some surface which is to them impermeable. I share this view completely, and I also believe that the expansive force of the gas arises from this motion. On the other hand, I am of opinion that this is not the only motion present.
>
> In the first place, the hypothesis of a rotatory as well as progressive motion of the molecules at once suggests itself; for at every impact of two bodies, unless the same happens to be central and rectilinear, a rotatory as well as a translator motion ensues.
>
> I am also of opinion that vibrations take place within the several masses in a state of progressive motion. (Clausius 1857b, p. 113)

Nevertheless, the role played by *vis viva* of a *rotatory motion* and *vibrations* is only particularly important in the case of gases, where "[...] each molecule consists of a great number of atoms." (Clausius 1857b, p. 114). Anyway, it is only the *vis viva* of the translation that is embodied inside Clausius' model. The motion of the molecules—continuously striking and rebounding from the surfaces of the container—is considered to determine the pressure of a gaseous system enclosed in a vessel. In fact,

> The pressure of the gas against a fixed surface is caused by the molecules in great number continually striking against and rebounding from the same. The force which must hence arise is, in the first place, by equal velocity of motion inversely proportional to the volume of the given quantity of gas; and secondly, by equal volume proportional to the *vis viva* of the translatory motion: the other motions do not here immediately come into consideration. (Clausius 1857b, p. 115)

By resting on the above premises, Clausius developed his mathematical model[18] from which he delivered the pivotal expression of the absolute temperature T of an ideal gas in terms of the *vis visa* of its constituent molecules. Below we retrace his main reasoning.

Firstly, Clausius obtained the mathematical expression for the pressure of a sample of gas exerted on the sides of the vessel in terms of the motion of the molecules. This mathematical derivation is based on a physical model that considers the sample of gas as constituted by n molecules with mass m moving with a velocity u in a vessel whose shape is specified as follows:

> As the shape of the vessel is indifferent, we will select that which is the most convenient for our purpose. We will assume the vessel to be very flat, and that two of its sides consist of parallel planes so close to one another that their distance asunder is infinitesimal when compared with other dimensions of the vessel. Hence, we need not consider the cases where the molecules strike against one of the strips of sides and we may assume that each move in a right line until it either strikes against another molecule or against one of the large parallel sides. In fact, to take the small sides into consideration would change nothing in the final result, and would only make the development more prolix. (Clausius 1857b, p. 126)

In particular, he assumed that the vessel is a parallelepiped made of two large parallel sides separated by an infinitesimal distance h. It is also assumed as a further simplification that, for each scattering molecule, the angle and velocity of reflection is equal to those of incidence, so that.

> According to these assumptions, it is evident, that, during the unit of time, each molecule will strike the side under consideration just as often as during that time it can, by following its peculiar direction, travel from the side in question to the other and back again. (Clausius 1857b, p. 127)

Thus he obtained the following expression for the quantity of motion imparted by one side of the vessel to all the molecules which strike against it during the unit of time:

$$\frac{mnu^2}{3h} \tag{11}$$

The physical meaning of this expression is then made to correspond to the pressure exerted by the gas in the unit of time through Clausius' following argument:

---

[18] We observe that the Newtonian modelling character (Pisano and Bussotti 2016, 2017b, 2022; Newton 1687, [1713] 1729) of this kinetic model of gases, through the inclusion of Newtonian quantities—and—dialectic (forces, velocities, etc.) in order to study a field of applicability of another thermodynamic. In other words, it works in the microscopic field of thermodynamics, while the latter is a macroscopic container, that is Carnot's theory. In fact, these Newtonian quantities were lacking in macroscopic thermodynamics (Gillispie and Pisano 2014; Thomson 1848–1849b). This is one of the reasons why, at that time, the kinetic statistical approach to thermodynamics was considered as a model and not as a theory. Still nowadays, this discipline is presented like a theory only when it included mechanics, i.e., statistical mechanics (Gillispie and Pisano 2014; Pisano and Capecchi 2013; Pisano et al. 2023).

> Since all forces, however, are measured by the quantity of motion they can produce in the unit of time, the above expression at once represents this force as well as the pressure exerted by the gas, the latter being equilibrated by the former. (Clausius 1857b, p. 129)

We underline the historical–epistemological relevance of this argument. Actually, it is the cognitive element bridging a macroscopic physical quantity (i.e., pressure) with a dynamical property (i.e., quantity of motion) of the microscopic constituents of the gas.

If α is the superficial area of the large side of the vessel, it follows that the pressure p per unit of surface is:

$$p = \frac{mnu^2}{3\alpha h} \tag{12}$$

Since the product αh geometrically represents the volume v of the vessel, we can re–write the foregoing Eq. (12) as follows:

$$p = \frac{mnu^2}{3v} \tag{13}$$

that can be algebraically rearranged as

$$\frac{3}{2} pv = \frac{mnu^2}{2} \tag{14}$$

This equation expresses the product pv as proportional to the *vis viva* of the *translatory motion* of all the molecules. Clausius assumed that the gas followed an ideal behaviour so that both Boyle–Mariotte's law (1662; 1676) and Charles–Gay-Lussac' law (1787; 1802) could be applied; that is to say that the product pv remains proportional to the absolute temperature T. In other words, from Eq. (14), it follows:

$$\frac{mnu^2}{2} = T \cdot C \tag{15}$$

where C is a constant. Hence, from Eq. (15), Clausius concludes that:

> [...] the *vis viva* of the translatory motion [of the constituent molecules] is proportional to the absolute temperature. (Clausius 1857b, p. 130)

This result remains a *chiave di volta* for all the subsequent developments of the early statistical mechanics. Both Clausius and Boltzmann, while attempting to find a proof of the Second Law of mechanical theory of heat, expressed macroscopic physical quantities such as work, heat and temperature in terms of the dynamical properties of a microscopic constituent. To this end both authors assumed—as the foundational principlethat the absolute temperature was proportional to the kinetic energy of the microscopic constituent elements. More specifically, as we have recalled in the foregoing sections while discussing Boltzmann's ideas and works, in 1866 (Boltzmann 1866) he delivered the expression

$$2\Sigma \log \int_{t_1}^{t_2} \frac{mc^2}{2} dt + C \tag{16}$$

that represents the following integral

$$\int \frac{dQ}{T} \tag{17}$$

in terms of the kinetic energy of the constituent particles. This result was obtained by starting from Clausius' assumption (1857) that "the *vis viva* of the translatory motion [of the constituent molecules] is proportional to the absolute temperature." (Clausius 1857b, p. 130). Clausius attained a similar result in his *Über die Zurückführung des zweiten Hauptsatzes der mechanischen Wärmetheorie auf allgemeine mechanische Prinzipien* (Clausius 1871; Fig. 12).

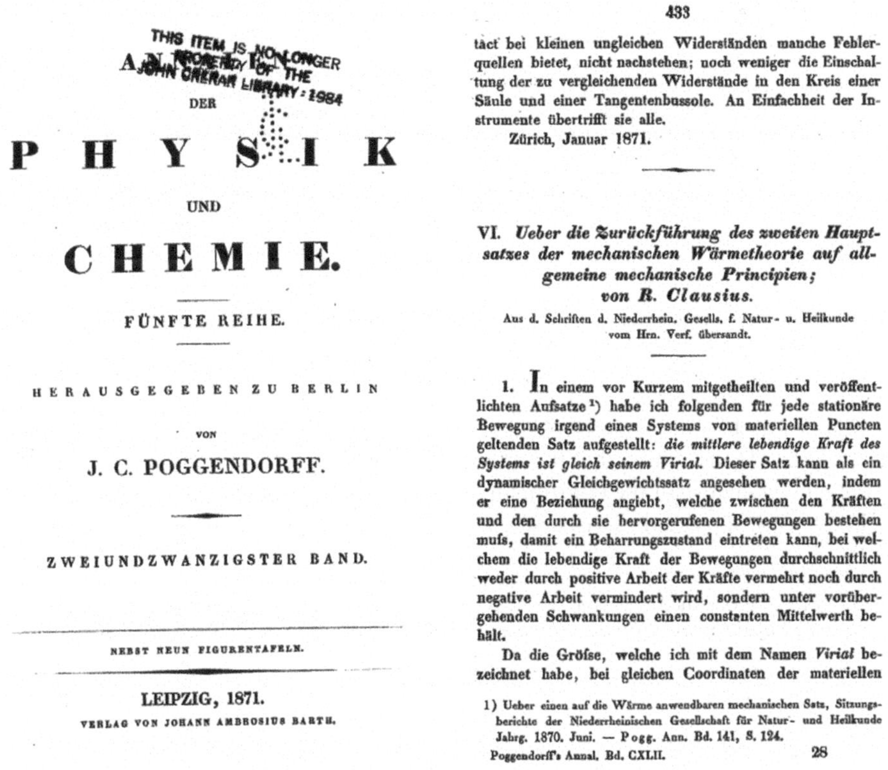

ANNALEN

DER

PHYSIK

UND

CHEMIE.

FÜNFTE REIHE.

HERAUSGEGEBEN ZU BERLIN

VON

J. C. POGGENDORFF.

ZWEIUNDZWANZIGSTER BAND.

NEBST NEUN FIGURENTAFELN.

LEIPZIG, 1871.

VERLAG VON JOHANN AMBROSIUS BARTH.

433

täct bei kleinen ungleichen Widerständen manche Fehlerquellen bietet, nicht nachstehen; noch weniger die Einschaltung der zu vergleichenden Widerstände in den Kreis einer Säule und einer Tangentenbussole. An Einfachheit der Instrumente übertrifft sie alle.

Zürich, Januar 1871.

VI. ***Ueber die Zurückführung des zweiten Hauptsatzes der mechanischen Wärmetheorie auf allgemeine mechanische Principien; von R. Clausius.***

Aus d. Schriften d. Niederrhein. Gesells. f. Natur- u. Heilkunde vom Hrn. Verf. übersandt.

1. In einem vor Kurzem mitgetheilten und veröffentlichten Aufsatze [1]) habe ich folgenden für jede stationäre Bewegung irgend eines Systems von materiellen Puncten geltenden Satz aufgestellt: *die mittlere lebendige Kraft des Systems ist gleich seinem Virial.* Dieser Satz kann als ein dynamischer Gleichgewichtssatz angesehen werden, indem er eine Beziehung angiebt, welche zwischen den Kräften und den durch sie hervorgerufenen Bewegungen bestehen mufs, damit ein Beharrungszustand eintreten kann, bei welchem die lebendige Kraft der Bewegungen durchschnittlich weder durch positive Arbeit der Kräfte vermehrt noch durch negative Arbeit vermindert wird, sondern unter vorübergehenden Schwankungen einen constanten Mittelwerth behält.

Da die Gröfse, welche ich mit dem Namen *Virial* bezeichnet habe, bei gleichen Coordinaten der materiellen

1) Ueber einen auf die Wärme anwendbaren mechanischen Satz, Sitzungsberichte der Niederrheinischen Gesellschaft für Natur- und Heilkunde Jahrg. 1870. Juni. — Pogg. Ann. Bd. 141, S. 124.

Poggendorff's Annal. Bd. CXLII. 28

Fig. 12 Clausius' kinetic model (Clausius 1871, frontispiece and p. 433) *Source* French National Library–Gallica. Public Domain

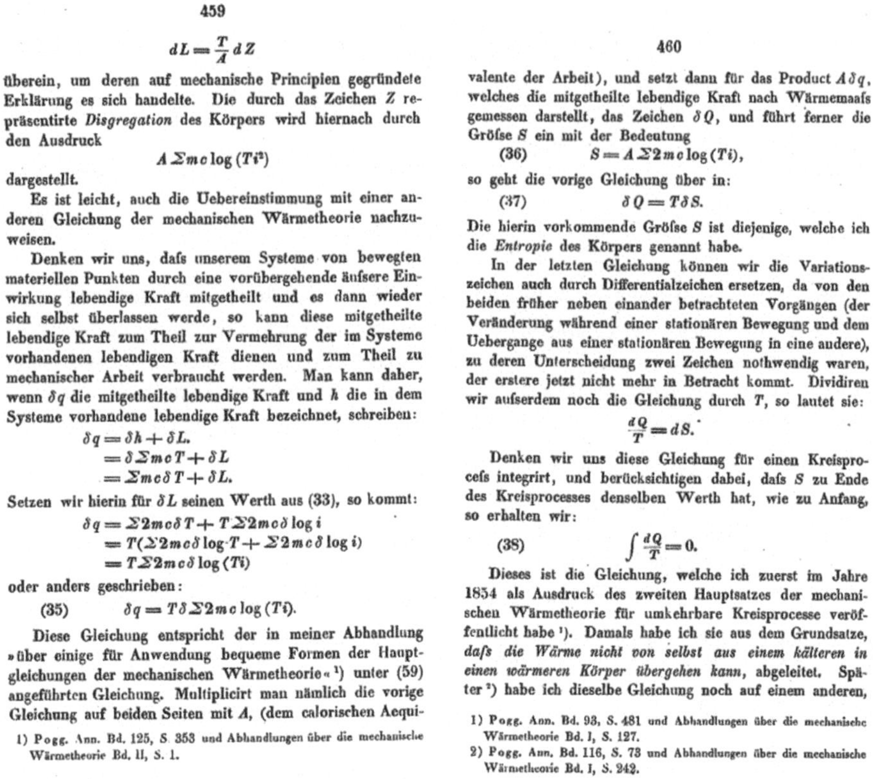

459

$$dL = \frac{T}{A}\, dZ$$

überein, um deren auf mechanische Principien gegründete Erklärung es sich handelte. Die durch das Zeichen Z repräsentirte *Disgregation* des Körpers wird hiernach durch den Ausdruck

$$A \Sigma mc \log (Ti^2)$$

dargestellt.

Es ist leicht, auch die Uebereinstimmung mit einer anderen Gleichung der mechanischen Wärmetheorie nachzuweisen.

Denken wir uns, dafs unserem Systeme von bewegten materiellen Punkten durch eine vorübergehende äufsere Einwirkung lebendige Kraft mitgetheilt und es dann wieder sich selbst überlassen werde, so kann diese mitgetheilte lebendige Kraft zum Theil zur Vermehrung der im Systeme vorhandenen lebendigen Kraft dienen und zum Theil zu mechanischer Arbeit verbraucht werden. Man kann daher, wenn δq die mitgetheilte lebendige Kraft und h die in dem Systeme vorhandene lebendige Kraft bezeichnet, schreiben:

$$\begin{aligned} \delta q &= \delta h + \delta L. \\ &= \delta \Sigma mcT + \delta L \\ &= \Sigma mc\delta T + \delta L. \end{aligned}$$

Setzen wir hierin für δL seinen Werth aus (33), so kommt:

$$\begin{aligned} \delta q &= \Sigma 2mc\delta T + T\Sigma 2mc\delta \log i \\ &= T(\Sigma 2mc\delta \log T + \Sigma 2mc\delta \log i) \\ &= T\Sigma 2mc\delta \log (Ti) \end{aligned}$$

oder anders geschrieben:

$$(35) \qquad \delta q = T\delta \Sigma 2mc \log (Ti).$$

Diese Gleichung entspricht der in meiner Abhandlung »über einige für Anwendung bequeme Formen der Hauptgleichungen der mechanischen Wärmetheorie« [1]) unter (59) angeführten Gleichung. Multiplicirt man nämlich die vorige Gleichung auf beiden Seiten mit A, (dem calorischen Aequi-

1) Pogg. Ann. Bd. 125, S. 353 und Abhandlungen über die mechanische Wärmetheorie Bd. II, S. 1.

460

valente der Arbeit), und setzt dann für das Product $A\delta q$, welches die mitgetheilte lebendige Kraft nach Wärmemaafs gemessen darstellt, das Zeichen δQ, und führt ferner die Gröfse S ein mit der Bedeutung

$$(36) \qquad S = A\Sigma 2mc \log (Ti),$$

so geht die vorige Gleichung über in:

$$(37) \qquad \delta Q = T\delta S.$$

Die hierin vorkommende Gröfse S ist diejenige, welche ich die *Entropie* des Körpers genannt habe.

In der letzten Gleichung können wir die Variationszeichen auch durch Differentialzeichen ersetzen, da von den beiden früher neben einander betrachteten Vorgängen (der Veränderung während einer stationären Bewegung und dem Uebergange aus einer stationären Bewegung in eine andere), zu deren Unterscheidung zwei Zeichen nothwendig waren, der erstere jetzt nicht mehr in Betracht kommt. Dividiren wir aufserdem noch die Gleichung durch T, so lautet sie:

$$\frac{dQ}{T} = dS.$$

Denken wir uns diese Gleichung für einen Kreisprocefs integrirt, und berücksichtigen dabei, dafs S zu Ende des Kreisprocesses denselben Werth hat, wie zu Anfang, so erhalten wir:

$$(38) \qquad \int \frac{dQ}{T} = 0.$$

Dieses ist die Gleichung, welche ich zuerst im Jahre 1854 als Ausdruck des zweiten Hauptsatzes der mechanischen Wärmetheorie für umkehrbare Kreisprocesse veröffentlicht habe [1]). Damals habe ich sie aus dem Grundsatze, *dafs die Wärme nicht von selbst aus einem kälteren in einen wärmeren Körper übergehen kann*, abgeleitet. Später [2]) habe ich dieselbe Gleichung noch auf einem anderen,

1) Pogg. Ann. Bd. 93, S. 481 und Abhandlungen über die mechanische Wärmetheorie Bd. I, S. 127.

2) Pogg. Ann. Bd. 116, S. 73 und Abhandlungen über die mechanische Wärmetheorie Bd. I, S. 242.

Fig. 13 Clausius' mathematical expression in terms of the kinetic energy (Clausius 1871, pp. 459–460) *Source* French National Library–Gallica. Public Domain

In this way he implemented a mathematical model based on the same principle, that is to say by assuming that the temperature of a gaseous system is proportional to the kinetic energy of the constituent molecules (Clausius 1871; Fig. 13).

In this way of reasoning we can identify two major steps (Clausius 1871, pp. 435–460):

(1) First, he obtained (Clausius 1871, p. 460) the mathematical expression—in terms of the kinetic energy of the constituent particles—of the following differential equation:

$$\frac{dQ}{T} = dS \tag{18}$$

where dQ represents the differential of heat, that in Clausius' notation is not written as δQ, even if it is treated as a partial differential. Conversely, dS is an exact differential that zeroes, by integrating it over a cyclic path.

(2) Secondly, he considered (Clausius 1871, p. 460) the particular case of cyclical processes hence obtaining, from Eq. (18) the following result:

$$\int \frac{dQ}{T} = 0 \tag{19}$$

We underline that Clausius used (Clausius 1871, p. 460) the same script to indicate both cyclical and not-cyclical integrals.

For sake of brevity we avoid tracing all the steps of Clausius' reasoning leading to the expression of Eq. (17) in terms of the kinetic energy (Clausius 1871, pp. 435–459). We focus on the conclusive part of the paper (Clausius 1871, pp. 457–460), by commenting on the following Eq. (20) obtained for a system made of a number of gas molecules, with mass m, moving with a velocity v, assuming that each molecule covers a closed trajectory in a time interval i:

$$\delta q = T\delta \sum 2mc \log(Ti) \tag{20}$$

To disclose the meaning of this equation, firstly we consider its right side. The conceptual key to understand it can be itemized in the following points.

- The sum is extended to all gas molecules contained in the sample;
- c is a constant and the script δ is the differential sign that in this context can be replaced with d.
- The letter T *has two meanings.*
- The script T before the sum is the absolute temperature of the gas.
- The second T, argument of the logarithm, corresponds to the *vis viva* of one molecule, according to the principle represented by Eq. (15).

The left side of the equation, i.e. δq, represents the differential of the heat, exchanged by the gas, expressed in mechanical units. Clausius comments this equation as follows:

> This equation corresponds to the equation given under (59) in my treatise "On some Convenient Forms of the main Equations of the mechanical Theory of Heat"(1).[19] (Clausius 1871, p. 459. Original quotation marks and footnote; Author's translation)

The Eq. "(59)" mentioned in the foregoing excerpt is in Clausius' 1865–work (Clausius 1865, p. 387; Fig. 14). It is the very famous differential expression of the Second Law for the general case (not-cyclical case) of reversible processes; that is to say:

$$\frac{dQ}{T} = dS \tag{21}$$

Now, we make explicit the correspondence between Eqs. (20) and (19) that Clausius stated in the foregoing excerpt.

At first, to express δq in unit Joule Clausius multiplied both members of the equation by the conversion factor A that is "the thermal equivalent of work" (Clausius

[19] "Diese Gleichung entspricht der in meiner Abhandlung "Über einige für Anwendung bequeme Formen der Hauptgleichungen der mechanischen Wärmetheorie"(1) unter (59) angeführten Gleichung.".

1871, p. 460). The product $A\ \delta q$ corresponds to the differential of the heat exchanged, expressed in thermal units, that is to say δQ. At this point of the reasoning, an artifice is outlined: Clausius defines a new function *S*, that for the moment does not have any physical meaning. It appears as an ad hoc mathematical inclusion in order to transform Eq. (20). In other words, to attribute a physical meaning to the kinetic expression given by Eq. (20), Clausius made use of a mathematical artifice: he (perhaps to be arbitrary) named as *S* the following quantity:

$$A\sum 2mc\log(Ti)$$

In this way, Eq. (20) becomes:

$$\delta Q = T\delta S$$

In his words:

> Namely, if one multiplies the previous equation on both sides by *A*, (that is the thermal equivalent of work), and then uses the notation δQ for the product $A\ \delta q$, which represents the aforementioned *vis viva* expressed in the thermal units of measure, and also introduces the quantity *S* defined as follows[20]
>
> $$S = A\sum 2mc\log(Ti) \qquad [22]$$
>
> then the previous equation becomes
>
> $$\delta Q = T\delta S \qquad [23]$$
>
> (Clausius 1871, pp. 459–460; Authors' translation)

---

20 "Multipliert man nämlich die vorige Gleichung auf beiden Seiten mit *A*, (der kalorischen Äquivalenter der Arbeit), und setzt dann für das Produkt $A\ \delta q$, welches die mitgeteilte lebendige Kraft nach Wärmemaaß gemessen darstellt, das Zeichen $A\ \delta q$, und führt ferner die Größe *S* ein mit der Bedeutung.

$$S = A\sum 2mc\log(Ti)$$

so geht die vorige Gleichung über in:

$$\delta Q = T\delta S''$$

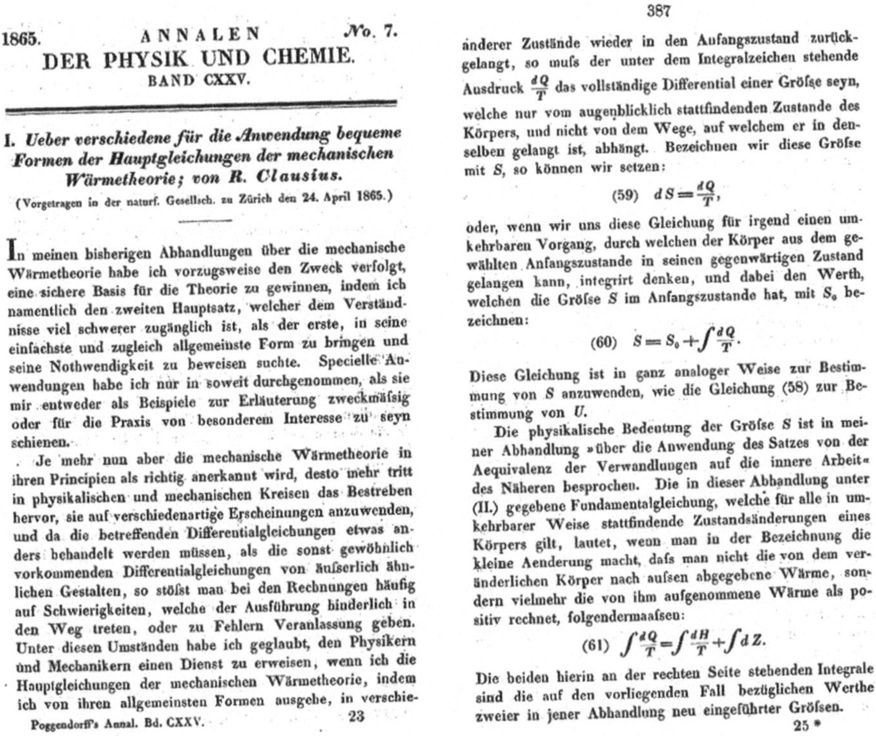
1865. ANNALEN No. 7.
DER PHYSIK UND CHEMIE.
BAND CXXV.

I. *Ueber verschiedene für die Anwendung bequeme Formen der Hauptgleichungen der mechanischen Wärmetheorie; von R. Clausius.*

(Vorgetragen in der naturf. Gesellsch. zu Zürich den 24. April 1865.)

In meinen bisherigen Abhandlungen über die mechanische Wärmetheorie habe ich vorzugsweise den Zweck verfolgt, eine sichere Basis für die Theorie zu gewinnen, indem ich namentlich den zweiten Hauptsatz, welcher dem Verständnisse viel schwerer zugänglich ist, als der erste, in seine einfachste und zugleich allgemeinste Form zu bringen und seine Nothwendigkeit zu beweisen suchte. Specielle Anwendungen habe ich nur in soweit durchgenommen, als sie mir entweder als Beispiele zur Erläuterung zweckmäfsig oder für die Praxis von besonderem Interesse zu seyn schienen.

Je mehr nun aber die mechanische Wärmetheorie in ihren Principien als richtig anerkannt wird, desto mehr tritt in physikalischen und mechanischen Kreisen das Bestreben hervor, sie auf verschiedenartige Erscheinungen anzuwenden, und da die betreffenden Differentialgleichungen etwas anders behandelt werden müssen, als die sonst gewöhnlich vorkommenden Differentialgleichungen von äufserlich ähnlichen Gestalten, so stöfst man bei den Rechnungen häufig auf Schwierigkeiten, welche der Ausführung hinderlich in den Weg treten, oder zu Fehlern Veranlassung geben. Unter diesen Umständen habe ich geglaubt, den Physikern und Mechanikern einen Dienst zu erweisen, wenn ich die Hauptgleichungen der mechanischen Wärmetheorie, indem ich von ihren allgemeinsten Formen ausgehe, in verschie-

Poggendorff's Annal. Bd. CXXV. 23

387

anderer Zustände wieder in den Anfangszustand zurückgelangt, so mufs der unter dem Integralzeichen stehende Ausdruck $\frac{dQ}{T}$ das vollständige Differential einer Gröfse seyn, welche nur von dem augenblicklich stattfindenden Zustande des Körpers, und nicht von dem Wege, auf welchem er in denselben gelangt ist, abhängt. Bezeichnen wir diese Gröfse mit S, so können wir setzen:

$$(59)\quad dS=\frac{dQ}{T},$$

oder, wenn wir uns diese Gleichung für irgend einen umkehrbaren Vorgang, durch welchen der Körper aus dem gewählten Anfangszustande in seinen gegenwärtigen Zustand gelangen kann, integrirt denken, und dabei den Werth, welchen die Gröfse S im Anfangszustande hat, mit S_0 bezeichnen:

$$(60)\quad S=S_0+\int\frac{dQ}{T}.$$

Diese Gleichung ist in ganz analoger Weise zur Bestimmung von S anzuwenden, wie die Gleichung (58) zur Bestimmung von U.

Die physikalische Bedeutung der Gröfse S ist in meiner Abhandlung »über die Anwendung des Satzes von der Aequivalenz der Verwandlungen auf die innere Arbeit« des Näheren besprochen. Die in dieser Abhandlung unter (II.) gegebene Fundamentalgleichung, welche für alle in umkehrbarer Weise stattfindende Zustandsänderungen eines Körpers gilt, lautet, wenn man in der Bezeichnung die kleine Aenderung macht, dafs man nicht die von dem veränderlichen Körper nach aufsen abgegebene Wärme, sondern vielmehr die von ihm aufgenommene Wärme als positiv rechnet, folgendermaafsen:

$$(61)\quad \int\frac{dQ}{T}=\int\frac{dH}{T}+\int dZ.$$

Die beiden hierin an der rechten Seite stehenden Integrale sind die auf den vorliegenden Fall bezüglichen Werthe zweier in jener Abhandlung neu eingeführter Gröfsen.

25 *

Fig. 14 Equation 59 in Clausius' 1865–paper (Clausius 1865, p. 353, p. 387) *Source* French National Library–Gallica. Public Domain

At this point all the theoretical elements are present to attribute the physical meaning of entropy to the function S. In fact, by replacing the operator δ with the differential d and dividing both members of Eq. (23) by the absolute temperature T we recognize, in the resulting equation, the mathematical expression of Clausius' Second Law obtained for the case of not-cyclical-reversible processes. The latter has a well-defined physical meaning wherein dS represents the differential of the entropy. Clausius sketched out this argument as follows:

> The quantity S occurring here is what I have called the entropy of the body. In the last equation we can also replace the variation [δ] signs by differential signs [d], since two signs were required to distinguish between the two processes previously considered side by side (the change during a stationary movement and the transition from one stationary movement to another), the former is no longer considered. If we also divide the equation by T, we read[21]:

[21] It is worth noticing that behind Clausius's choice to use the script S for entropy there is no hidden known meaning. In his works Clausius does not mention any reason for this choice. Very likely this latter is due to the fact that when writing his 1865 paper, almost all the other letters had already been used for other quantities. The rumour for which Clausius chose S in honour of Sadi (and why not C

$$\frac{dQ}{T} = dS \qquad [24]$$

(Clausius 1871, p. 460; Authors' translation)

The last step is to apply Eq. (24) to a cyclical process, that is to say to integrate both members of the equation over a closed curve.

Being dS path-independent, its cyclic integral is null. Clausius wrote:

> If we imagine integrating this equation for a cyclical process and take into account that S has the same value at the end of the cyclic process as at the beginning, we get:
>
> $$\int \frac{dQ}{T} = 0 \qquad [25]$$

(Clausius 1871, pp. 460–461; Authors' translation)

The resulting Eq. (25) is the expression of Clausius' Second Law for reversible cyclical process that was obtained in 1854 (Clausius 1854), starting from Carnot's principle, as we have discussed in the foregoing section. Clausius highlighted this result as follows:

> This is the equation I first published in 1854 as an expression of the second law of mechanical heat theory for reversible cycles$^{(1)}$. At that time, I derived it from the principle that heat cannot pass from a colder to a warmer body by itself. Later [Clausius 1850] I derived the same equation in another, very different way, namely from the principle that the work which heat can do in changing the arrangement of a body is proportional to the absolute temperature, in connection with the assumption that the heat actually contained in the body depends only on its temperature and not on the arrangement of its constituents [...]. The foregoing discussion now shows how [...] the second law of mechanical theory of heat can be traced back to general mechanical principles.[22] (Clausius 1871, pp. 460-461; Authors' translation)

---

as Carnot?), at this stage, has no historical-scientific evidences. "Die hierin vorkommende Grösser S ist diejenige, welche ich die Entropie des Körpers genannt habe. In der letzten Gleichung können wir die Variationszeichen auch durch Differentialzeichen ersetzen, da von den beiden früher neben einander befrachteten Vorgängen (der Veränderung während einer stationären Bewegung und dem Übergange aus einer stationären Bewegung in eine andere), zu deren Unterscheidung zwei Zeichen notwendig waren, der erstere jetzt nicht mehr in Betracht kommt. Dividieren wir aufserdem noch die Gleichung durch T, so lautet sie:

$$\frac{dQ}{T} = dS''$$

[22] "Denken wir uns diese Gleichung für einen Kreisprozess integriert, und berücksichtigen dabei, dass S zu Ende des Kreisprozesses denselben Werth hat, wie zu Anfang, so erhalten wir:

$$\int \frac{dQ}{T} = 0$$

Dieses ist die Gleichung, welche ich zuerst im Jahre 1854 als Ausdruck des zweiten Hauptsatzes der mechanischen Warme Theorie für umkehrbare Kreisprozesse veröffentlicht habe (1). Damals habe ich sie aus dem Grundsatze, dass die Wärme nicht von selbst aus einem kälteren in einen

Clausius succeeded in obtaining the same mathematical expression for the Second Law of the *Mechanical Theory of Heat* in two different ways.

- In 1854, he obtained it by starting from Carnot's principle.
- In 1871, he arrived at the same results though a different mathematical model, framed by considering the dynamical properties of the microscopic constituent particles of a gaseous system.

7 Boltzmann–Clausius: "*Zur Priorität Der Auffindung*"

In 1866, four years earlier than Clausius (Clausius 1871), Boltzmann obtained almost the same results (Boltzmann 1866). In 1871, just after Clausius' 1871–paper (Clausius 1871; see also Boltzmann 1866) Boltzmann also wrote a note to the *Wiener Akademie der Wissenschaften* entitled *Zur Priorität der Auffindung [On the Priority of Research] der Beziehung zwischen dem zweiten Hauptsatze der mechanischen Wärmetheorie und dem Prinzip der kleinsten Wirkung* (Boltzmann 1871) wherein he claimed the priority of the results (Boltzmann [1865–1874] 1909). In the incipit we read:

> On 7th November, 1870, in a lecture before the Lower Rhine Society for Nature and Medicine[23] and in a treatise – published in Pogg. Ann. 142, S. 433 – Mr. Clausius argued that the second law of the mechanical theory of heat could be proven through the principles of analytical mechanics and that the relevant calculations are identical to the calculations that are usually carried out when proving the principle of least action. I have already dealt with the same matter in a paper submitted to the Vienna Academy of Sciences on 8th February, 1866, which is included in the 53rd volume of its proceedings and is entitled "On the mechanical meaning of the second law of the theory of heat"; and I think that I can say that the fourth section of my paper, published more than four years earlier, is largely identical – as for as the results are concerned– to the aforementioned treatise of M. Clausius.[24] (Boltzmann [1865–1874] 1909, p. 228; Original quotation marks; Authors' translation)

---

wärmeren Körper übergehen kann, abgeleitet. Später (2) habe ich dieselbe Gleichung noch auf einem anderen, von jenem sehr verschiedenen Wege abgeleitet, nämlich aus dem oben angeführten Gesetze, dass die Arbeit, welche die Wärme bei einer Anordnungsänderung eines Körpers tun kann, der absoluten Temperatur proportional ist, in Verbindung mit der Annahme, dass die in einem Körper wirklich vorhandene Warme nur von seiner Temperatur und nicht von der Anordnung seiner Bestandteile Abhange."

Dabei betrachtete ich den Umstand, dass man auf diese Art zu der schon anderweitig bewiesenen Gleichung gelangen konnte, als eine Hauptstadt jenes Gesetzes. "Die vorstehende Auseinandersetzung zeigt nun, wie jenes Gesetz, und mit ihm der zweite Hauptsatz der mechanischen Wärme Theorie, sich auf allgemeine mechanische Prinzipien zurückfuhren last.".

[23] "niederrheinischen Gesellschaft für Natur und Heilkunde".

[24] "Hr. Clausius hat in einer am 7. November 1870 in der niederrheinischen Gesellschaft für Natur und Heilkunde vorgetragenen und in Pogg. Ann. 142, S. 433 abgedruckten Abhandlung nachgewiesen, dass sich der zweite Hauptsatz der mechanischen Wärmetheorie aus den Prinzipien der analytischen Mechanik beweisen lässt und dass die bezüglichen Rechnungen identisch sind mit den Rechnungen, die man beim Beweisendes Prinzips der kleinsten Wirkung auszuführen pflegt. Ich habe nun dieselbe Sache bereits in einer der Wiener Akademie der Wissenschaften am 8. Februar 1866 überreichten Abhandlung, die in den 53. Band ihrer Sitzungsberichte aufgenommen wurde (1) und den Titel hat "Über die mechanische Bedeutung des zweiten Hauptsatzes der Wärmetheorie",

Boltzmann defended the priority of his results by carrying out a detailed comparative analysis between the mathematical expressions that he obtained in his 1866–paper (Boltzmann 1866) and Clausius' results (Clausius 1871, 1872). He affirmed (Boltzmann [1865–1874], 1909, p. 229) that there were no differences between his approach/results and those of Clausius' other than the notation used in the mathematical expressions. More specifically, he established the correspondence between the mathematical expressions attained by Clausius and his own. In the following excerpt we find the logical key to Boltzmann's analysis:

The only difference concerns the notation used to designate the mechanical quantities. To represent the time that an atom takes to cover its closed trajectory I have used the expression $(t_2 - t_1)$ while Mr. Clausius used the letter *i*. I have indicated the average *vis viva* of an atom as

$$\frac{1}{t_2 - t_1}\int_{t_1}^{t_2} \frac{mc^2}{2} dt \qquad \text{while Clausius} \qquad \frac{m}{2}\overline{v^2} \text{ or } h$$

In order to conform the notations of my expressions with those of Mr. Clausius, it is necessary to place in the former ones:

$$i \text{ instead of } t_2 - t_1$$

$$i\frac{m}{2}\overline{v^2} \text{ or ih instead of } \int_{t_1}^{t_2} \frac{mc^2}{2} dt$$

$$i\overline{X\delta x + Y\delta y + Z\delta Z} \text{ instead of } \int_{t_1}^{t_2} dt\overline{X\delta x + Y\delta y + Z\delta Z}$$

(Boltzmann [1865–1874] 1909, p. 229; Authors' translation[25])

Below we present the final results that Boltzmann attained in this note, that is to say the expression of Clausius' integral in terms of the dynamical properties of the particles constituting the gaseous system. We read:

From this it can be seen that the integral

$$\iint \frac{\delta Q}{T} dk \qquad [26]$$

behandelt; und ich glaube behaupten zu können, dass der vierte Abschnitt meiner mehr als vier Jahre früher publizierten Abhandlung den Resultaten nach größtenteils identisch mit der ebengenannten Abhandlung des Hrn. Clausius ist.".

[25] "Nur in der Bezeichnung der mechanischen Größen herrscht eine Verschiedenheit. Die Zeit, welche ein Atom braucht, um seine geschlossene Bahn zu beschreiben, bezeichne ich mit $t_2 - t_1$, Hr. Clausius bezeichnet sie mit i, die mittlere lebendige Kraft eines Atoms bezeichne ich mit. Um also die Bezeichnung in meinen Formeln mit der des Hrn. Clausius konform zu machen, braucht man in den ersteren bloß zu setzen: [follow mathematical expressions]".

where one integration refers to the variable k, and which Clausius called the entropy of the body, has the value:

$$2\Sigma \log \int_{t_1}^{t_2} \frac{mc^2}{2} dt + C \qquad [27]$$

and it zeroes between equal limits if pressure and counter-pressure are always equal.[26] (Boltzmann [1865–1874], 1909, p. 28; Authors' translation)

We remark that:

- The sentence "[...] it zeroes between equal limits [...]" refers to Clausius' integral [Clausius (1854), p. 499] and means that the limits of integration coincide, i.e., the process is cyclical.
- The successive circumlocution "if pressure and counter-pressure are always equal" as equivalent to say "if the system is in equilibrium"; or, equivalently, "in case of reversible process".

In the end, Boltzmann obtained that:

- Eq. (27) mathematically expressed Clausius' integral (Clausius 1854, p. 499) in terms of kinetic energy of the constituent particles.
- The same Eq. (27) zeros in the case of cyclical-reversible processes.

Consequently, from scientific and historiographical standpoints, he found a proof that the equation

$$\int \frac{dQ}{T} = 0 \qquad (28)$$

holds true; that is to say that he has shown the validity of the Second Law for cyclical-reversible processes "[...] through the principles of analytical mechanics [...]" (Boltzmann [1865–1874] 1909, p. 228; Fig. 15). This was a significant step in the intellectual history of science, as well.

---

[26] "Hieraus ist ersichtlich, dass das Integral.

$$\iint \frac{\delta Q}{T} dk$$

wo sich die eine Integration auf das Zeichen 0 bezieht, und das von Clausius die Entropie des Körpers genannt wurde, den Wert.

$$2\Sigma \log \int_{t_1}^{t_2} \frac{mc^2}{2} dt + C$$

besitzt und zwischen gleichen Grenzen verschwindet, wenn Druck und Gegendruck immer gleich sind.".

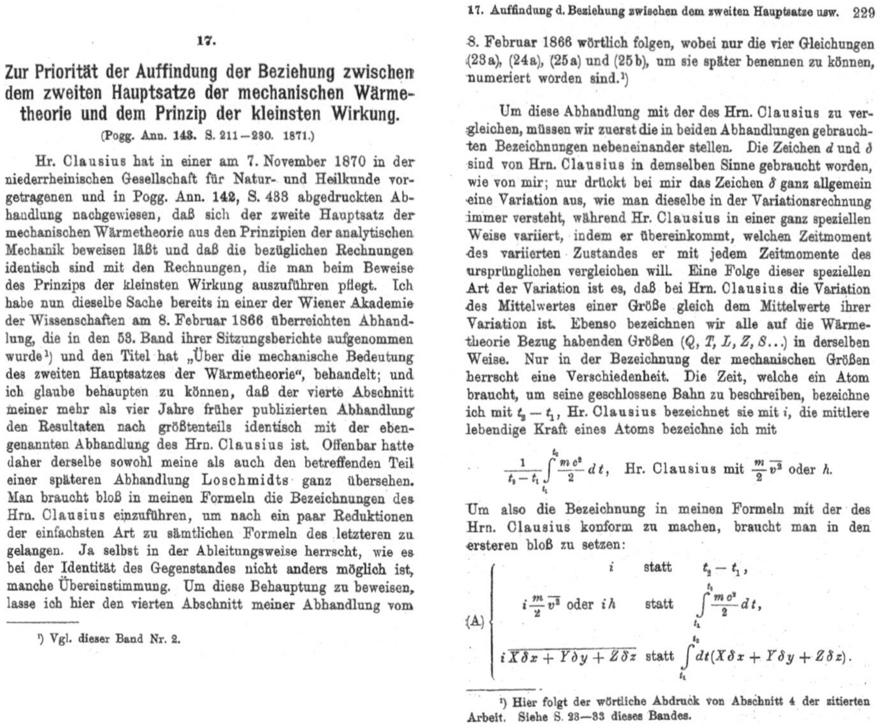

17.

Zur Priorität der Auffindung der Beziehung zwischen dem zweiten Hauptsatze der mechanischen Wärmetheorie und dem Prinzip der kleinsten Wirkung.

(Pogg. Ann. **143**. S. 211–230. 1871.)

Hr. Clausius hat in einer am 7. November 1870 in der niederrheinischen Gesellschaft für Natur- und Heilkunde vorgetragenen und in Pogg. Ann. **142**, S. 433 abgedruckten Abhandlung nachgewiesen, daß sich der zweite Hauptsatz der mechanischen Wärmetheorie aus den Prinzipien der analytischen Mechanik beweisen läßt und daß die bezüglichen Rechnungen identisch sind mit den Rechnungen, die man beim Beweise des Prinzips der kleinsten Wirkung auszuführen pflegt. Ich habe nun dieselbe Sache bereits in einer der Wiener Akademie der Wissenschaften am 8. Februar 1866 überreichten Abhandlung, die in den 53. Band ihrer Sitzungsberichte aufgenommen wurde[1]) und den Titel hat „Über die mechanische Bedeutung des zweiten Hauptsatzes der Wärmetheorie", behandelt; und ich glaube behaupten zu können, daß der vierte Abschnitt meiner mehr als vier Jahre früher publizierten Abhandlung den Resultaten nach größtenteils identisch mit der ebengenannten Abhandlung des Hrn. Clausius ist. Offenbar hatte daher derselbe sowohl meine als auch den betreffenden Teil einer späteren Abhandlung Loschmidts ganz übersehen. Man braucht bloß in meinen Formeln die Bezeichnungen des Hrn. Clausius einzuführen, um nach ein paar Reduktionen der einfachsten Art zu sämtlichen Formeln des letzteren zu gelangen. Ja selbst in der Ableitungsweise herrscht, wie es bei der Identität des Gegenstandes nicht anders möglich ist, manche Übereinstimmung. Um diese Behauptung zu beweisen, lasse ich hier den vierten Abschnitt meiner Abhandlung vom

[1]) Vgl. dieser Band Nr. 2.

17. Auffindung d. Beziehung zwischen dem zweiten Hauptsatze usw. 229

8. Februar 1866 wörtlich folgen, wobei nur die vier Gleichungen (23a), (24a), (25a) und (25b), um sie später benennen zu können, numeriert worden sind.[1])

Um diese Abhandlung mit der des Hrn. Clausius zu vergleichen, müssen wir zuerst die in beiden Abhandlungen gebrauchten Bezeichnungen nebeneinander stellen. Die Zeichen d und δ sind von Hrn. Clausius in demselben Sinne gebraucht worden, wie von mir; nur drückt bei mir das Zeichen δ ganz allgemein eine Variation aus, wie man dieselbe in der Variationsrechnung immer versteht, während Hr. Clausius in einer ganz speziellen Weise variiert, indem er übereinkommt, welchen Zeitmoment des variierten Zustandes er mit jedem Zeitmomente des ursprünglichen vergleichen will. Eine Folge dieser speziellen Art der Variation ist es, daß bei Hrn. Clausius die Variation des Mittelwertes einer Größe gleich dem Mittelwerte ihrer Variation ist. Ebenso bezeichnen wir alle auf die Wärmetheorie Bezug habenden Größen ($Q, T, L, Z, S \ldots$) in derselben Weise. Nur in der Bezeichnung der mechanischen Größen herrscht eine Verschiedenheit. Die Zeit, welche ein Atom braucht, um seine geschlossene Bahn zu beschreiben, bezeichne ich mit $t_2 - t_1$, Hr. Clausius bezeichnet sie mit i, die mittlere lebendige Kraft eines Atoms bezeichne ich mit

$$\frac{1}{t_2 - t_1}\int_{t_1}^{t_2} \frac{mc^2}{2}\,dt, \quad \text{Hr. Clausius mit } \frac{m}{2}\overline{v^2} \text{ oder } h.$$

Um also die Bezeichnung in meinen Formeln mit der des Hrn. Clausius konform zu machen, braucht man in den ersteren bloß zu setzen:

$$\text{(A)}\begin{cases} i & \text{statt} \quad t_2 - t_1, \\ i\frac{m}{2}\overline{v^2} \text{ oder } ih & \text{statt} \quad \int_{t_1}^{t_1} \frac{mc^2}{2}\,dt, \\ i\overline{X\delta x + Y\delta y + Z\delta z} & \text{statt} \quad \int_{t_1}^{t_2} dt(X\delta x + Y\delta y + Z\delta z). \end{cases}$$

[1]) Hier folgt der wörtliche Abdruck von Abschnitt 4 der zitierten Arbeit. Siehe S. 23–33 dieses Bandes.

Fig. 15 Boltzmann's note on the priority of his kinetic model (Boltzmann [1865–1874] 1909, pp. 228–229) *Source* Google Books, Public Domain

We remark that in Boltzmann's 1871–paper (Boltzmann [1865–1874] 1909, 1866) the mathematical steps that lead to showing the equivalence between Clausius'—and—Boltzmann's kinetic equations are not completely explicit. Thus, we have analysed and disclosed all the logical path-ways of the Second Law in its development, from Clausius, based on 1824–Carnot's statement (1854) to kinetic models by Boltzmann (1866) and Clausius (1871) respectively in the following comparative analysis (Table 3):

Table 3 Second Law in its historical critical development

Clausius (1854)	Boltzmann (1866)	Clausius (1871)
$\int \frac{dQ}{T} = 0$	$\Sigma \log \int_{t_1}^{t_2} \frac{mc^2}{2} dt + C = 0$	$A \sum 2mc \log(Ti) = 0$

Based on previous Table 3, below we comparatively analyse the development of the Second Law within Clausius (1854)–Boltzmann (1866)–Clausius(1871) apparatus.

We consider Boltzmann's (1866) following expression:

$$2\Sigma \log \int_{t_1}^{t_2} \frac{mc^2}{2} dt + C \tag{29}$$

We note that sum extended to all gases–atoms; C is a constant. To disclose the meaning of the logarithm we recall Boltzmann's definition of the temperature of an atom (Boltzmann [1865–1874] 1909, p. 14; above Fig. 5, left side) as in the following:

$$T = \frac{1}{t_2 - t_1} \int_{t_1}^{t_2} \frac{mc^2}{2} dt \tag{30}$$

where m is the mass of an atom, c its velocity and range $t_2 - t_I$ is the time that an atom takes to cover its closed trajectory.[27] Hence, Eq. (30) can be rearranged as follows:

$$\int_{t_1}^{t_2} \frac{mc^2}{2} dt = T(t_2 - t_1) \tag{31}$$

By replacing the integral of Eq. (29) with the expression given by Eq. (31), the foregoing Eq. (29) can be written as:

$$2\Sigma \log T(t_2 - t_1) + C \tag{32}$$

Clausius used (1871) the script i (Clausius 1871, pp. 438–439) for the time range that Boltzmann wrote as $(t_2 - t_I)$. Therefore, with this substitution, Eq. (32) becomes:

$$2\Sigma \log Ti + C \tag{33}$$

that, up to a constant, matches with Clausius' kinetic expression given by Eq. (22) (Clausius 1871, pp. 459–460).

Boltzmann's kinetic model, developed in 1866, deals with an expression of Clausius' integral for cyclical processes (1854) (Clausius 1854, p. 499)—in terms of the dynamical properties of the constituent particles—that is, another words,—it is equivalent to Clausius' results obtained in 1871 (Clausius 1871; Boltzmann [1865–1874] 1909) after a similar kinetic model. Finally, in claiming the priority of his model, Boltzmann sustained that the difference with Clausius' work is only formal, in the sense that the two models differ only for the notation used to designate the mechanical quantities.

[27] We let note how Newtonian mechanical paradigm performs the kinetic model (see above).

8 Entropy's Conceptual Frameworks (1865–1901)

In the following, we focus on two aspects of the long process (Pisano et al. 2019, 2021b) of conceptualization of entropy:

1. First, we itemize the mathematical-conceptual steps that lead Clausius to the former definition of entropy, starting form Carnot's principles, and to the mathematical expression whereby it can be calculated. Furthermore, we will outline the thermodynamic theory of J.W. Gibbs who—for the particular case of reversibility—reframed Clausius' equations inside a geometrical model, conceived by combining thermodynamic arguments with the methods of analytical geometry.
2. As second, we make explicit the intellectual path through which Planck could exploit Boltzmann's kinetic model to attain the statistical expression of entropy, known as Boltzmann's equation.

8.1 *Clausius (1850–1872) and Gibbs (1875): A Physical and Mathematical–Geometrical Entropy*

In 1865 Clausius coined the term entropy (Clausius 1865) in order to define the brand-new quantity S whose conceptualization was implicit in 1862 (Clausius 1862). He aimed to settle the complex issue of mathematically writing the Second fundamental theorem of the newborn *Mechanical Theory of heat* (for the special case of non-cyclical processes). In Clausius' words:

> We might call S the *transformational content* of the body, just as we termed the magnitude U its *thermal and ergonal content*. But as I hold it to be better to borrow terms for important magnitudes from the ancient language, so that they may be adopted unchanged in all modern languages, I propose to call the magnitude S the *entropy* of the body, from the Greek word τροπη, *transformation.* I have intentionally formed the word *entropy* so as to be as similar as possible to the word *energy*; for the two magnitudes to be noted but these words are so nearly allied in their physical meanings, that a certain similarity in designation appears to be desirable. (Clausius 1867, p. 357; Clausius' italics)

The former kinetic models delivered by Boltzmann (1866) and Clausius (1871)—that we have considered in the previous part of this work—opened the way to new conceptual frameworks in order to introduce entropy—as a *new* statistical function—inside the theoretical scaffolding of the emerging *Statistical Mechanics*. Actually, we did not find explicit traces of the definition of statistical entropy in the aforementioned Boltzmann and Clausius kinetic works (fl. 1866–1871). By the way, it cannot be. In fact, both models rely on the foundational assumption of cyclical processes that microscopically are represented by considering closed trajectories of the constituent particles; after a certain finite time, each particle comes back to its original position. Hence, in this special case of cyclical transformation the entropy variation zeroes because it is an integral path-independent function, or state function, and its cyclical integral zeroes. Nevertheless, from those former kinetic models stemmed the embryonic idea that a statistical expression, sometimes associated with Clausius' studies on entropy (1862–1865) (Clausius 1865, 1862) could be written as a function of the energy of the microscopic constituent elements of a physical system.

The intellectual scientific process leading Clausius to conceive entropy and mathematically express the *Second Law* took almost fifteen years (1850–1865). Here, we will focus on few main conceptual milestones of this topical research.[28]

In 1854 (Clausius 1854, 1867, pp. 136–207) Clausius obtained a definitive mathematical form of the Second Law for cyclical processes as he recalls in the incipit of the *Sixth Memoir* (Clausius 1867, pp. 215–250):

> In a memoir published in the year 1854 (Clausius 1854; 1867, pp. 136–207), wherein I sought to simplify to some extent the form of the developments I had previously published, I deduced, from my fundamental proposition that *heat cannot, by itself, pass from a colder into a warmer body*, a theorem which is closely allied to, but does not entirely coincide with, the one first deduced by S. Carnot from considerations of a different kind, based upon the older views of the nature of heat. (Clausius 1867, p. 215; Authors' italic)

He referred to the following expression[29]:

$$\int \frac{dQ}{T} \geq 0$$

where the integral is cyclical and the sign of equality holds true in the case of reversible processes.

In 1862 (Clausius 1862, 1867, pp. 215–250] he succeeded in delivering the former extension of his *Second Fundamental Theorem* to non-cyclical processes. More precisely, he attained the following mathematical expression that intertwines not only heat and temperature but also two other classical mathematical entities, that is to say internal heat (H) and the *Disgregation* (Z). In its differential form, this extension of the Second Law (Clausius 1867, p. 246) is:

$$\frac{dQ + dH}{T} + dZ \geq 0$$

where the equality holds true in case of reversibility.

We have already specified the meaning of H, *interior heat*. In the *Sixth Memoir* Clausius presented his *Disgregation*:

> By disgregation is represented [...] the degree of dispersion of the body. Thus, for example, the disgregation of a body is greater in the liquid state than in the solid, and greater in the aeriform than liquid state. (Clausius 1867, p. 226).

This concept clearly stemmed from an "atomistic" conception of matter, the same whereon Clausius based his new hypothesis on the nature of heat. Matter was modelled as made of microscopic particles whose freedom of motion was determined

---

[28] The reader can find more details on Clausius' mathematical model leading to entropy (Pellegrino and Cerruti 2015).

[29] In current notation this integral should be written as $\oint$ (Cfr. Pisano et al. 2019, 2021b).

by the state of aggregation. In the case of the gaseous systems, the constituent particles moved freely inside a vessel: the only constraints were the walls of the vessel and the interaction with the other particles, which, in Clausius' model, reduced to elastic collisions. This is the case in which *digregation* is the highest. Nevertheless, we have to wait until 1865 (Clausius 1865, 1867, pp. 327–365) to finally read the Second Law for non-cyclical processes in terms of entropy. In particular, we find the well-known expression in the *Ninth Memoir* (Clausius 1867, pp. 327–365) where Clausius deploys his reasoning, in the case of reversible processes:

> In fact if [...] the integral $\int \frac{dQ}{T}$ vanishes whenever the body, starting from any initial condition, returns thereto after its passage through any other conditions, then the expression $\frac{dQ}{T}$ under the sign of integration must be the complete differential of a magnitude which depends only on the present existing condition of the body, and not upon the way by which it reached the latter. Denoting this magnitude by S, we can write
>
> $$dS = \frac{dQ}{T}$$
>
> or, if we conceive this equation to be integrated for any reversible process whereby the body can pass from the selected initial condition to its present one, and denote at the same time by S_0 the value which the magnitude S has in that initial condition,
>
> $$S = S_0 + \int \frac{dQ}{T}.$$
>
> (Clausius 1867, pp. 354–355)

We conclude this short excursus on the story of Clausius' entropy historical frameworks by comparing Clausius' two formulations of the Second Law (for reversible processes) delivered in 1862 and 1865 as below presented in the Table 4:

Table 4 Second Law for cyclical and non-cyclical processes in its historical critic development

Clausius 1854 (Clausius 1854)	Clausius 1862 (Boltzmann 1866)	Clausius 1865 (Clausius 1871)
$\int \frac{dQ}{T} = 0$	$\frac{dQ+dH}{T} + dZ = 0$	$dS = \frac{dQ}{T}$

It is possible to obtain the 1865-formulation of the Second Law starting from the 1862 one (Pisano et al. 2019, 2021b): by considering the first true definition of entropy, in terms of *disgregation* and internal heat, we find it in the *Ninth Memoir* (Clausius 1867, p. 355):

$$dS = \frac{dH}{T} + dZ$$

In fact, by replacing the foregoing definition in the 1862-expression we obtain:

$$\frac{dQ}{T} + dS = 0$$

that, (paying attention to the sign) gives Clausius 1865s Law.

At the beginning, we cited the early physical meaning of the *transformational content* of the body that Clausius assigned to the conceptualization of entropy (Clausius 1867, p. 357). It was further developed by Josiah Willard Gibbs' (1839–1903) who embodied Clausius' equations inside a geometrical model by combining these thermodynamic arguments with the methods of analytical geometry. In particular, starting from the assumption that.

> […] the leading thermodynamic properties of a fluid are determined by the relations which exist between the volume, pressure, temperature, energy, and entropy of a given mass of the fluid in a state of thermodynamic equilibrium. (Gibbs 1906, p. 33)

Gibbs established (1875; Gibbs 1906) that the equilibrium state of an isolated thermodynamic system, undergoing a change at constant energy, corresponded to the maximum entropy state. After Gibbs' 1906–paper (Gibbs 1906) entropy becomes a truly *steering variable*, whose geometrical behavior serves to define the spontaneous direction of any thermodynamic transformation: it started playing the role of the fundamental thermodynamic state function (Pisano et al. 2019, 2021b).

8.2 Boltzmann–Planck (1865–1901): A Mathematical–Statistical Entropy

In the previous sections, we have retraced the process of conceptualization of temperature and heat inside the kinetic models developed both by Boltzmann and Clausius and disclosed some aspects of the well-known H-theorem by framing it inside Boltzmann's real interest area: analysing Clausius' Second Law proof for cyclic processes.

Boltzmann's immense scientific production remains one of the most important from the modern age until now. Among the numerous and distinguished hermeneutical works on Boltzmann, we focus on a particular one: Planck's idea to use the conceptual results of Boltzmann's kinetic model to obtain a statistical expression of entropy given by the famous equation $S = k \log W$, known as Boltzmann's equation.

During 1900–1901Max Planck developed (Planck 1901) the Boltzmannian early entropy idea (Clausius 1867, p. 357) and obtained an explicit functional form (Fig. 16) of the newborn statistical entropy, nowadays known as *Boltzmann' equation.* With reference to a system made of N resonators, he delivered the following definition:

> Now we state that the Entropy of the system S_N, is proportional – up to an additive constant – to the logarithm of the probability W for N resonators to have energy equal to U_N, that is to say:
>
> $$S_N = k \log W + \text{const.}$$

(Planck 1901, p. 556, Fig. 16; see also Planck 1914, pp. 140–141; Authors' translation)[30]

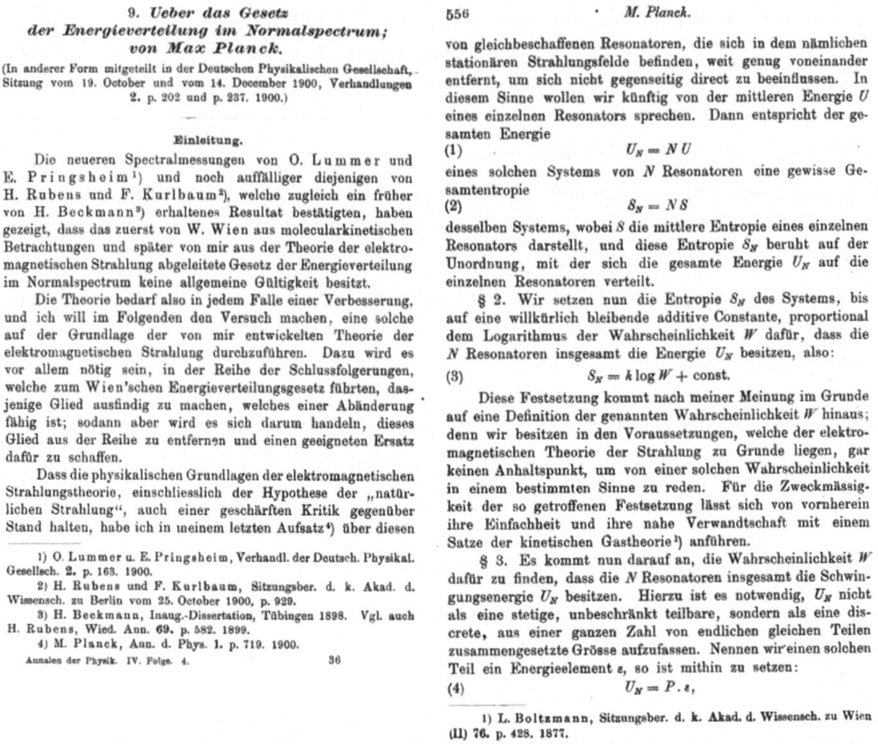

9. *Ueber das Gesetz der Energieverteilung im Normalspectrum;*
von Max Planck.

(In anderer Form mitgeteilt in der Deutschen Physikalischen Gesellschaft, Sitzung vom 19. October und vom 14. December 1900, Verhandlungen 2. p. 202 und p. 237. 1900.)

Einleitung.

Die neueren Spectralmessungen von O. Lummer und E. Pringsheim[1]) und noch auffälliger diejenigen von H. Rubens und F. Kurlbaum[2]), welche zugleich ein früher von H. Beckmann[3]) erhaltenes Resultat bestätigten, haben gezeigt, dass das zuerst von W. Wien aus molecularkinetischen Betrachtungen und später von mir aus der Theorie der elektromagnetischen Strahlung abgeleitete Gesetz der Energieverteilung im Normalspectrum keine allgemeine Gültigkeit besitzt.

Die Theorie bedarf also in jedem Falle einer Verbesserung, und ich will im Folgenden den Versuch machen, eine solche auf der Grundlage der von mir entwickelten Theorie der elektromagnetischen Strahlung durchzuführen. Dazu wird es vor allem nötig sein, in der Reihe der Schlussfolgerungen, welche zum Wien'schen Energieverteilungsgesetz führten, dasjenige Glied ausfindig zu machen, welches einer Abänderung fähig ist; sodann aber wird es sich darum handeln, dieses Glied aus der Reihe zu entfernen und einen geeigneten Ersatz dafür zu schaffen.

Dass die physikalischen Grundlagen der elektromagnetischen Strahlungstheorie, einschliesslich der Hypothese der „natürlichen Strahlung", auch einer geschärften Kritik gegenüber Stand halten, habe ich in meinem letzten Aufsatz[4]) über diesen

1) O. Lummer u. E. Pringsheim, Verhandl. der Deutsch. Physikal. Gesellsch. 2. p. 163. 1900.

2) H. Rubens und F. Kurlbaum, Sitzungsber. d. k. Akad. d. Wissensch. zu Berlin vom 25. October 1900, p. 929.

3) H. Beckmann, Inaug.-Dissertation, Tübingen 1898. Vgl. auch H. Rubens, Wied. Ann. 69. p. 582. 1899.

4) M. Planck, Ann. d. Phys. 1. p. 719. 1900.

Annalen der Physik. IV. Folge. 4. 36

556 *M. Planck.*

von gleichbeschaffenen Resonatoren, die sich in dem nämlichen stationären Strahlungsfelde befinden, weit genug voneinander entfernt, um sich nicht gegenseitig direct zu beeinflussen. In diesem Sinne wollen wir künftig von der mittleren Energie U eines einzelnen Resonators sprechen. Dann entspricht der gesamten Energie

(1) $$U_N = N U$$

eines solchen Systems von N Resonatoren eine gewisse Gesamtentropie

(2) $$S_N = N S$$

desselben Systems, wobei S die mittlere Entropie eines einzelnen Resonators darstellt, und diese Entropie S_N beruht auf der Unordnung, mit der sich die gesamte Energie U_N auf die einzelnen Resonatoren verteilt.

§ 2. Wir setzen nun die Entropie S_N des Systems, bis auf eine willkürlich bleibende additive Constante, proportional dem Logarithmus der Wahrscheinlichkeit W dafür, dass die N Resonatoren insgesamt die Energie U_N besitzen, also:

(3) $$S_N = k \log W + \text{const.}$$

Diese Festsetzung kommt nach meiner Meinung im Grunde auf eine Definition der genannten Wahrscheinlichkeit W hinaus; denn wir besitzen in den Voraussetzungen, welche der elektromagnetischen Theorie der Strahlung zu Grunde liegen, gar keinen Anhaltspunkt, um von einer solchen Wahrscheinlichkeit in einem bestimmten Sinne zu reden. Für die Zweckmässigkeit der so getroffenen Festsetzung lässt sich von vornherein ihre Einfachheit und ihre nahe Verwandtschaft mit einem Satze der kinetischen Gastheorie[1]) anführen.

§ 3. Es kommt nun darauf an, die Wahrscheinlichkeit W dafür zu finden, dass die N Resonatoren insgesamt die Schwingungsenergie U_N besitzen. Hierzu ist es notwendig, U_N nicht als eine stetige, unbeschränkt teilbare, sondern als eine discrete, aus einer ganzen Zahl von endlichen gleichen Teilen zusammengesetzte Grösse aufzufassen. Nennen wir einen solchen Teil ein Energieelement ε, so ist mithin zu setzen:

(4) $$U_N = P . \varepsilon,$$

1) L. Boltzmann, Sitzungsber. d. k. Akad. d. Wissensch. zu Wien (II) 76. p. 428. 1877.

Fig. 16 Boltzmann–Planck statistical entropy (Planck 1901, p. 553; p. 556) *Source* Google Books, Public Domain

In this section we made explicit Planck's reasoning that lead to this equation and its cognitive relationship with Boltzmann's kinetic theory by considering on the book *The theory of Heat radiation* (Planck 1917).

The foundational assumption of Planck's reasoning is that the entropy of a system—that is in a certain energy state—is a function expressing the probability that the system is in that energy state. More precisely, we read:

> Hence, we make the following proposition the foundation of our further discussion: *The entropy of a physical system in a definite state depends solely on the probability of this state.* (Planck 1917, p. 139, Author's italic)

The foregoing statement can be mathematically expressed as:

$$S = f(W)$$

[30] "Wir setzen nun die Entropie S_N des Systems, bis auf eine willkürlich bleibende additive Constante, proportional dem Logarithmus der Wahrscheinlichkeit W dafür, dass die N Resonatoren insgesamt die Energie U_N besitzen: $S_N = k \log W + \text{const.}$".

where $f(W)$ represents a general function of the argument W, that is the probability to find the system in a certain energy state. In the case of two independent systems, the probability to find the system—made of these two subsystems, in certain state is given by the product of the probabilities of the two systems, that is to say: $W = W_1 W_2$. More precisely, Planck argued that:

> […] the first system as any body whatever on the earth and […] the second system as a cavity containing radiation on Sirius, [such that] the probability that the terrestrial body be in a certain state 1 and that simultaneously the radiation in the cavity in a definite state 2 is
>
> $$W = W_1 W_2$$
>
> (Planck 1917, p. 140)

Let W_1 and W_2 be the respective probabilities of these systems. The entropies are respectively then given as:

$$S_1 = f(W_1)$$
$$S_2 = f(W_2).$$

> But, according to the second principle of thermodynamics, the total entropy of the two systems, which are independent […] of each other is $S = S_1 + S_2$ (Planck 1917, p. 140).

Consequently, we have:

$$f(W_1 W_2) = f(W_1) + f(W_2)$$

In order to determine the function f from the foregoing functional equation, Planck differentiates both sides of the equation with respect to W_1 and W_2, successively. Namely, by differentiating with respect to W_1, being W_2 constant, he obtains:

$$W_2 \dot{f}(W_1 W_2) = \dot{f}(W_1)$$

Then, by differentiating the foregoing result with respect to W_2, he obtained:

$$\dot{f}(W_1 W_2) + W_1 W_2 \ddot{f}(W_1 W_2) = 0$$

that is to say:

$$\dot{f}(W) + W \ddot{f}(W) = 0$$

The general solution of this second order differential equation was:

$$f(W) = k \log W + \text{ const.}$$

that, from the initial proposition $S = f(W)$, becomes:

$$S = k \log W + \text{const.}$$

This expression, intertwining entropy and probability, contains two constants, whose meaning and values are clarified as follows:

> The universal constant of integration k is the same for a terrestrial as for a cosmic system, and its value, having been determined for the former, will remain valid for the latter. The second additive constant of integration may, without any restriction as regards generality, be included as a constant multiplier in the quantity W which here has not yet been completely defined, so that the equation reduces to
>
> $$S = k \log W.$$

(Planck 1917, p. 141).

This is the currently well-known as Boltzmann's equation (Planck 1901, p. 556; Cfr. Planck 1914). But what is the ratio that links this result to Boltzmann's kinetic theory? Planck synthetically answers as follows:

> The logarithmic connection between entropy and probability was first stated by L. Boltzmann in his kinetic theory of gases. Nevertheless, our equation [i.e. $S = k \log W$], differs in its meaning from the corresponding one of Boltzmann in two essential points. Firstly, Boltzmann's equation lacks the factor k which is due to the fact that Boltzmann always used gram-molecules, not the molecules themselves, in his calculations. Secondly, and this is of greater consequence, Boltzmann leaves an additive constant undetermined in the entropy S as is done in the whole of classical thermodynamics, and accordingly there is a constant factor of proportionality, which remains undetermined in the value of the probability W. (Planck 1917, p. 141)

9 Concluding Remarks

In this paper, we have analysed three different approaches to the *Theory of Heat*, all three developed in the second half of the nineteenth century. We did it by analysing, interpreting and commenting on historical documents authored by Clausius and Boltzmann. In addition, we have focused on the process of the conceptualization of entropy, both in the *Mechanical Theory of Heat* and *Statistical Mechanics*, descending from Boltzmann-Clausius kinetic models.

To begin, we have considered Clausius' *Mechanical Theory of Heat*—developed between 1850 and 1865—where he built up a mathematical model starting from Carnot's principles. In particular, we have focused on how Clausius has settled the critical issue of the nature of heat, and we disclosed the intellectual process that allowed Clausius to open the way toward a conception of heat that is entirely associated with the motion of the microscopic constituent particles of matter. Clausius' implemented a new physical-mathematical scaffolding by modelling the microscopic properties of matter. More precisely, he conceptualized the nature of heat by splitting it into components: the first is the heat actually existing in the body that corresponds to what in German was named *latente Wärme* (H) whilst the second, named *freie Wärme* (Q), is the heat exchanged with the outside. We have made explicit how this conceptual duality was crucial in formulating Clausius' Laws of the Mechanical Theory of heat.

Then we detailed two approaches to the theory of heat, alternatives with respect to Clausius' Mechanical Theory of Heat that, as we have recalled, was implemented by starting from Carnot's principles. These alternatives were developed—by Boltzmann (1866) and Clausius (1871)—with reference to gaseous systems and were based on the idea that Laws of the Mechanical Theory of Heat could be obtained by applying the principles of mechanics to the microscopic particles, constituting the gaseous system. Boltzmann and Clausius independently developed two different models. After having detailed both, we have compared them with the support of a note that Boltzmann published, in 1871, to claim the priority of his own model. Based on an original epistemic reading of this note, we have made explicit that these two former kinetic models differ only for the notation used to designate the mechanical quantities. Both succeeded in obtaining the Second Law of the Mechanical theory of heat for reversible-cyclical process by applying the principle of analytical mechanics to the constituent particles of a gaseous system.

We have retraced the main steps of the long intellectual story of entropy. As first, we have analysed the main results of Clausius' decennial search for a general mathematical expression of the Second Fundamental Theorem. In particular, we focused on the former mathematical formulation of the Second Law in terms of heat, temperature, internal heat (H) and *Disgregation* (Z). Then we have commented on the 1865 formulation in terms of entropy. We have concluded by comparing the two expressions of the Second Law (1862 and 1865), hence making explicit the truly former definition of entropy in terms of *Disgregation* and internal heat. Secondly, we have deployed the conceptual path that lead to attaining the former statistical expression of entropy. More precisely, we have made explicit the reasoning that allowed Max Planck to exploit Boltzmann's kinetic achievements by obtaining a mathematical-statistical function, whose physical meaning is analogically related to the notion of entropy.

In conclusion, in this paper we dealt with never before examined aspects of the historical, fundamental—and—intellectual mechanisms of developments in the *Theory of heat*, in general, and of the concept of entropy, in particular. Briefly:

- Firstly, an original reading of Clausius *Memoirs* (Clausius 1867) (mainly the First, Forth, Sixth and Ninth) allowed us to analyse and discover previously overlooked mechanisms between the new conceptual idea of heat—intended as associated to the motion of the *ultimate particles*—and the definition of early-mathematical entities, such as *latente Wärme* and *innere Arbeit*. The latter mathematically allowed the expression of the *Principle of equivalence of heat and work* and the *First Law of Thermodynamics*. Furthermore, an historical–epistemological analysis of Clausius' *Sixth Memoir* (Clausius 1867), including the rarely commented Clausius' paper *Zur Geschichte der Mechanischen Wärmetheorie* (1872) (Clausius 1872), achieved to key-elements—till now unknown—that allowed sharpening the mathematical and conceptual role of entropy. We disclosed its interconnection with Clausius' atomistic conception of matter, through the antiquated entity framework named *Disgregation* (Z).

- Secondly, we analyzed the historical foundations of the former Boltzmann-Clausius kinetic models and, in particular:
  (a) 1866–1871. We disclosed the conceptual ring intertwining Boltzmann' kinetic model mechanism (1866) and Boltzmann's H-theorem (1871) framework;
  (b) 1871. We made explicit the intellectual path that Clausius followed to implement his kinetic model (1871);
  (c) 1866–1871. We compared Boltzmann and Clausius approaches supported by the historical-conceptual analysis of the rarely commented Boltzmann's note (Boltzmann 1871) *Zur Priorität der Auffindung der Beziehung zwischen dem zweiten Hauptsatze der mechanischen Wärmetheorie und dem Prinzip der kleinsten Wirkung*, wherein he claimed the priority of his kinetic model (1866) with respect to Clausius' one (1871).
  (d) 1901–1917. Starting from this intellectual framework, we also made explicit the mechanisms of physical–mathematical reasoning, which allowed Planck to deliver the statistical expression of entropy: that is to say the famous Boltzmann' equation

$$S = k \log W + \text{const.}$$

(Planck 1901, p. 556; Cfr. Planck 1914)

Acknowledgements We thank the anonymous reviewers for their careful reading of our manuscript and their many insightful comments and suggestions. We also express gratitude to Dr. Julie Robarts (University of Melbourne, Australia) for her observations on English language; eventual remaining English mistakes are our own.

References

Boltzmann L ([1865–1874] 1909) Wissenschaftliche Abhandlungen von Ludwig Boltzmann. I Band. Barth, Leipzig.

Boltzmann L (1866) Über die mechanische Bedeutung des zweiten Hauptsatzes der Wärmetheorie. Wiener Berichte 53:195–220.

Boltzmann L (1871) Zur Priorität der Auffindung der Beziehung zwischen dem zweiten Hauptsatze der mechanischen Wärmetheorie und dem Prinzip der kleinsten Wirkung. Annalen der Physik und Chemie 143:211–230.

Boltzmann L (1872) Weitere Studien über das Wärmegleichgewicht unter Gasmolekülen. Sitzungsberichte Akademie der Wissenschaften 66:275–370.

Boltzmann L (1877) Über die Beziehung zwischen dem zweiten Hauptsatze der mechanischen Wärmetheorie und der Wahrscheinlichkeitsrechnung resp. den Sätzen über das Wärmegleichgewicht. Wiener Berichte 76:373–435.

Boltzmann L (2003) Further Studies on the Thermal Equilibrium of Gas Molecules. The Kinetic Theory of Gases. History of Modern Physical Sciences 1:262–349.

Carathéodory C (1909) Untersuchungen über die Grundlagen der Thermodynamik. Math Ann 67:355–386.

Carnot S (1824) Réflexions sur la puissance motrice du feu et sur les machines propres à développer cette puissance. Bachelier, Paris.

Carnot S ([1824] 1978) Réflexions sur la puissance motrice du feu sur les machinés propre à développer cette puissance, édition critique par Fox Robert. Vrin J, Paris. [*Ivi*, Recherche d'une formule propre à représenter la puissance motrice de la vapeur d'eau, pp. 223–234].

Carnot S (1878a) Réflexions sur la puissance motrice du feu sur les machinés propre à développer cette puissance. [Archives of the Académie des Science–Institut de France, Paris].
Carnot S (1878b) Réflexions sur la puissance motrice du feu sur les machinés propre à développer cette puissance. Gauthier–Villars, Paris.
Carnot S (1986) Reflexions on the motive power of fire: a critical edition with the surviving scientific manuscripts translated and edited by Robert Fox. The Manchester University Press, Manchester.
Carnot S (s. d.) Notes sur les mathématiques, la physique et autres sujets. In: Carnot 1878b, pp. 89–102. The manuscript is conserved in Carnot 1878a.
Cercignani C (1998) Ludwig Boltzmann. The Man who Trusted Atoms. Oxford University Press, Oxford.
Clapeyron EBP (1834) Mémoire sur la puissance motrice du feu. Journal de l'École Royal Polytechnique XXIII/XIV:153–190. [Collections École polytechnique Z 5 (1834)].
Clausius R (1850) Über die bewegende Kraft der Wärme und die Gesetze, welche sich daraus für die Wärmelehre selbst ableiten lassen. Annalen der Physik und Chemie 79:368–397, 500–524. [English trans: *Id.*, On the Motive Power of Heat and on the Laws which can be deduced from it for the theory of heat itself. In: Mendoza E (1960) (ed) Reflections on the motive power of fire and other papers on the second law of thermodynamics by E Clapeyron and R Clausius. Dover, New York, pp 73–74, pp. 109–152].
Clausius R (1854) Über eine veränderte Form des zweiten Hauptsatzes der mechanischen Wärmetheorie. Annalen der Physik und Chemie 169/12:481–506.
Clausius R (1857a) Über die Art der Bewegung welche wir Wärme nennen. Annalen der Physik und Chemie 100/3:353–380.
Clausius R (1857b) The nature of the motion we call heat. Kinetic Theory. Selected Readings in Physics. II. Pergamon Press, Oxford, pp. 111–134.
Clausius R (1862) Über die Anwendung des Satzes von der Äquivalenz der Verwandlungen auf die innere Arbeit. Annalen der Physik und Chemie 116:73–112.
Clausius R (1865) Über verschiedene für die Anwendung bequeme Formen der Hauptgleichungen der mechanischen Wärmetheorie. Annalen der Physik und Chemie 75/7:353–400.
Clausius R (1867) The mechanical theory of heat with its applications to the steam-engine and the physical properties of bodies. Van Voorst, London.
Clausius R (1871) Über die Zurückführung des zweiten Hauptsatzes der mechanischen Wärmetheorie auf allgemeine mechanische Prinzipien. Annalen der Physik und Chemie 142:433–461.
Clausius R (1872) Zur Geschichte der Mechanischen Wärmetheorie. Annalen der Physik und Chemie 145:132–146.
D'Agostino S (2014) Maxwell's Dynamical Philosophy: An Early Solution to the Problem of a Discrepancy between Particles and Waves. Advances in Historical Studies 3:123–129.
Eddington AS ([1928] 1948) The Nature of the Physical World. Cambridge University Press, Cambridge.
Faraday M (1839–1855) Experimental Researches in Electricity. 3 Vols. Taylor, London.
Frigg R (2008) A field Guide to Recent Work on the Foundation of Statistical Mechanics. Ashgate, London.
Gabbey AW, Herivel JW (1966) Un manuscrit inédit de Sadi Carnot. Revue d'histoire des sciences et de leurs applications 19:151–166.
Gibbs JW (1892) Thermodynamischen Studien. Wilhelm Engelmann, Leipzig.
Gibbs JW (1899) Équilibre des Système Chimiques. Carré et Naud, Paris.
Gibbs JW (1902) Elementary Principles in Statistical Mechanics Developed with Especial Reference to the Rational Foundation of Thermodynamics. Charles Scribner's Sons, New York.
Gibbs JW (1906) On the Equilibrium of Heterogeneous Substances. The Scientific Papers of J. Willard Gibbs. Longmans, London, pp. 55–353.
Gillispie CC, Pisano R (2014) Lazare and Sadi Carnot. A Scientific and Filial Relationship. 2nd edn. Springer, Dordrecht.

Joule JP (1843) On the calorific effects of magneto-electricity, and on the mechanical value of heat. The London, Edinburgh, and Dublin Philosophical Magazine and Journal of Science. Conducted by Sir David Brewster, K.H. LL.D. F.R.S.L.&E. &c. Richard Taylor, F.L.S. G.S. Astr.S. Nat.H.Mosc. &c. Richard Phillips, F.R.S.L.& E. F.G.S. &c. Robert Kane, M.D. M.R.I.A., Richard and John E. Taylor, Red Lion Court, Fleet Street, Printers and Publishers to the University of London, London, New and United Series of the Philosophical Magazine, Annals of Philosophy, and Journal of Science XXIII: 263–276, 347–355, 435–443.

Joule JP (1845) On the changes of temperature produced by the rarefaction and condensation of air. By J. P. Joule, Esq. The London, Edinburgh, and Dublin Philosophical Magazine and Journal of Science [ser. 3, May 1845] 26/174:369–383. [*Id.*, (1884) The scientific papers of James Prescott Joule. Cambridge University Press, New York, pp.172–189].

Joule JP ([1849] 1850) On the mechanical equivalent of heat. By James Prescott Joule, F.C.S., Sec. Lit. and Phil. Society, Manchester, Cor. Mem. R.A., Turin, &c. Communicated by Michael Faraday, D.C.L., F.R.S., Foreign Associate of the Academy of Sciences, Paris, &c. &c. &c. Received June 6,-Read June 21, 1849. Philosophical Transactions of the Royal Society of London, 140:61–82.

Joule JP (1884) The scientific papers of James Prescott Joule. The Physical Society of London, London.

Joule JP (1887) Joint scientific papers of James Prescott Joule. The Physical Society of London. London.

Krönig A (1856) Grundzüge einer Theorie der Gase. Annalen der Physik und Chemie 175/10:315-322.

Lagrange JL (1788) Mécanique analytique. Desaint, Paris [See also: Lagrange JL (1973) Œuvres de Lagrange. Seconde Edition. Courcier (ed.), I–XIV vols. (in X) Paris: Gauthier–Villars].

Lavoisier AL (1789 [1937]) Traité élémentaire de Chimie. Gauthier–Villars, Paris.

Lavoisier AL (1862–1893) Œuvres de Lavoisier publiées par le soin de Son Excellence le Ministre de l'instruction publique et des cultes, 6 vols. Imprimerie Impériale, Paris.

Lavoisier AL, Laplace PS ([1780] 1784) Mémoires sur la Chaleur. In: Histoire de l'Académie Royale des Sciences. (First reading on 18th June 1783). Imprimerie Royale, Paris, pp. 355–408.

Maxwell JC (1864) [abstract] A dynamical theory of the Electromagnetic field [The paper was received in October 27, 1864, and read in December 8, 1864], The Proceedings of the Royal Society of London. From November 19, 1863, to December, 1861, inclusive. XIII:531–536.

Maxwell JC (1865) A Dynamical theory of the Electromagnetic field. Philosophical Transactions of The Royal Society of London CLV:459–512.

Maxwell JC (1871) Theory of heat. Longmann Green, Roberts & Green, London.

Mayer RJ (1842a) Bemerkungen über die Kräfte der unbelebten Natur. Weyrauch, Die Mechanik Warme, Stuttgart.

Mayer RJ (1842b) Bemerkungen über die Kräfte der unbelebten Natur. Annalen der Chemie und Pharmazie 42:233–240.

Newton I (1687) Philosophiae Naturalis Principia Mathematica. Autore IS. Newton, Trin. Coll. Cantab. Soc. Matheseos Professore Lucasiano, & Societatis Regalis Sodali. Imprimatur. S. Pepys, Reg. Soc. Præses. Julii 5. 1686. Londini, Jussu Societatis Regiæ ac Typis Josephi Streater. Prostat apud plures Bibliopolas. Anno MDCLXXXVII.

Newton I ([1713] 1729) The Mathematical principles of natural philosophy. Translated by Motte Andrew. Motte B–Middle Temple-Gate, London.

Ostwald W (1892) Klassiker der exakten Wissenschaften. Engelmann 37, Lipsia.

Pellegrino EM (2014) Entropy from Clausius to Kolmogorov: historical evolution of an open concept. In: Tucci P (ed). SISFA 2014. Pavia University Press, Pavia, pp. 139–147.

Pellegrino EM (2020) Lazare and Sadi Carnot: A Scientific and Filial Relationship by Charles Coulston Gillispie and Raffaele Pisano. The Mathematical Intelligencer 42/2:91–100.

Pellegrino EM, Cerruti L (2015) Clausius' disgregation: a conceptual relic that sheds light on the second law. Entropy 17/7:4500–4518.

Pellegrino EM Cerruti L (2016) From steam engines to chemical reactions: Gibbs' contribution to the extension of the second law. Entropy 18/5–162:1–27.

Pisano R (2010) On principles in Sadi Carnot's thermodynamics (1824). Epistemological Reflections. Almagest 2/1:128–179.

Pisano R (2013) History Reflections on Physics Mathematics Relationship in Electromagnetic Theory. In: Barbin E, Pisano R (eds). The Dialectic Relation between Physics and Mathematics in the 19th Century. Springer, Dordrecht, pp. 31–58.

Pisano R (2017) A Development of the Principle of Virtual Laws and its Framework in Lazare Carnot's Mechanics as Manifest Relationship between Physics and Mathematics. Transversal–International Journal for Historiography of science 2:166–203.

Pisano R (2021) An Intellectual History of Entropy as a Physics–Mathematics Dynamic Relationship. Speeches as Key Note Speaker at Science and Technologies Chemistry Department (October 29th), University of Palermo, Italy.

Pisano R, Bussotti P (2016) A Newtonian Tale Details on Notes and Proofs in Geneva Edition of Newton's Principia. Bulletin BSHM-Journal of the British Society for the History of Mathematics 31/3:160–178.

Pisano R, Bussotti P (2017a) On the Conceptualization of Force in Johannes Kep-ler's Corpus: an Interplay between Physics/Mathematics and Metaphysics. In: Pisano R, Agassi J, Drozdova D (eds). Hypothesis and Perspective within History and Philosophy of Science. Homage to Alexandre Koyré 1892–1964. Dordrecht, Springer, pp. 295–346.

Pisano R, Bussotti P (2017b) Newton's Principia Geneva edition: the action-and-reaction law. Historical and Nature of Science reflexions. Pavia University Press, Pavia, pp. 269–276.

Pisano R, Bussotti P (2022) Conceptual Frameworks on the Relationship between Physics–Mathematics in the Newton *Principia Geneva* Edition (1822). Foundations of Sciences (Springer) 27/3:1127–1182 [Recently: *Id.*, The Physics–Mathematics Interplay in Newton Principia Geneva Edition (1822): A New Case Study of the Three-body Problem, Proposition LXVI–Theorem XXVI. Submitted to Philosophical Transactions A].

Pisano R, Capecchi D (2013) Conceptual and Mathematical Structures of Mechanical Science in the Western Civilization around 18th Century. Almagest 4/2:86–121.

Pisano R, Marmottini D (2017) Nature–of–Science Teaching: Notes on the Lagrangian Methods in Maxwell's electromagnetic Theory. XXXVI SISFA Proceedings. Pavia University, Pavia, pp. 263–268.

Pisano R, Pellegrino EM, Anakkar A, Nagles M (2019) Thermodynamic foundations of physical chemistry: reversible processes and thermal equilibrium into the history. Foundation of Chemistry 21/3:297–323.

Pisano R, Vincent P, Mauricio P (2020) Introduction. Joule's Bicentenary: History of Science, Foundations and Nature of Science. Joule's Bicentenary Special Issue. Foundations of Science 26:531–551.

Pisano R, Coopersmith J, Peake M (2021a) Essay on Machines in General (1786) Text, Translations and Commentaries. Lazare Carnot's Mechanics. Volume 1. Springer, Dordrecht.

Pisano R, Pellegrino EM, Anakkar A, Nagles M (2021b) Conceptual polymorphism of entropy into the history: extensions of the second law of thermodynamics towards statistical physics and chemistry during nineteenth–twentieth centuries. Foundation of Chemistry 23/3:337–378.

Pisano R, Vincent P, Dolenc K, Ploj-Virtič M (2021c) Historical Foundations of Physics & Applied Technology as Dynamic Frameworks in Pre-Service STEM. Foundations of Science 26:727–756.

Pisano R, Marmottini D, Durlo A (2024) Lagrange's Method and Langrangian's Mechanism in Maxwell's A Treatise on Electricity and Magnetism (1873). In: Pisano R (ed). A History of Physics: Phenomena, Ideas & Mechanisms. Essays in Honor of Salvo D'Agostino. Springer, Dordrecht, pp. 183–261.

Planck M ([1897] 1903) Vorlesungen¨uber Thermodynamik. Walter de Gruyter, Berlin [Reprinted in English translation as Treatise on Thermodynamics. Dover, New York].

Planck M (1901) Ueber das Gesetz der Energieverteilung im Normalspectrum. Annalen der Physik, 309/3:553–563.

Planck M (1914) The Theory of Heat Radiation. Translation by Masius M. of the second German edition. Blakiston's Son & Co., Philadelphia.

Planck M (1917) Treatise on Thermodynamics. Translated with the Author's sanction by A. Ogg. Dover, New York.

Thompson B (1798) An inquiry concerning the source of the heat which is excited by friction. Philosophical Transactions of the Royal Society of London 88:80–102.

Thomson W (1848–1849a) An account of Carnot's theory of the motive power of heat with numerical results deduced from Regnault's experiments on steam. Transactions of the Edinburgh Royal Society XVI/5:541–574.

Thomson W (1848–1849b) On an absolute thermometric scale founded on Carnot's theory of the motive power of heat, and calculated from Regnault's observations. Proceedings of The Cambridge Philosophical Society 1/5:66–71.

Thomson W (1851a) On an absolute thermometric scale founded on Carnot's theory of the motive power of heat, and calculated from Regnault's observations. Mathematical and physical papers, I. Cambridge University Press, Cambridge, pp. 100–106.

Thomson W (1851b) On the dynamical theory of heat, with numerical results deduced from Mr Joule's equivalent of a thermal unit, and M. Regnault's observations on steam. Mathematical and physical papers, I, Cambridge University Press, Cambridge, pp. 175–183.

Thomson W (1882–1911) An account of Carnot's theory of the motive power of heat with numerical results deduced from Regnault's experiments on steam. Mathematical and physical papers by Sir William Thomson, I, Cambridge University Press, Cambridge, pp. 113–155.

Wien W (1894) Temperatur und Entropie der Strahlung. Annalen der Physik 52:132–165.

Chapter 15
Going Beyond the World of Atoms: Heinrich Hertz and the Electromagnetic Worldview

Jan Potters

Abstract In this paper, I argue that, contrary to what many historians have claimed, the hidden-masses mechanics introduced by Heinrich Hertz in his *Principles* did inspire reflections within fundamental physics concerning the role played by the ether. This occurred in particular in the context of the electromagnetic worldview, a program for the reduction of the whole of physics to electromagnetism that emerged a few years after *Principles* was published. This worldview was, however, short-lived, as it was soon more or less replaced by the theory of relativity. In the second part of the paper, I will discuss how Albert Einstein and Max Planck formulated the relativistic response to the electromagnetic worldview. This will then lead me to suggest that, in a sense, this relativistic response, even though it differed profoundly from the electromagnetic worldview, could equally well be characterized as Hertzian.

Keywords Electromagnetic worldview · Heinrich Hertz · Max Abraham · Max Planck · Theory of special relativity

1 Introduction

Heinrich Hertz's *Die Prinzipien der Mechanik* (Hertz 1894a) (published posthumously in 1894) famously consists of an introduction which provides philosophical reflections on the epistemology of physical theories, in particular the theory of mechanics. The rest of the book, more physical in nature, offers Hertz's own reconceptualization of that theory, and in particular of its central notion 'force', in terms of moving hidden masses rigidly connected to observable material masses.

Many historians (e.g., D'Agostino 1998; Lützen 2005 and Eisenthal 2018, 2021) have argued that these two parts should be seen as closely connected. At the same

J. Potters (✉)
Lichtenberg Group for History and Philosophy of Physics, Universität Bonn, Germany & Postdoctoral Fellow funded by Research Fund, Flanders (FWO), University of Antwerp, Antwerp, Belgium
e-mail: jan.potters@uantwerpen.be

R. Pisano (ed.), *A History of Physics: Phenomena, Ideas and Mechanisms*,
History of Mechanism and Machine Science 42,
https://doi.org/10.1007/978-3-031-26174-9_15

time, they have also pointed out that the two parts have not enjoyed an equal reception. The first part has inspired many philosophical reflections, both by philosophers and scientists, concerning the notion of representation (Eisenthal 2018, p. 45; Mulligan 1998; Majer 1998; Grasshoff 1998; Lützen 2005, Chap. 27 for overviews). At the same time, they all claim that the second part has not really been picked up by the physics community. D'Agostino, for example, has stated that "though widely read and commented upon, Hertz's proposal for a new form for the foundational axioms of mechanics did not find favor with his fellow physicists at the end of the century" (D'Agostino 1998, p. 96). Similarly, Eisenthal has claimed that.

> [...] even as it was admired for its elegance and scope Hertz's contemporaries could not find in it the kind of advances that they had hoped for. Indeed, there was a general sense of confusion regarding what *Principles* was supposed to have achieved. (Eisenthal 2018, p. 45)

And Joseph F. Mulligan has argued that.

> [...] when Hertz's *Prinzipien der Mechanik* was published in 1894, it was severely criticized by eminent theoretical physicists like Mach, Boltzmann, Lorentz, and Fitzgerald for its impracticality and its attempt to replace forces by the motions of even more obscure entities. (Mulligan 2001, p. 157)

That Hertz's *Principles* was seen as introducing obscurities is ironic, since as Hertz (1894a, p. 10) himself pointed out, his aim was precisely to remove a particular obscurity concerning how the three laws of Newtonian mechanics conceptualize the notion of force. According to Eisenthal (2018, p. 47), Hertz's *Principles* was seen as only introducing more obscurities because it was approached with the wrong expectations:

> [M]ost readers of *Principles*—both historical and contemporary—have regarded it as an attempt to lay the groundwork for some future ether mechanism, the details of which could be filled in later. But the inclination towards interpreting *Principles* this way has contributed to the dissatisfaction amongst Hertz's readers, for it ties the value of this project to the prospects of filling in these details. (Eisenthal 2018, p. 47)

While it seems that D'Agostino, Eisenthal and Mulligan are correct in claiming that Hertz's *Principles* did not really inspire concrete physical developments in mechanics, the aim of this paper is to further elaborate the study of Hertz's influence by showing that it did do so in a newly emerging approach to fundamental physics, namely the electromagnetic worldview (mainly elaborated in the first decade of the twentieth century). Two of its foremost contributors in particular, namely Wilhelm Wien and Max Abraham, explicitly presented their electromagnetic theories as accounting for, and going beyond, Hertz's hidden masses. And these theories were, moreover, primarily concerned with the issue of how to model the interaction between electron and ether. As such, contrary to the historians' claims above, Hertz's *Principles* did inspire physical reflection on the ether, just not in mechanical terms.

In the second part, I will then argue that the theory of relativity, which at the time was considered the foremost opponent of the electromagnetic worldview, was equally well elaborated in what one could call a Hertzian style (although its main adherents at the time, Max Planck and Albert Einstein, did not explicitly label it as such). Planck and Einstein argued, more specifically, that the electromagnetic worldview in fact maintained the traditional Newtonian framework of mechanics that it aimed to replace, and in this way also the obscurities that Hertz had attempted to overcome. In the case of the electromagnetic worldview, these obscurities concerned how to conceptualize the action-reaction principle for interactions between ether and (charged) matter. By means of their criticism, the adherents of the relativistic approach could then argue that the problem was that the electromagnetic worldview assumed that it could offer insight into the essence of the electron, while it in fact merely offered a possible representation, and that this essence had to be considered rather as hidden (in line with how Hertz had introduced a certain hiddenness).

This paper is structured as follows. After a short summary of Hertz's *Principles,* I will then discuss how this work inspired the attempts by Lorentz, Wien and Abraham to develop an electromagnetic worldview. I will show that, especially Wien and Abraham, explicitly saw their work as a further elaboration of Hertz's *Principles,* in the sense that they believed that their electron-theories could account for the emergence of both observable and hidden masses. In this way, they claimed, their theories showed how the electron's mass, and all mass more generally, was completely electromagnetic in nature. I will then turn to a discussion of the relativistic response to this claim, which took on the form, especially in Max Planck's work, of the claim that one could not infer any definite claims regarding the nature of mass from physical theories, and that, hence, the nature of mass was in a sense hidden. In this way, I will then conclude, we can distinguish at least two different ways in which Hertz's *Principles*, and in particular its notion of hiddenness, inspired reflections within fundamental physics.

2 Hertz's *Principles of Mechanics*

Hertz started his *Principles* by claiming that a scientific representation is successful when it displays what he called an essential correspondence ('eine wesentliche Übereinstimmung'), which means that it provides correct predictions: as he put it.

> [...] the necessary consequences of the images in thought are always again the images of the necessary consequences of the pictured objects. (Hertz 1894a, p. 1)[1]

The terms 'denknotwendig' and 'naturnotwendig' are quite difficult to translate adequately. They more or less mean that the consequences of the images are impossible to be, or to be thought, otherwise. Hertz immediately emphasized that regarding the nature of the entity represented, nothing further could be inferred from such success: on the contrary,

---

[1] "die denknotwendigen Folgen der Bilder stets wieder die Bilder seien von den naturnotwendigen Folgen der abgebildeten Gegenstände." (Hertz 1894a, p. 1; Translation from Eisenthal 2018, p. 51).

> [...] we do not know, nor have we any means of knowing, whether our images of things are in conformity with them in any other than this one fundamental respect. (Hertz 1894a, p. 2)[2]

Hertz (1894a, pp. 2–3) distinguished three criteria to evaluate different images of the same object. The first, correctness ('Richtigkeit'), concerns whether the images separately do indeed fulfil the essential correspondence-claim, i.e. whether their predictions are indeed successful. The second, permissibility ('Zuläßigkeit'), comes down to the claim that our images have to be logically consistent. The third, appropriateness ('Zweckmäßigkeit'), in turn concerns two subcriteria, distinctness ('Deutlichkeit') and simplicity ('Einfachheit'). Hertz conceptualized these last two subcriteria in terms of a distinction he had introduced earlier in his *Untersuchungen über die Ausbreitung der elektrischen Kraft* (Hertz 1894b), between indispensable ('unentbehrlich') and superfluous ('entbehrlich') content. He considered theoretical elements superfluous when "they cannot have any influence on [the derivation of] any possible phenomenon" (Hertz 1894b, p. 22),[3] and essential theoretical elements are therefore those that are necessary for the derivation of a phenomenon. An image is then more distinct than another if it allows for the derivation of more actual phenomena, i.e. for more essential correspondences, and it is simpler when it has fewer superfluous elements than the other.

In his *Electric Waves*, Hertz had then used these criteria to argue that representations of Maxwellian electrodynamics that conceptualized electricity in terms of action-at-a-distance ('Fernkräfte') were conceptually ambiguous. This had led him to elaborate an alternative representation, one that relied only on contact forces, in order to eliminate any superfluous and ambiguous elements. In *Principles*, Hertz's motivation was, according to Eisenthal (2021), very similar (at least partially), namely to formulate a conceptually clear notion of force that did not give rise to the obscurities underlying its Newtonian formulation.[4] On this Newtonian formulation, as Hertz put it, "force is introduced as the cause of movement existing before, and independently of, the motion" (Hertz 1894a, p. 5).[5] It is one of the theory's primitive notions, besides time, space and mass, and its relations to these other notions are described by means of the three laws of motions and d'Alembert's Principle. These elements together, Hertz then claimed, gave rise to a particular ambiguity, namely that whereas according to the first two laws, forces act on a body in a particular direction and bring about motion, according to the third law forces result out of motion and connect bodies. As Hertz put it:

---

[2] "Verschiedene Bilder derselben Gegenstände sind möglich und diese Bilder können sich nach verschiedenen Richtungen unterscheiden." (Hertz 1894a, p. 2).

[3] "sie [konnten] auf keine möglichen Erscheinungen einen Einfluss üben." (Hertz 1894b, p. 22).

[4] This was not Hertz's only motivation, however: he equally well had issues with the notion of energy. Lützen (Lützen 2005) offers an extensive discussion of the context in which Hertz was working. I do not doubt or dispute these other motivations. I rather choose to focus primarily on Hertz's issues with the Newtonian notion of force, in line with how Eisenthal (2021) discusses *Principles*, because, in this way a few interesting parallels with later discussions will emerge.

[5] "Die Kraft is dabei eingeführt als die vor der Bewegung und unabhängig von der Bewegung bestehende Ursache der Bewegung." (Hertz 1894a, p. 5).

> The force spoken of in the definition and in the first two laws acts upon a body in one definite direction. The sense of the third law is that forces always connect two bodies, and are directed from the first to the second as well as from the second to the first. It seems to me that the conception of force assumed and created in us by the third law on the one hand, and the first two laws on the other hand, are slightly different. This slight difference may be enough to produce the logical obscurity of which the consequences are manifest in the above example. (Hertz 1894a, pp. 7–8)[6]

Hertz then illustrated the issue by means of the example of a stone being swung around by means of a rope. Following the second law of mechanics, our hand exercises a force on the rock that puts it in motion. According to the third law, the stone at the same time also exercises a force on our hand, equal in value and opposite in direction. According to Hertz, however, this way of conceptualizing only introduced superfluous content by distinguishing the second force introduced—often called the centrifugal force—from the stone's inertial mass; Eisenthal (2021) extensively discusses this example:

> Is that what we now call swing force or centrifugal force something different from the stone's inertia? Can we take the effect of inertia into account twice, once as mass, once as force, without disturbing the clarity of our representations? In our laws of motion force was the cause of motion preceding the motion. Can we now, without confusing our concepts, speak all of a sudden of forces, which emerge out of motion, which are a consequence of motion? [...] All these questions are to be answered in the negative. (Hertz 1894a, p. 7)[7]

The example indicated, for Hertz, that there were conceptual issues underlying how the three laws of mechanics together conceptualized the relation between force, inertial mass, and motion. This was a purely conceptual issue: as Hertz (1894a, p. 10) pointed out, it solely concerned the appropriateness of Newtonian mechanics, not its correctness or permissibility. To overcome it, Hertz therefore set out to elaborate an alternative image of mechanics, which relied, in contrast to the other representations he discussed,[8] on only three primitive notions: space, time, and mass. The notion of force, on the other hand, became a derivative concept. He elaborated this concept by

[6] "Die Kraft, von welcher die Definition und die ersten beiden Gesetze reden, wirkt auf einen Körper in einseitig bestimmter Richtung. Der Sinn des dritten Gesetzes ist, daß die Kräfte stets zwei Körper verbinden und ebenso gut vom ersten zum zweiten, wie vom zweiten zum ersten gerichtet sind. Die Vorstellung der Kraft, welche dieses Gesetz und die Vorstellung, welche jene Gesetze voraussetzen und in uns erwecken, scheinen um ein Geringes verschieden, dieser geringe Unterschied aber reicht vielleicht aus, um die logische Trübung zu erzeugen, deren Folgen in unserem Beispiele zum Ausbruch kamen." (Hertz 1894a, pp. 7–8).

[7] "Ist das was wir jetzt Schwungkraft oder Centrifugalkraft nennen, etwas anderes als die Trägheit des Steines. Dürfen wir, ohne die Klarheit unserer Vorstellungen zu zerstören, die Wirkung der Trägheit doppelt in Rechnung stellen, nämlich einmal als Masse, zweitens als Kraft? In unseren Bewegungsgesetzen war die Kraft die vor der Bewegung vorhandene Ursache der Bewegung. Dürfen wir, ohne unsere Begriffe zu verwirren, jetzt auf einmal von Kräften reden, welche erst durch die Bewegung entstehen, welche eine Folge der Bewegung sind? [...]. Alle diese Frage sind offenbar zu verneinen." (Hertz 1894a, p. 7).

[8] Besides the Newtonian one, which had as its primitive concepts space, time, mass and force, connected by the three laws of mechanics and d'Alembert's Principle, Hertz also discussed the energeticist representation, which had as its primitive concepts space, time, mass and energy, connected by means of Hamilton's principle.

distinguishing two kinds of masses: the observable material masses we are all familiar with on the one hand, and hidden masses on the other. This idea of hiddenness Hertz summarized as follows:

> We can admit that a hiddenness is at play and still deny that this hiddenness is of a special kind of form. We are free to accept that the hidden is nothing other as, again, movement and mass, and precisely such movement and mass, that is not different from the observable, except in its relation to us and our normal means of observation. (Hertz 1894a, p. 30)[9]

Central to Hertz's conception of this hiddenness was that it in no way provided any kind of insight into any essences behind the phenomena: it was rather introduced solely to arrive at a conceptually clear formulation of the theory of mechanics. According to this formulation, with which the main body of *Principles* was concerned, the mechanics of moving masses could be conceptualized in terms of one fundamental law, according to which "each natural movement of an independent material system consists in the system following one of its straightest paths with constant speed" (Hertz 1894a, p. 33).[10] The idea behind it was that an unhindered single observable mass would travel in a straight path, and that the influence of forces on such a mass would then be conceptualized in terms of hidden masses, rigidly connected to it, that would constrain the observable system in such a way that its path was no longer straight but rather the straightest possible (Cfr. Lützen 2005; Eisenthal 2018, 2021).[11] Formulated in this way, Hertz (1894a, p. 33) then claimed, his fundamental law in combination with the hidden masses-hypothesis covered both the law of inertia and Gauss's Principle of Least Constraint, and hence it allowed for what he described as a purely deductive derivation of the content of mechanics. Moreover, it allowed for the formulation of a derivative notion of force that was not plagued by the conceptual obscurities underlying Newtonian mechanics. As Hertz put it:

> Force now no longer appears as something independent from us and alien to us, but rather as a mathematical auxiliary construction, whose properties we have completely under our control, and which in that way does not have anything mysterious to us. According to the fundamental law, wherever two bodies belong to the same system, the movement of the one must be co-determined by the movement of the other. The notion of force now arises from the fact that we find it appropriate ['zweckmässig'] to divide this determination of one movement by the other in two stages and to say: the movement of the first body first determines a force, which then only determines the movement of the second body. In this way, every force is always the cause of a movement, but at the same time, with the same

[9] "Wir können zugeben, das sein verborgenes Etwas mitwirke und doch leugnen, dass dieses Etwas einer besonderen Kategorie angehöre. Es steht uns frei anzunehmen, dass auch das Verborgene nichts anderes sei als wiederum Bewegung und Masse, und zwar solche Bewegung und Masse, welche sich von der sichtbaren nicht an sich underscheidet, sondern nur in Beziehung auf uns und auf unsere gewöhnlichen Mittel der Wahrnehmung." (Hertz 1894a, p. 30).

[10] "jede natürliche Bewegung eines selbständigen materiellen Systems bestehe darin, dass das System mit gleichbleibender Geschwindigkeit eine seiner geradesten Bahnen verfolge." (Hertz 1894a, p. 33).

[11] By means of these rigid connections, Hertz (1894a, p. 215) pointed out, he thus eliminated any form of action-at-a-distance from his mechanics, just as he had done in his reformulation of Maxwellian electrodynamics.

> right, it is also always the consequence of a movement; it becomes, to be precise, the merely imagined intermediary between two movements. (Hertz 1894a, pp. 33–34)[12]

As the quote indicates, Hertz's reformulation of Newtonian mechanics was to be evaluated not primarily on the basis of its correctness or permissibility—on these criteria, it scored equally well as other formulations—but rather with regards to its appropriateness: his reformulation of the notion of force, he claimed, allowed for the derivation of the same phenomena with less superfluous elements that were, moreover, conceptually clear. In line with his epistemology of scientific theories and his notion of hiddenness, he concluded by pointing out that this success should not be interpreted as providing any direct insight into the essence of mechanical concepts such as force. In fact, Hertz seems to have doubted whether such insight could ever be gained, since once his reformulation is achieved, "the question as to essences will not have been answered; but our minds, no longer vexed, will cease to ask unjustified questions" (Hertz 1894a, p. 9).[13] This shows how closely connected Hertz's physical and philosophical reflections were, for, as Eisenthal (2018, p. 54) puts it, according to Hertz's own epistemology of physical theories "the hypothesis of hidden masses *rules out* knowledge of the "fundamental" constituents of a system": it is a successful representation, but that only means that it provides us with successful predictions, and we should not infer from this that it in any way provides us with insight into the fundamental nature of reality. Hertz's only goal, as he put it, had been to clarify the logical-conceptual structure of the theory of mechanics, and whether forces were actually and essentially constituted by rigid connections between observable and hidden masses was a question that his theory did not purport to answer:

> In seeking the actual rigid connections we shall perhaps have to descend to the world of atoms. But such considerations are out of place here; they do not affect the question whether it is logically permissible to treat of fixed connections as independent of forces and precedent to them. (Hertz 1894a, p. 41)[14]

---

[12] "Aber die Kraft tritt nun nicht auf als etwas von uns unabhängiges und uns fremdes, sondern als eine mathematische Hilfskonstruktion, deren Eigenschaften wir völlig in unserer Gewalt haben, und welche also auch für uns nichts Rätselhaftes an sich haben kann. Nach dem Grundgesetze muss nämlich überall da, wo zwei Körper demselben System angehören, die Bewegung des einen durch die Bewegung des anderen mitbestimmt sein. Der Begriff der Kraft entsteht nun dadurch, dass wir es aus angebbaren Gründen zweckmässig finden, diese Bestimmung der einen Bewegung durch die andere in zwei Stadien zu zerlegen und uns zu sagen: die Bewegung des ersten Körpers bestimme zunächst eine Kraft, diese Kraft erst bestimme die Bewegung des zweiten Körpers. Auf diese Weise wird jede Kraft zwar stets Ursache einer Bewegung, mit gleichem Rechte aber zugleich auch stets Folge einer Bewegung; sie wird, genau gesprochen, das nur gedachte Mittelglied zwischen zwei Bewegungen." (Hertz 1894a, pp. 33–34).

[13] "Sind diese schmerzenden Widersprüche entfernt, so its zwar nicht die Frage nach dem Wesen beantwortet, aber der nicht mehr gequälte Geist hört auf, die für ihn unberechtigte Frage zu stellen." (Hertz 1894a, p. 9).

[14] "[B]ei der Suche nach den wirklichen starren Verbindungen wird sie vielleicht zur Welt der Atome hinabzusteigen haben, aber diese Erörterungen sind hier nicht am Platze, sie berühren nicht mehr die Frage, ob es logisch zulässig sei, feste Verbindungen unabhängig von und vor den Kräften zu behandeln." (Hertz 1894a, p. 41).

3 Going Beyond the World of Atoms

3.1 *Lorentz and Wien on the Electron's Electromagnetic Energy*

This last quote from Hertz, we will see later, became an explicit point of inspiration for Wilhelm Wien's and Max Abraham's formulation of the electromagnetic worldview. This view, which McCormmach describes as

> [...] a programmatic intent [... to focus] on problems whose solution promised to secure a universal physics based solely on electromagnetic laws and concepts. (McCormmach 1970, p. 459]

can be traced back at least to work by Wilhelm Weber from 1846 (McCormmach 1970, p. 472).[15] The first to elaborate it in Hertzian terms was Hendrik Antoon Lorentz who, in a 1899 lecture, conceptualized the program as the construction of images that display something resembling what Hertz called an essential correspondence:

> [...] as Hertz puts it, we construct inner mock images of external states of affairs in such a way that, what results out of these images according to the laws of our thinking, corresponds to what happens outside of us according to the laws of nature. (Lorentz 1899, p. 500)[16]

Lorentz distinguished three such images: a mechanical, an energeticist and an electromagnetic one. While he considered the last one to be the most promising, because it had already attained successes in, among other things, drawing chemistry closer to physics, it left one central question open: was it possible to formulate an electromagnetic account of gravitation? The most pressing issue in this regard, according to Lorentz, was that "while all electromagnetic forces propagate at the speed of light, astronomical observations teach us that the propagation of gravitational action, when it is not instantaneous, is in any case to be taken as much faster than that of light" (Lorentz 1899, p. 519).[17] Lorentz there did not elaborate on this issue any further.

---

[15] Besides Weber, McCormmach (1970) also lists Ottaviano Fabrizio Mossotti, Karl Friedrich Zöllner, Rudolf Clausius, Bernhard Riemann, Carl Neumann, and Emil Wiechert as predecessors. For historical discussions of the electromagnetic besides McCormmach's paper (McCormmach 1970; Jungnickel and McCormmach 1986, pp. 227–245; Kragh 2002, pp. 105–119; Seth 2004; Katzir 2005; Seth 2005; Staley 2008, pp. 260–293; Jungnickel and McCormmach 2007, pp. 364–373).

[16] "[W]ie Hertz es ausdruckt, wir machen uns 'innere Scheinbilder' der äusseren Gegenstande, und zwar so, dass das, was sich aus diesen Bildern nach den Gesetzen unseres Denkvermögens ergiebt, dem entspricht, was ausserhalb von uns nach den Naturgesetzen geschieht." (Lorentz 1899, p. 500).

[17] "Vorläufig stösst eine elektromagnetische Auffassung der allgemeinen Anziehungskraft auf die grosse Schwierigkeit, dass, soweit wir wissen, alle elektromagnetische Wirkungen sich mit der Geschwindigkeit des Lichts fortpflanzen, während die astronomische Erscheinungen lehren, dass, wenn überhaupt von einer nicht augenblicklichen Wirkung der Schwerkraft die Rede sein kann, ihre Fortpflanzungsgeschwindigkeit jedenfalls als viel grösser als die des Lichtes voraufgesetzt werden muss." (Lorentz 1899, p. 519).

A year later, however, Lorentz (1900) addressed this question—whether gravitation could be conceptualized as an electromagnetic action propagating at the speed of light—in more detail. He started from the assumption that all matter is completely made up of charged particles, which Lorentz called ions. He then ascribed the ions such charge values that the attractive force between opposite charges was a little bit bigger than the repulsive force between equally charged ions (Lorentz 1900, p. 566). Gravitation was then to be understood in terms of the difference between these two forces as follows:

> [...] the [charges] are acted on by the same force [...] in the same direction, which means that the force cannot be due to an electric field since that would move the charges oppositely. The force, Lorentz concluded, must then be gravitational. (McCormmach 1970, p. 477)

While this view, as Lorentz pointed out, led to predictions that did not agree with observations regarding Mercury's perihelion, it was also conceptually promising since

> What has been said is sufficient to show that gravitation may be attributed to actions which are propagated with no greater velocity than that of light. (Lorentz 1900, p. 573)[18]

Wien (1901) then further elaborated Lorentz's suggestion into a proposal for an electromagnetic worldview. He started not with gravitation, however, but rather with.

> [t]he inertia of matter, which besides gravitation provides the second independent definition of mass, [and which] can be derived without any additional hypothesis from the already often used concept of electromagnetic inertia. (Wien 1901, p. 507)[19]

The idea behind such inertia was that a charged body, when set in motion, "must [...] pass through its own electromagnetic field, with a consequent decrease in velocity—just as if it had gained mass" (Staley 2008, p. 219).[20] Wien conceptualized it as follows (where he called Lorentz's ions 'electrical quanta', m denoted their mass and E their electromagnetic energy):

> Accordingly, the mass defined by inertia would be constant only at low velocities and would increase with increasing velocity. Since inertia is proportional to the number of quanta out of which a body is composed, as well as to the gravitation emanating from this body, it follows that the mass defined by inertia must be proportional to the inertia determined by gravity. If we have a body with mass $m = 4/3\ Ec^2$ attract a body with mass M up to a distance r, then the electromagnetic energy supply of gravity is diminished with the value $\varepsilon\ 4/3\ E\ c^2\ M/r$, where ε denotes the gravitational constant. (Wien 1901, p. 508)[21]

---

[18] For a discussion of this attempt by Lorentz, see: McCormmach 1970, pp. 476–477.

[19] "Die Trägheit der Materie, welche neben der Gravitation die zweite unabhängige Definition der Masse giebt, lässt sich ohne weitere Hypothesen aus dem bereits vielfach benutzten Begriff der elektromagnetischen Trägheit folgern." (Wein 1901, p. 507).

[20] Wien was not the first to introduce this idea. It can already be found in the work of Riemann (1875), Thomson (1881, 1893), Heaviside (1889), Searle (1897), Wiechert (1894), Coudres (1899) and Lorentz (1899b).

[21] "Hiernach wäre die durch Trägheit definirte Masse nur bei kleinen Geschwindigkeiten constant und würde mit grösser werdender Geschwindigkeit zunehmen. Da die Trägheit der Anzahl der Quanten, aus denen sich ein Körper zusammensetzt, proportional ist, ebenso die von diesem Körper

The gravitational expression obtained in this way was quite close to Weber's, which, Wien (1901, p. 509) pointed out, provided quite good predictions regarding Mercury's perihelion. Moreover, it also opened up a new way to investigate gravitation experimentally, since the velocities required to make the change in mass observable could be attained by means of cathode rays (Wien 1901, pp. 509–510). Finally, as Wien then showed, it also indicated how to obtain electromagnetic analogues for the laws of mechanics. The first law could be reformulated in terms of the principle of conservation of electromagnetic energy, and the second law came down to the claim that the work exercised by a force on a body during a certain period was equal to the corresponding change in electromagnetic energy. The validity of the third law, the action-reaction principle, was, however, restricted, but Wien did not seem bothered by this (more on this below) (Wien 1901, p. 512).[22] Wien summarized the obtained results as follows:

> One can consider the reasoning sketch presented here as diametrically opposed to Hertz's. The rigid connections, which on Hertz's view are assumed in advance, show themselves here as the effect of entangled single forces. Likewise, the law of inertia is a relatively late consequence of the electromagnetic assumptions. While the Hertzian mechanics aims at providing the electromagnetic equations as consequences, here the relation is exactly the other way around. (Wien 1901, p. 512)[23]

Wien thus explicitly conceptualized the development of the electromagnetic worldview in terms of accounting for, and going beyond, Hertz's mechanics: whereas Hertz postulated his rigid connections between observable and hidden masses as (logical) primitives, Wien claimed that by means of the electron's velocity-dependent electromagnetic inertia, he could account for how mechanical mass and force, in the form of Hertz's rigid connections, came about.

As pointed out above, Wien's electromagnetic worldview entailed a restricted validity for Newton's third law, the action-reaction principle. The reason for this was the same as for Wien's source of inspiration, namely Lorentz's electromagnetic theory. Both theories took the ether as a substance that was absolutely at rest, to account for different experimental results concerning the possible influence of the

ausgehende Gravitation, so folgt, dass die durch die Trägheit definirte Masse der durch die Gravitation bestimmten proportional sein muss. Lassen wir einen Körper, dessen Masse $m = 4/3\ E\ c^2$ ist, bis in die Entfernung r von einem Körper von der Masse M anziehen, so ist die elektromagnetische Energievorrat der Gravitation um den Betrag $\varepsilon\ 4/3\ E\ c^2\ M/r$ vermindert, wo ε die Gravitationsconstante bezeichnet." (Wein 1901, p. 508).

[22] This was not the first discussion of the action-reaction principle by Wien. In a 1898 lecture (Wien 1898), he had already discussed how questions concerning its validity could arise out of different experimental results concerning the question whether matter could set the ether in motion.

[23] "Man kann die hier skizzirte Begründung der Mechanik als der Hertz'schen diametral entgegengesetzt bezeichnen. Die festen Verbindungen, welche bei Hertz zu den Voraussetzungen gehören, zeigen sich hier als Wirkung verwickelter Einzelkräfte. Ebenso ist das Gesetz der Trägheit eine verhältnismässig späte Consequenz aus den elektromagnetischen Voraussetzungen. Während die Hertz'sche Mechanik offenbar darauf abzielt, die elektromagnetischen Gleichungen als Folgerungen zu liefern, ist hier das Verhältnis gerade umgekehrt." (Wein 1901, p. 512).

earth's motion through the ether on optical and electromagnetic phenomena occurring in that same ether.[24] This absolute rest requirement entailed that

> [...] the ether acted on matter, but not the reverse; consequently, the violation of Newton's third law was built into Lorentz's electromagnetic theory. (Miller 1905, p. 42)

The issue concerned, more specifically, the calculation of the net force exerted on charged particles in a volume, which is expressed in terms of the Maxwell stress tensor—denoting the ponderomotive action of the electromagnetic fields on the charged bodies—and Poynting's energy flux vector—which denotes a flow of energy through the ether. When the charges are at rest, the net force inside the volume is zero. When the charges are in motion, however, this is not the case: rather, "Poynting's vector changes in time, the stresses do not vanish and must therefore move the ether as a whole" (McCormmach 1970, p. 469). The presence of these stresses indicated that the ether acted on the material bodies present. Given, however, that the ether was also taken as an unmovable substance, matter could not act back on it, which entailed that the action-reaction principle was not valid (Miller 1905, p. 43). Neither Lorentz nor Wien saw this, however, as a really pressing issue, since abandoning it could bring closer a completely electromagnetic worldview (Darrigol 1995, p. 22).

3.2 Poincaré, Abraham and Lorentz on the Ether's Electromagnetic Momentum

Henri Poincaré (Poincaré 1900), however, considered jettisoning the principle as unacceptable, because that also endangered e.g. the principle of relativity and different conservation principles (Miller 1905, p. 43; Goldberg 1967, p. 942). Poincaré only saw one solution, namely "to abandon the stricture of an absolute stationary ether, and to confer upon it some inertia and velocity" (Darrigol 1995, p. 18). This was, at the same time, a problem, since on his view, the action-reaction principle should apply only to material bodies, and not to the ether, which "if it existed, had to be a very elusive thing" (Darrigol 1995, p. 19).[25]

To address this issue, Poincaré proceeded as follows (here I follow: Miller 1905, pp. 41–44; Darrigol 1995, pp. 23–24). From the total electromagnetic force expression that entailed the violation of the action-reaction principle, Poincaré derived a conservation-expression for the system as a whole, i.e. consisting of both matter and ether (conceptualized as a fluid). This expression combined the material entity's velocity and mass density with Poynting's energy flux vector $c(E \times H)$, divided by c^2 (where E and H denote the electric and magnetic forces exercised by the ether on

---

[24] For some discussions of how these different experiments influenced the development of Lorentz's ether-model see (Hirosige 1969; McCormmach 1970; Hirosige 1976; Miller 1905, pp. 14–40; Darrigol 1995; Janssen 2009, pp. 29–31).

[25] The reason for this was that he believed that "Hertz's [electromagnetic] theory completely eliminated ether." (Darrigol 1995, p. 20).

unit charges). Poincaré then showed that if one ascribed the ether-fluid a particular mass density, the conservation equation entailed.

> [...] that the total momentum of matter and fluid is conserved and that the fluid was only fictitious, since it was created or destroyed at [a] rate [...] predicted by Poynting's theorem, and thus did not comply with the conservation of mass. (Darrigol 1995, p. 24).

The action-reaction principle could then be retained, at least in the form of the conservation of momentum. Moreover, in this way, one did not ascribe momentum or materiality to the ether in any physically substantial sense. According to Darrigol, however, Lorentz was not convinced:

> In [Lorentz's] opinion, the principle of reaction 'should not be regarded as a fundamental principle of physics'. Regarding the conservation of momentum, he congratulated Poincaré for his 'beautiful formula' [...] and suggested that the quantity $(E \times H)/c$ be regarded as 'equivalent to momentum'. He even speculated that the equivalence could some day become an identity, 'if we manage to consider ponderable matter as a modification of the ether'. (Darrigol 1995, pp. 29–30)[26]

As Darrigol (1995, p. 30) points out, it was Max Abraham who then turned this suggestion into one of the most elaborate formulations of the electromagnetic worldview. He did this in two articles from 1902, in which he elaborated a theory according to which the electromagnetic field completely determined the electron's equations of motion. Given that he took the electron to be the fundamental constituent of all matter, these equations could then be taken to provide an electromagnetic foundation for mechanics (Abraham 1902a, p. 20).

Abraham started by pointing out that, up until then, most electromagnetic electron-models conceptualized their mass in terms of electromagnetic energy (as, for example, Wien did). With regards to the electron's electromagnetic inertia, however, this meant that one could only obtain the electron's longitudinal mass, i.e. "the inertia opposing acceleration in the direction of its motion"[27]; not, however, their transverse mass, i.e. the "inertia at play in accelerations perpendicular to the direction of its path" (Abraham 1902a, p. 21),[28] since in the transverse case, there is no change in electromagnetic energy. This was a problem, Abraham pointed out, since the experiments on the velocity-dependency of the electron's mass carried out by Walter Kaufmann at that time indicated that it was primarily the transverse mass that was at play. To overcome this, Abraham (1902a, p. 25) therefore proposed to reconceptualize the electron's mass by means of the notion of electromagnetic momentum.[29]

---

[26] The quotes are from a letter by Lorentz to Poincaré from January 1901 (Miller 1980, pp. 70–71; Darrigol 1995, p. 30).

[27] "diejenige Trägheit welche sich einer Beschleunigung in der Bewegungsrichtung widersetzt." (Abraham 1902a, p. 21).

[28] "die bei Beschleunigung senkrecht zur Bahnrichtung in Betracht kommt." (Abraham 1902a, p. 21).

[29] According to Abraham (1902b, p. 110), it was Poincaré who had first introduced this notion. As Darrigol has pointed out, however, ascribing to Poincaré the specific idea of an electromagnetic momentum is not completely correct: "Poincaré is often regarded as the inventor of the notion of electromagnetic momentum. This is a double historical mistake: the notion already existed in

To conceptualize the electromagnetic field's influence on the electron's motion, it was essential to consider, according to Abraham (1902a, p. 23), that the electron's motion modifies this field, since a moving charge gives rise to a magnetic field. Hence, one could not represent the electron as a point particle: one had to take into account the charge distribution over its volume, to accommodate the fact that the electron's motion could give rise to an electromagnetic field that is, spatially speaking, distributed unequally. This vastly increased the complexity of the issue, however. To make it more feasible, Abraham followed an approach he borrowed from the study of conduction currents.[30] Such currents are called stationary when their strength is constant, in which case the magnetic field induced is determined purely by current strength. When the strength is not constant, this is strictly speaking no longer the case, but in low frequency cases, there are certain mathematical approximations to proceed as if the current is stationary, a state that is described as quasi-stationary. Abraham now proposed to proceed in a similar way: "the field will be assumed to be quasi-stationary, i.e. it will be determined completely by the instantaneous velocity of the electron" (Abraham 1902a, p. 23).[31]

Abraham (1902a, pp. 25–26) then conceptualized the field's electromagnetic momentum in terms of the Poynting energy flux vector, as Poincaré had done. In contrast to Poincaré, however, Abraham (1902a, p. 27) took it to be a real physical property of the ether, which acted as the source for the electron's electromagnetic mass, both in its longitudinal and transverse form. In fact, Abraham concluded his first paper, this was all there was to the electron's mass: if one applied his mass-expressions to Kaufmann's first results (Kaufmann 1901), the ether's electromagnetic momentum had to be responsible for the electron's entire mass, since any kind of added mechanical mass would lead to a velocity-dependency expression that deviated further from Kaufmann's. Hence, Abraham concluded, "[t]he electron's inertia is caused exclusively by its electromagnetic field" (Abraham 1902a, p. 40).[32]

In a second article from 1902 (Abraham 1902b), Abraham further elaborated his theory's foundations. These consisted of three elements: (a) a kinematical equation, which constrains the electron's freedom of motion; (b) field equations that govern the electromagnetic field produced by the electron; and (c) dynamical equations governing the electron's motion in an external electromagnetic field (Abraham 1902b, p. 108). While (b) and (c) were mostly taken over from earlier work—the field equations were borrowed from Lorentz's work, and the electron dynamics was mainly a further elaboration of Abraham's earlier (Abraham 1902a) paper—, the first part

Maxwell's theory, and it was only a fiction in Poincaré's analysis of Lorentz's theory." (Darrigol 1995, pp. 30–31).

[30] Abraham was quite familiar with this subject, since it was the topic of his PhD dissertation (supervised by Max Planck), which "was an examination of electrical oscillations in conductors and was the first in a series of studies in which Abraham applied his specialty, Maxwellian theory, to problems of wireless telegraphy and antenna theory" (Goldberg 1970, p. 8).

[31] "Das Feld soll weiterhin als quasistationär angenommen warden, d.h. es soll bestimmt sein durch Angabe der momentanen Geschwindigkeit des Electrons." (Abraham 1902a, p. 23).

[32] "Die Trägheit des Electrons ist ausschließlich durch sein electromagnetisches Feld verursacht." (Abraham 1902a, p. 40).

was new. Its central claim was that the electron's kinematics had to be considered as identical to the kinematics of the rigid body:

> [...] electricity adheres to the volume elements of the rigid body, in the same way as matter adheres to the volume elements of the rigid body. (Abraham 1902b, p. 108)[33]

The reason for this was that "[a] deformable electron would require the presence of non-electromagnetic forces in order to maintain its stability because of the mutual Coulomb repulsion between its constituent parts" (Miller 1973, p. 216). The resulting theory, according to Abraham, would look as follows:

> A thought similar to the one just presented may have guided Heinrich Hertz, when he allowed in his *Principles of Mechanics* only the existence of such kinematical connections that would give rise to neither the generation nor the destruction of kinetic energy. That was necessary because he wanted to trace back all energy to kinetic energy of moving masses, all forces to kinematic connections. To the objection that we find rigid connections in reality only approximately realized, Hertz (1894a, p. 41) replied as follows: 'In seeking the actual rigid connections our mechanics will possibly have to descend to the world of atoms'. Now, the electromagnetic mechanics will descend even further; in the atoms of negative electricity, these spheres whose radius is only the billionth part of a millimetre, it takes on the form of a rigid, unchanging arrangement of the electric charge. That it is permissible to speak of rigid connections before one speaks of forces, Hertz has shown convincingly. Our dynamics of the electron completely omits any talk of forces, that would aim to deform the electron. It speaks only of 'external forces', which are capable of endowing it with velocity or rotational velocity, and of 'internal forces' which, emerging from the electron's field, keep it in equilibrium. And these forces and rotational forces are only auxiliary concepts, defined through the kinematical and electromagnetic basic concepts. The same holds for words such as 'work', 'energy', 'momentum', which have only been chosen to let the analogy of the electromagnetic mechanics to the ordinary mechanics of material bodies clearly stand out. (Abraham 1902b, p. 109)[34]

---

33 "[W]ie die Materie an den Volumenelementen des starren Körpers, so haftet die Elektrizität an den Volumenelementen des starren Elektrons." (Abraham 1902b, p. 108).

34 "Eine der soeben angedeuteten verwandte Überlegung mag Heinrich Hertz geleitet haben, als er in seinen 'Prinzipien der Mechanik' nur solche kinematische Zusammenhänge zuließ, deren Bestehen weder Erzeugung noch Zerstörung kinetischer Energie bedingt. Das war notwendig, weil er alle Energie auf kinetische Energie bewegter Massen, alle Kräfte auf kinematische Verbindungen zurückführen wollte. Dem Einwande, daß wir starre Verbindungen in Wirklichkeit nur angenähert realisiert finden, begegnet Hertz (1894a, p. 41) mit den Worten: 'auf der Suche nach den wirklichen starren Verbindungen wird unsere Mechanik vielleicht zur Welt der Atome herabzusteigen haben'. Nun, die elektromagnetische Mechanik steigt noch weiter herab; in den Atomen der negativen Elektrizität, diesen Kugeln, deren Radius nur den billionten Teil eines Millimeters beträgt, nimmt sie eine starre, unveränderliche Anordnung der elektrischen Ladung an. Daß es zulässig ist, von starren Verbindungen zu reden, bevor man von Kräften gesprochen hat, das hat Hertz überzeugend dargetan. Unsere Dynamik des Elektrons unterläßt es überhaupt, von Kräften zu reden, die das Elektron zu deformieren bestrebt sind. Sie spricht nur von 'äußeren Kräften', die ihm eine Geschwindigkeit oder Drehgeschwindigkeit zu erteilen vermögen, und von 'inneren Kräften', die vom Felde des Elektrons herrührend, jenen das Gleichgewicht halten. Und auch diese 'Kräfte' und 'Drehkräfte' sind nur Hülfsbegriffe, die definiert werden durch die kinematischen und die elektromagnetischen Grundbegriffe. Das Gleiche gilt von Worten wie 'Arbeit', 'Energie', 'Bewegungsgröße', bei deren Wahl allerdings das Bestreben maßgebend war, die Analogie der elektromagnetischen Mechanik zur gewöhnlichen Mechanik materieller Körper deutlich hervortreten zu lassen." (Abraham 1902b, p. 109).

This quote shows how Hertz's *Principles* acted as a source of inspiration for Abraham's electromagnetic worldview in multiple ways. First of all, there is the claim, equally well to be found in the quote by Wien (see above), that an electromagnetic electron-dynamics could account for the emergence of Hertz's hidden masses-mechanics (Abraham even explicitly quotes Hertz's claim that the rigid connections are possibly to be found in the world of atoms, see above).

Second, Abraham's formulation of the electromagnetic worldview explicitly takes over an aspect of Hertz's mechanics, namely its rigid body-kinematics. The electron is to be seen as an extended material element that is constituted as a rigid body through the interaction between its charge and the surrounding electromagnetic field. No appeal to any forces is then really necessary, since this kinematic-electromagnetic equilibrium cannot in any way be disturbed or destroyed. Hence, Abraham could claim, in line with Hertz's approach, that in his theory as well, the notion of force was not a primitive but rather only a derivative notion, which could be obtained because the notion of electromagnetic momentum allowed one "to reduce the inner force to the 'impulse' and 'angular momentum' of the electron's electromagnetic field, and in this way it permits a simplified calculation of the electromagnetic mass and the electromagnetic moment of inertia" (Abraham 1902b, p. 110).[35]

Finally, Hertz's theory formed the explicit target for the construction of electromagnetic analogues for the laws of mechanics. As we have seen, Wien's claim that the notion of electromagnetic inertia could be used to account for the constitution of Hertz's rigid connections then led him to the formulation of electromagnetic replacements for the three laws of mechanics (with the third law in a restricted form). Abraham equally well elaborated such explicit analogues to the three laws, but he did so by means of the notion of electromagnetic momentum, which he conceptualized, as he had done earlier following Poincaré and Lorentz, in terms of the Poynting energy flux vector (Abraham 1902b, p. 125). This immediately provided Newton's third law, the action-reaction principle. It also allowed him to express equations of motion for the rigid electron, which led him to the following analogue for Newton's first law: "if the motion of the electron was, from the beginning, a uniform, purely translatory one with a velocity smaller than the speed of light, then to keep it uniform no external force or rotational force is required" (Abraham 1902b, pp. 142–143).[36] He then introduced a second conceptualization of electromagnetic momentum, in the form of a Lagrangian defined in terms of the difference between the electric and magnetic energy of the electron's electromagnetic field (Abraham 1902b, p. 143). After showing how to derive expressions for quasi-stationary acceleration and longitudinal and transverse mass from it (Abraham 1902b, pp. 149–151), he then used them

---

[35] "die Zurückführung der inneren Kräfte auf einen vom elektromagnetischen Felde des Elektrons abhängigen 'Impuls' und 'Drehimpuls', und gestattet so eine vereinfachte Berechnung der elektromagnetischen Masse und des elektromagnetischen Trägheitsmomentes." (Abraham 1902b, p. 110).

[36] "War die Bewegung des Elektrons von Anbeginn an eine gleichförmige, rein translatorische, und war die Geschwindigkeit kleiner als die Lichtgeschwindigkeit, so ist, um die Bewegung gleichförmig zu erhalten, keine äußere Kraft oder Drehkraft erforderlich." (Abraham 1902b, pp. 142–143).

to formulate an electromagnetic analogue of Newton's second law, which differed from the original as follows:

> The functional relation between force and acceleration is represented in the dynamics of the electron by a linear vector function of a more general kind than in ordinary mechanics. The electromagnetic mass, the coefficient system of the linear velocity function, is a tensor of rotational symmetry,[37] whose axis of symmetry is determined by the direction of motion of the electron. (Abraham 1902b, p. 153)[38]

In this way, Abraham saw his electron-theory as accounting for, and going beyond, the theory of mechanics (in the Hertzian formulation). His motivation for this commitment to the electromagnetic worldview again derived from Kaufmann's experiments, which showed, he claimed, that "the electron's mass is completely electromagnetic in nature" (Abraham 1902b, p. 107).[39] This claim was, in turn, strengthened by Kaufmann, who argued that new experimental results obtained in 1902 (Kaufmann 1902) and 1903 (Kaufmann 1903) were completely in line with Abraham's electron-theory, which brought Kaufmann to repeat Abraham's conclusion: "their mass is completely electromagnetic in nature" (Kaufmann 1903, p. 103).[40]

Abraham was not the only one to conceptualize the electron's mass in this way. Lorentz claimed that his 1904 elaboration of his electron-theory equally well entailed "that there is no other, no 'true' or 'material' mass" (Lorentz 1904, p. 821). His electrons were not rigid, however: they rather "have their dimensions changed by the effect of a translation" (Lorentz 1904, p. 818). This was needed to account for the failures to detect any effect of the earth's motion through the ether on optical and electromagnetic phenomena (the most famous such null-result being the 1887 Michelson-Morley experiment). After introducing the equations that governed how the electron's dimensions and its field equations would transform with such changes in motion, Lorentz (Lorentz 1904, p. 820) then pointed out that the transformation equations governing electromagnetic momentum would become very complex, unless one restricted the electron's motion, as Abraham had done, to quasi-stationary motion. One could then easily show that, with increasing velocity, a longitudinal and a transverse mass could be distinguished Lorentz (Lorentz 1904, p. 821). These differed, however, from Abraham's:

> The values [...] which I have found for the longitudinal and transverse masses of an electron, expressed in terms of its velocity, are not the same as those that have been formerly obtained by Abraham. The ground for this difference is solely to be sought in the circumstance that,

[37] To clarify to his audience what he meant by such a tensor, Abraham then referred to a recent article by him in the *Enzyklopädie der mathematischen Wissenschaften*, which provided "one of the first organized representations of vector analysis in Germany." (Goldberg 1970, p. 8).

[38] "Die funktionelle Beziehung zwischen Kraft und Beschleunigung wird in der Dynamik des Elektrons durch eine lineare Vektorfunktion allgemeiner Art, als in der gewöhnlichen Mechanik, dargestellt. Die elektromagnetische Masse, das Koeffizientensystem der linearen Vektorfunktion, ist ein Tensor von rotatorischer Symmetrie, dessen Symmetrieachse durch die Bewegungsrichtung des Elektrons bestimmt ist." (Abraham 1902b, p. 153).

[39] "Die Masse des Elektrons ist rein elektromagnetischer Natur." (Abraham 1902b, p. 107).

[40] "[d]eren [i.e. Elektronen] Masse rein elektromagnetischer Natur ist." (Kaufmann 1903, p. 103).

in his theory, the electrons are treated as spheres of invariable dimensions. Lorentz (Lorentz 1904, p. 826)

Lorentz (1904, p. 829) concluded by pointing out that, in spite of this difference, his account was equally well in line with Kaufmann's 1903 results. This surprised Abraham, since he had argued that only a rigid electron could be fully electromagnetic in nature. He therefore carried out a detailed comparison of the two theories, listing both the shared hypotheses as well as those on which they differed. This showed him that Lorentz's electron could not be completely electromagnetic in nature, since his deformation-hypothesis required the introduction of non-electromagnetic forces:[41]

> If one accelerates such an electron, its flattening is increased; thus work must be done against the electric forces. While for the undeformable electron the increase of the energy is equal to the work done by the external electric forces, this is not the case here anymore; the increase of energy accompanying an increase in velocity is greater than the work performed by the external forces. The consequence of this assumption is therefore that one includes, besides the inner electromagnetic forces, other, non-electromagnetic inner forces, which together determine the form of the electron. These would then do the required extra work during the contraction, which together with the work of the external forces would be equivalent to the increase in the electron's electromagnetic energy. As long as one does not specify by which law these forces are governed, the hypothesis system is incomplete. (Abraham 1904, p. 578)[42]

It was Poincaré (1906) who addressed this issue. He did this by first showing that Lorentz's transformation equations form a group, and that this fact could be used to establish the invariance of different physical properties over such transformations (Poincaré 1906, p. 146). After investigating how the electron's electromagnetic Lagrangian transformed under these equations, Poincaré turned to the question what would be required, given this transformation expression, to ensure the viability of a deformable electron. This led him to postulate an internal stress acting upon the electron, which made the difference in momentum found by Abraham (Cfr. Footnote 41) vanish. One advantage of this non-electromagnetic addition was that in this way the principle of relativity could be upheld (Poincaré 1906, pp. 164–165).

The principle of relativity was valid, on Lorentz's electron-theory, because the deformability of his electron allowed him to account for those experiments that

---

[41] Here, Abraham merely stipulated this claim. He elaborated it in more detail in his 1905 textbook on electromagnetism (Abraham 1905 59, pp. 201–208), by means of the two different expressions for electromagnetic momentum he had introduced in 1902 (in terms of the Poynting energy flux vector and in Lagrangian terms.

[42] "Beschleunigt man ein solches Elektron, so wird seine Abplattung vermehrt; es mus also gegen die elektrische Kräfte Arbeit geleistet werden. Während für das undeformierbare Elektron die Zunahme der Energie gleich der von den äusseren elektrischen Kräften geleisteten Arbeit ist, findet das hier nicht mehr statt; die Energiezunahme bei einer Geschwindigkeitsvermehrung ist grösser, als die Arbeit der äusseren Kräfte. Die konsequente Verfolgung der Hypothese zwingt also dazu, neben den inneren elektromagnetischen Kräften noch andere, nicht elektromagnetische, innere Kräfte anzunehmen, welche im Verein mit jenen die Form des Elektrons bestimmen. Diese würden dann bei der Kontraktion die erforderliche Arbeit leisten, die zusammen mit der Arbeit der äusseren Kräfte der Steigerung der elektromagnetischen Energie des Elektrons äquivalent ist. Solange man nicht angiebt, nach welchem Gesetz diese Kräfte wirken sollen, ist das Hypothesensystem unvollständig." (Abraham 1904, p. 578).

had failed to detect any influence of the earth's motion through the ether. Abraham's electron-model, on the other hand, could not account for these null-results: according to his theory, it should be possible to detect certain influences. He did not see this as a problem, however, since

> [t]he question, if and why an influence of the earth's motion on electrical and optical phenomena on the earth's surface cannot be detected, can at present not yet be answered decisively. (Abraham 1904, p. 579)[43]

He believed that more was to be gained from new electron-experiments, especially involving Becquerel-rays, since these could attain velocities almost up to that of light, and hence could provide more insight into the precise velocity-dependency of the electron's mass. Kaufmann was carrying out such experiments in 1905–1906. These came to be seen as potentially offering a decision between the electromagnetic world-view and the relativistic approach, because, by that time, it was generally accepted that Lorentz's electron was not completely electromagnetic, and because Albert Einstein, in his first relativity paper (Einstein 1905a), had obtained a velocity-dependency expression that was considered equal to Lorentz's. Kaufmann's conclusion was quite direct:

> These results clearly decide against the validity of the Lorentzian and hence also of the Einsteinian theory; insofar as one takes this as a refutation of these theories, one should also take the attempt, to base the whole of physics including electrodynamics and optics on the principle of relativity, as failed. [...] For the moment, we will rather maintain the assumption that the physical phenomena depend on motion relative to a completely determined reference system, which we denote as the absolutely resting ether. Even if up until now such an influence of the movement through the ether by electrodynamic or optical experiments has not yet been shown, this should not be taken to mean that it is impossible to detect it. (Kaufmann 1906, pp. 534–535)[44]

---

[43] "Die Frage, ob und wieso ein Einfluss der Erdbewegungen auf die elektrischen und optischen Erscheinungen an der Erdoberfläche sich nicht entdecken lässt, ist zur Zeit noch keineswegs spruchreif." (Abraham 1904, p. 579).

[44] "Die vorstehenden Ergebnisse sprechen entschieden gegen die Richtigkeit der Lorentzschen und somit auch der Einsteinschen Theorie; betrachtet man diese aber als widerlegt, so wäre damit auch der Versuch, die ganze Physik einschließlich der Elektrodynamik und der Optik auf das Prinzip der Relativbewegung zu gründen, einstweilen als mißglückt zu bezeichnen. [...] Wir werden vielmehr einstweilen bei der Annahme verbleiben müssen, daß die physikalischen Erscheinungen von der Bewegung relative zu einem ganz bestimmten Koordinatensystem abhängen, das wir als den absolut ruhenden Äther bezeichnen. Wenn es bis jetzt nicht gelungen ist, durch elektrodynamische oder optische Versuche einen derartigen Einfluß der Bewegung durch den Äther nachzuweisen, so darf daraus noch nicht auf die Unmöglichkeit eines solchen Nachweises geschlossen werden." (Kaufmann 1906, pp. 534–535).

4 The Construction of a Relativistic Response

Many saw this conclusion as a significant challenge for the validity of the relativity principle.[45] In addition, Einstein as well, in one of his few explicit discussions of the experiments,[46] admitted that, if there was no error underlying them, the results could mean "that the foundations of the theory of relativity do not correspond to the facts" (Einstein 1907, p. 439).[47] This was not an endorsement of the electromagnetic worldview, however, since, as a worldview, Einstein considered it too restricted:

> […] their basic assumptions concerning the dimensions of the moving electrons are not suggested by theoretical systems that encompass larger complexes of phenomena." (Einstein 1907, p. 439).[48]

This restrictedness-claim was based on how he had obtained expressions for the electron's longitudinal and transverse mass in his first 1905 relativity paper (Einstein 1905a). There he first elaborated a kinematics that was concerned with the behaviour of rigid rods and clocks constrained by the principle of relativity and the principle of the constancy of the velocity of light. Applying this kinematics to what he called the Maxwell-Hertz equations for empty space then provided him with transformation equations for the electromagnetic field and the electromotive forces they could exercise. From these, he then obtained expressions for the electron's longitudinal and transverse mass by reflecting on how the electron's equations of motions transformed between frames in relative motion (Einstein 1905a, p. 919). Einstein immediately emphasized two aspects of this derivation:

> Of course, with a different definition of force and acceleration we would obtain different numerical values for the masses; this shows that we must proceed with great caution when comparing different theories of the motion of the electron. It should be noted that these results concerning mass are also valid for ponderable material points, since a ponderable material point can be made into an electron (in our sense) by adding to it an *arbitrarily small* electric charge. (Einstein 1905a, p. 919)[49]

---

[45] See e.g. the reactions by Abraham (1908, p. 196), Sommerfeld et al. (2000, p. 251), Lorentz and Kox (2008, p. 203) and Poincaré (1906, p. 132). This aspect of Kaufmann's experiments has been discussed quite extensively (Cfr. Cushing 1981; Miller 1905; Janssen and Mecklenburg 2006; Staley 2008; Janssen 2009; Potters 2019; Giovanelli 2020).

[46] The only other place where he discussed the experiments in some way was in a small article from 1906 (Einstein 1906), in which he proposed an alternative way to measure the velocity-dependency of the electron's mass. Insofar as is known, these experiments were not carried out at the time (Miller 1905, p. 343).

[47] "daß die Grundlagen der Relativitätstheorie nicht den Tatsachen entsprechen." (Einstein 1907, p. 439). English translations of Einstein's papers and correspondence are freely consultable via the Einstein Papers Project: https://einsteinpapers.press.princeton.edu.

[48] "weil ihre die Maße des bewegten Elektrons betreffenden Grundannahmen nicht nahe gelegt werden durch theoretische Systeme, welche größere Komplexe von Erscheinungen umfassen." (Einstein 1907, p. 439).

[49] "Natürlich würde man bei anderer Definition der Kraft und der Beschleunigung andere Zahlen für die Massen erhalten; man ersieht daraus, daß man bei der Vergleichung verschiedener Theorien der Bewegung des Elektrons sehr vorsichtig verfahren muß. Wir bemerken, daß diese Resultate über

First, Einstein's minimal conception of an electron—nothing more than a negatively charged body—meant that his derivation concerned not merely the electron, but all material bodies in general. Second, depending on the specific mechanics (in the form of a definition of force and acceleration) used, one could obtain different expressions. The definition used by Einstein was the standard Newtonian one: "[n]umerical value of mass × numerical value of acceleration = numerical value of force" (Einstein 1905a, p. 919).[50] However, as Kaufmann (1906, pp. 530–531) was the first to point out, this provided expressions that differed slightly from Lorentz's (Miller 1905, p. 329) for a short discussion of this). One rather had to use a relativistic mechanics, which was first developed in 1906 by Max Planck (Planck 1906).

According to Planck, the relativity principle required that one replaced the ordinary Newtonian equations of motion for a free point particle ('eines freien Massenpunktes'), formulated in terms of mass and acceleration, with equations of motion formulated in terms of the time derivative of momentum, where the notion of time was to be understood in relativistic terms (Planck 1906, p. 139). In this way, Planck pointed out, one obtained a mechanics according to which the shape of entities such as the electron would change with velocity. While he acknowledged earlier discussions about how to conceptualize the work required to carry out such deformations (a reference, without mentioning any names, to Abraham), Planck did not see these as significant, since this work could be accounted for completely in terms of the electron's kinetic energy. Proceeding in this way had two advantages, according to Planck: first, there was no need to specify whether the electron's mass was completely electromagnetic or not; and second.

> [...] it is not necessary to ascribe to the electron a spherical shape, nor in fact any specific shape at all, in order to obtain a determinate dependency of inertia on velocity. (Planck 1906, p. 137)[51]

In this way, Planck's work indicated that, from a relativistic perspective, two of the main issues on which the electromagnetic worldview had focused—the electron's shape and the source of its inertial mass—were not really significant at all. In two articles from 1908, Planck then further elaborated these claims.

In the first (Planck 1908a), he argued that work by his former doctoral student Kurt von Mosengeil on a relativistic account of thermal radiation,[52] showed that there were many different ways in which mass could be velocity-dependent, depending on how one conceptualized a body's inertial mass (in this way broadening Einstein's claim

---

die Masse auch für die ponderablen materiellen Punkte gilt; denn ein ponderabler materieller Punkt kann durch Zufügen einer *beliebig kleinen* elektrischen Ladung zu einem Elektron (in unserem Sinne) gemacht werden." (Einstein 1905a, p. 919).

[50] "Massenzahl × Beschleunigungszahl = Kraftzahl." (Einstein 1905a, p. 919).

[51] "daß man dem Elektron weder Kugelgestalt noch überhaupt irgend eine bestimmte Form zuzuschreiben braucht, um zu einer bestimmten Abhängigkeit der Trägheit von der Geschwindigkeit zu gelangen." (Planck 1906, p. 137).

[52] Cfr. Goldberg 1976, pp. 133–137; Miller 1905, p. 360, for discussions of von Mosengeil's work. After his death in 1906, Planck had his work published as (Mosengeil 1907).

that it depended on the mechanics used).[53] This led Planck (1908a, p. 27) to question the significance of searching for the specific source of that velocity-dependency, since it became almost entirely an issue of definition ('Definitionssache'). In fact, the question no longer made much sense, Planck continued, since the notion of inertial mass,

> […] which in [classical] mechanics plays such a fundamental role, is reduced in [relativistic] dynamics to a secondary notion" (Planck 1908a, p. 27).[54]

From a relativistic perspective, Planck (1908a, p. 30) claimed, it made more sense to characterize physical bodies directly in terms of energy (and in a footnote, he then pointed out that this was, in essence, a generalization of the mass-energy equivalence obtained earlier by Einstein (Einstein 1905b)).

In the second article (Planck 1908b), Planck then reflected on what this all meant for one principle in particular, namely the action-reaction principle. He started by pointing out that the principle's universality had recently been threatened by Lorentz's electromagnetic theory. It was Abraham who saved the principle's generality, Planck (1908b, pp. 828–829) then argued, by introducing a second form of momentum, namely an electromagnetic one, besides the already existing mechanical one. Abraham argued for this, Planck (1908b, p. 829) claimed, by means of an analogy: just as the work-energy principle could only be upheld by introducing an electromagnetic form of energy, the action-reaction principle could only be maintained by introducing an electromagnetic form of momentum. According to Planck, however, there was an issue with this move:

> However, this comparison, which is in itself certainly indisputable, still leaves one essential difference untouched. For we already know a whole series of different kinds of energy: kinetic energy, gravity, elastic deformation energy, heat, chemical energy, and the addition of electromagnetic energy to these forms does not constitute a principled innovation. Regarding momentum, however, up until now we only know one kind: the mechanical one. While energy was already a universal physical concept beforehand, momentum was up until now a specifically mechanical concept, the reaction principle a specifically mechanical law; and because of that, the required extension has to be seen as also a fundamental change, which has turned the up until now relatively simple and uniform concept of momentum into quite a complicated one. (Planck 1908b, p. 829)[55]

---

[53] Planck (1908a, p. 28) distinguished, more specifically, between transverse mass, longitudinal isothermal-isochoric mass, longitudinal adiabatic-isochoric mass and longitudinal adiabatic-isobaric mass. For discussions of this work by Planck see: Miller 1905, pp. 361–362; Potters 2019, p. 79.

[54] "Diese Größe, welche in der reinen Mechanik eine so fundamentale Rolle spielt, sinkt in der allgemeinen Dynamik zu einem sekundären Begriff herab." (1908a, p. 27).

[55] "Indessen läßt dieser an sich gewiß unanfechtbare Vergleich doch noch einen wesentlichen Unterschied unberührt. Denn bei der Energie kennen wir ohnehin schon eine ganze Reihe verschiedener Arten: die kinetische Energie, die Gravitation, die elastische Deformationsenergie, die Wärme, die chemische Energie, und es bedeutet daher keine prinzipielle Neuerung, wenn man diesen verschiedenen Formen als eine weitere Form noch die elektromagnetische Energie angliedert. Dagegen bei der Bewegungsgröße kannte man bisher nur eine einzige: eben die mechanische. Während die Energie von vornherein schon einen universellen physikalischen Begriff darstellt, war die Bewegungsgröße bisher speziell ein mechanischer Begriff, das Reaktionsprinzip ein speziell

The theory of relativity offered a way to overcome this issue, Planck claimed, by conceptualizing momentum in terms of energy, more specifically as.

> [...] that vector, which denotes the flow of energy, not only Poynting's electromagnetic energy flux, but rather energy flow in general. (Planck 1908b, p. 829)[56]

This flow, Planck continued, could be brought about in many different ways, but by dividing the vector expressing it by c^2 one obtained a very general expression for momentum independent of the nature of the energy flow's source. Planck (1908b, p. 830) called this result the law of inertia of energy, and, as Miller put it, it in fact offered

> [...] a generalization of Einstein's mass-energy equivalence which included the flow of any sort of energy, and not just the total mechanical energy [and which] asserted that the effect of forces acting on a body was transmitted by a momentum density whose source was a flow of energy. (Miller 1905, p. 366)

In the last part of the paper, Planck then argued that in the case of electromagnetic momentum, this energy flow would take on the form of the Maxwell stress tensor. The significance of this result was that it addressed the issue that had plagued Lorentz's theory, namely how to conceptualize the action-reaction principle given an unmovable ether:

> It is remarkable how, through this law [of inertia of energy], the Maxwell stresses obtain a physical significance for the theory of the unmovable ether as well. For, as force of pressure these stresses have no real meaning in this theory, since one cannot really make any sense of a force, which acts on something absolutely unmovable. [Here, Planck refers in footnote to Lorentz's work] That the Maxwell stresses have nevertheless maintained themselves in the theory of the unmovable ether by the fact that they often proved to be a convenient mathematical aid for certain calculations, even though they were, so to speak, officially abolished, could suggest that they do play some significant physical role, which they also played in the theory of the unmovable ether. (Planck 1908b, p. 830)[57]

The struggle of Lorentz's theory with the action-reaction principle, we have seen, took on the form of the question how it could take the ether to be unmoveable,

---

mechanischer Satz, und daher mußte die als notwendig erkannte Erweiterung immerhin auch als eine Umwälzung prinzipieller Art empfunden werden, durch welche der bisher verhältnismäßig einfache und einheitliche Begriff der Bewegungsgröße einen erheblichen komplizierten Charakter erhält." (Planck 1908b, p. 829).

[56] "denjenigen Vektor [...], welcher die Energieströmung ausdrückt, aber nicht allein die Poyntingsche elektromagnetische Energieströmung, sondern die Energieströmung ganz im allgemeneinen." (Planck 1908b, p. 829).

[57] "Es ist bemerkenswert, wie durch diesen Satz die Maxwellschen Spannungen auch für die Theorie des ruhenden Äthers eine physikalische Bedeutung gewinnen. Denn als Druckkraft haben diese Spannungen in dieser Theorie keinen rechten Sinn, da man doch einer Kraft, die auf etwas absolut Unbewegliches wirkt, nicht wohl eine Bedeutung beimessen kann. Daß die Maxwellschen Spannungen sich dennoch, trotzdem sie sozusagen offiziel abgeschaffen waren, in der Theorie des ruhenden Äthers behauptet haben, indem sie sich eben für gewisse Rechnungen häufig als bequemes mathematisches Hilfsmittel erwiesen, konnte schon den Gedanken nahelegen, daß ihnen doch irgendeine besondere physikalische Rolle zufällt, durch die sie auch für den ruhenden Äther legitimiert werden." (Planck 1908b, p. 830).

while the theory's equations of motion for the electron at the same time entailed that a charge in motion should give rise to ether movements, in the form of Maxwell stresses and an energy flow in the ether (formally denoted in terms of variations of the Poynting energy vector with time). Abraham's solution was to erase the distinction between matter and ether underlying Lorentz's (and Wien's) account, and to claim that all inertia and momentum was electromagnetic in nature, regardless of whether one ascribed it to the electron, the ether or to matter (which were in any case all of the same nature).

According to Planck, however, Abraham's claim went too far. The Poynting vector in itself did not give any reasons to assume that the source of the energy flow was electromagnetic in nature, since it could denote the flow of any sort of energy. Moreover, if one conceptualized the electron's motion directly in terms of such energy flows, there was also no longer any reason to develop theories concerning the electron's precise shape. In this way, we come to see how Planck significantly elaborated and extended Einstein's criticism of the electromagnetic approach, that as a worldview it was too restricted, since it unnecessarily restrained itself only to electromagnetic interactions, and turned it into the claim that, from a relativistic perspective, the issues that were primordial for the electromagnetic worldview were either of no real significance (the electron's shape and the source of its inertial mass) or could be incorporated as secondary notions (inertial masses and the Maxwell stresses). This response, I will argue now, can be characterized as Hertzian in a specific kind of way.

5 Two Hertzian Responses

We have seen that Einstein obtained mass-expressions by conceptualizing the electron as a point particle endowed with an arbitrarily small electric charge. And Planck equally well claimed that specifying a body's shape was not required to obtain a definite velocity-dependency. Their approach hence differed from that of e.g. Abraham (as well as Wien and Lorentz), who claimed that one had to conceptualize the electron as extended in order to obtain any results. Moreover, one equally well had to introduce a quasi-stationary approximation, since otherwise the question would become too complex to resolve.

The importance of this approximation, however, lay not solely in the reduction of complexity. As Miller has shown, it also allowed them to elaborate electromagnetic analogues for the laws of mechanics: on Abraham's account.

> […] unambiguous identification of the electron's mass as the coefficient of its acceleration required restricting the particle to 'quasi-stationary acceleration'. (Miller 1905, p. 59)

This shows how the electromagnetic worldview, even though it proclaimed to replace the mechanical worldview, was still working within the traditional Newtonian framework in which forces are to be understood in terms of mass times acceleration. While Einstein also still employed this conceptualization, Planck soon pointed out that the theory of relativity in fact required a thorough reconceptualization of this framework, which led him to reduce the notion of inertial mass to only secondary importance, and to the claim that questions concerning the electron's shape or the velocity-dependency of its mass were of no real significance. In this way, Planck thus not solely elaborated a relativistic mechanics, but equally well questioned the electromagnetic worldview's methodological approach. As we have seen, both Wien and Abraham searched for laws that were almost exact electromagnetic analogues of the laws of mechanics. They believed that, in this way, they were elaborating a simple and unified electromagnetic worldview underlying these laws. In this way, however, they did not so much offer an alternative for mechanics, according to Planck, but rather just endowed the existing mechanical framework with an electromagnetic coating. They did not see, he argued, that elaborating a real alternative to the mechanical worldview meant thoroughly reconceptualising some of its fundamental concepts, such as force, mass, and energy. In this sense, the electromagnetic worldview did not go far enough methodologically speaking, according to Planck, since it retained the Newtonian conceptual framework, rather than radically reformulating our fundamental physical notions, as he and Einstein (and others) did in their relativistic reconceptualization of time, space, energy, inertial mass, etc.

At the same time, one could also say that, according to Planck, the electromagnetic worldview went too far, ontologically speaking, since there was no reason to assume that the mass of electrons and of all material bodies had to derive from one particular source, i.e. electromagnetism. Rather, as his law of the inertia of energy showed, energy flows could have different kinds of sources, and that was all: nothing more regarding the nature of e.g. inertial mass could be inferred from it than that it arose from an energy flow that could be captured by means of the Poynting energy vector. This criticism of the electromagnetic worldview was formulated very poignantly by Max von Laue[58]:

> [We often see] the mistaken belief, that the common behaviour of all forces under the Lorentz transformations points to a common origin, that all of them can be traced back to electrodynamic forces. The question, whether that is at all possible, is completely outside of our concerns. That commonality says nothing else than that the principle of relativity is valid in all areas of physics; and this we must assume, if this principle is to be more than a sometimes useful calculation rule. To conclude more from it would be as hasty as if one wanted to conclude from the general validity of the energy principle that all natural processes are in the end mechanical. (Laue 1911, p. 186)[59]

---

[58] Max von Laue made this claim in 1911 in what, according to Staley (2008, p. 334), was "the first and authorative textbook on relativity". In the textbook, von Laue (who worked in Berlin until 1909 as doctoral student and assistant of Planck) elaborated the relativistic dynamics developed by Einstein and Planck in Minkowskian terms (Goldberg 1976, p. 127). For extensive discussions of von Laue's work, see: Janssen (2009), Jassen and Mecklenburg (2006).

[59] "[...] der Irrtum, daß das gemeinsame Verhalten aller Kräfte gegen die Lorentz-Transformation auf einen gemeinsamen Ursprung von ihnen hinwiese, daß sich etwa alle auf elektrodynamische Kräfte zurückführen ließen. Die Frage, ob so etwas möglich ist, steht gänzlich außerhalb unserer Betrachtungen. Jene Gemeinsamkeit sagt nichts anderes aus, als daß das Relativitätsprinzip in allen Gebieten der Physik gilt; und dies müssen wir annehmen, wenn dies Prinzip mehr sein soll, als eine manchmal nützliche Rechnungsregel. Mehr daraus zu folgern, wäre so voreilig, als wenn man etwa aus der Allgemeingültigkeit des Energieprinzips schließen wollte, daß alle Naturvorgänge in letzter Linie mechanische sind." (Laue 1911, p. 186).

We have seen how some of the adherents of the electromagnetic worldview—in particular Wien and Abraham—explicitly presented their work as accounting for Hertz's *Principles*: their electron-theories, they believed, could account for how mass and force, and the Newtonian laws that described how these concepts were related, came about out of interactions between extended charged bodies and the ether. In this way, they also believed to be able to go beyond Hertz's mechanics, since while Hertz had claimed that the nature of force and mass was hidden, their theories showed that this nature was in fact electromagnetic. In this way, the electromagnetic worldview offers a first approach that was, one could say, inspired by Hertz's *Principles.*

The discussion of Planck's relativistic response to the electromagnetic worldview now suggests that we can delineate a second Hertzian approach as well (even though Planck never explicitly refers to Hertz). In line with Hertz's epistemological claim that from an image whose consequences are in conformity with the consequences of the things themselves, we cannot infer anything about the nature of the objects themselves beyond this conformity, Planck (and Laue) argued that, from the perspective of the theory of relativity, there was nothing to be inferred from the law of the inertia of energy regarding the nature, electromagnetic or otherwise, of the energy flows at play in different phenomena. In fact, one could say that Planck rendered the electromagnetic ether hidden in the same way as Hertz rendered the mechanical notion of force hidden. As we have seen, on Hertz's account, what was hidden was not of a special nature: like the observable parts of reality, it consisted of masses in motion, which were just not observable. Elaborating this theoretically then led him to a view according to which force was just a mathematical auxiliary construction. Similarly, on Planck's account, there was no special, underlying substance—the ether—underlying electromagnetic phenomena: rather, as we see above, it shows how, on his view, the stresses at play in the ether were just as well energy flows, and in this way, it becomes clear that the concept of an ether, insofar as it can play a role in electromagnetic theories, is merely a convenient mathematical aid, as the notion of force was for Hertz.

A final similarity between Hertz's work and the debate discussed here concerns the issue at play. As we have seen, Hertz was concerned with an obscurity at play in how the three laws of mechanics conceptualized force: whereas according to the first two laws, forces act on a body in a particular direction, according to the third law a force connects two bodies. Similarly, the issues with the action-reaction principle at play in the case discussed here emerged out of a tension between the principle's demand for a bidirectional interaction between ether and matter on the one hand, and their ether-model on the other, according to which the interaction between ether and matter could only be uni-directional: while the ether could act on matter, matter could not act back on the absolutely unmovable ether. In addition, similar to how Hertz overcame the issue by reconceptualising the notion of force in terms of hidden masses in motion, Planck was able to overcome the issue of his concern by reconceptualising the notion of electromagnetic momentum in terms of hidden energy in motion.

This suggests that Hertz's *Principles* did in some sense shape certain debates on the foundations of physics, even though, as in the case of Planck, this was not always explicit. At the moment, this paper has only suggested such similarities, however, and it primarily opens up many new questions: in how far was it really Hertz's *Principles* that inspired the work of Wien, Abraham and Planck, and in how far was their view of Hertz shaped by other authors? Was it really Hertz's *Principles* that shaped Planck's work, or was it rather another author that influenced both Hertz and Planck (e.g. Helmholtz)? In how far are there differences between the hiddenness at play in Hertz's and in Planck's work? All these are questions for the future.

6 Conclusion

In this paper, I have argued that, contrary to what many historians have argued, Heinrich Hertz's *Principles of Mechanics* did inspire reflections within fundamental physics, concerning, among other things, the role played by the ether. I have argued, more specifically, that we can distinguish two different forms of inspiration. The first concerned the development of the electromagnetic worldview by Wilhelm Wien and Max Abraham. They proclaimed that their electromagnetic electron-models, which were concerned with how the interaction between charge and ether gave rise to the electron's mass, could account for Hertz's hidden masses-mechanics. This allowed them, they then claimed, to formulate electromagnetic analogues for the laws of mechanics.

The second could be found in the relativistic response to the electromagnetic worldview, elaborated in particular by Max Planck. This response can be considered as a Hertzian response, I have argued, in the sense that Planck carried out a reconceptualization of the concepts underlying the theory of mechanics that led him to claim that, from a relativistic point of view, the electromagnetic assumption made by Wien and Abraham was unjustified: models of the electron could not show that the electron's mass was completely electromagnetic in nature. Insofar as one could speak of the electron's nature, it was rather in terms of a Hertzian kind of hiddenness, embodied in Planck's law of the inertia of energy. This suggests, I conclude, that Hertz's *Principles*, while they did not explicitly inspire the construction of mechanical ether-models at the time, still profoundly shaped the discussions that were going on in the foundations of physics.

Acknowledgements This research has been funded by the Research Fund—Flanders (FWO) and by the Lichtenberg Group for History and Philosophy of Physics (Universität Bonn). I would like to thank Joshua Eisenthal, Dennis Lehmkuhl and the members of Lichtenberg Group for their encouragement and comments and questions.

References

Abraham M (1902a) Dynamik des Elektrons. Nachrichten von der Gesellschaft der Wissenschaften zu Göttingen, Mathematisch-Physikalische Klasse: 20–41.

Abraham M (1902b) Prinzipien der Dynamik des Elektrons. Annalen der Physik 315(1):105–179.
Abraham M (1904) Die Grundhypothesen Der Elektronentheorie. Physikalische Zeitschrift 5(18):576–579.
Abraham M (1905) Theorie der Elektrizität. Teubner, Leipzig, B.G.
Abraham M (1908) Theorie der Elektrizität. Zweiter Band. Elektromagnetische Theorie der Strahlung Zweite Auflage. B.G. Teubner, Leipzig.
Coudres TD (1899) Handliche Vorrichtung zur Erzeugung Lenard'scher Strahlen und einige Versuche mit solchen Strahlen. Annalen der Physik 298:134–144.
Cushing JT (1981) Electromagnetic mass, relativity, and the kaufmann experiments. American Journal of Physics 49:1133–1149.
D'Agostino S (1998) Hertz's views on the method of physics: experiment and theory reconciled? In: Baird D, Hughes R, Nordmann A (eds) Heinrich Hertz: classical physicist, modern philosopher. Springer, Dordrecht, pp. 89–102.
Darrigol O (1995) Henri Poincaré's Criticism of Fin De Siècle Electrodynamics. Studies in the History and Philosophy of Modern Physics 26(1):1–44.
Ehrenfest P (1907) Die Translation deformierbarer Elektronen und der Flächensatz. Annalen der Physik 23:204–205.
Einstein A (1905a) Zur Elektrodynamik bewegter Körper. Annalen der Physik 17:891–921.
Einstein A (1905b) Ist die Trägheit eines Körpers von seinem Energieinhalt abhängig? Annalen der Physik 18:639–641.
Einstein A (1906) Über eine Methode zur Bestimmung des Verhältnisses der transversalen und longitudinalen Masse des Elektrons. Annalen der Physik 21:583–586.
Einstein A (1907a) Über das Relativitätsprinzip und die aus demselben gezogenen Folgerungen. Jahrbuch Der Radioaktivität Und Elektronik 4:411–462.
Einstein A (1907b) Bemerkungen zu der Notiz von Hrn. Paul Ehrenfest: Die Translation deformierbarer Elektronen und der Flächensatz. Annalen der Physik 23:206–208.
Eisenthal J (2018) Mechanics without mechanism. Studies in History and Philosophy of Modern Physics 62:45–55.
Eisenthal J (2021) Hertz's Mechanics and a unitary notion of force. Studies in History and Philosophy of Science 90:226–234.
Giovanelli M (2020) Like thermodynamics before boltzmann: on the emergence of einstein's distinction between constructive and principle theories. Studies in History and Philosophy of Modern Physics 71:118–157.
Goldberg S (1967) Henri Poincaré and Einstein's theory of relativity. American Journal of Physics 35:934–944.
Goldberg S (1970) The Abraham Theory of the electron: the symbiosis of experiment and theory. Archive for History of Exact Sciences 7(1):7–25.
Goldberg S (1976) Max Planck's philosophy of nature and his elaboration of the special theory of relativity. Historical Studies in the Physical Sciences 7:125–160.
Grasshoff G (1998) Hertz's philosophy of nature in Wittgenstein's Tractatus. In: Baird D, Hughes R, Nordmann A (eds) Heinrich Hertz: classical physicist, modern philosopher. Springer, Dordrecht, pp 243–268.
Heaviside O (1889) On the electromagnetic effects due to the motion of electrification through a dielectric. London, Edinburgh, Dublin Philosophical Magazine and Journal of Science 27:324–339.
Hertz H (1894a) Gesammelte Werke von Heinrich Hertz. Band III. Die Prinzipien der Mechanik. Johann Ambrosius Barth, Leipzig.
Hertz H (1894b) Gesammelte Werke von Heinrich Hertz: Band II Untersuchungen über die Ausbreitung der elektrischen Kraft. Johann Ambrosius Barth, Leipzig.
Hirosige T (1969) Origins of Lorentz' theory of electrons and the concept of the electromagnetic field. Historical Studies in the Physical Sciences 1:151–209.
Hirosige T (1976) The ether problem, the mechanistic worldview, and the origins of the theory of relativity. Historical Studies in the Physical Sciences 7:3–82.

Janssen M (2009) Drawing the line between kinematics and dynamics in special relativity. Studies in History and Philosophy of Modern Physics 40:26–52.
Janssen M, Mecklenburg M (2006) From classical to relativistic mechanics: electromagnetic models of the electron. In: Hendricks VF, Jørgensen KF, Lützen J, Pedersen SA (eds) Interactions: mathematics, physics and philosophy 1860–1930. Springer, Dordrecht, pp. 65–134.
Jungnickel C, McCormmach R (1986) Intellectual mastery of nature: theoretical physics from Ohm to Einstein, vol 2. University of Chicago Press, Chicago, The Now Mighty Theoretical Physics.
Jungnickel C, McCormmach R (2017) The second physicist: on the history of theoretical physics in Germany. Springer, Cham.
Katzir S (2005) On "the electromagnetic world-view": A comment on an article by Suman Seth. Historical Studies in the Physical and Biological Sciences 36(1):189–192.
Kaufmann W (1901) Die magnetische und electrische Ablenkbarkeit der Becquerelstrahlen und die scheinbare Masse der Elektronen. Nachrichten von der Gesellschaft der Wissenschaften zu Göttingen, Mathematisch-Physikalische Klasse, pp. 143–155.
Kaufmann W (1902) Ueber die elektromagnetische Masse des Elektrons. Nachrichten von der Gesellschaft der Wissenschaften zu Göttingen, Mathematisch-Physikalische Klasse, pp. 291–296.
Kaufmann W (1903) Ueber die "Elektromagnetische Masse" der Elektronen. Nachrichten von der Gesellschaft der Wissenschaften zu Göttingen, Mathematisch-Physikalische Klasse, pp. 90–103.
Kaufmann W (1906) Über die Konstitution des Elektrons. Annalen der Physik 19:487–553.
Kragh H (2002) Quantum generations: a history of physics in the twentieth century. Princeton University Press, Princeton.
Laue von M (1911) Das Relativitätsprinzip. Fried. Vieweg & Sohn, Braunschweig.
Lorentz HA (1899a) Elektromagnetische Theorien Physikalischer Erscheinungen. Physikalische Zeitschrift 1(498–501):514–519.
Lorentz HA (1899b) Simplified theory of electrical and optical phenomena in moving systems. Proceedings of the Royal Netherlands Academy of Arts and Sciences 1:427–442.
Lorentz HA (1900) Considerations on gravitation XE "Gravitation." KNAW Proceedings 2:559–574.
Lorentz HA (1904) Electromagnetic phenomena in a system moving with any velocity smaller than that of light. KNAW Proceedings: pp. 809–831.
Lorentz HA, Kox A (2008) The Scientific Correspondence of H.A. Lorentz, Vol 1. Springer, New York.
Lützen J (2005) Mechanistic images in geometric form: Heinrich Hertz's principles of mechanics. Oxford University Press, Oxford.
Majer U (1998) Heinrich Hertz's picture-conception of theories: its elaboration by Hilbert, Weyl, and Ramsey. In: Baird D, Hughes R, Nordmann A (eds) Heinrich Hertz: classical physicist, modern philosopher. Springer, Dordrecht, pp. 225–242.
McCormmach R (1970) H.A. Lorentz and the electromagnetic view of nature. Isis 61:459–497.
Miller AI (198) Albert Einstein's Special Theory of Relativity: Emergence (1905) and Early Interpretation (1905–1911). Addison-Wesley Publishing Company, Reading.
Miller AI (1973) A Study of Henri Poincaré's "Sur La Dynamique de l'Électron." Archive for History of Exact Sciences 10:207–328.
Miller A (1980) On some other approaches to electrodynamics in 1905. In: Woolf H (ed). Some Strangeness in the Proportion: A Centenial Symposium to Celebrate the Achievements of Albert Einstein. Addison-Wesley, Reading, pp. 66–91.
Mosengeil von K (1907) Theorie der stationären Strahlung in einem gleichförmig bewegten Hohlraum. Annalen der Physik 22:867–904.
Mulligan JF (1998) The reception of Heinrich Hertz's principles of mechanics by his contemporaries. In: Baird D, Hughes R, Nordmann A (eds) Heinrich Hertz: classical physicist, modern philosopher. Springer, Dordrecht, pp. 173–182.
Mulligan JF (2001) The Aether and Heinrich Hertz's The principles of mechanics presented in a new form. Physics in Perspective 3:136–164.

Planck M (1906) Das Prinzip der Relativität und die Grundgleichungen der Mechanik. Deutsche Physikalischen Gesellschaft. Verhandlungen 8:136–141.
Planck M (1908a) Zur Dynamik bewegter Systeme. Annalen der Physik 26(4):1–34.
Planck M (1908b) Bemerkungen zum Prinzip der Aktion und Reaktion in der allgemeinen Dynamik. Physikalische Zeitschrift 9:828–830.
Poincaré H (1900) La théorie de Lorentz et le principe de réaction. In: Bosscha J (ed). Recueil des travaux offerts par les auteurs à H.A. Lorentz à l'occassion du 25me anniversaire de son doctorat, Martinus Nijhoff, Den Haag, pp. 252–278.
Poincaré H (1906) Sur la dynamique de l'électron. Rendiconti del Circolo Matematico di Palermo 21:129–175.
Potters J (2019) Heuristics versus norms: on the relativistic responses to the Kaufmann experiments. Studies in History and Philosophy of Modern Physics 66:69–89.
Riemann B (1875) Schwere, Elektricität und Magnetismus. K. Hattendorff, Hannover.
Searle G (1897) On the steady motion of an electrified ellipsoid. London, Edinburgh, Dublin Philosophical Magazine and Journal of Sciencc 44:329–341.
Seth S (2004) Quantum theory and the electromagnetic world view. Historical Studies in the Physical and Biological Sciences 35(1):67–93.
Seth S (2005) Response to Shaul Katzir. On the electromagnetic world-view. Historical Studies in the Physical and Biological Sciences 36(1):193–196.
Sommerfeld A, Eckert M, Märker K (2000) Wissenschaftlicher Briefwechsel. Verlag für Geschichte der Naturwissenschaften und der Technik, Berlin.
Staley R (2008) Einstein's Generation: the origins of the relativity revolution. The University of Chicago Press, Chicago.
Thomson JJ (1881) On the electric and magnetic effects produced by the motion of electrified bodies. London, Edinburgh, Dublin Philosophical Magazine and Journal of Science 5(11):229–249.
Thomson JJ (1893) Notes on recent researches in electricity and magnetism. Clarendon Press, Oxford.
Wiechert E (1894) Bedeutung des Weltäthers. Physikalisch-Ökonomische Gesellschaft Zu Königsberg 35:4–11.
Wien W (1898) Ueber die Fragen, welche die translatorische Bewegung des Lichtäthers betreffen. Annalen der Physik 301:I–XVIII.
Wien W (1901) Ueber die Möglichkeit einer elektromagnetischen Begründung der Mechanik. Annalen der Physik 310(7):501–513.

Chapter 16
Maxwell's Electromagnetism: A Matter of Symmetry

Dante Roberto Salatino

Abstract The use of analogies contributed notably to the development of many significant physical theories. The purpose of this work is to propose that the analogies used by Maxwell respond to a kind of 'language' expressed through universal relational patterns, which reveal the 'symmetries' that link the fundamental elements that explain electromagnetism. The complementary method used to investigate conceptual relationships was Transcursive Logic, which considers concepts from the observer's point of view, thus enabling the possibility of evaluating various analogies as a source of knowledge. These 'universal patterns' are composed, structurally, by a Galois permutation group, and functionally, by a 'Galois connection', which allowed the relation of the fundamental equations of Maxwell, thus becoming a valid instrument to explain the phenomenon of electromagnetism and its relationship with light.

Keywords Models · Analogy · Maxwell's laws · Group theory · Transcurssive logic

1 Introduction

Kline Wrote:

> On the banks of the Rhine, a beautiful castle had been standing for centuries. In the cellar of the castle, an intricate network of cobwebs had been constructed by industrious spiders who lived there. One day a strong wind sprang up and destroyed the web. Frantically the spiders worked to repair the damage. They thought it was their webbing that was holding up the castle. (Kline 1980, p. 277)

The Maxwell's equations, which helped to unify the electric and magnetic fields, from their definition to the causes of their origin, constitute perhaps one of the most

---

D. R. Salatino (✉)
National University of Cuyo, Mendoza, Argentina
e-mail: dantesalatino@gmail.com

R. Pisano (ed.), *A History of Physics: Phenomena, Ideas and Mechanisms*,
History of Mechanism and Machine Science 42,
https://doi.org/10.1007/978-3-031-26174-9_16

remarkable achievements of the human mind. Assessing this achievement has been the task of not a few historians of science, ranging from dimensional analysis and its innovative ideas (D'Agostino 2001) to the physics-mathematics relationship in electromagnetic theory (Pisano 2013), just to mention a few. In this work, we will emphasize, above all, what could be the determinants of the emergence of a theory that relied on as varied as fertile analogies, which some have considered as a true contribution to the definition of "physical theory" (D'Agostino 2001, p. 77). As D'Agostino says, the analogical vision of a theory that Maxwell had, was based on investigating whether the mathematical laws in a known area of physics could have some correspondence with unknown physical laws, where he assumed that the mathematics was underlying. This vision is not far from Peirce's (1878) abductive proposal. That is, obtain the laws that govern unknown phenomena, through the laws that regulate and control the phenomena that we know. In *A Treatise on Electricity and Magnetism* (Maxwell 1881), Maxwell admits the existence of various ways of observing a phenomenon, being cautious when stating that a theory could be demonstrated by an experiment. He even raised the challenging possibility that an experiment could be valid without having to be tied to a particular theory. What Maxwell tried to tell us, as we see it here, is that there are different ways to explain the same phenomenon tested by a single experiment. It all depends on the point of view from which it is approached. From the Transcursive Logic (TL) (Salatino 2017c), a complementary research method of the traditional scientist, we propose that objective phenomena can also be approached from the observer's point of view. This allows that, regardless of the field of science to which a particular phenomenon belongs, it can be approached by strictly respecting the method accepted by the scientific community and give valid explanations of it. Philosophy and Science, as diverse branches of knowledge, harbor a logic that sustains them and gives the possibility that the first justifies the second. However, this is not the only coincidence or relationship, since in addition to trying to find a set of statements about reality, constitute a method. The methodological value provided by both philosophy and science is multifaceted. Basing an argument to reach the truth, is proper to philosophize. While to elaborate a hypothesis on a real phenomenon, to look for an element with which to measure it and to elaborate an experiment that allows corroborating the hypothesis, is a scientific task. Despite the apparent methodological differences, both pursue a common goal to explain the facts. The nature of this explanation, although it is shown in different ways according to the perspective from which it is analyzed, obeys a composition of similar basic elements. In this paper, we will try to base a useful methodology to approach the investigation of real phenomena, in search of an adequate explanation of them, elaborated from the perspective of the observer.

2 Explanation

Explanation is not so much to clear the facts of what prevents us from understanding, as an attempt to make them compatible with the beliefs, principles, and convictions that make up our knowledge. Given that reality manifests itself to us as a continuous, undefinable, and persistent process, explaining its facts will require making at least some of its fundamental aspects ostensible. Otherwise, we run the risk that Einstein's assertion when he jokingly analyzed quantum physics: "*Its math is brilliant, but its physics is a disaster*", will come true. Following the proposal by Aristotle (Aristóteles 2007, II, Chap. 3, *The causes*) we will start from the mere apprehension of the "what" to reach the first cause, to true knowledge, to an explanation of "why" the phenomenon is analyzed (Fig. 1). We could assimilate a real phenomenon by using a day-to-day frame of reference to the Cartesian system that we use to represent space. In its three dimensions, we will have three basic aspects that will allow us to characterize any phenomenon: "what" makes it manifest, "how" it does it and "when" it occurs.

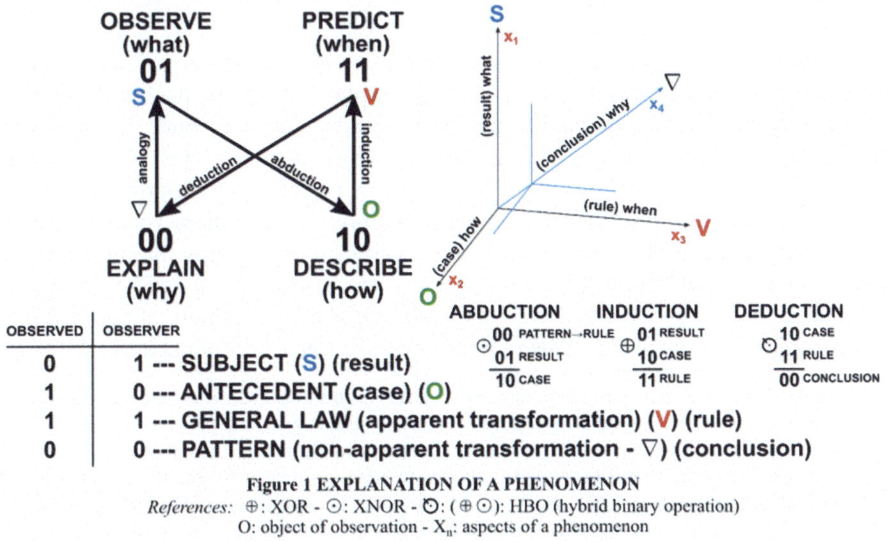

Fig. 1 Explanation of a phenomena *Source* Own production

To achieve our goal, we will turn to the argumentative options of the basic modes of reasoning provided by Philosophy (Aristóteles 2004), Logic (Peirce 1878) and Science (Samaja 2005). On the other hand, we will take with a certain license and as a guide, the nomological-deductive model of Hempel (1965).

In Fig. 1, we see a subject (S = 1D; unidimensional) observing the "what" of any phenomenon (1D). This observation is not made impartially but is conditioned by a "pattern" (Samaja 2005, p. 106) taken by analogy, as an "implicit universal individual" (Hegel 1982, II, p. 392). This way of "observing" will allow the observer, to extract from the "analogous rule" (known), the rule of the analogate (unknown) phenomenon. The "analogous rule" does not derive from observation itself, but from praxis. By an abductive mechanism (rule (pattern) + result → case), which contributes on existence and actuality, we can achieve the "how", that is, describe the observed phenomenon, thus obtaining the antecedent event demanded by the nomological-deductive model, as its *explanandum* or object (O = 2D; two-dimensional) of the observation (Hempel 1965, p. 246).

An inductive mechanism (result + case → rule) that tells us about generality allows us to predict the phenomenon ("when"). That is, to assign a general law, which is none other than the one provided by the "pattern", and so, based on experience (in a habit or custom) (Hume 1748–1894, p. 44) to expect in the future a succession of events similar to those that appeared in the past, in similar situations. We have reached the third dimension (3D) that characterizes our phenomenon, or the "general law" that in appearance, directly relates the observer subject and the observed object (V = 3D; three-dimensional).

Finally, a deductive mechanism (case + rule → conclusion) that promotes possibility and potentiality allows us to arrive at an explanation of the phenomenon that we are analyzing. The explanation obtained is independent of both the observer and the observed phenomenon, that is why it is "dimensionless" (∇ = 0D). One's "absence" in the Cartesian system tells us that it is projected in a fourth dimension (4D = coordinate axis inscribed in the figure). It represents an indirect (or qualitative) relationship between the observer and the observed, so it is impossible to detect it. Something like what Maxwell was able to discover when he tried to explain the 'lines of force' suggested by Faraday, who also believed in the usefulness of the analogy of mechanical models of magnetism, as he shows in one of his publications:

> It is not to be supposed for a moment that speculations of this kind are useless, or necessarily hurtful, in natural philosophy. They should ever be held as doubtful, and liable to error and to change; but they are wonderful aids in the hands of the experimentalist and mathematician. For not only are they useful in rendering the vague idea more clear for the time, giving it something like a definite shape, that it may be submitted to experiment and calculation; but they lead on, by deduction and correction, to the discovery of new phenomena, and so cause an increase and advance of real physical truth, which, unlike the hypothesis that led to it, becomes fundamental knowledge not subject to change. (Faraday 1855, p. 408)

3 Universal Pattern

In the previous point, we have verified that it is possible to move from the "what" to the Aristotelian "why" of a phenomenon, taking as a guide a pattern that provides the general law from which, by deduction, arise a conclusion, an explanation, or the cause (s) determinant (s) of the phenomenon that we are analyzing. This was possible by integrating inductive and deductive reasoning with reasoning by analogy (abduction). Such an elaboration would cease to be pure conjecture if we were able to find as a pattern an "implicit universal individual," inductively providing a general law based on a "sample" that encompasses the entire universe (ideal induction). In this way, it is possible to avoid the criteria of the critical approach that Hempel makes about induction, in his logical theory of confirmation (Hempel 1965, p. 5). Thus, we propose a complementary research method. The main foundation of this proposal lies in having discovered that the interrelationships maintained between the fundamental aspects that define any phenomenon (real or not), constitute a "universal pattern" based on a group of transformations. An invariant that respects the principle of symmetry and, therefore, adheres to the principle of "uniformity of nature" elaborated by Mill (1882, p. 400), as the foundation of induction. Principle confirmed by Emmy Noether (1918) when she demonstrated the conservation of natural laws over time, by relating in her theorem, the symmetry of a system with quantities that are conserved. The first outline of a possible universal pattern emerges when reading the "Poetics" of Aristotle:

> I explain the metaphor by analogy as what can happen when, of four things, the second remains in the same relation to the first as the fourth to the third; then you can talk about the fourth instead of the second, and the second instead of the fourth. (Aristóteles 2011, Chap. XXI, 1457b)

In symbols, it would be 2 : 1 :: 4 : 3. That is, *the second is to the first as the fourth is to the third*. This, in addition to expressing two oppositions by complementarity, represents the germ of all Aristotelian logic. It is curious, but the DNA molecule, the basis of the genetic code and life, turned out to be structured in the same way. Figure 2 shows the arrangement of its nitrogenous bases (cytosine, guanine, adenine, thymine) in opposite pairs. Both the decimal codes (which give their order or "relational valence") and the binary codes (which identify the pairs of opposite and complementary bases) assigned, have a solid foundation. The former is based on their electrochemical structure (number of nitrogen atoms with *sp2*[1] hybrid orbitals).

[1] *Orbital Hybridization*: is the mixing of atomic orbitals into new hybrid orbitals, with different energies, sharpes, etc., than the component atomic orbitals. The *sp2* hybridization is the mixing of one s and two p atomic orbitals, which involves the promotion of one electron in the *s* orbital to one of the *2p* atomic orbitals. The combination of these atomic orbitals creates three new hybrid orbitals equal in energy-level. The hybrid orbitals are higher in energy than the *s* orbital but lower in energy than the *p* orbitals, but they are closer in energy to the *p* orbitals. The new set of formed hybrid orbitals creates trigonal structures, configuring a molecular geometry of 120° (Petrucci 2003, p. 439), very similar to that acquired by the superficial (or quantitative) level of a PAU.

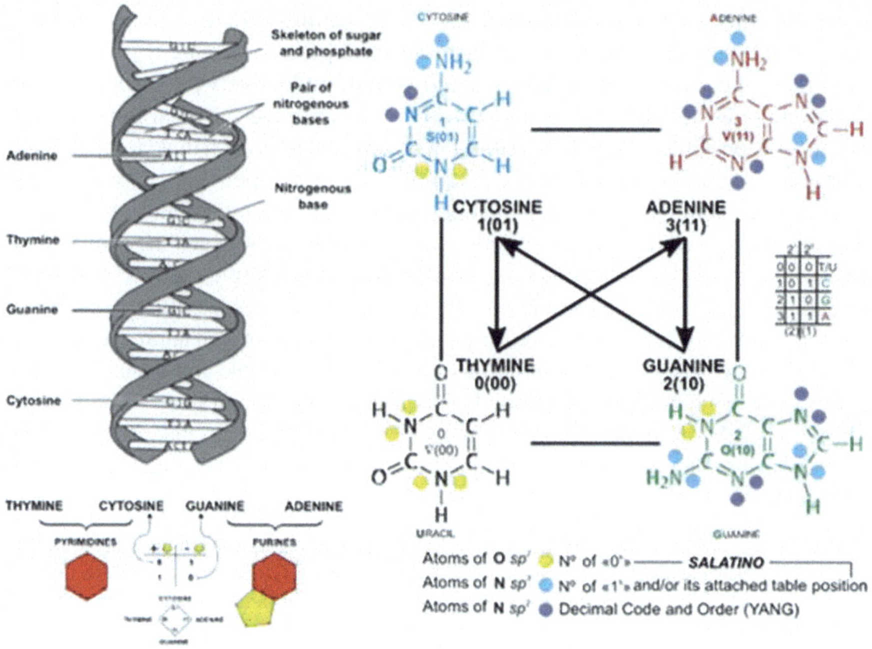

Fig. 2 DNA molecule (Salatino 2009) *Source* Own production

The assignment of binary codes to the basic components of the DNA molecule allowed us to demonstrate that this composition of abstract elements conforms to, structurally, a permutation group or Galois group. From the functional point of view, it also represents a "connection" of Galois, that is, the opposition of two aspects through another opposition.

> Different interests that lead to different procedures can gather evidence that can then materialize in an empirically grounded worldview. (Feyerabend 2001, p. 184)

The aphorism of Van Fraassen (Fraassen 1989) "Similar problems have similar solutions", and the fact that DNA is a "universal pattern", a series of practical examples were taken to demonstrate the relevance of the method of searching for a basic relational pattern as part of the solution of a given problem, as discussed below. In Transcursive Logic the key element to activate its basic principles is symmetry. Through it, analogies acquire the range of laws that may apply to different fields of study, although they are apparently dissimilar. Now, when we talk about symmetry, we are not referring to the intuitive idea that we have of it, such as the bilateral similarity of the human body and most animals, or the radial regularity of flowers, or the spatial harmony of numerous natural or man-made objects. Nor does the one suggested by Ampère, which exists, for example, between the parts of a circuit (D'Agostino 2001, p. 12), but rather the one whose laws were established with the help of the "invariants" of Emmy Noether (1918).

4 Generic Group (Properties): PAU (Universal Autonomous Pattern)

To address the different approaches, to which we should find a solution or an explanation, a tool based on a generic Galois group was built. Isolating the essential or relevant structure of a phenomenon is equivalent to defining a group of transformations that, once applied, leave the problem, essentially, in the same situation from where it started (invariance). These transformations are the "symmetries" of the problem (Noether 1918) With the essential aspects formed a group (a logical-structural arrangement) which we call PAU (Universal Autonomous Pattern), and the solution consists of a rule (a function) that depends only on those basic parameters (Fig. 3).

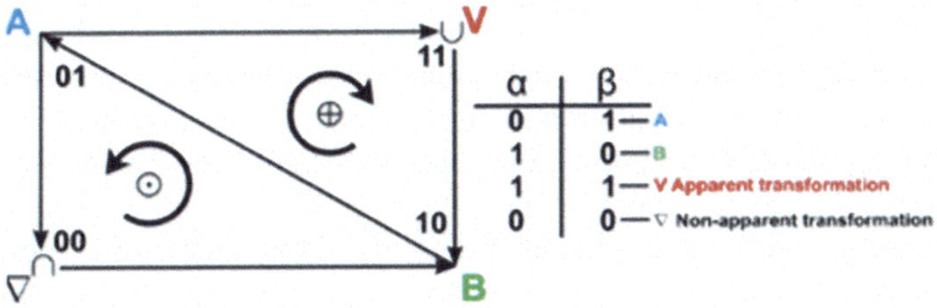

Fig. 3 Generic group (Salatino 2009) *Source* Own production

In the previous figure we see two essential aspects of any phenomenon (A and B) and the relationships that they keep with each other through two transformations: a superficial or apparent (V) and another hidden or profound (∇). The way we must identify each of these elements is through the individualization of some characteristic that is specific to each essential aspect (α and β). According to the resulting binary codes, A is identified as having the characteristic β, but not the α. On the contrary, B is identified as having the characteristic α, but not the β. This is telling us that A and B are opposites concerning the characteristics α and β, which are mutually exclusive. The transformations are distinguished because, the superficial (V) has both characteristics (α and β) of the aspects it relates (A and B), is what we call co-presence or organization, or what is evidenced in any phenomenon that is being observed. The profound transformation (∇), on the other hand, shows a total absence of the characteristics that identify the aspects it relates (co-absence or disorganization). As we can see, as with the essential aspects (A and B), here the transformations V and ∇ are opposite. Both pairs are also complementary, since added give the apparent unit (11), which is nothing but the evident or superficial transformation (the organization). It is thus formed an opposition mediated by another opposition (Galois connection). Kline's quote, in the epigraph of the paper, summarizes these two aspects. In appearance, the web of cobwebs supports the castle, according to the belief of the spiders. However, the wind revealed that the castle was always standing, because it was not supported by the web.

In the scheme of the figure, it is observed that between A and B there exists, in addition to the direct relationship, mediated by a transformation that we have described, an indirect relation between them shared by the two cycles that have been formed. A superficial one dextro-rotated (in the clockwise direction) that represents the symmetries of rotation of the system and constitutes the "objective or quantitative pole", and a deep one of left-handed rotation (in the anti-clockwise direction) as an evidence of reflection symmetries of our universal pattern and represents the 'subjective or qualitative pole'. The opposite turns are another way of complementarity that ensures the simultaneity in the operation of both cycles, where both the quantitative and the qualitative aspects of any real phenomenon or not must be present at the same time (heterarchy).

Finally, so that the whole system is not just an inert structure, there must be a rule or function that sets it in motion. This rule ($\oplus$ = XOR) allows to 'move' superficially to the right (quantitative or apparent transformation) each of the elements of the level to take the place of its successor in the sequence, without losing its own identity (symmetry of rotation), until the system returns to the beginning from where it started. With the previous operation, we assume that we have reached the solution of the problem posed by the observable aspects of the phenomenon. At a deeper level, and as it could not be otherwise, the rule or function used is the opposite of the superficial one ($\odot$ = XNOR or equivalence), which allows elements of the system to be moved to the left. Here, when the system returns to the original disposition, after the successive qualitative transformations, we obtain the complete solution to the problem (reflection symmetry). In short, it is this level, that aspects that characterize a phenomenon, the determinant of what the same phenomenon, shows us in appearance. What unites the two levels is the experience that arises from having solved, sometimes, a similar problem. The relational structure that we have used meets the characteristics that identify a permutation or Galois group.

(1) *Closure*: the application of a composition transformation (XOR) to a pair of its evident elements produces another element that belongs to the set.
(2) It has a neutral or *identity element* (00) such that composed with any other element it does not modify it.
(3) Each element of the set has its *inverse*, such that composed by a transformation, they give the neutral element.
(4) *Associative property*: all the compositions achieved through a certain transformation are independent of their grouping.
(5) The *closing of the conjugate*: there is an opposite transformation to the composition (XNOR) that, applied to the non-evident elements, produces another element of the set.

5 PAUs Library

To face the analysis of a great diversity of phenomena, which we consider as if they were systems, we have several types of PAUs. These relational structures are always shaped by the same generic and fundamental elements, only varies in each type, the sequence of their relationships (Fig. 4).

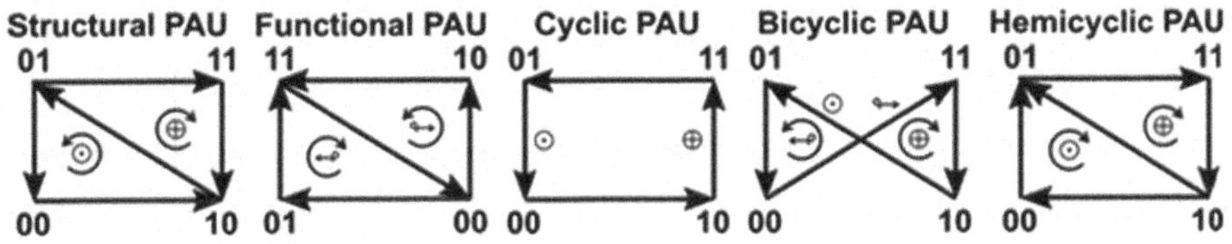

Fig. 4 PAUs library *Source* Own production

5.1 *Structural PAU*

This relational disposition is useful when we must analyze a system (composed of subsystems) that interacts with its immediate environment. Here, the function of the system depends, exclusively, on its relational structure. It handles the dynamics generated by only two variables (opposite and complementary) and does so through basic binary operations (XOR (⊕) and XNOR (⊙)). Application examples: Neurophysiology (Neuronal and cerebral functioning), Psychology (Psychic time, Psychic structure and function, Psychoanalytic theory, Subjective reality), Organic Chemistry (Bistable systems, DNA), Medicine (Immunology of pregnancy), Logic (Laws of equivalence, Dialogics), Mathematics (Theory of groups, Differential equations, Fundamental theorem of algebra, Statistics), Physics (Electromagnetism, Theory of relativity, Quantum physics, Particle physics, Automatic control, Electrical circuits), Art (Literature (tales of Borges), Painting (Escher's work), Music (language and musical composition), Creativity), Biology (Genetic code, Formation of the organs of flowers), Linguistics (Acquisition, understanding and production of natural language), Astronomy (Copernican Revolution), Economics (Quality control), Philosophy (Epistemology), Science (Scientific research, Methodology of research), etc. (Salatino 2009, 2012, 2014, 2015a, b, c, d, 2016, 2017a, b, c, d, e, 2018a, b, c, d, e, f; Salatino and Puglesi 2018).

5.2 *Functional PAU*

It allows address systems that represent an assembly of two or more structures to alternate in their functions. It can handle two or more variables through HBO (hybrid binary operations) (↫ (⊙⊕) / ↬ (⊕⊙)). Examples: Economics (Economic theory, Game theory, Nash equilibrium), Theory of Evolution (Adaptation, natural balance, readaptation), Philosophy (Classification of sciences, On knowledge according to Plato, Scientific research), Psychology (Behavior and conduct, psychoanalytic theory) (Salatino 2014, 2015a, b, c, d, 2017a, b, c, d, e).

5.3 *Cyclic PAU*

This PAU allows the analysis of systems where its two levels (superficial and profound) show an explicit heterarchy. That is, its dynamics are activated by a single operation or using both basic operations at the same time. This way of projecting the transformations causes the 'golden rule' of this approach to be violated: "The profound level of a PAU, always cycles in the opposite direction to the superficial level." Here, the profound level is "dragged" by the superficial level, so that a single cycle is formed. Examples: Physics (Maxwell's Law, Stepper Electric Motors, Electrical Circuits), Mathematics (Statistics, Logarithms) (Salatino 2016, 2017a, b, c, d, e).

5.4 *Bicyclic PAU*

The use of this PAU is convenient when the system analyzed shows a behavior that could be called "oscillatory," due to the cyclic variations that its dynamics show. However, it also allows us to study systems that show *hysteresis* in its different modalities. Thus, it allows describing both the tendency of a material to conserve some of its properties in the absence of the stimulus that generated it, and those systems, structurally stable, that tend to manifest discontinuity (abrupt changes in behavior) or divergence (positive feedback). It is possible to analyze systems where their current state depends on their previous history, but if the behaviors are reversed, never return to the initial state. Examples: Physics (Electromagnetism, Electrical circuits, Chemistry), Philosophy (Conceptual evolution), Psychology (Behavior and conduct), Sociology (Competitive behavior, Models of organizational change, Models of social evolution), Economy (Unemployment rate), Logic (Laws of implication), etc. (Salatino 2017a, b, c, d, e, 2018a, b, c, d, e, f; Salatino and Puglesi 2018).

5.5 Hemicyclic PAU

It admits two variants: (1) *Exogenous*: it starts as "external", and after reaching a certain threshold, it becomes "cyclical". (2) *Endogenous*: it begins as "internal", and when a certain threshold has been reached, it becomes "cyclical". We can use a hemicyclic PAU when our objective is not to analyze separately the elements that determine, in appearance, the functioning of a system, as we do with a structural PAU, but to try to separate appearance of reality, as in a functional PAU. But unlike the functional PAU, here the phenomenon is approached from the direct relation of its fundamental elements and not from its transformations. Therefore, the separation between appearance and reality is achieved regardless of the functional alternation to which the elements considered may be subject. Examples: Science (Methodology of research), Mathematics (Concept of limit, Derivatives), Physics (Changes of state of matter, Electromagnetism), Neurobiology (Neural action potential), Psychology (Psychic functioning), etc. (Salatino 2018a, b, c, d, e, f, 2019a, b, c, d, e, 2020a, b, c, d; Salatino and Puglesi 2018; Salatino and Tersoglio 2019; Salatino et al. 2019).

6 Evolution of the Argument of Symmetry

6.1 Noether

Noether's theorem (1918) relates many conservation laws to the symmetries of space, time, and 'internal' variables of a system. A homogeneous cylinder with its axis of symmetry, when it rotates about said axis, offers us the same image before and after the rotation. This is symmetry, although anyway unless we make a mark on the cylinder, we will not be able to capture the essence of that symmetry. To quantitatively define symmetry, we need to carry out some operation or transformation and see if we find any difference. If this difference is zero or infinitesimally small, we can affirm that this system is invariant under this transformation. The cylinder is invariant under rotation about its axis. The existence of invariance reveals the underlying symmetry. The transformation is a change of the reference frame, a new coordinate system. The momentum or energy of a system can be conserved, but not be, necessarily invariant. For example, one billiard ball approaching another. In the frame of reference of the billiard table, the moment of the system is different from zero, because, before the collision, one ball approaches another. But if we take the centre of mass of the ball as the frame of reference, the moment is zero. Since one frame of reference is nonzero and the other is equal to zero, we conclude that momentum is conserved within each frame, although it is not invariant between them.

Conservation: when there are no changes in a frame of reference.

Invariance: when there are no changes between reference frames.

Noether's theorem relates conservation to invariance, and therefore to symmetry, and determines that the transformations form a group or algebraic structure.

6.2 Van Fraassen

Van Fraassen (1989) defines symmetry as Noether did and proposes it as a guide to a scientific theory, since he considers it as the main key to understanding the world theoretically constructed. For example, we have a rule ***R*** to solve a certain problem. Either way, the problem has some 'input' data for the rule. The solution that the rule prescribes for him is the corresponding 'output', ***R (A)***. But also, we have a transformation class ***h***, which makes the problem essentially the same. If a specific problem with input ***A*** is modified by transformation ***h***, it becomes the problem with input ***h (A)***. We now have two inputs to the system: the original transformed output, and the output of the transformed problem. But, if the two problems are essentially the same, then the two solutions are too. Thus, the symmetry requirement is fulfilled: ***h (R (A)) = R (h (A))***. The problem remains the same after a solution is found. (Principle of invariance). His contributions focus on projecting the concept of symmetry beyond physics or from mathematics, trying to apply it to any scientific theory, suggesting that "similar problems have similar solutions." On the other hand, he establishes a method to individualize the features or relevant aspects of the solution, although he does not specify how to do it.

6.3 Salatino

In 2009 (Salatino 2009), we focused our interest on the group that involves symmetry and proposed several modifications, which allowed us to adapt it as a tool for the analysis of some social phenomena, such as language. Among the proposed modifications are the generation of a permutation group made up of two elements and two transformations arranged in opposition that represent, perfectly-identified, the essential elements of any problem. This makes it possible to identify two levels (or subgroups) with different frames of reference that demonstrate, in their temporal evolution, the existence of different symmetries, among them, some not originally proposed by Noether, such as running in the opposite direction of the levels, and the simultaneity or concurrence of the frames of reference, as happens, for example, when both billiard balls collide. In this work, we propose that it is possible to project this adaptation of the concept of symmetry beyond the Social Sciences and apply it to any scientific discipline, including Physics and Mathematics, and this with a double purpose; On the one hand, to validate this approach as a method of scientific research, and on the other, to provide a tool that can be useful for empirical science education.

7 Analogies

The use of the mathematical principles of Classical Mechanics defined by Newton to try to explain other unknown phenomena was a frequent practice in the middle of the XIX century. Maxwell set out to find a simple mathematical model that would behave analogously to the electric and magnetic fields and lines of force described by Faraday in the late eighteenth century. Thinking about electromagnetism as a mechanical phenomenon allowed him to discover the mathematical principles that explained it. Maxwell perceived that Faraday conceived the lines of force as a primary aspect, that is, as a fundamental factor in the explanation of the phenomenon that he intended to study. For this reason, he incorporated it into a mathematical theory of electromagnetism, but using a type of analogy like that used by Transcurssive Logic, that is, where only matter the relationships and not the related elements or aspects. In the four-part article, published in 1861–1862, "*On Physical Lines of Force*" Niven ([1890] 1965, p. 451), Maxwell begins his fertile analogical task. There, he understands that, according to what was observed in the experiments, the lines of force were something real that was beyond the mere resultant of two forces. Thus, in the first part of the mentioned article, by way of speculation, he dedicated himself to investigating the mechanical results of certain states of tension and motion in a medium, comparing this with the observed phenomena of magnetism and electricity. Using a particular type of mechanical illustration, as an aid to the imagination, but not to account for the phenomena, he succeeds in deducing the mathematical relationships between the electrotonic state, magnetism, electric currents, and the electromotive force. Incidentally, Faraday elaborated the metaphysical concept of "electrotonic state" and Maxwell mathematized it with the "vector potential", formulating the Maxwell-Faraday equation. Furthermore, he was able to show that all the forces acting between magnets, induced currents, and electric currents could be explained by presuming that in the surrounding environment, at each point, the pressures were different according to their direction, the direction of the lines of force being the one with the lowest pressure. Meanwhile, the difference between the maximum and minimum pressure was proportional to the square of the intensity of the force at that point. Because of the above, it could be said that the lines of force indicate the direction of minimum pressure at every point of the medium Niven ([1890] 1965, p. 467). In Part II, Maxwell asks himself: what is the mechanical cause of this pressure difference in different directions? The supposed answer lies in the suggested 'molecular vortices', whose axes of rotation are parallel to the lines of force. Now, the other questions are what makes them rotate, how are they distributed, what was related to the electric current, and what did the opposite rotation of the adjoining vortices mean? In this way, the 'layers of particles' arise between the vortices (Fig. 5).

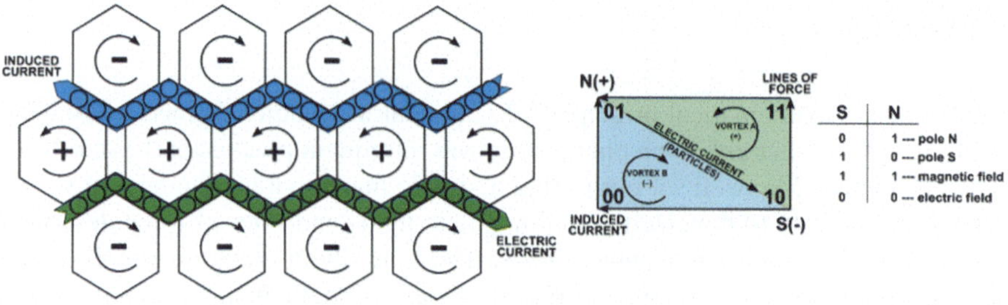

Fig. 5 Vortices, currents and lines of forces *Source* Own production

Figure 5 shows the essential elements of this fruitful analogy. On the left side a reproduction of the analogy used by Maxwell, while on the right side the interpretation made from the transcurssive logic. There we can see the poles of a magnet, opposed, as static elements. The lines of force described by Faraday, going from one to another, because of an electric current. It is thus proven that the variation of the generated magnetic field is what produces an induced current that runs in the opposite direction. The dynamic and opposing elements of the phenomenon can also be identified, represented by the electric and magnetic fields, which are generated simultaneously. All elements that make up the symmetries of an electromagnetic system.

Parts III and IV of the article, called *The Theory of Molecular Vortices applied to Statical Electricity*, and *to the Action of Magnetism on Polarized Light*, respectively, complete the analogy. In these writings, he indicated, in the first place, that if the particles linked to the vortices were endowed with a certain elasticity they would deform under the pressure of the electrical particles, which made it possible to explain the phenomena of attraction and repulsion of the bodies that they were electrically charged. Also, if the particles were elastic, the process was supposed to advance in the form of waves. Since the movement of the electric particles and the rotation of the vortices that described the lines of magnetic force were simultaneous, they were electromagnetic waves. Second, in Part IV, Maxwell used Faraday's experiment on the rotation of the plane of polarized light in the direction of propagation by the effect of a magnetic field, which, added to the rotation of the vortices, shows that the connection between electricity and magnetism has the same mathematical form as that between certain pairs of phenomena, of which one has a linear and the other a rotatory character; that is, both phenomena are opposite Niven ([1890] 1965, p. 503). Since both propagate in the same direction at a speed equivalent to that of light, allows him to conclude that light is an electromagnetic wave. These rich analogies lead to the well-known Maxwell equations.

In 1865, Maxwell proposed "*A Dynamical Theory of the Electromagnetic Field*", where he assumes that in the space that separates the electric charges there is matter in motion, by which the observed electromagnetic phenomena are produced. Therefore, the electromagnetic field is that part of space that contains and surrounds bodies that are in electrical or magnetic conditions. To elaborate the general equations of the electromagnetic field, he assumes that all the quantities that he could determine (twenty variables) have directions that are expressed in the three axes (x, y, z) of space. With these twenty quantities he was able to elaborate the following twenty equations (Maxwell 1865):

Three equations of	Magnetic force	B
Three equations of	Electric currents	C
Three equations of	Electromotive force	D
Three equations of	Electric elasticity	E
Three equations of	Electric resistence	F
Three equations of	Total currents	A
One equation of	Free electricity	G
One equation of	Continuity	H

After going through a series of reviews, carried out by Maxwell himself and other researchers, elemental Maxwell's equations for the electromagnetic field were obtained.

8 Symmetries

Let's see, with the help of transcurssive logic, some aspects of the symmetries that made it possible to determine the fundamental equations known as 'Maxwell's Laws'.

At any point in space, an electric or magnetic field has two properties: a magnitude and a direction. In general, these properties vary (continuously) from point to point. It is a convention to represent such fields in terms of their components measured with respect to a certain set of Cartesian axes, conveniently chosen (e.g., the standard axes: x, y, z). Of course, the orientation of these axes is arbitrary. In other words, different observers can choose different coordinate systems to describe the same field; consequently, the same electric or magnetic field can have different components according to different observers which highlights the importance that the transcurssive logic gives to the observer. As we can see, any description of an electric or magnetic field depends on two separate things. First, of the fundamental laws of physics that govern the behavior of electric or magnetic fields; and second of the arbitrary choice of the coordinate axes. It would be useful to be able to easily distinguish those elements of Maxwell's equations that depend on physics from those that only depend on coordinates. We can achieve this objective using a branch of mathematics called: "*Vector field theory*". This formalism allows Maxwell's equations to be written in such a way that they are completely independent of the choice of coordinate system. Also, these equations look much simpler when written in coordinate-free mode. In fact, instead of eight first-order partial differential equations, there are only four of these that are within the context of vector field theory (Fitzpatrick 2008).

8.1 Electromagnetic Field Quantities

According to Huray (2010) an electromagnetic field can be described by four field quantities:

$\vec{E}$ electric field intensity
$\vec{D}$ electric flux density or displacement
$\vec{H}$ magnetic field intensity
$\vec{B}$ magnetic flux density

The electric field intensity (***E***) is the field vector used in electrostatics when the charge is at rest in free space and is defined as the electric force in a unit test charge. The displacement or electric flux density (***D***) is a field vector used in the study of electric fields within material objects. Similarly, the intensity of the magnetic field (***H***) is a necessary vector to analyze magnetic phenomena. It is the field generated at a point in free space by constant or time-varying electric currents in a source; it is related to the magnetic force acting on a moving charge. Magnetic flux density (***B***) is useful in investigating magnetic fields within material objects where the material modifies the intensity of the field. When there are no variations in time of the field quantities, the electric field quantities (***E***, ***D***) are independent of the magnetic field quantities (***H***, ***B***). In time-dependent cases, however, the electric and magnetic fields are coupled; that is, the variation in time of (***E***, ***D***) gives rise to (***H***, ***B***) and vice versa. The electromagnetic properties of materials are governed by what is known as the 'constitutive relationships' between ***E*** and ***D***, and ***H*** and ***B***. The equations that represent these constitutive relationships [structure] are called Maxwell's equations. Taking into account all the above we have enough elements to approach these equations from Transcurssive Logic (TL). Since Maxwell's equations, as we have seen, represent the magnitudes of a vector field, we can establish the following relationships (Fig. 6).

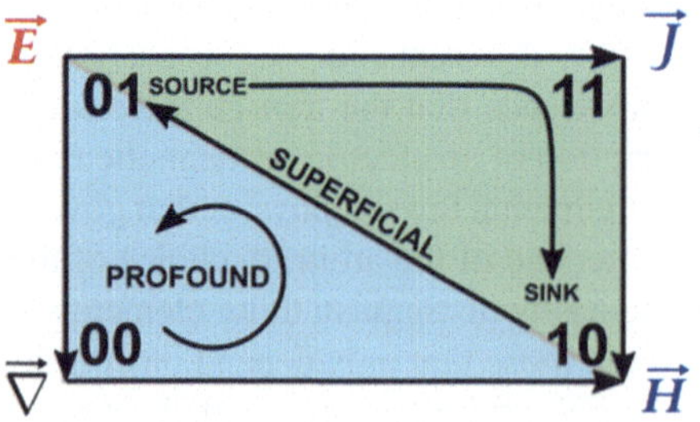

ELECTRIC FIELD	MAGNETIC FIELD
0	1 --- electric field intensity
1	0 --- magnetic field intensity
1	1 --- electric current density
0	0 --- Nabla

Fig. 6 PAU of electromagnetism *Source* Own production

The above diagram shows the fundamental relationships that exist between the electric and magnetic fields, mediated at the surface level (considering their intensities—the static) by ***J*** or electric current density, and at the deep level (if we consider their flux densities—the dynamic) by **∇** (nabla), the vector that summarizes as a differential operator the dynamic counterpart of this structure, as we will see later.

TL considers the above as a basic relational pattern or structure, which it calls PAU or universal autonomous pattern because it defines, as we have already seen, the 'structure' on which not only electromagnetism is based, but also all phenomena that can be observed in any area of reality. The binary codes of the **PAU** arise, as shown in the attached table, from considering its fundamental aspects in each field, which allows each participating element to be identified. This procedure makes it possible to verify that this set of quantities form, structurally speaking, a *Galois set*, and from a functional point of view, a *Galois connection*. These mathematical structures allow, for example, to demonstrate the presence of symmetry which certifies, in this case, the compliance and invariance of the laws and magnitudes during the temporary deployment of a physical system. Furthermore, from the pedagogical point of view, the PAU is a 'generator of knowledge' because it allows us to understand the relationships that characterize the fundamental concepts of any science, and how they are applied in practice. The justification for the binary code (**11**) of ***J*** is in the fact that the current density is closely related to ***D*** or displacement current, which is a quantity that, in turn, is related to a changing or varying electric field in time, and therefore generates a magnetic field. In other words, it relates to both fields.

Finally, we can say that Fig. 6 only warns us that there is an evident relationship (source → sink) between both fields, which will be expressed in one of the equations, and puts in evidence a force field and technically the presence of divergence of a Vector field.

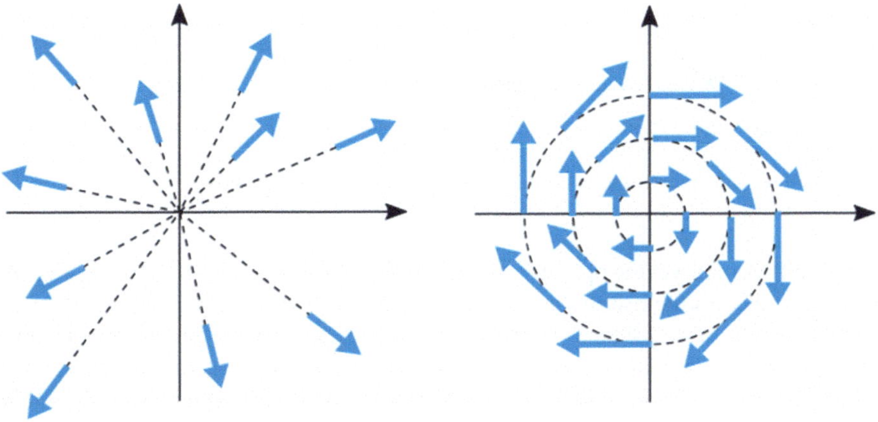

Fig. 7 Vectorial fields *Source* Own production

Figure 7 shows, on the left, an irrotational or conservative field (null rotational) with divergence or constant field source throughout the space, which occurs at the superficial level of the PAU. On the right, we see a solenoidal (null divergence) and rotational or vortex (non-zero rotational) field with a constant rotational throughout space. The latter is not evident since it belongs to the profound level of the PAU. To make it evident, that is, to analyze the dynamic aspect of this structure, we must 'unfold' this level that is represented by ∇.

8.2 Gradient of a Scalar Field

Taking into account the proposal by Huray (2010), we suppose we have a scalar quantity V that is a function of position, such that $V = V(x, y, z)$ in 3D. Such a function is called a scalar field and it has only one number associated with each point in space; that is, it has n^0 components (a tensor of rank 0). Examples of this are temperature, electric potential, density, or pressure. Said quantities are invariant in the face of a coordinate transformation. In other words, at a point in space, they are always the same even if they are expressed in Cartesian, cylindrical, or spherical coordinates.

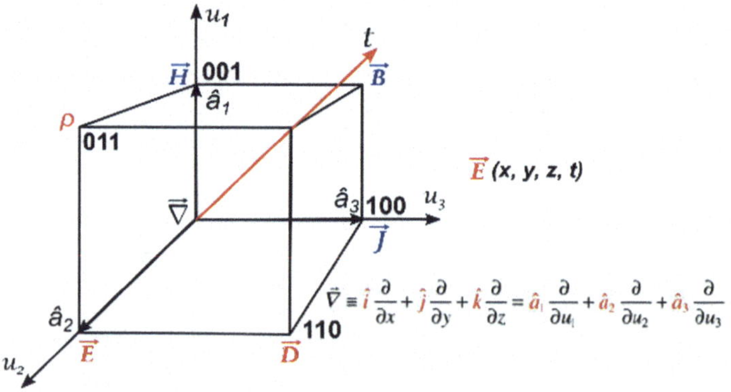

Fig. 8 Orthonormal curvilinear coordinate system *Source* Own production

The differential vector operator ∇ (nabla) is usually defined in 3D as in Fig. 8, where u_i indicates the use of curvilinear coordinates (cylindrical or spherical) such as those shown in said figure. The gradient of a scalar quantity is a vector quantity that represents both magnitude and direction; that is, it has three numbers associated with each point in 3D space and time. Examples are the Intensity of the Electric Field $E(x, y, z, t)$; the Intensity of the Magnetic Field $H(x, y, z, t)$, and the Electric Current Density $J(x, y, z, t)$. Such quantities are known as the vector field and typically, they vary under a coordinate transformation; that is, the components of a point in space are different when expressed in Cartesian, cylindrical, or spherical coordinates.

In Fig. 8, following the basic arrangement of the quantities of the electromagnetic field exposed in the PAU, those quantities that represent the dynamic aspect of this system are contrasted, that is, those that represent the flows and the electric charge density (***ρ***), which in 2D are 'represented' in **∇**. Thus, the intensity of the electric field (***E***) whose binary code would be **010** (x, y, z), is contrasted by the magnetic flux density (***B***) whose code is **101**[2] (the opposite), in each time. In the same way, the intensity of the magnetic field (***H***) of code **001** is contrasted with the electric flux density (***D***) with code **110** plus time. Finally, to the evident transformation that allows the divergence or the outward flow of a source and its orientation towards a 'sink', that is, the electric current density (***J***) of code **100**, and that gives a measure of the force of a radiation source (Fig. 7, left side), is contrasted with the profound transformation generated by another type of source, called a vortex source (Fig. 7, right side) that is described by a circulation of the vector field, in relation to the density volumetric electric charge (***ρ***) whose code is **011**. This 3D distribution of these quantities of electric and magnetic fields proposed by the TL allows to demonstrate by means of the symmetry of the formed group, for example, the conservation of the electric charge. The conservation of charge implies, like the conservation of mass, that at each point in space a continuity equation is satisfied that relates the derivative of the electric charge density with the divergence of the electric current density vector, said equation expresses that the net change in charge density within a preset volume is equal to the integral of the electrical current density over the surface enclosing the volume, which in turn is equal to the electrical current intensity. The curvature of a vector field gives a measure of the strength of the vortex source. Maxwell, by including the displacement current (D) in Ampere's law, gives the guideline for the behavior of fields that vary in time. The term $\partial \boldsymbol{D}/\partial \boldsymbol{t}$ introduced by Maxwell indicates the density of the displacement current.

[2] Binary codes indicate a possible location on the three axes of space in a given volume.

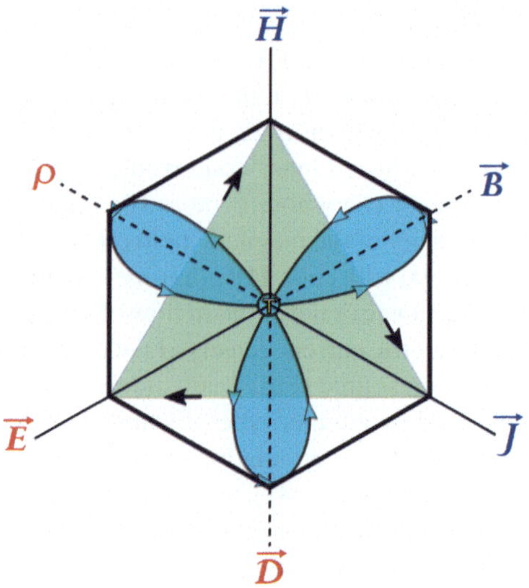

Fig. 9 Divergence and rotational *Source* Own production

Figure 9 shows another way in which the TL represents the static (triangle—divergence, linear) and dynamic (trefoil—rotational) relationships between the electric and magnetic fields and from where the four Maxwell equations can be deduced (Mundó 2008).

8.3 *First Equation*

Gauss's law for the electric field:

$$\nabla \cdot E = \frac{\rho}{\varepsilon_0}$$

Based on Coulomb's law, it expresses the link between electricity and one of the fundamental properties of nature: electric charge. Just as an example and to show that different degrees of aggregation can be approached through TL to try to explain a phenomenon, we will develop 'Coulomb's Law'.

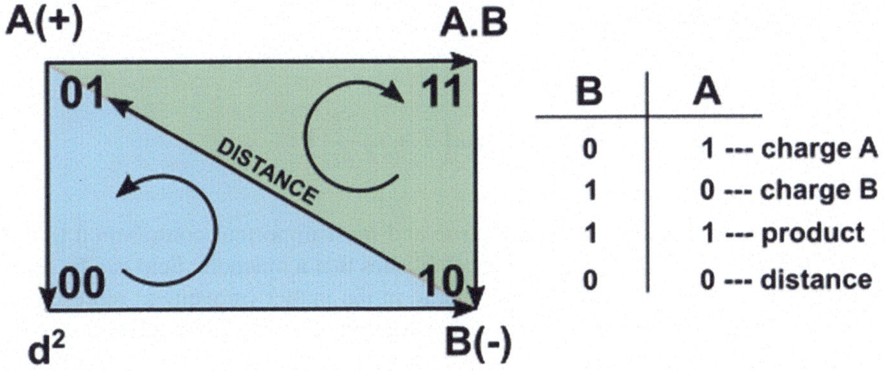

Fig. 10 PAU of Coulomb's law *Source* Own production

According to Fig. 10, the 'Coulomb's Law' that talks about static charges, tells us: "*Two electric charges* [**A** (**01**) and **B** (**10**)] *attract or repel with a force that is directly proportional to the product between them* (Huray 2010) (clockwise cycle), *and inversely proportional to the square of the distance that separates them*" [**00**]. (counterclockwise cycle). Code 11 tells us that both charges are linked by an evident transformation, while code 00 indicates that the distance exerts a non-evident transformation on them. From the previous PAU, the concept of electric field (***E***) that represents vectorially, the intensity of the electric force, of attraction or repulsion, arises in an intuitive way. This force is greater the closer we get to a given charge.

8.4 Second Equation

It expresses that there are no free magnetic monopoles.

$$\nabla.B = 0$$

8.5 Third Equation

$$\nabla \times E = -\frac{\partial}{\partial t}B$$

(Faraday's law) shows a relationship between magnetism and electricity: the variation in time of a magnetic field generates an electric field.

8.6 Fourth Equation

$$\nabla \times B = \mu_0 \left(J + \varepsilon_0 \frac{\partial E}{\partial t} \right)$$

(Ampere's law) represents Maxwell's true and most important contribution to the theory of the electromagnetic field. He establishes that a magnetic field can be originated both by movements of electric charge in the matter (movement represented by the current density vector ***J***) and by the variation of an electric field with respect to time.

9 Conclusion

Perhaps as a corollary of this work we could invoke Poynting's theorem (1884) which establishes the conservation of energy for electric and magnetic fields, showing that the energy flow vector, which carries the electromagnetic wave, depends only on the fields. We have been able to demonstrate the same through the analogy used by Maxwell, applying the symmetries of the electromagnetic phenomenon analyzed from Transcursive Logic, as we see below, thus confirming that the analogy is part of the scientific method.

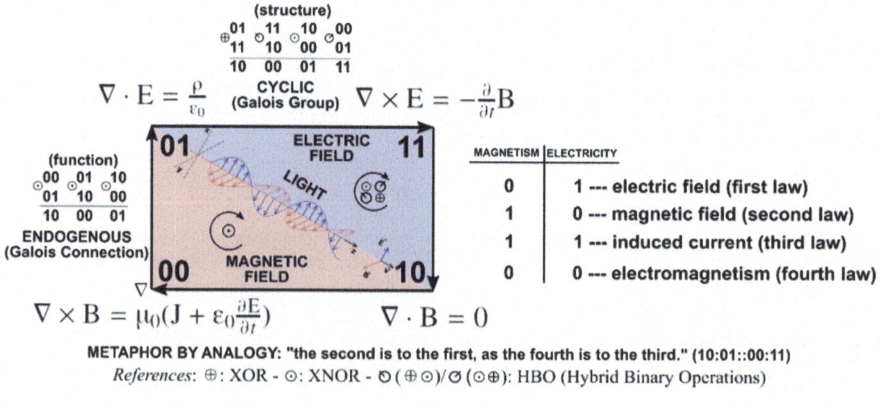

Fig. 11 PAU of elementary Maxwell's law *Source* Own production

Figure 11 summarizes the transcurssive approach of Maxwell's laws using an endogenous hemicyclic PAU (see above).

The *endogenous* element (determinant of the dynamic aspect or nabla (∇)) is represented by electric field generates a magnetic field (rotatory). The *cyclic* element produces the movement of the charges generating an induced current (linear) in the reverse direction and then produces an electromagnetic phenomenon. At the limit of both levels (apparent or electric field, and non-apparent or magnetic field): light as a wave with both perpendicular fields moving in one direction and at a constant speed. The fundamental elements of both fields are electric charges. Static (electrical), dynamic (magnetic). The cycle closes when we combine the first law with the fourth, and everything begins again. The binary operations (usual and hybrid), justify both the linear aspect or structure of the electromagnetism phenomenon expressed by a Galois group and the rotational aspect or function (dynamic) of said phenomenon (determined by nabla) expressed by a Galois connection. In this way, all the symmetries of electromagnetism that are expressed through a kind of 'universal language' are made explicit, and that they were found thanks to a 'metaphor by analogy', like the one used by Aristotle.

As a conclusion, we present the integration of the fundamental elements that support Maxwell's electromagnetic theory of light, starting from an analogy (Fig. 12). The diagram in the Fig. 12 explains Maxwell's electromagnetic theory of light, which shows that radiant energy propagates in the form of electromagnetic waves. Magnetic waves are indirectly evidenced, while electricity is directly evident. In this scheme, there is evidence of the four Maxwell equations that demonstrate that electricity, magnetism, and light are different manifestations of the same phenomenon, which can be assimilated to the existence of a common pattern (PAU) or a kind of 'universal language' that matches a Galois connection.

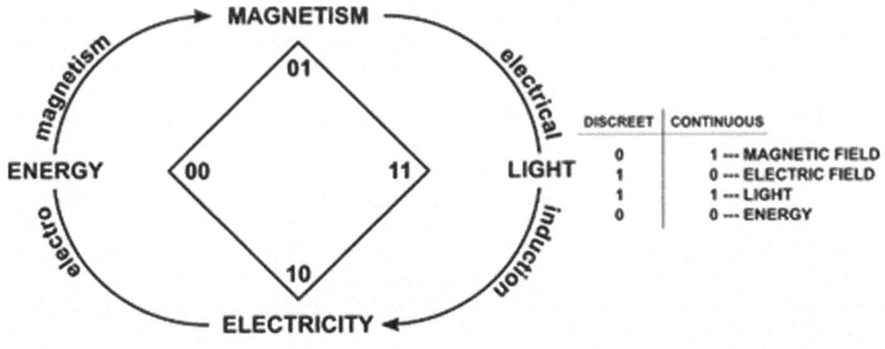

Fig. 12 PAU of elementary Maxwell's theory *Source* Own production

References

Aristóteles (2004) Tratados de lógica (el Organón). México, Editorial Porrúa.
Aristóteles (2007) Física. Barcelona, Editorial Gredos.
Aristóteles (2011) Poética. Valentín García Yebra (Traductor). Madrid, Editorial Gre-dos.

D'Agostino S (2001) A History of the Ideas of Theoretical Physics. Essays on the Nineteenth and Twentieth Century Physics. Kluwer Academic Publishers, Dordrecht.
Fraassen van BC (1989) Laws and Symmetry. Clarendon Press, Oxford.
Faraday M (1855) Experimental Researches in Electricity. Reprinted from the Philosophical Transactions of 1846–1852. Vol. III, Bernard Quaritch, London.
Feyerabend P (2001) La conquista de la abundancia. Paidós, Barcelona.
Fitzpatrick R (2008) Maxwell's Equations and the Principles of Electromagnetism. Infinity Science Press LLC, Massachusetts.
Hegel GWF (1982) Ciencia de la lógica. 2 Tomos. Ediciones Solar, Buenos Aires.
Hempel CG (1965) Scientific Explanation. Essay in the Philosophy of Science. The Free Press, NY.
Hume D (1748–1894) An Enquiry concerning the Human Understanding, and an Enquiry concerning the Principles of Morals. Clarendon Press, Oxford.
Huray PG (2010) Maxwell's Equations. John Wiley & Sons, Inc. New Jersey.
Kline M (1980) Mathematics. The Loss of Certainty. Oxford University Press, New York.
Maxwell JC (1865) A Dynamical Theory of the Electromagnetic Field. Philosophical Transaction of the Royal Society of London 155:459–512.
Maxwell JC (1881) A Treatise on Electricity and Magnetism. Second Edition. 2 Vol. Oxford University Press, Clarendon.
Mill JS (1882) A System of Logic, Ratiocinative and Inductive: being a Connected View of the Principles of Evidence and the Methods of Scientific Investigation. Harper & Brothers, New York.
Mundó J (2008) Las ecuaciones de Maxwell. Vida, pensamiento y obra – Colección grandes pensadores. Centro Editor PDA S. L., Madrid.
Niven WD ([1890] 1965) The Scientific Papers of James Clerk Maxwell – Two Vols. Dover Publications, New York.
Noether E (1918) Invariante Variationsprobleme, Nachrichten König. von der Gesellschaft der Wissenschaften zu Göttingen, Mathematisch-physikalische Klasse, s. 235–257.
Peirce CS (1878) Deduction, Induction, and Hypothesis. Popular Science Monthly 13:470–482.
Petrucci RH (2003) Química General. Pearson Educación, Madrid.
Pisano R (2013) Historical Reflexions on the Physics Mathematics Relationship in Electromagnetic Theory. In: Barbin E, Pisano R (eds). The Dialectic Relation Between Physics and Mathematics in the XIXth Century. Springer, Dordrecht, pp. 31–57.
Poynting JH (1884) On the transfer of energy in the electromagnetic field. The Royal Society, Volume 175, pp. 343–361.
Salatino DR (2009) Semiótica de los sistemas reales – Tesis Doctoral en Letras especialidad Psicolingüística por la Facultad de Filosofía y Letras de la Universidad Nacional de Cuyo, Mendoza.
Salatino DR (2012) Aspectos psico-bio-socioculturales del lenguaje natural humano. Introducción a la teoría psíquica del lenguaje. Desktop Publishing, Mendoza.
Salatino DR (2013) El proyecto de Freud. Psicoanálisis. Revista de la Asociación Psicoanalítica Colombiana 24(1/2):43–60.
Salatino DR (2014) Psyche – Structure and Function – First Desktop Publishing, Mendoza.
Salatino DR (2015a) Ficciones, solo ficciones In: Conceptos y Lenguajes en ciencia y tecnología. Cuadrado G, Redmond J, López R (eds). Serie: Selección de Textos, Volumen 3, Universidad de Valparaíso, Valparaíso, pp. 251–273.
Salatino DR (2015b) Las 2201 páginas de una teoría. La verdadera historia de la Lógica Transcursiva. 4 Vols. Autoedición, Mendoza.
Salatino DR (2015c) Bases neurológicas y psíquicas del lenguaje musical. Actas de ECCoM, La experiencia musical: cuerpo, tiempo y sonido en el escenario de nuestra mente 2/1:257–267.
Salatino DR (2015d) Sobre el inconsciente" – Psicoanálisis – Revista de la Asociación Psicoanalítica Colombiana XXVII(1):199–228.
Salatino DR (2016) La importancia de la simetría. Educación en Ciencias Empíricas en Facultades de Ingeniería. (ECEFI 2016). Universidad Tecnológica Nacional, Mendoza, pp. 19–49.

Salatino DR (2017a) Beyond the Decisions-Making: The Psychic Determinants of Conduct and Economic Behavior. International Journal and Methodology in Social Sciences 3/1:6–26.
Salatino DR (2017b) Beyond the Decisions-Making II: Methodological Aspects. International Journal and Methodology in Social Sciences 3/2:pp. 6–24.
Salatino DR (2017c) Tratado de Lógica Transcursiva. El origen evolutivo del sentido en la realidad subjetiva. Primera autoedición. Mendoza.
Salatino DR (2017d) Transcurssive Logic as Method. International Journal and Methodology in Social Sciences 3/3:5–18.
Salatino DR (2017e) Language production – First part. International Journal and Methodology in Social Sciences 3/4:5–21.
Salatino DR (2018a) Language production – Part 2. International Journal and Methodology in Social Sciences 4/1:5–17.
Salatino DR (2018b) Language acquisition – Part 1 of 3. International Journal and Methodology in Social Sciences 4/2:1–14 [with honourable mention].
Salatino DR (2018c) Language acquisition – Part 2 of 3. Genetic Typology of Human Natural Language. Hidden Colors Theory. International Journal and Methodology in Social Sciences 4/3:1–27.
Salatino DR (2018d) La 'geometría funcional' como fundamento de la estructura y la función del aparato psíquico. Revista de Psicopatología y Salud Mental del niño y del adolescente 32:77–94.
Salatino DR (2018e) Language acquisition – Part 3 of 3. Language Universals – Lexical Contextures – Sociocultural Heritage. International Journal and Methodology in Social Sciences 4/4:1–28 [with honourable mention].
Salatino DR (2018f) Tratado de Psicolingüística. Adquisición, Comprensión y Produc-ción del Lenguaje Natural Humano. Primera Autoedición, Mendoza.
Salatino DR (2019a) Treatise of Psycholinguistics. Acquisition, Comprehension, and Production of Human Natural Language. First Desktop Publishing, Mendoza.
Salatino DR (2019b) Los lenguajes de Freud. Enfoque subjetivo de la Teoría Psicoa-nalítica. Primera Autoedición, Mendoza.
Salatino DR (2019c) Fundamentals of a new research method. International Journal and Methodology in Social Sciences 5/1:52–73 [with honourable mention].
Salatino DR (2019d) The logic within the logic. Reinterpretation of Boolean algebra from Transcurssive Logic. International Journal and Methodology in Social Sciences, 5/1:83–99 [with honourable mention].
Salatino DR (2019e) The Mystery of Time. International Journal and Methodology in Social Sciences 5/3:1–29.
Salatino DR (2020a) The Intrinsic Beauty of the Penrose's Mosaics. International Journal and Methodology in Social Sciences 6/1:1–14.
Salatino DR (2020b) Subjective Universe. International Journal and Methodology in Social Sciences 6/1:1–12.
Salatino DR (2020c) Theory of Destination in Borges. A Transcurssive Approach. International Journal and Methodology in Social Sciences 6/1:13–21.
Salatino DR (2020d) New Nosological Model of Chronic Endometritis. International Journal and Methodology in Social Sciences 6/1:22–32.
Salatino DR, Puglesi, AE (2018) Neuron as an electric PAU. International Journal and Methodology in Social Sciences, 4/4:59–76 [with honourable mention].
Salatino DR, Tersoglio AE (2019) Risk of Autism in the use of assisted reproduction techniques. An analysis from the Transcurssive Logic. International Journal and Methodology in Social Sciences 5/1:1–16 [with honourable mention].
Salatino DR, Masera GA, Palma R (2019) Risks and critical infrastructures. An interpretation from the Transcurssive Logic. International Journal and Methodology in Social Sciences 5/2:17–25.
Samaja J (2005). Epistemología y Metodología. Elementos para una teoría de la investigación científica. Eudeba. Buenos Aires.

Part III
Essays in History of Physics—&—Mathematics

Chapter 17
Rethinking the Intellectual Genesis of Landauer's Principle

Javier Anta and Carl Hoefer

Abstract Our main aim in this paper is to evaluate from a critical historical-philosophical standpoint the intellectual origins of 'Landauer's principle', a thesis formulated in 1961 by the IBM researcher Rolf Landauer. He asserted that any physical implementation of a logically irreversible operation such as erasing a bit of information leads to an entropy increase in the system computing that operation. Contrary to the popular narratives from the thermodynamics of computation about the genesis of Landauer's principle, we argue that this idea was influenced, firstly, by Léon Brillouin's Negentropy principle, theoretically connecting information and entropy. Secondly, Landauer found in John von Neumann's ideas in 1949 the path to connect informational processes and thermal effects through the logical irreversibility of computational operations. Interestingly, both Brillouin and von Neumann also gave general and ill-defined formulations of this idea previous to the work of Landauer. Finally, alongside these two main intellectual sources one should include, complementing historiographic proposals such as Wright, the influence of the potential double-well model of the IBM researcher John Swanson in developing a concrete model to apply this thesis. Finally, we defend reconceiving the intellectual genesis of Landauer's principle as a direct historical heir to a rich multiplicity of physical, informational, and computational trends of thought during the 1950s.

Keywords Entropy · Information theory · Landauer's principle · Léon Brillouin · John von Neumann · Rolf Landauer

J. Anta (✉)
University of Seville, Seville, Spain
e-mail: antajav@gmail.com

C. Hoefer
ICREA, Universitat de Barcelona, Barcelona, Spain

R. Pisano (ed.), *A History of Physics: Phenomena, Ideas and Mechanisms*,
History of Mechanism and Machine Science 42,
https://doi.org/10.1007/978-3-031-26174-9_17

1 Introduction

It was in July 1961 when the German-born Rolf Landauer, Ph.D. from Harvard and worker at IBM since 1952, published a groundbreaking paper entitled 'Irreversibility and Heat Generation in the Computing Process' (Landauer 1961) in the IBM journal. His central thesis in this paper was:

> [...] It is argued that computing machines inevitably involve devices which perform logical functions that do not have a single-valued inverse. This logical irreversibility is associated with physical irreversibility and requires a minimal heat generation, per machine cycle, typically of the order of kT for each irreversible function. (Gleick 2011, p. 183)

The extremely impressive thing about this idea is that it can prima facie connect realms of such a seemingly different nature as logic, physics, and information. As can be inferred from the statistics of citations in this article (Wright 2016, p. 120), the attention that this thesis (later popularized as 'principle') attracted increased exponentially, not only within diverse sectors of the scientific community in the last decades, but also in the recent philosophy of physics literature. Illustratively, the philosopher Jos Uffink (2007) mentioned Landauer's principle (henceforth 'LP') as one of the central questions in the foundations of thermal physics. However, the main difference between how this idea was considered in the two domains is that, while LP was assumed as a fundamental theoretical physical principle by a large part of the scientific community, the philosophical community adopted a more critical perspective towards the validity of the idea (Earman and Norton 1999).

Despite the enormous interest aroused by LP in recent times, the analysis of the intellectual origins of this idea has been unfairly disregarded both by scientists and philosophers. Concerning the various scientific disciplinary areas (i.e., computer and information sciences or thermal physics), the historical origins of LP in particular have been simply subsumed as part of the natural development of a trend of physical thought in the mid-twentieth century to the present day centred on the informational processes, receiving the name of 'physics of information' (Anta 2021b; Bar-Hillel 1964, p. 69). From the recent philosophical reflection on this astonishing idea, the technical, conceptual, and interpretative origins of LP seem to be treated as a highly irrelevant task for the achievement of a deep understanding about its content. In this chapter we argue (i) that we should carefully analyse the manifold of factors involved in its intellectual genesis, and (ii) that illuminating its historical origins may enhance our understanding of this idea.

2 An Intellectual History of the Origins of Landauer's Principle

Landauer's 1961 paper went remarkably unnoticed during the 1960s and 1970s, being known only to his co-workers within the intellectual context of IBM, namely: Robert Keyes, James Woo (both collaborated with Landauer in later reformulations of the ideas of 1961), John Swanson and Charles Bennett. It was precisely the latter who popularized the modest thesis defended by Landauer (then raised to a fundamental principle of physics as 'Landauer's principle') beyond the walls of IBM with his foundational paper 'The Thermodynamics of Computation' (Bennett 1982).

In the 1980s, the promoters of the so-called 'thermodynamics of computation' (i.e., analysis of the physical costs of certain information processes) began to generate historical narratives wherein the intellectual origin of LP was inserted as a key element within the evolution of this emergent discipline. In particular, LP was popularized as saving the Second Law of Thermodynamics from Maxwell's demon. As paradigmatic examples of these popular historical narratives one could find:

> [...] Landauer tried in 1961 to prove von Neumann's formula for the cost of information processing and discovered that he could not. On the contrary, it seemed that most logical operations have no entropy cost at all [...] Only an irreversible operation, he argued, increases entropy. (Gleick 2011, p. 361)

or also

> [...] Any process that erases a bit in one place must transfer that same amount of information somewhere else. This is known as Landauer's principle, after Rolf Landauer, the pioneer of the physics of information, who discovered it in the early 1960s. (Lloyd 2006, p. 77)

Even the main recent criticisms of the theoretical validity of LP, notably that of Earman and Norton (1999), did not straightforwardly reject the 'ideological' (i.e., relative to this almost hegemonic informational trend) foundation underlying these types of historical narratives. One of the most referenced historical narratives in the literature is undoubtedly Leff and Rex's (1990, 2003) organized compilation of texts from the history of the thermodynamics of computation, wherein Landauer's 1961 paper was situated in contrast to Brillouin's proposal and solely by virtue of its later use in Bennett's (1982) exorcism.

The acceptance of these popular orthodox historiographies was typical within the literature, accounting not for the origins of LP but for its role in Bennett's (1982) arguments against Maxwell's demon. Mainly for this reason, the task of detailing the technical, and conceptual resources that historically generated the LP has been deeply neglected in the study of this idea, despite the enormous weight that this thesis has nowadays. Recently, this necessary historiographic task has been carried out by Wright in his 2016 'The Physics of Forgetting' (Wright 2016), who aimed to break with popular narratives and traced the conceptual changes that led to the genesis of computer thermodynamics from the rise of the Landauer's principle, from the initial influence of Swanson in 1959 in the elaboration of the famous 1961 paper to Bennett's role in establishing this thesis as a fundamental limit during 1970–1982. In this sense, each of these historical stages (i.e., 1950–1960, 1961–1969, 1970–1982) can be characterized by a central metaphysical statement, such as (i) the possibility in Swanson of defining Shannon information not in symbols but in units of physical volume, (ii) the physical consequences of implementing logically irreversible operations (underlying Landauer's principle itself), and (iii) the universal character of the previous thesis.

In this chapter we aim to show how LP historically emerged from a melting pot of diverse intellectual trends (Kline 2015). In this sense we also defend that the technical and conceptual bases underlying the historical origin of LP cannot be simply reduced to Swanson's intellectual contribution. Although Swanson's work played a role in conceptually shaping this thesis, it was limited to providing the practical conditions and the concrete model on which to assert its content. On the other hand, the origins of LP are inserted in an extremely rich intellectual context, where the author was decisively influenced by the statistical mechanical work of Léon Brillouin (his doctoral supervisor) in 1956 and by the proposals of von Neumann. Therefore, our aim of explaining the origins of LP will be focused on critically evaluating the complex intellectual context of the 1950s in which they generated the technical, conceptual, and interpretative conditions that eventually crystallized in this idea. Thus, our methodology does not rely on the popular narratives that one could usually finds within the literature, but foremost on the primary and secondary historical resources from the various fields involved in this historical development, such as books on statistical mechanics or talks on the connection between logic, information theory and thermodynamics. Furthermore, these will be critically analysed from the standpoint of the main arguments developed in the philosophy of thermal physics (Brillouin 1956). We begin now with a reassessment of the already-mentioned role of Swanson in this process.

3 John Swanson: The Physical Volume of Information

The computer scientists and physicist John Swanson was decisive in Landauer's intellectual formation, as Wright (Wright 2016) recently argued. It should be noted that the relationship between the two was not primarily academic, but of friendship, which went back to when they were both graduate students at Harvard in the late 1940s. This long-lasting friendship was pivotal for Landauer to join Swanson at the new IBM Research Laboratory in Poughkeepsie (New York) in 1952.

In the IBM research area of magnetic memories, where the immediate practical applicability of their theoretical results prevailed, both Landauer and Swanson were committed to the intellectual trend of applying Shannon's (1948) communication theory in various scientific fields. In Shannon's own words:

> Information theory has, in the last few years, become something of a scientific bandwagon [...] Our fellow scientists in many different fields, attracted by the fanfare and by the new avenues opened to scientific analysis, are using these ideas in their own problems. (Shannon 1956, p. 1)

Shannon's information theory served as a conceptual bridge between (i) the logical operations on which the computational architecture is based and (ii) the ferromagnetic processes that play the role of physical memories, as shown in Swanson's 1960 paper 'Physical versus Logical Coupling in Memory System' (Swanson 1960).

What is deeply original in Swanson's work was his application of von Neumann's idea concerning the cost of information processing in the particular case of magnetic memories, distinguishing between (a) the physical coupling of physical storage elements and (b) the logical coupling between these same elements. In the first case, the physical storage units are aligned by means of their physical-chemical properties; in the second, these are logically aligned by introducing redundancy between the values of said physical units. Then, each memory unit was assumed to be equivalent to a particle within a double potential well, where each of its two minima point represents two possible logical states '0' and '1'. Also, the physical coupling between said magnetic memory units n is manifested in the fact that the transition probability from one element n_0 to another possible element n_1 is homogeneously distributed. In this sense, the thermal effects linked to the molecular properties of the magnetic units (i.e., random transitions between elements) would prevent it from transiting in a controlled or non-homogenous manner from one certain logical state to another. Interestingly, this point has been recently (53 years later) defended by Norton (2013, pp. 1182–1192) in his critique of LP, arguing that molecular-based thermal fluctuations prevent reliable implementation of logical irreversible operations by thermodynamically reversible processes. However, it was precisely the logical coupling between magnetic memories that aimed to solve these implementation problems related to thermal and quantum effects. Additionally, the logical coupling between memory units n depends on introducing information-theoretical redundancy between these units (where only a subgroup k would be logically independent) so that the transition between their values can be carried out allowing the correction of possible errors linked to thermal and quantum effects. According to Swanson (1960), it was precisely the celebrated Shannon noisy-channel coding theorem (a main result in Shannon's theory) that determined the probability of occurrence of an error in the information states linked to the magnetic memory units. Of course, this somehow presupposed a physical interpretation of the noisy communication channel to which Shannon technically refers in that theorem[1] as the dynamic evolution of the magnetic memory apparatus. Additionally, the capacity of that 'physical' channel was reconceptualised via the informational capacity of that apparatus to store logical states in a time interval t_0–t with a certain degree of error.

In order to give coherence to the logical exploitability of magnetic memories required for the development of computers, Swanson (1960) intended to reconceptualise the theoretical structure of Shannon's proposal. As Wright (2016) correctly points out, Swanson did not explicitly mention (nor Landauer 1961) Shannon entropy, although it is already de facto implicit in his noisy channel theorem. Broadly speaking, Shannon's information-theoretical entropy was developed in message-transmission contexts to quantify the unexpectedness of sequences of symbols within a particular message regarding the frequency of each of these particular symbols. However, for Shannon's entropy to make sense in the context of magnetic memories,

[1] It should be remarked that Swanson's concept of 'noisy memory' was properly physical, precisely in the sense of being entirely constituted by physical properties (e.g., volume, magnetic direction, etc.) (Wright 2016, p. 128).

Swanson reconceptualised basic information-theoretical units beyond symbols and messages:

> [...] If information is stored at time $t = 0$ and read out at a later time t, we may regard the entire process as a (delayed) transmission of information, and may define a channel capacity in the sense of Shannon. However we define a capacity per unit volume rather than per symbol. (Swanson 1960, p. 307)

In this sense, Wright remarked that

> [...] by interchanging the informational concept "symbol" with the physical concept "volume" without other comment Swanson asserted their interchangeability. (Wright 2016, p. 128)

However, this necessary assumption in the IBM researcher's argument is extremely problematic for at least two reasons. First, in this 1960 paper Swanson tried to define the information stored in N elements from a unit of physical volume v, which is not only an unintentional attempt to interpretatively reify the information as if it were a substance in a container. Additionally, this assertion presupposed a misconception of the fact that the information of a sequence of symbols (or sequence of volumetric units, being charitable) can only be well-defined as a function on the probability distribution over the space of all possible sequences of symbols or messages, and in no way over an individual message or volume.

Also, Swanson was not able to give a value for the basic unit of volume v, simply stating that it has to be small enough to be physically optimal. He was not able to do so precisely because in the classical domain there is no fundamental physical volume unit (i.e., not based on conventional criteria) to partition space, while in the quantum domain one could find the so-called 'Planck length'. Interestingly, this fundamental unit of physical volume was used 12 years later by Jacob Bekenstein (1973) to quantify the amount of information-theoretical entropy of the event horizon of a black hole. However, the unit of volume v on which Swanson based his proto-physical information theory in the context of magnetic memories lacks (beyond its possible practical utility) any robust physical meaning.

Undoubtedly, Swanson's theoretical results influenced the genesis of LP (Wright 2016). However, it was not his reconceptualization of the notion 'information' via units of physical volume that influenced decisively the genesis of this idea, precisely because Landauer employed a substantially different conception of physical information in these fields. Otherwise, Landauer's major influences were, (i) the aim to apply Shannon's information-theoretical apparatus to the concrete field of IBM magnetic memories practices during the 1950s, and (ii) the concrete model of the double potential well to represent information bits to be processed in subsequent computational operations.

4 Leon Brillouin: The Physical Meaning of Information

One of the pivotal authors in Rolf Landauer's intellectual evolution during the period 1945–1961 was the French-American physicist Léon Brillouin (Segal 2003). Contrary to what is usually assumed in the historiographic literature,[2] the intellectual connection between the two physicists is by no means limited to the latter's occasional collaboration as an IBM consulting engineer in the 1950s and his subsequent appointment as director of Electronics Education. Actually, Brillouin was Landauer's Ph.D. supervisor at Harvard, and one of the greatest exponents of the informational theoretical reconceptualization of statistical mechanics in the 1950s. In fact, in this chapter we argue that the role of Brillouin's conceptual tools is explanatorily essential for obtaining a deep understanding of the intellectual genesis of LP, perhaps in a more substantial way than Swanson or other colleagues within the IBM research context. Notably, Landauer acknowledged this decisive influence in the introduction of his 1961 paper:

> An alternative way of anticipating our conclusions is to refer to the arguments by Brillouin and earlier authors, as summarized by Brillouin in his book, Science and Information Theory, to the effect that the measurement process requires a dissipation of the order of kT. (Landauer 1961, p. 183)

Interestingly, the book mentioned by Landauer in this quotation constituted one of the main systematic applications of information theory in the scientific field of thermal statistical physics. Within this line of research, Brillouin (1951) was known precisely for defending the thesis that the information that an observer can extract about a physical system corresponds to the degree of negative entropy (or 'negentropy', in his own terms) of the observed system, a thesis telegraphically synthesized in the expression 'information is negentropy'. This was called the 'Negentropic principle'. As to the informational part of this principle, Brillouin (1956) distinguished between two types of information concepts: on the one hand, 'free information', when the alternative units on which the informational measures are defined (i.e., the symbols of a message) have no specific physical meaning, and 'bound information', when each of these alternative units corresponds to what is known as a 'microstate' of the observed physical system. In statistical mechanics, the particular microstate of a physical system at a particular time corresponds to the state in which the values of position and velocity of all components of the system are determined.

Therefore, the bound information on which the Brillouin Negentropic principle is based was defined from sets of microstates, associated to observable properties of the system, such as pressure or temperature. However, these sets (also technically called 'macrostates') can be characterized by the limited observational capacity of agents, so that all microstates x and y within a particular macrostate M_1 are indistinguishable

[2] Wright recently suggested that: "Brillouin's work on information theory was published in English in 1956 [...] As a pioneer of both solid state physics and information theory, Landauer would have been familiar with his work." (Wright 2016, p. 131). Significantly, as we remark above, Landauer is so aware of (his own PhD's supervisor) Brillouin's work that he quotes it as a background in the first page of his 1961 paper.

from each other for the observer. So, bound information measures the (epistemic) incapacity of agents to distinguish the actual microstate *x* of the physical system among a set M_1 of macroscopically indistinguishable microstate. In this way, the basic unit on which Brillouin bases his notion of information was not the source of the symbols of the messages à la Shannon or a certain unit of physical volume à la Swanson, but rather the volume of possible (and equiprobable) microstates of the observed system contained within the same observationally-accessible macrostate M_1. Or equivalently, by the probabilistic measure μ (usually the Lebesgue measure) of the phase volume associated with a macrostate, i.e., $\mu(M_1)$.

According to Brillouin's Negentropy principle, the concept of bound information necessarily entails a reinterpretation of the statistical-mechanical concept of entropy (specifically in its Boltzmannian formulation), defined by the logarithm of measure the actual macrostate of the system, multiplied by Boltzmann constant *k*. Such informational interpretation of the statistical-mechanical entropy went back historically at least to suggestive statements such as that of Gilbert Lewis:

> [...] Gain in entropy always means loss of information (Lewis 1930, p. 577)

In this sense, Brillouin stated

> [...] entropy measures the lack of information about the actual structure of the system. This lack of information induces the possibility of a great variety of microscopically distinct structures, which we are in practice unable to distinguish from one another. (Brillouin 1956, p. 160)

Brillouin relied on his Negentropy principle to demonstrate what is known in the literature as 'Szilard's principle' (Fig. 1). The Hungarian physicist Leo Szilard aimed to show in 1929 that the acquisition of knowledge by an agent about the location of a molecule in a box divided into two volumes by means of a measurement operation would lead to a certain increase in the entropy of the medium (Bennett 1982; Brillouin 1956). Szilard's thermo-epistemic thesis was informationally reinterpreted in the 1950s by Brillouin with the technical and conceptual resources of Shannon's theory, so that the acquisition of 'knowledge' was assumed to be an acquisition of 'information' (in the precise sense of bound information) about the microscopic configuration of the observed gas, i.e.

> Let us suppose that we have additional information on the [micro]state of the gas. (Brillouin 1956, p. 157)

Therefore, an agent acquires bound information about the actual position and velocity of the molecular constituents of the observed system when the number of macroscopically-indistinguishable microstates decrease.

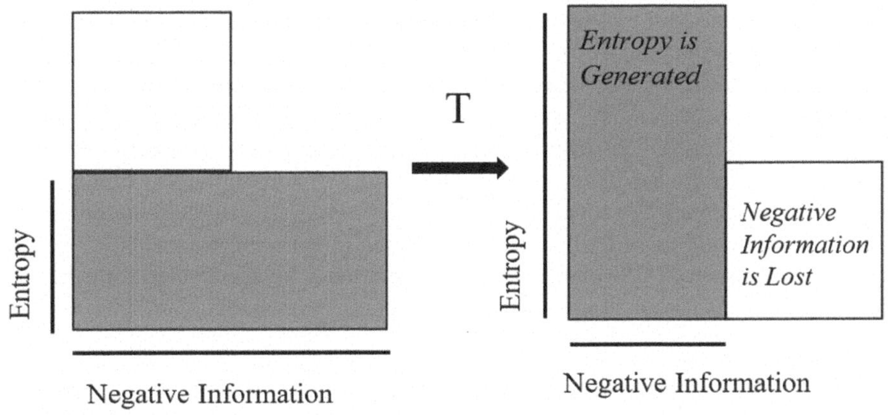

Fig. 1 Negentropy principle linking entropy and Information *Source* Author's images

In this phase diagram the macrostate (depicted as a rectangle) in which the actual microstate (depicted as a phase point) of the system is located is gray coloured, whose content encodes its 'bound information'. The projections of the phase volume of the macrostate on the vertical and horizontal axes correspond to the entropy and negative information of the system, respectively. The Negentropy principle relates the changes in entropy and negative information after performing a 'T' transformation in the system, where according to the Liouville theorem the reduction by half of the negative information (acquisition of information about the system) would be compensated by the duplication of the entropy of the system, generating $k \log 2 = 0.69\, k\text{T}$.

Due to Liouville's theorem of statistical mechanics, the phase volume of the actual macrostate M_1 remains constant during the system's dynamic evolution. Let us suppose that we divide the degrees of freedom of the system into (i) those relative (or projected on) the macroscopic knowledge of the observer and (ii) those relative only to the observed system. Liouville's theorem implies that, if obtaining information on the microscopic structure implies reducing the phase volume of the macrostate relative to the observer (i), then this decreased volume must be somehow compensated by an increase in phase volume elsewhere, which is none other than that relative to the observed system (ii). However, the increase in phase volume relative to the degrees of freedom of the system implies that the Boltzmann entropy relative to these degrees of freedom will increase accordingly. Therefore, due to Brillouin's Negentropy principle and Liouville's theorem, at the moment when the agent obtains bound information about the microscopic structure of the observed system (for example, halving the phase volume relative to the observer), the Boltzmann entropy of the system would increase proportionally by $k \log 2$ (doubling the phase volume relative to the observed system). This would be precisely Brillouin's (1956, p. 162) demonstration of his *sui generis* Szilard's principle: namely that the acquisition of one bit (i.e., halving the microstate relative to the observer) of information on the system would lead to an increase in the system entropy of $k \log 2$.

Up to this point, a clear demonstration of the thermal effects of information acquisition through measurements may seem a result. However, throughout the development of his proposal for informational thermophysics in his 1956 *Science and Information Theory*, Brillouin ambiguously used two substantially different (and somewhat problematic) concepts of 'information'. First, he exploited the notion of 'bound information' as a term of a strong semantic-epistemic character, closer to the everyday notion of information (e.g., 'information of an agent about something') than to Shannon's technical information concept, as Earman and Norton accurately pointed out:

> [...] Brillouin's labeling of the quantity [negentropy] as 'information' is intended, of course, to suggest our everyday notion of information as knowledge of a system. (Earman and Norton 1999, p. 112)

In this sense, an agent will possess more bound information ('epistemic information', 'knowledge' or even '*Können*', in Szilard's original German words) about the actual microstate of the system while fewer microscopic structures are macroscopically indistinguishable for the agent. But while this type of information was simply a new name for negative amounts of Boltzmann entropy, the concept of bound information adds nothing significant to standard statistical mechanical descriptions.

Beyond this sort of epistemic information, Brillouin also employed the term 'information' not as a semantic-epistemic notion (i.e., 'information as relevant knowledge') but in Shannon's structural-statistical sense (i.e., 'information as data'), grounded not on an agent's cognitive capacities but on the probability of occurrence of certain events relative to the observed system. In this sense, the observed system will encode more statistical information when more microscopic data are compatible with a certain macrostate, and therefore, when the volume of phase accessible to the observer is greater. Therefore, both senses of the term 'information' (i.e., epistemic information as microstatistical knowledge, and Shannon information as quantity of data or microstatistical bits) are intuitively presented as opposites. Brillouin was aware of this conceptual problem, however, his main strategy to solve it was to argue ad hoc that Shannon's informational entropy was identical to his notion of negentropy, except with the inverse sign:

> [...] The connection between entropy and information was rediscovered by Shannon, but he defined entropy with a sign just opposite to that of the standard thermodynamical definition. Hence what Shannon calls entropy of information actually represents negentropy. This can be seen clearly in two examples [...] where Shannon proves that in some irreversible processes (an irreversible transducer or a filter) his entropy of information is decreased. To obtain agreement with our conventions, reverse the sign and read negentropy. (Brillouin 1956, p. 161)

This apparent solution to eliminate the tension between the two different concepts of information was even more problematic. In any case, the key to the above quotation by Brillouin at the end of the chapter entitled 'The Negentropy Principle of Information' (Brillouin 1956, Chap. 12) was precisely the introduction of irreversible transformations between amounts of 'bound information' and negative amounts of 'Boltzmann entropy' as a conceptual trick to avoid the trivial physical meaning of the Negentropy principle as a mere 'relabelling'. That is, if the process that allows the negentropy (i.e., increasing entropy) of the system to be decreased after increasing the bound information of the agent is properly irreversible (in a thermodynamic sense, being charitable), then such a decrease in negentropy cannot be subsequently reversed by decreasing the bound information of the agent, according to the Second Law. Similarly, if by eliminating (via a thermodynamically irreversible operation) a certain amount of the agent's bound information we decrease the negentropy of the system, then this cannot be subsequently reversed by increasing the bound information, and therefore, this information about the system will have been lost forever:

> [...] Information can be changed into negentropy, and vice versa. If the transformation is reversible, there is no loss, but an irreversible transformation always corresponds to a loss. (Brillouin 1956, p. 184).

Therefore, by introducing the notion of irreversibility, the interchangeability between information and negentropy would no longer be trivial.

However, However, Brillouin introduced this idea in a sufficiently vague fashion to clearly delimit the nature of this irreversibility. For example, when Brillouin relied on Shannon's (1948) proto-notion of irreversibility it had a strictly information-theoretical meaning, i.e., after optimally encoding a message according to the frequency of its symbols, this encoding cannot be reversed. However, this kind of communication-theoretical irreversibility has no meaning within the context of the statistical mechanics of gases. In this sense, the fundamental theoretical differences between what one should call the Brillouin 'Irreversibility-based Negentropy principle' of 1956 and Landauer's principle of 1961 are (i) the substitution of the little-defined epistemic concept of 'bound information' by the technically sophisticated and extremely popularized Shannon's information and (ii) the transition from the undefined Brillouinian concept of irreversibility to a well-defined and computationally significant concept of 'logical irreversibility', proper to transitions between logical states and clearly distinguishable from the thermodynamic concept of thermal irreversibility (applicable only between physical states and not between epistemic states, such as agents' capacities to distinguish states). In order to understand the very complex intellectual process that allowed the generation of LP from the conceptual resources displayed in Brillouin's 'Irreversibility-based Negentropy principle', we will also need to give an explanatory account of how Rolf Landauer himself reached both (i) and (ii) through the rich context of scientific ideas in which he was involved during the 1950s. It is here that the famous von Neumann played a decisive role in the origin of this idea.

5 John von Neumann: The Physical Consequences of Logic

Apart from Brillouin and Swanson, the other pivotal historical author involved in the intellectual genesis of LP was the Hungarian-American mathematician John von Neumann, one of the main intellectual promoters of information theory as a conceptual mediator between thermal physics and computation. His influence ranged from the popularization of the Szilard principle as a thermal effect of the acquisition of microscopic knowledge to the choice of the name 'entropy' as a central concept in Shannon's theory:

> [...] Von Neumann told me [Shannon], 'You should call it entropy, for two reasons. In the first place your uncertainty function has been used in statistical mechanics under that name. In the second place, and more importantly, no one knows what entropy really is, so in a debate you will always have the advantage. (Tribus and McIrvine 2007, p. 180)

As far as the origin of LP is concerned, von Neumann fostered the exploitation of Shannon's theory beyond the domain of symbol transmission. Only one year after the publication of the 1948 paper and a few months after the publication of the famous 1949 book co-authored with Weaver (1949), von Neumann (1949) gave a talk in December 1949 entitled 'Theory and Organization of Complicated Automata' wherein he fervently defended a triple identity between the (a priori so disparate) concepts of Shannon's information theory, logic and thermodynamics:

> [...] I have been trying to justify the suspicion that a theory of information is needed and that very little of what is needed exists yet. Such small traces of it which do exist [...] indicate that, if found, it is likely to be similar to two of our existing theories: formal logics and thermodynamics. It is not surprising that it is likely to have a lot in common with thermodynamics. (Wright 2016, p. 62)

In this direction, von Neumann particularly emphasized that the formal similarity between the mathematical expression of (i) Gibbs' statistical mechanical entropy, i.e. $-k\mathrm{T}\ \Sigma\ p_i \ln p_i$, (ii) Shannon informational theoretical entropy, $-\Sigma\ p_i \log p_i$, (these first two differ only in the Boltzmann constant k and the natural logarithmic basis of the former) and (iii) certain expressions of inductive logic like $-\Sigma\ p_i \log p_i$, such as the one developed by Carnap (Carnap 1977), would suggest not a mere structural analogy, but a deep conceptual identification between all these notions. In particular, the von Neumannian connection between the first two concepts was already famously noted by Weaver in his comments to Shannon's theory:

> [...] Dr. Shannon's work roots back, as von Neumann has pointed out, to Boltzmann's observation, in some of his work on statistical physics. (Shannon and Weaver 1949, footnote 1)

His defense of the above (triple, or at least double) conceptual identity in this talk was deeply rooted within the scientific community during the 1950s, rejecting any criticism of this identification, such as the one that the philosophers Carnap and Bar-Hillel tried to make without much success.[3]

---

[3] An illustrative description of this historical episode is the following: "During one of my visits to him in Princeton, in 1952, von Neumann also came to see him [Carnap], and we started discussing

Based on this strong conceptual identification, von Neumann defended in that 1949 talk that every informational act would involve a minimum of entropy generation, reconceptualizing through Shannon's recently published theoretical informational tools the thesis (and numerical results) of Szilard two decades earlier in 1928:

> [...] He [von Neumann] then computed the thermodynamical minimum of energy per elementary act of information [...] For a binary act N = 2, and taking the temperature to be about 300 degrees absolute, he obtained [3 X 10^{-21} J] for the thermodynamical minimum. (Wright 2016, p. 66)

As the reader will appreciate, this proposal was intellectually decisive both for the informational reformulation of Szilard's principle by Brillouin and for the origins of LP. Thus, von Neumann opened a conceptual space of possibilities in which any information processing, however trivial it might be, has physical consequences in reality: in other words, any information process would entail a minimum thermophysical cost directly proportional to the number of bits involved in such a processing. However, both Szilard and Brillouin limited the conceptual space of this general von Neumann's thesis exclusively to the measurement-based acquisition of microscopic information, while Landauer delimited this same conceptual space into the domain of logically irreversible information processes (Maroney 2009).

Unlike Brillouin's vague concept of 'irreversible transformation', in his 1949 talk von Neumann accurately distinguished (in order to establish theoretical connections) in a context of computational analysis between (i) a thermodynamically irreversible process, in which physical states properly intervene, and (ii) a logically irreversible transformation, understood as an abstract map (usually many-to-one) between logical states, so that the information contained in the initial logical states cannot be inferred from the subsequent logical states. This distinction was central, firstly, to intellectually influence the distinction (already in Swanson) between (a) the physical coupling of magnetic memories, subject to thermodynamically reversible or irreversible processes, and (b) the logical coupling of the logical states (i.e., '0' or '1') represented by separate magnetic memories in an informationally redundant manner.

Therefore, von Neumann's suggestions (i.e., (i) informational processes have physical consequences, and that (ii) logical transformations are connected with thermodynamic processes) were historically determinant for Landauer to hypothesize in 1961 that logically irreversible transformations may involve thermodynamically irreversible and therefore entropically costly processes. In fact, LP intends precisely to show such a connection between the logical field of computing and computer physics

---

the talk I had heard von Neumann deliver [...] in which he had proclaimed, among other things, a triple identity between logic, information theory and thermodynamics [...] We were quite ready to agree that there existed a certain formal analogy [...] and had indeed ourselves shown [...] how entropy-like expressions, such as the famous $-\Sigma\, p_i \log p_i$, occurred in (inductive) logic, but could not see how stronger relationships could possibly be supposed to exist. We tried to convince von Neumann that this way of presenting the analogy as an identity must lead to confusion." (Bar-Hillel 1964, p. 11).

through theoretical information resources applied in the concrete model of the double well-potential, assuming *à la* von Neumann (1949) somehow the mediating role of information theory in the plausible interconnection of logic and thermodynamics. Interestingly, Brillouin mentioned in a marginal note of his book how von Neumann himself went so far as to formulate a thesis identical to LP, except that expressed in terms of an observer as a computer apparatus:

> [...] J. von Neumann also considers the case of the observer forgetting the information, and states that this process also means an increase of entropy. (Brillouin 1956, footnote 4)

In any case, the multiple quotes from von Neumann in his 1961 paper show how Landauer (1961, p. 185) not only knew the intellectual work of the Hungarian-American mathematician for being the greatest intellectual pioneer during the 1950s (using Shannon's information theory to evaluate the connection between logic and thermal physics) in the context of computing, but for having a real interest in his proposals.

6 Landauer at a Crossroad of Intellectual Trends

Up to this point in our argument, we defend that Landauer's statement in 1961 that 'any implementation of logically irreversible processes would necessarily entail a minimum entropic cost' was not discovering any idea that was not previously defended in general terms by von Neumann or in an interpretively inconsistent manner by Brillouin. The main theoretical merit of Landauer in 1961 lies precisely in (i) concretizing von Neumann's general abstract proposal (i.e., eliminating information has physical costs) by means of the double potential well model of concrete computation outlined by Swanson in 1960, and in (ii) eliminating the interpretative inconsistencies of Brillouin's Negentropic principle by defining the Shannon information measure of macrostate physical systems implementing logically irreversible computations. In this way, Swanson's influence in the intellectual origin of LP has been not only overestimated (in comparison with the role played by Brillouin and von Neumann) but also incorrectly evaluated by recent historiographic analyses of the genesis of this idea:

> [...] John Swanson made the transition from writing about the information stored per symbol in information theory to writing about information per volume of a computer memory device [...] These [...] shifts allowed him to use the tools and concepts of Claude Shannon's information theory in his study of physical computer memory. Swanson's work prompted Rolf Landauer to formulate what became known as Landauer's principles. (Wright 2016, p. 113)

Such a transition between symbols-based Shannon information and this same concept formulated from units of physical volume is not only deficient in the previously mentioned senses, but (i) neither did it serve to encourage Landauer's belief that informational processes would have physical effects (since this was done in a more theoretically robust way by von Neumann and Brillouin's proposals) (ii) nor was it adopted by Landauer to argue for the increase of entropy through the double-well model. On the contrary, Landauer reconceptualized the notion of Shannon information in a similar way as Brillouin defined his notion of 'bound information' (except for the substantial differences already noted between these two concepts of information (Anta 2021a)). That is, associating information values with the volume of a particular set of microstates associated with an observable property of the system implementing the computation, thus associating individual bits with phase regions of that system. Therefore, while for Swanson bits are encoded in physical system volume units v, for Landauer bits are encoded in $\mu(M_i)$, the measure of the phase volume associated with a macrostate:

> [...] Note that our argument here does not necessarily depend upon connections, frequently made in other writings, between entropy and information. We simply think of each bit as being located in a physical system, with perhaps a great many degrees of freedom, in addition to the relevant one. (Landauer 1961, p. 187)

We argue that the above quote is pivotal not only for understanding the intellectual genesis of that idea, but also for robustly grasping its meaning. While virtually all of the different proposals to apply information theory to the field of thermal physics during the 1950s were based on connecting different concepts of information and entropy (Anta 2021b, Chap. III), Landauer claimed to postulate something much more theoretically sound: a causal connection between a logically irreversible operation and a minimal increase in entropy. However, if we critically analyse the premises on which his argument was based, Landauer presupposed that every bit of information can indeed be encoded in individual macrostates M_i of physical systems. Furthermore, he presupposes (i) that the Boltzmann entropy of a physical system is proportional to the volume of the macrostate of the system at time t, and (ii) that the number of bits of the physical system (or 'statistical ensemble of bits', in the author's terms) depends on the number of binary partitions that are made on the macrostate containing the actual microstate.

> [...] Consider a statistical ensemble of bits in thermal equilibrium. If these are all reset to ONE, the number of states covered in the ensemble has been cut in half. The entropy therefore has been reduced by $k \log_e 2 = 0.6931\, k$ per bit. The entropy of a closed system, e.g., a computer with its own batteries, cannot decrease; hence this entropy must appear elsewhere as a heating effect, supplying $0.6931\, k\mathrm{T}$ per restored bit to the surrounding. (Landauer 1961, p. 187)

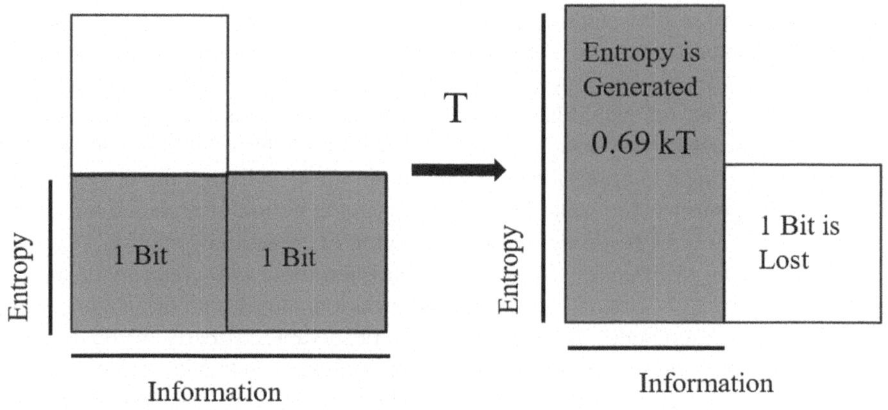

Fig. 2 Landauer's principle relating entropy and information *Source* Author's images

Therefore, although Landauer claimed that no direct connection between entropy (Boltzmann) and information (Shannon) was assumed, what is certain is that, since the value of both entropy and information was proportional to $\mu(M_i)$ the volume of the region of the phase space associated to the physical system, each individual bit ends up having a definite Boltzmann entropy value. Thus, the content of LP visually expressed via a phase diagram by (i) halving the phase volume of the macrostate of the system relative to the information it encodes after eliminating one of them, by (ii) doubling the phase volume of the macrostate of the system relative to its entropy (see Fig. 2), while the entropy of the macrostate of that closed system should be preserved during the dynamic evolution of the system

Note that the contents of the phase diagram of Brillouin's Negentropy principle (Fig. 1) and the phase diagram of Landauer's principle (Fig. 2) are identical except (i) for the concept of information used ('bound information' and 'Shannon entropy', respectively) and (ii) because the latter introduces the defined notion of implementation of logically irreversible operations. This equivalence was not merely a conventional fact, precisely because both were strictly identical insofar as they refer to the same thermo-informational phenomenon from two different theoretical perspectives. Although at first glance it may seem that Brillouin and Landauer contradict each other in stating that the increase in k log 2 entropy is linked to an increase in information and a decrease in information, respectively (as the literature has tried to show); The truth is that while Brillouin claimed that the increase in entropy k log 2 was linked to an increase (somehow irreversible) in the agent's knowledge of the actual microstate of the system, Landauer claimed that this increase was associated with a logically irreversible elimination of microstatistical data relating to the system.

In this direction, the main idiosyncratic element of Landauer's novel proposal as opposed to Brillouin's previous one was the introduction of a well-defined factor of asymmetry that allow to (a) re-establish a causal link between logical operations and thermal effects, and (b) reducing such increase of entropy by reversing the previously eliminated information bit. Thanks to von Neumann's famous indications, Landauer found this central asymmetric factor in the logical irreversibility of a computational operation, which when correctly implemented in the physical device would generate certain thermal effects:

> [...] We shall call a device logically irreversible if the output of a device does not uniquely define the inputs [...] Logical irreversibly, we believe, in turn implies physical irreversibility, and the latter is accompanied by dissipative effects. (Landauer 1961, p. 186)

Note that the amount of entropy generated 0.6931 kT (common to the analyses of Szilard, von Neumann and Brillouin) corresponding to the entropy produced after eliminating a bit is extremely difficult to measure due to the immeasurable technical complexity involved in detecting temperatures of the order of 1.38×10^{-23} K^{-1}. It is mainly for this reason that LP was first experimentally tested sixty-one years after it was originally formulated (Bérut et al. 2012).

7 Conclusion

In this paper we have argued that Landauer's principle did not emerge spontaneously in 1961 from the creative genius of a young IBM researcher, as it has been claimed in some of the popular narratives promoted from the main informationalist trend of scientific thought (Seife 2007); nor from the sole influence of co-workers like Swanson (Wright 2016). In fact, different versions of Landauer's principle were previously formulated by both Brillouin in 1956 and von Neumann in 1949, although in a certainly general and conceptually vague way. On the contrary, we have argued that when Landauer stated in his famous paper that every logically irreversible process has thermal costs, such a statement depended on the fact that Landauer was located in a historical confluence of a manifold of intellectual trends, i.e., the bandwagon of Shannon's theory after 1949, the informationalization of thermal physics by Brillouin during the 1950s (Anta 2021a, b, Chap. III), the evaluation of the computational implementation of logical structures by von Neumann, and the analysis of the physical limits of computers at the IBM. In this sense, the received view of Landauer in 1961 as an iconoclast popularized from mainstream narratives is not only historically incorrect but is a substantial impediment to a deeper understanding of the profound meaning of this idea. Therefore, the intellectual origins of Landauer's principle did not entail a break with the rich tradition of thermophysical, informational and computational theories in the 1950s, but a sophisticated confluence of these intellectual tendencies into a powerful idea developed in the practical context of the study of computational efficiency at IBM.

Acknowledgements This work has been funded by the Spanish Ministry of Education (FPU16/0774), and it was developed as part of the Research Group in Analytic Philosophy, LOGOS, the Barcelona Institute of Analytic Philosophy (BIAP) and the research project "Reassessing Scientific Objectivity" (PID2020-115114GB-I00).

References

Anta J (2021a) Can informational thermal physics explain the approach to equilibrium? Synthese 199(1/2):4015–4038.
Anta J (2021b) Historical and conceptual foundations of information physics. Dissertation, University of Barcelona, Spain.
Bar-Hillel Y (1964) Lenguage and information. Selected essays on their theory and application. Addison-Wesley, Massachusetts.
Bekenstein JD (1973) Black holes and entropy. Physical Review D 7:2333–2346.
Bennett C (1982) The thermodynamics of computation—a review. International Journal of Theoretical Physics 21:905–940.
Bérut A, Arakelyan A, Petrosyan A, Ciliberto S, Dillenschneider R, Lutz E (2012) Experimental verification of Landauer's principle linking information and thermodynamics. Nature 48:187–190.
Brillouin L (1951) Maxwell's demon cannot operate: information and entropy. International Journal of Applied Physics 22:334–337.
Brillouin L (1956) Science and information theory. Academic Press, New York.
Carnap R (1977) Two essays on entropy. In: Simony A (eds) University of California Press, Berkeley.
Earman J, Norton J (1999) Exorcist XIV: the wrath of Maxwell's demon. Part II. From Szilard to Landauer and beyond. History and Philosophy of Modern Physics 30(1):1–40.
Gleick J (2011) The information: a history, a theory, a flood. Pantheon Books.
Kline R (2015) The cybernetics moment: or why we call our age the age of information. Johns Hopkins University Press, Baltimore.
Landauer R (1961) Irreversibility and heat generation in the computing process. IBM Journal of Research and Development 5:183–191.
Leff H, Rex A (1990) Maxwell's demon: entropy, information, computing. Princeton University Press, New Jersey.
Leff H, Rex A (2003) Maxwell's demon 2: entropy, classical and quantum information, computing. Institute of Physics Publishing, Bristol-Philadelphia.
Lewis G (1930) The symmetry of time in physics. Science 71:568–577.
Lloyd S (2006) Programming the universe: a quantum computer scientist takes on the cosmos. Knopf, New York.
Maroney O (2009) Information processing and thermodynamic entropy. In: Zalta EN (ed) The Stanford encyclopaedia of philosophy. https://plato.stanford.edu/entries/information-entropy. Accessed 2021/12/22.
Neumann von J (1949) The role of high and of extremely high complication, fourth University of Illinois lecture. In: Burks AW (ed) Theory of self-reproducing automata (Reprinted in 1966). University of Illinois Press, Illinois.
Norton J (2013) The end of the thermodynamics of computation: a no go result. Philos Sci 80(5):1182–1192.
Segal J (2003) Le zéro et le un: Histoire de la notion scientifique d'information au 20e siècle. Éditions Syllepse, Paris.
Seife C (2007) Decoding the universe: how the new science of information is explaining everything in the cosmos from our brains to black holes. Penguin Group, London.
Shannon C (1948) A mathematical theory of communication. IEEE Press.
Shannon C, Weaver W (1949) the mathematical theory of communication. University of Illinois Press, Illinois.
Shannon C (1956) The Bandwagon. In: Collected papers. IEEE Press.
Swanson J (1960) Physical versus logical coupling in memory system. IBM Journal 4(3):305–310.
Tribus M, McIrvine E (2007) Compendium of the foundations of classical statistical physics. In: Butterfield J, Earman J (eds) Handbook for philosophy of physics. Elsevier, Amsterdam.

Uffink J (2007) Compendium of the foundations of classical statistical physics. In: Butterfield J, Earman J (eds) Handbook for philosophy of physics. Elsevier, Amsterdam.
Wright A (2016) The physics of forgetting: thermodynamics of information at IBM 1959–1982. Perspectives on Science 24(1):112–141.

Chapter 18
Foundations of the Operators Approach in Mathematics and Physics

Cécile Barbachoux and Joseph Kouneiher

Abstract The conception of the idea of operators has played a fruitful role in the evolution of mathematics and physics. The implementation of this idea made it possible to explore properties of solution spaces, the topology of certain spaces and to discover some underlying dualities, which made it possible to understand deep aspects of mathematical and physical theories. For instance, the equivalence between integral equations and differential equations has allowed the resolution of many essential problems in mathematics and physics. This is the case of what is called the Fredholm operator in mathematics and the formulation of wave mechanics by Schrödinger. Our aim in this paper is to study and discuss the origin and development of the idea of operators in mathematics and physics and the entanglement between the two, thanks to the properties of certain operators, which play a prominent role in old and recent developments in mathematics and physics.

Keywords Operators · Integral equation · Differential equations · Heat equation · Boole symbol-operator · Fredholm operator · Hilbert space · Born-Wiener approach · Operator spectrum · Quantum mechanics · Von Neumann · Non-commutative geometry

1 Introduction

The concept of operator is as old as symbolic algebra, for every symbol in classical algebra denotes a precept either to substitute a number or to perform an arithmetic operation. In the sixteenth century, from its origin, the algebraic language has been able to acquire a rating that has managed to abstract the operations of the objects to which they apply (or on which they carry) to make the objects of the thought. This

C. Barbachoux · J. Kouneiher (✉)
University of Côte d'Azur/INSPE, Nice, France
e-mail: joseph.kouneiher@univ-cotedazur.fr

C. Barbachoux
e-mail: cecile.barbachoux@univ-cotedazur.fr

R. Pisano (ed.), *A History of Physics: Phenomena, Ideas and Mechanisms*, History of Mechanism and Machine Science 42,
https://doi.org/10.1007/978-3-031-26174-9_18

language consists of symbolic systems that support algebraic—numbers, unknown, literal symbols—then, by generalizing, titles that designate geometric objects—vectors, transformations, tensors of logical entities—names, proposals, predicates, variables etc. These systems are organized in a real "conceptual writing" where the concepts are integrated in manipulative writing sometimes in the form of a formal and significant calculation (differential and integral calculation, calculation of proposals, predicates, modalities, etc.) that we often refer to the language name formalizes (first order languages, categories, diagrams, graphs, etc.). When the notations are specified, the thought emerges from the desire to verify, at each step of calculation, the first significance of the formal transformations operated on the symbolic arrangements which represent mathematical beings (algebraic, geometric, logical, etc.). By becoming formal, the rewrites reveal the types of abstracts operations apprehended from applied and more concrete operations; examination of the formal mechanism involved in the operation releases a new symbolic object more abstract; this is often designed—under the term operator. This acts on an operand to produce a result. While the operation is a concept that aims to capture the operand transformation process, the operator is being detachable from the operand and the operating process it generates. It can be thought in *itself* and for itself; it is intrinsically defined by its combinatorial latitudes and its possibilities, or constraints, arrangement with other operators, regardless of the actions that these operators trigger.

Several situations invite to define operations of operators independently of the objects on which they operate. The concept of an operator is used in a variety of practical and theoretical areas. Operators, and what they operate, i.e. their substrates, targets, or operands, have a wide variety of forms, functions, and properties. Operators have explicit philosophical significance. On the one hand, they represent important ontological issues of reality reflecting its dynamical nature when some objects act on other objects. The interactions which characterize our universe, for instance, can be decomposed of separate actions involving the structure of an operator. On the other hand, epistemological operators form the basic mechanism of cognition.

In mathematics, an operator is generally a mapping or function that acts on elements of a space to produce elements of another space (possibly the same space, sometimes required to be the same space). There is no general definition of an operator, but the term is often used in place of function when the domain is a set of functions or other structured objects (function that acts on functions to produce other functions, the sense in which Oliver Heaviside used the term). Also, the domain of an operator is often difficult to be explicitly characterized (for example in the case of an integral operator) and may be extended to related objects (an operator that acts on functions may act also on differential equations whose solutions are functions that satisfy the equation).

The most basic operators (in some sense) are linear maps, which act on vector spaces. However, when using *linear operator* instead of *linear map*, mathematicians often mean actions on vector spaces of functions, which also preserve other properties, such as continuity. For example, differentiation and indefinite integration are linear operators; operators that are built from them are called differential operators, integral operators. Operator is also used for denoting the symbol of a mathematical operation. This is related with the meaning of *operator* in computer programming. In general, the term *operator* in computer programming languages has the same meaning as in mathematics. This is particularly true in functional programming languages, where an operator is also a function. However, most programming languages distinguish between operators and functions in those operators are a special primitive part of the language, both syntactically and in terms of functionality. For example, most languages provide a '+' (addition) operator, which adds two numbers without making a function call.

In physics, an operator often takes on a more specialized meaning than in mathematics. Operators as observables are a key part of the theory of quantum mechanics. In that context operator often means a linear transformation from a Hilbert space to another, or (more abstractly) an element of a C*-algebra.

Our aim in this paper is to reflect about the historical and epistemological foundations of the concept of Operator in mathematics and physics and to understand what is common in physical objects called operators and abstract mathematical structures, with the name operator providing foundations for building a unified but flexible theory of operators. The paper treats also the functionalism and other philosophical aspects of the 'operation' of operators.

2 Historical and Epistemological Foundations

Humanity has always been trying to organize space to give him meaning and structures, by action, gesture and language. Mathematics, especially geometry, gave it the most stable of its conceptual restructuring of phenomenal space.

Greeks have introduced abstract reasoning to geometric shapes like straight lines, polygons and conic. They also introduced the axiomatic method that allowed Euclid to systematize and rigorously organize mathematics on the logic level. Before them and elsewhere, existed mainly algorithmic mathematics operating on numbers, gathered in "recipes". Greek mathematicians did not separate the geometry from arithmetic, so that many calculations on numbers have a meaning in terms of the properties of geometric figures, and vice versa. However, this had led to conceptual problems. For example, they ignored the existence of negative numbers, so some simple equations, of which they are the solutions, appeared as absurd for them. What could be the meaning of such numbers for measuring lengths, surfaces, volumes or angles? Centuries after Euclid, the mathematicians overcome these obstacles and found a geometric way of interpreting the negative numbers, as well as the strange calculation rule that the product of two negative numbers is positive (see for example Jean Dieudonne's book "*For the honour of the Human Spirit*"). The Greeks were perplexed too, but less, in front of the length of the diagonal of a square that could not be interpreted as the quotient of two integers. We know today that this problem involves irrational numbers.

2.1 Descartes and the Algebraization of Geometry

Although a prisoner, in a certain way, from the Greek conception of the relations between arithmetic and geometry, René Descartes will revolutionize the subject by discovering the analytical geometry that allowed him to translate more systematically algebra problems in geometric problems, and vice versa. After him, the points in the plan and the space became associated with couples and triplets of numbers, which entered in polynomial algebraic equations describing sets of straight lines, plans and even conic (circle and ellipse for example) or quadrics (sphere and ellipsoid for example) that can cut.

If algebra acquired a certain autonomy vis à vis geometry, the goal that was clearly set by its author René Descartes, from the first page of his Geometry, is the resolution of geometric problems, precisely those that he had met in the mathematical collection of Pappus, and who was presenting construction exercises. It is therefore not surprising that solving for Descartes means to build geometrically. Simply, between the geometric statement of the problem and its construction (also geometric), algebra has presented itself to him as an irreplaceable instrument. That algebra could be useful to geometry was far from obvious at the time, since algebra consisted in the manipulation of abstract magnitudes, devoid of any geometrical interpretation (which made it possible to consider powers greater than three), whose only *limit* was the respect of the rule of homogeneity—only algebraic terms of the same dimension can be added. This is the reason why the work of the sixteenth-century algebraists appears to Descartes as a confused set of formulas and recipes for calculations—"a confused and obscure art"—and not as a theory comparable to classical geometry. When we also know how to have "clear and distinct" ideas are one of the criteria of truth in Descartes, we can see that this independence of algebra from geometry was far from appearing to him as a progress. This is indeed what appears clearly when, before reorganizing the mathematical knowledge, Descartes endeavours to assess the existing mathematics, and the three *arts* which intervene there: formal logic, analysis of the surveyors and algebra of the moderns. If formal logic is autonomous and claims to govern all knowledge, one of its defects and the reason why "[...] the forms of the syllogism are of no help to perceive the truth of things [...]" (Descartes 1996, Rule XIV, p. 108), precisely because it is used alone, it is hollow, even dangerous: it turns empty.

Descartes therefore proposes to plunge the syllogism into a living intellectual activity, producing objects of knowledge: geometry. This new logic, alive, will be that of analysis, which: "[...] shows the true path by which a thing has been methodically invented, and shows how the effects depend on the causes [...]" (Descartes 1992, p. 279). As a general arithmetic, separated from geometry, algebra as it was at the time did not correspond either to the Cartesian project, and this is why Descartes will dissociate this algorithm and its notations, from the notion of number. Precisely, Descartes proposes to clarify the operations of algebra by giving to it a geometrical background: from the first pages of his essay, he thus makes correspond the fundamental operations of arithmetic (addition, multiplication, etc.) to geometric constructions of lines. Descartes's *algebraization* of geometry can, in fact, be analysed as an enterprise of re-founding the algebra then existing, again on the basis of geometry alone. In this new algebra, the algebraic symbols or letters that Descartes will use and on which the calculations and transformations will be carried out, designate no longer abstract numbers or quantities, but lengths of lines. Since on the other hand, in Descartes' Geometry, a line represents a dimension, we can say that Descartes' reduction of mathematical complexity is twofold: he begins by reducing the figures to lines, then designates these lines by letters. One line for each algebraic quantity, another representing the result of the operation performed. For addition or subtraction, it is therefore sufficient to postpone or delete on the same line a given segment. This being done, it is from geometry that Descartes will endeavour to reconstruct algebra, by attempting to explain in detail, and from the constructed figures, how "[...] to come to the equations that serve to solve the problems [...]" (Descartes 1992). Instead of naming a segment by these extremal points (eg AB), Descartes begins by designating a letter each of the lines (Descartes fixes as well as the last letters of the alphabet (x, y, z)) for the unknowns, the first ones (a, b, c) for the parameters, etc.), which will make it possible to examine their ratios, that is to say ratios of lengths, without the points of *attachment* (Extrémaux) need to be taken into account: it is therefore in fact all fold or crush on the same plane. From here, we can write the problem in the form of linear equations and solve the problem by respecting the geometric homogeneity. In doing so, all forms, regardless of their size, are designed as simple lines. In this case indeed, we can speak of symbols, but precisely because the relation is explicit as such, and that it is anchored in a geometrical reality of which it respects the rules. In addition, indeed, in a second time, and once the equations are solved, Descartes redeploys the solution in its geometrical form, and the problem is only really solved for him when one has reached a construction. If Deleuze (1988) could speak about the importance of the fold in Leibniz, it is also that this one, like Descartes, but this time vis à vis Pascal's pictorial integrals, proceeded to an identical work. Since the subject of the work remains the resolution of geometric problems, and algebra only plays the role of an instrument, we can say that Descartes' goal was not to eliminate geometry and representation, but on the contrary to give a real content to the current algebra, giving it an anchorage in the geometry. It was a question of clarifying things, and it appeared to Descartes that representing algebraic quantities by lines could be the simplest method to *visualize* the problem.

It is clear here that for Descartes, the geometry of the ancients remains the royal road towards the true mathematics and thus retains the primacy granted to it by the scholars of classical Greece: only in fact, its continuous magnitudes provide a content for symbolic operations. As a science of figures and lines to be constructed, it remains for him the criterion of the validity of the operations of algebra, since its object is what it considers to be the very essence of bodies: extension. One can then wonder why Descartes did not simply deny algebra any relevance, and instead sought to improve it. The content of geometric science is not questioned here, we will see that the algebraization of geometry can be analysed as a simple new formatting of geometry. It is indeed an algebra of lines that Descartes proposed to constitute, and not a geometry of coordinates (this last term will only appear with Leibniz): in fact, the real stake that constitutes today for us the large algebraic translation (the possibility of conceptualizing other types of geometries) did not appear immediately.

2.2 *Algebraization and Limits of the Representation*

The very essence of geometry combined for Descartes the best and the worst, depending on whether one considers its content or form. In the latter case, in fact, the defect of geometry is precisely due to being too constrained to the consideration of concrete figures, while the imagination (or representative capacity) is weak:

> The mind does not have it and it is difficult to form three lines in the brain and to draw them [...] but we have the greatest difficulty in imagining even a polygon of seven or eight sides. (Descartes 1981, p. 27)

Descartes thus recognizes that algebra has a great advantage: this "[...] kind of arithmetic relieves the imagination [...]" (Descartes 1981). The use of algebra will therefore make it possible to multiply the potentialities representative of its basic elements, the lines. We can see by this that Descartes' mathematics do not eliminate the representation: Descartes simply takes note of the limitation's representative of man, inventing a way not to go through them. Like Fermat, Descartes invokes for algebra its only convenience: the new writings will serve as a reminder and will allow to extend more clearly the traditional geometry of figures and curves—it is only in this way that we can understand his statement that he has completed mathematics. Do not be fooled by confusing form with matter: what we are thinking about here are representations. As Lon Brunschvicg says in his article *Mathématique et métaphysique chez Descartes, the originality of Cartesian geometry is that it applies an original method to problems that have been or could already have been solved by synthetic reasoning of the ancients.* (Brunschvicg 1927). According to Brunschvicg, Descartes thus substitutes for a spatial realism finding support in the imaginative representation of the extension and idealism of the relation, via algebra. By thus subordinating the examination of geometric curves to the classification of equations, Descartes would accomplish the intellectualization of extension. Geometry would thus become the preamble to the asceticism of the pure intelligence of which the metaphysical Meditations are the exercise. The whole problem is that Brunschwicg also regards formalization as the mark of an intellectualization of knowledge—and therefore, according to him, of a progress—whereas it marks rather the limitation of our imaginative power, which requires that we propose another method. After all, have we ever changed a method that works? As d'Alembert wrote in the wake of Descartes, in the article *Algebra* of the *Encyclopédie méthodique, the new algebra relieves memory and imagination*: it is then, at the time, what defines it primarily and is its raison d'être.

It is in the sixth *Meditation* that Descartes most clearly exposes the problem of representation in geometry, when he distinguishes imagination from pure intellection or conception. In the case of the triangle indeed, Descartes says that not only does he conceive of it, but that besides.

> [...] I consider these three lines as present by the force and inner application of my mind; and that's what I call imagining. (Descartes 1992, p. 173)
>
> But if I want to think of a chiliogone, I quite understand, indeed, that it is a figure composed of a thousand sides, as easily as I conceive that a triangle is a figure composed of only three sides; but I can not imagine the thousand sides of a chiliogone, as I do the three of a triangle, nor, as it were, look upon them as present with the eyes of my mind. And though [...] it happens that in designing a chiliogone I confusingly represent myself some figure, however it is very obvious that this figure is not a chiliogone, since it does not differ from what I will represent, if I thought of a myriogone. Therefore, I know clearly that I need a particular restraint of mind to imagine, of which I am not using myself to conceive, and this particular contention of mind obviously shows the difference that is between the imagination and the intellection or pure conception. (Descartes 1992, pp. 175)

It seems rather strange to us that this text could be analysed as showing the superiority of the conception over the imagination, a faculty of representation which for Descartes belongs to the soul insofar as it is closely related to the body. The chiliogone can only be conceived, failing to be able to represent it. Once again, we can see that we are not formalizing ourselves by a progress of the mind, but that formalization becomes necessary because of the limits of our representative faculty. On the other hand, we can still notice that the design of the chiliogone should give rise to a representation, which exists, but that our mind cannot represent itself. In other words, its content is not empty, even if we cannot see it. As we have seen, *distinctly conceiving* cannot be a confused representation of a chiliogone: we must find a way to preserve its spatial existence, so that we can represent it in our mind: is not this the whole purpose of the incredible representative condensation (and not elimination of representation) that algebra allows?

> As for that which does not require the immediate attention of the intellect, while being necessary to the conclusion, it is better to designate it by very short signs than by whole figures: thus, the memory will not be able to fail, and yet, during this time, thought will not be distracted from holding them back, while it applies to other inferences. (Descartes 1996, Rule XVI, p. 128)

The process is the same that led to the introduction of differential and integral calculus: Pascal created integral calculus in the form of a very complex pictorial language; Leibniz then used the power of Descartes' algebra to replace all these complicated figures with simple schemas. By this means, the differential formalism of Leibniz makes it possible to free oneself from many complications, like its "S" which represents integral calculus, and which will gradually be transformed until it acquires its present form. But if this symbol is certainly more practical, it contains exactly the same thing, and means as much as Pascal's drawings. In other words, it condenses. What this shows us is that formalized language, far from what formalist dogma can presuppose, underlies visual or spatial representations, the only difference being that they are not explicit as such. The same goes for mathematical physics: thus, in his book of analytical mechanics, Lagrange (Lagrange [1811] 1989) asserted that he does not contain figures; today, on the contrary, we see an extremely developed algebraic formalism.

2.3 Descartes and the Two Geometries

It is therefore first of all to solve practical problems, as we have seen, that equations have emerged. But over time, and according to a common interpretation in the history of science, algebra seems to have moved away from its original function, the signs and letters being deprived of any connection with reality, a link that we think we were at the time obvious. In fact, we will use the instrument (the medium) that was chosen for its ease of manipulation without necessarily performing the work of *redeployment*, which, moreover, would become too complex as the operations also made complexity. As we know, this use of letters, literal symbols, will free scientific thought from its anchoring in concrete reality. In addition, yet, from the moment when algebra is detached more and more from its geometric anchorage, by a slow *abstraction*, ranging from enumerated objects or measured quantities to abstract operations on letters denoting numbers, and later all kinds of objects, invariably arises again the problem of the reference. With the processes of empowering created objects, the question of their status has indeed been settled: do they represent anything other than symbolic forms or signs? This *history* of geometry could make us forget that the question of the foundational relations between geometry, mathematical objects and reality has always existed, and does not result from the axiomatization of geometry (which hides it rather than solves it), or of an "independence" that mathematical objects could have acquired vis à vis any geometric foundation. One can take as an example the question of the reality of imaginary numbers which one had to use to study the integers, but which were not interpretable geometrically. We began by calling them "impossible" (1629) (Girard 1629) then "imaginary"—Descartes describing the negative roots that are the source of "false"[1]—and finally of *Fictitious*, "between Being and non-Being" (Leibniz 2001).

To move to the notion of complex numbers and since, in practice, the imaginary number intervened in a pair of real numbers, Gauss had in 1799 the idea of taking the couple as a whole, calling it a complex number: the numbers Imaginaries lose their mystery by writing in the form: $a + ib$, (where i satisfies $i^2 = -1$); their ensemble will constitute what mathematicians will call an *algebraically closed body*. In other words, we pass here from the question of their reality to the fact of their existence as results of a definition that we pose (thus synthetics, the mathematical propositions eventually become analytic). The questioning of the real foundation of geometry thus began well before its axiomatization, especially since, as we shall see, it is from geometry that a real physical science came into being. We will try to see here, through the Cartesian position, the way we have for this reason, tried to anchor geometry in the real, distinguishing two types of geometry.

---

[1] *La Géométrie*, published in 1637, probably partly written in 1636 during the printing of *Les Météores*, is a "work of circumstances, hastily written". It finds its roots in the spirit of Descartes (among others) during his reflections on the problem of Pappus (1631). Before Descartes, it was understood that algebra and geometry were completely separate branches of mathematics with no connection between them. With Geometry Descartes wishes to reform algebra.

It has often been said that Descartes did not know how to draw in physics the benefit of his own work in geometry, a paradox that was emphasized by quoting his statement that *all my Physics is nothing else than Geometry.* If this were the case, in fact, we do not see why Descartes did not make full use of his geometry in order to formulate a true mathematical physics. But if we look more closely at the quoted passage, we realize that the apparent paradox is only the fact of a deeper thought that starts from considering the differences between physics and geometry. Thus, Descartes actually says exactly:

> [...] I do not want to study geometry anymore. But I have resolved to leave only abstract geometry, that is to say, the search for questions which only serve to exercise the mind; and this in order to have all the more leisure to cultivate another kind of geometry, which proposes for question the explanation of the phenomena of nature. Because [...] if it pleases to consider what I have written about salt, snow, rainbow, etc. [it] will be well known that all my Physics is nothing else than Geometry. (Descartes 1964–1974, Works 29, II, p. 268)

In other words, there are two types of geometry, one of which is merely formal, regulates and gives its method to the mind, and the other which aims above all to study natural phenomena. Just imagine yourself in the shoes of these two types of surveyors to realize that their work is totally different. But why, in this case, did Descartes finally say that his physics is nothing but his geometry? We think to find the answer in the Interview with Burman (Descartes 1981, pp. 72–74) where he writes:

> All demonstrations of mathematicians have true beings as their object, and the object of mathematics generally taken in their universality [...] is a real and true being, and possesses a true and real nature, no less than the object of physics itself. The difference consists only in the fact that this real and true being is the proper object of physics when we consider it as an act and existing as such, whereas mathematics considers it only as possible, and as a being. which in space probably does not exist in action but can nevertheless exist", we can notice here that Hilbert will also rely on such a conception of simply possible objects to assert the existence of mathematical objects in the absence of evidence of non-existence or contradiction. M. Fichant writes that" the difference between the possible and the existent is wholly relative to the nature of our mind and to the modalities of its operations. (Fourier [1822] 1988, p. 70)

Now, it does not seem to us that these two objects are distinguished only by the manner in which they are to be considered. In one case indeed, the ontological status of the object is assured because we see that it exists; in the other, the very real object of mathematics does not, however, receive confirmation of its existence: to speak as Leibniz, it is not passed to the act.

Finally, Ferdinand Alquié (1950) affirms in Descartes' metaphysical discovery of man that the thesis of the creation of eternal truths justifies the derealisation of physical objects; we would rather say that by giving mathematical objects the same status as physical objects, it is rather intended to give them a real content, as with the phenomena studied by the *philosophy of nature.* Let's consider for a moment the work of Descartes, which consisted first of all in an undertaking of geometric re-foundation of mathematics, then in a questioning of the objectivity of our senses and their legitimacy to ascertain the ontological value of objects they perceive (metaphysical Meditations), followed by a questioning of the testimony of the sight as being rather the fact of a *inspection of the mind* (example of the piece of wax). If Descartes seems to be opposed to a qualitative or metaphysical physics by proceeding exactly according to the Galilean method, the fact remains that, far from thinking that there is no difference in ontological status between mathematical objects and natural phenomena, Descartes has constantly started from this difference in order to reduce the gap, and this, by *derealizing* the real apprehended directly on one side and trying to provide a spatial representation, geometric, (which is nothing but an existence in action) of the mathematical objects of the other. We see it here, Descartes has ceaselessly reduced the ontological gap between these two types of objects, as he had done by including in the eternal truths freely created by God the mathematical truths in addition to the laws of nature—because this gap posed the whole problem of the status of the mathematical object—and the consistency of a physics, of a nature then essential.

3 Infinity and Infinitesimal

For nearly two millennia indeed, mathematicians, scalded by the famous paradoxes of Zenon, remained faithful to the essentially negative conception of the infinite brought by the Greek philosophical tradition. By recovering the Aristotelian distinction, it can be said that they are cautiously stored on the side of the finish and the *infinity in power*, and constantly rejected the *infinity in act* (perdition gulf for geometric evidence or even for reason). Thus, for example, Euclid did not point out: *There is an infinity of prime numbers*, [but] *the prime numbers are more numerous than any multiplicity given* (implied: finished) *of prime numbers.*

What brought the *wolf in the sheepfold*, in the first half of the seventeenth century, it was the Galilean problem of the geometrization of the movement. The question of the gradual passage of the movement and vice versa, the more general that of the indivisibility of the space, time and movement, seemed to make the *aporias* of *infinia* in act and the spectrum of paradoxes inexorably Zenon. The scholars of the time began to respond to the demanding requirement to think the infinite, as part of a geometric intelligibility.

While Pascal admitted in nature, underlining them, *the wonderful infinities it has proposed to men, not to conceive but to admire*; While Descartes advocated in his principles of philosophy (Cfr. § 26) *that we should not try to understand the infinite, but only to think that all what we do not find any terminals is undefined*; Leibniz, in his theory of abstract movement, explained that the movement is a continuous and that there are parties given in act in the continuation. "[...] These are infinite in act [...] indivisible or unexpected are given, without what neither the beginning nor the end of the movement and the body are conceivable".[2] It is on the grounds

[2] Later, Leibniz folded a little under the fire of criticism and considers these infinitesimals as "fiction".

of such philosophical debates, and in the extension of an intense activity around the calculations of tangents and areas, that Newton and Leibniz developed in the last decades of the 17th their *infinitesimal calculations.*[3]

Works where the future emphasizes to our eyes, those of the concepts of function, variable, continuity, limit, real number, but where confidence in a brand-new symbolism for Leibniz, the extraordinary arithmetic and algebraic skill for Newton and the unification of previous research for both have allowed a new step to be crossed and comes to fill these gaps. It is the very ambiguity of their positions, the gaps of their concepts that open a field of research specific to make them disappear in their turn (Raymond 1976).

In 1748 Leonhard Euler publishes his study *Introductio in Analytin Infinitorum* (Euler [1748] 1988) in which he wanted to consider the *duties of variables* as *object of the analysis*, by exhibiting various *species* of functions, although he then indicates a *means* uniform "to treat them. Starting from this first idea, the mathematicians mainly developed the study of the continuous, analytical, then elliptical functions and other categories arising from mathematical physics or the deepening of purely mathematical notions.

The idea of the representation of functions in the form of trigonometric series is due to Daniel Bernoulli,[4] about the theory of vibrating ropes. The work of Bernoulli will motivate, later, the works of Fourier on the heat where the vibrations will play a more crucial role than the propagation of heat. Then through those of Sturm and Liouville that led to the integral equations. The latter played an important role in the nineteenth century in electrostatics, (which gave the Mathematicians the problem of Dirichlet for example). We now realize that the technique of calculating the first variation of a functional is a kind of differentiation in a space of functions, and that a derivative in this context is a linear operator. While the early creators of variational calculus did not avail themselves of operators as abstractly conceived, they were implicitly using operators. Therefore, it is with the transforms of Pierre-Simon Laplace, Joseph Fourier and others, which to this day remain some of the most remarkable and most studied kinds of operators on spaces of functions. Integral operators were also implicit in the work of the self-taught British mathematician, George Green ([1828] 1871).

3.1 Infinitesimal Calculus

The infinitesimal calculation can find its origins in the efforts to express the instantaneous action of a *fluide* (Kepler) or a *force* (Newton) on the variable speed of the planets. Koyré (1961) shows how Johannes Kepler sought to explain the speed, variable depending on the distance to the sun of the planets on their orbit as the action of

[3] *Calculation of fluxions* and *differential calculation* respectively, quite different in spirits. If the calculation of Leibniz (and his successors the Bernoulli) makes the part of the algorithm, that of Newton invokes an analysis of the *first and last reasons*, ancestor of the limits.

[4] Cfr. *Réflexions et Éclaircissements sur les nouvelles vibrations des cordes* (1747 et 1748).

a fluid emitted isotropically by the sun. The idea was that each infinitesimal element of the orbit receives a quantity of fluid which propagates along the vector radius, and Kepler has tried at length to make the law of the laws quantitatively coincided with the idea that the fluid propagates isotropically. For this, he sought to express the areas swept by the vector radius as a sum of an infinite number of lengths and Koyré Alexandre very clearly shows that these attempts at Kepler are the beginning of the infinitesimal calculation. Newton's trial and trials, which led him to the calculation of fluxions, but it is historically indisputable that they are later than the celestial mechanics and inspired by the prior knowledge of the Kepler and Galilee works. This work is in continuity of the works of Apollonius from Perge and Archimedes from Syracuse on the cones.

3.2 *Analysis of the Eighteenth Century and Mastery of the Calculation*

The development of the power and operative coherence of the infinitesimal calculation in the eighteenth century is accompanied by a double movement of conquest of new territories downstream and elucidating its own foundations upstream. For a long time, indeed, we have tackling with infinity, as evidenced by the successive treaties cluttered with warnings, explanations and responses to objections about infinitesimal, much closer in this of the metaphysico-theological tradition than the great geometric style inherited from Greeks, as the Bishop G. Berkeley points out in his pamphlet *The Analyst* (Berkley 1734).

The new calculation wipes, to its great benefit, the critics of theologians, philosophers, and also and especially geometers. Here is, for example, the well-felt criticism of Maclaurin about the "infinite geometry" of its predecessors:

> By wanting to raise the geometry, it can be done well that we degrade it and that it is stripped of the character of its own, and which consists in the most perfect evidence. For such an arbitrary idea leaves in mind only obscure and imperfect knowledge, instead of clarity that results from truly geometric demonstrations. This is where the penchant has been bored in recent times to introduce mysteries into a science where everything must be demonstrated. (Raymond 1976, p. 3)

At the end of the century, the analysis noted the challenge and highly claims the non-recourse of the metaphysical infinite, as evidenced by the title of the work of Lagrange: theory of analytical functions containing the principles of the differential calculation, from any consideration of infinitely small or fading, limits or fluxions, and reduced to the algebraic analysis of the finished quantities (Lagrange 1797). On the side of the philosophers, it is Hegel (attentive reader of Lagrange) who took the measurement of the boldness of the infinitesimal calculation as a mathematical response to the problems of the infinite exceeding conventional metaphysical conceptions.

In a philosophical perspective, the infinite mathematics is important for this reason in fact it is the concept of the true infinite that is at its foundation, and that it stands much higher than what usually calls the metaphysical infinite, from which objections are made against the first. He concludes that it is to the philosophy that, however, has to elucidate the dialectic concept of the infinite that mathematics has cleared and implemented. The achievements of the infinitesimal calculation at the end of the eighteenth are so dazzling that Lagrange was able to write *that there is limited means to make great progress with the analysis in the current state where it is* (Euler [1748] 1988).

A period of stagnation followed in fact, where we realized that it was too good into account of the difficulties associated with infinitesimal. D'Alembert already noted that Euler wanted to *enlarge the building more than to enlighten the entrance.* If the progressive algebraization of the analysis, pushed by Lagrange, had evacuated the mysteries of infinity, the too formal nature of the symbolic operations of the infinitesimal calculation and the immoderate use of the series developments allowed a shadow on the foundations; and especially on the very faculty of a function of being developed in series of powers (its analyticity). By undertaking to enlighten *the entrance of the building,* Bolzano then Cauchy have been forced to think differently the notions of continuous and limit function (these limits that Lagrange was very to circumvent!). For them, a function (of a variable x) is continuous in a point x_0 if the absolute value of the difference $f(x) - f(x_0)$ decreases indefinitely with that of $(x - x_0)$. We can, as Weierstrass, express this condition even more explicit (or more pedantic): for any real number $\varepsilon > 0$, there is another δ such that $|x - x_0| < \delta$ involves $|f(x) - f(x_0)| < \varepsilon$.

Similarly, the notion of limit is laid, as transparent and non-metaphysical, and the integral is rigorously constructed (Cauchy, Riemann). To finish solidly building the conventional analysis framework, to complete what was called *the arithmetization of the analysis* and understand the continuum, there was no task anyway: to elucidate the notion of the real number. This task was carried out, a few 200 years after the infinitesimal calculation beginnings, by Weierstrass, Meray, Dedekind and Cantor.

Thus, after two and a half centuries, mathematicians, initially confronted with the requirement to think infinite on the problematic background of the continuous and movement, have produced, by refraining from forging a mathematical concept of the infinity, a calculation of infinity, calculation whose deepening has led them to a mathematical concept of the continuous (real numbers and continuous functions).

4 Operators Theory in Mathematics

The theory of operators today is an integral part of the various branches of analysis and mathematics. It plays an essential role in mathematical physics, in the variational calculation, the theory of processing and differential equations. Moreover, the history of these areas of mathematics is shaped by the emergence of the concept of operators.

If we define, however, the concept of an operator in a nontrivial manner as referring to the transformation of a function to another function, as it was defined by Born and Wiener[5] (Born and Wiener 1926), its origin can be traced back to the discovery of the differential calculus. In fact, it is often claimed that Leibniz (Leibniz 1710) was the first to think of the algebraic properties of the operations of calculus by treating differentiation and integration as operators when he stated, for instance, that the modus operandi of the differential operator $d^n(xy)$ shows a striking analogy to the expansion of $(x \pm y)^n$ and considers higher derivatives as successive operations we might write today as $D^n f(x)$.

Leibniz's approach was further elaborated by Lagrange (Lagrange 1797), who generalized it to a *new kind of calculus concerning differentiation and integration*. In this calculus, which may be regarded as the earliest operator calculus in history, Lagrange employed d/dx systematically as a fictitious quantity, subject to ordinary algebra as, for example, in the "Taylor expansion" $f(x + h) = \exp[h(d/dx)]f(x)$.

Lagrange's work was continued by Lorgna, Grüson, Arbogast, Francais, Lacroix, and, especially, Francois-Joseph Servois (1814), who was the first to pay attention to the formal aspects of symbolic operations and introduced the notions of *commutativity* and *associativity* of operators. The earliest detailed study of operator formalisms in *abstracto* was probably Robert Murphy's *First "memoir" on the theory of analytical operations*, published in 1837 (Murphy 1837). Distinguishing clearly between operand, operator, and result. Murphy stated:

> The elements of which every distinct analytical process is composed are three, namely, first the *Subject*, that is, the symbol on which a certain notified operation is to be performed; secondly, the *Operation* itself, represented by its own symbol; thirdly, the Result, which may be connected with the former two by the algebraic sign of equality. (Murphy 1837, p. 179; Author's italic)

The use of operators in pure mathematics for the solution of differential equations characterizes the second stage in the development of the operator conception. This phase was initiated by George Boole's paper "*On a general method in analysis*" (Boole 1844); it was given a detailed exposition in Boole's classic Treatise on Differential Equations. Boole's introductory remarks to his method in the Treatise are still interesting also from our modern point of view and deserve to be mentioned. *The question of the true value and proper place of symbolic methods,* wrote Boole, *is undoubtedly of great importance* (Boole 1844, p. 230). Their convenient simplicity, their condensed power, must ever constitute their first claim upon attention. I believe however that, in order to form a just estimate, we must consider them in another aspect, viz. as in some sort the visible manifestation of truths relating to the intimate and vital connexion of language with thought—truths of which it may be presumed that we do not yet see the entire scheme and connexion. Later, under the heading *Forms purely symbolical*, he declared:

---

[5] Max Jammer in his book "*The conceptual development of Quantum Mechnaics*" (Jammer 1967, Chap. 5.2), appears rather unambiguously that he would accord priority to the tandem Born-Wiener (Born and Wiener 1926). He writes that in work published 1925-6 *Born and Wiener, in generalizing matrix mechanics, introduced the operator calculus*. However, in 1903 Ludwig Silberstein published *Eine theorie der physikalischen operatoren* (Silberstein 1902) in *Otswald's Annalen* that might have been known by later authors.

> In any system in which thought is expressed by symbols, the laws of combination of the symbols are determined from the study of the corresponding operations in thought. But it may be that the latter are subject to conditions of possibility as well as to laws when possible. And thus, it may be that two systems of symbols, differing in interpretation, may agree as to their formal laws whenever they both express operations possible in thought, while at the same time there may exist combinations which really represent thought in the one but do not in the other. For instance, there exist forms of the functional symbol f, for which we can attach a meaning to the expression $f(m)$ but cannot directly attach a meaning to the symbol $f(d/dx)$. And the question arises: Does this difference restrict our freedom in the use of that principle which permits us to treat expressions of the form $f(d/dx)$ as if d/dx were a symbol of quantity? For instance, we can attach no direct meaning to the expression $\exp[h(d/dx)]\, f(x)$, but if we develop the exponential as if d/dx were quantitative, we have
>
> $$\exp\left(h\frac{d}{dx}\right)f(x) = \left(1 + h\frac{d}{dx} + \frac{1}{1.2}h^2\frac{d^2}{dx^2} + \text{etc.}\right)f(x) = f(x+h)$$
>
> by Taylor's theorem. Are we then permitted, on the above principle, to make use of symbolic language; always supposing that we can, by the continued application of the same principle, obtain a final result of interpretable form? Now all special instances point to the conclusion that this is permissible, and seem to indicate, as a general principle, that the mere processes of symbolical reasoning are independent of the conditions of their interpretation. (Boole 1844, p. 398)

Boole's principle of admitting symbolic reasoning, even if no immediate interpretation is in sight but has to be searched for, has proved extremely useful in the later development of quantum mechanics. The elaboration of symbolic techniques for the solution of differential equations, as, for example, by Hargreave, Gregory, Brouwin, Carmichael, Forsyth, became in the second half of the last century, especially among British mathematicians, a prolific subject of research.

Of special interest is the development which began when Charles James Hargreave presented a paper in June 1847 (Hargreave 1848), to the Royal Society in London on applications of the operator function $\varphi\ (d/dx)$ to the solution of differential equations, such as, for example, Riccati's equation, in the course of which questions concerning the interchange of symbolic operators began to play an important role.

Hargreave's work was soon generalized by Charles Graves (1847 [1847–1850]). In a series of papers read before the Royal Irish Academy, Graves, a mathematician, theologian, linguist, and archaeologist, developed an extensive and powerful operator formalism which, in one aspect at least, had an influence not only on Heaviside but even on Dirac at an important point of his work. Graves's paper *On the principles which regulate the interchange of symbols in certain symbolic equations* began as follows:

> Let π and ρ be two distributive symbols of operation, which combine according to the law expressed by the equation
>
> $$\rho\pi = \pi\rho + a$$
>
> a being a constant, or at least a symbol of distributive operation commutative with both ρ and π. (Graves 1847 [1847–1850], p. 144)

Graves then showed that in any symbolic equation $\varphi(\pi, \rho) = 0$, π may be exchanged with ρ and ρ with $-\pi$, yielding the "correlative equation" $\varphi(\rho, -\pi) = 0$. He also proved by induction that $\rho\pi^n = \rho\pi^n + n\pi^{n-1}$ and $\rho\varphi\pi = \varphi\pi\rho + \varphi\prime\pi$, where $\varphi\pi$ denotes any function of integral powers of π. It will have been recognized that Graves's reasoning and formal results recurred in the quantum-mechanical treatment of non-commutative quantities as, for example, in Dirac's q-number algorithm and its handling of the quantum conditions.

The third stage in the development of the operator conception may be characterized by the application of operators to problems in applied mathematics and, in particular, to the analysis of electric networks. The most important contribution was Oliver Heaviside's application of operators to electromagnetic problems (Heaviside 1893). To analyse networks of linear conductors with inductances, capacities, and resistances, Heaviside replaced the differential operator $D = d/dt$ by p and regarded the latter as subject to the ordinary laws of algebra. Applying this method also when partial differential equations were involved. Heaviside considers mathematics as an experimental science which explains the lack of habitual academic mathematical rigor in his approach. The widespread use of the delta function prior to its vindication by the theory of distributions shows that also the development of quantum mechanics was not exempt from Heaviside's antipuristic attitude.

The next and last stage in the development of the operator conception, before its use in quantum mechanics, is characterized by the effort to justify the previous methods by rigorous proofs, either by making use of the theory of functions of a complex variable, as, for example, in the work of T. J. I'A. Bromwich, or, more generally, by the use of integral transforms, on the historical development of which the reader is referred to S. Pincherle's excellent summary in the *Encyklopädie der mathematischen Wissenschaften.* It was in this connection that Wiener became interested in the operator calculus and introduced, together with Born, this powerful mathematical tool into quantum mechanics.

Other justifications were introduced in the mid-1920's using integrated equation techniques (as done by Carson) or Fourier processing (such as Norbert Wiener). It should be noted that the Polish mathematician Jan Mikusiński developed a different approach to the operational calculation in the 1930's using algebraic reasoning. In 1926, Norbert Wiener described the foundations of operator theory when reviewing the existential status of the operational calculation (Wiener 1926):

> The brilliant work of Heaviside is purely happy, devoid of the pretension of mathematical rigor. Its operators apply to voltages and electric currents, which can be discontinuous and certainly must not be analytical. For example, the preferred corpus vile on which he tries his operators is a function that disappears to the left of the origin and is 1 right. This excludes any direct application of Pincherle's methods. (Wiener 1926, p. 558)

4.1 Analytical Theory of Heat

The methods that Daniel Bernoulli and Leonhard Euler had tried successfully for the vibrant ropes or the membranes have indeed been taken up by Joseph Fourier, to treat this time the problem of the diffusion of heat by conduction in a material medium. It is on this occasion that Fourier will create the theory of Fourier series and thus prove the accuracy of the visions of Daniel Bernoulli on the universality of trigonometric solutions. The works of Fourier on the conduction of the heat started in the late 1790's. However, their presentation at the Academy dates from 1807, and it is the treaty published in 1822 (Fourier [1822] 1988) by the publisher Firmin Didot which will serve us as a reference.

To solve the problem of heat propagation, Fourier applies the method of separating the variables, which consists of looking for the solution as a trigonometric series. This method, applied to the equation of vibrating string was an invention of Daniel Bernoulli, but the latter did not possess the inversion formula, which precisely makes all the power of the process.

$$\frac{\partial^2 u}{\partial x^2} + \frac{\partial^2 u}{\partial y^2} + k^2 u = 0$$

The form of the solution was at the heart of the long discussion which took place in the previous century Fourier between Daniel Bernoulli, Jean d'Alembert, and Leonhard Euler, the first arguing that the solution had, for physical reasons, to be a trigonometric series, the other two offering a solution that they believed more general because they still did not know that all the periodic continuous functions are the sum of a trigonometric series. For our purpose the theory of vibrations is much more essential than that of heat. The theory of ropes or vibrating membranes is all the more essential (for our purpose here), that the equation of vibrating string above does not only describe the vibration of a membrane; it describes any vibration, as well as that of the ether. Thus, the acoustic or electromagnetic vibration modes are also described by this equation, which will later be called *Helmholtz equation*, in the context of electromagnetism (still non-existent in Fourier time) and acoustics.

Fourier considered the simplest and most typical problem is that of a rod, and it will be limited since it contains all the essential ideas. If X represents the abscissa along the rod and T the time, or $u(t,x)$ the temperature at the point x of the rod at the moment t. The evolution of the temperature over time is then described by the partial derivative equation

$$CD\frac{\partial u}{\partial t} = K\frac{\partial^2 u}{\partial x^2}$$

where C is the specific heat of the material whose rod is made, D the density, and K the conductivity of the material. The partial derivative equation is correct if the rod is perfectly isolated and if there is no heat dissipation in the surrounding medium, otherwise the second member must be decreased by C to take it into account.

4.2 The Problem of Sturm–Liouville

Fourier's work on heat inspired a very large number of contributions and it was in Paris at the beginning of the century the major subject, because electromagnetism was still poorly developed. On the way that will lead to the notion of Hilbert space, an essential step, recognized by all historians, is going to be the problem of Sturm–Liouville. The acknowledged goal of Charles Sturm (Sturm 1836) and Joseph Liouville (Liouville 1836) was to extend the Fourier method recalled above in case the coefficients C, D, or K of partial derivative equation depend on the X coordinate, that is to say in case the rod is inhomogeneous. The resolution of these *equations reveals what we will call integral equations.* Thus, an integral equation is equivalent to a differential equation with the first limits condition. Indeed, the integral equation is simply obtained by integrating the differential equation twice in a row while respecting the limits condition.

In 1836, Liouville presented the first memoir:

> When we want to determine the laws of heat movement in a heterogeneous bar, placed in a medium maintained at 0°, we come across the partial difference equation:
>
> $$g\frac{du}{dt} = \frac{d(k\frac{du}{dx})}{dx} - lu \qquad (1)$$
>
> In this equation, which must be used to determine the temperature u of each point as a function of the time t and the abscissa x of this point, the three letters g, k, l respectively represent the specific heat, the internal conductivity, and the emissive power; and, since the bar is heterogeneous, they must be regarded, not as constants, but as variable quantities given for each value of x. If the abscissas of the two extremities of the bar are x and X, we also have two conditions defined of the form
>
> $$\begin{cases} \frac{du}{dx} - hu = 0; \; pourx = x \\ \frac{du}{dx} + Hu = 0; \; pourx = X \end{cases} \qquad (2)$$
>
> h and H being constants which can have arbitrary values from 0 to $+\infty$. Finally, we must have
>
> $$u = f(x); \text{ pour } t = 0 \qquad (3)$$
>
> f(x) being an arbitrary function which represents the initial state of the temperatures and which satisfies the two conditions
>
> $$\frac{df(x)}{dx} - hf(x) = 0; \text{ pour} x = x$$
>
> $$\frac{df(x)}{dx} + Hf(x) = 0; \text{ pour} x = X$$

> Which are deduced from equations (2) by setting $t = 0$, which we have considered to hold for the general value of u of which $f(x)$ is only a particular case. (Liouville 1836, p. 253)

Then he exposes Liouville method of resolution which is exactly the method of Fourier (Liouville 1836, p. 254):

> To form the value of u which satisfies equation (1) and the defined conditions (2) and (3), we are led to expand the function $f(x)$ into a series whose successive terms differ from one of the other by a parameter r and have the property of satisfying both the general differential equation
>
> $$-rgV = \frac{d\left(k\frac{dV}{dx}\right)}{dx} - lV \tag{4}$$
>
> With the special conditions
>
> $$\frac{dV}{dx} - hV = 0; \text{ pour} x = \boldsymbol{x}$$
>
> $$\frac{dV}{dx} + hV = 0; \text{ pour} x = \boldsymbol{X}$$

The crucial novelty here is that the functions V_r are problem-specific functions, which change if the given functions *g(x)*, *k(x)*, *l(x)* change. If the given functions are constant, then the functions V_r will be trigonometric functions as with Fourier; but if the rod is heterogeneous, these will be new functions, for which there is no analytical formula which would express them using known functions. These V_r functions had been studied by his friend Charles-François Sturm for ten years when Liouville published these lines. Most of Sturm's work had been published the previous year in the same Journal, edited by Liouville (part of Sturm's original memoirs is lost).

What Strum Liouville shows is that for each particular case where the functions *g*, *k*, *l* are given, it is possible to find specific functions $V_n(x)$, which are analogous to the functions $v_n(x) = sin\left(\frac{\pi}{L}nx\right)$, and which make it possible to apply the same method as Fourier: instead of developing the function *f(x)* (the initial state of the rod) as a series of sines, it will be developed as a series of V_n. The sine functions are the V_n of the particular case where the functions *g*, *k*, *l* are constants.

The Eq. (4) obtained after separating the variables provides solutions V for a continuum of values of r but, just as in the Fourier problem, the corresponding functions V_r will satisfy the boundary conditions at $x = x$ and $x = X$ only for discrete values of r. Sturm and Liouville demonstrates "by the sole consideration of the differential equations in themselves", the useful properties of the functions V_n (the V_r for $r = r_n$): possibility of developing any function *f(x)* into a "series of Fourier" $\sum a_n V_n$, integral expression of the coefficients a_n, "orthogonality" of the V_n. Their demonstrations were not all absolutely rigorous, which caused quite a bit of criticism at the time.

Liouville then shows that Eq. (4) and the condition:

$$\frac{dV}{dx} - hV = 0; \; pour x = \mathrm{x}$$

[are satisfied by putting]

$$V = p_0 + \int_x^X \frac{dx}{k} \int_x^X (l - gr)V dx \tag{5}$$

the integral Eq. (5) is equivalent to the differential equation with the first boundary condition. Indeed, the integral equation is obtained simply by integrating the differential equation twice in a row, and the boundary condition at $x = \mathrm{x}$ is automatically satisfied by the choice of p_0 and the fact that we take the primitives which cancel at $x = \mathrm{x}$. And we thus obtain: $V = p_0 + p_1 + p_2 + \ldots$, where p_1, p_2, etc., are the iterates.

In the scientific literature of the nineteenth century, the series obtained by iteration from an integral equation (method of successive approximations), are often called Beer-Neumann series. Carl Neumann (Neumann 1877) seems to be the first to have provided a general and rigorous theory according to the mathematical criteria; A. Beer (Beer 1856) is known for having liberate the use in electromagnetism from *physicist reasoning*. The idea of using it (for the *Dirichlet problem*) had been proposed by Gauss. This literature seems to have ignored the work of Liouville in examining the texts of this intermediate period (ca. 1830–1860). Moreover, that electromagnetism had not yet taken the completed form that Maxwell gave him in his Treaty of 1873 (Maxwell [1873] 1989), and which is still taught today.

Historically, these electromagnetic applications played an important role because they led to the general equivalence between an integral equation and a differential problem with limits. The integral equation, as in Liouville's thesis, implicitly contains the limits conditions. However in this brief the use of the integral equation appeared as a particular technical process (to calculate the transcendent function), and it is the electromagnetism which will allow Gauss, Dirichlet, Riemann, Carl Neumann and to others, to formulate this equivalence in a general form and in an autonomous mathematical language and thus it has entered the tradition of *pure* mathematics (that which Dieudonné reports in history of functional analysis). These questions seem at first sight totally foreign to the future quantum mechanics, and of course that's what gives the feeling of the historic strangeness reported in the introduction. We will see the apparent gap still widens with the rest of the story, because from the moment when Gauss, Neumann, etc. gave the problem an autonomous mathematical life, the subsequent evolution should logically become even more independent. But this gap is only apparent: the quantum mechanics seems to have found by miracle the mathematical bases it needed; but it is simply because the quantum mechanics considers the wave properties of very small particles, and that all the waves (whether acoustic, electromagnetic, electronic) are mathematically described by the same equations, which are also those of a membrane. The hydrogen atom is physically similar to the electromagnetic cavity, which is itself physically similar to the membrane.

The study undertaken over a period that covers a century highlights a known of historical evolution, which is the specialization. Around 1900, it was still far from the extremes we observe today, but already visible. The work of Bernoulli and Euler on the membranes, those of Fourier, then from Sturm and Liouville, on the heat, those of Gauss on the electrostatic, were works of mathematical physics motivated by the natural philosophy; in doing so, they bequeathed to the posterity the problems resulting from this research, but made in a purely mathematical way that makes them autonomous in relation to their origin: the functions of Bessel, the Fourier series, the eigenvalues, the problem of Dirichlet, the integral equations, etc. These problems can then lead their own lives in the history of pure mathematics, as very clearly shows the work of Fredholm. However, these problems can keep over the time a physical content related to their origin, which for a long time will remain in the background, at the points of appearing in the eyes of contemporaries only as a case of application quite contingent or even forgotten. Until the day when, physics rediscovering the same phenomenon in another region of nature, the solutions made meanwhile by mathematicians will appear to be miraculous or pre-established harmony.

4.3 Operators and Matrices

From historical and algebraic perspective, we can consider the study of matrices[6] as the original model for operator theory. Nevertheless, we can find a trace of matrix methods long time before. For instance, Gauß elimination technics was found in a Chinese work, nine chapters of the Mathematical Art, from the Han Dynasty and even earlier, around 300 BC, the Babylonians worked with simultaneous linear equations. Likewise, Carl Friedrich Gauß[7] introduced the word "determinant[8]" in the 19th Century. Determinant had precursors some centuries before and were explicitly used since their 1683 by Takakazu Seki Kowa (Majima 2013), pp. 3–20) in Japan and Gottfried Leibniz in Europe.

At the beginning of the nineteenth century, Gauß and Hamilton on the one hand, Cayley[9] on the other hand, were to make a new leap in geometry, arithmetic and algebra. The Italian mathematicians of the sixteenth century had discovered, by

[6] The word *matrix* used the first time by James Joseph Sylvester in 1850. Sylvester's work was mainly algebraic. Sylvester and Arthur Cayley were close friends met around 1847 and they worked together.

[7] The term 'determinant' was first introduced by Gauss in *Disquisitiones arithmeticae* Gauss ([1801] 1965) while discussing quadratic forms. He used the term because the determinant determines the properties of the quadratic form. However the concept is not the same as that of our determinant. In the same work Gauss lays out the coefficients of his quadratic forms in rectangular arrays.

[8] It was Augustin Louis Cauchy in 1812 who used 'determinant' in its modern sense. Cauchy introduce in 1826 the ideas of eigenvalues and diagonalization in the process of finding normal forms for quadratic functions (Heisenberg 1925). Cauchy proved the spectral theorem for self-adjoint matrices, i.e. that every real, symmetric matrix is diagonable. The spectral theorem as generalized by John von Neumann is today the most important result of operator theory. In addition, Cauchy was the first to be systematic about determinants.

[9] In fact, Cayley's attention to matrix algebra is even slighter than his attention to group theory and is insignificant when compared to the large corpus he produced on invariant theory. Cayley's Memoir, which could have been a useful starting point for further developments, went largely ignored. His habit of instant publication and not waiting for maturation had the effect of making the idea available even if it was effectively shelved.

studying the solution of third-degree equations, new numbers so strange, involving square roots of negative numbers, that they were called *imaginary* and, later, complex numbers. Gauss succeeded, with others, in showing that they could be interpreted as points of the plane whereas the real numbers were associated with the points of a line. Hamilton searched soon after him to see if there were new numbers that would be points in space. This led him to discover quaternions.

Abstract algebra, in some sense, have been born with William Rowan Hamilton's discovery of quaternions in 1843, and Hermann Grassmann's introduction of exterior algebra the following year. Cauchy and Jean Claude de Saint–Venant also created abstract algebraic structures at about this time. Hamilton was motivated by the idea that quaternions give a better algebraic description of space and time, and for Grassmann the goal was geometric.

The quaternions presented an astonishing property: their product was no longer commutative. This non-commutativity also emerged from the calculations developed by Arthur Cayley in 1857 on tables of numbers that could be associated with systems of equations of lines or planes, as well as with conics and quadrics. These arrays, called matrices, could add or multiply as new numbers, this motivated the introduction of the idea of an algebra of matrices by Arthur Cayley (1858). In 1858 he showed that quaternions could be *represented* by matrices.

We owe to Camille Jordan the canonical form analysis of the matrices in his publication of 1870. This work is considered a prototype of the decomposition of the compact operators in the case of infinite dimension.

Giuseppe Peano, in his book *Calcolo Geometrico Secondo the Ausdehnungslehre di H. Grassmann Preceduto Dalle Operazioni Della Logica Deutiva* Published in 1888, develops an axiomatic formulation of linear spaces and announces the theorem that any operator defined on a vector space of finite dimension is a matrix. He develops the sum and the product of the linear operators giving premises to a form of a theory of operators.

4.4 Differential Equations and Spectral Theory

The subjects of operator theory and its most important subset, spectral theory,[10] came into focus rapidly after 1900. Today, many branches of analysis are inseparable from operator theory, notably variational calculus, transform theory, and differential equations.

[10] In mathematics, the spectral theory is an inclusive term for theories extending the theory of eigenvectors and the eigenvalues of a unique square matrix to a much wider theory of the operator structure in a variety of mathematical spaces. It is the result of linear algebra studies and solutions of linear equation systems and their generalizations. The theory is linked to that of analytical functions because the spectral properties of an operator are linked to the analytical functions of the spectral parameter. The name of spectral theory was introduced by David Hilbert in his original formulation of the Hilbert space theory, which was expressed in terms of quadratic forms in an infinity of variables.

Charles François Sturm (1836) and Joseph Liouville (1836) were the first to treat the question of eigenvalues in connection with differential equations. In contrast to Cauchy who treat the finite case, they considered the case of infinite dimensional, which allows phenomena that do not arise in the finite-dimensional case of linear algebra. For instance, the spectrum of the infinite-dimensional operators is continuous. This became evident with George Hill work on the theory of periodic Liouville equations in order to study the stability of the lunar orbit and where he introduced the infinite determinants.

The spectral theory of ordinary differential operators was originated from Sturm–Liouville works. The question of the eigenvalues of partial differential operators, particularly the Laplace operator was a subject of concern for the mathematician in the late Nineteenth century. One of the problems then was, what we call Dirichlet problem, to find a solution of Laplace's equation with specified boundary conditions. Working on this problem led the mathematicians to a better and more rigorous understanding of convergence of sequences of functions and the nature of what are now termed partial differential operators.

Today we recognize this as a question of topology of what we call the function spaces in which we treat the functions as points in sets. Grassmann, in 1862 and Salvatore Pincherle seem to have been the first to consider and write functions as abstract entities f, rather than $f(x)$. From this perspective the function is seeing as relations between domain and range values. The idea of a functions spaces is a central notion of Twentieth century analysis and was influenced by attempts to understand the Dirichlet problem, Fourier series and transforms, and the work of Vito Volterra and Ivar Fredholm on integral equations.

Olivier Heaviside had a great influence also (Heaviside 1893). Indeed, in his works between 1880 and 1887 he uses the differential operator d/dx, in novel ways which was efficient to solve differential equations despite a disdainful of mathematical rigor. His formal methods anticipate the developments of the pseudo–differential operators. Heaviside's operational calculus has continued to have a following among engineers and scientists to this day.

4.5 The Theory of Integral Equations

Dieudonné in his book History of Functional Analysis (Dieudonné 1981) refers the origin of the integral equations to the slow maturation of the ideas that begins with the vibrating string theory of Jean d'Alembert, Daniel Bernoulli, and Leonhard Euler (1745–1755), and continues with the analytical theory of heat (Joseph Fourier, 1807–1822), the extension to developments in series of arbitrary (orthogonal) functions and the discovery of *eigenvalues* by Charles Sturm and Joseph Liouville (1828–1837), the translation of these problems in terms of integral equations (Liouville) (Liouville 1836; Beer and Neumann; Beer 1856; Neumann 1877), the resolution of the integral equations by Fredholm (1900–1903) (Fredholm 1903), and finally the synthesis of all this by the Mathematicians of Göttingen, published by Hilbert under the title *Grundzüge Einen Allgemeinen Theory der Linearen Integraggleichungen* (1904–1907) (Hilbert 1909), and who will then make the essential material of one of the greatest classics, *Methoden der mathematischen Physik* of Richard Courant and David Hilbert (1924) (Courant and Hilbert 1924). Hilbert himself, in the introduction of Grundzüge, flying over this historic evolution of ideas, and highlights a certain logic of evolution. This is taken up and developed by Ernst Hellinger (Hellinger and Hilberts 1935) in a contribution included in Hilberts *Gesammelte Abhandlunden.*

The work of Dieudonné resumes the historical logic detected by Hilbert and Hellinger and considers the work of Sturm and Liouville as the beginning of functional analysis, probably because developments in series of arbitrary functions, which generalize the series developments. Trigonometric functions of Joseph Fourier are a first step towards a conception where the functions are considered as a generic element of a space, and no longer as the concrete data of a particular analytic expression. Unfortunately, the book of Dieudonné evacuates the fact that Fourier, Fish, Sturm, and Liouville sought to solve the problem of the heat propagation, just as the fact that the growing importance of the integral equations in the nineteenth century is motivated by electromagnetism. The trace left inside mathematics by the latter appears in the form of problems posed by Gauss. Thus, Dieudonné speaks about the problem of Dirichlet without explaining its origin. Similarly, the many works on the electrostatic, and using series developments or integral equations, are ignored. Finally, there is obviously no word about the relations between the spaces of Hilbert and the quantum mechanics, not even in the chapter *Applications of the functional analysis*.

Integral operator and Spectral theories. Since all these subjects predated operator theory as such by a century or two, it is no surprise that some of the earliest antecedents of operator theory are to be found in them. A major event was the appearance of Fredholm's theory of integral equations (Fredholm 1903) which arose as a new approach to the Dirichlet problem. In a preliminary report based on his dissertation published in 1900 and a landmark article in Acta Mathematica in 1903, Fredholm gave a complete analysis of an important class of integral equations, now known as Fredholm equations.[11]

It is the analogy between the systems of linear equations in finite dimension and the integral equations, which will allow that Fredholm or Volterra can identify the essential elements of the theory that carries Today the name of Fredholm's theory of equations. Fredholm proposes (1903) to solve the integral equation

$$\phi(x) + \int_0^1 f(x, y)\phi(x)dy = \psi(x) \tag{6}$$

[11] The index of a Fredholm operator is very important in physics. One way to define the Chern number of a topological insulator is in terms of the index of a Fredholm operator, as explained in (Bellissard et al. 1994). There is also the concept of an index of a pair of projections. This is seen a lot recently in physics paper (Akagi et al. 2017; Li and Mong 2019). That paper uses as a reference on the index of certain pairs of projections a paper in the Journal of Functional Analysis (Avron et al. 1994).

in a similar way to the resolution of systems of n linear equations with n unknowns by the method of Cramer determinants, and for this purpose introduces the functional equivalents of these determinants. Here is what Fredholm writes at the beginning of his article:

> [...] there exists a quantity D_f which plays with respect to the functional equation (6) the same role as the determinant plays with respect to a system of linear equations. (Fredholm 1903, p. 3; Author's translation)

Fredholm's method is easily understood if the interval [0,1] is discretized by the points $a_0 = 0$, $a_1 = n_1$, $a_2 = n_2$,... $a_n = 1$: the integral of (A) becomes a finite sum, so that the corresponding equation is a system of $n + 1$ equations with $n + 1$ unknowns whose matrix is $I + F$, I being the unit matrix and F the matrix coming from the discretization of f, whose elements are $F_{i,j} = f(a_i, a_j)$. All of Fredholm's formulas are then obtained as the limits when n tends to infinity of Cramer's resolution formulas. This aspect is completely overlooked in Fredholm's article but restored in the presentation that Hilbert will make shortly after (*Grundzüge einer allgemeinen Theorie der linearen Integralgleichungen*) (Hilbert 1909). In an effort to complement the work of Fredholm,[12] Hilbert reaches more general conception.[13]

$$\phi(x) - \lambda \int_0^1 f(x, y)\phi(x)dy = \psi(x)$$

In particular, he discovers that the success of the I. Fredholm method is based on the notion of *complete continuity* that it releases by formulating it for bilinear forms and that it studies in depth.

Hilbert spoke at that time on the meaning he intended to give the great synthesis of all these classic problems under the emblem of the integral equations: the unification of algebra and analysis. Indeed, the similarity between the Fredholm method for the integral equations and that of Cramer for finite dimensional systems must naturally lead to a single and universal method which was highlight by the discretization with passage to the limit (Hilbert 1909):

> This problem of determining an infinity of unknown from an inifited equations, because of its generality, seems to be ungratefully and inextricable; by wanting to deal with it is exposed to the danger of being lost in too complicated considerations, or too vague, or too vast, so that the efforts made to gain a deeper perception of the problems would be disproportionate. But if we do not let us turn away with such scruples, it will happen to us what happened in

[12] The parameter λ does not figure out in the Fredholm's paper it was introduced by Hilbert.

[13] Here we recognize an integral equation of a particular form with all its generality: knowing the functions ψ : [a,b]⟶R and f(x,y) (called the kernel), it is a question of finding the function(s) ϕ: [0,1]⟶R such that $\psi(x) = \phi(x) - \lambda \int_a^b f(x, y)\phi(y)dy$. In the case where the kernel is that of Dirichlet, the problem comes down to the Fourier transform, well known since the 1830s. But, except for a few particular kernels, the other cases remain insoluble. At the end of the nineteenth century, Hadamard, Lebesgue and Borel developed the notions of Lebesgue integral and topology.

> Siegfried, who saw the enchanting fire back before him and in reward we will see twinkle the wonderful treasure of a united formulation of algebra and analysis.[14]

This led Göttingen's mathematicians to the notion of Hilbert space on the one hand, their applications to partial derivative equations, the spectral theory of operators, etc. on the other hand. Integral operators[15] are involved in diffusion phenomena where classically integral equations are involved. The existence and the unicity of the solutions find solutions with the Fredholm alternative, when the latter is applicable, that is to say when the operator is compact. In a large number of cases in practice, there is already a complete study of the spectral analysis of the operator. It happens that such an operator admits an inverse which is also an integral operator. The kernel of the latter is then called the inverse kernel. Then just after that, E. Schmidt and Frechet deliberately introduce the language of Euclidean geometry in the Hilbert space. The concept of completely continuous linear application is found for the first time generally defined in the *mémoire* of Frigyes Riesz (1918). F. Riesz introduced the concept of an algebra of operators in a series of articles culminating in a 1913 book, where Riesz studied the algebra of bounded operators on the Hilbert space L^2. Riesz representation, orthogonal projectors, and spectral integrals made their first appearance in this work. In 1916 Riesz created the theory of what he called *completely continuous* operators, now more familiarly compact operators. Riesz's spectral theorem for compact operators made abstract, greatly extended, and largely supplanted Fredholm's work.

The definitive spectral theorem of self-adjoint, and more generally normal, operators, was the simultaneous discovery of Marshall Stone and John von Neumann in 1929–1932 (Neumann 1930). Although Stone is more readable today, von Neumann's contributions are somewhat more far-reaching. One of von Neumann's motivations was quantum mechanics, which had been discovered in 1926 (Heisenberg 1925), Schrödinger [1926] 1988) in two rather distinct forms by Erwin Schrödinger and Werner Heisenberg. It was von Neumann's insight that the natural language of quantum mechanics was that of self-adjoint operators on Hilbert space. This notion permeates modern physics. Von Neumann introduced or transformed many concepts which are at the core of operator theory.

---

[14] "Das genannte Problem der Bestimmung unendlichvieler Unbekannter aus unendlichvielen Gleichungen erscheint auf den ersten Blick wegen seiner Allgemeinheit undankbar und unzugänglich; bei einer Beschäftigung damit droht die Gefahr, daß wir uns in zu schwierige oder vage und weitschichtige Betrachtungsweisen verlieren ohne entsprechenden Gewinn für tiefere Probleme. Aber wenn wir uns durch solche Erwägungen nicht beirren lassen, geht es uns wie Siegfried, vor dem die Feuerzauber von selber zurückweichen, und als Lohn winkt uns entgegen der schöne Preis einer methodisch-einheitlichen Gestaltung von Algebra und Analysis." (Hilbert 1909, p. 59; Authors' translation).

[15] In mathematics, an integral operator or kernel operator is a linear operator defined using a parametric integral over some functional space. The image of a function by such an operator is therefore another function, whose domain can be very different. Such operators constitute fundamental objects in functional analysis, where they allow in particular transforming an equation to obtain a version a priori easier to solve. The first examples are the convolution and the Fourier or Laplace transforms, hence the name also encountered of integral transform.

Banach, in his turn, published a pushed study of the relations between a continuous linear application and its transposed, extended to standard spaces. This period witnessed the birth of a great theorems such as the theorem of the closed graph or the theorem of Banach-Steinhaus. The publication deals with Banach on linear operations marks the beginning of the adulthood of the functional analysis and the theory of operators.

The Hilbertian and banachic approaches have different readings of functional analysis. On one hand, the Hilbertian approach favours regularity, in particular through the study of Hilbertian products, whose classification was finished in 1935, which legitimized the Riesz representation theorem and other theorems in the same vein allowing to bring back numerous problems to studies on the same theme (Stone's theorem which gives the structure of so-called *normal* operators, Lax–Milgram theorem, Galerkin's method for differential equations, etc.). The usefulness of these results, especially in quantum physics, enhancing this work. On the other hand, the banachic approach favors *pathologies* by using the properties of approximation in order to determine the reasons for their possible everywhere continuous and nowhere differentiable, but also, for example, the definition of Banach spaces where no algebraic structure other than that of normed vector space intervenes.

In a series of articles from 1935, partly with Murray and von Neumann (1937, 1943) elaborated the theory of operator algebras, introduced by Riesz. They realized that the sets of operators that commute with an algebra was an important tool of analysis and classification and made many contributions to pure algebra as well as algebra.

The final seminal work that will be mentioned here is that of Israel Gelfand, who in a 1941 paper in Matematicheskii Sbornik extended the spectral theorem to elements of normed algebras. There are many problems in various fields of mathematics such as algebra, digital analysis, optimal control, spectral theory etc. are strongly linked to the notion of inverting elements. This is why some mathematicians thought of introducing new concepts of inverting that are useful for solutions to these problems. Among them, we quote von Neumann, Kaplansky, Nashed, Cardus, Koliha and many others.

In 1936, Von Neumann introduced the notion of general reverse for the elements of a ring, later and more precisely in 1948, Kaplansky gave an extention of this notion for algebras. Finally, the notion of pseudo-reverse was established independently by Moore, in 1922 and Roger Penrose in 1955, where the latter proved that any rectangular matrix (and more particularly any singular square matrix) has a unique kind of inverse called Moore–Penrose.

The theory of integral equations: the physical aspect.

In his 1856's paper Beer (1856) studied the electrostatic potential inside a conductive surface, if the charges that create it all are on this surface.[16] The symbol ∂ (see footnote below) here refers to the normal derivative on S. It is clear that if, in the first

[16] "Wenn nun erstlich der idioelektrische Körper außerhalb der Fläche S liegt, so verschwindet in der ganzen Ausdehnung des von ihr umschlossenen Raumes der Ausdruck ΔV, und man hat dem bekannten Green'schen Satze gemäß für jeden Punkt des erwähnten Raumes:

equation above, it is substituted for its given expression in the second, an integral equation is obtained. Beer proposes the resolution by iteration and proves the convergence of the process from physical arguments. In this context the integral equation is much more likely than the differential equation, because it directly expresses the feedback of the potential on the charges: the surface S being conductive, the charges (which cannot leave it, but can move along the long of this one) are influenced by the overall potential, which displaces them and modifies the potential. We recognized in this quote the problem of the cavity already mentioned above. But in this example Beer study only the electrostatic potential and not the vibrations. The problem of vibrations in a cavity has never received, along the nineteenth century an important importance to electrostatic problems (so-called *Dirichlet* or *de Neumann problems*).

Historically, these electromagnetic applications have played an important role because they have led to the general equivalence between an integral equation and a problem with limits. The integral equation, as in Liouille's thesis, implicitly contains the limits conditions. However, in this memo, the use of the integral equation appeared as a particular technical process (to calculate the transcendent function), and it is the electromagnetism which will allow Gauss, Dirichlet, Riemann, Carl Neumann and others, to formulate this equivalence in a general form and in an autonomous mathematical language. Thus, a differential problem with limits conditions is equivalent to an integral equation. A fairly simple calculation shows for example that the following two problems are equivalent:

(1) Find a function ϕ (x) on the interval [0, 1] such that:

$$\phi''(x) + q(x)\phi(x) = \chi(x)\phi(0) = 0 \ \phi(1) = 0 \tag{7}$$

(2) Find a function ϕ (x) on the interval [0, 1] such that

$$\phi(x) + \int_0^1 f(x, y)\phi(y)dy = \psi(x) \tag{8}$$

It is enough for that to take

$$f(x, y) = \kappa(x, y)q(y) \text{ and } \psi(x) = \int_0^1 \kappa(x, y)\chi(y)dy \tag{9}$$

$$V = \int \frac{\partial(\frac{V}{4\pi})}{r} dS + V'$$

$$V' = -\int V.\partial\left(\frac{1}{4\pi r}\right).dS$$

Die Function V ist offenbar selbst wiederum eine Potentialfunction, und der Ausdruck ΔV verschwindet allenthalben im Innern von S." (Beer 1856, p. 137).

With

$$\kappa(x, y) = \begin{cases} y(1-x)\, if\, 0 \leq y \leq x \leq 1 \\ x(1-y)\, if\, 0 \leq x \leq y \leq 1 \end{cases} \tag{10}$$

Indeed, this judicious choice of κ causes the integral Eq. (8), considering (9), is exactly equivalent to the differential equation plus the two conditions φ (0) = 0 and φ (1) = 0.

It can be seen that the integral Eq. (8) is exactly the shape of the integral equation studied by Fredholm (the case of any interval [a,b] instead of [0,1], there is reduced by elemental transformation). Notice that quantum mechanics considers the wavy properties of very small particles, and all the waves (whether acoustic, electromagnetic, electronic) are mathematically described by the same equations as those of a membrane. The hydrogen atom is physically similar to the electromagnetic cavity, which is itself physically similar to the membrane. Thus, quantum mechanics seems to have found by miracle the mathematical bases it needed.

Integral equation and Schrödinger wave mechanics.

Schrödinger in 1926 (Schrödinger [1926] 1988) had a different approach to Hilbert (unification of algebra and analysis):

> The analogy between the mechanics and the optics reported by Hamilton only concerns the geometric optics; Indeed, according to this analogy, to any trajectory of the representative point in the configuration space corresponds to a radius of light and this last notion can not be defined unambiguously than in geometric optics. [...]. This notion, and with it all classical mechanics, remain as approximations, valid only in the case where the dimensions of the trajectory are large with respect to the wavelength. On the contrary, for the case of "micromechanical movements", the fundamental equations of the mechanics are just as low that geometric optics for the treatment of di fraction problems; By analogy with what is happening in optics, the fundamental equations must be replaced by a single wave equation in the configuration space. (Schrödinger [1926] 1988, p. XIX)

It has been found that the atoms of the material, heated (for example in the stars), emit light only at the characteristic frequencies of each particular atom. This phenomenon is physically similar to the eigenten frequencies of an electromagnetic cavity, but this analogy has not been perceived as easily. It is precisely Erwin Schrödinger who draw the right consequences. In the atoms, the electrons have an undulating behaviour and, if they were "locked in a cavity", the electrons would have eigenteen frequencies as in the problem (4). It turns out that in the atoms the electrons are not locked in a cavity that would have a well-defined edge: they are confined by the Coulomb potential of the core. What Schrödinger showed in his memoirs is that if the problem of an electron is treated in the Coulombal field of a proton (hydrogen atom) in complete analogy with the problem (4), then we obtain for eigenvalues exactly Bohr levels, in perfect adequacy with the emission frequencies of this atom.

The analogy led Schrödinger to the wave equation (Servois 1814, pp. 2–3):

$$\frac{\partial^2 \psi}{\partial x^2} + \frac{\partial^2 \psi}{\partial y^2} + \frac{\partial^2 \psi}{\partial z^2} - \frac{2m}{K^2}\left(\mathrm{E} + \frac{e^2}{r}\right)\psi = 0$$

where $r = \sqrt{x^2 + y^2 + z^2}$, e being the charge and m the mass of the electron. K is the Planck constant divided by 2π and E the energy of the electron. Schrödinger indicates that:

> [...] If E is negative there is no solution, except for a discreet sequence of values of this constant. [...] The discreet spectrum corresponds to the Balmer's terms. (Schrödinger [1926] 1988, p. 4)

Moreover, the Courant-Hilbert Treaty (which had just appeared at Julius Springer) is frequently cited in these memories for the actual mathematics elements (especially sets in eigenfunctions, hermit polynomials, etc.). We find that Bernoulli or Schrödinger, it is always the problem of the membrane, that is to say that of the waves, which remains the main paradigm. Not only does the quantum mechanics do not exit (at least not in its first grind, sometimes called, to distinguish it, undulatory mechanics), but it even reinforced the uniformity: before, one processes the mechanics in a way different than electromagnetism; Around 1925, were discovered the common foundations underlying both.

The Schrödinger picture is the most frequently used formulation of quantum physics, but the Heisenberg picture is equivalent. In the Heisenberg picture it is the time dependence of the observables, operators, that describes the physical system. State vectors are of little consequence since they are evaluated at a fixed time. Of course, the possible results of a measurement, the eigenvalues, must be the same in both pictures, but the eigenvectors change with time in the Heisenberg picture. This may be shown by beginning with the eigenvalue equation for an operator $\hat{A}$ in the Schrödinger picture. The corresponding Heisenberg picture operator is $\hat{A}(t)$. As stated above, most problems in quantum physics are attacked using the Schrödinger picture. Why then even bother with the Heisenberg picture? Of course, there is the aesthetic beauty attendant to proving that the two pictures are equivalent. Another reason is that the Heisenberg picture makes the connection between classical and quantum physics clearer than does the Schrödinger picture. After all, wave functions and state vectors have no classical analogue. On the other hand, the observables in classical physics are usually time-dependent quantities so the Heisenberg equation of motion for an observable (Hermitian operator) is indeed analogous. Moreover, there is a close connection between this equation and Hamilton's equations of classical motion as well as a relationship between the Poisson brackets of classical physics and the commutator of quantum physics.

5 Operators in Physics

Essentially the use of operators in physics started with the birth of Quantum Mechanics. The mathematical methods used in quantum theory, being rather elementary at the very beginning, rapidly achieved a high level of sophistication. It was realized soon that relevant information on the operators may be collected in the objects called the spectral functions, which depend on the operator in question and on a real or complex parameter. In the context of quantum field theory (QFT) the most frequently used spectral function is the heat kernel. This function was first applied to the problems of quantum physics already in 1937, and then by Schwinger in his seminal paper (Schwinger 1951). In 1960's the heat kernel became a corner stone for his method of calculating quantum corrections. About 10 years later, Dowker and Critchley (1976) and Hawking (1977) suggested a regularization scheme of quantum field theory based on another prominent spectral function, the zeta-function. All these works determined for the rest of the century a landscape of methods of quantum field theory based on geometrical properties of operators. This has led to very interesting and important developments in practically all areas of QFT, ranging from quantum gravity, to anomalies, strings, and the Casimir effect. Apart form powerful technical tools this approach provided the physicists with a new rather adequate language to describe complicated phenomena, and with a new, very fruitful, point of view on quantum effects in general.

The operational viewpoint of physics is founded on the Heisenberg cut (Cfr. Heisenberg 1925), which divides the world into the system and an observer capable of making measurements. Any physicist longing to model a given physical phenomenon should have at his disposal a suitable set of observables, i.e. measurable quantities. These need not, in general, have definite numerical values, but could be associated with measurements, the outcomes of which are probabilistic, as indeed happens in quantum mechanics. The system, on the other hand, can find itself in different states. The statistical results of measurements of a given observable will depend on the state of the system. Note that the outcomes of any reasonable measurement in any state should be finite, hence the observables should be bounded in an appropriate sense.

The original concept of an operator, such as the differential operators d/dx, ∂/∂x, ∂/∂y, and ∂/∂z, Laplacian Δ, and Hamiltonian H, has been one of the most efficient and developed tools of theoretical physics. Even the basic arithmetic operations, such as $+$ and $-$, are also operators as it has become clear with the advent of computer programming. Such symbolic, mostly mathematical (Alexandrov et al. 1963; Cajori 1993), operators are natural operators that are studied by physics and describe physical concepts, reflecting their properties. For instance, mathematical operators are used as models of the term observable, which has become the standard concept in quantum mechanics as the counterpart of the term physical quantity or measurable quantity in classical physics. This term originated from the term observable quantity (*beobachtbare Grösse*) defined by Heisenberg in his groundbreaking work on matrix mechanics (Heisenberg 1925). His goal was to specify physical quantities by means of an operational definition. As a result, in contemporary quantum mechanics, the observables of a physical system are represented by self-adjoint operators acting in the Hilbert space H associated with the system. In general, operators play a fundamental role in physics. For instance, the physics of quantum mechanics was reduced to the mathematics of operators on Hilbert spaces (Jordan 1969; von Neumann 1932).

In general, an observable is a physical object or a property of such an object that some system (in a more restricted sense, people) can observe and/or measure. Thus, in the later development of quantum theory, several authors suggested the generalized representation of observables as positive operator measures (Davies and Lewis 1970;

States 1983; Günther 1964). This concept advanced the mathematical coherence and conceptual clarity at first of quantum mechanics and then of quantum field theory. From physics, mathematical operators were extended to quantum chemistry where they also have been successfully used to model different processes.

A paper that has had an influence in a specific way is a rather unknown paper by Ludwik Silberstein (Silberstein 1902), which anticipated to some extent the formal aspects of the operational approach in modern quantum mechanics. The study of symbolic integrals of the equations of the electromagnetic field suggested to Silberstein, who is known mostly for his writings on the theory of relativity, a theory of "physical operators," in terms of which he attempted to give a unified representation of such disparate phenomena as mechanical oscillations, heat conduction, and electrodynamic processes. Defining the "state" ("Zustand") of a physical system by a time-dependent function $\psi(t)$, Silberstein introduced what he called "chrono-operators" $\left\{ \begin{matrix} H \\ t \end{matrix} \right\}$ by means of which the state at time t = t can be determined from the knowledge of the state at time t = 0, in accordance with the equation $\psi(t) = \left\{ \begin{matrix} H \\ t \end{matrix} \right\} \psi(0)$. He defined the inverse operator $\left\{ \begin{matrix} H \\ t \end{matrix} \right\}^{-1}$ by $\left\{ \begin{matrix} H \\ -t \end{matrix} \right\}$ and showed that $\left\{ \begin{matrix} H \\ nt \end{matrix} \right\} = \left\{ \begin{matrix} H \\ t \end{matrix} \right\}^{n}$ and that $\left\{ \begin{matrix} H \\ t \end{matrix} \right\} e^{tF}$ where F is another operator, connected with H in a definite manner, but independent of t. Silberstein even spoke of the superposition of states and mentioned numerous other details which, including the notation, made their appearance in the operational formalism of quantum mechanics some twenty-five years later.

> Every class of physical phenomena [declared Silberstein in 1901] or at least those amenable to quantitative treatment, will be characterized by corresponding physical operators; and the scientific study of natural phenomena will proceed by detailed examinations of the properties of these operators, based on observation and nurtured by experiment. (Silberstein 1902, p. 375; see also Jammer 1967)

Whereas in Cambridge, Massachusetts, matrix mechanics was generalized into an operator calculus by Born and Wiener, in Cambridge, England, it was modified into an algebraic algorithm by Paul Adrien Maurice Dirac.

5.1 *Operators Theory in Mathematical Physics*

Often the study of the problems in physics[17] is reduced to the resolution of linear differential equations. In general, we can represent these equations by $Ly = f$, where y is the unknown function to look for, f a known function, and L a differential operator.

[17] Many physical or mechanical problems are often reduced to the resolution of an eigenvalue problem whose unknown is a function, or to a linear partial differential equation which can be studied with spectral methods.

For example, the heat equation can be written $Lu = q(x,t)$, where $L = \left(\partial_t - D\partial_x^2\right)$ and $q(x,t)$ is the source term. A priori, the search for solutions is the domain of the analysis. However, we can reduce the resolution of these equations in the field of matrix algebra of systems to N equations and n unknown of the genus $AX = B$. The very big advantage is that to make algebra, we have, in wholesale, that need for addition and multiplication. The Fourier and Laplace transforms are particular examples of tools well adapted to a certain class of equations that allowed us to reduce the analysis to algebra.

The birth of quantum physics dates from a century and this description of the physical phenomena, which has transformed our vision of the world, is still not called into question, which is exceptional for a scientific theory. His predictions have always been verified by experience with impressive accuracy. The fundamental concepts, such as probability amplitudes, linear states of states, which seem so strange for our intuition when we meet for the first time, still remain essential. However, an important evolution occurred in recent decades. The spectacular progress of observation techniques, methods of handling atoms, now make it possible to achieve such delicate experiments that they were considered only as "thought experiments" by the founding fathers of quantum mechanics. The existence of quantum "non-separable" correlations, which is at the base of the "paradox" of Einstein–Podolsky–Rosen (Einstein et al. 1935) and who violate the famous inequalities of Bell, could be confirmed experimentally with great precision. The "intricate" states of two systems, which show such quantum correlations, are better understood, and even used for concrete applications, such as quantum cryptography. The entanglement of a measuring device with its environment is an interesting track for a better understanding of the measurement process.

The formalization of the infinitely small of quantum mechanics is a miracle of abstraction. As much the development of classical physics (Newtonian mechanics, Lagrangian, Hamiltonian and relativity) is linked to that of geometry and tensor calculus, as much the emergence and development of quantum mechanics are related to the theory of Hilbert spaces and to the theory of measurement. As soon as an observable (which is accessible to an experimental measure) is represented by an operator on a Hilbert space, the mathematics are responsible for building probability measures on the spectrum of this operator and these probability measures become a physical reality in Born's transition laws.

At the end of the nineteenth century, physics rests on two pillars, Newtonian mechanics and Maxwell's electromagnetic theory. Four phenomena posed the problem in this theoretical framework:

- The experiments of Michelson-Morley could not highlight the speed of the earth in the repository of the ether (hypothetical physical support of electromagnetic waves). This problem will be solved by Einstein in 1905 by the introduction of the theory of restricted relativity;

- The experimental spectrum of the radiation of a black body did not coincide with the theory which provided for a divergence in the ultraviolet (divergence leading to a quantity of energy emitted by the body $\Longleftrightarrow$ the ultraviolet catastrophe);
- the atoms should not be stable, the electron orbits around the core to continually lose energy by electromagnetic radiation, it should be fragmented on the core over time;
- The observation of emission and absorption spectra of light by the material in the form of fine lines sequence was in contradiction with the theory that provided for continuous emission or absorption bands.
- To explain the last three problems, three competing theories were proposed: the theory of quantum, wave mechanics, the mechanics of matrices. This has highlighted the essential role of operators in quantum mechanics.

The turn to operators occurred in Heisenberg's 1925 paper (Heisenberg 1925), and it was a compromise between positions taken by Bohr (1988) and Born (1926). Bohr wanted minimal modifications to classical mechanics, like his quantization rules for the hydrogen atom, while Born wanted a new discrete mechanics governed by difference equations. Heisenberg's idea was to keep equations of motion more or less classical, but to reinterpret symbols in them. Those symbols represent observables (position, momentum, energy, etc.), and in Hamiltonian dynamics they are functions of positions x and momenta p, which form the phase space. Applying this idea Heisenberg discovered that his symbols would not commute, for example $xp - px = i\hbar$ with the Planck constant $\hbar$, so they could not be numbers. A mathematician friend of his pointed out that matrices do not commute, and could satisfy such relations, and this gave Heisenberg's proposal its name, matrix mechanics. Except that since electrons in an atom have infinitely many energy levels these *matrices* had to be of infinite size.

At about the same time Schrödinger was working from a different perspective, he wanted to reduce quantum effects to wave dynamics. So, he represented states of quantum systems by wave functions ψ on the phase space and looked for equations of motion. His idea was that quantum particles are wave packets, and he originally thought of $|\psi|^2$ as charge density, only later Born identified it with probability density. After matrix mechanics came out Schrödinger realized that he needed to introduce observables into his picture, but they could not be mere symbols. Since states are wave functions on the phase space rather than its points, observables can not be functions of them, they have to act on wave functions like matrices act on vectors. After some experimenting in 1926 he came up with representing position by a multiplication operator $\psi \mapsto x\psi$, and momentum by differentiation $\psi \mapsto -i\hbar\partial\psi/\partial x$. The big clue was that if we think of x, p as these operators[18] then $(xp - px)\psi = i\hbar\psi$. Then we can form other observables from these ones.

[18] Suppose that the commutation relation $[\mathrm{P}, \mathrm{Q}] = \frac{\hbar}{\mathrm{i}} 1$ is satisfied by operators P and Q acting on a Hilbert space H of finite dimension n (i.e. $\mathrm{H} \simeq \mathbb{C}^{\mathrm{n}}$). In this case, P and Q can be realized by $\mathrm{n} \times \mathrm{n}$ square matrices, the trace is a well-defined operation and we get the result.

$$0 = \mathrm{Tr}[\mathrm{P}, \mathrm{Q}] = \mathrm{Tr}\left(\frac{\hbar}{\mathrm{i}} 1_{\mathrm{n}}\right) = \frac{\hbar}{\mathrm{i}} \mathrm{n}.$$

We deduce that the Heisenberg relation cannot be realized on a finite dimensional Hilbert space. Quantum mechanics must therefore be formulated on an infinite-dimensional Hilbert space: on such a space, the trace is no longer a well-defined operation for all operators (in particular, the trace of

von Neumann gave an axiomatic formulation based on the notion of abstract Hilbert space that he introduced and established that matrix and wave mechanics were talking about the same thing in two different ways. In Schrodinger's picture the states became elements of a Hilbert space, namely the space L^2 of square–integrable (wave) functions on the position space. The observables became self-adjoint operators on it. Upon choice of an orthonormal basis in this space however the states become (infinite) vectors, and the operators become (infinite) matrices. This is the Heisenberg's picture.

Following discoveries in the field of mathematics in functional analysis and theory of operators in 1930 by Von Neumann, Paul Adrien Maurice Dirac conceptualized this picture in his textbook *Principles of Quantum Mechanics*, where he introduced the bra-ket notation among other things and proves that the three theories are actually three aspects of a single coherent theory, quantum mechanics, in which the Heisenberg matrices are generalized in operators who have for Schrödinger and spectrum vector own vector and for spectrum the suites of the quantum theory of Borh-Sommerfeld.

Thus, the mathematical formulation of quantum mechanics is based on the concept of an operator. Pure physical states in quantum mechanics are represented as unitary standard vectors (probabilities are standardized to one) in a special complex Hilbert space. The temporal evolution in this vector space is given by the application of the evolution operator. Any observable, that is, any greatness that can be measured in a physical experience, must be associated with a self-deputy linear operator. Operators must produce real eigenvalues because they are values that can result from the experiment. Mathematically, it means that operators must be Hermitens. The likelihood of each clean value is linked to the projection of the physical state on the subspace linked to this clean value.

The axiomatization of Quantum Mechanics by Von Neumann from 1927 to 1929, is a perfect illustration of Hilbert's formalist program. The results of this work are gathered in *Mathematische Grundlagen der Quantenmechanik*, published in 1932, a book that has become a classic treaty.

Von Neumann considers Hilbert's space as a mathematical framework for expressing quantum concepts about the nature of the physics measurements, Hilbert space operators corresponding to the *observables* of relativistic physics. Operators who introduce themselves are Hermitians but not limited, so are not liable for Hilbert's classical theory (see spectral theory). Von Neumann expanded the spectral theory to self-assistant non-bounded operators, while the general case led him to a detailed mathematical study of non-self-assistant symmetrical transformations. In an article of 1929, he introduces and describes in detail the operators' algebras called

operator 1 does not exist) and thus one can no longer deduce a contradiction from the Heisenberg commutation relation in the way indicated.

An inconsistency can still be deduced in another way on an infinite-dimensional Hilbert space by assuming that P and Q are both bounded operators; consequently, at least one of the two operators P and Q satisfying the Heisenberg relation must be unbounded and this fundamental relation cannot therefore be discussed without worrying about the domains of definition of the operators.

W*-Algebra or also *von Neumann algebra.* Attempts to use lattice in the classification of operator algebras factors led him to research on lattice theory; it must be in this area the two important concepts of continuous geometries and regular rings.

It is worth emphasizing that the most significant difference between von Neumann's mathematical axiomatization of quantum mechanics and Dirac's heuristic and systematic treatment of the same theory (Dirac 1925) was not so much the lack of mathematical rigour in the latter—although this point was stressed by von Neumann himself (Neumann 1932, p. 2), who was particularly annoyed with Dirac's δ-function and his closely related assumption that every self-adjoint operator can be diagonalized in the naive way of having a basis of eigenvectors—but the fact that Dirac's approach was relative to the choice of a (generalized) basis of a Hilbert space, whereas von Neumann's was absolute. In this sense, as a special case of his (and Jordan's) general transformation theory, Dirac showed that Heisenberg's matrix mechanics and Schrödinger's wave mechanics were related by a (unitary) transformation, whereas for von Neumann they were two different realizations of his abstract (separable) Hilbert space. In particular, von Neumann's approach a priori dispenses with a basis choice altogether; this is precisely the difference between an operator and a matrix Schmidt alluded to in the above quotation. Indeed, von Neumann's abstract approach (which as a co-founder of functional analysis he shared with Banach, but not with his mentor Hilbert) was remarkable even in mathematics; in physics it must have been dazzling. Von Neumann's abstract approach can deal with preferred observables and experimental contexts, too, though the formalism for doing so is more demanding.

Namely, for reasons ranging from quantum theory to ergodic theory via unitary group representations on Hilbert space, from 1930 onwards von Neumann developed his theory of "rings of operators" (nowadays called von Neumann algebras), partly in collaboration with his assistant Murray (Murray and Neumann 1937, 1943; Neumann 1932, 1938, 1940, 1949). For us, at least at the moment the point is that Dirac's diagonal observables are formalized by maximal commutative von Neumann algebras A on some Hilbert space. These often come naturally with some specific realization of a Hilbert space.

As early as 1927, Von Neumann gives a mathematical analysis of quantum mechanics in fully probabilistic terms: the concept of Von Neumann's statistical matrix, to simultaneously describe several systems that are not at the same quantum level, has become an essential tool. He uses these new notions to mathematically find the quantum thermodynamics. These are problems from the thermodynamics that led it, in 1931, almost simultaneously with Birkhoff, to elaborate the ergodic theory in its modern form. To put the work of Von Neumann in its innovative context, let us include Ulam:

> The ideas of nineteenth century physics, dominated mathematically by the differential and integral equations and the theory of analytical functions, had become insufficient. The new quantum theory needed a more general theoretical point of view, the primitive concepts themselves already including probabilistic distributions and functional spaces of infinite dimension. [...] The work of Von Neumann came at a time when all the complex of ideas

from the theory of Cantor sets and the algebraic work of the greatest mathematicians of the beginning of the century could be exploited for this purpose. (Ulam 1958, p. 40)

5.2 Non-Commutative Geometry Approach

In the 1930s, Norbert Wiener shows that a Banach of functions can be equipped with a commutative algebra structure by doing what is now called a Banach algebra. This makes it possible to study the spaces of functions that are Banach in the context of the groups and algebras of the Lie, by interviewing the topological duality, and the spectral theory: the systematic study of the operator algebras is launched. This theme is developed by the Russian school with mathematicians like Gelfand, Raikov, Chilov, Naimark. A particularly interesting case is that of the C*-algebra, meeting the spaces of Hilbert (1943 by Gelfald and Naimark) and properties found since 1932 by Von Neumann (Neumann 1932, 1938, 1940, 1949). The comprehension and classification of the C*-algebra is a work of several decades, punctuated by publications taking stock, such as those of Jacques Dixmier, and which is considered complete in 1978 following the work of Connes (Connes 1994). Despite this, the algebraization is not over: the homology and cohomology of the Banach algebras are developing on the inspiration of quantum physics; and Non-commutative geometry was designed both to meet mathematical needs and to address certain theoretical physics problems.[19]

In mathematics, it is a question of generalizing the tools of ordinary geometry which have been developed for more than a century: differentiable structures, metrics, group actions, fibrations, connections. These mathematical constructions are now widely used in theoretical physics, and most modern theories (the Standard Model of elementary particles, General Relativity, quantum field theory) are based on fine mathematical properties developed in this context. It is to Connes that we owe, in 1985 (Connes 1985), the first concrete avenues of research in the field of noncommutative geometry, by defining and studying cyclic cohomology. He thus showed that the notion of differential calculus on varieties admits a non-commutative equivalent, in the sense explained below.

In physics, from the middle of the 1950s, work on quantum field theories made it possible to bring out notions that should bring together geometric structures and algebraic structures: on the one hand operator algebras, on the other the gauge theories, that is to say the theory of connections on fibers. However, the strongest motivation still remains today the hope of writing a quantum theory of gravitation with this mathematics, since the founding principle of non-commutative geometry is to merge in the same conceptual framework the operational aspect of quantum mechanics and the geometric aspect of General Relativity (and gauge theories).

---

[19] The commutativity of Cartesian coordinates does not hold in the algebra of coordinates on the phase space of a microscopic system. What Heisenberg discovered was that quantum observables obey the rules of matrix mechanics and this led von Neumann to formalize quantum mechanics in terms of operators on Hilbert space.

The key idea of non-commutative geometry is first to characterize a very particular class of *geometric spaces* by an adapted type of function algebras, by equipping these algebras with appropriate algebraic tools. For example, it is possible to characterize a compact and separated topological space by its algebra of bounded continuous functions, and a measurable space by its algebra of classes of bounded measurable functions. In favourable cases, either these algebraic tools do not explicitly use the commutativity of function algebras, or more equivalent algebraic tools exist, which then makes it possible to study non-commutative algebras of the same type using of these algebraic tools, without having to mention an underlying space. This approach is widely encouraged by numerous mathematical results: characterizations of commutative von Neumann algebras, Gelfand-Naimark theorem on C^*-algebras, Serre-Swan theorem on bundles, cyclic cohomology of the algebra $C^\infty(M)$, relations between Dirac operators and Riemannian metrics, K-theory of C^*-algebras (Connes 1985). One of the strong motivations of noncommutative geometry is to obtain a coherent mathematical framework in which it would be possible to write quantum gravity. One of the avenues explored was to generalize the formalism used in the context of Einstein's General Relativity to operational situations. To do this, an essential first step consists first of all in considering the ordinary notions of Riemannian geometry in a more algebraic way, and then in considering doing without the commutativity of the algebra of functions C^∞. From all these works (Dubois-Violette et al. 1995; Madore et al. 1995; Madore 1999; Masson 2009), it emerges that the non-commutative constraints are so strong that the space of linear connections is very often reduced to a space of finite-dimensional parameters. This is often related to the fact that the center of these algebras is itself very small, a finite dimensional vector space. This situation is hardly encouraging in the hope of doing General Relativity in the *ordinary sense.*

Another possible approach consists in considering that the quantification of General Relativity induces a notion of non-commutative space, having as deformation parameter the Planck length. It is therefore not a question of rewriting General Relativity in one way or another directly, but of extracting possible quantum consequences from it and encoding them in a non-commutative space–time. Often, this non-commutative space–time is studied from the point of view of its *Riemannian geometry*, or it is the support of field theories whose properties are studied (Cfr. Maxwell ([1873] 1989).

Finally, another avenue of research for writing a theory of gravitation using noncommutative geometry was initiated and widely explored by Connes and Chamseddine (1997) for the first proposal, and (Chamseddine et al. 2007) for the latest version of this model and bibliographical references). The ambition of this approach, which leads to a fairly complete model in the last article cited, is to reconstruct the Standard Model of elementary particle physics coupled to gravitation, from principles derived from ideas of noncommutative geometry. This approach does not seek, unlike the previous ones, to consider using non-commutative geometry to write a *quantum gravity*, since the model it leads to is classical. The construction of Connes relies essentially on a spectral triple, consisting of a topological algebra, a Hilbert space on which this algebra is represented, and a Dirac operator on this Hilbert space.

These three objects satisfy compatibility relations, which are neither more nor less than what is necessary for the algebra of functions C^∞ on a compact manifold with spin, endowed with the Hilbert space obtained by completion of the L^2 sections of a spinor bundle and the natural Dirac operator, ie such a spectral triplet. This is the commutative model of this construction.

This notion of spectral triplet is quickly transformed into a spectral quintuplet when we add to it a possible Z_2-graduation and a notion of reality. These two concepts, formalized by two operators γ and j on Hilbert space, have their origins in the Standard Model of particles: γ is linked to chirality through the usual operator γ^5, and j is linked to charge conjugation (Connes 1994; Gracia-Bondía et al. 2001). In this approach, the Dirac operator represents the metric, and its (non-commutative) fluctuations are identified with gauge fields. We can, up to a certain point, understand these properties by the following heuristic argument.

The metric on a differentiable Riemannian manifold is given by the length invariant $ds^2 = g_{\mu\nu}\, dx^\mu dx^\nu$. It is well known that the natural action that one writes within the framework of the theory of relativity, for a free particle (Masson 2009), is

$$S[\gamma] = \int_\gamma d\tau$$

where γ is a path parametrized on the manifold, and $d\tau = \sqrt{ds^2}$ is here the infinitesimal proper time along this path (in a Minkowskian signature). Therefore, $S[\gamma]$ is the proper time of the object along the trajectory γ.

In relativistic quantum mechanics, we are led to replace the formalism of points and trajectories by the formalism of wave functions and their differential equations, using the "quantification procedure" $p_\mu \to -i\frac{\partial}{\partial x^\mu}$. In the space of momentum, the invariant corresponding to ds^2 is $m^2 = g_{\mu\nu}\, p_\mu p_\upsilon$, which, quantified, gives rise to the Laplacian Δ (generalized, in the sense that it can be the d'Alembertian). It is possible to reproduce the approach that leads to action in the space of positions by considering an operator D such that $D^2 = \Delta$, i.e. a Dirac operator. A natural action would then be of the form $\int D$, which only makes sense within the framework of the spectral theory of a certain class of operators. The spectral action proposed by Connes and Chamseddine is of the form

$$S[D] = \int f(D)$$

where f is a function defined on the spectrum of D.

Now imagine that we allow D to have fluctuations in a sufficiently *large* space, for example in such a way that the minimal couplings, which consist in "fluctuating" the p_μ in the form $p_\mu + A_\mu$ (A_μ being a Yang-Mills field), are possible. The action written above will then be an action containing the fields $g_{\mu\nu}$ and A_μ. The structure of this action requires that S[D] produce mathematical invariants from these fields. Also, the work of Chamseddine, Connes and Marcolli (Connes 1985) consisted in allowing the right fluctuations of D (by choosing, in particular, the algebra and the Hilbert space of the spectral triplet) in such a way that this action reproduces a version of the Standard Model of elementary particles (it is a model inspired by *see-saw* models, containing a mixture of neutrinos).

It is well known that this approach raises many problems. Technically, the Dirac operator that it is possible to consider in this approach has a compact resolvent, which effectively excludes Dirac operators from Minkowski space–time physics. Only Euclidean cases can give rise to such modelling, which makes the physical interest of this approach, in its current formalism, not very encouraging.[20] Some authors have started to explore the possibility of a non-elliptical operator. To do this, it has been suggested to further amplify the spectral tri(quintu)plet by adding an additional operator to it, a bit like a Krein space is a Hilbert space provided with a metric operator which connects the indefinite scalar product to the positive definite scalar product.

Another difficulty in the construction of Connes is that it requires some form of compactness for the noncommutative *space*. In Gayral et al. (2004), the definition of a non-compact spectral triple was proposed, and the example of a Moyal algebra (there are many possible definitions of *Moyal algebra*) is given there. This approach again requires extending the triple, by adding a non-unitary topological algebra. Accordingly, noncommutative geometry is an ideal tool for the physicist wishing to consider Yang-Mills-Higgs type models. Indeed, the differential geometry of gauge theories on SU(n) bundles is a sub-geometry of a noncommutative geometry of the algebra of endomorphism of E. This characterization of gauge theories makes it possible to better understand the origin and the place of the Higgs fields with respect to ordinary geometry, since they are interpreted, quite naturally, as the degrees of freedom in the purely noncommutative directions of the geometry of this algebra of endomorphisms. It is therefore a part not visible through the usual geometry. It should be noted that in the Standard Model elaborated by Connes mentioned above, the Higgs fields are also the components of the connections in purely noncommutative directions (the "finite geometry" of this model). This seems to answer both the physical question of the origin of the Higgs, and the mathematical question of the exact status of these scalar fields in spontaneous symmetry breaking gauge models, which until now had not been mathematically identified as gauge fields are today (as connections).

6 Conclusion

As we see, the motivations of research have suffered a considerable change since the time when Fourier, Sturm, and Liouville presented their memories from engineering problems. In the unified formulation of algebra and analysis, the fact that an equation

[20] Even if too many physicists think that a *simple* Wick rotation suffices to switch from one signature to another, what mathematics does not seem to support: how many theorems valid in the elliptical case are also valid in the case hyperbolic?

with edge conditions has a discreet family of eigenvalues is simply due to the fact that the zeros of an analytical function are isolated. The problem of the electromagnetic cavity can only have discrete vibration modes, is no more than a particular case subjected to the theorem of isolated zeros. For its part, the formalization of the infinitely small of quantum mechanics is a major achievement. As much as the development of classical physics (Newtonian, Lagrangian, Hamiltonian mechanics and relativities) is linked to that of geometry and tensor calculus, the emergence and development of quantum mechanics are linked to the theory of Hilbert spaces, the theory of operators and to measure theory. As soon as an observable (which is accessible to an experimental measurement) is represented by an operator on a Hilbert space, mathematics is responsible for constructing probability measures on the spectrum of this operator and these probability measures become a physical reality in Born's transition laws. Thus, in the world of ideas created in Göttingen, it is the general properties of the analytical functions that describe the change of sound of the string; But in the actual history of science, the synthesis made by the mathematicians of Göttingen consisted of gather what was common between problems of various origins and which already existed before. And if we find this as a particular case or an application of the isolated zeros theorem, it is because, among all the possible ways to find the mathematical analysis, the one held by Hilbert was the one we could find on the way Pythagoras, Bernoulli, d'Alembert, Euler, Fourier, etc. when they introduced the appropriate mathematical concepts for the study of the sound emitted by a string.

References

Akagi Y, Katsura H, Koma T (2017) A New Numerical Method for Topological Insulators with Strong Disorder. Journal of the Physical Society of Japan 86/12:123710.

Alexandrov A D, Kolmogorov A N, Lavrent'ev M A (1963) (eds) Mathematics. Its Content, Methods, and Meaning, in three volumes. MIT Press, Cambridge–MA.

Alquié F (1950) La découverte métaphysique de l'homme chez Descartes, Paris, PUF.

Ampère AM ([1823] 1990) Théorie mathématique des phénomènes électrodynamiques uniquement déduite de l'expérience. Gabay, Paris.

Avron J, Seiler R, Simon B (1994) The index of a pair of projections. Journal of Functional Analysis 120/1:220–237.

Beer A (1856) Allgemeine Methode zur Bestimmung der elektrischen und magnetischen Induction. Poggendorffer Annalen der Physik 98:137–142.

Bellissard J, Elst A van, Schulz-Baldes H (1994) The noncommutative geometry of the quantum Hall effect. Journal of Mathematical Physics 35/10:5373–5451.

Berkley G (1734) The Analyst, London edition, published Printed by and for S. Fuller at the Globe in Meath-street, Dublin.

Bohr N ([1935] 1988) Can Quantum-Mechanical Description of Physical Reality be Considered Complete? In: Faye J, Folse HJ (1998) (eds). The Philosophical Writings of Niels Bohr. Vol. IV: Causality and Complementarity. Ox Bow press, Woodbridge [see also: Physical Review:73–82].

Boole G (1844) On a general method in analysis, Philosophical Transactions of the Royal Society of London 134:227–282 [See also: *Id.* (1859), Treatise on Differential Equations (5th ed 1959), Chelsea Publishing Company New York].

Born M (1926) Göttingen Nachrichten. Mathematik und Physik Ki/1:146 [see also: *Id.*, (1968) My life and my views, Scribner's, NY, p. 55].
Born M, Wiener N (1926) A new formulation of the Laws of quantization of periodic and aperiodic phenomena. Journal of Mathematics and Physics 5:84–98 [*Id.*, (1925–1926) eine neue Formulierung des quantungesetze für periodische und nicht periodische Vorgäne. Z Phys 36:174–187.
Brunschvicg L (1927) Mathématique et métaphysique chez Descartes. Revue de métaphysique et de morale 34:277–324. [See also: *Id.* (1981) Descartes: Les étapes de la philosophie mathématique, Blanchard, Paris].
Cajori F (1993) A History of Mathematical Notations, Dover, NY.
Cayley A (1858) A memoir on the theory of matrices. Philosophical Transaction of the Royal Society of London 148:17–37.
Chamseddine AH, Connes A (1997) The spectral action principle. Communications in Mathematical Physics 186:731–750.
Chamseddine AH, Connes A, Marcolli M (2007) Gravity and the standard model with neutrino mixing. Advances in Theoretical and Mathematical Physics 11:991–1089.
Connes A (1985) Non-commutative differential geometry. Publications Mathématiques de l'IHES 62:41–144.
Connes A (1994) Noncommutative Geometry. Academic Press, London.
Courant R, Hilbert D (1924) Methoden der mathematischen Physik. Springer, Berlin.
Davies, EB, Lewis, JT (1970) An Operational Approach to Quantum Probability. Communications in Mathematical Physics 17:239–260.
Deleuze G (1988) Le pli, Leibniz et le Baroque. Les Éditions Minuit. Paris.
Descartes R (1964–1974) Œuvres de Descartes publiées par C. Adam et P. Tannery, Nouvelle présentation par B. Rochot et P. Costabel, 11 Vols., Vrin, Paris.
Descartes R (1981) Entretien avec Burman, Edition, traduction et annotation par J.-M. Beyssade. PUF, Paris.
Descartes R (1992) Méditations métaphysiques, Garnier-Flammarion, Paris.
Descartes R (1996) Règles pour la direction de l'esprit, Vrin, Paris.
Dieudonné J (1981) History of Functional Analysis. North-Holland, Amsterdam.
Dirac PAM (1925) The fundamental equations of quantum mechanics, Proceedings of the Royal Society of London A/109: 642–653.
Dorier JL (1996) Genèse des premiers espaces vectoriels de fonctions, Revue d'histoire des mathématiques 2:265–307.
Dowker JS, Critchley R (1976) Effective Lagrangian and energy momentum tensor in de Sitter space. Physical Review D/13:3224.
Dubois–Violette M, Madore J, Masson T, Mourad J (1995) Linear connections on the quantum plane. Letters in Mathematical Physics 35:351.
Dugac P (1978) Fondements de l'analyse, in Abrégé d'histoire des mathématiques. VI. Hermann, Paris.
Einstein A, Podolsky B, Rosen N (1935) Can Quantum Mechanical Description of Physical Reality be Considered Complete? Physical Review 47:777.
Euler L ([1748] 1988) Introductio in Analytin Infinitorum. *Introduction to analysis of the infinite, I*, Springer–Verlag, NY.
Fichant M (1998) Science et métaphysique dans Descartes et Leibniz. PUF, Paris.
Fourier J ([1822] 1988) Théorie analytique de la chaleur. Gabay, Paris.
Fredholm I (1903) Sur une Classe d'équations fonctionnelles. Acta Mathematica 27:365–390.
Gauss CF ([1801] 1965) Disquisitiones Arithmeticae. Yale University Press, London.
Gayral V, Gracia-Bondia JM, Iochum B, Schucker T, Varilly JC (2004) Moyal planes are spectral triples. Communications in Mathematical Physics 246:569–623.
Girard A (1629) Invention nouvelle en l'algèbre, ed. Jan Janssen, Amsterdam, p. 70.
Gracia-Bondía J, Várilly J, Figueroa H (2001) Elements of Noncommutative Geometry. Birkhaüser, Boston.

Graves C (1847 [1847–1850]) On the theory of linear differential equations. Proceedings of the Royal Irish Academy 4:88–91.
Green G ([1828] 1871) An Essay on the Application of Mathematical Analysis to the Theories of Electricity and Magnetism. In: Ferrers NM (ed). Mathematical Papers of the late George Green. Macmillan and Co, London.
Günther L (1964) Versuch einer axiomatischen Grundlegung der Quanten mechanik und allgemeinerer physikalischer Theorien. Zeitschrift für Physik 181:233–260.
Hargreave C J (1848) On the solution of linear differential equations. Philosphical Transactions of the Royal Society of London 138:31–54.
Hartong J (2000) Des Équations intégrales au Formalisme de la mécanique quantique, Philosophia Scientiæ 4/1:69–101.
Hawking SW (1977) Zeta function regularization of path integrals in curved space-time. Communications in Mathematical Physics 55:133.
Hawkings T (1975) Cauchy and the spectral theory of matrices. Historia Mathematica 2:1–29.
Heaviside O (1893) On operators in physical mathematics. Proceedings of the Royal Society of London A/52:504–529; 54:105–143. [see also: *Id.* (1899) and *Id.* (1950)].
Heisenberg W (1925) Über quantentheoretische Umdeutung kinematischer und mechanischer Beziehungen. Zeitschrift für Physik 33:879–893.
Hellinger E, Hilberts D (1935) Arbeiten über Integralgleichungen und unendliche Gleichungssysteme, in Hilberts gesammelten Abhandlungen. Springer, Berlin.
Hilbert D (1909) Wesen und Ziele einer Analysis der unendlichvielen unabhängigen Variablen. Rendiconti del Circolo matematico di Palermo 27:59–74. Hilberts gesammelten Abhandlungen Band III:56–72 [See also: *Id.* (1912) Grundzüge einer allgemeinen Theorie der linearen Integralgleichungen, Teubner, Leipzig].
Jammer M (1967) The conceptual development of Quantum Mechnaics, McGraw-Hill, NY.
Jordan TF (1969) Linear Operators for Quantum Mechanics; John Wiley & Sons, NY.
Klein F (1967) Vorlesungen über die Entwicklung der Mathematik im 19. Jahrhundert, NY.
Koyré A (1961) La Révolution astronomique. Hermann, Paris.
Kraus K States (1983) Effects and Operations. Springer, Berlin.
Lagrange JL (1797) Sur une nouvelle espèce de calcul relatif à la différentiation et à l'intégration des quantités variables. Œuvres III, pp. 441–476.
Lagrange JL ([1811] 1989) Mécanique Analytique, Editions Jacques Gabay, Paris.
Leibniz GW (1710) Symbolismus memorabilis calculi algebraici et infinitesimalis in comparatione potentiarum et differentiarum et de lege homogeneorum transcendentali, Miscellanea Berolinensia 1:160.
Leibniz GW (2001) The labyrinth of the continuum. Writings on the continuum problem, 1672–1686. Yale University Press, New Haven.
Liouville J (1836) Mémoire sur le développement des fonctions ou parties de fonctions en séries dont les divers termes sont assujettis à satisfaire une même équation différentielle du second ordre contenant un paramètre variable. Journal de Liouville 1:253–265. [see also: *Id.* (1837) Second mémoire sur le développement des fonctions ou parties de fonctions en séries dont les divers termes sont assujettis à satisfaire une même équation différentielle du second ordre contenant un paramètre variable. Journal de Liouville 2:16–35.
Madore J (1999) An Introduction to Noncommutative Differential Geometry and Its Physical Applications. 2nd edition. Cambridge University Press, Cambridge.
Madore J, Masson T, Mourad J (1995) Linear connections on matrix geometries. Journal of Classical and Quantum Gravity 12:1429.
Majima H (2013) Seki Takakazu, His Life and Bibliography. In: Knobloch E, Komatsu H, Liu D (eds). Seki, Founder of Modern Mathematics in Japan. A Commemoration on His Tercentenary. Dordrecht. Springer, pp. 3–20.
Masson T (2009) Quelques aspects de la géométrie non-commutative en liaison avec la géométrie différentielle. HDR Dissertation. University of Aix-Marseille, Marseille.
Maxwell JC ([1873] 1989) Traité d'électricité et de magnétisme. Gabay, Paris.

Murphy R (1837) First memoir on the theory of analytical operations. Philosophical Transactions of the Royal Society of London 127:179–210.

Murray FJ, Neumann von J (1936) On rings of operators. Annals of Mathematics 37:116–229.

Murray FJ, Neumann von J (1937) On rings of operators II. Transactions of the American Mathematical Society 41:208–248.

Murray FJ, Neumann von J (1943) On rings of operators IV. Annals of Mathematics 44:716–808.

Neumann CG (1877) Untersuchungen über das logarithmische und Newtonsche Potential. Teubner, Leipzig.

Neumann von J (1930) Zur Algebra der Funktionaloperationen und Theorie der normalen Operatoren. Mathematische Annalen 102:370–427.

Neumann von J (1931) Uber Funktionen von Funktionaloperatoren. Annals of Mathematics 32:191–226.

Neumann von J (1932) Mathematische Grundlagen der Quantenmechanik. Berlin: Springer–Verlag [English translation *Id.* (1955) Mathematical Foundations of Quantum Mechanics. Princeton University Press, NJ].

Neumann von J (1938) On infinite direct products. Compositio Mathematica 6:1–77.

Neumann von J (1940) On rings of operators III. Annals of Mathematics 41:94–161.

Neumann von J (1949) On rings of operators V. Reduction theory. Annals of Mathematics 50:401–485.

Raymond P (1976) La naissance du calcul infinitésimal. Philosophie et calcul de l'infini, Maspéro, Paris.

Riesz F (1918) Ueber lineare Funktionalgleichungen. Acta Mathematica 41:71–98 [See also: Œuvres complètes, II, pp. 1053–1080; *Id.* (1907) Sur les systèmes orthogonaux de fonctions et l'équation de Fredholm, Acta Mathematica 41:734–736; Œuvres complètes, I, pp. 382–385; *Id.* (1907) Sur une espèce de géométrie analytique des systèmes de fonctions sommables, Acta Mathematica 41:1409–141; Œuvres complètes, I, pp. 386–388].

Schrödinger E ([1926] 1988) Mémoires sur la Mécanique ondulatoire, Traduits en français par Al. Proca, Félix Alcan, Paris, 1933 et réédition Jacques Gabay, Paris.

Schwinger J (1951) On gauge invariance and vacuum polarization. Physical Review 82:664–679.

Servois FJ (1814) Essai sur un nouveau mode d'exposition des principes du calcul différentiel. Annales de Mathématique 5:93–104.

Silberstein L (1902) Versuch einer Theorie des physikalischen Operatoren. W. Osr-twald's Annalen der Natuphilosophie 2:201–254 [See also: *Id.* (1902) La théorie des opérateurs en physique – la connection des phénomènes dans le temps. Przeglad Filozoficzny 5:425–442; *Id.* (1901) Symbolische Integrale der electromahnetischen Gleichungen aus dem Anfangszustand des Feldes abgeleitet, nebst Andeutungen zu ziner allgemeinen Theorie physikalischer Operatoren. Annalen der Physik 6:373–397].

Sturm CF (1836) Mémoire sur les équations différentielles du second ordre, Journal de Liouville 1:106–168.

Sylvestre JJ ([1837–1853] 1904) The Collected Mathematical Papers of James Joseph Sylvester. Cambridge University Press, Cambridge.

Ulam S (1958) John von Neumann 1903–1957, Bulletin of American Mathematical. Society 64, p. 40.

Wiener N (1926) The operational calculus. Mathematische Annalen 95:557–584.

Zhi Li, Roger SK Mong (2019) A Local formula for the Z_2 invariant of topological insulators. Physical Review B 100:205101.

Chapter 19
Physics as a Fundamental Medical Humanity

Franco Blezza

Abstract Physics can be considered a Medical Humanity, and more generally a Humanity: for this reason, it can make many powerful contributions to training and professional practice of the doctor of medicine, starting with his beginning training. This fundamental science of nature constitutes an excellent paradigm of rigour, generality and research methodology, and it offers various reasons for interest also through the mediation of Philosophy. It is necessary to get out of the visions of Physics that are highly instrumental for the doctor, reasons that are also important and may require more in-depth training, to adopt an essential vision. In this paper, we bring some significant examples of issues in this sense, such as thermodynamics, blood circulation, the main sense organs, the foundation of scientific psychology, systems theory and the systemic approach, and we come to quantum and twentieth-century Physics regarding genetic mutations and molecular Biology. The historical dimension of the discourse turns out to be particularly valid and fruitful: in fact, some of thease great researchers were doctors and physicists together, and many relevant historical episodes have always been, and continue to be, straddling Physics and Medical Sciences. The professionalism of today's doctor has much to benefit from it.

Keywords History of Sciences · Humanities · M. D. Education · Medical science · Physics

F. Blezza (✉)
Department of Philosophical, Pedagogical and Economic-Quantitative Sciences, "Gabriele d'Annunzio" Chieti & Pescara University, Chieti, Italy
e-mail: francoblezza@email.it

R. Pisano (ed.), *A History of Physics: Phenomena, Ideas and Mechanisms*,
History of Mechanism and Machine Science 42,
https://doi.org/10.1007/978-3-031-26174-9_19

1 Introduction: Physics in the Beginning Training of Medical Doctors in Italy

Since 1969, the law "*Urgent measures for the University.*" (*Provvedimenti urgenti per l'Università*) n. 910 of December 11 signed by Tristano Codignola established that in the Italian University "the student can prepare a study plan different from those provided for by the educational systems in force, provided that within the disciplines actually taught and in the number of courses established." (*lo studente può predisporre un piano di studio diverso da quelli previsti dagli ordinamenti didattici in vigore, purchè nell'ambito delle discipline effettivamente insegnate e nel numero di insegnamenti stabilito*). Approval shall be granted taking into account "the "student's needs of cultural training and professional preparation" (*esigenze di formazione culturale e di preparazione professionale dello student*).

The ministerial curriculum of the *Corso di laurea in Medicina e Chirurgia*/ *Bachelor of Science in Medicine and Surgery* in Italy understandably had a very rigid system. For the first of the six years, it included, among others, an examination in Physics; the program for that examination never presupposed infinitesimal analysis, analytical geometry, statistical sciences or other mathematical prerequisites, so it was not characterized as a particularly demanding step; the discourse is more general, many doctors do not go beyond elementary mathematics. However, since the beginning of the law there were numerous attempts to replace that very limited subject, for instance by a History of medicine examination: It was argued that Physics was not consistent with the basic culture of the future medical doctor and that it educated to a different mentality. There were also serious attempts by the university collegial bodies to abolish the subject altogether from the ministerial curriculum, but they did not come true: the subject remained, renamed for example "Medical Physics" or "Applied Physics", with a very elementary level of mathematization, It must share 30 university educative credits with examinations of Applied Biology, General Psychology, Medical Statistics and Medical Genetics, out of a total of 360 credits in six years, and there continue to be no curricular courses even of mathematics institutions, nor optional in-depth courses. These subjects, as well as the basis for a Physics of actual university level and more professionalizing, would allow to rigorously approach a whole complex of medical problems that postulate greater Physics skills. Let's think of Epidemiology, and the relative mathematical elaborations for which physicists are often involved; to the analysis of traces from electrograms, more and more often left to an automatic analysis: to the functioning of the ear as an organ of hearing (logarithms, wave dynamics, Fourier transforms, harmonics, resonance, etc.) and as a detector of accelerations acting on the head (with the problem of the Coriolis accelerations, what they are and when they arise); Electrology (laser, ionizing and non-ionizing electromagnetic radiation); Physics as a basis for the diagnostic and therapeutic applications of radioactivity; or the always nuclear but different physical bases of the NMR; laser surgical too; to so much physiotherapeutic instruments; and are only the first subjects coming to mind by way of example, among the countless possibilities. In essence, that little bit of Physics that despite everything that is done at the beginning is justified by instrumental and non-essential considerations; it is believed despite everything that Physics is not part of the mentality or professionalism of the medical doctor and surgeon, but can offer him some practical contribution; and even this, given the very heavy limitation of the mathematical and cultural prerequisites, it is limited to classical physics and applications that are well represented by the physiological studies of the nineteenth century or earlier or, better, by a part of them, the simplest part. This is the problematic point: there are consistent reasons to go beyond pure instrumentality.

2 For an Essential Function

The essential and foundational importance, and not only instrumental and practical, of the Physics for the training of the medical doctor and surgeon should be evident. It is, if only we consider that all the parameters at stake, all the evidence, in his professional practice are physical or chemical or chemical-physical, and if we take into account how much human physiology, pathology and therapy refer properly to physical conceptualities in canonical form. This should be clearer in the homeland of Luigi Galvani (1737–1798), from whose name are derived various technical terms of Electrology and applied Physics [1, 2 anthological sections], or of Giovanni A. Borelli (1608–1679), one of the most eminent exponents of Iatrophysics or iatromechanics.

A very first level of study updated to the XXI century, and certainly not at the beginning of the nineteenth century or earlier, of the study and care of the cardiovascular system, the respiratory system, the main sense organs, the skeletal and muscular apparatus and related dynamics, just to bring some examples of greater evidence, can not be accomplished without an essential background of Physics. And instead, in Italy we continue to limit ourselves to elementary physics with very limited mathematical bases. Physics for the medic can be taught better as an operational tool; but it can be taught more and better as conceptuality, and there are very good reasons to do so.

3 Aim: Physics for Today's Doctor and for Medicine

The deepening of this background very strongly affects the possible professionalism of the doctor. On the other hand, the use of laser or ionizing radiation, bionics, tools such as Pacemakers, even leads to the exclusion that a single Physics examination with very elementary mathematical bases even approaches the cultural needs of the doctor. Many treatises on medical physics offer repertoires adequate for the purpose; but the problem is how they are treated; their lack of depth conditioned by an elementary Physic, in fact, pre-university Mathematic are neither adequate nor satisfactory.

What we will propose in this essay is a further goal and requires more advanced reflection. We will propose to pose the problem of a role of Physics as an essential component of the human culture of the surgeon, that is, as Medical Humanity at the level of a title, as education to rigour and continuous research, as a reconsideration of the nomothetic and idiographic dimensions, that is, of the essence of the scientificity of the medical-surgical matter, and as enrichment and greater qualification of the doctor, the professional who offers the patient as a first cure and remedy himself, his presence and his humanity. This humanity must be complete, in fact, also on the side of the Physics.

It is not only a matter of abandoning any absurd attempt to do without even that little elementary physics: It is necessary to carefully study the curriculum of the six-year degree course in medicine and surgery so that future doctors have adequate mathematical bases, in infinitesimal mathematics, in analytical geometry, in statistical science and more, on which to build a university-level physics preparation at least comparable to that of an industrial chemist or an architect. However, we intend to bring some bricks, canonically intuitive, to the construction of the idea of the Physics and Humanity in generally speaking, analogous to Philosophy and Literature, and we believe at the same time to honour the magisterium of Salvo D'Agostino, that is, a physicist who in the history of science has found an essential field of explanation precisely for his humanity.

4 What is Science?

We say, science without adjectives or further periphrastic specifications: Sciences in the strictest sense, or *stricto sensu.*

It is a cliché (lieu commun, τόπος) that the term "humanities" designates the study of subjects such as literature, philosophy, language or history, somehow conceptually opposed to sciences stricto sensu. It is a secular prejudice or preconception, not classic; the concept was foreign to the classical culture, both Greek and Latin, before and after the *Graecia capta* (Pidna battles, 168 and 148 b. C., destruction of Corinth, 146 b. C.), that is, since the scientific culture as the philosophical one also spread to Rome from Greece, and since the professions entered it starting from the medical *Ars medica*),

In classical Greece, there was not only Plato's distinction between ἐπιστήμη and δόξα science and opinion; the term ἐπιστήμη means ἐπί-ἵστημι, on-site "to be above", more properly mastery or something similar. Human knowledge had another diction, οί λόγοι, indicatively rational discourses even as that suffix -λογία that would be fixed to this day both for natural sciences (Biology, Cosmology, etc.) and for cultural sciences (Sociology, Anthropology, etc.) and different knowledge. The Romans, who originally were, if anything, great technicians, applied themselves to the sciences, id est after the *Graecia capta.*

In recent times, Kant's *Synthetische Urteile* (*Synthetic Judgements*) a priori was already in Newton's mechanics and the mathematics of his time; mathematical judgments are all synthetic a priori. It plays an important role in today's culture the reference to the German term *Wissenschaft*, with various meanings in the singular and plural *Wissenschaften*. However, it does not correspond to the English Science nor the Latin *scientia*, as it constitutes the abstract noun of the verb wissen, to know (with a suffix analogous to the English -ship); therefore, it is knowship, properly, although it is more immediate to translate as "science".

5 Some Appropriate Philosophical References

In fact, the distinction of principle between the *Naturwissenschaften* or Natural sciences and the variegated complex of the *Human- Sozial- Geistes-Wirtschafts-Wissenschaften* or *Social Sciences, Humanities, Economics* (and so on) has prevailed. In particular, the distinction from the Geisteswissenschaften ("spirit" Sciences) refers above all to Wilhelm Dilthey (1833–1911): it is a distinction based on the object of study, and possibly of application: but it does not prejudice and without foundation the question whether there are sectors of the second *Wissenschaften* which in turn constitute sciences in the strict sense (*stricto sensu*) such as the former, or not, and under what conditions.

Wilhelm Windelband (1848–1915) already rejected this distinction as metaphysical and unfounded in opposing "nature" to "spirit". He counter-proposed to replace the distinction on an object basis (nature and the human world) with a distinction of a method, precisely between nomothetic sciences, which seek to determine the general laws that express the regularity of phenomena, and idiographic sciences, i. e. descriptive, which place their attention to the individual phenomenon to understand its specificity and individuality. Physics is elective nomothetic science, History can be practiced as an idiographic science, Medicine is both.

Let us remember that, from a logical and formal point of view, a scientific law is a generalization of an implication and in a given domain. From the logical point of view:

$$\forall x \in D, f(x) \rightarrow g(x)$$

Well, not all sciences in the strict sense are nomothetic. For example, natural history or botanical and zoological systematics, or relevant areas of psychological sciences, are not; and often what seems like an application is actually a doubly hypothetical and personal imperative ("*if you want... then maybe* ..."). This applies to pedagogy (Blezza 2018, pp. 58–60, 4, pp. 123–126), but largely also to medicine and surgery, pharmacy and veterinary medicine.

Bringing down the wall between science in the strict sense and study and application to man and mankind was a first fundamental step forward, decisive for our discourse. Removing the postulation of nomoteticity was a second step, also of fundamental importance. It was over a century ago, when classical Physics had entirely consumed its irreparable crisis for decades and had already been alternated by the Einstein's theories of relativity (Einstein 1916) and quantum (Hutten 1967).

Meanwhile. from the branches of philosophical studies more attentive also to the social sciences came the fundamental contributions for today. Let us think of the concept of *future experience* and the *falsificationism* of the Pragmatists, concepts both taken from the Epistemology of Karl R. Popper (1902–1994) who established strictly *das Abgrenzungskriterium* i. e. *the demarcation criterion*, whose statute must be carefully considered:

> But I shall certainly admit a system as empirical or scientific only if it is capable of being tested by experience. These considerations suggest that not the verifiability but the falsifiability of a system is to be taken as a criterion of demarcation. In other words: I shall not require of a scientific system that it shall be capable of being singled out, once and for all, in a positive sense; but I shall require that its logical form shall be such that it can be singled out, by means of empirical tests, in a negative sense: it must be possible for an empirical scientific system to be refuted by experience.[1]

The arts in the strict sense such as literature or music or the figurative arts, the natural sciences, the social sciences, technology/technique and the logical-mathematical sciences are all domains in which the products of human creativity are classified, expressed in trying to solve man's problems: it is the method that distinguishes a subject from another one; and more generally there are some norms of method that allow distinguishing what is scientific from what is not scientific, such as logical coherence, and controllability with facts and the consequent asymmetric feedback.

There is therefore a scientific history, like the scientific histories of literature, philosophy, art(s), science(s); and there are non-scientific medicines such as many of the so-called "alternative" medicines and equally dogmatism applied to human health over the millennia.

Today the domain of Science stricto sensu is considered much wider and articulated than in the nineteenth century, in a unitary vision of the scientific method (Antiseri [1982] 2001).

It is true that historical Umanesimo / Humanism had its references in two great writers such as Francesco Petrarca (1304–1374) and Giovanni Boccaccio (1313–1375), who did not have a balanced culture on the scientific side as Dante Alighieri (1265–1321) had it, who gave evidence of it in the *Divina Commedia* too. But that cultural movement aimed to rediscover the Latin and Greek classics in their historical

---

[1] (Popper 1934–1935, A survey of some fundamental problems, p. 18). "Nun wollen wir aber doch nur ein solches System als empirisch anerkennen, das einer Nachprillung durch die "Erfahrung" fahig ist. Diese eberlegung legt den Gedanken nahe, als Abgrenzungs-kriterium nicht die Verifizierbarkeit, sondern die Falsifizierbarkeit des Systems vorzuschlagen; mit anderen Worten: Wir fordern zwar nicht, daß das System auf empirisch-methodischem Wege endgiiltig positiv ausgezeichnet werden kann, aber wir fordern, daß es die logische Form des Systems ermglicht, dieses auf dem Wege der methodischen Nachprüfung negativ auszuzeichnen: ein empirisch-wissenschaftliches System muß an der Erfahrung scheitern kännen. *Grundprobleme der Erkenntnislogik*, p. 12)".

and literal authenticity: It was, in short, an application of the same scientific spirit. Indeed, Humanism has expressed an original pedagogy *ante litteram* (Blezza 2020, p. 20) and didactics centered on dialogue, with a methodology that today we would consider *problem solving*. Recall that the term *paedagogy* appeared in Latin and soon in other European languages at the end of the fifteenth century; moreover, the Latin and German term "Humanismus" was coined in 1808 by the pedagogist Friedrich I. Niethammer (1766–1848), in order to enhance the studies of Greek and Latin within the curriculum studiorum. That said, and on this basis, Physics can be considered one of the humanistic subjects, and a humanistic subject of primary importance, especially if one considers it under its historical dimension (Hutten 1967; Geymonat 1970–1972; Rossi 1988–1989; Lindberg and Numbers 2003–2005 for the general discourse). This is general discourse: Physics is a Humanity, a human creation, operated in compliance with the norms of the method of science, to try to solve the problems of man (Blezza 2020, pp. 22–25 et sgg.; Popper 1934–1935, pp. 17–22); and in the laws that it establishes, always provisionally and interlocutorially, we read man, the same man who can be read in other humanistic disciplines such as literature, music or other arts, or technique and so on. The same can be said, in all evidence, for medical-surgical science.

6 Physical Science and Medical Science

6.1 An Exemplary Synthetic Review

In this section, we will exemplify some issues for which the importance of Physics as the essence of the human training of the medical surgeon seems to us to be better readable. Again, science, like Humanity, finds a very broad and particularly significant domain in medical science.

To begin the discourse in this sense, in our opinion, pays particular homage to the One to whom this collection is dedicated, as a master of the history of Physics and, more generally, of history of the natural sciences and science in the strict sense.

6.2 The Hypothetical, Provisional and Interlocutory Character of Science

Physics and Medicine are united in society by a widespread consideration of strong knowledge, producers of indisputable certainties and definitiveness. Those who have competence, and those who have historical competence in a particular way, are perfectly aware of the fallacious and dangerous character of these misunderstandings.

"Mathematics is not an opinion", "scientific certainties", "it is scientifically proven that [...]", "scientific truths", and so on listing by line and notes to all, are unsustainable myths that have hurt too much, and that have their ultimate origin in the first half of the nineteenth century, therefore in culturally distant times.

The whole Positivism was based on the strong deterministic vision of the Newtonian mechanical M that imposed itself from the last or precisely in the first half of the nineteenth century, so it is unfounded to study both that current of thought, and the philosophical dualism with Idealism, without adequate bases not only of Newtonian Mechanics and therefore of Mathematical Analysis, but Rational Mechanics and Analytical Mechanics. It's not just about having the necessary basis for a century or more of philosophical debate, it's an essential human component. We recall in particular Laplace's determinism.

6.3 Thermodynamics and Its Principles

About Thermodynamics, seen from a historical point of view, it is remarkable that it is a scientific theory formulated after the powerful affirmation of the corresponding technique. In addition, it was found to be compatible first with classical Mechanics and then with quantum Mechanics, but in this second case with less difficulty. The principle of conservation, increasing entropy, absolute zero are just three of his great conceptual tools. Among other things, here there is something to reflect for those who believe that technology is an applied science, and therefore that science and technology end up being two sides of the same medals: Among other things, this was one of the forcings to which the idealist philosophers minimized in the Italian school that little bit of science that remained, a domain of pure and simple practical value and not "true culture". The doctor knows that he also has practical tasks, and these are of fundamental importance, but that he fulfils these tasks with culture.

Entropy can only increase. Each human organism can be neghentropic locally, at the price of an increase in the overall entropy of the universe and this to be home to a flow of degrading energy. This aspect is no different from any machine and any device that consumes energy. As long as there is energy to degrade there is life, culture, history, science, technology and so on; without energy to degrade there is only the death of cosmic equilibrium, all atoms at the same temperature as a few Kelvin. Physicists do not listen to them, doctors who are listened to should be aware of these general theories of nature and evolution. Man, like any other living being, is inseparable from the environment with which he necessarily continuously exchanges matter, energy and information. Nor is it sensible to distinguish clearly the living from the environment in which he must live, nor to distinguish the molecules of the living from the molecules of the environment. Hence a rational ecology and a constructive attitude on the part of man towards the environment, which is so important in medical practice. In the medical field, an essential contribution can be made to the conscious and rationally founded respect for the environment, as a necessary condition for the survival of oneself and the species.

6.4 From the Nineteenth to the Twentieth Century

Secular Classic Mechanics entered into crisis in the second half of the nineteenth century, so the philosophical dispute between Positivism and Idealism continued without essential foundations. Beyond the empirical falsifications, on which we do not entertain, this crisis would have become irreparable with Maxwell's equations (since1865), that is, with the appearance of a general theory incompatible at the logical-mathematical level. Since 1824, that is, a few decades earlier, another general theory had appeared, precisely Thermodynamics, but it proved to be compatible with classical mechanics, it was said. In these crises, and the new theory, the foundations of quantum theory and theories of relativity will be found: Many twentieth-century discourses on relativism and indeterminism have spread despite being uncultivated and unfounded, due to the lack of Physics just like a Humanity. Moreover, for over a century we have been in a condition not describable with the conceptuality of "normal science" according to Kuhn (Kuhn [1962] 1970, pp. 10–42).

Einstein with the theories of relativity offers relevant teachings also on the level of method, on the relationship between scientific theory and related empiricism. His reflections were based on other scientific reflections, he was not an experimental physicist: but his theorizing was exposed to complex future experiments, which would corroborate it.

Enrico Fermi (1901–1954) with the steps of his scientific life testified well in sequence what they were or a theoretical physicist, an experimental physicist, and a nuclear engineer that is a technician. Physics like other empirical sciences has a theoretical branch; but every theorizingation has its foundation in other scientific theories that are controlled with facts, and above all offers infinite consequences that lend it to the scrutiny of future experience, so in Peirce (Hartshorne et al. 1931–1958) and in the other pragmatists (James, who was a doctor and psychologist (James 1890), and the pedagogist Dewey (1916, Chap. 11; 1938, *passim*), as in Popper and Critical Rationalism. For example, Pedagogy among the social sciences, has its theoretical branch in the philosophy of education: but its reflections are rooted in the organic experiential relationship with the reality of education, and above all, it must consider itself always open to the control of future experience through its consequences (Blezza 2020, Chaps. 2, 3, 5, pp. 127–131): Physics would be an excellent Humanity to prevent misunderstandings and contradictions to all these concerns (Popper 1934–1935; Bartley 1982–1983).

6.5 Quantum Physics, Molecular Biology, Mutations and Viral Variants

Mutations from one generation to the next are an essential component of the mechanism of the biological evolution of species by natural selection. In particular, virus mutations such as those plaguing the world during the COVID-19 pandemic that is still dramatic as we write, are consequences of quantum indeterminacy since they affect chemical reactions at the level of individual molecules, however macro they are. It would be appropriate to closely link the two concepts: not only to understand the unpredictable character of mutations, but to give rationality to the phenomenon, with a kind of historical study, a scientific idiographic but also very rigorous study, to the evolution of living beings and viruses themselves. With so much research and so many elaborations that meritoriously have been carried out in this global coronavirus tragedy, an attempt to connect with the Heisenberg Principle, which is the physical cause of mutations, that is, the most general, seems to be missing. We need conceptual and operational tools of quantum physics: but we understand that for this we need prerequisites of Mathematics and contemporary Physics that are still very far from the common way of feeling the training of doctors. On the other hand, in addition to spontaneous mutations, there may be those caused by ionizing radiation, which also affects the chromosomal level in a probabilistic way, and which can have natural or artificial origins.

In Italy, in the substantial shortage of civil nuclear power, diagnostics and therapeutics with radiation of measured doses is almost the only sector in which physicists can have a professional position and a career with roles similar to those of doctors. It is a question of carefully evaluating the cost/benefit ratio, given that ionizing radiations are susceptible to the most fearsome mutagenic effects. It is also one of the limited areas of professional exercise of Physics along with terrestrial Physics. It is no coincidence that the use of physicists in the study of meteorology and geological phenomena is required not for a particularly up-to-date Physics content, which is indeed classic, that dates back to about two centuries ago, nor to the use of tools or devices that is mostly operated by other professional figures, but to the sophistication of the mathematical tools to be used.

6.6 *Poiseuille, and other Physiological Problems*

Let us preface the reminiscence that blood circulation was discovered by William Harvey (1578–1657), and the discovery is not alien to the stay of the British scholar at the University of Padua in the period of Galileo, in a context of great integration with Physics at the origins of modern science.

The law of (Hagen) Poiseuille (1838) that studies the scope of a duct was originally formulated for blood circulation and was a few years later extended by George Stokes (1845) to the laminar motion of incompressible fluids. Recall that Jean L.M. Poiseuille (1799–1869) was a doctor who also won the qualification of physicist so much so that his name was named the unit of measurement of viscosity (1 Poiseuille = 1 Pa / s) (Geymonat 1970–1972; Lindberg and Numbers 2003–2005; Rossi 1988–1989). The formulation is mathematically simple, the result of empirical processing and not of a hypothetical-deductive one:

$$Q = \Delta P\, r^4 / 8\eta\, l$$

It is a very elementary formulation that presupposes the constant section of the duct, the lamellar flow, and that does not apply except in further simplified hypotheses such as the duct is not too short or too wide, and does not apply even to the terminal sections. The problem of mathematical prerequisites was the same then as it is today, about two centuries later. And yet, here is another testimony of how an integrated treatment of medicine and medical sciences highlights its humanistic character.

6.7 The Foundation of Scientific Psychology

A similar argument could be made intuitively for most of the human physiology if not for totality. But this discourse acquires a very particular meaning of close interdisciplinary integration between medical matter and Physics if it is referred to the sense organs, to the Physiology of perception and therefore to the foundations of scientific Psychology, and we are always talking about the nineteenth century, with organic reference to the culture of the time and the *Zeitgeist*. Here there are also consistent foundations for a science that would only be described later, namely Biophysics. Biophysics has strong contents and methods, but evanescent historiographical outlines, someone (Bartley III, p. 16.1) makes it go up to the studies on the flight of birds by Leonardo da Vinci (1452–1519). Instead, Psychology has been for millennia a branch of philosophical studies, the study of ψυχή that is a concept that can be roughly rendered with that of *soul*, from ancient Greece until the mid-nineteenth century.

In 1867, Wilhelm M. Wundt (1832–1920) began to speak about “Physiological Psychology”, in the context of the foundation of Physiology in the physical-sanitary sense that took place in that turning point; and created in 1880 in Leipzig the first laboratory of Experimental Psychology. Quantitative research on physiological Psychology and perception began to produce significant results in that same period, with prominent figures such as Ernest H. Weber (1795–1878), Gustav T. Fechner (1801–1887), Hermann von Helmholtz (1821–1894), Ernst Mach (1838–1916) and many others (Geymonat 1970–1972, IV, Chap. 17; Kuhn [1962] 1970, II; 11, VI). The close integration between the Physics and the Medicine allows the one to qualify the other, in absolute reciprocity, fully as Humanity, and as both generators and of scientific Psychology that would have a great development in the immediately following times. We note that at the center of this research was the Physiology of the inner ear too, with the great physicist and epistemologist empiriocritist Ernst Mach, together with Joseph Breuer (1842–1925), the same physician who collaborated so successfully with Freud (Breuer and Freud 1895), and the chemist Alexander Crum Brown (1838–1922).

6.8 About the Systemic Approach

The *systemic approach* is a synthetic approach without being able to be said to be global: in any case, it is opposed in principle to the *analytical approach.* It can be corrective to the excesses of the specialization of doctors or other professionals, a correction that refers precisely to the analytical and at the same time nomothetic character of the Physics, which is reconfirmed as a Humanity also in this specific regard in coherence with the others.

Modern Physics had among its foundations the possibility of breaking down each complex problem into a sequence of simpler sub-problems, to be studied and solved separately. It has been a powerful and successful application, but much less suited to a fruitful application to the Sciences of Culture and that, in fact, has attracted philosophical discredit towards science in general, even because of certain positivistic excesses.

The work of Ludwig von Bertalanffy (1901–1972), adhering to the Wiener Kreis, is twentieth century (Bertalanffy 1932, 1968). However, conceptuality is prior and more or less coeval to the analyticity of modern science, if it is studied in the applications of technical culture rather than of the scientific one; its main way of access to the scientific-naturalistic culture was that of biological and medical problems, consistent with the formation of the same. See Cornelius Drebber in the regulation of the temperature of an oven (1666), Christian Huygens in the regulation of the movement of the clock (1673), or James Watt and the regulation of the speed of the steam engine (1761–62). So, here we have another precedent of scientific theory formulated after technical application, as for the aforementioned Thermodynamics.

The conceptual rigour of Systems Theory is substantiated, as is known, in the two basic concepts of *state,* in particular, defined by a series of observable quantities, called in general *state variables,* and *Feedback* that allows the system itself to regulate, taking in real time a part (more or less relevant) of the product in order to acquire the information necessary for the regulation. The system may be susceptible to controlled transitions, or it may seek homeostasis. The systemic approach demonstrates its effectiveness in studying complexes such that certain modifications of some elements entail global modifications. It is not purely and simply a negation of the analyticity of nineteenth-century science, but it is understandably something else. The theory of systems can also have an interesting application to man, as an example of homeostasis, considering that the state of survival is characterized and controlled by complex of state variables, the first of which is body temperature, and which constitute the enormous domain of instrumental, laboratory and control examinations, which characterize a conspicuous part of medical science, also and the preventive one. Of course, it is a partial application, since there are no equations of state, but above all, because it is not conceivable for the human body other than evolution other than homeostasis, beyond which there is death. This is not a complication but a necessary differentiation, considering man in his particularity as the subject of cultural evolution. The only one; as always, considered in terms of Humanity. It is scientific deontology that emerges if a more advanced mathematical foundation would not allow a different elaboration and with other functions of allthese variables of state of man. Perhaps this conceptuality could be transferred to psychophysical development, and the related pedagogical considerations: and therefore in particular to the psychological and pedagogical sciences (Blezza 2018, Chap. 2); in fact, human life can be considered as a sequence of transitions from one state to succession or, from the simplest three-part scan (age of development—adulthood—third age), to more complex and realistic visions. Only one transition is organic, puberty: But all other transitions must be made the subject of medical practice in a contest of Humanities, among which the presence of Physics is evident and primary. If anything, we must warn against an extremely casual and uncultivated use of the term "system", to indicate an approach to man, the human body and the psyche that is not analytical as for modern sciences and contemporary sciences as much as, instead. synthetic: But those who know the sciences in a minimally adequate way certainly do not fall into such a trap. A vision in terms of Humanities leads to consider the so-called *narrative medicine*, in particular, as a place of acquisition of further evidence, such as the laboratory, the medical examination, the archive, the fever thermometer. All dualism with EBM therefore falls, for functional integration.

The discourse changes at the cellular level, for example when a cell encounters the transition of duplication or multiplication, and although there are descriptive attitudes that are not mathematically skillful and not quantified, it is immediately understood that it is consequently the increase in volume, which depends on the cube of its approximate radius, while the surface, i.e. the possibility of feeding and exchanging what is needed from the environment, increases according to the square of the radius, so that evolution can only be a multiplication or death. It is, however, a deep understanding first of Medicine and then of the biological Sciences, in their integration with Physics with a strong humanistic characterization. If this is the effect of mathematization, it can only surprise those who do not know Mathematics and the physical or chemical sciences, and those who know little about the life sciences. It is correct to say that in a system the whole is more than the sum of the parts, that the properties of the whole are not directly reproducible in the properties of the parts. However, the negative form of the sentence authorizes interpretations as reductive as uncultivated: a system can and must be studied with rigour exactly as in the analytical approach of the physical sciences of the modern age and the last centuries.

7 Some further Considerations about the Method

In this context, a comparison between the methodology of Physics as a science of nomothetic nature and the methodology of the medical clinic is appropriate. The result will reconfirm the character of Physics as Humanity, and in particular as Medical Humanity.

Physics constitutes a hypothetical-deductive system, in which hypotheses are not arbitrary but are created by man in compliance with the typical rules of any empirical science, there are the rules of internal or logical coherence and external coherence, that is, of control with *future experience.* Logically, the fundamental inference of internal coherence is the *modus* ponens:

$$(A \wedge (A \rightarrow B)) \rightarrow B$$

This formula is true only if the two premises are true, the underlying hypothesis and the first implication: this is as if to say that we can move with logical rules within a system of thought, if, in so far as and until it isn't falsified. And this brings us to the other logical basis, the *modus tollens* well treated by Popper but previously by Charles S. Peirce (Hartshorne et al. 1931–1958), the logician among the founders of Pragmatism:

$$((A \rightarrow B) \wedge -B) \rightarrow -A$$

This formula, on the other hand, is a tautology. From that the realistic sense of science as knowledge in motion, always hypothetical, fallible, devoid of certainties and definitiveness, and the idea of research as an evolutionary process without τέλος, no goal and no end.

Certain visions of Physics as an inductive empirical procedure, most recently those proposed by the logical Neopositivists of the Wiener Kreis, no longer have any epistemological and philosophical consistency, after the Pragmatists (Hartshorne et al. 1931–1958; James 1890; Dewey 1938; Bartley III) and after Popper. the critical Rationalists and other epistemologists of the '900 (Popper 1934–1935, Chaps. 3, 4; Kuhn [1962] 1970; Bartley III). It is essential in the formulation of physical laws, as well as the laws of any other nomothetic science, the universal quantifier within a given domain.

$$\forall x \in D, f(x) \rightarrow g(x)$$

For example, for each temperature difference between bodies in contact the heat flows from the body at the higher temperature to the one at the lower temperature; or in Ohm's first law for each circuit with certain characteristics $V = R\,I$ within certain limits, that is, in each circuit within a defined domain. We have seen, sometimes the reading is not so immediate: for example, *omne vivum e vivo* could be formulated as above, in the sense that if a subject is living, that is, for each subject belonging to the domain of the living beings, then its to be alive implies that it originated from other living beings. If you prefer, it is so for all living beings.

The way of proceeding of the medical clinic follows a reciprocal direction, from the particular to the general. Doctors are perfectly aware of treating the sick and not diseases, so they must start from the individual case considered with all its particularities and individual variability, which among other things is consistent in Pedagogy and Social Sciences (Blezza 2018, pp. 110–111, 2020, pp. 108–109) with the concept of "person". However, there would be neither medical science nor corresponding profession without the apparatus of general cases, that is, of diseases: and the doctor has the task of mediating from the particular case, the symptomatology of the patient, to the general case, the diagnosis of a disease, making use of his experience and professionalism (Fig. 1):

> **Data** disease A presents (or may present) phenomenology B
>
> **Medical science** The data relating to the B set of evidence of a particular case repeat aspects that can be considered and professionally tested to include references to the general case A. no other reasonable hypothesis also explains these evidence
>
> ---
>
> **Clinical diagnosis** the case we are studying, evidence B, is a special case of A

Fig. 1 Diagnosis of a disease *Source* Author

Pay attention: there are no "pure data" in which the Positivist and the logical Neopositivist too is believed, the data are impregnated with professional reflection, someone prefers to call the "facts", and with valid reasons. We must add what to a scientist is evident, that these data or facts are not made arbitrarily. The so-called "clinical eye", for example, is not a simple optical detector, but a sophisticated professional practice tool. It is old 'απαγωγή known by Aristotle (384–322 BC), used as a figure of speech in particular in the field of philosophy, logic and legal or juridical, approachable to *reductio ad absurdum* of Zeno from Elea (489–430 BC). This is used to justify the falsity of a statement underlining the absurdity of the consequences of its application.

The whole theoretical framework of practice for the profession of the pedagogist (Blezza 2018, pp. 47–56, 2020, pp. 78–82) refers to the classic Pragmatism or a Neopragmatism current perspective. Coherently, the concept of *abduction* is taken as it was formulated by Charles S. Peirce (1839–1914) (Hartshorne et al. 1931–1958; 7.218), as inference and not as an argument or a demonstration; it represents an alternative to induction and deduction, which provides an opportunity and a chance that we would call "corroboration" in accordance with the concept of the Critical Rationalism by Popper, with some similarity to the concept of "educated guess", but without any effect on the probability as shown by the well-known theorem of Pavel Tichý (1936–1994) (Tichý 1974, 1976; Svoboda et al. 2004) and consequences (Bartley III, I). Like the conceptual tools of *future experience* and *fallibilism*, this concept was theorised by Pragmatism (Hartshorne et al. 1931–1958; James 1890; Dewey 1916, 1938). This procedure, which is called "*abduction*" or "*retro-duction*" and is not a tautology, and therefore reproduces as an anthropological principle the fallibility of medical science, in analogy with the fallibility of any other science.

A firm mastery by the doctor in his initial and continuous training on the hypothetical-deductive method, also to be compared with the clinical method, has obvious human and professional implications. It would not only be about the Physics:

Even a theory of abstract mathematics such as Euclidean Geometry would be configured as a full-fledged Humanity, in the sense seen and in others; the difference is that scientific laws and theories are hypothetical and have continuous correlation by "future experience", which is not required of the axiomatic system that in Euclidean geometry or other branches of the mathematical sciences is assumed at the base. Which, among other things, allows us to fully understand the lack of meaning of the statement that mathematics is not an opinion.

8 Conclusion: Synthetic Reflection

Someone has considered Einstein's equation $E = m\ c^2$ even as the most beautiful poem of the twentieth century.

We are dealing here with aesthetic considerations, but rather with the character of human creation of a scientific theory, analogous to poetry, literature, the figurative arts, music; and on the other hand, also similarly to technique / technology, to the logical-formal sciences, to the human and social sciences. Undoubtedly, the foundations of reference and the choices of method change, and we have seen how the sciences properly called the Physical sciences and as the medical and life sciences are characterized, but these are articulations of the practices of a single human faculty. i. e. creativity.

Having said that, after discussing what Humanity is or how it integrates with medical sciences or other sciences, we should ask ourselves why a series of prejudices and distorted visions have arisen that have led us to the need to discuss these topics. Physics is a Humanity in its own right, and in particular, it is a Medical Humanity. In the training of the doctor, as in the training of various professionals, it can play a much more important role, and substantially different from that which it has played for a long time and still plays in recent years.

References

Antiseri D ([1982] 2001) Teoria unificata del metodo. Liviana, Padova, UTET, Torino.

Barbensi G (1967) (ed) Opere scelte di Luigi Galvani. UTET, Torino.

Bartley III, WW (1982–1983) (ed) Postscriptum to the Logic of Scientific Discovery by Karl R. Popper. 3 Vols. Hutchinson, London.

Bertalanffy von L (1932) Theoretische Biologie, 2 Vols. Gebruder Borntraeger, Berlin.

Bertalanffy von L (1968) General system theory–Foundations, development, applications. Penguin, London.

Blezza F (1983) Galvani e Volta—La polemica sull'elettricità. La Scuola, Brescia.

Blezza F (2018) La pedagogia professionale—Che cos'è, quali strumenti impiega e come si esercita. Libreria Universitaria, Limena PD.

Blezza F (2020) Il pedagogista. ETS, Pisa.

Breuer J, Freud S (1895) Studien über Hysterie. Franz Deuticke. Leipzig und Wien.

Dewey J (1916) Democracy and Education: An Introduction to the Philosophy of Education. The Macmillan Company, New York.
Dewey J (1938) Logic: The Theory of Inquiry. Henry Holt and Company, New York.
Einstein A (1916) Über die spezielle und allgemeine Relativitätstheorie (gemeinverständlich). Vieweg & Sohn a Braunschweig.
Geymonat L (1970–1972) (ed) Storia del pensiero filosofico e scientifico. 6 Vols. Garzanti, Milano.
Hartshorne C, Weiss P, Burks AW (1931–1958) (eds) Collected Papers of Charles Sanders Peirce. 8 Vols. Harvard University Press, Cambridge–Massachusetts.
Hutten E H (1967) The Ideas of Physics. Oliver & Boyd, Edinburgh & London.
James W (1890) Principles of Psychology. 2 Vols. Henry Holt and Company, New York.
Kuhn T S ([1962] 1970) The Structure of Scientific Revolutions. 2nd ed. University of Chicago Press, Chicago [with Postscript 1969].
Lindberg D C, Numbers R L (2003–2005) (eds) The Cambridge History of Science. 8 Vols. Cambridge University Press, Cambridge.
Popper K R (1934–1935) Logik der Forschung. Zur erkenntnistheorie der modernen naturwissenschaft, Springer–Verlag, Wien [English edition, translation prepared by the author, with the assistance of Julius Freed and Lan Freed The Logic of Scientific Discovery, Hutchinson & Co., London & New York 1959].
Rossi P (1988–1989) (ed) Storia della scienza. 3 Vols. 5 Tomes. UTET, Torino.
Svoboda V, Jespersen B, Cheyne C (2004) (eds) Pavel Tichý's collected papers in logic and philosophy. Prague and Otago University Press, Dunedin.
Tichý P (1974) On Popper's definition of verisimilitude. British journal for the philosophy of science 25:155–160.
Tichý P (1976) Verisimilitude redefined. British journal for the philosophy of science 27:25–42.
Vecli A (1975) Biofisica. Enciclopedia della Fisica. 2 Vols. ISEDI, Milano, p. 16.1.

Chapter 20
The Creator only may Destroy *Vis Viva*: Natural Philosophy in Early William Thomson

Mariano Bonaccorsi and Giuseppe Ruta

Abstract In this contribution we deal with an aspect that at first could not seem directly related with the discussions by Lord Kelvin and his colleagues on the fundamentals of the theory of heat. Indeed, we focus on William Thomson's profound spiritualism, which led him to found all of his natural philosophy on the immutability of some quantity, as established by the Creator himself. After his stay in Paris in 1845, where he worked at the laboratory of Henri-Victor Regnault, Lord Kelvin strongly supported the French scholars (e.g., Carnot and Clapeyron) who investigated thermal phenomena by introducing the caloric fluid. According to them, just like water produces mechanical power when falling through a difference in height, so does caloric when falling through a difference of temperatures. That is, caloric contains *vis viva*, a known, established concept in the mechanics of solid and fluid bodies. This entity must be conserved, since it is not possible to imagine a human process that can produce perpetual motion. Thus, a fundamental postulate of William Thomson's investigations on natural philosophy is that the Creator is the only supreme being that can create or annihilate *vis viva*. This originated a deep debate for at least two decades, in which Lord Kelvin investigated deeply on the possibility that heat can transform itself into *vis viva* and vice versa. We will analyse in some detail this conception and these investigations, with reference to their historical setting, providing some background of the English and European debate on the subject of the time.

Keywords Caloric · Spiritualism · Theory of heat · *Vis viva*

M. Bonaccorsi
University of Roma "La Sapienza", Roma, Italy

G. Ruta (✉)
Department of Structural and Geotechnical Engineering, University of Roma "La Sapienza", via Eudossiana 18, Roma, Italy
e-mail: giuseppe.ruta@uniroma1.it

R. Pisano (ed.), *A History of Physics: Phenomena, Ideas and Mechanisms*,
History of Mechanism and Machine Science 42,
https://doi.org/10.1007/978-3-031-26174-9_20

1 Introduction: Thermodynamics, a Theory of Everything

In this paper we describe how one of the fathers of classical thermodynamics was moved in his investigations by his humanity, between friendships, questions, doubts and his faith in the Creator, which led him to change his initial hypotheses on the essence of heat over time.

In history of science it is well established how asking questions is as important as looking to find answers. Questions about natural phenomena, the principles that rule the world, and the ways they can be investigated are strictly related to the cultural environment. Indeed, in history the men of science have been moved not only by empirics and rigour, but also by spiritual beliefs (often theology), guiding them to the ontological investigation of humankind and the analytic investigation of nature.

The early studies on thermal effects, so close to the ancestral connection of mankind with fire, are probably some of those offering one of the best outlooks of such close connections. Relying on a mechanistic approach to physical phenomena, dating back to Galileo, Euler, and Newton, thermal effects were first supposed to be due to an immaterial fluid called *caloric*. But the questions immediately arose: what is caloric, actually? Does it really exist? What is the true nature of heat? Can it be fully turned into mechanical effects? If not, is the remainder really lost? The approach by William Thomson (1824–1907) highlights how his cultural background led him to formulate his own statements on the principles of thermodynamics, showing that it is not always possible to draw a clear boundary between the figures of a scientist and that of a man of faith.

Thermodynamics is a classic topic in the informative literature and didactic textbooks; it always manages to arouse a particular interest in the reader, perhaps dictated by the direct daily experience that people have in life with heat. Extensive essays have been produced that closely retrace the main historical stages of the theory of heat, e.g. those by Truesdell (1980), Truesdell and Bharatha (1989). These works present the fundamental theorems as well as their fields of application and introduce to all the evolutions achieved in the innumerable related fields of investigation. Indeed, it is not a gamble to say that life, as we know it now, is possible only thanks to the intimate nature of heat. This intuition was certainly immediately clear to primitive men, whose discovery and domination of fire allowed the survival of the evolving human species. Any kind of animated and still life is supported by thermodynamics that, regardless of human understanding, continues to regulate physical phenomena and dictate their direction from the beginning of time. In this regard, in some notes Einstein stated that thermodynamics

> [...] is the only physical theory of universal content concerning which I am convinced that, within the framework of the applicability of its basic concepts, it will never be overthrown. (Einstein 1949, p. 12)

That is, thermodynamics is like a theory of everything that silently rules and governs every corner of the universe.

The historical narration about those who paved the way to a deeper understanding of thermodynamics sometimes neglects some points that look crucial to us. Indeed, we remark some overlooking of the vast amount of philosophical ideas, common thoughts and, most of all, set of cultural and even religious values that supported the fathers of thermodynamics in coming to enunciate their principles. These principles, well known today, are taken as factual data, tacitly accepted because they passed the scrutiny of severe experimental verification over times. The scope of the present historiographical dissertation, based on the works by Smith (1976, 1977), Smith and Norton Wise (1989), and by Thompson (1910), is twofold: on the one hand, it closely follows the evolution of the scientific thought of William Thomson (better known as Lord Kelvin) in his investigations into the question of heat and the resulting mechanical effects; on the other hand, it stresses how this evolution was indissolubly tied with his philosophy and religious belief. This evolution is framed within a historical context of great dynamism among scientists of the time.

Indeed, if it is true, as mentioned, that thermodynamics embraces every aspect of the cosmos, then it is not possible to try to fully understand the results by William Thomson without involving all the facets of his person, which have defined and accompanied him in this process. In particular, we will point out his spiritualism and we will show how this dimension was one of his compasses along the way.

2 Science in the Victorian Age

It was not a coincidence that the most flowery process of defining thermodynamics took place in the late Victorian age that, from the second half of the nineteenth century, saw the United Kingdom and the rest of Western Europe to be the cradle of the second industrial revolution, through strong technological innovation and the use of new raw materials. The research fields were open on several fronts and all connected to each other, often carried out by the same minds, which delighted in ranging from one problem to another in physics. The British physicist Sir Oliver Joseph Lodge (1851–1940) in his work *My philosophy* (1933) presents an interesting perspective of what was the scientific landscape of the time, in which the possibility to unify knowledge under the same laws was constantly pursued (Wilson 1971). Moreover, in those years the originality of physical research found a powerful thrust in the seeming compatibility between science and (Christian) religion. This was nourished by the spirituality of the same men of science, who saw the two areas as the sides of the same coin: it was possible to know the truth of reality only by resorting to both.

William Thomson himself considered unthinkable to be able to answer the scientific questions of the world exclusively through the scientific method: it is also necessary to rely on the philosophical and theological method in order to understand and define with logic what is still hidden and not directly visible or measurable, yet knowable by the human mind. Furthermore, this feeling was not something relegated to remain in a private dimension: rather, Lord Kelvin's faith was always formalised and presented publicly as a starting point in his approaching the things of the material world. As a chairperson of the Christian Evidence Society, an apologetic organisation founded in 1870 to counter atheism, William Thomson stated on May 23rd, 1889:

> My primary reason for accepting the invitation to preside was that I wished to show sympathy with this great Society which has been established for the purpose of defending Christianity as a Divine Revelation. I also thought something was due from Science. I have long felt that there was a general impression in the non-scientific world [that] believes Science has discovered ways of explaining all the facts of nature without adopting any definite belief in a Creator. I have never doubted that impression was utterly groundless. It seems to me that when a scientific man says – as it has been said from time to time – that there is no God, he does not express his own ideas clearly. He is, perhaps, struggling with difficulties; but when he says that he does not believe in a creative power I am convinced he does not faithfully express what is in his mind. He is out of this depth. [...] Peter speaks of scoffers who said that ‘all things continue as they were from the beginning,’ but the Apostle affirms himself that ‘all these things shall be dissolved.’ It seems to me that even physical science absolutely demonstrates the scientific truth of these words. (On the Miracle of Life by Lord Kelvin, retrieved via: https://zapatopi.net/kelvin/papers/on_the_miracle_of_life.html)

These feelings were not uncommon among scientists at the time and even before: we have the remarkable examples, just to quote a couple of well-known scholars, of: the French mathematician Augustin-Louis Cauchy (1789–1857), an openly rigorous Catholic, member of the Society of Saint Vincent de Paul, and having links to the Society of Jesus; the Italian mathematician Gabrio Piola (1794–1850), deeply influenced by spiritualism, who taught religion in a parish in Milan throughout his life and put his ideas on what he felt as close connections between science and Catholicism in his work (Piola 1825). As a further example, in 1875 the British geophysicist Balfour Stewart (1828–1887) and the mathematician Peter Guthrie Tait (1831–1901, co-author with Lord Kelvin of the *Treatise on Natural Philosophy* in 1867), manifested together in their work (Stewart and Tait 1875) the need to demonstrate that the alleged incompatibility between science and religion in reality does not exist, despite the accusations of being gullible orthodox. Indeed, for the authors it is not a question of subordinating material evidence to the dogmas of faith, nor of limiting Christianity and the immortality of the soul to well-established scientific conclusions, but to show that it is precisely by virtue of the visible world that one is brought to accept the existence of an invisible world. In history, Tait and Stewart’s attempt is not clearly the first: the philosopher and theologian St. Thomas Aquinas’ (1225–1274) *Quinque viæ* (Five Ways) in his *Summa Theologica* may represent another well-known reasoning. Stewart and Tait argued that mankind is led

> [...] to believe that there exists now an invisible order of things intimately connected with the present, and capable of acting energetically upon it-for, in truth, the energy of the present system is to be looked upon as originally derived from the invisible universe. (Stewart and Tait 1875, p. 158)

However, this idea was not common to many of their contemporaries: Stewart and Tait accused the drift of science of their time to separate the two dimensions, science and religion, thus doing harm not only to spirituality but to science itself.

The same outlook on science and faith was shared by William Thomson and his older brother James Thomson (1822–1892), plus James Prescott Joule (1818–1889), James Clerk Maxwell (1831–1879) and others. They all collected the legacy of their direct ancestors, such as Newton, Faraday and Boyle, the studies and scientific thoughts of whom, albeit with different expressions, were the fruits of a profound desire to contemplate and read the complex material world as governed by the only one divine intelligence, that of the Creator.

It was not everyone's prerogative to approach the study of thermodynamics with the great spiritual questions: one instance is the German physicist Rudolph Clausius (1822–1888). Although we have not many sources capable of documenting Clausius' gradual studies on heat and energy transformations, it seems that he never expressed the need to link his claims to those of a religious nature. Despite being son of a Protestant pastor and having middle name Emanuel, which means "God with us", he came to postulate his thermodynamic principles as mere acknowledgments of the world. To him, all of his hypotheses on heat were reasonable only because they were confirmed by natural processes, thus limiting himself simply to the direct experience of reality. The question then arises, why did men of science on English soil show themselves to be so tied to the science-faith question as to merge them together? In order to try to answer this question, we need to take a step back.

In November 1660, King Charles II established The Royal Society of London for Improving Natural Knowledge, thus institutionalising science and starting a slow process of scientific development in all fields in England. In the seventeenth century, science was the expression of Protestant ethics, which strongly influenced scientists in their investigations. Man's work was placed as the privileged path for his salvation by God and the studies in natural philosophy made by men of science were the proper way to access the state of divine grace. Furthermore, the purpose of natural philosophy was that of expanding man's control over nature and of glorifying God. Thus, science became soon seen at the service of both society and divinity.

According to the American sociologist Robert Merton (1910–2003), in those years

> [...] Science was to be fostered and nurtured as leading to the improvement of man's lot on earth by facilitating technologic invention. (Merton 1938, p. 446)

Furthermore, many of the early members of the Royal Society were Puritans, whose ethics was even more rigid than the Protestants', and saw a crucial purpose in science: the use of reason was based on the conception that rationality could be an effective resource for restraining and distracting from the impulses of passions, thus favouring all those activities that particularly required the use of rigorous reasoning. Merton also stated that

> [...] the puritan ethic, as an ideal-typical expression of the value-attitudes basic to ascetic Protestantism generally, so canalized the interests of seventeenth century Englishmen as to constitute one important element in the enhanced cultivation of science. The deep-rooted religious interests of the day demanded in their forceful implications the systematic, rational and empirical study of nature for the glorification of God in His works and for the control of the corrupt world [...] if nature is the manifestation of His power, then nothing in nature is too mean for scientific study. (Merton 1936, p. 1)

According to Merton's thesis, Puritan ethics promoted the juxtaposition of rationalism and empiricism, leading to the investigation of nature mainly by an experimental approach. Indeed, experiments were seen as the best scientific expression of the practical Puritan attitude of working. On the other hand, as we read in Shapin (1989) (and in the famous preface of Newton's *Principia*), in seventeenth-century England a rational scientific thinking was quite distinct from the practical activity in a laboratory. Natural philosophers left laboratory activity to technicians, since practice was considered less noble and misleading than reasoning and validating results. These last, in any case, were associated to the name of the natural philosopher only, by virtue of his authority in proposing and carrying on the experimental campaign. Thus, knowledge was based not only on the genius of great scientists, but also on the invisible work of many men who made the laboratory research part of their daily life. Still, for (Christian) English scientists, science was an investigation on the perfection of divine creation: the Anglo-Irish natural philosopher, chemist and physicist Robert Boyle (1627–1691), member of the first council of the Royal Society, as reported in Merton (1938), claimed that

> [...] the knowledge of the Works of God proportions our Admiration of them, they participating and disclosing so much of the inexhausted Perfections of their Author, that the further we contemplate them, the more Footsteps and Impressions we discover of the Perfections of their Creator; and our utmost Science can but give us a juster veneration of his Omniscience. (Merton 1938, p. 462)

In the thirties of the nineteenth century, the Royal Society went through a period of crisis, losing part of its prestige. This gave way to the nascent British Association, soon to become the main stage of the men of science in England. Here, step by step, the new scientific materialism, marked by Darwin's evolutionism, began to take the place of the traditional view of the British Christianity on knowledge, which for centuries had left an important imprint on science until then.

3 The Encounter Between William and Heat

In a passage of the letter that James Thomson wrote to his brother William on August 13th, 1863, the former recalled an experience of the past common to both:

> Even you and I at Walsall (in 1842 I think) when watching the consumption of power in the flow of water into a canal lock were speculating on what became of the power as we could not suppose the water worn and therefore altered like as solids might be supposed to be when power is consumed in their friction. (Smith 1977, p. 233)

Born in Belfast on 26 June 1824 to James Sr. and Margaret Gardner, second of seven children, William was just eighteen years old in 1842, the year referred to in the previous quote. At that time, he had enrolled at the University of Cambridge and the previous year he already published his first work, entitled *On Fourier's expansions of functions in trigonometric series*, in the Cambridge Mathematical Journal. He used the pseudonym P.Q.R., and it was apparent that he had read and accepted the ideas of the French physicist Jean Baptiste Joseph Fourier (1768–1830) with emphasis. The latter, in his book *La Théorie Analytique de la Chaleur* (1822), studied the possibility of describing and approximating a generic mathematical function with a series of sines and cosines, and proved the particular convenience of this new tool for solving the differential equations that describe the problem of heat propagation.

It was precisely in writing this first paper, where he defended Fourier's ideas from the attacks of British mathematicians, that William began to consolidate all of his interest in the question of the heat flow between bodies and in the possibility of harnessing it to obtain certain amounts of mechanical effects from suitable machines to support human labour. In order to grasp the subsequent developments of William's thought and approach to thermodynamics, and how his spiritual dimension is intimately intertwined in his investigations, it is necessary to take a look at the context of his intellectual formation and growth. Indeed, as it will be apparent, these gave him the tools that were central to his vision and study of the world and the universe.

4 William's Father Role in His Life

A decisive influence on Lord Kelvin's vision of science came from the family sphere, especially from the father, during the years of his childhood and early adolescence. James Thomson Sr. was born and raised in a family of Protestant Presbyterian tradition of the province of Ulster, which in the nineteenth century became a ground of close encounter between the newly arrived evangelicalism, a movement that grew up in that century in eastern Scotland and the British colonies of North America, led by the figure of Reverend Henry Cooke in Ireland, and the Orthodox Protestants of the traditional Presbyterian Church of Ireland (Holmes 2011).

Born in 1786, when still a child, James Thomson Sr. was deeply affected by the experience of the Irish rebellion of 1798, which also broke out in his hometown of County Down. In that uprising against the political and ideal domination of the English in the Irish lands, he saw a brutal clash in a complex political and religious framework between the local Presbyterians, the British Anglican Church, and the Roman Catholics supported by the French. The latter had the ambitions to take advantage of the moment to weaken the English kingdom, in a succession of close war events at the turn of the seventeenth and eighteenth centuries. In 1798 France was ruled by the Executive Directory, a committee of five members originated at the end of the French Revolution, a member of which was Sadi Carnot's father Lazare (1753–1823). There is a curious intertwining of stories "at a distance" between James Sr. and Lazare, of a sort of passing of the baton between their sons: the first studies of the heat carried out by William Thomson are indeed right on the trail of those by Sadi Carnot (1796–1832).

The Irish rebellion of 1798 was an episode of utter violence that brought James Sr. deeply close to the ideals of freedom and equality, which led him to distance himself from all those forms of political and religious intolerance capable of generating divisions and useless armed clashes.

Since attending Trinity College in Dublin was barred for Presbyterian students and ministers, in 1810, aged 24, James Sr. enrolled in the Scottish University of Glasgow, wishing to become the ministry of his local Presbyterian church. He received a typical Scottish higher education: liberal, practical, and focused on the importance of applied science to industry, an approach he would later bequeath to his sons William and James.

Central to education at the University of Glasgow were the teachings of Moral philosophy and of Natural philosophy; the chair of the latter had been given in 1803 to William Meikleham (1771–1846), who held it until his death. After completing his studies, in 1814 James Sr. became a teacher in the arithmetic school of the newly inaugurated Royal Belfast Academical Institution. This was born in the wake of the 1798 rebellion, willing to strengthen the Presbyterian identity in Northern Ireland, with an education based on Bacon's cognitive method and an inductive scientific approach.

The following year 1815 James Sr. was elected professor in the college and the studies of mathematical science under his teaching were the pivot of a convinced liberal education for all attending students. The offered course was extremely practical and its approach stood in a broader context that characterized the early nineteenth century at Belfast: studying science and natural history was seen as the tool for a more aware and strong Presbyterian class.

Natural history was different from Natural philosophy since the latter introduced the concept of movement, linked to the evolution in time of all the physical systems under investigation. For James Sr., the possibility to approach and systemize knowledge in a scientific apparatus was to be considered a divine gift to mankind, one talent among the others received in God's creation:

> The all-wise Author of our nature has given us minds as varied structure and powers as our bodies and has implanted in them principles of which it is doubtless his intention that we should exercise, just as he has endowed us with passions and propensities, the well-regulated operation of which intended that we should suspend. (Holmes 2011, p. 115)

A peculiarity of James Sr.'s scientific teaching, which will also be a distinctive element in his son William's personal approach to science, was that he never included principles of a theological nature. Indeed, James Sr. believed that the research in mathematical sciences while investigating our world had to be limited to studying and describing the matter and the laws of nature, without going back to explanations on divine causes and how God created the universe. As a matter of fact, had it been possible otherwise, the creative power of the supreme being would have been limited.

During his teaching at the Royal Belfast Academical Institution, James Sr. reunited with Reverend Samuel Edgar (1767–1826), Seceder's professor of Divinity.[1] The friendship between the two dated back to James Sr.'s childhood years in the hometown of Ballynahinch, where the reverend prepared him to join the college. Although Samuel Edgar was a minister of an evangelical congregation, an offshoot of the Presbyterian Church of Scotland, their friendship represented for James Sr. the closest point of contact with evangelicalism of those years. As a matter of fact, he always declared to be an Orthodox Protestant, adhering to the liberal principles and the given importance to science typical of the Ulster Orthodox Presbyterians, opposed to the religious and political movement headed by Reverend Henry Cooke (1788–1868). William Thomson will remain in the same position of his father.

In 1832, James Sr. was given the chair of mathematics at the University of Glasgow, which forced the family to move there the following year. This implied the interruption of the studies begun by William at the Royal Belfast Academical Institute, which were then continued in Glasgow.

5 Natural Philosophy to Investigate the World

Thus, in 1834, at the age of just ten, William was enrolled with his two years older brother James in the courses for brilliant young people organised by the University of Glasgow. Here he came in contact with Natural philosophy, still held in those years by William Meikleham, whose teaching we know through the notes of William and his father James. In a note taken in class by the latter it is possible to appreciate which was the sector of investigation of Natural philosophy according to Meikleham:

> [Natural philosophy is] that knowledge which we acquire of the material world by the exercise of our reason [...] the material world affords subjects of consideration so very different in their nature, and furnishes conclusions so various, and each of them so important, it is usual to separate the study of matter into several parts, and to regard each of these as a distinct department of knowledge. (Smith 1976, p. 294)

Thus, besides Natural history or Chemistry, a new branch of science is identified as a part of Natural philosophy, which finds definition in the continuation of James' notes:

> The more peculiar object of Natural Philosophy [...] is the consideration of matter when subject to the influence of any power capable of producing motion; and this department of general physics is distinguished by the name of Mechanical Philosophy [whose] chief object is the theory of motion. (Smith 1976, p. 294)

The concept of movement of matter and the study of acting causes were already dominant in every field of investigation of nature. From astronomy to magnetism, the theory of motion was the link that conceptually united the different fields of

[1] "Seceders" were those of evangelical view in the Presbyterian Church of Ireland of the time and subscripted to the Westminster Confession of Faith of 1646.

knowledge. This peculiar vision will characterise the approach to the study of the world of young William: in 1839 he used to attend with attention the classes of Meikleham, who emphasised in his lessons [as reported by Smith (1976)] the need to distinguish knowledge in two major branches of nature:

> [...] natural history or the history of natural events, which teaches us to observe and discriminate objects, but which does not imply the concept of motion, and natural philosophy or the discovery of the laws of nature and the investigation of the way in which objects act on one another, which does imply motion. [...] Natural philosophy may be defined as the science of the pressure of motion of bodies, or the science of the forces. (Smith 1976, p. 295)

William took the chair of Natural philosophy at the University of Glasgow in 1846 (Fig. 1), after the death of Meikleham, fallen ill in 1839, during the course attended by William. He took up the legacy of the recently deceased professor by writing in his introductory lecture to the course that 'the fundamental subject of Natural Philosophy is Dynamics, or the science of force'. Thus, William transposed the role of dynamics, central to Meikleham's concept of mechanics, into the heart of natural philosophy. It was during Meikleham's lessons that young William became passionate about the works of the well-known French mathematical physicists Lagrange and Laplace; and it was the substitute for the sick Meikleham, the Scottish astronomer John P. Nichol (1804–1859), who introduced him to the study of Fourier's texts.

Fig. 1 Young William Thomson, professor at Glasgow in 1846 *Source* https://www.gutenberg.org/Public Domain

6 Evidence of the Existence of God

While attending Natural philosophy during the academic year 1839–40 in Glasgow, the Thomson brothers also attended the lectures of Moral philosophy by William Fleming (1794–1866). From the notes by William Thomson, quoted in Smith (1976), we know that a brief consideration on the proofs of the existence of God was initially addressed, which can be classified as a priori and a posteriori. Among the arguments supporting the a priori evidence there are those of Samuel Clarke (1675–1729), who claimed in his *A Demonstration of the Being and Attributes of God* (1705) that 'something must have existed from eternity for something now to exist' and of Descartes, whereby our clear and distinct idea of perfection cannot have any other source than from the perfection of God himself. The physical arguments (that is, those attributable to the study of phenomena in nature), presented by W. Fleming[2] and reported in the original William's notes, of the a posteriori proofs are two: the first, is that 'the universe exhibits proofs of active power', and the second is based in favour of a divine design because 'when we see as well as mere motion means tending towards an end' (Smith 1976, p. 298).

From the same notes the reflection emerges on how it is not possible to know something about first causes, except for the effects that belong to the visible and finite world. Thus, 'God can be known for His activity in the world and his Eternity implies the immutability of the world and His desire and design must remain the same forever with His Goodness, love and will, through which He generated the life of men and guides events in the created universe'.[3] The notes continue by referring to the three main perfections belonging strictly to God: omnipresence, omniscience and omnipotence.

The first is "an attribute peculiarly beyond our reach" and "the continual presence of God is necessary to preserve as to create", in fact 'the same power requisite to create any thing is requisite to preserve it in being'. The second puts the accent on the visible harmony in the creation that cannot but be thought and desired, 'if the beauty and order prove that these works are not the products of chance but of an intelligent designing cause, the prolongation of that order proves'. Finally, on omnipotence, the notes refer to the reflection on the concept of impossible and on the verifiability or otherwise of miracles, concluding that 'impossibilities are merely things which God had determined shall not be' (Smith 1976, pp. 299–300). And faced with the question of whether God and nature can come to coincide, the answer that William will mature will be negative, according to the vision that the laws of nature are posthumous to creation and therefore cannot control the divine will, and it is the immutability of God that preserves them over time.

It appears clear that, in the eyes of young William nature was increasingly governed by God and that in reality 'the distinction between great things and little

---

[2] Extended discussion in William Fleming, *Moral Philosophy*, Part II, Book III: Theistic ethics, or natural theology.

[3] "Now, the first and most general proof which we have of the Goodness of God appears in His having called us into existence." (*Ivi*, p. 350).

do not exist in divine government' (Smith 1976, p. 300). Nothing, therefore, can escape the influence of the divine active providence that pervades the creation. This is why all created things must be supported and there is a difference in saying that man 'makes' when he simply gives a different arrangement. God 'makes' in another sense. The term 'makes', used in William's notes and associated first to man and then to God, clearly assumes a different meaning when we want to refer to the divine capacity to 'create' from nothing, which is not up to man as he can only give different shapes to created things. Therefore, nature is supported by the absolute power of God, *potentia absoluta*, who created the universe, and His ordering power, *potentia ordinata*, which maintains the framework of absolute and immutable laws. This vision of a profound relationship between God and nature was also characteristic of Meikleham, according to whom, as reported by William's notes, one of the reasons for studying natural philosophy was because 'above all it leads us to view the Creator as the Great First Cause, and as the maintaining the energies of nature' (Smith 1976, p. 301). This motivation will be resumed and extended with emphasis in the introduction drawn up by William in 1846, once he became a teacher of Natural philosophy, which reads:

> When we look back upon knowledge of nature from earliest times, and on the progress which has been made by the human mind in the discovery of truth, we feel that the power of investigating the laws established by the Creator for maintaining the harmony and permanence of His works is the noblest privilege which He has granted to our intellectual state. (Thompson 1910, p. 245)

7 Carnot-Clapeyron's Theory of Heat

In 1841, William went to university studies at Cambridge. After completing his degree course in mathematics there, at the beginning of 1845 William chose to settle in Paris for a few months. In doing this, he followed the advice of his father, to gain laboratory experience and thus complement his previous purely theoretical training. While in Paris William took part in a campaign of research experiments involving extremely accurate measurements of temperatures, coefficients of expansion and specific and latent heats of materials at the laboratory of the French chemist Henri-Victor Regnault (1810–1878). In this period, William was reading the work of Benoît-Paul-Émile Clapeyron (1799–1864) entitled *Mémoire sur la puissance motrice de la chaleur* (1834) and came to know about Sadi Carnot's treatise on the motive power of heat, which he could not find until the end of 1848.[4]

At the time, heat was the term used to describe the invisible agent responsible for heating a body. The sensory perception of a body more or less hot than another led in history to define the notion of temperature as a measure of this intrinsic property of each object, with the consequent attempts to measure such a quantity

[4] For a description of the editorial difficulties of Sadi Carnot's work one may refer, e.g., to Gillispie and Pisano, *Lazare and Sadi Carnot* (Gillispie and Pisano 2014) and Pisano, *Essay on Machines in General (1786)* (Pisano et al. 2021).

by an instrument. Galileo Galilei (1564–1642) and his natural heir, the *Accademia del Cimento*, had already taken the first steps in the realization of a thermometer. However, one must wait until 1714, when the Polish-Dutch physicist Daniel Gabriel Fahrenheit (1686–1736) presented an instrument based on liquid mercury in a closed capillary and a relevant temperature scale (the centigrade one was introduced in 1742 by the Swedish astronomer and physicist Anders Celsius (1701–1744)).

The modern conception of heat begins to mature in the nineteenth century, in which William found himself surrounded by prominent personalities, with whom he would contribute significantly to understanding the nature of heat and its exploitation to produce mechanical effects. Such a possibility had been thought of even in ancient times; indeed, the first thermal machine known to us is an invention of Hellenistic culture: the Aeolipile of Heron of Alexandria (Fig. 2), dated the first century AD, could convert heat into work, but disappeared in the following centuries. The development of thermal machines in the eighteenth century will originate the need to know more and more about heat, beginning with describing its physical passage from one body to another.

Fig. 2 Graphic representation of Heron's aeolipile *Source* https://en.wikipedia.org/wiki/Public_domain

In this regard, the French scientist (today appointed as father of chemistry) Antoine-Laurent Lavoisier (1743–1794) in his *Traité Élémentaire de Chimie* (1789) introduced the concept of caloric. This was imagined as a material and invisible fluid substance which, with its lower or higher concentration, was responsible for the temperature of the bodies. Taking two bodies and putting them in contact, this hypothetical fluid would leave the body at a higher temperature, where it was more concentrated, to enter that at a lower concentration, thus causing the temperature variations necessary to reach the final equilibrium condition. This doctrine of considering heat of material nature was added the further notion that the caloric is preserved: it cannot be created or destroyed but can only move from one body to another. However, this was not the only vision on heat at that time: on the wake of the English philosopher Francis Bacon (1561–1626), some well-known scientists from earlier times, among which Robert Boyle (1627–1691) and Robert Hooke (1635–1703), had presented a profoundly different vision of the heat contained in a body. According to this, heat should be attributed to a kind of movement of minute particles, a vision resumed with experimental demonstrations at the very beginning of the nineteenth century by the physicist Benjamin Thompson (Count Rumford, 1753–1814) and by the chemist Humphry Davy (1778–1829). Both produced heat from the friction of two bodies, thus concluding that the amount of heat produced in this way was equal to the amount of work mechanically employed to maintain the movement against the resistance to friction. These results later inspired William, who at first firmly shared the idea of heat as due to the amount of caloric.

A new starting point for William's investigations on heat was Sadi Carnot's work of 1824 entitled *Réflexions sur la puissance motrice du feu.* There, strongly influenced by his father Lazare's investigations on machines[5] and from his recent studies on steam engines, he discussed the possibility of obtaining mechanical work in an engine by the driving force of heat, seen as caloric in the same point of view defined by Lavoisier.

Sadi Carnot had focused his reflections on the action of steam in a heat engine in analogy to water whirlpools. Indeed, the amount of work that a wheel can do when driven by a given quantity of water depends on the prevalence of the flow, i.e., the difference in height between the supply channel at the top and the tail one at the bottom. In a similar way, the work done by a heat engine is due to the flow of a given quantity of caloric fluid from a hot source (high density of caloric) to a cold one (state of low density): it depends on a prevalence-like quantity, i.e., the difference in temperature between the boiler and the condenser (or of the steam discharge). Experience teaches that the steam driving the engine always comes out from the cylinder colder than at its entrance, and that heat-caloric tends to pass from a warmer body to a colder one. Then, assuming that caloric is indestructible, Carnot stated that the production of motive power in a steam engine is not due to an actual consumption of caloric, but only to its move from a hot body to a cold one: actually, steam is only a means of conveying the quantity of caloric involved.

Developing these conclusions further, he came to propose his ideal cycle of a heat engine, now dubbed after him. By this, one can obtain the maximum extractable motive power from two sources at different temperatures only by reversible transformations and a continuous flow of caloric between the two sources. Most of all, this result does not depend on the type of fluid used to convey caloric between the two sources. Sadi Carnot's work was resumed, analysed and formalized mathematically by Clapeyron (1834) who provided a precise representation of Carnot's cycle and

---

[5] For details on Lazare Carnot's influential work on machines one may refer, e.g., to Pisano, *Essay on Machines in General (1786)* (Pisano et al. 2021).

underlined the essential role played by the supposed indestructibility of caloric with much more awareness than Carnot himself.

Independently, in 1838 (aged 19) the Scotsman J. P. Joule in Manchester was trying to improve the electromagnetic motor invented by the English physicist William Sturgeon (1783–1850). This experience made Joule reflect on the production of heat by an electric current: he perceived the possibility of generating heat from the mechanical work produced by the engine, thus contradicting Carnot's supposed indestructibility (and, consequently, non generability) of caloric. Further, he began to think of the possibility of transforming heat into work and vice versa, thus starting the first estimates on the quantification of the mechanical equivalent of heat. Maturing further reflections on the issue, Joule (1844) began criticising the material vision of heat-caloric by Carnot and Clapeyron (Drago and Pisano 2005). Following conclusions were that heat-caloric can actually be consumed, since the steam expanding in the cylinder of an engine loses heat, converting it into mechanical work. Heat-caloric is thus not fully conserved moving from the hot to the cold source, since the portion converted into motive power abandons the steam when it reaches the condenser:

> [...] we may infer that the steam, while expanding in the cylinder, loses heat in quantity exactly proportional to the mechanical force which it communicates by means of the piston; and that on the condensation of the steam, the heat thus converted into power is *not* given back. Supposing no loss of heat by radiation, &c., the theory here advanced demands that the heat given out in the condenser shall be less than that communicated to the boiler from the furnace, in exact proportion to the equivalent of mechanical power expended. (Joule 1845, p. 383)

In June 1847, Joule brought the efforts of his work in front of the British Association in Oxford, describing the perfected apparatus of rotating blades in a fluid driven by the descent of a weight, through which he had defined the measurement of the mechanical equivalent of heat. Young William Thomson was among the attendance and met Joule, discussing these remarkable results and being so impressed that on his way back he wrote an emphatic letter to his father, reporting about this experience. William shared Joule's results with his father and his brother James, stating that he thought Joule was somehow wrong but looked 'to have discovered some facts of extreme importance, such as, for example, that heat is developed by friction within moving fluids' (Smith and Norton Wise 1989, p. 243). This seemed to give a definitive answer to the question involving the Thomson brothers on the banks of a river lock at Walsall in 1842 (see the quotation in Sect. 3). Indeed, basing on a thorough investigation of Carnot-Clapeyron's theory, supported by a dense correspondence with his brother James after returning from his stay in Paris, until his meeting with Joule William was still well founded on the approach by the French authors on caloric. He was so much in favour of it that he could not accept the possibility of transforming heat into mechanical motive power (thus denying the conservation of caloric), since valid demonstrative evidence did not exist yet to support this hypothesis. Thus, William was impressed by Joule's physical reading and an intense correspondence between the two was born, together with their friendship.

8 Joule and the Conservativeness of *Vis Viva*

Joule's founding pivot of his discussion on heat transfer in a thermal machine, i.e., on Carnot's principle, was the concept of *vis viva*, well known in dynamics. Historically, the term was proposed by the German polymath (somewhere even dubbed 'universal genius') Wilhelm Gottfried Leibniz (1646–1716). In the essay *Specimen Dynamicum* (1695) (Leibniz 1989) he defined *vis viva* (living force) a quantity coinciding with the product of mass and the square of the intensity of the velocity of a single body (its sum for more bodies). This quantity, meaningful in dynamics, represented the capacity possessed by moving bodies to cause effects in a collision; its conservation was known to hold in many mechanical systems and its sense is close to the modern one of kinetic energy, equal to one half of the *vis viva*. Leibniz opposed *vis viva* to *vis mortua* (dead force), an expression indicating the force possessed by a body immediately before getting to move, close to the current idea of potential energy, and related to the meaning given at the time to what we call inertia. Thus, the concept of *vis viva* was born in a context of moving bodies, i.e., of what we would now call 'cold', or pure, mechanics. However, some observations (such as that of Count Rumford in 1798 on the production of heat during the process of drilling cannon holes) brought a new awareness in the eighteenth century: the heat generated by some form of mechanical motion was actually another form of *vis viva*. Indeed, since this quantity, pertaining to moving bodies, was assumed to be conserved, the generation of heat by motion should imply that form of heat to be *vis viva*. In addition, indeed, in one work dated 1843 (Joule 1843) Joule attributed the nature of heat to the vibratory movement of the dense atmospheres of electricity and magnetism surrounding all atoms. Thus, heat to him was not a substance but a dynamic state of vibration: it was then natural to think of the mutual convertibility of mechanical work and heat, both being of dynamic nature. Then, it shall be possible to obtain heat by pure mechanical actions (as in friction) and the opposite as well, that is, mechanical work through heat. It was already clear to the Thomson brothers what was the role of *vis viva* within Carnot-Clapeyron's vision of a steam engine: as water falling from a height, so does the caloric acquire *vis viva* through the fall of temperature from the boiler to the condenser. Joule's vision of *vis viva*, however, began to weaken William Thomson's certainties up to that moment on the nature of heat-caloric (that is, a 'material' vision of heat). In fact, the principle of conservativeness of *vis viva* led Joule to state that

> [Mr. Clapeyron] agrees with Mr. Carnot in referring power to *vis viva* developed by caloric contained in the vapour, in its passage from the temperature of the boiler to that of the condenser. I conceive that this theory, however ingenious, is opposed to the recognised principles of philosophy, because it leads to the conclusion that *vis viva* may be destroyed by an improper disposition of the apparatus: thus Mr. Clapeyron draws the inference, that "the temperature of the fire being from 1000° (C.) to 2000° (C.) higher than that of the boiler, there is an enormous loss of *vis viva* in the passage of the heat from the furnace into the boiler". Believing that the power to destroy belongs to the Creator alone, I entirely coincide with Roget and Faraday in the opinion, that any theory which, when carried out, demands the annihilation of [living] force, is necessarily erroneous. (Joule 1845, pp. 382–383)

In 1848, one year after their meeting, William wrote to Joule after having analysed the latter's writings, showing himself well aware of the importance of the objections raised to Carnot-Clapeyron's theory about the destruction of *vis viva*. Indeed, William had thought that the lost quantity of caloric is due to an imperfect disposition of the thermal machine: a part of caloric flows by conduction through the solid walls of the machine from the boiler to the condenser and does not participate in the production of mechanical work. However, such an explanation did not solve the matter, since William also admitted that the quantity of caloric lost in such an irreversible process cannot be later turned into mechanical work again: hence, *vis viva* is actually lost, and Carnot-Clapeyron's theory does not provide any convincing answer on the question.

William supported Joule in the consideration that the power to destroy or annihilate *vis viva* is an exclusive privilege of God. We infer that this adhesion of thought reflects the importance of natural philosophy and theology in William's education as a student at the University of Glasgow. Thus, he began to look for a new way to deepen and improve Carnot-Clapeyron's theory: a different, new thought was necessary to better understand the transfer of caloric in a heat engine. Indeed, at that time caloric remained for William the most accredited physical hypothesis to describe heat transfer; however, due to the first doubts arose after his meeting with Joule, he began not to exclude the possibility to totally remove the idea of caloric in the future.

Thanks to Joule's experiments and papers, the concept of *vis viva* began to be extended from mechanics to all phenomena of nature, with the additional ingredients of convertibility and indestructibility. Thus, those who stuck to the older theory needed to try reconciling Joule's and Clapeyron's models of heat; a brave, yet still immature, attempt was made by William's brother James. He tried to disengage from the analogy of water fall between two heights and compared the caloric drop to the variation between two different speeds, well known in dynamics. Thus, analogously to how momentum is conserved, so James Thomson posed of heat, regardless of its being substantial or not. James' theory of heat foresaw the new idea of *vis calida* (hot force), defined as the product of the heat capacity of the body by the square of the temperature. *Vis calida* did not necessarily have to be conserved under all circumstances, accepting Joule's criticism of Clapeyron and trying to hint at a possible mutual convertibility between thermal and mechanical effect, that is, *vis calida* into *vis viva*.

A further element that kept the Thomson brothers stuck to Carnot's theory was the prediction of the variation of the freezing point of water with outer pressure in a pure theoretical way. It was known that freezing water expands: thus, it was possible to imagine an experiment with water put into a cylinder in contact with a cold source. If the cylinder is closed by a piston, the expansion of water during its freezing would push the piston against the outer pressure, which, however, is unspecified. In particular, pressure could even increase in correspondence to the same expansion of freezing water, thus getting greater mechanical work without increased expense. This is clearly not possible, since it would imply a perpetual motion using only one single heat source; therefore, the freezing point has to drop as pressure increases. James Thomson provided approximated mathematical relations for this, while the first experimental verification came from William only in 1850. The question led him to wonder about how to measure variations in temperatures below the freezing point of water at a given pressure.

9 An Absolute Thermometric Scale

An example of the attempt by the Thomson brothers to reconcile Carnot-Clapeyron's and Joule's theories, which remained at the centre of their correspondences throughout 1848, is in a passage from a letter from James to William:

> Perhaps Joule would say that if a hot pound of water loses a degree of heat to a cold one, the cold one may receive a greater absolute amount of heat than that lost by the hot one; the increase being due to the mechanical effect which might have been produced during the fall of heat from the high temperature to the low one. (Smith and Norton Wise 1989, p. 242)

Strengthened by his laboratory experience in Paris, William focused on temperature measurement in his paper of 1848 (Thomson 1848), dealing with the need of a measure instrument independent of the thermometric substance. A perfect thermometer would indicate always the same variations of temperature in correspondence of the same supplied heat, and its scale would serve as a reference for all the other measuring instruments. However, it was apparent that manufacturing a thermometer under these conditions was impossible: all the thermometric substances in use had proven to have different specific heats for different temperature ranges. That included the air thermometer, which was, however, recognized to be the best: its scale was based on the fact that, keeping pressure constant, equal expansions of gas imply an equal increase in temperature. This linearity between volume and temperature of a gas was measured in the experiments of Joseph-Louis Gay-Lussac (1778–1850) in 1802 and previously discovered by Jacques Charles (1746–1823) and Alessandro Volta (1745–1827).

William wondered if 'there is a reason on which an absolute thermometric scale can be based' (Thomson 1848, p. 102), with the further advantage of assuming only positive values. Indeed, in the common thermometers there were two references on the graduated scale, linked to freezing and boiling of known fluids.[6] This turned out uncomfortable in laboratory, whenever it was necessary to measure temperatures outside that range. He found a positive answer to the question by noting that:

> The relation between motive power and heat, as established by Carnot, is such that quantities of heat, and intervals of temperature, are involved as the sole elements in the expression for the amount of mechanical effect to be obtained through the agency of heat; and since we have, independently, a definite system for the measurement of quantities of heat, we are thus furnished with a measure for intervals according to which absolute differences of temperature may be estimated. (Thomson 1848, p. 102)

[6] Celsius, e.g., designed the scale of the thermometer by labelling 0 the boiling point of water and 100 its freezing point; after his death the scale was reversed in 1745 by Linnaeus.

In other words, Carnot's theory made it possible to establish the value of a degree associated with an absolute thermometric scale by independent measures of mechanical work and heat, since the cycle itself is independent of the type of substance used for transferring caloric from the hot to the cold source. About Clapeyron's experiments on calculating the quantities of mechanical effect due to a unit of heat falling one degree and measured with an air thermometer in various parts of the scale, William stated that:

> In M. Clapeyron's paper various experimental data, confessedly very imperfect, are brought forward, and the amounts of mechanical effect due to a unit of heat descending a degree of the air-thermometer, in various parts of the scale, are calculated from them, according to Carnot's expressions. The results so obtained indicate very decidedly, that what we may with much propriety call the value of a degree (estimated by the mechanical effect to be obtained from the descent of a unit of heat through it) of the air-thermometer depends on the part of the scale in which it is taken, being less for high than for low temperatures. (Thomson 1848, p. 103)

In support of this claim, he stated in a note that:

> This is what we might anticipate, when we reflect that infinite cold must correspond to a finite number of degrees of the air-thermometer below zero; since if we push the strict principle of graduation, stated above, sufficiently far, we should arrive at a point corresponding to the volume of air being reduced to nothing, which would be marked as −273° of the scale (−100/.366, if .366 be the coefficient of expansion); and therefore −273° of the air-thermometer is a point which cannot be reached at any finite temperature, however low. (Thomson 1848, p. 104)

William based on the experimental results of Charles and Gay-Lussac, from which it was known that the trends of gases at different pressures, represented in the plane volume–temperature, are lines converging to the abscissa corresponding to −273.15 °C. This, however, corresponds to zero gas volume, beyond which it is not possible to go: indeed, this condition is attained only removing infinite quantities of heat. In addition, this implies a greater value of a single degree about this limit temperature, small but finite. On the other hand, when increasing temperatures, there is no upper limit to the volume that a gas can occupy, so, he proposed his own thermometric scale in which:

> […] all degrees have the same value; that is, that a unit of heat descending from a body A at the temperature T° of this scale, to a body B at the temperature (T-1)°, would give out the same mechanical effect, whatever be the number T. This may justly be termed an absolute scale, since its characteristic is quite independent of the physical properties of any specific substance. (Thomson 1848, p. 104)

Having clarified the meaning, in an absolute sense, of the value of a degree of temperature, William pointed out that not even the scale of air thermometers could be defined in this way. Indeed, besides being referred to a particular type of gas at a given pressure, they still remained tied to two standard references for the scale. Thus, William paved the way to subsequent developments in the field of thermometry, in which only by resorting to the perfect gas model, with specific heat independent of temperature, it is possible to establish an absolute thermometric scale, always positive and originating from a single standard reference for any thermodynamic system.

10 William's Account of Carnot's Theory

Towards the end of 1848, after writing on the absolute thermometric scale, William managed to get hold of a copy of the original text of Carnot, given to him by his mentor, professor Lewis D. B. Gordon (1815–1876), and not yet found around despite numerous research attempts; once he finished reading it, he immediately went back to writing to Joule, taking up the conflict that he considered most central: the problem of conduction.

He conceived two mental experiment, related with Joule's objections to Carnot and Clapeyron. They are sketched in Fig. 3 and deal with the fall of caloric through a given temperature difference for a gas with initial volume V_0. In one experiment, the gas follows an adiabatic compression *A* first, then an isothermal expansion *B*, thus producing perfectly reversible work. The other mental experiment is a pure thermal conduction *C* between the same temperature fall, keeping constant volume: here, neither *vis viva* nor work results from the transfer of caloric, that is, no mechanical effect is observed.

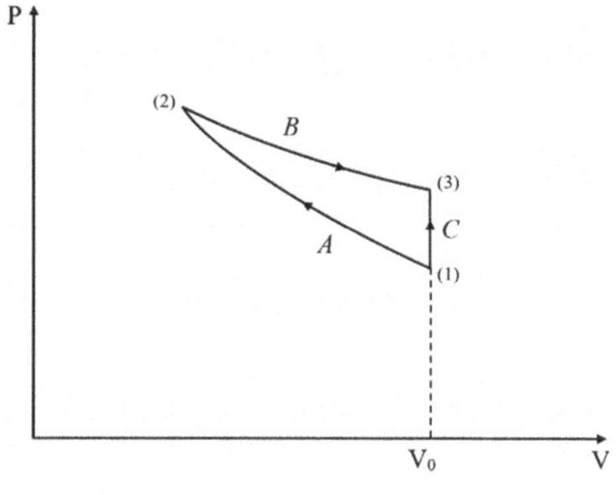

Fig. 3 The representation of William's mental experiments, where two different transformations occur between (1) and (3): the first along an adiabatic compression *A* and an isothermal expansion *B*, the second along the conduction path *C* at constant volume V_0 *Source* Redrawn from Smith and Norton Wise (1989, p. 313)

Thus, the work produced in the first experiment is not produced in the second and could seem somehow lost, although the same amount of caloric is exchanged. Then, William began to realise that the conduction of heat through a solid can produce some invisible dynamic effects, thus anticipating that he will later see conduction through solids as an irreversible transformation. In addition, it comes to light that William began considering the quantity of heat flowing from one body to another as a function of the physical conditions of the bodies. Such a function, characteristic of the state of matter composing the bodies in those physical conditions, should depend only on the initial and final conditions and not on the kind of transformations performed between them: thus, under the same conditions the quantities of heat exchanged are the same.

It is easy to see that at this point William began to move away from the idea of caloric as a substance and transposed the model he already had in mind for electricity to heat: conduction heat flow could match the electric flow in a conductor, where the effect is to change the state of charge of the body in which it flows. In this way, William was convinced to combine Carnot's caloric axiom with the hypothesis that the amount of heat flowing through a body depends exclusively on the change in its physical state. A problem arose: in a cyclic transformation the whole system returns to its initial state and it should not be possible to obtain useful mechanical effects. However, this holds for the system as a whole; caloric flows only from the hot to the cold source, though, and this unique variation remains the sole cause of the *vis viva* acquired by the system, still supporting Carnot-Clapeyron's theory.

One could ask why William focused on seeing heat and electricity as states of matter rather than properties of matter itself. The answer can still be found in the Scottish tradition of natural philosophy, which limited to investigating motions of extended bodies: it was the "applied" science of chemistry (not prone to pure and abstract speculations but related to actual experiments) that was devoted to everything linked to infinitesimal molecular motions. Thus, heat conduction changes the state of matter, but is it possible to recover this unexpressed mechanical effect anyway, since caloric cannot be destroyed? Reconciliation with Joule's mutual convertibility of heat into work was not yet within William's reach, but the laws of heat conduction, which he had learned and studied so much from Fourier, proved to be for him much more than mathematical tools with which to seek the solution of physical problems. William still could not think of heat in a dynamic form: *vis viva* that came in the system as heat and turned into mechanical *vis viva* cannot came out again as a *vis viva* of heat.

These considerations were noted and exposed for the first time the following year 1849 in Thomson (1849). In this work he investigated how, through a more detailed explanation of the reasons provided by Carnot, it can be possible to resolve the following issues:

> What is the precise nature of the thermal action by means of which mechanical effect must be produced, without any other type of effects? How can one estimate the amount of this thermal action required to perform a given amount of work? (Thomson 1849, p. 114)

On this purpose, he tried to get numerical responses using the experimental results on steam provided by Regnault's experimental campaign. The use of the term *mechanical effect* instead of mechanical work was inherited by the same professor

L.B.D. Gordon, who introduced it in the University of Glasgow, translating the German *mechanische Wirkung*. William referred to some discoveries made by Joule, according to which the generation of heat was actually possible, because it developed inside a closed conductor in the vicinity of a magnet, and in the friction of moving fluids, thus going against the common opinion that heat could only emerge from a source where it was present in a sensible or latent form. But, again, in a note of the text, William reports his perplexities about the role of conduction, in an attempt to define a perfect engine:

> When "thermal agency" is thus spent in conducting heat through a solid, what becomes of the mechanical effect which it might produce? Nothing can be lost in the operations of nature – no energy can be destroyed. What effect, then, is produced in place of the mechanical effect which is lost? (Thomson 1849, p. 118)

Here he also began to use the word energy, still far from the modern meaning, but close to the world of natural philosophy. As we read in a note of Smith (1977), as early as 1846 in a letter to his father, William said he was reading *A Course of Lectures on Natural Philosophy and the Mechanical Arts* (1807) by Thomas Young (1773–1829), in which the term energy was used to express the *vis viva*.

In Carnot, William was able to reflect himself, not only in the common questions on heat, but also in the spiritual approach that guided Carnot by his religious sensitivity. Recognising man under the loving eyes of God means recognising life not as a case of nature, therefore all the phenomena of the world cannot be left to chance and require an in-depth study to understand their causes. As reported in the appendix of the work edited by Thurston (1897) among some of unpublished Carnot's statements about his religious sentiment, we read

> Men attribute to chance those events of the causes of which they are ignorant. If they succeed in diving these causes, chance disappears. To say that a thing has happened by chance, is to say that we have not been able to foresee it. I do not myself believe that any other acceptation can be given to this word. What to an ignorant man is chance, cannot be chance to one better instructed. (Thurston 1897, p. 215–216)

In presenting his treatment, starting from Carnot's results, William confirmed that the maximum efficiency of a Carnot thermal machine must be a function of the temperature of the two sources only, supported by the idea of heat as a function of state:

$$\text{efficiency} = \frac{M}{Q} = f(T_1, T_2) = \int_{T_1}^{T_2} \mu dT \tag{1}$$

In (1) M is the mechanical effect, Q the provided quantity of heat, T the temperature, and μ, which he calls Carnot's coefficient, is characterized by greater values for low temperatures and vice versa for low temperatures.

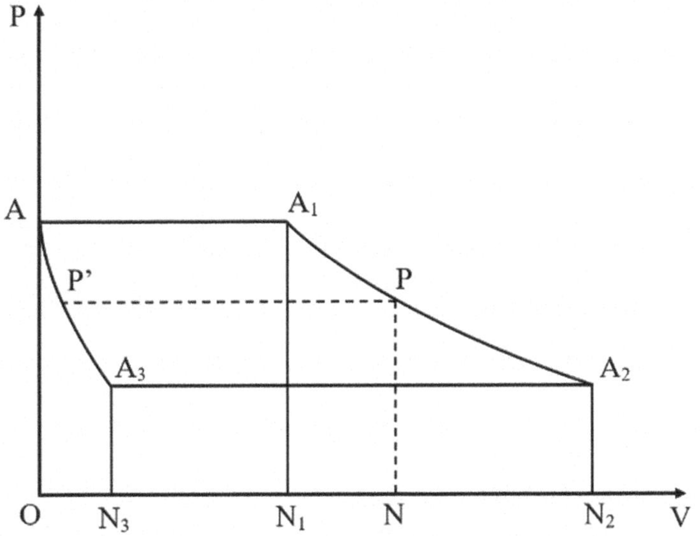

Fig. 4 William's scheme of the 4 phases of Carnot's cycle on the pressure–volume plane; since we are in the field of transition liquid + vapour, isobaric transformations are also isothermal *Source* Redrawn from Thomson (1849, p. 124)

Starting from the graphic representation (Fig. 4) on the plane (P, V) of the two isothermal transformations of the cycle, in the passage of the liquid–vapor phase of water, William quantified the mechanical work obtainable from the cycle as:

$$M = (1 - \sigma) \cdot \frac{dp}{dt} \cdot \frac{H}{k} \cdot (S - T) \tag{2}$$

equal to the area of the cycle. Here S is the hot temperature of saturated steam along A–A_1, T the cold condensation temperature along A_2–A_3, H the total latent heat units inside the cylinder along A_1–A_2, k the latent heat of a unit of saturated steam at temperature t, dp/dt the rate of increase of pressure with temperature at constant volume. Supposing that the quantity H of caloric flown is the same under the same temperature gap between the two adiabatic transformations, the difference of the volumes of steam at the same pressure between the second and fourth operations equals the ratio H/k. Then, if σ is the ratio of steam density to that of water, the volume actually occupied by the steam subtracts the quantity $\sigma \cdot H/k$, equal to the volume of water from which the steam was formed and added as a volume available to be occupied by the steam itself in the cylinder. Under the further hypothesis of extremely small intervals t of the hot and cold sources, the quantities σ and k, functions of t, can be considered constant and taken out of the integral between the lower and upper pressure of the cycle, finally obtaining the quantification of the mechanical effect in Eq. (2).

The coefficient in front of $H \cdot (S - T)$ is precisely Carnot's μ, which, according to Carnot, must be the same for all the saturated vapours of different liquids at the same temperature and for gaseous masses, in which μ has the form

$$\frac{E p_0 v_0}{v \frac{dq}{dv}} \tag{3}$$

valid for atmospheric machines.[7] William, however, claimed to be well aware of the fact that, in reality, this coefficient turned out to be a function of temperature and that it had to be experimentally evaluated at different altitudes of the same thermometric scale, to ultimately estimate the mechanical effect:

$$M = H \cdot \int_S^T \mu dt \tag{4}$$

Several numerical attempts close William's work (Thomson 1849).

William's formulation to evaluate the mechanical work of the cycle, which led him to define μ, was strongly influenced by Carnot's use of adiabatic transformations within his cycle. Carnot's mathematical approach was probably not fully known to William, who limited to analyse the four transformations from a purely mechanical point of view. As reported in the essay of Drago and Pisano (Drago and Pisano 2005), the efficiency evaluated by Carnot, proportional to the thermal difference between the sources, apparently does not account for the adiabatic in the cycle. For Carnot, in fact, adiabatic transformations were neutral transformations from the point of view of the production of work, because they did not exchange heat with the outside world. Indeed, as an inheritance of the principle of virtual work of his father Lazare, where a displacement along the lines of action of forces is necessary to produce mechanical work, in the thermal case a passage of heat through a temperature gap between the sources is necessary. Thus, in Carnot's calculation of the efficiency, according to the synthetic approach of his father Lazare, adiabatic transformations had the same role of dummy infinitesimal transformations and were eventually suppressed, analogously to virtual displacements. This allowed Carnot to neglect the real nature of heat and was the starting point for William.

11 Towards a New Theory of Heat

At the end of the same year, William received a letter from Joule, who wished to bring his attention on considerations about the value of Carnot's coefficient: for the Scottish, μ was nothing more than the inverse of the absolute temperature at which

---

[7] William started from Gay-Lussac's results on ideal gases and proceeded to study the pressure increase dp with an infinitesimal variation dv of the gas, due to an entering quantity of heat dq, provided E is the expansion coefficient of the gas and p_0, v_0 the gas pressure and volume at 0 °C.

heat is received. In the following months the exchange between the two proceeded, having learned of the results of the German physician, chemist and physicist Julius Robert von Mayer (1814–1878). The latter already in 1840 conceived the idea of a link between heat and work observing the different colouring of blood in organisms in hot climates, during a trip to the Dutch Indies. As told by Joule (Cfr. Smith 1977, p. 251), Mayer in 1842 claimed that the heat produced by compressing air was equivalent to the force used, although he did not make experiments to prove it.

Meanwhile, the Scottish engineer and physicist William John Macquorn Rankine (1820–1872) published his investigations on steam engines, started as early as 1842, in his *On the Mechanical Action of Heat, especially in Gases and Vapours* (1850). Rankine was convinced, like Joule, of the mutual convertibility of heat and work and exhibited a peculiar mechanical theory of molecular vortices, according to which there is a form of elasticity due to heat, deriving from the centrifugal force of the revolutions and oscillations of particles in the atmospheres around nuclei. In his essay, Rankine (who got the chair of engineering at Glasgow in 1872, after professor Gordon's death) stressed that it is possible to have a mechanical effect only if the amount of heat emitted to bring the body back to the starting volume is less than that received during the expansion, in contrast with William's view of heat as a function of state of the body. Thus, following the principle of conservativeness of *vis viva*, Rankine tried to derive a state function U depending only on molecular forces. Yet he recognized that U can be easily determined only for perfect gases within a Carnot cycle, where one can evaluate the variations of heat in a body starting from the variations of volume and temperature.

At the same time, the work of the German physicist-mathematician Rudolf Clausius (1822–1888) began to influence the panorama of thermodynamics. Starting from Carnot, Clausius was working on a more virtuous mathematical formulation of the theory of heat, detaching from the idea of caloric and relative to the most probable hypothesis of heat as a state of motion of molecules. William always remained hostile to Clausius' approach, since the latter did not support his thermodynamic assumptions by any solid theological arguments. Inspired by Rankine's mechanical interpretation of heat and by Davy's results, which led to the evaluation of heat as movement between the particles of a body, William definitively gave up the idea of caloric as a substance and function of state, accepting convertibility of heat into work and reconciling with the vision of his friend Joule. Yet he did not accept Rankine's mechanical hypothesis of heat linked with molecular interactions: he still thought the molecular repulsive and attractive forces as responsible for inner work and explained molecular motion in terms of *vis viva*. That is, the amount of heat represents the *vis viva* of molecular motion and the mechanical effect obtained by a thermal action is caused by an external resistance reducing such motion. Conduction and propagation of heat merge in the transmission of *vis viva* among contiguous molecules, and the 'loss' of mechanical effect due to conduction is actually intrinsic in the directionality of some phenomena:

> I believe the tendency in the material world is for motion to become diffused, and that as a whole the reverse of concentration is gradually going on. (Smith 1977, p. 282)

12 Outcome: The dynamic Theory of Heat

On 17th March 1851, William Thomson read at the Royal Society of Edinburgh the first parts of a series of contributions, completed only in 1855 for a total of seven, later collected in Thomson (1851). He exposed numerous considerations on the now established vision of the dynamic nature of heat and the most varied consequences in the field of application, passing through the two fundamental laws of thermodynamics expressed in his personal mathematical-physical language.

The intent is clear: he did not want to present new personal discoveries, but to deliver a new organic work that could collect and present the new dynamic theory of heat, as can be seen from his words of introduction:

> Sir Humphry Davy, by his experiment of melting two pieces of ice by rubbing them together, established the following proposition: – "The phenomena of repulsion are not dependent on a peculiar elastic fluid for their existence, or caloric does not exist". And he concludes that heat consists of a motion excited among the particles of bodies. "To distinguish this motion from others, and to signify the cause of our sensation of heat," and of the expansion or expansive pressure produced in matter by heat, "the name *repulsive* motion has been adopted." [...] The recent discoveries made by Mayer and Joule, of the generation of heat through the friction of fluids in motion, and by the magneto-electric excitation of galvanic currents, would either of them be sufficient to demonstrate the immateriality of heat; and would so afford, if required, a perfect confirmation of Sir Humphry Davy's views. (Thomson 1851, pp. 174–175)

According to the caloric theory, the increase of the aforesaid *forces of repulsion* among the particles of bodies had to be attributed to the increase of this invisible fluid around the molecules. As reported in Pisano (Pisano et al. 2021), a distinction was made between free and combined caloric; the first indicated that part of caloric moving between ponderable particles in form of thermal radiation, regardless of the molecular state of motion. Free caloric was measurable by thermometers and, according to Laplace's definition, temperature coincided with the density of the aforementioned thermal radiation. Combined caloric represented the portion surrounding molecules, responsible for the repulsive forces. The absorption of caloric, therefore, was divided between the two, generating a thermal expansion of the body and an increase in temperature. At a given point, all the supplied heat is absorbed only as combined caloric, leading to an increase in repulsive forces and a consequent change of state. Heat becomes latent (i.e., it stays hidden), since no temperature variation is measured, and the free caloric remains constant.

William attributed to his writing the following purposes:

> (1) To show what modifications of the conclusions arrived at by Carnot, and by others who have followed his peculiar mode of reasoning regarding the motive power of heat, must be made when the hypothesis of the dynamical theory, contrary as it is to Carnot's fundamental hypothesis, is adopted.
>
> (2) To point out the significance in the dynamical theory, of the numerical results deduced from Regnault's observations on steam, and communicated about two years ago to the Society, with an account of Carnot's theory, by the author of the present paper; and to show that by taking these numbers (subject to correction when accurate experimental data regarding the density of saturated steam shall have been afforded), in connection with Joule's

mechanical equivalent of a thermal unit, a complete theory of the motive power of heat, within the temperature limits of the experimental data, is obtained.

(3) To point out some remarkable relations connecting the physical properties of all substances, established by reasoning analogous to that of Carnot, but founded in part on the contrary principle of the dynamical theory. (Thomson 1851, pp. 176–177)

He exposed the whole theory of the motive power of heat basing on the two known propositions of thermodynamics:

Prop. I. (Joule) – When equal quantities of mechanical effect are produced by any means whatever from purely thermal sources, or lost in purely thermal effects, equal quantities of heat are put out of existence or are generated.

Prop. II. (Carnot and Clausius) – If an engine be such that, when it is worked backwards, the physical and mechanical agencies in every part of its motions are all reversed, it produces as much mechanical effect as can be produced by any thermo-dynamic engine, with the same temperatures of source and refrigerator, from a given quantity of heat. (Thomson 1851, p. 178)

Regarding the first proposition, William stated that, according to the dynamic theory of heat, the temperature of a substance can be made to increase only by working on the substance itself. In such a way, one produces an increase in thermal motions inside it, plus those modifications of mutual distances and distribution of particles that can be accompanied by a variation in temperature.

Furthermore, William exposed his statement of the second principle, elaborated at the same time as Clausius, as:

It is impossible, by means of inanimate material agency, to derive mechanical effect from any portion of matter by cooling it below the temperature of the coldest of the surrounding objects. (Thomson 1851, p. 179)

The use of the term *inanimate* is all typical of William's sensibility, wishing to limit his claims to the material world and not relevant to life itself, a divine miracle that cannot be understood by the investigation of science alone:

We are all out of our depth when we approach the subject of life. The scientific man, in looking at a piece of dead matter, thinking over the results of certain combinations which he can impose upon it, is himself a living miracle, proving that there is something beyond that mass of dead matter of which he is thinking. His very thought is in itself a contradiction to the idea that there is nothing in existence but dead matter. Science can do little positively towards the objects of this society. But it can do something, and that something is vital and fundamental. It is to show that what we see in the world of dead matter and of life around us is not a result of the fortuitous concourse of atoms. (Excerpt from an address at the annual meeting of Christian Evidence Society on May 23rd 1889, free access at the link https://zapatopi.net/kelvin/papers/on_the_miracle_of_life.html)

William realised that by the hypothesis of mutual convertibility of heat and mechanical work it is not necessary to put all of Carnot's results aside, as highlighted by Truesdell:

Carnot's expression, relating to the mechanical effect deriving from a certain amount of heat by means of a perfect machine operating in an infinitely small temperature range, truly

> expresses the greatest possible effect that can be obtained in this case; and even if, in reality, it is only an infinitely small part of the entire mechanical equivalent of the heat supplied, the remaining part being irremediably lost to man, that is, 'dissipated', even if not annihilated. (Truesdell 1980, p. 333)

This passage is crucial to understand the extreme clarity reached by William in thinking that there is no absolute loss in the material world: any loss is due irremediably to the hands of man, who, with his creativity, has the only power to arrange machines so to minimise this losses, because 'the *vis viva* of the motion of heat in the whole must remain constant' (Smith and Norton Wise 1989, p. 263).

In the second part of the work, in quantifying the mechanical effect in a Carnot's cycle between two finite temperatures, he adopted the explanation of the concept of mutual reversibility between heat and mechanical power, writing:

$$W = J \cdot (H - R) \tag{5}$$

with H and R the quantities of heat respectively coming from the hot source and leaving towards the cold source; he obtained R again as a function of Carnot's coefficient

$$R = H \cdot e^{-\frac{1}{J}\int_T^S \mu\, dt} \tag{6}$$

introducing the mechanical equivalent of heat J, indicated with the initial of his friend Joule. After obtaining various measurements starting from experimental values of μ, he concluded with the suggestion received by Joule on the real value that this coefficient could assume, that is, equal to the inverse of the absolute temperature. Without developing further calculations, William limited to showing the new quantification of the work obtainable from the cycle:

$$W = J \cdot \left(\frac{S - T}{\frac{1}{E} + S} \right) \tag{7}$$

with E coefficient of expansion of the gas, the inverse of which is added to the temperature S and gives precisely the absolute temperature of the hot source.

In May 1854, William (Fig. 5) published the sixth part of his monography expressing the first law for a closed cycle in the new form, well known today:

$$W + J(H_t + H_{t'} + \cdots + H_{t^{(n-1)}} + H_{t^{(n)}}) = 0 \tag{8}$$

where W indicates the total amount of work consumed in producing the required operations and J indicates the mechanical equivalent of the thermal unit, thus definitively abandoning the Carnot's coefficient.

These ideas were then taken up by him twenty-three years later, in his publication of 1875 *The Kinetic Theory of the Dissipation of Energy*, in which he explicitly stated that it will never be possible for man to restore the heat that was produced by

Fig. 5 William Thomson (1824–1907). In recognition of his scientific merits, in 1892 he was elected at the House of Lords as Baron Kelvin of Largs, after the River Kelvin at the University of Glasgow. He was buried at Westminster Abbey, together with Isaac Newton *Source* https://en.wikipedia.org/wiki/Public_domain

irreversible phenomena, and thus regaining the forces that produced it; all his considerations on irreversibility would become a great reason for reflection for Clausius, who would come to express the concept of dissipation by introducing the term of *entropy* (from the Greek ἐν, 'inside', and τροπή, 'transformation').

13 Conclusions

In this contribution, we have exposed in some detail the complex spiritual and cultural environment that surely influenced the speculations of the future Lord Kelvin on the very nature of heat and its possible applications to human activities.

Indeed, investigating something that cannot yet be seen in detail leads to grasp to the few fixed points of knowledge already achieved, which become the origin for supporting further research process. This was the case of William Thomson who, in his investigations on the nature of heat, had to deal with his conception of the immanent world, with the vision of heat inherited from Carnot, and with his laboratory experience. In the effort to come to a realistic and formal vision of heat, William was encouraged and inspired by all those relationships, in a special way those with Joule and his brother James Jr., who were able to turn on new points of view, leading him to the new dynamical vision of heat in the end. We have presented the historical, social, philosophical and religious background in which Lord Kelvin took his first steps in science, showing how it was fundamental for his approach to the analysis of physical phenomena concerning heat and heat transformations.

What basically remained constant in William's research on the true entity of heat was the awareness of man's limitation in interacting and governing things of the world, since in his spiritual conception only the transcendent creator of matter really had this privilege. Thus, all his steps and changes of perspective were embraced as long as they did not enter into conflict with this own primitive act of faith. This was also the springboard towards another kind of act of faith, namely, that of believing in a new description of heat, far from the idea of caloric, without yet having direct access to its intimate nature, but sensing from its visible effects on the world as ruled by the immutable laws of a divine Creator.

Acknowledgements Giuseppe Ruta acknowledges the support of the institutional grant RM12017294D1B7EF of the university of Rome "La Sapienza" and of the PRIN grant 20177TTP3S_006 of the Italian Ministry of Research.

References

Clapeyron BEP (1834) Mémoire sur la Puissance Motrice de la Chaleur, Journal de l'École Royale Polytechnique. Paris, Imprimerie Royale. Cahier 23/Tome XIV:153–190.
Drago A, Pisano R (2005) La nota matematica nelle Réflexions sur la Puissance motrice du feu di Sadi Carnot: interpretazione del calcolo con il metodo sintetico. Quaderni di Storia della Fisica, Giornale di Fisica (13):37–58.
Einstein A (1949) Notes for an autobiography. The Saturday Review of Literature, 26th November, p. 12.
Gillispie CC, Pisano R (2014) Lazare and Sadi Carnot. 2nd. ed. Springer, Dordrecht.
Holmes AR (2011) Professor James Thomson Sr. and Lord Kelvin: Religion, Science, and Liberal Unionism in Ulster and Scotland. Journal of British Studies 50:100–124.
Joule JP (1843) On the calorific effects of magneto-electricity, and on the mechanical value of heat, The London, Edinburgh, and Dublin Philosophical Magazine and Journal of Science 23(152):263–276.
Joule JP (1844) On the Changes of Temperature produced by the Rarefaction and Condensation of Air. Proceedings of the Royal Society of London 5:517–518.
Joule JP (1845) On the Changes of Temperature produced by the Rarefaction and Condensation of Air, Philosophical Magazine ser. 3/26:369–383.
Leibniz GW (1989) Specimen Dynamicum. In: Loemker LE (ed). Philosophical Papers and Letters. Texts and Studies in the History of Philosophy. Vol 2. Springer, Dordrecht.
Merton R (1936) Puritanism, Pietism, and Science. The Sociological Review 28:1–30.
Merton R (1938) Science, Technology and Society in Seventeenth Century England. Osiris 4:360–632.
Piola G (1825) Lettere scientifiche di Evasio ad Uranio. Soliani, Modena (in Italian).
Pisano R, Coppersmith J, Peake M (2021) Essay on Machines in General (1786). Text, Translations and Commentaries. Lazare Carnot's Mechanics. Springer, Dordrecht.
Shapin S (1989) The Invisible Technician. American Scientist 77/6:554–563.
Smith CW (1976) Natural Philosophy and Thermodynamics: William Thomson and 'The Dynamical Theory of Heat'. The British Journal for the History Science (9):293–319.
Smith CW (1977) William Thomson and the Creation of Thermodynamics: 1840–1855. Archive of Rational Mechanics 16:231–288.
Smith CW, Norton Wise M (1989) Energy and Empire. A biographical study of Lord Kelvin. Cambridge University Press.

Stewart B, Tait P G (1875) The Unseen Universe. Physical Speculations on a Future State. Cambridge University Press.
Thomson W (1848) On an Absolute Thermometric Scale founded on Carnot's Theory of the Motive Power of Heat and calculated from Regnault's Observations. Mathematical and Physical Papers 1:100–106.
Thomson W (1849) An Account of Carnot's Theory of the Motive Power of Heat; with Numerical Results deduced from Regnault's Experiments on Steam. Mathematical and Physical Papers 1:113–155.
Thomson W (1851) On the Dynamical Theory of Heat, with numerical results deduced from Mr. Joule's Equivalent of a Thermal Unit, and M. Regnault's Observations on Steam. Mathematical and Physical Papers 1:174–332.
Thurston RH (1897) Reflections on the Motive Power of Heat by Sady Carnot accompanied by Sir William Thomson (Lord Kelvin). New York, John Wiley & Sons.
Thompson SP (1910) The Life of William Thomson, Baron Kelvin of Largs. vol. 1. Cambridge University Press.
Truesdell CA (1980) The tragicomical history of thermodynamics 1822–1854. Springer, New York.
Truesdell CA, Bharatha S (1989) The concepts and logic of classical thermodynamics as a theory of heat engines. Springer, Berlin.
Wilson DB (1971) The Thought of Late Victorian Physicists: Oliver Lodge's Ethereal Body. Victorian Studies 15:29–48.

Chapter 21
Details in Galileo's Mechanism of *Inertia* as New Intellectual Historical Insights

Paolo Bussotti and Raffaele Pisano

Abstract In the twentieth century, although evident and remarkable works correctly studied Galileo's Inertia, giving fundamental insights, they did not grasp the nuances of Galileo's concept, idea and mechanism of Inertia in its intellectual subtlety. Galileo offered crucial insights on the concept of inertia, and there are many studies on this subject. We think that much investigation remains before the processus of Galileo's scientific thought is complete, and we can fully appreciate both his uncertainties and understandings in the field. We think that a full understanding of some aspects of Galileo's concept of inertia is lacking. We try to fulfil this gap. This paper deals with Galileo's conceptualisation of Inertia as presented in the First Two Days of *Dialogo sopra i due massimi sistemi del mondo tolemaico e copernicano* (1632) and Last Two Days of *Discorsi e dimostrazioni matematiche intorno a due nuove scienze* (1638). Our work consists in three main parts: *scientific*, *historical*, and *historiographical*. (1) After introducing Inertia scientifically, we present it *into* historical research from its embryonic statements respectively as Impetus, laws and incorporated into *Principle of virtual laws*. (2) We historically analyse Galileo's inertial paths, showing how, in specific physical–geometrical circumstances, his ideas confront both circular and rectilinear Inertia. (3) We examine and discuss some of the most influential historiographical contributions in the range of Galileo's Inertia—and—his mechanics. Our paper intends to offer an additional analysis on the subject within the history of physics—and—intellectual history of ideas, including both the relationship between physics–mathematics and the current historiographical debate.

Keywords Circular inertia · Drake · Frova · Galileo · History of physics · Impetus · Intellectual history of ideas · Koyré · Mass · Momentun · Perpetual motion · Physics and mathematics relationship · Principle of virtual laws · Rectilinear inertia · Weight

P. Bussotti
DIUM, University of Udine, Udine, Italy
e-mail: paolo.bussotti@uniud.it

R. Pisano (✉)
HOPAST at IEMN, Lille University, Villeneuve d'Ascq, Lille, France
e-mail: raffaele.pisano@univ-lille.fr

R. Pisano (ed.), *A History of Physics: Phenomena, Ideas and Mechanisms*, History of Mechanism and Machine Science 42,
https://doi.org/10.1007/978-3-031-26174-9_21

1 An Outline

Generally speaking, since ancient times, Inertia was conceived within the following crucial research questions:

Is motion possible in void space? And what kind of motion?

- *If yes, how can we scientifically describe this motion?*
- *If not, how can we scientifically explain the resistance to the motion?*

Can resistance induce a body to rest?

- *If yes, how can we scientifically explain a body at rest?*

In addition, one should consider all the consequences and difficulties of defining, measuring and calculating these quantities connected to Inertia, according to the various disciplinary conventions including physics and mathematics.

Here one might take the position:

a. Inertia is an invention. It does not exist. It should be not be a citizen in a scientific measurable Physics. But, it was used to describe the behaviour of certain physical phenomena. Thus, it became a necessary element in Science. Nevertheless, it is not possible to measure or calculate it; one must calculate something correlated to it. Inertia came to be a *necessary property* of an object in order to justify certain behaviours and physical laws.

Here one might take another position:

b. Inertia does not exist, but acceptance is what is required for a concept to become a component of scientific Knowledge: timely, convenient and reasonable. In other words, it was necessary to construct a functioning (dynamical) physics. Even if it was not commonly accepted (see below).

Taking into account Gillispie's thesis, as well:

> Inertial motion, it will be recalled, had entered science as a physical consequence of Descartes' geometrization of space-matter, combined with the immutability of God. (Gillispie 1960, pp. 367–368)

A question arises: what should one scientifically discuss to answer the questions above?

1.1 Defining Inertia

Nowadays we define Inertia a property *á la* Newton by an a priori definition as *Vis insita* (Figs. 1 and 2). However, sometimes Inertia is also (ambiguously) defined as the resistance of an object in motion:

> *The resistance offered by an object in motion to change (vectorally) its velocity.*

The latter produces confusion between Inertia, as *property*, and the *friction resisting forces*, such as the forces which act against the motion. Thus, in order to justify Newton's First Law, one only considers the propensity (*property*) of an object [Cfr. 34]. But this can produce a misunderstanding, as well. In fact,

- *In order to explain Inertia as presented in Newton's First Law, one should introduce Newton's Second Law.*

  In fact,

- *Inertial forces are any force involved in the validity of Newton's Second Law of motion (rotating, accelerating at a constant rate). In other words, they arise also because the inertial tendency of a body cannot be actualized (e.g., the bucket experiment).*

But

- *The dynamic forces (producing acceleration) as presented by Newton's Second Law are not the same forces presented by Newton's First Law (rest or uniform motion).*

Thus

- *If Inertia is a property of an object in motion or at rest, then which forces are the cause of the Inertia (property) of this object? Which of Newton's Laws should we eventually (or necessarily) consider?*

Finally, in calculus, starting from Newton's First Law and in the absence of friction (ideal case), one should adopt the following features (in order, in any case, to avoid the above misunderstanding):

- A straight-line motion;
- At a constant velocity;
- Within an infinite absolute space—and—time
- The vectorial sum of the external forces ($\neq mg$ as *Force-Weight*) acted upon can be:

  $\sum_{i=1}^{n} F_i^{ext} = 0$ The object (i.e., an isolated physical body) is at rest or moving at a constant speed in a straight line; it will remain at rest or keep moving in a straight line at constant speed unless it is acted upon by a force. In modern terms, this state is represented by an *Inertial Frame of Reference.*

  $\sum_{i=1}^{n} F_i^{ext} \neq 0$ This causes a change of its Inertial state.

This also means that the external forces, which act upon the object, exist and their vectorial sum is zero; and not that the forces do not act because they do not exist. In other words, this phenomenon is kinetically indistinguishable from the absence of forces. This point is very important to discuss both the epistemological role played by *Vis insita* as defined by Newton (Fig. 1) and his rectilinear Inertia applied to his First Law (Fig. 2).

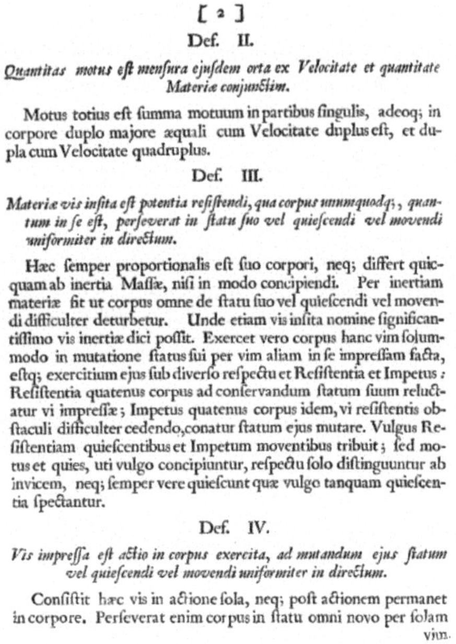

[2]

Def. II.

Quantitas motus eſt menſura ejuſdem orta ex Velocitate et quantitate Materiæ conjunctim.

Motus totius eſt ſumma motuum in partibus ſingulis, adeoq; in corpore duplo majore æquali cum Velocitate duplus eſt, et dupla cum Velocitate quadruplus.

Def. III.

Materiæ vis inſita eſt potentia reſiſtendi, qua corpus unumquodq;, quantum in ſe eſt, perſeverat in ſtatu ſuo vel quieſcendi vel movendi uniformiter in directum.

Hæc ſemper proportionalis eſt ſuo corpori, neq; differt quicquam ab inertia Maſſæ, niſi in modo concipiendi. Per inertiam materiæ fit ut corpus omne de ſtatu ſuo vel quieſcendi vel movendi difficulter deturbetur. Unde etiam vis inſita nomine ſignificantiſſimo vis inertiæ dici poſſit. Exercet vero corpus hanc vim ſolummodo in mutatione ſtatus ſui per vim aliam in ſe impreſſam facta, eſtq; exercitium ejus ſub diverſo reſpectu et Reſiſtentia et Impetus: Reſiſtentia quatenus corpus ad conſervandum ſtatum ſuum reluctatur vi impreſſæ; Impetus quatenus corpus idem, vi reſiſtentis obſtaculi difficulter cedendo, conatur ſtatum ejus mutare. Vulgus Reſiſtentiam quieſcentibus et Impetum moventibus tribuit; ſed motus et quies, uti vulgo concipiuntur, reſpectu ſolo diſtinguuntur ab invicem, neq; ſemper vere quieſcunt quæ vulgo tanquam quieſcentia ſpectantur.

Def. IV.

Vis impreſſa eſt actio in corpus exercita, ad mutandum ejus ſtatum vel quieſcendi vel movendi uniformiter in directum.

Conſiſtit hæc vis in actione ſola, neq; poſt actionem permanet in corpore. Perſeverat enim corpus in ſtatu omni novo per ſolam

vim.

DEFINITION III.

The Vis insita, *or Innate Force of Matter, is a power of resisting by which every body, as much as in it lies, endeavours to persevere in its present state, whether it be of rest or of moving uniformly forward in a right line.*

(Newton [1713] 1729, Def. III, I, p. 2; Author's Italic style and capital letters)

Fig. 1 Newton's *Vis insita* (Newton 1687, Def. III, I, p 2) *Source* French National Library–Gallica. Public Domain

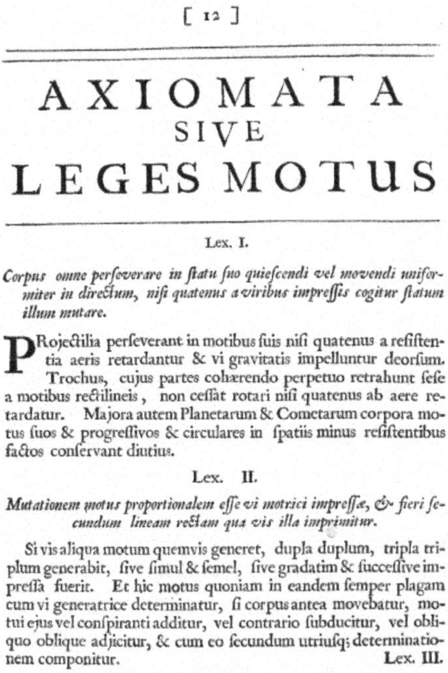

[12]

AXIOMATA
SIVE
LEGES MOTUS

Lex. I.

Corpus omne perſeverare in ſtatu ſuo quieſcendi vel movendi uniformiter in directum, niſi quatenus a viribus impreſſis cogitur ſtatum illum mutare.

PRojectilia perſeverant in motibus ſuis niſi quatenus a reſiſtentia aeris retardantur & vi gravitatis impelluntur deorſum. Trochus, cujus partes cohærendo perpetuo retrahunt ſeſe a motibus rectilineis, non ceſſat rotari niſi quatenus ab aere retardatur. Majora autem Planetarum & Cometarum corpora motus ſuos & progreſſivos & circulares in ſpatiis minus reſiſtentibus factos conſervant diutius.

Lex. II.

Mutationem motus proportionalem eſſe vi motrici impreſſæ, & fieri ſecundum lineam rectam qua vis illa imprimitur.

Si vis aliqua motum quemvis generet, dupla duplum, tripla triplum generabit, ſive ſimul & ſemel, ſive gradatim & ſucceſſive impreſſa fuerit. Et hic motus quoniam in eandem ſemper plagam cum vi generatrice determinatur, ſi corpus antea movebatur, motui ejus vel conſpiranti additur, vel contrario ſubducitur, vel oblíquo oblique adjicitur, & cum eo ſecundum utriuſq; determinationem componitur. Lex. III.

Axioms or Laws of Motion.

Law I.

Every body perseveres in its state of rest, or of uniform motion in a right line, unless it is compelled to change that state by forces impressed thereon.

Law II.

The alteration of motion is ever proportional to the motive force impressed; and is made in the direction of the right line in which that force is impressed.

Law III.

To every Action there is always opposed an equal Reaction: or the mutual actions of two bodies upon each other are always equal, and directed to contrary parts.

(Newton [1713] 1729, I, pp. 19-20; Author's Italic style and capital letters)

Fig. 2 Newton's Axioms or Law of Motion (Newton 1687, I, pp. 12) *Source* French National Library–Gallica. Public Domain

Both the versions of Newton's First Law represent Inertia as a property of an object in motion or—at—rest. A scientific and epistemological problem arises: Newton's First Law of Inertia states that rest and rectilinear uniform motion are *equivalent*. In other words, Newton's statement deals with the two cases as if they were the same thing within the same frame of reference: "[...] in its state of rest, or of uniform motion" (Newton [1713] 1729, I, p. 19; Pisano 2014). But what does *equivalent* mean in this case and in physics? In addition, we are unable, e.g.:

- To find an *Inertial* frame of reference.
- To find a scientific criterion for establishing when forces do not act upon a body (*vectorial sum* = *0*) because in the First Law we do not have an adequate quantitative definition of impressed force.

Nevertheless, in spite of its *unjustifiability*, even today, in classical mechanics we define—&—teach Inertia according to Isaac Newton (1642–1726) in his (nowadays didactically adapted) 1687–*Philosophiæ Naturalis Principia Mathematica*[1] (Newton 1687, [1713] 1729).

Imagining a parallel universe fiction story, we could conceive that Newtonian Inertia conceptualization (Def. III & First Law) might be the maximum level of understanding at which Galileo could have arrived by means of the physical and mathematical sciences of his time? But, in reality he did not.

In this paper, we analyse the texts and discuss Galileo's concept of Inertia. We enrich these topics with new historical/epistemological and historiographical insights. Below we show how Galileo gradually conceptualized his Inertia.

The following (Table 1) is the starting–point of our research:

[1] Since 2014, for Oxford University Press, we have been engaged in a critical translation (into English) of this whole apparatus as expounded in the Newton Geneva Edition (GE). No translation exists of the notes–commentaries of the GE. Besides the translation of the notes, the variants of the notes existing in the 1739–1742 and 1760 editions will be highlighted. Therefore, 5 volumes are expected within 2025. This will clarify the nature, purposes and structure of all the notes/commentaries (apparatus) added to Newton's text by the commentators, so providing a useful guide to understand the relevance of Newton's mathematical methods within his physics (Bussotti and Pisano 2013; Bussotti and Pisano 2014a, 2014b; Pisano and Bussotti 2016; Pisano and Bussotti 2017a, 2017b, 2017c; Pisano and Bussotti 2020; Pisano and Bussotti 2022; Pisano 2014).

Table 1 Our Interpretation of Galileo's Motion—&—Inertia [Cfr. 12–17]

Kind of motion	Inertial	Non inertial
Uniform circular motion of the planets around the Sun and of the satellites around the planets;	Yes	
Uniform circular rotation of a celestial object around its axis;	Yes	
Uniform circular movement of an object (e.g. a ship) which moves on an ocean, a see or a lake, after having received an initial impulse;	Yes	
Circular uniform movement, in proximity to the Earth's surface, for example that of a sling;		Yes
Uniform rectilinear motion;	Yes	
Accelerated rectilinear and circular motions		Yes

Particularly:

- The First and Second Days of the *Dialogo sopra i due massimi sistemi del mondo tolemaico e copernicano* (Galilei 1632, 1638) present much room for discussing the intellelectual mechanism of Galileo's Inertia (Pisano 2024-pre print, Galilei 1623).
- Third and the Fourth Days (*Della forza della percossa*) of *Discorsi e dimostrazioni intorno a due nuove scienze* (1638) (Galilei 1914, 1967; see also Pisano 2024-pre print) present interesting material, as well.

These two masterpieces will be, hence, our main sources, but we will also draw some argumentation from other works.

1.2 *Avvertimento*

At first glance, the historical picture we are tracing of Galileian Inertia might appear odd due to uniform circular motion being Inertial as presented in the first three physical situations above (Table 1). Whereas we argue that in the fourth situation, the one of a circular uniform motion in proximity of the Earth's surface, Galileo did not deal with such a motion as Inertial (Galilei 1632, II; Galilei 1967). This also means that, from a modern perspective, there is a patent contradiction in Galileo's view of Inertia (*Ibidem*). Generally speaking, due to an object's vectorial velocity, vector is constantly changing direction, the moving object is submitted to an acceleration by a centripetal force in the direction of the center of rotation. Thus for circular motion a centripetal force is needed; and also in the case of a uniform circular motion (Cfr. Galilei 1890–1909, XIV, pp. 109–146). As mentioned above, Galileo's ideas and assumptions were developed on the borderline of Newtonian modern science, that is, in a pre-Newtonian mechanical context. More precisely, Galileo's studies were crossed/influenced by Ancient/Archimedean science (Archimedes 1503, 1544, 1558, 1565, 1792, 1880–1881, 1881, 1897, 1912, 1987, 2002a, 2002b; Aristotle

[1936] 1955a, 1955b; Pisano 2009b). Early Modern Science/Fortifications/Machines (Clagett 1959, 1964–1984, 1976, 1978; Pisano and Bussotti, 2014, 2015a, 2015b, 2015c, 2017d; Pisano 2007, 2016, 2020, 2009b, 2017a; Pisano and Capecchi 2010), Science of Weights (Pisano and Bussotti 2017d; Pisano and Capecchi 2016; Capecchi and Pisano 2008) and moving towards New Modern Science (Pisano 2009a, 2009c, 2009d, 2011; Pisano and Bussotti 2017b; Bussotti et al. 2017). Thence, it is not to be taken as given that he considered uniform circular motion to describe a single phenomenon. In fact, Galileo did not work this out. He considered (Galilei 1632, I-II; Galilei 1967) there to be two kinds of inertial circular motion:

(1) The motion of celestial objects;
(2) The motion of terrestrial objects along a large portion of Earth's surface.

He did not consider *inertial* the *local motion*[2] in proximity to the Earth's surface, such as that of a sling. By *local circulal motion* Galileo meant the change of position of a body in proximity to the Earth's surface along a trajectory whose length has to be considered negligible in respect to the Earth's radius or circumference (Galilei 1638, III-IV; Galilei 1914). It is interesting to show how the conceptualisation of local motion was a subtle matter to deal with:

> THE PUBLISHER TO THE READER.
>
> […]. But the divine and natural gifts of this man are shown to best advantage in the present work where he is seen to have discovered, though not without many labors and long vigils, two entirely new sciences and to have demonstrated them in a rigid, that is, geometric, manner: and what is even more remarkable in this work is the fact that one of the two sciences deals with a subject of never-ending interest, perhaps the most important in nature, one which has engaged the minds of all the great philosophers and one concerning which an extraordinary number of books have been written. I refer to motion [*moto locale*], a phenomenon exhibiting very many wonderful properties, none of which has hitherto been discovered or demonstrated by anyone. The other science which he has also developed from its very foundations deals with the resistance which solid bodies offer to fracture by external forces [per *violenza*], a subject of great utility, especially in the sciences and mechanical arts, and one also abounding in properties and theorems not hitherto observed.[3] (Galilei 1914, pp. XX-XXI).

---

[2] Hereafter we will use the term *local motion* from this Galilean perspective. This is confirmed by the complete title of Galileo's masterpiece *Discorsi e dimostrazioni matematiche intorno a due nuove scienze attinenti alla meccanica e i movimenti locali* (Galilei 1638).

[3] "LO STAMPATORE A I LETTORI. […]. Ma molto più si fa manifesta la grazia concedutagli da Dio e dalla natura (per mezzo però di molte fatiche e vigilie) nella presente opera, nella quale si vede, lui essere stato ritrovatore di due intere scienzie nuove, e da i loro primi principii e fondamenti concludentemente, cioè geometricamente, dimostrate: e, quello che deve rendere più maravigliosa questa opera, una delle due scienzie è intorno a un suggetto eterno, principalissimo in natura, speculato da tutti i gran filosofi, e sopra il quale ci sono moltissimi volumi scritti; parlo del moto locale, materia d'infiniti accidenti ammirandi, nessuno de' quali è sin qui stato trovato, non che dimostrato, da alcuno: l'altra scienzia, pure da i suoi principii dimostrata, è intorno alla resistenza che fanno i corpi solidi all'essere per violenza spezzati; notizia di grande utilità, e massime nelle scienzie ed arti mecaniche, ed essa ancora piena d'accidenti e proposizioni sin qui non osservate. Di queste due nuove scienzie, piene di proposizioni che in infinito saranno accresciute col progresso del tempo dagl'ingegni specolativi, in questo libro si aprono le prime porte, e con non piccolo numero di proposizioni dimonstrate si addita il progresso e trapasso ad altre infinite, sì come da gl'intelligenti sarà facilmente inteso e riconosciuto." (Galilei 1890, p 45).

Thus his reasoning implies no contradiction. In fact, in his view, according to different physical circumstances, the uniform circular motions might be described as different phenomena.

1.3 The Structure of This Paper

In the following, we present the main structure of this paper:

- Introduction;
- Inertia: circular uniform motion of celestial bodies and along a portion of the Earth's surface;
- Inertia: rectilinear uniform motion
- The Historiography of Galileo's Concept of Inertia
- Concluding Remarks

2 Introduction: From Ancient Times to 1803

The captivating Inertia conceptualisation – defined first within natural and violent motions ideas, then as Impetus from the Renaissance until the early nineteenth century—has always been an object of vast scientific, historical, epistemological and philosophical inquiring. For the sake of brevity, in this section we only present an ad hoc introduction composed of the main issues in relation to the aim of our paper.

Since ancient times, Inertia–or what they used to refer to as a *property*, denoted something related to a diverse set of things: heavy bodies, a kind of resistance, friction, velocity and motion, displacements, work.

In his *Physics* Aristotle (384–322 B.C.) discussed *void space*. In his words:

> Let us explain again that there is no void existing separately, as some maintain. If each of the simple bodies has a natural locomotion, e.g. fire upward and earth downward and towards the middle of the universe, it is clear that it cannot be the void that is the condition of locomotion. What, then, will the void be the condition of? It is thought to be the condition of movement in respect of place, and it is not the condition of this. Again, if void is a sort of place deprived of body, when there is a void where will a body placed in it move to? It certainly cannot move into the whole of the void. The same argument applies as against those who think that place is something separate, into which things are carried; viz. how will what is placed in it move, or rest? Much the same argument will apply to the void as to the 'up' and 'down' in place, as is natural enough since those who maintain the existence of the void make it a place. And in what way will things be present either in place—or in the void?
>
> [...]
>
> Further, no one could say why a thing once set in motion should stop anywhere; for why should it stop here rather than here? So that a thing will either be at rest or must be moved ad infinitum, unless something more powerful get in its way. Further, things are now thought to move into the void because it yields; but in a void this quality is present equally everywhere,

> so that things should move in all directions. Further, the truth of what we assert is plain from the following considerations. We see the same weight or body moving faster than another for two reasons, either because there is a difference in what it moves through, as between water, air, and earth, or because, other things being equal, the moving body differs from the other owing to excess of weight or of lightness.
>
> [...]
>
> It is clear, then, from these considerations that there is no separate void. (Aristotle 1930, IV)

In his *Mechanical problems* (Aristotle [1936] 1955) he—or what currently we define as the Aristotelian School (hereafter simply Aristotle) argued regarding the lever and an embroyonic conceptualisation of force as the cause of motion; including heaviness (something like gravity, but not really) and its correlated *Principle of virtual velocities* (or also *Principle of virtual displacements/work* in Nemorarius' account):

> [On the lever; small forces can move great weights. Problem] 3. Why is it that small forces can move great weights by means of a lever, as was said at the beginning of the treatise, seeing that one naturally adds the weight of the lever? For surely the smaller weight is easier to move, and it is smaller without the lever. Is the lever the reason, being equivalent to a beam with its cord attached below, and divided into two equal parts? For the fulcrum acts as the attached cord: for both these remain stationary, and act as a centre. But since under the impulse of the same weight the greater radius from the centre moves the more rapidly, and there are three elements in the lever, the fulcrum, that is the cord or centre, and the two weights, the one which causes the movement, and the one that is moved; now the ratio of the weight moved to the weight moving it is the inverse ratio of the distances from the centre. Now the greater the distance from the fulcrum, the more easily it will move. The reason has been given before that the point further from the centre describes the greater circle, so that by the use of the same force, when the motive force is farther from the lever, it wil cause a greater movement. (Aristotle [1936] 1955, 85ab, p. 353)
>
> [...]
>
> [On the moving object and its stopping; Problem] 33. Why, again, does a body travel at all except by its own motion, when the discharging force does not follow and continue to push it? Surely it is clear that the initial impulse given causes it to push something else in the first instance, while this in turn pushes something else; it stops when the force which is pushing the travelling object has no longer power to push it along, and when the weight of the travelling object pulls it down more than the power of the pushing force can drive it forwards. (Aristotle [1936] 1955, 858a, p. 407)

Taking into account both his inaccurate mechanism of motion—and—rest (i.e., object moves only as long as cause constantly continues to be applied) and his *Principle of virtual velocities* (Pisano and Capecchi 2016; Pisano 2017b), in other places, Aristotle described this sustained motion of a projectile (in modern terms as external ballistic) as a necessary condition for the projectile continuing to move in the air. The main Aristotelian *simplicity hypothesis*, concerning virtual work and the correlated (modern) *Principle of Least Action* was based on an actual movement of a heavy moving body as a *natural motion*, tending to minimize the motion of a particular material body, i.e., an early Aristotelian Inertia conceptualisation (*Ibidem*).

This complex mechanism between virtual velocities and Inertia in Jordanus de Nemore's (fl. 12th/13th) mechanism (De Nemore 1533, 1565; Koyré 1960; see also De Nemore fl. 13th) became the *Principle of virtual displacements*: a physical system (e.g., masses subjected to forces) is in a state of equilibrium if and only if the (forces–) weights are inversely proportional to their virtual displacements.

In his words:

> If two weights descend along diversely inclined planes, then, if the inclinations are directly proportional to the weights, they will be of equal force when descending [idem force – equilibrium].[4]

Idem in Tartaglia (see also below):

> QUESTION. XV. DEFINITION XII.
>
> A body is said to be more or less heavy in descent than another when the straightness, obliquity, or pendency of the place or space where it descends makes it descend more or less heavily than the other, and similarly more or less rapidly than the other, though both were simply equal in heaviness.
>
> QUESTION. XVI. DEFINITION XIII.
>
> A body is said to be positionally more or less heavy than another when the quality of the place where it rests and is located makes it heavier [or less heavy] than the other, even though both are simply equal in heaviness.[5]

Johannes Philoponus (fl. 6th) proposed a very interesting dissimilarity from Aristotle on the subject. He mounted a serious criticism of Aristotle's mechanism of motion—and—heavy moving body in the range of void space considered as a "[…] place deprived of body […]". In Aristotle's words:

> As a step towards settling which view is true, we must determine the meaning of the name. The void is thought to be place with nothing in it. The reason for this is that people take what exists to be body, and hold that while every body is in place, void is place in which there is no body, so that where there is no body, there must be void. Every body, again, they suppose to be tangible; and of this nature is whatever has weight or lightness. Hence, by a syllogism, what has nothing heavy or light in it, is void. This result, then, as I have said, is reached by syllogism. It would be absurd to suppose that the point is void; for the void must be place which has in it an interval in tangible body. But at all events we observe then that in one way the void is described as what is not full of body perceptible to touch; and what has heaviness and lightness is perceptible to touch. So we would raise the question: what would they say of an interval that has colour or sound-is it void or not? Clearly they would reply that if it could receive what is tangible it was void, and if not, not. In another way void

---

[4] "Quaestio decima. Si per diversarum obliquitatum vias duo pondera descendant, fueritque declinationum, & ponderum una proportio eodem ordine sump[t]a, una erit eritutriusque in discendendo [idem force – equilibrium]." (De Nemore 1565, 7r; see also Pisano 2016, p. 413).

[5] "QUESITO XV. DIFFINITIONE XII. Un corpo se dice essere piu, over men grave d'un'altro nel descendere, quando che la rettitudine, obliquita, over dependentia del luoco, over spacio dove descende lo fa descendere piu, over men grave dell'altro, e similmente piu, over men veloce dell'altro, anchor che siano ambidui simplicemente eguali in gravita. QUESITO XVI. DIFFINITIONE XIII. Un corpo si dice essere piu grave, over men grave d'un'altro, secondo il luoco, over sito, quando che la qualita del luoco dove che lui se riposa, e giace, lo fa essere piu grave dell'altro anchor che fusseno simplicemente egualmente gravi." (Tartaglia [1554] 1959, VIII, Q. XVI, Def. XIII, 84r; see also Pisano and Capecchi 2016, p. 293).

> is that in which there is no 'this' or corporeal substance. So some say that the void is the matter of the body (they identify the place, too, with this), and in this they speak incorrectly; for the matter is not separable from the things, but they are inquiring about the void as about something separable. Since we have determined the nature of place, and void must, if it exists, be place deprived of body, and we have stated both in what sense place exists and in what sense it does not, it is plain that on this showing void does not exist, either unseparated or separated; the void is meant to be, not body but rather an interval in body. This is why the void is thought to be something, viz. Because place is, and for the same reasons. For the fact of motion in respect of place comes to the aid both of those who maintain that place is something over and above the bodies that come to occupy it, and of those who maintain that the void is something. They state that the void is the condition of movement in the sense of that in which movement takes place; and this would be the kind of thing that some say place is. But there is no necessity for there being a void if there is movement. It is not in the least needed as a condition of movement in general, for a reason which, incidentally, escaped Melissus; viz. that the full can suffer qualitative change. But not even movement in respect of place involves a void; for bodies may simultaneously make room for one another, though there is no interval separate and apart from the bodies that are in movement. And this is plain even in the rotation of continuous things, as in that of liquids. And things can also be compressed not into a void but because they squeeze out what is contained in them (as, for instance, when water is compressed the air within it is squeezed out); and things can increase in size not only by the entrance of something but also by qualitative change; e.g. if water were to be transformed into air. In general, both the argument about increase of size and that about water poured on to the ashes get in their own way. For either not any and every part of the body is increased, or bodies may be increased otherwise than by the addition of body, or there may be two bodies in the same place (in which case they are claiming to solve a quite general difficulty, but are not proving the existence of void), or the whole body must be void, if it is increased in every part and is increased by means of void. The same argument applies to the ashes. It is evident, then, that it is easy to refute the arguments by which they prove the existence of the void. (Aristotle 1930, IV, 7.214a, pp. 17-20; Cfr. Grant 1964, pp. 265-268)

In his *Commentary on the Physics* (Grant 1964) Philoponus argued that the Aristotelian (air–) medium does not play any action in maintaining a body in motion, while, in contrast he suggested the presence of a property assigned to the moving body (one might think of projectiles). The emergent mechanism is not so far from a primitive idea of Inertia. In fact, Philoponus also debated that a motion in void space depends on velocity, which has a direct proportionality to the "weight" (mass) of the moving body (Cfr. Grant 1964, p. 266); and that the Aristotelian medium only retards the body in the void. (Of course, if a body was in a medium, then it was not in the void). Philoponus discussed these arguments up to *Physics*' Sect. 9 (Aristotle 1930, IV, Sect. 7–9). Finally, he focused on the role played by the Aristotelian medium, which seems to have a double feature:

(a) To be able to warrant motion to a moving body;
(b) To be able to impede motion in a moving body.

It is known that the late Aristotelian School strongly discussed this duality.

Ibn Rushd, in Latin also known as Averroes (1126–1198), was one of the scholastic philosophers who moved against the above criticism-theses. Later on, by means of another primeval form of Inertia/Impetus, Persian Pūr Sina' (Avicenna) suggested that a moving body in a void space continues its path unless acted upon. In other words, he added an external action to warrant Inertia as a property of a moving body.

In the Middle ages, during the Aristotelian schools' speculations—and—Scholastic's opposing arguments, Johannes Buridanus (below also Buridan, Buridani) (ca. 1300–ca. 1360; fl. 14th) *adapted* it by means of an ad hoc theory named Impetus[6] (Buridan 1513, 1942; Pisano and Capecchi 2016: Pisano 2017b) presented in his *Subtilissimae Quaestiones super octo Physicorum libros Aristotelis* (1509) (Buridani 1509). The main features are:

- *Impetus changes–proportionally to the velocity of a moving body and its body (mass): the more velocity impressed by the (Aristotelian) engine on the moving object, the stronger the transmitted Impetus.* This mechanism of Impetus (weight[mass] · velocity) might be comparable (considered cautiously) to the concept of *Momentum*, as well.
- *Impetus is a permanent property different and distinct from the motion and the mass of the projectile.* A characteristic of permanence of Impetus might be weakened by motion, i.e., by air resistance and the degree of inclination of the launch.
- *The resistance of the air and the weight (nowadays mass) of a moving body are the causes of the arrest of the body.* Both are opposite elements to his Impetus. Therefore, he excluded both the possibility of a physical phenomenon in which the motion of a moving body can produce a property (Impetus) and that it can naturally degenerate.
- *Buridanus convinced himself that his Impetus could be naturally both rectlinear and circular; for, celestial bodies can have a motion.* This last aspect of his Impetus' theory is very important for our aim because it fits well with the arguments presented in the following section.

The resemblances to modern Inertia are evident. But we should remark that Buridanus only aimed to revise Aristotle's idea on this specific subject. In fact, he conserved other of the Aristotelian School's convictions, e.g., such as the difference between a moving object and the same at rest.

Below we present the main differences between the Aristotleian engine and Buridanus' Impetus. As usual, it will be the case–study of a projectile.

Aristotle (IV B.C.)	Buridanus (14th)
First Path In surrounding middle (air) Impetus appears (straight line) after receiving an initial boost (e.g., cannonball)	First Path Impetus is a necessary property in order to start its (straight line) trajectory. Heavyness (gravity) does not intervene

(continued)

[6] The historical genesis of Impetus—later applied to the motion of projectiles—is quite complex and varied. Aside from Aristotle's initial theory among the scholars who dealt with the topic, we note: Pūr Sina' (Persian) son of Sina called Avicenna (980–1037), Roger Bacon (1214–1292), Thomas Aquinas (1225–1274), Pierre Jean Olivi (1248–1298), Nicole Oresme (ca. 1320/1325–1382), Francesco of Marchia or of Esculo, or of Ascoli (fl. 14th), William of Ockham (ca. 1280–ca. 1349), and for some considerations, Jordanus de Nemore (fl. 12th/13th | also Nemorarius), too (Pisano and Capecchi 2016; Pisano 2017b).

(continued)

Aristotle (IV B.C.)	Buridanus (14th)
Second Path The moving body decreases its velocity. (In Aristotle's view it can be also included in the First Path above)	Second Path Air resistance opposes the moving body, which decreases its velocity. Heavyness (gravity) intervenes for the moving body trajectory downwards such as a part of circumference to be completed
Third Path After a certain trajectory, the aristotelian engine force is off and moving body trajectory tends downwards	Third Path Impetus is off Heavyness (gravity) completes its intervenance

On the subject, we specify that:

(a) Aristotelian/Buridanus Impetus and Momentum cases. If we consider:

$$\text{Impetus} = \text{Weight [Mass]} \cdot \text{Velocity}.$$

We can clearly see that Impetus is presented as similar to (e.g., Galilean) Momentum (Galluzzi 1979; Galluzzi and Torrini 1975–1984). Taking it into account the motion, in both cases, will change only when an external cause (e.g., a force) acts on the moving body.

(b) Aristotelian/Buridanus Impetus and Inertia cases. If we consider:

$$\text{Impetus} = \text{Weight [Mass]} \cdot \text{Velocity}.$$

But we also see that it is different from Inertia. In fact, for the latter (uniform rectilinear) motion—and—rest are equivalent. But as above announced:

- This equivalent term should be better specified. In fact, in the case of an object at rest, Impetus does not intervene.
- Buridanus' Impetus theory dealt both with rectilinear and circular motions. This is an important assumption in the Impetus' theory because it is an intellectual mechanism–speculation, which will persist until Galileo, as we will show in the next sections.

Finally, Buridanus attempted a more accurate reasoning of Impetus[7] without, however, producing any formal or mathematical language to describe it. It is quite probable that before 1607, Galileo had not yet clarified the theory on the composition of vertical and horizontal motion (Pisano 2024-pre print; Pisano and Capecchi 2016). This non–clarification was mostly likely due to a lack of sufficient *sensate esperienze*, but then he wrote his notes in the famous codex *Ms. 72* (Galilei fl. 17th, 116v; Hill

[7] Clagett (Clagett 1959) and in turn reproduced by Maier (Maier 1509) which, in turn, includes – with some modifications—the Parisian edition from *Subtilissimae Quaestiones* (Buridani 1509, 1513, 1942). Clagett dated Buridan's manuscript around 1357. It is archived at the Vatican Library in Roma (Vat. Lat. 2136, 1r.).

1986; Pisano et al. 2020; see also Codex *Ms 71*) in which, instead he outlined the solution. We also remark that at Galileo's time, typically Aristotelian motion was supported by the dichotomy of violent and natural motion.

Niccolò Fontana also known as Tartaglia (1499?/1500?–1557) (Pisano and Capecchi 2016) presented contributions to the art of warfare in *Nova scientia* (Tartaglia 1537, Books I–II) and *Quesiti et invention diverse* (Tartaglia 1546; Tartaglia [1554] 1959, Books I–III; Pisano and Capecchi 2016). The arguments regarded military artillery, cannonballs, gunpowder linked to the problem of natural—and—violent motion; hence Impetus was included, as well. One of his major results was the general law of the trajectory of a cannonball and its maximum range, for any given degree–measure (Tartaglia 1537, I, 4r; Tartaglia [1554] 1959, I, Q. II, 9r; see also: 78, I, Qs. I–II–III–VI-VII, 5rv–16rv). Tartaglia was not able to formulate a modern theory of projectiles by means of, e.g., a correct mathematical interpretation (nowadays) of a parabolic trajectory: the path was clearly curved but not parabolic. In 1609, Galileo formulated it (Naylor 1976; Naylor and Drake 1983; see also Galilei fl. 16th, Ms A, Ms B, Ms m; Pisano et al. 2020). This involves a case–study on the trajectory of projectiles which Tartaglia had not yet sufficiently theoretically developed: an Aristotelian rectilinear path and a final curvilinear one (Tartaglia [1554] 1959, I, Q. VII, 16rv; Pisano and Capecchi 2016). In addition, his studies were, in any cases—more or less based on the medieval theory of Impetus. As cited above, this medieval theory of Impetus included embryonic reasoning on resistance and the ancient concept of Inertia. In Tartaglia's law of elevation at 45 degrees, the angle refers to one (although partial) final curved projection (a sort of semi–parabolic curve) of the projectile. For, his form of Inertia was a *property* of an object in motion caused by both a *virtù impressa* and *natural one*, which had the *property* to overcome the initial force that, in time, became weaker and allowed the projectile to fall. We remark that the problem of the physical and mathematical knowledge of the projectile trajectory was complicated at that time. In fact, it was clear that a cannon ball in the air proceeded in a *non–rectilinear*[8] way. However, the idea that an equation (of motion) of the second degree could indeed mathematically interpret the physical motion of a cannon shot along the trajectory was barely mature for sixteenth century. Taking into account both Aristotleian speculations concerning improbable differences of motions in everyday and celestial spaces, and Copernicus' heliocentric assumptions (1543), Galileo Galilei (1564–1642)—in developing his forthcoming astronomical Copernican model—echoed, at first, the Aristotelian school. Neverthless, we remark that Galileo is a turning point with regard to the concept of Inertia. In his writings there is no explicit formulation of it *à la* Descartes, or, even less, *à la* Newton; the term Inertia itself is not used by Galileo. This notwithstanding, he was the scholar who first problematized this question, though he did not use a specific term to denote inertia and did not offer a synthetic description of such principle. We will specifically discuss this in the following sections on Galileo's Inertia below.

[8] For example, see Leonardo da Vinci, 1504: Codex Atlanticus, Mortar' model, 33r (Pisano 2016; Pisano 2009b).

The term Inertia was presented, perhaps for the first time, by Johannes Kepler (1571–1630) in his *Epitomes Astronomiae Copernicanae usitata forma quaestionum et responsionum conscriptae* (1618–1621)[9] (Kepler 1618–1621, VII, IV; see also Pisano and Bussotti 2017a). In this work he also produced and offered an extensive exposition of his concept of *copia materiae*, which may be discussed as mass (see below; Pisano and Bussotti 2017a).

In the history of the concept of Inertia, Kepler is an important author. His characterization of Inertia is far from comparable with Galilean or Newtonian ones. In fact, his concept of Inertia was essentially presented as the resistance to a motion; but the latter and rest were studied as separated phenomena as well. Particularly, rest was considered a *natural state*. In detail, in his metaphysical view, Kepler's Inertia was introduced to explain the opposition to a motion, so the Aristotelian *violent* and *natural motions* were also considered. The conceptualization of gravity also represents an access key to Kepler's ideas on his *Inertia*. As above cited, in his opinion, if there is no external action, each body stays at rest. We can read this in the *Introduction* to his *Astronomia Nova* (1609) (Pisano et al. 2017a):

> Each corporeal substance, as far as it is corporeal, has been created so to be suitable to stay at rest in every place in which it is located by itself, outside the sphere of influence of a cognate body.[10]

This conception of Inertia is further developed and clarified in *Epitomes Astronomiae Copernicanae usitata forma quaestionum et responsionum conscriptae*, where Kepler wrote:

> The entire Earth, considered as a whole, and in respect to its matter, is naturally subject to no motion. For, inertia is proper to the matter of which most part of the Earth is composed. Opposition to motion due to inertia is the stronger the more copia materiae compressed in a narrow space exists.[11]

At first glance, Kepler's concept of Inertia might, thence, appear similar to the Aristotelian. But, this is not the case because of two reasons:

(1) Kepler argued that our Earth was not the only centre of Inertia of the universe. Instead Aristotle adopted this idea.

---

[9] In his *Epitome Astronomiae Copernicanae*, since "*Liber Primus / Pars Quarta* (Kepler 1618–1621, I, IV, p. 79) Kepler dealt with several types of "inertia", "inertia naturalis", "inertia materiae globi", "interia materialis globi", "inertia materiae movendae", "inertia corporis Lunaris", etc. (Kepler 1618–1621). We remark that independently from Latin genitive (*Epitomes* of *Epitome*) the reader can read the title of this Kepler's opera both as *Epitome Astronomiae Copernicanae usitata forma quaestionum et responsionum conscriptae* [...] and *Epitomes Astronomiae Copernicanae usitata forma quaestionum et responsionum conscriptae* [...]: Books I-III (1618, *Epitome* and *Epitomes*), Book IV (1620, *Epitomes*) and Books V-VII (1621, *Epitomes*). Hereafter, we cite it as *Epitomes*.

[10] "Omnis substantia corporea, quatenus corporea, apta nata est quiescere omni loco, in quo solitaria ponitur, extra orbem virtutis cognate corporis." (Kepler 1618–1621, III, p. 25, lines 9–10; translation: Kepler 1992, p. 55).

[11] "Terra tota, quatenus tota, et respectu suae materiae, motum planè nullum habet naturaliter: materiae enim, qua plurima Terra constat, propria est inertia, repugnans motui, eaque tanto fortior, quanto major est copia materiae in angstum coacta spacium." (Kepler 1618–1621, VII, p. 88, lines 9-13; Authors' translation).

(2) Each body has its own Inertia.

In Kepler's words:

> Though a celestial globe is not heavy in the same manner in which a stone is told to be heavy on the Earth, and it is not light as for us fire is, nonetheless it holds, in relation to its own matter, a natural lack of power to move from one place to another. It has a natural inertia or stillness, because of which it stays at rest everywhere it is collocated alone. Thus, in order to remove such a body from its place and its rest, a power [force-cause] is necessary. It has to be bigger than its matter and its body for it to prevail on the body's natural inertia.[12]

Thus, according to Kepler, each body has its own Inertia: it naturally stays at rest, *unless* a force moves it (Pisano and Bussotti 2017a). Motion is a physical condition opposed to the lack of motion. This confirms that—in Kepler's conception—the forces have to explain not only the change of movement, but the origin of movement itself. However, no privileged centre of inertia exists.

There is a further noteworthy insight connected, for Kepler, with Inertia: it is the concept of *copia materiae*. Let us re-analyse the last part of the first previous quotation:

> Opposition to motion due to inertia is the stronger the more *copia materiae* compressed in a narrow space exists. [See above]

What is the "*copia materiae* compressed in a narrow space"?

This locution *copia materiae* can probably be interpreted as *mass*,[13] so that *copia materiae* compressed in a given volume is probably equivalent to density. This seems to us a plausible interpretation of the notion of *copia materiae*. Probably Kepler guessed the existence of bodies' property different from weight. In other sections of Kepler's work too, *copia materiae* seems identifiable with mass, as it is the case of this interesting passage:

> If the volume of Saturn, 10, is multiplied by its density, one obtains a *copia materiae* of 50, as if the volume of Jupiter, 5, is multiplied by its density, 10.[14]

Kepler seems close to Newton's concept of mass. But this impression is in contradiction with this further assertion:

---

[12] "Etsi globus aliquis coelestis non est sic gravis, vt aliquod in Terra saxum grave dicitur, nec sic levis, vt penes nos ignis: habet tamen ratione suae materiae naturalem *adynamian* transeundi de loco in locum, habet naturalem inertiam seu quietem, qua qui'escit in omni loco, vbi solitarius collocatur. Inde verò ex situ et quiete sua vt emoveatur, opus est illi potentia aliqua, quae sit amplius quippiam, quam sua materia et corpus nudum, quaeque inertiam hanc ejus naturalem vincat." (Kepler 1618–1621 VII, IV/II, p. 296, lines 30–36).

[13] The mass also has an influence on the mean motion because the more massive a planet, the bigger its resistance to the action of the *solar virtus* (Cfr. Pisano and Bussotti 2017a) is, due to the natural Inertia of the masses.

[14] "Multiplicata enim mole Saturni 10, in densitatem 5, prodiret copia materiae 50, tantundem scilicet, quantum, si molem Iovis 5 in densitatem ejus 10 multiplicasses". (Kepler 1618–1621, VII, p. 283, lines 33–35; Authors' translation).

> There are four causes which contribute to the length of periodical time. The first one is the path's length, the second one is the weight or the *copia materiae* to move, the third one is the motive force's power, the fourth one is the volume [*molis*] or the space in which the matter to move is located.[15]

Thus, a global interpretation of Kepler's *copia materiae* cannot be definitive because, according to the different circumstances we have referred to in our quotations, it seems to mean both:

(1) A notion similar to that of Newtonian *mass*;

(2) *Weight* (in the Renaissance common sense of it (Pisano and Robarts 2022)).

We are inclined to interpret such a locution as similar to mass in the embryonic sense of quantity of matter, but the interpretation is neither trivial nor granted.

Evangelista Torricelli (1608–1647), in his *Opera geometrica* (1644) (Torricelli 1644), presented his known principle in mechanics according to which two heavy bodies linked together cannot move by themselves unless their common centre of gravity descends. At beginning of *Liber primus* of *De motu gravium naturaliter descendentium* (Torricelli 1644, pp. 97–153) he expressed this *Praemittamus* – linked both to motion/rest and intervening forces as in the following:

> *Premise*: Two heavy bodies joined together cannot move by themselves unless their common centre of gravity descends.[16]

The *Praemittamus* (premise) is nowadays called *Torricelli's Principle* in mechanics. Torricelli resorted to it in the proofs of Proposition I and Proposition II (Torricelli 1644, pp. 99–100). The *Praemittamus* is considered a specific–basic–early expression of the *Principle of virtual work* (Pisano 2017b; Pisano and Bussotti 2024; Capecchi and Pisano 2007, 2010). In these proofs, *Torricelli's Principle* is used by means of a *reductio ad absurdum*. In a recent paper (Pisano and Bussotti 2024) we showed that Torricelli used this principle to prove the intriguing Proposition XVI of the *De motu ac mometis varia*.[17] In the literature, important authors, such as Ernst Mach (1838–1916), tried to reduce Torricelli's mechanical principle to the ones expounded or, anyway, used by Galileo in his *Le Mecaniche*[18] (Pisano 2024-pre

---

[15] "Causae ad constituendam temporis periodici longitudinem, concurrunt quatuor. Prima est itineris longitudo, secunda pondus seu copia materiae transportandae, tertia fortitudo virtutis motricis, quarta moles seu spacium in quod explicatur materia vehenda." (Kepler 1618–1621, VII, p. 306, lines 37–40; Authors' translation).

[16] "*Praemittamus*: Duo gravia simul coniuncta ex se moveri non posse, nisi centrum commune gravitatis ipsorum discenda" (Torricelli 1644, p. 99; 101, p. 105; Torricelli's italic; Authors' translation). On Torricelli see our recent *forthcoming* paper *Historical and Methodological Details on the De motu gravium naturaliter descendentium in Torricelli's Opera Gometrica (1644)* (Pisano and Bussotti 2024) and book *380th Anniversary of Torricelli's Opera Geometrica (1644). Text, Transcription, Commentaries and Selected Essays as New Historical Insights* (Pisano et al. 2024a; Cfr. Capecchi and Pisano 2007, 2010).

[17] This proposition is missing in 1644–edition; see *Opere di Evangelista Torricelli by Loria and Vassura* (Torricelli 1919–1944, pp. 231–249, Proposition XVI, pp. 243–246).

[18] An extensive study on the two *Le Mecaniche* including the complete translation and transcription side–by–side two *Le Mecaniche* (Galilei [1598–1599] 1623) texts into English is *forthcoming*

print, II, pp. 147–191) and other works. In his *Opera geometrica*, Torricelli suggested that this supposition may be *proved* beginning with a "theorem" according to which "the Momentum of equal bodies on planes unequally inclined are to each other as the perpendicular lines of equal parts of the same planes" (Torricelli 1644, p. 99); and, of course, it is linked with why and how it can happen in void space—and—in (air) middle. Torricelli assumed that Galileo's theorem (Galilei 1890–1909, VIII, III, p. 205) had not yet been demonstrated. Galileo wrote:

> [Sagr] [...] [it] seems that up to the present we have established the definition of uniformly accelerated motion which is expressed as follows: A motion is said to be equally or uniformly accelerated when, starting from rest, its Momentum [*celeritatis momenta*] receives equal increments in equal times. Salv. This definition established, the Author makes a single assumption, namely. The speeds acquired by one and the same body moving down planes of different inclinations are equal when the height of these planes are equal.[19]

Torricelli named Galileo's theorem "*Supposit. Galilei* [...] *Galileus principium* [...]" and claimed:

> To afford the naturally accelerated Motion, Galileo assumes hypothetically a principle which he himself does not consider completely evident [≠ he considers it experimentally evident], because he tries to prove it with the not very exact experiment of the pendulum, that is: [...].[20]
>
> Because we did not previously encounter this Theorem, first we will confirm it with the same proof, then we will arrive to show what in Galileo is a principle or a petition.[21]

Later on, he added an important *Lemma*, as well:

> The total *Momentum* of a heavy body is to the *Momentum* the body assumes on an inclined plane as the length of the inclined plane [its hypotenuse] is to the perpendicular.[22]

(Pisano 2024-pre print). In this work for the first time short—and—long versions of *Le Mecaniche* (manuscripts 1594/fl. 17th; 1623, 1634) are discussed and published in English.

[19] "Salv. Fermata cotal definizione [*motus aequaliter, seu uniformiter acceleratum dicimus eum, qui, a quiete recedens, temporibus aequalibus, aequalis celeritatis momenta sibi superaddit*] un solo principio domanda e suppone per vero l'Autore, cioè: *Accipio, Gradus velocitatis eiusdem mobilis super diversas planorum inclinationes acquisitos tunc esse aequales, cum eorumdem planorum elevation aequales sint*. Ex hac petitione dependet quasi universa illius doctrina de motu tùm accelerato, tum proiectorum." (Galilei 1890–1909, VIII, III, p. 205; Galileo's italic. Translation retrieved from Galilei 1954, p. 169).

[20] "*Acturus de Motu naturalitèr accelerato Galileus principium supponit, quod & ipse non admodum euidens putat, dum illud parum exacto penduli experimento nititur comprobare. hoc est.*" (Torricelli 1644, p. 98; Authors' translation; Torricelli's italic).

[21] "*Nos quia in huiusmodi Theorema non incidimus, hoc primum aliqua demonstratione confirmabimus: protinus ad ostedendum* [we are going to demonstrate by means of] *id quod Galileo principium sive petitio est, accedemus*." (Torricelli 1644, pp. 98–99; Authors' translation; Torricelli's italic).

[22] "*Lemma*. Momentum totale gravius ad Momentum quod habet in plano inclinato, est ut longitudo ispius plani inclinati ad perpendiculum" (Torricelli 1644, p. 101; authors' translation). [...]. Hic [without ad absurdum proof] Propositio secunda iterum. Momenta gravium aequalium super planis inaequaliter inclinatis sunt in reciproca ratione cum longitudinibus planorum [...]." (Torricelli 1644, p. 102; Authors' translation; Torricelli's italic).

René Descartes (1596–1650) has been accredited as (one of) the first authors to have stated Inertia as a principle. Between the 20s and the 30s of the seventeenth century he dealt with Inertia in his *Le Monde ou Traité de la lumière*[23]—hereafter *Le Monde* – fl. 1629–1633, then published posthumously in 1664 (Descartes 1897–1913, XI; see also 110, IX-2, IV, §§20–27, pp. 133–183; 109). In the following, we present his law of Inerta from *Le Monde*:

> The first [Law of Inertia] is: That each individual part of matter always continues to remain in the same state unless collision with others constrains it to change that state. That is to say: if it has a certain size, it will never become smaller unless others divide it; if it is round or square, it will never change this shape without others compelling it to; if it is at rest in any place, it will never leave it except if others drive it out; and if it once begins to move, it will continue always with the same force until others bring it to rest or slow it down.[24]

In 1644, in *Principia Philosophiae* Descartes presented his mechanistic conception (Descartes 1897–1913, VIII; see also Descartes 1964–1974). An attempt to unify the entire physical world on the basis of a few principles and rules is present; the treatment was not mathematized and accurate definitions are missing. The geometric model proposed provided weak indications about the positions of the bodies in the function of time. Important laws are introduced such as the laws of Inertia (Descartes 1897–1913, VIII-1, II, XXXVII, pp. 62–63), conservation of motion and rectilinear motion (Descartes 1897–1913, VIII-I, II, XXXIX, XL, pp. 63–65); the laws are immutable. In his Words:

1. XXXVII. The First Law of [intertia] Nature: That any object, in and of itself, always perseveres in the same state; and thus what is moved once always continues to be moved.[25]

[23] Descartes 1897–1913, VI. *Discours de la Méthode*, pp. 1–79. It also includes *La Dioptrique*, pp. 80–228; *Les Météores*: pp. 231–366; *La Géométrie*, pp. 367–485; *Le Monde*, XI-1, pp. 3–118.

[24] "La peremiere [Loi d'Inertie] est: Que chaque partie de la matière, en particulier, continue toujours d'être en un même état, pendant que la rencontre des autres ne la contraint point de le changer. C'est à dire que: si elle a quelque grosseur, elle ne deviendra jamais plus petite, sinon que les autres la divisent; si elle est ronde ou carrée, elle ne changera jamais cette figure sans que les autres l'y contraignent; si elle est arrêtée en quelque lieu, elle n'en partira jamais que les autres ne l'en chassent; & si elle a une fois commencé à se mouvoir, elle continuera toujours avec une égale force jusques à ce que les autres l'arrêtent ou la retardent." (Descartes 1897–1913, XI, VII, p. 38 (pagination 435–436). Translation from Descartes [1979] 1995; Gaukroger 2004). He also wrote two more laws/rules on motion: 2nd. "Je suppose pour seconde Règle: Que quand un corps en pousse un autre, il ne saurait lui donner aucun mouvement, qu'il n'en perde en même temps autant du sien; ny luy en oster, que le sien ne s'augmente d'autant. Cette Règle jointe avec la précédente, se rapporte fort bien à toutes les expériences, dans lesquelles nous voyons qu'un corps commence ou cesse de se mouvoir, parce qu'il est poussé ou arrêté par quelque autre" (Descartes 1897–1913, XI, VII, p. 41 (pagination 437–438)). 3rd. "J'ajouterai pour la troisième: Que lorsqu'un corps se meut, encore que son mouvement se fasse le plus souvent en ligne courbe & qu'il ne s'en puisse jamais faire aucun, qui ne soit en quelque façon circulaire, ainsi qu'il a été dit ci-dessus, toutefois chacune de ses parties en particulier tend toujours à continuer le sien en ligne droite. Et ainsi leur action, c'est à dire l'inclination qu'elles ont à se mouvoir, est différente de leur movement." (Descartes 1897–1913, XI, VII, pp. 43–44 (pagination 439–440)).

[25] "XXXVII. Prima lex naturae: quòd unaquaeque res quantum in se est, semper in eode statu perseveret; sicque quod semel movetur, semper moveri pergat." (Descartes 1897–1913, VIII-1,

2. XXXXIX. The Second Law [of Nature]: That every motion of itself is rectilinear; and hence what is moved circularly tends always to recede from the center of the circle it describes.[26]
3. XL. The Third Law [of Nature]: That a body, in colliding with another larger one, loses nothing of its motion; but, in colliding with a smaller one, loses as much as it transfers to that one.[27]

Since a mathematical and definitional apparatus is missing, it is challenging—and probably in part wrong—to interpret the Cartesian conception of Inertia and quantity of motion as part of Newtonian and post-Newtonian physics. Nevertheless, it is necessary to point out that in other cases of complex phenomena, such as refraction and the rainbow, Descartes is faithful to the mathematical approach of his own work. Pierre Boutroux (1880–1922) after having eulogized Descartes for the introduction of the Inertia principle and the conservation of the quantity of motion principle, wrote regarding the collision:

> Unfortunately, Descartes makes a very serious mistake that is surprising from his part.[28]

The inaccuracy consists in the fact that Descartes did not catch the vectorial nature of the quantity of motion. The mistakes in the collision rules are due, according to Boutroux, to this misconception. Still in *Principia Philosophiae*, space cannot be distinguished from *res extensa*. It is always relative (Descartes 1897–1913, VIII-I, II, X, p. 45). Time is also relative (Descartes 1897–1913, VIII-I, I, LVI-LVII, pp. 26–27). It is known that a scientific (i.e., by magnitudes) distinction between mass and weight was problematic at that time. In addition, in *Principia Philosophiae*, there is an insufficient connection between theory and experience. In some cases—as in the one of the movements of the planets—there are intrinsic difficulties in connecting theory and experience because no provisional model is supplied, but only a descriptive one (Cfr. Bussotti and Pisano 2013; see also: Blackwell 1966; Schuster 2000, 2013a, b). In other cases, such as the collision rules, experience is not consistent with the theoretical rules (Descartes 1897–1913, VIII-I, II, XLVI-LII, pp. 68–70). Descartes ignored experience, claiming that many conditions can influence experiments and experience (Descartes 1897–1913 VIII-I, II, LIII, p. 70; see also Bussotti and Pisano 2013; Cohen 1964).

II, XXXVII, pp. p. 62; pp. 62–63 (pagination 54–55); Author's capital letters; Translation from Descartes [1979] 1995; see also Blackwell 1966).

[26] "XXXIX. Altera lex naturae: quòd omnis motus ex se ipso sit rectus, et ideò quae circulariter moventur, tendere semper, ut recedant à centro circuli quem decribunt." (Descartes 1897–1913, VIII-1, II, XXXIX, p. 63; pp. 63–65 (pagination 55–56); Author's capital letters; Translation from Descartes [1979] 1995; see also Blackwell 1966).

[27] "XL. Tertia lex: quòd unum corpus alteri fortiori occurrendo, nihil amittat de suo motu; occurrendo verò minùs forti, tantum amittat, quantum in illud transfert." (Descartes 1897–1913, VIII-1, II, XL, p. 65 (pagination 57–58); Author's capital letters; Translation from Descartes [1979] 1995; see also Blackwell 1966).

[28] "Malheureusement, Descartes commet une erreur très grave et qui est bien surprenante da sa part." (Boutroux 1921, p. 677; Authors' translation).

In the sections above, we briefly introduced Newton's Inertia. But, what about after Newton? (Pisano 2014; Pisano 2017b). In the history of science, after Newton's laws, other assumptions and developments in Inertia case–studies were proposed.

Lazare Nicolas Marguérite Carnot (1753–1823) (Gillispie 1960) introduced his *Première Hypothèse* as the Inertia law (Pisano 2014, 2017b; Pisano et al. 2021; Gillispie and Pisano 2014) in opposition to Newton's law. Lazare Carnot had special consideration for force, space and *void* (Pisano et al. 2021; Gillispie and Pisano 2014). For our aims we only refer to his *Essai sur les machines en général* (Carnot [1783] 1786) (hereafter *Essai*) and *Principes fondamentaux de l'équilibre et du mouvement* (1803, here after *Principes*) (Carnot 1803; see also Coulomb 1784; Oliveira and Pisano 2022; Drago 1991).

The *Essai* adopted a very simple level of mathematics combined with a simplified geometrical modelling of machines/mechanics and advanced studies such as the machines in motion. The particular apparatus of machineries within machines (equilibrium and in motion) were systematically deduced from his *Hypothèses* (laws) of mechanics. Carnot rigorously used geometry and trigonometry (vector calculus) rather than the advanced mathematics of his contemporaries. Combining *Geometric motion, Moment of activity–concept of Work* with the *moment of quantity of motion*, he found the invariants of the *communication of motion* and established the maximum effect produced by mechanical machines, e.g., the continuity in the transmission of power. The latter is the apparatus where Inertia was included and discussed together with the *Principle of virtual laws*. He was convinced that an innovative general theory of machines—an applied science of machines—was of crucial necessity for the economic development of a society. From this derived the importance of dealing with Inertia and the *Principle of virtual laws*. His *Essai* came to be of interest to technicians by the 1780's as well. But, the current principles and related theories at that time were insufficient to perform motion and equilibrium for all machines. This was one of the main challenges of his *machines en général*:

> Given the virtual motion of any system of bodies (i.e., that which each of the bodies would describe if it were free), find the real motion it will assume in the next instant in consequence of the mutual interaction of the bodies considered as they exist in nature, i.e., endowed with the Inertia that is common to all the parts of matter. (Galilei 1890–1909, X, p. 21)[29]

What about bodies and motion? In order to answer this question in Carnot's account, we come back to *Principle of virtual laws* and Inertia.

The bodies with infinite mass (that is constraints only) were not considered by Carnot because they do not produce *Work* (Carnot [1783] 1786, pp. 58–59). In fact, to produce work, other intermediary mechanisms are necessary. As in the previous sections, the discussion dealt with Inertia and *Principle of virtual laws*, in the following the reader finds scientific arguments about Carnot's physical—and—mathematical roots devoted to the concept of Inertia and *Work—Principle of virtual laws*.

The *Principle of virtual laws* (also *law, works, work, displacement, displacements*, depending on specific cases) in *Mechanics* throughout history encountered

[29] Hereafter for Carnot's translation: (Pisano et al. 2021).

several intellectual difficulties, both for fundamental and applied research. For one example, see *rational mechanics*, which did not allow indeterminable situations. In the *mechanics of rigid bodies*, the application of the *principle of virtual law* solves certain problems related to *vincula.* The latter *vincula* are related to *constraint reactions*, such as *auxiliary unknowns*, which are then eliminated by substitution during the solution of single static problems; idem by *friction constraints* (Coulomb 1784, 1785, 1799) one can formulate certain *constitutive laws*. On the contrary, in *continuum mechanics,* the *Principle of virtual laws* is very important. In fact, one can see some particular values of the mathematical expression of the *principle of virtual law*, e.g., by using the *method of finite elements.*

The *principle of virtual works* states that the *total virtual work* performed by all the forces acting on a system in static equilibrium is zero for a set of infinitesimal *virtual displacements.* The infinitesimal displacements are virtual because they need not be obtained by a displacement that actually occurs in the physical system. The virtual work is the work performed during the virtual displacements, which can be arbitrary, and are consistent with the constraints of the system. Its common mathematical expression is:

$$\delta W = \sum_i F_i^{(a)} \delta s_i = 0$$

The theory of mechanical machines may be based on the *principle of virtual work*, and thought of as a consequence of the principle of the impossibility of perpetual motion, e.g., applied to machines and constraints: *it is impossible that the reactions of the constraints on the actions of the bodies, which make up the machine, produce positive work*. In other words, it is impossible for forces of constraints to produce work:

$$\sum_i R_i ds_i \leq 0$$

The generalization of Lazare Carnot's *Principle of virtual work* is historically very important because it precedes Lagrange's approach[30] on the subject. Carnot

[30] In 1762, in *Application de la méthode exposée dans la mémoire précédente à la solution des différents problèmes de dyna mique* (Lagrange 1762, 1764, 1788, 1797, 1806, 1813, 1870–1873, 1881, 1892, 1973a, b) Lagrange published the results of his research on the *Principle of Least action.* He perfected its formulation and substantiated it with a convincing proof. Nevertheless, before this Lagrange conceived of a more general principle than that of least action. In a letter to Euler on 24th November 1759, Lagrange wrote about having composed elements of differential calculus and mechanics and developing the true metaphysics of its principle. Euler's *Correspondence with Joseph Louis de Lagrange* (*Opera Omnia.* Series IV, Vol. I (Lagrange 1892, Tome XIV, pp. 170–174; see also: Lagrange 1973a, pp. 137–145; Euler [1730–1731] 1738, [1773] 1774, 1736, 1749, [1750] 1752, 1751, 1757, 1775; Bernoulli 1727). Generally speaking, one can also see: Johann Bernoulli, *Theoremata selecta pro conservatione virium vivarum demonstranda et esperimenta confirmanda* (Bernoulli 1727), *Remarques sur le principe de la conservation des forces vives pris dans un sens général* Bernoulli (Bernoulli [1748] 1750). D'Alembert's *Traité de dynamique* (D'Alembert [1743]

began by stating his principles, which he also referred to as laws, to underline their empirical content. In his *Principes* (1803) Carnot wrote:

> Any motion that, when imparted to a system of bodies, has no effect on the intensity of the actions that they exert or can exert on each other in the course of any other motions imparted to them, will be named geometric. (Coulomb 1784, §136, p. 108).

On the subject, some years before, in *Essai* (1786) he wrote:

> First law. *The reaction is always equal and opposite to the action.*
>
> Second law. *When two hard bodies act on each other by collision or pressure—that is to say, they act on each other in virtue of their impenetrability—their relative speed, immediately after the interaction, is always null.* (Carnot [1783] 1786, pp. 21–22. Carnot's *italic*).

Generally speaking, the first law dealt with the fact that all bodies change their state of rest—or—motion always due to the action of another body. All bodies resist this change of state. Carnot often used the term "force d'inertie".[31] On this subtle concept Carnot was distant, e.g., from Euler's formulation who considered it as a theorem (theorem 7; see also theorem 6) by means of activity (force) and passivity (Inertia); typically of applied mechanics. In detail, Euler's theorem founding principles of dynamics strictly follows those assumed in *Mechanica sive motus scientia analytice exposita* (1736) where he tried to reformulate Newtonian mechanics in a more systematic form.

> PROPOSITION 7. THEOREM. 56. A body remains in a state of absolute rest, unless it is disturbed to move by some external cause. (Euler 1736, p. 21; Cfr. Euler 1749, [1750] 1752, 1751, 1757, 1775; Bernoulli 1727).

Thus, the *Axioms or Laws of motion à la* Newton are now presented as theorems which can be demonstrated. This methodological difference assumed a great importance for the development of mechanics and successive applied mechanics. It means that in certain cases, the principle of Inertia is not a principle but a mere theorem; so it can be proved within a certain mathematics. Euler also assumed that laws of mechanics can be deduced without any recourse to experiments. On can also see D'Alembert ([1743] 1758) who called it "loi" (law) and not a theorem, we read:

> II. Law. 6. A body once put in motion by whichever cause, must persevere uniformly and in straight line, unless a new cause, different from that has caused the motion, will act on it. (D'Alembert [1743] 1758, p. 4).

Coming back to Carnot, his Inertia force is a result of two combined motions. In his words:

> A body that one forces to change its state of rest or of motion resists (XI) the agent which produces the change, & it is this resistance which one calls the force of Inertia. To evaluate

1758, 1751–1780, 1767) and *Lagrange's Recherches sur la libration de la Lune* (Lagrange 1764; Cfr. Pisano et al. 2021).

[31] Carnot [1783] 1786, pp. 21–22; see also p. 20, p. 60, p. 61, pp. 62–63, pp. 68–69, p. 77, p. 78, p. 81, p. 86, p. 97.

> this force it is necessary then to decompose the actual motion of the body into two parts, one of which is [the motion] that it will have the instant afterwards. [because] The other [part of the motion] will evidently be that which must be destroyed in order to force the body's change of state, that is to say, the resistance with which it opposes this change or its force of Inertia, from which it is easy to conclude, that the force of Inertia of a body is the resultant of its actual motion & the motion equal and opposite to that which it must have the following instant. (Carnot [1783] 1786, footnote, p. 61).

The second law in *Essai* concerns hard (or completely soft) bodies. Carnot thought that the content of this law allowed moving the attention to the hard bodies only. Thus, the elastic bodies can be re–conducted to hard bodies. But this assumption was clearly a forced justification. By applying his principles to a system of free hard bodies (Pisano et al. 2021, see endotes), Carnot formulated a first principle of mechanics as in the following:

$$\sum mVU \cos Z = 0$$

It is the "first fundamental equation of mechanics (E)" (Carnot [1783] 1786, p. 27): m is the mass of the corpuscles of the system, V the true velocity after the impact, U the lost velocity (such that $W = V + U$ is the velocity the mass would have before the impact) and Z the angle between V and U. At this point Carnot introduced the concept of *geometric motion*.

> XVI. [...] *if a system of bodies moves from a given position, with an arbitrary motion, but such that it was possible also to make another absolutely equal and directly opposite motion, then each of these motions will be called a geometric motion* [...]. (Carnot [1783] 1786, p. 28. Carnot's italic; see also *Ivi*, pp. 29–34, pp. 41–45)

In *Principes* (1803) (Carnot 1803) the definition was formulated in a slightly different manner:

> DEFINITIONS. 136. *Any motion will be called geometric if, when it is impressed upon a system of bodies, it has no effect on the intensity of the actions that they do or can exert on each other when any other motion is impressed upon them.* (Carnot 1803, p. 108; Carnot's italic)

The first one is purely geometric: *geometric motions* are reversible and congruent with constraints. The second one, by means of "action", alluded to force or Work, as well. In *Essai*, geometric motions could be also infinitesimal. These *speeches of thoughts* move directly to *Principes*, too (Carnot 1803, Theorem IX, p. 130). In effect, even if Carnot was not interested in infinitesimal cases, the finite—or—infinitesimal nature of geometrical motion, in the end made no difference for Carnot's formulations. In fact, he used a quantity u associated to a single geometric motion. He called it *geometric velocity,* which, in modern terms is not so far from *virtual velocity* or *virtual displacement*. Geometric motion was one of his main contributions to mechanics. By means of geometric motion, he wrote his first fundamental equation of motion when the true velocity V—after the impact – substituted the geometric velocity u, as in the following:

$$\sum muU \cos z = 0 \quad (z \text{ is the angle between } u \text{ and } U).$$

Carnot called this equation the "second fundamental equation of mechanics (F)" (Carnot [1783] 1786, p. 32); by changing u among all possible geometric motions, one can obtain all equations needed to find the lost motions U of all masses implicated. In this way, the following problem is solved: given the initial velocities V of a system of masses to find the final velocities W after the impact. Given U, the final velocities are:

$$W = U + V$$

By taking into account the discussion on hard bodies as above cited, this result could be uncorroborated simply because hard bodies are ideal bodies, so far from the practical use of mechanics within machines theory. Carnot tried to solve this correlated problem by moving towards real cases, avoiding any a priori assumptions on Inertia or absolute space and time. Instead, to consider the motion lost in the impact, he used the motion lost by imperceptible degrees and identifies mU (the lost motion) with a force F. We can then write the second fundamental equation "F" expressed by the following form:

$$\sum Fu \cos z = 0$$

In *Essai*, Carnot could have alluded to the *Principle of virtual work* most by means of the application of his *geometric motion*, i.e., *virtual velocities* (Carnot [1783] 1786, XXXIV, pp. 68–70). As above cited, Carnot's theory of *geometric motions*, coincided with velocities and not with displacements. Thus, Carnot avoided a formulation of the *Principle of virtual work* by infinitesimal displacements. The latter could have produced some scientific embarrassment with respect to his projects (Pisano et al. 2021; Galluzzi 1979). Furthermore, by applying the principle of virtual velocity to mechanical machines, the (forces–)weight that balance each other are reciprocal to their virtual velocities. Incidentally, the two conceptually different approaches/formulations could be mathematically equivalent. In particular, Lazare Carnot in applying the principle of virtual work also discussed the role played by forces applied to bodies. In his words:

> Fundamental Theorem
>
> General Principle of equilibrium & of motion in Machines.
>
> XXXIV. *Whatever the state of rest or of motion found in any system of forces applied to a Machine, if we suddenly give it* [the system] *an arbitrary geometric motion, without changing these forces,* [then] *the sum of the products of each force and each* [initial] *speed* [when taken from the point of application, and in the direction of the force] *will be equal to zero.*

That is to say, therefore, that in naming by F each of these forces (1), u the velocity at the first point of application, & z the angle between the directions of F & u, and if one imparts a geometric motion to the Machine, then it must be proved that for the whole system, $[\Sigma]F u \cos z=0$. But, this equation is precisely the equation (AA) [$\Sigma F u \cos Z=0$ (Carnot [1783] 1786, p. 63] found (XXX) [*Ivi*, pp. 60–63] which is nothing else at root but the same [second] fundamental equation (F) [$\Sigma m u U \cos z=0$ (*Ivi*, p. 32] presented in a different form.
It is easy to perceive that this general principle is only, properly speaking, that *Descartes* to which one gives a sufficient generalisation, for it to include not only all the conditions between two forces, but as well all those of equilibrium & motion in a system composed of any number of forces. Also, the first consequence of this theorem will be the principle of *Descartes*, rendered complete by the conditions that we have seen were omitted by him (V). (Carnot [1783] 1786, § XXXIV, pp. 68–70 and footnote "(I)". Carnot's italic and Capital letters)

By working on these subjects, Carnot aimed to obtain a mathematical expression of its invariant with respect to all possible kinds of working substances. Consequently, to find invariants with regard to the efficiency (and reversibility) of a mechanical machine was necessary. In fact, the *principle of virtual work* establishes the condition of *equilibrium* of the forces that act on the bodies to produce work. Coming back to his *first* equation "E" (Carnot [1783] 1786, p. 27) and *second* one "F" (Carnot [1783] 1786, p. 32) and taking into account $\sum F u \cos z = 0$, one can generalize to multi–body systems and write:

$\sum mVU \cos(\langle \vec{U}, \vec{V} \rangle) = 0$ (E) $\sum muU \cos(\langle \vec{U}, \vec{u} \rangle) = 0$ (F)	m = mass of the body W = velocity before interaction V = velocity after interaction $U = W–V$ u = *arbitrary geometric motion*

In short:

- The mass of the parts of a machine.
- Global magnitudes, abstracting from the mass of the mechanism.
- Kinematics first, then dynamics, and statics is a special case of dynamics.
- A theory of machines concerns a theory of the *communication of motions*.
- A machine is a connected system of (hard) bodies.
- The connections between the bodies constrain the *communication* of motion of the bodies.
- The theory of interaction–collisions by means of insensible degrees (Cfr. Carnot 1803, §293, pp. 261–262) as the result of a sequence of infinitesimally small percussions.

Taking into account the paralleling between *Essai* ([1783] 1786) and *Principes* (1803) about the *Principle of virtual work*, we can write down the *Law of conservation* for *hard* (plastic) *bodies* in the following way:

$\sum_i m\vec{U}_i = 0$	m_i = mass of i–th body [for isolate system] U_i = velocity lost (by that body) during the collision W_i = velocity before interaction V_i = velocity after interaction $\vec{U}_i = \vec{W}_i - \vec{V}_i$

Therefore, using the hypotheses of *parfaitement élastiques* bodies, e.g., in *Principes* (Carnot 1803, p. 105) we can write:

$\sum_i m_i\vec{U}_i\vec{V}_i = 0$	V_i = velocity after interaction is the same for all bodies; Generalization for all bodies using n–elasticity index

Lazare Carnot proposed an attempt at generalization from plastic bodies to all bodies both in *Essai* (Carnot [1783] 1786, pp. 15–22) then in *Principes* (Carnot 1803, pp. 103–106, pp. 131–146). But in these two books, the generalizations are differently presented. Precisely, the first has an inverse procedure with respect to second one. At this stage, the *law of conservation of kinetic energy* for soft bodies can be formulated as in the following:

$$\sum_i m_i\vec{W}_i^2 = \sum_i m_i V_i^2 = 0$$

By introducing *geometric motions* (Carnot [1783] 1786) and in regard with the above *Law of conservation,* we have:

$\sum_i m_i\vec{U}_i \cdot \vec{u}_i = 0$	m_i = mass of the i–th body U_i = velocity lost (by that body) during the collision u_i = velocity called "mouvement géométrique"

u is the velocity of any geometric $motion$.

However, mathematically u is an indeterminate variable. Its specification produces an equation applicable to the physical system considered. For example, given

$$\vec{u}_i = \text{const}$$

we have an extension of the *Principle of virtual velocity* to the collision of several bodies and we can write:

$$\sum_i m_i\vec{U}_i \cdot \vec{u}_i = 0 \rightarrow \vec{u}\sum_i m_i\vec{U}_i = 0$$

Instead, given.

$$u - arbitrariness.$$

then, the equation becomes

$$\sum_i m_i \vec{U}_i = 0$$

$$\vec{U}_i = \vec{W}_i - \vec{V}_i$$

Therefore, we have the following situation:

- A theory of interacting bodies by means of collisions.
- A collision is a basic phenomenon. Continuously accelerated motion is a limiting case of a system driven by a series of pulses.
- Newton's second law is replaced by Carnot's second fundamental equation (F) for a system of *n*–bodies.
- Due to the arbitrariness of $u_{i,}$ it can be a constant, as well. Thus, geometric uniform motions applied to all bodies follows.
- In addition, by considering another *ad hoc* geometric motion

$$\vec{u}_i = \vec{\omega} \times \vec{r}_i$$

(e.g., the rotation of the system with angular velocity around a fixed axis) and by using the properties of the triple product and the arbitrariness of "$\vec{\omega}$", we have the following *Laws of conservation* as *invariants* of motion:

$\sum_i m_i \vec{W}_i = \sum_i m_i \vec{V}_i$	*Law of conservation of the total–quantity–of–motion.*
$\sum_i m_i \vec{r}_i \times \vec{W}_i = \sum_i m_i \vec{r}_i \times \vec{V}_i$	*Law of conservation of the total–angular–Momentum.*

3 Inertia of Uniform Circular Motion and the Platonic Myth

Almost at the beginning of the first Day of the *Dialogo sopra i due massimi sistemi del mondo tolemaico e copernicano* (Galilei 1632, 1967) Salviati, in agreement with Simplicio and Sagredo, points out the perfection of the universal order: all the positions and movements of celestial bodies are as perfect as possible. Rectilinear motion is infinite in its nature because an object moves continuously away from the place where it is. This means that it cannot be the natural motion of celestial bodies (Fig. 3):

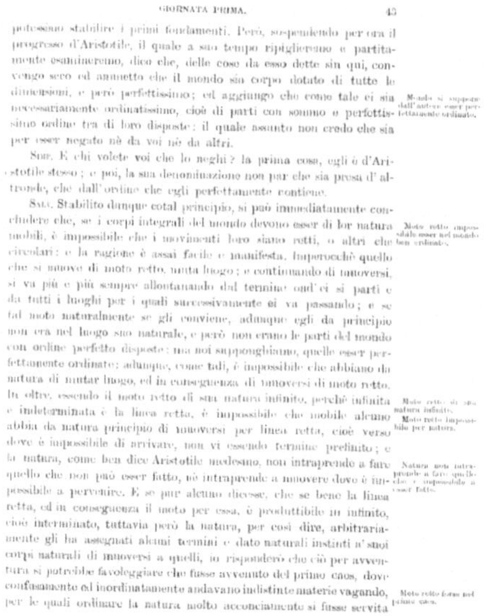
GIORNATA PRIMA. 43

potessimo stabilire i primi fondamenti. Però, sospendendo per ora il progresso d'Aristotile, il quale a suo tempo ripiglieremo e partitamente esamineremo, dico che, delle cose da esso dette sin qui, convengo seco ed ammetto che il mondo sia corpo dotato di tutte le dimensioni, e però perfettissimo; ed aggiungo che come tale ei sia necessariamente ordinatissimo, cioè di parti con sommo e perfettissimo ordine tra di loro disposte: il quale assunto non credo che sia per esser negato nè da voi nè da altri. Mondo si suppone dall'autore esser perfettamente ordinato.

Sagr. E chi volete voi che lo neghi? la prima cosa, egli è d'Aristotile stesso; e poi, la sua denominazione non par che sia presa d'altronde, che dall'ordine che egli perfettamente contiene.

Salv. Stabilito dunque cotal principio, si può immediatamente concludere che, se i corpi integrali del mondo devono esser di lor natura mobili, è impossibile che i movimenti loro siano retti, o altri che circolari: e la ragione è assai facile e manifesta. Imperocchè quello che si muove di moto retto, muta luogo; e continuando di muoversi, si va più e più sempre allontanando dal termine ond'ei si partì e da tutti i luoghi per i quali successivamente ei va passando; e se tal moto naturalmente se gli conviene, adunque egli da principio non era nel luogo suo naturale, e però non erano le parti del mondo con ordine perfetto disposte: ma noi supponghiamo, quelle esser perfettamente ordinate; adunque, come tali, è impossibile che abbiano da natura di mutar luogo, ed in conseguenza di muoversi di moto retto. Moto retto impossibile esser nel mondo ben ordinato. In oltre, essendo il moto retto di sua natura infinito, perchè infinita e indeterminata è la linea retta, è impossibile che mobile alcuno abbia da natura principio di muoversi per linea retta, cioè verso dove è impossibile di arrivare, non vi essendo termine prefinito; e la natura, come ben dice Aristotile medesimo, non intraprende a fare quello che non può esser fatto, nè intraprende a muovere dove è impossibile a pervenire. Moto retto di sua natura infinito. Moto retto impossibile per natura. Natura non intraprende a fare quello che è impossibile a esser fatto. E se pur alcuno dicesse, che se bene la linea retta, ed in conseguenza il moto per essa, è produttibile in infinito, cioè interminato, tuttavia però la natura, per così dire, arbitrariamente gli ha assegnati alcuni termini e dato naturali instinti a' suoi corpi naturali di muoversi a quelli, io risponderò che ciò per avventura si potrebbe favoleggiare che fusse avvenuto del primo caos, dove confusamente ed inordinatamente andavano indistinte materie vagando, per le quali ordinare la natura molto acconciamente si fusse servita Moto retto forse nel primo caos.

[Salv] [...] if all integral bodies in the world are by nature movable, it is impossible that their motions should be straight, or anything else but circular; and the reason is very plain and obvious. For whatever moves straight changes place and, continuing to move, goes ever farther from its starting point and from every place through which it successively passes. If that were the motion which naturally suited it, then at the beginning it was not in its proper place. So then the parts of the world were not disposed in perfect order. But we are assuming them to be perfectly in order; and in that case, it is impossible that it should be their nature to change place, and consequently to move in a straight line. Besides, straight motion being by nature infinite (because a straight line is infinite and indeterminate), it is impossible that anything should have by nature the principle of moving in a straight line; or, in other words, toward a place where it is impossible to arrive, there being no finite end. For nature, as Aristotle well says himself, never undertakes to do that which cannot be done, nor endeavours to move whither it is impossible to arrive.

Fig. 3 On the Motion and its Cause (Galilei 1632, I, p. 43, left side; Galilei 1967, I, p. 19, right side) *Source* French National Library–Gallica. Public Domain

Galileo then, further clarifies the cosmological role of rectilinear motion (Fig. 4):

Galileo's opinion can be expressed like this: Rectilinear motion is not the proper condition of bodies when they are set in their natural order. It has, instead, a modifying character since it produces a translation which can create disorder from order as well as order from disorder. When celestial bodies, after their creation, were moved from the place in which they were created in the primitive chaos and located in their natural place, rectilinear motion created order from a previous chaotic condition. As soon as celestial bodies reached their natural position, their rectilinear motion ended and the uniform circular one began.[32] It must be noted that in the quoted passages Galileo regarded rectilinear motion as the kind of motion able to modify a physical state. However, he did not specify its nature, by which we mean if and how velocities changed. For, a question arises: *is it an accelerated or a uniform*

[32] Galileo did not believe in the ellipticity of the orbits. However, he was perfectly aware that the motion of the planets around the Sun is not regular. The irregularities of the planetary motions were known since ancient times. While speaking of uniform circular motion of the planets, he was, thence, referring to their mean motion.

44 DIALOGO SOPRA I DUE MASSIMI SISTEMI DEL MONDO.

de i movimenti retti, i quali, sì come movendo i corpi ben costituiti gli disordinano, così sono acconci a ben ordinare i pravamente disposti; ma dopo l'ottima distribuzione e collocazione è impossibile che in loro resti naturale inclinazione di più muoversi di moto retto, dal quale ora solo ne seguirebbe il rimuoversi dal proprio e natural luogo, cioè il disordinarsi. Possiamo dunque dire, il moto retto servire a condur le materie per fabbricar l'opera, ma fabbricata ch'ell'è, o restare immobile, o, se mobile, muoversi solo circolarmente; se però noi non volessimo dir con Platone, che anco i corpi mondani, dopo l'essere stati fabbricati e del tutto stabiliti, furon per alcun tempo dal suo Fattore mossi di moto retto, ma che dopo l'esser pervenuti in certi e determinati luoghi, furon rivolti a uno a uno in giro, passando dal moto retto al circolare, dove poi si son mantenuti e tuttavia si conservano: pensiero altissimo e degno ben di Platone, intorno al quale mi sovviene aver sentito discorrere il nostro comune amico Accademico Linceo; e se ben mi ricorda, il discorso fu tale. Ogni corpo costituito per qualsivoglia causa in istato di quiete, ma che per sua natura sia mobile, posto in libertà si moverà, tutta volta però ch'egli abbia da natura inclinazione a qualche luogo particolare; chè quando e' fusse indifferente a tutti, resterebbe nella sua quiete, non avendo maggior ragione di muoversi a questo che a quello. Dall'aver questa inclinazione ne nasce necessariamente che egli nel suo moto si anderà continuamente accelerando; e cominciando con moto tardissimo, non acquisterà grado alcuno di velocità, che prima e' non sia passato per tutti i gradi di velocità minori, o vogliamo dire di tardità maggiori: perchè, partendosi dallo stato della quiete (che è il grado di infinita tardità di moto), non ci è ragione nissuna per la quale e' debba entrare in un tal determinato grado di velocità, prima che entrare in un minore, ed in un altro ancor minore prima che in quello; anzi par molto ben ragionevole passar prima per i gradi più vicini a quello donde ei si parte, e da quelli a i più remoti; ma il grado di dove il mobile piglia a muoversi è quello della somma tardità, cioè della quiete. Ora, questa accelerazion di moto non si farà se non quando il mobile nel muoversi acquista; nè altro è l'acquisto suo se non l'avvicinarsi al luogo desiderato, cioè dove l'inclinazion naturale lo tira; e là si condurrà egli per la più breve, cioè per linea retta. Possiamo dunque ragio-

But someone might say nevertheless that although a straight line (and consequently the motion along it) can be extended in infinitum (that is to say, is unending), still nature has, so to speak, arbitrarily assigned to it some terminus, and has given her natural bodies natural instincts to move toward that. And I shall reply that this might perhaps be fabled to have occurred in primordial chaos, where vague substances wandered confusedly in disorder, to regulate which nature would very properly have used straight motions. By means of these, just as well-arranged bodies would become disordered in moving, so those which were previously badly disposed might be arranged in order. But after their optimum distribution and arrangement it is impossible that there should remain in them natural inclinations to move any more in straight motions, from which nothing would now follow but their removal from their proper and natural places; which is to say, their disordering. We may therefore say that straight motion serves to transport materials for the construction of a work; but this, once constructed, is to rest immovable—or, if movable, is to move only circularly.

Fig. 4 On the Effects of Rectilinear Motion (Galilei 1632, I, pp. 43–44, left side; Galilei 1967, I, pp. 19–20, right side) *Source* French National Library–Gallica. Public Domain

motion? This initially missing reference is not casual. Galileo is distinguishing the properties of rectilinear motion from those of circular motion, regardless of whether the rectilinear motion is accelerated or uniform. Nonetheless, we will see that the distinction between uniform and accelerated rectilinear motion was highlighted as soon as Galileo specified how celestial bodies reached their natural place and began to rotate uniformly. Galileo referred to Plato's myth (Fig. 5):

44 DIALOGO SOPRA I DUE MASSIMI SISTEMI DEL MONDO.

de i movimenti retti, i quali, sì come movendo i corpi ben costituti gli disordinano, così sono accomci a ben ordinare i pravamente disposti; ma dopo l'ottima distribuzione e collocazione è impossibile che in loro resti naturale inclinazione di più muoversi di moto retto, dal quale ora solo ne seguirebbe il rimuoversi dal proprio e natural luogo, cioè il disordinarsi. Possiamo dunque dire, il moto retto servire a condur le materie per fabbricar l'opera, ma fabbricata ch'ell'è, o restare immobile, o, se mobile, muoversi solo circolarmente; se però noi non volessimo dir con Platone, che anco i corpi mondani, dopo l'essere stati fabbricati e del tutto stabiliti, furon per alcun tempo dal suo Fattore mossi di moto retto, ma che dopo l'esser pervenuti in certi e determinati luoghi, furon rivolti a uno a uno in giro, passando dal moto retto al circolare, dove poi si son mantenuti e tuttavia si conservano: pensiero altissimo e degno ben di Platone, intorno al quale mi sovviene aver sentito discorrere il nostro comune amico Accademico Linceo; e se ben mi ricorda, il discorso fu tale. Ogni corpo costituito per qualsivoglia causa in istato di quiete, ma che per sua natura sia mobile, posto in libertà si moverà, tutta volta però ch'egli abbia da natura inclinazione a qualche luogo particolare; chè quando e' fusse indifferente a tutti, resterebbe nella sua quiete, non avendo maggior ragione di muoversi a questo che a quello. Dall'aver questa inclinazione ne nasce necessariamente che egli nel suo moto si anderà continuamente accelerando; e cominciando con moto tardissimo, non acquisterà grado alcuno di velocità, che prima e' non sia passato per tutti i gradi di velocità minori, o vogliamo dire di tardità maggiori: perchè, partendosi dallo stato della quiete (che è il grado di infinita tardità di moto), non ci è ragione nissuna per la quale e' debba entrare in un tal determinato grado di velocità, prima che entrare in un minore, ed in un altro ancor minore prima che in quello; anzi par molto ben ragionevole passar prima per i gradi più vicini a quello donde ei si parte, e da quelli a i più remoti; ma il grado di dove il mobile piglia a muoversi è quello della somma tardità, cioè della quiete. Ora, questa accelerazion di moto non si farà se non quando il mobile nel muoversi acquista; nè altro è l'acquisto suo se non l'avvicinarsi al luogo desiderato, cioè dove l'inclinazion naturale lo tira; e là si condurrà egli per la più breve, cioè per linea retta. Possiamo dunque ragio-

[…] these world bodies, after their creation and the establishment of the whole, were for a certain time set in straight motion by their Maker. Then later, reaching certain definite places, they were set in rotation one by one, passing from straight to circular motion, and have ever since been preserved and maintained in this. […] Every body constituted in a state of rest but naturally capable of motion will move when set at liberty only if it has a natural tendency toward some particular place; for if it were indifferent to all places it would remain at rest, having no more cause to move one way than another. Having such a tendency, it naturally follows that in its motion it will be continually accelerating. Beginning with the slowest motion, it will never acquire any degree of speed (velocità) without first having passed through all the gradations of lesser speed—or should I say of greater slowness? For, leaving a state of rest, which is the infinite degree of slowness, there is no way whatever for it to enter into a definite degree of speed before having entered into a lesser, and another still less before that. It seems much more reasonable for it to pass first through those degrees nearest to that from which it set out, and from this to those farther on. But the degree from which the movable body began to move was that of most extreme slowness; that is to say, from rest.

Fig. 5 On Celestial Bodies Nature, Motion and Plato's myth (Galilei 1632, I, pp. 44–45, left side; Galilei 1967, I, p. 21, right side) *Source* French National Library–Gallica. Public Domain

God used rectilinear motion only initially, to carry the bodies towards their final natural places and to impart them their speeds. After this initial phase, he converted the rectilinear motion into a uniform circular one, which is maintained without further interventions (Galilei 1632, I, p. 45; Galilei 1967, I, p. 21). The rectilinear motion used by God in this operation is uniformly accelerated. Galileo clarified that God imparted to the celestial bodies a rectilinear motion passing through all degrees of speed or slowness ("gradi di velocità o tardità" (Galilei 1632, I, p. 44; Galilei 1967, I, p. 21) until each of them reaches its own place at a certain speed. This motion is accelerated because a body cannot instantaneously reach the speed v with which it will move circularly. This means it is impossible that each body acquires its speed v without passing through all the degrees of speed from 0 to v. According to Galileo, God contrived the mechanism of creation and positioning of celestial bodies in their natural place, so that when a planet reached the final speed v, God converted the rectilinear accelerated motion into a circular uniform one whose speed is equal to that reached by the body in the last instant of its previous rectilinear motion. This means that:

The natural, Inertial motion of celestial bodies is the circular uniform one.

One might object that, remaining in Galileo's way of reasoning, a further possibility might subsist: God might have accelerated a body until a certain velocity v, after which a uniform rectilinear motion might have begun and the planet could have reached its natural place under the action of such a uniform motion. Probably Galileo did not contemplate this hypothesis because he considered the mechanism he devised simpler and more harmonious.

At the end of his reasoning Galileo offered a synthesis of his whole argument (Fig. 6):

GIORNATA PRIMA. 45

nevolmente dire che la natura, per conferire in un mobile, prima costituito in quiete, una determinata velocità, si serva del farlo muover, per alcun tempo e per qualche spazio, di moto retto. Stante questo discorso, figuriamoci aver Iddio creato il corpo, v. g., di Giove, al quale abbia determinato di voler conferire una tal velocità, la quale egli poi debba conservar perpetuamente uniforme: potremo con Platone dire che gli desse di muoversi da principio di moto retto ed accelerato, e che poi, giunto a quel tal grado di velocità, convertisse il suo moto retto in circolare, del quale poi la velocità naturalmente convien esser uniforme.

SAGR. Io sento con gran gusto questo discorso, e maggiore credo che sarà doppo che mi abbiate rimossa una difficultà: la quale è, che io non resto ben capace come di necessità convenga che un mobile, partendosi dalla quiete ed entrando in un moto al quale egli abbia inclinazion naturale, passi per tutti i gradi di tardità precedenti, che sono tra qualsivoglia segnato grado di velocità e lo stato di quiete, li quali gradi sono infiniti; sì che non abbia potuto la natura contribuire al corpo di Giove, subito creato, il suo moto circolare, con tale e tanta velocità.

SALV. Io non ho detto, nè ardirei di dire, che alla natura e a Dio fusse impossibile il conferir quella velocità, che voi dite, immediatamente; ma dirò bene che *de facto* la natura non lo fa; talchè il farlo verrebbe ad esser operazione fuora del corso naturale, e però miracolosa (1).

SAGR. Adunque voi credete che un sasso, partendosi dalla quiete, ed entrando nel suo moto naturale verso il centro della Terra, passi

(1) Nell'esemplare dell'edizione originale posseduto dalla Biblioteca del Seminario di Padova (intorno al qual esemplare veggasi l'Avvertimento), e precisamente sul margine inferiore della pag. 33, la quale termina con le parole « e però miracolosa », si legge, scritto di mano di GALILEO, quanto appresso:

« Muovasi con qual si voglia velocità qual si sia poderosissimo mobile, ed incontri qual si voglia corpo costituito in quiete, ben che debolissimo e di minima resistenza; quel mobile, incontrandolo, già mai non gli conferirà immediatamente la sua velocità: segno evidente di che ne è il sentirsi il suono della percossa, il quale non si sentirebbe, o per dir meglio non sarebbe, se il corpo che stava in quiete ricevesse, nell'arrivo del mobile, la medesima velocità di quello. »

This assumed, let us suppose God to have created the planet Jupiter, for example, upon which He had determined to confer such-and-such a velocity, to be kept perpetually uniform forever after. We may say with Plato that at the beginning He gave it a straight and accelerated motion; and later, when it had arrived at that degree of velocity, converted its straight motion into circular motion whose speed thereafter was naturally uniform.

Fig. 6 On God and Motion (Galilei 1632, I, pp. 44–45, left side; Galilei 1967, I, p. 21, right side) *Source* French National Library–Gallica. Public Domain

It is, thus, clear that for Galileo the final movement of the planets, their circular uniform one, is their real Inertial motion.[33] The planets also rotate around their axes. In particular, Galileo addressed the diurnal rotation of the Earth in the second Day of the *Dialogo*. In the final section of this Day, he claimed that, since both the annual and the diurnal rotations take place from West to East (counterclockwise), they were originated by the same principle. Hence, the reason that determines the eternity and Inertiality of the annual rotation is the same as that which determines the diurnal rotation (Fig. 7):

[33] In the Fourth Day of the *Discorsi e dimostrazoni intorno a due nuove scienze* (Galilei 1638) Galileo re-proposed a synthetic version of the platonic myth as an annotation to the Theorem III, Proposition III (Galilei 1638, IV, pp. 281–282; Galilei 1914, IV, pp. 258–259). The essence of the myth is the same as in the *Dialogo*.

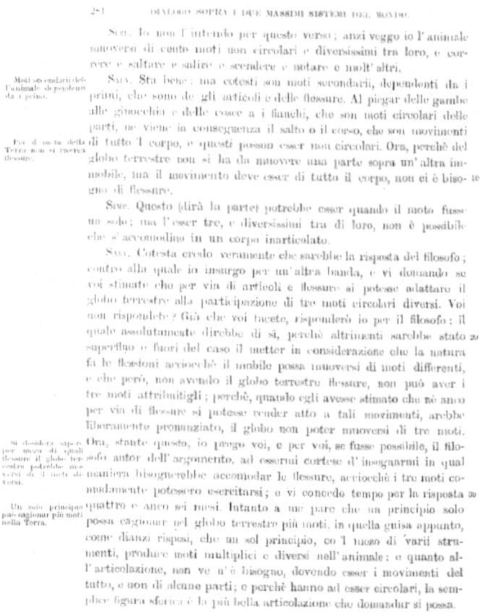
284 DIALOGO SOPRA I DUE MASSIMI SISTEMI DEL MONDO.

SIMP. Io non l'intendo per questo verso; anzi veggo io l'animale muoversi di cento moti non circolari e diversissimi tra loro, e correre e saltare e salire e scendere e notare e molt' altri.

Moti secondarii dell'animale dependenti da i primi.

SALV. Sta bene: ma cotesti son moti secondarii, dependenti da i primi, che sono de gli articoli e delle flessure. Al piegar delle gambe alle ginocchia e delle cosce a i fianchi, che son moti circolari delle parti, ne viene in conseguenza il salto o il corso, che son movimenti di tutto 'l corpo, e questi posson esser non circolari. Ora, perchè del globo terrestre non si ha da muovere una parte sopra un' altra immobile, ma il movimento deve esser di tutto il corpo, non ci è bisogno di flessure.

Nel moto della Terra non si ricerca flessure.

SIMP. Questo (dirà la parte) potrebbe esser quando il moto fusse un solo; ma l'esser tre, e diversissimi tra di loro, non è possibile che s'accomodino in un corpo inarticolato.

SALV. Cotesta credo veramente che sarebbe la risposta del filosofo; contro alla quale io insurgo per un'altra banda, e vi domando se voi stimate che per via di articoli e flessure si potesse adattare il globo terrestre alla participazione di tre moti circolari diversi. Voi non rispondete? Già che voi tacete, risponderò io per il filosofo: il quale assolutamente direbbe di sì, perchè altrimenti sarebbe stato superfluo e fuori del caso il metter in considerazione che la natura fa le flessioni acciocchè il mobile possa muoversi di moti differenti, e che però, non avendo il globo terrestre flessure, non può aver i tre moti attribuitigli; perchè, quando egli avesse stimato che nè anco per via di flessure si potesse render atto a tali movimenti, arebbe liberamente pronunziato, il globo non poter muoversi di tre moti. Ora, stante questo, io prego voi, e per voi, se fusse possibile, il filosofo autor dell'argomento, ad essermi cortese d'insegnarmi in qual maniera bisognerebbe accomodar le flessure, acciocchè i tre moti comodamente potessero esercitarsi; e vi concedo tempo per la risposta quattro e anco sei mesi. Intanto a me pare che un principio solo possa cagionar nel globo terrestre più moti, in quella guisa appunto, come dianzi risposi, che un sol principio, co 'l mezo di varii strumenti, produce moti multiplici e diversi nell'animale: e quanto all'articolazione, non ve n'è bisogno, dovendo esser i movimenti del tutto, e non di alcune parti; e perchè hanno ad esser circolari, la semplice figura sferica è la più bella articolazione che domandar si possa.

Un solo principio può cagionar più moti nella Terra.

In the meantime it seems to me that a single principle can cause more than one motion of the earth in exactly the same way in which I have just told you that a single principle, by means of various instruments, produces many and diverse motions in animals. As to joints, there is no need for them, the required movements being of the whole and not of some parts only. And since they must be circular, the simpler spherical shape is the most beautiful joint which one can ask for.

Fig. 7 Relations between Motions and Principles of the Motion (Galilei 1632, p. 284, left side; Galilei 1914, p. 260, right side) *Source* French National Library–Gallica. Public Domain

Thus, according to Galileo, uniform circular motion is Inertial in the case of both planetary orbits and a spherical body rotating around an axis passing through its centre.

4 Motion on an Extended Portion of Earth's Surface

Let us suppose that an impulse be given to a ship on an ocean or sea. How should the motion of the ship continue if the surface of the water were perfectly smooth and no continent or island were interposed? According to Galileo, such motion would be circular around a great circle of the Earth; it would be eternal and the crew of the ship would feel no tendency to recede along the tangent. The motion would be horizontal and Inertial.

In *Istoria e dimostrazioni intorno alle macchie solari* (1613) Galileo wrote:

> Finally, they [the heavy bodies] are indifferent to some movements—as are the same heavy bodies to horizontal motion—to which they have neither inclination, because it is not toward the center of the Earth, nor aversion, because it is not away from the same center. And therefore, with all the external impediments removed, a heavy body on the spherical surface concentric to the Earth will be indifferent to rest and to movement toward any part of the horizon, and it will remain in the state in which it has been put; that is, if it has been put in a state of rest it will remain in it, and if it has been put in motion, toward the west, for example, it will maintain the same state. And thus a ship, for example, having received one single time some impetus, would move continuously through a quiet sea around our globe without ever stopping, and if one were to bring it gently to rest it would perpetually remain at rest, provided that in the first case all the extrinsic impediments could be removed, and in

the second one that no external mobile cause came upon it.[34] (Reeves and Van Helden 2010, p. 125)

Galileo expressed the same train of thought in the Second Day of his *Dialogo* (Fig. 8):

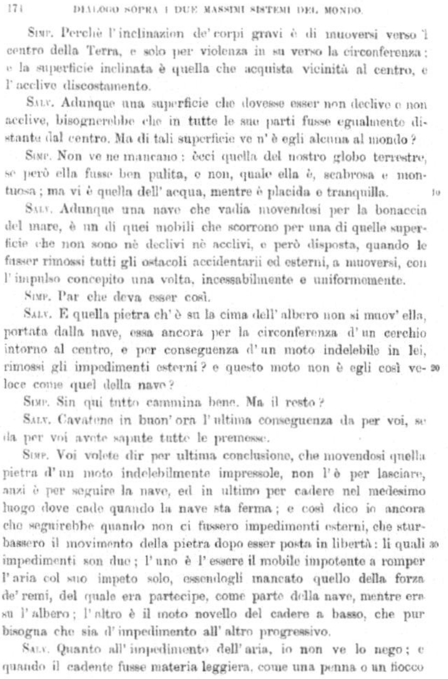

174 DIALOGO SOPRA I DUE MASSIMI SISTEMI DEL MONDO.

SIMP. Perchè l'inclinazion de' corpi gravi è di muoversi verso 'l centro della Terra, e solo per violenza in su verso la circonferenza: e la superficie inclinata è quella che acquista vicinità al centro, e l'acclive discostamento.

SALV. Adunque una superficie che dovesse esser non declive e non acclive, bisognerebbe che in tutte le sue parti fusse egualmente distante dal centro. Ma di tali superficie ve n'è egli alcuna al mondo?

SIMP. Non ve ne mancano: ècci quella del nostro globo terrestre, se però ella fusse ben pulita, e non, quale ella è, scabrosa e montuosa; ma vi è quella dell' acqua, mentre è placida e tranquilla.

SALV. Adunque una nave che vadia movendosi per la bonaccia del mare, è un di quei mobili che scorrono per una di quelle superficie che non sono nè declivi nè acclivi, e però disposta, quando le fusser rimossi tutti gli ostacoli accidentarii ed esterni, a muoversi, con l'impulso concepito una volta, incessabilmente e uniformemente.

SIMP. Par che deva esser così.

SALV. E quella pietra ch'è su la cima dell' albero non si muov' ella, portata dalla nave, essa ancora per la circonferenza d'un cerchio intorno al centro, e per conseguenza d'un moto indelebile in lei, rimossi gli impedimenti esterni? e questo moto non è egli così veloce come quel della nave?

SIMP. Sin qui tutto cammina bene. Ma il resto?

SALV. Cavatene in buon' ora l'ultima conseguenza da per voi, se da per voi avete sapute tutte le premesse.

SIMP. Voi volete dir per ultima conclusione, che movendosi quella pietra d'un moto indelebilmente impressole, non l'è per lasciare, anzi è per seguire la nave, ed in ultimo per cadere nel medesimo luogo dove cade quando la nave sta ferma; e così dico io ancora che seguirebbe quando non ci fussero impedimenti esterni, che sturbassero il movimento della pietra dopo esser posta in libertà: li quali impedimenti son due; l'uno è l'essere il mobile impotente a romper l'aria col suo impeto solo, essendogli mancato quello della forza de' remi, del quale era partecipe, come parte della nave, mentre era su l'albero; l'altro è il moto novello del cadere a basso, che pur bisogna che sia d'impedimento all' altro progressivo.

SALV. Quanto all' impedimento dell' aria, io non ve lo nego; e quando il cadente fusse materia leggiera, come una penna o un fiocco

Salv. Then in order for a surface to be neither downward nor upward, all its parts must be equally distant from the center. Are there any such surfaces in the world?

Simp. Plenty of them; such would be the surface of our terrestrial globe if it were smooth, and not rough and mountainous as it is. But there is that of the water, when it is placid and tranquil.

Salv. Then a ship, when it moves over a calm sea, is one of these movables which courses over a surface that is tilted neither up nor down, and if all external and accidental obstacles were removed, it would thus be disposed to move incessantly and uniformly from an impulse once received?

Simp. It seems that it ought to be.

Fig. 8 Galileo's Concept of Horizontal Plan? (Galilei 1632, II, p. 174, left side; Galilei 1917, II, p. 148, right side) *Source* French National Library–Gallica. Public Domain

This passage confirms that this kind of motion is considered Inertial by Galileo.

It is to be noted that, while speaking of this kind of motion, Galileo never mentioned the tendency to recede along the tangent. As we will see in the next section, he was perfectly aware of this feature connoting circular motion. However, when he referred to this property, he inserted his observations in a context in which

[34] "Finalmente, ad alcuni movimenti [i corpi pesanti] si trovano indifferenti, come pur gl'istessi gravi al movimento orizontale, al quale non hanno inclinazione, poi che ei non è verso il centro della terra, né repugnanza, non si allontanando dal medesimo centro: e però, rimossi tutti gl'impedimenti esterni, un grave nella superficie sferica e concentrica alla terra sarà indifferente alla quiete e ai movimenti verso qualunque parte dell'orizonte, e in quello stato si conserverà nel qual una volta sarà stato posto; cioè se sarà messo in stato di quiete, quello conserverà, e se sarà posto in movimento, v.g. verso occidente, nell'istesso si manterrà: e così una nave, per essempio, avendo una sol volta ricevuto qualche impeto per il mar tranquillo, si moverebbe continuamente intorno al nostro globo senza cessar mai, e postavi con quiete, perpetuamente quieterebbe, se nel primo caso si potessero rimuovere tutti gl'impedimenti estrinseci, e nel secondo qualche causa motrice esterna non gli sopragiungesse." (Galilei 1613, V, pp. 134–135).

he answered the objection of the anti-Copernicans according to which, if the Earth rotated, bodies would be projected into space by the rotation. Therefore, the context is that in which he analysed the behaviour of the bodies located on a rotating surface, rather that of bodies animated by a proper circular motion. So far, we have to conclude that, according to Galileo, the (approximately) uniform circular motions of the planets around the Sun and of the satellites around the planets are Inertial. Thus:

- The rotation of planet around its axis is Inertial;
- The motion on a great circle of the Earth it is also Inertial, provided that the ground produces no friction.

5 Local Circular Uniform Movement

Let us see how Galileo considered local circular motion on the Earth.

In this case, the situation is completely different from those so far analysed, for Galileo recognised that a continuous action is necessary for circular motion to be maintained. Otherwise, the tendency to recede along the tangent would carry the body along a rectilinear path with a uniform motion. Galileo clearly stated in the Second Day of the *Dialogo*:

> Circular motion has the property of casting off, scattering, and driving away from its center the parts of the moving body, whenever the motion is not sufficiently slow or the parts not too solidly attached together.[35] (Galilei 1967, II, p. 132).

The recognition that circular motion has the property of casting off the parts of the moving body is a fundamental step towards the understanding that uniform circular motion is not Inertial, whereas rectilinear uniform is indeed Inertial.

In other passages of the *Dialogo*, Galileo is explicit on the tendency of the object animated by a circular motion to recede along the tangent. In the following passage of the Second Day, we read:

> Up to this point you knew all by yourself that the circular motion of the projector impresses an impetus upon the projectile to move, when they separate, along the straight line tangent to the circle of motion at the point of separation, and that continuing with this motion, it travels ever farther from the thrower. And you have said that the projectile would continue to move along that line if it were not inclined downward by its own weight from which fact the line of motion derives its curvature.[36] (Galilei 1967, II, pp. 193–194).

[35] "Il moto circolare ha facoltà di estrudere dissipare e scacciar dal suo centro le parti del corpo che si muove, qualunque volta o 'l moto non sia assai tardo o esse parti non sian molto saldamente attaccate insieme." (Galilei 1632, II, p. 158).

[36] "Sin qui avete per voi stesso saputo che il moto circolare del proiciente imprime nel proietto impeto di muoversi (quando avviene ch'e' si separino) per la retta tangente il cerchio del moto nel punto della separazione, e, continuando per essa il moto, vien sempre allontanandosi dal proiciente; ed avete detto che per tal linea retta continuerebbe il proietto di muoversi, quando dal proprio peso non gli fusse aggiunta inclinazione all'in giù, dalla quale deriva l'incurvazione della linea del moto." (Galilei 1632, II, p. 220).

This is a perfect description of what happens. An Inertia principle is not explicitly formulated, but it is applied correctly, which is even more important. Circular motion is not considered Inertial, while uniform rectilinear motion is. If there were no impediments, motion along the tangent would continue ad infinitum. These are the basic properties of Inertial motions.

The situation is, thus, rather intricate because, as we have pointed out in our Outline and Introduction above, Galileo considered uniform circular motion to be Inertial in some circumstances and not-Inertial in other circumstances. This is a litmus paper indicating that he did not hold a unitary and complete view of this phenomenon. In order to highlight the difficulties encountered by Galileo in providing a coherent conception of circular motion, we analyse the following example drawn from the Second Day of the *Dialogo*.

Galileo is going to prove that, whatever the diurnal rotation speed of the Earth might be, terrestrial bodies would never be thrown into the sky and begin to orbit around the Earth or fly far from the Earth itself, but, after a certain local flight, they would land on the Earth's surface. Obviously, this is false. We will mention the whole passage by Galileo because, in this way, it is easier to explain our analysis (Fig. 9):

> So let us mark thus a perpendicular line toward the center, AC, and let the horizontal line AB be at right angles to this, along which the motion of projection is made, and which the projectile would continue to follow with uniform motion if its weight did not bend it downward. Now suppose a straight line AE to be drawn from A, making any desired angle with AB, and let us mark off on AB some equal spaces AF, FH, and HK, drawing from these the perpendiculars FG, HI, and KL, down as far as AE. And since as we have remarked on other occasions the falling body starting from rest acquires always a greater degree of speed as time goes on, according to the time elapsed, we can picture the spaces AF, FH, and HK as representing equal times, and the perpendiculars FG, HI, and KL as representing the degrees of speed acquired in the said times. Thus the degree of speed acquired in the whole time AK will, by the line KL, be represented relatively to the degree HI acquired in the time AH, and to the degree FG acquired in the time AF; which degrees KL, HI, and FG obviously have the same ratios as the times KA, HA, and FA. And if other perpendiculars are drawn from arbitrary points marked on the line FA, smaller and smaller degrees will be found ad infinitum, always proceeding toward the point A, which represents the first instant of time and the original state of rest. This withdrawal toward A represents the infinite diminution of the original tendency toward downward motion with the approach of the moving body to the original state of rest, which approach is infinitely augmentable. Now let us find that other diminution of speed which can be made ad infinitum by decreasing the weight of the body. This will be represented by drawing another line from the point A, making a smaller angle than BAE; let this be AD. This, cutting the parallels KL, HI, and FG in the points M, N, and O, shows us the degrees FO, HN, and KM acquired in the times AF, AH, and AK to be less than the other degrees FG, HI, and KL acquired in the same times but by a heavier body, this being a lighter one. And it is obvious that by withdrawal of the line EA toward AB, restricting the angle EAB (which can be done ad infinitum, just as weight can be infinitely decreased), the speed of the falling body and consequently the cause that impeded its projection comes likewise to be diminished ad infinitum. Hence it appears that from a combined diminution ad infinitum of the two causes counter to it, projection cannot be impeded. Reducing the whole argument to a few words, let us say; By restricting the angle EAB, the degrees of speed LK, IH, and GF are diminished. By also withdrawing the parallels KL, HI, and FG toward the angle A, these same degrees are diminished, and both diminutions may proceed ad infinitum. Therefore the downward speed of motion can indeed be diminished so much (admitting of a twofold diminution ad infinitum) that it no

longer suffices to restore the moving body to the surface of the wheel, and consequently to impede its projection or prevent it. On the other hand, then, in order to prevent projection taking place it is necessary that those spaces through which the projectile has to descend in order to get back to the wheel must be made so short and close that however slow the descent of the moving body may be, even if infinitely diminished, it still suffices to take it back there. Hence it would be necessary to find a diminution of these spaces which was not merely infinite, but of an infinity such as to overcome the double infinity accomplished in decreasing the downward speed of the body [...]. The degrees of speed, infinitely diminished by the decrease of the weight of the moving body and by the approach to the first point of motion (the state of rest), are always determinate. They correspond proportionately to the parallels included between the two straight lines meeting in an angle such as the angle BAE, or BAD, or some other angle infinitely acute but still rectilinear. But the diminution of the spaces through which the moving body must go to return to the surface of the wheel is proportional to another sort of diminution included between lines which contain an angle infinitely narrower and more acute than any rectilinear angle whatever.[37] (Galilei 1967, II, pp. 199–200).

---

[37] "Segniamo dunque una linea perpendicolare verso il centro, e sia questa AC, ed ad essa sia ad angoli retti la orizontale AB, sopra la quale si farebbe il moto della proiezione e vi continuerebbe d'andare il proietto con movimento equabile, quando la gravità non lo inclinasse a basso. Intendasi ora dal punto A prodotta una linea retta, la quale con la AB contenga qualsivoglia angolo, e sia questa AE, e notiamo sopra la AB alcuni spazii eguali AF, FH, HK, e da essi tiriamo le perpendicolari FG, HI, KL sino alla AE. E perchè, come altra volta si è detto, il grave cadente, partendosi dalla quiete, va acquistando sempre maggior grado di velocità di tempo in tempo, secondo che l'istesso tempo va crescendo, possiamo figurarci gli spazii AF, FH, HK rappresentarci tempi eguali, e le perpendicolari FG, HI, KL gradi di velocità acquistati in detti tempi, sì che il grado di velocità acquistato in tutto il tempo AK sia come la linea KL rispetto al grado HI acquistato nel tempo AH, e'l grado FG nel tempo AF, li quali gradi KL, HI, FG hanno (come è manifesto) la medesima proporzione che i tempi KA, HA, FA; e se altre perpendicolari si tireranno da i punti ad arbitrio notati nella linea FA, sempre si troverranno gradi minori e minori in infinito, procedendo verso il punto A, rappresentante il primo instante del tempo e il primo stato di quiete: e questo ritiramento verso A ci rappresenta la prima propensione al moto in giù, diminuita in infinito per l'avvicinamento del mobile al primo stato di quiete, il quale avvicinamento è agumentabile in infinito. Troveremo adesso l'altra diminuzion di velocità, che pure si può fare in infinito per la diminuzion della gravità del mobile; e questo si rappresenterà col produrre altre linee dal punto A, le quali contengano angoli minori dell'angolo BAE, qual sarebbe questa AD, la quale, segando le parallele KL, HI, FG ne' punti M, N, O, ci figura i gradi FO, HN, KM acquistati ne i tempi AF, AH, AK, minori de gli altri gradi FG, HI, KL acquistati ne i medesimi tempi, ma questi come da un mobile più grave, e quelli da un più leggiero. Ed è manifesto che col ritirar la linea EA verso AB, ristrignendo l'angolo EAB (il che si può fare in infinito, sì come la gravità in infinito si può diminuire), si vien parimente a diminuire in infinito la velocità del cadente, ed in conseguenza la causa che impediva la proiezione: e però pare che dall'unione di queste due ragioni contro alla proiezione, diminuite in infinito, non possa ella esser impedita. E riducendo tutto l'argomento in brevi parole, diremo: Col ristrigner l'angolo EAB si diminuiscono i gradi di velocità LK, IH, GF; ed in oltre col ritirar le parallele KL, HI, FG verso l'angolo A si diminuiscono pure i medesimi gradi, e l'una e l'altra diminuzione si estende in infinito: adunque la velocità del moto in giù si potrà ben diminuir tanto e tanto (potendosi doppiamente diminuire in infinito), che ella non basti per restituire il mobile sopra la circonferenza della ruota, e per fare, in conseguenza, che la proiezione venga impedita e tolta. All'incontro poi, per far che la proiezion non segua, bisogna che gli spazii per i quali il proietto deve scendere per riunirsi alla ruota, si facciano così brevi ed angusti, che per tarda, anzi pur diminuita in infinito, che sia la scesa del mobile, ella pur basti a ricondurvelo; e però bisognerebbe che si trovasse una diminuzione di essi spazii non solo fatta in infinito, ma di una infinità tale che superasse la doppia infinità che si fa nella diminuzion della velocità del cadente in giù [...]. I gradi della velocità diminuiti in infinito, sì per la diminuzion della gravità del mobile sì per l'avvicinamento al primo termine del moto,

Below we present the figure used by Galileo to describe and exaplain his geometrical reasoning on motion and surfaces.

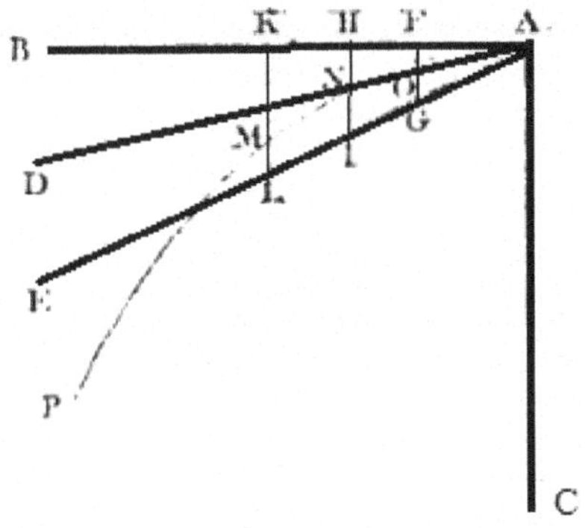

The letter C represents the centre of a circumference whose radius is AC. It indicates a section of the Earth carried out with a plane passing through the Earth's centre. The reference to the angle which is narrower than any acute angle is relative to the angle of contingence between a circumference and a tangent at its point.

Fig. 9 The Figure Used by Galileo to explain his Reasoning in his *Dialogo Source* French National Library–Gallica. Public Domain

Below we specify the explanation in his Dialogo (Galilei 1632):

(1) Figure 9 (1632) represents a great circle, which is a section of the terrestrial sphere, though Galileo spoke of a wheel.
(2) If AF, AH, AK represent the times of a body in free fall, then FI, HG, KL represent the fall-speeds because speeds are as times.
(3) Galileo conceded the most favourable conditions to those who claim that if the rotation speed exceeds a certain amount, the objects will fly away forever from the Earth's surface. Thus, he admited (*Ibidem*) that a body falls slower the lighter it is. Obviously, he himself had proved this not to be the case.
(4) At the nascent state of motion, the fall-speed can be reduced twice ad infinitum because it is possible to choose an arbitrarily light body and because the nascent state of motion can, in fact, be considered.
(5) The possibility that a body escapes along the tangent as a consequence of the speed of the Earth's diurnal rotation depends inversely on the fall speed. Galileo does not specify what kind of inverse dependence he was referring to, but from the whole argument we guess that he was thinking of a simple inverse relation.
(6) Thus, according to Galileo, the fall's speeds can be reduced to an infinitesimal quantity representable by an arbitrarily small rectilinear angle.
(7) From a point K, approaching along the tangent the point A of the Earth's surface, the decrement of the spaces between a point on the tangent and one on the Earth's surface will no longer be inverse to a rectilinear angle, but to the angle of contingence

cioè allo stato di quiete, sempre son determinati, e proporzionatamente rispondono alle parallele comprese tra due linee rette concorrenti in un angolo, conforme all'angolo BAE o BAD o altro in infinito più acuto, ma però sempre rettilineo; ma la diminuzione degli spazii per li quali il mobile ha da ricondursi sopra la circonferenza della ruota è proporzionata ad un'altra sorte di diminuzione, compresa dentro a linee che contengono un angolo infinitamente più stretto ed acuto di qualsivoglia acuto rettilineo, quale sarà questo." (Galilei 1632, II, pp. 225–227).

between the tangent and the circumference. Using modern language, such an angle is an infinitesimal of the second order in respect to a rectilinear angle assumed as a first order infinitesimal. Referring to the language of that time it is enough to claim that the angle of contingence is infinitely smaller than any rectilinear angle.

(8) This implies that whatever the rotation speed, it will never be sufficient for a body to fly away from the Earth or to enter into orbit around our planet because it is inversely as an infinitesimal rectilinear angle. Instead, the spaces between the tangent to the Earth surface and the surface itself, calculated on the perpendicular to the tangent, decrease quicker approaching at A since they decrease as the inverse of the angle of contingence.

It seems to us that the weaknesses of Galileo's argument are two:

(A) At the beginning of his reasoning, Galileo admits abstractly that the rotation speed of the Earth influences the possibility that a body does not return to the terrestrial surface. In the case of the wheel this is equivalent to saying that it does not remain constrained at the wheel. However, this presupposition disappears from Galileo's analysis and he does not identify any further a quantitative relation between the effects on the speed due to the free fall and those due to the rotation of the Earth around its axis. This means, considering the accelerations, to compare the effects due to gravity and those due to the diurnal motion of the Earth around its axis. Galileo, finally, does not consider the centrifugal effects due to the terrestrial rotation and, thus, he constructs a figure which does not represent the whole situation, an incomplete figure.

(B) Galileo compares two angles: A rectilinear angle and the angle of contingence. The former represents velocities, the latter spaces. It is, hence, incorrect to draw Galileo's conclusion: It is not legitimate from a physical standpoint. If the spaces were considered as represented respectively by the areas of the triangles and of the sector of the plane included between the tangent and the circumference, then the two areas would have a finite ratio and the whole of Galileo's reasoning would, anyway, collapse.[38]

6 Inertia and Rectilinear Uniform Motion

The problem of rectilinear Inertia is difficult because Newton's formulation is very abstract, being referred to ideal bodies on which no force acts. Therefore, one might think a clear idea of rectilinear Inertia to be impossible without distinguishing between mass and weight. It seems to us that such a distinction was not clear to

[38] While commenting on this passage by Galileo, Koyré (Koyré 1939, pp. 262–268) claims that Galileo makes evident errors. However, he does not specify what these errors are. Koyré also writes that such mistakes depend on the fact that Galileo did not develop a dynamical reasoning, but only a geometrical one, which is not sufficient in this case. This observation is true, but does not supply any clarification about the internal logic of Galileo's reasoning. For, a priori, geometry might also have been enough. Anyway, it is necessary to specify why it is not. Drake (Drake 1992, pp. 207–208) treats Galileo's reasoning synthetically and defends its validity. However, it seems to us that, in this case, Drake does not offer convincing interpretative elements. We point out that we did not have the possibility to check the original English version (1990) of Drakes' work and we refer to the Italian translation. The reader can find another example of Galileo's ingenuity, but at the same time, of his difficulties in dealing with circular motion a few pages after the example we have examined (Galilei 1632, VII, pp. 242–243). For a commentary of this passage see Paolo Bussotti (Bussotti 2017, pp. 93–96).

Galileo. Thus, he could not conceive of rectilinear Inertia. Furthermore, the Inertia principle implies, at least theoretically, the possibility that the motion be continued ad infinitum (Pisano 2024-pre print, 2009c, 2009d; Pisano and Cioci 2020). Since Galileo did not have an infinitary conception of universe, he could not conceive an Inertia principle. These are the two general objections addressed by Koyré against the idea that Galileo was the discoverer of such a fundamental law of nature.[39]

Let us examine the first objection: before Newton, no one had a clear idea of the distinction between mass and weight; rather the use of weight, heaviness etc., abounded (Pisano 2007, 2009a, 2009b, 2009c, 2016, 2020; Clagett 1959; Pisano and Capecchi 2016; Capecchi and Pisano 2007). Thus, from a purely logical point of view, the Inertia principle could not be formulated. However, the history of physics, though having its inner logic, did not proceed only on the basis of logical coherence. It is absolutely possible that in its initial phase—of which Galileo was one of the protagonists—some concepts were introduced whose meaning was not yet completely clear and defined. This does not mean these concepts were useless or that their use, if restricted to certain classes of problems, implied mistakes. Furthermore, mutual coherence among the propositions of a theory is attained only when its concepts and principles are explicitly defined. Actually, some inconsistencies do not necessarily undermine the value of all the obtained results. Therefore, it makes perfect sense that Galileo, though not posing the distinction mass–weight, wondered what happens to a body moving in uniform rectilinear motion. Indeed, Galileo's assertion that a rotating body tends to recede along its instantaneous tangent and to continue its motion in that direction with the same speed it held in the last instant of its circular motion shows that he had guessed the main features of rectilinear Inertia.

What might be called "Galileo's bucket experiment" is evidence that his ideas in this regard were rather clear. For we read in the Second Day of the *Dialogo*:

> […] how true it is that heavy bodies, whirled quickly around a fixed center, acquire an impetus to move away from that center even when they have a natural tendency to go toward it. Tie one end of a cord to a bottle containing water and, holding the other end firmly in your hand (making your arm and the cord the radius, and your shoulder knot the center), cause the vessel to go around swiftly so that it describes the circumference of a circle. Whether this is parallel to the horizon, or vertical, or slanted in any other way, the water will not spill out of the bottle in any event; rather, he who swings it will always feel the cord pull forcibly to get farther away from his shoulder. And if a hole is made in the bottom of the bottle, the water will be seen to spurt forth no less toward the sky than laterally or toward the ground. And if in place of water you put pebbles in the bottle, upon your turning it in the same manner it will be felt to exert the same force against the cord.[40] (Galilei 1967, II, p. 190)

[39] This is the general thesis expressed by Koyré (Koyré 1939; see also Pisano 2024-pre print).

[40] "[…] come sia vero che i corpi gravi, girati con velocità intorno a un centro stabile, acquistano impeto di muoversi allontanandosi da quel centro, quando anco e' sieno in stato di aver propensione di andarvi naturalmente. Leghisi in capo di una corda un secchiello, dentrovi dell'acqua, e tenendo forte in mano l'altro capo, e fatto semidiametro la corda e'l braccio, e centro la snodatura della spalla, facciasi andare intorno velocemente il vaso, sí che egli descriva la circunferenza di un cerchio; il quale o sia parallelo all'orizonte, o siagli eretto, o in qualsivoglia modo inclinato, in tutti i casi seguirà che l'acqua non cascherà fuori del vaso, anzi colui che lo gira sentirà sempre tirar la corda e far forza per allontanarsi piú dalla spalla; e se nel fondo del secchiello si farà un foro, si vedrà

This is a perfect description of what really happens. Here the principle of Inertia is correctly applied. It is true that Galileo denied the possibility of a rectilinear motion to prosecute ad infinitum, eternally (Galilei 1632, II, p. 161). However, this simply depends on the fact that an external action will sooner or later intervene to modify such a kinetic state. The argument that a body will never leave the Earth (which is seen as a sort of black hole for matter, so to say), though being wrong, is not evidence against the possibility that Galileo conceived rectilinear Inertia. In fact, such an argument can be read like this: gravity is always preponderant so that no body leaves the Earth. If gravity did not exist, every body would leave a rotating Earth because of the tendency to recede along the tangent.

Let us now analyse Koyré's (Koyré 1939; Pisano and Bussotti 2017a; see also Koyré 1957, 1960) second objection against the possibility that a scientist who believes in a finite universe could conceive rectilinear Inertia: in Newton's First Law (Pisano 2014) and in the following brief explanation there is no reference to the universe's infinity. The attention is focused on the impressed forces which, in fact, modify after a certain time (which is, in general, relatively brief) the Inertial state of a body. Newton's interest is to understand what happens when a body enters the gravitational field of another body with a certain Inertial velocity, what effects (apparent forces) depend on Inertia in non-uniform rectilinear motions, and so on. These are concrete physical problems. Newton adopted the infinity of the universe, but a scientist who had a different opinion would not have been prevented from envisioning rectilinear Inertia. For, while being an abstract principle, it has very concrete consequences. It is true that, in abstract and general terms, an Inertia principle implies that a uniform rectilinear motion perseveres to infinity in the absence of external action. However, as a principle of scientific praxis it can be accepted independently of the concept of universe that one holds.

Instead, there is a different problem with Inertia: it is referred to within an abstract physical–mathematical world because concretely impressed forces always exist. Neither does this difficulty prevent, nonetheless, a correct idea of rectilinear Inertia. In Galileo's case, he had an *intuitive* and *local* conception of Inertia.

Intuitive (local): though Galileo considered the body to be always heavy (as Koyré claims), this did not prevent him from intuiting rectilinear Inertia. This also depends on the fact that science was in a nascent phase and that an apparatus of definitions, axioms and theorems did not yet exist. When the concepts are not well defined and the deductive connections among them are not explicitly stated, it is absolutely normal that ideas, which a deeper analysis could show to be mutually contradictory, cohabit.

Terrestrial (local): when Galileo employs the principle of rectilinear Inertia, this use is confined to local motion on the Earth. However, these motions are not always horizontal (as "Galileo's bucket experiment" shows), though in most cases they are.

---

l'acqua zampillar fuori non meno verso il cielo che lateralmente e verso la terra; e se in cambio d'acqua si metteranno pietruzze, girando nell'istesso modo, si sentirà far loro l'istessa forza contro alla corda". Drake translates "secchiello" and "vaso" by "bottle". Only once he translates "vaso" as "vessel". This, and not "bottle", is the correct translation for "vaso". The correct translation for "secchiello" is "little bucket". (Galilei 1632, II, p. 43; Galileo's quotation marks).

There are passages in Galileo's works in which Inertia is even more clearly stated. At the beginning of the Fourth Day of the *Discorsi e dimostrazioni matematiche* (Galilei 1638) we read:

> Imagine any particle projected along a horizontal plane without friction; then we know, from what has been more fully explained in the preceding pages, that this particle will move along this same plane with a motion which is uniform and perpetual, provided the plane has no limits. But if the plane is limited and elevated, then the moving particle, which we imagine to be a heavy one, will on passing over the edge of the plane acquire, in addition to its previous uniform and perpetual motion, a downward propensity due to its own weight; so that the resulting motion which I call projection [projectio], is compounded of one which is uniform and horizontal and of another which is vertical and naturally accelerated.[41] (Galilei 1914, IV, p. 244)

Two different physical situations are described in this brief but complex passage.

First physical framework.

At the beginning Galileo describes uniform rectilinear motion independent of gravity. Such a motion is envisioned in an abstract physical–mathematical space (Pisano and Bussotti 2022; see also Pisano and Bussotti 2017a; Pisano 2011, 2017b). In this case it proceeds potentially to infinity, if no external action intervenes. Galileo wrote of a "horizontal plane" (Galilei 1914, IV, p. 244). From our point of view, this is an assumption which should be discussed. In fact, a plane conceived as belonging to an abstract space does not have a position that could be defined "horizontal" unless a reference frame is established. Thus, Galileo's assumption is not evident. Nonetheless, Galileo's idea is clear: in an abstract space gravity is not an ontological property of bodies. The use of the term "horizontal" indicates that Galileo is no longer considering the behaviour of bodies on Earth, but is not yet fully capable of conceiving the principles of physics as referred to an abstract physical–mathematical space. Despite these limitations, the essence of the Inertia principle is clearly formulated in the following sentence (see also above):

> [...] this particle will move along this same plane with a motion which is uniform and perpetual, provided the plane has no limits. (Galilei 1914, IV, p. 244).

Second physical framework.

The description of the second physical situation begins with these words (see also above):

> But if the plane is limited and elevated [...]. (Galilei 1914, IV, p. 244).

Here Galileo does not describe an ideal condition, rather the concrete physics on the Earth, for he adds that now the mobile is thought of as endowed with gravity. It

---

[41] "Mobile quoddam super planum horizontale proiectum mente concipio, omni secluso impedimento: iam constat, ex his quae fusius alibi dita sunt, illius motum aequabilem et perpetuum super ipso plano futurum esse, si planum in infinitum extendatur; si vero terminatum est et in sublimi positum intellegamus, mobile, quod gravitate praeditum concipio, ad plani terminum delatum, ulterius progrediens, aequabili atque indelebili priori lationi superaddet illam quam a propria gravitate habet deorsum propensionem, indeque motus quidam emerget compositum ex aequabili horizontali et ex deorsum naturaliter accelerato, quem projectionem voco." (Galilei 1638, IV, p. 268).

is evident that in the previous sentences he had considered it as devoid of gravity; by the way, this ought to induce one to question whether Koyré's idea that Galileo was unable to conceive of bodies if not as heavy bodies is always correct.

Galileo, thus, guessed the main features of rectilinear Inertia, though he did not develop a system of physical causality based on the Inertia law. This immense step was taken only by Newton in his *Principia*. It should also be added that Galileo considered the Inertiality of rectilinear uniform motion also in cases in which motion is not horizontal, as we can see in the following passage drawn from *Della forza della percossa*. Galileo here imagines two heavy bodies in equilibrium on a pulley. If an impulsive force downwards is given to one of the two bodies, it will move forever with a rectilinear uniform motion, if no impediment intervenes. We quote the entire passage as it is particularly significant to clarify Galileo's views on Inertia:

> To all appearances, what happens here is exactly what happens to a heavy and perfectly round mobile, for, if it is located on a very clean and inclined plane, it will descend naturally, always acquiring an increasing speed. In contrast to this, if you want to move that body upwards from below, it will be necessary to impart to it an impetus, which progressively diminishes and finally is annihilated. However, if the plane is not inclined, but horizontal, the round body on it will do what you want. If it is posed on the plane at rest, it will remain at rest. If we give it an impetus in a direction, it will move in that direction, always maintaining the same speed received by our hand, for there is no action that might increase or decrease such a speed, as the plane is neither downward nor upward sloping. Analogously, if two bodies of the same weight are tied to the ends of a rope, they will remain in equilibrium. If an impetus downwards is imparted to one of them, the movement of that body will maintain constantly uniform. This would follow if all the external and accidental impediments were removed. I am referring to the rope's hardness and weight, to swivels and to frictions when the rope wraps around an axis. Further accidents could also exist.[42]

We point out three elements included in this passage: (a) the context is concrete because Galileo speaks of pulleys, ropes, impacts, frictions, and so on; (b) in these situations he approximates the spherical Earth's surface with a plane in which the motion is rectilinear uniform; (c) a uniform rectilinear motion is created through an impulsive force directed downwards which breaks the equilibrium between the two heavy bodies. This is a completely correct description of what happens.

---

[42] "E qui pare che accada per appunto quello che accade ad un mobile grave e perfettamente rotondo, il quale, se si porrà sopra un piano pulitissimo ed alquanto inclinato, da per sé stesso naturalmente vi scende, acquistando sempre velocità maggiore: ma se, per l'opposito, dalla parte bassa si vorrà quello cacciare su, ci bisognerà conferirgli impeto, il quale si anderà sempre diminuendo e finalmente annichilando; ma se il piano sarà non inclinato, ma orizontale, tal solido rotondo, postovi sopra, farà quello che piacerà a noi, cioè, se ve lo metteremo in quiete, in quiete si conserverà, e dandogli impeto verso qualche parte, verso quella si moverà, conservando sempre l'istessa velocità che dalla nostra mano averà ricevuta, non avendo azione né di accrescerla né di scemarla, non essendo in tal piano né declività né acclività: ed in simile guisa i due pesi eguali, pendenti da' due capi della corda, ponendogliene in bilancio, si quieteranno, e se ad uno si darà impeto all'in giù, quello si andrà conservando equabile sempre. E qui si dee avvertire che tutte queste cose seguirebbero quando si movessero tutti gli esterni ed accidentari impedimenti, dico di asprezza e gravità di corda, di girelle e di stropicciamenti nel volgersi intorno al suo asse, ed altri che ve ne potessero essere". (Galilei 1638, pp. 336–337; Authors' translation). For the context of this passage, we refer to Vergara Caffarelli (Vergara Caffarelli 2007).

Therefore, it seems to us legitimate to conclude that Galileo understood many aspects of rectilinear Inertia, although he never elevated the Inertia principle to the rank of an explicit axiom of his theory. He intuited the validity of this principle also for bodies on which gravity does not act. He did not restrict his analysis to horizontal motions.

7 The Historiography of Galileo's Concept of Inertia

Until the studies of Alexandre Koyré (1892–1964) it was generally accepted that Galileo was the discoverer of the Inertia principle (Cfr. Koyré 1939). One of the most influential authors in this respect was Ernst Mach (1838–1916) who, in his *Science of mechanics* paid close attention to the theme of Inertia in Galileo. He pointed out how Galileo developed his idea of rectilinear Inertia, stressing that, though Galileo had the merit of discovering the Inertia principle, he did not use it extensively in his physics (Mach 1919, pp. 140–141).

Koyré proposed a completely different interpretation: Galileo did not believe in rectilinear Inertia. He conceived only Inertia of uniform circular motion. Koyré's reading is linked to his general theses on the scientific revolution which he illustrated in his celebrated work *From the Closed World to the Infinite Universe* (Koyré 1957):

(1) The scientific revolution was possible because the idea of an ordered and finite cosmos was replaced by that of the infinite universe;
(2) Mathematics was systematically applied to physics;
(3) Physicists believed that the laws they were discovering were inherent to the universe;
(4) Technological and economical factors had scarce influence on the development of theoretical science;
(5) The experimental aspect was far less important than normally thought of in the birth of modern science.

Points (2–5) represent the Platonic tendency shared by many scientists who lived in the sixteenth-seventeenth centuries. Obviously, Koyré does not claim that all scientists believed in every idea expressed by these five points. Rather, he thinks that they contributed to the progress of science to the extent that they supposed in such convictions. (Cfr. Bussotti 2017, pp. 59–63; see also Pisano, Agassi et al. 2017).

As for Galileo's concept of Inertia, Koyré thought that he believed only in circular Inertia. Galileo was impressed by the natural persistency of circular motion, confirmed by the daily observation of the planets which rotate around the Sun in agreement with his Copernican convictions (Koyré 1939, p. 206). According to Koyré, the Platonic myth is the most significant text in Galileo's works to understand his view of Inertia.

Koyré also claims that, while referring to motion along a horizontal plane on the Earth's surface, Galileo was not thinking of a geometrical plane, but of a spherical surface. Inertial motion on a vast part of the Earth's surface is, therefore, circular, not rectilinear (Koyré 1939, pp. 207–208). Galileo did not reach the concept of rectilinear

Inertia and remained anchored to circular Inertia because, unlike Descartes, he did not accept the consequences of a complete geometrization of the universe; namely he did not accept the infinity of the universe and the destruction of the ordered cosmos (Koyré 1939, p. 211). Koyré also highlights that Galileo thought every body to be intrinsically heavy, which prevented him from reaching the concept of rectilinear Inertia (Koyré 1939, pp. 239–276; in particular p. 258). As we have hinted, he stresses that the Inertia principle is based on the distinction between mass and weight. We have already addressed this objection in our previous analysis.

More generally, Koyré's position is surprising from an historical perspective: He held that Galileo did not have a clear concept of rectilinear Inertia, while other physicists, such as Descartes, had. Koyré should have noted, though, that in Descartes the distinction between mass and weight is not clearer than in Galileo. We agree with Burns' opinion that "the distinction between weight and mass emerged only at the end of the seventeenth century" (Burns 2001, Entry "Force", p. 105). Contrary to Koyré, Galileo seems to us to be the scientists who, in the period 1620–1640, had the clearest concept of rectilinear Inertia, though with the limitations we have pointed out. No scientific notion appears suddenly in its perfect form, but it is developed in genetic phases. This is particularly true for a difficult and abstract concept as rectilinear Inertia. In the phase during which this idea was formed, Galileo was the physicist who had the most satisfying view. It was far from being perfect, but was, nonetheless, basically correct.

While commenting on the numerous passages where Galileo seems to refer to rectilinear Inertia, Koyré claims that the Italian scientist came close to conceiving it, but refused to make the final step (Koyré 1939, p. 262).

Koyré's analyses of Galileo's concept of Inertia are unsatisfactory from a conceptual and methodological standpoint. First, Koyré overestimated the importance that Galileo ascribed to circular Inertia. As we have explained, we are convinced that Galileo believed in circular Inertia and that this concept is part of his cosmology. However, circular Inertia plays no role in the development of Galileo's arguments in the *Dialogo* and in the *Discorsi*. Therefore, though he believed in circular Inertia, it is misleading to regard such a conception as a milestone to interpret Galileo's physical view and work. The most unsatisfactory aspect of Koyré's interpretation is the thesis that Galileo could not reach the concept of rectilinear Inertia because he did not factor in the infinity of the universe. Now, if one analyses the *Platonic myth*, Galileo explicitly claims the potentially infinite character of rectilinear motion (be it indifferently uniform or accelerated motion, since he asserts the infinitary character of rectilinear motion before stating that such motion is accelerated). In this context, it is appropriate to stress a distinction that Koyré seems to have overlooked: The phrase "natural motion" is not exactly identifiable with what we call "Inertial motion". The term "natural" means "appropriate" in a meaning which is closer to the order of thinking of Aristotelian metaphysics than to modern physics. Uniform circular motion is natural for celestial bodies because it is appropriate for them according to a metaphysical meaning. Moreover, it is also Inertial because, when it begins, it continues eternally. Rectilinear motion, be it uniform or accelerated, is not ontologically natural for celestial bodies, but this does not prevent uniform rectilinear motion from being Inertial in the physical meaning of this term.

Furthermore, as we have outlined in the previous sections, it is rather clear that the infinity of the world has nothing to do with rectilinear Inertia because Inertia is a "limit-concept" in the sense that every physicist—and Galileo, too—is perfectly aware that no motion is properly Inertial. At best, a motion is approximately Inertial. In other terms:

It can be considered Inertial, granted some circumstances. The principle of Inertia has a general theoretical validity, but, in fact, a motion can be Inertial only locally, though the notion of "local" includes many motions in proximity of the Earth's surface.

When Galileo claims that circular motion has the tendency to extrude the rotating objects along the instantaneous tangent of their trajectories, he is not interested in analysing what happens at infinity. Rather he is interested in stating that:

(1) The sling's circular motion does not continue without an external action;
(2) If the circular motion is interrupted, the rotating body moves with rectilinear uniform motion along its instantaneous tangent. What the body will do at infinity is irrelevant.

The principle of Inertia is a regulatory idea concerning local motions. According to Koyré's views coherently developed, a physicist who nowadays believes that the universe is finite—and no one knows whether it is or not—should refuse to admit–for instance—that the potential of an electric field is null at infinity. But this phrase "at infinity" means "very far from the sources of the electric field"; it has nothing to do with the infinity of the universe. The same is true for Inertia. On the other hand, Newton himself who believed in the infinity of the universe, did not connect the Inertia principle with the assumption of the infinity of the universe. In the bucket experiment, the concept of Inertia plays a fundamental role, but there is no reference to the infinity of the universe. Despite these objections, Koyré had the merit of stimulating further studies on Inertia in Galileo, many of which supported or rejected his ideas.

Eduard Jan Dijksterhuis (1892–1965) shared Koyré's view in substance: according to Galileo, the real Inertial motion is the circular uniform one. Like Koyrè, Dijksterhuis claims that, while referring to a horizontal surface, Galileo has in mind a portion of the spherical terrestrial surface. This author recognizes that in the Fourth Day of *Discorsi e dimostrazioni matematiche* Galileo arrived almost at conceiving rectilinear Inertia. However, as in the *Dialogo*, his idea of a *planar horizontal surface* is tied to the Earth's sphere, so that the notion of Inertia expressed in the two great works is, as a matter of fact, the same (Dijksterhuis [1950] 1961, pp. 351–352). However, Dijksterhuis ascribed a great merit to Galileo: he first guessed (Dijksterhuis [1950] 1961, p 351, pp. 353–354) that in any motion there is a component that tends to maintain its kinetic state, whereas another component is produced by an external action. Furthermore, according to Dijksterhuis, Galileo first clarified the concept of the participation of a body in the Inertial system to which it belongs (Dijksterhuis [1950] 1961, p. 355). This means that Dijksterhuis separates this issue from the problem of Inertia: Galileo did not fully grasp the concept of rectilinear Inertia, but—to paraphrase Dijksterhuis' expression—in "intuitively Inertial conditions" a body participates in the motion of the system to which it belongs (Dijksterhuis [1950] 1961, pp. 349–359).

Stillman Drake (1910–1993) brilliantly devoted much thought to the issue of Inertia in Galileo. In 1964, he claimed that Galileo was the discoverer of the inertial principle. Drake wrote:

> In my opinion the essential core of the inertia concept lies in the ideas, explicitly stated above, of a body's indifference to motion or to rest and its continuance in the state it is once given. This idea is, to the best of my knowledge, original with Galileo. It is not derived from, or even compatible with, impetus theory, which assumed a natural tendency of every body to come to rest. (Drake 1964, pp. 606)

According to Drake, Galileo was aware of the principle of rectilinear Inertia, though restricted to horizontal planes (Drake 1964, pp. 606–607); while dealing with horizontal motions, he was not referring to portions of the Earth's surface (*Ibidem*); but to planes which approximate a small portion of such a surface with a negligible error (*Ibidem*). As for the *Platonic myth*, following Drake, it was simply a concession to the Aristotelians, a sort of literary expedient to captivate the sympathies of the adversaries (Drake 1964, pp. 607–608).

On our side, we argue that there is no scientific reason either to overestimate or to underestimate the value of the *Platonic myth*. Since Galileo also referred to it in the *Discorsi*, we can assume that he considered it to be scientifically factual. In 1972 Drake re-examined (Drake 1972) the problem of Inertia and proposed that the main question is not to decide whether Galileo considered either rectilinear or circular Inertia rather, it is to analyse his concept of neutral motion (Drake 1972, p. 249); a notion already pointed out by Coffa (Coffa 1968). This is a motion which does not begin naturally, but that, once begun, does not stop without an external action. Uniform rectilinear motion as well as the motion along a portion of Earth's surface are neutral. Drake also discussed that neutral motion is a fundamental step towards the idea of rectilinear Inertia (Drake 1972, pp. 248–253). In this context Drake moved an entirely acceptable criticism of Koyré: He had by far overestimated the importance of a possible philosophy to which Galileo adhered on Galileo's scientific conceptions. Scientific ideas and praxes are not strictly related to scientist being a Platonic or a Pythagorean or something else.[43] Since Drake was aware that the Inertia affair in Galileo is (rightly) a complicated matter, maybe he adopted a sceptical position. That is, to recognize that Galileo left the door open to both interpretations: i.e. that he conceived either rectilinear or circular Inertia, so avoiding a final decision. Insofar as his concrete physics was concerned, Galileo applied rectilinear Inertia correctly (Drake 1972, p. 270).

> But we do not really have to decide for one view or the other; if anything, we should avoid a final decision. What we must recognize is that Galileo's discussion leaves open the *possibility* of either circular or rectilinear continuation of uniform motion in the case of free

[43] A very interesting discussion concerning the ovestimation of the philosophical convictions of a seventeenth century scientist on his science is developed by Drake in the Introduction of (Drake 1972, pp. 1–17).

> fall (a "natural" motion), whereas he unambiguously specifies *rectilinear* continuation for the "violent" component in the motion of cannonballs and terrestrial objects released from slings or flung from rapidly rotating wheels. (Drake 1972, p. 270; Author's Italic and capital letters)

This sceptical position seems to us to be extreme in Drake's next work (Drake 1992) where he proposed that asking whether Galileo reached a correct concept of Inertia makes no sense because Inertia is a dynamical concept whereas Galileo's is only a kinematic physics. Finally, in Drake's words:

> Inertia being a dynamic concept, it did not enter into Galileo's physics at all.[44]

We think that the problem of Inertia in Galileo's mechanics can be faced in a different manner. On the other hand, sometimes secondary literature simply avoids facing it. It is true that Galileo's physics was essentially kinematics, but it makes perfect sense to ask whether a motion can continue on its own without any external action.

A further important and more recent contribution to the concept of Inertia in Galileo is due to Vergara Caffarelli who has a clear purpose in his studies:

(1) To prove, contrary to Koyré, that the experimental aspect is fundamental in Galileo;
(2) To demonstrate that Galileo clearly reached the concept of rectilinear Inertia and refused that of circular Inertia.

Vergara Caffarelli traces an evolution in Galileo's thought: by analysing some passages of *De motu*, *Le Mecaniche*, *Istoria e dimsotrazioni intorno alle macchie solari*, the *Dialogo* and *Della forza della percossa* Vergara Caffarelli highlighted that Galileo guessed rectilinear Inertia from his initial works, that this concept was progressively refined until *Della forza della percossa* and that he reached, substantially, a modern notion of Inertia (Vergara Caffarelli 2007). On our side, we disagree with his ideas on circular Inertia, as we have clarified in the previous sections. In particular, with regard to Galileo's idea that the motion on the Earth's surface is conserved, Vergara Caffarelli ascribed to Galileo the discovery of the principle of conservation of the angular Momentum (Vergara Caffarelli 2007, p. 76). We think that idea of circular Inertia is sufficient to explain Galileo's remarks on this topic.

Boccaletti in his otherwise excellent book (Bussotti and Lotti 2022) also claimed that Galileo did not consider circular Inertia (Boccaletti 2016, Chap. 4).

As we have pointed out, our argument is that Galileo considered both circular and rectilinear Inertia, but that rectilinear Inertia played a role in Galileo's physics, whereas circular Inertia did not. However, we do not use this argument to modernize our analysis or Galileo's wording.

In a stimulating paper, Andrea Frova (Frova 2006) doubted that Galileo understood all the consequences of the principle of rectilinear Inertia. Frova analysed the passage from the Second Day of the *Dialogo* (Galilei 1632) where Galileo argued about the

---

[44] We point out that this is from original English version *Galileo, Pioneer Scientist* (Drake 1972, p. 163). In the Italian translation we read: "Essendo l'inerzia un concetto dinamico, essa non entra affatto nella fisica di Galileo." (Drake 1992, p. 171; Cfr. Oliveira and Pisano 2022).

"imberciatori d'uccelli" ["hunters of birds"] (Bussotti and Lotti 2022, p. 205). The problem is: *where should the hunters point their harquebus to hit a flying bird?* Galileo considered the right answer, but he rejected it and reasoned according to the following rule: *the harquebus has to be pointed directly at the bird*. Frova stressed (Frova 2006, p. 44) that this solution is manifestly wrong because, if r indicates the length of the harquebus, R indicates the altitude of the bird, the ratio R/r is much bigger than 1, so that the tangential velocity V of the bird is far superior to the tangential velocity v of the projectile. Therefore, if one follows Galileo's above mentioned rule, the projectile would not reach the bird and would not hit it. This is an evident proof that Galileo did not have a clear concept of the kinematical magnitudes connoting uniform circular motion. Frova concludes:

> This is a simple case of application of the Inertia principle and Galileo makes a trivial mistake as if he were a beginner! It's a good consolation for us common mortals and a warning to doubt even our strongest certainties.[45]

8 Concluding Remarks

With regard to Galileo's conception of Inertia, our main purpose in this paper has been to prove the validity of the interpretative scheme presented in the Outline and Introduction. The case study we have analysed is, furthermore, an interesting example which shows a general and paradigmatic (though not universally valid) way in which scientific ideas develop: for, they have a gestational phase and progressively they are better defined and refined. At the beginning not all the consequences of the concepts are guessed; in the course of time, such consequences are extended and better understood, so as to compose a part of a whole scientific theory. The case of Inertia in emblematic: the problems connected to motion have always interested philosophers and scientists. From the late sixteenth century mathematicians tried to deal with them more systematically than in past. In this context, Galileo, because of his broad views and the profound results he obtained, can be considered one of the fathers of modern physics. With regard to the principle of Inertia, the contributions made by Galileo cannot even be compared to those of the preceding and contemporary scientists, excluding Descartes. Galileo's activity as a physicist is connected to a first expression of the Inertia principle. In his time the conception of which motion had to be regarded as Inertial was not formulated in a clear manner and the notion itself of Inertia was not clear. Galileo ascribed the property we may call Inertia to two motions:

(1) The uniform rectilinear;
(2) The uniform circular as in the circumstances we have specified. As to this latter movement, sometimes he considers the tendency to recede along the tangent, whereas other times does not.

[45] "Si tratta di un semplice esercizio di applicazione del principio di inerzia e Galileo ci casca come un principiante! Una bella consolazione per noi comuni mortali, e un monito a dubitare sempre anche delle nostre certezze più salde." (Frova 2006, p. 44; Authors' translation).

Thus, principles and conclusions, which, from our point of view are mutually contradictory, cohabit in Galileo. But for him there was no contradiction. This depends on the fact that many of Galileo's conceptions on uniform circular motion were imprecise and he did not have a general view of this phenomenon. This is also probably connected to the fact that the mathematical means available to Galileo were not satisfactory to study circular motion. In particular, it seems to us that without the idea of velocity and acceleration being vectorial quantities[46] it is difficult to grasp many features of circular motion. When the ideas are not precise enough, one can reach conclusions which a more attentive analysis reveals to be contradictory. Galileo had the great merit to develop, at least in outline and with many uncertainties, the kinematics of circular motion. There were three possibilities for Inertial motion:

- The uniform circular
- The uniform rectilinear or both of them
- The rectilinear one.

The process through which we have understood the third option to be the correct one includes many physicists and philosophers, from ancient and medieval times until Galileo, Torricelli and Descartes; but also Baliani, Gassendi, Huygens, until reaching Newton, who ascribes only to Galileo the merit for the discovery of the concept of Inertia.

From intellectual history of ideas and epistemological standpoints, a statement becomes meaningful when it is part of a theoretical context where its applications show a series of consequences impossible to draw otherwise. A principle or an intuition formulated as an isolated statement is not a law, so it is not very significant. In this perspective, the first to fully understand the physical consequences of Inertia as law (but also as Axiom) was Isaac Newton. Therefore, in a sense, he was the discoverer. Does this mean that Galileo's contribution to the understanding of rectilinear Inertia is not noteworthy? Not at all. It is noteworthy. For, Galileo, as we have seen, refers on numerous occasions to the Inertial character of uniform rectilinear motions. Furthermore, with regard to the parabolic motion his distinction between a component which varies and which is uniformly accelerated and a component that is rectilinear uniform and that maintains its state Inertially is fundamental. Therefore, in the comprehension of the Inertia principle Galileo, despite his errors, deserves second place behind Newton.

Acknowledgements We thank the anonymous reviewers for their careful reading of our manuscript and their many insightful comments and suggestions. We also express gratitude to Dr. Julie Robarts (University of Melbourne, Australia) for her observations on English language; eventual remaining English mistakes are our own.

[46] Of course, we are not claiming that an abstract notion of vector, which was acquired only in the second half of the nineteenth century (Cfr. Crowe), is necessary to have a clear idea of velocity and acceleration. However, the distinction between the scalar aspect (more intuitively, we might say the aspect connected to the intensity of a magnitude) and that connected to directions and sense is.

References

Archimedes (1503) Tetragonismus idest circuli quadratura per Campanum Archimedem Syracusanum atque Boetium mathematicae perspicacissimos adinuenta. [Gaurico L (ed)] Impressum Venetiis: per Ioan. Bapti. Sessa, 1503 die 28 august.
Archimedes (1544) Opera quae quidem extant omnia. Basel, Gechauff.
Archimedes (1558) Archimedis opera non nulla a Federico Commandino Urbinate nuper in latinum conversa et commentariis illustrata. apud Paulum Manutium, Aldi F. Venetiis.
Archimedes (1565) Archimedis De iis quae vehuntur in aqua libri duo. A F. Commandino U. in pristinum nitorem restituti, et commentariis illustrati. Ex Offcina Benacii, Bononiae.
Archimedes (1792) Archimedis quae supersunt omnia, cum Eutocii Ascalonitae commentariis, ex recensione Josephii Torelli Veronensi cum nova versione latina (accedunt versiones variantes ex codd. Mediceo et Parisiensibus). Robertson. Oxford.
Archimedes (1880) Archimedis Opera omnia cum Commentariis Eutocii e codice Florentino recensuit, Latine vertit, notisque illustravit J.L. Heiberg. Stutgardiae, in aedibus B.G. Teubneri.
Archimedes (1880–1881) Archimedis opera omnia com commentariis Eutocii e codice Florentino recensuit, edidit Iohann Ludvig Heiberg, 3 vols., Lipsiae, in aedibus B.G. Teubneri.
Archimedes (1881) Opera Omnia cum commentari Eutocii. Heiberg JL (ed), Teubneri, Lipsiae.
Archimedes (1897) On the Equilibrium of Planes. In Heath 2002, pp. 189–220.
Archimedes (1912) The Method of Archimedes recently discoverd by Heiberg. Edited by Heath, T., L. The Cambridge University Press. Cambridge.
Archimedes (1987) The Method of Mechanical Theorems. Dijksterhuis EJ (ed). Translation by C. Dickshoorn. Princeton University Press, Princeton.
Archimedes (2002a) On the equilibrium of planes. In: Archimedes 2002b, pp. 189–220.
Archimedes (2002b) The works of Archimedes. Edited by Heath TL. Dover Publications Inc. New York.
Aristotle (1930) Physics. Translation by Hardie RP and Gaye RK. Clarendon Press, Oxford.
Aristotle ([1936] 1955a) Minor Works. Translated by Hett WS. Cambridge MA –The Harvard University Press—Heinemann W LTD, London.
Aristotle ([1936] 1955b) Mechanical problems. In: Hett WS (ed) Aristotle. Minor works. William Heinemann, Cambridge, pp. 328–414.
Bernoulli Johann (1727) Theoremata selecta pro conservatione virium vivarum demonstranda et esperimenta confirmanda. Novi commentarii Academiae Scientiarum Imperialis Petropolitanae 2, pp. 200–207.
Bernoulli D ([1748] 1750) Remarques sur le principe de la conservation des forces vives pris dans un sens général. Mémoire de l'Académie Royale des Sciences et des Lettres de Berlin, vol IV, pp. 356–364.
Blackwell RJ (1966) Descartes' Laws of Motion. Isis 57(2):220–234.
Boutroux P (1921) L'histoire des principes de la dynamique avant Newton. Revue De Métaphysique Et De Morale 28:657–688.
Boccaletti D (2016) Galileo and the Equations of Motion. Springer, Dordrecht.
Buridani J (1509) Subtilissimae Quaestiones super octo Physicorum libros Aristotelis, Johanne Dullaert, Paris. [Reprint: *Id*. (1964) Kommentar zur Aristotelischen Physik. Minerva, Frankfurt].
Buridan J (1513) Quaestiones super decem libros Ethicorum Aristotelis ad Nicomachum, Paris. [Reprint: *Id*. (1968) Super decem libros Ethicorum. Minerva, Frankfurt].
Buridan J (1942) Iohannis Buridani Quaestiones super libris quattuor De caelo et mundo. In: Moody EA (ed). Medieval Academy of America. Studies and documents 6/40.
Burns W E (2001) The Scientific Revolution. An Encyclopedia. ABC-Clio, Santa Barbara.
Bussotti P (2017) Galileo e il suo concetto di inerzia. In: Neri A (ed). Là dove il pensiero incontra l'esperienza. ETS, Pisa, pp. 51–110.
Bussotti P, Lotti B (2022) Cosmology in the Early Modern Age: a Web of Ideas. Springer, Dordrecht.
Bussotti P, Pisano R (2013) On the Conceptual Frames in René Descartes' *Physical Works*. Advances in Historical Studies 2(3):106–125.

Bussotti P, Pisano R (2014a) Newton's Philosophiae Naturalis Principia Mathematica "Jesuit" Edition: The Tenor of a Huge Work. Accademia Nazionale Lincei-Matematica e Applicazioni 25/4:413–444.

Bussotti, P, Pisano, R (2014b) On the Jesuit Edition of Newton's Principia. Science and Advanced Researches in the Western Civilization. In Pisano R (ed). Isaac Newton and his Scientific Heritage: New Studies in the History and Historical Epistemology of Science. Special Issue. Advances in Historical Studies 3/1:33–55.

Bussotti P, Pisano R (2017) Historical and Philosophical Details on Leibniz's Planetary Movements as Physical-Structural Model. In: Fichant M, Bussotti P, Oliveira ARE (eds) Pisano R. College Publications, London, pp. 49–92.

Capecchi D, Pisano R (2007) La teoria dei baricentri di Torricelli come fondamento della statica. Physis XLIV/1:1–29.

Capecchi D, Pisano R (2008) La meccanica in Italia nei primi anni del Cinquecento. Il contributo di Niccolò Tartaglia. In: Tucci P (ed). Proceedings of XXV SISFA Congress. University of Milano Press, Milano, C17.1–C17.6.

Capecchi D, Pisano R (2010) Reflections on Torricelli's principle in mechanics. Organon 42:81–98.

Carnot L ([1783] 1786) Essai sur les machines en général. Defay, Dijon.

Carnot L (1803) Principes fondamentaux de l'équilibre et du mouvement. Deterville, Paris. [see also Id. (1803) Géométrie de position à l'usage ceux qui se destinent à mesurer les terrains. Duprat, Paris].

Carnot L (1813) Réflexions sur la métaphysique du calcul infinitésimal. [2nd edition of 1797] Courcier, Paris.

Chalmers A (1991) Galilean Relativity and Galileo's Relativity. In: French S, Kamminga H (eds). Correspondence, Invariance and Heuristics: Essays in Honour of Heinz Post. Kluwer Academic Publishers, Dordrecht.

Clagett M (1959) The Science of Mechanics in the Middle Age. University of Winsconsin Press, Madison.

Clagett M (1964–1984) Archimedes in the Middle Ages. Madison-Philadelphia, Memoirs of the American Philosophical Society. 5 Vols., 10 Tomes. The Clarendon University Press, Oxford.

Clagett M (1976) Archimedes in the Middle Ages. Philadelphia, American Philosophical Society, II.

Clagett M (1978) Archimedes in the Middle Ages. Philadelphia, American Philosophical Society, III.

Cohen BI (1964) 'Quantum in Se Est': Newton's Concept of Inertia in Relation to Descartes and Lucretius. Notes and Records of the Royal Society of London 19(2):131-155.

Coulomb CA (1784) Observations théoriques et expérimentales sur l'effet des moulins à vent,&sur la figure de leurs ailes. In: Histoire de l'Académie Royale des Sciences avec les Mémoires de Mathématique&de Physique, pour lamême année 1781. Imprimerie Royale, Paris, pp. 65–81.

Coulomb CA (1785) Théorie des machines simples en ayant égard au frottement de leurs parties et à la roideur du cordage. Mémoire de mathématique et de physique présentes à l'Académie Royal des Sciences par divers savants. Bachelier, Paris.

Coulomb CA (1799) Résultat de plusieurs expériences destinées a déterminer la quantité d'action que les hommes peuvent fournir par leur travail journalier suivant les différentes manières dont ils emploient leurs forces, vol II. Mémoires de l'Institut National des sciences et arts, Paris, pp. 340–428.

Coffa JA (1968) Galileo's concept of Inertia. Physis X(4):261–281.

Crowe MJ (1967) A History of Vector Analysis. The Evolution of the Idea of a Vectorial System. Dover, New York.

D'Alembert JLR ([1743] 1758) Traité de dynamique. Nouvelle édition, revue et fort augmentée par l'Auteur Chez David Libraire, Paris.

D'Alembert JLR (1751–1780) Encyclopédie ou dictionnaire raisonné des sciences, des arts et des métiers, Diderot & D'Alembert. Briasson–David–Le Breton–Durand, Paris.

D'Alembert JLR (1767) Mélanges de littérature, d'Histoire et de philosophie. Tome V. Chatelain & Fils. Amsterdam, pp. 3–274.
De Nemore I (1533) Liber Iordani Nemorarii viri clarissimi, de ponderibus propositiones XIII & earundem demonstrationes, multarumque rerum rationes sane pulcherrimas complectens. [edited by Apianus] Ioh Petreium, Norimbergae.
De Nemore I (1565) Iordani Opusculum de Ponderositate, Nicolai Tartaleae Studio Correctum Novisqve Figuris avctum [Tartaglia N (ed)]. Cum Privilegio Traiano Curtio, Venetiis, Apud Curtium Troianum. MDLXV.
De Nemore I (fl. 13th) Ms. Elementa Jordani super demonstratione de ponderibus. Oxford, Bodleian Library, Ms. Auct. F.5.28. Folia 125v–133r.
Descartes R ([1632–1633] 1664) Le monde de M Descartes ou le Traité de la lumière et des autres principaux objets des Sens. Girard, Paris.
Descartes R (1897–1913) Oeuvres de Descartes. Par Charles Adams et Paul Tannery, [Discours de la méthode et Essais, Specimina philosophiae, Vol. VI; Physico– mathematica, Vol. X, Le Monde ou Traité de la lumière, Vol. XI], 12 Vols. Vrin, Paris.
Descartes R (1964–1974). Oeuvres. Nouvelle présentation par Rochet E, et Costabel P, 11 Vols. Vrin, Paris.
Descartes R ([1979] 1995) Le Monde ou Traité de la lumière. Translation and introduction by Michale Sean Mahoney. Abaris Books, NY.
Dijksterhuis E J ([1950] 1961) The mechanization of the world picture. Clarendon Press, Oxford.
Drago A (1991) Le due opzioni. La meridiana. Molfetta.
Drake S (1964) Galileo and the Law of Inertia. American Journal of Physics 32:601–608.
Drake S (1972) Galileo Studies. University of Michigan Press, Ann Arbor.
Drake S (1992) Galileo Galilei pioniere della scienza. Muzzio, Padova [*Id.*, Galileo: Pioneer Scientist. University of Toronto Press, Toronto, 1990].
Euler L ([1730–1731] 1738) De communicatione motus in collision corporum. Commentarii academiae scientiarum Petropolitanae 5:159–168.
Euler L (1736) Mechanica sive motus scientia analytice exposita. Auctore Leonhardo Eulero academiae imper. scientiarum membro et matheseos sublimioris professore. Tomus I. Instar supplementi ad commentar. acad. scient. imper. Petropoli. Ex typographia academiae scientarum.
Euler L (1749) Recherches sur les mouvements des corps célestes en général (1747). Mémoire de l'Académie Royale des Sciences et des Lettres de Berlin, Vol. I, pp. 93–143.
Euler L (1750] 1752) Découverte d'un nouveau principe de Mécanique. Mémoires de l'académie des sciences de Berlin. Vol. VI, 1752, pp. 185–217.
Euler L (1751) De machinis in genere Novi Commentarii Academiae Scientarum Imperialis Petro–Politanae. Petropoli Typis Academiae Scientiarum.
Euler L (1757) Principes généraux du mouvement des fluides. Mémoires de l'Academie des Sciences de Berlin 11:274–315.
Euler L ([1773] 1774) De pressione ponderis in planum in cui incumbit. Novi Commentarii academiae scientiarum Petropolitanae, vol XVIII, pp. 289–329.
Euler L (1775) Nova methodus motum corporum rigidorum determinandi. Novi Commentarii academiae scientiarum.
Frova A (2006) Anche Galileo può errare. In Atti di "Pianeta Galileo 2006". Regione Toscana, Firenze, pp. 35–47.
Galilei G (1613) Istoria e dimostrazioni intorno alle macchie solari e loro accidenti. In: Galilei 1890–1909, V, pp. 71–249.
Galilei G ([1598–1599] 1623) Le Mecaniche. In: Galilei 1890–1909, II, pp. 147–191.
Galilei G (1632) Dialogo sopra i due massimi sistemi del mondo tolemaico e copernicano. In: Galilei 1890–1909, VII, pp. 21–520.
Galilei G (1638) Discorsi e dimostrazioni matematiche intorno a due nuove scienze. In: Galilei 1890–1909, VIII, pp. 49–448.

Galilei G (1890–1909) Le opere di Galileo Galilei. Edizione Nazionale sotto gli auspici di Sua Maestà il Re d'Italia. Favaro. A (ed). Tipografia di G. Barbèra, Firenze.

Galilei G (1914) Dialogues Concerning Two New Sciences. Macmillan, New York.

Galilei G (1954) Dialogues Concerning Two New Sciences, translated by Henry Crew & Alfonso de Salvio, with an introduction by Antonio Favaro, Dover Publications, Inc., New York [Originally published in 1904 by the MacMillan company].

Galilei G (1967) Dialogue Concerning the Two Chief World Systems: Ptolemaic and Copernican, translation and with notes, 2nd edn. University of California Press, Berkeley-Los Angeles.

Galilei G (fl. 16th) Ms. A—Trattato di fortificazioni e modo d'espugnare la città con disegni e piante di fortificazioni. Biblioteca Ambrosiana. D: 328 Inf—38640.

Galilei G (fl. 16th) Ms. B—Trattato delle fortificazioni, con disegni diversi. Biblioteca Ambrosiana. D: 296 Inf–38499.

Galilei G (fl. 16th) Ms. m—Trattato di fortificazioni, con disegni e figure (acephalous). Biblioteca Ambrosiana. N: 281 Sup–2– 80277.

Galilei G (fl. 17th) Ms. 72, 33rv–196v. Biblioteca Nazionale Centrale, Florence Istituto e Museo di Storia della Scienza, Florence–Max Planck Institute for the History of Science, Berlin.

Galluzzi P (1979) Momento: Studi galileiani. Ateneo & Bizzarri, Roma.

Galluzzi P, Torrini M (1975–1984) (eds) The works of Galileo Galilei's disciples. Correspondence (in Italian). Giunti–Barbera, Firenze, 2 Vols.

Gaukroger S (2004) Descartes an intellectual biography. Oxford University Press, Oxford.

Gillispie CC (1960) The Edge of Objectivity: An Essay in the History of Scientific Ideas. Princeton University Press, Princeton–NJ.

Gillispie CC, R Pisano (2014) Lazare and Sadi Carnot. A Scientific and Filial Relationship, 2nd edition. Dordrecht.

Grant E (1964) Motion in the Void and the Principle of Inertia in the Middle Ages. Isis 55(3):265–292.

Hill DK (1986) Galileo's Work on 116v: A New Analysis. Isis 77:283–291.

Kepler J (1618–1621) Epitome ASTRONOMIAE Copernicanae Usitata forma Quaestionum et Responsionum conscripta inqu;[e] VII. libros digesta, quorum TRES hi priores sunt de DOCTRINA SPHAERICA. HABES, AMICE LECTOR, HAC PRIMA parte praeter physicam accuratam explicationem motus terrae diurnae ortusque ex eo circulorum sphaerae totam doctrinam sphaericam nova et concinniori methodo auctiorem, additis exemplis omnis generis computationum astronomicarum et geographicarum, quae integrarum praeceptionum vim sunt complexa. AUTHORE JOANNE KEPPLERO IMPERATORIS CAESARIS MATTTHIAE, Ordd[inumque] q; ill[ium] Archiducatus Austriae supra Onasum, Mathematico. Cum Privilegio Caesareo ad Annos XV. Lentijs ad Danubium, excudebat. Johannes Plancus ANNO MDCXVIII [Books I-III, 1618]; — Epitomes ASTRONOMIAE Copernicanae Usitata forma Quaestionum et Responsionum conscriptae, Liber Quartus, doctrinae Theoricae Primus : Quo PHYSICA CAELESTIS, Hoc Est, OMNNIUM IN CAELO MAGNITUDInum, motuum proportionumque causae vel Naturales vel Archetypicae explicantur, Et Sic PRINCIPIA DOCTRINAE Theoricae demonstrantur. QUI QUOD VICE SUPPLEMENTi librorum Aristotelis de Caelo esset, certo consilio seorsim est editus. AUTHORE JOANNE KEPPLERO. Cum Privilegio Caesareo ad Annos XV. Lentijs ad Danubium, excudebat Johannes Plancus, ANNO M.DC.XX [1620, Book IV; [see also: impensis Godefridi Tampachij, 1622]; — Epitomes ASTRONOMIAE Copernicanae usitata forma quaestionum et responsionum conscriptae, Libri V. VI. VII. Quibus proprie DOCTRINA THEORICA (post principia libro IV. praemissa) comprehenditur. AUTHORE IOANNE KEPLERO. Cum privilegio Caesareo ad Annos XV. Francoforti, Sumptibus Godefridi Tampachij. ANNO M.DC.XXI [Books V-VII, 1621]. In: KGW, VII.

Kepler J (1937–2012) Johannes Kepler, Gesammelte Werke (KGW). Van Dyck W, Caspar M. et al. (eds). Revised April 2013. X Vols. Deutsche Forschungsgemeinschaft und Bayerische Akademie der Wissenschaften. Beck'sche Verlagsbuchhandlung, München.

Kepler J (1992) New Astronomy. Translation of Astronomia Nova. Cambridge University Press, Cambridge.

Koyré A (1939) Études Galiléennes. Hermann, Paris. [English translation: Galileo Studies. Humanities Press, New Jersey, 1978].
Koyré A (1957) From the Closed World to the Infinite Universe. John Hopkins Press, Baltimore.
Koyré A (1960) La dynamique de Nicolo Tartaglia (1957). In: La science au seizième siècle. Colloque international de Royaumont, 1–4 juillet 1957. Hermann, Paris, pp 91–116 [see also: Koyré A (1966) La dynamique de Niccolo Tartaglia. Études d'histoire de la pensée scientifique:101–121].
Lagrange JL (1762) Application de la méthode exposée dans le mémoire précédente à la solution des différents problèmes de dynamique. In: Lagrange 1870–1873, I, pp. 363–468.
Lagrange JL (1764) Recherches sur la libration de la Lune dans lesquelles on tache de résoudre la question proposes par l'Académie Royale des Sciences, pour le Prix de l'année 1764. Prix de l'académie Royale des Sciences de Paris, IX (1777), pp. 1–50. [See also Lagrange 1870–1873, Tome VI, pp. 5–61].
Lagrange JL (1788) Mécanique Analytique. Desaint, Paris.
Lagrange JL (1797) Théorie des fonctions analytiques contenant les principes du calcul différentiel, dégagés de toute considération d'infiniment petits, de limites et de fluxions, et réduits à l'analyse algébrique des quantités finies. Journal de l'École Polytechnique III(9):91–93.
Lagrange JL (1806) Leçons sur le calcul des fonctions, Nouvelle. Courcier, Paris.
Lagrange JL (1813) Théorie des fonctions analytiques. Courcier, Paris.
Lagrange JL (1870–1873) Œuvres de Lagrange. Seconde édition, Courcier (ed), I–XIV vols., (in X) Paris Gauthier–Villars, Paris.
Lagrange JL (1881) Théorie des fonctions analytiques contenant les principes du calcul différentiel, dégagés de toute considération d'infiniment petits, de limites et de fluxions, et réduits à l'analyse algébrique des quantités finies. In: Serret JA (ed) Œuvres de Lagrange. Cinquième édition, vol. IX, Chapter VII. Gauthier–Villars, Paris, pp. 409–410.
Lagrange JL (1892) [Lagrange's letter to Euler on November 24th 1759. In: Œuvres de Lagrange. Tome XIV. Darboux G (ed). Gauthier–Villars, Paris, pp. 170–174.
Lagrange JL (1973a) Œuvres de Lagrange. Seconde édition, Courcier (ed), I–XIV vols., (in X) Paris. Gauthier–Villars, Paris.
Lagrange JL (1973b) Théorie des fonctions analytiques. In: Lagrange (1973a), vol. IX.
Mach E (1919) The science of mechanics. A critical and historical account of its development. The Open Court Publishing Co., Chicago-London [English translation of the fourth edition of Mach's Die Mechanik in ihrer Entwickelung historisch-kritisch dargestellt, first edition 1883].
Maier A (1509) Zwei Grundprobleme der scholastichen Naturphilosphie, (3° ed.) Edizioni Storia e Letteratura, Roma, 1968 [which, in turn, includes—with some modifications—The Parisian edition from 1509].
Naylor RH (1976) Galileo: the Search for the Parabolic Trajectory. Annals of Science 33:153–172.
Naylor RH, Drake S (1983) Discussion on Galileo's early experiment on projectile trajectories. Annales of Science 40:391–396.
Newton I (1687) Philosophiae Naturalis Principia Mathematica. Autore IS. Newton, Trin. Coll. Cantab. Soc. Matheseos Professore Lucasiano, & Societatis Regalis Sodali. Imprimatur. S. Pepys, Reg. Soc. Præses. Julii 5. 1686. Londini, Jussu Societatis Regiæ ac Typis Josephi Streater. Prostat apud plures Bibliopolas. Anno MDCLXXXVII.
Newton I ([1713] 1729) The Mathematical principles of natural philosophy. Translated by Motte Andrew. Motte B–Middle Temple-Gate, London.
Oliveira ARE, Pisano R (2022) Historical Details on Coulomb's Torsional Forces by Metal Wires. In: Ceccarelli M, López-García R (eds) Explorations in the History and Heritage of Machines and Mechanisms, Vol 40. Springer, Dordrecht, pp. 241–254.
Palmieri P (1988) Re-examining Galileo's Theory of Tides. Archive for History of Exact Sciences 53:223–375.
Pisano R (2007) Brief history of centre of gravity theory. Epistemological notes. In Kokowski M (Ed). Proceedings of the 2nd ESHS Congress, Krakow, Poland, pp. 934–941.

Pisano R (2009a) Il ruolo della scienza archimedea nei lavori di meccanica di Galilei e di Torricelli. In: Giannetto E, Giannini G, Capecchi D, Pisano R, (eds). Da Archimede a Majorana: La fisica nel suo divenire. Guaraldi Editore, Rimini, pp. 65–74.

Pisano R (2009b) Continuity and discontinuity. On method in Leonardo da Vinci' mechanics. Organon 41:165–182.

Pisano R (2009c) On method in Galileo Galilei' mechanics. In Hunger H (ed) Proceedings of ESHS 3rd Conférence. Austrian Academy of Science, Vienna, pp. 147–186.

Pisano R (2009d) Galileo Galileo. Riflessioni epistemologiche sulla resistenza dei corpi. In Giannetto E, Giannini G, Toscano M (eds). Relatività, Quanti Chaos e altre Rivoluzioni della Fisica. Guaraldi Editore Rimini, pp. 61–72.

Pisano R (2011) Physics–Mathematics Relationship. Historical and Epistemological notes. In Barbin E, Kronfellner M, Tzanakis C (eds). European Summer University History And Epistemology In Mathematics, Verlag Holzhausen GmbH–Holzhausen Publishing Ltd., Vienna, pp. 457–472.

Pisano R (2014) Historical Reflections on Newton's First Law and Carnot's Première Hypothèse. In: Gianna Katsiampoura G (ed). International Conference of the European Society for the History of Science. Athens (GR) 1–3 November 2012. National Hellenic Research Foundation/Institute of Historical Research/ Section of Neohellenic Research/ Programme of History, Philosophy and Didactics of Science and Technology, Athens, pp. 214–220.

Pisano R (2016) Details on the mathematical interplay between Leonardo da Vinci and Luca Pacioli. Journal of the British Society for the History of Mathematics 31(2):1–8.

Pisano R (2017a) Introduction. 1564–2014. In: Pisano and Bussotti 2017a, pp. 5–13.

Pisano R (2017b) A Development of the Principle of Virtual Laws and its Framework in Lazare Carnot's Mechanics as Manifest Relationship between Physics and Mathematics. Transversal-International Journal for Historiography of Science 2:166–203.

Pisano R (2020) A Tale of Tartaglia's Libro Sesto & Gionta in Quesiti et Inventioni diverse (1546–1554). Exploring the Historical and Cultural Foundations. Foundations of Science 25/2:477–505.

Pisano R (2024-pre-print) Galileo Galilei: The Two Manuscripts On Mechanics. Translations, Text and Commentaries. Dordrecht, Springer, *forthcoming*.

Pisano R, Bussotti P (2014) Historical and Philosophical Reflections on the Culture of Machines around the Renaissance. How Science and Technique Work? Acta Baltica Historiae et Philosophiae Scientiarum 2/2:20–42.

Pisano R, Bussotti P (2015a) Fibonacci and the Abacus Schools in Italy. Mathematical Conceptual Streams, Science Education and its Changing Relationship with Society. Almagest 6/2:126–165.

Pisano R, Bussotti P (2015b) Historical and Philosophical Reflections on the Culture of Machines around the Renaissance: Machines, Machineries and Perpetual Motion. Acta Baltica Historiae Et Philosophiae Scientiarum 3/1:69–87.

Pisano R, Bussotti P (2015c) Galileo in Padua: architecture, fortifications, mathematics and « practical » science. Lettera Matematica Pristem International 2/4:209–221.

Pisano R, Bussotti P (2016) A Newtonian Tale Details on Notes and Proofs in Geneva Edition of Newton's Principia. Bulletin BSHM-Journal of the British Society for the History of Mathematics 31/3:160–178.

Pisano R, Bussotti P (2017a) On the Conceptualization of Force in Johannes Kepler's Corpus: an Interplay between Physics/Mathematics and Metaphysics. In: Pisano R, Agassi J, Drozdova D (eds) Alexandre Koyré 1892–1964. Springer, Dordrecht, pp. 295–346.

Pisano R, Bussotti P (2017b) The Fiction of Infinitesimals in Newton's Works. On the Metaphoric Use of the Infinitesimals in Newton. In Marcacci F, Rossi MG (eds). Isonomia-Epistemologica. Reasoning, Metaphor and Science 9:141–160.

Pisano R, Bussotti P (2017c) Newton's Principia Geneva edition: the action-and-reaction law. Historical and Nature of Science reflexions. Pavia University Press, Pavia, pp. 269–276.

Pisano R, Bussotti P (2017d) (eds) Homage to Galileo Galilei 1564–2014. Reading Iuvenilia Galilean Works within History and Historical Epistemology of Science. Special Issue Philosophia Scientiae 21/1.

Pisano R, Bussotti P (2020) Newton's Geneva Edition: the Notes on Integral Calculus. In: La Rana A, Rossi P (eds): International 39th Congress of the Italian Society for the History of Physics and Astronomy (Pisa/VIRGO 2019, 9–12 Sept.). Pavia, Pavia University Press, pp. 127–134.

Pisano R, Bussotti P (2022) Conceptual Frameworks on the Relationship between Physics–Mathematics in the Newton *Principia Geneva* Edition (1822). Foundations of Sciences (Springer) 27/3:1127–1182.

Pisano R, Bussotti P (2024) Historical and Methodological Details on the *De motu gravium naturaliter descendentium* in Torricelli's *Opera Gometrica* (1644). In: Pisano Dhombres, Radelet de Grave and Bussotti 2024a.

Pisano R, Cioci V (2020) Galileo's Free fall into History Physics and Nature of Science Teaching. In: International 38th Congress for The SISFA–Italian Society of Historians of Physics and Astronomy. Messina 3–6, October, Italy, Pavia University Press, pp. 271–278.

Pisano R, Robarts J (2022) A Short Comparative Historiography of Science & Technology of Mechanics into Engineering—and—Architecture Literatures during the Renaissance. In: Ceccarelli M, López-García R (eds) Explorations in the History and Heritage of Machines and Mechanisms, vol 40. Springer, Dordrecht, pp. 227–240.

Pisano R, Capecchi D (2010) Galileo Galilei: Notes on Trattato di Fortificazione. In Altamore A, Antonini G (eds), Galileo and the renaissance scientific discourse, Roma Edizioni nuova cultura, Roma, pp. 28–41.

Pisano R, Capecchi D (2016) Tartaglia's Science of Weights and Mechanics in the Sixteenth Century. Selections from Quesiti et inventioni diverse: Books VII–VIII. Springer, Dordrecht.

Pisano R, Agassi J, Drozdova D (2017) (eds) Hypothesis and Perspective within History and Philosophy of Science. Homage to Alexandre Koyré 1892–1964. Springer, Dordrecht.

Pisano R, Cooppersmith J, Peake M (2021) Essay on Machines in General (1786). Text, Translations and Commentaries. Lazare Carnot's Mechanics–Vol. 1. Springer, Dordrecht.

Pisano R, Dhombres J, Radelet de Grave P, Bussotti P (2024) 380th Anniversary of Torricelli's Opera Geometrica (1644). Springer, Text, Transcription, Commentaries and Selected Essays as New Historical Insights. Springer, Cham.

Reeves E, Van Helden A (2010) (eds) On Sunspots. Galileo Galilei and Christoph Scheiner. Chicago University Press, Chicago and London.

Schuster JA (2000) Descartes opticien: the construction of the law of refraction and the manufacture of its physical rationales, 1618–1629. In: Gaukroger S, Schuster J, Sutton J (2000) (eds). Descartes' natural philosophy. Routledge London and New York, pp. 258–312.

Schuster JA (2013a) Descartes–Agonistes. Physico–mathematics, Method and Corpuscular–Mechanism 1618–1633. Springer, Dordrecht.

Schuster JA (2013b) Cartesian Physics. In: Oxford T (ed) Buchwald Jz and Fox R. The Oxford University Press, Oxford, Handbook of the History of Physics, pp 56–95.

Tartaglia N (1537) Nova scientia inventa da Nicolo Tartalea. B. Disciplinae mathematicae loquuntur[.] Qui cupitis rerum varias cognoscere causas disate nos cunctis hac patet una via. In Vinegia per Stefano Nicolini da Sabio. Ad instantia di Nicolo Tartalea Brisciano il qual habita a San Saluador. MDXXXVII.

Tartaglia N (1546) Quesiti et inventioni diverse de Nicolo Tartalea Brisciano. Stampata in Venetia per Venturino Ruffinelli ad instantia et requisitione, & a proprie spese de Nicolo Tartalea Brisciano Autore. Nel mese di luio L'anno di nostra salute. M.D.XLVI.

Tartaglia N ([1554] 1959) La nuova edizione dell'opera "Quesiti et inventioni diverse de Nicolo Tartaglia brisciano, Riproduzione in facsimile dell'edizione del 1554 [with Gionta al sesto Libro] Masotti A (ed). Commentari dell'Ateneo di Brescia, Tipografia La Nuova Cartografica, Brescia.

Torricelli E (1644) Opera geometrica. Masse & de Landis, Florentia.

Torricelli E (1919–1944) Opere di Evangelista Torricelli. Loria G, Vassura G (eds). 4 Vols., Montanari, Faenza.

Vergara Caffarelli R (2007) Il principio d'inerzia negli ultimi scritti di Galilei. Cronos 10:63–88.

Chapter 22
Deciphering Enigma

Thomas Cuff and Jean-Marc Ginoux

Abstract The idea of secret writing (cryptography), whether in terms of codes or ciphers, has a long history. Written communication exhibits a number of phenomena with respect to the frequency of occurrence of letters of the alphabet in words. The aforementioned phenomena are true both for single letters and combinations of two letters (digraphs) that appear in words. The science of cryptography seeks to suppress these phenomena in order to frustrate the cryptanalyst's efforts to convert secret writing back into clear text. The mechanism by which the idea of secret writing has been realized evolved from manual to electro-mechanical operations. Perhaps the most famous cryptographic artifact used to mechanize the process of encryption and decryption is the Enigma machine, which is the subject of this chapter. We will reverse engineer the Enigma machine so the reader can see for themselves the mechanical and electrical mechanisms used by this portable, battery powered device. To make this exposition more understandable, extensive use will be made of annotated photographs of actual Enigma machines, drawings from patents, drawings from instruction manuals, other visual aids, etc.

Keywords Enigma machine · Cryptography · Mono-alphabetic Substitution cipher · Poly-alphabetic Substitution Cipher · Electro-mechanical device · Rotor

1 Introduction

The German Navy's submarine service attacked the lone ships and convoys of ships carrying supplies and materiel to Great Britain during WWII. Eventually organized into Wolf Packs, these submarines (U-boats) threatened to starve Britain into submission. Key to the success of the Wolf Pack strategy was the use of central control

T. Cuff
Frederick Community College, Frederick, USA

J.-M. Ginoux (✉)
Department of Mechanical Engineering, Université de Toulon, Université de Lorraine, France & Archives Henri Poincaré, Nancy, France
e-mail: ginoux@univ-tln.fr

R. Pisano (ed.), *A History of Physics: Phenomena, Ideas and Mechanisms*,
History of Mechanism and Machine Science 42,
https://doi.org/10.1007/978-3-031-26174-9_22

via radio communication, and the contents (orders and information) conveyed in the radio telegraphy messages was insured by encrypting the message content by first the M3 type Enigma machine, and later the more cryptographically secure M4 Enigma machine. In fact, all three of the German armed services: army, navy, and air force, eventually employed the Enigma machine in 1930, 1934, and 1935, respectively (Calvocoressi 1980, p 26). At Bletchley Park (BP), Alan M. Turing and his cohort, building on the pioneering work of three Polish mathematicians (Marian Rejewski, Jerzy Różycki, and Henryk Zygalski), designed and commissioned the building of automatic machines, Bombes or Bomby, to help with the decryption of the messages being set from and to the Wolf Packs. Eventually, BP succeeded in breaking the encryption of the M3 in near real time. But in early 1942, the German Navy finished updating their M3 Enigma machines to the M4 version, and BP was no longer able to read the messages. The M4 required a more complicated Bombe organized into larger groups of machines. Due to the ongoing war effort, Britain did not have sufficient spare design and manufacturing capacity to create reliable new Bombes, and the task was taken over by the Americans. This document will concentrate on the describing and explaining the inner structure and function of the M3. Although the military Enigma secret was first made public in the early 1970s, many people still have misconceptions about how the Enigma machines actually worked from the inside. While the machines were quite complicated, they are still amenable to elucidation with the help of photographs, mechanical drawings, and electrical schematic diagrams of the actual machines.

2 Radio (Wireless) Networks, A Problem in Search of a Solution

Modern warfare depends upon communication for everything from tactics, to strategy, to logistics, to personnel matters, to intelligence, etc. Under most circumstances, the aforementioned information is communicated over radio networks, which are susceptible to eavesdropping by the enemy. For example, the British could easily intercept German military's wireless traffic originating from Europe, or even Russia, at interception stations in England using large antenna arrays, commercially available radio receivers manned by operators—usually women—proficient at Morse code, and equipped with a pencil and a specially ruled pad of paper, or a typewriter. For places like the North African desert and Asia, mobile interception stations mounted in trucks could be employed. Note, the interception stations were known as the Y Service, while the main station for decryption, Bletchley Park, was known as Station X.

Radio (wireless) communications utilized two information modalities: Morse code, and voice. These two modalities refer to the type of information being transmitted, but in to order project these baseband signals through empty space or the air over great distances, they must be imposed on radio waves. For example, you can shout out of an open window in your house or apartment, but the range of this type of information transmission is quite limited. On the other hand, radio waves, depending on their frequency, can travel hundreds, and even thousands of miles, for example, VHF (Very High Frequency) radio waves, 144 MHz, have been bounced off the Moon by amateur radio operators in 1953, and the echoes picked up using suitable receivers on Earth (Blakeslee 1972). Information, such as Morse code or voice, can be imposed on radio waves by the process of modulation. During WWII, there were three main methods of modulation of the RF (Radio Frequency) carrier used to transmit the information through space: CW (Continuous Wave), which was used for Morse code transmission, also called radio telegraphy, and AM (Amplitude Modulation) or FM (Frequency Modulation), which were used to transmit voice and music, also called radio telephony. Physically a radio transmitter is connected to an antenna, which serves to couple the RF energy to the surrounding space in an efficient manner. At the receiving station, another antenna couples the incoming, modulated radio waves to a receiver. Inside the receiver, the intercepted RF energy is discriminated by its center frequency, i.e., the receiver is tuned by varying a knob on its front panel, the desired signal is then demodulated to recover the original baseband signal carrying the information: Morse code, or voice.

Everyone has probably seen a movie or TV series, set during WWII, wherein a spy is equipped with a radio wave transmitter and receiver in a suitcase. Once they are in a secure room/structure, or a secluded area outdoors, they opened the suitcase, put up an aerial (antenna), and began transmitting and/or receiving information on a prearranged channel (center frequency). This bulky apparatus in the suitcase, with its battery powered transceiver (trans[mitter-re]ceiver), was the forerunner of the modern smart phone. The modern smart phone besides containing three transceivers (GSM, Wi-Fi, and Bluetooth), has a high energy density, lithium-ion battery, and the ability to transmit and/or receive data, voice, music, and video information.[1]

For a number of practical reasons, CW modulation and Morse code was the dominant scheme used in radio (wireless) communication in WWII. Unfortunately, because of their very nature, radio waves were reasonably easy to intercept, and anyone who knew how to convert the dots and dashes of Morse code back to letters and numbers would be able to read the original message's content. The problem was that Morse code was not a secret code. Before we continue, let us distinguish between a code and a cipher. For our purposes, we shall explain the difference by example instead of by definition. Morse code is an example of a code because it replaces each letter, decimal digit, or punctuation mark by a series of dots (•) and dashes (-) . For example, the international distress call is given by the letters 'SOS', which in Morse code becomes •••_ –-_ •••, where the underscore (_) simply represents a long silence/pause to delimit one Morse character from the next, where 'S' •••, 'O' –-, and _ serves as the delimiter/separator between each character. Thus, in a code, different symbols and different numbers of symbols are employed. In a cipher, 'SOS' would be replaced by, for example, 'VTV', if the substitution was mono-alphabetic,

[1] GSM (Global System for Mobile Communications) is a second generation, digital, cellular telephone protocol that has been adopted worldwide.

or 'VZX' if the substitution was poly-alphabetic. Therefore, in a cipher the same symbols and same number of symbols are employed.

Thus, all Enigma machines perform a substitution encryption in which a given letter of clear text is replaced with different letter of the Modern European alphabet in the cipher text; the term 'plain text' is reserved for the character set out of which the unencrypted message is written, the clear text, and the encrypted message is written, the cipher text. If each letter of the clear text was replaced with a single letter in the cipher text, then we would have what is known as a mono-alphabetic substitution cipher, e.g., see the following, where we have used the monospace font Courier New,

```
Clear Text Letters: ABCDEFGHIJKLMNOPQRSTUVWXYZ
Cipher Text Letters: KSGJTDAYOBXHEPWMIQCVNRFZUL
```

In the above example, the clear text letter 'C' is always replaced with the letter 'G' in the cipher text. For example, the clear text word 'FOOD' would be rendered as 'DWWJ' in the cipher text. In contrast, the Enigma machines employed a poly-alphabetic substitution cipher, wherein a given letter in the clear text was substituted by more than one letter in the cipher text. For example, in the case of a certain setting of the rotors of the Enigma machine, the clear text word 'FOOD' would be rendered as 'HYWI' in the cipher text. In other words, each time the same key of a particular clear text letter was pressed on the Enigma machine keyboard a different letter would appear as the cipher text letter in the lighted display.

The Enigma machine was and is a marvel of technology. How it came to be, and what it tells us about human ingenuity both from the prospective of the cryptographer (maker of codes and ciphers), and the cryptanalyst (breaker of codes and ciphers) is well worth knowing. Cryptology (the study of the making and breaking of codes and ciphers) has been with us for centuries, but it was only in the first half of the twentieth century that it became mechanized, commercialized, and indispensible.

3 The Enigma Machine

It is time to actually gaze upon a real Enigma machine (Fig. 1). Note, all the photographs of Enigma machines in this document will be of actual WWII vintage Enigma machines, not reproductions.

Fig. 1 A Real M3 Type of German Military Enigma Machine *Source* Photograph by Tom Cuff; On 5/24/13 at the National Cryptographic Museum, Fort Meade, Maryland

Starting at the bottom of the photograph (Fig. 1), we first encounter the plug board, which are sets of electrical contacts that can be connected via a set of patch cords; two spare patch cords can be seen the wooden lid at the top of the photograph. Next comes the input device, the 26-key keyboard on which the operator either enters clear text for the purpose of encryption, or enters cipher text for the purpose of decryption in order to recover the original clear text. Above the keyboard is the output device, a display of 26 letters each able to be independently backlit by one of 26 low voltage, incandescent bulbs/lamps. In response to a single key being pressed, one and only one letter will light. Above the display, there are three rotors, invisible except for three raise areas in the top cover. To the left of each raised area, there is a window allowing the operator to see the position of the rotor. To the right of the three rotors there is the power selector switch, and then two terminals for supplying external power, if the need arises. Finally, there is the wooden lid containing spare patch cords; spare low voltage, incandescent bulbs/lamps; display filter to improve contrast; and a plaque with the maintenance and care instructions written on it.

Let us begin our discussion of the Enigma machine with an overview of how it was used in the field by, say, the German Army. Assume a message needed to be sent from a forward field unit describing the results of their latest operation. Typically, a team of three people would be required to send the message. The first person would write out the message on a special form that grouped the alphabetic characters of the message in groups of five letters. An example will make this procedure clear. Suppose the clear text message was: "Sighted enemy tank column. We will engage and destroy." The resultant message in groups of five letters would be: "SIGHT EDENE MYTAN KCOLU MNWEW ILLWN GAGEA NDDES TROYX". The message was solely in uppercase alphabetic letters because the keyboard of the Enigma machine only had 26 keys, one for each letter of the alphabet; there were no lowercase letters, punctuation marks, space character, or decimal digits. If procedure allowed it, an infrequently letter such as 'X' could be used to separate words since the Enigma keyboard lacked a space character, otherwise, the words of the message could just be run together with the 'X' character added at the end of the message to pad (make complete) the last five letter group. After filling out the form, the first person of the team would slowly dictate each letter of the message to the second person of the team, who was sitting in front of the Enigma machine. The second person had three jobs: (1) setting up the Enigma machine, which is a complicated task we shall discuss presently; (2) typing each letter of the message that needed to be encrypted; and (3) and calling out the encrypted letter that appeared in the lighted display located directly above the keyboard. The third person of the team, the radio operator, would write the encrypted letters being called out by the second person on a similar form to that used by the first person of the team so that the letters were in groups of five.

The message itself would contain three parts: (1) a header/preamble describing the radio network to which the message was being sent, like the forward and return addresses on the envelope of a letter, which was not encrypted; (2) a portion of the cryptographic key, specifically the initial position of the three rotors of the Enigma machine, which might be partially encrypted; and (3) the body of the message, i.e., the content of the message, which was encrypted. Before encryption of the body of the message, the third person of the team would add the header/preamble to the form and the partially encrypted portion of the cryptographic key. Once the first two parts of the message were completed on the form, then the first team member would begin feeding the clear text letters of the message body, to the second team member operating the Enigma machine, who, in turn, would call out the encrypted letters to the third team member. If the message did not completely fill the last five positions in the form, the second team member would add padding characters such as the letter 'X', suitably encrypted. Once the body of the message was encrypted, the third person, the radio operator, would transmit the message using Morse code.

Because there would be many German radio operators on the same radio network, the header/preamble would indicate whether the message was meant for them. If the message was not meant for them, then they would ignore the remaining two parts of the message. If the message was meant for the receiving radio operator, then he would fill in his form with the five letter groups. Once the form was filled out properly, then radio operator would convene a three person team to perform the decryption of the message to recover the clear text. The radio operator would then present the partially encrypted key representing the initial positions of the three rotors to the second team member. Once the second team member had setup the Enigma machine, the decryption of the message body would begin, and the Enigma machine operator would type in the encrypted letters one-at-a-time and call out the decrypted letters appearing in the lighted display of the Enigma machine to the third team member so that he could fill out his form. Note, the third team member's form would contain all three parts of the message, and so contain an integer number of five letter groups.

The previously mentioned processes of encryption and decryption might appear to be both cumbersome and time consuming, and they were, but the saving grace was that the machine generated encryption and decryption operation was less error prone than the older manual methods. Exactly how time consuming was the encryption and decryption process, for example? Consider the following example in which the USMC (United States Marine Corps) compared the speed of a Hagelin model M209 pin and lug machine, a mechanical rotor cipher machine similar to the electro-mechanical Enigma machine, to a Native American Code Talker sending the same message (Miller and Fields 2011). Like the Enigma machine, the Hagelin machine's cipher was transmitted via Morse code, while the Navajo, Choctaw, or Comanche Code Talkers spoke their respective native languages, languages with no written form and only spoken by members of these three Native American tribes, over a radio telephony link. Under battlefield conditions, the Code Talkers could encode a three line message from English into, say, Navajo at the transmission end and decode the message from Navajo back into English at the receiving end in 20 s compared with 30 min to compete the same task using the state of the art encryption machine such as the Hagelin, model M 209, Converter[2] (Simpson 2017; Paul 1973; McClain 1994). This speed benefit and the fact that the Code Talkers' code was never broken explained why the use of Code Talkers continued for so long after the introduction of mechanical and electro-mechanical encryption/decryption machines. Besides being used in the Pacific campaign during WWII, the Code Talkers were employed during the Korean Conflict and the Vietnam War, and their very existence remained a closely guarded secret until about 1968, according to Simon Singh (Garliński 1979, pp. 152–162). Other sources, such as David Kahn, provide evidence that the existence of Code Talkers was known much earlier than 1968. Kahn's book has a copyright of 1967 and the entries in the Notes section for pp. 549–550, wherein the Code Talkers are

[2] Windtalkers (June 14, 2002), screenplay: John Rice & Joe Batteer, Lion Rock Productions, 134 min.

discussed, give as sources various newspaper articles from 1921, 1938, 1940, 1941, and 1945. Note, the Index section of Kahn's book only has an entry for the term 'Navahos [sic]' and nothing for the terms: 'Choctaws', 'Code Talkers', 'Codetalkers [Kahn's term]', or 'Comanches' (Kahn 1967).

Let us clarify the purpose of some of the procedures we have mentioned previously. The header/preamble of the message served two functions: (1) it contained the forward address so that radio operators, listening on a certain center frequency of the radio network, could tell if the message was meant for them; and (2) the return address in the header/preamble allowed the intended recipient to send an acknowledgment message back to the sender indicating the message was received without any garbling or dropped letters, otherwise the intended recipient would send the sender a request for the message to be resent. The grouping of letters in groups of five, for example, was done to help detect and prevent translational errors due to a dropped letter in the message. Translational errors were especially a problem for the person receiving the cipher text, but they also created a problem for the sender, who would invariably be asked by the receiver to resend the message. Suppose the radio operator transmitting the cipher text drops one letter of the cipher text in the middle of the message. If the radio operator receiving the cipher text did not use a form which grouped the letters in groups of five letters, he probably would not realize that the message was missing a single letter. As the corrupted cipher message was read out to the Enigma machine operator, he, in turn, would call out the clear text letter appearing in the machine's display to the person writing down the clear text. The clear text would only be understandable up to the missing letter after which point it would become gibberish because the poly-alphabetic substitution, which required that the enciphering Enigma machine and the deciphering Enigma both start with the same configuration, would fall out of synchronization if even a single letter were dropped by either the transmitting or receiving radio operator. In other words, the two Enigma machines must start in the same state, and advance to the next states in lockstep with the letters of the original message. A loss of synchronization means that if the nth encrypted letter of the message is missing, the next encrypted letter will be paired with the wrong state of the deciphering Enigma machine (Cfr. p. 2) (sheet 1, lines 20–40), from the US patent for the commercial Enigma machine (Cfr. Patents 2).

The inventor indicated in the US patent a counter could be added to the cipher machine so that, for example, a bell might ding to indicate that a predetermined number of characters had been enciphered.

The Enigma machine was a battery powered, electromechanical device that enciphered clear text into an apparently meaningless string of letters, cipher text. It had to be battery powered because it was meant to be portable, and used in the field. However, like the smart phone, which is also battery powered, the Enigma machine had a pair of terminals to allow it to be powered by either an external battery or from an AC outlet by means of what used to be called a battery eliminator, i.e., an AC to DC power supply similar to the AC adapter provided with your smart phone. However, unlike your smart phone, which is an electronic device, i.e., it uses ICs (Integrated Circuits), and semiconductor piece parts (transistors, diodes, etc.) to control the flow of electricity, and unlike the WWII transceiver in a suitcase, which utilized electron vacuum tubes (or as the British referred to them, valves or lamps) to control the flow of electricity, the Enigma machine is purely an electrical–mechanical device instead of an electronic device. Electronic circuits utilize active devices (electron vacuum tube, transistors, Ics), which, when presented with an input signal, can produce an output signal with a power gain above the power of the input signal, i.e., active devices allow for the process of amplification; in contrast, circuits that exclusively contain passive devices (mechanical switches, resistors, capacitors, inductors, transformers, etc.) cannot produce amplification. The electrical circuits in the Enigma machine are, in their essence, no more sophisticated than the circuit employed in an electric doorbell, and contain no active components. A traditional doorbell circuit consists of a SPST (Single Pole Single Throw), spring loaded, momentary- contact switch connected via two wires to an electric buzzer/chime/bell in series with an electrical power source (usually a step-down transformer powered from the AC mains. The Enigma machine has a large number of mechanical switches; a set of 26 low voltage, incandescent bulbs/lamps; and a battery as its power source. There are no electron vacuum tubes, transistors, or Ics inside the Enigma machine.

We previously mentioned that the second part of the message contained a portion of the cryptographic key, i.e., the initial positions of the rotors, but what about the rest of the cryptographic key, and why must there be a cryptographic key, i.e., what is its purpose? In order for the two different Enigma machines to form a complementary encryption and decryption pair, they must be configured to be the same, but this configuration must be both amenable to change, and contain many members. Like all modern cryptography machines, the Enigma machine relies on the concept of a cryptographic key, i.e., the selection of an initial machine configuration from a large space, called the key space, of initial machine configurations. The cryptographic key prevents someone who has, for example, captured the cryptographic machine from simply using it to read cipher messages created on it.

Think of the lock to the front door of your house or apartment; the lock and its key form a complementary pair in the sense that the lock is locked or unlocked by the key, and the pattern cut into the key together with the complementary tumbler settings in the lock provides a deterrent to anyone simply opening your door with their own key. Of course, theoretically one could try every possible key from the key space, but, if the key space is sufficiently large, this approach—the brute force method—would take too long to be operationally effective, i.e., most secret messages have a finite shelf life, a time when the secret message must remain secret, beyond which it does not matter. There is also the problem that, just by coincidence, some of the keys used in the brute force approach may yield fragments of clear text having nothing to do with the actual clear text message, which can cause the cryptanalyst to erroneously think that, for example, the clear text is a mixture of actual words from the source language and code book entries that look like gibberish or that the message was garbled in either transmission or interception (reception by the opposition) or both. When, first, the Poles, and later the British, and later still the Americans, utilized the Bombe/Bomba machines, which were used to help break the encryption of the Enigma machine, it is important to realize that the brute force method was not and could not be used because the key space of the Enigma machines was simply too large, i.e., to build the Bombe/Bomba, the wiring of the rotors, entrance cylinder, and reflecting cylinder all had to be known. Instead of the brute force approach, the British and American Bombe/Bomba machines had to be provided with clues or cribs about what the body of the message contained.

Now, we must ask what elements of the Enigma machine make up its key space? The answer, which has many times been given incorrectly, is that there are five elements to the key space. We shall list these five elements in the order in which the operator of the Enigma machine would have to approach them. Note, when discussing Enigma machines, we shall initially use the verbose terminology "3 rotor slot or position Enigma machine" or "4 rotor slot or position Enigma machine" to hopefully avoid confusion when also talking about the number of rotors supplied with each Enigma machine. For example, a 3-rotor slot or position Enigma machine might be supplied with 5 rotors, but only 3 of the 5 rotors can be installed in the machine at the same time because there are only 3 slots or positions available.

Element #1, select from the available set of five rotors, as directed, three rotors, (Figs. 5 and 6).

Element #2, set the so-called ring setting of each the three selected rotors as directed; we will explain what the term 'ring setting' means later (Figs. 9 and 10). This portion of the key space is sometimes erroneously omitted from the description.

Element #3, place the chosen three rotors in a given order, from left-to-right, on their common shaft/spindle as directed, and place the populated shaft/spindle inside the Enigma machine (Figs. 2, 5 and 6).

Element #4, plug the provided patch cables into the plug board on the front of the Enigma machine in the specified number and manner, i.e., as directed (Fig. 12).

And lastly, Element #5, set the initial position of the three rotors, i.e., rotate each rotor until the proper number or letter appears in the respective window of the top panel (Figs. 2 and 11). We mentioned that Elements #1–4 were set "as directed", but not Element #5. Why? The answer to this question has to do with the subject known as key distribution.

Enigma machine operators must know ahead of time how to configure their respective machines, and this is codified by providing each operator with a document containing Elements #1–4 for a given period of time, say, six months. That is what we mean by the term 'key distribution'. As the host of the BBC, TV series The Secret War, William Woollard mentioned in Episode 6 – Still Secret, the key setting (configuration) was changed at midnight (probably Berlin time as opposed to local time, although, he does not say) every day, the key setting for a specific day had to be looked up in a document, and that different groups within, say the German Army, were in different Enigma radio networks and so had different key setting documents.[3] Element #5, as mentioned earlier, was included as the second part of the message. Element #5 was left up to the Enigma machine operator, although, certain services

[3] The Secret War (1977), screenplay by Brian Johnson, based on the book *Most Secret War* (1978) by Reginald Victor Jones, BBC, 300 min (in particular Episode 6: Still Secret, directed by Fisher Dilke).

such as the German Navy had very stringent requirements on how to choose the initial positions of the rotors, not unlike the modern rules for creating a password, which stress good password hygiene. Allowing the operator to choose Element #5 was meant to provide an added measure of unpredictability to the key space in order to further frustrate the efforts of the cryptoanalyst. Thus, Element #5 is distributed in real time as part of the message.

We have mentioned that Elements #1–5 of the cryptographic key could be changed, but there are also two other parts of the encryption/decryption stack (entrance cylinder, three rotors, and reflecting cylinder), which could be used in the encryption process: (1) the entrance cylinder, and (2) the reflecting cylinder (Fig. 4). Normally, these other two parts of the encryption/decryption stack are fixed, and so if an enemy were able to capture an Enigma machine, they would know how the entrance and reflecting cylinders were wired. Of course, the last sentence is a generalization, which is not completely correct since the reflecting cylinder went through some iteration: A, B, etc. Note, the wiring of the entrance cylinder stymied the British completely early in the war, and later in the war, the German Navy's utilization of an easily reconfigurable reflecting cylinder and set of eight rotors with their M4 type Enigma machine proved too much for the British Bombes constructed for use against the M3 type of Enigma machine.

4 The Innards of the M3 Enigma Machine

Having presented a photograph of a M3 type Enigma machine (Fig. 2), it is time to finally discuss how it works in detail. To accomplish this exposition, we will need photographs, mechanical drawings, and electrical schematic diagrams. If all these visuals aids seem overwhelming, the reader should take heart in that the writers were no less intimidated when they first encountered these same aids.

4.1 Where Encryption Occurs: the Stack and the Plug Board

The first question we need to answer is where does the encryption occur within the Enigma machine? The answer is that the main source of encryption is the stack of rotating and fixed contacts made up of the entrance cylinder, three movable rotors, and the reflecting cylinder. Encryption also takes place through the plug board. Note, not all Enigma machines were equipped with a plug board, for example, the Enigma machine used by the German railways did not have a plug board. To see the encryption/decryption stack, we must lift up the top cover of the machine (Fig. 2).

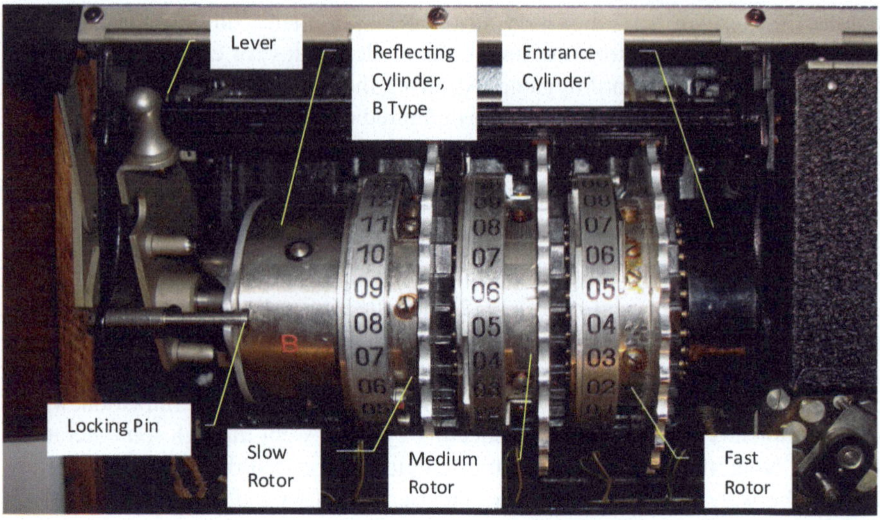

Fig. 2 The Encryption/Decryption stack of fixed and rotating electrical contacts *Source* Photograph by Tom Cuff

Table 1 Rotor Mappings for the Railway Enigma Machine (Rocket)[4]

Rotor #	ABCDEFGHIJKLMNOPQRSTUVWXYZ	Date introduced	Model Name & Number
I	JGDQOXUSCAMIFRVTPNEWKBLZYH	7 February 1941	German Railway (Rocket)
II	NTZPSFBOKMWRCJDIVLAEYUXHGQ	7 February 1941	German Railway (Rocket)
III	JVIUBHTCDYAKEQZPOSGXNRMWFL	7 February 1941	German Railway (Rocket)

Source http://en.wikipedia.org/wiki/Enigma_rotor_details, circa February 2013

Starting from the right, we have the black entrance cylinder, followed by three silver rotors, and at the left the reflecting cylinder with the letter 'B' embossed on it. The entrance cylinder has 26 fixed, electrical contacts, one for each letter of the keyboard. All of the electrical signals going in to and out of the encryption/decryption stack pass through the entrance cylinder. Each of the three rotors has two set of 26 rotating, electrical contacts, one set on the right face and one set on the left side. The wiring between the 26 contacts on the right face and the 26 contacts on the left

[4] Let's note that the entries in this table were originally obtained from a treatise penned by Alan M. Turing, see the following URL: http://cryptocellar.org/Turing/index.html and they contain some errors as detailed in a blog by Frode Weierud, http://cryptocellartales.blogspot.com/2013/05/railway-enigma-and-other-special.html.

face of a given rotor is such that one contact on the right side is connected to one-and-only-one contact on the left side. If we chose to simulate an Enigma machine, then to make side-by-side comparisons of the simulator outputs possible a specific Enigma machine such as the Railway Enigma machine known by the codename Rocket would be selected (Table 1).

Thus, the wiring of Rotor I was as follows (the monospace font Courier New is used below):

```
Right Side: ABCDEFGHIJKLMNOPQRSTUVWXYZ
Left Side: JGDQOXUSCAMIFRVTPNEWKBLZYH
```

In other words, the 'A' position on the right side of the rotor is connected to the 'J' position on left side, and the 'A' position on the left side of the rotor is connected to the 'J' position on the right side. The reflecting cylinder at the left end of the stack has 26 fixed, electrical contacts. It is called a reflecting cylinder because each of its 26 contacts is connected in pairs, 13 pairs in all. In other words, the electrical signal entering the entrance cylinder passes through the three rotors until it reaches the reflecting cylinder at which point it is turned around and fed back through the stack in the opposite direction but through a different path back to the entrance cylinder (Fig. 3).

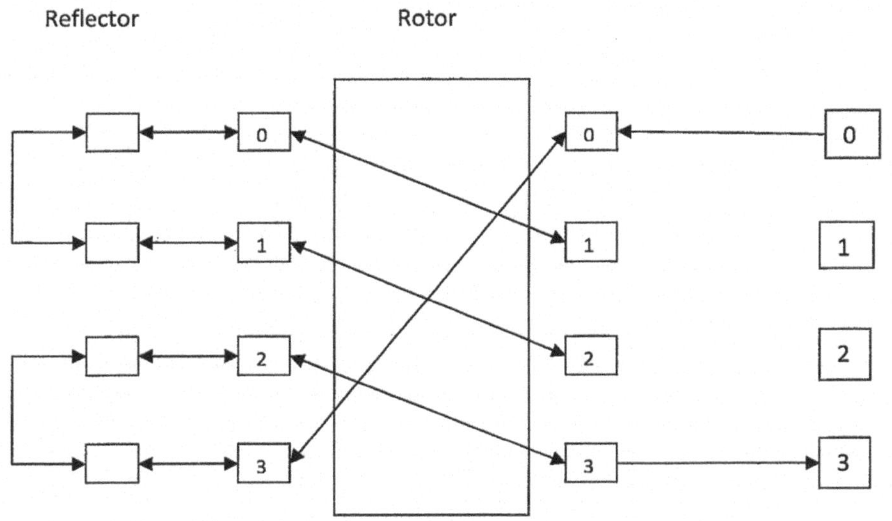

Fig. 3 Minimal Enigma Encryption/Decryption Stack with a 4-Letter Alphabet[5]

There are two consequences of the reflecting cylinder: (1) no letter is mapped to itself by either the encryption, or decryption process; and (2) the Enigma machine is self-decrypting. The self-decrypting property means that the Enigma machine does

[5] Note, the indices start with 0, while a real Enigma machine's indices usually begin with 01.

not have a two-position switch marked Encryption and Decryption, i.e., the Enigma machine's operator performs an encryption (clear text to cipher text) the exact same way as they perform a decryption (cipher text to clear text). Of course, the two machines must have the same configuration (key), but there is no need to change anything else about the machines, like flipping a switch to go from encryption to decryption mode. As to who invented the reflecting cylinder/reflector/half-cylinder, there seems to be intense disagreement. Arthur Scherbius is given the honor by Józef Garliński[6] (Garliński 1979), and David Kahn (Kahn 1967); Hugo A. Koch is championed by Louis Kruh and Cipher Deavours (Kruh and Deavours 2002); and Willi Korn is firmly given the nod by Friedrich L. Bauer (Bauer 2002). If one looks at the US patents of the three candidates, then it is clear from the drawings and text that Scherbius' commercial Enigma machine passes the electrical signal through the stack only in one direction; the decryption process passes the electrical signal through the stack in the opposite direction (lines 121–130, page 4 and lines 1–5, page 5). There is no reflecting cylinder. The US patent of Koch performs encryption by directing the motive force (air, water, oil, electricity, or optical energy) through the stack from left-to-right, while decryption is realized by directing the motive force through the stack from right-to-left via a changeover valve (Figs. 6 and 7, p. 2). Thus, again, there is no reflecting cylinder. Only the US patents of Korn explicitly depicts something that can be readily recognized as a reflecting cylinder (Fig. 2, p. 1), of US patent 1,705,641, for example see Patents 1–2 & 5–7. At the far left of the encryption/decryption stack is a silver colored lever with a bulbous handle. In its position as shown in Fig. 2, the lever locks in place the smooth shaft/spindle on which rides the three rotors between the entrance cylinder and the reflecting cylinder; in its position as shown in the following Fig. 4, the lever is in the unlock position. The following Fig. 4, shows the Enigma machine with the shaft/spindle and three rotors removed, note the position of the lever. When the lever is moved to the upright (vertical) position it does three main things: (1) the lobe associated with the cam that is part of the lever pushes the roller and pawl assembly towards the rear of the Enigma, and so disengages them from the rotors; (2) the reflecting cylinder shifts to the left more than enough so that all the spring loaded electrical contacts are no longer in contact with their opposing non-spring load contacts; and (3) the leftward motion of the reflecting cylinder frees the shaft/spindle allowing it and the rotors to be lifted out as a unit. To get an idea of how much leftward travel the reflecting cylinder does in response to the lever going from its locked position, Fig. 2, to its unlocked position, Fig. 4, compare how far the locking pin overlaps the reflecting cylinder in these two figures. Note, the locking pin is fixed to the frame of the Enigma machine so it is a good datum (reference point). Note, the square box, abelled '2', near the entrance cylinder is the battery. Also note the bank of 26 low voltage, incandescent bulbs that forms the display.

[6] At page 9, Garliński claims (Garliński 1979, p. 9) that Arthur Scherbius invented the reflector and at page 209 he provides a selected bibliography section with a citation to Kahn's 1967 book.

4.2 Poly-Alphabetic Substitution Cipher

Now we shall give an explanation of how the Enigma machine implements its poly-alphabetic substitution cipher. If the three rotors were stationary, i.e., did not rotate, the Enigma machine would implement a mono-alphabetic substitution cipher. An example of a mono-alphabetic substitution cipher would render the clear text word 'WOODCHUCK' into the cipher text word 'XCCEDZYDA', where, for example, the clear text letters 'O' and 'C' are always mapped to the cipher text letters 'C' and 'D'. In a poly-alphabetic substitution cipher, in contrast, repeated occurrences of the same clear text letter are mapped to different cipher text letters, up to the cycle length of the machine, i.e., the number of rotor rotations before the initial setting is repeated, which is approximately 17,576 or 263. However, in the Enigma machine, the three rotors perform a rotate/step motion, which gives rise to a poly-alphabetic substitution. Very roughly, the rotation/stepping goes something like what happens in a mechanical odometer, sometimes referred to as cyclometric stepping. An odometer measures the number of miles a car has traveled. The right-most cylindrical display indicates mileage in 0.1 mile increments from 0 to 0.9 miles, the next cylindrical display indicates mileage in 1 mile increments from 0 to 9 miles, etc. When the right-most display increments from 0.9 back to 0, then the next display increments from 0 to 1 mile, etc. In cyclometric stepping, the right-most rotor steps one index position each time any key on the keyboard is depressed. Note, each rotor has 26 index positions, corresponding to each letter of the alphabet, marked on a ring as either letters ('A', 'B', 'C', etc.) or numbers (01, 02, 03, etc.) (Fig. 4). After the right-most or fast rotor completes one full rotation (26 index positions), the medium rotor advances one index position. In like manner, the left-most or slow rotor advances one index position after the medium rotor completes one full rotation. When the odometer returns to its initial setting of all zeros, it is said to have "rolled over", which corresponds to its cycle length. Note, what we have just stated is a simplification because we have not mentioned the role of the turnover notch, which we shall review in the next several paragraphs.

Fig. 4 Top: M3 Type Enigma Machine with the Top Cover Lifted Up and the Rotors + Shaft/Spindle Removed, Bottom: Detail of the Pawls from a Different Viewpoint *Sources* Photographs by Tom Cuff

To understand how cyclometric stepping happens, we need Fig. 4 and the following two figures, Figs. 4 and 5, showing the rotors. The rotation of the rotors is due to the interaction amongst three things:

(1) The movement of the pawls;
(2) The ratchet gears on the right side of the rotors, which is an asymmetrically shaped gear meant to be driven in only one direction; and
(3) The turnover notche(s) on the left side of the rotors.

In Fig. 4, there are three silver colored, V-shaped arms located between the entrance and the reflecting cylinder. At the top of these arms are U-shaped tongues meant to engage with a ratchet gear. These pivoted V-shaped arms with their U-shaped tongues are known as pawls. In a somewhat analogous manner, the combination of the stepping rotors, their pawls, and their stop members (to be discussed later) can be likened to the escapement mechanism of a mechanical timepiece, or clock. The main difference is that in the escapement mechanism, the escape wheel is subject to a constant force (actually, a constant torque) due to either hanging weights, or a main spring, while the stepping rotors are subject to an impulsive force (torque), communicated by the pawls, due to how sharply a key on the keyboard is struck (Robinson and Ward 1964, pp. 933–937; Lenoir 1964, pp. 401–403). Every time a key on the keyboard is depressed, all three pawls move in such a way so as to try and engage with the ratchet gear on the right side of the rotors. We say try because only the right-most or fast rotor rotates by one index position for each and every key that is pressed, and for this reason, it is called the fast rotor. The other two rotors are only turned by their respective pawls if a certain condition is satisfied. Consider the medium rotor. It will be turned by one index position only if the turnover notch on the left side of the fast rotor is at the same position as the middle pawl, which allows the middle pawl to engage with the ratchet gear on the right side of the medium rotor. If you have not already done so go back to Fig. 4 and observe that the middle and left-most tongues at the end of the pawls are about twice the width of the tongue of the right-most pawl. The added width of the middle and left-most tongues is necessary so that the tongues can engage with both the turnover notch of the previous rotor and the ratchet gear on the succeeding rotor. If the turnover notch is not at the same position as the tongue, then the shoulder of the metal ring, into which the turnover notch is cut, will not allow the tongue to engage the ratchet gear, and there will be no stepping of the succeeding rotor.

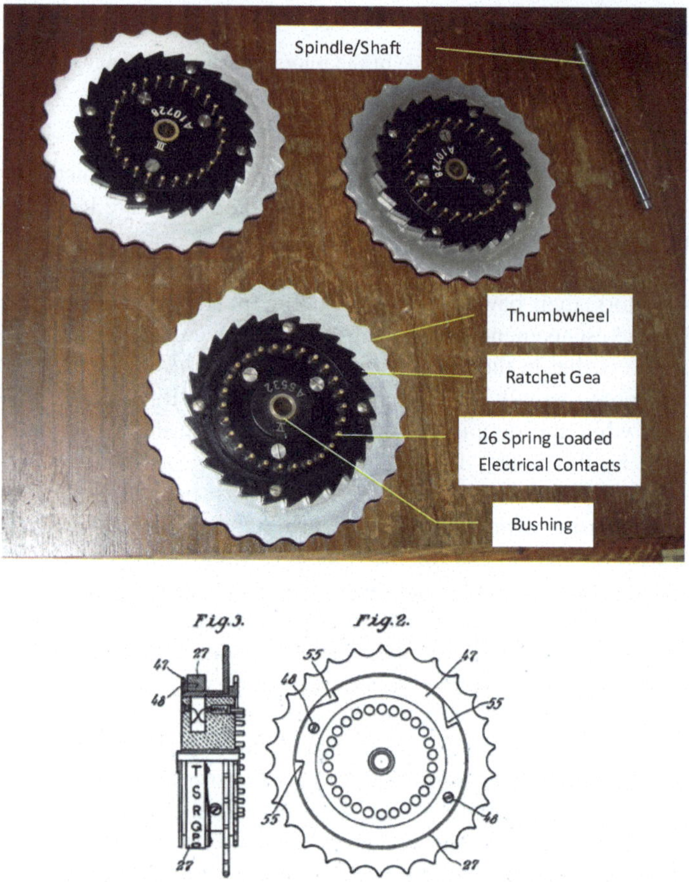

Fig. 5 Top: Right Side of the Rotors, Bottom: Drawing of Rotor *Sources* Photograph by Tom Cuff and Drawings from Patents 6

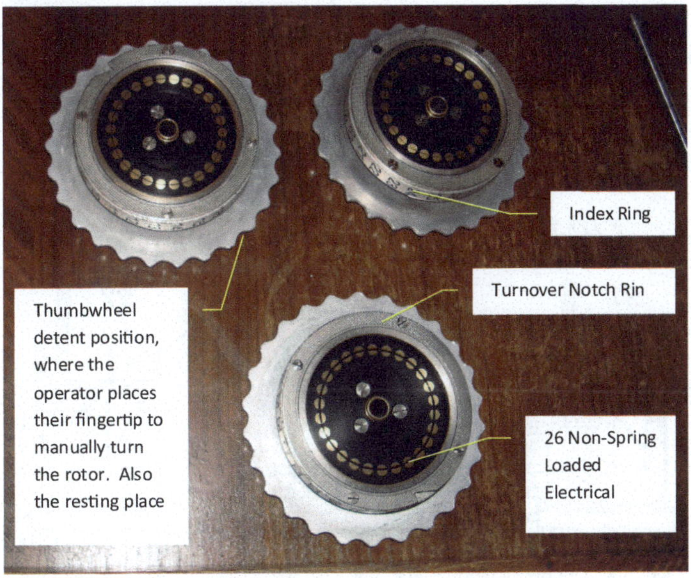

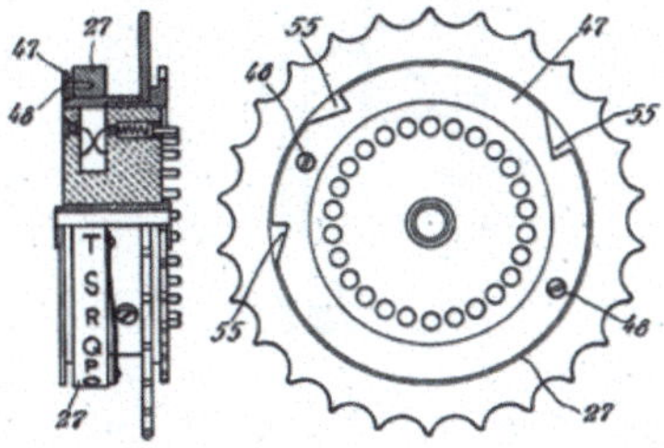

Fig. 6 Top: Left Side of the Rotors, Bottom: Drawing of Rotor *Sources* Photograph by Tom Cuff and Drawings from Patent 6

Looking at Figs. 4, 5 and 6, we notice something interesting about the four interfaces between electrical contacts in the encryption/decryption stack. Starting at the far left-hand end of the stack, we note that spring loaded contacts of the reflecting cylinder face the non-spring loaded contacts on the left-hand side of the slow rotor (Fig. 7). Similarly, the electrical contact interface between the slow and the medium rotors is comprised of the spring loaded contacts on the right-hand side of the slow rotor facing the non-spring loaded contacts on the left-hand side of the medium rotor. This arrangement continues to the last electrical contact interface where the spring loaded contacts on the right-hand side of the fast rotor face the non-spring loaded contact of the entrance cylinder. The pressure from the spring loaded electrical contacts on the facing non-spring loaded contacts together with the relative rotation of these same contacts with respect to one another break through any oxide layer that might form at the surface of the contacts.

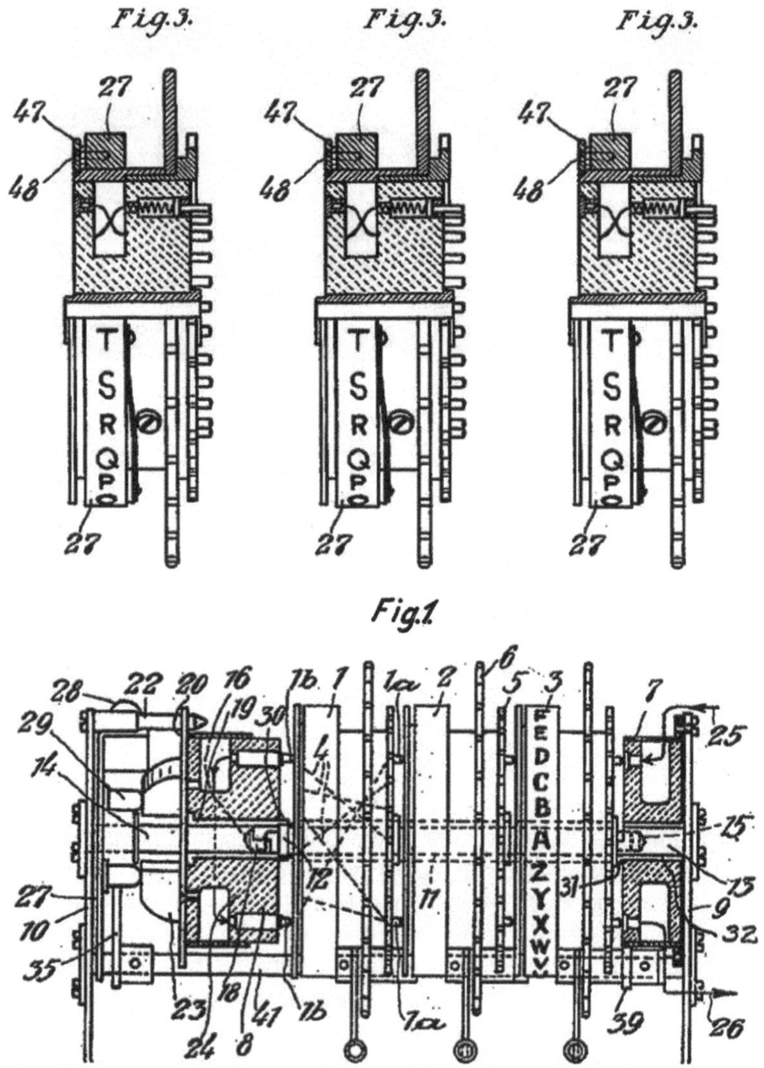

Fig. 7 Top: Orientation of the Stepping Rotors of the Encryption/Decryption Stack Showing the Alternating Interface Layers of Spring-Loaded and Fixed Contacts. Bottom: Full Stack Comprised from Left to Right of the Reflecting Cylinder (8), Slow Rotor (1), Medium Rotor (2), Fast Rotor (3), and Entrance Cylinder (7) *Sources* Top, composite of a drawing from Patents 6, and Bottom, drawing from Patents (5)

Let us look more closely at Figs. 5 and 6. In Fig. 5, notice the Roman Numerals (II, III, and V), which identify these rotors as members of a set of, at least, five rotors, i.e., Element #1 of the cryptographic key, with each rotor having a different wiring between its right and left electrical contacts. The shaft/spindle shown in Fig. 5 is smooth, and meant to pass through the bushing in the center of the rotor. There are no splines or mechanical keying present on the shaft/spindle, which means that the rotors are free to rotate on the shaft/spindle. When the three rotors are slipped on the shaft/spindle, and placed inside the Enigma machine with the lever in the locked position, each of the rotors are prevented from rotating by a spring load roller pressing into one of the detents of the silver colored thumbwheel shown in both Figs. 5 and 6. The following figure, Fig. 8, shows the location of the spring loaded rollers, and how they mate with the detents of the rotor thumbwheels. These thumbwheels protrude through the top cover of the Enigma machine, and allow the operator to set Element #5 of the cryptographic key, i.e., the initial index position of the three rotors, by manually rotating the three thumbwheels to the desired starting points.

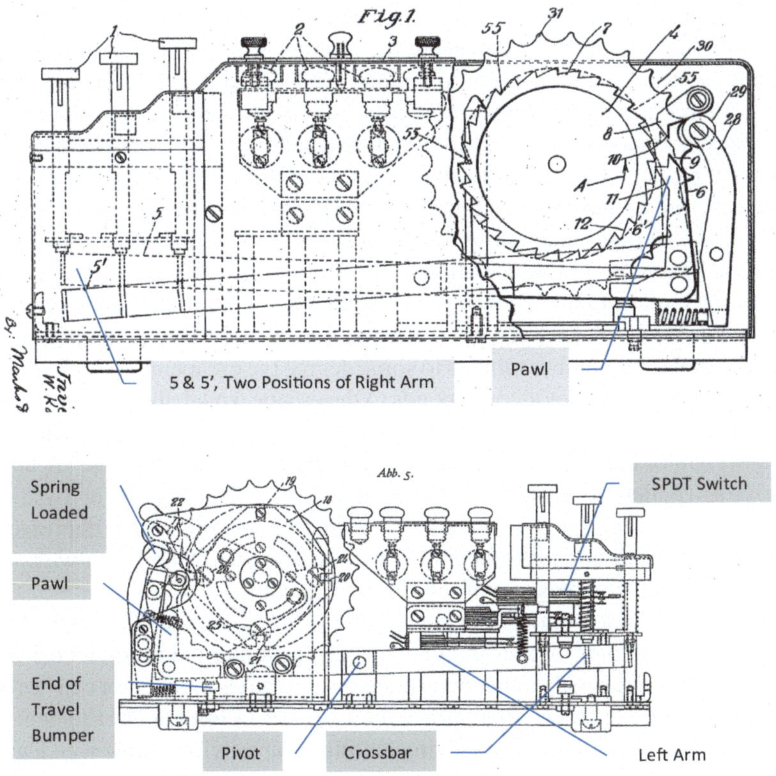

Fig. 8 Top: Right-Hand View, Bottom: Left-Hand View *Sources* Top Drawing from Patents 6 and Bottom Drawing from Patents 3

The mechanical drawings in Fig. 8 also reveal how depressing a key on the keyboard results in the rotors rotating by one index position. Starting at the front of the Enigma machine, we see that below the keyboard there are two arms on either side, and that these arms are linked by three crossbars, one for each row of keys, see 5 in the top drawing of Fig. 8. When a key is depressed, its plunger does two things:

(1) Due to a linkage with the key's plunger, the center contact of the SDPT (Single Pole Double Throw) switch associated with the key, which is normally in contact with the upper terminal, is shifted downwards so it breaks contact with the upper terminal and makes contact with the lower terminal; and
(2) The key's plunger presses down on one of the three crossbars, depending on the row of the key.

Pressing down on any of the crossbars forces the left and right arms, attached to the crossbars, to move downward, while the portion of the arms behind the pivot move upward like a seesaw. The upward motion of the arms at the back of the Enigma machine raises the pawl assembly so that they can engage with the ratchet gears on all three rotors, see 5' in the top drawing of Fig. 8. As mentioned earlier, the fast rotor always turns by one index position each time any key is depressed, but the motions of the medium and slow rotors depends on the alignment of the turnover notch on the previous rotor. One other thing that needs mentioning is the purpose of the three items called out as Stop Members in the top photo of Fig. 4, and as 8 in the top drawing of Fig. 8. Item 8 is what prevents rotors from rotating more than one index position due to the rotational inertia of a rotor when responding to, say, a particularly sharp and forceful pressing of a key of the keyboard. In other words, when a pawl engages with a ratchet gear, and gives it a push, depending on the strength of the push, the rotor could rotate more than one index position by simply lifting the tongue of the pawl by the next tooth of the ratchet gear. Item 8 momentarily presses down on the tongue of the engaged pawl, and so prevents the next tooth of the ratchet gear from lifting the tongue of the engaged pawl. As mentioned earlier, the combination of the rotor, pawl, and stop member mimics to some degree the escapement mechanism in a timepiece, or clock. The rotor corresponds to the escape wheel, the pawl corresponds to the source of impulsive force (actually torque) similar to the main gear with its hanging weight or main spring, and the stop member corresponds to the anchor with its two arms (pallets) used to make sure the escape wheel only turns by a single increment (Dale 1992, pp. 37–43). As far as the implications of the depressed key and the SPDT switch are concerned, they will be discussed later when we consider the electrical schematic diagram of the Enigma machine.

There is however something that cannot be readily discerned from Figs. 4 and 5. Namely, that if one picks up an individual rotor, gripping the thumbwheel by one hand and the index ring in the other hand, these two portions of the rotor can be rotated with respect to one another. This allows the operator to implement Element #2 of the cryptographic key, the so-called ring setting. The previous sentences leave out something, though. The two parts of the rotor are normally prevented from rotating with respect to one another by a locking mechanism, either a spring-loaded pin-and-socket or a ratchet. The rotors in Figs. 4 and 5 utilize the spring-loaded pin-and-socket mechanism so when gripping the two parts of the rotor in both hands, one fingernail has to be used to lift the pin from its socket to allow the two parts to be rotated with respect to one another (Fig. 9), and for an exploded view of a rotor (Fig. 10).

In the following, Fig. 9, there is another view of the encryption/decryption stack. Notice what is in the valley or groove between the index ring and the thumbwheel, i.e., of the left-most rotor – the slow rotor. Direct your attention to index marks 06 and above, and see the holes/sockets. These holes/sockets accommodate the pin that is currently next to index mark 05. The pin currently at index mark 05 and the T-shaped protuberance/handle currently between index marks 04 and 03 are both attached to the same curved, flat piece of spring steel, which is anchored with machine screws currently at index marks 26 and 25. Note, we use the word 'currently' because the flat piece of spring steel is actually anchored to the part of the rotor associated with the thumbwheel and the ratchet gear. The T-shaped handle must be lifted to allow the two parts of the rotor to rotate with respect to one another.

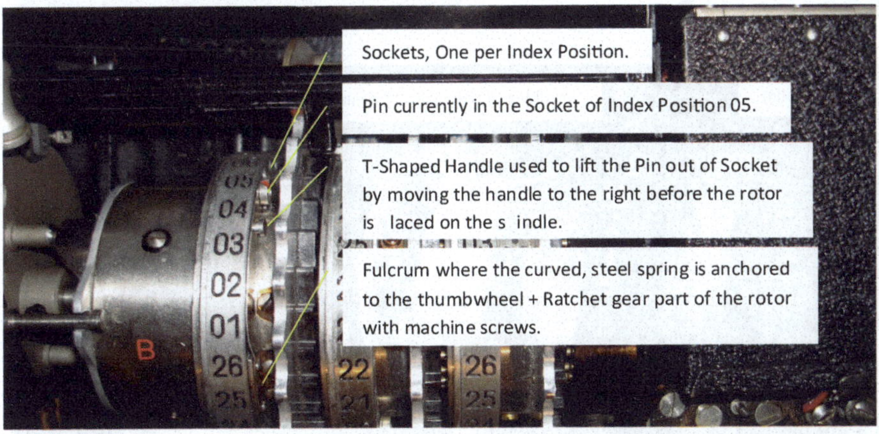

Fig. 9 Another View of the Encryption/Decryption Stack *Source* Photograph by Tom Cuff

Fig. 10 Two Views of a Disassembled/Exploded Rotor *Sources* Photograph by Tom Cuff

Figure 10, above, shows two rotors. The rotor at the far left is fully assembled, while rotor occupying the rest of the photograph is a disassembled/exploded rotor.

Let us try and explain how the various pieces in Fig. 10 go together. Starting at the far right of the top photo, the Insulating Ring with 26 holes is placed over the spring loaded electrical contacts of the Left & Right Electrical Contact Assembly, and held in place by three machine screws screwed into three threaded holes in the hub of the right side of the Assembly (Fig. 5). The non-spring loaded electrical contacts on the left side of the Assembly are likewise held in place by three machine screws screwed into three threaded holes in the hub of the left side of the Assembly, see Fig. 5. This fixes the orientation the two sets of electrical contacts with respect to one another.

Note, the metal bushing, which is not visible in either photo in Fig. 10, is held captive inside the Assembly. Now, we turn our attention to the pieces on the left of the top photo of Fig. 10. The Turnover Notch Ring is attached to the left side of the Index Ring with four machine screws (Fig. 6). And then, the Collar, minus the Guide for Index Ring and the Pin & T-Shaped Handle on a Curved Spring, slides through the already assembled Turnover Notch Ring plus Index Ring until the Lip of the Collar rests on the Bearing Surface, located at the inside diameter of the Groove in the left face of the Index Ring. The far right end of the Collar slides over the Shoulder of the Contact Assembly. The Collar is attached to the Shoulder of the Contact Assembly via machine screws screwed into radial, tapped holes in the Shoulder, not visible in either photo in Fig. 10. Lastly, the Guide for Index Ring and the fulcrum of the Pin & T-Shaped Handle on a Curved Spring are attached to the Shoulder of the Contact Assembly by way of machine screws traversing through-holes in the Collar and screwing into radial, tapped holes in the Shoulder (Figs. 2 and 9).

But we can summarize the important points about how the rotor is assembled as follows. The silver colored ring with the turnover notch and the silver colored, index ring are held together with screws, and form one part/unit of the rotor, while all the other components, including parts that are not visible such as the central bushing, are fixed to the thumbwheel and ratchet gear part/unit. Thus, when one rotates the part/unit comprised of the ring with the turnover notch and index ring relative to the rest of the rotor, you are changing the mapping of the rotor but not in a random way. Obviously, when you turn the part/unit comprised of the turnover notch and index ring 26 index positions the original mapping between the right and left electrical contacts is restored, i.e., rotating the index ring creates a different cyclic permutation of the mapping of the two sets of electrical contacts, as well as the turnover position.

4.3 *Difference Between the Turnover Notch Position and the Turnover Position*

One last thing we should like to mention is the difference between the turnover notch position and the turnover position. Each of the three rotors, when installed in the Enigma machine, displays its index position in a window in the top cover (Fig. 11). In Fig. 11, one can see the outline of the corner of a trapdoor built-in to the top cover that protects the thumbwheels from inadvertent disturbance; to set the initial positions of the rotors the trapdoor must be raised to gain access to the thumbwheels. Now, we mentioned that each rotor had a turnover notch, and the position of this turnover notch can be read off the index ring. But suppose that you wanted to set the initial positions of some or all rotors to their respective turnover notch positions?

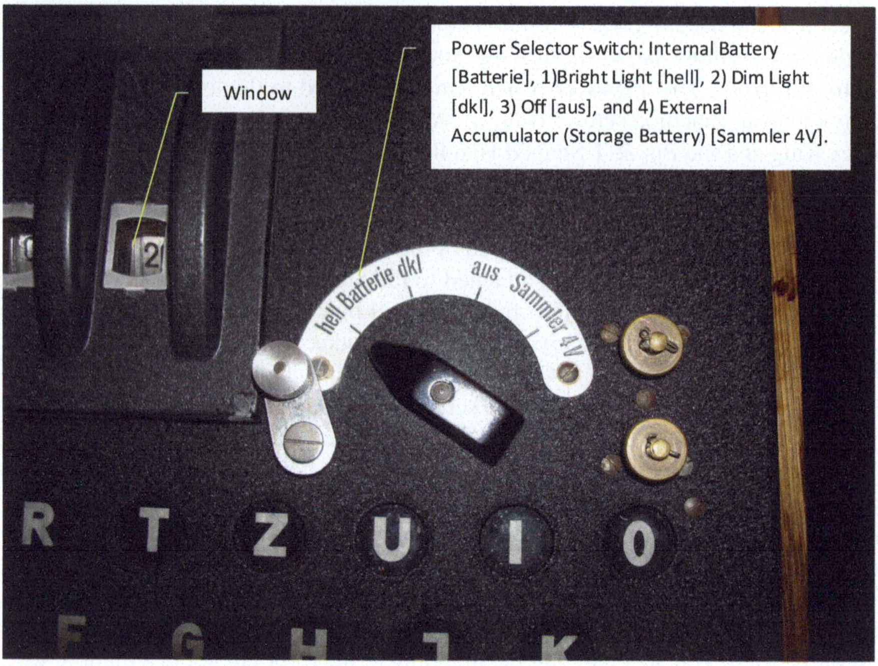

Fig. 11 Close Up View of the Top Cover *Sources* Photograph by Tom Cuff

The turnover notch position for each rotor can be determined by simply looking at the left side of the rotor before it is placed on the spindle and noting the index position or letter—depending on how the index ring is marked—that is next to the turnover notch (Fig. 8) where the turnover notch position is next to the index number 06. For example, suppose the turnover notch position is 06. How is this related to the index number that appears in the window in the top cover of the Enigma machine, i.e., the turnover position (Fig. 11)? Is the turnover position, the number displayed in the window, the same as the turnover notch position, the index position next to the notch on the rotor?

It turns out that due to the how the Enigma machine is constructed, the turnover position seen in the window is eight index positions ahead of the turnover notch position on the rotor. If you were to lift the top cover on the Enigma machine, you would notice that the letter 'B' embossed on the reflecting cylinder is coincident with the indexes showing through the windows in the top cover (Fig. 4), Bottom. That the turnover notch position on the rotor is behind the turnover position shown in the top cover window should become obvious when both the positions of the pawls in Fig. 4, Bottom, which engage the turnover notches, and the direction of rotation of the rotors (curved arrow) in Fig. 8, Top, are considered.

4.4 Electrical Portion of the Enigma Machine

We mentioned earlier that both the stack and the plug board take part in the encryption process. It is now time to look at the electrical schematic diagram of the Enigma machine so that we may see how the plug board worked its obfuscation. In many ways, the plug board is very similar to the old style telephone switch board, except in the case of the Enigma machine, the connections are made with the intention of rendering the communication indecipherable. In fact, an early reproduction of the Enigma machine built by the Polish intelligence service employed telephone jacks and plugs; the plugs had two electrically isolated terminals called the tip and the ring as did the jacks (Calvocoressi 1980, pp. 54–55; Johnson 1978, p. 313). Andrew Hodges in his biography of Alan M. Turing also utilized the telephone switch board analogy when referring to the plug board (Hodges 1983, p. 169). It will be important, when looking at both the Enigma machine's electrical schematic diagrams, to realize two things:

(1) These are electrical not electronic schematic diagrams, i.e., the only components are switches, electrical contacts, wires, a battery, and incandescent light bulbs, there are no electron vacuum tubes, transistors, or IC's;

(2) The motive power used to rotate the rotors of the stack comes solely from the operator's finger pressure as it depresses a key on the keyboard, i.e., there are no devices that convert electrical energy into mechanical effect, such as electric motors, relays, or solenoids.

Figure 12 shows that in a real Enigma machine there are 52 terminals, 26 pairs of terminals, on the plug board. For example, there are two terminals marked 'A', a larger diameter jack/socket and a smaller diameter jack/socket, with one terminal going to the entrance cylinder from the keyboard, and the other terminal going to the display from the entrance cylinder. The patch cords have pairs of plugs/prongs, one larger diameter and one smaller diameter, on both ends with one insulated wire of the cable connecting together the larger plugs/prongs and the other insulated wire of the same cable connecting together the smaller plugs/prongs. Having jacks/sockets and plugs/prongs of different diameters essentially polarizes the connection, and, in so doing, prevents the patch cord from being plugged into the plug board the wrong way. This does not, of course, explain how, when no patch cords are plugged into the patch board, there are no interchanges of connections. For that we need a more detailed electrical schematic diagram such as in Fig. 13.

Fig. 12 Top: Close Up of the Plug Board Showing 26 Pairs of Terminals (52 Terminals, Total), Middle & Bottom: Left & Right Cable Testing Jacks/Sockets, Respectively *Sources* Photograph by Tom Cuff

Figure 13 contains two detailed electrical schematic diagrams of the Enigma machine. To keep the wiring diagrams from looking like an explosion in a spaghetti factory, the keyboard of the left-hand diagram is shown with only four keys and four display lights, and the right-hand diagram is shown with only two keys and two display lights. Though we do not have the space to review the history of the development of the modern rotor based cipher machines that preceded the Enigma, and then the various commercial and military Enigma machines, let us, at least, mention the names of those whose work anticipated, paralleled, and directly contributed to the Enigma machine: Theo van Hengel and Rudolf Spengler in the Dutch East Indies in 1915, Edward H. Hebern in the USA in 1917 and onwards, Arthur Scherbius in Germany in February 1918, Arvid G. Damm in Sweden in October 1919, and Hugo A. Koch in the Netherlands also in October 1919, and Willi Korn in Germany circa 1926 (filing year for one of his German patents) and onwards.[7] Note, Korn's German patents did not include the plug board, which remained a closely guarded secret throughout WWII, see the right-hand diagram in Fig. 13, below. Ralph Simpson, in one of his PowerPoint slide presentations, mentions that even very famous propaganda photographs of the Enigma machine being used by the German armed forces always hid the existence of the plug board by obliterating or covering them or photographing the Enigma machine in such a way as to obscure the view of the plug board.[8] For example, who has not seen the iconic photograph of German Army General Heinz Guderian and his team of Enigma machine operators during field operations during the *Blitzkrieg* around Sedan, France in 1940 (Bauer 2002, p. 86; 19, p. 81; 20, p. 21; 21, pp. 226–227; the Enigma Machine was cropped out this photograph; Lunde 2012, p. 116). The Enigma machine in this picture has the front face of the wooden case closed and some type of white material - or was the photo simply retouched - covering the top of the plug board so it cannot be seen. Also, it is not, even clear after all these years, who exactly invented the plug board.

Now, it is time to examine the actual wiring diagram of the Enigma machine. To this end, the next figure, Fig. 13, depicts two such diagrams: the left-hand one taken from the Internet, and the right-hand one taken from one of Willi Korn's German patents. Note, the left-hand detailed diagram can also found, with some minor differences in the callouts and signal current arrows, first in a 1979 non-fiction book about the Enigma machine by Józef Garliński 1979, p. 21). Today, this same wiring diagram can also be found in a reproduction of the 1937 German Army instruction manual for the Enigma machine, which can be purchased from online retailers[9] (Lunde 2012, p. 11). The right-hand detailed diagram comes from a 1928 German patent by Willi Korn, the designer of the military version of the Enigma machine. Note that the right-hand detailed diagram in Fig. 13 does not have a plug board, which as we have already mentioned was, at the time and during WWII, a closely held military secret (Patent 3).

---

[7] Cfr. Ralph Simpson, Ralph@CiperMachines.com (11/9/2016), The History and Technology of the Enigma Cipher Machine. CipherMachines.com, p. 7 and Ralph Simpson, Ralph@CiperHistory.com (June 12, 2018) The Race to Crack the Enigma Machine, San Francisco Navy League, CipherHistory.com, p 6. See also Patents 3–4 and (Simpson 2017; Leeuw 2003).

[8] Ralph Simpson, Ralph@CiperMachines.com (11/9/2016), The History and Technology of the Enigma Cipher Machine. CipherMachines.com, pp. 17–18 & 22.

[9] Note, this manual was authorized by ("I. A." ≡ im Auftrage ≡ by order of) General Fritz Erich Fellgiebel, who would be executed for the part he played in the failed assassination attempt on Hitler, code named Operation Valkyrie, on July 20, 1944.

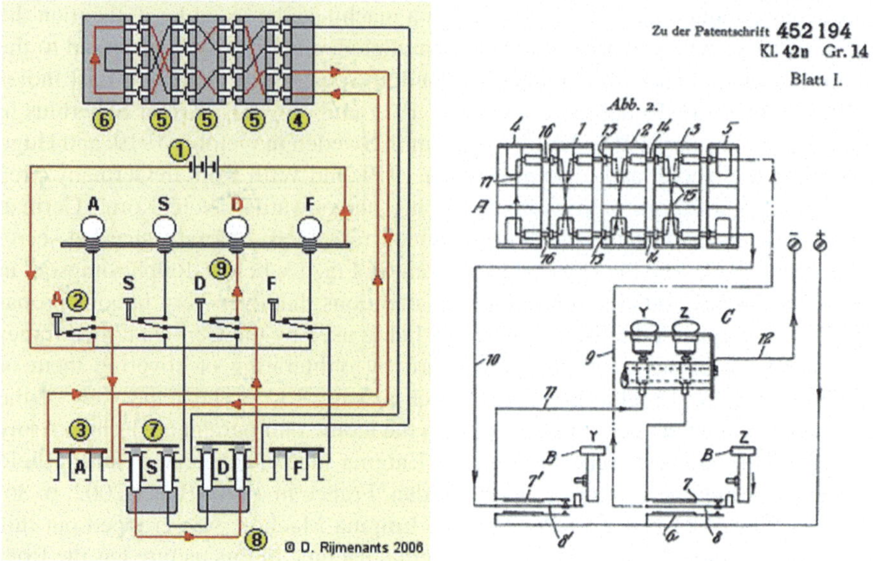

Fig. 13 Detailed Electrical Schematic Diagrams of the M3 Type of Enigma Machine *Sources* Left: http://users.telenet.be/d.rijmenants/en/enigmatech.htm with the permission of Dirk Rijmenants; Right: drawing from Patents 3

The first thing to note about the left-hand diagram (Fig. 13) is the pairs of electrical contacts/jacks/sockets of the plug board, 3 and 7. Behind each pair of electrical contacts/jacks/sockets there is an electrical shorting bar, which electrically shorts the two contacts together when no patch cord is plugged into the contacts/jacks/sockets. Thus, the shorting bars allow the plug board to function in absence of a patch cord, 3, except there is no interchange of connections. When a patch cord, 8, is plugged into the contacts/jacks/sockets, S & D, the plugs/prongs push against the insulating cylinders at the bottom of the jacks lifting the shorting bar, 7, and allowing the electrical signals to travel through the patch cord cable and interchange the connections of a pair of letters.

Finally, the detailed electrical schematic diagrams also imply that no light in the display is lit if no key of the keyboard is depressed and held down. This behavior can be verified by the following observations of, for example, the left-hand diagram. In order for a light to be turned on, there must be a closed circuit between the battery, shown as a series connection of three cells (callout #1), and any bulb in the display. The threaded portions of the bulb bases are always connected to the positive (+) terminal of the battery; the long line is the plus (+) terminal of the cell, and the shorter, thicker parallel line is the minus (-) terminal. Now, if no key of the keyboard is depressed this means that all the SPDT (Single Pole Double Throw) switches (callout #2) connected to the keys have their moveable contact contacting the upper fixed contact. However, in this switch configuration, the negative (-) terminal of the battery cannot reach any of the center terminals of the bulb bases in the display. This means that power is only drawn from the battery for the second or two when a key is held down, thus, extending the battery life.

Although this might seem like restating the obvious, it still needs to be done. The detailed electrical schematic diagrams depicted in Fig. 13 clearly show that the Enigma machine does not contain any electronic devices such as electron vacuum tubes, transistors, or ICs. In addition, as mentioned earlier, there are no electric motors, relays, or solenoids, i.e., no devices wherein electric energy is transformed into mechanical energy. All the mechanical motions, such the rotation of the rotors, are derived from the mechanical pressure of the operator's fingers on the keys of the keyboard. In other words, each time a key is depressed two things happen, one of which is mechanical and conveyed by linkages resulting in the rotation of the rotors, and the other which is electrical and annunciates, via the display, the result of the encryption or decryption by the encryption/decryption stack and the plug board. To drive home this point, we present the following figure, which is a photograph taken from the left-hand side of the Enigma machine after it has been removed from its wooden carrying case; compare this with the drawings in Figs. 8 and 14.

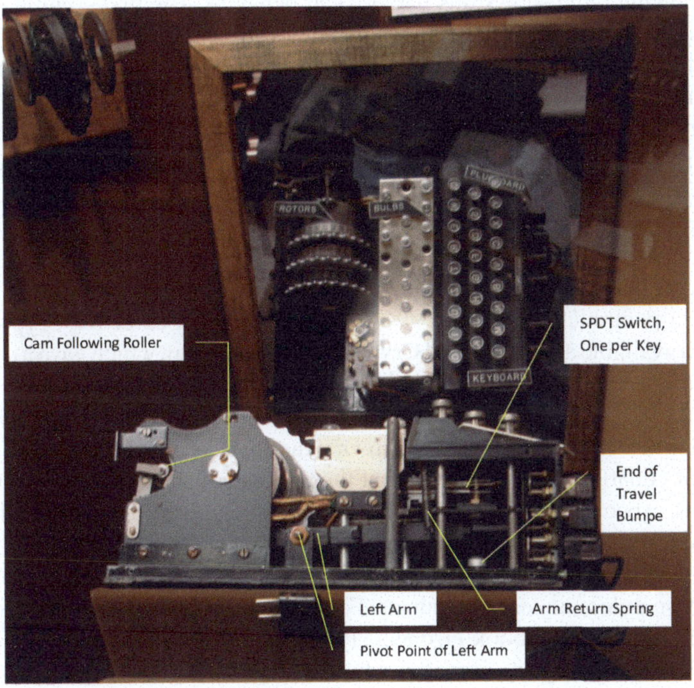

Fig. 14 Left-Hand, Side View of an Enigma Machine without its Case *Source* Photograph by Tom Cuff

What is not obvious from the left-hand diagram in Fig. 13 is whether or not one needs to populate the plug board with the maximum allowable number of patch cords, i.e., 13 cords. The maximum number of patch cords is a direct consequence of the fact that the plug board has 26 pairs of terminals (52 terminals in all; Fig. 12), and each patch cord connects four terminals, therefore, $52/4 = 13$ is the maximum number of patch cords. Or let us put it a different way. With each patch cord added to the plug board, the encryption increases. Think about it this way. When there are no patch cords plugged into the plug board, there is only one way to interchange the connections, the identity interchange, {(A,A), (B,B), (C,C), …, (Z,Z)}, which is essentially no interchange, but with a single patch cord plugged into the plug board, the sets (collections) of interconnections increases to 325. But does all this mean that the maximum amount of encryption corresponds to the maximum number of patch cords, 13, that can be plugged into the plug board? The answer, which is counterintuitive, is, no; the maximum encryption via the plug board occurs with 11 patch cords plugged into the plug board. The first time we were consciously aware of this paradox was May 2013 at the National Cryptographic Museum when Dr. Hamer mentioned it to one of us (TC) during his demonstration of the inner workings of the M3 type Enigma machine.

The following table shows this paradoxical behavior. The first column is simply the number of patch cords plugged into the plug board, while the second column is the possible number of sets (collections) of legal pairs of letters that can be swapped by this number of patch cords. The term 'legal pairs' means: (1) order of the swapping is unimportant, i.e., the pair (A,B) is equivalent to the pair (B,A); and (2) the pairings in a set of pairings must be unique with respect to each member of a pair, i.e., (A,B) [or equivalently (B,A)] cannot appear more than once. In other words, the letter interchanges or swaps must be combinations, not permutations.

Surprisingly, even though this combinatory paradox is mentioned in many books on the Enigma machine, one almost never sees the complete derivation of the formula. There was a pamphlet printed in 1996 and 2001 by the Center for Cryptographic History and available at the National Cryptographic Museum, which gave a derivation of the combinatory formula of the plug board, but a well-known cryptographer, upon reviewing the document, discovered errors, and the document was withdrawn sometime around 2013 or earlier. Our own examination of this pamphlet revealed misprints of equations on pp. 9 & 12, but the document, as a whole, has interesting information, especially in its Notes and References sections; there is, in addition, a 1995 journal article with the same title by the same author (Miller [1996] 2001; Miller 1995). Even people who worked at Bletchley Park were reluctant to grind through this derivation unless under great duress. For example, the non-mathematician Peter Calvocoressi in his 1980 book erroneously mentioned that the greatest number of combinations of letter pairs occurred when 10 patch cords were plugged into the plug board; note #6, on p. 17, in the Notes section of the pamphlet by A. Ray Miller mentioned that by 1941 the German military had standardized the number of patch cords to 10 for all traffic, which may be why Calvocoressi thought this number of cords gave the highest encryption (Calvocoressi 1980, p. 26). In contrast, W. Gordon Welchman (Welchman 1982, pp. 51–52), who worked with Alan Turing on the British Bombes, mentioned in his 1982 book that paradoxical behavior did indeed happen when around 10 or 11 patch cables were plugged into the plug board, but he appeared unwilling to share the rigors of deriving the formula even though he was a trained mathematician.[10] Andrew Hodges, another mathematician, said in 1983 that the maximum occurred with 11 patch cords plugged into the plug board, though instead of providing an algebraic formula, he simply showed what the formula looked like after the relevant numbers were put into it; leaving it to the bemused reader to reverse engineer the actual algebraic formula (Johnson 1978). It is not clear from his Notes section for the chapter in question, Chap. 4—The Relay Race, where or if he found the derivation of the formula somewhere else.

The mathematical formula used to calculate the numbers in the second column can be expressed in a number of different ways. For example, the following formula gives the number of possible sets/collections of pairs of letter swaps for n patch cords,

$$S(n) = \frac{26!}{2^n \times n! \times (26 - 2 \times n)!} \tag{1}$$

Some authors prefer simply writing the various sets/collections of combinations explicitly as a product after cancellation by the terms in the denominator:

$$S(0) = 1 \tag{2a}$$

$$S(1) = 25 \times 13 \tag{2b}$$

$$S(2) = 25 \times 23 \times 13 \times 6 \tag{2c}$$

$$S(3) = 25 \times 23 \times 22 \times 21 \times 13 \tag{2d}$$

$$S(4) = 25 \times 23 \times 21 \times 19 \times 13 \times 11 \times 5 \tag{2e}$$

$$\vdots \tag{2f}$$

$$S(13) = 25 \times 23 \times 21 \times 19 \times 17 \times 15 \times 13 \times 11 \times 9 \times 7 \times 5 \times 3 \tag{2g}$$

[10] Let's note that Welchman's terminology about the plug board "steckerboard", and patch cords, "stecker pairs" is different than what we have employed; Stecker is the German noun for plug. With regard to the number of patch cords necessary for the maximum number of combinations of letter pairs, Welchman expressed this as either six or four self-steckered letters. By which he meant three or two pairs of letters which were mapped to themselves, i.e., not mapped to another letter, which could be achieved with either 10 or 11 patch cords, respectively.

However, neither Eq. (1) nor Eqs. (2) are particularly helpful in helping the reader understand how Eq. (1) was derived. The following expression, on the other hand, with some textual hints, at least, provides the reader with some clue as to how to fill-in the steps of the derivation,

$$S(n) = \prod_{i=1}^{n} \frac{C(26 - 2 \times (i - 1), 2)}{i} \tag{3a}$$

$$S(n) = \frac{C(26,2)}{1} \times \frac{C(24,2)}{2} \times \frac{C(22,2)}{3} \times \cdots \times \frac{C(26 - 2 \times (n-1), 2)}{n} \tag{3b}$$

where $C(m,l)$ is the standard formula for the number of combinations of m objects (letters) grouped (connected together) l at a time,

$$C(m,l) = \binom{m}{l} = \frac{m!}{l! \times (m-l)!} \tag{4}$$

The expression for $S(n)$, Eq. (3b), contains n products of fractions. The first fraction represents the possible combinations of pairs of letters taken from a pool of 26 letters due to the first patch cord; the denominator is 1 because there are no duplicate pairs. The second fraction represents the possible combinations of pairs of letters taken from a pool of 24 letters due to the second patch cord (2 of the letters from the pool have already been used with the first patch cord); the denominator is 2 because there are 2 times the number of duplicate pairs. The third fraction represents the possible combinations of pairs of letters taken from a pool of 22 letters due to the third patch cord (4 of the letters from the pool have already been used with the first and second patch cords); the denominator is 3 because there are 3 times the number of duplicate pairs. Etc., etc., etc. The product of the numerators in Eq. (3a) increases as the number of patch cords increases, but the product of the denominators also increases with the number of patch cords and this rate of increase counteracts the slowing rate of increase of the product of the numerators somewhere before 11 patch cords causing the maximum observed in Table 2.

Note, the solution for $S(13)$ is also the solution to the question of how many possible reflecting cylinders it is possible to construct? In a reflecting cylinder, there are 13 jumpers connecting 13 pairs of distinct letters, which is exactly the case for the plug board with 13 patch cords plugged into it. Because both $S(0)$ and $S(13)$ are boundary conditions, their behavior is different than the corresponding intermediate values for $1, 2, 3, \ldots, 12$ patch cords as enumerated in Eqs. (2). In fact, the expression corresponding to the case of 13 patch cords can be recast as follows,

$$S(13) = \frac{26!}{2^{13} \times 13!} = (26-1)!! \tag{5a}$$

$$S(13) = 25 \times 23 \times 21 \times 19 \times 17 \times 15 \times 13 \times 11 \times 9 \times 7 \times 5 \times 3 \times 1 \tag{5b}$$

Table 2 Mapping between the number of patch cords plugged into the plug board of an Enigma Machine and the number of possible sets/collections of legal pairs (Possible Sets/Collections of Pairs of Swapped Letters)

Number of Patch Cables	Number of Sets/Collections of Possible Legal Pairs
0	1
1	325
2	44,850
3	3,453,450
4	164,038,875
5	5,019,589,575
6	100,391,791,500
7	1,305,093,289,500
8	10,767,019,638,375
9	53,835,098,191,875
10	150,738,274,937,250
11	205,552,193,096,250
12	102,776,096,548,125
13	7,905,853,580,625

where we have used the notation for what is called the double factorial or semi-factorial, which is simply the factorial of the number n formed by multiplying n by all the positive integers less than it and of the same parity (odd or even) as n. Note, a one line proof of the identity in the first line of Eqs. (5) can be found on the first page of the 2012 journal article by Gould and Quaintance (Gould and Quaintance 2012). Occasionally, one finds journal articles or books about the Enigma machine that use the double factorial notation with respect to the reflecting cylinder and/or the plug board with 13 patch cords installed, though, they usually do not explain what the notation means (Bauer 2002, p. 46; Miller 1995, p. 71). There are, of course, combinatorial expressions for the possible number of different entrance cylinders, and the possible number of stepping rotors, but they are simpler and much easier to derive than the ones for the plug board and the reflecting cylinder. As usual, there is still something missing from what we have said about even the detailed electrical schematics. The reader may have noticed, when comparing the photograph of the Enigma machine in Figs. 1 and 3 to the detailed electrical schematic diagrams in Fig. 13, that the detailed schematics do not contain the power selector switch and the pair of terminals for external power (Fig. 10); the bulb testing & cable testing sockets, see Fig. 4 (Top); nor cable testing jacks/sockets, see Fig. 12 (Middle & Bottom). The following are explanations of these items.

The 4-position power switch, Fig. 10, allows the operator to completely disconnect the internal dry cell battery from the Enigma machine's circuitry [*aus*]; directly connect the internal dry cell battery to the circuitry so that the display lights have high intensity when they are lit [*hell*]; indirectly connect the internal dry cell battery to the circuitry through a resistor so that the display lights have low intensity when they are lit [*dkl*]; or connect an external accumulator (storage battery) through a pair of terminals [*Sammler* 4 V]. As an aid to the Enigma machine operator, there are also two diagnostic features built into the machine. At the far right hand side of the middle row of display lights, there is an empty socket labeled *Lampenprüfung* [lamp/bulb testing] (Fig. 4), Top. Anytime electrical power is applied to the Enigma machine's circuitry, the operator can test whether a light bulb is good or bad by simply screwing the bulb in question into the electrically 'hot' *Lampenprüfung* socket. If the lamp glows, then it is a good lamp, otherwise it is bad. Lastly, at the far left hand side of the middle row of display lights, there is an empty socket labeled *Kabelprüfung* [cable testing], (Fig. 4), Top, which allows the operator to test for electrical continuity of the two insulated wires comprising each patch cord by using the two raised jacks/sockets at either end of the middle row of the plug board. Note, these two jacks/sockets, besides being raised, are not in large and small diameter pairs like the others on the plug board. To use this diagnostic feature, first, simply screw a known good lamp into the *Kabelprüfung* socket. Second, using the two raised jacks/sockets at both ends of the middle row of the plug board (Fig. 12) Middle & Bottom, plug in or touch the large [small] diameter plug/prongs of the patch cord. If the lamps glows, then the wire associated with either the large or small diameter plugs/prongs of the patch cord are good, i.e., they have electrical continuity.

We have told the reader much about what goes on inside the Enigma machine but not everything. But for the sake of full disclosure, we shall briefly disclose and discuss these other behaviors. Many Enigma machine behaviors can best, and sometimes only, be seen in a movie or a video of its actual operation. For example, we stated that the rotation of the rotors behaved like an odometer, i.e., they underwent cyclometric rotation. In fact, the easiest way to induce non-cyclometric rotation was to add a second turnover notch to one or more of the rotors. But even without adding a second turnover notch, the Enigma machine still exhibits what is called Enigma stepping, though it is hard to see this behavior without a movie or video and/or a very detailed textual explanation. For the interested reader, we shall provide a citation (Hamer 1997).

In the Spring semester of 2013, one of us (TC), was enrolled in a class on the programming language C + +, wherein the class project was to simulate an Enigma machine in software. One of the students (Chris) mentioned that, in response to the pressing of a key on the keyboard of the Enigma machine, the fast stepping rotor moves first and then the light lights on the display, i.e., this is similar to the difference between the pre-increment, + + i, and post-increment, i + +, operations in Java and C + +. This is an important fact about the operation of the Enigma machine, which is rarely talked about. One can easily confirm that the military version of the Enigma machine does indeed employ the pre-increment operation as highlighted in the DVD of the movie Enigma, Scene 20—Decoding the Intercepts. In this scene the protagonists have obtained copies of some intercepts, which they try to decode using an Enigma machine. With the front panel raised, the lead character, Tom Jericho (Dougray Scott), demonstrates what happens when a key on the keyboard is pressed. One can see the key being pressed, the fast stepping (right-most) rotor turning, and then the light bulb of the light panel illuminating; viewing this scene frame-by-frame will more clearly show the pre-incrementing process.[11]

It should be noted that the military version of the Enigma machine used in the movie is, in fact, an actual Enigma machine loaned to the movie company by one of the movie's producers, the singer Mick Jagger of the rock group the Rolling Stones, who also appears in the movie for a few seconds in Scene 10 – Happiest Moment of His Life.

5 Conclusion

While the Enigma machine was not the first nor the last rotor-based, electro-mechanical cryptographic machine, its importance during WWII made, at least, its name familiar to the general public. We have tried to explain in detail its inner workings. With the help of annotated visual aids and text, the Enigma machine has been deconstructed into simpler mechanical and electrical subsystems from which one may abstract the function of and interaction amongst these subsystems. Through clever design, the machine's mechanical subsystems get their power from the keystrokes of the machine operator, and its battery's stored electrical energy is only called upon for the instant a key on the keyboard is depressed to the end of its downward travel. Its low power demands, and small size and weight made the Enigma machine a very portable and rugged cryptographic system. Although an enormous amount of material has been written about the military version of the Enigma machine since its existence was first revealed to the public in the mid 1970s, many of the details of the machine have been omitted in the various expositions. With this document, we hope to provide the interested layperson or professional with enough background, including citations to original source material, so that they can discover for themselves both the anatomy (structure) and physiology (function) of this man made mechanism.

Acknowledgements First, we would like to thank Mr. Joshua McDonald, Adjunct Professor at Frederick Community College, whose course Introduction to C++, CIS-208, Spring Semester 2013, introduced one of us (TC) to the need to understand the inner workings of the Enigma machine as part of a course project to simulate either the Enigma machine or the Bombe in software—all the student teams opted to try to simulate the Enigma machine. Note, all the photographs taken in May 2013 were done with the help of Dr. David H. Hamer who, with the permission of the staff of the National Cryptographic Museum (NCM), disassembled the single, publicly accessible Enigma machine on display so that one of us (TC) might benefit from seeing and photographing the innards of the machine. Thank you, to both Mr. McDonald and Dr. Hamer for all their help. Lastly, we would like to express our appreciation to the staff for the exhibits and signage at the NCM without which this document could not have been written; we should also like to compliment them on their library. Thanks for the memories and the explanations.

[11] Enigma (2001), screenplay: Tom Stoppard; Based on the book "Enigma" by Robert Harris, Manhattan Pictures International, LLC, 119 min.

References

Bauer FL (2002) Decrypted Secrets. Springer-Verlag, Berlin, Methods and Maxims of Cryptology.

Blakeslee EP (1972) The Radio Amateur's VHF Manual. American Radio Relay League, Newington, Connecticut.

Calvocoressi P (1980) Top Secret Ultra. Pantheon Books, New York.

Dale R (1992) Timekeeping. The British Library, London.

de Leeuw K (2003) The Dutch invention of the rotor machine, 1915–1923. Cryptologia 27(1):73–94.

Der Chef des Oberkommandos der Wehrmacht, Berlin (12.1.1937) Geheim! Gebrauchsanleitung für die Chiffriermaschine Enigma, Vom 12.1.1937, Prüf-Nr. 965 [Secret! Instructions of Use for the Encryption Machine Enigma from January 12, 1937, Test No. 965]. Books on Demand, Norderstedt, ©2020 by Thomas Heise.

Garliński J (1979) The Enigma War. Charles Scribner's Sons, New York.

Gould H, Quaintance J (2012) Double fun with double factorials. Mathematics Magazine 85(3):177–192.

Hamer D (1997) Enigma: Actions involved in the "Double Stepping" of the middle rotor. Cryptologia 21(10):47–50.

Hodges A (1983) Alan Turing: The Enigma. Simon and Schuster Inc., New York.

Johnson B (1978) The Secret War. Methuen Inc., New York.

Kahn D (1967) The Codebreakers, The Story of Secret Writing. The Macmillan Company, New York.

Keegan J (2001) World War II. A Visual Encyclopedia. PRC Publishing Ltd., London.

Kruh L, Deavours C (2002) The Commercial Enigma: Beginnings of Machine Cryptography. Cryptologia 26(1):1–16.

Lenoir E (1964) Watch. In: Encyclopædia Britannica. Vol. 23, William Benton, Publisher, Chicago, Illinois.

Lewin R (1978) Ultra Goes to War, The First Account of World War II's Greatest Secret Based on Official Documents. McGraw-Hill Book Company, New York.

Lunde P (2012) The Secret of Codes. Weldon Owen Inc., San Francisco, Califorina.

McClain S (1994) The Navajo Weapon. Books Beyond Borders, Boulder, Colorado.

Miller AR (1995) The Cryptographic Mathematics of Enigma. Cryptologia 19(1):65–80.

Miller AR ([1996] 2001) The Cryptographic Mathematics of Enigma. Published by the Center for Cryptographic History, Fort Meade, Maryland.

Miller R, Fields W (2011) The "American" Enigma. Officer Review 51(1):8–11.

Paul DA (1973) The Navajo Code Talkers. Dorrance Publishing Company, Pittsburgh, Pennsylvania.

Robinson RC, Ward ED (1964) Clock. In: Encyclopædia Britannica. Vol. 5, William Benton, Publisher, Chicago, Illinois.

Simpson R (2017) Crypto Wars, 2000 Years of Cipher Evolution. CreateSpace Independent Publishing Platform, Middletown, Delaware.

Singh S (2001) The Code Book. Delacorte Press, New York.

Welchman G (1982) The Hut Six Story, Breaking the Enigma Codes. McGraw-Hill Book Company, New York.

Patents

1. Koch, Assignor/Inventor, Naamlooze Vennootschap Ingenieursbureau "Securitas" of Amsterdam, Netherlands, Assignee (April 14, 1925) Coding and Decoding Machine. US Patent 1,533,252, 11 pp., 3 pp. (diagrams) & 8 pp. (text).

2. Korn, Assignor/Inventor, Chiffriermaschinen Aktiengesellschaft in Berlin, Assignee (November 14, 1928) Elektrische Vorrichtung zum Chiffrieren und Dechiffrieren [Electrical apparatus for encryption and decryption]. Reichspatentschrift 452,194, Klasse 42n, Gruppe 14, 8 pp., 3 pp. (text) & 5 pp. (diagrams).
3. Korn, Assignor/Inventor, Chiffriermaschinen Aktiengesellschaft in Berlin, Assignee (June 1, 1929a) Chiffriervorrichtung zur Verwendung bei Chiffriermaschinen [Encryption apparatus for application in cipher machines]. Reichspatentschrift 460,457, Klasse 42n, Gruppe 14, 5 pp., 4 pp. (text) & 1 pp. (diagrams).
4. Korn, Assignor/Inventor, N/A [Not Applicable], Assignee (March 19, 1929b) Permutating Device for Use in Coding Machines. US Patent 1,705,641, 5 pp., 1 pp. (diagrams) & 4 pp. (text). [Note the Crypto Museum website says that this US patent is identical to the German Patent 460,457.]
5. Korn, Assignor/Inventor, N/A Assignee (April 25, 1933a) Coding Machine. US Patent 1,905,593, 7 pp., 3 pp. (diagrams) & 4 pp. (text).
6. Korn, Assignor/Inventor, Chiffriermaschinen Aktiengesellschaft [Cipher Machines, Joint-Stock Company], Berlin, Germany, Assignee (December 5, 1933b) Electrical Coding and De-Coding Devices. US Patent 1,938,028, 11 pp., 6 pp. (diagrams) & 5 pp. (text).
7. Scherbius, Assignor/Inventor, Chiffriermaschinen Aktiengesellschaft [Cipher Machines, Joint-Stock Company], Berlin, Assignee (January 24, 1928) Ciphering Machine. US Patent 1,657,411, 7 pp.: 1 p. (diagram) & 6 pp. (text).

Chapter 23
Between Natural Philosophy and Mathematics: Euler's Investigation of Planetary Perturbations in the 1740s

Giovanni Ferraro

Abstract In the late 1740s Euler wrote two memoirs devoted to the investigation of celestial perturbations in general and Saturn perturbation in particular. In these memoirs, he raised doubts over the validity of the inverse-square law but, at the same time, he developed analytical methods for determining planetary perturbations that were of great importance in celestial mechanics.

Keywords Euler · Inverse-Square law · Celestial perturbations · Newton's second law · Trigonometric series

1 Introduction

On March 26, 1746 the Paris Academy of Sciences announced the topic of the 1748 competition:

> A theory of Jupiter and Saturn explicating the inequalities that these planets appear to cause in each other's motions, especially about the time of their conjunction.[1] (Wilson 2007, p. 124)

Euler, who had a remarkable interest in astronomical problems and had already published some memoirs and a book on the motion of celestial bodies (Euler 1744, 1746) took part in the competition and won the prize with a long memoir that bears the title *Recherches sur la question des inégalités du mouvement de Saturne et de Jupiter* (Euler 1749b). The memoir, which despite the title only deals with Saturn's perturbations, was preceded by another paper, *Recherches sur le mouvement des corps célestes en general* (Euler 1749a), submitted to the Berlin Academy, on June

[1] "Une théorie de Saturne et de Jupiter par laquelle on puisse expliquer la cause physique des inégalités qu'on remarque dans les mouvemens de ces deux Planetes, principalement dans le temps de leur conjunction." (AA.VV. 1746, p. 87; Translation from Wilson 2007).

G. Ferraro (✉)
University of Molise, Campobasso, Italy
e-mail: gferraro@libero.it

R. Pisano (ed.), *A History of Physics: Phenomena, Ideas and Mechanisms*,
History of Mechanism and Machine Science 42,
https://doi.org/10.1007/978-3-031-26174-9_23

8, 1747, where Euler derived the differential equations used in the investigation of Saturn's orbit.

In his (Euler 1749b) Euler did not provide a satisfactory explanation for the motion of Saturn (for this reason, the same topic was later (1750 and 1752) proposed by the Paris Academy of Science as the subject for competitions (Maheu 1966, p. 214); however, these two memoirs are of remarkable interest in the history of different fields of exact sciences:

- In the history of astronomy, since in *Recherches sur la question des inégalités* and in *Recherches sur le mouvement des corps* Euler raised doubts over the validity of the inverse-square law (even if he based his memoirs on the inverse-square law) and used analytical methods for determining the perturbative effects caused by the mutual actions of the celestial bodies;
- In the history of mechanics, since in *Recherches sur le mouvement des corps* Euler, for the first time in a published paper[2] expressed the so-called Newton's second law of motion in the form $F_x = ma_x$, $F_y = ma_y$, $F_z = ma_z$[3];
- In the history of mathematical analysis and statistics, since these memoirs contain the attempt to introduce trigonometric series and a statistical technique to fit parameters to observations.

Several features of Euler's mathematical activity converge in *Recherches sur la question des inégalités* and in *Recherches sur le mouvement des corps*. But the most interesting aspect of these memoirs is the concept of mathematics emerging from them. Indeed, at the middle of eighteenth century, Euler was pursuing a project aiming to rebuilding mathematics around analysis. In his conception, analysis was the most general and abstract mathematical discipline and any other mathematics disciplines had to apply the findings of analysis (Ferraro 2020). The two *Recherches* are one of the first attempts to apply analysis to celestial mechanics and are the starting point of the process leading to analytical mechanics.

In this paper,[4] I will not discuss Euler's contribution to the study of the planetary perturbations in all its aspect, but I limit myself to some points. Specifically, in Sect. 2, I will highlight Euler's doubts on the inverse-square law. Then, in Sect. 3, I will outline the historical path that led Euler to write Newton's second law in the form $F_x = ma_x$, $F_y = ma_y$, $F_z = ma_z$. Finally, in Sect. 4, I will illustrate Euler's methods to investigate Saturn's perturbations.

[2] In his (Verdun 2013), Verdun analyzed Euler's unpublished manuscript and showed that Euler applied equations of motion in three dimensions already in 1744.

[3] This is the modern notation and not that used by Euler; see also footnote 28 below.

[4] An attempt to reconstruct the sources and evolution of Euler's contributions to celestial mechanics is made in (Ferraro 2010, 2012). The role played by (Euler 1749a, b) in the context of Euler's reorganization of mathematical sciences is scrutinized in (Ferraro 2020). The article is partly based on (Ferraro 2012, 2020).

2 Inverse-Square Law

In his *Recherches sur le mouvement des corps célestes en general* (Euler 1749a, pp. 93–97), Euler *began* his investigation by observing that planetary perturbations could be arise from different reasons: an irregular distribution of mass in the Sun, the presence of forces other than the attractive one of the Sun (such as the attractive forces of other planets), and the approximate validity of the inverse-square law.[5]

Euler's doubts had a twofold origin: the philosophical implications of the concept of *action-at-a-distance and* empirical evidence that called into question the accuracy of the inverse-square law. As regards the philosophical implications of *action-at-a-distance*, Euler seems not to share Newton's famous *dictum* "Hypotheses non fingo". In the *General Scholium* of the second edition of the *Principia*, Newton had used this phrase to answer those who criticized the lack of an explanation for the qualitative nature and causes of gravity and to claim that he did not make any hypothesis or conjecture concerning such causes.[6] In effect, the criticism to Newton's law of attraction lasted decades (Cfr. Gingras 2001). For instance, in *Nouvelles pensées sur le système de M. Descartes* (Bernoulli 1730), a paper that won the prize of the Academy of Science of Paris, Johann Bernoulli still tried to show the mathematical compatibility of the vortex theory with the elliptic orbits of planets. He attributed the elliptical orbit to the combination of a circular motion of the planet about the Sun and a radial oscillation about an equilibrium position, taking the planet in turn into layers of greater and smaller density, while the line of apsides rotated owing to a lack of synchronia between the circulation and radial oscillation. A few years later, Johann Bernoulli published a new paper, *Essai d'une nouvelle physique céleste* (Bernoulli 1735), where he partially changed his approach and attempted to form a new system by combining elements of Descartes's and Newton's conceptions. In this paper, J. Bernoulli accepted gravitation as a fact and admitted that planets moved toward each other, but he also stated that the principle of gravitation was unintelligible and rejected the idea that attraction was the cause of the motion or that it was a physical principle. According to Bernoulli, it was necessary to give an explanation for gravitation: Abandoning the Cartesian hypothesis that planets were carried by the vortices, he tried to base this explanation on the assumption that a perpetual stream of matter tended from all the sides to the geometrical centre.

Johann Bernoulli's memoir was one of the two that won the prize of the Paris Academy in 1734: the other was Daniel Bernoulli's Recherches physiques

---

[5] *As concerns this last point, he stated:* "Car quelques recherches & réflexion que j'ai faites, tant sur l'origine de ces forces, que sur les dérangements, qu'on remarque dans le mouvement de la Lune, & des Planètes supérieures, m'ont porté à croire, que les forces, dont on soutient, que les Planètes, s'attirent les unes les autres, ne suivent pas exactement la raison réciproque des quarrés des distances." (Euler 1749a, pp. 95–96). *A similar statement is found in* (Euler 1749b*; see below*).

[6] "Rationem verò harum gravitatis proprietatum ex phænomenis nondum potui deducere, et hypotheses non fingo. Quicquid enim ex phænomenis non deducitur, *hypothesis* vocanda est; et hypotheses seu metaphysicæ, seu physicæ, seu qualitatum occultarum, seu mechanicæ, in *philosophiâ experimentali* locum non habent. In hâc philosophiâ Propositiones deducuntur ex phænomenis, et redduntur generales per inductionem." (Newton 1713, p. 484).

et astronomiques (Bernoulli 1734). In this memoir, Daniel Bernoulli replaced the ethereal vortices by a solar atmosphere. However, he still believed that the idea of vortices had to be maintained: the cause of gravity might be found in a complicated system of independent vortices moving through each other without mutual interference. Ever later Daniel Bernoulli stated that the theory of vortices could not be definitely rejected (Bernoulli 1741).

Following the opinion of the Bernoullian school, Euler was critical of the idea of action-at-a-distance. In his *Lettres à une princesse d'Allemagne* (Euler 1768b), Euler asserted that the problem of action-at-a-distance was more philosophical than mathematical and thus one could not pretend to afford it with absolute success

> It now remain that we inquire into the cause of these attractive forces; but this research belongs rather to the province of metaphysics than of mathematics. I dare not therefore myself with the prospect of absolute success in the prosecution of it.[7] (Euler 1833, p. 230)

In his papers Euler, however, advanced "[...] a non-Newtonian demand of explanation of the nature and causes of forces" (Maronne and Panza 2014, p. 14): According to him, the notion of force could not be primitive and a mathematical theory about forces could not be separated from their causes. Like Descartes, Euler required that the basic concepts of mechanics had to be derived from matter and motion.[8] Euler also claimed that the whole space was filled with subtle, though invisible, matter, and that it was more reasonable to ascribe the gravitational attraction to an action which the ether exercised upon the bodies (though how the ether acted was unknown), rather than to have recourse to an unintelligible property:

> As we know that the whole space which separates the heavenly bodies is filled with a subtle matter, called ether, it seems more reasonable to ascribe the mutual attraction of bodies to an action which the ether exercises upon them, though its manner of acting may be unknown to us, rather than to have recourse to an unintelligible property.[9]

---

[7] "Il s'agit à présent d'approfondir la véritable source de ces forces attractives, ce qui appartient plutôt à la Metaphysique qu'aux Mathematiques; & je ne saurois me atter d'y reussir aussi heureusement." (Euler 1768b, p. 265; Translation from 32).

[8] For instance, see (Euler 1738, Sect. 2). On this question, see (Gaukroger 1982).

[9] "Puisque nous savons donc que tout l'espace entre les corps célestes est rempli d'une manière subtile qu'on nomme l'éther, il semble plus raisonnable d'attribuer l'attraction mutuelle des corps, â une action que l'éther y exerce, quoique la maniere nous soit inconnue, que de recourir a une qualité inintelligible." (Euler 1768b, p. 268). In his *Lettres à une princesse d'Allemagne, Euler wrote:* "As you see nothing that impels (small bits of iron and steel] toward the loadstone), we say that the loadstone attracts them; and this phenomenon we call attraction. It cannot be doubted, however, that there is very subtle, though invisible, matter, which produces this effect by actually impelling the iron towards the lodestone [...]. Though this phenomenon be peculiar to the loadstone and iron, it is perfectly adapted to convey an idea of the signification of the word attraction, which philosophers so frequently employ. They allege then, that all bodies, in general, are endowed with a property similar to that of the loadstone, and that they mutually attract." (Maronne and Panza 2014, p. 13); Translation, *ibidem*). "Comme on ne voit rien, qui les (de petits morceaux de fer ou d'acier) pousse vers l'aimant, on dit que l'aimant les attire, & l'action même, se nomme *attraction*. On ne sauroit douter cependant qu'il n'y ait quelque matiere très subtile, quoiqu'invisible, qui produise cet effet, en poussant effectivement le fer vers l'aiman; (...) Quoique ce phénomène soit particulier à l'aimant, & au fer, il est très propre à éclaircir le terme d'attraction, dont les Philosophes modernes

According to Euler (Euler 1768b, p. 283), occult qualities and imaginary causes were to be rejected (Euler 1740, p. 237) and only two possible explanations for gravity were admissible: Either there was a continual flux of subtle matter into the central sun or planet, bearing down with it all bodies it passed through, or there was a vortex about the sun or planet, the centrifugal force of which caused gross bodies to accelerate towards the centre. The first hypothesis entailed a force moving ether to the centre and then a mechanism for evacuating it from the centre: so it was implausible. Therefore, no other cause of these forces was left except one consisting in centrifugal force (Wilson 1992, p. 404). The vortices had to be so disposed as to yield an inverse-square force toward the centre. However, if one accepted this model, then it seemed improbable that the law of gravitation, whatever its mathematical form was, could be valid at any distance (Cfr. Wilson 1995).

Apart from these philosophical and a priori doubts regarding *action-at-a-distance*, Euler had other doubts originated from empirical evidence, in particular the evidence regarding the lunar motion.[10] Indeed, according to him, if one supposed that the forces acting on the Moon from both the Earth and the Sun were perfectly proportional reciprocally to the square of the distances, then calculations showed that the motion of the lunar apogee to be almost two times slower than the observations made it (Waff 1975, pp. 37–38). In his *Recherches sur la question des inégalité*, Euler states:

> I compared the observations of the Moon very carefully with the theory and found that the distance of the Moon from the Earth is not so great as it should be, according to the theory. Hence it follows that the gravity of the Moon towards the Earth is a little less than according to the inverse ratio of the squares of the distances. Some small irregularities in the movement of the Moon, which one could not explain by this theory, confirmed me still more in this feeling. It therefore seems to me that the Newtonian proportion according to the squares of the distances is only true approximately in the forces of celestial bodies, and that perhaps it deviates all the more from the truth as the distances are large. In this case, therefore, there would be no reason to be surprised if the motion of Saturn were subject to still other inequalities than those caused by the action of Jupiter: and it seems to me very likely that the very action of Jupiter on the body of Saturn, deviates considerably from the inverse ratio of the squares of the distances.[11]

---

se servent si frequemment. Ils disent donc, qu'une propriété semblable à celle de l'aimant, convient à tous les corps, en general, & que tous les corps au monde s'attirent mutuellement." (Euler 1768b, pp. 219–220).

[10] It is interesting to observe that the discrepancy between theory (based on the inverse-square law) and observation of the motion of the lunar apse was first stated by Newton in his *Principia* (Newton 1713, I, Prop. XLV, Probl. XXXI).

[11] "Ayant comparé fort soigneusement les observations de la Lune avec la théorie, j'ai trouvé que la distance de la Lune à la Terre n'est pas si grande qu'elle devroit être, selon la théorie: d'où il s'ensuit que la gravité de La lune vers la Terre est un peu moindre, que selon la raison inverse des quarrés des distances: et quelques petites irrégularités dans le mouvement de la Lune, qu'on ne sçauroit expliquer par cette théorie, m'ont encore davantage confirmé dans ce sentiment. Il me semble donc que la proportion Newtonienne selon les quarrés des distances, n'est vraie qu'à peu près dans les forces des corps célestes, et que peut-être elle s'écarte d'autant plus de la vérité que les distance sont grandes. Dans ce cas, il n'y auroit donc pas lieu de s'étonner si le mouvement de Saturne étoit assujetti encore à d'autre inégalités qu'à celles qui sont causées par l'action de Jupiter: et il me paroît fort vraisemblable que l'action même de Jupiter sur le corps de Saturne, s'écarte considérablement de la raison inverse des quarrés des distances." (Euler 1749b, p. 49).

In the mid-eighteenth century, the most important French mathematicians, *Jean-*Baptiste Le Rond d'Alembert and Alexis Clairaut, undertook to develop general methods for investigating the problem of the moon's orbit and the three-body problem.[12] Like Euler, both d'Alembert and Clairaut found disagreements between observations and the results derived from the inverse-square law and hence doubted the validity of inverse-square law.

D'Alembert and Clairaut worked in secret and, in order to have priority in case of new discoveries, deposited uncompleted portions of their works at the Paris Academy in the form of sealed notes (Hankins 1970; Maheu 1966, pp. 32–34). In one of these sealed notes, deposited on September 6, 1747, Clairaut gave the numerical determination of the equations for the Moon's orbit and noted a discrepancy between the theory and the observed motion of the lunar apse. Two months later, d'Alembert came to the same conclusion; he also deposited a sealed note on this problem at the Academy (Hankins 1970, pp. 31–34).

In September 1747, d'Alembert and Clairaut also learnt of Euler's doubts about the accuracy of the inverse-square law (they were two of the judges of the 1748 prize, Maheu 1966, p. 214) and had the possibility of reading Euler's prize essay). Both of them discussed the question with the Swiss mathematician.[13] Thus, on September 30, 1747, Euler wrote to Clairaut that he was able to give several proofs that the forces which acted on the moon did not exactly follow the Newton's gravitational law, and the one Clairaut drew from the movement of the apse was the most striking. Since the errors could not be attributed to the observations, Euler did not doubt that a certain derangement of the forces supposed in the theory was the cause. This circumstance made Euler think that the vortices or some other material cause of forces was very probable, since it was then easy to understand that these forces ought to be altered when they were transmitted by some other vortex (Euler 1980, pp. 175–177).

At the public meeting of the Paris Academy of 15 November 1747, Clairaut gave a lecture on this subject (Maheu 1966, pp. 220–222; Clairaut 1745a). He observed that the discrepancy was of the same order as the discrepancy with the far older supposition of a Moon rotating around the Earth on a circular orbit. Clairaut hypothesized that the inverse-square law was an approximation and suggested the addition of an extra term to the mathematical expression for gravitational attraction, one that decreased with distance more rapidly than the basic inverse-square relationship expressed in the first term (Clairaut 1745a, pp. 358–359).

Clairaut's announcement of the falsehood of the inverse-square law created a sensation and provoked controversy. In particular, George-Louis *Leclerc de* Buffon attacked Clairaut on the basis of metaphysical considerations concerning the simplicity of the inverse-square law:

---

[12] On the role of the investigation of the motion of the lunar apogee in the establishment and reception of Newton's inverse-square law, see (Waff 1975; Wepster 2010; Schroeder 2007).

[13] Euler's correspondence is in (Euler 1980). For an extensive investigation of the relationship between Euler and d'Alembert, see Ferraro (2008a).

> Every time the expression of a law is not represented by a single term, the simplicity and unity of the expression, which forms the essence of the law, no longer remains and, as a consequence, there is no longer a physical law.[14] (Lyon and Sloan 1981)

According to Buffon, a term in a mathematical expression governing some basic kind of physical phenomenon had to correspond to some specific thing in the world that could be measured: the value of the term varied as the corresponding measurement did (Cfr. Dear 2005).

> By whatever means, we thus would suppose that a physical quality could vary, since this quality is unitary, its variation will be simple and always expressible by a single term, which will be the measure of it. But as soon as one would wish to employ two terms, the unity of physical quality will be destroyed because these two terms will be represent two terms are, in fact, two measures, two separate and unequal variable if two terms are allowed to represent the effect of the central force of a star, it is necessary to admit that in place of one force there are two [...] [translation from Lyon and Sloan 1981].[15] (Lyon and Sloan 1981)

In other words, if a law of nature had more than one term in its analytical expression, such as $\frac{a}{r^2} + \frac{1}{mr^4}$ (Leclerc and Buffon 1745b, pp. 581–582), then the law expressed the influence of more than one causes (Cfr. Dhombres 2002).

The controversy between Buffon and Clairaut caused excitement in d'Alembert, who, however, thought that the discrepancy did not necessarily involve the rejection of inverse-square law. D'Alembert suggested two hypotheses that might explain the deviation from the gravitational law: A radial asymmetric distribution of the mass of the Moon or an additional non-gravitational force acting solely between the Earth and the Moon (Euler 1980, pp. 276–286).

After a short time, d'Alembert changed his mind: indeed, he noted errors in his calculations (Euler 1980, pp. 22–23). After having corrected the errors, there were still disagreements between empirical data and the inverse-square law; they, however, are no longer so evident. In the same days, Clairaut also came to the same conclusion: He improved his analysis by considering more terms in certain series expansions (Hankins 1970, p. 35) and realized that it was the lack of these terms that caused the discrepancy. Thus, on May 17, 1749, Clairaut made another announcement in the Paris Academy: this time he announced that the inverse-square law was correct. (Clairaut 1745b, pp. 577–588).

Euler, who had shared Clairaut's previous opinion on the need of adding some terms to inverse-square law, maintained his critical standpoint on Newton's law of attraction since his calculations still showed disagreements between data and

---

[14] "Toutes les fois que l'échelle d'une loi ne sera pas représentée par un seul terme, cette simplicité & cette unité d'échelle, qui fait l'essence de la loi, ne subsiste plus, & par conséquent il n'y a plus aucune loi physique." (Leclerc and Buffon 1745a, pp. 497–498).

[15] "De quelque façon que nous puissions donc supposer qu'une qualité physique puisse varier, comme cette qualité est une, sa variation sera simple & toujours exprimable par un seul terme qui en sera la mesure; & dès qu'on voudra employer deux termes, on détruira l'unité de la qualité physique [...] deux termes sont en effet deux mesures, toutes deux variables & inégalement variables [...] si on admet deux termes pour représenter l'effet de la force centrale d'un astre, il est nécessaire d'avouer qu'au lieu d'une force il y en a deux [...]" (Leclerc and Buffon 1745a, p. 498; Translation from Lyon and Sloan 1981).

theory (Ferraro 2008a, pp. 278–279). Euler was convinced only when he saw the mathematical details of Clairaut's announced solution.[16] It is worthwhile noting that, on the one hand, Euler's acceptation of Newton's law did not imply a rejection of certain Cartesian principles (see the above quotations from Euler 1768b); on the other hand, these Cartesian principles did not impede him to consider a mathematical science of motion to be "[...] perferctly possible also in the ignorance of the actual causes of attractive forces [...]" (Maronne and Panza 2014, p. 14.[17] Indeed (Euler 1749b) grounded his investigation of Saturn's perturbations on Newton's law of attraction.

3 Euler and the Second Law of Mechanics

Euler's investigation of Saturn's perturbations presented important methodological novelties, some of which were highlighted in his paper:

(1) Euler described the motion by means of equations expressing the relation between time and space.[18]

(2) Euler used the angles into calculations instead of using their sine and cosine. This means that the sine and cosine were no longer conceived of as geometric entities but as analytical functions.[19]

(3) Euler did not study the "curvature" of the trajectory, since the consideration of the trajectory would have led to very complicate quantities, mainly if the motions were not plane. In practice, this mean that forces and other physical magnitudes were expressed in rectangular Cartesian co-ordinates and investigated by considering their components in a Cartesian coordinate system (Euler 1749b, pp. 101–102; Cfr. Verdun 2013).

---

[16] In 1750 the subject prize competition sponsored by the St. Petersburg Academy was: to demonstrate whether all the inequalities observed in lunar motion are in accordance with Newtonian (gravitational) theory, and if they are not, to demonstrate the true theory behind all these inequalities, so that the exact position of the moon at any time can be computed with the aid of this theory (Maheu 1966) Clairaut won the prize with his *Théorie de la Lune déduite du seul principe de l'attraction réciproquement proportionnelle aux quarrés des distances* (Clairaut 1752).

[17] According to Maronne and Panza, "[...] (for Euler) the only way to ensure that some forces are caused by some reasons is to show that the consideration of these reasons leads to the well-known mathematical laws of motion. These laws—rather than any possible mechanical model—are thus finally understood as the only sure expression of the reality of the universe." (Maronne and Panza 2014, p. 14).

[18] Until about 1742–1744 Euler used the equations of motion parametrized by length in order to have direct access to the geometry of the orbit or trajectory. In his (Verdun 2013) Verdun showed that Euler's investigation of lunar theory played a crucial role in understating the importance of the parametrization by time.

[19] In the 1740s, the calculus of the trigonometric functions is a novelty and, thus, in his *Recherches sur la question des inégalités* (Euler 1749b, pp. 99–100), Euler explicitly warns his readers that he will use such new analytical instruments.

Euler's methodology made clear what his main objective was: The elaboration of the analytical tools that enabled the complete understanding of the phenomenon of the planetary perturbations.[20]

Euler began his investigation by considering the following problem:

> If a body is subjected to a force, determine the instantaneous change that this force produces in the movement of the body. (Euler 1749b, p. 102)

To solve this problem, (Euler 1749b, pp. 102–103) termed the coordinates of M as x, y, z and stated that the forces acting on the body in M could be always reduced to three forces parallel to the directions of the coordinates (namely, he considered the components of the force acting in the directions of the x-, y-, and z-axes).[21] He denoted the "absolute" or "moving forces" by the letters X, Y, and Z; therefore, the "accelerating" forces were equal to $\frac{X}{m}, \frac{Y}{m}, \frac{Z}{m}$ (m is the mass of the body[22]). Taken dt as a constant, Euler stated that the "instantaneous change" of the motion was given by the equations:

$$2\frac{d^2x}{dt^2} = \frac{X}{m}, \; 2\frac{d^2y}{dt^2} = \frac{X}{m}, \; 2\frac{d^2z}{dt^2} = \frac{X}{m}. \tag{1}$$

This is the first time that Eq. (1) are given in an explicit and general form in a published paper.[23] In a scholium, (Euler 1749b, p. 104) asserted that Eq. (1) were nothing but a reformulation of Newton's second law, which he wrote as

$$du = pdt \tag{2}$$

($u = \frac{ds}{dt}$ is the velocity and p is the "accelerating" power). From (2) Euler derived

$$du = \frac{d^2s}{dt} \text{and} \frac{d^2s}{dt^2} = p. \tag{3}$$

At this point, Euler claimed that even if the foundation of Eq. (1) was nothing else than the analytical representation of Newton's second law, some reflexion was

---

[20] Euler's reformulation of some basic notions of Newton's *Principia* is subject-matter of *Euler and the structure of mathematics* (Ferraro 2020, pp. 15–21). I refer to this paper for details and clarifications.

[21] AT that time, the three axes were not given a priori but one only axis was represented in an explicit way. Therefore, in order to introduce the coordinate of the point M, Euler considered a fixed plane and, from the point M, drew a segment MQ perpendicular to this plane. Then Euler took a straight line on the plane (the axis of the abscissas) and a point C on the line. He drew a segment QP perpendicular to the straight line: the segments CP = x, QP = y, MQ = z were the perpendicular coordinates of M. Such a way of introducing a coordinate system did not reduce the (physical and geometric) space to R^3 and had a clear meaning as concerns the conception of analytical geometry and the reducibility of the whole geometry to analytical geometry (Cfr. Ferraro 2007a, pp. 50–55).

[22] Euler denoted the mass by the same symbol M that denoted the point.

[23] The coefficient 2 is a factor of normalization.

necessary before one could see that this law held equally for each partial motion into which the true motion was thought of as reduced (in modern terms, some reflexion was necessary before one realized that forces were vectors and that it was possible to reduce the investigation of forces to that of their components). In Euler's opinion, Eq. (1) included all the principles ordinarily used in the determination of curvilinear motions (Euler 1749a, p. 104). According to Truesdell:

> Euler's statement is misleading in its modesty. All of Newtons's brilliant work on the problem of three bodies was done by the aid of inequalities, groping approximations, and physical insight alone, *without the equations of motion*. To get results by such means required the genius of a Newton. His disciples, who might reasonably have been expected to build upon his foundation, did not raise the structure an inch higher. Real further progress came only after the equations of motion had been discovered. (Truesdell 1960, p. 252)

In effect, in the *Principia*, Newton (Newton 1687, p. 12) had formulated the second law by stating: "Mutationem motus proportionanalem esse vi motrici impressae, et fieri secundum lineam rectam quas vis illa imprimatur". In a modern symbolism, this statement can be rendered as $F = \Delta(mv)$, where mv is the motion (*momentum*). The expression "*mutatio motus*" is not univocal in Newton and, elsewhere, Newton stated that a centripetal force is proportional to the motion that it generates in a given time: this "sounds as" $F = ma$ (to use an expression of Maltese 2002). However, it has to be strongly emphasized that Newton's formulation cannot be translated in terms of equations, as Euler did, without a radical change in the conception of mechanics,[24] since this translation implies the overcoming of the Newtonian conception of mechanics as a geometric investigation of the trajectories and the transition to a mechanics that used analysis (and, in particular, differential equations) as its main instrument. In Newton's *Principia*, the investigation of the motion of bodies was intrinsically geometric. The trajectories[25] were supposed to be geometric curves, and the physical magnitudes associated with the motion at any point of the curve were indirectly represented as segments proportional to the values of speeds or forces at that point[26] (Newton understood the motion as a continuous composition of rectilinear and uniform motions, which, at any point, expressed the motion the body would have if speed and forces acting in that point did not change). This geometric representation allowed Newton to compose forces according to the law of parallelogram and to geometrize mechanics by interpreting motion as a geometric magnitude (Panza 2002, p. 262). Of course, the idea of continuously changing punctual motions could

[24] Maltese stated: "[...] it was Euler who built what we now call the "Newtonian tradition" in mechanics, grounded on the laws of linear and angular momentum (which Euler was the first to consider as principles general and applicable to each part of every macroscopic system), on the concept that forces are vectors, on the idea of reference frame and of rectangular Cartesian coordinates, and finally, on the notion of relativity motion." (Maltese 2000, pp. 319–320; Author's translation).

[25] Non-punctual bodies were assumed to behave with respect to attractive forces as if their mass were concentrated in their centre of gravity.

[26] Given two velocities or two forces, and, the segments A and B that represented them had to satisfy the proportion = A:B.

not be dealt with by classical geometrical methods and Newton added the geometric method of prime and last ratios to classic geometry (Panza 2002, pp. 262–263).

Immediately after the publication of the *Principia*, Leibniz attempted to use his differential and integral calculus to rewrite Newton's results. Leibniz generalized (Leibniz 1689a) the law governing the rectilinear motion by stating the following principle: the elements of space are in compound ratio of the velocities and the elements of the time, namely *ds* is proportional to *vdt*, where *s*, *v*, and *t* denote the space, velocity, and time, respectively. Leibniz represented (Leibniz 1689b) accelerated motion as "[...] a series of infinitesimal uniform rectilinear motions interrupted by impulses [...]" (Bertoloni Meli 1993, p. 91).

This approach to mechanics was still widely geometric. Indeed, physical magnitudes were, first, represented geometrically and, then, the concepts and symbols of Leibniz's calculus were applied. Thus, for example (Leibniz 1689a), instantaneous velocity was represented by means of infinitesimal segments: the differential symbolism *ds*, *dt*, *dv*, ..., translated the geometric representation into an analytical one, but the latter was understandable only by referring to the former (Blay 1992, [1993] 1998).

The Bernoullis, Varignon, and Hermann achieved many results that contributed to the transformation of mechanics. From a conceptual viewpoint, Varignon's contribution is very interesting (Cfr. Varignon 1700a, b). In his *Manière générale de déterminer les forces, les vitesses, les espaces et les temps* (Varignon 1700a), first, Varignon geometrically represented space, time, force and velocity and associated algebraic symbols to these magnitudes (Varignon 1700a, p. 23); then, he assumed the speed of a body as uniform during each instant (Fig. 1).

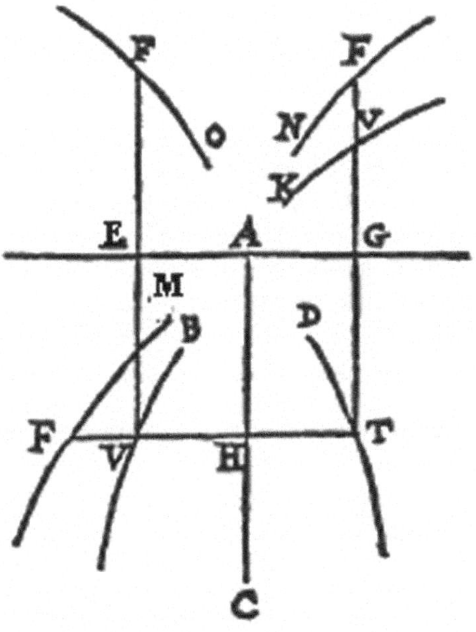

Fig. 1 Varignon's representation of physical quantities *Source* Varignon 1700a, p. 22

In this way, he could express the speed v at each instant as $v = \frac{dx}{dt}$ (x and t are space and time, respectively), the variation of the velocity as $dv = \frac{d^2x}{dt}$, and the accelerating force y at each instant as $y = \frac{dv}{dt}$ or $y = \frac{d^2x}{dt^2}$ (Varignon 1700a, p. 23).

At that time the consideration of ratios such as space/time was problematic. Varignon tackled the question explicitly in 1707 and stated that space and time being heterogeneous, they could not be compared in the strict sense of the term (Varignon 1707, p. 223). However, space and time could be expressed by homogeneous magnitudes (two lines, or two numbers, or any two other such homogeneous magnitudes); it was these homogeneous magnitudes that could be compared. In other terms, it was the geometrical representation that enabled to compare space and time (and other heterogeneous magnitudes).

It is to be highlighted that Varignon and other components of Leibnizian and Bernoullian school used the language of differential calculus to speak of mechanical configurations represented by appropriate geometric diagrams and the rules of differential calculus were "[…] applied in order to get the relevant quantitative relations between the elements depicted in these diagrams [...]" (Maronne and Panza 2014, p. 15). In this way, the investigation of a mechanical problem relied on the diagram that represented that specific problem; this caused a fragmentation of mechanics in many geometrically different problems (a specific problem corresponded to a specific diagram and had a specific solution). This was the situation when Euler started his activity.[27] In 1736, he published an innovative treatise on kinematics and dynamics of a point-mass, *Mechanica sive motus scientia analytice exposita* (Euler 1736), where he attempted to overcome the above-described fragmentation by applying analytic methods to mechanics in a systematic way. In this treatise (Euler 1736, I, Sects. 154–155), Euler formulated the Newton's second law of motion by means of the differential equation

$$dc = \frac{npdt}{A} \tag{4}$$

where A is the body and its mass, p is the force, c is the velocity, t is the time, n is a constant of proportionality depending on the unity of measure (Euler sets $n = ½$, see footnote n. 26). From the last equation, Euler was able to derive all differential equations necessary to describe the motion of a point-mass. The novelty of Euler's approach was understood and criticized, for instance, by Benjamin Robins, who wrote:

> I have no design to charge this author [Euler] with haste or negligence on account of these errors; but I consider them solely, as the effect of that inaccuracy in conception, to which the differential calculus is disposed to betray its admirers [...]. In the beginning of the third chapter, which treats of right-lined motion, Mister Euler has given Galileo's theory of falling

[27] It should be mentioned that the equation of motion as used by Euler was first published in Leibnizian notation by Jacob Hermann in his *Phoronomia* (Hermann 1716, p. 57). However, in his work, Hermann used this notation to refer to mechanical configurations represented by appropriate geometric diagrams. In this way, the investigation of mechanics lost generality. In the preface of *Mechanica* (Euler 1736), Euler criticized *Phoronomia*.

> bodies, in its own nature no difficult subject; but it is here so compounded with differential computations, that this subject may be much better learned from what has been writ in a more simple manner by others. (Robins 1761, pp. 203, p. 205)

In his *Mechanica* (Euler 1736) Euler used intrinsic coordinates systems: he decomposed speeds and forces according to directions that depended upon the intrinsic nature of the problem (Cfr. Maronne and Panza 2014, p. 16): this limited the generality of procedures. The novelty of *Recherches sur le mouvement des corps célestes en général* (Euler 1749a) consisted of using an extrinsic references frame and considering Eq. (1) as the bases of mechanics. This was made clearer in 1750 when Euler devoted a paper (Euler 1750) to his new approach to mechanics (Ferraro 2020).

4 Saturn's Perturbations

In his *Recherches sur la question des inégalités du mouvement de Saturne et de Jupiter* (Euler 1749b), Euler investigates the motion of Saturn with respect to an observer placed in the Sun by using the plane of Jupiter as the plane of reference.[28]

Using the results of his (Euler 1749a), Euler first determines the four differential equations of Saturn's motion[29]:

$$ddz - zd\varphi^2 = -a^3 d\zeta^2 \left(\frac{(1+v)\cos\psi^3}{z^2} + \frac{nz}{v^3} + \frac{n\cos\omega}{y^2} - \frac{ny\cos\omega}{v^3} \right) \qquad (5_{\mathrm{I}})$$

$$2dzd\varphi + zdd\varphi = -na^3 d\zeta^2 \sin\omega \left(\frac{1}{y^2} - \frac{y}{v^3} \right) \qquad (5_{\mathrm{II}})$$

$$d\pi = \frac{na^3 d\zeta^2 \sin(\varphi - \pi)\sin(\theta - \pi)}{zd\varphi} \left(\frac{1}{y^2} - \frac{y}{v^3} \right) \qquad (5_{\mathrm{III}})$$

$$d(\ln\tan\rho) = \frac{-na^3 d\zeta^2 \cos(\varphi - \pi)\sin(\theta - \pi)}{zd\varphi} \left(\frac{1}{y^2} - \frac{y}{v^3} \right) \qquad (5_{\mathrm{IV}})$$

where φ is the longitude of Saturn, ψ is the latitude of Saturn, z is the shortened distance from Saturn to the Sun, π is the longitude of the ascending node, θ is the longitude of Jupiter, a is the mean distance between Jupiter and the Sun, ζ is the mean anomaly of Jupiter, y is the distance between Jupiter and the Sun, $\nu = 1/3021$ is the ratio between the masses of Saturn and Sun, $n = 1/1067$ is the ratio between the masses of Jupiter and Sun, $\omega = \theta - \varphi$ is the elongation of Saturn and Jupiter,

$$\mathrm{v} = \sqrt{y^2 + z^2 / \cos^2\psi - 2yz\cos\omega}$$

[28] For the sake of simplicity, I introduce some modern notations that do not exist in Euler's papers.

[29] On the derivation of Eq. (5), see Wilson 2007.

is the distance between Jupiter and Saturn, and ρ is the inclination of the orbital planes. In the calculation the element dζ of the mean anomaly of Jupiter is taken as a constant.

To simplify the problem Euler assumes the following hypotheses. First, the motions of Saturn and Jupiter lie on the same plane. This implies that one has to take only the first two equations into account, which can be written in the form (since the latitude ψ of Saturn (with respect the plane of Jupiter's orbit) is zero and then *cos* ψ = 1):

$$ddz - zd\varphi^2 = -a^3 d\zeta^2 \left(\frac{1+v}{z^2} + \frac{nz}{v^3} + \frac{n\cos\omega}{y^2} - \frac{ny\cos\omega}{v^3} \right) \tag{6_{I}}$$

$$2dzd\varphi + zdd\varphi = -na^3 d\zeta^2 \sin\omega \left(\frac{1}{y^2} - \frac{y}{v^3} \right) \tag{6_{II}}$$

where

$$\mathrm{v} = \sqrt{y^2 + z^2 - 2yz\cos\omega}$$

(Euler 1749b, p. 58).

Second, Jupiter's orbit is supposed to be circular. Consequently, $y = a$, $\theta = \zeta$, and $d\omega = d\theta – d\varphi = d\zeta – d\varphi$ and Eqs. (6_{I}) and (6_{II}) become:

$$ddz - zd\varphi^2 = -a^3 d\zeta^2 \left(\frac{1+v}{z^2} + \frac{nz}{v^3} + \frac{n\cos\omega}{a^2} - \frac{na\cos\omega}{v^3} \right) \tag{7_{I}}$$

$$2dzd\varphi + zdd\varphi = -na^3 d\zeta^2 \sin\omega \left(\frac{1}{a^2} - \frac{a}{v^3} \right) \tag{7_{II}}$$

where

$$\mathrm{v} = \sqrt{a^2 + z^2 - 2az\cos\omega}$$

(Euler 1749b, pp. 58–59).

Third, the unperturbed orbit of Saturn is supposed to be circular. Moreover, the attraction due to Jupiter causes small change in the circular motion of Saturn and, therefore, the differential dφ of the longitude of Saturn can be expressed by the formula

$$d\varphi = md\zeta + ndx,$$

where *m* is an appropriate coefficient and the term *ndx* is very small with respect to *m*dζ and only depends on ω. The following approximate equalities hold

$$d\varphi = md\,\varsigma \text{and } d\omega = (1-m)d\varsigma$$

Since $d\zeta$ is constant, from $d\varphi = md\,\zeta + ndx$, it follows that:

$$d^2\varphi = nd^2x \text{ and } d\varphi^2 = m^2d\zeta^2 + 2mnd\zeta dx$$

(n^2dx^2 is neglected). If we denote the mean distance between Saturn and the Sun by f, *in the absence of Jupiter's* perturbing influence, we have $f = z$, instead in the presence of Jupiter, it is $z = f(1 + nr)$, where nr is very small and r depends solely on ω. Moreover,

$$dz = nfdr \text{ and } d^2z = nfd^2r$$

At this point Euler sets $f = \lambda a$, $g = \frac{2\lambda}{1+\lambda^2} = \frac{2fa}{a^2+f^2}$, $h = \lambda(1+\lambda^2)^{3/2}$ and approximates $(1+nr)^{-2}$ by $1\text{-}2nr$ (since—he says—the powers of nr can certainly be neglected in the calculation). By appropriate substitutions, Euler (Euler 1749b, pp. 59–60) obtains:

$$m^2d\zeta + 2mndx + m^2nrd\zeta - n\frac{d^2r}{d\zeta} = \frac{1+v}{\lambda^3}d\zeta - \frac{2nr}{\lambda^3}d\zeta + \frac{n\cos\omega}{\lambda}d\zeta + \frac{n(\lambda-\cos\omega)}{h(1-g\cos\omega)^{\frac{3}{2}}}d\zeta \quad (8_{\mathrm{I}})$$

$$2mdr + \frac{d^2x}{d\zeta} = -\frac{\sin\omega}{\lambda}d\zeta + \frac{\sin\omega}{h(1-g\cos\omega)^{\frac{3}{2}}}d\zeta \quad (8_{\mathrm{II}})$$

The greatest difficulty in solving this system of differential equations is that of calculating the integral.

$$\int (1-g\cos\omega)^{-\frac{3}{2}}d\omega.$$

Since it is not possible to give a closed form for it, Euler is forced to investigate in detail the question of the integration of $(1-g\cos\omega)^{-\mu}$. The Swiss mathematician tries to integrate this function by series, but he soon realizes that this traditional technique is of no help. Indeed, the expansion of the function $(1-g\cos\omega)^{-\mu}$ by the binomial theorem gives rises to the series

$$(1-g\cos\omega)^{-\mu} = 1 + \frac{\mu}{1}g\cos\omega + \frac{\mu(1+\mu)}{1\cdot 2}g^2\cos^2\omega + \ldots$$

which is not suitable to determine the planetary motion. Indeed, Euler states that the quantity $(1-g\cos\omega)^{-\mu}$ cannot be developed into a convergent sequence, since the value of g is approximately $= 4/5$ and also that the series is not sufficiently convergent.[30] In effect the series is convergent; however, the speed of convergence depends on the parameter g (in its turn, g depends on a and f). In order to obtain an approximate value < 0,001, without any assumption about $cos\ \omega$, it is necessary to calculate 30 terms of the series at least. Therefore, *it is not convergent for practical uses*.[31]

Euler observed that the function can be approximately integrated by evaluating the area under the curve $(1 - gcos\omega)^{-\mu}$; however, this technique cannot be practised when the eccentricity of the orbits is to be taken into account.

The problem arises from the fact that the series contains powers of $cos\omega$; however, one can remedy this disadvantage by reducing the powers of the cosine of the angle ω, to the cosines of multiples of the angle. Since

$$2cos^2\omega = cos2\omega + 1,$$

$$4cos^3\omega = cos3\omega + 3cos\omega,$$

$$8cos^4\omega = cos4\omega + 4cos2\omega + 3, \ldots,$$

Euler stated that $(1 - gcos\omega)^{-\mu}$ has an expansion of the type

$$\sum_{i=0}^{\infty} a_i \cos i\omega$$

To calculate the coefficients a_i, Euler first expresses these coefficients by means of the following infinite series

$$\begin{aligned} a_0 = 1 &+ \frac{2}{4}\frac{\mu(\mu+1)}{1\cdot 2}g^2 + \frac{4\cdot 3}{4\cdot 8}\frac{\mu(\mu+1)(\mu+2)(\mu+3)}{1\cdot 2\cdot 3\cdot 4}g^4 \\ &+ \frac{6\cdot 5\cdot 4}{4\cdot 8\cdot 12}\frac{\pi(\mu+1)\ldots(\mu+5)}{1\cdot 2\cdot 3\cdot\ldots\cdot 6}g^6 \\ +\cdots \end{aligned} \tag{9}$$

$$a_1 = \frac{\mu}{1}g\left(1 + \frac{3}{4}\frac{(\mu+1)(\mu+2)}{2\cdot 3}g^2 + \frac{5\cdot 4}{4\cdot 8}\frac{(\mu+1)(\mu+2)(\mu+3)(\mu+4)}{2\cdot 3\cdot 4\cdot 5}g^4 + \ldots\right) \tag{10}$$

However, Euler thinks that the determination of the approximate values of the coefficients by means of infinite series is not appropriate. Thus he tries to find a recursive formula for the coefficients a_i ($i > 1$). He sets

$$s = \sum_{i=0}^{\infty} a_i \cos i\omega = (1 - g\cos\omega)^{-\mu}$$

[30] "Ne se peut résoudre dans une suite convergente, vû que la valeur de g est environ = 4/5 [...] tant parce qu'elle n'est pas assez convergente." (Euler 1749b, p. 60).

[31] On the eighteenth-century theory of series see (Ferraro 2007b, 2008b).

and calculates the differential

$$\frac{ds}{s} = \frac{-\mu g \sin \omega d\omega}{1 - g \cos \omega} \text{of } log\ s$$

Hence,

$$\sum_{i=0}^{\infty} (ia_i \sin i\omega - iga_i \cos \omega \sin i\omega - \mu ga_i \sin \omega \cos i\omega) = 0 \tag{11}$$

Since $2cosn\omega\ sin\omega = sin(n + 1)\ \omega\text{-}sin(n - 1)\ \omega$, Eq. (11) can be written in the form

$$\sum_{i=1}^{\infty} \left(-ia_i + \frac{g(\mu + i - 1)a_{i-1}}{2} + \frac{g(1 + i - \mu)a_{i+1}}{2} \right) \sin i\omega = 0$$

Using the method of indeterminate coefficients Euler obtains.

$$a_{i+1} = \frac{2ia_i - g(\mu + i - 1)a_{i-1}}{g(1 + i - \mu)} \text{ for } i \geq 1$$

Euler has still to determine the coefficients a_0 and a_1. He might calculate them by means of (9) and (10), but he prefers to use the approximate formulas:

$$a_0 = \frac{1}{2k} \sum_{i=0}^{k-1} \left(\left(1 - g \sin \frac{2i+1}{2k} \frac{\pi}{2} \right)^{-\mu} + \left(1 + g \sin \frac{2i+1}{2k} \frac{\pi}{2} \right)^{-\mu} \right) k = 1, 2, \ldots \tag{12}$$

$$\frac{1}{2} a_1 = \frac{1}{2k} \sum_{i=0}^{k-1} \left(\sin \frac{2i+1}{2k} \frac{\pi}{2} \left(1 - g \sin \frac{2i+1}{2k} \frac{\pi}{2} \right)^{-\mu} - \sin \frac{2i+1}{2k} \frac{\pi}{2} \left(1 + g \sin \frac{2i+1}{2k} \frac{\pi}{2} \right)^{-\mu} \right) k = 1, 2, \ldots \tag{13}$$

These approximations can be derived by observing that[32]

$$(1 - g \sin \omega)^{-\mu} + (1 + g \sin \omega)^{-\mu} = 2 \sum_{i=0}^{\infty} (-1)^i a_{2i} \cos 2i\omega \tag{14}$$

If we set $\omega = \frac{\pi}{4} = \frac{1}{2}\frac{\pi}{2}$, the value of (14) is

$$(1 - g \sin \frac{\pi}{4})^{-\mu} + (1 + g \sin \frac{\pi}{4})^{-\mu} = 2 \sum_{i=0}^{\infty} (-1)^i a_{4i}$$

---

[32] By replacing by into $(1 - g \cos \omega)^{-\mu} = \sum_{i=0}^{\infty} a_i \cos i\omega$, one has $1 - g \cos(\omega + \pi))^{-\mu} = (1 + g \cos \omega)^{-\mu} = \sum_{i=0}^{\infty} (-1)^i a_i \cos i\omega$. It follows that $(1 - g \cos \omega)^{-\mu} + (1 + g \cos \omega)^{-\mu} = 2 \sum_{i=0}^{\infty} a_{2i} \cos 2i\omega$. By replacing by one has Eq. (14).

Hence

$$a_0 = \frac{1}{2}\left(\left(1 - g\sin\frac{\pi}{4}\right)^{-\mu} + \left(1 + g\sin\frac{\pi}{4}\right)^{-\mu}\right) - \sum_{i=1}^{\infty}(-1)^i a_{4i} \qquad (15)$$

If we neglect the series $\sum_{i=1}^{\infty}(-1)^i a_{4i}$ in Eq. (15), we obtain the first approximation for a_0:

$$a_0 = \frac{1}{2}\left((1 - g\sin\frac{\pi}{4})^{-\mu} + (1 + g\sin\frac{\pi}{4})^{-\mu}\right)$$

The error is about equal to a_4. Now let us evaluate (14) for $\omega = \frac{\pi}{8} = \frac{1}{4}\frac{\pi}{2}$ and $\omega = \frac{3\pi}{8} = \frac{3}{4}\frac{\pi}{2}$. By summing the two obtained values, we have:

$$a_0 = \frac{1}{4}\left(\begin{array}{l}\left(1 - g\sin\frac{1}{4}\frac{\pi}{2}\right)^{-\mu} + \left(1 - g\sin\frac{3}{4}\frac{\pi}{2}\right)^{-\mu} \\ +\left(1 + g\sin\frac{1}{4}\frac{\pi}{2}\right)^{-\mu} + \left(1 + g\sin\frac{3}{4}\frac{\pi}{2}\right)^{-\mu}\end{array}\right) - \sum_{i=1}^{\infty}(-1)^i a_{8i}$$

If we neglect the series $\sum_{i=1}^{\infty}(-1)^i a_{8i}$ we obtain the second approximation for a_0, with an error about equal to a_8. This procedure can be repeated by taking more small subdivisions of π/2 and so obtaining improved approximations, which provide Eq. (12).

In a similar way, one obtains Eq. (13) for a_1. By applying this procedure and using the values 0.8405 and 3/2 for g and μ, (Euler 1749b, p. 72) obtains the following values[33]

$$a_0 = 3,21789,\ a_1 = 4,70357,\ a_2 = 3,07731 a_3 = 3,92413 \qquad (16)$$

$$a_4 = 3,18601,\ a_5 = 3,75144$$

Euler is therefore able to calculate the coefficients of trigonometric series. Thus, (Euler 1749b, p. 66) writes Eqs. (8_{I}) and (8_{II}) in the form

$$m^2 d\zeta + 2mndx + m^2 nrd\zeta - n\frac{d^2r}{d\zeta} = \frac{1+v}{\lambda^3}d\zeta - \frac{2nr}{\lambda^3}d\zeta + \frac{n\cos\omega}{\lambda(1-m)}d\omega$$
$$+\frac{nd\omega\cos\omega}{\lambda(1-m)} + \frac{nd\omega}{h(1-m)}(\lambda - \cos\omega)\sum_{j=0}^{\infty}a_j\cos j\omega \qquad (17_{\mathrm{I}})$$

[33] Euler did not say what value of k he uses in his calculation; but it is probably that he takes $k = 5$. L. A. Golland and R. W. Golland (Golland and Golland 1993, p. 62) have calculated the value of a_0 as 3.217598 (following Euler's formula and taking $k = 10$). Using the integral $\frac{1}{2\pi}\int_{-\pi}^{\pi}(1 - g\cos\omega)^{-3/2}d\omega$, the value of a_0 is 3.217598.

$$2mdr + \frac{d^2x}{d\zeta} = -\frac{d\omega \sin\omega}{\lambda(1-m)} + \frac{d\omega \sin\omega}{h(1-m)} \sum_{j=0}^{\infty} a_j \cos j\omega \qquad (17_{\mathrm{II}})$$

To integrate this system, Euler again applies the formula.

$$2\cos n\omega \sin\omega = \sin(n+1)\omega - \sin(n-1)\omega$$

The second equation becomes

$$2mdr + \frac{d^2x}{d\zeta} = -\frac{d\omega \sin\omega}{\lambda(1-m)} + \frac{d\omega}{2h(1-m)}\left((2a_0 - a_3)\sin\omega + \sum_{j=2}^{\infty}(a_j - a_{j+2})\sin j\omega\right)$$

By integrating term-by-term, he obtains[34]

$$2mr + \frac{dx}{d\zeta} = \frac{\cos\omega}{\lambda(1-m)} - \frac{1}{2h(1-m)}\left((2a_0 - a_3)\cos\omega + \sum_{j=2}^{\infty}\frac{1}{j}(a_j - a_{j+2})\cos j\omega\right) \qquad (18)$$

Euler states that, the quantity $\frac{dx}{d\zeta}$ in Eq. (18) is finite (namely, it does not contain differentials). By replacing it into Eq. (17I), he finds the following expression for r:

$$r = \sum_{i=1}^{\infty} C_i \cos i\omega \qquad (19)$$

where

$$C_i = \frac{\mu^2 B_i}{m^2(i^2 - \mu^2)},\ B_i = \frac{\mu(2a_0 - a_2)}{ih} + \frac{2\lambda a_1 - 2a_0 - a_2}{2h} - \frac{2\mu - 1}{\lambda},$$
$$B_i = \frac{\mu(2a_{i-1} - a_{i+1})}{ih} + \frac{2\lambda a_i - a_{i-1} - a_{i+1}}{2h}(i = 2, 3, \ldots)$$

By substituting (19) into (18) and integrating, Euler derives

$$x = \left(\frac{\mu^2 - 2\mu C_1}{\lambda m^2} - \frac{\mu^2(2a_0 - a_2)}{2m^2 h}\right)\sin\omega + \sum_{i=2}^{\infty}\left(-\frac{2}{i}\mu C_i - \frac{\mu^2(a_{i-1} - a_{i-1})}{2i^2 m^2 h}\right)\sin i\omega, \qquad (20)$$

where $\mu = \frac{m}{1-m}$. Using (16), (19), and (20) Euler is able to calculate the shortened distance z and the longitude φ of Saturn:[35]

$$z = f(1 + 0.0008542\cos\omega + 0.0001449\cos 2\omega + 0.0000335\cos 3\omega + 0.0000105\cos 4\omega + \ldots)$$

---

[34] Euler asserted that there is no constant of integration since x and r depend solely on and did not contain a constant part (Euler 1749b, p. 67).

[35] η is the mean longitude of Saturn.

$$\varphi = \eta + 4'' \sin\omega - 32'' \sin 2\omega - 7'' \sin 3\omega - 2'' sin4\omega - ...$$

In the following parts of his memoir, Euler improves the above result by considering first Saturn's eccentricity, then, Jupiter's one. In both the cases, he however supposes that the orbit of the other planet is circular (he also neglects Saturn's attraction on Jupiter). After a long sequel of calculations, he finds that the following values for z and φ[36]:

$$\frac{z}{f} = 1 + 0.05700372 \cos q + 0.0008542 \cos\omega + 0.0001449 \cos 2\omega + 0.0000335 \cos 3\omega + \dots$$
$$- 0.0000429 \cos(\omega + q) - 0.0003752 \cos(\omega - q) - 0.0000168 \cos p$$
$$+ 0.000030 \cos(\omega + q) - 0.000030 \cos(\omega - q) + 0.0000127 \cos(2\omega + p) +$$
$$0.0000050 p \sin(\omega - p) + 0.0005719 \cos(2\omega - p)$$
$$\varphi = \eta + (-23515'' sin\ q + 167'' \sin\ \rightarrow 2q + 4'' \sin\ \omega - 32'' \sin 2\omega - 7'' \sin 3\omega$$
$$- 2'' \sin 4\omega + 13'' \sin(\omega + q) - 257'' \sin(\omega - q) + 3'' \sin\ p - 3'' \sin p. - 3'' \sin(\omega + p)$$
$$- 3'' \sin(2\omega + p) - 0.00001 p\ \cos(\omega - p) - 3'' \sin(\omega - p) - 243'' \sin(2\omega - p))$$

Note the last formula contains the terms:

$$-257'' sin(\omega - q) - 243'' \sin(2\omega - p)$$

They are wrong because of sign on p. 80 (the correct values are $-182''$ and $413''$[37]). Apart this slip of the pen, Euler's calculations contains a serious problem. It is true that they allow one to obtain many terms in the longitude of Saturn that are correct; however, Euler neglects the terms that are proportional to the third dimension in the products and powers of the eccentricities (and inclinations—see below) of the two planets. These products and powers are small but, in reality, because of the presence of small divisors, they became observationally sizable (Wilson 2007, p. 130).

In his (Euler 1749b) Euler also investigates the motion of nodes and the variation in the inclination of Saturn's orbit with respect to Jupiter. In order to do this, the Swiss mathematician eliminates the hypothesis that Jupiter and Saturn are on the same plane and takes Eqs. (5_{III}) and (5_{IV}) into account (Euler 1749b, p. 102). Euler again assumes the orbits of both Saturn and Jupiter be circular (Euler 1749b, p. 104). Since ψ is smaller than $1°15'20''$, Euler (Euler 1749b, p. 103) sets $cos\psi \approx 1$ and

$$\text{v} = \sqrt{y^2 + z^2 / \cos^2\psi - 2yz\cos\omega} = \sqrt{y^2 + z^2 - 2yz\cos\omega}$$

Under these hypotheses, Euler writes Eqs. (5_{III}) and (5_{IV}) in the form:

$$d\pi = \frac{n}{mh} d\zeta \sin(\varphi - \pi) \sin(\theta - \pi) \left(\alpha - \sum_{j=1}^{\infty} a_j \cos j\omega\right)$$
$$d(\ln\tan\rho) = \frac{n}{mh} d\zeta \cos(\varphi - \pi) \sin(\theta - \pi) \left(\alpha - \sum_{j=1}^{\infty} a_j \cos j\omega\right)$$

---

[36] P and q are the eccentric anomalies of Jupiter and Saturn, respectively.

[37] These agree with Laplace's values (Cfr. Wilson 1995, p. 100).

where $\frac{h}{\lambda} - a_0 = \alpha$. By neglecting several terms, Euler (Euler 1749b, pp. 105–106) finds:

$$\pi = \mathrm{C} - \frac{n}{4m^2h}a_1\varphi + \frac{n}{mh}\left(\frac{\alpha}{2(1-m)}\sin\omega - \frac{\alpha}{2(1+m)}\sin(\theta+\varphi-2\pi) - \frac{a_1}{8(1-m)}\sin 2\omega + \frac{a_1}{8m}\sin 2(\varphi-\pi) + \frac{a_1}{8}\sin 2(\theta-\pi)\right) \qquad (21_{\mathrm{I}})$$

$$\mathrm{In}\tan\rho = \mathrm{D} - \frac{n}{mh}\left(\frac{\alpha}{2(1-m)}\cos\omega + \frac{\alpha}{2(1+m)}\cos(\theta+\varphi-2\pi) - \frac{a_1}{8(1-m)}\cos 2\omega + \frac{a_1}{8m}\cos 2(\varphi-\pi) + \frac{a_1}{8}\cos 2(\theta-\pi)\right) \qquad (21_{\mathrm{II}})$$

where C and D are constants of integrations. By replacing the numerical values into Eqs. (2_{II}) and (2_{III}), he obtains

$$\pi = Const. - 0.0004066\varphi + 141\text{''}sin\omega - 28\text{''}sin\ 2\omega + 42\text{''}\sin\ 2(\varphi-\pi) + 17\text{''}\cos 2(\theta-\pi) - 60\text{''}\sin(\theta+\varphi-2\pi)$$

$$\mathrm{In}\tan\rho = Const. - 141\text{''}\cos\ \omega + 28\text{''}\cos\ 2\omega - 42\text{''}\cos\ 2(\varphi-\pi) - 17\text{''}\cos 2(\theta-\pi) - 60\text{''}\cos(\theta+\varphi-2\pi)$$

These equations imply that the longitude of nodes of Saturn is variable (Euler 1749b, p. 106).

In the years that followed, Euler dealt with the motion of celestial bodies in other papers.[38] I limit myself to noting that he realized that the node of any planet perturbed by another planet retrogresses on the orbital plane of the perturbing planet and undertook a systematic investigation of the *motion of the Ecliptic* in a memoir published in 1756 (Euler 1754).[39]

5 Conclusion

Around 1750 Euler doubted the strict validity of the inverse-square law, however, in *Recherches sur la question des inégalités du mouvement de Saturne et de Jupiter* (Euler 1749b), he assumed the strict accuracy of this law and developed analytical methods for investigating the perturbation of Saturn due to Jupiter following from that assumption. The memoir was submitted in the contest of Paris Academy's prize for the 1748. Despite (Euler 1749b) failed to account for the motion of Saturn, the memoir was highly innovative and was awarded the prize.

[38] As concerns planetary perturbations, e.g., Euler see Euler 1753, 1758, 1766, 1767, 1768a.

[39] On the motion of Ecliptic, see Wilson 2007, 130, 1992.

The methodology of *Recherches sur la question des inégalités du mouvement de Saturne et de Jupiter* was the result of an evolution which had had a crucial step with the publication of *Mechanica* (Euler 1736), where Euler applied analytical methods to mechanics in a systematic way and criticized Newton's geometric methodology. Later, to improve the generality of procedures, Euler introduced a new approach, which was published, first, in *Recherches sur le mouvement des corps célestes en général* (Euler 1749a) and, then, in *Découverte d'un nouveau principe de Mécanique* (Euler 1750). The novelty of new approach consisted of using an extrinsic references frame and considering equations $F_x = ma_x$, $F_y = ma_y$, $F_z = ma_z$. as the bases of mechanics. Basing on this approach, Euler determined the differential equations of Saturn's motion by which he coped with Saturn's perturbations.

References

AA.VV (1746) Procès-verbaux Académie royale des Sciences. Vol. 65. Retrieved via: https://gallicabnffr/ark:/12148/bpt6k55741qf89, last accessed 2021/10/21

Bernoulli J (1730) Nouvelles pensées sur le système de M Descartes et la manière d'en déduire les Orbites & les Aphélies des Planètes. Claude Jombert, Paris. In: Bernoulli J Opera, III, pp. 132–173.

Bernoulli D (1734) Recherches physiques et astronomiques. Pièces qui ont remporté le prix de l'Académie Royale des Sciences 3:93–122.

Bernoulli J (1735) Essai d'une nouvelle physique céleste Servant à expliquer les principaux Phénomènes du Ciel, & en particulier la cause physique de l'inclinaison des Orbites des Planètes par rapport au plan de l'Equateur du Soleil, G Martin et al, Paris. In Bernoulli J, Opera, III, pp. 261–364.

Bernoulli D (1741) Traité sur le flux et reflux de la mer. Pièces qui ont remporté le prix de l'Académie Royale des Sciences 4:56–57.

Bernoulli J (1742) Opera Omnia, Marc-Michel Bousquet. Lausannae & Genevae.

Bertoloni Meli D (1993) Equivalence and Priority: Newton Versus Leibniz, Including Leibniz's Unpublished Manuscripts on the "Principia". Oxford University Press, Oxford.

Blay M (1992) La naissance de la mécanique analytique : La science du mouvement au tournant des xviie et xviiie siecles. PUF, Paris.

Blay M ([1993] 1998) Reasoning with the infinite From the Closed World to the Mathematical Universe. University of Chicago Press, Chicago.

Clairaut AC (1745a) Du système du monde dans les principes de la gravitation universelle. Mémoires de l'Académie Royale des Sciences, pp. 329–364.

Clairaut AC (1745b) Avertissement de M Clairaut, au sujet des Mémoires qu'il a donnez en 1747 et 1748, sur le système du monde dans les principes de l'Attraction. Mémoires de l'Académie Royale des Sciences, pp. 577–578.

Clairaut AC (1752) Théorie de la Lune déduite du seul principe de l'attraction réciproquement proportionnelle aux quarrés des distances, Pièce qui a remporté le prix de l'Académie impériale des sciences de Saint-Pétersbourg en 1750, de l'Imprimerie de l'Académie Impériale des Science, Saint-Pétersbourg.

Dear P (2005) What Is the History of Science the History of Early Modern Roots of the Ideology of Modern Science. Isis 96:390–406.

Dhombres J (2002) The Mathematics Implied in the Laws of Nature and Realism, or the Role of Functions around 1750 In: Cerrai P, Freguglia P, Pellegrini C (eds). The Application of Mathematics to the Sciences of Nature: Critical Moments and Aspects. Kluwer, New York, 2002, pp. 207–222.

Euler L (1736) Mechanica sive motus scientia analytice exposita. I, ex typ Academiae Scientiarum, Petropoli. In Euler, Opera Omnia, I, ser 2.

Euler L (1738) Dissertatio de igne, in qua eius natura et proprietates explicantur. Pièces qui ont remporté le prix de l'académie royal des sciences de Paris, pp. 1–19.

Euler L (1740) Inquisitio physica in causam fluxus ac refluxus mari. Pièces qui ont remporté le prix de l'académie royale des sciences de Paris, pp. 235–350.

Euler L (1744) Theoria motuum planetarum et cometarum Continens methodum facilem ex aliquot observationibus orbitas cum planetarum tum cometarum determinandi Una cum calculo, quo cometae, qui annis 1680 et 1681 itemque ejus, qui nuper est visus, motus verus investigatur, sumtibus Ambrosii Haude, Berolini In Euler, Opera Omnia, 28, ser 2, pp. 105–251.

Euler L (1746) Tabulae astronomicae solis et lunae, in Opuscula varii argumenti, Berolini, Sumtibus A Haude & JC Speneri, I, pp. 137–168. In Euler, Opera Omnia, 23, ser 2, pp. 1–10.

Euler L (1749a) Recherches sur le mouvement des corps célestes en général. Mémoires de l'Académie des sciences de Berlin 3:93–143.

Euler L (1749b) Recherches sur la question des inégalités du mouvement de Saturne et de Jupiter, sujet proposé pour le prix de l'année 1748, par l'académie royale des sciences de Paris Pièce qui a remporté le prix de l'Académie Royale des Sciences, Martin, Coignard et Guerin, Paris. In Euler Opera Omnia, 25, ser 2, pp. 45–157.

Euler L (1750) Découverte d'un nouveau principe de Mécanique, Mémoires de l'académie des sciences de Berlin 6:185–217. In Euler, Opera Omnia, 5, ser 2, pp. 81–108.

Euler L (1753) De perturbatione motus planetarum ab eorum figura non sphaerica oriunda, Novi Commentarii academiae scientiarum Petropolitanae 3:235–253. In Euler, Opera Omnia, 25, ser 2, pp. 158–174.

Euler L (1754) De la variation de la latitude des étoiles fixes et de l'obliquité de l'écliptique, Mémoires de l'académie des sciences de Berlin 10:296–336. In Euler, Opera Omnia, 29, ser 2, pp. 125–159.

Euler L (1758) De motu corporum coelestium a viribus quibuscunque perturbato, Novi Commentarii academiae scientiarum Petropolitanae 4:161–196. In Euler, Opera Omnia, 25, ser 2, pp. 175–209.

Euler L (1766) Considerationes de motu corporum coelestium, Novi Commentarii academiae scientiarum Petropolitanae 10:544–558. In Euler, Opera Omnia, 25, ser 2, pp. 246–257.

Euler L (1767) De motu rectilineo trium corporum se mutuo attrahentium, Novi Commentarii academiae scientiarum Petropolitanae 11:144–151 In: Euler, Opera Omnia, 25, ser 2, pp. 281–289.

Euler L (1768a) Methodus facilis motus corporum coelestium utcunque perturbatos ad rationem calculi astronomici revocandi, Novi Commentarii academiae scientiarum Petropolitanae, 12:129–165. In Euler, Opera Omnia, 25, ser 2, pp. 305–331.

Euler L (1768b) Lettres à une princesse d'Allemagne sur divers sujets de physique & de philosophie, Imprimerie de l'Académie Impériale des Sciences, I. Saint-Pétersbourg.

Euler L (1833) Letters of Euler on Different Subjects in Natural Philosophy addressed to a German Princess. I. Harper, New York.

Euler L (1911–) Opera omnia. Birkhäuser, Basel.

Euler L (1980) Correspondance de Leonhard Euler avec AC Clairaut, J d'Alembert et JL Lagrange. Juskevic AP, Taton R (eds). Birkhäuser, Basel.

Ferraro G (2007a) Euler's treatises on infinitesimal analysis: Introductio in analysin infinitorum, Institutiones calculi differentialis, Institutionum calculi integralis. In: Baker R (ed) Euler Reconsidered Tercentenary Essays. Kendrick Press, Heber City, pp. 39–101.

Ferraro G (2007b) Convergence and formal manipulation in the theory of series from 1730 to 1815. Historia Mathematica 34:62-88.

Ferraro G (2008a) D'Alembert visto da Eulero. Bollettino di Storia delle Scienze Matematiche 28:257-275.
Ferraro G (2008b) The Rise and Development of the Theory of Series up to the Early 1820s. Springer, New York.
Ferraro G (2010) Mathematics and Natural Philosophy in Euler's Investigation of Saturn's Perturbations. In: Badolati E (ed) First International Meeting on Cultural Astronomy. Loffredo, Napoli, pp. 125–157.
Ferraro G (2012) Mathematical facets of Euler's astronomical works: In: Badolati E (ed). Proceedings of the third International Meeting on Cultural Astronomy. Libellula Edizioni, Tricase, pp. 49–75.
Ferraro G (2020) Euler and the structure of mathematics. Historia Mathematica 50:2-24.
Gaukroger S (1982) The Metaphysics of Impenetrability: Euler's Conception of Force. The British Journal for the History of Science 15:132–154.
Gingras Y (2001) What did Mathematics do to physics? History of Science 39:383-416.
Golland, LA, Golland RW (1993) Euler's troublesome series: An early example of the use of trigonometric series. Historia Mathematica 20:54–67.
Hankins TL (1970) Jean d'Alembert Science and the Enlightenment. Clarendon Press, Oxford.
Hermann J (1716) Phoronomia, sive De viribus et motibus corporum solidorum et fluidorum Wetstenios, Amstelædami.
Juškevič A, Taton R (1980) Introduction to Correpondance d'Euler avec J D'Alembert. In: Euler, Opera Omnia, 5, ser IVA, pp. 1–63.
Leclerc GL Buffon comte de (1745a) Réflexions sur la loi de l'attraction. Histoire de l'Académie royale des sciences avec les mémoires de mathématique et de physique, pp. 493–500.
Leclerc GL Buffon comte de (1745b) Second addition au Mémoire qui a pour titre: Réflexions sur la loi de l'attraction. Histoire de l'Académie royale des sciences avec les mémoires de mathématique et de physique, pp. 580–583.
Leibniz GW (1689a) Schediasma de resistentia medii et motu projectorum gravium in medio resistante. Acta Eruditorum 1689a, pp. 38–47 [See also: Leibniz GW (1860) Mathematische Schriften, VI, Halle, pp.135–144].
Leibniz GW (1689b) Tentamen de Motuum Coelestium Causis, Acta Eruditorum 1689, pp. 82–96 [See also: Leibniz GW Mathematische Schriften, VI, pp. 144–160].
Lyon J, Sloan PR (1981) From natural history to the history of nature. University of Notre Dame Press, Notre Dame and London.
Maheu G (1966) La vie scientifique au milieu du XVIIIe siècle Introduction à la publication des lettres de Bouguer à Euler. Revue d'histoire des sciences et de leurs applications19:206–224.
Maltese G (2000) On the Relativity of motion in Leonhard Euler's Science. Archive for History of Exact Sciences 54:319-348.
Maltese G (2002) On the Changing Fortune of the Newtonian Tradition in Mechanics In: Williams K (ed). Two Cultures Essays in Honour of David Speiser. Birkhäuser, Basel, pp. 97–113.
Maronne S, Panza M (2014) Euler, reader of Newton: mechanics and algebraic analysis. Advances in Historical Studies 3:12–21.
Newton I (1687) Philosophiae Naturalis Principia Mathematica, Londini, Josephi Streater.
Newton I (1713) Philosophiæ naturalis principia mathematica Auctore Isaaco Newtono, Equite Aurato Editio secunda auctior et emendatior Cantabrigiæ.
Panza M (2002) Mathematisation of the science of motion and the birth of analytical mechanics A historiographical note. In: Cerrai P, Freguglia P, Pellegrini C (eds) The Application of Mathematics to the Sciences of Nature: Critical Moments and Aspects. Kluwer, New York, pp. 253–271.
Robins B (1761) Remarks on Mr Euler's Treatise of Motion in Mathematical Tracts of the late Robins. Benjamin WJ (ed). Nourse, London, pp. 197–221.
Schroeder p. (2007) La loi de la gravitation universelle Newton, Euler et Laplace. Le cheminement d'une révolution scientifique vers une science normale. Springer Verlag, Paris.

Shea WR (1987) The quest for Scientific rationality: Some historical considerations. In: Pitt J, Pera M (eds) Rational Changes in Science Essays on Scientific Reasoning. Reidel, Boston, pp. 155–176.
Truesdell CA (1960) The rational mechanics of flexible or elastic bodies 1638–1788. Introduction to Euler L Opera omnia, Vols X-XI, ser. 2, Orell Füslli, Zürich.
Varignon p. (1700a) Manière générale de déterminer les forces, les vitesses, les espaces et les temps, une seule de ces quatre choses étant donnée dans toutes sortes de mouvements rectilignes variés à discrétion, Mémoires de mathématique et de physique de l'Académie Royale des Sciences, pp. 22–27.
Varignon p. (1700b) Du mouvement en générale par toute sortes de courbes; et des forces, tant centrifuges que centripètes, nécessaires auc cops qui les décrivent, Mémoires de mathématique et de physique de l'Académie Royale des Sciences, 83–101.
Varignon p. (1707) Des Mouvemens variés à volonté, comparés entr'eux & avec les uniformes, Histoire de l'Académie royale des sciences avec les mémoires de mathématique et de physique pour la même année tirée des registres de cette académie, Mémoires de l'Académie Royale des Sciences, pp. 222–276.
Verdun A (2013) Leonhard Euler's early lunar theories 1725–1752. Archive for Histo-ry of Exact Sciences 67:235–303, 477–551.
Waff CB (1975) Universal gravitation and the motion of the moon's apogee: the establishment and reception of Newton's inverse-square law, 1687–1749. PhD Dissertation. Johns Hopkins University, Baltimore.
Waff CB (1992) Clairaut and the motion of the lunar apse: the inverse-square law undergoes a test. In: Harman PM, Shapiro AE (eds) The Investigation of Difficult Things. University Press, Cambridge, pp. 35–46.
Wepster SA (2010) Between theory and observations Tobias Mayer's explorations of lunar motion, 1751–1755. Springer, New York.
Wilson C (1985) The great inequality of Jupiter and Saturn – from Kepler to Laplace. Archive for History of Exact Sciences 33:15-290.
Wilson C (1992) Euler on action-at-a-distance and Fundamental Equations in Continuum mechanics. In: Harman PM, Shapiro AE (eds) The Investigation of Difficult Things. University Press, Cambridge, pp. 399–420.
Wilson C (1995) The problem of perturbation analytically treated: Euler, Clairaut, d'Alembert. In: Taton R, Wilson C (eds) Planetary Astronomy from the Renaissance to the Rise of Astrophysics Part B: The eighteenth and nineteenth centuries. University Press, Cambridge, pp. 89–107.
Wilson C (2007) Euler and Applications of Analytical Mathematics to Astronomy Bradley RE, Sandifer CE (eds). Leonhard Euler: life, work and legacy. Elsevier, New York, pp. 121–145.

Chapter 24
Physics Between the Wars: Mobility and Exchange, Lisbon Physics Laboratories and Institutions Beyond the Iberian Peninsula

Quintino Lopes

Abstract This paper provides a contribution to the history of physics in the period between the two world wars, examining relations between the physics laboratories of the University of Lisbon Faculty of Science and the Instituto Superior Técnico in Lisbon on the one hand and the Laboratoire Curie, the Collège de France and other prestigious science institutions on the other hand. The location of the two Lisbon laboratories on the so-called European periphery did not mean that they did not produce knowledge. While the latter laboratory was equipped with the most advanced scientific instruments of the era, such as a spectrograph and a Zeiss microphotometer, at the former a research school with international standing was established from 1929 to 1947. Analysis is carried out in accordance with the historiographical methodology known as 'following the money trail', involving the study of funding, conducted in depth, from an impartial standpoint, which has the potential for producing new findings and leading to the reanalysis of established interpretations. In this paper, we examine the funding granted to these two Portuguese physics institutions by the Junta de Educação Nacional (National Education Board), established in 1929, which enabled Portugal to play a part in the European and American movement for the creation of scientific planning and funding institutions following World War I. This focus on the support provided by the dictatorial regime in Portugal to the two Lisbon laboratories also reveals the existence of resentment among scientists and resistance within these institutions, which limited the findings that they produced.

Keywords Funding for science · Physics laboratories · Scientific centres and peripheries · Circulation of knowledge · International science networks · Resistance in the field of science

Q. Lopes (✉)
Institute of Contemporary History, NOVA University of Lisbon, University of Évora Unit, Lisbon, Portugal
e-mail: quintinolopes1@gmail.com

R. Pisano (ed.), *A History of Physics: Phenomena, Ideas and Mechanisms*, History of Mechanism and Machine Science 42,
https://doi.org/10.1007/978-3-031-26174-9_24

1 Introduction

In this paper, we analyse research in the field of physics during the period between the wars, examining the work of the first Portuguese institution which organised and funded science, the Junta de Educação Nacional (JEN)—National Education Board (1929–1936) (Lopes 2017a). We focus on the support that this state institution, and its successor, the Instituto para a Alta Cultura (IAC)—Institute for High Culture (1936–1952), provided to the physics laboratories of the Instituto Superior Técnico (IST)—Higher Technical Institute—and the Faculdade de Ciências de Lisboa (FCL)—University of Lisbon Faculty of Science. We assess how such support fostered the circulation of both knowledge and physicists between Portugal and Europe beyond the Pyrenees, and gauge its impact on the development of this field of knowledge. We also assess whether during the period under analysis research practices introduced to Portugal by physicists who had trained abroad gave rise to support or resistance to such within these institutions.

Exchanges with the Laboratoire Curie and the Collège de France are highlighted, not from a traditional perspective, according to which scientific peripheries are regarded as playing a passive role as being exclusively receptive of ideas originating at the centre, but from a perspective of the circulation of both knowledge and physicists in a Europe in which, despite the presence of nationalist tendencies, authoritarian political regimes and autarchy, international science networks were established during the period between the wars (Schroeder-Gudehus 2020; Crawford 1992; Roberts 2009).

2 The National Education Board: Science Planning and Funding in Portugal between the Wars

The outbreak of World War I had a profound impact on scientific activity at the international level, notably with the mobilisation of academia to aid the war effort. The profound scientific and technological nature of the conflict aroused the interest of the military authorities to the potential of specialists, particularly physicists, chemists, engineers, doctors, veterinarians, geologists and geographers for providing a contribution in this sphere. Associated with the academic world, they served as professors, researchers and technicians, and were now expected to respond to military needs through their work in the laboratory (Salgueiro 2015, pp. 83–86; Crawford 1988, pp. 163–201; Macleod 1993, pp. 455–481).

The predominant role of science in war-making and the consequent perception of its influence on the realignment of states in the post-war period led to the creation of numerous bodies, usually as part of state systems whose aims were to plan and promote scientific research in the countries in which they operated, coordinating this with economic development (Edgerton 2013, pp. 759–776; Vanpaemel 2013, pp. 33–48; Hull 1999). Meanwhile, science was also funded by organisations in the private sector (Siegmund-Schultze 2001).

One of the combatants in the war, Portugal enlisted the support of many in society, including the mobilisation of students, lecturers and professors, which had a marked impact on the academic world, while there was also a scarcity of funds for university science laboratories and institutes, which stunted their development and impacted their work. At the same time, the Portuguese authorities sporadically resorted to the use of science facilities to try to resolve issues associated with military aeronautics and armaments, and there were also developments in knowledge in fields such as surgery and neurology (Salgueiro 2015, pp. 81–92).

The growing recognition of the importance of the role of scientists in meeting the needs arising from the war, associated with the difficulties experienced by Portuguese universities in the context of World War I, explains how the reform of higher education of 1918 came about, which enhanced the role of universities in the field of science and led to a proposal for setting up a state body to promote scientific research and teaching reform (Salgueiro 2015, pp. 92–100). In the event, the reform proposal was shelved, and the JEN (1929–1936) only came into existence 11 years later. It should be noted that the initial period of its operation coincided with the Great Depression, which had a considerable impact in financial and political terms, while the Portuguese government was faced with widespread opposition, particularly in the academic sector. Thus, the creation of this innovative institution may have represented an attempt to appease critics at the universities, and for the first time meant that there was a specific state body whose role was to plan the organisation and development of scientific research in Portugal (Fitas and Videira 2004, pp. 34–35).

It is notable that in Portugal a series of different political regimes operated in the early to mid-twentieth century, ranging from the Monarchy (up until 1910) to the First Republic (1910–1926), the Military Dictatorship (1926–1933) and the Estado Novo—New State (1933–1974), the last of these being an authoritarian, nationalist, conservative regime. Thus, although the coming of the Military Dictatorship and the New State marked the establishment of a new form of political organisation in contrast with the First Republic, the creation of the JEN by the Military Dictatorship, and its subsequent maintenance by the Estado Novo, meant that in terms of science policy, the aspirations of the First Republic came to fruition during the dictatorship. It should also be noted that the JEN was strongly influenced by the operational model of two similar international institutions: Spain's Junta para Ampliación de Estudios e Investigaciones Científicas (1907–1939), whose main objective was the Europeanisation of science and teaching in Spain, and Belgium's Fonds National de la Recherche Scientifique, set up in 1928, which played a vital role in the development and internationalisation of science in the post-World War I period (Lopes 2017b; Sánchez Ron 2007, pp. 102–113; Halleux and Xhayet 2007).

Influenced by these institutions, about whose work the leaders of the JEN were keenly aware, especially Augusto Celestino da Costa, president of the board from 1934, and Luís Simões Raposo, its first secretary, the board implemented a science policy based on the awarding of scholarships abroad to Portuguese researchers, aimed at enabling them to receive specialist training in the most advanced research methods and techniques of the time. These long-term scholarship holders, who spent an average of thirteen months abroad, mainly chose to study in France, Germany and England, followed by Spain, Belgium, Switzerland, Italy, Austria, Sweden, USA, Yugoslavia, Scotland, the Netherlands, Denmark and Poland. Having carried out such training abroad, researchers would then obtain scholarships in Portugal. With domestic study centres being provided with funding, these scientists were thus enabled to continue their studies in Portugal. Aiming at the internationalisation of scientific knowledge produced in Portugal, the JEN also funded scientific publications and encouraged cultural development and academic exchange, enabling the active participation of Portuguese researchers at international scientific meetings and the holding of congresses in Portugal. With regard to fields of knowledge to be funded, the board was guided in its work by the following policy: preference was to be shown for research in fields with a less established tradition of research or fields with a lower level of development in Portugal, combined with other criteria: priority for the pure sciences, followed by studies in applied science; the improvement of teaching methods at all levels of education; priority for research plans with defined, concrete themes, whose findings obtained abroad could be used or supplemented by means of continued research in Portugal; finally, in association with this criterion, without constituting a central concern, there was the promotion of the acquisition of knowledge and techniques for the economic development of the home country and the colonies (Lopes 2017b).

Bearing in mind that physics is the field of knowledge of interest in focus in this paper, it should be mentioned that the combination of the above-mentioned principles led to the following resolution being adopted: support for studies in the field of physics and chemistry, which Augusto Celestino da Costa advocated on several occasions, establishing pure sciences as a priority field, based on the assumption that providing support in this field would enhance the development of the applied sciences (Costa 1930, pp. 13–17, 1939, pp. 13–14). It should be noted that Costa was fully aware of how scientific and cultural institutions in Belgium operated, particularly the Fonds National de la Recherche Scientifique, whose establishment was closely associated with the speech made by King Albert I on 1st October 1927, in which he stressed that applied science is necessarily dependent on pure science (Halleux and Xhayet 2007, pp. 11–15; Wullus-Rudiger 1948). The influence of this body on the JEN in determining the importance conferred of scientific and technical-scientific disciplines is thus clear. However, Costa stressed the differences between physics and chemistry in Portugal, regarding the latter as a discipline with a more established tradition which, overall, enjoyed better facilities at faculties of science (Costa 1930, pp. 15–16). According to the first criterion stated above, this pointed to the need for a greater degree of support for physics. A comparative analysis of support provided by the board to physics laboratories (225,250 escudos) and chemistry laboratories (66,250 escudos) reveals how this principle was applied in practice (Lopes 2017b, p. 48).

3 The Microphotometer of the Instituto Superior Técnico Physics Laboratory: "A Marvel of Mechanical and Optical Construction"

The physics laboratory of the IST was one of the first laboratories of its kind in Portugal to receive substantial state funding in the early 1930s. As we shall see, this support, provided by the JEN, enabled the facility to be equipped to a standard similar to that of the most advanced physics centres of the time. In order to provide an understanding of how its head, António da Silveira, used these funds, we need to examine the career path he followed.

Funded by the board, Silveira carried out research on the Raman effect at the Collège de France in Paris from 1930 to 1932, publishing four articles in *Comptes Rendus de l'Académie des Sciences de Paris*. In August 1932 he applied for an extension to the scholarship, as in his opinion there were not the adequate facilities for the continuation of his studies in Portugal. As his internship came to an end, anticipating the establishment of such facilities on Portugal, he requested funding for continuing his work in Portugal.[1] While still at the Collège de France, he strategically advised the board of the large sum that would be required in order that full advantage might be gained from his proposed internship, describing the equipment which would be essential in order for his work to be carried out and its cost. Thus, the following are required:

(1) A spectrograph with a high degree of luminosity and great dispersive potential [...] perfectly adapted and custom-built for me by Mr. [Paul] Langevin [...] cost: 20,000 francs.

(2) Two quartz mercury vapour lamps [...] These are easier to obtain, although the two cost 2,000 francs.

(3) A good reflector [which] costs 300 francs.

(4) A collection of quartz or Pyrex tubes [...] The cost of these tubes is usually estimated at 40 francs each and as about 12 are required this means an approximate total of 500 francs.

(5) A microphotometer, an extremely expensive device that the laboratory did not have at the time I started there. [...] The cost of this equipment is about 60,000 francs. This marvel of mechanical and optical construction is produced at the Zeiss workshops. [...]

> I would like to stress that what has been spent on my work at the Collège de France covers studies of the Raman effect on liquids, solutions and gases in particular. [...] There were liquids [...] that cost upwards of 1,000 francs per kilo.[2]

---

[1] His application for a scholarship in Portugal was made on 21st October 1932 (Arquivo do Camões, Instituto da Cooperação e da Língua, I.P. Caixa 1339, Processo 7, Documentos 34 e 37).

[2] "Assim, com efeito, exige-se:

1º) Um espectrógrafo de grande luminosidade e de grande poder dispersivo [...] Um perfeitamente adaptado e mandado construir expressamente para mim pelo Sr. [Paul] Langevin [...] custou 20.000 fr.

2º) Duas lâmpadas de vapor de mercúrio de quartzo [...] Isto é mais fácil de obter se bem que as duas tenham custado 2.000 fr.

As a consequence, the pressure on the JEN was notable: if the intention was to ensure that full advantage should be drawn from the investment made, and any loss minimised, the laboratory Silveira headed would need to be funded to the tune of at least 83,800 francs, that is, 67,040 escudos.[3] Believing in the "capital mission" of higher technical schools and its role in the encouragement of Portuguese industry and agriculture, which involved expanding the scope of basic research, the board awarded funding from 1932 to 1936 to the Instituto Superior Técnico physics laboratory which was even greater than that indicated by its head: 75,000 escudos. Thanks to this huge investment the laboratory was able to purchase the desired research material, especially the most expensive items, including the Zeiss spectrograph in 1933–34, soon after Silveira had completed his internship in Paris, allowing him, as a scholarship-holder in Portugal, to address the president of the JEN as follows:

> António da Silveira, professor of physics at the IST, a JEN fellow, continued at his laboratory in Portugal the work he had started in Paris [...] on Raman's effect on saline solutions. He was vehemently urged to defend his thesis at the UP [University of Paris] before returning, but although his findings thus far fully justified this, he did not want to do this, claiming that he wished to complement his work with results he hoped to obtain in Portugal. [...]. Some of the findings, which I expected, I have just obtained thanks to the excellent facilities I have managed to provide here at my laboratory, but I can only complete them in Paris, because I do not have access to equipment like a microphotometer, for example, here.[4]

While Silveira's return to the French capital in the summer of 1933 was funded with a new grant from the JEN, improvements in research facilities in the field of physics in Portugal—through the purchase of the latest and most advanced instruments—induced Francisco de Paula Leite Pinto and Herculano de Carvalho, two important leaders of the board, to present the following proposal in July 1935 for consideration by the executive committee:

3º) Um bom reflector [que] custou 300 fr.

4º) Uma colecção de tubos de quartzo ou de pírex [...] O custo destes tubos orçou sempre à roda de 40 fr. cada um e como foram necessários cerca de doze isto dá um total aproximado de 500 fr.

5º) Um microfotómetro, aparelho extremamente caro que não existia à data em que entrei para o laboratório. [...] O preço deste aparelho anda à roda de 60.000 fr. É uma maravilha de construção mecânica e óptica executada pelas oficinas Zeiss. [...].

Pretendendo sobretudo salientar o que se despendeu comigo no Collège de France direi ainda que estudei especialmente o efeito de Raman nos líquidos, nas soluções e nos gases. [...] Ora líquidos houve [...] que custaram para cima de 1.000 fr. o quilo". This information is contained in a report submitted to the board in 1930 (Arquivo do Camões, Instituto da Cooperação e da Língua, I.P. Caixa 1339, Processo 7).

[3] Given the average exchange rate in 1930 (1 French franc = 0.800 escudos), 83,800 francs was worth 67,040 escudos (http://www.bportugal.pt/estatisticasweb/(S(gsreikym5y0jdn450wyygm45)/SeriesChronologicas.aspx).

[4] "António da Silveira, professor de Física no IST, bolseiro da JEN, prosseguiu em Portugal, no seu laboratório, os trabalhos que tinha iniciado em Paris [...] sobre o efeito de Raman nas soluções salinas. Foi veementemente instado para defender tese na U.P. antes de regressar, mas embora os resultados até então obtidos a justificassem plenamente não quis, alegando que desejava completá-la com os resultados que esperava obter em Portugal. [...]. Parte desses resultados, que antevia, acabo de os obter, mercê da boa instalação que consegui realizar no meu laboratório, mas só os posso completar em Paris, por não dispor aqui, por exemplo, dum microfotómetro". (Arquivo do Camões, Instituto da Cooperação e da Língua, I.P. Caixa 1308, Processo 2, Documento 1).

> No laboratory, however modest the work carried out there, can do without a microphotometer today, especially if its work is in the field of spectroscopy. [...] Some research work is already being conducted now in Portugal that requires the use of a microphotometer. We should not forget that the board has already funded training and is continuing to provide for the training of researchers in the fields of spectroscopy and nuclear physics: these researchers must have access to equipment like this and will be obliged to move abroad if there are no such facilities in Portugal. [...] The results of a survey of the main suppliers carried out by the secretary general of the board show that Zeiss, which builds the best microphotometers, has great interest in supplying the board with such equipment. [...] Therefore, we propose:
>
> a) that the board should purchase a Zeiss register microphotometer;
>
> b) that this device should be housed at physics laboratory of the Instituto Superior Técnico, whose head [...] is today one of the leading Portuguese researchers in the field of physics;
>
> c) that only researchers who are qualified to use this apparatus may be authorised to do so.[5]

Approving the application, the Executive Committee immediately granted funds to the tune of 20,000 escudos to the IST physics laboratory and in 1936 made a grant of a further 30,000 escudos. Thus, the state was able to save on future costs associated with the funding of internships abroad and the head of the laboratory received a coveted Zeiss microphotometer, describing it as a "marvel of mechanical and optical construction".[6]

Science practice in Portugal, however, proved to be more complex than board leaders would have liked. While in 1939 António da Silveira informed the president of the IAC that he was in the process of obtaining new findings associated with the Raman effect, and while in the early 1940s Francisco Mendes used the microphotometer of the IST with the authorisation of António da Silveira, at the same time, Manuel Valadares, another scientist who was recognised as possessing the skills required to operate the instrument, had to carry out an internship in Italy, with IAC

---

[5] "Nenhum laboratório, por muito modestos que sejam os trabalhos que nele se realizam, pode hoje dispensar um microfotómetro registador, sobretudo se esses trabalhos são de espectroscopia. [...] Ora alguns trabalhos de investigação se fazem já hoje em Portugal que exigem um microfotómetro. Não devemos esquecer que a Junta já preparou e está continuando a preparar investigadores nos campos da espectroscopia e da física nuclear, investigadores esses que necessitam absolutamente de tal aparelho e se verão obrigados a deslocarem-se fora do país se ele não existir em Portugal. [...] De um inquérito feito junto dos principais construtores pelo Secretário Geral da Junta conclui-se que a Casa Zeiss, que é precisamente a que constrói o mais perfeito microfotómetro, tem o maior interesse em fornecer a Junta. [...] Nestas condições propomos:

a) Que a Junta adquira um microfotómetro registador Zeiss;

b) Que este aparelho fique à guarda do Laboratório de Física do Instituto Superior Técnico, cujo director [...] é hoje um dos melhores investigadores portugueses no campo da Física;

c) Que deste aparelho só se possam servir os investigadores com competência para isso". In 1935 Leite Pinto was Secretary-General of the JEN, and Herculano de Carvalho was a member of the board (Arquivo do Camões, Instituto da Cooperação e da Língua, I.P. Caixa. 1320, Processo 2, Documento 1).

[6] "Uma maravilha de construção mecânica e óptica. The cost of the instrument was 60,000 escudos, of which the board provided 50,000 escudos, requiring that Portuguese physics laboratories should make up the shortfall." (Arquivo do Camões, Instituto da Cooperação e da Língua, I.P. Caixa 1320, Processo 2).

funding, in order to gain access to a microphotometer (Gaspar 2008, p. 132; Valadares 1950, p. 97).[7]

This internship took place in 1940–41, soon after the dissolution of the Núcleo de Matemática, Física e Química—Mathematics, Physics and Chemistry Unit, from which time relations between two of its main figures, António da Silveira and Manuel Valadares, deteriorated (Gil 2003, pp. 77–92; Gaspar 2008, pp. 70–75).[8] From 1940, the latter regularly asked the IAC to purchase a microphotometer for the Physics Studies Centre of the University of Lisbon Faculty of Science, while recognising, in December 1945, that there were reasons that went beyond his will preventing the processing of the spectrograms he then studied by means of a microphotometer.[9]

Assuming that it was the deterioration of relations with Silveira that effectively denied Valadares access to a piece of scientific instrument which was unique in Portugal, it should be highlighted how the present case study suggests that, paradoxically, the Portuguese dictatorship contributed to keeping several academics who were trained in the field of physics up to date with the latest developments during the period between the wars, while a degree of enmity and resentment among some of them meant that the impact of this was limited.

4 The Physics Laboratory of the Lisbon Faculty of Science: A Research School Created Through the JEN and the IAC

As a result of the support provided by the National Education Board and the Institute for High Culture, the University of Lisbon Faculty of Science physics laboratory was the only successful experimental physics research school in Portugal from 1929 to 1947 (Simões 2011, p. 16; Gaspar 2008; Fitas 2006, pp. 1–22). The argument that supports this thesis, based on the criteria defined by Gerald Geison, consists of the idea that the laboratory had a research programme focussing on X-ray spectrography, radioactivity and nuclear physics, it trained disciples, sending some abroad and attracting others, both scientists from Portugal and abroad, from outside and within the faculty, it assured scientists that they could publish works both in Portugal and abroad and take part in international networks, and it created new specialist journals. Among these was *Portugaliae Physica*, which enabled the dissemination at the international level from 1942–43 of works produced in the laboratory, among

[7] It should be mentioned that this internship also had other aims (Arquivo do Camões, Instituto da Cooperação e da Língua, I.P. Caixa 0974, Processo 6; Caixa 1480, Processo 11).

[8] Founded in 1936 by some of the JEN's former associates who had held scholarships abroad, and supported by other researchers, the Mathematics, Physics and Chemistry Unit was formally dissolved on 5th November 1939 due to disagreements among its members having arisen.

[9] Arquivo do Camões, Instituto da Cooperação e da Língua, I.P. Caixa 0488, Processo 9; Caixa 1480, Processo 11.

others (Simões 2011, p. 19; Gaspar and Simões 2011, pp. 303–343; Geison 1981, pp. 20–40).

To enable this, in accordance with its scientific policy, from 1929 the JEN sent researchers from the FCL physics laboratory to prestigious international study centres: for example, Herculano Amorim Ferreira, Manuel Valadares, Aurélio Marques da Silva and Manuel Teles Antunes, in addition to Amaro Monteiro, Francisco Mendes and Armando Gibert, who were later funded by the IAC (Gaspar and Simões 2011, p. 320). The level of competence achieved by these grantees may be verified by the number of doctorates completed, particularly abroad. Having begun his internship at the Radium Institut Suisse in 1929, Valadares transferred to the Laboratoire Curie, completing his PhD in Physics at the University of Paris in December 1933 under the guidance of Marie Curie. In December 1938, after a 56-month internship at the Laboratoire Curie, Silva, funded by the JEN and the IAC, also received a PhD in Physics at the University of Paris under the guidance of Frédéric Joliot. After serving an internship from 1932 to 1936 with Miguel A. Catalán at the Instituto Nacional de Física y Química, Antunes completed his PhD in Physics at the Central University of Madrid (Lopes 2017b, pp. 276–290; Boudia 2001).[10] Finally, financed by the IAC from 1942 to 1946, Gibert completed his PhD in Zurich, under the guidance of Paul Scherrer (Serra 2010, pp. 1–14; Fitas and Videira 2004, pp. 147–148, 227–228). These internships were complemented with the publication of articles in highly regarded journals and active participation at international scientific congresses, in addition to the publication of two works by Valadares and Silva as part of the prestigious collection Actualités Scientifiques et Industrielles (Valadares 1935).

Meanwhile, the return of scholarship holders to the FCL physics laboratory was marked by the implementation of the practice of research, funded once more by the JEN. In this regard, in accordance with the board's *modus operandi*, described above, it was concerned to grant scholarships in Portugal to the above-mentioned former external scholarship-holders, Ferreira, Valadares, Silva and Antunes, which enabled them to continue the research they had begun abroad. By way of example, let us consider the case of Manuel Valadares, who was awarded a 12-month scholarship at the FCL physics laboratory, subsequently suspended by the minister of public instruction and then reinstated in May 1935, effective from the date of its suspension.[11] After outlining in his application for funding the essential nature of a scholarship for the continuation of the research he had begun in Geneva and Paris, Valadares received notification of the approval of his application from Professor Achilles Machado, whose statement anticipated the future development of the FCL physics laboratory:

> If a large amount of money were to be spent on a four-year scholarship abroad and no further advantage was derived from this before the scholar in question had forgotten all that he had

[10] Archives de l'Institut du Radium de l'Université de Paris. Laboratoire Curie. Pièces 1695, 1747, 1750, 1923, 1929, 2095, 2123, 2125, 3360, 3368, 3373, 3378 and 3380.

[11] Arquivo do Camões, Instituto da Cooperação e da Língua, I.P. Caixa 1215, Processo 3, Documentos 4, 8, 17 e 19.

> learnt, the investment would have proven to be completely unproductive. Working in this country has the advantage of the fact that the equipment purchased and the techniques used may be employed to the advantage of others who will no longer need to seek training abroad in order to carry out research work of the same nature.[12]

This concern with the training of new researchers whose requirement for specialist studies did not necessarily involve carrying out an extended internship abroad establishes a connection between the reinstatement of Valadares' grant and the award of a grant enabling him to purchase bibliography and research instruments. Having agreed that this grant should revert to the institution to which he was attached as a scientist, Manuel Valadares had the following to say to the Secretary General of the JEN on its importance:

> I am sending you an explanatory note on the use of the amount of 2,526.95 escudos that the National Education Board has deigned to grant me this year for the purchase of material and books. I would like to take this opportunity to inform you that this grant has been of great use to me in carrying out the experiments I am conducting.[13]

At the time, in October 1935, Valadares was engaged in X-ray spectrography studies, for which he still benefited from the funding which the board had provided from 1930 with a view to improving research facilities at the FCL physics laboratory (Table 1).

According to Augusto Celestino da Costa, in 1930 the greatest shortcoming of the FCL physics laboratory was the scarcity of books and journals, and during the initial period from 1930 to 1932 funds were earmarked for the purchase of bibliography, enabling Herculano Amorim Ferreira to state in December 1932:

> The grants made by the JEN over the last two years have led to a great expansion of the laboratory library, especially regarding specialist monographs on modern physics subjects. There are subscriptions to the following journals at the laboratory:
>
> - Anales de la Sociedad Española de Física y Química (Madrid);
> - Le Journal de Physique et le Radium (Paris);
> - Revue d'Optique (Paris);
> - The Proceedings of the Physical Society (London);
> - Science Abstracts (London);
> - Journal of Scientific Instruments (London);
> - The Physical Review (New York);

[12] "Se se gastaram alguns contos de réis com 4 anos de uma bolsa no estrangeiro para não mais aproveitar, antes para esquecer o que o bolsista aprendeu, a despesa feita foi completamente improdutiva. Trabalhando no País tem a vantagem dos aparelhos adquiridos e a técnica empregada poderem aproveitar a outros que, para efectuar trabalhos de investigação da mesma natureza, não terão necessidade de ir buscar ao estrangeiro o ensino necessário". Report by Achilles Machado, dated 5th April 1935 (Arquivo do Camões, Instituto da Cooperação e da Língua, I.P. Caixa 1215, Processo 3, Documento 15).

[13] "Junto envio a V. Ex.ª nota explicativa do emprego da verba de Esc. 2.526$95 que, para aquisição de material e livros, a Junta de Educação Nacional se dignou conceder-me no corrente ano. Aproveito a ocasião para informar V. Ex.ª que este subsídio me foi de grande utilidade na realização das experiências que estou executando". Letter dated 19th October 1935 (Arquivo do Camões, Instituto da Cooperação e da Língua, I.P. Caixa 1215, Processo 3, Documento 25).

Table 1 Funding of the Physics Laboratory of the University of Lisbon Faculty of Science (JEN/IAC: 1929–1947; figures in escudos)

Financial year	Sum granted	Use to which funding was put
1930–31	28,500	Publications, purchase of bibliography and research material
1931–32	3,000	Publications and purchase of bibliography
1934–35	26,000	Purchase of research material
1936	43,000	Radiotorium hire and radium salt purchase
1937	19,000	Transformation and installation of radium salt, and purchase of research material
1938	5,000	Purchase of research material
1939	10,200	Purchase of research material
1940	42,700	Purchase of bibliography and research material
1942	15,000	Publication of journal *Portugaliae Physica* and purchase of research material
1943	23,000	Publication of journal *Portugaliae Physica* and purchase of research material
1944	30,000	Publication of journal *Portugaliae Physica* and purchase of research material
1945	29,000	Publication of journal *Portugaliae Physica*, purchase of bibliography and research material
1946	99,000	Publication of journal *Portugaliae Physica,* purchase of bibliography and research material
1947	59,451.25	Purchase of bibliography and research material
Total	432,851.25	

Sources Arquivo do Camões, Instituto da Cooperação e da Língua, I.P. Caixa 1212, Processo 24; Caixa 1214, Processo 21; Caixa 1254, Processo 6; Caixa 1319, Processo 5; Caixa 1320, Processos 7 e 11; Caixa 0974, Processo 5; Caixa 3248, Processo 4; Caixa 3249, Processos 1 e 5; Caixa 3250, Processo 3; Livro de Actas da Comissão Executiva da JEN 1929–35; Livro de Actas da Comissão Executiva da JEN e Direcção do IAC 1935–42; Livro de Actas da Direcção do IAC 1942–48

- Journal of the Optical Society of America (Ithaca);
- The Review of Scientific Instruments (Ithaca). […]

It would be useful to be able to continue what, with JEN's support, has begun. In these matters: books, journals and publications, any interruption would be highly damaging.[14]

[14] "Os subsídios concedidos pela JEN nos dois últimos anos permitiram aumentar muito a biblioteca do Laboratório, especialmente em monografias especializadas de assuntos de Física moderna. Assinam-se actualmente neste Laboratório as seguintes revistas:

Anales de la Sociedad Española de Física y Química (Madrid);
Le Journal de Physique et le Radium (Paris);
Revue d'Optique (Paris);
The Proceedings of the Physical Society (Londres);
Science Abstracts (Londres);
Journal of Scientific Instruments (Londres);

The importance of this testimony is clear when one considers that, in 1930, before the intervention of the board, the library of the physics laboratory could not afford to subscribe to any periodical publications (Gibert 1950, p. 88). In addition, the head of the laboratory, Cirilo Soares, earmarked JEN and IAC funds for the purchase of research material essential for the work of grantees Ferreira, Valadares and Silva, among others (Valadares 1950, p. 102).

The purchase of an electrometer and the payment of the first instalment on the purchase of a spectrograph in financial year 1930–31, instruments that Manuel Valadares indicated in 1933 as being essential for the continuation of his studies in Lisbon, shows how his return to Portugal was planned for from the beginning of his internship abroad through the concerted effort of the JEN and Cirilo Soares.[15] In the event, the spectrograph was not immediately purchased, neither was the other material sought by Valadares in September 1933—the purchase or hire of radioactive substances made available, due to Soares's commitment to other grantees. With regard to this Soares' stance, the following extract from his letter sent to the chairman of the board in August 1934 should be noted:

> On 28th September 1933 this Laboratory [...] submitted to the National Education Board a request for 17,000 escudos intended for the installation of a gamma radiation spectrograph at the laboratory [...] which lecturer M. Valadares would use in the continuation of his studies after returning from abroad. [...]. However, Professor Amorim Ferreira who has just returned from abroad would like to be able to continue at this laboratory the work he has carried out in London. [...] The total amount requested by this laboratory is therefore 33,000 escudos. However, you were kind enough to inform me that the board could not make a grant like this to this laboratory, so I am obliged to disregard one of the applications made this financial year. As lecturer M. Valadares will be able to continue working this year with X-ray equipment that has been loaned to him by the chemistry laboratory of this faculty, I am therefore giving priority to the application for 16,000 escudos required for Prof. Amorim Ferreira's facility. [...]. I must also inform you that what I have just set out is the result of a comprehensive agreement between those concerned as regards the work of this laboratory.[16]

The Physical Review (Nova York);
Journal of the Optical Society of America (Ithaca);
The Review of Scientific Instruments (Ithaca). [...].
Seria interessante poder continuar o que, com o apoio da JEN, se iniciou. Qualquer dos assuntos (livros, revistas e publicações) é daqueles em que uma interrupção seria altamente prejudicial.". Letter from Herculano Amorim Ferreira, dated 8th December 1932, to the President of the JEN (Arquivo do Camões, Instituto da Cooperação e da Língua, I.P. Caixa 1214, Processo 21, Documento 7).

[15] Arquivo do Camões, Instituto da Cooperação e da Língua, I.P. Caixa 1254, Processo 6, Documento 1.

[16] "Em 28 de Setembro de 1933 este Laboratório [...] apresentou à Junta de Educação Nacional um pedido de Esc. 17.000$00 que se destinavam à montagem duma instalação de espectrografia de radiação gama [...] com a qual prosseguiria, após regresso do estrangeiro, os seus estudos o assistente M. Valadares. [...]. Sucede porém que acaba de regressar do estrangeiro o Professor Amorim Ferreira que desejaria poder continuar neste Laboratório os trabalhos que sobre Termiónica realizou em Londres. [...] O total dos pedidos feitos por este Laboratório atinge pois a quantia de Esc. 33.000$00. Teve porém V. Ex.ª a amabilidade de me informar que a Junta não poderia pôr à disposição deste Laboratório uma tal verba, pelo que me vejo obrigado a desistir, neste ano económico, dum dos pedidos formulados. Como o Assistente M. Valadares poderá continuar a

Once this grant had been made and Valadares' requests, particularly for the hire of radium and the purchase of radium salt, had been put on hold, in the event he was prevented from carrying out research in the field of radioactivity, probably denying him the opportunity to be the first to discover the element with atomic number 85 (Gaspar 2008, p. 81). At the same time, with the equipment purchased as a result of funding provided by the board in mid-1935—X-ray and spectrographic instruments—Valadares was able to replace the instruments he had adapted and reused since December 1933, resulting in his obtaining photographs of the satellite band between stripes α_1 and α_2 on the lead L spectrum, and he published his findings in *Comptes Rendus de l'Académie des Sciences de Paris* in 1938 (Valadares and Mendes 1938, pp. 744–746; Gaspar and Simões 2011, p. 321).

While Manuel Valadares' specialist training abroad and the subsequent award of a scholarship in Portugal, in association with the facilities at the FCL physics laboratory, enabled him, once he had returned to Portugal, to continue to disseminate the results of his research by publishing articles in a reputable international journal, equally important was the fact that these circumstances encouraged and benefited other researchers, eventually resulting in the development of a new geographical configuration of doctoral training.

On the circumstances which attracted scientists, of note is a case that, while it did not lead to the scientist in question completing his PhD at the laboratory, provides an indication of the international recognition of the high standard of research produced at the laboratory. The minutes of a faculty board meeting in May 1943 attest to this:

> Prof. C. Soares, hereby informs the council that at the congress [the Luso-Spanish Congress for the Progress of Science] held in Porto the Spanish professor Julio Palacios of the Instituto Nacional de Física y Química in Madrid showed an interest in visiting the physics laboratory of our faculty, especially to learn about the facilities available for the study of X-ray spectrography, and inquired about the possibility of scientists from his laboratory coming here to work. Professor Palacios' associate arrived in Lisbon a few days ago and is already working at our laboratory with Dr. Valadares. On behalf of the council the head of the faculty welcomes the fact that it is of great importance for our faculty.[17]

---

trabalhar este ano com uma instalação de Raios X que lhe foi emprestada pelo Laboratório de Química desta Faculdade, colocarei assim em 1.º lugar o pedido de Esc. 16.000$00 necessário para a instalação do Prof. Amorim Ferreira. [...]. Cumpre-me ainda comunicar a V. Ex.ª que o que acabo de expor resulta duma perfeita concordância entre os interessados no respeitante à marcha dos trabalhos deste Laboratório". Letter from Cirilo Soares dated 2nd August 1934 (Arquivo do Camões, Instituto da Cooperação e da Língua, I.P. Caixa 1254, Processo 6, Documento 4).

[17] "O prof. C. Soares dá conhecimento ao Conselho que no Congresso [Luso-Espanhol para o Progresso das Ciências] do Porto o prof. espanhol Julio Palacios do Instituto Nacional de Física e Química de Madrid mostrou interesse em visitar o Laboratório de Física da nossa Faculdade, especialmente para tomar conhecimento do material de que dispomos para o estudo da espectrografia dos Raios X e inquiriu da possibilidade de aqui vir algum dos seus colaboradores. Esse colaborador chegou há poucos dias a Lisboa e já está a trabalhar no nosso Laboratório, junto do Dr. Valadares. O Sr. director em nome do Conselho congratula-se com o facto referido que é de grande importância para a nossa Faculdade". Livro de Actas do Conselho Escolar da Faculdade de Ciências da Universidade de Lisboa 1933–44, Acta da sessão de 7 de Maio de 1943, pp. 159–163 (http://ciuhct.org/media/default/online/docs/1441c.pdf).

Table 2 Scientific production at the Physics Laboratory of the University of Lisbon Faculty of Science (1930–1946)

Years	Published works			Doctorates		
	Domestic journals	International journals	Total	Portugal	Abroad	Total
1930–34	9	6	15	1	1	2
1935–38	8	10	18	0	2	2
1939–42	16	11	27	1	0	1
1943–46	25	5	30	4	1	5
Total	58	32	90	6	4	10

Adapted from Fitas (2006, p. 14)

When one considers the development of physics and chemistry in Spain thanks to the funding provided by the Junta para Ampliación de Estudios—which led to the establishment of the Instituto Nacional de Física y Química in Madrid in 1932 through Rockefeller philanthropy (Sánchez Ron 2009, pp. 47–74; Fernández Terán 2014)—while a Portuguese laboratory, which did not receive any private funding, constituted in the early 1940s the chosen institution for one of its researchers, we can be assured of the high standard of the research produced at the latter and, ultimately, the importance of the state institutions that provided support for it: the JEN and the IAC (Table 2).

With regard to the recruitment of new researchers whose work resulted in the completion of doctorates, it should be noted that while research in physics was practically non-existent at Portuguese faculties of science from 1911 to 1930—which made it impossible to award doctorates in this field of science, in keeping with the legislation of the First Republic that required a year of experimental work (Gaspar and Simões 2011, p. 317)[18]—the impact of the work of the JEN and the IAC can be seen in the enabling of such research at this laboratory and the consequences in terms of allowing it to offer doctoral training. As confirmation of this, there are the cases of Lídia Salgueiro, Marieta da Silveira, Carlos Braga and José Sarmento, who completed their doctorates between 1943 and 1946 based on experimental work carried out at the FCL physics laboratory, known at the time as the Physics Studies Centre (Fitas 2006, pp. 13–14). Regarding this new stage in the life of the laboratory, Carlos Braga stated:

> When I thought about doing experimental physics work for my PhD in 1942, I soon realised that, unless I went abroad, only in Lisbon could I find the necessary means to start my work with the likelihood of a successful outcome. Indeed, the working environment at the study centre attached to the physics laboratory was encouraging.[19] (Braga 1950, p. 90).

[18] The PhD that is recorded, in Portugal, in the period between 1930–34 is that of Amorim Ferreira, who carried out the experimental part of his PhD training in London (Gaspar 2008, p. 71).

[19] "Quando em 1942 pensei em realizar um trabalho de Física experimental para o meu doutoramento depressa conclui que, a não ser saindo do país, só em Lisboa encontrava os meios necessários para o iniciar com algumas probabilidades de êxito. Efectivamente no Centro de Estudos, anexo ao Laboratório de Física, o ambiente de trabalho era animador".

In other words, in contrast to the circumstances experienced by their tutors, Manuel Valadares and Aurélio Marques da Silva, in the 1940s, the four above-mentioned new PhD students had not been obliged to move abroad to achieve the status of qualified scientists, and this was due to the work of an institution that organised and funded scientific research in Portugal.

It is interesting to note that, although it came late in the day, the granting of physics doctorates by the University of Lisbon Faculty of Science involving experimental work carried out at the faculty was opposed by some members of the faculty board. Already analysed in detail by other authors, this question of scientific policy (Gaspar and Simões 2011) illustrates the difficulty encountered by the university administration in introducing a new model of doctoral training, in accordance with the stance set out by Victor de Lemos at the faculty board session of 18th November 1944, at which he argued that "[...] a doctor is a super-graduate and therefore training for a doctorate is a lengthy process, because one must necessarily review one's entire graduate training and make improvements to that which has already been achieved.".[20] In stressing this, and also due to the encyclopaedic nature of the requirements for achieving success in doctoral examinations at the time, in the Lemos' view a PhD holder could be distinguished from a graduate by virtue of the former possessing a "higher cultural level", which assumed that they had carried out a lengthy period of training from the completion of their degree to that of their doctorate.[21]

Cirilo Soares, Torre de Assunção and Flávio Resende held a distinguished position in this respect, arguing that a PhD holder is a researcher who has specialised in a particular field and produces new findings, and in practice the University of Lisbon Faculty of Science recognised Lídia Salgueiro, Marieta da Silveira, Carlos Braga and José Sarmento as having achieved this.[22]

On the work produced by Lídia Salgueiro during her doctoral training it is interesting to read what Cirilo Soares had to say to the president of the IAC in May 1944:

> Another fact that I am pleased to point out to you, which shows the success not only of *Portugaliae Physica*, but also, through it, the Studies Centre, of which I have the honour of acting as head, is that the article published in the second issue [of] January 1944 by scholarship-holder Lídia Salgueiro establishes priority for discovering new sources of radiation in the gamma spectrum of Radium D, a study that has been pursued by several researchers, and has only since been confirmed by the work presented in March of the current year to the

[20] "Um doutor é um super-licenciado e que, portanto, a preparação para o doutoramento exige tempo, pois há que rever toda a licenciatura e melhorar a preparação já conseguida". Livro de Actas do Conselho Escolar da Faculdade de Ciências da Universidade de Lisboa 1944–52, Acta da sessão de 18 de novembro de 1944, pp. 1–5 (http://ciuhct.org/media/default/online/docs/1442c.pdf).

[21] Livro de Actas do Conselho Escolar da Faculdade de Ciências da Universidade de Lisboa 1944–52, Acta da sessão de 18 de novembro de 1944, pp. 1–5 (http://ciuhct.org/media/default/online/docs/1442c.pdf).

[22] Livro de Actas do Conselho Escolar da Faculdade de Ciências da Universidade de Lisboa 1944–52, Acta da sessão de 18 de novembro de 1944, pp. 1–5 (http://ciuhct.org/media/default/online/docs/1442c.pdf). Having carried out experimental work at the Physics Studies Centre of the University of Lisbon, Carlos Braga and José Sarmento received their PhDs at the University of Porto in 1944 and 1946, respectively (Gaspar 2008, p. 145).

> Paris Academy of Sciences by authorised scientist Marcel Frilley, "maître de recherches" at the Laboratoire Curie.[23]

While the evaluation of the work of the JEN and IAC assumes a comparison being made with the situation when they began operations, the case study under analysis reveals how the JEN and the IAC, by awarding scholarships to the most competent candidates and providing funding to the FCL physics laboratory while making grants for the international dissemination of their work, in particular funding *Portugaliae Physica,* contributed to transforming a Portuguese laboratory which did not carry out practical research and was unable to offer doctorate courses, into an internationally recognised research school.[24]

5 Conclusion

In the post-World War I period, Portugal was part of the international movement for the creation of state bodies for the planning and funding of scientific activity. Current political historiography stresses the rivalry between European states, nationalism and economic autarchy, while, although the scope of action of these science institutions was domestic, they promoted the circulation of ideas and scientists among different countries. At the same time, some of them provided a model for the creation of similar institutions in other countries, such as Portugal, where the science policy of the National Education Board was strongly influenced by the Spanish Junta para Ampliación de Estudios e Investigaciones Científicas and the Fonds National Belge de la Recherche Scientifique.

By considering as essential investment in basic science and fields with a less marked tradition or those which were less developed in Portugal, the National Education Board and its successor, the Institute for High Culture, provided significant support for physics. Regarding support for the physics laboratory of the Instituto Superior Técnico we find that they enabled the facility to be equipped with some of the most advanced research instruments of the time, similar to those found at the Collège de France, where the head of the laboratory, António da Silveira, had carried out research.

---

[23] "Outro facto que eu tenho o prazer de apontar a V. Ex. e que representa um êxito não só da *Portugaliae Physica*, mas por seu intermédio, do Centro de Estudos que tenho a honra de dirigir é que o artigo publicado no segundo fascículo [de] Janeiro de 1944 da autoria da bolseira D. Lídia Salgueiro marca a prioridade da descoberta de novas radiações no espectro gama do rádio D, estudo que tem sido prosseguido por vários investigadores, e só posteriormente se acha confirmado pelo trabalho apresentado em Março do ano corrente à Academia de Ciências de Paris pelo autorizado cientista Marcel Frilley, "maître de recherches" no Laboratório Curie". Letter dated 18th May 1944 (Arquivo do Camões, Instituto da Cooperação e da Língua, I.P. Caixa 3249, Processo 1, Documento 29).

[24] On the favourable reaction of various international scientific institutions to the work published in *Portugaliae Physica*, see Gaspar (2008, pp. 152–153).

The science policy of the JEN and IAC had a profound impact on the physics laboratory of the University of Lisbon Faculty of Science. The several years during which its members carried out research at the Laboratoire Curie, the Radium Institut Suisse and the Instituto Nacional de Física y Química, among other scientific centres abroad, led to radical changes in academia in Portugal, among them the enhanced role assigned to the physics laboratory of the University of Lisbon Faculty of Science enabling scientists to complete their doctorates in Physics in Portugal. The practical consequence of this was that from the 1940s a new generation of Portuguese PhD Physics students no longer had to move abroad to complete their studies and thus achieve the status of physicist.

However, these changes did not take place with the full agreement of all scientists and some expressed reservations. Different conceptions of the role of universities—that they should either focus mainly on teaching or developing research activities that would lead to researchers specialising in a particular field, in keeping with the aspirations of the physics associates who were funded by the JEN and the IAC—, caused a degree of discord among the members of the board of the University of Lisbon Faculty of Science. Meanwhile resistance and resentment are also evident among institutions or alternatively may have stemmed from personal relations between scientists. Manuel Valadares' difficulty in gaining access to the microphotometer at the physics laboratory of the Instituto Superior Técnico, while his assistant, Francisco Mendes, was authorised by António da Silveira to use what was then a uniquely prized instrument in Portugal for the purposes of carrying out research, shows how research in physics in Portugal between the wars depended heavily on peer relations.

Regardless of the difficulties which arose, thanks to the work of the institutions which organised and funded research in Portugal, from 1929 to 1947 the physics laboratory of the University of Lisbon Faculty of Science was an internationally recognised centre for scientific research. This not only provides an example of how institutions on the European periphery were not merely uncritical, passive consumers of knowledge produced by scientific centres, but also evidences how the latter recognised the knowledge produced on the periphery and benefited from it (Hauswedell et al. 2019).

Acknowledgements The author would like to thank Michael Lewis for the translation work. The IHC receives national funding from the FCT—Foundation for Science and Technology, I.P. under projects UIDB/04209/2020, UIDP/04209/2020, and LA/P/0132/2020. Research for this paper was also funded by the FCT under project 2021.00906.CEECIND.

References

Boudia S (2001) Marie Curie et son laboratoire: sciences et industrie de la radioactivité en France. Éditions des Archives Contemporaines, Paris.

Braga C (1950) À memória do Professor A. Cyrillo Soares. Gazeta de Física II(4):90–92.

Costa A (1930) Relatório do Vice-presidente do ramo de ciências sobre as necessidades da investigação científica em Portugal. In: Junta de Educação Nacional. Relatórios, propostas e projecto de orçamento para o ano económico de 1930–1931. Lisboa, pp. 9–43.
Costa A (1939) O Problema da Investigação Científica em Portugal. Coimbra.
Crawford E (1988) Internationalism in science as a casualty of the First World War: relations between German and Allied scientists as reflected in nominations for the Nobel prizes in physics and chemistry. Social Science Information 27(2):163–201.
Crawford E (1992) Nationalism and Internationalism in Science, 1880–1939: Four Studies of the Nobel Population. Cambridge University Press, Cambridge.
Edgerton D (2013) Science in the United Kingdom: a Study in the Nationalization of Science. In: Krige J, Pestre D (eds). Science in the Twentieth Century. Routledge, London–New York, pp. 759–776.
Fernández Terán R (2014) El profesorado del 'Instituto Nacional de Física y Química' ante la Guerra Civil, el proceso de depuración y el drama del exilio. (PhD thesis, Universidad Complutense de Madrid).
Fitas A (2006) A Ciência em Portugal ao longo do séc. XX (cenas exemplares de um percurso incompleto). Prelo 3:1–22.
Fitas A, Videira A (2004) Cartas entre Guido Beck e cientistas portugueses. Instituto Piaget, Lisboa.
Gaspar J (2008) A Investigação no Laboratório de Física da Universidade de Lisboa (1929–1947). Master thesis in History and Philosophy of Science. University of Lisbon.
Gaspar J, Simões A (2011) Physics on the Periphery: A Research School at the University of Lisbon under Salazar's Dictatorship. Historical Studies in the Natural Sciences 41(3):303–343.
Geison G (1981) Scientific Change, Emerging Specialties and Research Schools. History of Science 19:20–40.
Gibert A (1950) O Centro de Estudos de Física do Instituto para a Alta Cultura Anexo à Faculdade de Ciências de Lisboa. Gazeta de Física II(4):86–89.
Gil F (2003) Núcleo de Matemática, Física e Química: uma contribuição efémera para o movimento científico português. Boletim da Sociedade Portuguesa de Matemática 49:77–92.
Halleux R, Xhayet G (2007) La Liberté de Chercher: Histoire du Fonds National Belge de la Recherche Scientifique. Editions de l'Université de Liège, Liège.
Hauswedell T, Körner A, Tiedau U (eds) (2019) Re-Mapping Centre and Periphery. Asymmetrical Encounters in European and Global Contexts. University College London, London.
Hull A (1999) War of words: the public science of the British scientific community and the origins of the Department of Scientific and Industrial Research, 1914–16. British Journal for the History of Science 32:461–481.
Lopes Q (2017a) The National Education Board (1929/36) and scientific research in Portugal. Portuguese Journal of Social Science 16(1):71–85.
Lopes Q (2017b) A europeização de Portugal entre guerras. A Junta de Educação Nacional e a investigação científica. Caleidoscópio, Casal de Cambra.
Macleod R (1993) The Chemists go to War: The Mobilization of Civilian Chemists and the British War Effort, 1914–1918. Annals of Science 50:455–481.
Roberts L (2009) Situating Science in Global History: Local Exchanges and Networks of Circulation. Itinerario 33(1):9–30.
Salgueiro Â (2015) Ciência e Universidade na I República. Ph.D dissertation in Contemporary History, New University of Lisbon Faculty of Social and Human Sciences.
Sánchez Ron J (2007) Las ciencias Físicas y Químicas en la JAE. In: Puig-Samper Mulero M (ed). Tiempos de investigación. JAE-CSIC, cien años de ciencia en España. Consejo Superior de Investigaciones Científicas, Madrid, pp. 102–113.
Sánchez Ron J (2009) Del Laboratorio de Investigaciones Físicas al Instituto Nacional de Física y Química. In: González Ibáñez C, Santamaría García A (eds). Física y Química en la Colina de los Chopos. 75 años de investigación en el edificio Rockefeller del CSIC (1932–2007). Consejo Superior de Investigaciones Científicas, Madrid, pp. 47–74.

Schroeder-Gudehus B (2020) International Science from the Franco-Prussian War to World War Two: An Era of Organization. In: Slotten H, Numbers R, Livingstone D (eds). The Cambridge History of Science: Modern Science in National, Transnational, and Global Context, Vol. 8. Cambridge University Press, Cambridge, pp. 43–59.

Serra I (2010) Armando Gibert um físico português em luta pela ciência. Travaux de Laboratoire L(II):1–14.

Siegmund-Schultze R (2001) Rockefeller and the internationalization of Mathematics between the Two World Wars: documents and studies for the Social History of Mathematics in the 20th Century. Birkhauser Verlag, Basel.

Simões A (2011) O ano 1947 e o Laboratório de Física da Faculdade de Ciências de Lisboa. Gazeta de Física 34(2):16–21.

Valadares M (1935) Transmutation des éléments par des particules accélérées artificiellement. Hermann & Cie, Paris.

Valadares M (1950) O Laboratório de Física da Faculdade de Ciências de Lisboa, sob a direcção do Prof. Dr. A. Cyrillo Soares (1930–1947), e a investigação científica. Gazeta de Física II(4):93–106.

Valadares M, Mendes F (1938) Étude des satellites Lα, de l'élément 82 (Pb). Comptes Rendus de l'Académie des Sciences de Paris 206:744–746.

Vanpaemel G (2013) To be or not to be: Belgian science policy in the interwar period. In: Fitas A, Príncipe J, Nunes F, Bustamante M (eds). A Junta de Educação Nacional e a Investigação Científica em Portugal no Período entre Guerras. Caleidoscópio, Casal de Cambra, pp. 33–48.

Wullus-Rudiger J (1948) La science au service de la nation et de l'humanité: la Fondation Universitaire et le Fonds National de la Recherche Scientifique. Service d'information et de documentation du Ministère de la Défense nationale, Bruxelles.

Chapter 25
Physics as Science Proper: A New Historical Justification

Peeter Müürsepp

Abstract Since its very beginning, physics has enjoyed the position of the only science proper among the different branches of sciences, even for social ones. There is a good justification for this. The method of modern science, worked out by Galileo, is strongly based on physics (mechanics). Actually, science started as mechanics. It is not clear, however, whether being a firm foundation for the whole of science would make the best model for any kind of science. There is science that is physics-like and has to be such and science that is not physics-like and perhaps does not have to be like that. It is interesting that there is also a science that is partly physics-like and partly not physics-like, namely chemistry. It is the former as far as it looks for the laws of nature, i.e. the law of periodicity, and it is the latter as long as it deals with substances (or stuff). The current essay takes a deeper look into the similarity and difference between physics and chemistry from the historical and philosophical perspective. The analysis of the dual character of chemistry from the historical perspective allows proposing a more solid historical justification to the special position of physics as the only science proper.

Keywords Chemistry · Constructive-hypothetico-deductive · Physics · Physics-like science · Science proper · φ-science

1 Introduction

There is a well justified tradition to call physics science proper. Modern science was started by Galileo. The method that later became classified as scientific was initially the method of what we now call classical mechanics. Thus, classical mechanics became the very foundation of any kind of science. In order to be scientific, a field had to be developed along the lines of the method of classical mechanics. Put briefly, this is the accepted historical justification of physics being the science proper.

P. Müürsepp (✉)
Tallinn University of Technology, Tallinn, Estonia
e-mail: peeter.muursepp@taltech.ee

R. Pisano (ed.), *A History of Physics: Phenomena, Ideas and Mechanisms*,
History of Mechanism and Machine Science 42,
https://doi.org/10.1007/978-3-031-26174-9_25

There is also the philosophical justification of physics being science proper. I shall address this, as well as the historical one, in more detail below. At the core of this justification is the idea that the method of physics is hypothetico-deductive in its essence. Some even say that it is actually constructive-hypothetico-deductive. This means that a physicist does not approach reality as it is but rather constructs the research object for herself, adjusting reality to a considerable extent. According to this approach, the physicist deals with her constructed research object, not with reality directly. This move enables the scientist to follow the method for more or less properly starting from a testable hypothesis and moving towards results by means of measuring and, experimenting and in the end, formulating the results in the language of mathematics. Therefore, following the method becomes a convenient endeavor for a scientist. This is how we get the idea that science has to be a field, or rather a practical activity, that is adjustable to this convenient method.

As mentioned above, I shall take a more detailed look into the justification of the position of physics as science proper below. The core of the current analysis, however, is historical rather than philosophical. I shall briefly go over the three stages of the development of physics: classical, nonclassical and post-nonclassical, to show that although shifting quite far away from the initial classical ideal in the case of quantum mechanics or self-organization theory for instance, physics still retains its core feature as the model for anything scientific in its most proper, although narrow sense. In principle, history of physics will still need the philosophical eye to keep the focus and follow the main thread properly. This idea will become clearer below as well.

2 Hypothetico–Deductive or Constructive–Hypothetico–Deductive: The Philosophical Justification

We could ask, where does the hypothetical nature of science come from if even Isaac Newton himself did not recognize the need to hypothesize, at least seemingly? Newton explained his famous attitude to hypotheses as follows:

> I have not as yet been able to discover the reason for these properties of gravity from phenomena, and I do not feign hypotheses. For whatever is not deduced from the phenomena must be called a hypothesis; and hypotheses, whether metaphysical or physical, or based on occult qualities, or mechanical, have no place in experimental philosophy. In this philosophy particular propositions are inferred from the phenomena, and afterwards rendered general by induction. (Newton ([1726] 1999, p. 943)

According to Newton, experimental philosophy, which was his term for what we call science today, should be based on particular propositions inferred from the phenomena. Those propositions are the same as what we tend to call hypotheses these days. Thus, Newton quite clearly advocated the approach of a typical inductivist, or perhaps we should say verificationist. His rejecting of hypotheses rather meant resistance to speculations, rather than what we understand as hypotheses today.

By all evidence, William Whewell had a similar attitude as Newton but placed the emphasis quite differently:

> What is requisite is, that the hypothesis should be close to the facts, and not connected with them by other arbitrary and untried facts; and that the philosopher should be ready to resign it as soon as the facts refuse to confirm it. (Whewell 1840, p. 438)

In principle, Whewell says the same as Newton. He just calls the particular propositions of Newton hypotheses. Whewell's hypotheses also have to be connected to facts, just like Newton's particular propositions. However, Whewell's approach could be called falsificationist rather than verificationist. The scientist, by the way, is still called the philosopher in Whewell's days. The unfortunate split between philosophy and science that had been caused by Newton and Descartes already in Maxwell (2017, pp. 51–57) did not yet reflect in terminology in 1840. In addition, we may claim that here we have the first declaration of a method that we know as hypothetico-deductive. As far as it is known, the term itself belongs to Whewell as well, although it does not directly appear in this particular quote given above.

It is a shame that William Whewell, who obviously still kept philosophy and science together as we can see from the terminology he used, also became the first well-known thinker who introduced the split in a way. Whewell was thinking about science and did science but was definitely better versed with the former activity rather than the latter.

The hypothetico-deductive method in which the hypothesis is based on facts does not necessarily put physics in a special position. Any science, never mind natural or social, fits in. There are facts everywhere. So, why the special position of physics after all?

Let us remember how Galileo worked. He proposed hypotheses and tried to prove them by means of experiments. But his experiments were not about what was really going on in reality. He constructed the experimental situation according to his goals. Starting from Galileo, experimenting has been "[...] a purposeful and critical theory-guided constructive, as well as manipulative, material interference with nature [...]" (Vihalemm and Lõhkivi 2012, p. 3). Modern physicists are constructing their objects of research. They are doing this on the basis of the empirical experience they have obtained, on the basis of facts, but it is still a construction. This construction helps physicists to follow the scientific method properly. Such kind of constructing the object of research in a pure form is not possible in any other branch of science except for physics. Only a part of chemistry is of the same type—the part that deals with the laws of nature. The story of finding the law of periodicity, for instance, is exactly an example of the constructive-hypothetico-deductive type of approach. The law is about chemical elements. Then, what is a chemical element? Rein Vihalemm has put it like this: "[a] fundamental idealisation substantiated by experimental chemistry—namely, a definite position in the periodical system based on the periodic law." (Vihalemm 2015, p. 111). We see that the chemical element is a construction of the scientist rather than a pure piece of reality.

Now we have the choice between physics and chemistry over which of them is more proper science. However, physics still has an advantage. It is for, more or less fully of the constructive-hypothetico-deductive type, while chemistry is just partly that. The part of chemistry that is dealing with stuff (or substances) is rather of the natural philosophy type of research. Therefore, it is still physics that is science proper. According to this approach, any other science has to be constantly striving for being as physics like as possible if its developers are interested in seeing it as science.

Rein Vihalemm has introduced a further terminology that helps to clarify the position of physics. He called mathematized exact science φ-science and any other type of science non-φ-science. The most typical φ-science is classical physics, of course. Obviously, φ-science is based on the constructive-hypothetico-deductive type of cognition and non-φ-science does not have that basis. Rein Vihalemm called the basis of the latter classifying-historico-descriptive type of cognition (Müürsepp et al. 2019).

Why not speak just about physics-likeness? This is an important question for our further analysis. The issue is that physics is a developing science that keeps changing. Non-classical physics, quantum mechanics and relativity theories, was definitely not pure φ-science, but it was still very close to it. Statistical approach came in but there was still no consideration of irreversibility as something objective. Time did not have much of a meaning except for the context of relativity.

The real shift away from φ-science happened in the closing decades of the twentieth century, when self-organisation theory, synergetics and chaos theory emerged. All of these, self-organisation theory in particular, recognised irreversibility. For these approaches, the unique direction of time is an objective phenomenon, not just a subjective empirical appearance. The arrow of time became real. This approach no longer fits strictly into the model of φ-science. The role of the experiment changed. Strictly speaking, the experiments could no longer be repeated.

The decisive change concerning the inclusion of time into physics was first introduced by Ludwig Boltzmann. An important contributor was Henri Poincaré. More on this will follow below in the historical analysis.

Irreversibility of phenomena is more visible in experimental chemistry rather than in physics, as in the case of the experiments of Henri Bénard (Bénard's convection) or Belousov-Zhabotinsky. We can repeat these experiments but the pattern we can observe never develops in the same way. As a result of this kind of studies, chemistry seemingly started to undermine the leading position of physics as science proper, becoming a more innovative traditions-breaking field. However, as I shall claim below, chemistry actually rather began to offer additional support to the model of φ-science as a necessary foundation for anything properly scientific.

The real champion of irreversibility in science was Ilya Prigogine. He declared that irreversibility is not subjective but a real theoretical fact. The explanation and justification has been provided, for instance by Leo Näpinen and Peeter Müürsepp:

> By accepting the considering of irreversibility (the distinction between past and future) as the basic element of description of physical world, by referring to the starting position of the mathematical description of "large" physical systems that lies in the infinity of a

> number of degrees of freedom, by acknowledging the central role of Poincaré resonances, nonintegrability and quantum chaos in modelling the physical world, we have by the same token referred to the fact that, for the adequate evaluation of Ilya Prigogine's approach, the impossibility to measure the initial conditions precisely enough cannot be the heart of the matter in principle. In the 'new science' a totally different approach becomes necessary. (Näpinen and Müürsepp 2002, p. 477)

The core of the necessary totally different approach was making a clear and constant distinction between the past and the future. The 'arrow of time' had to be permanently included into science. This has not been the case neither in the classical physics nor quantum mechanics. Irreversibility is the key concept in the 'new science' that was introduced.

Looking at the new situation from the perspective of the conception of φ-science, we can clearly see that the classical requirement for the repeatability of the experiments has to be dropped. The other basic requirements, testable hypotheses and mathematical expression, however, are retained. And in some sense, as we just saw above, the experiments still play an important, although an altered, role. It can be argued whether the new science requires the object of research to be constructed. If it does not then here is another characteristic feature, showing that physics today is no longer entirely φ-science. Thus, the model is badly needed.

What is the role of chemistry here? In a way, contemporary physics has got closer to chemistry as the latter has always been on both sides of the demarcation line, being partly φ-science and partly not. It's partly physics-like and partly natural philosophy like.

The next question would be: why not take chemistry as the model of science as it has a hand in both major kinds of science and more clearly so than physics. The easy answer would be that being on both sides is exactly the reason why we cannot make chemistry the model. There is a need for the clearly defined model of science. Even physics itself is not good enough, but obviously it is still much closer to the model of φ-science than chemistry. However, in a way, the existence of chemistry even supports the special status of physics.

3 The Historical Justification. How Has Chemistry Been Able to Support the Position of Physics?

3.1 Before the Arrow of Time

The first physicist was Aristotle. His physics was qualitative. Our modern physics is quantitative. This does not mean, however, that Aristotle is fully obsolete from the modern point of view. Just the opposite is the case. There is a lot to learn from Aristotle, and physicists today should pay much more attention to his legacy than they are normally used to do (Müürsepp 2003). After all, these are the qualitative results, rather than exact quantitative measurements, that normally interest us, humans. It is important to realize that in the ancient times already, (Aristotelian) physics occupied the central stage in the pool of knowledge about the world. Perhaps even more importantly, Aristotle (or his followers) retained a tight connection between basic philosophy (*Metaphysica*) and the study of nature (*Physica*). The terms used indicate that, actually the study of nature may have been even more important than pure philosophy, although there are also arguments that work in the opposite direction in the context of Aristotelian legacy.

The four-cause theory of the Stagirite deserves special attention. Modern science lost the grasp that each and every process or phenomenon has its material, formal, efficient and final aspect. Galileo started modern physics but, at the same time, committed a crime of different meaning (Mumford 1970). This is an assessment used by Lewis Mumford, who considered introducing a strict separation of the objective and the subjective spheres of human cognition no less than a criminal act. The philosophy of Francis Bacon and the science of Galileo killed the Aristotelian legacy. Generally, this has been considered a progressive move. This is so in the short perspective, though perhaps a mistaken attitude in the long run.

The crime of Galileo has yet another aspect, even a more significant one. In the process of objectifying science, he cancelled the human observer. Since this bold move by Galileo, the objective method of science has been in place. Surprisingly, while cancelling out the human observer, no one noticed that the method was put even under stronger human control. The researcher began to construct the object of research for herself, shifting away from reality. Here we have the beginning of the realism-constructivism debate about the essence of modern science that will never end by all evidence.

Why and how did this problematic situation around physics enable it to win and retain the position of the only science proper? It appears that chemistry helped. As we know, we can speak not just about physics but about physics-like science. The only physics-like science in addition to physics itself is chemistry. Moreover, even chemistry is not fully physics-like. Chemistry is physics-like when chemists are looking for the laws of nature. To be precise, we should actually make a distinction between the laws of physics and the laws of chemistry, the law of periodicity being the model example of the latter. As we saw above, finding out the law of periodicity was really a physics-like activity based on the construction of the notion of the chemical element. Still the law itself is not a law of physics, it's a law of chemistry.

For taking a historical position on chemistry we need to take at least a brief look into alchemy. There may be the question whether alchemy plays the same role for modern chemistry as Aristotelian physics did in the case of physics. There is a certain similarity here but only when looking from a very broad perspective. After all, Aristotelian physics was not just physics, but the study of inert nature, kind of a φ-science of the Antiquity but having a very different essence. In Aristotle's philosophy, we can see the basic structure of thought not concerning physics only but concerning chemistry as well. First of all, it is the four-cause theory that lays down the guidelines for cognition and thinking for an approach to any kind of processes and phenomena in the world.

To some extent, alchemy can be called qualitative chemistry. On the other hand, however, it was a cultural phenomenon of its time with certain visibility in society, which can hardly be said about the qualitative physics of the Stagirite. There is an interesting move in the thinking of Aristotle that has to be remembered here. It concerns bringing in the so-called fifth element, the "aether" or "quintessence" as the Scholastics called it. The alchemists derived the basis of their knowledge directly from the main elements and basic movements that were present in Aristotle's qualitative physics. The fifth element was absolutely necessary for putting the full picture together.

Alchemy might well have played even a greater role in the mediaeval society than science does today. People are more interested in fields that are not fully rationally confirmed but rather include some emotional excitement, not to speak of magic. By all evidence, the alchemists enjoyed the position of somewhat special people who were supposed to be dealing with something not too far from magic but retaining its rational core. Most importantly, however, alchemy was special because of its prospective practical value, never mind the expectations were almost never ever met. Still, alchemy strongly contributed to chemistry becoming a practical science (Müürsepp 2016). Interestingly, and unfortunately, the attitude to scientists today is not necessarily very different from the general attitude to alchemy in mediaeval times. Fortunately, modern scientists still have the upper hand. Differently from the alchemists, they are able to deliver. But this is often not enough for the general public. People expect miracles.

Quite typically, Rein Vihalemm calls the chapter dedicated to alchemy in his book about the history of chemistry 'Mysterious Alchemy' (Vihalemm 2019, pp. 37–57). What was the mystery there? One thing is that alchemists were taken to be mysterious, as they were believed to be able to make gold from precious metals or produce the Elixir of Youth or of the Wise, also called the Philosopher's Stone. In reality, alchemists were not able to achieve anything of special value but the procedures they worked out and the laboratory equipment they developed was a valuable contribution to the scientific chemistry of modern times.

The mystery also connects to Aristotle's qualitative physics. Actually, in the context of that it is not inconceivable at all to produce a metal from other metals. But still it remained something slightly mysterious. Perhaps because it could never be realised.

Was alchemy even chemistry? Did it provide us with scientific knowledge or rather with knowledge that belongs to non-science (Müürsepp 2011)? By all evidence, the latter is true. Alchemy was a cultural phenomenon of its time. It provided a lot for modern scientific chemistry. However, this legacy had to be put into the framework of the modern scientific method provided by the physicists starting from Galileo. This is another important moment where chemistry has actually helped physics to strengthen its role as science proper. Physics has obviously provided the methodological basis for chemistry in its development into a science.

After the Middle Ages and Renaissance, the stage was set for Galileo, Kepler, Newton, etc. Physics remained the only real science for many for quite a while. Still, there was also Robert Boyle. Thanks to him, chemistry was not out of the picture.

More than that, it may be that, without Boyle, the scientific method would never have been established. It is interesting, however, that without physics being there, Boyle might never have conceived to developing chemistry into a science similar to physics. Boyle had a sceptical-critical mind, similar to contemporary scientists. At the same time, he was an alchemist who believed in the possibility of the transmutation of metals. It appears that alchemy triggered some of the greatest minds that are called straightforward scientists today, such as Robert Boyle and Isaac Newton. It is no secret that the great Newton was also keen on alchemy.

Robert Boyle defined chemistry as a science on the model of physics and therefore gave significant support to physics being recognized as science proper. This is even more remarkable because chemistry developed into a science of dual character. It became a physics-like science looking for lawlike relationships and at the same time a natural philosophy type of field of knowledge that deals with stuff (substances).

The conception of the twofold character of chemistry is an observation that has been discussed in quite several publications. We do not need to review them all here. The idea has been explained by Rein Vihalemm in many papers and by the author of this article in some. A concise insight can be found, for instance in Müürsepp (Müürsepp et al. 2019). The authors write:

> What does the dual character of chemistry actually mean? Based on what has been said above, it is perhaps obvious that the idea has to be that chemistry is partly a φ-science and partly a non-φ-science. Chemistry is physics-like to some extent beyond any reasonable doubt. Just like physics, chemistry also aims at finding out regulations in nature that can be called laws. However, this is not what the whole chemistry is about. Chemistry also deals with substance or stuff. This raises the question whether the physics-like part of chemistry is not just physics-like but actually belongs to physics. We are going to show, however, that this would not be an appropriate interpretation of chemistry. (Müürsepp et al. 2019, p. 10).

An important observation of the authors is that even the laws that a part of chemistry is looking for, are different from the laws that physics is dealing with. Both the laws of physics and the laws of chemistry qualify as the laws of nature. Still, they are not of the same kind. Again, a good example is the law of periodicity of chemical elements here. Studying its unfolding clearly reveals that the law is specifically chemical and cannot really fit into physics, although the constructive aspect is present there as we saw above. Here we have another supporting fact to physics being science proper here. The laws of physics do adhere more strictly to the method of science, being clearly based on constructive type of thinking about the objects of research. The law of periodicity is similar in that respect but still different as it's objects, the chemical elements, do not belong to physics. As a rule, the other laws of chemistry deviate even more from the similarity to the laws of physics.

3.2 The Arrow of Time Making Its Presence in Science

An important development started in late nineteenth early twentieth century. These changes were mostly connected to Ludwig Boltzmann and Henri Poincaré. Obviously, the arrow of time connects with irreversibility. The nineteenth century left us with the dilemma: the laws of nature, such as Newton's law, which describe the time-reversible universe, and an evolutionary description associated with entropy. The latter is an essential part of thermodynamics, the science that deals specifically with irreversible, time-oriented processes. In contrast to time-reversible processes, such as the motion of a frictionless pendulum, where future and past play the same role, irreversible processes have a direction in time. For instance, a radioactive substance prepared in the past will disappear in the future.

We know already that the primordial role of the direction of time is evident in the processes we study at the macroscopic level, such as chemical reactions. Nature involves both time-reversible and time-irreversible processes. However, irreversible processes are clearly the rule and reversible processes the exception. The latter correspond to idealizations. We have to ignore friction to make the pendulum move reversibly. Such idealizations are problematic as there is no absolute void in nature. Time-reversible processes are described by equations of motion like Newton's equation in classical mechanics or Schrödinger's equation in quantum mechanics. For irreversible processes we need a description that breaks time symmetry.

The distinction between reversible and irreversible processes was introduced through the concept of entropy associated with the second law of thermodynamics. Entropy was defined by Rudolf Julius Clausius in 1865. According to the second law, irreversible processes produce entropy, reversible processes leave it constant. Clausius's statement of the second law was the first formulation of an evolutionary view of the universe. It was Arthur Stanley Eddington, however, who called entropy the "arrow of time".

For Ludwig Boltzmann, the nineteenth century was the century of Charles Darwin. It is the tragedy of Boltzmann that, for most physicists, he is today associated with a conclusion quite opposite to that of Darwin. Both Boltzmann and Darwin replaced the study of individuals with the study of populations and showed that slight variations taking place over a long period of time can generate evolution on a collective level. But while Darwin attempted to explain the appearance of new species, Boltzmann described an evolution toward equilibrium and uniformity. Boltzmann's interpretation of irreversibility succumbed to its critics and he was forced to retreat. Boltzmann was convinced that in order to understand nature we have to include evolutionary features and that irreversibility, as defined by the second law, was a decisive step in this direction. But he also realized that it stood in the way of his attempt to give a microscopic meaning to the arrow of time. Unfortunately, Boltzmann's attempt was leading to a failure. Explaining irreversibility in terms of trajectories that are time-reversible processes, however numerous, is simply a logical error. This was noted by Henri Poincaré. In case we invert the sign of the velocity of all molecules, the system would go into its own past. If entropy was increasing before velocity inversion, it would now decrease. When faced with severe criticism, Boltzmann replaced his microscopic interpretation of the second law with a probabilistic interpretation based upon our lack of information.

Boltzmann's probability-based interpretation makes the macroscopic character of our observations responsible for the irreversibility we observe. If we could follow the individual motion of the molecules, we would see a time-reversible system in which each molecule follows the laws of Newtonian physics. Therefore, according to the probabilistic interpretation, irreversibility is not a basic law of nature. It is merely a consequence of the approximate, macroscopic character of our observations.

In the wake of Boltzmann's failure, several arguments were put forward, which imply that it is our own ignorance that leads to the second law. For such approach, a well-informed observer like Laplace's demon would see the world as perfectly time reversible.

It was Max Planck who first opposed the idea of incomplete information in his *Treatise on Thermodynamics* (Planck 2016). Planck observed correctly that the validity of the second law does not depend in any way on the skill of the physicist or chemist in observing or experimenting. The law asserts that there exists in nature a quantity which always changes in one and the same direction in any natural process. Propositions stated in such general form may be true or false. But whichever they may be, it will remain so, irrespective of whether there is an intelligent being in the universe able to perform measurements and experiments. According to Max Planck, the limits of the second law, if any, must lie in the observed nature itself, not in the observer.

Planck's views were to remain isolated, even rejected. The tables turned, however, when Ilya Prigogine took the stage. Referring to developments in nonequilibrium physics and chemistry in the last decades of the twentieth century, Prigogine gives the arrow of time the role of a source of order. This is shown by quite simple experiments, i.e., thermal diffusion. The constructive role of irreversibility becomes even clearer in far-from-equilibrium situations where nonequilibrium leads to new forms of coherence. In *From Being to Becoming* Prigogine formulates the following basic conclusions:

- irreversible processes are as real as reversible processes described by the fundamental laws of physics; they do not correspond to approximations added to the basic laws;
- irreversible processes play a fundamental constructive role in nature (Prigogine 1980).

The theory of microscopic irreversibility, created by Prigogine and his colleagues, is a part of a new mathematical natural science which involves irreversibility and chance. This science formulates the new types of laws, which are at least to some extent in accordance with the open, evolving world including human beings. Prigogine postulated the second law as a fundamental theoretical fact, as a principle of selection leading to the breaking of time symmetry. Treatment of time as an operator by Prigogine has made it possible to explain the inner breaking of time symmetry simultaneously in classical and quantum mechanics as well as relativity theories.

From our current perspective, it is important to note that Ilya Prigogine can be considered a chemist and a physicist at the same time. He received the Nobel Prize in chemistry. Thus, chemistry is again, directly involved in the development of exact natural science—not countering it—but rather supporting. Once again, we see the need for a model of science, the φ-science. Still, it is physics that is the closest science to the model. Even in the 'Prigoginian science', there are hypotheses, measurements, experiments (although not strictly repeatable) (Müürsepp 2013), expressions in the language of mathematics and, most importantly, looking for lawlike regularities. The new physics of Prigogine is not fully φ-science, but it is close enough to the model. The model of φ-science is needed more than ever before to keep science on track, preventing it from diverging into an alchemy-type of cultural phenomenon.

4 Conclusion

The core idea of the current essay is to justify the position of physics as the only science proper from the historical perspective of the development of physics in conjunction with the development of chemistry, the only other truly physics-like science, although not entirely such.

Comparing physics and chemistry on philosophical grounds and following their parallel historical development we can see that physics has normally been a step or two ahead, except for in the most recent decades. Chemistry is an especially useful field of comparison as it is partly physics-like and partly not. We have good grounds for retaining the position of science proper for physics for different reasons. The method of modern science was coined on the basis of physics. The development of physics (classical mechanics) enabled chemistry to become a science, mostly as a result of the efforts of Robert Boyle. However, Boyle's success in making chemistry a hypothetico-deductive science emphasized the position of physics as the basis of everything scientific in the modern sense.

The objective nature of the arrow of time seemingly turned the tables of most contemporary science towards chemistry as the model for a science proper. However, let us not forget that the phenomenon of irreversibility was discovered and studied by physicists first. Thus, even the most recent developments in these sciences strengthen the leading role of physics. Chemistry is lifted to a special position as well but in a different sense. In a way, we can even call chemistry the basic science (Müürsepp et al. 2021), while physics still retains its position as science proper.

Acknowledgements The author thanks Estonian Research Council for the research Grant PRG462.

References

Maxwell N (2017) In Praise of Natural Philosophy. A Revolution for Thought and Life. McGill-Queen's University Press Montreal & Kingston, London, Chicago.
Mumford L (1970) The Myth of the Machine. The Pentagon of Power. Harcourt Brace Jovanovich, Inc., New York.
Müürsepp P (2003) The Need for Aristotle in Contemporary Science. Problemos 64:111–123.
Müürsepp P (2011) Knowledge in Science and Non-Science. Baltic Journal of European Studies 1(9):61–73.
Müürsepp P (2013) The Changing Role of Scientific Experiment. Studia Philosophica Estonica 5(2):152–166.
Müürsepp P (2016) Chemistry as a Practical Science (Edward Caldin Revisited). Foundations of Chemistry 18(3):213–223.
Müürsepp P, Nurysheva G, Bekenova Zh, Usenov G (2019) From the Dual Character of Chemistry to Practical Realism and Back Again: Philosophy of Science of Rein Vihalemm. Problemos 96:107–120.
Müürsepp P, Nurysheva G, Ramazanova A, Amirkulova Zh (2021) Chemistry as the Basic Science. Foundations of Chemistry 23:69–83.
Näpinen L, Müürsepp P (2002) The Concept of Chaos in Contemporary Science: On Jean Bricmont's Critique of Ilya Prigogine's Ideas. Foundations of Science 7:465–479.
Newton I ([1726] 1999) Philosophiae Naturalis Principia Mathematica, General Scholium. Third edition. In: The Principia: The Authoritative Translation and Guide: Mathematical Principles of Natural Philosophy. I. Bernard Cohen and Anne Whitman's 1999 translation. University of California Press, Berkeley.
Planck M (2016) Treatise on Thermodynamics. The Project Gutenberg EBook. The Project Gutenberg eBook: Treatise on Thermodynamics, via https://www.gutenberg.org/ebooks/50880. Last accessed 2021/12/29.
Prigogine I (1980) From Being to Becoming: Time and Complexity in the Physical Sciences. W.H. Freeman and Company, San Francisco.
Vihalemm R (2015) Philosophy of Chemistry against Standard Scientific Realism and Anti-Realism. Philosophia Scientiae 19(1): 99–113.
Vihalemm R (2019) A Story of a Science: On the Evolution of Chemistry. Special Issue Acta Baltica Historiae et Philosophiae Scientiarum 7(2).
Vihalemm R, Lõhkivi E (2012) Guest Editorial. Philosophy of Science in Practice and Practical Realism. Studia Philosophica Estonica 5(2):1–6.
Whewell W (1840) The Philosophy of the Inductive Sciences. London.

Chapter 26
Giambattista Suardi's Machine for Drawing Plane Curves and other Instruments for Cycloids

Nicla Palladino

Abstract Giambattista Suardi's "Geometric Pen" was an instrument based on wires and gears that can be used to draw a variety of curves generated by combining different rotary motions. Suardi described his machine in 1752 and explained that its mechanism was inspired by Ptolemaic theory. The instrument allowed him to draw many curves based on the combination of three basic parameters: the angular velocity of the wheels, the ratio of their radii or their number of teeth, and the possible directions of the motions. This article describes some of the complex mechanical artifices that Suardi used to draw curves, as well as correlations between Suardi's machine and some later instruments. His machine was probably a precursor of the "Spirograph," a geometric drawing device that produces mathematical roulette curves as hypotrochoids and epitrochoids. The well-known toy version was developed by British engineer Denys Fisher, who exhibited his Spirograph at the Nuremberg International Toy Fair in 1965. The invention is often attributed to Abdank-Abakanowicz, the inventor and designer of integraphs, a mechanical device for plotting logarithmic spirals; I show here where the error comes from.

Keywords Intergraph · Mechanical machines · Spirograph · Trochoids

1 Introduction: Suardi's Treatise on Instruments

1.1 Suardi's Curves

In 1752, the Italian mathematician Giambattista Suardi (1711–1767) published an essay on drawing and mathematical tools: *Nuovi istromenti per la descrizione di diverse curve antiche e moderne e di molte altre che servir possono alla speculazione de' geometri ed all'uso de' pratici: col progetto di due nuove macchine per la nautica*

N. Palladino (✉)
Dipartimento di Medicina e Scienze della Salute, Università del Molise, Campobasso, Italy
e-mail: nicla.palladino@unina.it

R. Pisano (ed.), *A History of Physics: Phenomena, Ideas and Mechanisms*,
History of Mechanism and Machine Science 42,
https://doi.org/10.1007/978-3-031-26174-9_26

ed una per la meccanica, e con alcune osservazioni sopra de' poligoni rettilinei regolari (Schilling 1911). He came from a noble family and moved to Padua at the age of 20 to study law, but also attended mathematics classes under the supervision of the famous physicist and mathematician Giovanni Poleni (1683–1761), who sparked his interest in practical applications and mathematical tools.

In his 1752 treatise, Suardi described the construction of many well-known curves, thanks also to some instruments that he considered original: instruments for drawing conchoids, the cissoid of Diocles and the curve of Mr. Carré, quadratrix, conic sections, cycloids, Cartesian ovals, curves generated by the combination of two rotary motions, loxodromes and regular polygons.

The ninth instrument, which Suardi called *Penna geometrica* (*Geometric pen*) is particularly interesting and original; it allows him to draw.

> Curves generated by the motion composed of two circles, i.e. straight line, ellipse, circle, geometric flowers, cycloids with a circular base, spirals and particularly the spiral of Archimedes.[1] (Schilling 1911, p. 85; Author's translation)

The author divided the description into four parts: description of curves by points, construction of the mechanism, use of the instrument, use of a similar instrument made of gears. At the end of the volume there are many tables with drawings of the instrument and the curves that can be drawn. Suardi presented two versions of his Geometric Pen: the first consists of cylinders connected by wires; the second is equipped with gears that remedy some of the drawbacks of the first version. The instrument allows to draw many curves, but also plane figures such as polygons, based on the combination of three basic parameters: the angular velocity of the cylinders or wheels, the ratio of their radii or the number of teeth, and the relative directions of the motions (clockwise and counter clockwise).

Suardi was not interested in the theoretical treatment of curves, which he limited to what he called organic explanation, but rather, in the development of methods and combinations of parameters that made it possible to create curves graphically; in fact, he seems to have proceeded experimentally to describe the results of his work. In particular, the version with cylinders and wires allows Suardi to draw the Archimedean spiral and the elliptical spiral, using accessories in the form of cylinders and cones. The gear instrument can be used to draw epitrochoids and hypotrochoids, figures that can be easily drawn today with a well-known device sold under the name of *Spirograph*, a geometric drawing device that generates mathematical roulette curves.

Suardi wrote that he was inspired by the Ptolemaic system and named the two main parts of the device *Primo Mobile* (*First Moved*) and *Epiciclo* (*Epicycle*) because they move around each other. The Primo Mobile rotates around itself, while the epicycle as it rotates around itself also moves along the Primo Mobile. In Ptolemaic systems, planets are assumed to move in a small circle, the epicycle, which in turn moves on a larger circle, the deferent. Both circles rotate clockwise and are roughly parallel to

---

[1] "Curve generate dal moto composto di due circoli, ossia linea retta, ellisse, circolo, fiori geometrici, cicloidi di base circolare, spirali e particolarmente la spirale di Archimede".

the plane of the Sun's orbit. Although the system is considered geocentric, neither circle was exactly concentric with the Earth. More specifically, the motion of each planet was centered at a planet-specific point slightly away from the Earth called the eccentric.

With his instrument, Suardi was able to describe orbits of planets according to the geocentric model. He wrote that the curve displayed in Fig. 1a is an example of the planetary orbit resolving around the epicycle, whose centre in turn rotates on the Primo mobile circumference. To draw the curve in Fig. 1b, one must set the angular velocities equal and the motion in the opposite direction, while the radii of the two circles can be set arbitrarily. The construction of the curve was described point by point: the radius of the Primo mobile is *cn*, while the epicycle *a126* has centre *n* and radius *na*. He divided the peripheries of the two wheels into a number of parts equal to the reciprocal of the velocities; in this case, since the velocities are equal, both peripheries are divided into 8 parts. From the centre *c* of the Primo mobile, the radii *co*, *cp*, etc. are drawn and extended. With the centres in *o*, in *p*, etc. and distance *na*, circumferences corresponding to the epicycle are drawn.

When the radius *cn* moves to the left, i.e. to the points *o*, *p*, etc., it drags its epicycle with it; when *cn* goes to position *co*, the radius *na* would coincide with *ov* if it had not moved either to point *3* (in the same direction) or to *b* (in the opposite direction, as in this case). Moreover, when the radius *cn* goes to position *co*, *cn* has already covered the arc *no*; thus the part *a1* of the periphery of the epicycle, that runs from *v* to *b*, is drawn by a compass with aperture na. If the radius *cn* has already covered two sections up to *cp*, then two parts *a2* of the periphery of the epicycle from *x* to *d* are drawn and so on. The points *a*, *b*, *d* belong to the curve and in this way other points are obtained connecting with a line as it was suggested. Finally, it is worth noting that in this way the ratio between the angles *vob* and *nco* are in the same relation as the velocities; as here the velocities are equal, *vob* and *nco* are also equal.

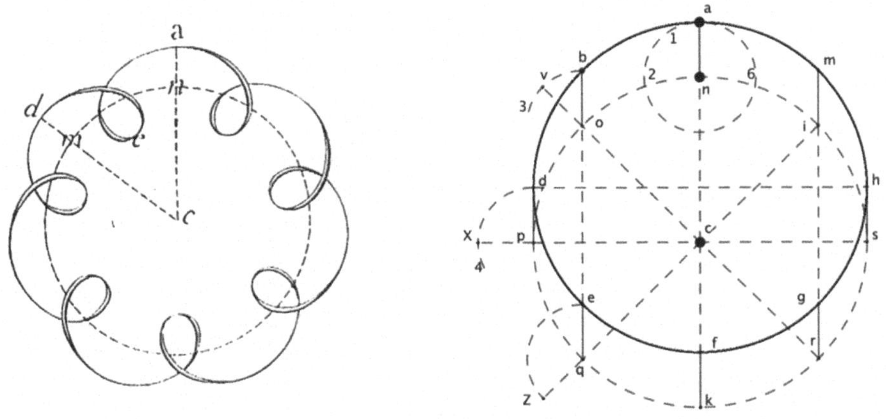

(a) **(b)**

Fig. 1 **a, b** Construction of a curve (Schilling 1911, Tav. XV) *Source* Google Books, Public Domain

Suardi reported that by calculating the combinations that can be obtained by varying angular velocity of the cylinders or wheels, the ratio of their radii and the possible directions of the motions, 1273 different curves can be drawn.

1.2 How Geometric Pen Works

As mentioned above, Suardi designed two versions of his Geometric pen; the first one is based on cylinders, rods and wires (Fig. 2).

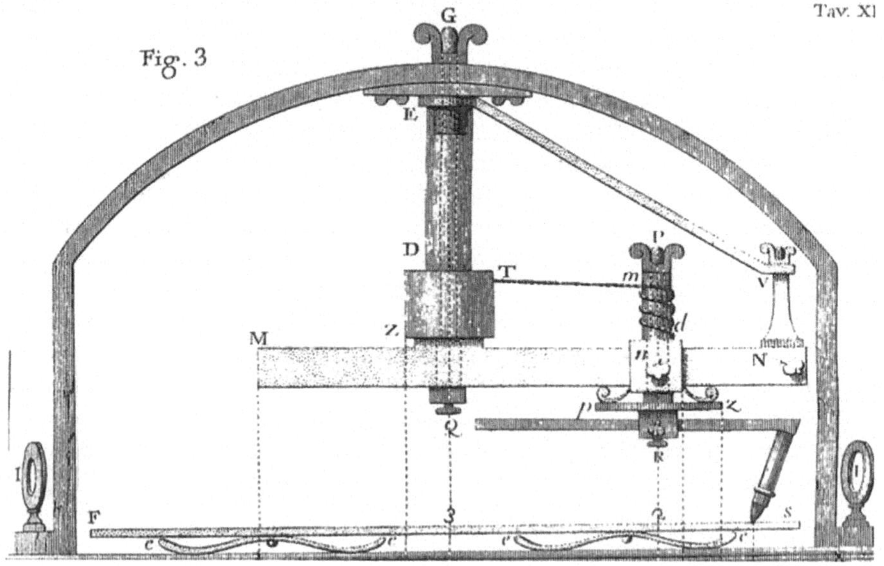

Fig. 2 The first version of Suardi's Geometric pen (Schilling 1911, Tav. XI) *Source* Google Books, Public Domain

The device consists of a central cylinder *TQZ*, whose center *Q* coincides with the center of the Primo mobile. The rod *MN* is attached to the cylinder. The second cylinder *mrd* with center *r* is attached to *MN* such that *Qr* is the radius of the Primo mobile. Connected to the cylinder *mrd* is another rod of variable length ending with a pen at point *S* so that *rS* is the radius of the epicycle. The pen is used to draw curves. Suardi inserted a string that starts from the cylinder *TQZ* and wraps around the cylinder *mrd*. The string that winds up or unwinds on the central cylinder winds up or unwinds on the smaller cylinder at the same time; this is how the movements of the deferent and the epicycle are generated.

We describe how it is possible to vary the three parameters to define the different curves. To vary the lengths of the radii, Suardi varied *Qr* and *rS*. Note that the angular velocities are inversely proportional to the diameters of the two cylinders.

For example, if we assume that the diameter of *mrd* is in the ratio 1:3 to the diameter of *TQZ*, the two circumferences are in the same ratio. Therefore, to change the ratio between the two circumferences, the smaller cylinder must be replaced. To make the angular velocities of the two circles different, he designed twelve cylinders with diameters in the ratio 1:2, 1:3, 1:4 and so on up to 1:12. If we assume in Fig. 2 that the radius of the Primo mobile has travelled 1/3 of the circumference (arriving at point 3), then the cylinder is also rotated by 1/3 and the point *T* reaches *Z*. The circumference *mrd* is equal to 1/3 of *TQZ*, then the small cylinder (thanks to the wire connecting the two cylinders) has completed one revolution. So in the time that *TQZ* makes one complete turn, *mrd* makes 3. To make the cylinders rotate in the same direction, Suardi planned to attach the wire to the alternate parts of the two cylinders (*Tm*); to make them rotate in opposite directions, the wire is attached to the same side instead (*Td*).

Suardi, who admitted that he had not tested the construction of the instrument, imagined at least two potential problems. In the first place, he thought that the string does not always remain taut in the same way: the tension may vary according to the force that the pen must overcome to draw the curve. The second problem is related to the dimensions: the string has its own thickness, which must therefore be added to the diameter of the cylinders. Not entirely satisfied, Suardi improved the instrument by replacing wires and cylinders with gears. To vary the angular velocities, Suardi designed a series of gears with a certain number of teeth in a certain ratio. For example, to achieve a ratio of 1:2, the first gear has 8 teeth, the second 16. Assuming that the size of the teeth is fixed, he divided the circumference of a wheel with 16 teeth into 16 parts, each containing one tooth. In this case, he also used proportionality: he made 12 gears with a number of teeth in the same proportion to the diameters of the two replaced cylinders. Assuming that the first wheel has 8 teeth, the second must have 16 teeth, the third 24, and so he proceeded until the twelfth wheel (96 teeth). Each arc of circumference corresponding to a tooth is $d\pi/16$ long; therefore, a wheel with 8 teeth has a circumference of $d\pi/2$. He reproduced exactly the same situation as with the cylinders.

Finally, if the wheel of the epicycle is directly stuck in the wheel of the Primo mobile (i.e. there are only two gears in the geometric pen), then the Primo mobile and the epicycle have the same direction of rotation. To allow motions in opposite directions, Suardi inserted a third gear with any number of teeth between the two gears (Fig. 3).

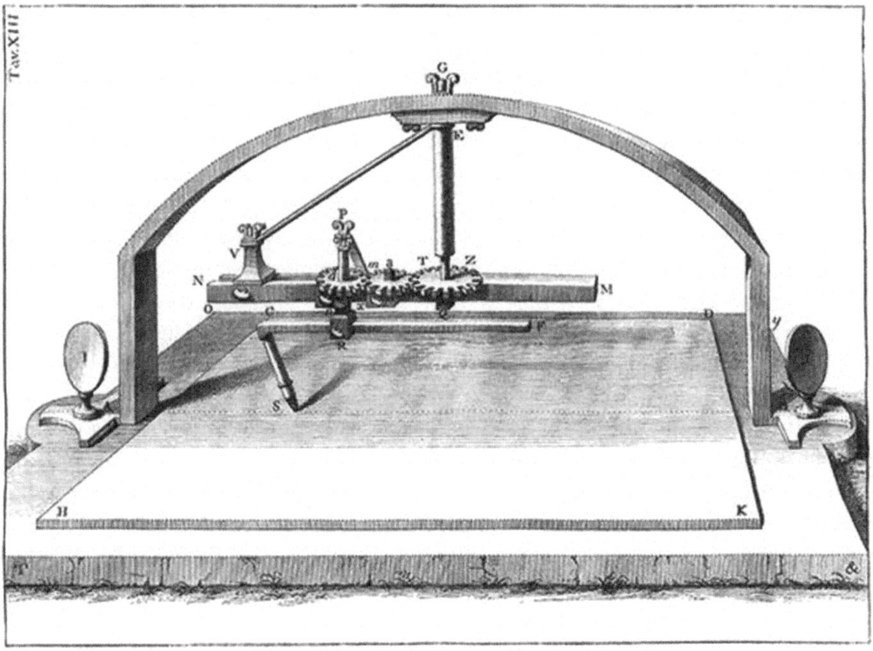

Fig. 3 The second version of Suardi's Geometric pen (Schilling 1911, Tav. XIII) *Source* Google Books, Public Domain

2 Methods for Drawing Hypotrochoids and Epitrochoids

2.1 Suardi's Trochoids

The notion of trochoid refers to the curve described by a point linked to a disk rolling without slipping on a line. Suardi used his Geometric pen to draw curves that have the name of hypotrochoids and epitrochoid. A hypotrochoid (from the Greek *hupo* - *under* and *trokhos* - *wheel*) is the curve described by a point attached to a circle of radius r rolling without slipping around the inside of a fixed circle of radius R, where the point is a distance O from the center of the rolling circle. The parametric equations for a hypotrochoid are:

$$\begin{cases} x = (R - r)\cos t - O * \cos\left(\dfrac{R-r}{r}t\right) \\ y = (R - r)\sin t - O * \sin\left(\dfrac{R-r}{r}t\right) \end{cases}$$

When $O = r$ we have a hypocycloid.

An epitrochoid (from the Greek *epi - over*) is the curve traced by a point attached to a circle of radius r rolling without slipping around the outside of a fixed circle of radius R, where the point is at a distance O from the center of the rolling circle. The parametric equations for an epitrochoid are:

$$\begin{cases} x = (R+r)\cos t - O * \cos\left(\frac{R+r}{r}t\right) \\ y = (R+r)\sin t - O * \sin\left(\frac{R+r}{r}t\right) \end{cases}$$

When $O = r$ we have an epiclycloid.

Suardi's instrument, as conceived by him, could only draw epitrochoids but he managed to set it in such a way as to draw hypotrochoids as well, transforming them into epitrochoids. In the Geometric pen, the rolling circle has its center on the fixed circle, so the equation of the curve becomes:

$$\begin{cases} x = ((R-r)+r)\cos t - O * \cos\left(\frac{(R-r)+r}{r}t\right) \\ y = ((R-r)+r)\sin t - O * \sin\left(\frac{(R-r)+r}{r}t\right) \end{cases}$$

Following Fig. 4, Suardi showed that a hypocycloid with fixed circle *dok* and rolling circle *dmx* can be drawn as epicycloid with fixed circle *enp* and rolling circle *dmw*.

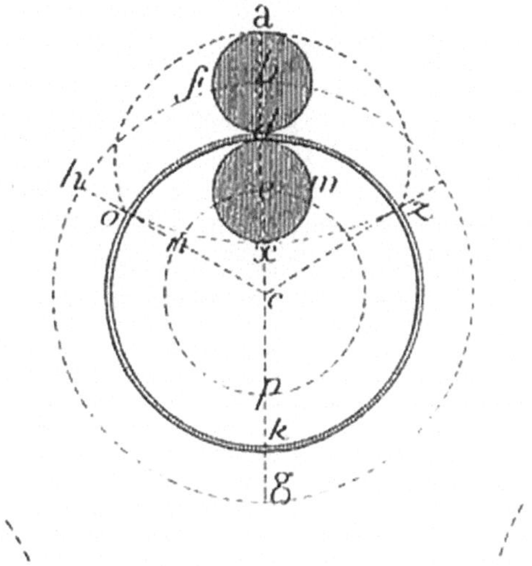

Fig. 4 Construction of hypotrochoids (Schilling 1911, Tav. XIX) *Source* Google Books, Public Domain

In fact, because of his transformation, Suardi used a fixed circle of radius R-r and the equation of the resulting epicycloid is:

$$\begin{cases} x = ((R-r)+r)\cos t - O * \cos\left(\frac{(R-r)+r}{r}t\right) \\ y = ((R-r)+r)\sin t - O * \sin\left(\frac{(R-r)+r}{r}t\right) \end{cases}$$

Replacing *R* by *R-r* (because of the rolling circle has its center on the fixed circle), the equation becomes:

$$\begin{cases} x = ((R-r-r)+r)\cos t - O * \cos\left(\frac{(R-r-r)+r}{r}t\right) \\ y = ((R-r-)+r)\sin t - O * \sin\left(\frac{(R-r-r)+r}{r}t\right) \end{cases}$$
$$= \begin{cases} x = (R-r)\cos t - O * \cos\left(\frac{R-r}{r}t\right) \\ y = (R-r)\sin t - O * \sin\left(\frac{R-r}{r}t\right) \end{cases}$$

This is exactly the equation of the original hypocycloid.

Starting from paragraph 5 of Part three, Suardi showed settings for his instrument to draw various figures that can be constructed as hypo- and epitrochoids. He described how to obtain a segment, an ellipse, and polygons, but the ones that seem particularly interesting for the purposes of this article are the constructions of "geometric flowers" ("fiori geometrici"). Suardi learned them from Guido Grandi (Grandi 1729). Since he had constructed only a few gears, he could not draw any epi- and hypotrochoid, but only those with the same circles as the wheels he built. Among the various geometric flowers he could build, Suardi drew the figure he called "ettafoglio" (Fig. 1a). To obtain this figure, the ratio between the radii and velocities in the same direction of rotation of the Primo mobile and the epicycle is 7:2 and 1:7, respectively. When the radii of the wheels are the same and the ratio between velocity of the Primo mobile and the epicycle is 5:8, he drew the figures called "ottofoglio" (Fig. 5a if the directions of rotation are the same, Fig. 5b otherwise).

Fig. 5 Two "ottofoglio" (Schilling 1911, Tav. XVII) *Source* Google Books, Public Domain

2.2 Spirograph

Suardi's machine was probably a precursor of the "spirograph", a geometric drawing device that produces mathematical roulette curves as hypotrochoids and epitrochoids (Fig. 6). The well-known toy version of the spirograph was developed by British engineer Denys Fisher who exhibited his tool at the 1965 Nuremberg International Toy Fair.

Born in 1918, Fisher was brought up in Leeds. He was interested in the infinitesimal calculus; he was captivated and a life-long passion for logic and mathematics ensued. Academia was not to his liking and, after dropping out of Leeds University, he joined the family firm Kingfisher (Lubrification) Ltd. In 1960 Fisher left business to establish Denys Fisher Engineering in Leeds. His success in business allowed him to pursue a project about the idea of using gears and cogs to create patterns. Fisher spent the Christmas of 1962 building various drawing machines from Meccano[2] pieces but finally he imagined a new device made by a series of perforated plastic cogwheels and racks. He hurried to his workshop where he created his prototype Spirograph from a Meccano cog and a geared ring. He had originally intended to create a draftsman's tool but he decided to market the spirograph as a toy (Suardi 1752). The box read "Pattern drawing by revolving stencils—a new invention for everyone".

---

[2] Meccano is a brand of model construction system created in 1898 by Frank Hornby in Liverpool, United Kingdom.

Fig. 6 A modern spirograph
Source Author's own made

The Spirograph was named "Toy of the Year" by the British Association of Toy Retailers. At the same time, the toy topped the bestseller list in the United States for two consecutive years. In 1827, Peter Hubert Desvignes, an English architect and engineer born in Greece, developed and advertised a "speiragraph," a device for making elaborate spiral drawings. A man named J. Jopling soon claimed to have invented a similar method previously. While working in Vienna between 1845 and 1848, Desvignes constructed a version of the machine to prevent banknote counterfeiting, as the nearly endless variations of roulette patterns that the device could produce were extremely difficult to replicate.

The original US-released Spirograph consisted of two differently sized plastic rings (or stators), with gear teeth on both the inside and outside of their circumferences. Once either of these rings were held in place (either by pins, with an adhesive, or by hand) any of several provided gearwheels (or rotors)—each having holes for a ballpoint pen—could be spun around the ring to draw geometric shapes. One of the discs is chosen and a coloured pen is placed through one of the holes. As the disc is rolled along the inside or outside of the ring or bar, a trochoid curve is drawn. Later, the Super-Spirograph introduced additional shapes such as rings, triangles, and straight bars. All edges of each piece have teeth to engage any other piece; smaller gears fit inside the larger rings, but they also can rotate along the rings' outside edge or even around each other. Gears can be combined in many different arrangements.

US distribution rights were acquired by Kenner, Inc., which introduced it to the United States market in 1966 and promoted it as a creative children's toy. With a hit toy on its hands, Kenner introduced a variety of Spirograph-related offerings to the market, including Super Spirograph (with more gears than the original), Spirotot (aimed at the kindergarten crowd), Magnetic Spirograph (using magnets to hold shapes in place on a page), and a variety of refill sets. In 1968, Kenner's Spiroman also appeared on the scene. Spirotot was the first spin-off of Spirograph; it was a simpler drawing toy for children aged three to seven. According to the box it "Does everything Spirograph does—and a whole lot more!".

Spirograph continues to be sold today. The name has been a registered trademark of Hasbro Inc. since 1998; Hasbro has marketed a variety of sets over the years.

Drawing toys based on gears have been around since at least 1908, when The Marvelous Wondergraph was advertised in the Sears catalog. An article describing how to make a Wondergraph drawing machine appeared in the Boys Mechanic publication in 1913.

All figures produced by Suardi's Geometric Pen can be considered hypotrochoids or epitrochoids. However, the Spirograph has the limitation of not being able to draw figures with cusps. This is because the instrument has physical limitations due to its structure and, unlike the Geometric Pen, it is not possible to use the pen exactly on the periphery of the circle.

3 Methods to Draw Spirals

3.1 *Spirals in Suardi's Nuovi Istromenti*

In paragraph 10, Suardi studied spirals and stated that also spirals can be drawn with his Geometric pen as they are generated by a simultaneous rectilinear and circular motion.

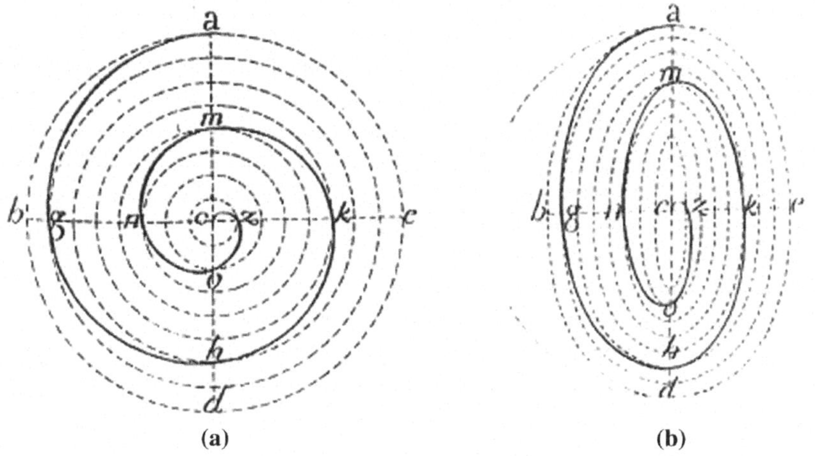

Fig. 7 Archimedean spiral and elliptical spiral (Schilling 1911, Tav. XVIII) *Source* Google Books, Public Domain

Therefore, he presented the description of a point moving with uniform rectilinear motion on a straight line while the latter rotates with uniform circular motion, that is an *Archimedean spiral* (Fig. 7a). In the second figure (Fig. 7b), an *elliptical spiral* is displayed, composed of a rectilinear motion and an elliptical motion. To create the curves, Suardi used the geometric pen made of wires and cylinders (Fig. 2). He placed the pen *S* between *Q* and *R* (otherwise the pen would not be able to draw the center of the spiral) so that once the wire is completely wrapped around the cylinder *ZT* the pen is exactly at point *Q*. He tightens the screw *P* so that the cylinder *md* no longer rolls and then loosens the screw *n* so that the cylinder *mn* can slide along the rod *MN*. The rectilinear motion is due to the cylinder *mn* along *MN* and the circular motion is due to *MN* around *QZD*.

Suardi prepared four pieces (Fig. 8): the first piece is a right circular cylinder, the second an elliptical cylinder, the third a circular cone and the fourth an elliptic cone. Insert the pieces in place of the cylinder ZT.

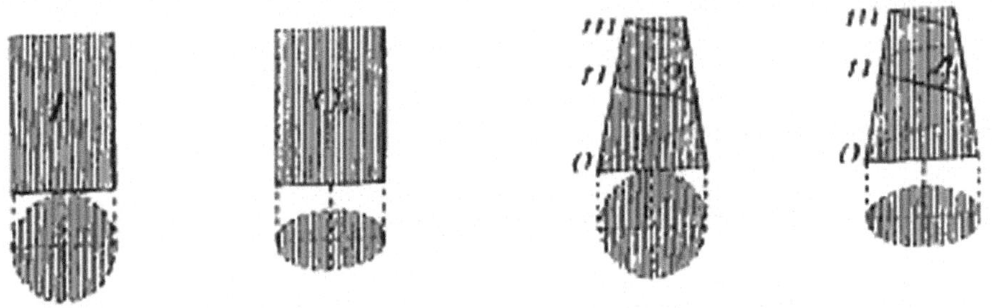

Fig. 8 Pieces for spirals (Schilling 1911, Tav. XI) *Source* Google Books, Public Domain

Through the first and second pieces, Suardi draws an Archimedean and an elliptical spiral; through the third and the fourth, he draws figures similar to the first two, but with spaces between the coils increasing or decreasing towards the center depending on whether he fixes the wire at the vertex *m* or at the base *o* of the piece. When *MN* has completed a full revolution, the distance between one coil and the other is equal to the circumference of the cylinder *ZT*. Using the first cylinder, the S pen starts at the point *a*. Considering a wire of twice the length of the circumference of *ZT*, after a quarter turn of *MN* the pen is at *g*; after one full turn, at *m*; finally, after two revolutions, it arrives at the center *c* which corresponds to *Q*. In practice, the Geometric pen drew an Archimedean spiral.

Suardi also designed an instrument for the logarithmic spiral which is a modification of Giovanni Poleni's instrument; it is presented as instrument 5 in the first paragraph (Schilling 1911, p. 27) and its picture is in Tav. IV.

3.2 Abdank-Abakanowicz's Integraphs

The invention of the spirograph is often attributed to Bruno Abdank-Abakanowicz, famous inventor and builder of integraphs. Abdank-Abakanowicz (1852–1900) was a Polish mathematician, inventor, and electrical engineer. Abakanowicz was born in 1852 in the Russian Empire (now Lithuania). After graduating from the Riga Technical University, Abakanowicz passed his habilitation and began an assistantship at the Technical University of Lwów. In 1881, he moved to France where he purchased a villa in Parc St. Maur on the outskirts of Paris. Earlier he invented the *integraph*, a form of the integrator, which was patented in 1880, and was henceforth produced by the Swiss firm Coradi. His other patents included devices to draw parabolas and spirals, the electric bell used in trains, and an electric arc lamp of his own design.

Abakanowic's famous treatise is *Les intégraphes. La courbe intégrale et ses applications.* The input to the integraph is a tracing point that is the guiding point that traces the differential curve (Abdank-Abakanowicz 1886). The output is defined by the path a disk that rolls along the paper without slipping takes. The mechanism sets the angle of the output disk based on the position of the input curve: if the input is zero, the disk is angled to roll straight, parallel to the x axis on the Cartesian plane. If the input is above zero the disk is angled slightly toward the positive y direction, such that the y value of its position increases as it rolls in that direction. If the input is below zero, the disk is angled the other way such that its y position decreases as it rolls.

The hardware consists of a rectangular carriage which moves left to right on rollers. Two sides of the carriage run parallel to the x axis. The other two sides are parallel to the y axis. Along the trailing vertical (y axis) rail slides a smaller carriage holding a tracing point. Along the leading vertical rail slides a second smaller carriage to which is affixed a small, sharp disc, which rests and rolls (but does not slide) on the graphing paper. The trailing carriage is connected both with a point in the center of the carriage and the disc on the leading rail by a system of sliding crossheads and wires, such that the tracing point must follow the disc's tangential path. Given the differential curve *abdfg*, the integraph draws the integral curve *ABCDE* (Fig. 9). Abdank-Abakanowicz started from the observation that to trace the integral curve of a differential curve means to draw a curve such that the generic ordinate *BK* (where *B* is the corresponding point of *b* on the integral curve and *K* is the corresponding point of *B* and *b* on the line parallel to *x* axis passing for A) is proportional to the area bounded by the differential curve on the interval [a;c] (similarly CF is proportional to the area bounded by the differential curve on the interval [a;e] and so on) (Palladino 1985).

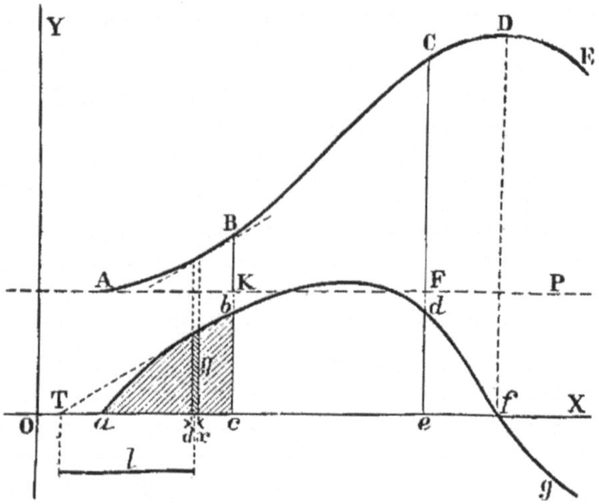

Fig. 9 Integral and differential curves (Abdank-Abakanowicz 1886, p. 2) *Source* Google Books, Public Domain

3.3 Abakanowicz's Spirograph

In the appendix X of his treatise on integraphs, Abakanowicz presented a device that allowed him to trace spirals as an application of the wheel (*roulette*) described in the previous pages: "Application de la roulette au tracé de quelques courbes". He designed it to plot integrals of curves.

The fundamental property of the roulette is to always move on the same plan and Abakanowicz's instrument draws the curves for which the inclination of the tangent with the axes (in case of Cartesian coordinates) or with the vector radius (for polar coordinates) is given. The trace of the integral curve is only a particular case in which the inclination of the tangent follows the law $\mathrm{tg}\alpha = y$, where α is the angle between the tangent of the integral curve and the x axis.

Abakanowicz's instrument for drawing the logarithmic spiral is an immediate application of the principle of the continuous screw. He called logarithmic spiral interesting because of its applications. In polar coordinates the curve has equation $r = ae^{k\varphi}$ and its tangent has always the same angle with the vector *r*. He used this latter property to build his instrument.

Abakanowicz built a carriage that can move along a straight line and turn around a wheel; the angle between the wheel and the frame can assume any value. he fixed the roulette bracket so that the angle was constant and then gave a rotational movement of the frame around its fixed point. The carriage moves along the tracks and the wheel describes the logarithmic spiral on the paper determined by the initial angle (Fig. 10):

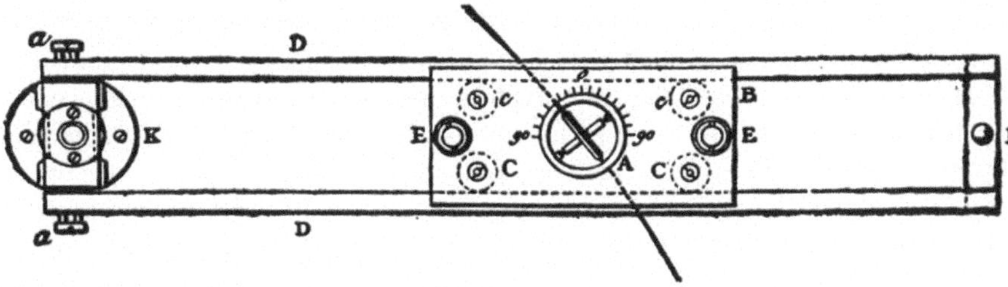

Fig. 10 Abakanowicz's spirograph (Abdank-Abakanowicz 1886, p. 137) *Source* Google Books, Public Domain

The wheel is knife-edged and movable in a carriage along a straight line. The angle of inclination of the wheel can be adjusted with the help of scales from –90° to +90°. The instrument can be turned around the center of the plate K, the pole of the instrument; let design the angle of rotation of the whole instrument, counted clockwise from an initial position. The carriage is moved completely to the left and the inclination of the wheel is fixed. If now the instrument is turned clockwise by its handle L (i.e. with φ permanently increasing from its initial position $\varphi = 0°$), the knife-edge of the wheel sinks into the paper and forces the wheel to keep its direction. As the direction of the wheel is oblique to the circle which it would describe if allowed to glide, there results an evasive movement, made possible by the mobility of the carriage along the straight line. The carriage will move towards L, its distance from K being proportional to the sine of the angle of inclination α, i.e. proportional to $\sin\alpha$, and of course also proportional to the arc s. To determine the trace of the knife-edged wheel one has to consider small ("infinitesimal") changes dr and $d\varphi$. If the wheel is a distance r away from K and if the instrument is turned by a small angle $d\varphi$, the length s of the arc mentioned above will be $rd\varphi$. Consequently, the displacement of the carriage will be proportional to both s and $\sin\alpha$, that is, proportional to $r\sin\alpha d\varphi$. As α is fixed (and by that also $\sin\alpha$), it is constant, and one has $dr = krd\varphi$, k a constant. This is a differential equation, which is written (after separating the variables, i.e. separate grouping of terms containing r and φ, respectively, which is possible in this case) $dr/r = k\, d\varphi$. Integrating both sides now gives $\ln r = k\varphi + C$, or $r = r(\varphi) = e^{k\varphi+C} = ae^{k\varphi}$ (C is a constant of integration, and $a = e^C$). If the pen is placed at *E*, for example, lying on the straight line and having distance b from the point of contact of the wheel, it will describe a curve given by $r(\varphi) = e^{k\varphi+C} + b = ae^{k\varphi} + b$.

4 Conclusions

Suardi's Geometric pen actually seems to us, as he himself declared, an original instrument and the first to be designed to construct curves that are epitrochoids and hypotrochoids. The curves he designed had already been known and studied but no device seems to have been designed to draw them. Through the two versions of his device, Suardi described how to draw 1273 different curves including polygons and geometric flowers that he explicitly declared to have studied by Guido Grandi. The attribution of the invention of the spirograph to Abakanowicz seems to us a mistake: the Polish inventor created a machine for spirals which he presented in the appendix to his most famous treatise and which he needed for the integration of some curves.

Suardi's cogwheels instrument is probably precursor of other devices spreading in the following century. One of the most prominent producers of models for commercial use was the firm of Ludwig Brill of Darmstadt, Germany. In 1899 Martin Schilling took over and expanded the business from Brill. Originally located in Halle, Germany, the firm moved to Leipzig. The firm produced numerous types of mathematical models and instruments, including twelve kinematic models. The kinematic models were designed by German mathematician Frederick Schilling (1868–1950), professor of mathematics at Gottingen. The construction and the study of these plaster models were especially popular in Germany, in France and in the United Kingdom and they were used in several filed of mathematics (for research and didactics). In the catalogue edit by Brill and Schilling (Schilling 1911) twelve devices to draw curves are sold. Some of these tools were used to trace epitrochoids and hypotrochoids and they consist of toothed metal ring and cogged wheels; for the mode of operation (Shell-Gellasch 2015).

One of the first to tackle the prickly topic of straight-line motion was Scottish engineer James Watt (1736–1819), often credited with inventing the steam engine. One of the pressing questions Watt needed to answer was how to convert circular motion produced by the steam engine to the linear motion needed to drive the wheels. The answer was Linkage, a set of metal or wooden bars linked at pivot points that can be articulated, creating the straight-line motion so useful in engineering. Drawing toys based on gears have been around since at least 1908, when The Marvelous Wondegraph was advertised in the Sears catalogue. An article describing how to make a Wondergraph drawing machine appeared in the Boys Mechanic publication in 1913. To a piece of wide board, three grooved circular disks are fastened with screws so as to revolve freely about the centers, guide wheel and driver with attached handle.

Even the wondergraph was then taken over by other manufacturing companies that still now market different versions of the toy.

Acknowledgements I would like to thank Paolo Sgammini student of History of Mathematics at University of Perugia.

References

Abdank-Abakanowicz B (1886) Les intégraphes, la courbe intégrale et ses applications: étude sur un nouveau système d'intégrateurs mécaniques. Gauthier-Villars, Paris.

Grandi G (1729) I fiori geometrici. F. Marescandoli, Lucca.

Palladino F (1985) Ernesto Pascal e gli integrafi per equazioni differenziali. Annali dell'Istituto e Museo di Storia della Scienza di Firenze (X):67–76.
Schilling M (1911) Catalog mathematischer Modelle für den höheren mathematischen Unterricht. Verlag, Leipzig.
Shell-Gellasch A (2015) Schilling kinematic models at the Smithsonian. Journal of Humanistic Mathematics 5/1:167–179 via: scholarship.claremont.edu/jhm/vol5/iss1/9.
SN (2002) Denys Fisher obituary. The Times, 26 October.
Suardi GB (1752) Nuovi istromenti per la descrizione di diverse curve antiche e moderne, e di molte altre, che servir possono alla speculazione de' Geometri, ed all'uso de' Pratici. Col progetto di due nuove macchine per la nautica ed una per la meccanica, e con alcune Osservazioni sopra de' Poligoni rettilinei regolari. Brescia.

Chapter 27
The Laws of Collision at the Royal Society, 1668–9: A Case Study in the Organization of Inquiry in Early Classical Physics

John A. Schuster

Abstract In the late 1660s the Royal Society of London investigated the laws of collision. The contributions by Wren, Wallis and Huygens have long been studied by internalist historians of physics as a telling moment in the gestation of classical physics, between the earlier work of Galileo and Descartes and the later synthesis of Newton. This study examines the institutional and cognitive dynamics of this episode at the intersection of two novel historiographical dimensions: (1) The rules of collision project is analysed by applying techniques and insights first developed by Alan H. B. Taylor and then by Taylor and Schuster for understanding the Royal Society's organizational patterns of management of investigations. (2) Classical mechanics is viewed as a domain that in this period was in the process of diverging from the larger field of natural philosophising and crystallizing as the hegemonic discipline in the emerging ecology of new scientific fields. The rules of collision project displays the processes and tensions involved in this larger trajectory. It will be shown that (1) and (2) are closely related and jointly frame an historiographically innovative account of the rules of collision project.

Keywords Classical mechanics · Historiography of physics and natural philosophy · Laws of collision · Natural philosophy · Royal Society—organization of inquiry

1 Introduction: The Historiographical Challenge

On January 11 1669 [NS] the *Philosophical Transactions of the Royal Society* (hereafter *Phil Trans*) published two papers on the laws of collision of bodies by Christopher Wren and John Wallis. On April 12 was published a similar but shorter and

J. A. Schuster (✉)
Campion College, Sydney, Australia
e-mail: drjaschuster@gmail.com

R. Pisano (ed.), *A History of Physics: Phenomena, Ideas and Mechanisms*,
History of Mechanism and Machine Science 42,
https://doi.org/10.1007/978-3-031-26174-9_27

very lucid paper by the great Christiaan Huygens.[1] The contributions of Wren and Huygens deal with perfectly elastic collisions; that of Wallis with perfectly inelastic collisions. Internalist historians of physics have always acknowledged that the rules of impact offered in these papers are essentially correct—albeit unconventionally presented by post-Newtonian standards (Dijksterhuis 1961; Dugas 1955; Hall 1966; Murray et al. 2011). Similarly, internalist history of physics has long recognized these publications as a telling moment in the gestation of classical physics, between the earlier work of Galileo and Descartes and the later synthesis of Newton. The present study envisions no change or addition to the standard internalist history of these matters. Rather, in the manner of more recent historiographical approaches to the Scientific Revolution, it is concerned with two intertwined axes of analysis that are often ignored in linear internalist historiography of early classical mechanics and other emergent sciences of the day.

Firstly, as explained in Sect. 2 below, this study does not treat early classical mechanics as a smoothly maturing, self-contained and autonomous field. Rather, it is viewed as a domain of theory and practice deeply entangled in the dynamics of the larger and supervening field of natural philosophy, which in the second and third generation of the seventeenth century was marked by unprecedented competition and proliferation of approaches. Only late in the century did such natural philosophical turbulence subside, as embryonic classical mechanics (and other emerging fields like physical optics) began to crystallize into recognizably modern forms and to separate from the field of natural philosophy, which itself began a long process of dissolution. Secondly, we need to place the larger trajectory of emergence of classical mechanics in relation to the specific locations and interactions through which that trajectory was played out. Obviously, the process of maturation of classical mechanics and its effective separation from overt natural philosophical contestation did not drop into lived history from some ideal realm. We are dealing with the concatenated outcomes of sequences of specific encounters and debates, often inside the new institutions that appeared in the 1660s. Accordingly, this paper looks at how the laws of collision project was conducted at the Royal Society. In particular, as explained below in Sect. 4, this study mobilizes and extends techniques and insights first developed by Dr. Alan H. B. Taylor, and subsequently by Taylor and the present author, for understanding the Royal Society's patterns of management of investigations; its handling of internal theoretical disputation (including over natural philosophical matters); and its policies for reporting results to the public through the *Phil Trans*.

Our two axes of inquiry are interrelated: On the one hand, we better comprehend the crystallization of classical mechanics away from natural philosophy, if we examine the local settings in which occurred moments in that trajectory. No textbook or definitive pedagogy suddenly emerged in 1669 to show that a classical mechanics discipline now existed and that it could be demarcated from natural philosophizing.

[1] Wren's paper had been read to the Society 17 December 1668, Birch (1756, p. 335). The paper of Wallis had been produced, but not read, 26 November 1668, Birch (1756, p. 328). On the prepublication history of Huygens' paper see below in Sect. 5. English translations by Curtis Wilson of all three papers are usefully presented in Murray (Murray et al. 2011), along with valuable commentaries.

The laws of collision project is one significant episode where that disciplinary character and demarcation were in the act of being debated and shaped. On the other hand, we better understand the early Royal Society itself, if we study how the institution and its leading actors managed significant events that occurred there—events bearing on the longer processes of maturation of new, more modern-looking disciplines and the concomitant fragmentation of natural philosophy that complemented those processes.

2 The Field of Natural Philosophy and Emergence of Classical Mechanics

I have argued elsewhere that we cannot understand the Scientific Revolution, including the histories of such specialist disciplines as astronomy, optics, mechanics, physiology, hydrostatics and aerostatics, unless we understand the category 'natural philosophy'—its nature, dynamics and fate in the period. This section offers an introduction to these issues and to the relations between the field of natural philosophy and the emergent discipline of classical mechanics.

In my historiography 'natural philosophy' does not simply denote the new mechanical philosophy of the mid to late seventeenth century, nor the still institutionally entrenched neo-Scholastic Aristotelianism.[2] Rather, I model natural philosophy as a wider institution, sub-culture and disciplinary field in all its variety, contention and change: All natural philosophers and natural philosophies constituted one subculture in *dynamic process over time*. With the late sixteenth century revival of alternative natural philosophies, such as atomism, Stoicism and neo-Platonism, the field became more dynamic and contested than it was within the confines of the neo-Scholastic Aristotelianism of the universities. This contestation reached a peak roughly in the years 1620–1660, a phase of the Scientific Revolution I have termed the 'Critical' Stage. Then, from around 1650 there began a long, heterogeneous and complex process during which the field of natural philosophy tended to fragment, as new sciences, as well as radically altered versions of the sciences inherited from antiquity, began to crystallize out of it.

Despite these developments, neo-Scholastic Aristotelianism was *hegemonic* in the field, meaning that including all its sub-varieties, it was the biggest and most widespread genus of natural philosophizing, and that it supplied what I term the *template* even for alternative approaches to natural philosophy. As a 'template' for all genera of natural philosophizing, neo-Scholastic Aristotelianism provided educated men with a general framework for the proper content and aims of a natural

[2] Material in this and the following two paragraphs is treated more extensively in Schuster (2002, 2012, 2013a, Chap. 2; 2013c, 1990; Schuster and Watchirs 1990). The thesis is extensively developed in my current work in progress, provisionally titled: *Natural Philosophy and its Fate in the Scientific Revolution 1550–1700: An Historiographical Inquiry into the Birth of Modern Sciences in Early Modern Europe*. On the persistence of Aristotelian categories in the field of natural philosophizing, see also Grant (1978).

philosophy. For example, when one 'natural philosophized' one was constructing claims regarding one or more of the following dimensions of natural philosophical discourse: the nature of matter, the cosmological structuring of that matter, the principles of causation and the methodology for (supposedly) acquiring or justifying such natural knowledge.[3] A claim within the field of natural philosophy could be as small as a single proposition or as large as a treatise encompassing arguably 'systematically' interrelated claims about all four dimensions. Additionally, as part of its 'template' status, Scholastic Aristotelianism entrained an entire geography of knowledge: the way in which other disciplines were conceived, and related to each other, and to natural philosophizing. Some disciplines were considered superior to natural philosophy (such as theology); others cognate with it (such as mathematics); or subordinate to it. These latter included sciences inherited from classical antiquity: the subordinate mixed mathematical sciences, such as geometrical astronomy, geometrical optics, mechanics, statics, and music theory, as well as the bio-medical domains, such as anatomy, medical theorizing and physiology in the manner of Galen. Any variety of natural philosophy had to include a selection and weighting of some of these disciplines, which I term its *entourage*. The core matter and cause categories of that natural philosophy had to be, in my terms, '*articulated*' to that set of entourage disciplines.[4]

This way of conceptualizing natural philosophizing in the Scientific Revolution undercuts the myth that after about 1660 'natural philosophy' died and was replaced 'Modern Science'.[5] The 'Scientific Revolution, in its most turbulent, 'Critical' phase, in the early and mid-Seventeenth century, was a set of transformations, a civil war, inside the seething, contested culture of natural philosophizing. As natural philosophical culture evolved into the later seventeenth century, there developed a loose consensus in favor of an experimentally oriented corpuscular-mechanical approach to natural philosophy, linked to acceptance of versions of realist Copernican cosmology. This consensus, entangled with the rise of the new scientific institutions, tended to mute natural philosophical competition and debate. In the Royal Society and *Accademia del Cimento*, for example, we find the muting in public of internal theoretical controversy (which still existed) and the routine recourse to experimental

[3] The common method training, allowing of course for the unending technical debates about method, its meaning, contents, scope etc., is of the upmost importance (Maclean 2007). This is not because knowledge was actually discovered and demonstrated by method (Schuster 1986, 1993, 2013a). Rather, method discourse provided universally understood packaging and rhetorical framing for claims of natural philosophical type, and by means of the tools of logic provided natural philosophical players, as subjective agents, the technical capability for reflexively criticizing, comparing, overthrowing and radically reworking the claims of others and of themselves.

[4] My term of art 'articulation' means in this connection the ways in which the higher registers of theory in the subordinate sciences contained or were shaped by matter and cause categories drawn from some favored genre of natural philosophy. For example, the neo-Platonic categories conditioning Kepler's notions of celestial forces and structure; or in Ptolemaic astronomy the underlying commitment to geo-centrism, uniform circular motions and earth/heavens differentia, all derived from and enforced by Aristotelian natural philosophy. Schuster (2013a, pp. 50–53).

[5] Schuster (2002; 2013a, pp. 77–88; 2013b, pp. 16–21), Schuster and Taylor (1996, 1997). First mooted in Schuster (1990), Schuster and Watchirs (1990).

evidence (even if that evidence was difficult to produce and inconclusive in its implications).[6] The more narrow and subordinate 'entourage' sciences, were transformed through these processes. At first, in the heightened natural philosophical turmoil of the second and third generation of the seventeenth century, these fields were imbued with novel doctrines of matter and cause and they became more closely linked to the field of natural philosophy than was the case under the neo-Scholastic Aristotelianism. Then, from the 1650s, as natural philosophical contention was muted, and attention in the new institutions tended to shift to specialist inquiries, these newly minted versions of the traditional fields began respectively to evolve into increasingly autonomous forms, crystallizing out of, and away from, the field of natural philosophy—as pointed out above—thus initiating a long period of disintegration of the latter.[7]

The genealogy of the embryonic classical mechanics of the 1660s and 70s shows exactly this pattern. At the turn of the seventeenth century mechanics was still one of the classical mixed mathematical sciences, consisting of statics, hydrostatics and the study of the simple machines. Key texts included the recovered statical and hydrostatical works of Archimedes, the texts of the so-called Medieval Science of Weights, and the recovered text of the pseudo-Aristotelian *Mechanical Questions*, or *Mechanica*. During the later sixteenth century, there had been novel attempts to promote mechanics into natural philosophy or up to its status level. Some sixteenth-century commentators on the *Mechanical Questions* tried to collapse the distinction between physical explanations of natural phenomena and geometrical explanations of machines, thus inviting mathematics into discourse concerning physical causation (Hattab 2005; Laird 1986; Rose and Drake 1971). Other attempts to make natural philosophical capital out of the older mechanics tradition took the form of taking a dynamical approach to statics and the simple machines, following the lead of the *Mechanical Questions*.[8] Figures like Benedetti and the young Galileo tried to attack

[6] For the Royal Society this is broadly the well known finding of Shapin (1994). However, as shown below, he misunderstood the continued role of theorizing inside the projects of the Society—matters addressed by Schuster and Taylor (1997). Boschiero (2007) presents a more accurate depiction of the delicate presence of theorizing, theoretical debate and natural philosophical agendas inside the *Accademia del Cimento*.

[7] Other new fields also began to emerge and similarly crystallize away from the (itself fragmenting) trunk of natural philosophy, for example, first aerostatics in the later seventeenth century, and later in the early to mid-eighteenth century electricity and magnetism, as well as heat theory. This line of interpretation was foreshadowed in Schuster and Watchirs (1990) and emerged in Schuster and Taylor (1996) and then Schuster (2002). The matter is treated extensively in the work in progress cited in Note 2 above.

[8] In the terminology developed in Gaukroger and Schuster (2002), and Schuster (2012; 2013a, pp. 56–59 and Chaps. 3, 4; 2013b, pp. 57–61; 2017b), these were 'physico-mathematical' gambits. They involved radical attempts to infuse mixed mathematical categories with natural philosophical ones concerning matter and cause, thus bringing the 'target' mixed science closer to natural philosophy than was the case inside neo-Scholastic Aristotelianism. Such moves amounted to attempts to 'physicalize' mixed mathematics, rather than mathematicize natural philosophy. These bids also usually involved early commitment to realist Copernicanism that raised such 'mechanical' questions and stimulated the ideas of Kepler, Beeckman and the young Descartes concerning the mechanics of planetary motion. They are termed 'radical' because they had an anti-Aristotelian agenda, whereas

Scholastic Aristotelianism and promote realist Copernicanism, by using concepts and arguments in statics and hydrostatics to criticize Aristotelian principles such as heaviness and lightness and the nature of fall. These were early symptoms of the substantial entanglement of 'mechanics' with innovative natural philosophical agendas in the Critical period of natural philosophical contention, in the work of the younger Descartes, his mentor Beeckman, Kepler and the mature Galileo.

Kepler's bold work to bring optics and mechanics into intimate relations with natural philosophy was done under the aegis of his own species of non-Aristotelian, neo-Platonic natural philosophy. His 'celestial physics' is concerned with the study of the mathematical laws of the physical *causes* of the motions of the planets and in particular the elliptical planetary orbits around the sun. In his natural philosophy such causes can only be those of an immaterial neo-Platonic kind, operating in simple and elegant mathematical manners, the effects manifesting themselves in the simple, lawlike behavior of the gross matter making up planets, as determined—one hoped—'sufficiently accurately' in observational data. Kepler's claims about celestial forces, and his rather traditional conception of 'inertia' all needed 'correction' or excision under the regime of Newtonian mechanics and celestial mechanics. But this is not the point of a critically informed and categorically correct historiography of natural philosophy and emergent classical mechanics: Kepler's claims and gambits became part of the ongoing weave of contestation and changing structure and dynamics of natural philosophizing and its entourage subjects, such as mechanics.

At first glance Galileo's work in mechanics does not appear to be thus entangled in the natural philosophical struggles of his day. Galileo's mathematical kinematics in the *Two New Science*s (1638) is limited to terrestrial phenomena across a narrow scope of problems—falling bodies, projectile motion, and the fall of bodies along chords and arcs of circles, aiming at what Galileo thought happens with pendula (Gaukroger 1978). This kinematics was largely autonomous of natural philosophical systematics and contestations. However, the genealogy of Galileo's kinematics was implicated in natural philosophical conflict, growing as it did from the agendas of sixteenth century anti-Aristotelian and pro-realist Copernican devotees of the science of the mechanics, including the young Galileo, as noted above. In his *de Motu* of 1590, Galileo had tried to extract a dynamics out of the *Mechanical Questions*, replacing the Aristotelian categories of motion and cause residing behind that text in the service of an anti-Aristotelian campaign in natural philosophy.[9] Failure of that gambit helped shape his mature mechanics, thus rendering the constrained new kinematics of the *Two New Sciences* an unintended culmination of the sixteenth century search for an anti-Aristotelian 'new science' of motion and mechanics. Indeed, the very existence of a mathematical science of local motion—even if published without exposing its natural philosophical overtones—constituted an affront to Aristotelian natural philosophy. Another dimension of Galileo's mechanics, more engaged in natural philosophical contestation, was his deployment of concepts from his mechanics inside his

there had been moderate, even timid attempts along these lines in some Aristotelian circles, as brilliantly delineated by Dear (1995).

[9] Gaukroger (2000, pp. 63–68), Gaukroger and Schuster (2002, pp. 547–549).

Dialogues Concerning the Two Chief World Systems (1632) as weapons in his battle against Aristotelian natural philosophy and for realist Copernicanism. First, he used his conceptions of relativity of motion and 'inertia' not to prove the Earth spins, but to defend against arguments that if the Earth were rotating we would observe resulting phenomena such as twisting trajectories of projectiles and deflection of falling bodies. Then, in the supposedly triumphal concluding 'Day Four' of the *Two World Systems*, he played his trump card—a very problematical one—his theory of the tides, a strategy grounded in his mechanics, aimed at contesting Aristotelian natural philosophy and promoting realist Copernicanism.

The rise of the corpuscular-mechanical philosophy of nature in the Critical period marked a decisive stage in the interweaving of new ideas in mechanics with natural philosophizing. Some version of a mechanics was needed to provide the causal account of corpuscular motions. All corpuscular-mechanists were also realist Copernicans, so that novel ideas about mechanics also had to be articulated with the Copernican cosmology, thus constituting forms of proto-celestial mechanics. From around 1610 the visionary Isaac Beeckman developed a fully formed corpuscular-mechanical philosophy of nature.[10] The innovative core of Beeckman's corpuscular-mechanism resided not in his atomistic matter theory, but in the causal dimension of the natural philosophy, where it was expected that a theory of mechanics would run the corpuscular proceedings. Promoting any type of mechanics into the nucleus of a natural philosophy was an epochal strategy. Beeckman sought this in a '*Mechanica*' inspired dynamical interpretation of the balance beam as a model for collision, macro and corpuscular.[11]

Beeckman was Descartes' mentor for his early forays into corpuscular-mechanical thinking and attempts to relate mixed mathematical disciplines like hydrostatics and optics to this species of natural philosophizing. After a long and complex trajectory—especially involving his discovery of the law of refraction of light and attempts to render physical rationales for it—Descartes finally devoted himself to systematic construction of corpuscular-mechanism, first in his *Le Monde* and then in the *Principia philosophiae*.[12] The causal dimension of this natural philosophy consisted in what Stephen Gaukroger and I have termed a 'punctiform dynamics' of corpuscles.[13] Ultimately deriving from attempts at understanding the physical rationale for his law of refraction, Descartes' new kind of dynamics applied instant to instant to the motions, impacts and tendencies to motion of his corpuscles. In his two natural philosophy treatises, Descartes teaches that bodies in motion, or tending to motion,

---

[10] On Beeckman, see Berkel (2012), Schuster (2013a, pp. 108–112; 2019, 2022).

[11] Gaukroger and Schuster (2002, pp. 555–6), Schuster (2013a, pp. 110–1).

[12] This trajectory also importantly included the collapse of Descartes' methodological dreams, embodied in his *Regulae ad directionem ingenii*, thus pushing him in the direction of more traditional forms of natural philosophizing. Schuster (2013a, Chaps. 7, 8). *Le Monde* composed between 1629 and 1633 remained unpublished in his lifetime. The *Principia philosophiae* appeared in Latin 1644 and French 1647.

[13] Gaukroger and Schuster (2002, pp. 568–569), Schuster (2013a, pp. 169–172, pp. 369–371, 2013b, pp. 62–63).

are characterised from moment to moment by the possession of two sorts of dynamical quantity: (1) the absolute quantity of the 'force of motion' and (2) the directional modes of that quantity of force, the directional components along which the force or parts of the force act.[14] It is these Descartes termed actions, tendencies, or most often determinations.[15] As corpuscles undergo instantaneous collisions with each other, their quantities of force of motion and determinations are adjusted according to certain universal laws of nature. Applied to the problem of circular motion, first in the case of a stone in a sling, these concepts were then deployed in Descartes' non–trivial vortex theory of planetary motion.[16] Descartes' dynamics of corpuscles was not mathematically articulated. It was qualitative and discursive like the rest of his systematic natural philosophizing. Still, it offers a good example of the larger historical dialectic we have in view: Descartes forged concepts in 'mechanics' for his natural philosophy in four areas: inertia, so-called; the laws of collision; the problem of circular motion; and fluid dynamics/vortical motion. His answers were not permanent, often seen as problematical and ultimately as largely incorrect. However, they both stimulated and demanded radical revision inside the later attempts at forging versions of the emerging classical mechanics by Huygens and Newton and Leibniz (Schuster 2013b, pp. 82–86).

To summarize: The mixed mathematical field of mechanics had been launched on its unintended trajectory of maturation by being variously brought into the natural philosophical field from the late sixteenth century and especially in the heightened natural philosophical contestation of the Critical stage.[17] The eventual crystallization of classical mechanics resulted in large part from a concatenation of such moves by

[14] Schuster (2000, pp. 258–261; 2013a, pp. 169–173). '(1)' is defined by the first law of nature in *Le Monde* and the *Principia philosophiae*, '(2)' is defined by the third law of nature in *Le Monde* and second law of nature in the *Principia philosophiae*.

[15] The understanding of determination used here develops work of Sabra (1967, pp. 118–121), Knudsen and Pedersen (1968, pp. 183–186), Prendergast (1975, pp. 453–462); as well as Gabbey (1980), Gaukroger (1995), McLaughlin (2000), Mahoney (1973), Schuster (2000, 2005, 2013a).

[16] On the serious and non–trivial conceptual and systematic construction of the vortex theory, see Schuster (2005; 2013a, pp. 453–477; 2013b, pp. 73–78), Schuster and Brody (2013).

[17] My stress on the emergence of embryonic classical mechanics within the natural philosophical conflict of the Critical Period might be contested based on our previous remarks about the non-natural philosophical character of Galileo's mathematical kinematics. Our previous remarks should be taken in conjunction with the following supplemental observations, now that we have met the mechanical philosophers: The truth of the matter is that the mechanical philosophers who embraced or tried to encounter Galileo's kinematics could do so only rhetorically. Neither Gassendi, Descartes nor Hobbes could find convincing matter/cause accounts through which to integrate into their respective systems of natural philosophy the mathematical law of continuous acceleration in free fall. This was not for lack of trying. Descartes went through this experience during his 1618 work with Beeckman on the physico-mathematics of free fall. Schuster (2013a, pp. 139–153). In my account they were seeking not only a mathematical law of fall but, as physico-mathematicians, its causal rationale. They were faced by what I have termed 'too many possible laws, too many possible corpuscular causal regimes', with no possibility of meaningful measurement. On Gassendi's extended struggles with the problem of reconciling a mechanical impact cause of fall with the continuity of the law, see the subtle analysis of Palmerino (2004); on similar problems in Hobbes see Brandt (1928, pp. 316–320) and Leijenhorst (2004). That one element in the eventual emergence of classical mechanics passed through the period down to Huygens and Newton largely outside of the realm of natural

major players of that generation, until there was enough density of embryonic 'classical mechanical' problems, concepts, exemplary solutions, tools, and techniques for actors themselves to start acting upon and within that framework.[18] By the time of the Royal Society's inquiry into the rules of impact in the late 1660s, some experts were practicing this new type of mechanics, proto-classical and piecemeal. That is to say it was on the verge of being recognized as a discipline autonomous of natural philosophy, meaning that it did not necessarily need to be practiced for natural philosophical reasons.[19] The situation, however, remained fluid, because many practitioners, committed at least loosely to corpuscular-mechanical natural philosophy, hoped that findings in the mechanics of macroscopic systems might reveal truths about the mechanics of the underlying corpuscular world. Additionally, since a full-blown discipline of classical mechanics did not exist, the content and boundaries of the fledgling field had to be made out on the run and in the course of inquiry and debate. It was exactly this tentative emergence of classical mechanics that framed what happened at the Royal Society, conditioning how the expert participants and their natural philosophizing critics behaved.

3 Embryonic Classical Mechanics at the Time of the Royal Society Inquiry

E. J. Dijksterhuis, doyen of mid-twentieth century internalist historians of early mechanics, captured the state of the embryonic classical mechanics at the time of the Royal Society inquiry about the rules of collision this way: In this field

philosophical contention does not negate the larger patterning. The exception proves the rule in Ciceronian fashion: Had all parts of all physico-mathematical disciplines arisen so remotely from natural philosophizing, we would indeed be faced with a very different history than the one we are sketching here.

[18] It is not being argued that the early maturation of classical mechanics circa 1600–1670 was entirely encapsulated in the relations of the young field to the dynamics of natural philosophizing. Rather, we are emphasizing this set of relations that followed from the physico-mathematical gambits of some natural philosophers, and by extension, the questions that thereby arose concerning recognition of the relative autonomy of mechanics and its working boundaries with natural philosophy—since these latter were clearly in play and being managed during the Royal Society inquiry into the rules of collision.

[19] This idea of a complex process of emergence of classical mechanics through the heightened natural philosophical conflict of the early and mid seventeenth century, first came to light in informal discussions with Stephen Gaukroger (circa 2002–3) about how to think through the birth of classical mechanics. As I noted in Schuster (2005, p. 77, Note 82) '*It would seem, as Stephen Gaukroger has expressed to me in discussion of themes of this and related work, that the long term genealogy of classical mechanics should be written, at least in part, in terms of the concatenation of unintended conceptual windfalls bequeathed to the emerging discipline by this and other nodes in the natural philosophical turbulence of the early-and mid- seventeenth century.*'Gaukroger's magisterial (2006a) contains what is likely to remain for a long time the best technically informed, detailed account of interactions between natural philosophizing and classical–mechanics–emerging in the seventeenth century.

sui generis new concepts such as Galilean relativity and Torricelli's axiom (to which we might add Cartesian inertia), were deployed and interrelated in a growing tool kit of problem solving schemas.[20] Compared to the ancestral mixed mathematical field of mechanics, the new domain was more difficult, specialised and esoteric in mathematical techniques and conceptual innovations. Alan Gabbey has a felicitous description of what the new mechanics had become by this stage: '...a quite general mapping of the natural world', referencing in this connection John Wallis' 800 page treatise of 1670, *Mechanics, or On Motion by Geometrical Means* (*Mechanica, sive De Motu, Tractatus Geometricus*).[21] Since Aristotle motion had been the key subject matter of natural philosophy. Now 'Mechanics' transcended and encompassed traditional statics, hydrostatics and the study of the simple machines, becoming the study of mathematically described, idealized versions of otherwise potentially realizable macroscopic systems of bodies in motion or equilibrium. Thus far the accrued concepts and problem-solving techniques were loosely assembled, but were increasing over time in number and scope.[22]

We arguably first see a fully fledged classical mechanics in Newton's *Principia* (1687). By the later 1660s on the verge of Wallis' publication, the embryonic new mechanics was more fragmentary and just beginning to hove into view for interested parties. A counter-factual conceit helps elucidate this: Imagine that Huygens had at some stage between 1670 and 1687 published his own *Treatise of Mechanics* with chapters on collision, centrifugal force, pendula, but that Newton had held back his *Principia* until the 1720s. Would we not look back on Huygens and Wallis (along with the likes of Borelli and Mariotte) as *per force* the pioneers of the newly hatched discipline of classical mechanics, born in the 1670s and 80s, while we would judge the situation in 1668 as still a bit disjointed and preparatory?[23]

Two other dimensions of the embryonic discipline need to be taken into account before our case study: The changing meaning of 'theory' at the time, and the related

---

[20] Dijksterhuis (1961, pp. 362–636, p. 380): '*Before the beginning of the seventeenth century the mathematical knowledge that formed part of every scientific (sic) education had generally been sufficient, and anyone who had received such an education had been qualified to participate in a discussion about phenomena of motion and equilibrium. Now this was already becoming a matter for specialists, while the finding of really new methods was reserved to mathematically highly gifted scientists such as Huygens*'.

[21] Gabbey (2004, p. 22). Wallis' book was made known at a Royal Society meeting 30 April 1668, Birch (1756, p. 282). The Fellows urged its speedy completion and publication.

[22] It is of course necessary to include other contemporary Continental pioneers, such as Borelli and Marci in this picture. By far the most illuminating conceptualization of the process of development of mechanics in the seventeenth century is contained in Bertoloni Meli (2006, p. 309). The reader should consult in particular his schematic diagram and accompanying discussion in his concluding chapter entitled 'Mapping the transformations of Mechanics'. Of course my emphasis here has been, and will continue to be on natural philosophy/emergent mechanics relations.

[23] Imagine the same scenario, but add Hooke also publishing between 1670 and 1687 his work on circular motion, orbits, pendula and elasticity in a third 'early classical mechanics' monograph alongside Wallis and Huygens. In principle, in this imagined world an arguably mature classical mechanics would have materialized before and independently of Newton, but still not as early as the Royal Society inquiry into collision.

issue of new categories and concepts being constructed to meet newly explored experimental situations—concepts that arguably had no pedigree reaching back into the history of natural philosophy in the period. David Wootton has shown that at this time the term 'theory' begins to acquire its modern meaning: an evidentially based conceptual scheme that accounts for ranges of old and new evidence. Previously, 'theory' applied in astronomy to geometrical planetary models and, in practical mathematics, to geometrical schemes for instrumental calculation and prediction.[24] In the new usage, theory did not signify deep natural philosophical concepts of matter and cause, nor was it merely a name for correlation of appearances.[25] The *Phil Trans* calls Wren's paper on collision rules, a 'theory'—the same term Oldenburg as editor applied earlier to Wallis' mechanical explanation of the tides and which is also used in the index to the volume.[26] For these savants, the mechanical analysis of macroscopic systems involved 'theories' consisting mainly in what Boyle was to call 'intermediate' concepts. These were categories that did not belong either to ordinary language descriptions of experience, or to deep natural philosophical discourse about matter and cause (in one or another genre of natural philosophizing). They were constructed in the course of experimental inquiries and related to (but not naive reflections of) various experimental hardwares and outcomes. In general, in any domain where attention began to focus on related sets of experimental hardwares and manipulations, there tended to emerge a conceptual language (which is termed a 'dialect' in my model of the process), the make-up of which owed less to inherited natural philosophical categories and more to the ongoing theory construction by the experts in that domain.[27] Alan Chalmers has illuminated this process in the case of the maturation of hydrostatics. He shows that the new concept of 'hydrostatic pressure' marked a break with both commonsensical notions of pressure and natural philosophically derived conceptual 'obstacles' to its formulation. Forged in the work of Boyle, with debts to Pascal and subsequent refinement by Newton, this concept became the lynchpin of mathematical articulation of specific sequences of experiments that constituted the

---

24 Bennet (1998). An alternative wording for theory in practical mathematics might be rendered as: a logistical/instrumental problem solving protocol.

25 Stephen Gaukroger has attempted to characterize the core of the new 'experimental philosophy' of the later seventeenth century in terms of actors' prioritizing of the correlation of phenomenal appearances displayed in the working of privileged experimental hardwares, such as Boyle's air pump or Newton's prismatic spectrum. Gaukroger (2006a, pp. 354–356, pp. 397–399; 2010, pp. 153–157). My monograph (Note 2 above) will critique this approach to understanding experimental hardwares, actors' accounts of 'results' and the general state of experimental natural philosophizing in the period, showing it not to be a fruitful or viable historiographical way forward. Schuster and Watchirs (1990) and Schuster (2002, 2018, pp. 415–417) foreshadowed my present model, which is inconsistent with the *correlation of appearances as exemplar* approach to which I have given the name 'one-hardware phenomenalism'.

26 See Wootton (2015, p. 396); *Philosophical Transactions* vol. I, pp. 263, p. 378, p. 403; vol. III. p. 867. On Wallis' theory of the tides, see also Hooper (2004).

27 Schuster and Watchirs (1990), Schuster (2018, pp. 415–417) and in more detail in my monograph in progress (Note 2 above) which develops my full set of historiographical terms of art for investigating these developments, and places these emergences in the context of a wider model of the dynamics, evolution and fate of the field of natural philosophizing itself.

new, modern science of hydrostatics. In collision studies in mechanics at this stage, the following 'intermediate' concepts were explicitly or tacitly in play: inertia, quantity of motion (scalar or vector-like), force of motion, conservation of momentum, hardness, elasticity or spring, and—via Torricelli's axiom about the motion of the center of gravity of two bodies—what was later in the century called vis viva. Thus, the reminting of the term 'theory' and the emergence of 'intermediate' mathematicised concepts keyed to specific sets of experimental hardwares and outcomes are good indicators that the new mechanics had begun the process of detaching from natural philosophizing.

Nevertheless, for those partial to corpuscular-mechanical natural philosophy, the possibility remained that the new mechanics might throw light on the laws governing the behaviours of ultimate particles. The new mechanics was not totally autonomous of natural philosophizing: Not in the minds of practitioners of that mechanics, and certainly not in the minds their critics at the Royal Society. For example, we shall see that the category 'elasticity', that Hooke and others were happy to explore at the 'intermediate' and experimental level inside the emerging mechanics, could also be attributed to corpuscles (as of course so could the other categories listed in the previous paragraph). Those prioritizing natural philosophical inquiry preferred to elucidate elasticity first at the corpuscular level, while those focussed on the new mechanics and macroscopic systems saw the order of study as quite opposite. At the Royal Society during the rules of collision inquiry, there were still open questions about who would decide such issues, and how they would do it, because there still was no canonical disciplinary structure—social and conceptual—for the new mechanics. All of which brings us to the question of how exactly the early Royal Society organized its inquiries.

4 A Model for the Study of the Organization of Inquiry at the Early Royal Society

This examination of the inquiry about the rules of collision at the Royal Society constitutes a new case study in a program originally devised by Dr. Alan Taylor and later jointly pursued by Taylor and the present author, focussing on the organisation of experimental inquiries at the early Royal Society of London.[28] Taylor had pioneered a new technique of investigation: The delineation of what he termed routine patterns of 'decision-making' and 'action-taking', leading to analysis of how these patterns shaped inquiries inside the institution, as well as representations of those inquiries to outsiders. Later, superimposing upon the study of decision-action patterns, the

[28] Dr. Alan B. H. Taylor (b 1938) served as a Royal Marine (battlefield tactician) and latterly was an early senior corporate IT strategist in Australia. In later life, he turned himself into an extraordinary historian of scientific organisation and practice, under the supervision of Professor Roderick Home in the HPS Department at the University of Melbourne (Taylor 1989). He then worked as a postdoctoral research fellow with the present author, leading to several collaborative works, before he decided to retire from active engagement in the field of HPS around 2001.

present author brought into play more attention to the role of theory—including natural philosophical categories—in experimentation and debate. Taylor and I were also motivated by claims that modern experimental 'Science' was born at the early Royal Society.[29]

In conceptual terms the decision/action patterns we identified were present above and beyond specific passages of activity. However, they only existed for the actors concerned in so far as those actors continually reproduced and sustained them. From an actor's standpoint the patterns were constraints and resources—they might be played differently in terms of actors' skills and agendas.[30] It is important to identify the decision/action patterns, but also to follow different actors as they play within and upon them. We found that the initiation and subsequent leadership of projects was left to those who chose to be active; but, once initiated, a course of in-house investigation was subject to routine patterns, routinely ordered by the Fellows—chiefly the repetition, re-testing and witnessing of experiments as a condition for their acceptance. This situation imposed a frequently observed pattern on the actual meetings: When material emerged at the Society's meetings outside the sphere of competence of most of the Fellows, they were left with little opportunity to participate. But, for the few Fellows who had the skills required for a given project, a window of opportunity opened for them to participate fully and to introduce activities and views that favoured their preferred beliefs, skills and interests. Hence, for an active member intent upon advancing his own intellectual agenda, the best policy was to keep the remaining Fellows engaged by feeding routine decisions their way, and to manage any spin off committees or actions the Fellows might collectively call for, such as the writing of reports. In this way the patterns the Society followed facilitated the contributions of active players and provided activity for most of the other Fellows engaged at meetings. The existence of lead players and of routine patterns were interdefining organisational dimensions at the Royal Society. As to reports about inquiries communicated to the public, they routinely failed to reveal the Society's in-house processes of the construction of knowledge claims. The Society systematically created the impression that it was merely the impartial reporter of news supplied by others. The contrast between what was reported and what happened in-house was endemic, and no accident, but rather a consequence of deeply embedded policies.[31]

The issue of selection and self-selection for participation in a project makes expertise and its distribution a crucial issue in each case study. Where expertise went, so did experts' assumptions, techniques, and of course, their theories. As mentioned,

[29] Implausible—indeed impossible—claims, historically and epistemologically speaking, that we dealt with in extensive review essays: Schuster and Taylor (1996, 1997).

[30] The theoretical inspiration for this and the two preceding sentences is generally from Giddens (1976) on structures and 'structuration'.

[31] These policies both enabled Henry Oldenburg (Secretary of the Society 1662–1677) to carry out the day to day provision of news to those involved, and curtailed his actions. We found that Oldenburg, as editor of the *Philosophical Transactions*, was not completely free to publish what he pleased. He followed the policies of the Society and met the day to day directions of the Council and Fellows: Taylor (1989) and several of our case studies to appear in the monograph mentioned above (Note 2).

the meaning of the term 'theory' was evolving at this time. Theories of various sorts were involved in the projects we studied, but what were not in play were the atheoretical 'matters of fact' that mythically reside at the base of Shapin's (1994) widely cited account of 'the experimental life' at the early Royal Society. Moreover, our emphasis on expertise runs counter to Shapin's picture of self-consciously non-expert 'gents' running the show.[32] We found that theoretical commitments were endemically present, lurking behind decisions and actions, but that conflict about theories was managed in-house, and occluded in official publications, if any, resulting from a project.

Some of Taylor's own early work aimed to uncover the 'corporate policy settings' of the Royal Society as set down in its Charters and Statutes and ongoingly interpreted by the Council. In Taylor's interpretative grid, shaped by his own high level organizational roles and experience, the 'product' of the Society was natural knowledge; its 'aim' the improvement of natural knowledge.[33] Taylor insisted that his most important finding about deep policy settings was this: The Royal Society's corporate policies did not include a detailed day to day method, or procedure for investigating nature by means of experiments. 'Corporate Policy' encouraged the use of disparate views on approaches to experiment. This claim obviously runs counter to the various unproven assertions in the traditional secondary literature that the Royal Society was either 'Baconian'; or 'Boylean'; or later 'Newtonian'. Taylor insisted—and all our cases confirm—that the early Royal Society was open textured by design and functioning.[34] In our case studies the decision/action patterns involved in experimental investigations allowed for a flexible style of inquiry—differences of approach and interest of the Fellows could be mobilized in order to achieve the Society's aims.[35]

---

[32] These problems will be addressed in Sect. 10. See also our review essay on Shapin, Schuster and Taylor (1997, pp. 512–513 and Note 8 thereto).

[33] Royal Society (1940, pp. 250–263, pp. 287–300), Taylor (1989, pp. 18, pp. 153–157).

[34] Here we negatively heuristically denote 'Baconianism' as the virtually unattainable ideal of a more or less systematically articulated natural philosophy as recoverable from the published texts of Bacon. The Royal Society was in many ways accountably 'Baconian' in a range of looser, more generic senses of the term, consistent with Rees (2000, p. 69) correct—rather post-Quentin Skinnerian—statement that '*There is no such thing* [as Baconianism]'. The term denotes any one of a plethora of interpretations/articulations of resources derivable from Bacon's corpus of writings. Essentializing Baconianism, and making 'it' the controlling ideology is an empty historiographical pastime. One should observe the decisions and actions, rather than pre-judge them under confected contextual determinations, social or intellectual. If one is considering the natural philosophical interests and agendas of many Fellows, as one should, then rather than invoking a non-specifiable Baconianism, one should follow scholars like Taylor (1989), Rattansi (1968) and Hoppen (1976) who explored the natural philosophical diversity inside the Royal Society, intent on showing that the members were not all Baconians but rather espoused a variety of natural philosophical positions whose rough, clashing edges were filtered and smoothed inside the Society.

[35] It should be added, not just 'mobilized' but, as we are about to see, contained without conflict. Shapin (1994) therefore overdramatized the nature of theoretical conflict at the early Royal Society: you could have divergence without conflict and mere differences did not necessarily entail conflict—you can have different types of 'decorum', including a decorum for containing and shaping—rather than completely quelling—theoretical conflict. See below Sect. 10 for further conclusions of this study differing from the well known claims of Shapin.

Our project originally had three cases[36]: [1] An investigation into May-Dew in 1664, led and managed by Fellow Thomas Henshaw. This case was investigated by Taylor (1994) and published by him as the first example of his approach. [2] A project on constitution of metals from 1680, led and managed by Robert Hooke. [3] The first round of debate about Newton's new theory of light and colours from 1672, managed on a technical basis by Hooke and organizationally by Oldenburg.[37] Four features mark the cases: [a] Experiments were endemically called for and some were performed, and relevances of their outcomes debated. [b] Some actors had clear natural philosophical agendas. These resided along a spectrum from those with systematizing commitments to those who held particular natural philosophical positions in a piecemeal manner. Thus, in some instances, participants were committed to one or another explicit, detailed philosophy of nature—a deep doctrine of matter theory and of causation, for example Cartesian, Gassendist or van Helmontian. This motivated them; they expected results advantaging their own natural philosophy. But, the same issues of motivation and goals also applied to those whose underlying natural philosophical commitments were more of a piecemeal character. [c] Theories were also embraced that did not originate in the available archive of natural philosophical concepts of the day.[38] Finally [d], theoretical commitments and agendas, of whatever character, natural philosophical or not, were always muted and 'managed' in-house and not permitted to dominate published results, although hints and clues sometimes passed into print.

The rules of collision project will present the following features: For the leading players the paramount aim was not to articulate natural philosophical theories of matter and cause. The issue was a problem in mechanics as they now understood it—the mathematical analysis of motion in macroscopic but idealized systems. The wider debate was shaped by the fact, discussed above, that for the practitioners of new mechanics, the discipline was in the process of separating from the field of natural philosophy. However, some Fellows, not central practitioners of the new mechanics, still believed that natural philosophical issues needed to be raised within the laws of collision inquiry. We shall see such opinions muted and managed in-house and kept out of the official publications. Moreover, the main players, convinced corpuscular-mechanists in natural philosophy, continued to hope that the rules of collision found for macro systems might ultimately transfer to the micro world of corpuscles. Finally,

[36] The suite of three cases were variously presented at seminars and conferences in Australia, the US and Europe, and are brought together with the rules of collision case in my manuscript in progress (Note 2 above). One of our seminar presentations, University of Ghent 2008, is available at descartes-agonistes.com/2020/01/02/schuster-organizing-the-experimental-life-at-the-early-royal-society-ghent-2008/.

[37] Like mechanics, optics had entered the seventeenth century as a traditional mixed mathematical field. Starting with the work of Kepler and Descartes, entangled in their respective natural philosophical agendas, optics was evolving, even before Newton, into a new form as a more modern looking 'physico-mathematical' discipline. On this conception see Schuster (2012, 2013a). We had an additional case in Schuster and Taylor (1997, pp. 524–527), not located in the Royal Society, but in the Boyle corpus: A critique of Shapin's (1994, pp. 247–258) treatment of Boyle's worries about icebergs, buoyancy and witness reports by sea-captains.

[38] Theories, that is, consisting of the sort of 'intermediate concepts' discussed above in Sect. 3.

one must remember that this was classical mechanics in its early stages, so that theoretical claims were not particularly open to precise experimental exploration. Despite this, and in accord with the dominating corporate policy rhetoric prioritizing experimentation, issues were expected to be referred to evidence from experiment. Throughout the project experiments were called for and some were performed. Yet gaps are evident everywhere between the realm of experiment and the contents of the papers, published and unpublished.

5 Genesis of the Project: 'Consensus' and 'Carve Out'

It seems that Hooke sparked the 1668 rules of collision project on 22 October 1668, when he called for experiments to be tried to 'to state at last the nature and laws of motion'. Interestingly, the President, Brouncker immediately reminded everyone that Wren and Huygens had worked on the matter effectively to closure (Birch 1756, p. 315). As we shall see, in 1661 Wren had done well known experiments on the topic for the Society, in some of which Huygens took part. The latter's work in the area was well known to some. Wallis was only asked into this group later at the 12 November meeting, when the Fellows suggested that John Collins peruse and report upon the relevant work of Descartes, Borelli and Marcus Marci and that Oldenburg consult Wallis about these matters.[39] As already noted in Sect. 3, Wallis had his big *Mechanica* in process, as the Society had been well aware from at least 30 April prior. The papers of Wren and Wallis appeared together in January 1669. Huygens' paper appeared three months later in April 1669.[40] A month later, in May 1669, Oldenburg would tell Huygens the debate was over and that almost everyone agreed (Oldenburg 1965–1985, vol. 5, p. 582). In addition, as was almost always the case with Royal Society publications, the three papers were offered to the reading public as though the Society were just neutrally reporting findings submitted to it, with no hint of any in-house organization or debate about the topics. Thus, at first sight the enterprise looks straight forward, neatly executed and closed. One would be tempted

[39] Birch (1756, p. 320). Hall (1966, p. 28) suggested that Borelli's *de vi percussionis* arrived in London to be discussed in July, and that it might have inspired Hooke's October call to action. However, from Birch (1756, pp. 306–307) we learn only that on 16 July Moray informed the meeting of a letter from Wallis dated two days before that discussed an experiment reported in Borelli. Hall (1966, p. 28) does correctly report that Borelli was never reviewed in the *Phil Trans*. On Borelli see Bertoloni (2006, pp. 229–231).

[40] There had been a bit of a *contretemps* with Oldenburg when the latter told Huygens the Society did not need to publish his paper because it agreed with Wren. Huygens, rightly, prided himself on the depth and scope (and priority) of his research in mechanics. Displeased, Huygens placed his paper, already aired in January 1668 at the Parisian *Académie des Sciences* (Bertoloni Meli 2006, p. 234) in the *Journal des Sçavans*, 8 March 1669. That reminded the Society to show some respect and maintain decorum with such a key player, so that, as noted, three months later the *Phil Trans* published Huygens' paper. This was the occasion of Oldenburg adding a prefatory essay to the publication of Huygens paper, 12 April 1669, *Phil Trans,* vol. IV, pp. 925–928. A useful translation by Curtis Wilson exists in Murray et al. (2011, pp. 154–156).

to conclude that complete consensus and effectively complete closure of the matter reigned unopposed. However, a closer look at the actors and events involved suggests that the issue was not so straightforward. We shall discover that the proffered closure on the occasion of publishing the three papers was a nicely managed outcome, and that the publicly represented consensus occluded much internal contestation requiring expert management.

First let us note that work on the rules of collision was done very early in the life of the Royal Society. In 1661 Wren and Lawrence Rooke experimented with differently weighted bobs on colliding pendula. Huygens visited London in April 1661 and was shown some of this work. By the mid 1650s he had already worked out post-Cartesian rules of collision (Bertoloni Meli 2006, p. 231). Meeting with Wren, Rooke and others, Huygens quickly and correctly calculated—algebraically, not geometrically—the results of certain trials. When offering his predictions, Huygens blotted out his calculations and did not state his own rules. The event was well remembered by members during the ensuing decade. In 1665 Moray reminded Oldenburg of the event, claiming that Wallis too recalled the matter in detail.[41] Similarly in November 1668 Huygens recalled it, in responding to Oldenburg's request for his work on motion.[42] And at the December 17 1668 meeting when Wren produced his theory of collision, he recalled the previous episode with Rooke (Birch 1756, p. 335). Others at that meeting, including President Brouncker and Sir Paul Neile, also recalled the 1661 events as eye witnesses. Clearly we have here a long corporate memory and tacit acceptance by insiders of these studies of perfectly elastic collision. Hooke on 22 October 1668 certainly also knew about all of this history, although his general call to arms to investigate the laws of collision had to be immediately specified by Brouncker back to the Wren/Huygens history, which in turn immediately prompted a directive to Oldenburg to find out what Wren and Huygens now possessed on the subject and the publication status of that work (Birch 1756, p. 315). However, there is little sign of concerted action on collision studies as such until Hooke's call and this quick mobilization in October 1668, seven years after the Wren, Rooke and Huygens work.

In the preceding years one can find periodic but largely unsystematized work on the collision problem as well as a range of what we shall call 'cognate problems' in mechanics. For example, in October 1666 experiments concerning 'the propagation of motion' had been conducted using pendula, with no larger principles of mechanics or natural philosophy apparently having been discussed.[43] The following January Oldenburg tried to rekindle the Society's interest in the problem of general principles of motion, asserting that the Council now saw fit that '*the experiments for making out a theory of the laws of motion formerly begun by Dr. Wren, Dr Croune and Mr.*

---

[41] Oldenburg (1965–1985, vol. 2, pp. 624–5) 27 Nov 1665. Also Hall (1966, p. 27). Generally speaking Hall provides the earliest full secondary account of the Society's pre-1668 dealings with collision studies. See also generally on the events mentioned in this and the next paragraph: Bertoloni (2006, pp. 234–237) and Hyslop (2015, pp. 248–252, pp. 257–258).

[42] Oldenburg (1965–1985, vol. 5, p. 127) 3 Nov 1668.

[43] Birch (1756, pp. 116, p. 117). Meetings of 17 and 24 October 1666.

Hooke…should be prosecuted' (Birch 1756, p. 140). But we hear nothing about this until Hooke's call nineteenth months later on 22 October 1668. In the meantime, one of the Fellows, William Neile, whom we shall see would loom large in the in-house debate after October 1668, had sent a paper about motion via Oldenburg in anonymous form to Wallis in 1667. Wallis had been critical and the paper was sidelined only to appear later.[44] Before October 1668 Hooke and Wren pursued numerous studies on pendula and the phenomena of spring and elasticity.[45]

Next, turning to the Society's activities in the period immediately following 22 October 1668, one finds a great deal of cognate inquiry in mechanics, some of it potentially bearing on the eventual content of the three published papers, but that was never allowed to impinge upon their content and publication. For example, from October 1668 through January 1668/9, Hooke reported on experiments on communication of motion through three linked pendula; upon fall in a purported void; and upon the question of whether the strength of a falling body in impact is in duplicate proportion to its velocity. In addition various work was discussed and debate conducted, centering on Hooke, concerning the 'springiness' or elasticity of bodies, particularly the relation of rebound to the latter.[46]

Thus, the long lead up to Hooke's call, as well as its immediate context and sequels, places the swift and decisive recruitment, review and publication of Wren, Wallis and Huygens in a interesting light, and raises the question of how this came about. Our cursory review of recent activities of the Society in mechanics and their pre-history indicate that nothing was predestined, or obvious, about Hooke's October 1668 call. In each of our other case studies (Sect. 4) we found a clear in-house leader of the inquiry. In the rules of collision case it seems that because of the well established history of successful work, several Fellows with expert investments were able to work cooperatively together—if mainly tacitly—to produce the outcome: The three papers, and only the three papers, were published, despite all other cognate and arguably relevant matters raised, explored and debated. Given that Hooke was involved in many of these cognate matters at the very same time, he certainly was displaying vision and leadership; grasping the nettle of pushing for collision studies to be presented, reviewed and quite possibly published. By that point he had vast experience with related issues of pendulum motion and elasticity or spring. We have just seen that Hooke was central to much of the 'cognate' mechanics inquiry then going on at the Society. But, he recognized the separable character of the basic collision rules studies, sharing with some others the hope for deeper natural philosophical insight ultimately flowing from them. Thus, although he did not turn out to be a direct,

44 Kemeny (2017, p. 184) citing Wallis (2005, vol. II, pp. 323–324). This document is likely the immediate ancestor of Neile's paper that came into play early in 1669 (See Sect. 6).

45 Patterson (1952, pp. 288–296) especially for Wren and Hooke interactions in pendulum studies in 1666–7.

46 Communication of motion through pendula: 5 November, Birch (1756, p. 318). Fall in a void: 5 November (Birch 1756, p. 318). Strength of impact of a falling body: 7 January 1668/9 (Birch 1756, pp. 338–339). Spring or elasticity: 29 October (Birch 1756, p. 316); 12 November (Birch 1756, pp. 320); 26 November (Birch 1756, p. 328); 3 December 1668 (Birch 1756, p. 332); 10 December (Birch 1756, p. 333); 17 December (Birch 1756, p. 335); 18 February 1668/9 (Birch 1756, p. 347).

published contributor on the question of basic rules of collision, he had the insight to request that the problem of collision be focussed upon at that moment, and he had the confidence to push the idea. However, others were clearly just as involved in keeping the project for publication focussed. This is apparent in the reported behaviour of Brouncker and Oldenburg, not to mention Wren. He had much the same recent research profile as Hooke. He delivered his paper quickly and always supported the direct advancement of the project.[47] It was therefore a cooperative tactic by these leaders to excise this specific question from the wider field of cognate investigations, and it deserves to be denominated as such. We shall term this the 'carving out' of the collision project from everything else then underway and arguably relevant to it. This again indicates that non-trivial decisions and actions were involved in the genesis of the project.

However, it was one thing for several leaders to display the *nous* to call for an inquiry into the rules of collision to be carved out of the wider organizational field of activity. It was another for the inquiry in fact to be quickly executed, closed and published. Debate and dissidence were prominent—as we shall see in Sects. 6, 8 and 9—and they needed to be managed so as to preserve the viability and circumscribed nature of the collision inquiry. We shall discover that there were two interacting dimensions of debate at the Society about the collision project: one dimension concerned fundamental natural philosophical issues, while the other was located at the technical/conceptual level in macro mechanics. We shall encounter people who wanted the inquiry directly grounded in, begun and ended with considerations of ultimate matter theory and conceptions of natural philosophical causation. Even the lead players in macro mechanics held out the possibility of ultimate natural philosophical relevances. So our axis of tension about classical mechanics emerging from natural

[47] The importance of Hooke in focusing the project while conducting most of the cognate studies that were left out for its benefit should be noted. In the Taylor/Schuster case studies, Hooke is a clear and unopposed leader and director of two of them: the 'metals project' of 1680, and the Newton optical work controversy 1672. Here we are dealing with our earliest case. Of course Hooke's status grew, so that, for example, in our case study of the project on metals (1680) one should not be surprised to see Hooke in total command. He certainly was not in total control in the present case, given that it was always intended to consist in the available work of others. Still, one gets the impression he was a leader in this project, exercising a subtle directing role, as a result of giving sustained thought to the agenda of the Society in mechanics and how to advance it. The traditional perspective on Hooke as a social inferior to the bulk of the Fellows and a kind of under worker first to Boyle and then to the Society (which he was at first) has tended to be corrected by recognition of his many leading contributions both to late seventeenth century natural philosophy, and we might add, to leading organizational roles. Essay reviews by Gaukroger (2006b) and Boschiero (2005) pointed the way to these more fruitful and accurate directions. I thank Dr. Adrian Wilson of the HPS Programme, University of Leeds for drawing to my attention these two penetrating reviews. I suggest we learn more about the natural philosophy and sciences of the day, the workings of the Royal Society and the historical standing of Hooke by following him around, 'in action' at the Society over time, rather than freezing his image and surrounding it with equally frozen contextual determinations about large gauge social status. Hooke became overwhelmingly significant and wielded very considerable power, emerging into view even in this episode. Remember, at the time of the rules of collision inquiry he was already the author of the popular and impressive *Micrographia* (1666).

philosophizing was obviously in play. On the level of macro mechanics per se, we shall see in Sect. 9 that mechanics experts, including the three authors, continued to disagree about technical details, and that others who advanced serious objections were constantly told to reconcile their own views with those of Wren, Wallis and Huygens and to provide experimental evidence for those reconciled claims. In point of fact when in May 1669 Oldenburg told Huygens the project was all over and closed, both natural philosophical and technical critics were still in full flight (Oldenburg 1965–1985, vol. 5, p. 582). As for the outside reading public, the *Phil Trans* of course presented only consensus.

There are two likely reasons why the leaders' suggested carve out was readily accepted and the resulting papers pushed as a consensus to which all competing views and queries would have to be reconciled with experimental support. First of all, as we have seen, at least something about the Wren and Huygens results had long been known, having echoed impressively inside the institution. They therefore were probably presumed to be solid, experimentally backed and ready to be presented. But something deeper also seems to have been in play, to wit, a desire to establish and publish some putatively quite basic laws of mechanics—basic in the sense that they probably also applied at the corpuscular level. In other words, Hooke, Wren and some others grasped that regarding rules of collision, some central elements of the new discipline were at stake, likely to help consolidate the field, at the same time that some fundamental corpuscular laws of motion might be uncovered.[48] Interestingly, a similar dynamic later played out at the Paris *Académie des Sciences*.[49]

We obtain a flavor for these intricate reasons behind the carve out (and the difficulties of sticking to it) through events at the critical early meetings. For example, as shown, at the meetings in late October and November, Hooke was deeply involved in several experiments (some quite inconclusive) and debates about mechanics issues potentially related to the rules of collision. This work and the debates were focussed on spring, rebound and elasticity as well as pendula of various types. Not only were the experiments difficult to interpret, but debate erupted about, yes, the possible corpuscular bases of elasticity.[50] This seems to have been dragging Hooke himself away from the clear cut aims of his original 26 October suggestion. Indeed on 26

[48] See Sprat's (1667, p. 311) encomium of Wren's paper, cited below Note 62 and Hooke's concern with possible corpuscular insights discussed in Sect. 6. Hunter and Wood (1986, p. 67) note that Hooke's call had been 'Prompted by his desire to secure and elaborate the foundations of the mechanical philosophy [...].'

[49] In 1675 when Minister Colbert told the Parisian Academicians to concentrate on useful applications of mechanics, they replied that they first needed to establish principles of the discipline, or 'the natural principles of mechanics'. Huygens and Mariotte were assigned to lead this task but in the end Colbert vetoed the work on mechanical principles and only a volume on mechanical inventions eventuated. I thank Dr. Luciano Boschiero for this point, and the primary reference: *Registres des procès-verbaux de l'académie des sciences de Paris*, Vol. 8, 44–6. He first mentioned it to me in correspondence in 2017 about the origins of the present paper. Later he presented the material in a paper titled titled 'Machines, Motion and Mechanics in Mariotte's *Traitté de la percussion* (1673)' for the *Workshop on Themes from the Work of Carla Rita Palmerino on Natural Philosophy*, February 2019, organized by Peter Anstey at the University of Sydney.

[50] See the instances cited in the first paragraph of Sect. 6 below.

November Croone punctuated a debate about Hooke's work on spring and reflexion by asking whether the Society might investigate motion, setting apart the issues about spring. This micro event has not been much noted, but it appears to be rather telling: It seems likely that Croone too was grasping the idea that if one really did wish to proceed from fundamental and simple macroscopic rules, those putatively also applying at the corpuscular level, perhaps the Society should bracket off the tricky experimental investigation of spring, elasticity and their degrees and causes, and establish first of all the rules of macroscopic, as near as possible to perfectly elastic collisions. For some inside the Society this is exactly what the early Wren/Rooke/Huygens work was believed to have done. We see here that when Hooke called for deep conclusions about the rules of collision, he had been on the same wavelength that Croone was to display a month later. Arguably, they had both glimpsed the forest beyond the individual trees. But, by late November, Hooke and the Society were so enmeshed in detailed work on spring, elasticity and less than perfectly elastic collision (and the underlying causes of elasticity), that even Hooke had to be reminded, along with the Society, about the call on the table for Wren's and Huygens' work on the simpler (perhaps fundamental) case of in effect perfectly elastic macro collisions. This in turn perhaps also renders explicable the quick acceptance and publication of Wallis' contrasting exercise in the rules of perfectly inelastic collision, that was produced and registered, without being read, at that same November 26 meeting. Wallis' quick work arrived at a timely moment, it would seem.[51]

6 Did Mechanics Presuppose a Natural Philosophical Basis at the Early Royal Society?

Our findings in Sects. 2 and 3 naturally lead us to pose the question of whether and how the emerging domain of classical mechanics was related to natural philosophizing activities inside the Royal Society rules of collision project. Nobody involved explicitly insisted that the new mechanics had nothing whatsoever to do with natural philosophical issues. Even those most concerned with the project, being corpuscular-mechanists of one sort or another, held out the in principle possibility that finding basic laws in macro mechanics might eventually lead to insight about the 'mechanics' of the corpuscular realm. Because of this lurking possibility of natural philosophical relevance, the raising of issues at the macro level had potential resonances at the corpuscular level. For example, Hooke was several times involved in discussions

[51] There is more to be said about the dynamics of the 26 November 1668 meeting (Birch 1756, p. 328). We see in the minutes the following three entries in sequence: The opening item in the minutes is the report on Hooke's experiments with both 'springy and not springy' balls intended to show that 'reflection of motion results from springiness'. The experiments are reported as not satisfactory. Right then Croone suggests that the Society might study the *'business of motion [...] without taking in the notion of the springiness of bodies'* Indeed! The next item in the series of three is Oldenburg's production of Wallis' paper. In fourteen lines much of the tenor of debate, and contours of the final product, had been aired, and pretty well resolved.

about the deep cause of 'springyness', sometimes initiated by himself, sometimes urged by other members. On 29 October a week after Hooke's call for work on the rules of collision, he told the Fellows that experiments were needed to determine whether bodies rebound due to springy particles or whether it is the case the bodies are springy because of the presence of air (presumably the presence of small volumes of air trapped in the substance of springy bodies)—a flagging of Boyle's spring of the air.[52] The former explanation clearly appeals to the existence of springy micro particles, the latter inserts Boyle's 'intermediate cause' of spring of the air into hypothetically existing small interstitial volumes distributed inside a material. On 17 December Hooke was asked to submit what he thought caused springyness and to experiment upon the question of whether more springyness caused more rebound—a matter much discussed at this stage.[53] Here there seemed to be no problem raising issues about how macroscopic investigation and results in mechanics might illuminate the corpuscular world and basic laws of nature. There was also some impatience with this sort of thing, as shown by Croone's intervention on 26 November cited above. All this was quite different from the strategy, redolent of traditional systematic contestation in natural philosophy, that one should start the inquiry about the rules of collision at the natural philosophical level, and move to macroscopic mechanics, and possible confirmatory evidence later. The differences of opinion and emphasis that emerged about issues at the macro level only will be addressed later in Sect. 9, when we look at the way such 'intra-mechanics' differences were managed in the interest of consensus and limited publication of results. What concerns us for the moment are the problems and tensions introduced by those advocating in-house a 'natural philosophy first' approach to the rules of collision problem.

William Neile was the chief of those Fellows for whom it remained the case that fundamentals about motion at the corpuscular level, and related issues in matter theory, needed to be addressed prior to asserting anything in the new mechanics about the principles of motion at the macro level. He was the son of the distinguished foundation Royal Society member Sir Paul Neile. The younger Neile, thirty-one years old in late 1668, was an accomplished pure mathematician, but not so much a student of the new, emergent mechanics. He was intelligent, engaging, diplomatic and well regarded. Elected a Fellow in January 1663, he unfortunately died young, in July 1670. We have already mentioned how Wallis sidelined a paper on motion that Neile submitted in 1667. That paper very likely was the basis of another one he finally tabled at the 29 April 1669 meeting of the Society, during the long course of a series of letters between himself and Wallis, mediated by Oldenburg. These documents

[52] Birch (1756, p. 316): '*Mr. Hooke moved, that experiments might be made to see, whether all hard bodies, that rebound, do not so upon the account of having springy particles in them; and that it might be inquired into, whether there be any body springy upon any other force, than that it has air in it.*' Referencing Boyle (1660) and Cfr. Chalmers (2017, pp. 119–122).

[53] Birch (1756, p. 335) for the request to Hooke. Other discussions at the macroscopic level of the problem of elasticity and rebound occurred on Nov 12 1668 (Birch 1756, p. 320); 17 December (Birch 1756, p. 334); 18 Feb 1668/9 (Birch 1756, p. 347). On the latter occasion some members suggested that experiments should be carried out with the most springy bodies—expecting that these would produce the most close approximation to Wren's rules.

reveal the foundation of Neile's position: He held that macroscopic bodies consist of micro particles which are in continuous small scale motion, in all directions. If a particular direction predominates amongst these particles, the macro body translates in that direction. If no direction predominates, the body is at rest. Two 'rules' of particle collision control micro events: If one micro particle impacts another particle at rest, both move off in the direction and at the speed of the incoming particle, an 'interesting' proposition as Huygens later commented. Secondly, if two particles collide from opposite directions, regardless of their initial speeds, they come to rest.[54] From speculative accounts of such micro outcomes, he tried to argue for types of macro results.

This was deep natural philosophical theorizing, seeking a matter theory and theory of causation. It echoed Pierre Gassendi's version of atomism, most likely sourced from Walter Charleton's *Physiologia Epicuro-Gassendo-Charletoniana* (1654) a widely read translation and synthesis of Gassendi's atomist natural philosophy. The Gassendist program had always threatened to place a barrier between micro matter theory and results in mechanics (Dijksterhuis 1961, pp. 429–30): It rejected the main line corpuscular-mechanists' hope, originally articulated clearly by Descartes, in simple transdiction down from the macro to the micro. Descartes had famously and incisively claimed that there are no rules of mechanics that do not pertain to physics (natural philosophy); that is, apply at the micro corpuscular level.[55] Gassendist atomism, often classed as just another species of corpuscular-mechanism, fails this test; always did and continued to do so in the hands of Neile, who embraced Gassendi's not-quite-mechanistic-atomism—or perhaps stumbled into Gassendi's pitfall, to wit, a disjunction between macro mechanics and the declaratory mechanics of atoms.

Twenty-five years earlier, it would have been perfectly feasible for Gassendi to debate this way against Descartes. But this Royal Society debate was not going to be about natural philosophical systems. Indeed outright debate about systems was never tolerated. Additionally, the emergent classical mechanics had advanced beyond Descartes and Galileo. The lead players in this project were doing macro mechanics, hoping perhaps sometime in the future to be able to transdict down to their micro world of corpuscles. In contrast Neile embraced Gassendi's theory of matter, hoping to work up from that in order to correct findings in macro mechanics. In the last analysis, Neile's views and his strategy of explanation were not going to be accepted

[54] Kemeny (2017, pp. 184–185) for a lucid exposition of these principles. See also Jalobeanu (2011, pp. 117–119). Both accounts are based largely on Neile's 7 May 1669 letter to Oldenburg (1965–1985, vol. 5, pp. 517–578, 534, 526). Huygens had been sent a copy of Neile's work on 18 June 1669. Six weeks later, 31 July, Huygens replied with his 'interesting' comment, adding that Neile's claimed rules erroneously maintain that objects offer no resistance to motion and that rest is a state different from motion (which violated Huygens' brilliant deployment of Galilean relativity of motion to the problem of collision). Huygens also claimed Neile's rules conduce to a range of absurd physical predictions—predictions in macro mechanics. Oldenburg (1965–1985, vol. 6, pp. 161–164).

[55] Descartes, *Principia philosophiae*, Part IV, Art. 203 '*Et sane nullae sunt in Mechanica rationes, quae non etiam ad Physicam, cuius parts vel species est, pertineant*'.

inside this project, let alone published, as indicated by the swift selection and—as observed above—carve out of the three key papers. Still, this sort of behaviour needed to be managed and finessed inside the Society, because Neile was a respected Fellow, an intellectual and gentlemanly peer. The organizational questions are, How exactly was that done and what does it show us about the early Royal Society? Those are issues for Sect. 8 below and our Conclusion. For the moment we need to look more closely at Neile's claims, and his manner of proceeding.

Neile correctly judged himself to be raising fundamental issues about atomistic matter theory. Unfortunately, in this Royal Society project and among the mechanics experts and other leaders, nobody was buying the relevance to the new mechanics of the dynamical and kinematic aspects of Gassendi's matter theory. Tellingly, Neile constantly questioned—legitimately from his standpoint—how his adversaries, mechanics experts like Wren, could dare to assert things about ultimate reality, i.e. matter theory, on the basis of appearances (that is the rules of macro collision found in up to date mechanics). He objected to Wren's work proceeding incorrectly from macro experimental results, asserted as 'axioms', whereas to behave 'philosophically' demanded one start from considerations of ultimate matter and cause at the particulate level.[56] Of course the mechanics experts could proceed from macro mechanics to the possibility of micro insight because they were true corpuscular-mechanists and believed macro to micro transdiction would perhaps someday be possible, whereas immediate elevation of putative matter theoretical principles into rules of macro mechanics did not seem possible or fruitful—especially the odd neo-Gassendist principles of matter theory asserted by Neile. Similarly, during the late 1668 debates at the Society about elasticity, Neile had also asked in private correspondence via Oldenburg about hardness, offering his own explanation in terms of micro-particulate motion, rather than rigid entwinement.[57] Hence a perfectly hard macro body, like a diamond, was for Neile a perfectly elastic one—the opposite of what Wallis and Hooke thought. Again, Neile considered that from his 'dynamics' of micro particles he could derive or at least rationalize macroscopic mechanical rules. But clearly that could only be done by beginning with necessarily speculative, qualitative, natural philosophical discourse. His opponents held a different view of the relation of macroscopic mechanics to natural philosophy: One starts with problems in mechanics at the idealized macroscopic level, while the solutions might eventually prove to have the potential to be related to natural philosophical issues.

---

[56] For example in late January 1668/9 Neile wrote via Oldenburg, '*Dr. Wren I think he assumes his axiome a great deale sooner then he need to doe for if it be possible that the nature of motion be really no otherwise then it seemes to be by experiment to conclude that the aparence is the reality and that the aparence must not be denied to be really true under pretence that it is an axiome meethinks is not very philosophical*' Oldenburg (1965–1985, vol. 5, p. 363). Cited in Jalobeanu (2011, p. 116). Neile's early paper on motion set aside by Wallis as referee in 1667 was mentioned above in Sect. 5. Neile's excursion into the collision project began in earnest when he sent to Oldenburg a series of deep (i.e. natural philosophical) queries about motion that Oldenburg forwarded to Wallis, who duly responded with answers read and entered in the *Letter Book* 10 December 1668 (Birch 1756, p. 333; Jalobeanu 2011, pp. 113–114; Kemeny 2017, p. 183, p. 188; Oldenburg 1965–1985, vol. 5, pp. 220–221).

[57] 18 December 1668 (Oldenburg 1965–1985, vol. 5, p. 264; Jalobeanu et al. 2011, p. 116).

The crescendo of Neile's criticisms came following the publication of Wallis and Wren's papers in January 1668/9. In a flurry of correspondence carried out via Oldenburg over several months, Neile launched some tricky questions at Wallis—queries that touch on the macro world as well[58]: Is motion ever lost or gained; will the cosmos unwind to zero motion? But ultimately the answers, and the discussion, were taken to reside in speculative theorizing about his Gassendist particles that make up macroscopic bodies. Surely this explains why in April and May 1669 the upholders of the new consensus at the Royal Society about the rules of collision in mechanics kept asking Neile to 'reconcile' his 'hypothesis'—his matter theory that is—with the findings of Wren and Huygens. They comprehended that in their construal, the inquiry into of the rules of collision of bodies was a problem in the new mechanics, not in the matter and cause deep structure of this or that philosophy of nature.

Eventually, as we shall see, Neile threw up his hands at these requests, and I suggest, at the entire tenor of his treatment—or rather 'management'—by the project insiders at the Society. Moreover—and I do not think any analyst has yet made this point—even if Neile had somehow managed to 'reconcile' his matter theory to the rules of Wren and Huygens (to *their* satisfaction) do we really think the Royal Society would have published his Gassendist, qualitative, discursive matter theoretical claims? Moreover, what 'experiments' could possibly have illuminated that kind of natural philosophical imaginings? We today can see this latter point, but surely at the time Wren, Huygens, Wallis, Hooke and Oldenburg also grasped it and arguably were acting upon it.

7 Accurate and Fruitful Historiography Depends upon Choice and Design of Categories

Historians of any type will correctly tell you that periodization is the indispensable armature of historical inquiry. Indeed as Sect. 2 has briefly mentioned, this study is guided by interpretative constructions about the main stages in the Scientific Revolution that I have detailed in other publications. But as I have also insisted elsewhere, that is only half the story. Historical understanding also requires conceptualization of the entities and structures in play. Only this allows serious narration *and* explanation to be slung across a periodization armature. There is no point telling stories and giving explanations about entities and processes whose nature and dynamics have not been carefully thought out. These categories or models are heuristic and revisable, but they are ignored, or frozen, at the peril of one's life as a serious historian. In history of science, history of philosophy and intellectual history generally, these categories

[58] Jalobeanu (2011) and Kemeny (2017) offer excellent and detailed accounts of this correspondence. In Sect. 7 we shall have occasion to discuss the wider theses of the former paper. Here we note only that our concern is not with the blow by blow accounting of this correspondence, but with Neile's general position and strategy in the debate and the ways he was managed to produce and 'carve out' the consensus.

pertain to the knowledge-seeking fields, domains, disciplines or traditions taken to be in play. Only by constructing models of their structure and dynamics of change can historians hope adequately to describe and explain actors' decisions and actions within those domains. In addition, models of this type also allow for a disciplined and fruitful approach to the question of how forces and structures 'external' to a given discipline can and should be taken to have effects and consequences within it.[59] In the present study, we have tried to work within a model of how the field of natural philosophizing was structured and changing in the period, including as part of that the ways the traditional mixed mathematical disciplines were being transformed and eventually crystallizing out of the larger field. Additionally, we have taken the view that the Royal Society was not simply the site of this or that methodological regime, or the summed total of the sociologically characterized traits of its members. We have worked within the Taylor–Schuster modelling of how the institution organized its typical products—inquiries into natural knowledge and (selective) publication of chosen results. Any satisfactory historiographical approach to the history of science requires this sort of work. Of course, different historical specialisms, or particular branches within them, tend to inculcate differing descriptive/explanatory categorical frameworks. We can illustrate this in our present case, and thereby underscore the categories we have chosen to articulate, by looking at an important and telling contrasting approach to the Royal Society collision rules project, coming from the field of history of early modern philosophy.

In an important article Dana Jalobeanu (Jalobeanu 2011) tells us the these events involved the competition of two natural philosophical teams, Cartesian and Gassendist. Yes, Neile was (at least in part) a Gassendist playing for high natural philosophical stakes, and he was marginalized for such overt natural philosophical partisanship. But, Wallis, Wren and Huygens were not a Cartesian team. They were corpuscular-mechanists, with ultimate aims in that domain. But in the first instance in their Royal Society collision work they were practitioners of the emergent new mechanics, as described throughout the present study. I suggest that in order to construe Wren, Wallis and Huygens as Cartesian natural philosophers, Jalobeanu in effect conflates two versions of Cartesianism: what one might term a 'hard' version, in Descartes own 'project for a mathematical physics', and a 'soft' version amounting to the idea that practicing mechanics in the 1660s was being a Cartesian. Let us look at both.[60]

The 'Cartesian project for a mathematical physics' is a scholars' illusion promoted by some historians of philosophy since the nineteenth century. It was created by amalgamating two projects of Descartes: his early, pre–mechanical philosophy text, the *Regulae*, and some very general statements about ontology in his much later *Principles*. The *Regulae* speak of a mathematical science of relations amongst measurable,

[59] Schuster (2019; 2013a, pp. 65–70; 2013c, pp. 26–28; 2022, pp 31–34, pp. 38–42).

[60] Jalobeanu (2011, pp. 107–110). At one point the systematizing Cartesian project is termed 'bad' Cartesianism, in the eyes of the Royal Society 'Cartesians', and their own softer version looking for mathematical laws of motion, is termed 'good'. The supposedly good Cartesianism, is just doing mechanics and does not need to be classed as a species of Cartesianism, that is, within the same genus as the 'bad'.

macroscopic physical dimensions. (I have written about this extensively and shown how and why Descartes abandoned his project of the *Regulae* in 1629 and only then turned to systematic corpuscular-mechanical natural philosophizing (Schuster 1986; 1993; 2013a, Chap. 7). This defunct project of the *Regulae* is glued to Descartes' later posturing about matter being mere geometrical extension and to his related method rhetoric about strictly mathematical elaboration of statements about such matter-extension bodies and their motions. But this latter description is also misleading, as we have seen in Sect. 2: Descartes' natural philosophizing was not mathematical: there is not an equation or geometrical proportion in sight. His *Le Monde* and *Principia philosophiae* are entirely qualitative and discursive. Moreover, these texts are not really about Descartes' famous matter-extension; their explanatory detail is de facto corpuscular, as Thomas Kuhn long ago argued and I have shown in considerable detail in numerous publications. In short there was no hard Cartesian project for a mathematical physics for anyone to follow.[61]

Jalobeanu's soft or 'good' Cartesianism is just doing the new mechanics in the 1660s, meaning—as we have reiterated throughout—the geometrical and experimentally mediated study of idealized systems of macro bodies. Experts in the new mechanics, such as Wallis, Wren and Huygens certainly hoped that macro findings in mechanics might prove able to be translated down into being the laws of the micro world. They had what might be called natural philosophical investments, as many did at the time, but they were not focally pursuing basic natural philosophical questions of matter and cause, let alone systematically and certainly not part of any 'team' gambits. The supposed 'Cartesians' such as Wren, Wallis and Huygens, as we already have learned, were working away at problems in mechanics. There is no need to make them into a Cartesian party in natural philosophy per se, when their focus was not on promoting and extending any kind of 'full-service' natural philosophy of Cartesian or any other sort. We have alluded to Wallis' forthcoming 800 page tome, titled "*Mechanics, or On Motion by Geometrical Means*". Sprat was later to give a similar gloss to Wren's collision studies.[62] Huygens' complex, fruitful relation to Descartes did not amount to supporting the system.[63] Huygens' brilliant achievements in mechanics owed little if anything in point of technical, conceptual and mathematical detail to Descartes' system of natural philosophy or his supposed 'project for a mathematical physics'. So arguably there was no Cartesian team at the Royal Society.

As for Gassendists, only Neile is discussed at length. It is not proven that other dissidents like Willughby or Croone were Gassendists. We have seen that Neile wished to revivify Gassendist matter theory, and we shall soon see that the leaders

---

61 See items cited above Sect. 2, Notes 12–16, Kuhn (1959, p. 240).

62 Sprat (1667, p. 311) praises Wren's work on the rules of collision this way: Wren's '*doctrine of motion*' [sic] is '*the most considerable of all for establishing the first principles of* [natural] *philosophy*' That is, this macroscopic work in mechanics may well help provide (new) principles for natural philosophizing itself. Not vice versa—one does not start with Descartes'—or anybody else's—corpuscular rules of motion and collision and work up to the macro world.

63 Cf. the remarks of Westfall (1971, pp. 184–185) on Huygens' relation to Cartesian systematics as opposed to a broad corpuscular-mechanistic orientation.

of the Society organized quite predictable tactics for managing this gentlemanly systematist amongst their fellowship. Jalobeanu approves of the fact that Neile, who indeed was quite Gassendist, raised issues about fundamental causes, such as the definition of elasticity. But we have seen that the inquiry leaders were not impressed by Neile's urging a 'deep natural philosophizing first' strategy inside a set of investigations that per force concerned 'intermediate' level rules and concepts in mechanics. When the leaders ignored Neile on some points and criticised him on others, and managed his interventions, they were not staging a 'Cartesian' fight back. The putative members of the Cartesian team were just doing high level mechanics. Indeed the main evidence for Jalobeanu about the Gassendi grouping comes from the accounts of the Royal Society by Sorbière, Glanville and Sprat that date from slightly before this project and deal with micro-politically aimed accusations about the existence, or not, of natural philosophical sects inside the Society.[64] Jalobeanu was right to ask what we can learn from studying such conflicts over institutional descriptions and accounting. The problem is that she has moved too far toward hypostatizing such political accounting rhetorics.

Let us additionally note the following, in terms of putative team membership: As we shall see: at a late stage of his correspondence with Wallis via Oldenburg, Neile interestingly invokes ideas about the structure of explanation in natural philosophy that ultimately could only have come from latter portions of Descartes *Principles*. Therefore, was Neile Gassendist *and* 'Cartesian' in terms of natural philosophical genealogy? As for Wallis, in his 1664 mechanical 'theory' of the tides, as the *Phil Trans* index dubbed it, he uses a Galilean notion of circular inertia for celestial mechanical systems, the furthest thing from Descartes' approach or of anyone else, such as Huygens or Newton who saw that circular motion is forced motion, requiring analysis of its dynamical conditions. Shall we therefore say Wallis was not part of the putative 'Cartesian' team? Or should we say that he was simply making conceptual choices inside mechanics, in its early and somewhat inchoate state?

The conclusion is forced upon us that the creation of these feuding natural philosophical 'teams' at the early Royal Society is not a plausible or fruitful categorical construction in historiography. Amongst other things, it makes little allowance for how the meta-accounting and rhetoric put forward by an actor might or might not match up to the actual language and commitments that are more organically involved in his specialist, technical activities and actions. Thus, how Wren or Huygens worked upon the 'mathematical laws of collision' has everything to do with what they respectively thought mechanics might be, or become—where the latter did indeed include the possibility of applying to the micro corpuscular realm in which they both believed. However, for them technical work and debate in mechanics—and let us add the organization of this inquiry at the Royal Society—already had strictly nothing to do with any meta-posturing, rhetorically or programmatically, as some kind of 'Cartesians',

---

[64] Jalobeanu (2011, pp. 103–105). Actors' accounting rhetoric is not a sound guide to historians' category construction, unless additional evidence and argument is provided for proceeding in that manner. Cf. our observations about Peter Anstey's claims regarding 'experimental philosophy', below, Sect. 10.

that might be imputed to them by historians—meta-posturing that, we concede, was objectively possible for them to have done, but for which in fact we lack any evidence, quite apart from the aforementioned irrelevancy to technical work in mechanics.

8 End Game Organizational Manoeuvres: May–June 1669

William Neile was a Fellow and an insider of the Royal Society, a gentleman, an accomplished mathematician and no crank. Wallis indulged him, exchanging numerous letters in the first half of 1669. Neile was taken seriously—up to a point—and patiently managed mostly outside of official business by this exchange of letters. Oldenburg was the pivot. Wallis and Neile mainly corresponded through him. In organizational terms Wallis managed Neile and Oldenburg managed the encounter for the Society by keeping the interaction outside the institution. But, Oldenburg and others eventually lost patience: On 29 April 1669 Neile's 'Hypothesis on Motion', apparently a version of his original anonymous submission to Wallis in 1667, was read at a Royal Society meeting. Neile was 'instructed' to complete it; to consider how experimentally to verify it, and to accommodate it to the rules of Dr. Wren and M. Huygens.[65] What we have here is a textbook example of the gentlemanly management of a respected Fellow who was not pursuing the right sort of approach to the problem of collision (inside the emergent mechanics). Expert insiders present at the meeting knew no experiment could deal with Neile's micro particulate speculations; they knew that Neile's claims could not be accommodated to the now 'consensus' findings in mechanics. Moreover, he was asked to do this immediately. At the next meeting, 6 May, Neile was asked again for his completed theory applied to the rules of Wren and Huygens.[66] One does not have to have spent much time contemplating the culture of English clubs, then or now, to imagine what was happening: This was a gentlemanly way of telling Neile that so far as the managers of this matter were concerned, his work in their area was over inside the precincts of the Royal Society.

The next day, May 7, Neile wrote to Oldenburg—but not about the 'tasks' the Society had just set him (Oldenburg 1965–1985, vol. 5, pp. 517–518). Neile pontificated about the conditions of explanation in natural philosophy, the role of experiment and principles derived from reason. What he says is redolent of Descartes' late views on the matter at the end of his *Principia philosophiae*, to wit, corpuscular-mechanical explanations consist in the imaginative design of models aimed at explaining the widest range of known relevant facts, without meeting negative evidence, provided that the corpuscular-mechanical model in question asserts nothing inconsistent with the 'truths' of metaphysics previously established or immediate deductions from

[65] Birch (1756, vol. 2, p. 361). Amongst those present were Hooke, Boyle, Oldenburg and Aubrey. We also know that Neile's paper was ordered to be put in the *Register Book*. See Oldenburg to Wallis 8 May 1669 (Wallis 2005, vol. 3, p. 171).

[66] Birch (1756, p. 362). Neile apparently 'intimated' that he had started this work and would produce it as soon as he finished. Hooke, Wren, Boyle, Croone and Moray amongst others were present on this second occasion. On Croone at this meeting, see Note 69 below.

them (Schuster 2017a). Neile has a very similar view but uses the terminology 'principles derived from reason' instead of Descartes' speaking of principles derived from metaphysical truths. In both accounts such principles control the construction of particular corpuscular explanations—with Neile more consistently using 'experiment' to denote the realm of possible confirmatory/non-confirmatory evidence. Clearly Neile had studied Descartes carefully.[67] Hence our earlier remark that, if you please, he might be termed both Gassendist and Cartesian in natural philosophy. Jalobeanu (Sect. 7) does not talk about this management of Neile. For her Neile's letter is a creative change of topic. But, in its organizational and inter-personal contexts, it seems preferable to interpret it as Neile's rather lame and embarrassed attempt to change the subject to abstract theory of knowledge and structure of explanation in natural philosophy. His digression had little to do with the now months long flurry of letters concerning motion and collision. To be clear, yes of course, the two topics were connected, or able to be connected at least for some contemporary actors, provided that like Neile, one was in a very Cartesian fashion now insisting on prior sorting out of the metaphysics and higher level principles that supposedly control or constrain natural philosophical theorizing. But, in the context of how the debate had unfolded, and was being managed, at the Society, this change of subject was arguably more silly (and desperate) than 'creative'. Presumably Neile wanted to convey that this was a much more important and fundamental issue than what had transpired concerning him at the Society. He aimed to occlude the fact that just the day before the Society had more or less told him to stop pursuing his project and queries—at least in their environs and with expectations of acceptance and publication by them. Neile's gentlemanly colleagues had painted him in a corner, finally tired of his twenty-year out of date attempt to build up a mechanics from micro-matter theory. They did this in a diplomatic but crystal clear way, manifesting gentlemanly decorum, because Neile was one of them. And Neile certainly grasped all this; hence his elaborate, and arguably embarrassed natural philosophical backchat to Oldenburg about natural philosophical foundations that had nothing immediately to do with the specific jobs he had been given a week before, and that had been reiterated to him as recently as twenty-four hours previously.[68]

While Neile was being sidelined and diverting himself by taking up privately with Oldenburg his new grand theme of 'proper explanation in natural philosophy', the management of affairs ground on inside the Royal Society. When Oldenburg

[67] Kemeny (2017, pp. 186–187) has an excellent discussion of the import of the letter and the relation to Descartes' mature views on natural philosophical explanation.

[68] Kemeny (2017, pp. 190–191) provides a very useful account of the latter stages of Neile's campaign. He had virtually given up his agenda by October 1669, but not without retreating to yet another position, proposing reforms to how the Royal Society should operate—a not unusual pastime at the time as Kemeny, following Hunter and Wood (1986) points out. Neile proposed creating a committee on causes that would carefully 'indagate' causes from experiments, equally carefully carried out by the experts at the Society. In a way Neile was more correct about what procedure in so-called 'experimental philosophy' required, than were some of his contemporaries, or present day historians, who tended to downplay the omnipresence of theoretical considerations in all their work. Cf. above Sect. 4 and below Sect. 10 discussion of the modern historiographical category of 'Experimental Philosophy'.

wrote Huygens that the discussion of collision at the Royal Society was over; that everybody, almost, agreed with the published findings, this was only three weeks after Neile had been asked to accommodate his views to those of Wren and Huygens by way of experiment. The Society's leaders had no expectation Neile could do any such thing, and in any case they had long since made up their minds.

9 Achieving 'Consensus' in Early Physics at the Royal Society—A Close, Organizational View

We have seen William Neile carefully managed both before and after the publication of the three 'consensus' papers—January and April 1669—and even after May when Oldenburg told Huygens that almost everybody at the Society agreed with the published papers. When Oldenburg and Wallis were dealing intensively with Neile in early 1669—outside the formal precincts of the Society—the issue was an outright challenge concerning the presumed priority of natural philosophy over the new mechanics, where the published papers were clearly instances of the latter posed autonomously of natural philosophizing. The diplomatic but firm management and marginalization of Neile was a case of necessary local boundary drawing between natural philosophizing and practice of the new mechanics. Well honed practices for internal management were drawn upon, since Neile was a peer-Fellow and respected gentleman. What concerns us now is the existence of what may be called 'intra-mechanics' debate and dissent, which similarly required management both before and after the publications of the papers. Even if Neile had never surfaced with his interventions requiring management, there still would have occurred management of the consensus with regard to these 'intra-mechanics' challenges.

Consider first these differences amongst the authors of the three papers: Wren and Huygens disagreed on what was needed to support their rules. Wren, coming in part from the English practical mathematics tradition, was happy with piling up experimental confirming instances and appealing to the utility of the results. Huygens, being Huygens, relied on argument from fundamental principles to the rules: Principles not stated in his paper, such as relativity of motion and conservation of the motion of the centre of gravity of a system. Additionally, their modes of mathematical analysis and expression differed. Wallis had his own idiosyncratic but effective symbolism; Huygens formally worked in geometry but as we saw calculated privately in algebra if needed. Wren's solid and practical approach eschewed proof and relied on intuitions of symmetry. These sorts of issues, potentially quite discussable in relation to the emerging field of mechanics, were not published of course, let alone raised as points of further inquiry. Wider intra-mechanics issues were also potentially on the table amongst the authors. For example, Wallis differed from Wren and Huygens in his willingness to raise issues about elasticity and hardness that as we have seen were also deeply concerning Hooke at the time. Potentially, this could have opened up questions about matter theory; hence, perhaps, Wallis' willingness to engage with

Neile—at least in correspondence outside the actual meetings. But more to our point here, again these cognate issues, arguably quite germane to the study of the rules of collision, while mentioned or discussed, were not allowed to disturb the selection and publication of just those three papers. The Royal Society left all of these issues untouched in its public presentation: the three papers accepted and published thus represented the result of the inquiry.

There were other debated intra-mechanics issues, including those advanced by Fellows other than the three authors. These tended to emerge after the publication of the papers and just around the time in May 1669 when Oldenburg proclaimed the closure to Huygens, but some were present all along. For example, Croone—who as we have seen objected in November 1668 to Hooke and Wallis that one need not problematize the notion of elasticity in order to work on collision rules—submitted a paper, late in the game in terms of publication policy, in January 1668/9. For his trouble Croone kept being asked to present experiments that would reconcile his views to those of Wren and Huygens. He responded by continually promising to deliver them, yet it appears that nothing was forthcoming, despite the curious fact that Croone continued to come to meetings to discuss other matters.[69] Similarly late in the process, shortly after Oldenburg proclaimed the consensus in May 1669, Francis Willughby, the distinguished naturalist, objected to the idea that motion could, according to Wren's rules, be increased or decreased in collision. Wren retorted that this was not an anomaly or problem but the actual state of affairs, given how one understood motion and that his rules had been confirmed by experiment. In fact Willughby's treatment of some of Wren's cases did not show conservation of motion, because of Willughby's own construal of motion as in effect a scalar quantity. In any case Willughby seems to have been given short shrift and quick closure, being sent a copy of the original report on Wren's and Huygens' papers to peruse for his trouble.[70]

---

[69] Croone was formally asked to submit his paper ('hypothesis of motion') 14 January 1668/9 (Birch 1756), p. 339). The paper was submitted, read and registered at the next meeting and ordered to be compared by an expert committee (probably including Pell, Wallis, Wren, Hooke and others) to 'other theories already given in' (Birch 1756, p. 340). On 6 May (Birch 1756, p. 362) Croone told the Society the he hoped shortly to be give an account of the 'experiments of motion' through which he 'had begun to accommodate' his principles to the rules of Huygens and Wren. (This entry immediately follows the 6 May reiteration of the solicitation of Neile for experiments accommodating his views to those of Wren and Huygens.) On June 3 (Birch 1756, p. 377) Croone was asked about his experiments on motion (as well as on transfusion). He replied that he (still) needed the help of the operator, who was ordered to help him as soon as it was convenient to Croone. It seems that Croone then continued attending meetings to discuss other matters, despite the lack of any further mention of his mechanics work. On Croone's paper see Hall (1966, pp. 35–36).

[70] Willughby's first letter dated 29 May 1669 was presented by Oldenburg at the 10 June 1669 meeting, with the name withheld. It was entered in the *Register*, Birch (1756, p. 381). That was the occasion of Wren's retort. Then a second letter was sent to Oldenburg dated 21 June and read at the Society meeting 8 July (Birch 1756, p. 392). The Society's answer was to order that Willughby be sent a copy of the report on the original experiments by Wren and Rooke (in 1661), a move that rather confirms our observations about the foundational status of those events in the history of this affair, as viewed by most of the Fellows. Kemeny (2017, p. 188) briefly discusses the objections of Croone and Willughby. Wardhaugh (2016, pp. 133–139) has by the far the most detailed, and best contextualized, discussion of Willughby's intervention. He observes that Willughby's absence

So, quite apart from the height of the Neile affair, but in the same time period in early and mid 1669, the consensus was maintained regarding intra-mechanics debate. By April 1669 all three papers had been published. Nothing prior to or after that date was allowed to change or undercut the three papers or to signal to outside readers anything about the content or extent of ongoing discussion of the project inside the Society, let alone discussion of 'cognate' issues under study (cf. Sect. 5 above). All such wider cognate questions were institutionally shunted into different avenues for investigation and, at best, possible future publication. Hooke, Wren, Boyle and others of course could continue on with studies of spring, elasticity, hardness, pendula etc., a line having been drawn, in public, under the basic study of the rules of collision.[71] Clearly these cognate studies by leaders were seen as potentially fruitful and publication worthy. As for the likes of Neile, Croone and Willughby, there was much polite calling for papers and/or experiments by them to reconcile their animadversions with the already published papers. Some papers were forthcoming from them, as we have seen, but at best only registered and not published, while most of the calls for experiment were unanswered. It is arguably not drawing too long a bow to judge that neither the diplomatically requested 'reconciliation' with the published papers, nor the production of experiments to confirm such an achievement, were viewed by the managers as likely to eventuate.

Reflecting on these organizational moves, one is tempted to say that there were in fact two phases in the Society's 'carve out' and closure of the project. The first obviously was the decision to publish the three papers. But, a second post-publication phase surely was in play, punctuated by Oldenburg's claim to Huygens in May 1669 that the discussion was over and almost everybody agreed with the papers. In the months before this and for a while afterward nothing more was selected for publication despite much debate inside and outside the Society, as well as solicitations of papers and experiments. Neile, Croone and Willughby all continued to be 'organized', so that the three papers and only the three papers continued to constitute the public results of the project. Perhaps one might say that 'closure' and 'carve out' were also organizational processes, not single events.

---

from the crucial meetings and the resulting delays in communication contributed greatly to the lateness and resulting isolation of his rather interesting and competent contribution. The incident, Wardhaugh continues, shows the difficulty of being in the conversation unless physically present; or on Oldenburg's list of preferred correspondents for a given topic (Wardhaugh 2016, p. 138)—a very shrewd and apt observation. This seems to the present author to be the logical corollary of the policy based facts that the Royal Society most valued in-house contributions and actions (while occluding them from the public), at the same time that Oldenburg actively managed the communication side of projects—cf. Sect. 4 above.

[71] In modern usage the analogous policy would be giving each of these topics a different 'research area coding', all of which would differ from that of the now closed rules of collision project.

10 Conclusions about the Historiographical Models Deployed and Reflections on Alternatives

The Royal Society was managing the rules of collision project and its boundaries in the following two senses: [1] New, foundational work in the emerging mechanics was not allowed to be undercut or diverted by claims about the need for *prior* natural philosophical clarification of fundamentals. This policy was reinforced by the hope that foundational work in mechanics potentially might illuminate the laws of the corpuscular realm. [2] Within the emerging field of classical mechanics, cognate and potentially fruitful work (by leading insiders) was treated differently from the animadversions of other Fellows whose criticisms and alternatives were judged both unlikely to be fruitful in the long run, as well as disruptive of the published consensus. The field of classical mechanics was still somewhat fragmented, loosely defined both internally and in relation to the still possibly pertinent realm of natural philosophizing. Hence the organizational ways and means of the Royal Society had to be exercised by inquiry leaders, especially Wren, Wallis, Hooke and Oldenburg, to manage events and outcomes, in order not only to impose closure, but also to 'carve out' the project.

Allowing, as usual, for the specificity of the subject matter, this case study confirms the general contours of the findings in the other cases in the Taylor/Schuster program of study of the early Royal Society (Cf. Sect. 4 above): We find an open textured attitude toward 'method'—meaning possible technical modes of inquiry, not some mythical unique, transferable and efficacious method—with the corresponding reliance upon the specific expertise profiles of the inquiry leaders. There is the usual disjunction between publicly available results, offered seemingly as neutral reporting of incoming news, compared to the rough and tumble of internal debate, managed via organizational decision/action patterns.

Nevertheless, specific differences from our other case studies are also to be expected. Unlike the other cases Taylor and I have examined, the present case was embedded in an already existing expert area, so that we did not find one leader guiding other Fellows through the stages of an internal process.[72] The expert question about rules of collision was well entangled in other questions in theoretical and practical mechanics, so it did not have that one-off, separable feel of the 'May-dew' or 'metals cases'. Indeed for those reasons it had to be 'carved out' as a discrete enterprise. Nor did the rules of collision case contain a revolutionary challenge from outside the actual meetings such as posed by Newton's new theory of light and colors. Still, the internal conduct of business was the same as in other cases, as were the of handling 'non-consensual' views by peer-Fellows and the publication choices and filtering. Similarly, as in our other cases there were repeated calls for experiments in the form of reports, or in-house witnessed repetitions, etc. But in this case one might say that the ratio of repeated calls for experiments, pressuring different players at different times, compared to actual deliveries, was very high. Regarding the rules of collision, it is as though the leading Fellows had seen it all back in 1661, or heard about that

[72] As Henshaw did with May-dew and Hooke did in the case of the 'metals project'.

subsequently. This relative lack of delivery of experiments contrasts with our other three cases, in which experimental work when requested was generally delivered.

Our findings prompt comments on two lines of leading scholarship concerning the nature and organization of inquiry at the early Royal Society, pioneered respectively by Stephen Shapin and Peter Anstey. First, as to Shapin this case study confirms little regarding Shapin's claims about the Royal Society, except for the organizational suppression of overt natural philosophizing. Shapin's larger claims about the role of trust amongst gentlemanly virtuosi, and their decorous exchange of atheoretical matters of fact, are not confirmed in this case or any of our other ones. The Royal Society's gentlemanly decorum was not a mutual trust fest, where putatively atheoretical facts were found and communicated. Knowledge production at the early Royal Society revolved around relevant expertises, and good, fruitful theorizing (Schuster and Taylor 1997). Neither in this case, nor in our others, do we find a single claimed 'matter of fact' that is without theory-loading of some sort. Neile's opponents specifically did not expect that he would tell them 'the truth'. He was a gentleman and a Fellow, but they *distrusted* his claims and his proclivities as wrong and wrong headed. It is true that if we follow Shapin's lead and attend generally to the organization of decorum, we will penetrate further into these matters than if we remain content with Jalobeanu's somewhat abstract 'history of philosophy' categories of analysis. But what was being managed were disciplinary and theory-related claims and conflicts, not some ideal world of precious expertise-avoiding gents finding and exchanging (non-existent) atheoretical matters of fact.[73]

Finally, we must comment on the relation of our findings to Peter Anstey's claim that, even prior to Newton, English 'science' at the Royal Society consisted of action under the umbrella of a new 'Experimental Philosophy'. In a number of pioneering studies Anstey has established that the language of 'Experimental Philosophy' surpassing and supplanting 'Speculative Philosophy', previously associated almost entirely with pronouncements of Isaac Newton, was in fact being deployed in English, later British, natural inquiry well before Newton, by the third generation of the seventeenth century (Anstey 2005, 2014). This is a substantial historical finding. But what exactly is its significance? Anstey himself (and his followers) take these

[73] Shapin may be said to have fallen into two fundamental historiographical and epistemological errors, as Taylor and I pointed out many years ago (Schuster and Taylor 1997). These errors were [1] claiming that theory-less 'facts' exist and [2] claiming that at the Royal Society processes and results were actually accomplished in this supposedly completely theory and natural philosophy free space. The open espousal of systems and ongoing overt contention about them certainly was not organizationally allowed. But, middle level theoretical categories, as well as deeper natural philosophical commitments, categories and even tacit agendas continued to exist and play roles in projects and processes. Moreover, the work of inquiry at the Society was not the collection, re-testing and exchange of atheoretical matters of fact. It consisted, as just shown in this case, in organized courses of inquiry, led by assertive experts, who, taking advantage of the routine decision/action patterns, worked to channel, shape and in the end close inquiry by marginalizing some gambits and pushing the completion and publication of others. In short we see an embryonic foreshadowing of what scientific research eventually began to look like across the board in its classic era of the later nineteenth and early twentieth centuries.

categories as substantive, as constitutively descriptive of actual practices and activities: 'Speculative philosophy' was the almost entirely verbal and imagination driven systematized natural philosophy of earlier figures, notably Descartes and Hobbes, who supposedly were mainly concerned with bolstering system aims with speculative fantasizing of deep causes, ignoring or downplaying empirical evidence.[74] 'Experimental philosophy' was a progressive, successful, fruitful endeavor of unified technique and approach. It broke with speculation and paved the way, in some cases immediately, for the breakout of true experimental science.

Now, it is indeed clear from all our case studies that calls for experimental support for claims were in fact the preferred and endemic tactic inside the Royal Society. Experimental support was called for any paper submitted and for any claim made regarding the standing or consistency of any already accepted paper. Almost all the actors at the Royal Society operated within a discourse that preferred and privileged claims grounded in or supported by experiment. But, one must be very careful about how all this is understood: The key point is that neither Anstey nor any of his supporters has shown how this experimental philosophizing actually functioned *qua* an experimentally grounded protocol of 'fruitful, productive and meritorious' research.[75] That is to say—and this is really quite surprizing—Anstey offers no detailed dissection of 'hardware/theory' relations upon which one might reasonably expect the use of the term 'experimental' to depend.[76] It therefore must be realized that what we have here are merely labeling or accounting *topoi* used by some actors. It is a very good thing that these actors' terms have been located and traced in careful scholarship. But one must understand what it means to have identified a new genre of accounting rhetoric in the historical record: These terms were used by actors as resources of meta-discourse and accounting; that is, they were used broadly to categorise, describe and evaluate the work, outputs and attitudes of themselves, allies and opponents. These are not terms that *taken by themselves and directly applied* can accurately describe the actual practices—material, discursive and organizational—that produced concrete outcomes (hardwares, texts, claims) in natural philosophy and the more narrow domains of inquiry.[77] Moreover, it must be said that as counters

---

[74] Elsewhere I have shown Descartes' serious attention to empirical detail, not only in isolated instances, but as conditioned by specifically recognized system-articulating moves. I argue that systems such as that of Descartes, derided later in the century as 'speculative', reveal to the careful modern analyst very intricate interweavings of appeals to evidence, system driven detailed conceptual constructions and interesting architectonic theory manoeuvres that mere 'piecemeal' research would not have elicited. Schuster (2013a, Chaps. 11, 12; Schuster and Brody 2013).

[75] To invoke Chalmers' (2017) felicitous list of the virtues of actual scientific theories.

[76] On deployment of the categories 'hardware' and the 'discourse' invested in it, see Schuster and Watchirs (1990; Schuster 1995, Chap. 14, pp. 145–146; Schuster 2018, pp. 415–417); and my forthcoming monograph (Note 2 above) Chaps. 4, 11, and 14.

[77] I have articulated elsewhere insights coming from Bachelard (1949, 1975a, b), Kuhn (1970), Collins (1975), Pinch (1985), Barnes (1982), and Shapin (1982); as well as more recently Chalmers (2017), to create a model of the what I term 'the phenomenology of experiment and experimental hardware use'. (In the monograph forthcoming Note 2 above; foreshadowed in Schuster (1995, pp. 145–146, 2018, pp. 415–417). Pinch (1985) is a very good place to start for anyone innocent of the complexities of hardware/discourse or hardware/theory relations in serious scientific practice.

available to actors for rhetorical purposes of *presentation of self* and *description of others*, these terms play similar rhetorical and communicative roles to those attributed to methodological claims or 'method-talk', as I and others have described it over the past two generations.[78] In no single instance in any of the Taylor/Schuster case studies does the verbal content of the category 'experimental philosophy' describe, let alone explain anything of serious technical, practical worth.[79]

Despite these obvious historical and epistemological facts, one can still charitably conclude, with Anstey, that the constant demand for experimental support for this or that claim; or, for 'reconciliation' of one line of claims with another by means of experiment, shows something important: That the early Royal Society, or almost all its members, de facto operated under the shared meta-accounting cloak of what some

---

Nobody considering this material would reasonably retreat to the idea that the purported discursive content of 'experimental philosophy' did and can account for how humans make, use and debate the outputs of experimental hardwares.

[78] Schuster and Yeo (1986), Schuster (1984, 1986, 1993, 2013a), Mulkay and Gilbert (1981), Miller (1986) and Moyer (1992). We lose a lot if—as Anstey does with the rhetorical resources he has uncovered—we *essentialise* that rhetoric. Like all method or accounting rhetoric, the discourse of 'experimental' versus 'speculative' philosophy was fluid and could not constitute a genuinely efficacious *modus operandi* for actually *doing* any specific course of experimental science work. Yes, for a time many English natural philosophers called themselves experimental and berated so-called speculative ones. This proves nothing much about the macro history of experimental natural philosophy débouching into emergent sciences. (Cf. Schuster and Watchirs 1990; Schuster 2002; 2018, pp. 415–417). This rhetorical machinery simply was overlayed on parts of any contemporary players' activities, as binary meta–categorisations of an underlying spectrum of types of actual moves in experimental natural philosophy. Who really was 'speculative' and who really 'experimental': do we let Boyle determine, or should Newton? Should Boyle be charged with speculation for having natural philosophical lexicon materials in his natural history protocol: A point to be demonstrated in Chap. 14 of my monograph signaled above in Note 2. Accounting rhetorics are fluid and political in their use—that is why they are called accounting rhetorics.

[79] Anstey does offer something that superficially looks like a workable guide to actual experimental practices inside his 'experimental philosophy' structure. It is his category of 'natural history' (Anstey and Hunter 2008). That this construal of 'natural history' does not offer a unique, general, repeatable 'method' or *modus operandi* has already been demonstrated by a card carrying supporter of Anstey, Vanzo (2016) who shows, despite himself, that there was no protocol for a complete natural history and that judgement and interpretation were required on a case by case basis to decide if a 'sufficient' empirical base had been attained for any given explanatory/experimental project in hand. (And, let it be added, that even if a real 'method' of natural history had existed, that still would not have constituted a unitary, efficacious, transferable 'method' of doing experiments.) Vanzo also argues correctly for the interpenetration of corpuscular-mechanical categories into 'experimental philosophy', something Anstey, Gaukroger (2006a, 2010), Chalmers (2017) and others have imagined to have been largely banned from the realms of the emergent more modern looking experimental philosophy (or 'experimental science' in Chalmers' case). In my monograph in process (Note 2 above) I review Vanzo's most enlightening findings and also show that Anstey and Hunter's construction of Boyle's natural history for service in experimental philosophy does not acknowledge the obvious continued use by Boyle in that locus of categories and conceits drawn from the wide 'lexicon' (as I term it) of natural philosophical discourse. In short the world of 'speculative' philosophy, that is, actual, living natural philosophizing (without the trumped up rhetorical tag), had not been transcended and displaced. In later seventeenth natural philosophy, there was more use of and appeal to experiment, but no substantive, unified procedural entity 'experimental philosophy' existed in anything beyond certain actors' preferred rhetorical repertoire.

leaders formally termed 'experimental philosophy'. That this cloak was rhetorical (Schuster and Yeo 1986; Schuster 1993; Schuster 1986)—not a royal road to technical understanding of experimentation—is reinforced by recalling a fundamental matter concerning rhetorical regimes. These rhetorical labels—experimental versus speculative philosopher—by their nature may be deployed in a fluid manner. For example, William Neile accepted the demands for experiments, called for them himself, or hoped for them in his own cause. He could not do otherwise in the organizational culture of the Royal Society, so that Neile, too, could have self-described as an experimental philosopher—'one of the boys' as it were—particularly had Hobbes or a reincarnated Descartes strolled through the door thus drawing fire from the Fellows as 'speculative philosophers'. Indeed, which of our lead characters, including Neile, Croone and Willughby, would not have claimed to be an 'experimental philosopher' given the right social situation and pressures?[80] But on the other hand, who, being on the receiving end of the consensus and carve out, like Neile, would not have been labelled 'speculative' by his critics?[81]

To conclude, then, historians should not essentialize actors' rhetorical accounting resources[82]; rather we should study how actors deployed such resources and contested each others gambits—while simultaneously we historians should study experimentation and experimental hardware in their full complexity. That is part of the burden of the historiography of the modern experimental sciences. In this case study we have not had to do much of that, but we have had to attend to other two requisite dimensions of such an historiography: the (evolving) structure and dynamics of the relevant expert disciplines—mechanics and natural philosophy—as located in and further shaped by, the organizational structure and dynamics of the institutional settings of such inquiry.

References

Anstey P (2005) Experimental versus Speculative Natural Philosophy. In: Anstey P, Schuster JA (eds). The Science of Nature in the Seventeenth Century: Changing Patterns of Early Modern Natural Philosophy. Springer, Dordrecht, pp. 215–242.

Anstey Peter (2014) Philosophy of Experiment in Early Modern England: The Case of Bacon, Boyle and Hooke. Early Science and Medicine 19:103–132.

---

[80] Such a counter-factual 'Neile' could have maintained that he was 'only slightly re-negotiating' what 'experimental philosopher' means, recognizing as his peers did not—at least in their own accounting rhetoric—the actually unavoidable role of theory in experimental investigations. Cf the remarks about Neile's end game gambits in Note 68.

[81] The classic study of the rhetorical/political fluidity of methodological labels is Mulkay and Gilbert (1981). Anyone tempted to believe in a substantive late seventeenth 'experimental philosophy' should read that paper along with Pinch (1985) and then, like Newton, 'think hard' on the matter.

[82] Jalobeanu (2011) too was doing that, first by using *ex post facto* accounts by other actors who had not been directly engaged in the activities, and then, by hypostatising their terms of accounting and rhetorical aims.

Anstey P, Hunter M (2008) Robert Boyle's 'Designe about Natural History', Early Science and Medicine 13:83–126.
Bachelard G (1949) Le rationalisme appliqué. Presses Universitaires de France, Paris.
Bachelard G (1975a) Le Nouvel Esprit Scientifique. 13e edition. Presses Universitaires de France, Paris.
Bachelard G (1975b) La Formation de l'Esprit Scientifique. 9e edition. Vrin, Paris.
Barnes B (1982) T. S. Kuhn and Social Science. MacMillan, London.
Bennett J (1998) Practical Geometry and Operative Knowledge. Configurations 6(2):195–222.
Berkel van K (2012) Isaac Beeckman on Matter and Motion: Mechanical Philosophy in the Making. Johns Hopkins University Press, Baltimore.
Bertoloni Meli D (2006) Thinking with Objects: The Transformation of Mechanics in the Seventeenth Century. Johns Hopkins University Press, Baltimore.
Birch T (1756) The History of the Royal Society of London. Vol II. A. Millar, London.
Boschiero L (2005) Stories about the Rise of Modern Science. Minerva 43(3):311–318.
Boschiero L (2007) Experiment and Natural Philosophy in Seventeenth Century Tuscany: The History of the Accademia del Cimento. Springer, Dordrecht.
Boyle R (1660) New Experiments Physico-Mechanicall, Touching the Spring of the Air and Its Effects. Hall, Oxford.
Brandt F (1928) Thomas Hobbes' Mechanical Conception of Nature. Levin & Munksgaard, Copenhagen.
Chalmers A (2017) One Hundred Years of Pressure: Hydrostatics from Stevin to Newton. Springer, Dordrecht.
Collins HM (1975) The seven sexes: a study in the sociology of a phenomenon. Sociology 9:205–224.
Dear P (1995) Discipline and Experience: The Mathematical Way in the Scientific Revolution. Chicago University Press, Chicago.
Dijksterhuis EJ (1961) The Mechanization of the World Picture. Clarendon Press, Oxford.
Dugas R (1955) A History of Mechanics. Tr J.R. Maddox. Dover, New York.
Gabbey A (1980) Force and Inertia in the Seventeenth Century: Descartes and Newton. In: Gaukroger S (ed). Descartes: Philosophy, Mathematics and Physics. Harvester, Sussex, pp. 230–320.
Gabbey A (2004) What was 'Mechanical' about 'The Mechanical Philosophy? In Palmerino CR, Thijssen JMMH (eds). The Reception of the Galilean Science of Motion in Seventeenth–Century Europe, Kluwer, Dordrecht, pp. 11–23.
Gaukroger S (1978) Explanatory Structures. Harvester, Hassocks.
Gaukroger S (1995) Descartes: An Intellectual Biography. Oxford University Press, Oxford.
Gaukroger S (2000) The Foundational Role of Hydrostatics and Statics. In: Gaukroger S, Schuster JA, Sutton J (eds). Descartes' Natural Philosophy, Routledge, London, pp. 60–80.
Gaukroger S (2006a) The Emergence of a Scientific Culture: Science and the Shaping of Modernity 1210—1685. Clarendon Press, Oxford.
Gaukroger S (2006b) Review of Robert Hooke Tercentennial Studies. Aestimatio 3:52–5.
Gaukroger S (2010) The Collapse of Mechanism and the Rise of Sensibility 1680–1760. Clarendon Press, Oxford.
Gaukroger S, Schuster JA (2002) The Hydrostatic Paradox and the Origins of Cartesian Dynamics. Studies in the History and Philosophy of Science 33:535–572.
Giddens A (1976) New Rules of Sociological Method. Hutchinson, London.
Grant E (1978) Aristotelianism and the Longevity of the Medieval World View. History of Science 16:93–106.
Hall AR (1966) Mechanics and the Royal Society. British Journal for the History of Science 3 (9):24–38.
Hattab H (2005) From Mechanics to Mechanism: The Quaestiones Mechanicae and Descartes' Physics. In: Anstey P, Schuster JA (eds). The Science of Nature in the Seventeenth Century: Changing Patterns of Early Modern Natural Philosophy. Springer, Dordrecht, pp. 99–129.

Hooper W (2004) Seventeenth-Century Theories of the Tides as a Gauge of Scientific Change. In: Palmerino CR, Thijssen JMMH (eds). The Reception of the Galilean Science of Motion in Seventeenth–Century Europe. Kluwer, Dordrecht. pp. 199–242.
Hoppen KT (1976) The Nature of the Early Royal Society. Pt 1. British Journal for the History of Science 9:1–24.
Hunter M, Wood P (1986) Towards Solomon's House: Rival Strategies for Reforming the Early Royal Society. History of Science 24(1):49–108.
Hyslop S (2015) The Mathematics of Collision and the Collision of Mathematics in the Seventeenth Century, Ph.D. Dissertation, Department of History and Philosophy of Science, University of Indiana.
Jalobeanu D (2011) The Cartesians of the Royal Society: The Debate Over Collisions and the Nature of Body (1668–1670). In Anstey P, Jalobeanu D (eds). Vanishing Matter and the Laws of Nature: Descartes and Beyond, Routledge, London, pp. 103–129.
Kemeny M (2017) What Motion Is: William Neile and the laws of motion. Annals of Science 74 (3):179–191.
Knudsen O, Pedersen KM (1968) The Link between 'Determination' and Conservation of Motion in Descartes' Dynamics. Centaurus 13:183–186.
Kuhn TS (1959) The Copernican Revolution. Vintage, New York.
Kuhn TS (1970) The Structure of Scientific Revolutions. 2nd ed. University of Chicago Press, Chicago.
Laird WR (1986) The Scope of Renaissance Mechanics. Osiris 2:43–68.
Leijenhorst C (2004) Hobbes and the Galilean Law of Free Fall. In: Palmerino CR, Thijssen JMMH (eds). The Reception of the Galilean Science of Motion in Seventeenth–Century Europe. Kluwer, Dordrecht, pp. 165–184.
Maclean I (2007) Logic, Signs and Nature in the Renaissance. CUP, Cambridge.
Mahoney M (1973) The Mathematical Career of Pierre de Fermat 1601–1665. Princeton University Press, Princeton.
McLaughlin P (2000) Force, Determination and Impact. In: Gaukroger S, Schuster JA, Sutton J (eds). Descartes' Natural Philosophy. Routledge, London, pp. 81–112.
Miller DP (1986) Method and the 'Micropolitics' of Science: The Early Years of the Geological and Astronomical Societies of London. In: Schuster JA, Yeo RR (eds). The Politics and Rhetoric of Scientific Method, Reidel, Dordrecht, pp. 227–257.
Moyer AE (1992) A Scientist's Voice in American Culture: Simon Newcomb and the Rhetoric of Scientific Method. University of California Press, Oakland.
Mulkay M, Gilbert N (1981) Putting Philosophy to Work: Sir Karl Popper's Influence on Scientific Practice. Philosophy of the Social Sciences 11:389–407.
Murray G, Harper W, Wilson C (2011) Huygens, Wren, Wallis and Newton on Rules of Impact and Reflection. In: Anstey P, Jalobeanu D (eds). Vanishing Matter and the Laws of Nature: Descartes and Beyond. Taylor and Francis, London, pp. 153–191.
Oldenburg H (1965–1985) The Correspondence of Henry Oldenburg. Hall AR, Hall MB (eds). University of Wisconsin Press, Madison.
Palmerino CR (2004) Galileo's Theories of Free Fall and Projectile Motion as Interpreted by Pierre Gassendi. In: Palmerino CR, Thijssen JMMH (eds). The Reception of the Galilean Science of Motion in Seventeenth–Century Europe. Kluwer, Dordrecht, pp. 137–164.
Patterson LD (1952) Pendulums of Wren and Hooke. Osiris 10:277–321.
Pinch T (1985) Towards an analysis of scientific observation: the externality and evidential significance of observational reports in physics. Social Studies of Science 15:3–36.
Prendergast TL (1975) Motion, Action and Tendency in Descartes' Physics. Journal of the History of Philosophy 13:453–62.
Rattansi PM (1968) The Intellectual Origins of the Royal Society. Notes and Records of the Royal Society 23:129–43.
Rees G (2000) Baconianism. In: Applebaum W (ed). Encyclopedia of the Scientific Revolution, pp. 69–71. Garland, New York.

Rose PL, Drake S (1971) The Pseudo-Aristotelian 'Questions of Mechanics' in Renaissance Culture. Studies in the Renaissance 18:65–104.

Royal Society (1940) The Record of the Royal Society of London. 4th edition. Morrison and Gibb, London.

Sabra AI (1967) Theories of Light from Descartes to Newton. Oldbourne, London.

Schuster JA (1984) Methodologies as Mythic Structures: A Preface to the Future Historiography of Method. Metascience: Review of the Australasian Assoc. for the History, Philosophy and Social Studies of Science 1–2:15–36.

Schuster JA (1986) Cartesian Method as Mythic Speech: A Diachronic and Structural Analysis. In: Schuster JA, Yeo RR (eds). The Politics and Rhetoric of Scientific Method, pp. 33–95. D. Reidel, Dordrecht.

Schuster JA (1990) The Scientific Revolution. In: Olby RC, Cantor GN, Christie JRR, Hodge MJS (eds). The Companion to the History of Modern Science, pp. 217–242. Routledge, London.

Schuster JA (1993) Whatever Should We Do with Cartesian Method: Reclaiming Descartes for the History of Science. In: Voss S (ed). Essays on the Philosophy and Science of René Descartes. OUP, Oxford, pp. 195–223.

Schuster JA (1995) The Scientific Revolution: An Introduction to the History and Philosophy of Science. Open Learning Australia. Located at http://descartes-agonistes.com/.

Schuster JA (2000) Descartes Opticien: The Construction of the Law of Refraction and the Manufacture of its Physical and Methodological Rationales 1618–1629. In: Gaukroger S, Schuster JA, Sutton J (eds). Descartes' Natural Philosophy. Routledge, London, pp. 258–312.

Schuster JA (2002) L'Aristotelismo e le sue Alternative. In: Garber D (ed). La Rivoluzione Scientifica, Instituto della Enciclopedia Italiana, Rome, pp. 337–357. [An English translation is available, with the title 'Aristotelianism and its Alternatives' on the website http://descartes-agonistes.com/].

Schuster JA (2005) 'Waterworld': Descartes' Vortical Celestial Mechanics—A Gambit in the Natural Philosophical Agon of the Early Seventeenth Century. In: Anstey P, Schuster JA (eds). The Science of Nature in the Seventeenth Century: Patterns of Change in Early Modern Natural Philosophy. Springer, Dordrecht, pp. 35–79.

Schuster JA (2012) Physico–mathematics and the Search for Causes in Descartes' Optics—1619–37. Synthèse 185:467–499.

Schuster JA (2013a) Descartes–Agonistes: Physico–Mathematics, Method and Corpuscular–Mechanism, 1619–1633. Springer, Dordrecht.

Schuster JA (2013b) Cartesian Physics. In: Buchwald JZ, Fox R (eds). The Oxford Handbook of the History of Physics. Oxford University Press, Oxford, pp. 56–95.

Schuster JA (2013c) What was the relation of Baroque Culture to the Trajectory of Early Modern Natural Philosophy. In: Gal O, Chen-Morris R (eds). Science in the Age of Baroque, Springer, Dordrecht, pp. 13–45.

Schuster JA (2017a) Did Descartes Teach a 'Philosophy of Science' or Implement 'Strategies of Natural Philosophical Explanation'. In: Gaukroger S, Wilson C (eds). Descartes and Cartesianism: Essays in Honour of Desmond Clarke. Oxford University Press, Oxford, pp. 3–25.

Schuster JA (2017b) Consuming and Appropriating Practical Mathematics and the Mixed Mathematical Fields, or Being 'Influenced' by Them: The Case of the Young Descartes. In: Cormack L, Walton, S, Schuster JA (eds). Mathematical Practitioners and the Transformation of Natural Knowledge in Early Modern Europe. Springer, Dordrecht, pp. 37–65.

Schuster JA (2018) The Pitfalls and Possibilities of Following Koyré: The Younger Tom Kuhn, 'Critical Historian', on Tradition Dynamics and Big History. In: Pisano R, Agassi J, Drozdova, D (eds). Hypotheses and Perspectives in the History and Philosophy. Springer, Dordrecht, pp. 391–420.

Schuster JA (2019) Pitfalls and Opportunities of Contextual Explanation: The Case of Isaac Beeckman's Invention of the Mechanical Philosophy. Isis 110(2):308–311.

Schuster JA (2022) Isaac Beeckman in the Context of the Scientific Revolution. In: van Berkel K, Clement, A, van Dixhoorn A (eds). Knowledge and Culture in the Early Dutch Republic: Beeckman in Context. University of Amsterdam Press, Amsterdam, pp. 37–55.

Schuster JA, Yeo RR (eds) (1986) Introduction. In: Schuster JA, Yeo RR (eds). The Politics and Rhetoric of Scientific Method. D. Reidel, Dordrecht, pp. ix–xxxvii.

Schuster JA, Watchirs G (1990) Natural Philosophy, Experiment and Discourse in the Eighteenth Century: Beyond the Kuhn/Bachelard Problematic. In: Le Grand HE (ed). Experimental Inquiries: Historical, Philosophical and Social Studies of Experimentation in Science, pp. 1–47. Kluwer, Dordrecht.

Schuster JA, Taylor ABH (1996) Seized by the Spirit of Modern Science. Metascience 9:9–26.

Schuster JA, Taylor ABH (1997) Blind Trust: The Gentlemanly Origins of Experimental Science. Social Studies of Science 27:503–536.

Schuster JA, Brody J (2013) Descartes and Sunspots: Matters of Fact and Systematising Strategies in the Principia Philosophiae. Annals of Science 70(1):1–45.

Shapin S (1982) History of science and its sociological reconstructions. History of Science 20: 157–211.

Shapin S (1994) A Social History of Truth: Civility and Science in Seventeenth Century England. University of Chicago Press, Chicago.

Sprat T (1667) The History of the Royal Society of London. London.

Taylor ABH (1989) The Early Royal Society of London. Ph.D. Thesis, University of Melbourne.

Taylor ABH (1994) An Episode with May–dew. History of Science 32:163–84.

Vanzo A (2016) Experiment and Speculation in Seventeenth-Century Italy: The Case of Geminiano Montanari'. Studies in History and Philosophy of Science 56:52–61.

Wallis J ([2012] 2005) Correspondence of John Wallis. Vol II (1660–1668) and III (October 1668–1671). Beeley PC, Scriba C (eds). Oxford University Press, Oxford.

Wardhaugh B (2016) Willughby's Mathematics. In: Birkhead T (ed). The Virtuoso by Nature: The Scientific Worlds of Francis Willoughby FRS 1635–1672. Brill, Leiden, pp. 122–141.

Westfall R (1971) Force in Newton's Physics: The Science of Dynamics in the Seventeenth Century. MacDonald, London.

Wootton D (2015) The Invention of Science: A New History of the Scientific Revolution. Allen Lane, London.

Chapter 28
Historia Matheseos: From the History of Phenomena to the History of Ideas. Early Formation of a History of Mathematics

Galina I. Sinkevich

Abstract A mathematical phenomenon or event is the publication of a mathematical document—letters, manuscripts, articles, books. The foundation of the history of mathematics is a scientific biography and bibliography. For two millennia, the history of mathematics presented as a chronologically ordered sequence of mathematical phenomena: "who is after whom." The belief reigned that mathematics is an unshakable body of knowledge, given to us by a Greek miracle. The history of mathematics reached a scientific level in the Age of Enlightenment. A doctrine appeared about historical documents, about the signs of their reliability, ways to distinguish falsification, about ancient written instruments and materials, about styles. For the first time, an understanding of the progress of mathematics arose. The history of mathematics began to be presented by professional mathematicians as the history of concepts, ideas, theories. The article focuses on the evolvement of the history of mathematics as a science and the development of its methodology from the fourth century B.C. to the age of Enlightenment. We consider works of Eudemus of Rhodes, Pappus, Thabit ibn Qurra, Johannes Buteo, Pierre de la Ramée, Joseph Justus Scaliger, Dionisius Petavius, Bernardino Baldi, Joseph Blancanus, Gerardus Johann Voss, Claude-Francois de Chales, Jean Mabillon, John Wallis, Edward Bernard, Joseph Raphson, Christian von Wolff and Johann Heilbronner.

Keywords Historia Matheseos · History of mathematics

G. I. Sinkevich (✉)
Saint Petersburg State University of Architecture and Civil Engineering, St. Petersburg, Russian Federation
e-mail: sinkevich.g.i@lan.spbgasu.ru

R. Pisano (ed.), *A History of Physics: Phenomena, Ideas and Mechanisms*,
History of Mechanism and Machine Science 42,
https://doi.org/10.1007/978-3-031-26174-9_28

1 Introduction

The history of mathematics was evolving together with the mathematics itself. Its methodology gradually established. Having started with short excursuses in the history of certain issues and biographical details of scientists, the history of mathematics ended up with research using historical, textual, and mathematical methods, and achieved significant results. One would find wealth of diverse literature devoted to the history of mathematics—from popular literature to rigorous research. People in all countries studied the history of mathematics; it is included in education courses; it is interesting for all lovers of mathematics. The fruitful period after 1758 began with the appearance of *Histoire des mathématiques* by J.-É. Montucla, it is very well studied. In the eighteenth and nineteenth centuries, the main scientific approaches to the history of mathematics were formed, they represent a move to a more systematic history of mathematics. It was a stage for later works. The new rigorous approaches to historiography that developed in the nineteenth century in general to some extent affected history of science as well. But before the age of Enlightenment the history of mathematics was in its formative phase, which was less studied. Our goal will be to give an overview of the historiography of the history of mathematics in the early period before 1758.

2 4th Century B.C., Eudemus of Rhodes

Mathematics was laid down as a science in Ancient Greece, and the first famous work devoted to the history of mathematics was the *History of Geometry* (Γεωμετρικὴ ἱστορία) by Eudemus of Rhodes,[1] Aristotle's student. This work was repeated in *The Philosophical and Mathematical Commentaries of Proclus on the First Book of Euclid's Elements* by Proclus (The Catalogue of Geometers; Diadochus 2013, pp. 82–86). We do not know the historiographical style of Eudemus. Undoubtedly, he has undergone changes in the retelling of Proclus. As J. Mansfeld wrote,

> The ideas of earlier philosophers were used and interpreted in many ways, and, more often than not, served merely as springboards [...]. Surveys of earlier philosophers and philosophies and even anthologies containing purple passages were also composed for the delectation of a more general public, but the doctrinal contents of such works as well as the selections that were made, though containing mostly traditional material, were often updates and reflects the interest and predilections of their times, which as a rule were indebted to those of the professional philosophers. The transmission of the views of the early Greek philosophers (the so-called *physikoi*) therefore is not only quite fragmentary but also often coloured or even biased. (Mansfeld 1999, pp. 22–23)

Eudemus is known to have also written *The History of Arithmetic* and *The History of Astronomy,* none of which has survived to our days and are only mentioned by

[1] *Geometry and Geometers before Euclid* with a fragment from Eudemus of Rhodes was first published in Russian in a journal of A.I. Goldenberg, "Mathematical Leaflet", v. I, 1879–1880.

other authors. He concluded that arithmetic resulted from trade activities of the Phoenicians, and geometry, from geodetic activities of Egyptians. He described a period of three centuries in Greek mathematics in his works. According to L.Ya. Zhmud', researcher of the ancient world, Eudemus laid emphasis on three aspects: he specified the "ground breaker", looked for a faithful mathematical proof, and compared works of several scientists who studied the same issue (Zhmud 1997), p. 277).

> The main question he raised in his *Histories*, was 'who discovered and what? (Zhmud 1997, p. 289).

Eudemus arranged his materials chronologically—from discoveries of the teacher to discoveries of disciples. Proclus began classifying mathematical literature by goal: "subject of investigation and for students" (Diadochus 2013, Introduction, Chap. 7, p. 87). Therefore, he compared teachers' merits as well.

Whereas the attitude to discovery of mathematical results was heuristic, meaning that ancient mathematicians 'discovered', not 'created', the history of mathematics was presented as a history of discoveries, i.e. as a linear flow 'who follows whom'.

Works of ancient mathematicians was normally preceded by a historical approach, for example, in Archimedes' works (third century B.C.). Starting with Diogenes Laertius (2nd–third century), biographical compendiums began to appear, i.e. summary versions of classic authors with comments which also included elements of history of mathematics. *Mathematical Collection* by Pappus of Alexandria (3rd–fourth century) were a most complete compendium translated into Latin by Federico Commandino in 1588. It is thanks to Pappus that we now know many ancient problems. Let me also mention a compendium of Eutocius of Ascalon (5th–sixth century) which along with fragments from *The History of Geometry* by Eudemus of Rhodes contains solutions of the problem of cube duplication (Delosian problem) by ancient mathematicians; and Simplicius' philosophical works (5th–sixth century). The genre of collections of rendered works (compendiums) became predominant in the Late Antiquity and Middle Ages up until the early Modern Age. Their subject was not only Greek mathematics, but Byzantine (e.g. treatises of a Byzantian, Maximus Planudes (thirteenth century)), and a little of Indian mathematics as well. Since the chronological framework of such works was not broad, and all the characters lived in the same region, the sequence of events was indicated according to the principle "who follows whom", or the year of the Olympiad was called, or the name of the ruler.

Thabit ibn Qurra[2] (836–901), a mathematician and astronomer from Bagdade, translated works of Archimedes, Apollonius, Euclid, Ptolemy, and other ancient writers into Arabic. We can only read his translations of Archimedes' Treatises *Book of the Construction of the Circle Divided into Seven Equal Parts, On Tangent Circles*, as well as Books V–VII of *The Conics* by Apollonius. In Arabic mathematical literature of 8th–thirteenth centuries, Arabic translations of ancient classic authors were accompanied by comments and supplemental information intended

[2] Sometimes, in Russian literature his name is spelled as Qurra.

to systematise the knowledge of Greece, India, and the Arab world. Mathematicians' biographies were arranged chronologically in the Collections with extensive catalogues of manuscripts accompanied by historical comments. For example, in his *Introduction on Superiority of the Science of History* (*Prolegomena* or *Muqaddimah, Introduction to History*), an Arab historian Ibn Khaldun (1332–1406) reviewed the development of Greek and Islamic mathematics, dating its stages by periods of reign of tsars and khalifs.[3] The treatise of Kâtib Çelebi (Katib Çelebi 1835–1858), Turkish historian of the seventeenth century, *The Removal of Doubt from the Names of Books and the Arts,* containing 14,500 book titles and 10,000 names of authors and scholiasts in alphabetic order, which was published in Leipzig in seven volumes in Latin (Kâtib Çelebi 1835–1858), is a most vivid example of bibliographic encyclopaedic genre.

As of the twelfth century, translations from Arab and Greek, or renderings of ancient mathematicians, and comments containing historical data, began to appear in Europe. Euclid was first translated or rendered by Adelard of Bath (twelfth century), an English scholastic philosopher who also translated al-Khwarizmi's astronomical tables and renderings of Euclid's Elements in 15 books by Giovanni Campano (Campanus, Campani, deceased in 1296). Campano himself did not know Arabic. He made his renderings with comments based on earlier translations.

With advent of book printing in late fifteenth century, Greek and Latin texts of Archimedes, Euclid, and other ancient authors started to appear. Scientists began to compare translations, analyse translators' and scholiasts' (relaters') errors.

3 1559, Johannes Buteo

In 1559, a book of J. Buteo[4] (Buteo, Jean Borrel, 1492–between 1564 and 1572), *Quadrature of the Circle,* was published in two books where he invalidated many quadratures and sheltered Archimedes from hostile criticism; he also provided a list of errors made by Campani, Zamberti, Finé, and Peletier in their interpretations of Euclid in Latin (Buteo 1559). Buteo analysed the errors of these translators of Euclid and Archimedes; provided approximate calculations of Bryson of Heraclea, Archimedes, and Ptolemy; and criticised the common misconception originating from Zamberti to the effect that the author of the demonstrations in Euclid's elements was Theon of Alexandria.[5] *Buteo has confidently mastered Archimedes' method and*

---

[3] This book is available in French (Les Prolégomènes d'Ibn Khaldoun 1863).

[4] This is the Buteo which in 1559 calculated the capacity of Noah's Ark.

[5] Campano was the author of one of the first renderings of Euclid into Latin (*The Elements* in 15 books). Italian Zamberti (Zamberti, Zambertus, 1473 – after 1543) was the first to publish a printed translation of Euclid from Greek in 1505 (The Elements and other books). Zamberti corrected the mistakes in the medieval Campano's version in Latin. However, Zamberti was not a mathematician. Therefore, Luca Pacioli criticised him for his assaults upon Campano. In 1543, Tartaglia published his first translation of Euclid considering the text of Campano and Zamberti. Zamberti believed that Theon was the true author of the proof, while those were only the definitions and statements which

provided a summary of its use in the ancient world and Middle Ages (Beckmann (1971–2015, p. 97).

4 1567, Pierre de la Ramée

Buteo's book, as well as the book of Ramus (Pierre de la Ramée, 1515–1572) contains no dates, adhering to the principle 'who follows whom'. In 1567, Ramée published his *Introduction to Mathematics* divided into three books (Ramée 1567) in Latin which formed a preamble to his large work, *Thirty-One Books of Mathematical Essays* (Ramée 1569). In the sixteenth century, the idea of comparing, researching and verifying the information communicated by predecessors arose, and a new criterion of scientific knowledge was gradually forming, connected with observation and practice. Poor historical material, which Ramée had, serves him only as a springboard for attack to the old methods of teaching. The fact that the exposition of mathematics has not changed in two thousand years is, for Ramée, proof of the "complexity and ambiguity" of the subject, although, of course, it does not diminish the value of the "Elements". But, despite the high authority of Euclid, it is necessary to subject them to research, removing the "non-methodical" shortcomings: introducing definitions of mathematical concepts before they become necessary, teaching first geometry, and then algebra. According to Ramée, it is not knowledge that develops, but only the way of its presentation, teaching.

This was a summary of prior discoveries in mathematics divided into three periods: from Adam to Abraham (Chaldean); from Abraham (Egyptian period); from Thales to Proclus and Theon of Alexandria (Greco-Roman period); and the fourth, modern period, from Theon (fifth century) to Copernicus, Regiomontanus, and Cardano. The first book of the Introduction (pp. 1–39) describes the first three periods and lists 65 names of Greek mathematicians, and the whole of Greek and Roman science is presented on 36 pages. The second book – classification of mathematical sciences (described only arithmetic and geometry as a mathematical science; as to astronomy, optics, and music, he assigned them to physics) and their development in various European countries (pp. 39–71). He listed around 30 names of sixteenth century Reformers in his second book. They were mostly theologians, translators and scholiasts. Of mathematicians and astronomers, he mentioned, inter alia, Regiomontanus (but young Tycho Brahe is absent, Ramée got to know him two years later); sixteenth century: Herlinus, Copernicus, Rheticus (Copernicus' student), Rheinold,

belonged to Euclid. A French mathematician and cartographer Orontius (Oronce Finé, Orontius Finnæus, or Finæus, French Oronce Finé; 1494–1555) was Buteo's teacher. In 1532, he published his Protomathesis (Introduction in Mathematics) in Paris, where he explained the main notions used in The Elements by Euclid and the calculation of areas of planar figures as provided in Archimedes' works. He also described in this book his method of solving the circle quadrature problem which was later criticized by his student Johannes Buteo. Buteo criticized all the above authors including his teacher.

Santbecus, Leovitius (Cyprian Karasek Lvovicky), Dasypodius, Clavius, Landtgravius, Morshemius, Grunius, Xylander. No mention of Stiefel. Perhaps this is due to Ramée's negative attitude towards hypotheses as a scientific tool, as well as to Stiefel's attempts to combine the "scientific with the miraculous" (Cantor 1857, p. 364).

Ramée noted that the copies of ancient manuscripts were preserved in Florence thanks to the Medici family,[6] which evidences of his good knowledge of Italian historical literature. He set forth his speculations on the changes in methods of teaching mathematics in Christian Europe, giving preference to teaching in Protestant universities and criticising Aristotle. He mentioned Latin translations of Euclid and dissemination of information on Greek mathematics in Christian Europe. However, he mentioned nothing of the development of mathematics in the Islamic East. As in early works devoted to history of mathematics, Ramée considered mathematics as a combination of ancient Greek achievements remaining unchanged until the seventeenth century to be looked up to, sometimes criticising teaching methods. He told nothing about Kepler's, Cardano's, or Tartaglia's results, although mentioned their names. There was no evolution dynamics of mathematics or its contents in Ramée's. In the emerging Age of Enlightenment, the notion of the progress of science had not yet become a historical category.

The third book described the development of teaching mathematical methods in European universities. The purpose of mathematics was described as practical application in trade, physics, architecture, astronomy, and other areas (Ramée 1569, pp. 71–107). The book contained almost no historical information—only rhetoric on the way ancient classic mathematicians should be presented in educational institutions. The book mentioned sixteenth century mathematicians, such as Cardano, Maurolico, Piccolomini, Commandino, Tartaglia, and Dürer. Ramus sets out his main views on teaching. According him, first of all, the general formulation of the problem should be given, after which definitions of basic concepts should be introduced; then the problem is divided into its component parts, each of them is defined, and, finally, explanations are made using illustrative examples. Ramus sharply criticized not only Aristotle, but also Euclid, reproaching him in the absence of a method. He called science arts: for example, arithmetic is the art of counting well, geometry is the art of measuring well. New methods should be based on the conscious assimilation of knowledge and promote the development of students' independent thinking. In the textbooks of Ramus himself, only the most necessary theoretical information is given, the main content is given to a large number of examples. He consistently pursues the idea that the main source of scientific knowledge is practice, it and only it verifies the authenticity of the theories. It is necessary to revise all the sciences, to abandon the system of disputes. Mathematicians of 16–17 centuries recognized for the textbooks of Ramus "Arithmetic", "Algebra" and "Geometry" high methodological advantages. This created the glory of Rama "great enlightener", but gave rise to many enemies.

---

[6] Ramus' book was devoted to his patroness Catherine de Medici, the Queen of France. However, it did not help to retrieve the Huguenot from ruin on the St. Bartholomew's night.

5 New Chronology

Chronology was an important problem in the history of mathematics as well as in general history. Each culture had its own chronology, and history of each culture did not correlate with other cultures; social time was described in each culture regardless of others. In Greece, they dated events by Olympiad; in the Arab world, from Hegira and by khalif; in Rome, time keeping was 'from the founding of Rome' (ab urbe condita); in Byzantium, 'from Adam', i.e. 'since the creation of the world'. Social time in different cultures was autonomous. The errors were significant. In 1582, the Pope, Gregory XIII published a bulla, *Inter gravissimas,* with an invitation to switch to a new calendar. First, some Catholic countries began gradually switching to the Gregorian calendar; thereafter, over a period of 17th—eighteenth centuries, Protestant countries, including Great Britain in 1752, were switching to it. Russia switched to the new calendar in 1918. It was Dionysius Exiguus who in the sixth century suggested keeping time 'from the year of our Lord' (ab Anno Christi, ab inscriptione, Anno Domini). In Europe, this way of time keeping spread in late Middle Ages. In 725, in addition to the time keeping by Olympiad and by emperor, Bede the Venerable introduced absolute chronology for the first time. Beginning Chapter Two of Book One of his *Ecclesiastic History of the English People* (Historia ecclesiastica gentis anglorum), Bede wrote in 731: "ante incarnationis dominicae tempus" (before the incarnation of our Lord). This was the first time ever that the time countdown was mentioned. This is not to say that the countdown scale—B.C. and A.D.—developed in a prompt and natural way. Before the sixteenth century, along with 'A.D.' time keeping, they used 'anno mundi' and many other ways of reckoning.

The historical research space has expanded not only geographically, but also chronologically. There was a need for a unified time scale.

In the period from 1583 to 1629, books of Joseph Justus Scaliger (1540–1609) were published. He was a connoisseur of ancient culture and ancient calendars and the founder of modern chronology as an auxiliary science of history. Scaliger found ways of conversion between the systems of Ancient Rome, Ancient Greece, East Asia, and Mexico, using the method of astronomic dating of events by eclipse. This enabled him to correlate scientific discoveries in various cultures in the course of time.

In 1627, Dionisius Petavius (Denis Pétau, 1583–1652), a French scholar, suggested a 'before Christ' (ante Christum, B.C., century before Christian Era) system of counting down dates.[7] This system was universally recognized by the end of the eighteenth century.

[7] It should be noted that it's in these years that a new understanding of a number scale comes into existence too: minus (negative) numbers, as numbers that are less than zero (according to Stifel), are located to the left of (behind) zero. In 1629, A. Girard wrote about negative solutions of equations: "Solution using minus is explained in geometry as reversion, and minus retreats where plus goes further." (Descartes 1637, p. 228).

6 Bernardino Baldi

Bernardino Baldi (1553–1617), an Italian poet and mathematician, Commandino's student, was the first to try and use the new chronology in combination with former traditions of presenting history as a chronicle. It took him 12 years to create a *Chronicle of Mathematicians including their curriculum vitae* (Baldi 1707) as basis for a more comprehensive work. Among other preparatory materials for this work, his writings about Pythagor, Ctesibius, Hero of Alexandria, and Copernicus were preserved. The *Chronicle of Mathematicians* was written in Italian and contained around 200 life histories and an index of names. The book was written as a popular Who's Who. Names were arranged in a chronological order; in the left field, Greek chronology by Olympiad; in the right field, years before Christ or after Christ. E.g. about Euclid: 122 (122nd Olympiad) in the left field, 290 (anni avanti Christo – 290 before Christ) in the right field; and a text as follows:

> *Euclid.* There is some evidence that the most esteemed mathematician from the City of Gela in Sicily studied in Alexandria and probably in Athens. We has written a lot, that is to say a book entitled *The Elements of Geometry* in which he surpassed all those who wrote before him, and he was so glorious that he was named στοιχειωτής – Stichiota[8] (Sic! – G.S.). In addition to the *Elements,* he wrote a book entitled *Data, Porisms* in three volumes about perspective projection, about mirrors, a book about phenomena, and, to the best of my knowledge, a book entitled *Conics*, instead of the book about basics of music erroneously assigned to him. There is another apocryphal book about division of surfaces assigned to him by Mohammed from Bagdad. There is also a Plato section of Euclid which, according to Proclus, prepares the use of *The Elements* for the purposes of plotting Platonic solids. (Baldi 1707, pp. 22–23)

Baldi also mentioned Arab and Persian mathematicians, as well as mathematicians from North Europe; many astronomers; some philosophers and theologians (e.g. Thomas Bradwardine, Nicholas of Cusa, Abraham Zacuto). Notably, there was no article devoted to Girolamo Cardano, although he mentioned his name in the articles devoted to Swineshead and Tartaglia. The timeframe covered by this book was from 545 B.C. to 1596 A.D. Since Baldi included Muslim mathematicians in his chronicle, he needed a unified time scale.

7 1615, Joseph Blancanus

In 1615, in his *Thesis on the Nature of Mathematics,* Joseph Blancanus (Giuseppe Biancani, 1566–1624), Italian mathematician and astronomer, represented the history of European and Asian mathematics as a history of discoveries in accordance with the new chronology, *Chronology of famous mathematicians* (Blancanus 1615), in Latin.

[8] The Author of "The Elements", Euclid the stoicheiotes.

Although it contained certain inaccuracies,[9] it was more complete than Ramée's work and included Islamic mathematicians. This was an attempt to use a unified time scale representing the European and Arab history of mathematical discoveries, so Biancani uses the new timeline.

8 1650, Gerardus Johann Voss

The book which enriched the history of mathematics with new methods was written by a Dutch historian and philologist, Gerardus Johann Voss (Vossius, 1577–1649). His materials on the history of literature were such extensive that they included information on the development of mathematics as well. He collected these materials in his work On the Nature and Structure of All Mathematical Sciences Supplemented by Mathematicians' Chronology published posthumously (Voss 1650) and republished as a part of the book entitled *On Four Main Arts, On Philology and Mathematical Sciences, Supplemented by Mathematicians' Chronology*, issued in three books (Voss 1660). Voss was not a mathematician and, at times, used inaccurate information, for which he was reasonably blamed by researchers. However, he was the first to use philological and source study methods in his historical and mathematical review.

Voss began with the history of alphabetical and digital numbers and symbols, systematised his presentation by section (geometry, arithmetic, optics, music, mechanics, logistics, geodesy, astronomy, calendar, chronology). Having mentioned Greek mathematicians, among others, he mentioned such mathematicians as Boethius, Alcuin, al-Farghani, Ibn al-Haytham, Sacrobosco, Nicholas of Cusa, Regiomontanus, Zakuto, Dürer, Copernicus, Maurolico, Cardano, Gemma, Commandino, Mercator, Ramée, Clavius, Viète, Ludolph van Ceulen, Tycho Brahe, Neper, van Roomen, Grégoire de Saint-Vincent, Stifel, Mersenne, Snellius, J. Golius, Cavalieri. He placed emphasis on translations of Greek classic authors into Arabic and thereafter, from Arabic into Latin (Voss 1660, p. 55); addressed works of Arabian historians. In the section devoted to the history of Alfonsine tables, Voss mentioned the notion of a progress of science for the first time—"progress of astronomy after Greeks" (Voss 1660, p. 146). For him, publishing a book, including a translation, or annotation, was a mathematical event. Voss' book was followed by an index rerum & verborum, i.e. an index of objects and words, an index of names and subjects with page numbers, and in addition, a list of printing mistakes. All this suggested a new sample form of a historical and mathematical research.

[9] E.g. Thābit ibn Qurra (836–901) was described as a thirteenth century scientist; Roger Bacon (thirteenth century), as a fourteenth century scientist; Leonardus Pisanus (Fibonacci, early thirteenth century), as a fifteenth century scientist.

9 1674, Claude-François De Chales

French mathematician, Jesuit professor de Chales (Dechales, 1621–1678)[10] was the first in historiography to express his consciousness of the advance of mathematics in his *Treatise on the Advance of Mathematics and on Famous Mathematicians* which made part of Volume One of his three-volume edition of *The Course and World of Mathematics* (Chales 1674), a pansophy which contained information from mathematics, physics, astronomy, astrology, and architecture. De Chales translated Euclid, and this translation was popular in France although it was worse than Roberval's translation. D. E. Smith wrote that, although de Chales published Euclid, his own contribution in the subject was more than modest (Smith 1951, p. 386).

10 1681, Jean Mabillon

In 1681, a book of J. Mabillon (1632–1707) was published. Historian, founder of the discipline of palaeography, historical criticism, and chronology, wrote a book entitled *Diplomacy in Six Books* (Mabillon 1681). Mabillon understood diplomacy as a science addressing historical documents, evidence of their accuracy, methods of identifying forgery, ancient written instruments and materials, styles. Mabillon's book contained engraving plates with examples of ancient writing. Voss' and Mabillon's works influenced subsequent researchers and J. Wallis, in the first place.

11 1685, John Wallis

His *Treatise of algebra both Historical and Practical. Shewing, the Original, Progress and Advancement thereof, from time to time; and by what Steps it hath attained to the Heighth at which now it is. With some additional Treatises. London. M.DC.LXXXV* (Wallis 1685) was published in 1685. Historians of mathematics (Cajori, Bobynin,[11] Popov) blamed him of nationalism and lack of personal modesty which consisted in attributing discoveries of other mathematicians to himself (or his compatriots). This was a just reproach which, however, did not make his treatise less interesting, as it was written by an outstanding mathematician. The history of mathematics (algebra) here was for the first time presented as a history of ideas. Speaking of works of ancient classic authors, Wallis gave names of translators and publishers.

[10] Not to be confused with geometer Michel Chasles (1793–1880), French mathematician (geometer) and historian of mathematics.

[11] Victor V. Bobynin (1849–1919), Russian scientist, teacher, historian of mathematics, professor at Moscow University; one of the authors of the Encyclopedic Dictionary of the Brockhaus and Efron Publishing Society; publisher of the first Russian journals on the history of mathematics (1884–1905).

But it appears from the following example that the Arab history (chronology) had still existed apart from the European history. Wallis believed that Arabs knew algebra:

> After Diophantus (if not before, also) this Learning was pursued by *Arabic* Authors (but little known in *Europe* for a long time) [...] Divers writers (is said) there are of *Algebra* in that Language, and from them (I suppose) the Denominations of *Diophantus* (if from him they learned it) came to be changed; and (beside the Denominations of Root, Square, and Cube) that of Sursolide (first, second, thirds, etc.) introduced. But I rather think the *Arabs*, either of themselves, or from some others, had it long before *Diophantus*, and think this reckoning of *Powers* (by Sursolids, etc.) different from Diophantus. (Wallis 1685, p. 5; Author's Italics)

According to Wallis, in England, algebra started to develop in the twelfth and thirteenth centuries—earlier than in Europe—thanks to the fact that English scholastics knew Arabic. Englishmen used to visit Spain and bring many mathematical manuscripts with them. For example, in 1180, J. Morley (Morlacus, Morley, around 1140—around 1210), mathematician and astronomer, studied Arab mathematical manuscripts in Toledo and brought a valuable set to England. Wallis believes that the English were the first to translate Greek mathematical texts from Arabic. For example, in 1130, Adelard was the first to translate Euclid's Elements. Wallis mentioned an English theologian and historian, monk Bede Venerable (Saint Beda, Beda Venerabilis, late 7th—early thirteenth century), who wrote the history of English people; and then Alcuin (Alcuinus, around 735–804). Wallis erroneously called him Bede's student.[12] Wallis told in detail which Arab translations of ancient authors were brought to Oxford (including Merton College) and translated into Latin and English. In fact, this was a rendering of Voss' history of mathematics in the context of English history. He further told about the numerical values which originated from Moors and Arabs (Wallis 1685, p. 7) and Maximus Planudes. He also considered other number notations—Roman and Greek literal and number notations worldwide. He admitted that, although Arabic figures came from Saracens and Arabs, they originated from India. He compared Sacrobosco's (Johannes Sacrobosco, Sacrobosco, John of Holywood, around 1195–around 1256) numbering, who described fundamentals of Indo-Arabian numbering and arithmetic in his treatise *Algorithm* (*Algorismus de integris*): operations of addition, subtraction, averaging, duplication, multiplication, division, summing up arithmetical progressions, rooting, and cube-root extraction. Wallis believed that it was Luca Pacioli who was the first to bring the new notation to Europe (Luca de Burgo, Luca Pacioli, Summa de Arithmetica, Geometria, Proportioni, et Proportionalita, 1494). Thereafter, Wallis divided his narration into topics, which tells the difference between him and his predecessors.

Further Wallis' synopsis:

Astronomic tables: Ptolemy, Copernicus. Decimal fractions appeared; then, logarithms. Archimedes' methods, including the method of using big numbers (with

[12] Alcuin was born after Bede had died, and was in tutelage of Archbishop Egbert, Bede's student.

the help of 60-ary fractions, e.g. ¼ = 15′). Operations on them. On decimal fractions and use thereof in certain branches of arithmetic; on antiquity of decimal fractions. Described works of Briggs, Oughtred,[13] Gellibrand[14] (*Trigonometria Britannica*, 1633), Regiomontan (1464), Ramus (1560), Schoener[15] (1585), Record (1550), Stevin (1585). Reducing fractions, or proportions, to a smaller number of characters with the nearest approximation to the real value: the development line from Archimedes: Van Gulen, Snellius.[16] Chapter 11: application of the same in relation to the proportion of the diameter and length of circumference (from Archimedes).

In Chap. 12: *On Logarithms*, Wallis wrote about Neper, Briggs, Kepler, Rudolphian tables (1627), Mercator's *Logarithmotechnia* (1668), however, he did not mention logarithmic tables of J. Speidell[17] (Speidell 1619, 1622; Hobson 1614, p. 43) or the logarithmic rule invented by the English (astronomers Gunter and Wingate, mathematician Oughtred).

In Chap. 13, Wallis spoke about algebra:

> From the Arabs or Saracens, together with their Algorism by the Numeral Figures, (and other parts of Mathematical Learnings) we received also our *Algebra*, brought into *Europe*, partly be the way of *Greece* (as may seem by what we have of *Maximus Planudes,*) and partly by the *Moors* in *Spain.* Whither I find, that divers of our *English* Mathematicians (about the Twelfth Century) did resort, on purpose there to learn from the *Moors*, not only the *Arabic* Language, but especially the Astronomical and other Mathematical Learning. And this (no doubt) of Algebra amongst the rest; though I have not yet seen any thing of Algebra in any Ancient Manuscripts. (Wallis 1685, p. 61; Author's italics and orthography)

About Leonardo Pisano, Luca Pacioli, Cardano, Tartaglia, Nunes, Bombelli, and other authors of *Algebra* before Viète. He told a lot of good things about Pacioli (Pacioli's own books and his translation of Euclid).[18] In Part Five of his *Sum*, Pacioli provided the basic materials on arithmetic as provided by ancient authors and his contemporaries. Subsequently, Wallis rendered Voss' story of Leonardo Pisano. Lucas de Burgo was the first to describe the abacus method, i.e. that of Luca Pacioli. In page 62, he mentioned Stifel, his *Arithmetic* of 1544, Rudolph, Cardano's *Arithmetic* and his *Great Art* (*Ars magna*) of 1545. Cardano's rule of solving a cubic equation which, according to Cardano, was found by Tartaglia too. Cardano's student Luigi Ferrari also added a bit.[19] In 1567, Pedro Nunes published *Algebra* in Spanish.[20]

[13] William (Guilelm) Oughtred, 1575–1660.

[14] Henry Gellibrand, 1597–1636, Professor of Astronomy in Oxford who completed Briggs' unfinished work.

[15] Lasarus Schonerus, Schoener, Schöner, 1543–1607, Ramée's publisher, scholiast, and partly a contributing author. He taught mathematics in Neustadt; was Provost in Schmalkalden, Thuringia; and taught mathematics in Corbach high school.

[16] Willebrord Snellius, 1580–1626, Dutch mathematician and astronomer.

[17] John Speidell, 1600–1634, teacher of mathematics in London, drafter of logarithmic tables.

[18] Wallis called this period the Italian period. However, in effect, this was a revised publication of the Latin translation of Euclid made by Campano, where Pacioli had corrected numerous errors.

[19] This is what Wallis wrote. However, this 'a bit' was a solution of a 4th-degree equation.

[20] Wallis was mistaken here. It was in Portuguese.

In 1579, Rafael Bombelli published the treatise *Algebra*[21] in Italian. In this book, he published the rule of solving a cubic and biquadratic equation as Tartaglia and Cardano had done before (Wallis 1685, p. 63).

We believe that Bobynin's observation to the effect that Wallis took the credit for the solution of the irreducible case of a cubic equation (Popov 1920, p. 149) is groundless. On pages 173–174, Wallis spoke of the similarity of Cardano's and Harriot's methods. Further, in Chapters 46–48 (p. 175–181), he presented his own method of cube-root extraction from a binomial in the form.

$$a + \sqrt{b}, \text{ where } b \text{ may have any sign.}$$

He found this method in 1647–1648, and it was really similar to Bombelli's method published in 1572, and Wallis was familiar with Bombelli's book.

Then, Pierre de la Ramée (1570) was mentioned. The book was published by Schoener who also wrote books (*Numerical Geometry*). Leonard Digges[22] was one of our (i.e. Wallis') compatriots who wrote a book in 1579 entitled *Stratioticos* (military message deliverer). Another was Robert Recorde, 1552.[23] Chapter 14 was devoted to François Viète and his *symbolic arithmetic*.[24]

Chapters 15–29 were devoted to Oughtred (Wallis was his disciple) and his book *The Key to Mathematics* (*Clavis Mathematicae*; Oughtred 1631), a textbook of arithmetic which was republished three times in Oughtred's lifetime and was thereafter used even in the eighteenth century. Wallis rendered it in the tiniest detail. Beginning with Chap. 30, Wallis wrote about Harriot's *Algebra* (Harriot 1631), rendered it in detail, and asserted that Descartes adhered to Harriot. In this book, Harriot showed the way algebraic equations were laid down by multiplying linear binomials for the first time. Harriot rule (as Euler set it forth with reference to Harriot[25]) was as follows: each equation has as many positive roots as many variations of sign it contains, or as many negative roots as many repeat signs it contains. This only applies to those equations in which all roots are real (Euler 1949, p. 468).

Now we refer to this rule as Descartes rule. Descartes himself claimed that, although he had Harriot's book (1631) at home, he only read it after he had finished his *Geometry* (1637). In Chapter 53, Wallis accused Descartes of borrowings from Harriot. In particular, Wallis highly appreciated the innovation of Harriot who suggested that all members of an equation be written as one member of the equality, setting them to zero.[26]

---

[21] This is the second publication. The first one was published in 1572.

[22] Leonard Digges (around 1515—around 1559), an English mathematician and topographer.

[23] "If I be not misinformed"—Wallis' note.

[24] Wallis meant *Francisci Vietae – in artem analyticem isagoge.*

[25] Wallis' book was in Euler's library in St. Petersburg.

[26] It should be noted that Harriot did not select this form of notation as the final one. In his records, the constant term is more often on the right side. One can come across an equation written with a zero in the right side as far back as in Stifel's works.

Wallis was right about the priority of Harriot. Descartes had Harriot's treatise when he was writing his *Geometry*. Many ideas which were thoroughly set forth and systematised by Descartes, had been first uttered by Harriot. It should be also noted that a book of A. Girard, *Invention nouvelle en l'Algèbre*, was published in Amsterdam in 1629 to formulate the basic theorem in algebra eight years before Descartes. However, Wallis did not mention this fact. Further, in Chapter 55 (Wallis 1685, p. 208), Wallis repeated that Descartes' reasoning was based on Harriot's *Algebra* published in 1631, while Descartes' *Geometry* was published in 1637 in French and thereafter, in 1649 and in 1659, in Latin. Wallis demonstrated that, although used by Viète and Bombelli, many procedures could be asserted much easier based on Harriot's *Algebra*. This, certainly, does not derogate Descartes' role, who, unlike the English, did not proceed from geometry to algebra. Instead, developing algebra and generalizing the notion of a number, he set the analytical direction in the development of geometry.

Wallis listed Harriot's achievements (up to 200-th page): symbols, terms, generation of equations by multiplying binomials, rule of signs (the number of positive and negative roots), methods of determining the number of real and imaginary roots, research of a quadratic equation, dividing an equation by a binomial, simplifying a cubic equation. Wallis acknowledged that almost all of these discoveries were made by Harriot, although some of them had been discovered earlier by Viète.

Wallis highly appreciated the role of Leonardo Pisano who reproduced Arab rules and symbols without resort to Diophantus who remained unknown in Europe until 1572.[27] According to Wallis, Stiefel was a good author who had never moved beyond quadratic equations (Wallis 1685, sheet a3). Scipione del Ferro, Cardano, Tartaglia, and other developed a solution of a cubic equation. Bombelli took it a step further, solving biquadratic equations (with the help of cubic ones,[28] reducing them to two quadratic equations). Nunes, Ramus, Schoener, Salignac,[29] Clavius, Record, T. Digges,[30] and some of our men (i.e. Englishmen—G.S.), were developing this subject in the last century. However, by and large, they had failed to take it a step further than the quadratic equations. At the same time, thanks to Xilander,[31] Diophantus became known in Latin, and thereafter, thanks to Bachet, in Greek and Latin[32]; all

[27] Bombelli found a manuscript of Diophantus in the library of Vatican and published 143 problems in his *Algebra*. Wallis was mistaken about symbols. Leonardus Pisanus had no symbols. Wallis had not seen Pisanus' works. He learnt about them in Pacioli's works. (Thanks to J. Cesiano for this remark).

[28] This is not true! It was Ferrari who created a formula to solve a biquadratic equation! However, Bombelli uttered nothing about Ferrari, although Ferreri's formula had been set forth in Cardanus' *Ars magna* – G.S.

[29] Johannes Salignacus, Scottish.

[30] Thomas Digges (1546–1595), son of Leonardo Digges, English mathematician and astronomer, one of the first partisans and promoters of the heliocentric world.

[31] Xilander published *Diophanti Alexandrini Rerum Arithmeticarum libri sex* in 1575 in Basel.

[32] Bachet de Méziriac published *Diophanti Alexandrini Arithmeticorum libri sex; et de Numeris multangulis liber unus. Nunc primum graece et latine editi, atque absolutissimis commentariis illustrati* in1621 in Paris.

his methods differed from Arab methods (followed by others). In particular, the procedure for naming exponents (Wallis 1685, sheet a3 verso): using new symbols and figures were an important step in algebra. The next major step in the development of algebra was made by François Viète in 1590 in his "specious arithmetic".[33] Wallis expressed his appreciation of this step. He noted that, unlike the preceding authors, in designating exponents, Viète adhered to Diophantus, not to Arabs.

Wallis' historical sketch was structured by subject matters. As a prominent mathematician, Wallis was by far more knowledgeable about mathematical information than other authors. He gave an unbiased account of the development of methods and discoveries in algebra and emerging analysis, although sometimes, one could, of course, blame him in subjective assessment. Unlike preceding authors, his narration was not a discrete set of biographies or discoveries. He showed mathematics as a continuous development of ideas and algebra in the first place. He demonstrated its internal relations and their continuity, genesis of mathematical knowledge, the creativity of mathematicians, but not their heuristic. He discerned algebraic and geometrical methods and distinguished the inception of an analytical method, i.e. *Differential Calculation* method, in works of his contemporaries and, first of all, of Newton. The Wallis book, perhaps, is the first historical and mathematical book, which outlines the history of mathematical thought.

12 1704, Edward Bernard

Edward Bernard (1638–1697) was a Savilian Professor[34] of astronomy in Oxford. He was connoisseur of ancient manuscripts; studied many manuscripts of Apollonius of Perga; worked in Bodleian Library (Oxford) with Arab manuscripts brought from Spain, Morocco, Syria, Arab countries, and Turkey, which, to a large extent, replenished its collection. Edward Bernard found an Arab text of Apollonius entitled *Determinate Section*; tried to recover those fragments that had been lost and translate them into Latin; he edited Josephus Flavius. Most of Edward Bernard's work consisted of annotating books from Bodleian Library: his work *On Ancient Weights and Measures (De mensuris et ponderibus antiquis*, 1688) was enclosed with a work of E. Pococke (1604–1691), an Orientalist scholar from Oxford. *Bernard's Catalogue* (Bernard 1697) comprised manuscripts from British and Irish libraries and served as a basic tool of scientists of that time. Many works of Bernard were not completed, which made his colleagues joke.[35] After Bernard 's death, his colleagues published a book about him (Huntington 1704, Sect. 9, pp. 1–78) which included Bernard's work

[33] So Wallis calls Viiet's "species logistic" this way.

[34] In 1619, Sir Henry Savile, mathematician, custodian of Merton College in Oxford and provost of Eton College, complaining of the "poor condition of mathematical research in England", constituted two positions to be funded at his own expense: professor of geometry and professor of astronomy, which are in existence to the present day. The first professor of geometry was Henry Briggs.

[35] E.g. epigram of Cl. Barksdale (1609–1687): "Savilian Bernard's a right learned man;/Josephus he will finish when he can".

entitled *A Short List of Ancient Greek, Latin, and Arab Mathematicians, prepared by Dr. Eduardo Bernardo, the most honoured and educated man* (Bernard 1704, Sect. 11, p. 1–44)—annotated plan of republishing oeuvres of classic authors kept in European archives and libraries, on circa 44 pages. The adaptations and translations of Apollonius' *Conics*[36] Bernard had made were subsequently used by Edmond Halley (1656–1742) in the 1710 publication of Apollonius' works.

13 1715, Joseph Raphson

In 1715, a small posthumous edition of Newton's disciple, Joseph Raphson,[37] was published. It was *History of Fluxions* (Raphson 1715), and the goal of this publication was to assert Newton's priority in the discovery of differential calculation. Newton allowed Raphson to look through his works and his correspondence with Leibnitz, and the respective presentment of this correspondence in Raphson's book provided a strong support to Newton's position in this dispute.

14 1741. Christian Von Wolff

Christian von Wolff (1679–1754), German philosopher, lawyer, and professor of mathematics, published a *Report on additions to mathematical sciences over one century* in Halley in 1707; *Mathematical Vocabulary* in Leipzig in 1716 (Wolff [1716] [1734] 1747)—a dictionary of mathematics in German, the best of those available by that time although not the first one, on 788 pages, the list of sources alone was on 54 pages; and an article entitled *A Summary of the Most Renowned Mathematical Works* in Volume V of *Elementary Fundamentals of Mathematical Sciences* (Wolfius 1741, pp. 3–168). In Chapter One (pp. 5–28), Wolff reviewed books, beginning with Euclid and finishing with publications of Academia Petropolitana until 1731. He, inter alia, mentioned works of young Euler. He devoted a paragraph with a summary to each of the above books. Those were books of French, English, and Dutch authors. Chapter Two (pp. 29–32) was devoted to the history of arithmetic from Nicomachus to Neper. Chapter Three (pp. 32–50), Geometry: Euclid and his translators, publishers, and annotators, European geometers, finishing with the year 1699. Chapter Four (pp. 51–69): analytical works from the ancient world to the inception of differential calculation. Much prominence was given to the dispute regarding

[36] *Conics, a* fundamental treatise of Apollonius of Perga consisted of eight books. The Greek text of four of them has been preserved; three other books have survived translated into Arabic; the eighth book was renovated in the eighteenth century by E. Halley who published Apollonius' works (Oxford, 1710).

[37] Joseph Raphson, an English mathematician, Newton's disciple, died before 1715. We know very little about his life. He was the author of the most appropriate statement of Newton's approximation approach.

the priority of Newton and Leibnitz. And Wolff, professional lawyer and Leibnitz' friend, gave preference to Leibnitz in this matter, arranging their correspondence and publications chronologically, without getting involved in the mathematical substance. Chapter Five (pp. 71–77): trigonometry from Ptolemy to Ozanam. Chapters Six to Thirteen were devoted to statics, mechanics (up to Euler), hydrostatics, aerometry, hydraulics, optics, catoptric, dioptric, perspective projection, astronomy, chronology, geography, gnomonics, civil architecture, pyrotechnics, and military architecture—well-established sections of the eighteenth century mathematics. The book contains an index of names.

15 1742, Johann H. Heilbronner

The last book in the early period of historiography (before Montucla) was published in 1742. This was *The History of Mathematics at Large—from the Creation to the 16th Century A.D. including life stories of famous mathematicians, their doctrines, works, and manuscripts; in addition, a summary of main mathematical collections and works, and history of arithmetic to the present day* (Heilbronner 1742) in Latin. Its author was a German theologian and mathematician Johann Heilbronner (1706–1745/1747). The book contained an index of names. It was a bulky book—924 pages. Montucla called this Heilbronner's work "chaos" (Popov 1920, pp. 152–153). The author paid much attention to philosophical issues and described the structure of mathematics. He presented the sequence of certain names and discoveries in mathematics in considerable detail, although not free from errors. He listed famous manuscripts and published books. This *Historia matheseos* compared favourably with the previous books thanks to the two special features. First, the author added modest information from the history of Arab and Chinese mathematics (names and discoveries) to the European history; second, all these different national histories were reduced to a single time scale. Heilbronner used the achievements in chronology of the last century and dated each event in mathematics in several ways: mentioned the eclipses which happened at that time or other celestial events, specifying their characteristics from astronomical tables (of Ptolemy and other astronomers), year anno mundi (ad annum Mundi), year from the founding of the City of Rome (ab urbe condita), year B.C. (ante Christum natum, ante Christi nativitatem), or year A.D. (ab Anno Christi). This presentation was not infallible. E.g. on page 353, Thang-Heng,[38] Chinese mathematician and astronomer, was dated to the year 164. Heilbronner carried Michael Psellos (eleventh century) back to the ninth century (p. 410), while Al-Farabi (872–950) and Ibn Musa (al-Khwarizmi, around 820) were dated to the tenth century. However, regardless of numerous disadvantages, it was in Heilbronner's book that the image of the history of world mathematics appeared for the

[38] Heilbronner used to be very reliant on the letters about the history of Chinese astronomy of Antoine Gaubil, cartographer and missionary in China, which were published in European journals as of 1729.

first time ever, combining histories of different cultures. After Heilbronner's death, his library was purchased by Kästner who wrote a history of mathematics of his own. However, this is a topic for another article.

16 Coda

The list of the books provided can be supplemented. A good review can be found in Popov's book (Popov 1920), although one could blame him of some inaccuracies and gaps. However, the author was very scrupulous writing his book, he had read all the books he wrote about. A book devoted to historical development of historiography and mathematics in various countries was published in 2002 (Dauben and Scriba 2002). However, the period till 1750 was illustrated in it all too briefly. Thus, in the first two millennia of its existence, the history of mathematics began developing scientific methodology: performing scientific analysis of works, sources (original, translated, renderings, and annotations); distinguishing between facts and interpretations; drafting catalogues and reference books; issues relating to individual and collective authorship (of national school), analysing the application and teaching of mathematical methods; chronology. In addition, textual analysis was emerging, the purpose of writing of mathematical works was not identified (research, teaching). Authors did not consider the role and reciprocal influence of ancient civilizations—as a rule, they began the history of mathematics from Greeks and considered it mostly in Latin culture. Arab manuscripts began to become the custom; Chinese manuscripts were hardly mentioned; and Indian manuscripts were almost unknown. The issues of national priorities were solved quite simply—each historian knew mathematical literature of his own country pretty well, giving preference to his compatriots (as, for example, Wallis or Wolff). This was the way the historical and mathematical memory of the nation consolidated and its mentality shaped. Absolute chronology was evolving up to the seventeenth century. Therefore, mathematical achievements in the social time of different civilisations only began to be compared. They did not distinguish between the development stages, periods of boom and bust, they did not give prominence to the trends in the evolution of mathematics, they did not emphasise its independence or the degree of its dependence on the needs of that time. Gradually, the presentation of the history of mathematics was undergoing stages – from describing discoveries and biographies by type of chronicles to genesis of ideas and understanding of mathematical progress.

17 Conclusions

As a science, the history of mathematics was evolving together with the evolution of mathematics itself. Ancient works described the sequence of discoveries in mathematics based on the principle 'who discovered what', 'who taught whom', 'who followed whom'. Eudemus of Rhodes compared works that were topically related. Proclus began distinguishing research from educational works. With Dionysius de Laerte, biographies appeared in the history of mathematics. However, the history of mathematics was confined in a single culture.

The phenomenon of scientific translation as an art and its systematisation appeared in Arab culture of the 8th–twelfth centuries. Thanks to Arabic translations, the ancient heritage was preserved. This tradition was carried on in the Christian medieval period: Greek, Byzantine, and Arabic texts were translated into Latin; compendiums came into being—those were abridged renderings of classic authors which contained historical information. In Muslim culture of the 14th–seventeenth centuries, great importance was attached to accounting and classifying manuscripts, describing thereof providing information from their authors' biographies; first catalogues began to be created. As of the twelfth century, Englishmen began collecting manuscripts brought to England from the East; research libraries emerged, e.g. the Bodleian library (fourteenth century). The advent of book printing (fifteenth century) gave an impulse to dissemination of works of ancient classic authors; they were annotated and discussed, which included criticism as well. They began accounting each issue of a book as an individual scientific event. However, until the sixteenth century, the entire body of mathematical knowledge appeared to be static, lacking development (Ramus), although it was consistently establishing in time. Only the development of teaching methods was considered. It was the seventeenth century that a notion of progress in mathematics appeared for the first time in Voss' works. The history of creation gradually replaced the history of discoveries.

Chronologically, research works were arranged in an ordered fashion inside each culture: the chronology by Olympiad, since the creation of Rome; Muslim chronology—from Hegira; Chinese, by dynasty; Christian, Anno Domini, etc. Events, including mathematical events, which happened in various cultures were not related to one another; the story of each culture was stated independently. Wallis, for example, believed that Arabs knew algebra: "After Diophantus, not to say before him, the notion of exponentiation was studied by Arab authors. However, they remained unaware of it in Europe for a rather long time." (Wallis 1685, pp. 4–5). This also resulted in "national shortsightedness", when discoveries of compatriots seemed closer and more important (e.g., Wallis and Wolff).

Thanks to the works of Scaliger and Petavius, the chronology reform of the 16th–seventeenth centuries made it possible to bring historical events in compliance with astronomical phenomena and reduce them to a single scale which had a starting point and a direct reading-scale and a countdown-scale (A.D., B.C.). It should be noted that by this time, zero began to be apprehended as a reference point and a negative number, as a possibility of a countdown of steps, time, temperature (Wallis in the seventeenth century, Celsius in the eighteenth century). Authors began to include not only Christian mathematicians in the history of mathematics, Muslim mathematicians were mentioned as well (e.g., Baldi and Blankanus).

Methods used in other historical sciences began to be used in the history of mathematics. Those were methods of paleography, historical criticism, chronology, the doctrine of historical sources, evidence of their authenticity, ways to identify forgery, ancient written instruments and materials, styles. In the eighteenth century, name indices began to appear in books.

Authors began to try and present the history of mathematics not as a chronicle but as a history of ideas (Buteo in the sixteenth century, then Wallis in the seventeenth century).

All this made the history of mathematics ready for its next fledging period which began in 1758 when Montucla's History of Mathematics appeared.

References

Baldi B (1707) Cronica de' matematici: overo Epitome dell' istoria delle vite loro. Angelo Antonio Monticelli, Urbino.

Beckmann P (1971–2015) A History of Pi. Golem Press, New York.

Bernard E (1697) Catalogi librorum manuscriptorum Angliae et Hiberniae in unum collecti cum indice alphabetico. E Theatro Sheldoniano, Oxford.

Bernard E (1704) Epistolae et Veterum Mathematicorum, Graecorum, Latinorum, & Arabum, Synopsis. Robert Huntington, Edward Bernard, Thomas Smith. Epistolae. Typis G. Bowyer, impensis A. & J. Churchill, London. [Separate pagination]. Part 11:1–44.

Blancanus J (1615) De mathematicarum Natura dissertatio. Una cum Clarorum mathematicorum chronologia. Apud Bartholomaeum Cochium, Bologna.

Buteo J (1559) De quadratura circuli libri duo: vbi multorum quadraturae confutantur & ab omnium impugnatione defenditur Archimedes; eiusdem Annotationum opuscula in errores Campani, Zamberti, Orontij, Peletarij, Io. Penae interpretum Euclidis. Apud Gulielmum Rouillium, Lyon.

Cantor M (1857) Petrus Ramus, Michael Stifel, Hieronymus Cardanus, drei mathematische Charakterbilder aus dem 16. Jahrhundert. Zeitschrift für Mathematik und Physik. Literaturzeitung 2:353–376.

Chales de CF (1674) Cursus seu Mundus Mathematicus. Ex officina Annisoniana, Lyon.

Dauben JW, Scriba CJ (2002) (eds) Writing the History of Mathematics: Its Historical Development. Birkhäuser Verlag, Basel-Boston-Berlin.

Descartes R (1637) Geometria à Renato Des Cartes. Apud Ludovicum&Danielem Elzevirios, Amsterdami. [In English: (1979) English translation by Michael Mahoney. Dover, New York; In Russian: (1938) Nauchtekhizdat, Moscow-Leningrad].

Diadochus P (2013) Proclus's Commentary on Euclid, Book I. [In Russian] Aristej. Vestnik klassicheskoj filologii i antichnoj istorii: Appendix III:82–86.

Euler L (1755–1949) Institutiones Calculi Differentialis cum eius vsu in Analysi Finitopum. [In Russian: (1949) Gos. izd-vo tekhniko-tekhnicheskoj literatury, Moscow-Leningrad].

Harriot T (1631) Artis Analyticae Praxis ad Aequationes Algebraicas Resolvendas. Apud Robertum Barker, London. [In English: (2007) Thomas Harriot's Artis analyticae praxis. Springer, New York].

Heilbronner JC (1742) Historia matheseos universae a mundo condito ad saeculum post Christ. nat. XVI. Accedit recensio elementorum, compendiorum et operum mathematicorum atque historia arithmetices ad nostra tempora. Gleditsch, Leipzig.

Hobson EW (1914–2012) John Napier and the Invention of Logarithms, 1614. Cambridge University Press.

Huntington R (1704) Edward Bernard, Thomas Smith. Epistolae: Et Veterum Mathematicorum, Graecorum, Latinorum, & Arabum, Synopsis. Typis G. Bowyer, impensis A. & J. Churchill, London. [Separate pagination]. Part 9:1–78.

Ibn K ([1863] 1938) Les Prolégomènes d'Ibn Khaldoun. Troisième partie. 2nd ed. Librairie orientaliste Paul Geuthner, Paris.

Katib Çelebi MBA (1835–1858) Lexicon bibliographicum et encyclopaedicum, a Mustafa Ben Abdallah Katib Jelebi dicto et nomine Haji Khalfa celebrato compositum. Flügel (ed). Gustav Leberecht, Leipzig.

Mabillon J (1681) De re diplomatica, libri VI. Louis Billaine, Paris.

Mansfeld J (1999) Sources. The Cambridge Companion to Early Greek Philosophy. Long AA (ed). Cambridge University Press, Berkeley, pp. 22–44.

Oughtred W (1631) Clavis Mathematicae. Typis Lichfieldianis, Oxoniae [In English: Id., (1694) Key of the Mathematics] Salusburn, London.

Popov GN (1920) History of mathematics. In Russian. Tipo-Litografiya Moskovskogo Kartoizdatelskogo Otdela korpusa Voyennoy Topografii, Moscow.

Ramée [Ramus] P (1567) Proemium mathematicum in tres libros distributum. Wechel, Paris.

Ramée [Ramus] P (1569) Scholarum mathematicarum libri unus et triginta. Eusebius Episcopius, Basel.

Raphson J (1715) Historia fluxionum: sive tractatus originem & progressum peregregiæ istius methodi brevissimo compendio (et quasi synopticè) exhibens. Per Josephum Raphsonum. Printed by William Pearson, London.

Smith DE (1951) History of Mathematics. Vol. 1. Dover Publications Inc., New York.

Speidell J (1619) New Logarithmes, London.

Speidell J (1622) New Logarithmes (2nd ed). Printed by Edward Allde, London.

Voss [Vossius] GJ (1650) De universae matheseos natura et constitutione liber; cui subjungitur chronologia mathematicorum. Ex typogr. J. Blaeu, Amsterdam.

Voss [Vossius] GJ (1660) De quatuor artibus popularibus, de philologia, et scientiis mathematicis. Cui operi subjungitur Chronologia Mathematicorum. Libri tres. Ex typographeio I. Blaeu, Amsterdam.

Wallis J (1685) Treatise of algebra both Historical and Practical. Richard Davis, London.

Wolff C ([1716] [1734] 1747) Vollständiges mathematisches Lexicon. Gleditsch, Leipzig.

Wolfius C (1741) Elementa matheseos Universae. Tomus V. Renger, Halle.

Zhmud L (1997) History of mathematics by Eudemus of Rhodes. [In Russian] Hyperboreus 3:274–297.

Part IV
In Memoriam

Chapter 29
In Memoriam: Prof. Joseph Agassi (1927–2023)

Michael Segre

Abstract *In Memoriam of Prof. Joseph Agassi (1927–2023)* I offer a brief scientific biography of Joseph. Particularly, I refer to his distinguished contributions to his science & social sciences, metaphysics, logics, analysis of scientific method and politics.

Keywords Joesph Agassi · Popper · Foundations of Science · Physics · History of Science · Logic

Joseph

It is quite a challenge to write about a master such as the late Joseph Agassi. Not only because of his remarkable personality and his broad intellectual contribution, but also because he was outstanding in the literal meaning of the term: able to detach himself from circumstances, examine them from an unbiased, rational point of view, and reach striking conclusions—excellence in the proper meaning of the concept.

Born into a religious, poor, though cultured Jerusalem family, JosephAgassi (Joske) was brought up as a pious Jew. Despite the considerable knowledge and methodology he gained from Talmudic studies, he did not find his place in the religious sphere. A crisis led him to seek his fate elsewhere, in the Israeli paratrooper forces and in the study of physics at the Hebrew University of Jerusalem. There he met his wife Judith Buber (1924–2018), future acclaimed sociologist and granddaughter of philosopher Martin Buber, who is known for his philosophy of dialogue.

Agassi found a congenial intellectual environment at the London School of Economics and Political Science as a Ph.D. student of Karl Popper, an interlude he describes in his autobiographical *A Philosopher's Apprentice: In Karl Popper's Workshop* thus:

M. Segre (✉)
University of Chieti "Gabriele d'Annunzio", Chieti, Italy
e-mail: michael.segre@outlook.it

R. Pisano (ed.), *A History of Physics: Phenomena, Ideas and Mechanisms*,
History of Mechanism and Machine Science 42,
https://doi.org/10.1007/978-3-031-26174-9_29

> A melancholy account both of my apprenticeship with Sir Karl Popper, the greatest living philosopher, and of my way of having resigned myself to his rejection of myself. (Agassi 1993, p. ix)

Popper's critical rationalism was the main avenue, and this included boldly and respectfully criticizing Popper. Agassi's doctoral dissertation in the philosophy of science under Popper's supervision, entitled *The Function of Interpretation in Physics* (1956), led him to research the history of science in a broad context, delving into works of scientists including Bacon, Boyle, Galileo, and Faraday (Agassi 1971a, b, 1977, 2013; Cfr. Segre 2022). It would be the milestone of his intellectual contribution. Agassi's early research led him to criticize, inter alia, traditional history of science as ritualistic, white and black, and publish the ground-breaking *Towards an Historiography of Science* (1963), suggesting a more critical rational approach (Agassi [1936] 1967). The first step in his academic career was at the University of Hong Kong, followed by the universities of Boston and Tel Aviv and York University in Toronto.

Though criticizing plain "professorial" erudition, Agassi himself accumulated an enviable, encyclopaedical erudition spanning philosophy, the sciences, logic, technology, the humanities, and social sciences from sociology and political science to anthropology and psychology—including psychotherapy. In hundreds of publications, including dozens of books written in or translated into several languages, he offered brilliant discussions and fresh solutions to open questions alongside criticizing past and current views and pinpointing flaws.[1] The last book published in his life, *The Philosophy of Practical Affairs* (Agassi 2023) is a masterpiece of wisdom (another book was published post-mortem: Weiss and Agassi 2023).

Agassi insisted on dialogue and critical rational analysis in philosophy and other fields and on the importance of metaphysics, and applied these principles in all his studies. He was particularly sensitive to the interplay between metaphysics and empirical science: "There is no theory-free observation." (Agassi 2023, p. 81). He became deeply involved in areas including the Israeli political discussion, calling for a separation between the state and the religious establishment, lamenting the lack of discussion in Israel about the need of a constitution, and distinguishing between the Israeli nation and the Jewish communities overseas (Agassi 1999).

Above all, those who knew Joske closely recall his open, generous, and empathic personality (indeed, the volume of essays in honour of his 90th birthday is entitled *Encouraging Openness*; Bar-Am and Gattei 2017). His house was open to everybody, and he was always available to discuss any issue and offer openings and solutions, from private, reserved matters up to basic philosophical, scientific, or social questions. Dialoguing with him was a thrilling experience, and his many students, colleagues, and admirers will miss him greatly.

[1] Joseph Agassi's list of publications: https://www.tau.ac.il/~agass/pub.html, accessed June 1, 2023.

References

Agassi J ([1936] 1967) Towards an Historiography of Science. Wesleyan University Press, Connecticut Wesleyan.

Agassi J (1971a) On Explaining the Trial of Galileo. Organon 8/138–166.

Agassi J (1971b) Faraday as a Natural Philosopher. Chicago University Press, Chicago.

Agassi J (1977) Who Discovered Boyle's Law? Studies in History and Philosophy of Science 8:189–250.

Agassi J (1993) A Philosopher's Apprentice: In Karl Popper's Workshop. Rodopi, Amsterdam

Agassi J (1999) Liberal Nationalism for Israel: Towards an Israeli National Identity. Gefen, Jerusalem.

Agassi J (2013) The Very Idea of Modern Science. Springer, Dordrecht.

Agassi J (2023) The Philosophy of Practical Affairs: An introduction. Rowman and Littlefield, London.

Bar-Am N, Gattei S (2017) Encouraging Openness: Essays for Joseph Agassi on the Occasion of his 90th Birthday. Springer, Cham.

Segre M (2022) Agassi's Contribution to the History of Science. Philosophy of the Social Sciences 52/6:372–379 [See also: https://www.academia.edu/91843438/Agassis_Contribution_to_the_History_of_Science].

Weiss U, Agassi J (2023) Games to Play and Games not to Play: Strategic Decisions via Extensions of Game Theory. Springer, Cham.

Chapter 30
In Memoriam: Prof. Dante Roberto Salatino (1949–2022)

Guillermo Alberto Cuadrado

Abstract In *Memoriam of Prof. Dante Roberto Salatino (1949–2022)* I offer a brief scientific biography of Dante. Particularly, I refer to his distinguished contributions to Transcursive Logic.

Keywords Dante Roberto Salatino · Medicine & Physics · History of Science · Logic

Dante

Dante Salatino was born in Mendoza and studied at the National University of Cuyo, where he graduated as a physician and then as a specialist in cardiology, a profession he practiced for almost 40 years. He was also a systems analyst and senior programmer at NCR and specialized in artificial intelligence, which he applied to various areas of Medical Informatics, such as cardiac emergencies and assisted toxicology. He promoted Medical Informatics in Postgraduate studies in Cardiology and Hospital Administration. Salatino also cultivated a taste for the Humanities. His inclinations for Linguistics, Psycholinguistics and Artificial Intelligence led him to join the Institute of Linguistics (Faculty of Philosophy and Letters, UNCuyo) in 2005. It was there that he obtained his Ph.D. in 2011 as *Doctor of Philosophy in Linguistics*.

Salatino's doctoral thesis refers to certain psycho-bio-socio-cultural aspects of human natural language (as he particularly considers in his work) and the psychological theory of language. In order to prove his thesis he developed the *Transcursive Logic*, a creation that shows his facets, as a scientist and as humanist as well, which is why it can be applied in both fields and even combine them with each other. This is demonstrated by his work and the work of those who adopted it to explain very diverse matters, such as literary analysis or electronic circuits. It is a complementary scientific method that considers the perspective of the observer of a phenomenon, allows to explain it as it happens and records what affects the person

G. A. Cuadrado (✉)
National University of Cuyo, Mendoza, Argentina
e-mail: gacuadrado@gmail.com

R. Pisano (ed.), *A History of Physics: Phenomena, Ideas and Mechanisms*,
History of Mechanism and Machine Science 42,
https://doi.org/10.1007/978-3-031-26174-9_30

who observes in that process of evolution of the event. It is based on the existence of basic characteristic elements in any phenomenon, which has the structure of a group functionally defined by a Galois connection. This system of relations forms the syntax of a universal language, whose operative unit is a universal autonomous pattern composed of two triads: a superficial one, which represents the appearance of the real fact, and a deep and recondite one that explains it. Some of the possibilities offered by the *Transcursive Logic* are: to relate the concrete existence to the abstract one, to link sensations with ideas or to explore realities that affect the subject, it also allows to take the first step of an investigation in areas with little theory, because it allows to detect the determining factors that explain a phenomenon.

Dr. Salatino was a true university man in every sense, loved and admired by his students and colleagues, always ready to discuss the results of a work, suggesting other points of view or contributing with new ideas. Furthermore, he was a distinguished member of several academic societies, advisor to scientific journals, member of the editorial board of specialized publications. Salatino organized the *Forums on Creativity, Research and Transcursive Logic* at the Facultad Regional Mendoza, Universidad Tecnológica Nacional from 2017 to 2020, whose selection of papers are reflected in four volumes, with a quite extensive variety of applications.

In 2021 he was declared illustrious citizen by the Municipality of Godoy Cruz, the department where he lived. Moreover, in addition to his scientific publications, which exceed half a hundred, there are some historical novels. Almost all of these contributions were generously shared in his blog (https://aprend3r.blogspot.com/), which is currently managed by his son Diego.

Undoubtedly, Dante's departure leaves a void difficult to fill in the lives of those of us who knew him and in the intellectual history of the City of Mendoza.

Part V
Salvo D’Agostino’s Publications

Chapter 31
Salvo D'Agostino's Publications: 1964 to 2019–[Posthumous 2022]: 237 Works

Raffaele Pisano

Abstract By means of his finest intellectual vision of history of physics and science, Salvo D'Agostino continuously published for more than fifty-five years, crossing two centuries. Other several works are unpublished or remain in *pre-print* (see below); and others could be unknown, as well. Below the huge list of Salvo Dagostino's References has been collected by Raffaele Pisano in collaboration with Pasquale Tucci, Arcangelo Rossi and Caterina D'Agostino. The reader will also find previously unseen images of Salvo D'Agostino (Figs. 1, 2, 3 and 4). They are a kind courtesy—and—authorised by Caterina D'Agostino, Salvo's daughter.

R. Pisano (✉)
HOPAST at IEMN, Lille University, Villeneuve d'Ascq, France
e-mail: raffaele.pisano@univ-lille.fr

R. Pisano (ed.), *A History of Physics: Phenomena, Ideas and Mechanisms*,
History of Mechanism and Machine Science 42,
https://doi.org/10.1007/978-3-031-26174-9_31

Fig. 1 Salvo with his fellow high school teachers in Viterbo, early 1950s. Salvo is standing, second from the right *Source* Courtesy D'Agostino Family Private Archive

1. D'Agostino S (1964) Un insegnamento non dogmatico della Scienza. Scuola e Città:514–515.
2. D'Agostino S (1965) Il contributo di Newton allo sviluppo dell'ottica. Giornale di Fisica 1965 VI/3:188–196.
3. D'Agostino S (1965) Traduzione di Zadou-Naisky Georges. L'insegnamento delle scienze fisico-matematiche. La Nuova Italia, Firenze.
4. D'Agostino S (1966) Alcune considerazioni sul calcolo dei campi magnetici nei regimi quasi stazionari. Giornale di Fisica 7/3:179–184.
5. D'Agostino S (1967) Maxwell e la scoperta della Teoria Elettromagnetica della luce. Archimede 4/5:276–282.
6. D'Agostino S (1967) Alcune osservazioni su teorie elastiche e prime teorie del campo elettromagnetico nella prima metà dell'Ottocento. Scientia 102:166–171.
7. D'Agostino S (1968) Il pensiero scientifico di Maxwell e lo sviluppo della teoria del campo elettromagnetico nella memoria "On Faraday's lines of force". Scientia 103:291–301.

Fig. 2 "Enrico Fermi" International School of Physics, History of Twentieth Century Physics, Varenna (Como), 31 July–12 August 1972 *Source* Courtesy D'Agostino Family Private Archive

8. D'Agostino S (1968) I vortici dell'etere nella teoria del campo elettromagnetico di Maxwell: la funzione del modello nella costruzione della teoria. Physis X/3:188–202.
9. D'Agostino S (1968) Per la preparazione all'insegnamento della fisica nella Scuola Media. Le Scienze 1:33–36.
10. D'Agostino S, Cametti C (1970) Esperienze didattiche di diffrazione con microonde. Giornale di Fisica 11:36–48.
11. D'Agostino S (1971) Hertz e Helmholtz sulle onde elettromagnetiche. Scientia 106:622–636.
12. D'Agostino S (1973) L'elettrodinamica di Lorentz sino alla soglia della teoria speciale della relatività di Einstein. Physis XV:260–279.
13. D'Agostino S (1973) L'elettrodinamica di Ampère e l'elettromagnetismo di Maxwell: due esempi dell'uso di considerazioni di simmetria nella fisica dell'ottocento. In: Agazzi E (ed). Seminari Interdisciplinari di Venezia. La Simmetria. Il Mulino, Bologna, pp. 87–120.
14. D'Agostino S (1974) L'opera elettromagnetica di Maxwell nel saggio di C.W.F. Everitt alla voce 'Maxwell' del Dictionary of Scientific Biography. Physis XVI:267–70.

Fig. 3 Reading a rare original copy of *Sidereus Nuncius* (1610), 1958–1959, at the University of Oklahoma Librairy, History of Science Collection, USA *Source* Courtesy D'Agostino Family Private Archive

15. D'Agostino S (1974) Heinrich Hertz e la verifica della teoria elettromagnetica di Maxwell. Giornale di Fisica 15(3):230–237.
16. D'Agostino S (1975) Hertz's Researches on Electromagnetic Waves. Historical studies in the physical sciences 6:261–323.
17. D'Agostino S (1975) L'elettromagnetismo classico. Sansoni, Firenze.
18. D'Agostino S (1976) La scoperta di una velocità quasi uguale alla velocità della luce nell'elettrodinamica di Wilhelm Weber (1804–1891). Physis XVIII(3/4):297–318.
19. D'Agostino S (1976) Paradigmi storici e insegnamento scientifico. La Fisica nella scuola 9(3–4):150–153.
20. D'Agostino S (1976) Il problema del moto assoluto e della distinzione tra forze impresse e forze insite nel "Philosophiae naturalis principia mathematica" di Isaac Newton. Giornale di astronomia—Rivista di informazione, cultura e didattica della Società astronomica italiana 4(3):275–290.
21. D'Agostino S, Ianniello MG (1977) Il Deutsches Museum tra storia e scienza. La Fisica nella Scuola 4:179–181.

Fig. 4 International conference: *Albert Einstein and Hermann Weyl, Fifty Years after their Death: Unsolved Epistemological Issues*, Presicce (Lecce), April 22–23, 2005. From left to right: Arcangelo Rossi, Salvo D'Agostino, Dario Antiseri, Luigi Binanti, Gabriella Sava, Mario Castellana and Demetrio Ria *Source* Courtesy D'Agostino Family Private Archive

22. D'Agostino S (1977) La fondazione newtoniana della scienza empirica. Memorie e rendiconti, Accademia di scienze, lettere e belle arti degli Zelanti e dei Dafnici di Acireale VII(II):219- 235.
23. D'Agostino S (1977) Maxwell's Experimental Researches with the Electrical Standard Committee and the Development of the Electromagnetic Theory of Light. Communication at 15th International Union of the History and Philosophy of Science Congress. Abstracts of Scientific Section Papers, Edinburgh University, Edinburgh p. 92.
24. D'Agostino S (1978) Esperimento e teoria nell'opera di Maxwell: Le misure per le unità assolute elettromagnetiche e la velocità della luce. Scientia 113/ 5–8:453–468.
25. D'Agostino S (1978) Il rinnovamento dei musei scientifici. A.N.M.S.—Notiziario dell'Associazione Nazionale Musei Scientifici 5:2–3
26. D'Agostino S (1979) Modelli ed oggetti della teoria fisica. La nuova critica––Rivista di scienze dell'uomo e di filosofia delle scienze XIII(49):5–19.
27. D'Agostino S, Ianniello MG (1979) Catalogo della collezione di strumenti elettrici e magnetici del Museo dell'Istituto Fisico dell'Università di Roma. Università di Roma, Roma.

28. D'Agostino S (1979) Storia della strumentazione scientifica e sviluppo della scienza: i problemi posti dall'uso conoscitivo della techne. Medicina nei secoli 1:13–29.
29. D'Agostino S, Ianniello MG (1979) Catalogo strumenti. Catalogo della collezione di strumenti elettrici e magnetici. Museo dell'Istituto di fisica dell'Università di Roma, Roma.
30. D'Agostino S, Fabbri B, Ianniello MG, La Forgia M, Tarsitani C, Veronese L, Violini G (1979) La termodinamica nella secondaria superiore. Riforma della Scuola 5:51–56.
31. D'Agostino S, Ianniello MG (1979) Linee di sviluppo degli strumenti elettromagnetici dell''800 e problemi di storia della strumentazione. Nota Interna n. 721. Istituto di Fisica dell'Università di Roma, University of Roma La Sapeinza, Roma.
32. D'Agostino S (1979) The absolute system of units and the Nineteenth Century Discovery of the velocity of light within Electricity. In: 15th International Union of the History and Philosophy of Science Congress. Travaux de la Metrologie historique, Edinburgh 1977. Deutsches Museum Publishing, pp. 100–114.
33. D'Agostino S (1979) Maxwell's Experimental Researches with the Electrical Standard Committee and the Development of the Electromagnetic Theory of Light. In: 15th International Union of the History and Philosophy of Science Congress. Travaux de la Metrologie historique, Edinburgh 1977. Deutsches Museum Publishing, pp. 78–99.
34. D'Agostino S (1979–1980) Attività scientifica e didattica del gruppo di storia della fisica dell'Istituto di fisica, 1974–1979. Bollettino della Facoltà di Scienze matematiche, fisiche e naturali 4:5–25.
35. D'Agostino S, Ianniello MG, Violini G (1980) L'esperimento storico nella didattica della fisica, iniziative e proposte. La Fisica nella scuola XIII(1):19–21.
36. D'Agostino S (1980) Maxwell ed Einstein come esempi della forza dei processi sintetici nella crescita del pensiero fisico. La Fisica nella scuola XIII(2):49–55.
37. D'Agostino S, Ianniello MG (1980) Elettrometri e galvanometri dell'Ottocento. Evoluzione strumentale e contesti teorici. Annali dell'Istituto di Storia della Scienza di Firenze 5(2):69–81.
38. D'Agostino S, Tarsitani C, La Forgia M (1980) J.C. Maxwell: filosofia, metodo ed equazioni. Istituto di Fisica G. Marconi, Università di Roma, Roma [typescript document by 143 pages].
39. D'Agostino S (1980) Quale museo per gli strumenti scientifici? Risultati e proposte di una ricerca sulla storia degli strumenti scientifici. In: D'Agostino S, Ianniello MG (eds). Storia della scienza e della tecnica: problemi di ricerca storica e didattica nella scuola e nei musei scientifici. Atti del Seminario di studio "Storia della scienza e della tecnica e insegnamento scientifico". Università di Roma—Centro Europeo dell'Educazione, Villa Falconieri, Frascati, 3–7 novembre. Centro Europeo dell'Educazione, Frascati, pp. 183–191.

40. D'Agostino S (1980) H. Hertz's Experiments on Electromagnetic Waves. In: Bericht uber das Kolloquium des Forschunginstitut der Deutsches Museum in Sommer 1978. Deutsches Museum Publishing, Munchen, p. 34.
41. D'Agostino S, Ianniello MG (1970s–1980s) Raccolta fotografica Strumenti del Museo di Roma Contiene fotografie di strumenti, prevalentemente di elettricità e magnetismo, conservati al Museo di fisica dell'Università degli studi La Sapienza di Roma. Le fotografie, alcune delle quali realizzate nel 1980 per un articolo mai pubblicato di Salvo D'Agostino e Maria Grazia GIanniello, furono scattate presumibilmente negli anni '70 del XX secolo. Via: https://bibdig.museogalileo.it/Teca/Viewer?an=000001016940&lng=en
42. D'Agostino S (1981) Weber and Maxwell on the Discovery of the Velocity of light in 19th Century Electrodynamics. In: Grmek MD, Cohen RS, Cimino G (eds). On Scientific Discovery: The Erice Lectures 1977. Reidel, Dordrecht, pp. 281–293.
43. D'Agostino S (1981) Book review [Hackmann, Willem D, Electricity from glass: the History of the frictional electrical machine, 1600–1850. Sijthoff & Noordhoff, Alphen aan den Rijn]. Annali dell'Istituto di Storia della Scienza di Firenze 1:261–263.
44. D'Agostino S (1981) Nineteenth Century Electro-mechanical Instruments and the role in the Developments of Electrodynamics. Proceedings of 16th International Union of the History and Philosophy of Science Congress. Under the Auspices of UNESCO and the Academy of the Socialist Republic of Romania. Publishing House of the Academy of the Socialist Republic of Romania, Bucharest, p. 127 et suc.
45. D'Agostino S (1982) Sintesi della relazione: Electromechanical instruments and mechanicism in the history of nineteenth century electrodynamics. Seminario Storico Epistemologico. Ottobre.
46. D'Agostino S (1982) La presentazione del fondo per la storia della fisica quantica in occasione dell'inaugurazione del Centro per la storia della scienza e dei Quaranta presso l'Accademia Nazionale delle Scienze. Physis XXIV(4):577–578.
47. D'Agostino S (1982) Vi sono asserti metafisici nelle teorie fisiche? A proposito della teoria del "monopolo" di Dirac. Medicina nei secoli XIX(3):239–247.
48. D'Agostino S (1982) Empirismo e strumenti scientifici. Physis XXIV(2):281–283.
49. D'Agostino S (1982) Book review [Cantor GN, Hodge MS (eds). Conceptions of ether: studies in the history of ether theories, 1740–1900. Cambridge University Press, Cambridge]. Archives Internationales d'Histoire des Sciences 32(109):306–307.
50. D'Agostino S (1982) Book review [Woolf H (ed), Some strangeness in the proportion: a centennial Symposium to celebrate the achievements of Albert Einstein. Addison-Wesley, London]. Archives Internationales d'Histoire des Sciences 32(109):312–313.

51. D'Agostino S (1983) Strumenti scientifici e teorie fisiche: considerazioni storico-critiche sulla storia della strumentazione scientifica. In: Tarozzi G (ed). Gli strumenti nella storia e nella filosofia della scienza. Istituto per i Beni Artistici Culturali Ambientali della Regione Emilia Romagna, Bologna, pp. 173–182.
52. D'Agostino S (1983) On the problem of the redundancy of absolute motion in Newton's Principia. Physis XXV(1):167–169.
53. D'Agostino S (1983) Die Bedeutung wissenschaftlicher Instrumente für die Erkenntnis – eine philosophische und historische Betrachtung. Kultur & Technik 7(4):228–234.
54. D'Agostino S (1983) Symmetry in Ampère electrodynamics. In: Doncel MG, Hermann A, Michel L, Pais A (eds). Symmetries in physics (1600–1980). Proceedings 1st International Meeting on the History of Scientific Ideas, Sant Feliu de Guixols, Catalonia, Spain, September 20–26. World Scientific Publishing, Singapore, pp. 93–103.
55. D'Agostino S (1983) Conceptions and concepts in physics teaching. In: Kennedy PJ (ed). Proceedings of the International Conference on Using History of Physics in Innovatory Physics Education. La Goliardica Pavese, Pavia, pp. 173–182.
56. D'Agostino S (1983) Le ricerche sui raggi cosmici in Italia negli anni trenta: Bruno Rossi e la scuola fiorentina. In: Tucci P (ed). Atti del IV Congresso nazionale di Storia della Fisica. Centro Volta, Villa Olmo, Como, 3–5 novembre. CLUED, Milano, pp. 220–228.
57. D'Agostino S (1983) L'importanza della storia delle scienze nel rinnovamento dell'insegnamento scientifico. In: AA.VV. Parte seconda. Atti del Convegno Internazionale presso la Fondazione Cini. Fondazione Cini, Venezia, p. 263.
58. D'Agostino S (1983) On the problem of the presumed redundancy of absolute motion in Newton's principia. Physis XXV(1):167–169.
59. D'Agostino S (1984) Leopoldo Nobili e la fisica degli osservabili. Giornale di Fisica 25(3):231–235.
60. D'Agostino S (1984) Velocità della luce in elettrodinamica: dalla scoperta di un rapporto di unità a una costante fondamentale. In: Grmek MD, Somenzi V (eds). La scoperta scientifica: aspetti logici, psicologici e sociali. Armando editore, Roma, pp. 241–253.
61. D'Agostino S (1984) Alcune considerazioni sull'opera di Bruno Rossi e della scuola fiorentina di fisica nelle ricerche sui raggi cosmici. Annali dell'Istituto di Storia della Scienza di Firenze 1984 IX(2): 69–83.
62. D'Agostino S (1984) Alcune annotazioni sulla posizione del pensiero scientifico nel pensiero contemporaneo. Cultura e scuola 90:239–243.

63. D'Agostino S (1985) Metrology and Dimensional Theories in Maxwell's Treatise. Communication at 18th International Union of the History and Philosophy of Science Congress. Abstracts of Scientific Section Papers, Berkeley University of California, Berkeley, Act. I, Pf.
64. D'Agostino S (1985) La matematica e la fisica nelle idee di P.A.M. Dirac. Memorie di scienze fisiche e naturali. Istituto Poligrafico e Zecca dello Stato, Roma. Rendiconti della Accademia nazionale delle scienze detta dei XL, Serie V, IX(II):287–290.
65. D'Agostino S (1985) Maxwell's Dimensional Approach to the Velocity of Light: Rise and Fall of a Paradigma. Mathematical Models and Physical Theories. Memorie di scienze fisiche e naturali. Rendiconti della Accademia nazionale delle scienze detta dei XL, Serie V, IX(2):147–167.
66. D'Agostino S (1985) Introduction. In: D'Agostino S, Petruccioli S (eds). Mathematical models and physical theories. Atti del V Convegno Nazionale di Storia della Fisica, Roma 29–31 Ottobre1984. Memorie di scienze fisiche e naturali. Rendiconti della Accademia nazionale delle scienze detta dei XL, Serie V, IX(2):67–69.
67. D'Agostino S (1985) Strumenti elettromeccanici e concezioni meccaniciste nella storia dell'elettrodinamica. Epistemologia 8:119–138.
68. D'Agostino S (1985) The problem of the link between correspondence and complementarity in Niels Bohr's papers: 1925–1927. Proceedings of the International symposium on Niels Bohr, Rome, February 25–27. Special Issue—Rivista di Storia della Scienza 2(3):369–390.
69. D'Agostino S (1986) Leopoldo Nobili e la fisica degli osservabili. In: Tarozzi G (ed). Leopoldo Nobili e la cultura scientifica del suo tempo. Istituto per i beni artistici, culturali, naturali della Regione Emilia-Romagna. Atti del Convegno tenuto a Reggio Emilia, 1985. Nuova alfa Editoriale, Bologna, pp. 97–103.
70. D'Agostino S (1986) Maxwell's Dimensional Approach to the Velocity of Light. Centaurus 29:178–204.
71. D'Agostino S (1986) La formazione dello storico della scienza: problemi e prospettive. Il Nuovo Saggiatore 3:8–10.
72. D'Agostino S (1986) Scienza e cultura nella Germania dell'Ottocento. Le basi filosofiche della fisica matematica nell'opera di Helmholtz. Cultura e Scuola 99:265–282.
73. D'Agostino S (1986) Dall'occhio alle teorie della luce e ritorno: la luce come problematica innovatrice nella storia dell'ottica ottocentesca. In: Tracce, immagini, numeri. Atti dei seminari e conferenze dell'A.S.T., Roma, marzo-dicembre 1983. Il bagatto, Roma, pp. 109–117.

74. D'Agostino S (1986) Book review [Hendry J, James Clerk Maxwell and the Theory of the Electromagnetic Field. Adam Hilger, Bristol]. Physics Bulletin 37(12):500.
75. D'Agostino S (1986) Strumenti scientifici e beni culturali: alcune considerazioni. Insegnare II:10–12.
76. D'Agostino S (1986–1987) Federico Enriques (1871–1946) and History of Science in Italy in the Pre-War Years. Organon 7/22–23:17–24.
77. D'Agostino S (1987) Il principio di indeterminazione e la transizione dall'ontologia della fisica classica a quella della meccanica quantistica. In: Petruccioli S, La Forgia M (eds). Rappresentazione e oggetto: dalla fisica alle altre scienze. Theoria, Roma, pp. 23–34.
78. D'Agostino S (1987) Le relazioni di indeterminazione e la fondazione della meccanica quantistica nell'opera di Bohr ed Heisenberg, 1925–1927. Atti del VII Congresso nazionale di Storia della fisica, Padova, 2–7 ottobre 1986. Gruppo nazionale di coordinamento per la storia della fisica del C.N.R. Litografia Abbiatese, Milano, pp. 63–85.
79. D'Agostino S (1988) Concetto e metafora nel dialogo critico fra Einstein e Bohr, 1927–1935. Atti del VIII Congresso nazionale di Storia della fisica, Napoli, 12–17 ottobre 1987. Gruppo nazionale di coordinamento per la storia della fisica del C.N.R. Grafiche GV, Milano, pp. 77–89.
80. D'Agostino S (1988) Cento anni fa Hertz fabbricava in laboratorio le onde elettromagnetiche. Il Nuovo Saggiatore 4(2):54–59.
81. D'Agostino S (1988) Boscovich's reception of Newton's legacy. In: Bossi M, Tucci P (eds). Proceedings of the Bicentennial commemoration of R. G. Boscovich, Milano, September 15–18, 1987. Edizioni Unicopli, Milano, pp. 27–45.
82. D'Agostino S (1988) History of science at the service of a new scientific pedagogy. In: Blondel Ch, Brouzeng P (eds). Enseignement Scientifiqueet histoire de la physique, 21–25 novembre, 1988, Paris. Cité des Sciences et de l'industrie. Université de Orsay, Université de Paris-XI, Paris, pp. 41–50.
83. D'Agostino S (1989) Pourquoi Hertz et non pas Maxwell a-t-il Découvert les Ondes Électriques? In: Cazenobe, J (ed). L'électricité il y a cent ans. Édition de l'École des hautes études en sciences sociales, Paris, pp. 71–80.
84. D'Agostino S (1989) Pourquoi Hertz et non pas Maxwell a-t-il découvert les ondes électriques? Centaurus. 32:66–76.
85. D'Agostino S (1989) La prima tappa dello sviluppo dell'elettromagnetismo: l'abolizione del fluido magnetico. Cultura e Scuola 110:253–260.
86. D'Agostino S (1989) La Classificazione degli strumenti scientifici. Nuncius 4(1):87–92.
87. D'Agostino S (1989) Fisica e filosofia in James Clerk Maxwell. Special issue—De anima. Nuova Civiltà delle macchine 7(2):83–91.

88. D'Agostino S (1989) From Boskovic's notes on the work of Benedict Stay: a criticism of one of Newton's alleged proofs of absolute motion. Atti del Congresso Astronomy after Newton through the Nineteenth century Cagliari, 4–5 November 1988. Memorie della Società Astronomica Italiana 60(4):759–768.
89. D'Agostino S (1990) From Bošković's supplements to the poem of Benedikt Stay: A criticism of one of Newton's alleged proofs of absolute motion. [Iz Boškovićevih dopuna spjevu Benedikta Staya: kritika Newtonova navodnog dokaza za apsolutno gibanje]. Synthesis Philosophica: International Edition of the Journal Filozofska Istraživanja 9(5–6):1605–1610.
90. D'Agostino S (1990) Insegnamento scientifico e rinnovamento culturale. Atti e memorie della Accademia Petrarca di lettere, arti e scienze LII:131–141.
91. D'Agostino S (1990) Introduzione all'edizione Italiana di: Krafft F, La fisica e il suo divenire: sull'esperienza storica della conoscenza scientifica [*Id.*, Das Selbstverstandnis der Physik im Wandel der Zeit]. Armando editore, Roma, pp. 7–10.
92. D'Agostino S (1990) Boltzmann and Hertz on the Bild-Conception of Physical Theory. History of Science 28:380–398.
93. D'Agostino S (1990) Maxwell: Metodo e programma di ricerca. In: Costantini D, Galavotti MC (eds). Atti del Congresso Nuovi problemi della logica e della filosofia della scienza. Viareggio, 8–13 Maggio. CLUEB, Bologna, pp. 179–185.
94. D'Agostino S (1990) Matematica ed elettrodinamica nell'opera di Ampère: lo sviluppo delle teorie dell'elettromagnetismo. Cultura e Scuola 116:261–271.
95. D'Agostino S (1990) Da Maxwell a Boltzmann: la costituzione della fisica teorica. In: Ballio A, Paoloni L (eds). Scritti di storia della scienza in onore di Giovanni Battista Marini Bettolo nel 75mo compleanno. Memorie di scienze fisiche e naturali. Rendiconti della Accademia nazionale delle scienze detta dei XL, Serie V, XVI(2):343–357.
96. D'Agostino S (1991) Teorie sistematiche e strumenti precisi. Per una storia degli strumenti dell'elettromagnetismo. In: Erdas F, Dragoni G (ed). Instrumenta. Il patrimonio storico scientifico italiano: Una realtà straordinaria. Soprintendenza per i Beni Culturali della regione Emilia Romagn. Grafis, Bologna, pp. 247–254.
97. D'Agostino S (1989) Gli strumenti dell'elettromagnetismo. Teaching Speeches. Università di Roma La Sapienza, Roma.
98. D'Agostino S (1991) The bild-conception of physical theory and the foundation of Boltzmann's mechanics. Atti del X Congresso nazionale di Storia della Fisica, Cagliari, 28 settembre/2 ottobre 1989. Gruppo nazionale di coordinamento per la storia della fisica del CNR. Grafiche GV, Milano, pp. 95–110.
99. D'Agostino S, Orlando L (1992) Il criterio di corrispondenza e la genesi della teoria gravitazionale einsteiniana. Atti del XI Congresso Nazionale di Storia della Fisica, Trento, 8–13 ottobre 1990. Gruppo Nazionale di Coordinamento per la Storia della Fisica. Grafiche CV, Milano, pp. 111–121.

100. D'Agostino S (1991) Ampère e l'elettrodinamica: leggi matematiche e ipotesi fisiche. Giornale di Fisica 32(2–3):127–135.
101. D'Agostino S (1991) Erwin Schrödinger: la teoria fisica come "bild" e l'integrazione di scienza e cultura. Nuova Civiltà delle Macchine 9(34):151–157.
102. D'Agostino S (1991) Book reviews [Walter M, Schrödinger. Life and Thought. Cambridge University Press; Kilmister CW (ed), Centenary Celebration of a Polymath, Cambridge University Press]. Physis. Nuova Serie XXVIII(3):994–997.
103. D'Agostino S (1991) Immagini della scienza per una cultura della riforma. Quaderni del Centro di metodologia delle scienze sociali Serie 1-Saggi, 3–32. L.U.I.S.S., Roma.
104. D'Agostino S (1991) Einstein e Kant: la fisica teorica contro il logico ma vuoto schema dei concetti. In: Petruccioli S (ed). Immagini, linguaggi, concetti. Theoria, Roma, pp. 197–220.
105. D'Agostino S (1991) Un programma di ricerca e di attività per il reperimento, catalogazione e valorizzazione degli strumenti di interesse storico; La collezione di strumenti di misura elettromagnetici del Museo del Dipartimento di Fisica; Sistemi teorici e strumenti di precisione: per una storia degli strumenti dell'elettromagnetismo. In: D'Agostino S, Angelini, Alessandra; Chierego, Nedda (eds). Gli strumenti dell'elettromagnetismo. Collezioni del Museo del Dipartimento di Fisica de "La Sapienza". Istituto Tecnico Nautico Colonna, Roma, pp. 5–6, 7–8, 15–28.
106. D'Agostino S (1991) Il secondo Schrödinger. La riformulazione del concetto di teoria fisica. In: Cattaneo G, Rossi A (eds). I Fondamenti della Meccanica Quantistica. Analisi storica e problemi aperti. EditEI, Commenda di Rende, pp. 91–103.
107. D'Agostino S (1992) Schrödinger on continuity and completeness in physical theory. Physis. Nuova Serie XXIX(2):539–561.
108. D'Agostino S (1992) Continuity and completeness in physical theory: Schrödinger's return to the wave interpretazion of quantum mechanics in the 1950's. In: Bitbol M, Darrigol O (eds). Erwin Schrödinger, Philosophy and the Birth of quantum mechanics. Editions Frontières, Gif-sur-Yvette, pp. 339–362.
109. D'Agostino S (1993) Hertz's Researches and their Place in nineteenth Century theoretical Physics. Centaurus 36:46–82.
110. D'Agostino S (1993) Gli strumenti scientifici e il nuovo museo della scienza. In: Borchi E, Macii F, Vetrano, F (eds). Strumenti di fisica e cultura scientifica nell'Ottocento in Italia. Atti del Convegno nazionale su Strumenti di fisica e cultura scientifica nell'Ottocento in Italia. Villa Bottini e Liceo Machiavelli, Lucca, 8–9 maggio 1991. Litografia I.P., Firenze, pp. 123–131.
111. D'Agostino S (1993) Hertz on energetism and energy conservation. I beni culturali scientifici nella Storia e Didattica. Atti del Convegno del 14–15 dicembre 1990. Università degli studi di Pavia, Pavia, pp. 149–160.

112. D'Agostino S (1993) Boltzmann and Hertz on the Bild-conception of physical theory. In: Battimelli G, Ianniello MG, Kresten O (eds). Proceedings of the International symposium on Ludwig Boltzmann, Rome, February 9–11, 1989. Verlag der Österreichischen Akademie der Wissenschaften, Wien, pp. 75–94.
113. D'Agostino S (1993) Maxwell: La teoria del campo elettromagnetico e la nuova fisica. Cultura e Scuola 126:193–209.
114. D'Agostino S (1993) Guglielmo Marconi: l'invenzione della telegrafia senza fili sullo sfondo della fisica e della tecnologia del suo tempo. Rendiconti dell'Accademia Nazionale delle Scienze detta dei XL. Memorie di Scienze Fisiche e Naturali, Serie V, XXVII(2):159–175.
115. D'Agostino S (1993) Boscovich's physical theory of space and matter. In: Bursill HP (ed). R.J. Boscovich: vita e attività scientifica. Atti del Convegno, Roma 23–27 maggio 1988. Istituto della Enciclopedia italiana, Roma, pp. 41–48.
116. D'Agostino S (1993) A consideration of the rise of theoretical thysics in Europe and of its interaction with the philosophical tradition. First European Physical Society Conference on History of Physics in Europe in the Nineteenth and Twentieth Centuries. Edizioni Compositori, Bologna, pp. 5–28.
117. D'Agostino S (1993) Book review [Gallino L, Incerta alleanza. Einaudi, Torino]. Rivista di Storia della Scienza 2(1):160–165.
118. D'Agostino S (1993) Heinrich Hertz (1857–1894). [unpublished; scheduled for Museoscienza].
119. D'Agostino S (1993) Il Corso di Storia della Fisica I presso il Dipartimento di Fisica dell'Università di Roma. [Intervento tenuto nel corso dell'Incontro Nazionale dei Docenti di Storia della Fisica dal titolo "Venti anni di didattica universitaria di storia della fisica", Napoli, 14–15 aprile 1989] In: Atti del XII Congresso Nazionale di Storia della Fisica, L'Aquila, 30 settembre - 5 ottobre 1991. Gruppo Nazionale di Coordinamento per la Storia della Fisica del CNR. Grafiche GV, Milano, Appendix Section, pp. 323–326.
120. D'Agostino S (1994) L'opera di Marconi, 1895–1902: teorie elettromagnetiche e contesto della scoperta. Reprint.
121. D'Agostino S (1994) The fundamental role of absolute instruments in the evolution of classical electrodynamics: Thomson's absolute electrometer. Dragoni G, McConnell A, Turner G (eds). Proceedings of the eleventh International scientific instrument Symposium: Bologna University, Italy, 9–14 September 1991. Grafis, Bologna, pp. 153–155.
122. D'Agostino S (1994) Heinrich Hertz (1857–1894). La scoperta delle onde elettromagnetiche e i contributi alla teoria di Maxwell. Cultura e Scuola 33/130:191–203.
123. D'Agostino S (1994) Erwin Schrödinger: Scienza e Wissenchaft. In: Bertotti B, Curi U (eds). Erwin Schrödinger: Scienziato e filosofo. Il Poligrafo, Padova, pp. 101–110.
124. D'Agostino S (1994) Alcune osservazioni sulle ricerche italiane sui raggi cosmici e la fisica nucleare (1930–40). Giornale di Fisica XXXV(1/2):95–100.

125. D'Agostino S (1994) La scienza di Maxwell. In: Petruccioli S (ed). Dagli atomi di elettricità alle particelle atomiche. Problemi di storia e filosofia della fisica tra Ottocento e Novecento. Istituto della Enciclopedia Italiana, Roma, pp. 29–47.
126. D'Agostino S (1994) Book review [Sanzo U (ed), J.H. Poincaré, Scritti di fisica-matematica. UTET, Torino]. Physis. Nuova Serie XXXI(2):676–679.
127. D'Agostino S, Orlando L (1994) Il principio di corrispondenza e la genesi della teoria gravitazionale einsteiniana. Rivista di Storia della Scienza 2:51–74.
128. D'Agostino S (1995) Alcune considerazioni critiche sulla storiografia dell'universo della precisione di Koyré. In: Vinti C (ed). Alexandre Koyré. L'avventura intellettuale. Edizioni Scientifiche Italiane, Napoli, pp. 617–634.
129. D'Agostino S (1995) Einstein's Life-Long Doubts on the Physical Foundations of the General Relativity and Unified Field Theories. Garola C, Rossi A (eds). The Foundations of Quantum Mechanics – Historical Analysis and Open Questions. Kluwer Academic Publishers, Dordrecht, pp. 167–178.
130. D'Agostino S, Rossi A (1995) Enrico Fermi e l'origine della fisica teorica in Italia. In: Rossi A (ed). Atti del XIV e del XV Congresso nazionale di storia della fisica, Udine, 27 settembre-2 ottobre 1993; Lecce, 26 settembre-1 ottobre 1994. Conte, Lecce, pp. 467–474.
131. D'Agostino S (1995) L'archivio scientifico e la dimensione 'personale' nella storia della scienza. In: A.A.V.V. Gli archivi per la storia della scienza e della tecnica. Atti del convegno Gli archivi per la storia della scienza e della tecnica, Desenzano del Garda, 4–8 giugno 1991. Ministero per i beni culturali e ambientali. Ufficio centrale per i beni archivistici. Vol. 1. Roma, pp. 135–140.
132. D'Agostino S (1995) Alcune osservazioni su Guglielmo Marconi e il problema della trasmissione agli antipodi. In: Rossi A (ed). Atti del XIII Congresso Nazionale di Storia della Fisica, Pavia, 5–10 ottobre 1992. Conte, Lecce, pp. 401–413.
133. D'Agostino S (1995) Il problema delle basi empiriche della relatività generale di Einstein. Considerazioni sulle recenti ricerche storico-critiche. Rivista di Storia della Scienza 3(2):191–208.
134. D'Agostino S (1996) James Clerk Maxwell e la prima fondazione matematica delle teorie di campo. In: D'Agostino S (ed). Una teoria dinamica del campo elettromagnetico. Traduzione di James Clerk Maxwell, 'A dynamical theory of the electromagnetic field'. Teknos, Roma, pp. VII-XXXV.
135. D'Agostino S (1996) Absolute systems of unites and dimensions of physical quantities: a link between Weber's electrodynamics and Maxwell's electromagnetic theory of light. Physis XXXIII(1/3):5–51.
136. D'Agostino S (1996) The physics of the effects: a gate to atomism. In: Hoffmann D, Stuewer RH (eds). The emergence of modern physics: Proceedings of a conference commemorating a century of physics, Berlin 22–24 March 1995. La Goliardica pavese, Pavia, pp. 27–43.
137. D'Agostino S (1996) Fisica e matematica nell'opera di Dirac: le teorie del positrone e del monopolo. Cultura e Scuola 139/140:211–221.

138. D'Agostino S, Rossi A (1996) Entry: "Fermi, Enrico". Dizionario biografico degli italiani. Istituto della Enciclopedia Italiana, Vol. 46. Roma, pp. 329–343.
139. D'Agostino S (1996) Book review [Miller A, Insights of genius. Copernicus, NY]. Rivista di Soria della Scienza 4(2):169–171.
140. D'Agostino S (1997) Book review [Galison P, How experiments end, University of Chicago Press, Chicago]. Physis XXXIII(1/3):396–401.
141. D'Agostino S (1997) Giovanni Giorgi, ingegnere elettrotecnico, nel suo tempo. In: Tucci P (ed). Atti del XVI Congresso nazionale di Storia della Fisica e dell'Astronomia, Centro Volta, Villa Olmo, Como 24–25 Maggio 1996. Commissione di Studio per la Storia della Fisica e dell'Astronomia. Amminsitrazione Provinciale di Como—Camera di Commercio di Como—Centro di Cultura A. Volta, Como, pp. 345–353.
142. D'Agostino S (1997) Note per una storiografia dewlla visione microscopica. Marconi S (ed). Scritti in onore di Corrado Maltese. Edizioni Quasar, Roma, pp. 129–134.
143. D'Agostino S (1997) Heinrich Hertz and electric waves: An essay review. Centaurus. 39(3):267–272.
144. D'agostino S (1997) Book review [Buchwald J, The Creation of Scientific Effects. Heinrich Hertz and Electric Waves. Chicago University Press, Chicago]. Centaurus 39/3:267–272.
145. D'Agostino S (1997) From Maxwell's Electromagnetism to Ferraris' Electrotechnics. Galileò Ferraris and the conversion of energy developments of electrical engineering over a century. Proceedings of the International symposium, 27–29 October 1997. SIREA, Torino, pp. 77–98.
146. D'Agostino S (1997–1998) La corrispondenza di Vittorio Fossombroni con l'Accademia delle scienze detta dei Quaranta. Atti e memorie della Accademia Petrarca di lettere, arti e scienze 59–60:83–99.
147. D'Agostino S (1998) Heinrich Hertz on the methods of theoretical physics: Holism and Bild-conception of theory. In: Arbusci VM (ed). Prospettive della logica e della filosofia della scienza: Atti del Convegno triennale (1996–1998) della Società Italiana di Logica e Filosofia delle Scienze, Roma, 3–5 gennaio. Edizioni ETS, Pisa, pp. 389–401.
148. D'Agostino S (1998) La relatività generale nel dibattito degli anni Venti fra neo-kantiani ed empiristi logici: annotazioni su recenti studi einsteiniani. Physis. Nuova Serie XXXIV(3):643–658.
149. D'Agostino S (1998) Hertz's Views on the Methods of Physics: Experiment and Theory Reconciled? In: Baird D, Hughes RIG, Nordmann A (eds). Heinrich Hertz: Classical physicist, modern philosopher. Kluwer Academic Publishers, Dordrecht, pp. 89–102.
150. D'Agostino S (1998) Maxwell ed Einstein: metodo ed equazioni. Tucci P (ed). Atti del XVII Congresso nazionale di Storia della Fisica e dell'Astronomia, Centro Volta, Villa Olmo, Como 22–22 Maggio 1997. Commissione di Studio per la Storia della Fisica e dell'Astronomia. Relazioni generali, Milano, pp. 22–36.

151. D'Agostino S (1998) J.J. Thomson al tempo della scoperta dell'elettrone. Il metodo di Thomson spiegato in una lettera inedita di Thomson a Righi. Memorie di scienze fisiche e naturali. Rendiconti della Accademia nazionale delle scienze detta dei XL, Serie V, XXII(2):317–336.
152. D'Agostino S, Rossi A (1999) (eds) Galileo Ferraris e il suo tempo. Special issue Physis XXXV(2):209–478.
153. D'Agostino S (1999) Alcune considerazioni sulle teorie dello spazio e sulle geometrie da Descartes ad Einstein. In: Conti L, Mamone CM (eds). La scienza e i vortici del dubbio. Edizioni scientifiche italiane, Napoli, pp. 194–211.
154. D'Agostino S (1999) Book review [Monti D, Equazione di Dirac. Bollati Boringhieri, Torino]. Physis. Nuova Serie XXXVI(1):299–304 [Two copies are available in the *Fondo D'Agostino*; one of them it is an autograph–annotated copy].
155. D'Agostino S, Morando AP (1999) La nascita dell'ingegneria elettrica: dal mutuo induttore maxwelliano al trasformatore di Ferraris e di Steinmetz. Atti del XVIII Congresso nazionale di Storia della Fisica e dell'Astronomia, Centro Volta, Villa Olmo, Como 15–16 Maggio 1998. Commissione di Studio per la Storia della Fisica e dell'Astronomia. Universitá di Milano, Milano, pp. 135–150.
156. D'Agostino A (2000) A Controversial Role for Correspondence in Theoretical Physics and Quantum Mechanicas, in: Carola G, Rossi A (eds). The Foundation of Quantum Mechanics. Atti del Congresso di Lecce (1998) World Scientific Press, Singapore.
157. D'Agostino S (2000) Galileo Ferraris and the post-Maxwellian Electrical Engineering. Memorie di scienze fisiche e naturali. Rendiconti della Accademia nazionale delle scienze detta dei XL, Serie V, XXIV(5):117–129.
158. D'Agostino S, Leva S, Morando AP (2000) Per una storia della teoria delle reti elettriche. In: Tucci P (ed). Atti del XVIII Congresso nazionale di Storia della Fisica e dell'Astronomia, Centro Volta, Villa Olmo, Como 28–29 Maggio 1999. Commissione di Studio per la Storia della Fisica e dell'Astronomia. Universitá di Milano, Milano, pp. 229–252.
159. D'Agostino S (2000) On the Difficulties of the Transition from Maxwell's and Hertz's Pure-Field Theories to Lorentz's Electron. Physics in Perspective 2:398–410.
160. D'Agostino S (2000) A history of the Ideas of Theoretical Physics: Essays on the Nineteenth and Twentieth Century Physics. Kluwer Academic Publishers, Dordrecht.
161. Borgo S, D'Agostino S (2000) Book review [Chevalley C (ed), Werner Heisenberg, Philosophie: le manuscript de 1942. Sources du savoir Seuil Paris]. Physis. Nuova Serie XXXVII(1):256–262.
162. D'Agostino S (2001) The Bild-Conception of Physical Theories from Helmholtz to Hertz. In: Albertazzi L (ed). The Dawn of Cognitive Science Early European Contributors. Springer, Dordrecht, pp. 151–166.

163. D'Agostino S, Rossi A (2001) (eds). Enrico Fermi e l'Enciclopedia Italiana. Istituto della Enciclopedia Italiana Treccani, Roma.
164. D'Agostino S, Orlando L (2001) Il principio di corrispondenza e la genesi della teoria gravitazionale Einsteiniana. In: Fano V, Tarozzi G, Stanzione M (eds). Prospettive della logica e della filosofia della scienza: Atti del Convegno triennale della Società italiana di logica e filosofia delle scienze, Cesena e Urbino, 15–19 febbraio 1999. Rubbettino, Soveria Mannelli, pp. 51–74.
165. D'Agostino S (2002) From Rational Numbers to Dirac's Bra and Ket: Symbolic Representation of Physical Laws. Physics in Perspective 4:216–229.
166. D'Agostino S (2002) Esperimento e teoria in Heinrich Hertz. Αρχ´η IV: 217–225.
167. D'Agostino S (2002) Book review [Auletta G (ed), Foundations and Interpretation of Quantum Mechanics: In the Light of a Critical-Historical Analysis of the Problems and a Synthesis of the Results. World Scientific Press, Singapore]. Physis. Nuova Serie XXXIX(1):373–376.
168. D'Agostino S (2002) Book review [Badino M, L'epistemologia di Planck nel suo contesto storico. Edizioni scientifiche italiane, Roma]. Physis. Nuova Serie XXXIX(1):368–372.
169. D'Agostino S (2008) Dirac: la ragionevole potenza della matematica (la matematizzazione della fisica nell'opera di Dirac). Atti del XXI Congresso di Storia della Fisica e dell'Astronomia, Università della Calabria, Arcavacata di Rende (CS), 6–8 giugno 2001. Online–SISFA pp. 79–91.
170. D'Agostino S (2003) Dirac: la ragionevole potenza della matematica (la matematizzazione della fisica nell'opera di Dirac). Nuova Civiltá delle Macchine 21:92.
171. D'Agostino S. (2003) Dirac: la ragionevole potenza della matematica. Nuova Civiltà delle Macchine XXI(3):92–105.
172. Rossi A, D'Agostino S, Morando AP (2003) Giovanni Giorgi e la tradizione dell'elettrotecnica italiana. Atti del XXII Congresso Nazionale di Storia della Fisica e dell'Astronomia, Università degli Studi di Genova, Genova-Chiavari, 6–8 giugno 2002. Stampa Microart's S. p. A., Recco, pp. 314–328.
173. D'Agostino S, Leva S, Morando AP (2003) From Volta's Pile to Ferraris' Motor: A History of Instruments and Theories. In Giannetto E (ed). Volta and the History of Electricity. Hoeply, Milano, pp. 309–305.
174. D'Agostino S (2003–2004) La riflessione critica in Europa e la fisica teorica. Arché (1/2)5:217–230.
175. D'Agostino S, Leva S, Morando AP, Rossi A (2004) Il rapporto tra strumenti e teorie nell'ingegneria elettromagnetica in Italia e nel contesto internazionale tra Ottocento e Novecento. Dal Trasformatore e dal Teorema di Poynting alla teoria delle macchine elettriche e delle reti di Giovanni Giorgi ed Ercole Bottani. In: Tucci P, Garuccio A, Nigro M (eds). Atti del XXIII Congresso nazionale di Storia della Fisica e dell'Astronomia, Università degli Studi di Bari, Dipartimento di Fisica, Bari, 5–7 Giugno 2003. Progedit, Bari, pp. 145–153.

176. D'Agostino S (2004) Sostanza o funzione? Il problema dell'oggettivazione nella fisica post-einsteniana. Physis. Nuova Serie XLI(1):219–240.
177. D'Agostino S (2004) Book review [Oreskes N, Le Grand H (eds), Plate tectonics. An insider's history of the modern theory of the earth. Westview press, Boulder]. Physis. Nuova Serie XLI(1):276–280.
178. D'Agostino S (2004) Book review [Castellana M, Razionalismi senza dogmi. Per un'epistemologia della fisica matematica. Rubbettino, Soveria Mannelli]. Physis. Nuova Serie XLI(1):285–287.
179. D'Agostino S (2004) The bild conception of physical theory: Helmholtz, Hertz and Schrödinger. Physics in perspective 6:372–389.
180. D'Agostino S, Rossi A, (2004) Introduzione. Atti del Convegno Giovanni Giorgi nella realtà del suo tempo. Elettrotecnica, scienza e cultura nell'Italia del '900'. Istituto Treccani of the Enciclopedia Italiana. Physis. Nuova Serie XLI(2):309–317.
181. D'Agostino S (2004) Un problema delle teorie classiche e quantistiche: esiste una logica della strumentazione? In: Tassani I (ed). Quanti Copenaghen? Bohr, Heisenberg e le interpretazioni della meccanica quantistica. Il ponte vecchio, Cesena, pp. 135–148.
182. D'Agostino S (2004) La formazione umanistica è indispensabile per un efficace insegnamento della scienza? Nuova Civiltà delle Macchine XXII(1):144–146.
183. D'Agostino S (2004s) Considerazioni sui metodi d'insegnamento della fisica. Fisicamente via: https://fisicamente.blog/2021/03/12/considerazioni-sui-metodidinsegnamento-della-fisica
184. D'Agostino S (2005) Book review [Galison P, Gli orologi di Einstein, Le mappe di Poincaré. Imperi del tempo. Raffaello Cortina, Milano]. Physis. Nuova Serie XLII(2):556–559.
185. D'Agostino S (2005) Certezze e limiti nella scienza moderna. Atti della Accademia roveretana degli Agiati di scienze, lettere e arti. Contributi della Classe di scienze matematiche, fisiche e naturali 255(8B):33–45.
186. D'Agostino S (2005) Il difficile ricupero dell'Anschaulichkeit di Goethe nell'opera di Helmholtz. Nuncius 20(2):401–414.
187. D'Agostino S (2005) Gli strumenti scientifici e la scienza. Barbieri editore, Manduria [One autograph–annotated copy is available in the *Fondo D'Agostino*].
188. D'Agostino S, Leva S, Morando AP, Rogante G, Rossi A (2005) Reminiscenze elettriche in una città marchigiana: Fermo e l'elettrotecnica italiana. Atti del XXV Congresso Nazionale di Storia della Fisica e dell'Astronomia, Milano, 10–12 novembre 2005. Sezione di Storia della fisica, Istituto di fisica generale applicate. Università degli Studi di Milano, Milano, Online SISFA, pp. C19.1-C19.6.

189. D'Agostino S, Leva S, Morando AP, Rossi A (2005) Adolfo Bartoli e la pressione di radiazione. Atti del XXV Congresso Nazionale di Storia della Fisica e dell'Astronomia, Milano, 10–12 novembre 2005. Sezione di Storia della fisica. Istituto di fisica generale applicate. Università degli Studi di Milano, Milano, pp. C20.1-C20.4.
190. D'Agostino S (2006) Theory-Laden instruments and quantum mechanics. In: Garola C, Rossi A, Sozzo S (eds). The Foundations of Quantum Mechanics, Historical Analysis and Open Questions, 4–9 Ocober 2004. World Scientific Press, Singapore, pp. 130–141.
191. D'Agostino S, Salciccia E (2006) Einstein 1905. Nascita della teoria della relatività. Le scienze naturali nella scuola XV 27(1):61–66.
192. D'Agostino S (2007) (PhD Supervsor) Giacinto Cestoni, i rapporti con Redi e le scienze della vita nel XVII secolo. PhD Candidate: Tanga M., University of Pisa deg. Dottorato nteruniversitario degli Atenei Toscani.
193. D'Agostino S (2007) Book reviews [Boncinelli E, L'etica della vita. Siamo uomini o embrioni? Rizzoli, Milano; L'anima della tecnica, Rizzoli, Milano]. Physis. Nuova Serie XLIV(2):557–560.
194. D'Agostino S (2007) Completezza e sovradeterminazione. Le teorie di campo nella filosofia di Einstein in Autobiographical notes e Replay to criticism. In: Minazzi F, Tassani I (eds). Albert Einstein filosofo e metodologo. Atti del Convegno di Cesena, 25–26 Nov. 2005. Special issue. Il Protagora XXXV(9):101–113.
195. D'Agostino S (2007) Le teorie della fisica sono generalizzabili? Note su storiografia e storia della fisica Nuova Civiltà delle macchine 25(2):109–122.
196. D'Agostino S (2007) Book review [Carrea M, L'uomo asimmetrico e i limiti del neo-darwinismo. Borla, Roma]. Physis. Nuova Serie XLIV(2):560–561.
197. D'Agostino S (2007) Book review [Habermas J, Fra Scienza e fede. Laterza, Bari]. Physis. Nuova Serie 44(2):562–566.
198. D'Agostino S (2007) Book review [Buttaro C, Rossi A, Franco Rasetti. Una Biografia scientifica. Aracne, Roma]. Physis. Nuova Serie XLIV(2):566–568.
199. D'Agostino S, Leva S, Morando AP, Rossi A (2007) Una lettura post-maxwelliana dell'elettromagnetismo classico: Charles Proteus Steinmetz e la nascita della General Electric. Atti del XXIV Congresso Nazionale di Storia della Fisica e dell'Astronomia (Napoli-Avellino 3–6 Giugno 2004). Bibliopolis, Napoli, pp. 141–146.
200. D'Agostino S (2008) Scienza e Religione. Per una nuova alleanza. In: Pisano R, Pascual R (eds). Giornata di studio sulla relazione Scienze e Religioni. Il ruolo delle scienze naturali. Ipotesi di studio e prospettive, UNI Service, Trento, pp. 107–112.
201. D'Agostino S (2008) I contributi di Planck e di Einstein alla riflessione critica sulla filosofia teorica. Studi germanici XLVI(2):329–342.
202. D'Agostino S (2008) I contributi di Planck e di Einstein alla riflessione critica sulla fisica teorica. Nuova civiltà della macchine 26(4):143–153.

203. D'Agostino S (2009) Einstein: la scienza come autocritica. In: Giannetto E; Giannini G, Capecchi D, Pisano R (eds). Da Archimede a Majorana: la fisica nel suo divenire. Atti del XXVI Congresso Nazionale di Storia della Fisica e dell'Astronomia Roma. Guaraldi, Bergamo, pp. 233–246.
204. Morando AP, D'Agostino S, Rossi A (2009) Gian Antonio Maggi. In: Giannetto E, Giannini G, Capecchi D, Pisano R (eds). Da Archimede a Majorana: la fisica nel suo divenire. Atti del XXVI Congresso Nazionale di Storia della Fisica e dell'Astronomia, 2006 Roma. Guaraldi, Bergamo, pp. 195–200.
205. D'Agostino S (2009) La trasformazione dei modelli nella relatività di Einstein e nelle teorie quantistiche. Fisica e metafisica: XI Scuola estiva di filosofia della fisica. Special issue—Giornale di Fisica 50(1):87–95.
206. D'Agostino S (2009) Il 'gatto' di Schroedinger in una lettura del testo originale del celebre saggio. Physis. Nuova Serie 46(1/2):235–252.
207. D'Agostino S (2009) Fra storia e memoria. Ricerca ed insegnamento nella fisica italiana. Barbieri Selvaggi Manduria.
208. D'Agostino S (2009) La storia della fisica è utile solo nell'insegnamento secondario? Nuova civiltà delle macchine 27(2):107–116.
209. D'Agostino S (2009) Book review [Hermann von Helmholtz Meulders M, Dal secolo dei lumi alle neuroscienze, Bollati Boringhieri, Torino]. Physis. Nuova Serie XLVI(1/2):351–356.
210. D'Agostino S (2010) Basi filosofiche della fenomenologia matematica nell'opera di Helmholtz. Franco G, (ed). Sentieri aperti della ragione: verità, metodo, scienza: scritti in onore di Dario Antiseri nel suo 70 compleanno. La biblioteca Pensa multimedia, Lecce, pp. 175–188.
211. D'Agostino S (2010) Schrödinger: una veduta unitaria fra nuove statistiche, seconda quantizzazione e destino del gatto. Nuova Civiltà delle Macchine 27(4):135–153.
212. D'Agostino S, Morando AP (2010) La strumentazione dell'elettromagnetismo ottocentesco. In: Giannetto E, Giannini G, Toscano M (eds). Atti del XXVII Convegno Nazionale di Storia della Fisica e dell'Astronomia, Università degli Studi di Bergamo, Facoltà di Scienze della Formazione, sede di S. Agostino, Bergamo, 20–23 giugno 2007. Guaraldi, Rimini, pp. 115–126.
213. D'Agostino S (2011) Fisici e storici della scienza. Nuova Civiltà delle Macchine 28(4):161–172.
214. D'Agostino S (2011) Hermann von Helmholtz e il problema delle due culture. Nuova civiltà delle machine 27(2):207–218.
215. D'Agostino S (2011s) Gli strumenti nello sviluppo della scienza: da ausiliari a protagonisti. Fisicamente via: https://web.archive.org/web/20180924050539/http://www.fisicamente.net/FISICA/index-1505.htm#_ftnref1
216. D'Agostino S (2012) Newton, Ampère, Maxwell, Einstein: sulla deduzione dei fenomeni. Special Issue Complessità e riduzionismo. Isonomia. pp. 47–56.
217. D'Agostino S (2012) Strumentaria: filosofia e storia della strumentazione fisica. In: Mantovani R (ed). Atti del XXX Congresso Nazionale di Storia della Fisica e dell'Astronomia. Argalia Editore Urbino, pp. 255–264.

218. D'Agostino S (2013–2014) Book review [Newton. Un filosofo della natura e il sistema del mondo, Le scienze, Milano]. Physis. Nuova Serie XLIX(1/2):396–397.
219. D'Agostino S, Morando AP, Rossi A (2012) James Clerk Maxwell e la cultura italiana. In: mantovani R (ed). Atti del XXX Convegno Nazionale di Storia della Fisica e dell'Astronomia, Università degli Studi di Urbino, 09–11 Aprile 2010. Argaglia, Urbino, pp. 167–176.
220. D'Agostino S (2013) La scienza dell'elettricità da Volta ad Einstein. Una prospettiva storico-didattica. Booksprint Editore, Romagnano al Monte.
221. D'Agostino S (2013–2014) Fenomenologia e matematica nell'opera di Helmholtz e di Kirchhoff. Physis. Nuova Serie XLIX(1/2):243–258.
222. D'Agostino S (2014) Maxwell's Dynamical Philosophy: An Early Solution to the Problem of a Discrepancy between Particles and Waves. Advances in Historical Studies 3/3:123–129.
223. D'Agostino S (2014) Il celebre esperimento mentale del gatto. In: Giannetto E, Ricciardo S, Antonello E, Mazoni M (eds). Cielo e terra Fisica e Astronomia Un antico legame. Atti dei Congressi SISFA e della SIA 14–18 settembre 2009. Aracne, Roma, pp. 207–2014.
224. D'Agostino S (2015) Il sorriso del babbuino. I problemi dell'evoluzione visti da un fisico. Physis. Nuova Serie L(1/2):357–364.
225. D'Agostino S (2015) A Good Chance for the Cat's Life: Erwin Schroedinger's New Statistics and a Second Quantization Theory of Quantum. Advances in Historical Studies 4/1:1–7.
226. D'Agostino S (2015) Einstein risponde a Reichenbach, Born e Schrödinger critici della relatività. Il Protagora 23-XLII 24(1/2):19–36.
227. D'Agostino S (2016a) An Overview of the Nineteenth and Twentieth Centuries' Theories of Light, Ether, and Electromagnetic Waves. Advances in Historical Studies 5/1:1–11.
228. D'Agostino S (2016b) Equivalence Principle and Ether: Two Revolutionary Kernels of Einstein's General Relativity. Advances in Historical Studies 5/5:240–253.
229. D'Agostino S (2017) The Centennial Recurrence of Einstein's General Relativity. LLAP, Balti.
230. D'Agostino S (2018) An asymmetric evaluation of Einstein's epistemologic contributions. Lambert Academic Publishing, Saarbrücken.
231. D'Agostino S (2018) Letture di Storia della fisica. Da Galileo a Newton: Edizioni Accademiche Italiane, Beau Bassin–Berlin [A Book review of this publication is available: Pisano R (2021) Seventeen Lectures in the History of Physics. Metascience 30:207–210].
232. D'Agostino S (2019) Le ricerche sui raggi cosmici a Firenze e a Roma. La nuova scienza astronomica. Giornale di Astronomia—Rivista di informazione, cultura e didattica della Società astronomica italiana 45(3):27–30.

233. D'Agostino S (2018–2019) Winter Notebooks more—or—less scientific for Friends and Family. The fog at the high hills. 11 pages of 1/2 A4 format. Content: 1. Gravitational waves explained without mathematics, 2. Science as culture, 3. Development of theories of elasticity: elastic bodies as models for the electromagnetic field, 4. Field theories as a subversion of reductionist methods, 5. The general equations of elasticity, 6. The problem of the Two Cultures in Germany: Hermann Von Helmholtz; 6. (Sic!). Heinrich Hertz and the verification of Maxwell's electromagnetic theory, 7. Ludwig Boltzmann. 8. Niels Bohr the "Principle of Correspondence". Planck: reconciliation of tradition and innovation, 9. The life of Boltzmann. At the end of the 7. "END" is handwritten and the pages of the following paragraphs are crossed out. In another Notebook with the same title (63 pages of 1/2 A4 format) the crossed-out paragraphs have been deleted and a Preface and a chapter "The Miracle of Celano" have been added. In the Preface D'Agostino writes that this is the third of his notebooks. It is not clear what the second notebook is [unpolished]. In summary the Notebooks are three but the third notebook is a reprint of the second with the corrections that had been made by hand and with 2 chapters added (Preface and The Miracle of Celano) [unpublished].
234. D'Agostino S ([submitted end 2019, published posthumous] (2022) Schrodinger 1950: The Cat's Situation on the Background of the New Statistics and Second Quantization. Advances in Historical Studies 11/2:81–85.
235. D'Agostino S (s.d.) Forces, Fluxes, Quantities and Dimensions: the Geometrization of Physics in Maxwell's Treatise.
236. D'Agostino S (s.d.) Osservazione, esperimento e linguaggio scientifico (per una ricerca sull'impiego del laboratorio nella formazione del linguaggio scientifico)
237. D'Agostino S (s.d.) Il successo della teoria delle forze centrali nel campo dell'elettricità "frizionale" e del magnetismo.

References

AA.VV. (2021) SISFA: 40 Years of history of physics in Italy. Percorsi 37(1–2):37–48.

Abdank-Abakanowicz B (1886) Les intégraphes, la courbe intégrale et ses applications: étude sur un nouveau système d'intégrateurs mécaniques. Gauthier-Villars, Paris.

Abraham M (1902a) Dynamik des Elektrons. Nachrichten von der Gesellschaft der Wissenschaften zu Göttingen, Mathematisch-Physikalische Klasse: 20–41.

Abraham M (1902b) Prinzipien der Dynamik des Elektrons. Annalen der Physik 315(1):105–179.

Abraham M (1904) Die Grundhypothesen der Elektronentheorie. Physikalische Zeitschrift 5(18):576–579.

Abraham M (1905) Theorie der Elektrizität. B.G. Teubner, Leipzig.

Abraham M (1908) Theorie der Elektrizität. Zweiter Band. Elektromagnetische Theorie der Strahlung Zweite Auflage. B.G. Teubner, Leipzig.

Agassi J (1964) The Nature of Scientific Problems and Their Roots in Metaphysics. In: Bunge M (1964) (ed). The Critical Approach Essays in Honor of Karl Popper. Free Press of Glencoe, NY., pp. 189–211.

Agassi J (1975) Determinism: Metaphysical versus Scientific. Memoirs, 5th International Congress of Logic, Methodology and Philosophy of Science. Kluwer, Dordrecht, pp. V 1–4.

Agassi J (1977) Towards a Rational Philosophical Anthropology. Nijhoff. The Hague.

Amsterdamski S (1975) Between experience and metaphysics/Philosophical problems of the evolution of science. Reidel, Dordrecht–Boston.

Anstey P (2005) Experimental versus Speculative Natural Philosophy. In: Anstey P, Schuster JA (eds). The Science of Nature in the Seventeenth Century: Changing Patterns of Early Modern Natural Philosophy. Springer, Dordrecht, pp. 215–42.

Anstey P, Hunter M (2008) Robert Boyle's'Designe about Natural History', Early Science and Medicine 13:83–126.

Anstey, Peter (2014) Philosophy of Experiment in Early Modern England: The Case of Bacon, Boyle and Hooke. Early Science and Medicine 19:103–132.

Anta J (2021a) Can Informational Thermal Physics explain the Approach to Equilibrium? Synthese 199(1/2):4015–4038.

Anta J (2021b) Historical and Conceptual Foundations of Information Physics. Dissertation, University of Barcelona, Spain.

Antiseri D ([1982] 2001) Teoria unificata del metodo. Liviana, Padova, UTET, Torino.

Antiseri D, Tagliagambe S (2008) Storia della Filosofia. Filosofi Italiani contemporanei, vol. 14. RCS Libri, Bompiani, Milano, pp. 138–151.

R. Pisano (ed.), *A History of Physics: Phenomena, Ideas and Mechanisms*, History of Mechanism and Machine Science 42,
https://doi.org/10.1007/978-3-031-26174-9

Archimedes (1503) Tetragonismus idest circuli quadratura per Campanum Archimedem Syracusanum atque Boetium mathematicae perspicacissimos adinuenta. [Gaurico L (ed)] Impressum Venetiis: per Ioan. Bapti. Sessa, 1503 die 28 august.

Archimedes (1544) Opera quae quidem extant omnia. Basel, Gechauff.

Archimedes (1558) Archimedis opera non nulla a Federico Commandino Urbinate nuper in latinum conversa et commentariis illustrata. apud Paulum Manutium, Aldi F. Venetiis.

Archimedes (1565) Archimedis De iis quae vehuntur in aqua libri duo. A F. Commandino U. in pristinum nitorem restituti, et commentariis illustrati. Ex Offcina Benacii, Bononiae.

Archimedes (1792) Archimedis quae supersunt omnia, cum Eutocii Ascalonitae commentariis, ex recensione Josephii Torelli Veronensi cum nova versione latina (accedunt versiones variantes ex codd. Mediceo et Parisiensibus). Robertson. Oxford.

Archimedes (1880–1881) Archimedis opera omnia com commentariis Eutocii e codice Florentino recensuit, edidit Iohann Ludvig Heiberg, 3 vols., Lipsiae, in aedibus B.G. Teubneri.

Archimedes (1912) The Method of Archimedes recently discoverd by Heiberg. Edited by Heath, T., L. The Cambridge University Press. Cambridge.

Archimedes (1987) The Method of Mechanical Theorems. Dijksterhuis EJ (ed). Translation by C. Dickshoorn. Princeton University Press, Princeton.

Archimedes (2002a) On the equilibrium of planes. Archimedes 2002b, pp 189–220.

Archimedes (2002b) The works of Archimedes. Edited by Heath TL. Dover Publications Inc. New York.

Aristóteles (2004) Tratados de lógica (el Organón). México, Editorial Porrúa.

Aristóteles (2007) Física. Barcelona, Editorial Gredos.

Aristóteles (2011) Poética. Valentín García Yebra (Traductor). Madrid, Editorial Gredos.

Aristotle (1930) Physics. Translation by Hardie RP and Gaye RK. Clarendon Press, Oxford.

Aristotle ([1936] 1955) Mechanical problems. In: Hett WS (ed) Aristotle. Minor works. William Heinemann, Cambridge.

Arntzenius F (2012) Space, Time, and Stuff. Oxford University Press, New York.

Asimov I (1951) Foundation Series. Gnome Press, NY.

ATLAS Collaboration (2012) Observation of a new particle in the search for the standard model Higgs Boson with ATLAS detector at the LHC. Physics Letter B 716/1:1–29. http://www.sciencedirect.com/science/article/pii/S037026931200857X

Atten von M, Kennedy J (2003) On the philosophical development of Kurt Gdel. The Bulletin of Symbolic Logic 9:425–476.

Avishai Y, Ekstein H (1974) Einstein's Equivalence Principle and its Derivation from Special Relativity Presymmetrey. Communications in Mathematical Physics 37:193.

Baaquie BE (2004) Quantum Finance: Path Integrals and Hamiltonians for Options and Interest Rates. Cambridge University Press, New York.

Bachelard G (1949) Le rationalisme appliqué. Presses Universitaires de France, Paris.

Bachelard G (1975a) La Formation de l'Esprit Scientifique. 9e edition. Vrin, Paris.

Bachelard G (1975b) Le Nouvel Esprit Scientifique. 13e edition. Presses Universitaires de France, Paris.

Bacry H, Lévy-Leblond JM (1968) Possible Kinematics. Journal of Mathematical Physics 9:1605.

Baldi B (1707) Cronica de' matematici: overo Epitome dell' istoria delle vite loro. Angelo Antonio Monticelli, Urbino.

Baracca A, Rossi A (1974) 1789: prassi e organizzazione della scienza. Sapere 775:46–52.

Baracca A, Rossi A (1976) Marxismo e scienze naturali. De Donato, Bari.

Baracca A, Rossi A (1977) Scienza e produzione nel '700. Guaraldi, Firenze.

Baracca A, Rossi A (1978) Materia e energia: antologia di testi. Feltrinelli, Milano.

Baracca A, Ruffo S, Russo A (1979) Scienza e industria, 1848–1915. Laterza, Bari.

Barbensi G ed (1967) Opere scelte di Luigi Galvani. UTET, Torino.

Barbour J, Pfister H (eds) (1995) Mach's Principle: From Newton's Bucket to Quantum Gravity. Birkhauser, Boston.

Bar-Hillel Y (1964) Lenguage and Information. Selected Essays on Their Theory and Application. Addison-Wesley. Massachusetts.
Barnes B (1982) T.S. Kuhn and Social Science. MacMillan, London.
Barrett MD, Chiaverini J, Schaetz T, Britton J, Itano WM, Jost JD, Knill E, Langer C, Leibfried D, Ozeri R, Wineland DJ (2004) Deterministic quantum teleportation of atomic qubits. Nature 429:737–739.
Barrow JD, Tsagas CG (2004) Dynamics and stability of the Gödel universe. Retrieved 26 January 2022 from Cornell University Library Web site: http://arxiv.org/abs/gr-qc/0308067
Bartley III, WW (1982–198) (ed) Postscriptum to the Logic of Scientific Discovery by Karl R. Popper. 3 Vols. Hutchinson, London.
Barut AO (1974) On the formulations of electrodynamics from a single principle. Foundations of Physics 24(4):477–485.
Battimelli G, De Maria M, Rossi A (1984) (eds) La Ristrutturazione delle scienze fra le due guerre mondiali, Atti del Congresso, Firenze e Roma, 1980. Editrice Universitaria La Goliardica, Roma, vol. 1 e 2.
Bauer FL (2002) Decrypted Secrets, Methods and Maxims of Cryptology. Springer-Verlag, Berlin.
Beckmann P (1971–2015) A History of Pi. Golem Press, New York.
Bekenstein J D (1973) Black Holes and Entropy. Physical Review D 7:2333–2346.
Belfer I (2014) Informing Physics: Jacob Bekenstein and the Informational Turn in Theoretical Physics. Physics in Perspective 16:69–97.
Bell JS (1964) On Einstein Podolsky Rosen paradox. Physics 1:195–200.
Belot G (2005) Dust, time, and symmetry. British Journal for the Philosophy of Science 56:255–291.
Bennett C (1982) The thermodynamics of computation – a review. International Journal of Theoretical Physics 21:905–940.
Bennett CH (1993) Teleporting an unknown quantum state via dual classical and Einstein-Podolsky-Rosen channels. Physical Review Letters 70:1895–1899.
Bennett J (1998) Practical Geometry and Operative Knowledge. Configurations 6(2):195–222.
Berkel von K (2012) Isaac Beeckman on Matter and Motion: Mechanical Philosophy in the Making. Johns Hopkins University Press, Baltimore.
Bernard E (1697) Catalogi librorum manuscriptorum Angliae et Hiberniae in unum collecti cum indice alphabetico. E Theatro Sheldoniano, Oxford.
Bernard E (1704) Epistolae et Veterum Mathematicorum, Graecorum, Latinorum, & Arabum, Synopsis. Robert Huntington, Edward Bernard, Thomas Smith. Epistolae. Typis G. Bowyer, impensis A. & J. Churchill, London. [Separate pagination]. Part 11:1–44.
Bernoulli D ([1748] 1750) Remarques sur le principe de la conservation des forces vives pris dans un sens général. Mémoire de l'Académie Royale des Sciences et des Lettres de Berlin, vol IV, pp. 356–364.
Bernoulli Johann (1727) Theoremata selecta pro conservatione virium vivarum demonstranda et esperimenta confirmanda. Novi commentarii Academiae Scientiarum Imperialis Petropolitanae 2, pp. 200–207.
Bertalanffy von L (1932) Theoretische Biologie, 2 Vols. Gebruder Borntraeger, Berlin.
Bertalanffy von L (1968) General system theory–Foundations, development, applications. Penguin, London.
Bertoloni MD (2006) Thinking with Objects: The Transformation of Mechanics in the Seventeenth Century. Johns Hopkins University Press, Baltimore.
Bérut A, Arakelyan A, Petrosyan A, Ciliberto S, Dillenschneider R, Lutz E (2012) Experimental verification of Landauer's principle linking information and thermodynamics. Nature 48:187–190.
Berzi V, Gorini V (1968) Reciprocity Principle and the Lorentz Transformation. Journal of Mathematical Physics 10:1518 [These authors offer a very large list of the previous works, beyond the ones quoted here].
Birch T (1756) The History of the Royal Society of London. Vol II. A. Millar, London.

Birkhoff GD (1944) Flat Space-time and Gravitation. Proceedings of the National Academy of Sciences of the United States of America 30:324.
Bishop E (1967) Foundations of Constructive Mathematics. Mc Graw-Hill, New York.
Blackwell RJ (1966) Descartes' Laws of Motion. Isis 57(2):220–234.
Blakeslee EP (1972) The Radio Amateur's VHF Manual. American Radio Relay League, Newington, Connecticut.
Blancanus J (1615) De mathematicarum Natura dissertatio. Una cum Clarorum mathematicorum chronologia. Apud Bartholomaeum Cochium, Bologna.
Blezza F (1983) Galvani e Volta – La polemica sull'elettricità. La Scuola, Brescia.
Blezza F (2018) La pedagogia professionale – Che cos'è, quali strumenti impiega e come si esercita. Libreria Universitaria, Limena PD.
Blezza F (2020) Il pedagogista. ETS, Pisa.
Boccaletti D (2016) Galileo and the Equations of Motion. Springer, Dordrecht.
Boltzmann L ([1865–1874] 1909) Wissenschaftliche Abhandlungen von Ludwig Boltzmann. I Band. Barth, Leipzig.
Boltzmann L (1866) Über die mechanische Bedeutung des zweiten Hauptsatzes der Wärmetheorie. Wiener Berichte 53:195–220.
Boltzmann L (1871) Zur Priorität der Auffindung der Beziehung zwischen dem zweiten Hauptsatze der mechanischen Wärmetheorie und dem Prinzip der kleinsten Wirkung. Annalen der Physik und Chemie 143:211–230.
Boltzmann L (1872) Weitere Studien über das Wärmegleichgewicht unter Gasmolekülen. Sitzungsberichte Akademie der Wissenschaften 66:275–370.
Boltzmann L (1877) Über die Beziehung zwischen dem zweiten Hauptsatze der mechanischen Wärmetheorie und der Wahrscheinlichkeitsrechnung resp. den Sätzen über das Wärmegleichgewicht. Wiener Berichte 76:373–435.
Boltzmann L (2003) Further Studies on the Thermal Equilibrium of Gas Molecules. The Kinetic Theory of Gases. History of Modern Physical Sciences 1:262–349.
Boschiero L (2005) Stories about the Rise of Modern Science. Minerva 43(3):311–318.
Boschiero L (2007) Experiment and Natural Philosophy in Seventeenth Century Tuscany: The History of the Accademia del Cimento. Springer, Dordrecht.
Boudia S (2001) Marie Curie et son laboratoire: sciences et industrie de la radioactivité en France. Éditions des Archives Contemporaines, Paris.
Boutroux P (1921) L'histoire des principes de la dynamique avant Newton. Revue de Métaphysique et de Morale 28:657–688.
Boyle R (1660) New Experiments Physico-Mechanicall, Touching the Spring of the Air and Its Effects. Hall, Oxford.
Braga C (1950) À memória do Professor A. Cyrillo Soares. Gazeta de Física II(4):90–92.
Brandt F (1928) Thomas Hobbes' Mechanical Conception of Nature. Levin & Munksgaard, Copenhagen.
Brassard G (1996) Teleportation as a quantum computation. Retrieved 26 January 2022 from Cornell University Library Web site: https://arxiv.org/abs/quant-ph/9605035
Breuer J, Freud S (1895) Studien über Hysterie. Franz Deuticke. Leipzig und Wien.
Brillouin L (1951) Maxwell's Demon Cannot Operate: Information and Entropy. International Journal of Applied Physics 22:334–337.
Brillouin L (1956) Science and Information Theory. Academic Press, New York.
Buridani J (1509) Subtilissimae Quaestiones super octo Physicorum libros Aristotelis, Johanne Dullaert, Paris. [Reprint: *Id.* (1964) Kommentar zur Aristotelischen Physik. Minerva, Frankfurt].
Buridan J (1513) Quaestiones super decem libros Ethicorum Aristotelis ad Nicomachum, Paris. [Reprint: *Id.*, (1968) Super decem libros Ethicorum. Minerva, Frankfurt]
Buridan J (1942) Iohannis Buridani Quaestiones super libris quattuor De caelo et mundo. In: Moody EA (ed). Medieval Academy of America. Studies and documents 6/40.
Burns WE (2001) The Scientific Revolution. An Encyclopedia. ABC-Clio, Santa Barbara.

Burtt EA ([1924, 1932] 2003) The Metaphysical foundations of Modern Science. Doubleday. Dover, New York.
Busenmeyer JR, Bruza OD (2012) Quantum Models of Cognition and Decision. Cambridge University Press, New York.
Bussotti P (2017) Galileo e il suo concetto di inerzia. In: Neri A (ed). Là dove il pensiero incontra l'esperienza. ETS, Pisa, pp. 51–110.
Bussotti P, Lotti B (2022) Cosmology in the Early Modern Age: a Web of Ideas. Springer. Dordrecht.
Bussotti P, Pisano R (2013) On the Conceptual Frames in René Descartes' Physical Works. Advances in Historical Studies 2/3: 106–125.
Bussotti P, Pisano R (2014a) Newton's Philosophiae Naturalis Principia Mathematica "Jesuit" Edition: The Tenor of a Huge Work. Accademia Nazionale Lincei–Matematica e Applicazioni 25/4:413–444.
Bussotti P, Pisano R (2014b) On the Jesuit Edition of Newton's Principia. Science and Advanced Researches in the Western Civilization. In Pisano, R (ed). Isaac Newton and his Scientific Heritage: New Studies in the History and Historical Epistemology of Science. Special Issue. Advances in Historical Studies 3/1:33–55
Bussotti P, Pisano R (2017) Historical and Philosophical Details on Leibniz's Planetary Movements as Physical-Structural Model. In: Pisano R, Fichant M, Bussotti P, Oliveira ARE (eds). College Publications, London, pp. 49–92.
Buteo J (1559) De quadratura circuli libri duo: vbi multorum quadraturae confutantur & ab omnium impugnatione defenditur Archimedes; eiusdem Annotationum opuscula in errores Campani, Zamberti, Orontij, Peletarij, Io. Penae interpretum Euclidis. Apud Gulielmum Rouillium, Lyon.
Calvocoressi P (1980) Top Secret Ultra. Pantheon Books, New York.
Cantor M (1857) Petrus Ramus, Michael Stifel, Hieronymus Cardanus, drei mathematische Charakterbilder aus dem 16. Jahrhundert. Zeitschrift für Mathematik und Physik. Literaturzeitung 2:353–376.
Capanna M (1988) Formidabili quegli anni. Rizzoli, Milano.
Capecchi D, Pisano R (2007) La teoria dei baricentri di Torricelli come fondamento della statica. Physis XLIV/1:1–29.
Capecchi D, Pisano R (2008) La meccanica in Italia nei primi anni del Cinquecento. Il contributo di Niccolò Tartaglia. In: Tucci P (ed). Proceedings of XXV SISFA Congress. University of Milano Press, Milano, C17.1–C17.6.
Capecchi D, Pisano R (2010) Reflections on Torricelli's principle in mechanics. Organon 42:81–98.
Capocaccia A, Eco U, Geymonat L (1965) Verso una nuova cultura. Esigenze, aspetti e problemi di una cultura unitaria. Civiltà delle Macchine XIII(1).
Carathéodory C (1909) Untersuchungen über die Grundlagen der Thermodynamik. Mathematische Annalen 67:355–386.
Carnap R (1977) Two Essays on Entropy. Edited by Abner Simony. University of California Press, Berkeley.
Carnot L ([1783] 1786) Essai sur les machines en général. Defay, Dijon.
Carnot L (1803) Principes fondamentaux de l'équilibre et du mouvement. Deterville, Paris. [see also *Id*. (1803) Géométrie de position à l'usage ceux qui se destinent à mesurer les terrains. Duprat, Paris].
Carnot L (1813) Réflexions sur la métaphysique du calcul infinitésimal. [2nd edition of 1797] Courcier, Paris.
Carnot S (1824) Réflexions sur la puissance motrice du feu et sur les machines propres à développer cette puissance. Bachelier, Paris.
Carnot S ([1824] 1978) Réflexions sur la puissance motrice du feu sur les machinés propre à développer cette puissance, édition critique par Fox Robert. Vrin J, Paris. [*Ivi*, Recherche d'une formule propre à représenter la puissance motrice de la vapeur d'eau, pp. 223–234].
Carnot S (1878a) Réflexions sur la puissance motrice du feu sur les machinés propre à développer cette puissance. [Archives of the Académie des Science–Institut de France, Paris].

Carnot S (1878b) Réflexions sur la puissance motrice du feu sur les machinés propre à développer cette puissance. Gauthier–Villars, Paris.
Carnot S (1986) Reflexions on the motive power of fire: a critical edition with the surviving scientific manuscripts translated and edited by Robert Fox. The Manchester University Press, Manchester.
Carnot S (s. d.) Notes sur les mathématiques, la physique et autres sujets. In: Carnot 1878b, pp 89–102. The manuscript is conserved In: Carnot 1878a.
Cassirer E (1970) Determinismo e indeterminismo nella fisica moderna. La Nuova Italia, Firenze.
Cercignani C (1998) Ludwig Boltzmann. The Man who Trusted Atoms. Oxford University Press, Oxford.
Cerroni U (1970) Tecnica e libertà. De Donato, Bari.
Chales de CF (1674) Cursus seu Mundus Mathematicus. Ex officina Annisoniana, Lyon.
Chalmers A (1991) Galilean Relativity and Galileo's Relativity. In: French S, Kamminga H (eds). Correspondence, Invariance and Heuristics: Essays in Honour of Heinz Post. Kluwer Academic Publishers, Dordrecht.
Chalmers A (2017) One Hundred Years of Pressure: Hydrostatics from Stevin to Newton. Springer, Dordrecht.
Chandrasekhar S, Wright JP (1961) The geodesics in Gödel's universe. Proceedings of the National Academy of the Sciences of the USA 47:341–347.
Ciccotti G, Cini M, De Maria M, Jona Lasinio G (1976) L'ape e l'architetto. Paradigmi scientifici e materialismo storico. Feltrinelli, Milano.
Ciccotti G, Cini M, De Maria M, Jona-Lasinio G, Donini E ([1976]2011). L'Ape e l'Architetto. Paradigmi Scientifici e Materialismo Storico. Feltrinelli, Milano.
Cini M, Maccacaro GA (1975) (eds) Scienza e Politica. Scienza e potere, AA. VV. Feltrinelli, Milano.
Ciufolini I, Wheeler JA (1995) Gravitation and Inertia. Princeton University Press, Princeton.
Clagett M (1959) The Science of Mechanics in the Middle Age. University of Winsconsin Press, Madison.
Clagett M (1964–1984) Archimedes in the Middle Ages. Madison-Philadelphia, Memoirs of the American Philosophical Society. 5 Vols., 10 Tomes. The Clarendon University Press, Oxford.
Clagett M (1976) Archimedes in the Middle Ages, II, Philadelphia, American Philosophical Society.
Clagett M (1978) Archimedes in the Middle Ages, III, Philadelphia, American Philosophical Society.
Clapeyron EBP (1834) Mémoire sur la puissance motrice du feu. Journal de l'École Royal Polytechnique XXIII/XIV:153–190. [Collections École polytechnique Z 5 (1834)].
Clausius R (1850) Über die bewegende Kraft der Wärme und die Gesetze, welche sich daraus für die Wärmelehre selbst ableiten lassen. Annalen der Physik und Chemie 79:368–397. 500–524. [English trans: *Id.*, On the Motive Power of Heat and on the Laws which can be deduced from it for the theory of heat itself. In: Mendoza E (1960) (ed) Reflections on the motive power of fire and other papers on the second law of thermodynamics by E Clapeyron and R Clausius. Dover, New York, pp 73–74, 109–152].
Clausius R (1854) Über eine veränderte Form des zweiten Hauptsatzes der mechanischen Wärmetheorie. Annalen der Physik und Chemie 169/12:481–506.
Clausius R (1857a) The nature of the motion we call heat. [English translation: Brush (1965) (ed) Kinetic Theory. Selected Readings in Physics. II. Pergamon Press, Oxford, pp. 111–134.
Clausius R (1857b) Über die Art der Bewegung welche wir Wärme nennen. Annalen der Physik und Chemie 100/3:353–380.
Clausius R (1862) Über die Anwendung des Satzes von der Äquivalenz der Verwandlungen auf die innere Arbeit. Annalen der Physik und Chemie 116:73–112.
Clausius R (1865) Über verschiedene für die Anwendung bequeme Formen der Hauptgleichungen der mechanischen Wärmetheorie. Annalen der Physik und Chemie 75/7:353–400.
Clausius R (1867) The mechanical theory of heat with its applications to the steam-engine and the physical properties of bodies. Van Voorst, London.

Clausius R (1871) Über die Zurückführung des zweiten Hauptsatzes der mechanischen Wärmetheorie auf allgemeine mechanische Prinzipien. Annalen der Physik und Chemie 142:433–461.
Clausius R (1872) Zur Geschichte der Mechanischen Wärmetheorie. Annalen der Physik und Chemie 145:132–146.
Clifton T, Barrow JD (2006) The existence of Gödel, Einstein, and de Sitter universes. Retrieved 27 January 2022 from Cornell University Library Web site: https://arxiv.org/abs/gr-qc/0511076
CMS Collaboration (2012) Observation of a new boson at a mass of 125 GeV with the CMS experiment at the LHC. Physics Letter B 716:30–61. http://www.sciencedirect.com/science/article/pii/S0370269312008581
Coffa JA (1968) Galileo's concept of Inertia. Physis X(4):261–281.
Cohen BI (1964) 'Quantum in Se Est': Newton's Concept of Inertia in Relation to Descartes and Lucretius. Notes and Records of the Royal Society of London 19(2):131–155.
Coleman S, Hartle JB, Piran T, Weinberg, S (1991) (eds) Quantum Cosmology and Baby Universes. World Scientific. River Edge, NJ.
Collins HM (1975) The seven sexes: a study in the sociology of a phenomenon. Sociology 9:205–224.
Comte C (1986) Leibniz aurait-il pu découvrir la relativité? European Journal of Physics 7:225.
Contessa G (2010) Modal truthmakers and two varieties of actualism. Synthese 174:34–353.
Costa A (1930) Relatório do Vice-presidente do ramo de ciências sobre as necessidades da investigação científica em Portugal. In: Junta de Educação Nacional. Relatórios, propostas e projecto de orçamento para o ano económico de 1930–1931. Lisboa, pp. 9–43.
Costa A (1939) O Problema da Investigação Científica em Portugal. Coimbra.
Coudres Td (1899) Handliche Vorrichtung zur Erzeugung Lenard'scher Strahlen und einige Versuche mit solchen Strahlen. Annalen der Physik 298:134–144.
Coulomb CA (1784) Observations théoriques et expérimentales sur l'effet des moulins à vent,&sur la figure de leurs ailes. In: Histoire de l'Académie Royale des Sciences avec les Mémoires de Mathématique&de Physique, pour lamême année 1781. Imprimerie Royale, Paris, pp. 65–81
Coulomb CA (1785) Théorie des machines simples en ayant égard au frottement de leurs parties et à la roideur du cordage. Mémoire de mathématique et de physique présentes à l'Académie Royal des Sciences par divers savants. Bachelier, Paris.
Coulomb CA (1799) Résultat de plusieurs expériences destinées a déterminer la quantité d'action que les hommes peuvent fournir par leur travail journalier suivant les différentes manières dont ils emploient leurs forces, vol II. Mémoires de l'Institut National des sciences et arts, Paris, pp. 340–428.
Crawford E (1988) Internationalism in science as a casualty of the First World War: relations between German and Allied scientists as reflected in nominations for the Nobel prizes in physics and chemistry. Social Science Information 27(2):163–201.
Crawford E (1992) Nationalism and Internationalism in Science, 1880-1939: Four Studies of the Nobel Population. Cambridge University Press, Cambridge.
Crowe MJ (1967) A History of Vector Analysis. The Evolution of the Idea of a Vectorial System. Dover, New York.
Cushing JT (1981) Electromagnetic Mass, Relativity, and the Kaufmann Experiments. American Journal of Physics 49:1133–1149.
D'Agostino S (1964) Un insegnamento non dogmatico della Scienza. Scuola e Città, pp. 514–515.
D'Agostino S (1965) Il contributo di Newton allo sviluppo dell'ottica. Giornale di Fisica VI (3):188–196.
D'Agostino S (1968a) Il pensiero scientifico di Maxwell e lo sviluppo della teoria del campo elettromagnetico nella memoria "On Faraday's lines of force". Scientia 103:291–301.
D'Agostino S (1968b) Per la preparazione all'insegnamento della fisica nella Scuola Media. Le Scienze 1:33–36.
D'Agostino S (1968c) I vortici dell'etere nella teoria del campo elettromagnetico di Maxwell: la funzione del modello nella costruzione della teoria. Physis 10:188–202.

D'Agostino S (1971) Hertz e Helmholtz sulle onde elettromagnetiche. Scientia. Rivista di Scienza 106:622–636.
D'Agostino S (1973) L'elettrodinamica di Lorentz sino alla soglia della teoria speciale della relatività di Einstein. Physis 15:260–279.
D'Agostino S (1975a) Hertz's researches on electromagnetic waves. Historical studies in the physical sciences 6:261–323.
D'Agostino S (1975b) L'elettromagnetismo classico. Sansoni, Firenze.
D'Agostino S (1975c) L'elettromagnetismo classico (with an anthology of texts, translated by Maria Grazia Ianniello). Sansoni, Firenze.
D'Agostino S (1976a) La scoperta di una velocità quasi uguale alla velocità della luce nell'elettrodinamica di Wilhelm Weber (1804-1891). Physis XVIII (3–4):297–318.
D'Agostino S (1976b) Il problema del moto assoluto e della distinzione tra forze impresse e forze insite nel "Philosophiae naturalis principia mathematica" di Isaac Newton. Giornale di Astronomia, Rivista di informazione cultura e didattica della Società Astronomica Italiana 4:275–290.
D'Agostino S (1978) Esperimento e teoria nell'opera di Maxwell: Le misure per le unità assolute elettromagnetiche e la velocità della luce. Scientia 113(5–8):453–468.
D'Agostino S (1979a) Il rinnovamento dei musei scientifici. A.N.M.S., Notiziario dell'Associazione Nazionale Musei Scientifici.
D'Agostino S (1979b) Modelli ed oggetti della teoria fisica. La nuova critica. Rivista di scienze dell'uomo e di filosofia delle scienze XIII(49):5–19.
D'Agostino S (1979c) Storia della strumentazione scientifica e sviluppo della scienza: i problemi posti dall'uso conoscitivo della techne. Medicina nei secoli 1:13–29.
D'Agostino S (1979–1980) Attività scientifica e didattica del gruppo di storia della fisica dell'Istituto di fisica, 1974–1979. Bollettino della Facoltà di Scienze matematiche, fisiche e naturali 1979–1980 4:5–25.
D'Agostino S (1980) Quale museo per gli strumenti scientifici? risultati e proposte di una ricerca sulla storia degli strumenti scientifici. In: D'Agostino S, Ianniello MG (eds) Storia della scienza e della tecnica: problemi di ricerca storica e didattica nella scuola e nei musei scientifici. Atti del Seminario di studio "Storia della scienza e della tecnica e insegnamento scientifico" Università di Roma, Centro Europeo dell'Educazione, Villa Falconieri, Frascati, 3–7 novembre 1980, Centro Europeo dell'Educazione, Frascati. pp. 183–191.
D'Agostino S (1981) Weber and Maxwell on the discovery of the velocity of light in 19th century electrodynamics. In: Grmek MD, Cohen RS, Cimino G (eds) On Scientific Discovery, The Erice Lectures 1977. Reidel, Dordrecht, pp. 281–293.
D'Agostino S (1982) Empirismo e strumenti scientifici. Physis XXIV(2):281–283.
D'Agostino S (1983a) Conceptions and concepts in physics teaching. In: Kennedy PL (ed). Using history of physics in innovatory physics education. Atti del Seminario Internazionale Centro Studi per la Didattica, Pavia, pp. 173–182.
D'Agostino S (1983b) L'importanza della storia delle scienze nel rinnovamento dell'insegnamento scientifico, in AA. VV. L'insegnamento della matematica e delle scienze sperimentali nella scuola secondaria superiore. Parte seconda, Atti del Convegno Internazionale presso la Fondazione Cini, Fondazione Cini, Venezia, p. 263.
D'Agostino S (1983c) On the problem of the presumed redundancy of absolute motion in Newton's principia. Physis XXV(1):167–169.
D'Agostino S (1983d) Symmetry in Ampère electrodynamics. In: Doncel MG, Hermann A, Michel L, Pais A (eds) Symmetries in physics (1600–1980). Proceedings, 1st International Meeting on the History of Scientific Ideas, Sant Feliu de Guixols, Catalonia, Spain, September 20–26. World Scientific Publishing, Singapore, pp. 93–103.
D'Agostino S (1983e) Strumenti scientifici e teorie fisiche: considerazioni storico-critiche sulla storia della strumentazione scientifica. In: Tarozzi G (ed) Gli strumenti nella storia e nella filosofia della scienza. Istituto Beni Artistici Culturali Ambientali Regione Emilia Romagna, Bologna, pp. 173–182.

D'Agostino S (1984) Alcune annotazioni sulla posizione del pensiero scientifico nel pensiero contemporaneo. Cultura e scuola 90:239–243.

D'Agostino S (1985a) La matematica e la fisica nelle idee di P. A. M. Dirac. Memorie di scienze fisiche e naturali. Rendiconti della Accademia nazionale delle scienze detta dei XL, serie V, IX(II):287–290.

D'Agostino S (1985b) The problem of the link between correspondence and complementarity in Niels Bohr's papers: 1925–1927. In: Proceedings of the International symposium on Niels Bohr. Università "La Sapienza", Rome. Rivista di Storia della Scienza 2(3):369–390.

D'Agostino S (1985c) Strumenti elettromeccanici e concezioni meccaniciste nella storia dell'elettrodinamica, Epistemologia 8:119–138.

D'Agostino S (1986a) La formazione dello storico della scienza: problemi e prospettive. Il Nuovo Saggiatore 3:8–10.

D'Agostino S (1986b) Maxwell's dimensional approach to the velocity of light. Centaurus. International Journal of the History of Science and its Cultural aspects 29:178–204.

D'Agostino S (1986c) Scienza e cultura nella Germania dell'800. Le basi filosofiche della fisica-matematica nell'opera di Helmholtz. Cultura e Scuola 99:265–282.

D'Agostino S (1986d) Strumenti scientifici e beni culturali: alcune considerazioni. Insegnare II:10–12.

D'Agostino S (1986–1987) Federico Enriques (1871–1946) and History of Science in Italy in the Pre-War Years. Organon 22–23:17–24.

D'Agostino S (1987a) Le relazioni di indeterminazione e la fondazione della meccanica quantistica nell'opera di Bohr ed Heisenberg, 1925–1927. Atti del VII Congresso nazionale di Storia della fisica - Padova, 2–7 ottobre 1986. Gruppo nazionale di coordinamento per la storia della fisica del C.N.R, Milano, pp. 63–85.

D'Agostino S (1987b) Il principio d'indeterminazione e la transizione dall'ontologia della fisica classica a quella della meccanica quantistica. La Forgia M, Petruccioli S (eds) Rappresentazione e oggetto. Dalla fisica alle altre scienze. Theoria, Roma, pp. 23–34.

D'Agostino S (1989a) Pourquoi Hertz et non pas Maxwell a-t-il découvert les ondes électriques? Centaurus. International Journal of the History of Science and its Cultural aspects 32:66–76.

D'Agostino S (1989b) La Classificazione degli strumenti scientifici. Nuncius 4(1):87–92.

D'Agostino S (1989c) The bild-conception of physical theory and the foundation of Boltzmann's mechanics. X SISFA Proceedings, Cagliari, 28 settembre - 2 ottobre 1989b. Gruppo nazionale di coordinamento per la storia della fisica del CNR, Milano, pp. 95–110.

D'Agostino S (1989d) (ed) Gli strumenti dell'elettromagnetismo. Università La Sapienza di Roma Press, Roma.

D'Agostino S (1990a) Teorie sistematiche e strumenti precisi: per una storia degli strumenti dell'elettromagnetismo. In: Dragoni G (ed) Instrumenta, il patrimonio storico-scientifico italiano: una realtà straordinaria. Soprintendenza per i Beni Culturali della regione Emilia Romagna, Bologna. pp. 247–254.

D'Agostino S (1990b) Boltzmann and Hertz on the Bild-Conception of Physical Theory. History of Science 28:380–398.

D'Agostino S (1990c) Matematica ed elettrodinamica nell'opera di Ampère: lo sviluppo delle teorie dell'elettromagnetismo. Cultura e Scuola 116:261–271.

D'Agostino S (1990d) Da Maxwell a Boltzmann: la costituzione della fisica teorica. Memorie di scienze fisiche e naturali, Rendiconti della Accademia nazionale delle scienze detta dei XL, serie V, XVI(2):343–357.

D'Agostino S (1991a) Einstein e Kant: la fisica teorica contro il "logico ma vuoto schema dei concetti. Petruccioli S (ed). Immagini Linguaggi Concetti. Theoria, Roma, pp. 195–217.

D'Agostino S (1991b) Gli strumenti dell'elettromagnetismo. Collezioni del Museo del Dipartimento di Fisica, "La Sapienza". Istituto Tecnico Nautico "Colonna", Roma. Università degli studi di Roma "La Sapienza", Roma.

D'Agostino S (1991c) Immagini della scienza. Per una cultura della riforma. Quaderni del Centro di Metodologia delle Scienze Sociali, Serie I – Saggi. L.U.I.S.S., Roma.

D'Agostino S (1991d) Ampère e l'elettrodinamica: leggi matematiche e ipotesi fisiche. Giornale di fisica 32 (2–3):127–135.
D'Agostino S (1991e) Erwin Schrödinger: la teoria fisica come "bild" e l'integrazione di scienza e cultura. Nuova Civiltà delle Macchine 9(34):151–157.
D'Agostino S (1991f) Il secondo Schrödinger. La riformulazione del concetto di teoria fisica. In: Cattaneo G, Rossi A (eds) I Fondamenti della Meccanica Quantistica. Analisi storica e problemi aperti, EditEI, Commenda di Rende, pp. 91–103.
D'Agostino S (1992a) Continuity and completeness in physical theory: Schrödinger's return to the wave interpretation of quantum mechanics in the 1950's. In: Bitbol M, Darrigol O (eds) Erwin Schrödinger, Philosophy and the Birth of quantum mechanics. Editions Frontières, Gif-sur-Yvette, pp. 339–362.
D'Agostino S (1992b) Schrödinger on continuity and completeness in physical theory. Physis. Nuova Serie 29(2):539–561.
D'Agostino S (1993) Hertz's researches and their Place in nineteenth Century theoretical Physics. Centaurus. International Journal of the History of Science and its Cultural aspects 36: 46–82.
D'Agostino S (1995) Einstein's life-long doubts on the physical foundations of the general relativity and unified field theories. In: Garola C, Rossi A (eds) The Foundations of Quantum Mechanics - Historical Analysis and Open Questions, Kluwer Academic Publishers, Dordrecht, pp.167–178.
D'Agostino S (1996) Absolute systems of units and dimensions of physical quantities: a link between Weber's electrodynamics and Maxwell's electromagnetic theory of light. Physis 33(1–3):5–51.
D'Agostino S (1998a) J. J. Thomson al tempo della scoperta dell'elettrone. Il metodo di Thomson spiegato in una lettera inedita di Thomson a Righi. Memorie di scienze fisiche e naturali, Rendiconti della Accademia nazionale delle scienze detta dei XL, serie V, XXII(2):317–336.
D'Agostino S (1998b) La relatività generale nel dibattito degli anni venti fra neo-kantiani ed empiristi logici: annotazioni su recenti studi einsteiniani, Physis. Nuova Serie, 34(3):643–658.
D'Agostino S (1998c) Hertz's Views on the Method of Physics: Experiment and Theory Reconciled? In: Baird D, Hughes R, Nordmann A (eds). Heinrich Hertz: Classical Physicist, Modern Philosopher. Springer, Dordrecht, pp. 89–102.
D'Agostino S (1999) Alcune considerazioni sulle teorie dello spazio e sulle geometrie da Descartes ad Einstein. Conti L, Mamone Capria M (eds). La scienza e i vortici del dubbio. Edizioni scientifiche Italiane, Napoli, pp. 193–211.
D'Agostino S (2000a) A History of the Ideas of Theoretical Physics. Essays on the Nineteenth and Twentieth Century Physics, Boston Studies in the Philosophy of Science. Kluwer, Dordrecht.
D'Agostino S (2000b) Ch. 10. Einstein's Correspondence Criterium and the construction of General Relativity. Ch. 11. Einstein's life-long doubts on the physical foundations of the General Relativity and Unified Field Theories. A History of the Ideas of Theoretical Physics. Essays on the Nineteenth and Twentieth Century Physics, Boston Studies in the Philosophy of Science. Kluwer, Dordrecht, pp. 219–252.
D'Agostino S (2000c) Correspondence and Complementarity in Niels Bohr's papers 1925–1927. A History of the Ideas of Theoretical Physics. Essays on the Nineteenth and Twentieth Century Physics, Boston Studies in the Philosophy of Science. Kluwer, Dordrecht, pp. 253–271.
D'Agostino S (2000d) From the 1926 wave mechanics to a second-quantisation theory: Schrödinger's new interpretation of wave mechanics and microphysics in the 1950's. A History of the Ideas of Theoretical Physics. Essays on the Nineteenth and Twentieth Century Physics, Boston Studies in the Philosophy of Science. Kluwer, Dordrecht, pp. 273–302.
D'Agostino S (2000e) A history of the ideas of theoretical physics. Essays on the nineteenth and twentieth century physics, Boston studies in the philosophy of science n. 213. Kluwer Academic Publishers, Dordrecht.
D'Agostino S (2000f) A History of the Ideas of Theoretical Physics. Essays on the Nineteenth and Twentieth Century Physics. Kluwer, Boston.
D'Agostino S (2000g) On the difficulties of the transition from Maxwell's and Hertz's pure-field theories to Lorentz's electron. Physics in Perspective 2:398–410.

D'Agostino S (2002) From rational numbers to Dirac's bra and ket: symbolic representation of physical laws. Physics in Perspective 4:216–229.

D'Agostino S (2003–2004) La riflessione critica in Europa e la fisica teorica. Arché, Rivista di Filosofia 5:217–230.

D'Agostino S (2004a) La formazione umanistica è indispensabile per un efficace insegnamento della scienza? Nuova Civiltà delle Macchine XXII(1):144–146.

D'Agostino S (2004b) The bild conception of physical theory: Helmholtz, Hertz and Schrödinger. Physics in perspective 6:372–389.

D'Agostino S (2004c) Theory-Laden instruments and quantum mechanics. In Garola C, Rossi A, Sozzo S (eds). The Foundations of Quantum Mechanics/Historical Analysis and Open Questions-Cesena 2004. World Scientific Press, Singapore, pp. 130–141.

D'Agostino S (2004d) Un problema delle teorie classiche e quantistiche: esiste una logica della strumentazione? In: Tassani I (ed). Quanti Copenaghen? Bohr, Heisenberg e le interpretazioni della meccanica quantistica: Atti del Convegno, Cesena, 17–19 ottobre 2002. Il ponte vecchio, Cesena, pp. 135–148.

D'Agostino S (2005a) Gli strumenti scientifici e la scienza. Barbieri Selvaggi editore, Manduria.

D'Agostino S (2005b) Il difficile recupero dell'Anschaulichkeit di Goethe nell'opera di Helmholtz. Nuncius 20(2):401–414.

D'Agostino S (2005c) Gli strumenti scientifici e la scienza. Barbieri, Manduria.

D'Agostino S (2009a) Fra storia e memoria. Ricerca e insegnamento nella fisica italiana. Barbieri Selvaggi Editori, Manduria.

D'Agostino S (2009b) Il "gatto" di Schroedinger in una lettura del testo originale del celebre saggio. Physis. Nuova Serie 46(1–2):235–252.

D'Agostino S (2011) Hermann von Helmholtz e il problema delle due culture. Nuova civiltà delle macchine 27(2): 207–218.

D'Agostino S (2013) La scienza dell'elettricità da Volta ad Einstein. Una prospettiva storico-didattica. Booksprint Editore, Romagnano al Monte (SA).

D'Agostino S (2013–2014) Fenomenologia e matematica nell'opera di Helmholtz e di Kirchhoff. Physis. Nuova Serie 49(1/2):243–258.

D'Agostino S (2014) Maxwell's Dynamical Philosophy: An Early Solution to the Problem of a Discrepancy between Particles and Waves. Advances in Historical Studies 3:123–129

D'Agostino S (2017) The centennial recurrence of Einstein's general relativity. Lambert Academic Publishing, Saarbrücken.

D'Agostino S (2018a) An asymmetric evaluation of Einstein's epistemologic contributions. Lambert Academic Publishing, Saarbrücken.

D'Agostino S (2018b) Letture di storia della fisica. Da Galileo a Newton. Edizioni Accademiche Italiane, Beau Bassin.

D'Agostino S (2018c) The Centennial Recurrence of Einstein's General Relativity. Lambert, Beau Bassin.

D'Agostino S (2018d) An Asymmetric Evaluation of Einstein's Epistemological Contribution. Lambert, Beau Bassin.

D'Agostino S (2019a) A Piedi dalle Alpi verso il mare. In: Complementi ai quaderni, Unpublished.

D'Agostino S (2019b) La cultura scientifica. In: Quaderni più o meno scientifici per amiche, amici e conoscenti. Unpublished.

D'Agostino S, Cametti C (1970) Esperienze didattiche di diffrazione con microonde. Giornale di Fisica 11:36–48.

D'Agostino S, Ianniello MG (1977) Il Deutsches Museum tra storia e scienza. La Fisica nella Scuola 4:179–181.

D'Agostino S, Ianniello MG (1979) Catalogo strumenti. Catalogo della collezione di strumenti elettrici e magnetici. Museo dell'Istituto di fisica dell'Università di Roma, Roma.

D'Agostino S, Ianniello MG (1980) Elettrometri e galvanometri dell'Ottocento. Evoluzione strumentale e contesti teorici. Annali dell'Istituto di Storia della Scienza di Firenze 5(2):69–81.

D'Agostino S, Ianniello MG, Violini G (1980) L'esperimento storico nella didattica della fisica, iniziative e proposte. Fisica nella scuola XIII(1):19–21.

D'Agostino S, Rossi A (1995) Enrico Fermi e l'origine della fisica teorica in Italia. In: Rossi, A (ed) Atti del XIV e del XV Congresso nazionale di storia della fisica, Udine, 27 settembre - 2 ottobre 1993–Lecce, 26 settembre - 1 ottobre 1994. Conte, Lecce, pp. 467–474.

D'Agostino S, Rossi A (1996) Fermi, Enrico In: Dizionario biografico degli italiani, Istituto della Enciclopedia Italiana, Roma. Vol. 46, pp. 329–343.

D'Agostino S, Rossi A (1998) (eds) Galileo Ferraris e il suo tempo. Physis XXXV(2).

D'Agostino S, Rossi A (2001) (eds) Enrico Fermi e l'Enciclopedia Italiana. Istituto dell'Enciclopedia Italiana, Roma.

D'Agostino S, Rossi A (2004) (eds) Giovanni Giorgi nella realtà del suo tempo. Physis XLI(2).

D'Agostino S, Fabbri B, Ianniello MG, La Forgia M, Tarsitani C, Veronese L, Violini G (1979) La termodinamica nella secondaria superiore. Riforma della Scuola 5:51–56.

Dale R (1992) Timekeeping. The British Library, London.

D'Alembert JLR ([1743] 1758) Traité de dynamique. Nouvelle édition, revue et fort augmentée par l'Auteur Chez David Libraire, Paris.

D'Alembert JLR (1751–1780) Encyclopédie ou dictionnaire raisonné des sciences, des arts et des métiers, Diderot & D'Alembert. Briasson–David–Le Breton–Durand, Paris.

D'Alembert JLR (1767) Mélanges de littérature, d'Histoire et de philosophie. Tome V. Chatelain & Fils. Amsterdam, pp. 3–274.

Darrigol O (1995) Henri Poincaré's Criticism of Fin De Siècle Electrodynamics. Studies in the History and Philosophy of Modern Physics 26(1):1–44.

Darrigol O (2000) Electrodynamics from Ampère to Einstein. Oxford University Press, Oxford.

Darrigol O (2022) Relativity Principles and Theories from Galileo to Einstein. Oxford University Press, Oxford.

Dauben JW, Scriba CJ (2002) (eds) Writing the History of Mathematics: Its Historical Development. Birkhäuser Verlag, Basel-Boston-Berlin.

Davidon WC (1975) Consequences of the Inertial Equivalence of Energy. Foundations of Physics 5:525.

Davies EB (2002) The Role of Astronomy in the History of Science. arXiv:physics/0207043.

De Leeuw K (2003) The Dutch invention of the rotor machine, 1915-1923. Cryptologia 27(1):73–94.

De Luca (2019) An unpublished approach to electrodynamics by Richard Feynman. European Journal of Physics 40(6):1–15.

De Nemore I (1533) Liber Iordani Nemorarii viri clarissimi, de ponderibus propositiones XIII & earundem demonstrationes, multarumque rerum rationes sane pulcherrimas complectens. [edited by Apianus] Ioh Petreium, Norimbergae.

De Nemore I (1565) Iordani Opusculum de Ponderositate, Nicolai Tartaleae Studio Correctum Novisqve Figuris avctum [Tartaglia N (ed)]. Cum Privilegio Traiano Curtio, Venetiis, Apud Curtium Troianum. MDLXV.

De Nemore I (fl. 13th) Ms. Elementa Jordani super demonstratione de ponderibus. Oxford, Bodleian Library, Ms. Auct. F.5.28. Folia 125v–133r.

Dear P (1995) Discipline and Experience: The Mathematical Way in the Scientific Revolution. Chicago University Press, Chicago.

Denbigh, KG, & Denbigh, JS (1985) Entropy in Relation to Incomplete Knowledge. Cambridge University Press, Massachusetts.

Der Chef des Oberkommandos der Wehrmacht, Berlin (12.1.1937) Geheim! Gebrauchsanleitung für die Chiffriermaschine Enigma, Vom 12.1.1937, Prüf-Nr. 965 [Secret! Instructions of Use for the Encryption Machine Enigma from January 12, 1937, Test No. 965]. Books on Demand, Norderstedt, ©2020 by Thomas Heise.

Descartes R ([1632–1633] 1664) Le monde de M Descartes ou le Traité de la lumière et des autres principaux objets des Sens. Girard, Paris.

Descartes R (1637) Geometria à Renato Des Cartes. Apud Ludovicum&Danielem Elzevirios, Amsterdami. [In English: (1979) English translation by Michael Mahoney. Dover, New York; In Russian: (1938) Nauchtekhizdat, Moscow-Leningrad].
Descartes R ([1644] 1983) Principles of Philosophy. Translation by Miller VR and Miller RP. Reidel, Dordrecht.
Descartes R (1897–1913) Oeuvres de Descartes. Par Charles Adams et Paul Tannery, [Discours de la méthode et Essais, Specimina philosophiae, Vol. VI; Physico– mathematica, Vol. X, Le Monde ou Traité de la lumière, Vol. XI], 12 Vols. Vrin, Paris.
Descartes R (1964–1974) Oeuvres. Nouvelle présentation par Rochet E, et Costabel P, 11 Vols. Vrin, Paris.
Descartes R ([1979] 1995) Le Monde ou Traité de la lumière. Translation and introduction by Michale Sean Mahoney. Abaris Books, NY.
Deser S (1970) Self-Interaction and Gauge Invariance. General Relativity and Gravitation 1:9.
Dewey J (1916) Democracy and Education: An Introduction to the Philosophy of Education, The Macmillan Company, New York.
Dewey J (1938) Logic: The Theory of Inquiry. Henry Holt and Company, New York.
Di Mauro M (2021) Some insights into Feynman approach to electromagnetism. European Journal of Physics 42(2) 025206.
Di Mauro M, Naddeo A (2021) A Road Map for Feynman Adventures in the Land of Gravitation. arXiv:2102.11220v2.
Diadochus P (2013) Proclus's Commentary on Euclid, Book I. In Russian. Aristej. Vestnik klassicheskoj filologii i antichnoj istorii: Appendix III:82–86.
Dick PhK (1962) The Man in the High Castle. Putnam's Sons, New York [numerous editions following; in this remarkable example of counterfactual history, the plot takes place in America, twenty years after it has lost Second world war and is occupied by Nazi Germany and imperial Japan].
Diener G (2013) Obtaining Maxwell equations heuristically. American Journal of Physics 81(2):120–123.
Dijksterhuis E J ([1950] 1961) The mechanization of the world picture. Clarendon Press, Oxford.
Disalle R (2020) Space and time: Inertial Frames. Stanford Encyclopedia of Philosophy, Zalta EN (ed). Retrieved 23 February 2022 from https://plato.stanford.edu/entries/spacetime-iframes/.
Dorato M (2002) On becoming, cosmic time and rotating universes. Royal Institute of Philosophy Supplement 50:253–276.
Drago A (1987) An effective definition of incommensurability. VIII LMPS, Moscow, 4, pt.1, 159–162 [see also: Cellucci C (1988) (ed) Temi e prospettive della logica e della filosofia della scienza contemporanea. II, CLUEB, Bologna, pp. 117–120].
Drago A (1988) A Characterization of Newtonian Paradigm. In: Scheurer PB, Debrock G (eds). Newton's Scientific and Philosophical Legacy. Kluwer Academy Press, Dordrecht, pp. 239–252.
Drago A (1991) Le due opzioni. La meridiana. Molfetta.
Drago A (2003) Volta and the strange history of electromagnetism. In: Giannetto EA (ed) Volta and the history of Electricity. Hoepli, Milano, pp. 97–111.
Drago A (2011) The birth of the non-Euclidean geometries As the more significant crisis in the foundations of modern Mathematics, Logic and Philosophy of Science 9(1):103–110.
Drago A (2012) Pluralism in Logic. The Square of opposition, Leibniz's principle and Markov's principle. In: Béziau J-Y, Jacquette D (eds) Around and Beyond the Square of Opposition. Birkhaueser, Basel, pp. 175–189.
Drago A (2013) The emergence of two options from Einstein's first paper on Quanta. In: Pisano R, Capecchi D, Lukesova A (eds). Physics, Astronomy and Engineering. Critical Problems in the History of Science and Society. Scientia Socialis, Siauliai, pp. 227–234.
Drago A (2017) Dalla storia della Fisica alla scoperta dei fondamenti della Scienza. Aracne, Roma.
Drago A, Pisano R (2000) Interpretazione e ricostruzione delle Réflexions di Sadi Carnot mediante la logica non classica, Giornale di Fisica 41:195–215 [English translation: *Id.,* (2004) Atti Fondazione Ronchi 59:615–644]

Drake S (1964) Galileo and the Law of Inertia. American Journal of Physics 32:601–608.

Drake S (1972) Galileo Studies. University of Michigan Press, Ann Arbor.

Drake S (1992) Galileo Galilei pioniere della scienza. Muzzio, Padova [*Id.*, Galileo: Pioneer Scientist. University of Toronto Press, Toronto, 1990].

Dugas R (1955) A History of Mechanics. Tr J.R. Maddox. Dover, New York.

Dummet M (2011) Metaphysical Necessity. In: DeVidi D, Hallett M, Clarke P (2011) (eds) Logic, Mathematics, Philosophy, Vintage Enthusiasms: Essays in Honour of John L. Bell, Springer, Dordrecht, pp. 19–34.

Earman J (1995) Bangs, Crunches, Whimpers, and Shrieks. Singularities and Acausalities in Relativistic Spacetimes. Oxford University Press, New York.

Earman J, Norton J (1999) Exorcist XIV: the wrath of Maxwell's Demon. Part II. From Szilard to Landauer and beyond. Studies in History and Philosophy of Modern Physics 30(1):1–40.

Earman J, Smeenk C, Wüthrich C (2009) Do the laws of physics forbid the operation of time machines? Synthese 169:91–124.

Eddington A (1933) The Expanding Universe. Macmillan, New York.

Eddington AS ([1928] 1948) The Nature of the Physical World. Cambridge University Press, Cambridge.

Eddington AS (1939) The Philosophy of Physical Science. Cambridge University Press, Cambridge.

Edgerton D (2013) Science in the United Kingdom: a Study in the Nationalization of Science. In: Krige J, Pestre D (eds). Science in the Twentieth Century. Routledge, London; New York, pp. 759–776.

Ehrenfest P (1907) Die Translation deformierbarer Elektronen und der Flächensatz. Annalen der Physik 23:204–205.

Einstein A (1905a) Ist die Trägheit eines Körpers von seinem Energieinhalt abhängig?. Annalen der Physik 18:639–641.

Einstein A (1905b) Ueber einen die Erzeugung der Verwandlung des Lichtes betreffenden heuristisch Gesichtpunkt, Annalen der Physik 17:132–148 [reprinted: Stachel J (1989) (ed) Collected Papers of Albert Einstein. Princeton University Press, Princeton, II, pp. 149–16].

Einstein A (1905c) Zur Elektrodynamik bewegter Koerper. Annalen der Physik 17:891–921.

Einstein A (1906) Über eine Methode zur Bestimmung des Verhältnisses der transversalen und longitudinalen Masse des Elektrons. Annalen der Physik 21:583–586.

Einstein A (1907a) Bemerkungen zu der Notiz von Hrn. Paul Ehrenfest: Die Translation deformierbarer Elektronen und der Flächensatz. Annalen der Physik 23:206–208.

Einstein A (1907b) Über das Relativitätsprinzip und die aus demselben gezogenen Folgerungen. Jahrbuch der Radioaktivität und Elektronik 4:411–462.

Einstein A (1914) Die formale Grundlagen der allgemeinen Relativitätstheorie. Königlich Preussische Akademie der Wissenschaften Berlin. Sitzungsberichte:1030–1085. Reprinted as Document 9 [Einstein 1996, pp. 72–130].

Einstein A (1916a) Die grundlage der allgemeinen relativitätstheorie. Annalen der Physik 49:769–822. Document 30 [Einstein 1996, pp. 283–339].

Einstein A (1916b) Über die spezielle und allgemeine Relativitätstheorie (gemeinverständlich). Vieweg & Sohn a Braunschweig.

Einstein A (1947) Paul Langevin, La Pensée 12:13 [see also: Einstein A (1993) Œuvres choisies. Le Seuil-CNRS, Paris, Vol. 4, p. 183].

Einstein A (1996) The Collected Papers of Albert Einstein. Volume 6: The Berlin Years: Writings, 1914–1917. In: Klein MJ, Kox AJ, Schulman R (eds). Princeton University Press, Princeton.

Einstein A, Podolsky B, Rosen N (1935) Can quantum-mechanical description of physical reality be considered complete? Physical Review 47:777–780.

Eisenthal J (2018) Mechanics without mechanism. Studies in History and Philosophy of Modern Physics 62:45–55.

Eisenthal J (2021) Hertz's Mechanics and a unitary notion of force. Studies in History and Philosophy of Science 90:226–234.

Ellis GFR (2001) Contributions of K. Gödel to relativity and cosmology. In: Hájek (ed). Gödel '96. Logical Foundations of Mathematics, Computer Science, and Physics—Kurt Gödel's Legacy. AK Peters, Natick, pp. 34–49.

Emery N, Markosian N, Sullivan M (2020) Time. Stanford Encyclopedia of Philosophy. Zalta EN (ed). Retrieved 8 February 2022 from https://plato.stanford.edu/entries/time/.

Esclangon E (1936) Sur les formules de Lorentz et la théorie de la relativité. Compte Rendu de Académie des sciences 202:1492.

Euler L ([1730–1731] 1738) De communicatione motus in collision corporum. Commentarii academiae scientiarum Petropolitanae 5:159–168.

Euler L (1736) Mechanica sive motus scientia analytice exposita. Auctore Leonhardo Eulero academiae imper. scientiarum membro et matheseos sublimioris professore. Tomus I. Instar supplementi ad commentar. acad. scient. imper. Petropoli. Ex typographia academiae scientarum.

Euler L (1749) Recherches sur les mouvements des corps célestes en général (1747). Mémoire de l'Académie Royale des Sciences et des Lettres de Berlin, Vol. I, pp. 93–143.

Euler L (1750] 1752) Découverte d'un nouveau principe de Mécanique. Mémoires de l'académie des sciences de Berlin. Vol. VI, 1752, pp. 185–217.

Euler L (1751) De machinis in genere Novi Commentarii Academiae Scientarum Imperialis Petro–Politanae. Petropoli Typis Academiae Scientiarum.

Euler L (1755–1949) Institutiones Calculi Differentialis cum eius vsu in Analysi Finitopum. [In Russian: (1949) Gos. izd-vo tekhniko-tekhnicheskoj literatury, Moscow-Leningrad].

Euler L (1757) Principes généraux du mouvement des fluides. Mémoires de l'Academie des Sciences de Berlin 11:274–315.

Euler L ([1773] 1774) De pressione ponderis in planum in cui incumbit. Novi Commentarii academiae scientiarum Petropolitanae, vol XVIII, pp. 289–329.

Euler L (1775) Nova methodus motum corporum rigidorum determinandi. Novi Commentarii academiae scientiarum.

Faraday M (1839–1855) Experimental Researches in Electricity. 3 Vols. Taylor, London.

Faraday M (1855) Experimental Researches in Electricity. Reprinted from the Philosophical Transactions of 1846–1852. Vol. III, Bernard Quaritch, London.

Feferman S (1984) Kurt Gödel: Conviction and caution. Philosophia Naturalis 21:546–562.

Feferman S (1986) Gödel's life and work. [Gödel 1986, pp. 1–36].

Fermi E (1932) Quantum theory of radiation. Reviews of Modern Physics 4:87–132.

Fernández Terán R (2014) El profesorado del 'Instituto Nacional de Física y Química' ante la Guerra Civil, el proceso de depuración y el drama del exilio. (PhD thesis, Universidad Complutense de Madrid).

Feyerabend P (2001) La conquista de la abundancia. Paidós, Barcelona.

Feynman R, Leighton R, Sands M (2005) The Feynman Lectures on Physics. Addison Wesley, Boston, Massachusetts.

Feynman RP, Morinigo FB, Wagner WG (1995) Feynman Lectures on Gravitation. Addison-Wesley, Massachusetts.

Finkelstein AM, Kreinovich V (1989) Can Science Attains its Goals? Informatika Technical Reports, St. Petersburg, Russia (in Russian)

Fitas A (2006) A Ciência em Portugal ao longo do séc. XX (cenas exemplares de um percurso incompleto). Prelo 3:1–22.

Fitas A, Videira A (2004) Cartas entre Guido Beck e cientistas portugueses. Instituto Piaget, Lisboa.

Fitzpatrick R (2008) Maxwell's Equations and the Principles of Electromagnetism. Infinity Science Press LLC, Massachusetts.

Fox D, Schmidt M, Koshelev M, Kreinovich V, Longpre L, Kuhn J (1998) We must choose the simplest physical theory: Levin-Li-Vitanyi theorem and its potential physical applications. In: Erickson GJ, Rychert JT, Smith CR (eds). Maximum Entropy and Bayesian Methods, Kluwer, Dordrecht, pp. 238–251.

Fraassen van BC (1989) Laws and Symmetry. Clarendon Press, Oxford.

Frank PH, Rothe H (1911) Uber die Transformation der Raumzeitkoordinaten von ruhenden auf bewegte Systeme. Annalen der Physik 34:825; *Id.*, (1912) Zur Herleitung der Lorentz transformation, Physik Z 13:750.
Frege G (1891) Funktion und begriff. Reprinted [Frege 1967, pp. 125–142].
Frege G (1892) Über sinn und bedeutung. Zeitschrift für Philosophie und philosophische Kritik (n.s.) 100:25–50. Reprinted [Frege 1967, pp. 143–162].
Frege G (1967) Kleine Schriften. Angellini (ed.). Georg Olms, Hildesheim.
Frigg R (2008) A field Guide to Recent Work on the Foundation of Statistical Mechanics. Ashgate, London.
Frova A (2006) Anche Galileo può errare. "In Atti di Pianeta Galileo 2006". Regione Toscana, Firenze, pp. 35–47.
Gabbey A (1980) Force and Inertia in the Seventeenth Century: Descartes and Newton. In: Gaukroger S (ed). Descartes: Philosophy, Mathematics and Physics. Sussex, Harvester, Sussex, pp. 230–320.
Gabbey A (2004) What was 'Mechanical' about 'The Mechanical Philosophy?' In Palmerino CR, Thijssen JMMH (eds). The Reception of the Galilean Science of Motion in Seventeenth–Century Europe, Kluwer, Dordrecht, pp. 11–23.
Gabbey AW, Herivel JW (1966) Un manuscrit inédit de Sadi Carnot. Revue d'histoire des sciences et de leurs applications 19:151–166.
Galilei G ([1598–1599] 1623) Le Mecaniche. In: Galilei 1890–1909, II, pp. 147–191.
Galilei G (1613) Istoria e dimostrazioni intorno alle macchie solari e loro accidenti. In: Galilei 1890–1909, V, pp. 71–249.
Galilei G (1632) Dialogo sopra i due massimi sistemi del mondo tolemaico e copernicano. In: Galilei 1890–1909, VII, pp. 21–520.
Galilei G (1638) Discorsi e dimostrazioni matematiche intorno a due nuove scienze. In: Galilei 1890–1909, VIII, pp. 49–448.
Galilei G (1890–1909) Le opere di Galileo Galilei. Edizione Nazionale sotto gli auspici di Sua Maestà il Re d'Italia. Favaro. A (ed). Tipografia di G. Barbèra, Firenze.
Galilei G (1914) Dialogues Concerning Two New Sciences. Macmillan, New York.
Galilei G (1954) Dialogues Concerning Two New Sciences, translated by Henry Crew & Alfonso de Salvio, with an introduction by Antonio Favaro, Dover Publications, Inc., New York [Originally published in 1904 by the MacMillan company].
Galilei G (1967) Dialogue Concerning the Two Chief World Systems: Ptolemaic and Copernican, translation and with notes. 2nd ed. University of California Press, Berkeley–Los Angeles.
Galilei G (fl. 16th) Ms. A – Trattato di fortificazioni e modo d'espugnare la città con disegni e piante di fortificazioni. Biblioteca Ambrosiana. D: 328 Inf.– 38640.
Galilei G (fl. 16th) Ms. B – Trattato delle fortificazioni, con disegni diversi. Biblioteca Ambrosiana. D: 296 Inf.– 38499
Galilei G (fl. 16th) Ms. m – Trattato di fortificazioni, con disegni e figure (acephalous). Biblioteca Ambrosiana. N: 281 Sup.– 2– 80277.
Galilei G (fl. 17th) Ms. 72, 33rv–196v. Biblioteca Nazionale Centrale, Florence Istituto e Museo di Storia della Scienza, Florence–Max Planck Institute for the History of Science, Berlin.
Gallino L (1992) L'incerta alleanza. Modelli di relazioni tra scienze umane e della natura. Einaudi paper backs, Torino.
Galluzzi P (1979) Momento: Studi galileiani. Ateneo & Bizzarri, Roma.
Galluzzi P, Torrini M (1975–1984) (eds) The works of Galileo Galilei's disciples. Correspondence (in Italian). Giunti–Barbera, Firenze, 2 Vols.
Gamow G (1946) Rotating Universe? Nature 158:549.
Gannett JW (2007) Nothing but relativity redux. European Journal of Physics 28:1145.
Garliński J (1979) The Enigma War. Charles Scribner's Sons, New York.
Gaspar J (2008) A Investigação no Laboratório de Física da Universidade de Lisboa (1929–1947) A Investigação no Laboratório de Física da Universidade de Lisboa (1929–1947). Master thesis in History and Philosophy of Science. University of Lisbon.

Gaspar J, Simões A (2011) Physics on the Periphery: A Research School at the University of Lisbon under Salazar's Dictatorship. Historical Studies in the Natural Sciences 41(3):303–343.
Gaukroger S (1978) Explanatory Structures. Harvester, Hassocks.
Gaukroger S (1995) Descartes: An Intellectual Biography. Oxford University Press, Oxford.
Gaukroger S (2000) The Foundational Role of Hydrostatics and Statics. In: Gaukroger S, Schuster JA, Sutton J (eds). Descartes' Natural Philosophy, Routledge, London, pp. 60–80.
Gaukroger S (2004) Descartes an intellectual biography. Oxford University Press, Oxford.
Gaukroger S (2006a) The Emergence of a Scientific Culture: Science and the Shaping of Modernity 1210—1685. Clarendon Press, Oxford.
Gaukroger S (2006b) Review of Robert Hooke Tercentennial Studies. Aestimatio 3:52–5.
Gaukroger S (2010) The Collapse of Mechanism and the Rise of Sensibility 1680-1760. Clarendon Press, Oxford.
Gaukroger S, Schuster JA (2002) The Hydrostatic Paradox and the Origins of Cartesian Dynamics. Studies in the History and Philosophy of Science 33:535–572.
Geison G (1981) Scientific Change, Emerging Specialties and Research Schools. History of Science 19:20–40.
Geymonat L (1960) Filosofia e Filosofia della scienza. Feltrinelli, Milano.
Geymonat L ed (1970–1972) Storia del pensiero filosofico e scientifico. 6 Vols. Garzanti, Milano.
Giannetto E (2019) Something is rotten in the state of Denmarke: Søren Kierkegaard, Niels Bohr e la nascita della fisica quantistica, Altre Modernità 22:179–197 via: https://riviste.unimi.it/index.php/AMonline/article/view/12152
Giannetto E, Giannini G, Capecchi D, Pisano R (2009) (eds) Da Archimede a Majorana: La fisica nel suo divenire. Proceedings of XXVI SISFA Congress (Roma, 15–17 June, 2006). Guaraldi Editore, Rimini.
Giannetto E, Ricciardo S, Antonello E, Mazzoni M (2014) (eds) Cielo e Terra: Fisica e Astronomia, un antico legame. Proceedings of the XXIX SISFA Congress, (Firenze, 14–18 September, 2009). Rimini, Guaraldi.
Gibbs JW (1892) Thermodynamischen Studien. Wilhelm Engelmann, Leipzig.
Gibbs JW (1899) Équilibre des Système Chimiques. Carré et Naud, Paris.
Gibbs JW (1902) Elementary Principles in Statistical Mechanics Developed with Especial Reference to the Rational Foundation of Thermodynamics. Charles Scribner's Sons, New York.
Gibbs JW (1906) On the Equilibrium of Heterogeneous Substances. The Scientific Papers of J. Willard Gibbs. Longmans, London, pp. 55–353.
Gibert A (1950) O Centro de Estudos de Física do Instituto para a Alta Cultura Anexo à Faculdade de Ciências de Lisboa. Gazeta de Física II(4):86–89.
Giddens A (1976) New Rules of Sociological Method. Hutchinson, London.
Gil F (2003) Núcleo de Matemática, Física e Química: uma contribuição efémera para o movimento científico português. Boletim da Sociedade Portuguesa de Matemática 49:77–92.
Gillispie CC (1960) The Edge of Objectivity: An Essay in the History of Scientific Ideas. Princeton University Press, Princeton–NJ.
Gillispie CC, Pisano R (2014) Lazare and Sadi Carnot. A Scientific and Filial Relationship. 2nd edn. Springer, Dordrecht.
Giovanelli M (2020) Like Thermodynamics before Boltzmann: On the Emergence of Einstein's Distinction between Constructive and Principle Theories. Studies in History and Philosophy of Modern Physics 71:118–157.
Gleick J (2011) The Information: A History, a Theory, a Flood. Pantheon Books.
Gödel K (1930) The completeness of the axioms of the functional calculus of logic. Translation in: Gödel 1986, pp. 100–123.
Gödel K (1931) On formally undecidable propositions of the Principia Mathematica and related systems I. Translation in: Gödel 1986, pp. 144–195.
Gödel K (1938) The consistency of the axiom of choice and of the generalized continuum hypothesis. Reprinted in: Gödel 1990, pp. 26–27.
Gödel K (1944) Russell's mathematical logic. 3rd ed. Reprinted in: Gödel 1990, pp. 102–141.

Gödel K (1947) What is Cantor's continuum problem? American Mathematical Monthly 54:515–525; errata 55:151. Reprinted in: Gödel 1990, pp. 176–187.
Gödel K (1949a) An example of a new type of cosmological solutions of Einstein's field equations of gravitation. Reviews of Modern Physics 21:497–450. Reprinted in: Gödel 1990, pp. 189–198.
Gödel K (1949b) A remark about the relationship between relativity theory and idealistic philosophy. Reprinted in: Gödel 1990, pp. 199–207.
Gödel K (1949c) Lecture on Rotating Universes. Reprinted in: Gödel 1995, pp. 261–289.
Gödel K (1951) Some basic theorems on the foundations of mathematics and their implications. The Gibbs lecture in: Gödel 1995, pp. 290–323.
Gödel K (1952) Rotating universes in general relativity theory. Proceedings of the International Congress of Mathematicians: Cambridge, Massachusetts, U.S.A. August 30–September 6, 1950, I. American Mathematical Society, Providence, pp. 175–181. Reprinted in: Gödel 1990, pp. 208–216.
Gödel K (1964) Revised and expanded version of Gödel 1947. In: Gödel 1990, pp. 254–270.
Gödel K (1972) On an extension of finitary mathematics which has not yet been used. In: Gödel 1990, pp. 271–280.
Gödel K (1975) Letter to Grandjean of 19 August 1975 and Grandjean's questionnaire. In: Gödel 2003, pp. 443–449.
Gödel K (1986) Collected Works, Volume I: Publications 1929–1936. Feferman S, et al. (eds). Oxford University Press, New York.
Gödel K (1995) Collected Works, Volume III: Unpublished Essays and Lectures. Feferman S, et al. (eds). Oxford University Press, New York.
Gödel K (2003) Collected Works, Volume IV: Correspondence A-G. Feferman S, et al. (eds). Oxford University Press, New York.
Goldberg S (1967) Henri Poincaré and Einstein's Theory of Relativity. American Journal of Physics 35:934–944.
Goldberg S (1970) The Abraham Theory of the Electron: The Symbiosis of Experiment and Theory. Archive for History of Exact Sciences 7(1):7–25.
Goldberg S (1976) Max Planck's Philosophy of Nature and his Elaboration of the Special Theory of Relativity. Historical Studies in the Physical Sciences 7:125–160.
Gott III JR (1991) Closed timelike curves produced by pairs of moving cosmic strings: exact solutions. Physical Review Letters 66:1126–1129.
Gould H, Quaintance J (2012) Double fun with double factorials. Mathematics Magazine 85(3):177–192.
Gould SJ (1996) Full House. Harmony Books, New York.
Grandi G (1729) I fiori geometrici. F. Marescandoli, Lucca.
Grant E (1964) Motion in the Void and the Principle of Inertia in the Middle Ages. Isis 55/3:265–292.
Grant E (1978) Aristotelianism and the Longevity of the Medieval World View. History of Science 16:93–106.
Grasshoff G (1998) Hertz's Philosophy of Nature in Wittgenstein's Tractatus. In: Baird D, Hughes R, Nordmann A (eds) Heinrich Hertz: Classical Physicist, Modern Philosopher. Springer, Dordrecht, pp. 243–268.
Guerraggio A (2010) (ed) Il '68 e la scienza in Italia. Pristem–Storia, Note di Matematica, Storia, Cultura 27–28:1–167.
Gupta S N (1954) Gravitation and electromagnetism Physical Review 96:1683; *Id.*, (1957) Einstein and Other Theories of Gravitation. Reviews of Modern Physics 29:334.
Hahn E (1913) Grundlagen zu einer Theorie der Lorentztransformationen. Archiv der Mathematik und Physik III:21/1.
Hall AR (1966) Mechanics and the Royal Society. British Journal for the History of Science 3(9):24–38.
Hamer D (January 1997) Enigma: Actions involved in the "Double Stepping" of the middle rotor. Cryptologia 21(10):47–50.

Hanson NR (1958) Pattern of Discovery/An inquiry into the conceptual foundations of science. Cambridge University Press, Cambridge.
Harriot T (1631) Artis Analyticae Praxis ad Aequationes Algebraicas Resolvendas. Apud Robertum Barker, London. [In English: (2007) Thomas Harriot's Artis analyticae praxis. Springer, New York].
Hartshorne C, Weiss P, Burks AW (1931–1958) (eds) Collected Papers of Charles Sanders Peirce. 8 Vols. Harvard University Press, Cambridge–Massachusetts.
Hattab H (2005) From Mechanics to Mechanism: The Quaestiones Mechanicae and Descartes' Physics. In: Anstey P, Schuster JA (eds). The Science of Nature in the Seventeenth Century: Changing Patterns of Early Modern Natural Philosophy. Springer, Dordrecht, pp. 99–129.
Hauswedell T, Körner A, Tiedau U (eds) (2019) Re-Mapping Centre and Periphery. Asymmetrical Encounters in European and Global Contexts. University College London, London.
Haven E, Khrennikov A (2013) Quantum Social Science, Cambridge University Press, Cambridge, UK.
Hawking SV (1967) The occurrence of singularities in cosmology. III. Causality and singularities. Proceedings of the Royal Society of London. Series A 300:187–201.
Hawking SV, Ellis GFR (1973) The Large-Scale Structure of Space-Time. Cambridge University Press, Cambridge.
Hawking SV, Penrose R (1970) The singularities of gravitational collapse and cosmology. Proceedings of the Royal Society of London. Series A 314:529–548.
Heaviside O (1889) On the Electromagnetic Effects due to the Motion of Electrification through a Dielectric. The London, Edinburgh, and Dublin Philosophical Magazine and Journal of Science 27:324–339.
Heckmann O, Schücking E (1955) Bemerkungen zur Newtonschen Kosmologie. I. Zeitschrift für Astrophysik, 38:95–109.
Heckmann O, Schücking E (1956) Bemerkungen zur Newtonschen Kosmologie. II. Zeitschrift für Astrophysik, 40:81–92.
Hegel GWF (1982) Ciencia de la lógica. 2 Tomos. Ediciones Solar, Buenos Aires.
Hegerfeldt G (1994) Causality problems for Fermi's two atom system. Physical Review Letters 72:596–599.
Hegerfeldt G (1997) Problems about causality in Fermi's two-atom model and possible resolution. Retrieved 27 January 2022, from Cornell University Library Web site:
Heilbronner JC (1742) Historia matheseos universae a mundo condito ad saeculum post Christ. nat. XVI. Accedit recensio elementorum, compendiorum et operum mathematicorum atque historia arithmetices ad nostra tempora. Gleditsch, Leipzig.
Hempel CG (1965) Scientific Explanation. Essay in the Philosophy of Science. The Free Press, NY.
Hertz H (1890) Über die Grundgleichungen der Elektrodynamik für ruhende Korper. Göttinger Nachrichten, March, and Annaler der Physik 40 [Über die Grundgleichungen der Elektrodynamik für bewegte Korper. Annaler der Physik 41].
Hertz H (1893) Electric Waves. MacMillan, London.
Hertz H (1894a) Gesammelte Werke von Heinrich Hertz. Band III. Die Prinzipien der Mechanik. Johann Ambrosius Barth, Leipzig.
Hertz H (1894b) Gesammelte Werke von Heinrich Hertz: Band II Untersuchungen über die Ausbreitung der elektrischen Kraft. Johann Ambrosius Barth, Leipzig.
Hesse M (1965) Forces and Fields. Littlefield & Adams, London.
Hessen B ([1903] 2017) Le radici sociali ed economiche della meccanica di Newton. Saggio introduttivo di Gerardo Ienna e Giulia Rispoli. Castelvecchi, Roma.
Higgs PW (1964) Broken symmetries and the masses of Gauge Bosons. Physics Review Letter 13:508–509.
Higgs PW (1966) Spontaneous symmetry breakdown without Mass less Bosons. Physics Review Letter 145:1156–1163.
Hill DK (1986) Galileo's Work on 116v: A New Analysis. Isis 77:283–291.
Hintikka J (1998) On Gödel's philosophical assumptions. Synthese 114:13–23.

Hintikka J (2000) On Gödel. Wadsworth, Belmont.
Hirosige T (1969) Origins of Lorentz' Theory of Electrons and the Concept of the Electromagnetic Field. Historical Studies in the Physical Sciences 1:151–209.
Hirosige T (1976) The Ether Problem, the Mechanistic Worldview, and the Origins of the Theory of Relativity. Historical Studies in the Physical Sciences 7:3–82.
Hobson EW (1914–2012) John Napier and the Invention of Logarithms, 1614. Cambridge University Press.
Hodges A (1983) Alan Turing: The Enigma. Simon and Schuster Inc., New York.
Hoefer C (2000) Kant's hands and Earman's pions: Chirality arguments for substantival space. International Studies in the Philosophy of Science 14:237–256.
Hoefer C, Huggett N, Read J (2021) Absolute and relational space and motion: Classical theories. Stanford Encyclopedia of Philosophy. Zalta EN (ed). Retrieved 26 January 2022 from https://plato.stanford.edu/entries/spacetime-theories-classical/
Holton G (1973), Thematic origins of scientific thought, Kepler to Einstein. Harvard University Press, Harvard.
Hooper W (2004) Seventeenth-Century Theories of the Tides as a Gauge of Scientific Change. In: Palmerino CR, Thijssen JMMH (eds). The Reception of the Galilean Science of Motion in Seventeenth–Century Europe. Kluwer, Dordrecht. pp. 199–242.
Hoppen KT (1976) The Nature of the Early Royal Society. Pt 1. British Journal for the History of Science 9:1–24.
Hubble E (1934) The distribution of extra-galactic nebulae. Astrophysical Journal 79:8–76.
Huggett N, Hoefer C, Read J (2021) Absolute and relational space and motion: Post-Newtonian theories. Stanford Encyclopedia of Philosophy. Zalta EN (ed). Retrieved 26 January 2022 from https://plato.stanford.edu/entries/spacetime-theories/
Hull A (1999) War of words: the public science of the British scientific community and the origins of the Department of Scientific and Industrial Research, 1914–16. British Journal for the History of Science 32:461–481.
Hume D (1748–1894) An Enquiry concerning the Human Understanding, and an Enquiry concerning the Principles of Morals. Clarendon Press, Oxford.
Hunter M, Wood P (1986) Towards Solomon's House: Rival Strategies for Reforming the Early Royal Society. History of Science 24(1):49–108.
Huntington R (1704) Edward Bernard, Thomas Smith. Epistolae: Et Veterum Mathematicorum, Graecorum, Latinorum, & Arabum, Synopsis. Typis G. Bowyer, impensis A. & J. Churchill, London. [Separate pagination]. Part 9:1–78.
Huray PG (2010) Maxwell's Equations. John Wiley & Sons, Inc. New Jersey.
Hutten E H (1967) The Ideas of Physics. Oliver & Boyd, Edinburgh & London.
Hyslop S (2015) The Mathematics of Collision and the Collision of Mathematics in the Seventeenth Century, Ph.D. Dissertation, Department of History and Philosophy of Science, University of Indiana.
Ibn K ([1863] 1938) Les Prolégomènes d'Ibn Khaldoun. Troisième partie. 2nd ed. Librairie orientaliste Paul Geuthner, Paris.
Ignatowsky WV (1910–1911) Das Relativitätsprinzip. Archiv der Mathematik und Physik III:17:1–18; Physik Z 11:972 & 12:776–779.
Israel G (2008) Chi sono i nemici della scienza? Lindau s.r.l., Torino.
Jalobeanu D (2011) The Cartesians of the Royal Society: The Debate Over Collisions and the Nature of Body (1668–1670). In Anstey P, Jalobeanu D (eds). Vanishing Matter and the Laws of Nature: Descartes and Beyond, Routledge, London, pp.103–129.
James W (1890) Principles of Psychology. 2 Vols. Henry Holt and Company, New York.
Jammer D, Stachel J (1980) If Maxwell had worked between Ampere and Faraday. An historical fable with a pedagogical moral, American Journal of Physics 48:5–7.
Janssen M (2004) Relativity. New Dictionary of the History of Ideas. Horowitz MC et al (eds). Charles Scribner's Sons, New York.

Janssen M (2009) Drawing the Line between Kinematics and Dynamics in Special Relativity. Studies in History and Philosophy of Modern Physics 40:26–52.
Janssen M, Mecklenburg M (2006) From Classical to Relativistic Mechanics: Electromagnetic Models of the Electron. In: Hendricks VF, Jørgensen KF, Lützen J, Pedersen SA (eds). Interactions: Mathematics, Physics and Philosophy 1860-1930. Springer, Dordrecht, pp. 65–134.
Jenkin F (1873) Electricity and Magnetism. Appleton, New York.
Johnson B (1978) The Secret War. Methuen, Inc., New York.
Joule JP (1843) On the calorific effects of magneto-electricity, and on the mechanical value of heat. The London, Edinburgh, and Dublin Philosophical Magazine and Journal of Science. Conducted by Sir David Brewster, K.H. LL.D. F.R.S.L.&E. &c. Richard Taylor, F.L.S. G.S. Astr.S. Nat.H.Mosc. &c. Richard Phillips, F.R.S.L.&E. F.G.S. &c. Robert Kane, M.D. M.R.I.A., Richard and John E. Taylor, Red Lion Court, Fleet Street, Printers and Publishers to the University of London, London, New and United Series of the Philosophical Magazine, Annals of Philosophy, and Journal of Science XXIII: 263–276, 347–355, 435–443.
Joule JP (1845) On the changes of temperature produced by the rarefaction and condensation of air. By J. P. Joule, Esq. The London, Edinburgh, and Dublin Philosophical Magazine and Journal of Science [ser. 3, May 1845] 26/174:369–383. [*Id.*, (1884) The scientific papers of James Prescott Joule. Cambridge University Press, New York, pp.172–189].
Joule JP ([1849] 1850) On the mechanical equivalent of heat. By James Prescott Joule, F.C.S., Sec. Lit. and Phil. Society, Manchester, Cor. Mem. R.A., Turin, &c. Communicated by Michael Faraday, D.C.L., F.R.S., Foreign Associate of the Academy of Sciences, Paris, &c. &c. &c. Received June 6,-Read June 21, 1849. Philosophical Transactions of the Royal Society of London, 140:61–82.
Joule JP (1884) The scientific papers of James Prescott Joule. The Physical Society of London, London.
Joule JP (1887) Joint scientific papers of James Prescott Joule. The Physical Society of London. London.
Jungnickel C, McCormmach R (1986) Intellectual Mastery of Nature: Theoretical Physics from Ohm to Einstein. Volume 2. The Now Mighty Theoretical Physics. University of Chicago Press, Chicago.
Jungnickel C, McCormmach R (2017) The Second Physicist: On the History of Theoretical Physics in Germany. Springer, Cham.
Kahn D (1967) The Codebreakers, The Story of Secret Writing. The Macmillan Company, New York.
Kant I ([1781–1787] 1956) Kritik der reinen Vernunft. Felix Meiner, Hamburg.
Katib Çelebi MBA (1835–1858) Lexicon bibliographicum et encyclopaedicum, a Mustafa Ben Abdallah Katib Jelebi dicto et nomine Haji Khalfa celebrato compositum. Flügel (ed). Gustav Leberecht, Leipzig.
Kaufmann W (1901) Die magnetische und electrische Ablenkbarkeit der Becquerelstrahlen und die scheinbare Masse der Elektronen. Nachrichten von der Gesellschaft der Wissenschaften zu Göttingen, Mathematisch-Physikalische Klasse, pp. 143–155.
Kaufmann W (1902) Ueber die elektromagnetische Masse des Elektrons. Nachrichten von der Gesellschaft der Wissenschaften zu Göttingen, Mathematisch-Physikalische Klasse, pp. 291–296.
Kaufmann W (1903) Ueber die "Elektromagnetische Masse" der Elektronen. Nachrichten von der Gesellschaft der Wissenschaften zu Göttingen, Mathematisch-Physikalische Klasse, pp. 90–103.
Kaufmann W (1906) Über die Konstitution des Elektrons. Annalen der Physik 19:487–553.
Keegan J (2001) World War II, A Visual Encyclopedia. PRC Publishing Ltd., London.
Kemeny M (2017) What Motion Is: William Neile and the laws of motion. Annals of Science 74 (3):179–191.

Kepler J (1618–1621) Epitome ASTRONOMIAE Copernicanae Usitata forma Quaestionum et Responsionum conscripta inqu;[e] VII. libros digesta, quorum TRES hi priores sunt de DOCTRINA SPHAERICA. HABES, AMICE LECTOR, HAC PRIMA parte praeter physicam accuratam explicationem motus terrae diurnae ortusque ex eo circulorum sphaerae totam doctrinam sphaericam nova et concinniori methodo auctiorem, additis exemplis omnis generis computationum astronomicarum et geographicarum, quae integrarum praeceptionum vim sunt complexa. AUTHORE JOANNE KEPPLERO IMPERATORIS CAESARIS MATTTHIAE, Ordd[inumque] q; ill[ium] Archiducatus Austriae supra Onasum, Mathematico. Cum Privilegio Caesareo ad Annos XV. Lentijs ad Danubium, excudebat. Johannes Plancus ANNO MDCXVIII [Books I-III, 1618]; — Epitomes ASTRONOMIAE Copernicanae Usitata forma Quaestionum et Responsionum conscriptae, Liber Quartus, doctrinae Theoricae Primus : Quo PHYSICA CAELESTIS, Hoc Est, OMNNIUM IN CAELO MAGNITUDInum, motuum proportionumque causae vel Naturales vel Archetypicae explicantur, Et Sic PRINCIPIA DOCTRINAE Theoricae demonstrantur. QUI QUOD VICE SUPPLEMENTi librorum Aristotelis de Caelo esset, certo consilio seorsim est editus. AUTHORE JOANNE KEPPLERO. Cum Privilegio Caesareo ad Annos XV. Lentijs ad Danubium, excudebat Johannes Plancus, ANNO M.DC.XX [1620, Book IV; [see also: impensis Godefridi Tampachij, 1622]; — Epitomes ASTRONOMIAE Copernicanae usitata forma quaestionum et responsionum conscriptae, Libri V. VI. VII. Quibus proprie DOCTRINA THEORICA (post principia libro IV. praemissa) comprehenditur. AUTHORE IOANNE KEPLERO. Cum privilegio Caesareo ad Annos XV. Francoforti, Sumptibus Godefridi Tampachij. ANNO M.DC.XXI [Books V-VII, 1621]. In: KGW, VII.

Kepler J (1937–2012) Johannes Kepler, Gesammelte Werke (KGW). Van Dyck W, Caspar M. et al. (eds). Revised April 2013. X Vols. Deutsche Forschungsgemeinschaft und Bayerische Akademie der Wissenschaften. Beck'sche Verlagsbuchhandlung, München.

Kepler J (1992) New Astronomy. Translation of Astronomia Nova. Cambridge University Press, Cambridge.

Kline M (1980) Mathematics. The Loss of Certainty. Oxford University Press, New York.

Kline R (2015) The Cybernetics Moment: Or Why We Call Our Age the Age of Information. Johns Hopkins University Press, Baltimore.

Knudsen O, Pedersen KM (1968) The Link between 'Determination' and Conservation of Motion in Descartes' Dynamics. Centaurus 13:183–186.

Kosheleva O, Kreinovich V (2020) Relativistic effects can be used to achieve a universal square-root (or even faster) computation speedup. In: Blass A, Cegielsky P, Dershowitz N, Droste M, Finkbeiner B (eds). Fields of Logic and Computation III, Springer, Cham, Switzerland, pp. 179–189.

Kosheleva O, Zakharevich M, Kreinovich V (2014) If many physicists are right and no physical theory is perfect, then by using physical observations, we can feasibly solve almost all instances of each NP-complete problem. Mathematical Structures and Modeling 31:4–17.

Koyré A (1939) Études Galiléennes. Hermann, Paris. [English translation: Galileo Studies. Humanities Press, New Jersey, 1978].

Koyré A (1957) From the Closed World to the Infinite Universe. John Hopkins Press, Baltimore.

Koyré A (1960) La dynamique de Nicolo Tartaglia (1957). In: La science au seizième siècle. Colloque international de Royaumont, 1–4 juillet 1957. Hermann, Paris, pp 91–116 [see also: Koyré A (1966) La dynamique de Niccolo Tartaglia. Études d'histoire de la pensée scientifique:101–121].

Kragh H (2002) Quantum Generations: A History of Physics in the Twentieth Century. Princeton University Press, Princeton.

Kreinovich V, Kosheleva O (2021) Joule's 19th Century Energy Conservation Meta-Law and the 20th Century Physics (Quantum Mechanics and Relativity): 21st Century Analysis. Foundations of Science 26:703–725.

Kreinovich V, Kosheleva O (2023) Why quantum techniques are a good first approximation to economic phenomena, and what next. In: Nguyen Hoang Phuong, Kreinovich V (eds) Biomedical and Other Applications of Soft Computing. Springer, Cham, pp. 57–66.

Kreinovich V, Araiza R, Ferret J (2011) Analysis of Information and Computation in Physics Explains Cognitive Paradigms: from Full Cognition to Laplace Determinism to Statistical Determinism to Modern Approach. In: Dodig-Crnkovic G, Burgin M (eds). Information and Computation, World Scientific, Singapore, pp. 203–223.

Kreinovich V, Lakeyev A, Rohn J, Kahl P (1998) Computational Complexity and Feasibility of Data Processing and Interval Computations. Kluwer, Dordrecht.

Kreisel G (1980) Kurt Gödel, 28 April 1906–14 January 1978. Biographical Memoirs of Fellows of the Royal Society 26:148–224. Corrections 27:697; 28:718.

Kripke S ([1972] 1980) Naming and Necessity. Harvard University Press, Cambridge.

Krönig A (1856) Grundzüge einer Theorie der Gase. Annalen der Physik und Chemie 175/10:315–322.

Kruh L, Deavours C (2002) The Commercial Enigma: Beginnings of Machine Cryptography. Cryptologia 26(1):1–16.

Kuhn TS (1959) The Copernican Revolution. Vintage, New York.

Kuhn T (1962) The structure of scientific revolutions. University of Chicago Press, Chicago.

Kuhn T ([1962] 1969) La struttura delle rivoluzioni scientifiche. Einaudi, Torino.

Kuhn TS ([1962] 1970) The Structure of Scientific Revolutions. 2nd ed. University of Chicago Press, Chicago [with Postscript 1969].

Kuhn TS (1969) The structure of the scientific revolutions. Chicago University Press, Chicago.

Lagrange JL (1762) Application de la méthode exposée dans le mémoire précédente à la solution des différents problèmes de dynamique. In: Lagrange 1870–1873, I, pp. 363–468.

Lagrange JL (1764) Recherches sur la libration de la Lune dans lesquelles on tache de résoudre la question proposes par l'Académie Royale des Sciences, pour le Prix de l'année 1764. Prix de l'académie Royale des Sciences de Paris, IX (1777), pp. 1–50. [See also: Lagrange 1870–1873, Tome VI, pp. 5–61].

Lagrange JL (1788) Mécanique analytique. Desaint, Paris [See also: Lagrange J.L. (1973) Œuvres de Lagrange. Seconde Edition. Courcier (ed.), I–XIV vols. (in X) Paris: Gauthier–Villars].

Lagrange JL (1797) Théorie des fonctions analytiques contenant les principes du calcul différentiel, dégagés de toute considération d'infiniment petits, de limites et de fluxions, et réduits à l'analyse algébrique des quantités finies. Journal de l'École Polytechnique III(9):91–93.

Lagrange JL (1806) Leçons sur le calcul des fonctions, Nouvelle edn. Courcier, Paris.

Lagrange JL (1813) Théorie des fonctions analytiques. Courcier, Paris.

Lagrange JL (1870–1873) Œuvres de Lagrange. Seconde édition, Courcier (ed), I–XIV vols. (in X) Paris Gauthier–Villars, Paris.

Lagrange JL (1881) Théorie des fonctions analytiques contenant les principes du calcul différentiel, dégagés de toute considération d'infiniment petits, de limites et de fluxions, et réduits à l'analyse algébrique des quantités finies. In: Serret JA (ed) Œuvres de Lagrange. Cinquième édition, vol IX, Chapter VII. Gauthier–Villars, Paris, pp. 409–410.

Lagrange JL (1892) [Lagrange's letter to Euler on November 24th 1759 In: Œuvres de Lagrange. Tome XIV. Darboux G (ed). Gauthier–Villars, Paris, pp. 170–174.

Lagrange JL (1973a) Œuvres de Lagrange. Seconde édition, Courcier (ed), I–XIV vols, (in X) Paris. Gauthier–Villars, Paris.

Lagrange JL (1973b) Théorie des fonctions analytiques. In: Lagrange 1973a, vol IX.

Lalan V (1936–1937) Compte Rendu de Académie des sciences 203:1491 [see also Bulletin des Sciences Mathématiques 65:83].

Landau LD, Lifschitz EM (1975) The Classical Theory of Fields, III. Pergamon, London.

Landauer R (1961) Irreversibility and Heat Generation in the Computing Process. IBM Journal of Research and Development 5:183–191.

Langevin P (1911) L'évolution de l'espace et du temps. Scientia 10:34.

Lavoisier AL (1789 [1937]) Traité élémentaire de Chimie. Gauthier–Villars, Paris.

Lavoisier AL (1862–1893) Œuvres de Lavoisier publiées par le soin de Son Excellence le Ministre de l'instruction publique et des cultes, 6 vols. Imprimerie Impériale, Paris.

Lavoisier AL, Laplace PS ([1780] 1784) Mémoires sur la Chaleur. In: Histoire de l'Académie Royale des Sciences. (First reading on 18th June 1783). Imprimerie Royale, Paris, pp 355–408.
Laue von M (1911) Das Relativitätsprinzip. Fried. Vieweg & Sohn, Braunschweig.
Le Roy E (1936) Sur les postulats qui sont à la base des cinématiques. Compte Rendu de Académie des sciences 202:794.
Lecardonnel JP (1979) Un autre point de vue pour présenter les transformations de Lorentz. Bulletin de l'union des physiciens 615:1171.
Lee AR, Kalotas TM (1975) Lorentz Transformations from the First Postulate. American Journal of Physics 43:434.
Leff H, Rex A (1990) Maxwell's Demon: Entropy, Information, Computing. Princeton University Press, New Jersey.
Leff H, Rex A (2003) Maxwell's Demon 2: Entropy, Classical and Quantum Information, Computing. Institute of Physics Publishing, Bristol–Philadelphia.
Leibniz GW (1698) Essai de dynamique sur les lois du mouvement. MS, VI, pp. 215–221.
Leibniz GW (1969) Philosophical Papers and Letters, Loemker LE (ed). 2nd ed. Reidel, Boston.
Leijenhorst C (2004) Hobbes and the Galilean Law of Free Fall. In: Palmerino CR, Thijssen JMMH (eds). The Reception of the Galilean Science of Motion in Seventeenth–Century Europe, pp. 165–184. Dordrecht, Kluwer, Dordrecht.
Lenoir E (1964) Watch. In: Encyclopædia Britannica. Vol. 23, William Benton, Publisher, Chicago, Illinois.
Lévy-Leblond JM (1976a) One More Derivation of the Lorentz Transformations. American Journal of Physics 44:271.
Lévy-Leblond JM (1976b) What is So Special about Relativity? Group-Theoretical Methods in Physics. Janner A & al (eds). Lecture Notes in Physics n. 50. Springer-Verlag, New York.
Lévy-Leblond JM (1980) Speed(s). American Journal of Physics 48:345.
Lévy-Leblond JM (2001) Science's fiction. Nature 413:573.
Lévy-Leblond JM (2003) What if Einstein had not been there? In: Gazeau J-P & al (eds). Proceedings of the 24th International Colloquium on Group Theoretical Methods in Physics, Institute of Physics, London.
Lèvy-Leblond JM (2007) La Velocità dell'ombra. Ai limiti della scienza. Codice Edizioni, Torino.
Lévy-Leblond JM, Provost JP (1979) Additivity, Rapidity, Relativity. American Journal of Physics 47:1045.
Lewin R (1978) Ultra Goes to War, The First Account of World War II's Greatest Secret Based on Official Documents. McGraw-Hill Book Company, New York.
Lewis D (1968) Counterpart theory and quantified modal logic. The Journal of Philosophy, 65:113–126.
Lewis D (1986) On the Plurality of Worlds. Blackwell, New York.
Lewis D (1991) Parts of Classes. With an appendix by Burgess J, Hazen A, Lewis D. Basil Blackwell, Oxford.
Lewis G (1930) The symmetry of time in physics. Science 71:568–577.
Lindberg D C, Numbers R L (2003–2005) (eds) The Cambridge History of Science. 8 Vols. Cambridge University Press, Cambridge.
Lloyd S (2006) Programming the Universe: A Quantum Computer Scientist Takes On the Cosmos. Knopf, New York.
Lombardo Radice L (1962) L'educazione della mente. Editori Riuniti, Roma.
Lombardo Radice L (1981) L'Infinito. Numeri, mondi, Dio, spazio, tempo. Da Pitagora a Cantor itinerari filosofici e matematici di un concetto di base. Editori Riuniti, Roma.
Lopes Q (2017a) The National Education Board (1929/36) and scientific research in Portugal. Portuguese Journal of Social Science 16(1):71–85.
Lopes Q (2017b) A europeização de Portugal entre guerras. A Junta de Educação Nacional e a investigação científica. Caleidoscópio, Casal de Cambra.
Lorentz HA (1899a) Elektromagnetische Theorien physikalischer Erscheinungen. Physikalische Zeitschrift 1:498–501, 514–519.

Lorentz HA (1899b) Simplified Theory of Electrical and Optical Phenomena in Moving Systems. Proceedings of the Royal Netherlands Academy of Arts and Sciences 1:427–442.

Lorentz HA (1900) Considerations on Gravitation. KNAW, Proceedings 2:559–574.

Lorentz HA (1904) Electromagnetic Phenomena in a System Moving with Any Velocity Smaller than that of Light. KNAW, Proceedings, pp. 809–831.

Lorentz HA, Kox A (2008) The Scientific Correspondence of H.A. Lorentz, Volume 1. Springer, New York.

Lunde P (2012) The Secret of Codes. Weldon Owen, Inc., San Francisco, Califorina.

Lützen J (2005) Mechanistic Images in Geometric Form: Heinrich Hertz's Principles of Mechanics. Oxford University Press, Oxford.

Mabillon J (1681) De re diplomatica, libri VI. Louis Billaine, Paris.

Mach E (1919) The science of mechanics. A critical and historical account of its development. The Open Court Publishing Co., Chicago-London [English translation of the fourth edition of Mach's Die Mechanik in ihrer Entwickclung historisch-kritisch dargestellt, first edition 1883].

Maclean I (2007) Logic, Signs and Nature in the Renaissance. CUP, Cambridge.

Macleod R (1993) The Chemists go to War: The Mobilization of Civilian Chemists and the British War Effort, 1914–1918. Annals of Science 50:455–481.

Maddox J (1994) Time machines still over horizon. Nature 367:509.

Mahoney M (1973) The Mathematical Career of Pierre de Fermat 1601-1665. Princeton University Press, Princeton.

Maier A (1509) Zwei Grundprobleme der scholastichen Naturphilosphie, (3° ed.) Edizioni Storia e Letteratura, Roma, 1968 [which, in turn, includes – with some modifications – The Parisian edition from 1509].

Majer U (1998) Heinrich Hertz's Picture-Conception of Theories: Its Elaboration by Hilbert, Weyl, and Ramsey. In: Baird D, Hughes R, Nordmann A (eds). Heinrich Hertz: Classical Physicist, Modern Philosopher. Springer, Dordrecht, pp. 225–242.

Malament D (1977) The class of continuous timelike curves determines the topology of spacetime. Journal of Mathematical Physics 18:1399–1404.

Malament D (1985) Time travel in the Gödel universe. In Asquith PD, Kitcher P (eds). PSA: Proceedings of the 1984 Biennial Meeting of the Philosophy of Science Association. Philosophy of Science Association, East Lansing, pp. 91–100.

Mansfeld J (1999) Sources. The Cambridge Companion to Early Greek Philosophy. Long AA (ed). Cambridge University Press, Berkeley, pp. 22–44.

Maroney O (2009) Information Processing and Thermodynamic Entropy, The Stanford Encyclopaedia of Philosophy, Edward N. Zalta (ed). Via: https://plato.stanford.edu/entries/information-entropy. Retrieved 2021/12/22.

Martin DA (2005) Gödel's conceptual realism. The Bulletin of Symbolic Logic 11:207–213.

Masterman M (1970) The Nature of a Paradigm. In: Lakatos I, Musgrave A (eds). Criticism and the Growth of Knowledge. Cambridge University Press, London, pp. 59–89.

Matvievskaya GP (1981) Ramus. 1515–1572. Nauka, Moscow.

Maxwell JC (1864) [abstract] A dynamical theory of the Electromagnetic field [The paper was received in October 27, 1864, and read in December 8, 1864], The Proceedings of the Royal Society of London. From November 19, 1863, to December, 1861, inclusive. XIII:531–536.

Maxwell JC (1865) A Dynamical theory of the Electromagnetic field. Philosophical Transactions of The Royal Society of London CLV:459–512.

Maxwell JC (1871) Theory of heat. Longmann Green, Roberts & Green, London.

Maxwell JC (1873) A Treatise on Electricity and Magnetism. Oxford University Press, Oxford.

Maxwell JC (1881) A Treatise on Electricity and Magnetism. Second Edition. 2 Vol. Oxford University Press, Clarendon.

Maxwell N (2017) In Praise of Natural Philosophy. A Revolution for Thought and Life. McGill-Queen's University Press Montreal & Kingston, London, Chicago.

Mayer RJ (1842a) Bemerkungen über die Kräfte der unbelebten Natur. Weyrauch, Die Mechanik Warme, Stuttgart.

Mayer RJ (1842b) Bemerkungen über die Kräfte der unbelebten Natur. Annalen der Chemie und Pharmazie 42:233–240.

McClain S (1994) The Navajo Weapon. Books Beyond Borders, Boulder, Colorado.

McCormmach R (1970) H.A. Lorentz and the electromagnetic view of nature. Isis 61:459–497.

McLaughlin P (2000) Force, Determination and Impact. In: Gaukroger S, Schuster JA, Sutton J (eds). Descartes' Natural Philosophy. Routledge, London, pp. 81–112.

McTaggart JME (1908) The unreality of time. Mind (n.s.) 17:457–474.

Mermin ND (1984) Relativity without light. American Journal of Physics 52:120.

Mill JS (1882) A System of Logic, Ratiocinative and Inductive: being a Connected View of the Principles of Evidence and the Methods of Scientific Investigation. Harper & Brothers, New York.

Miller AI (1973) A Study of Henri Poincaré's'Sur La Dynamique de l'Électron'. Archive for History of Exact Sciences 10:207–328.

Miller A (1980) On Some Other Approaches to Electrodynamics in 1905. In: Woolf H (ed). Some Strangeness in the Proportion: A Centenial Symposium to Celebrate the Achievements of Albert Einstein. Addison-Wesley, Reading, pp. 66–91.

Miller AI (1981) Albert Einstein's Special Theory of Relativity: Emergence (1905) and Early Interpretation (1905–1911). Addison-Wesley Publishing Company, Reading.

Miller AR (1995) The Cryptographic Mathematics of Enigma. Cryptologia 19(1):65–80.

Miller AR ([1996] 2001) The Cryptographic Mathematics of Enigma. Published by the Center for Cryptographic History. Fort Meade, Maryland.

Miller DP (1986) Method and the 'Micropolitics' of Science: The Early Years of the Geological and Astronomical Societies of London. In: Schuster JA, Yeo RR (eds). The Politics and Rhetoric of Scientific Method, Reidel, Dordrecht, pp. 227–257.

Miller R, Fields W (2011) The "American" Enigma. Officer Review 51(1):8–11.

Misner C (1969) Mixmaster universe. Physical Review Letters 22:1071–1074.

Misner CW, Thorne KS, Wheeler JA (1973) Gravitation, Freeman & Company, NY, p. 424.

Morando AP (1998) Galileo Ferraris e la nascita dell'ingegneria elettrica moderna. Physis XXXV(2):291–339.

Morando AP (2004) Giovanni Giorgi, fisico matematico ed ingegnere elettrico. Physis XLI(2):391–447.

Morgenstein D, Kreinovich V (1995) Which algorithms are feasible and which are not depends on the geometry of space-time. Geombinatorics 4(3):80–97.

Morris M, Thorne K, Yurtsever U (1988) Wormholes, time machines, and the weak energy condition. Physical Review Letters 61:1446–1449.

Mosengeil von K (1907) Theorie der stationären Strahlung in einem gleichförmig bewegten Hohlraum. Annalen der Physik 22:867–904.

Moyer AE (1992) A Scientist's Voice in American Culture: Simon Newcomb and the Rhetoric of Scientific Method. University of California Press, Oakland.

Mulkay M, Gilbert N (1981) Putting Philosophy to Work: Sir Karl Popper's Influence on Scientific Practice. Philosophy of the Social Sciences 11:389–407.

Mulligan JF (1998) The Reception of Heinrich Hertz's Principles of Mechanics by his Contemporaries. In: Baird D, Hughes R, Nordmann A (eds). Heinrich Hertz: Classical Physicist, Modern Philosopher. Springer, Dordrecht, pp. 173–182.

Mulligan JF (2001) The Aether and Heinrich Hertz's The Principles of Mechanics Presented in a New Form. Physics in Perspective 3:136–164.

Mumford L (1970) The Myth of the Machine. The Pentagon of Power. Harcourt Brace Jovanovich, Inc., New York.

Mundó J (2008) Las ecuaciones de Maxwell. Vida, pensamiento y obra – Colección grandes pensadores. Centro Editor PDA S. L., Madrid.

Murray G, Harper W, Wilson C (2010) Huygens, Wren, Wallis and Newton on Rules of Impact and Reflection. In Anstey P, Jalobeanu D (eds). Vanishing Matter and the Laws of Nature: Descartes and Beyond. London: Taylor and Francis, London, pp. 153–191.

Müürsepp P (2003) The Need for Aristotle in Contemporary Science. Problemos 64:111–123.

Müürsepp P (2011) Knowledge in Science and Non-Science. Baltic Journal of European Studies 1(9):61–73.

Müürsepp P (2013) The Changing Role of Scientific Experiment. Studia Philosophica Estonica 5(2):152–166.

Müürsepp P (2016) Chemistry as a Practical Science (Edward Caldin Revisited). Foundations of Chemistry 18(3):213–223.

Müürsepp P, Nurysheva G, Bekenova Zh, Usenov G (2019) From the Dual Character of Chemistry to Practical Realism and Back Again: Philosophy of Science of Rein Vihalemm. Problemos 96:107–120.

Müürsepp P, Nurysheva G, Ramazanova A, Amirkulova Zh (2021) Chemistry as the Basic Science. Foundations of Chemistry 23:69–83.

Näpinen L, Müürsepp P (2002) The Concept of Chaos in Contemporary Science: On Jean Bricmont's Critique of Ilya Prigogine's Ideas. Foundations of Science 7:465–479.

Naylor RH (1976) Galileo: the Search for the Parabolic Trajectory. Annals of Science 33:153–172.

Naylor RH, Drake S (1983) Discussion on Galileo's early experiment on projectile trajectories. Annales of Science 40:391–396.

Neumann von J (1949) The Role of High and of Extremely High Complication, Fourth University of Illinois lecture. In: Burks AW (ed). Theory of Self-Reproducing Automata, [Reprinted in 1966, University of Illinois Press, Illinois].

Newton I (1687) Philosophiae Naturalis Principia Mathematica. Autore IS. Newton, Trin. Coll. Cantab. Soc. Matheseos Professore Lucasiano, & Societatis Regalis Sodali. Imprimatur. S. Pepys, Reg. Soc. Præses. Julii 5. 1686. Londini, Jussu Societatis Regiæ ac Typis Josephi Streater. Prostat apud plures Bibliopolas. Anno MDCLXXXVII.

Newton I ([1713] 1729) The Mathematical principles of natural philosophy. Translated by Motte Andrew. Motte B–Middle Temple-Gate, London.

Newton I ([1726] 1999) Philosophiae Naturalis Principia Mathematica, General Scholium. Third edition. In: The Principia: The Authoritative Translation and Guide: Mathematical Principles of Natural Philosophy. I. Bernard Cohen and Anne Whitman's 1999 translation. University of California Press, Berkeley.

Nielsen M, Chuang I (2010) Quantum Computation and Quantum Information. Cambridge University Press, New York.

Niven WD ([1890] 1965) The Scientific Papers of James Clerk Maxwell – Two volumes. Dover Publications, New York.

Noether E (1918) Invariante Variationsprobleme, Nachrichten König. von der Gesellschaft der Wissenschaften zu Göttingen, Mathematisch-physikalische Klasse, s. 235–257

Norton J (2013) The End of the Thermodynamics of Computation: A No Go Result. Philosophy of Science 80(5):1182–1192.

Norton JD (2019) The hole argument. Stanford Encyclopedia of Philosophy, Zalta EN (ed). Retrieved 26 January 2022 from https://plato.stanford.edu/entries/spacetime-holearg/

Nyquist H (1924) Certain Factors Affecting Telegraph Speed. Bell System Technical Journal 3:324–346.

Oldenburg H (1965–1985) The Correspondence of Henry Oldenburg. Hall AR, Hall MB (eds). University of Wisconsin Press, Madison.

Oliveira ARE, Pisano R (2022) Historical Details on Coulomb's Torsional Forces by Metal Wires. In: Ceccarelli M, López-García R (eds). Explorations in the History and Heritage of Machines and Mechanisms. Vol. 40. Dordrecht, Springer, pp. 241–254.

Ostwald W (1892) Klassiker der exakten Wissenschaften. Engelmann 37, Lipsia.

Oughtred W (1631) Clavis Mathematicae. Typis Lichfieldianis, Oxoniae. [In English: *Id.*, (1694) Key of the Mathematics] Salusburn, London.

Page N (1912) Derivation of the fundamental relations of electrodynamics from those of electrostatics. American Journal of Science 34:57–68.

Pal PB (2003) Nothing but relativity. European Journal of Physics 24:135.

Palladino F (1985) Ernesto Pascal e gli integrafi per equazioni differenziali. Annali dell'Istituto e Museo di Storia della Scienza di Firenze (X):67–76.
Palmieri P (1988) Re-examining Galileo's Theory of Tides. Archive for History of Exact Sciences 53:223–375.
Papadimitriou CH (1994) Computational Complexity. Pearson, Upper Saddle City, New Jersey.
Pars LA (1921), The Lorentz transformation, Philosophical Magazine 42:249.
Parsons C (1964) Infinity and Kant's conception of the "possibility of experience". [Reprinted] Mathematics in Philosophy: Selected Essays. Cornell University Press, Ithaca, pp. 95–109.
Parsons C (1995) Platonism and mathematical intuition in Kurt Gödel's thought. The Bulletin of Symbolic Logic 1:44–74.
Parsons C (2010) Gödel and philosophical idealism. Philosophia Mathematica (III) 18:166–192.
Pascual R, Pisano R (2008) (eds) Scienze e Religioni. Il ruolo delle scienze naturali, ipotesi di studio e prospettive. Uniservice, Trento.
Patterson LD (1952) Pendulums of Wren and Hooke. Osiris 10:277–321.
Paul DA (1973) The Navajo Code Talkers. Dorrance Publishing Company, Pittsburgh, Pennsylvania.
Peirce CS (1878) Deduction, Induction, and Hypothesis. Popular Science Monthly 13:470–482.
Pellegrino EM (2014) Entropy from Clausius to Kolmogorov: historical evolution of an open concept. In: Tucci P (ed). SISFA 2014. Pavia University Press, Pavia, pp. 139–147.
Pellegrino EM (2020) [Book review] Lazare and Sadi Carnot: A Scientific and Filial Relationship by Charles Coulston Gillispie and Raffaele Pisano. The Mathematical Intelligencer 42/2:91–100.
Pellegrino EM, Cerruti L (2015) Clausius' disgregation: a conceptual relic that sheds light on the second law. Entropy 17/7:4500–4518
Pellegrino EM Cerruti L (2016) From steam engines to chemical reactions: Gibbs' contribution to the extension of the second law. Entropy 18/5–162:1–27.
Penrose R (1965) Gravitational collapse and space-time singularities. Physical Review Letters, 14:57–59.
Peter Bowler PJ (2013) Darwin Deleted. Imagining a World without Darwin. Chicago Press, Chicago.
Petrucci RH (2003) Química General. Pearson Educación, Madrid.
Petruccioli S (1988) Atomi Metafore Paradossi - Niels Bohr e la costruzione di una nuova fisica. Theoria, Roma.
Pinch T (1985) Towards an analysis of scientific observation: the externality and evidential significance of observational reports in physics. Social Studies of Science 15:3–36.
Pisano R (2007) Brief history of centre of gravity theory. Epistemological notes. In Kokowski M (ed). Proceedings of the 2nd ESHS Congress, Krakow, Poland, pp. 934–941.
Pisano R (2009a) Il ruolo della scienza archimedea nei lavori di meccanica di Galilei e di Torricelli. In: Giannetto E, Giannini G, Capecchi D, Pisano R, (eds). Da Archimede a Majorana: La fisica nel suo divenire. Guaraldi Editore, Rimini, pp. 65–74.
Pisano R (2009b) Continuity and discontinuity. On method in Leonardo da Vinci' mechanics. Organon 41:165–182.
Pisano R (2009c) On method in Galileo Galilei' mechanics. In Hunger H (ed) Proceedings of ESHS 3rd Conférence. Austrian Academy of Science, Vienna, pp. 147–186.
Pisano R (2009d) Galileo Galileo. Riflessioni epistemologiche sulla resistenza dei corpi. In Giannetto E, Giannini G, Toscano M (eds). Relatività, Quanti Chaos e altre Rivoluzioni della Fisica. Guaraldi Editore Rimini, pp. 61–72.
Pisano R (2010) On principles in Sadi Carnot's thermodynamics (1824). Epistemological Reflections. Almagest 2/1:128–179.
Pisano R (2011) Physics–Mathematics Relationship. Historical and Epistemological notes. In Barbin E, Kronfellner M, Tzanakis C (eds). European Summer University History And Epistemology In Mathematics, Verlag Holzhausen GmbH–Holzhausen Publishing Ltd., Vienna, pp. 457–472.

Pisano R (2013a) Historical Reflexions on the Physics Mathematics Relationship in Electromagnetic Theory. In: Barbin E, Pisano R (eds). The Dialectic Relation Between Physics and Mathematics in the XIXth Century. Springer, Dordrecht, pp. 31–57.
Pisano R (2013b) Introduction to the International 32nd Congress SISFA. In: Pisano R, Capecchi D, Lukešová A (eds). Physics, Astronomy and Engineering. Critical Problems in the History of Science and Society. International 32nd Congress for The SISFA–Italian Society of Historians of Physics and Astronomy (Roma, 27-29 September, 2012) The Scientia Socialis UAB & Scientific Methodical Centre Scientia Educologica Press, Šiauliai University, Lithuania, pp. xi–xxi.
Pisano R (2013c) On the New Boson Higgs's Studies at the CERN-ATLAS Experiment. The Emergency of a Historical Discovery. AHS 2/1:3–5.
Pisano R (2014a) Historical Reflections on Newton's First Law and Carnot's Première Hypothèse. In: Gianna Katsiampoura G (ed). International Conference of the European Society for the History of Science. Athens (GR) 1–3 November 2012. National Hellenic Research Foundation/Institute of Historical Research/ Section of Neohellenic Research/ Programme of History, Philosophy and Didactics of Science and Technology, Athens, pp. 214–220.
Pisano R (2014b) On Lagrangian in Maxwell's electromagnetic theory. Scientiarum VI. The Federate University of Rio de Janeiro Press, Rio de Janeiro, pp. 44–59.
Pisano R (2016) Details on the mathematical interplay between Leonardo da Vinci and Luca Pacioli. Journal of the British Society for the History of Mathematics 31(2):1–8.
Pisano R (2017a) Introduction. 1564–2014. In: Pisano and Bussotti 2017b, pp. 5–13.
Pisano R (2017b) A Development of the Principle of Virtual Laws and its Framework in Lazare Carnot's Mechanics as Manifest Relationship between Physics and Mathematics. Transversal–International Journal for Historiography of science 2:166–203.
Pisano R (2020) A Tale of Tartaglia's Libro Sesto & Gionta in Quesiti et Inventioni diverse (1546–1554). Exploring the Historical and Cultural Foundations. Foundations of Science 25/2:477–505.
Pisano R (2021) An Intellectual History of Entropy as a Physics–Mathematics Dynamic Relationship. Speeches as Key Note Speaker at Science and Technologies Chemistry Department (October 29th), University of Palermo, Italy.
Pisano R (2024) Briefs on Symmetries in the History of Physics–Mathematics: Maxwell (1865–1873), Noether (1915–1918) and Einstein (1905–1926), Institute of Physics Publ. (UK), Forthcoming.
Pisano R (2024-pre print) Galileo Galilei: The Two Manuscripts On Mechanics. Translations, Text and Commentaries. Dordrecht, Springer, forthcoming.
Pisano R, Bussotti (2014) Historical and Philosophical Reflections on the Culture of Machines around the Renaissance. How Science and Technique Work? Acta Baltica Historiae et Philosophiae Scientiarum 2/2:20–42.
Pisano R, Bussotti (2015a) Fibonacci and the Abacus Schools in Italy. Mathematical Conceptual Streams, Science Education and its Changing Relationship with Society. Almagest 6/2:126–165.
Pisano R, Bussotti (2015b) Historical and Philosophical Reflections on the Culture of Machines around the Renaissance: Machines, Machineries and Perpetual Motion. Acta Baltica Historiae et Philosophiae Scientiarum 3/1:69–87.
Pisano R, Bussotti P (2015c) Galileo in Padua: architecture, fortifications, mathematics and « practical » science, Lettera Matematica Pristem International 2/4:209–221.
Pisano R, Bussotti P (2016) A Newtonian Tale Details on Notes and Proofs in Geneva Edition of Newton's Principia. Bulletin BSHM-Journal of the British Society for the History of Mathematics 31/3:160–178.
Pisano R, Bussotti P (2017a) (eds) Homage to Galileo Galilei 1564–2014. Reading Iuvenilia Galilean Works within History and Historical Epistemology of Science. Special Issue Philosophia Scientiae 21/1.
Pisano R, Bussotti P (2017b) Newton's Principia Geneva edition: the action-and-reaction law. Historical and Nature of Science reflexions. Pavia University Press, Pavia, pp. 269–276.

Pisano R, Bussotti P (2017c) On the Conceptualization of Force in Johannes Kepler's Corpus: an Interplay between Physics/Mathematics and Metaphysics. In: Pisano R, Agassi J, Drozdova D (eds). Hypothesis and Perspective within History and Philosophy of Science. Homage to Alexandre Koyré 1892–1964. Dordrecht, Springer, pp. 295–346.

Pisano R, Bussotti P (2017d) The Fiction of Infinitesimals in Newton's Works. On the Metaphoric Use of the Infinitesimals in Newton. In Marcacci F, Rossi MG (eds). Isonomia-Epistemologica. Reasoning, Metaphor and Science 9:141–160.

Pisano R, Bussotti P (2020) Newton's Geneva Edition: the Notes on Integral Calculus. In: La Rana A, Rossi P (eds): International 39th Congress of the Italian Society for the History of Physics and Astronomy (Pisa/VIRGO 2019, 9–12 Sept.). Pavia, Pavia University Press, pp. 127–134.

Pisano R, Bussotti P (2022) Conceptual Frameworks on the Relationship between Physics–Mathematics in the Newton *Principia Geneva Edition* (1822). Foundations of Sciences (Springer) 27/3:1127–1182.

Pisano R, Bussotti P (2024) Historical and Methodological Details on the *De motu gravium naturaliter descendentium in Torricelli's Opera Gometrica (1644).* In: Pisano Dhombres, Radelet de Grave and Bussotti 2024a.

Pisano R, Capecchi D (2010) Galileo Galilei: Notes on Trattato di Fortificazione. In Altamore A, Antonini G (eds), Galileo and the renaissance scientific discourse, Roma Edizioni nuova cultura, Roma, pp. 28–41.

Pisano R, Capecchi D (2013) Conceptual and Mathematical Structures of Mechanical Science in the Western Civilization around 18th Century. Almagest 4/2:86–121.

Pisano R, Capecchi D (2016) Tartaglia's Science of Weights and Mechanics in the Sixteenth Century. Selections from Quesiti et inventioni diverse: Books VII–VIII. Springer, Dordrecht.

Pisano R, Cioci V (2020) Galileo's Free fall into History Physics and Nature of Science Teaching. In: International 38th Congress for The SISFA–Italian Society of Historians of Physics and Astronomy. Messina 3–6, October, Italy, Pavia University Press, pp. 271–278.

Pisano R, Marmottini D (2017) Nature–of–Science Teaching: Notes on the Lagrangian Methods in Maxwell's electromagnetic Theory. XXXVI SISFA Proceedings. Pavia University, Pavia, pp. 263–268.

Pisano R, Robarts J (2022) A Short Comparative Historiography of Science & Technology of Mechanics into Engineering—and—Architecture Literatures during the Renaissance. In: Ceccarelli M, López-García R (eds). Explorations in the History and Heritage of Machines and Mechanisms. Vol. 40. Dordrecht, Springer, pp. 227–240.

Pisano R, Agassi J, Drozdova D (2017) (eds) Hypothesis and Perspective within History and Philosophy of Science. Homage to Alexandre Koyré 1892–1964. Dordrecht, Springer.

Pisano R, Coopersmith J, Peake M (2021a) Essay on Machines in General (1786) Text, Translations and Commentaries. Lazare Carnot's Mechanics. Volume 1. Springer, Dordrecht.

Pisano R, Cooppersmith J, Peake M (2021b) Essay on Machines in General (1786). Text, Translations and Commentaries. Lazare Carnot's Mechanics – Vol. 1. Springer, Dordrecht.

Pisano R, Dhombres J, Radelet de Grave P, Bussotti P (2024) 380th Anniversary of Torricelli's *Opera Geometrica (1644).* Text, Transcription, Commentaries and Selected Essays as New Historical Insights. Springer. Springer, Cham.

Pisano R, Pellegrino EM, Anakkar A, Nagles M (2019) Thermodynamic foundations of physical chemistry: reversible processes and thermal equilibrium into the history. Foundation of Chemistry 21/3:297–323.

Pisano R, Pellegrino EM, Anakkar A, Nagles M (2021) Conceptual polymorphism of entropy into the history: extensions of the second law of thermodynamics towards statistical physics and chemistry during nineteenth–twentieth centuries. Foundation of Chemistry 23/3:337–378.

Pisano R, Vincent P, Dolenc K, Ploj-Virtič M (2021) Historical Foundations of Physics & Applied Technology as Dynamic Frameworks in Pre-Service STEM. Foundations of Science 26:727–756.

Pisano R, Vincent P, Mauricio P (2020) Introduction. Joule's Bicentenary: History of Science, Foundations and Nature of Science. Joule's Bicentenary Special Issue. Foundations of Science 26:531–551.

Planck M ([1897] 1903) Vorlesungen¨uber Thermodynamik. Walter de Gruyter, Berlin [Reprinted in English translation as Treatise on Thermodynamics. Dover, New York].

Planck M (1901) Ueber das Gesetz der Energieverteilung im Normalspectrum. Annalen der Physik, 309/3:553–563.

Planck M (1906) Das Prinzip der Relativität und die Grundgleichungen der Mechanik. Deutsche Physikalischen Gesellschaft. Verhandlungen 8:136–141.

Planck M (1908a) Bemerkungen zum Prinzip der Aktion und Reaktion in der allgemeinen Dynamik. Physikalische Zeitschrift 9:828–830.

Planck M (1908b) Zur Dynamik bewegter Systeme. Annalen der Physik 26(4):1–34.

Planck M (1914) The Theory of Heat Radiation. Translation by Masius M. of the second German edition. Blakiston's Son & Co., Philadelphia.

Planck M (1917) Treatise on Thermodynamics. Translated with the Author's sanction by A. Ogg. Dover, New York.

Planck M (2016) Treatise on Thermodynamics. The Project Gutenberg EBook. The Project Gutenberg eBook #50880: Treatise on Thermodynamics. Last accessed 2021/12/29.

Poincaré H (1900) La théorie de Lorentz et le principe de réaction. In: Bosscha J (ed). Recueil des travaux offerts par les auteurs à H.A. Lorentz à l'occassion du 25me anniversaire de son doctorat, Martinus Nijhoff, Den Haag, pp. 252–278.

Poincaré H ([1902] 1905) Science and Hypothesis. Walter Scott Publishing Co., NY.

Poincaré H (1906) Sur la dynamique de l'électron. Rendiconti del Circolo Matematico di Palermo 21:129–175.

Popławski NJ (2010) Radial motion into an Einstein-Rosen bridge. Physics Letters B 687:110–113.

Popov GN (1920) History of mathematics. In Russian. Tipo-Litografiya Moskovskogo Kartoizdatelskogo Otdela korpusa Voyennoy Topografii, Moscow.

Popper K (1935) Logik de Forschung. Springer, Berlin.

Popper K (1945) The Open Society and Its Enemies. Routledge, London.

Popper K (1950) Indeterminism in Quantum Physics and in Classical Physics. British Journal for Philosophy of Science 1:117–195.

Popper K (1952) The Nature of Philosophical Problems and their Roots in Science [Republished in Popper 1963].

Popper K (1957) The Poverty of Historicism. Routledge, London.

Popper K (1959) The Logic of Scientific Discovery. Routledge, London.

Popper K (1963) Conjectures and Refutations. Routledge, London.

Popper K (1982) The Open Universe, An Argument for Indeterminism: From the Postscript to The Logic of Scientific Discovery. Routledge, London.

Popper K R (1934–1935) Logik der Forschung. Zur erkenntnistheorie der modernen naturwissenschaft, Springer–Verlag, Wien [English edition, translation prepared by the author, with the assistance of Julius Freed and Lan Freed The Logic of Scientific Discovery, Hutchinson & Co., London & New York 1959].

Popper KR (1970) Logica della scoperta scientifica. Einaudi, Torino.

Popper KR (1972) Conoscenza oggettiva, un punto di vista evoluzionistico. Rossi A (ed), Armando Editore, Roma.

Potters J (2019) Heuristics Versus Norms: On the Relativistic Responses to the Kaufmann Experiments. Studies in History and Philosophy of Modern Physics 66:69–89.

Pour-El MB, Richards JI (2017) Computability in Analysis and Physics. Cambridge University Press, Cambridge.

Poynting JH (1884) On the transfer of energy in the electromagnetic field. The Royal Society, Volume 175, pp. 343–361.

Prendergast TL (1975) Motion, Action and Tendency in Descartes' Physics. Journal of the History of Philosophy 13:453–62.

Prigogine I (1980) From Being to Becoming: Time and Complexity in the Physical Sciences. W.H. Freeman and Company, San Francisco.
Putnam H (1959) Review of Pattern of Discovery/An inquiry into the conceptual foundations of science. Science 129(3364):1666–1667.
Putnam H (1967) Time and physical geometry. The Journal of Philosophy 64:240–247.
Ramée [Ramus] P (1567) Proemium mathematicum in tres libros distributum. Wechel, Paris.
Ramée [Ramus] P (1569) Scholarum mathematicarum libri unus et triginta. Eusebius Episcopius, Basel.
Raphson J (1715) Historia fluxionum: sive tractatus originem & progressum peregregiæ istius methodi brevissimo compendio (et quasi synopticè) exhibens. Per Josephum Raphsonum. Printed by William Pearson, London.
Rattansi PM (1968) The Intellectual Origins of the Royal Society. Notes and Records of the Royal Society 23:129–43.
Raychaudhuri A (1955) Relativistic cosmology. I. Physical Review 98:1123–1126.
Raychaudhuri A (1957) Relativistic and Newtonian cosmology. Zeitschrift für Astrophysik 43:161–164.
Raychaudhuri A (1979) Theoretical Cosmology. Clarendon, Cambridge.
Rees G (2000) Baconianism. In: Applebaum W (ed). Encyclopedia of the Scientific Revolution, pp. 69–71. Garland, New York.
Reeves E, Van Helden A (2010) (eds) On Sunspots. Galileo Galilei and Christoph Scheiner. Chicago University Press, Chicago and London.
Reichenbach H ([1928] 1958) The Philosophy of Space and Time. Translation by Reichenbach M and Freund J. Dover, New York.
Riebe A, Häffner H, Roos CR, Hänsel W, Benhelm J, Lancaster GP, Körber TW, Becher C, Schmidt-Kaler F, James DF, Blatt F (2004) Deterministic quantum teleportation with atoms. Nature 429:734–747.
Riemann B (1875) Schwere, Elektricität und Magnetismus. K. Hattendorff, Hannover.
Rijn van Alkemade van AC (1912) Über die Transformationsgleichungen der Relativitätstheorie. Annalen der Physik 38:1033.
Rindler W (2009) Gödel, Einstein, Mach, Gamow, and Lanczos: Gödel's remarkable excursion into cosmology. American Journal of Physics 77(6):498–510.
Robb A ([1911] 2015) Optical Geometry of Motion, a New View of the Theory of Relativity. Heffner & sons. Cambridge 9. Andesite Press, Warsaw.
Roberts L (2009) Situating Science in Global History: Local Exchanges and Networks of Circulation. Itinerario 33(1):9–30.
Robinson RC, Ward ED (1964) Clock. In: Encyclopædia Britannica. Vol. 5, William Benton, Publisher, Chicago, Illinois.
Rodwell G (2013) Whose History? Engaging history students through historical fiction University of Adelaide Press, Adelaide. Chapter 7, Counterfactual Histories and the Nature of History, pp. 81–98.
Rose PL, Drake S (1971) The Pseudo-Aristotelian Questions of Mechanics in Renaissance Culture. Studies in the Renaissance 18:65–104.
Rossi A (2019) Ricordo di Salvo D'Agostino. Physis. Nuova Serie LIV(1–2):491–493.
Rossi B (1957) Optics. Addison Wesley, New York [Italian Translation: *Id.* (1979) Ottica. Masson Italia, Milano].
Rossi P (1986) Italo-marxismo, scienza società. I ragni e le formiche. Un'apologia della storia della scienza. Il Mulino, Bologna, pp. 229–247, p. 236.
Rossi P (1988–1989) (ed) Storia della scienza. 3 Vols. 5 Tomes. UTET, Torino.
Roth Ph (2004) The Plot Against America. Houghton Mifflin, Boston.
Royal Society (1940) The Record of the Royal Society of London. 4th edition. Morrison and Gibb, London.
Ryckman TA (2011) What does history matter to philosophy of physics? Journal of the Philosophy of History 5:496–512.

Ryckman TA (2021) Early philosophical interpretations of general relativity. Stanford Encyclopedia of Philosophy. Zalta EN (ed). Retrieved 29 January 2022 from https://plato.stanford.edu/archives/fall2021/entries/genrel-early/

Rynasiewicz R (2011) Newton's views on space, time, and motion. Stanford Encyclopedia of Philosophy. Zalta EN (ed). Retrieved 26 January 2022 from https://plato.stanford.edu/entries/newton-stm/

Rynasiewicz R (2018) Newton's Scholium on time, space, place and motion. [Preprint]. Retrieved 11 March 2022 from http://philsci-archive.pitt.edu/id/eprint/15525

Sabra AI (1967) Theories of Light from Descartes to Newton. Oldbourne, London.

Salatino DR (2009) Semiótica de los sistemas reales – Tesis Doctoral en Letras especialidad Psicolingüística por la Facultad de Filosofía y Letras de la Universidad Nacional de Cuyo, Mendoza.

Salatino DR (2012) Aspectos psico-bio-socioculturales del lenguaje natural humano. Introducción a la teoría psíquica del lenguaje. Desktop Publishing, Mendoza.

Salatino DR (2013) El proyecto de Freud. Psicoanálisis. Revista de la Asociación Psicoanalítica Colombiana 24(1/2):43–60.

Salatino DR (2014) Psyche – Structure and Function – First Desktop Publishing, Mendoza.

Salatino DR (2015a) Ficciones, solo ficciones In: Conceptos y Lenguajes en ciencia y tecnología. Cuadrado G, Redmond J, López R (eds). Serie: Selección de Textos, Volumen 3, Universidad de Valparaíso, Valparaíso, pp. 251–273.

Salatino DR (2015b) Las 2201 páginas de una teoría. La verdadera historia de la Lógica Transcursiva. 4 Vols. Autoedición, Mendoza.

Salatino DR (2015c) Bases neurológicas y psíquicas del lenguaje musical. Actas de ECCoM, La experiencia musical: cuerpo, tiempo y sonido en el escenario de nuestra mente 2/1:257–267.

Salatino DR (2015d) Sobre el inconsciente" – Psicoanálisis – Revista de la Asociación Psicoanalítica Colombiana XXVII(1):199–228.

Salatino DR (2016) La importancia de la simetría. Educación en Ciencias Empíricas en Facultades de Ingeniería. (ECEFI 2016). Universidad Tecnológica Nacional, Mendoza, pp. 19–49.

Salatino DR (2017a) Beyond the Decisions-Making: The Psychic Determinants of Conduct and Economic Behavior. International Journal and Methodology in Social Sciences 3/1:6–26.

Salatino DR (2017b) Beyond the Decisions-Making II: Methodological Aspects. International Journal and Methodology in Social Sciences 3/2:pp. 6–24.

Salatino DR (2017c) Tratado de Lógica Transcursiva. El origen evolutivo del sentido en la realidad subjetiva. Primera autoedición. Mendoza.

Salatino DR (2017d) Transcurssive Logic as Method. International Journal and Methodology in Social Sciences 3/3:5–18.

Salatino DR (2017e) Language production – First part. International Journal and Methodology in Social Sciences 3/4:5–21.

Salatino DR (2018a) Language production – Part 2. International Journal and Methodology in Social Sciences 4/1:5–17.

Salatino DR (2018b) Language acquisition – Part 1 of 3. International Journal and Methodology in Social Sciences 4/2:1–14 [With honourable mention].

Salatino DR (2018c) Language acquisition – Part 2 of 3. Genetic Typology of Human Natural Language. Hidden Colors Theory. International Journal and Methodology in Social Sciences 4/3:1–27.

Salatino DR (2018d) La 'geometría funcional' como fundamento de la estructura y la función del aparato psíquico. Revista de Psicopatología y Salud Mental del niño y del adolescente 32:77–94.

Salatino DR (2018e) Language acquisition – Part 3 of 3. Language Universals – Lexical Contextures – Sociocultural Heritage. International Journal and Methodology in Social Sciences 4/4:1–28 [With honourable mention].

Salatino DR (2018f) Tratado de Psicolingüística. Adquisición, Comprensión y Producción del Lenguaje Natural Humano. Primera Autoedición, Mendoza.

Salatino DR (2019a) Treatise of Psycholinguistics. Acquisition, Comprehension, and Production of Human Natural Language. First Desktop Publishing, Mendoza.
Salatino DR (2019b) Los lenguajes de Freud. Enfoque subjetivo de la Teoría Psicoanalítica. Primera Autoedición, Mendoza.
Salatino DR (2019c) Fundamentals of a new research method. International Journal and Methodology in Social Sciences 5/1:52–73 [With honourable mention].
Salatino DR (2019d) The logic within the logic. Reinterpretation of Boolean algebra from Transcurssive Logic. International Journal and Methodology in Social Sciences, 5/1:83–99 [With honourable mention].
Salatino DR (2019e) The Mystery of Time. International Journal and Methodology in Social Sciences 5/3:1–29.
Salatino DR (2020a) The Intrinsic Beauty of the Penrose's Mosaics. International Journal and Methodology in Social Sciences 6/1:1–14.
Salatino DR (2020b) Subjective Universe. International Journal and Methodology in Social Sciences 6/1:1–12.
Salatino DR (2020c) Theory of Destination in Borges. A Transcurssive Approach. International Journal and Methodology in Social Sciences 6/1:13–21.
Salatino DR (2020d) New Nosological Model of Chronic Endometritis. International Journal and Methodology in Social Sciences 6/1:22–32.
Salatino DR, Puglesi, AE (2018) Neuron as an electric PAU. International Journal and Methodology in Social Sciences, 4/4:59–76 [With honourable mention]
Salatino DR, Tersoglio AE (2019) Risk of Autism in the use of assisted reproduction techniques. An analysis from the Transcurssive Logic. International Journal and Methodology in Social Sciences 5/1:1–16 [With honourable mention].
Salatino DR, Masera GA, Palma R (2019) Risks and critical infrastructures. An interpretation from the Transcurssive Logic. International Journal and Methodology in Social Sciences 5/2:17–25.
Salgueiro Â (2015) Ciência e Universidade na I República. Ph.D dissertation in Contemporary History, New University of Lisbon Faculty of Social and Human Sciences.
Samaja J (2005) Epistemología y Metodología. Elementos para una teoría de la investigación científica. Eudeba. Buenos Aires.
Sánchez Ron J (2007) Las ciencias Físicas y Químicas en la JAE. In: Puig-Samper Mulero M (ed). Tiempos de investigación. JAE-CSIC, cien años de ciencia en España. Consejo Superior de Investigaciones Científicas, Madrid, pp. 102–113.
Sánchez Ron J (2009) Del Laboratorio de Investigaciones Físicas al Instituto Nacional de Física y Química. In: González Ibáñez C, Santamaría García A (eds). Física y Química en la Colina de los Chopos. 75 años de investigación en el edificio Rockefeller del CSIC (1932–2007). Consejo Superior de Investigaciones Científicas, Madrid, pp. 47–74.
Savitt S (1994) The replacement of time. Australasian Journal of Philosophy 72:463–474.
Savitt S (2000) There's no time like the present (in Minkowski spacetime). Philosophy of Science 67–Proceedings:S563–S574.
Schaffer J (2016) The metaphysics of causation. Stanford Encyclopedia of Philosophy. Zalta EN (ed). Retrieved 3 February 2022 from https://plato.stanford.edu/entries/causation-metaphysics/
Schilling M (1911) Catalog mathematischer Modelle für den höheren mathematischen Unterricht. Verlag, Leipzig.
Schilpp PA (ed.) (1949) Albert Einstein, Philosopher-Scientist. Library of Living Philosophers, Evanston.
Schroeder-Gudehus B (2020) International Science from the Franco-Prussian War to World War Two: An Era of Organization. In: Slotten H, Numbers R, Livingstone D (eds). The Cambridge History of Science: Modern Science in National, Transnational, and Global Context, Vol. 8. Cambridge University Press, Cambridge, pp. 43–59.
Schuster JA (1984) Methodologies as Mythic Structures: A Preface to the Future Historiography of Method. Metascience: Review of the Australasian Assoc. for the History, Philosophy and Social Studies of Science 1–2:15–36.

Schuster JA (1986) Cartesian Method as Mythic Speech: A Diachronic and Structural Analysis. In: Schuster JA, Yeo RR (eds). The Politics and Rhetoric of Scientific Method, pp. 33–95. D. Reidel, Dordrecht.

Schuster JA (1990) The Scientific Revolution. In: Olby RC, Cantor GN, Christie JRR, Hodge MJS (eds). The Companion to the History of Modern Science, pp. 217–242. Routledge, London.

Schuster JA (1993) Whatever Should We Do with Cartesian Method: Reclaiming Descartes for the History of Science. In: Voss S (ed). Essays on the Philosophy and Science of René Descartes. OUP, Oxford, pp. 195–223.

Schuster JA (1995) The Scientific Revolution: An Introduction to the History and Philosophy of Science. Open Learning Australia. Located at http://descartes-agonistes.com/

Schuster JA (2000) Descartes opticien: the construction of the law of refraction and the manufacture of its physical rationales, 1618–1629. In: Gaukroger S, Schuster J, Sutton J (2000) (eds). Descartes' natural philosophy. Routledge London and New York, pp. 258–312.

Schuster JA (2002) L'Aristotelismo e le sue Alternative. In: Garber D (ed). La Rivoluzione Scientifica, Instituto della Enciclopedia Italiana, Rome, pp. 337–357. [An English translation is available, with the title 'Aristotelianism and its Alternatives' on the website http://descartes-agonistes.com/]

Schuster JA (2005) 'Waterworld': Descartes' Vortical Celestial Mechanics—A Gambit in the Natural Philosophical Agon of the Early Seventeenth Century. In: Anstey P, Schuster JA (eds). The Science of Nature in the Seventeenth Century: Patterns of Change in Early Modern Natural Philosophy. Springer, Dordrecht, pp. 35–79.

Schuster JA (2012) Physico–mathematics and the Search for Causes in Descartes' Optics—1619-37. Synthèse 185:467–499.

Schuster JA (2013a) Descartes–Agonistes: Physico–Mathematics, Method and Corpuscular–Mechanism, 1619–1633. Springer, Dordrecht.

Schuster JA (2013b) Cartesian Physics. In: Buchwald Jz and Fox R (eds). The Oxford Handbook of the History of Physics, The Oxford University Press, Oxford, pp. 56–95.

Schuster JA (2013c) What was the relation of Baroque Culture to the Trajectory of Early Modern Natural Philosophy. In: Gal O, Chen-Morris R (eds). Science in the Age of Baroque, Springer, Dordrecht, pp. 13–45.

Schuster JA (2017a) Did Descartes Teach a 'Philosophy of Science' or Implement 'Strategies of Natural Philosophical Explanation'. In: Gaukroger S, Wilson C (eds). Descartes and Cartesianism: Essays in Honour of Desmond Clarke. Oxford University Press, Oxford, pp. 3–25.

Schuster JA (2017b) Consuming and Appropriating Practical Mathematics and the Mixed Mathematical Fields, or Being 'Influenced' by Them: The Case of the Young Descartes. In: Cormack L, Walton, S, Schuster JA (eds). Mathematical Practitioners and the Transformation of Natural Knowledge in Early Modern Europe. Springer, Dordrecht, pp. 37–65.

Schuster JA (2018) The Pitfalls and Possibilities of Following Koyré: The Younger Tom Kuhn, 'Critical Historian', on Tradition Dynamics and Big History. In: Pisano R, Agassi J, Drozdova, D (eds). Hypotheses and Perspectives in the History and Philosophy. Springer, Dordrecht, pp. 391–420.

Schuster JA (2019) Pitfalls and Opportunities of Contextual Explanation: The Case of Isaac Beeckman's Invention of the Mechanical Philosophy. Isis 110(2):308–311.

Schuster JA (2022) Isaac Beeckman in the Context of the Scientific Revolution. In: van Berkel K, Clement, A, van Dixhoorn A (eds). Knowledge and Culture in the Early Dutch Republic: Beeckman in Context. University of Amsterdam Press, Amsterdam, pp. 37–55.

Schuster JA, Brody J (2013) Descartes and Sunspots: Matters of Fact and Systematising Strategies in the Principia Philosophiae. Annals of Science 70(1):1–45.

Schuster JA, Taylor ABH (1996) Seized by the Spirit of Modern Science. Metascience 9:9–26

Schuster JA, Taylor ABH (1997) Blind Trust: The Gentlemanly Origins of Experimental Science. Social Studies of Science 27:503–536.

Schuster JA, Watchirs G (1990) Natural Philosophy, Experiment and Discourse in the Eighteenth Century: Beyond the Kuhn/Bachelard Problematic. In: Le Grand HE (ed). Experimental Inquiries: Historical, Philosophical and Social Studies of Experimentation in Science, pp. 1–47. Kluwer, Dordrecht.

Schuster JA, Yeo RR (eds) (1986) Introduction. In: Schuster JA, Yeo RR (eds) The Politics and Rhetoric of Scientific Method, D. Reidel, Dordrecht, pp. ix–xxxvii.

Schütz I (1897) Prinzip der absoluten Erhaltung der Energie. Königlich Preussische Akademie der Wissenschaften, Nachrichten, pp. 110–123.

Schwinger J S (1976) Gravitons and photons: The methodological unification of source theory General Relativity and Gravitation 7:251–256.

Searle G (1897) On the Steady Motion of an Electrified Ellipsoid. The London, Edinburgh, and Dublin Philosophical Magazine and Journal of Science 44:329–341.

Segal J (2003) Le zéro et le un: Histoire de la notion scientifique d'information au 20e siècle. Éditions Syllepse, Paris.

Seife C (2007) Decoding the Universe: How the New Science of Information is Explaining Everything in the Cosmos from our Brains to Black Holes. Penguin Group, London.

Serra I (2010) Armando Gibert um físico português em luta pela ciência. Travaux de Laboratoire L(II):1–14.

Seth S (2004) Quantum theory and the electromagnetic world view. Historical Studies in the Physical and Biological Sciences 35(1):67–93.

Seth S (2005) Response to Shaul Katzir. "On the electromagnetic world-view". Historical Studies in the Physical and Biological Sciences 36(1):193–196.

Sexl RU (1967) Theories of Gravitation. Fortschritte der Physik 15:269.

Shannon C (1948) A Mathematical Theory of Communication. IEEE Press.

Shannon C (1956) The Bandwagon. In Collected Papers. IEEE Press.

Shannon C, Weaver W (1949) The Mathematical Theory of Communication. University of Illinois Press, Illinois.

Shapin S (1982) History of science and its sociological reconstructions. History of Science 20: 157–211.

Shapin S (1994) A Social History of Truth: Civility and Science in Seventeenth Century England. University of Chicago Press, Chicago.

Shell-Gellasch A (2015) Schilling Kinematic Models at the Smithsonian. Journal of Humanistic Mathematics 5/1:167–179 via: scholarship.claremont.edu/jhm/vol5/iss1/9

Siegmund-Schultze R (2001) Rockefeller and the internationalization of Mathematics between the Two World Wars: documents and studies for the Social History of Mathematics in the 20th Century. Birkhauser Verlag, Basel.

Silverberg R (2003) Roma Æterna. Eos Books, New York.

Simões A (2011) O ano 1947 e o Laboratório de Física da Faculdade de Ciências de Lisboa. Gazeta de Física 34(2):16–21.

Simpson R (2017) Crypto Wars, 2000 Years of Cipher Evolution, CreateSpace Independent Publishing Platform, Middletown, Delaware.

Singaravélou P & Deluermoz Q (2016) Pour une histoire des possibles. Analyses contrafactuelles et futurs non advenus, Seuil, Paris.

Singh S (2001) The Code Book. Delacorte Press, New York.

Sinkevich GI (2015) On the history of number line. Antiquitates Mathematicae. 9/1:83–92.

Smeenk C, Wüthrich C (2011) Time travel and time machines. In Callender (ed). Oxford Handbook of Philosophy of Time. Oxford University Press, New York, pp. 577–630.

Smith DE (1951) History of Mathematics. Vol. 1. Dover Publications Inc., New York.

SN (2002) Denys Fisher obituary. The Times, 26 October.

Snow CP (1962) The two cultures and the scientific evolution. Feltrinelli, Milano.

Soler L, Trizio E, Pickering A (2015) (eds) Science as it Could Have Been: Discussing the Contingency/Inevitability Problem. University of Pittsburgh Press, Pittsburgh.

Sommerfeld A (1952) Electrodynamics, Lectures on Theoretical Physics. Vol. 111, Academic Press, New York.
Sommerfeld A, Eckert M, Märker K (2000) Wissenschaftlicher Briefwechsel. Verlag für Geschichte der Naturwissenschaften und der Technik, Berlin.
Speidell J (1619) New Logarithmes, London.
Speidell J (1622) New Logarithmes (2nd ed). Printed by Edward Allde, London.
Spinrad N (1972) The Iron Dream. Avon Books, New York.
Sprat T (1667) The History of the Royal Society of London. London.
Stachel J (2005) Structure, individuality, and quantum gravity. Retrieved 27 January 2022 from Cornell University Library Web site: https://arxiv.org/abs/gr-qc/0507078
Stachel J (2007) Review of Yourgrau, A World Without Time: The Forgotten Legacy of Gödel and Einstein. Notices of the American Mathematical Society 54:861–868.
Stachel J (2014) The hole argument and some physical and philosophical implications. Living Reviews in Relativity 17(1). Retrieved 26 January 2022 from https://link.springer.com/article/https://doi.org/10.12942/lrr-2014-1
Staley R (2008) Einstein's Generation: The Origins of the Relativity Revolution. The University of Chicago Press, Chicago.
Stein H (1968) On Einstein-Minkowski space-time. The Journal of Philosophy 65:5–23.
Stein H (1970) On the paradoxical time-structures of Gödel. Philosophy of Science 37:589–601.
Stein H (1991) On relativity theory and the openness of the future. Philosophy of Science 58:147–167.
Stein H (2009) "Definability," "conventionality," and simultaneity in Einstein-Minkowski Space-Time. In Myrvold WC, Christian J (eds). Quantum Reality, Relativistic Causality, and Closing the Epistemic Circle (Essays in Honor of Abner Shimony). Springer, pp. 403–442.
Suardi GB (1752) Nuovi istromenti per la descrizione di diverse curve antiche e moderne, e di molte altre, che servir possono alla speculazione de' Geometri, ed all'uso de' Pratici. Col progetto di due nuove macchine per la nautica ed una per la meccanica, e con alcune Osservazioni sopra de' Poligoni rettilinei regolari. Brescia.
Svoboda V, Jespersen B, Cheyne C (2004) (eds) Pavel Tichý's collected papers in logic and philosophy. Prague and Otago University Press, Dunedin.
Svozil K (2011) Physical unknowables. In Baaz M, et al (eds). Kurt Gödel and the Foundations of Mathematics. Cambridge University Press, Cambridge, pp. 213–251.
Swanson J (1960) Physical versus Logical Coupling in Memory System. IBM Journal 4(3):305–310.
Tartaglia N (1537) Nova scientia inventa da Nicolo Tartalea. B. Disciplinae mathematicae loquuntur[.] Qui cupitis rerum varias cognoscere causas disate nos cunctis hac patet una via. In Vinegia per Stefano Nicolini da Sabio. Ad instantia di Nicolo Tartalea Brisciano il qual habita a San Saluador. MDXXXVII.
Tartaglia N (1546) Quesiti et inventioni diverse de Nicolo Tartalea Brisciano. Stampata in Venetia per Venturino Ruffinelli ad instantia et requisitione, & a proprie spese de Nicolo Tartalea Brisciano Autore. Nel mese di luio L'anno di nostra salute. M.D.XLVI.
Tartaglia N ([1554] 1959) La nuova edizione dell'opera "Quesiti et inventioni diverse de Nicolo Tartaglia brisciano, Riproduzione in facsimile dell'edizione del 1554 [with Gionta al sesto Libro] Masotti A (ed). Commentari dell'Ateneo di Brescia, Tipografia La Nuova Cartografica, Brescia.
Taylor ABH (1989) The Early Royal Society of London. Ph.D. Thesis, University of Melbourne.
Taylor ABH (1994) An Episode with May–dew. History of Science 32:163–84.
Thirring W (1961) An Alternative Approach to the Theory of Gravitation. Annales of Physics 16:96.
Thompson B (1798) An inquiry concerning the source of the heat which is excited by friction. Philosophical Transactions of the Royal Society of London 88:80–102.
Thomson JJ (1881) On the Electric and Magnetic Effects Produced by the Motion of Electrified Bodies. The London, Edinburgh, and Dublin Philosophical Magazine and Journal of Science 5(11):229–249.
Thomson JJ (1893) Notes on Recent Researches in Electricity and Magnetism. Clarendon Press, Oxford.

Thomson W (1848–1849a) An account of Carnot's theory of the motive power of heat with numerical results deduced from Regnault's experiments on steam. Transactions of the Edinburgh Royal Society XVI/5:541–574.

Thomson W (1848–1849b) On an absolute thermometric scale founded on Carnot's theory of the motive power of heat, and calculated from Regnault's observations. Proceedings of The Cambridge Philosophical Society 1/5:66–71.

Thomson W (1851a) On an absolute thermometric scale founded on Carnot's theory of the motive power of heat, and calculated from Regnault's observations. Mathematical and physical papers, I. Cambridge University Press, Cambridge, pp. 100–106.

Thomson W (1851b) On the dynamical theory of heat, with numerical results deduced from Mr Joule's equivalent of a thermal unit, and M. Regnault's observations on steam. Mathematical and physical papers, I, Cambridge University Press, Cambridge, pp. 175–183.

Thomson W (1882–1911) An account of Carnot's theory of the motive power of heat with numerical results deduced from Regnault's experiments on steam. Mathematical and physical papers by Sir William Thomson, I, Cambridge University Press, Cambridge, pp. 113–155.

Thorne KS (1994) Black Holes and Time Warps: Einstein's Outrageous Legacy. Norton, New York.

Thorne KS, Blandford RD (2017) Modern Classical Physics: Optics, Fluids, Plasmas, Elasticity, Relativity, and Statistical Physics. Princeton University Press, Princeton, New Jersey.

Tichý P (1974) On Popper's definition of verisimilitude. British journal for the philosophy of science 25:155–160.

Tichý P (1976) Verisimilitude redefined. British journal for the philosophy of science 27:25–42.

Torricelli E (1644) Opera geometrica. Masse & de Landis, Florentia.

Torricelli E (1919–1944) Opere di Evangelista Torricelli. Loria G, Vassura G (eds). 4 Vols., Montanari, Faenza.

Tribus M, McIrvine E (2007) Compendium of the foundations of classical statistical physics. In: Butterfield J, Earman J (eds). Handbook for philosophy of physics. Elsevier, Amsterdam.

Uffink, J (2007) Compendium of the foundations of classical statistical physics. In: Butterfield J, Earman J (eds) Handbook for philosophy of physics. Elsevier, Amsterdam.

Ursin R, Jennewein T, Aspelmeyer M, Kaltenbaek R, Lindenthal M, Walther P, Zeilinger A (2004) Quantum teleportation across the Danube. Nature 430:849.

Valadares M (1935) Transmutation des éléments par des particules accélérées artificiellement. Hermann & Cie, Paris.

Valadares M (1950) O Laboratório de Física da Faculdade de Ciências de Lisboa, sob a direcção do Prof. Dr. A. Cyrillo Soares (1930–1947), e a investigação científica. Gazeta de Física II(4):93–106.

Valadares M, Mendes F (1938) Étude des satellites Lα, de l'élément 82 (Pb). Comptes Rendus de l'Académie des Sciences de Paris 206:744–746.

Vanpaemel G (2013) To be or not to be: Belgian science policy in the interwar period. In: Fitas A, Príncipe J, Nunes F, Bustamante M (eds). A Junta de Educação Nacional e a Investigação Científica em Portugal no Período entre Guerras. Caleidoscópio, Casal de Cambra, pp. 33–48.

Vanzo A (2016) Experiment and Speculation in Seventeenth-Century Italy: The Case of Geminiano Montanari'. Studies in History and Philosophy of Science 56:52–61.

Varicak Vladimir (1910) Anwendung der Lobatschefskijschen Geometrie in der Relativtheorie. Physikalische Zeitschrift 11:93–96

Vecli A (1975) Biofisica. In Fieschi R. ed. Enciclopedia della Fisica. 2 Vols. ISEDI, Milano, p. 16.1.

Vergara Caffarelli R (2007) Il principio d'inerzia negli ultimi scritti di Galilei. Cronos 10:63–88.

Vihalemm R (2015) Philosophy of Chemistry against Standard Scientific Realism and Anti-Realism. Philosophia Scientiae 19(1): 99–113.

Vihalemm R (2019) A Story of a Science: On the Evolution of Chemistry. Special Issue Acta Baltica Historiae et Philosophiae Scientiarum 7(2).

Vihalemm R, Lõhkivi E (2012) Guest Editorial. Philosophy of Science in Practice and Practical Realism. Studia Philosophica Estonica 5(2):1–6.

Voss [Vossius] GJ (1650) De universae matheseos natura et constitutione liber; cui subjungitur chronologia mathematicorum. Ex typogr. J. Blaeu, Amsterdam.

Voss [Vossius] GJ (1660) De quatuor artibus popularibus, de philologia, et scientiis mathematicis. Cui operi subjungitur Chronologia Mathematicorum. Libri tres. Ex typographeio I. Blaeu, Amsterdam.

Wallis J (1685) Treatise of algebra both Historical and Practical. Richard Davis, London.

Wallis J ([2012] 2005) Correspondence of John Wallis. Vol II (1660–1668) and III (October 1668-1671). Beeley PC, Scriba C (eds). Oxford University Press, Oxford.

Wang H (1974) From Mathematics to Philosophy. Humanities, New York.

Wang H (1995) Time in philosophy and physics: From Kant and Einstein to Gödel. Synthese 102:215–234.

Wardhaugh B (2016) Willughby's Mathematics. In: Birkhead T (ed). The Virtuoso by Nature: The Scientific Worlds of Francis Willoughby FRS 1635–1672. Brill, Leiden, pp. 122–141.

Watkins J (1970) Imperfect Rationality. In: Borger R, Cioffi F (1970) (eds) Explanation in the Behavioral Sciences. Cambridge University Press, Cambridge, pp. 167–218.

Weiner C (1977) (ed) History of 20th century physics, Atti della LVII Scuola Internazionale di Fisica Enrico Fermi, Varenna, 1972. Academic Press, New York.

Welchman G (1982) The Hut Six Story, Breaking the Enigma Codes. McGraw-Hill Book Company, New York.

Westfall R (1971) Force in Newton's Physics: The Science of Dynamics in the Seventeenth Century. MacDonald, London.

Whewell W (1840) The Philosophy of the Inductive Sciences. London.

Wiechert E (1894) Bedeutung des Weltäthers. Physikalisch-ökonomische Gesellschaft zu Königsberg 35:4–11.

Wieleitner H (1923) Die Geschichte der Mathematik von Descartes bis zum Hálfie des 19 Jahrhunderts. De Gruyter. [In Russian: (1960) Istoria matematiki ot Dekarta do serediny XIX stoletiya. Gosudarstvennoe izdatel'stvo fiziko-matematicheskoj literatury, Moscow].

Wien W (1894) Temperatur und Entropie der Strahlung. Annalen der Physik 52:132–165.

Wien W (1898) Ueber die Fragen, welche die translatorische Bewegung des Lichtäthers betreffen. Annalen der Physik 301:I–XVIII.

Wien W (1901) Ueber die Möglichkeit einer elektromagnetischen Begründung der Mechanik. Annalen der Physik 310(7):501–513.

Wittgenstein L ([1921] 1922) Tractatus Logico-Philosophicus. Routledge, London.

Wolff C ([1716] [1734] 1747) Vollständiges mathematisches Lexicon. Gleditsch, Leipzig.

Wolfius C (1741) Elementa matheseos Universae. Tomus V. Renger, Halle.

Wootton D (2015) The Invention of Science: A New History of the Scientific Revolution. Allen Lane, London.

Wright A (2016) The Physics of Forgetting: Thermodynamics of Information at IBM 1959–1982. Perspectives on Science 24(1):112–141.

Wullus-Rudiger J (1948) La science au service de la nation et de l'humanité: la Fondation Universitaire et le Fonds National de la Recherche Scientifique. Service d'information et de documentation du Ministère de la Défense nationale, Bruxelles.

Yourgrau P (1999) Gödel Meets Einstein. Time Travel in the Gödel Universe. Expanded ed. Open Court, Chicago.

Zhmud L (1997) History of mathematics by Eudemus of Rhodes. [In Russian] Hyperboreus 3:274–297.

Ziman J (2003) Emerging out of Nature into History: the Plurality of the Sciences. Philosophical Transactions of the Royal Society of London A 361:1617–1633.

Ziman J, Midgley M (2001) Pluralism in Science: a Statement. Interdisciplinary Science Reviews 26:153.

Names Index

R. Pisano (ed.), *A History of Physics: Phenomena, Ideas and Mechanisms*, History of Mechanism and Machine Science 42,
https://doi.org/10.1007/978-3-031-26174-9

Subject Index

R. Pisano (ed.), *A History of Physics: Phenomena, Ideas and Mechanisms*, History of Mechanism and Machine Science 42,
https://doi.org/10.1007/978-3-031-26174-9

D

I

J

O

P

T

U

V

The manufacturer's authorised representative in the EU is Springer Nature Customer Service Centre GmbH, Europaplatz 3, 69115 Heidelberg, Germany. If you have any concerns regarding our products, please contact ProductSafety@springernature.com

Printed and bound by CPI Group (UK) Ltd, Croydon, CR0 4YY
01/10/2026
02225857-0019